BASIC OP AMP MODULES

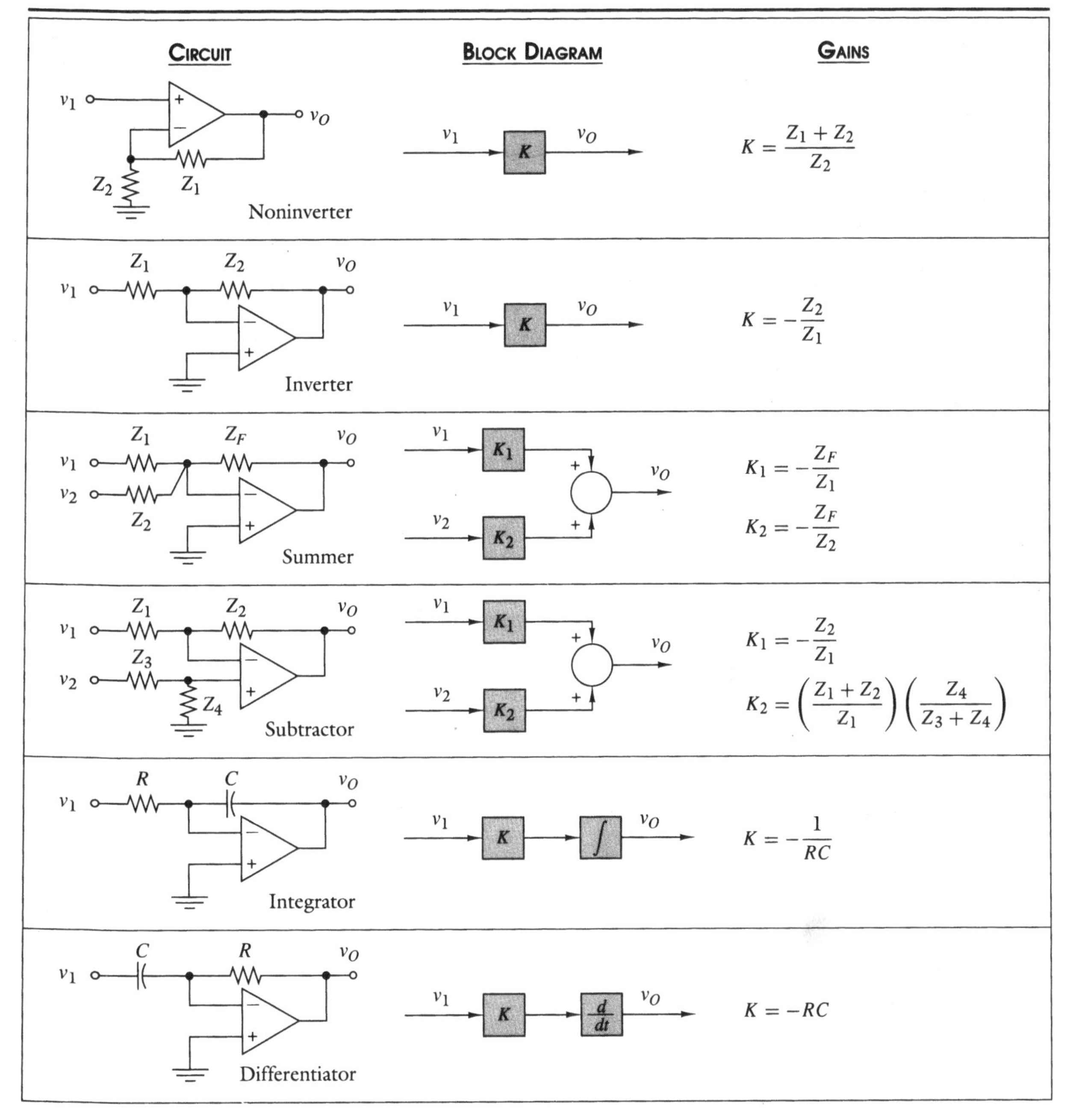
CIRCUIT
BLOCK DIAGRAM
GAINS
Noninverter
$K = \frac{Z_1 + Z_2}{Z_2}$
Inverter
$K = -\frac{Z_2}{Z_1}$
Summer
$K_1 = -\frac{Z_F}{Z_1}$
$K_2 = -\frac{Z_F}{Z_2}$
Subtractor
$K_1 = -\frac{Z_2}{Z_1}$
$K_2 = \left(\frac{Z_1 + Z_2}{Z_1}\right)\left(\frac{Z_4}{Z_3 + Z_4}\right)$
Integrator
$K = -\frac{1}{RC}$
Differentiator
$K = -RC$

THE ANALYSIS AND DESIGN OF LINEAR CIRCUITS

FIFTH EDITION

THE ANALYSIS AND DESIGN OF LINEAR CIRCUITS

ROLAND E. THOMAS
Professor Emeritus
United States Air Force Academy

ALBERT J. ROSA
Professor of Engineering
University of Denver

JOHN WILEY & SONS, INC.

Sr. Acquisitions Editor & Product Manager Catherine Fields Shultz
Associate Publisher Daniel Sayre
Associate Editor Kelly Boyle
Production Editor Valerie Vargas
Marketing Manager Phyllis Cerys
Design Director Harry Nolan
Illustration Editor Sigmund Malinowski
Media Editor Stefanie Liebman
Production Management Services Suzanne Ingrao/Ingrao Associates
Cover Photo ©MEHAU KULYK/Photo Researchers, Inc.

This book was set in 10/12 Times Ten Roman by Techbooks and printed and bound by R.R. Donnelley and Sons. The cover was printed by Lehigh Press, Inc.

This book is printed on acid free paper. ∞

ISBN-13 978-0-471-76095-5
ISBN-10 0-471-76095-1

Printed in the United States of America

10 9 8 7 6 5 4 3 2 1

To our wives
Juanita and Kathleen

PREFACE

Why this Text

The approach to circuits in this text differs from others. It recognizes that studying circuits can be rewarding and useful, even for students who are not majoring in electrical or computer engineering. Most students who pursue engineering studies are looking for opportunities to be creative and design things. The longer it takes for them to encounter such opportunities, the more likely it is that they will become disillusioned or even change to a different major. The authors have long believed that an early introduction to design and design evaluation raises the excitement level and greatly increases student interest in their chosen discipline. More than thirty years of teaching experience at several institutions have only served to strengthen our belief. This new edition furthers our philosophy by adding more *Design* and *Evaluation* examples, exercises, and homework problems, as well as additional real-world applications.

The fifth edition of *The Analysis and Design of Linear Circuits* combines the two versions of the fourth edition to provide greater flexibility to present users of either version. A later section discusses how to use the *Fifth Edition* to pursue either a traditional *Phasor First* approach or the *Laplace Early* approach. The fifth edition assumes the same student prerequisites as in past editions and is aimed at introductory circuit analysis courses that may include some students in other engineering disciplines. The fifth edition continues the Authors' thirty-year commitment to providing a modern and innovative approach to teaching circuit analysis and design.

Continuing Features

Objectives

This text is structured around a sequence of carefully defined learning objectives and related evaluation tools based on *Bloom's Taxonomy of Educational Objectives*. The initial learning objectives focus on enabling skills at the *Knowledge*, *Comprehension,* and *Application* levels of the Taxonomy. As students demonstrate mastery of these lower levels, they are introduced to higher-level objectives involving *Analysis*, *Synthesis* (Design), and *Evaluation*. Each learning objective is explicitly stated in terms of expected student proficiency, and each is supported by at least ten homework problems specifically designed to evaluate student mastery of the objective. This framework has been a standard feature of all five editions of this book and has allowed us to maintain a consistent level of expected student performance over the years. A new feature of this edition is that the objectives are listed in the chapter openers to orient the student to the expected outcomes.

Circuit Analysis and Design

Our experience convinces us that a student's grasp of circuit analysis fundamentals is reinforced by an interweaving of analysis and design topics. Early involvement in design provides motivation as students apply their newly acquired analysis tools to practical situations. Using computer simulation software to verify their designs provides an early degree of confidence that they have actually created a design that meets a specification. A supporting laboratory program where they actually build and test their designs provides the final confirmation that they can create useful products.

Design Evaluation

Realistic design problems do not have unique solutions, so it is natural for students to wonder how their design compares with those of other students. Using judgment to compare alternative solutions is a fundamental trait of good engineering. The evaluation of alternative designs gives the students an early introduction to real-world engineering practices. Including design and the evaluation of design in an introductory course helps to convince students that circuits perform useful functions and that circuit courses are not simply vehicles for teaching routine skills like node-voltage and mesh-current analyses. Although this philosophy is present in all previous editions, this edition offers expanded coverage of design and evaluation both in the worked examples and homework problems.

The OP AMP

An early introduction and integrated treatment of the OP AMP continues to be a central feature of this text. The modular form of OP AMP circuits simplifies analog circuit analysis and design by minimizing the effects of loading and allowing an interconnection of simple building blocks to produce complex signal processing functions. The close agreement between theory, simulation, and hardware allows students to analyze, design, and successfully build useful OP AMP circuits in the laboratory. The text covers numerous OP AMP applications such as digital-to-analog conversion, transducer interface circuits, comparator circuits, block diagram realization, first-order filters, and multiple-pole active filters. These applications are especially useful to students from other engineering disciplines that require knowledge of instrumentation, filtering, or biomedical signal processing.

Laplace Transforms

In electrical engineering, Laplace transforms can be used to treat important concepts such as zero-state and zero-input responses, impulse and step responses, convolution, frequency response, and filter design. An important question is "when should Laplace transforms be taught"—in the

Circuits course, the Signals and System course, or both, or elsewhere? The traditional answer is to first teach phasors and use them to study ac circuit analysis, steady-state ac power, polyphase circuit analysis, magnetically coupled circuits, and frequency response. This extended treatment of phasor analysis means that Laplace transforms are often delayed to the last weeks of the second semester and treated as an advanced topic along with Fourier methods and two-ports networks.

The Authors have long advocated an *early Laplace* approach—one in which Laplace transforms are introduced and applied to circuit analysis before introducing phasors. The advantage of treating Laplace-based circuit analysis first is that once mastered, it makes learning phasor-based analysis relatively easy. Students quickly make the connection between phasor analysis and the concepts of network functions, transient response, and sinusoidal steady-state response developed through s-domain circuit analysis. We do not claim that Laplace analysis is more fundamental or even more important than phasor analysis. We do claim that the learning effort needed to master both phasor analysis *and* Laplace analysis is not a zero sum game. Our experience is that less classroom time is needed to mastering *both* methods of analysis when Laplace transform analysis is treated before phasor analysis. Emphasizing transform methods in the circuits course also better prepares the students to handle the profusion of transforms they will encounter in subsequent signals and systems courses.

SIGNALS PROCESSING

Our treatment of dynamic circuits begins with a separate chapter on waveforms and signal characteristics. This chapter gives students an early familiarity with all of the important input and output signals encountered in the study of linear circuits. Introducing signals at the beginning lets students become comfortable with time-varying signals without simultaneously dealing with new concepts like differential equations, phasors, or Laplace transforms. Additional emphasis on signal processing and systems is achieved through use of block diagrams, input–output relationships, and transforms methods. The ultimate goal is for students to understand that time-domain waveforms and frequency-domain transforms are simply alternative ways to characterize signals and signal processing. Viewing signals in both domains leads naturally to discussions of important concepts like signal bandwidth, signal sampling, and reciprocal spreading.

COMPUTER TOOLS

Our philosophy recognizes that students come to the circuits course already knowing how to use several useful computer tools. Our goal is to help them learn *when* to use these tools and *how to interpret* the results they supply. Three types of computer programs are used to illustrate computer-aided circuit analysis, namely spreadsheets (Excel®), math solvers (Mathcad® & MATLAB®), and circuit simulators (Orcad® and Electronics WorkBench®). Examples of computer-aided circuit analysis are integrated into all chapters beginning with Chapter 2. The purpose of these examples is to help students develop a problem-solving style that includes the intelligent use of the productivity tools routinely used by practicing engineers. Although all of the homework problems can be solved using a scientific calculator, a computer icon ( ) is used to identify those problems where the use of a computer tool offers a significant learning advantage.

WEB APPENDICES

The fifth edition continues the chapter length treatment of *Fourier Transforms* and *Two-port Networks* as Web appendices. These appendices are fully integrated into the text with index references and answers to selected homework problems. These appendices are available at:

http://www.wiley.com/college/thomas

NEW FEATURES FOR THE FIFTH EDITION

CHAPTER OPENERS

The chapter openers have been redesigned to add new features. Two new brief introductory paragraphs give the student some historical background for the chapter and explain why the chapter remains important today. The chapter learning objectives are listed together in the openers so the teachers and students can see the expected outcomes up front.

CIRCUIT DESIGN AND EVALUATION

New design and evaluation examples have been added in Chapters 3, 4, 7, 11, 12, and 14. More design and evaluation homework problems have been added where appropriate. A sample listing of some new evaluation examples is:

Page	Example	Topic
189	4-22	Temperature Sensor Interface Circuits
299	7-11	First-order *RC* Circuits
568	12-7	First-order High-pass Filters
607	12-24	Piezoelectric Pressure Transducers

APPLICATION EXAMPLES

New application examples have been added in Chapters 2, 3, 4, 6, 11, 13, and 14. These applications emphasize computer engineering, instrumentation, signal processing, and biomedical engineering. A sample listing of application examples is:

Page	Example	Subject
45	2-19	Binary Current Divider
238	5-19	ECG Waveform
542	11-25	Digital Clock Delay Skew
641	13-13	Signal Sampling

OP AMPS

The treatment of OP AMPs in Chapter 4 has been rewritten to further emphasize *Design* and *Evaluation*. A new section called *OP AMP Circuit Applications* collects the applications contained in several previous sections in one area and expands their number.

WHY STUDY CLASSICAL SOLUTIONS

The final section in Chapter 7 has been rewritten to give the student a rationale for studying classical methods of solving differential equations even though our ultimate goal is to use Laplace transform methods for circuit analysis and design.

PHASORS AND NETWORK FUNCTIONS

Chapter 11 on *Network Functions* has been modified to provide a bridge between sinusoidal steady-state analysis in the s-domain and phasor circuit analysis in Chapter 8. This change provides flexibility to those wanting to use a *Laplace Early* approach by delaying the treatment of phasor-based ac circuits until after Chapter 11. (See separate section on Combined Edition below.) The section on transfer function design has been improved to place added emphasis on evaluating alternative design solutions.

FREQUENCY RESPONSE AND ACTIVE FILTERS

Chapter 12 on Frequency Response and Chapter 14 on Active Filters have been rewritten. Reviewers of previous editions indicated that they use Chapter 12 but may not have time for Chapter 14. Accordingly, Chapter 12 has been revised to be a stand-alone chapter on frequency response, Bode plots, first-order filter design, and second-order *RLC* circuit filters. Users also report that Chapter 14 is often useful to students in subsequent design courses where active filters may be needed. As a result, Chapter 14 has been re-titled *Active Filter Design* and rewritten to concentrate on multiple-pole active filters. Revised discussion of Sallen-Key filters, active notch filters, and Chebyshev high-pass filters make this chapter a better introduction and reference for active filter design. Both chapters have expanded design and evaluation examples as well as homework problems. A new Appendix D has been added to describe procedures for finding the poles of Butterworth and Chebyshev filters.

USING THE FIFTH EDITION

The fourth edition was made available in a *Traditional* version and a *Laplace Early* version. In the *Traditional* version, phasor analysis (Chapter 8) comes before the study of Laplace transform techniques (Chapters 9–14). In the *Laplace Early* version the topic flow goes directly from first- and second-order circuit transient response into the study of Laplace transform techniques. Phasors are introduced later as an alternative approach to sinusoidal steady-state analysis. The primary difference between the two versions is the location of the chapter introducing phasor analysis and the approach taken in the introductory sections of that chapter.

The fifth edition is designed to accommodate users of either version of the fourth edition. Those using the *Traditional* version can follow fifth edition chapter organization through Chapter 8 on phasor analysis, possibly delaying of Chapter 7 until the second semester. Those using the *Laplace Early* version can follow the same chapter organization through Chapter 7, skip Chapter 8, and proceed directly to the Laplace chapters. The fifth edition adds an introduction to phasor analysis in section 11-5 dealing with the sinusoidal steady state. As a result, Laplace early users can use the phasor analysis in Chapter 8 anytime after Chapter 11.

The next table shows suggested chapter sequencing for the traditional and Laplace early approaches for three different subject matter emphases. The second author uses the Traditional-Electronics sequence at the USAF Academy and has used the Laplace Early-Systems sequence at the University of Denver. Enough material is available in the printed text and in the Web appendices to construct other topic sequences.

	Semester 1							Semester 2							
Traditional	**1**	**2**	**3**	**4**	**5**	**6**	**8/7**								
Power								7/8	15	16	9	10	11	12	13
Systems								7/8	9	10	11	12	13	14	W1
Electronics								7/8	9	10	11	12	14	15	W2
Laplace Early	**1**	**2**	**3**	**4**	**5**	**6**	**7**								
Power								9	10	11	12	13	8	15	16
Systems								9	10	11	12	13	14	8	W1
Electronics								9	10	11	12	14	8	15	W2

ACKNOWLEDGMENTS

Individuals at John Wiley & Sons, Inc. involved with this edition include Associate Editor Kelly Boyle, Production Services Manager Jeanine Furino, and Marketing Manager Phyllis Cerys. The support of our former Executive Editor Bill Zobrist allowed us to continue to present our ideas about teaching circuits. It is a great pleasure to acknowldege the meticulous work of our Developmental Editor Karen Osborn. It has been our great good fortune to have Suzanne Ingrao as our Production Manager for all three Wiley editions of this book. This edition would not have been possible without her professionalism and honesty.

Over the years the following individuals have helped shape this work in many ways: Robert M. Anderson, Iowa State University; Doran J. Baker, Utah State University; James A. Barby, University of Waterloo; William E. Bennett, United States Naval Academy; Maqsood A. Chaudhry, California State University at Fullerton; Micheal Chier, Milwaukee School of Engineering; Don E. Cottrell, University of Denver; Robert Curtis, Ohio University; Micheal L. Daley, University of Memphis; Ronald R. Delyser, University of Denver; Prasad Enjeti, Texas A&M University; John C. Getty, University of Denver; James G. Gottling, Ohio State University; Frank Gross, Florida State University; Robert Kotiuga, Boston University; Hans H. Kuehl, University of Southern California; K.S.P. Kumar, University of Minnesota; Nicholas Kyriakopoulos, George Washington University; Michael Lightner, University of Colorado at Boulder; Jerry I. Lubell, Jaycor; Reinhold Ludwig, Worcester Polytechnic Institute; Lloyd W. Massengill, Vanderbilt University; Frank L. Merat, Case Western Reserve University; Richard L. Moat, Motorola; Gene Moriarty, San Jose State University; Dudley Outcalt, Milwaukee School of Engineering; Anil Pahwa, Kansas State University; Michael Polis, Oakland

University; Pradeep Ramuhalli, Michigan State University; William Rison, New Mexico Institute of Mining and Technology; Martin S. Roden, California State University at Los Angeles; Pat Sannuti, the State University of New Jersey; Alan Schneider, University of California at San Diego; Ali O. Shaban, California Polytechnic State University; Jacob Shekel, Northeastern University; Kadagattur Srinidhi, Northeastern University; Peter J. Tabolt, University of Massachusetts at Boston; Len Trombetta, University of Houston; David Voltmer, Rose-Hulman Institute of Technology; and Bruce F. Wollenberg, University of Minnesota. Special thanks go to James K. Kang of California State University at Pomona for his meticulous editing of the homework solutions for both the fourth and fifth editions.

The first author expresses his indebtedness to his wife Juanita whose proofreading and constructive comments are an integral part of this book.

CONTENTS

CHAPTER 1 • INTRODUCTION 1

1-1 About This Book 2
1-2 Symbols and Units 3
1-3 Circuit Variables 5
SUMMARY 10
PROBLEMS 11
INTEGRATING PROBLEMS 13

CHAPTER 2 • BASIC CIRCUIT ANALYSIS 14

2-1 Element Constraints 15
2-2 Connection Constraints 19
2-3 Combined Constraints 26
2-4 Equivalent Circuits 32
2-5 Voltage and Current Division 39
2-6 Circuit Reduction 45
2-7 Computer-aided Circuit Analysis 51
SUMMARY 56
PROBLEMS 57
INTEGRATING PROBLEMS 64

CHAPTER 3 • CIRCUIT ANALYSIS TECHNIQUES 66

3-1 Node-voltage Analysis 67
3-2 Mesh-current Analysis 81
3-3 Linearity Properties 90
3-4 Thévenin and Norton Equivalent Circuits 98
3-5 Maximum Signal Transfer 111
3-6 Interface Circuit Design 114
SUMMARY 125
PROBLEMS 126
INTEGRATING PROBLEMS 135

CHAPTER 4 • ACTIVE CIRCUITS 137

4-1 Linear Dependent Sources 138
4-2 Analysis of Circuits with Dependent Sources 141

4-3 The Transistor 152
4-4 The Operational Amplifier 157
4-5 OP AMP Circuit Analysis 164
4-6 OP AMP Circuit Design 180
4-7 OP AMP Circuit Applications 184
Summary 195
Problems 196
Integrating Problems 204

CHAPTER 5 • SIGNAL WAVEFORMS 207
5-1 Introduction 208
5-2 The Step Waveform 209
5-3 The Exponential Waveform 214
5-4 The Sinusoidal Waveform 218
5-5 Composite Waveforms 226
5-6 Waveform Partial Descriptors 231
Summary 239
Problems 240
Integrating Problems 243

CHAPTER 6 • CAPACITANCE AND INDUCTANCE 245
6-1 The Capacitor 246
6-2 The Inductor 253
6-3 Dynamic OP AMP Circuits 258
6-4 Equivalent Capacitance and Inductance 266
Summary 270
Problems 270
Integrating Problems 274

CHAPTER 7 • FIRST- AND SECOND-ORDER CIRCUITS 276
7-1 *RC* and *RL* Circuits 277
7-2 First-order Circuit Step Response 287
7-3 Initial and Final Conditions 295
7-4 First-order Circuit Sinusoidal Response 301
7-5 The Series *RLC* Circuit 308
7-6 The Parallel *RLC* Circuit 318
7-7 Second-order Circuit Step Response 323
7-8 Other Second-order Circuits 331
Summary 333
Problems 334
Integrating Problems 340

CHAPTER 8 • SINUSOIDAL STEADY-STATE RESPONSE 343
8-1 Sinusoids and Phasors 344
8-2 Phasor Circuit Analysis 349
8-3 Basic Circuit Analysis with Phasors 353
8-4 Circuit Theorems with Phasors 365
8-5 General Circuit Analysis with Phasors 375
8-6 Energy and Power 384
SUMMARY 390
PROBLEMS 390
INTEGRATING PROBLEMS 397

CHAPTER 9 • LAPLACE TRANSFORMS 399
9-1 Signal Waveforms and Transforms 400
9-2 Basic Properties and Pairs 404
9-3 Pole-zero Diagrams 412
9-4 Inverse Laplace Transforms 416
9-5 Some Special Cases 421
9-6 Circuit Response Using Laplace Transforms 426
9-7 Initial Value and Final Value Properties 434
SUMMARY 439
PROBLEMS 439
INTEGRATING PROBLEMS 443

CHAPTER 10 • *s*-DOMAIN CIRCUIT ANALYSIS 444
10-1 Transformed Circuits 445
10-2 Basic Circuit Analysis in the *s* Domain 454
10-3 Circuit Theorems in the *s* Domain 458
10-4 Node-voltage Analysis in the *s* Domain 467
10-5 Mesh-current Analysis in the *s* Domain 475
10-6 Summary of *s*-Domain Circuit Analysis 481
SUMMARY 486
PROBLEMS 487
INTEGRATING PROBLEMS 493

CHAPTER 11 • NETWORK FUNCTIONS 495
11-1 Definition of a Network Function 496
11-2 Network Functions of One- and Two-port Circuits 499
11-3 Network Functions and Impulse Response 510
11-4 Network Functions and Step Response 513
11-5 Network Functions and Sinusoidal Steady-state Response 518

11-6 Impulse Response and Convolution 523
11-7 Network Function Design 529
Summary 544
Problems 545
Integrating Problems 550

Chapter 12 • Frequency Response 552
12-1 Frequency-response Descriptors 553
12-2 Bode Plots 554
12-3 First-order Low-Pass and High-Pass Responses 557
12-4 Bandpass and Bandstop Responses 570
12-5 The Frequency Response of *RLC* Circuits 576
12-6 Bode Diagrams with Real Poles and Zeros 586
12-7 Bode Diagrams with Complex Poles and Zeros 597
12-8 Frequency Response and Step Response 604
Summary 608
Problems 609
Integrating Problems 614

Chapter 13 • Fourier Series 616
13-1 Overview of Fourier Analysis 617
13-2 Fourier Coefficients 618
13-3 Waveform Symmetries 627
13-4 Circuit Analysis Using the Fourier Series 630
13-5 RMS Value and Average Power 636
Summary 643
Problems 644
Integrating Problems 648

Chapter 14 • Active Filter Design 650
14-1 Active Filter 651
14-2 Second-order Low-pass and High-pass Filters 652
14-3 Second-order Bandpass and Bandstop Filters 660
14-4 Low-Pass Filter Design 665
14-5 Low-pass Filter Evaluation 681
14-6 High-Pass Filter Design 683
14-7 Bandpass and Bandstop Filters 690
Summary 694
Problems 695
Integrating Problems 698

CHAPTER 15 • MUTUAL INDUCTANCE 700
15-1 Coupled Inductors 701
15-2 The Dot Convention 703
15-3 Energy Analysis 706
15-4 The Ideal Transformer 708
15-5 Transformers in the Sinusoidal Steady State 714
15-6 Transformer Equivalent Circuits 719
Summary 721
Problems 722
Integrating Problems 725

CHAPTER 16 • POWER IN THE SINUSOIDAL STEADY STATE 726
16-1 Average and Reactive Power 727
16-2 Complex Power 729
16-3 AC Power Analysis 733
16-4 Load-flow Analysis 737
16-5 Three-phase Circuits 743
16-6 Three-phase AC Power Analysis 747
Summary 758
Problems 759
Integrating Problems 763

APPENDIX A—STANDARD VALUES A-1
APPENDIX B—SOLUTION OF LINEAR EQUATIONS A-2
APPENDIX C—COMPLEX NUMBERS A-12
APPENDIX D—BUTTERWORTH AND CHEBYCHEV POLES A-16

ANSWERS TO SELECTED PROBLEMS A-20
INDEX I-1

WEB APPENDICES
APPENDIX W1—FOURIER TRANSFORMS W-1
W1-1 Definition of Fourier Transforms W-1
W1-2 Laplace Transforms and Fourier Transforms W-7
W1-3 Basic Fourier Transform Properties and Pairs W-10
W1-4 Circuit Analysis Using Fourier Transforms W-18
W1-5 Impulse Response and Convolution W-22
W1-6 Parseval's Theorem W-26

SUMMARY W-32
PROBLEMS W-33

APPENDIX W2—TWO-PORT CIRCUITS W-37
W2-1 Introduction W-37
W2-2 Impedance Parameters W-39
W2-3 Admittance Parameters W-41
W2-4 Hybrid Parameters W-43
W2-5 Transmission Parameters W-46
W2-6 Two-Port Conversion W-49
W2-7 Two-Port Connections W-50
SUMMARY W-53
PROBLEMS W-54

CHAPTER 1

INTRODUCTION

The electromotive action manifests itself in the form of two effects which I believe must be distinguished from the beginning by a precise definition. I will call the first of these "electric tension," the second "electric current."

André-Marie Ampère, 1820,
French Mathematician/Physicist

Some History Behind This Chapter

André Ampère (1775–1836) was the first to recognize the importance of distinguishing between the electrical effects we now call voltage and current. He also invented the galvanometer, the forerunner of today's voltmeter and ammeter. A natural genius, he had mastered all the then-known mathematics by age 12. He is best known for defining the mathematical relationship between electric current and magnetism, now known as Ampère's law.

Why This Chapter Is Important Today

Welcome to the study of Linear Circuits. In this chapter you are introduced to the lexicon of electrical engineering. You will learn both the terminology and the variables that will be used throughout the book. Important concepts introduced here are voltage and current, the reference marks used to define them, and a voltage benchmark called ground.

Chapter Sections

1–1 About This Book

1–2 Symbols and Units

1–3 Circuit Variables

Chapter Learning Objectives

1-1 Electrical Symbols and Units (Sect. 1-2)

Given an electrical quantity described in terms of words, scientific notation, or decimal prefix notation, convert the quantity to an alternate description.

1-2 Circuit Variables (Sect. 1-3)

Given any two of the three signal variables (i, v, p) or the two basic variables (q, w), find the magnitude and direction (sign) of the unspecified variables.

1–1 About This Book

The basic purpose of this book is to introduce the analysis and design of linear circuits. Circuits are important in electrical engineering because they process electrical signals that carry energy and information. For the present we can define a **circuit** as an interconnection of electrical devices and a **signal** as a time-varying electrical entity. For example, the information stored on a compact disc is recovered in the CD-ROM player as electronic signals that are processed by circuits to generate audio and video outputs. In an electrical power system some form of stored energy is converted to electrical form and transferred to loads, where the energy is converted into the form required by the customer. The CD-ROM player and the electrical power system both involve circuits that process and transfer electrical signals carrying energy and information.

In this text we are primarily interested in **linear circuits**. An important feature of a linear circuit is that the amplitude of the output signal is proportional to the input signal amplitude. The proportionality property of linear circuits greatly simplifies the process of circuit analysis and design. Most circuits are only linear within a restricted range of signal levels. When driven outside this range they become nonlinear, and proportionality no longer applies. Although we will treat a few examples of nonlinear circuits, our attention is focused on circuits operating within their linear range.

Our study also deals with interface circuits. For the purposes of this book, we define an **interface** as a pair of accessible terminals at which signals may be observed or specified. The interface idea is particularly important with integrated circuit (IC) technology. Integrated circuits involve many thousands of interconnections, but only a small number are accessible to the user. Designing systems using integrated circuits involves interconnecting complex circuits that have only a few accessible terminals. This often involves relatively simple circuits whose purpose is to change signal levels or formats. Such interface circuits are intentionally introduced to ensure that the appropriate signal conditions exist at the connections between complex integrated circuits.

Course Objectives

This book is designed to help you develop the knowledge and application skills needed to solve three types of circuit problems: analysis, design, and evaluation. An **analysis** problem involves finding the output signals of a given circuit with known input signals. Circuit analysis is the foundation for understanding the interaction of signals and circuits. A **design** problem involves devising one or more circuits that perform a given signal-processing function. There usually are several possible solutions to a design problem. This leads to an **evaluation** problem which involves picking the best solution from among several candidates using factors such as cost, power consumption, and part counts. In real life the engineer's role is a blend of analysis, design, and evaluation, and in practice the boundaries between these categories are often blurred.

This text contains many worked examples to help you develop your problem-solving skills. The **examples** include a problem statement and provide the intermediate steps needed to obtain the final answer. The examples often treat analysis problems, although design and evaluation examples are included. This text also contains a number of **exercises** that include only the problem statement and the final answer. You should use the exercises to test your understanding of the circuit concepts discussed in the preceding section.

CHAPTER OBJECTIVES

At the end of each chapter we provide a carefully structured set of enabling skills called **chapter objectives**. Collectively, these objectives define the basic knowledge and understanding needed to master the topics covered in each chapter. The objectives explicitly state the expected behavior and are followed by a graduated set of homework problems designed to help you assess your level of achievement. Each objective also lists worked examples and exercises in the text that help you work the related homework problems. Once you understand all of the chapter objectives, you can move on to the integrating problems at the very end of the chapter. These problems require mastery of several chapter objectives and provide an opportunity to test your ability to deal with comprehensive problems.

1–2 SYMBOLS AND UNITS

Throughout this text we will use the international system (SI) of units. The SI system includes six fundamental units: meter (m), kilogram (kg), second (s), ampere (A), kelvin (K), and candela (cd). All the other units of measure can be derived from these six.

Like all disciplines, electrical engineering has its own terminology and symbology. The symbols used to represent some of the more important physical quantities and their units are listed in Table 1–1. It is not our purpose to define these quantities here, nor to offer this list as an item for memorization. Rather, the purpose of this table is merely to list in one place all the electrical quantities used in this book.

Numerical values in engineering range over many orders of magnitude. Consequently, the system of standard decimal prefixes in Table 1–2 is used. These prefixes on a unit abbreviation symbol indicate the power of 10 that is applied to the numerical value of the quantity.

Exercise 1–1

Given the pattern in the statement 1 kΩ = 1 kilohm = 1×10^3 ohms, fill in the blanks in the following statements using the standard decimal prefixes.

(a) _______ = _______ = 5×10^{-3} watts
(b) 10.0 dB = _______ = _______
(c) 3.6 ps = _______ = _______
(d) _______ = 0.03 microfarads = _______
(e) _______ = _______ gigahertz = 6.6×10^9 Hertz

Answers:

(a) 5.0 mW = 5 milliwatts
(b) 10.0 decibels = 1.0 bel
(c) 3.6 picoseconds = 3.6×10^{-12} seconds
(d) 30 nF or 0.03 μF = 30.0×10^{-9} Farads
(e) 6.6 GHz = 6.6 gigahertz

TABLE 1–1 SOME IMPORTANT QUANTITIES, THEIR SYMBOLS, AND UNIT ABBREVIATIONS

Quantity	Symbol	Unit	Unit Abbreviation
Time	t	second	s
Frequency	f	hertz	Hz
Radian frequency	ω	radian/second	rad/s
Phase angle	θ,ϕ	degree or radian	° or rad
Energy	w	joule	J
Power	p	watt	W
Charge	q	coulomb	C
Current	i	ampere	A
Electric field	$\mathscr{E}$	volt/meter	V/m
Voltage	v	volt	V
Impedance	Z	ohm	Ω
Admittance	Y	siemens	S
Resistance	R	ohm	Ω
Conductance	G	siemens	S
Reactance	X	ohm	Ω
Susceptance	B	siemens	S
Inductance, self	L	henry	H
Inductance, mutual	M	henry	H
Capacitance	C	farad	F
Magnetic flux	ϕ	weber	wb
Flux linkages	λ	weber-turns	wb-t
Power ratio	P	bel	B

TABLE 1–2 STANDARD DECIMAL PREFIXES

Multiplier	Prefix	Abbreviation
10^{18}	exa	E
10^{15}	peta	P
10^{12}	tera	T
10^{9}	giga	G
10^{6}	mega	M
10^{3}	kilo	k
10^{-1}	deci	d
10^{-2}	centi	c
10^{-3}	milli	m
10^{-6}	micro	μ
10^{-9}	nano	n
10^{-12}	pico	p
10^{-15}	femto	f
10^{-18}	atto	a

1–3 CIRCUIT VARIABLES

The underlying physical variables in the study of electronic systems are **charge** and **energy**. The idea of electrical charge explains the very strong electrical forces that occur in nature. To explain both attraction and repulsion, we say that there are two kinds of charge—positive and negative. Like charges repel, while unlike charges attract one another. The symbol q is used to represent charge. If the amount of charge is varying with time, we emphasize the fact by writing $q(t)$. In the international system (SI), charge is measured in **coulombs** (abbreviated C). The smallest quantity of charge in nature is an electron's charge ($q_E = 1.6 \times 10^{-19}$C). Thus, there are $1/q_E = 6.25 \times 10^{18}$ electrons in 1 coulomb of charge.

Electrical charge is a rather cumbersome variable to measure in practice. Moreover, in most situations the charges are moving, so we find it more convenient to measure the amount of charge passing a given point per unit time. If $q(t)$ is the cumulative charge passing through a point, we define a signal variable i called **current** as follows:

$$i = \frac{dq}{dt} \tag{1–1}$$

Current is a measure of the flow of electrical charge. It is the time rate of change of charge passing a given point in a circuit. The physical dimensions of current are coulombs per second. In the SI system, the unit of current is the **ampere** (abbreviated A). That is,

$$1 \text{ coulomb/second} = 1 \text{ ampere}$$

Since there are two types of electrical charge, there is a bookkeeping problem associated with the direction assigned to the current. In engineering it is customary to define the direction of current as the direction of the net flow of positive charge.

A second signal variable called **voltage** is related to the change in energy that would be experienced by a charge as it passes through a circuit. The symbol w is commonly used to represent energy. In the SI system of units, energy carries the units of **joules** (abbreviated J). If a small charge dq were to experience a change in energy dw in passing from point A to point B in a circuit, then the voltage v between A and B is defined as the change in energy per unit charge. We can express this definition in differential form as

$$v = \frac{dw}{dq} \tag{1–2}$$

Voltage does not depend on the path followed by the charge dq in moving from point A to point B. Furthermore, there can be a voltage between two points even if there is no charge motion, since voltage is a measure of how much energy dw would be involved if a charge dq was moved. The dimensions of voltage are joules per coulomb. The unit of voltage in the SI system is the **volt** (abbreviated V). That is,

$$1 \text{ joule/coulomb} = 1 \text{ volt}$$

The general definition of physical variable called **power** is the time rate of change of energy:

$$p = \frac{dw}{dt} \tag{1–3}$$

The dimensions of power are joules per second, which in the SI system is called a **watt** (abbreviated W). In electrical circuits it is useful to relate power to the signal variables current and voltage. Using the chain rule, Eq. (1–3) can be written as

$$p = \left(\frac{dw}{dq}\right)\left(\frac{dq}{dt}\right) \tag{1–4}$$

Now using Eqs. (1–1) and (1–2), we obtain

$$p = vi \tag{1–5}$$

The electrical power associated with a situation is determined by the product of voltage and current. The total energy transferred during the period from t_1 to t_2 is found by solving for dw in Eq. (1–3) and then integrating

$$w_{\mathrm{T}} = \int_{w_1}^{w_2} dw = \int_{t_1}^{t_2} p\,dt \tag{1–6}$$

EXAMPLE 1–1

The electron beam in the cathode-ray tube shown in Figure 1–1 carries 10^{14} electrons per second and is accelerated by a voltage of 50 kV. Find the power in the electron beam.

FIGURE 1–1

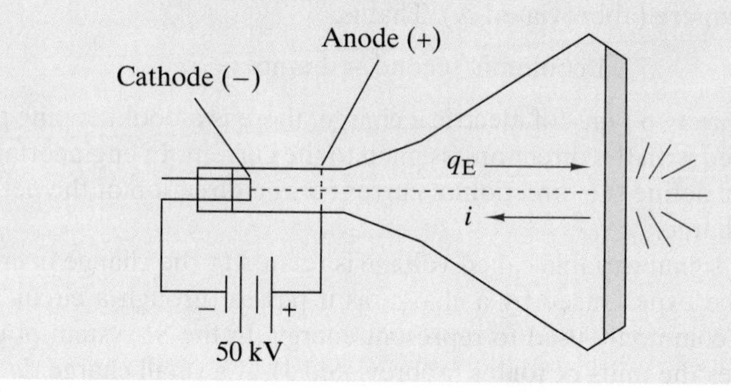

SOLUTION:

Since current is the rate of charge flow, we can find the net current by multiplying the charge of the electron q_{E} by the rate of electron flow dn_{E}/dt.

$$i = q_{\mathrm{E}}\frac{dn_{\mathrm{E}}}{dt} = (1.6 \times 10^{-19})\,(10^{14}) = 1.6 \times 10^{-5}\ \mathrm{A}$$

Therefore, the beam power is

$$p = vi = (50 \times 10^{3})(1.6 \times 10^{-5}) = 0.8\ \mathrm{W}$$ ■

EXAMPLE 1–2

The current through a circuit element is 50 mA. Find the total charge and the number of electrons transferred during a period of 100 ns.

SOLUTION:
The relationship between current and charge is given in Eq. (1–1) as

$$i = \frac{dq}{dt}$$

Since the current i is given, we calculate the charge transferred by solving this equation for dq and then integrating

$$q_T = \int_{q_1}^{q_2} dq = \int_0^{10^{-7}} i\,dt$$

$$= \int_0^{10^{-7}} 50 \times 10^{-3}\,dt = 50 \times 10^{-10}\text{C} = 5\text{ nC}$$

There are $1/q_E = 6.25 \times 10^{18}$ electrons/coulomb, so the number of electrons transferred is

$$n_E = (5 \times 10^{-9}\text{ C})(6.25 \times 10^{18}\text{ electrons/C}) = 31.2 \times 10^9\text{ electrons} \quad ■$$

Exercise 1–2

A device dissipates 100 W of power. How much energy is delivered to it in 10 seconds?

Answer: 1 kJ

Exercise 1–3

The graph in Figure 1–2(a) shows the charge $q(t)$ flowing past a point in a wire as a function of time.

(a) Find the current $i(t)$ at t = 1, 2.5, 3.5, 4.5, and 5.5 ms.
(b) Sketch the variation of $i(t)$ versus time.

Answers:

(a) −10 nA, +40 nA, 0 nA, −20 nA, 0 nA.
(b) The variations in $i(t)$ are shown in Figure 1–2(b).

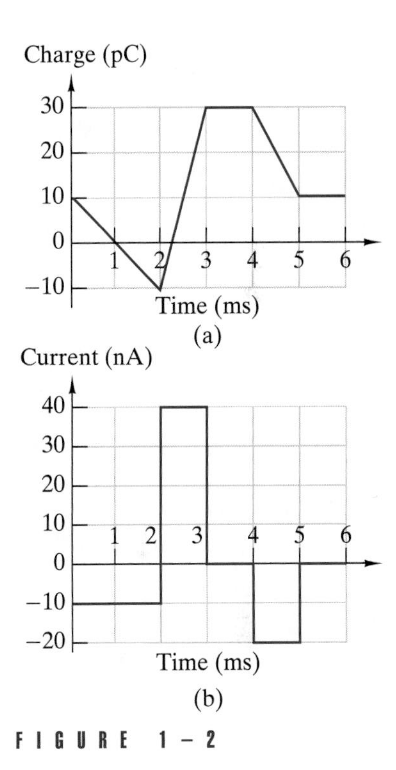

FIGURE 1–2

THE PASSIVE SIGN CONVENTION

We have defined three circuit variables (current, voltage, and power) using two basic variables (charge and energy). Charge and energy, like mass, length, and time, are basic concepts of physics that provide the scientific foundation for electrical engineering. However, engineering problems rarely involve charge and energy directly, but are usually stated in terms of voltage, current, and power. The reason for this is simple: The circuit variables are much easier to measure and therefore are the most useful working variables in engineering practice.

At this point, it is important to stress the physical differences between current and voltage variables. Current is a measure of the time rate of charge passing a point in a circuit. We think of current as a *through variable,* since it describes the flow of electrical charge through a point in a circuit. On the other hand, voltage is not measured at a single point, but rather between two points or across an electrical device. Consequently,

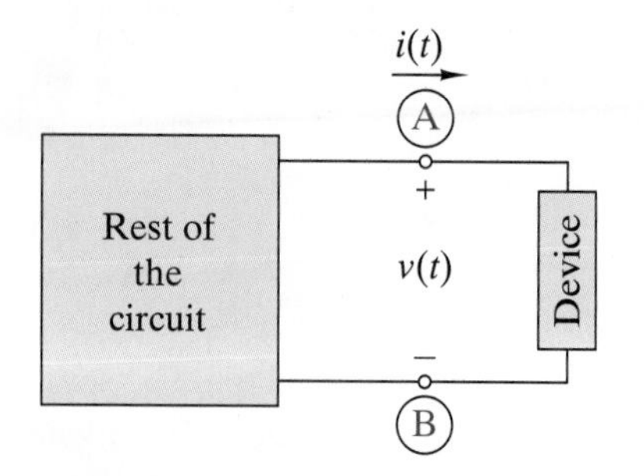

FIGURE 1–3 *Voltage and current reference marks for a two-terminal device.*

we think of voltage as an *across variable* that inherently involves two points.

The arrow and the plus and minus symbols in Figure 1–3 are *reference marks* that define the positive directions for the current and voltage associated with an electrical device. These reference marks do not represent an assertion about what is happening physically in the circuit. The response of an electrical circuit is determined by physical laws, not by the reference marks assigned to the circuit variables.

The reference marks are benchmarks assigned at the beginning of the analysis. When the actual direction and reference direction agree, the answers found by circuit analysis will have positive algebraic signs. When they disagree, the algebraic signs of the answers will be negative. For example, if circuit analysis reveals that the current variable in Figure 1–3 is positive [i.e., $i(t) > 0$], then the sign of this answer, together with the assigned reference direction, indicates that the current passes through point A in Figure 1–3 from left to right. Conversely, when analysis reveals that the current variable is negative, then this result, combined with the assigned reference direction, tells us that the current passes through point A from right to left. In summary, the algebraic sign of the answer together with arbitrarily assigned reference marks tell us the actual directions of a voltage or current variable.

In Figure 1–3, the current reference arrow enters the device at the terminal marked with the plus voltage reference mark. This orientation is called the **passive sign convention**. Under this convention, the power $p(t)$ is positive when the device absorbs power and is negative when it delivers power to the rest of the circuit. Since $p(t) = v(t) \times i(t)$, a device absorbs power when the voltage and current variables have the same algebraic sign and delivers power when they have opposite signs. Certain devices, such as heaters (a toaster, for example), can only absorb power. The voltage and current variables associated with these devices must always have the same algebraic sign. On the other hand, a battery absorbs power [$p(t) > 0$] when it is being charged and delivers power [$p(t) < 0$] when it is discharging. Thus, the voltage and current variables for a battery can have the same or opposite algebraic signs.

In a circuit some devices absorb power and others deliver power, but the sum of the power in all of the devices in the circuit is zero. This is more than just a conservation-of-energy concept. When electrical devices are interconnected to form a circuit, the only way that power can enter or leave the circuit is via the currents and voltages at device terminals. The existence of a power balance in a circuit is one method of checking calculations.

The passive sign convention is used throughout this book. It is also the convention used by circuit simulation computer programs.[1] To interpret correctly the results of circuit analysis, it is important to remember that the reference marks (arrows and plus/minus signs) are reference directions, not indications of the circuit response. The actual direction of a response is determined by comparing its reference direction with the algebraic signs of the result predicted by circuit analysis based on physical laws.

---

1 We discuss computer-aided circuit analysis in subsequent chapters.

GROUND

Since voltage is defined between two points, it is often useful to define a common voltage reference point called **ground**. The voltages at all other points in a circuit are then defined with respect to this common reference point. We indicate the voltage reference point using the ground symbol shown in Figure 1–4. Under this convention we sometimes refer to the variables $v_A(t)$, $v_B(t)$, and $v_C(t)$ as the voltages at points A, B, and C, respectively. This terminology appears to contradict the fact that voltage is an across variable that involves two points. However, the terminology means that the variables $v_A(t)$, $v_B(t)$, and $v_C(t)$ are the voltages defined between points A, B, and C, and the common voltage reference point at point G.

Using a common reference point for across variables is not an idea unique to electrical circuits. For example, the elevation of a mountain is the number of feet or meters between the top of the mountain and a common reference point at sea level. If a geographic point lies below sea level, then its elevation is assigned a negative algebraic sign. So it is with voltages. If circuit analysis reveals that the voltage variable at point A is negative [i.e., $v_A(t) < 0$], then this fact together with the reference marks in Figure 1–4 indicate that the potential at point A is less than the ground potential.

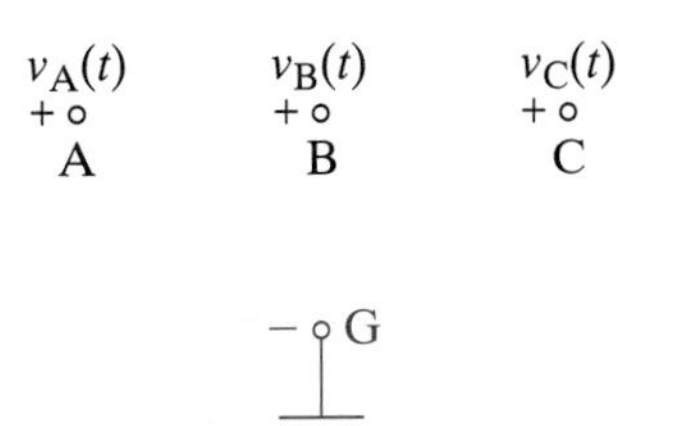

FIGURE 1–4 *Ground symbol indicates a common voltage reference point.*

EXAMPLE 1–3

Figure 1–5 shows a circuit formed by interconnecting five devices, each of which has two terminals. A voltage and current variable has been assigned to each device using the passive sign convention. The working variables for each device are observed to be as follows:

	DEVICE 1	DEVICE 2	DEVICE 3	DEVICE 4	DEVICE 5
v	+100 V	?	+25 V	+75 V	–75 V
i	?	+5 mA	+5 mA	?	+5 mA
p	–1 W	+0.5 W	?	0.75 W	?

(a) Find the missing variable for each device and state whether the device is absorbing or delivering power.
(b) Check your work by showing that the sum of the device powers is zero.

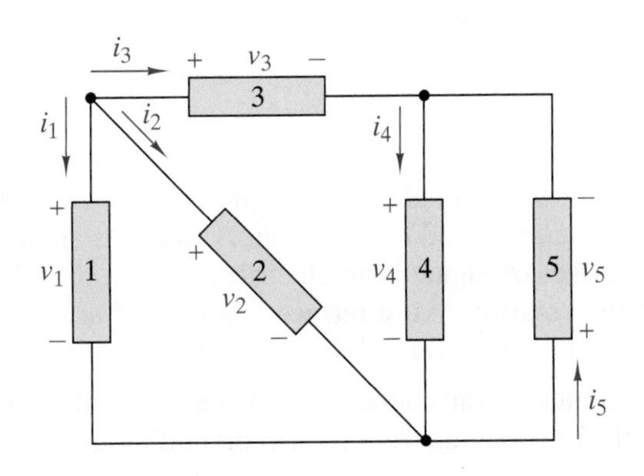

FIGURE 1–5

SOLUTION:

(a) We use $p = vi$ to solve for the missing variable since two of the three circuit variables are given for each device.

Device 1: $i_1 = p_1/v_1 = -1/100 = -10$ mA [$p(t) < 0$, delivering power]

Device 2: $v_2 = p_2/i_2 = 0.5/0.005 = 100$ V [$p(t) > 0$, absorbing power]

Device 3: $p_3 = v_3 i_3 = 25 \times 0.005 = 0.125$ W [$p(t) > 0$, absorbing power]

Device 4: $i_4 = p_4/v_4 = 0.75/75 = 10$ mA [$p(t) > 0$, absorbing power]

Device 5: $p_5 = v_5 i_5 = -75 \times 0.005 = -0.375$ W [$p(t) < 0$, delivering power]

(b) Summing the device powers yields

$$p_1 + p_2 + p_3 + p_4 + p_5 = -1 + 0.5 + 0.125 + 0.75 - 0.375$$
$$= +1.375 - 1.375 = 0$$

This example shows that the sum of the power absorbed by devices is equal in magnitude to the sum of the power supplied by devices. A power balance always exists in the types of circuits treated in this book and can be used as an overall check of circuit analysis calculations. ■

Exercise 1–4

The working variables of a set of two-terminal electrical devices are observed to be as follows:

	DEVICE 1	DEVICE 2	DEVICE 3	DEVICE 4	DEVICE 5
v	+10 V	?	−15 V	+5 V	?
i	−3 A	−3 A	+10 mA	?	−12 mA
p	?	+40 W	?	+10 mW	−120 mW

Using the passive sign convention, find the magnitude and sign of the unknown variable and state whether the device is absorbing or delivering power.

Answers:

Device 1: $p = -30$ W (delivering power)
Device 2: $v = -13.3$ V (absorbing power)
Device 3: $p = -150$ mW (delivering power)
Device 4: $i = +2$ mA (absorbing power)
Device 5: $v = +10$ V (delivering power)

SUMMARY

- Circuits are important in electrical engineering because they process signals that carry energy and information. A **circuit** is an interconnection of electrical devices. A **signal** is an electrical current or voltage that carries energy or information. An **interface** is a pair of accessible terminals at which signals may be observed or specified.
- This book defines overall course objectives at the analysis, design, and evaluation levels. In **circuit analysis** the circuit and input signals are given and

the object is to find the output signals. The object of **circuit design** is to devise one or more circuits that produce prescribed output signals for given input signals. The **evaluation** problem involves appraising alternative circuit designs using criteria such as cost, power consumption, and parts count.

- Charge (q) and energy (w) are the basic physical variables involved in electrical phenomena. Current (i), voltage (v), and power (p) are the derived variables used in circuit analysis and design. In the SI system, charge is measured in coulombs (C), energy in joules (J), current in amperes (A), voltage in volts (V), and power in watts (W).
- **Current** is defined as dq/dt and is a measure of the flow of electrical charge. **Voltage** is defined as dw/dq and is a measure of the energy required to move a small charge from one point to another. **Power** is defined as dw/dt and is a measure of the rate at which energy is being transferred. Power is related to current and voltage as $p = vi$.
- The **reference marks** (arrows and plus/minus signs) assigned to a device are reference directions, not indications of the way a circuit responds. The actual direction of the response is determined by comparing the reference direction and the algebraic sign of the answer found by circuit analysis using physical laws.
- Under the **passive sign convention**, the current reference arrow is directed toward the terminal with the positive voltage reference mark. Under this convention, the device power is positive when it absorbs power and is negative when it delivers power. When current and voltage have the same (opposite) algebraic signs, the device is absorbing (delivering) power.

PROBLEMS

OBJECTIVE 1–1 ELECTRICAL SYMBOLS AND UNITS (SECT. 1–2)

Given an electrical quantity described in terms of words, scientific notation, or decimal prefix notation, convert the quantity to an alternate description.
See Exercise 1–1

1–1 Express the following quantities to the nearest standard prefix using no more than three digits.
(a) 52,000 W
(b) 92.5×10^4 Hz
(c) 200×10^{-7} s
(d) 2500 V

1–2 Express the following quantities to the nearest standard prefix using no more than three digits.
(a) 0.000110 A
(b) 200×10^5 C
(c) 2.02×10^8 J
(d) 3,200,000 Ω

1–3 An ampere-hour (Ah) meter measures the time-integral of the current in a conductor. During an 8-hour period a certain meter records 3300 Ah. Find the number of coulombs that flowed through the meter during the recording period.

1–4 Electric power companies measure energy consumption in kilowatt-hours, denoted kWh. One kilowatt-hour is the amount of energy transferred by 1 kW of power in a time period of 1 hour. A power company billing statement reports a user's total energy usage to be 1489 kWh. Find the number of joules used during the billing period.

1–5 Fill in the blanks in the following statements.
(a) To convert capacitance from picofarads to microfarads, multiply by ________.
(b) To convert resistance from megohms to kilohms, multiply by ________.
(c) To convert voltage from millivolts to volts, multiply by ________.
(d) To convert energy from megajoules to joules, multiply by ________.

OBJECTIVE 1–2 CIRCUIT VARIABLES (SECT. 1–3)

Given any two of the three signal variables (i, v, p) or the two basic variables (q, w), find the magnitude and direction (sign) of the unspecified variables.
See Examples 1–1, 1–2, 1–3 and Exercises 1–1, 1–2, 1–3

1–6 A wire carries a constant current of 120 mA. How many coulombs flow through the wire in 3 s?

1–7 The net positive charge flowing through a device is $q(t) = 10 + 3t$ mC. Find the current through the device.

1–8 Figure P1–8 shows a plot of the net positive charge flowing in a wire versus time. Sketch the corresponding current during the same time period.

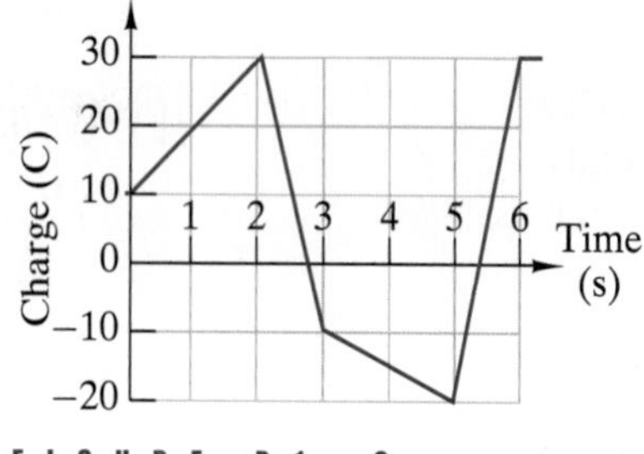

FIGURE P1–8

1–9 The net positive charge flowing through a device varies as $q(t) = 3t^2$ C. Find the current through the device at $t = 0$ s, $t = 1$ s, and $t = 3$ s.

1–10 The current through a device is given by $i(t) = 0.05t$ A. How many coulombs flow through the device between time $t = 0$ s and $t = 5$ s?

1–11 For $t \leq 5$ s, the current through a device is $i(t) = 10$ A. For $5 < t \leq 10$ s, the current is $i(t) = 20 - 2t$ A, and $i(t) = 0$ for $t > 10$. Sketch $i(t)$ versus time and find the total charge through the device between $t = 0$ s and $t = 10$ s.

1–12 The charge flowing through a device is $q(t) = 1 - e^{-500t}$ μC. Sketch the current through the device versus time for $t > 0$.

1–13 The 12-V automobile battery in Figure Pl–13 has an output capacity of 100 ampere-hours (Ah) when connected to a head lamp that absorbs 200 watts of power. Assume the battery voltage is constant.

(a) Find the current supplied by the battery.

(b) How long can the battery power the headlight?

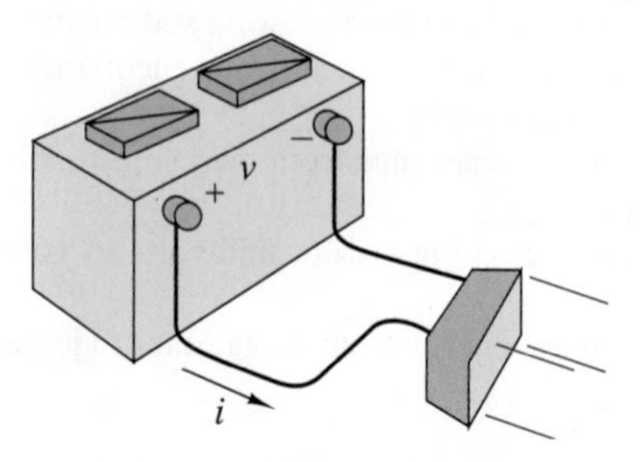

FIGURE P1–13

1–14 An incandescent lamp absorbs 60 W when connected to a 120-V source.

(a) Find the current through the lamp.

(b) Find the cost of operating the lamp for 24 hours when electricity costs 6.8 cents/kWh.

1–15 The current through a device is zero for $t < 0$ and is $i(t) = 3e^{-2t}$ A for $t \geq 0$. Find the charge $q(t)$ flowing through the device for $t \geq 0$.

1–16 The current through and voltage across a two-terminal device are $i(t) = 10 - 5\sin(10t)$ mA and $v = 15$ V. Find the maximum and minimum power delivered to the device.

1–17 When illuminated, the $i - v$ relationship for a photocell is $i = e^v - 10$A. For $v = -2, 2,$ and 3 V, find the device power and state whether it is absorbing or delivering power.

1–18 A new 1.5-V AA battery delivers 40 kJ of energy during its lifetime. How long will the battery last in an application that draws 10 mA continuously. Assume the battery voltage is constant.

1–19 The maximum power the device can dissipate is 0.5 W. Determine the maximum current allowed by the device power rating when the voltage is 15 V.

1–20 A constant current of 2 A charges a battery for 4 hours. During the charging period, the battery voltage is $v(t) = 12 - 2e^{-t}$, where t is in hours. Determine the energy stored in the battery.

1–21 Two electrical devices are connected as shown in Figure P1–21. Using the reference marks shown in the figure, find the power transferred and state whether the power is transferred from A to B or B to A when

(a) $v = +33$ V and $i = -2.2$ A

(b) $v = -12$ V and $i = -1.2$ mA

(c) $v = +37.5$ V and $i = +40$ mA

(d) $v = -15$ V and $i = -43$ mA

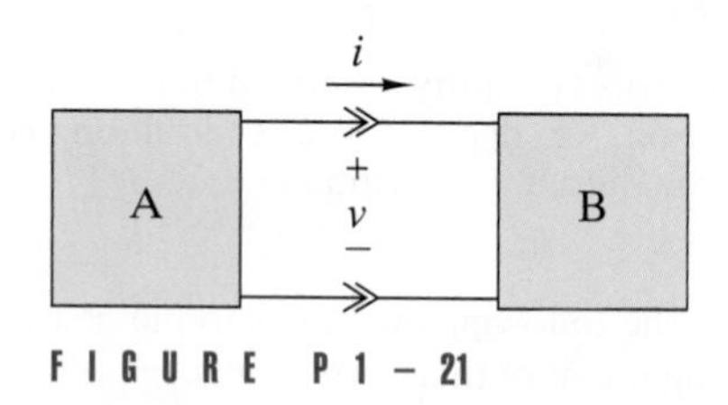

FIGURE P1–21

1–22 Figure P1–22 shows an electric circuit with a voltage and a current variable assigned to each of the six devices. The device voltages and currents are observed to be

	v(V)	i(A)
Device 1	15	−1
Device 2	5	1
Device 3	10	2
Device 4	−10	−1
Device 5	20	−3
Device 6	20	2

Find the power associated with each device and state whether the device is absorbing or delivering power. Use the power balance to check your work.

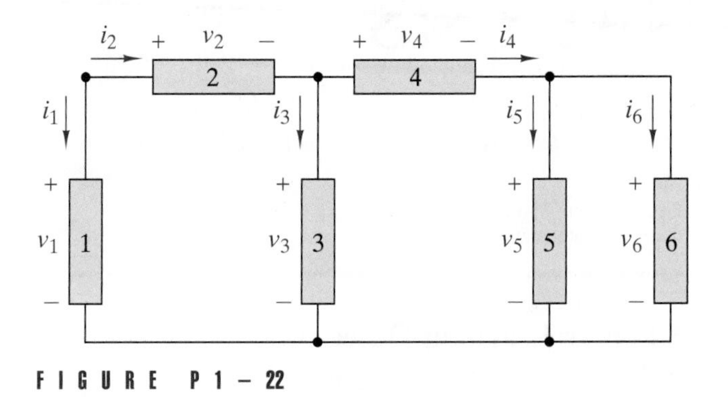

FIGURE P1–22

1–23 Figure Pl–22 shows an electric circuit with a voltage and a current variable assigned to each of the six devices. Use power balance to find v_4 when $v_1 = 20$ V, i_1 −2 A, $p_2 = 20$ W, $p_3 = 10$ W, $i_4 = 1$ A, and $p_5 = p_6 = 2.5$ W. Is device 4 absorbing or delivering power?

1–24 Using the passive sign convention, the voltage across a device is $v(t) = 5\cos(10t)$ V and the current through the device $i(t) = 0.5\sin(10t)$ A. Calculate the device power at $t = 0.2$ s and $t = 0.4$ s and state whether the device is absorbing or delivering power.

1–25 For $t \geq 0$, the voltage across and power absorbed by a two-terminal device are $v(t) = 2e^{-t}$ V and $p(t) = 40e^{-2t}$ mW. Find the total charge delivered to the device for $t \geq 0$.

INTEGRATING PROBLEMS

1–26 Power Ratio in dB
In complete darkness the voltage across and current through a two-terminal light detector are +5.6 V and +8 nA. In full sunlight the voltage and current are +0.9 V and +4 mA. Express the light/dark power ratio of the device in decibels (dB), where the power ratio in dB is

$$PR_{dB} = 10\log_{10}(p_2/p_1)$$

1–27 AC to DC Converter
A manufacturer's data sheet for the converter in Figure P1–27 states that the output voltage is $v_{dc} = 5$ V when the input voltage $v_{ac} = 120$ V. When the load draws a current $i_{dc} = 40$ A, the input power $p_{ac} = 300$ W. Find the efficiency of the converter.

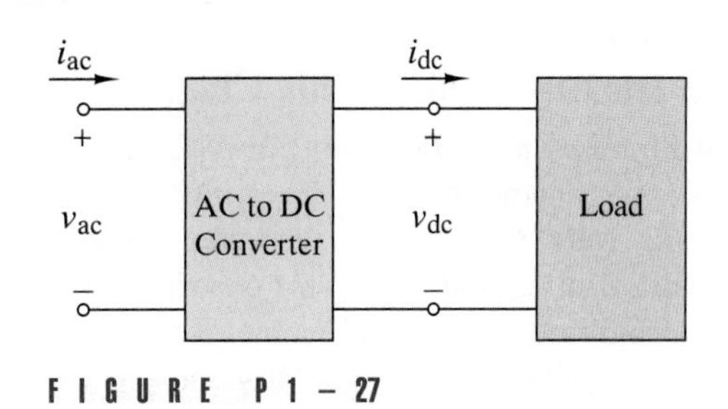

FIGURE P1–27

1–28 Charge Storage Device
A capacitor is a two-terminal device that can store electric charge. In a linear capacitor the amount of charge stored is proportional to the voltage across the device. For a particular device the proportionality is $q(t) = 10^{-7}v(t)$. If $v(t) = 0$ for $t < 0$ and $v(t) = 10(1 - e^{-5000t})$ for $t \geq 0$, find the energy stored in the device at $t = 200$ μs.

CHAPTER 2

BASIC CIRCUIT ANALYSIS

The equation S = A/L shows that the current of a voltaic circuit is subject to a change, by each variation originating either in the magnitude of a tension or in the reduced length of a part, which latter is itself again determined, both by the actual length of the part as well as its conductivity and its section.

Georg Simon Ohm, 1827,
German Mathematician/Physicist

Some History Behind This Chapter

Georg Simon Ohm (1789–1854) discovered the law that now bears his name in 1827. His results drew heavy criticism and were not generally accepted for many years. Fortunately, the importance of his contribution was eventually recognized during his lifetime. He was honored by the Royal Society of England in 1841 and appointed a Professor of Physics at the University of Munich in 1849.

Why This Chapter Is Important Today

A circuit is an interconnection of electric devices that performs a useful function. This chapter introduces some basic tools you will need to analyze and design electric circuits. You will also be introduced to several important electric devices that control currents and voltages in a circuit. These devices range from everyday things like batteries to special integrated circuits that meter out predetermined voltages or currents.

Chapter Sections

2–1 Element Constraints

2–2 Connection Constraints

2–3 Combined Constraints

2–4 Equivalent Circuits

2–5 Voltage and Current Division

2–6 Circuit Reduction

2–7 Computer-Aided Circuit Analysis

Chapter Learning Objectives

2-1 Element Constraints (Sect. 2-1)

Given a two-terminal element with one or more electrical variables specified, use the element $i-v$ constraint to find the magnitude and direction of the unknown variables.

2-2 Connection Constraints (Sect. 2-2)

Given a circuit composed of two-terminal elements:

(a) Identify nodes and loops in the circuit.
(b) Identify elements connected in series and in parallel.
(c) Use Kirchhoff's laws (KCL and KVL) to find selected signal variables.

2-3 Combined Constraints (Sect. 2-3)

Given a linear resistance circuit, use the element constraints and connection constraints to find selected signal variables.

2-4 Equivalent Circuits (Sect. 2-4)

Given a linear resistance circuit, find an equivalent circuit at a specified pair of terminals.

2-5 Voltage and Current Division (Sect. 2-5)

(a) Given a linear resistance circuit with elements connected in series or in parallel, use voltage or current division to find specified voltages or currents.
(b) Design a voltage or current divider that delivers specified output signals.

2-6 Circuit Reduction (Sect. 2-6)

Given a linear resistance circuit, find selected signal variables using successive application of series and parallel equivalence, source transformations, and voltage and current division.

2–1 ELEMENT CONSTRAINTS

A **circuit** is a collection of interconnected electrical devices. An electrical **device** is a component that is treated as a separate entity. The rectangular box in Figure 2–1 is used to represent any one of the two-terminal devices used to form circuits. A two-terminal device is described by its ***i–v* characteristic**; that is, by the relationship between the voltage across and current through the device. In most cases the relationship is complicated and nonlinear, so we use a linear model that approximates the dominant features of a device.

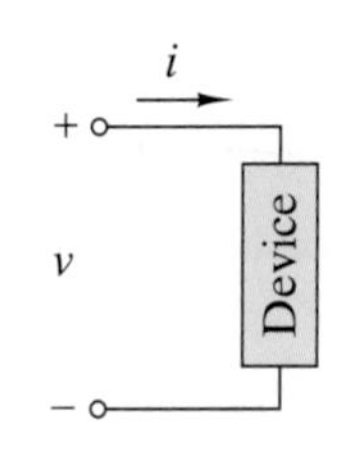

FIGURE 2–1 *Voltage and current reference marks for a two-terminal device.*

To distinguish between a device (the real thing) and its model (an approximate stand-in), we call the model a circuit **element**. Thus, a device is an article of hardware described in manufacturers' catalogs and parts specifications. An element is a model described in textbooks on circuit analysis. This book is no exception, and a catalog of circuit elements will be introduced as we go on. A discussion of real devices and their models is contained in Appendix A.

THE LINEAR RESISTOR

The first element in our catalog is a linear model of the device described in Figure 2–2. The actual *i–v* characteristic of this device is shown in Figure 2–2(b). To model this curve accurately across the full operating range shown in the figure would require at least a cubic equation. However, the graph in Figure 2–2(b) shows that a straight line is a good approximation to the *i–v* characteristic if we operate the device within its linear range. The power rating of the device limits the range over which the *i–v* characteristics can be represented by a straight line through the origin.

For the passive sign convention used in Figure 2–2(a), the equations describing the *linear resistor* element are

$$v = Ri \quad \text{or} \quad i = Gv \tag{2–1}$$

where R and G are positive constants that are reciprocally related.

$$G = \frac{1}{R} \tag{2–2}$$

The relationships in Eq. (2–1) are collectively known as **Ohm's law**. The parameter R is called **resistance** and has the unit **ohms**, Ω. The parameter G is called **conductance**, with the unit **siemens**, S. In earlier times the unit of conductance was cleverly called the mho, with a unit abbreviation symbol ℧ (ohm spelled backward and the ohm symbol upside down). Note that Ohm's law presumes that the passive sign convention is used to assign the reference marks to voltage and current.

The Ohm's law model is represented graphically by the black straight line in Figure 2–2(b). The *i–v* characteristic for the Ohm's law model defines a circuit element that is said to be linear and bilateral. **Linear** means that the defining characteristic is a straight line through the origin. Elements whose characteristics do not pass through the origin or are not a straight line are said to be **nonlinear**. **Bilateral** means that the *i–v* characteristic curve has odd symmetry about the origin.[1] With a bilateral resistor, reversing the polarity of the applied voltage reverses the direction but not the

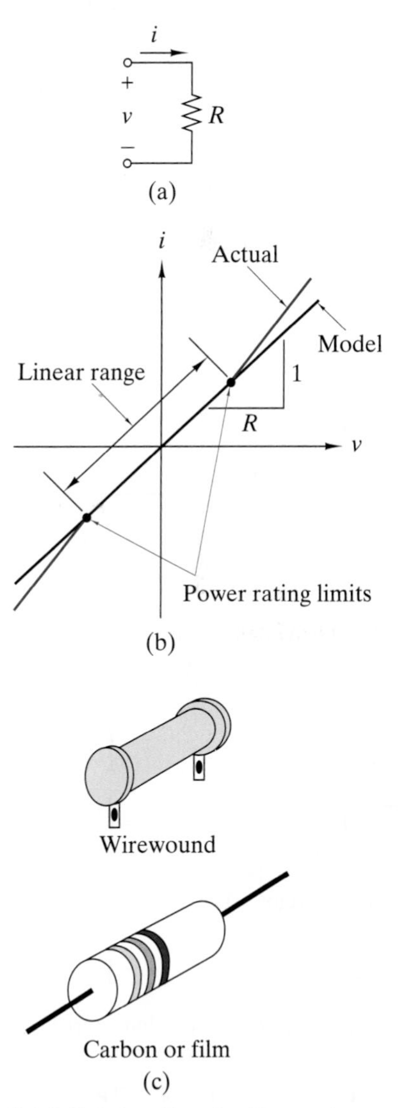

FIGURE 2–2 *The resistor: (a) Circuit symbol. (b)* i–v *characteristics. (c) Some actual devices.*

---

1 A curve $i = f(v)$ has odd symmetry if $f(-v) = -f(v)$.

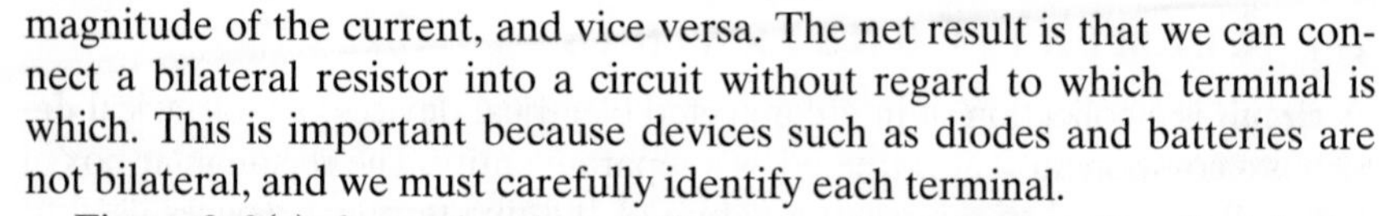

magnitude of the current, and vice versa. The net result is that we can connect a bilateral resistor into a circuit without regard to which terminal is which. This is important because devices such as diodes and batteries are not bilateral, and we must carefully identify each terminal.

Figure 2–2(c) shows sketches of discrete resistor devices. Detailed device characteristics and fabrication techniques are discussed in Appendix A.

The power associated with the resistor can be found from $p = vi$. Using Eq. (2–1) to eliminate v from this relationship yields

$$p = i^2R \tag{2–3}$$

or using the same equations to eliminate i yields

$$p = v^2G = \frac{v^2}{R} \tag{2–4}$$

Since the parameter R is positive, these equations tell us that the power is always nonnegative. Under the passive sign convention, this means that the resistor **always absorbs power**.

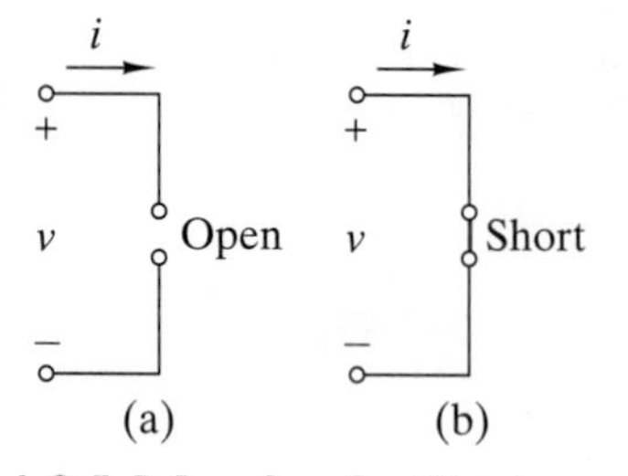

FIGURE 2–3 *Circuit symbols: (a) Open-circuit symbol. (b) Short-circuit symbol.*

EXAMPLE 2–1

A resistor operates as a linear element as long as the voltage and current are within the limits defined by its power rating. Suppose we have a 47-kΩ resistor with a power rating of 0.25 W. Determine the maximum current and voltage that can be applied to the resistor and remain within its linear operating range.

SOLUTION:

Using Eq. (2–3) to relate power and current, we obtain

$$I_{MAX} = \sqrt{\frac{P_{MAX}}{R}} = \sqrt{\frac{0.25}{47 \times 10^3}} = 2.31 \text{ mA}$$

Similarly, using Eq. (2–4) to relate power and voltage, we obtain

$$V_{MAX} = \sqrt{RP_{MAX}} = \sqrt{47 \times 10^3 \times 0.25} = 108 \text{ V}$$

■

OPEN AND SHORT CIRCUITS

The next two circuit elements can be thought of as limiting cases of the linear resistor. Consider a resistor R with a voltage v applied across it. Let's calculate the current i through the resistor for different values of resistance. If $v = 10$ V and $R = 1$ Ω, using Ohm's law we readily find that $i = 10$ A. If we increase the resistance to 100 Ω, we find i has decreased to 0.1 A or 100 mA. If we continue to increase R to 1 MΩ, i becomes a very small 10 μA. Continuing this process, we arrive at a condition where R is very nearly infinite and i just about zero. When the current $i = 0$, we call the special value of resistance (i.e., $R = \infty$ Ω) an **open circuit**. Similarly, if we reduce R until it approaches zero, we find that the voltage is very nearly zero. When $v = 0$, we call the special value of resistance (i.e., $R = 0$ Ω), a **short circuit**. The circuit symbols for these two elements are shown in Figure 2–3. In circuit analysis the elements in a circuit model are assumed to be interconnected by zero-resistance wire (that is, by short circuits).

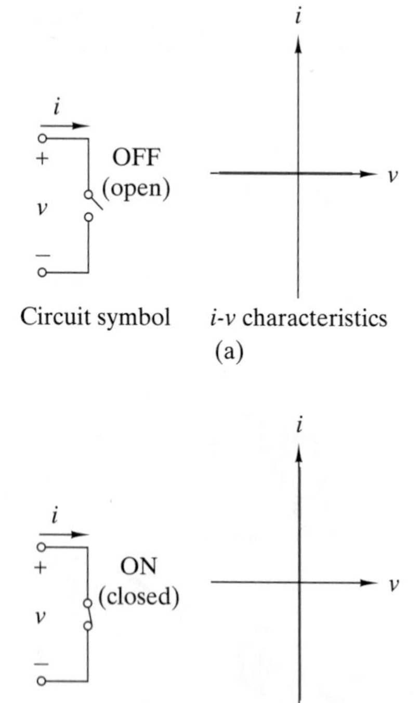

FIGURE 2–4 *The circuit symbol and i–v characteristics of an ideal switch: (a) Switch OFF. (b) Switch ON.*

THE IDEAL SWITCH

A switch is a familiar device with many applications in electrical engineering. The **ideal switch** can be modeled as a combination of an open- and a short-circuit element. Figure 2–4 shows the circuit symbol and the i–v characteristic of an ideal switch. When the switch is closed,

$$v = 0 \quad \text{and} \quad i = \text{any value} \tag{2–5a}$$

and when it is open,

$$i = 0 \quad \text{and} \quad v = \text{any value} \tag{2–5b}$$

When the switch is closed, the voltage across the element is zero and the element will pass any current that may result. When open, the current is zero and the element will withstand any voltage across its terminals. The power is always zero for the ideal switch, since the product $vi = 0$ when the switch is either open ($i = 0$) or closed ($v = 0$). Actual switch devices have limitations, such as the maximum current they can safely carry when closed and the maximum voltage they can withstand when open. The switch is operated (opened or closed) by some external influence, such as a mechanical motion, temperature, pressure, or an electrical signal.

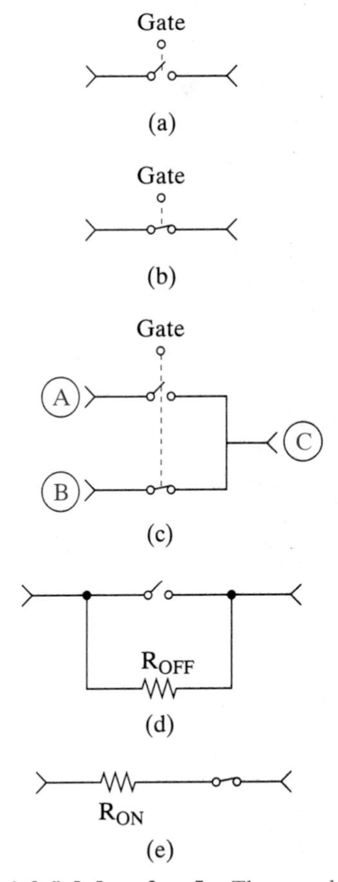

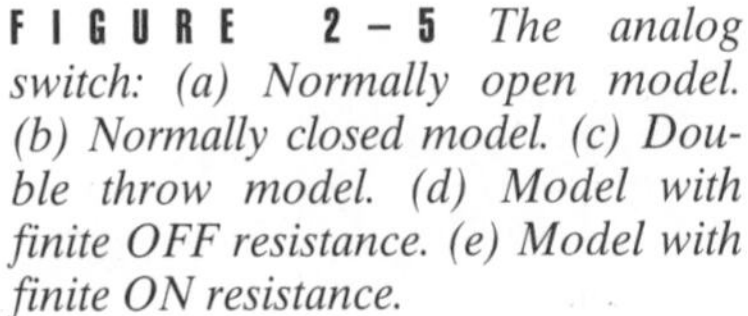

FIGURE 2–5 *The analog switch: (a) Normally open model. (b) Normally closed model. (c) Double throw model. (d) Model with finite OFF resistance. (e) Model with finite ON resistance.*

APPLICATION EXAMPLE 2–2

The **analog switch** is an important device found in analog-to-digital interfaces. Figures 2–5(a) and 2–5(b) show the two basic versions of the device. In either type the switch is actuated by applying a voltage to the terminal labeled "gate." The switch in Figure 2–5(a) is said to be *normally open* because it is open when no voltage is applied to the gate terminal and closes when voltage is applied. The switch in Figure 2–5(b) is said to be *normally closed* because it is closed when no voltage is applied to the controlling gate and opens when voltage is applied.

Figure 2–5(c) shows an application in which complementary analog switches are controlled by the same gate. When gate voltage is applied, the upper switch closes and the lower opens so that point A is connected to point C. Conversely, when no gate voltage is applied, the upper switch opens and the lower switch closes to connect point B to point C. In the analog world this arrangement is called a double throw switch since point C can be connected to two other points. In the digital world it is called a two-to-one multiplexer (or MUX) because it allows you to select the analog input at point A or point B under control of the digital signal applied to the gate.

In many applications an analog switch can be treated as an ideal switch. In other cases, it may be necessary to account for their nonideal characteristics. When the switch is open an analog switch acts like a very large resistance (R_{OFF}), as suggested in Figure 2–5(d). This resistance is negligible because it ranges from perhaps 10^9 to 10^{11} Ω. When the switch is closed it acts like a small resistor (R_{ON}), as suggested in Figure 2–5(e). Depending on other circuit resistances, it may be necessary to account for R_{ON} because it ranges from perhaps 20 to 200 Ω.

This example illustrates how ideal switches and resistors can be combined to model another electrical device. It also suggests that no single model can serve in all applications. It is up to the engineer to select a model that adequately represents the actual device in each application.

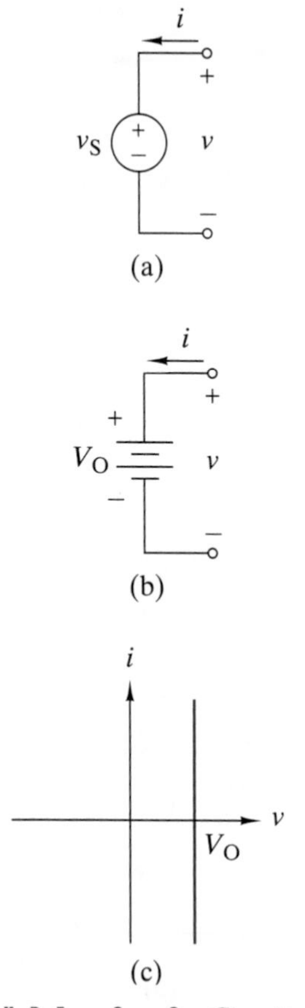

FIGURE 2–6 *Circuit symbols and i–v characteristic of an ideal independent voltage source: (a) Time-varying. (b) Constant (Battery). (c) Constant source i–v characteristics.*

IDEAL SOURCES

The signal and power sources required to operate electronic circuits are modeled using two elements: **voltage sources** and **current sources**. These sources can produce either constant or time-varying signals. The circuit symbols and the i–v characteristic of an ideal voltage source are shown in Figure 2–6, while the circuit symbol and i–v characteristic of an ideal current source are shown in Figure 2–7. The symbol in Figure 2–6(a) represents either a time-varying or constant voltage source. The battery symbol in Figure 2–6(b) is used exclusively for a constant voltage source. There is no separate symbol for a constant current source.

The i–v characteristic of an **ideal voltage source** in Figure 2–6(c) is described by the following element equations:

$$v = v_S \quad \text{and} \quad i = \text{any value} \tag{2–6}$$

The element equations mean that the ideal voltage source produces v_S volts across its terminals and will supply whatever current may be required by the circuit to which it is connected.

The i–v characteristic of an **ideal current source** in Figure 2–7(b) is described by the following element equations:

$$i = i_S \quad \text{and} \quad v = \text{any value} \tag{2–7}$$

The ideal current source produces i_S amperes in the direction of its arrow symbol and will furnish whatever voltage is required by the circuit to which it is connected.

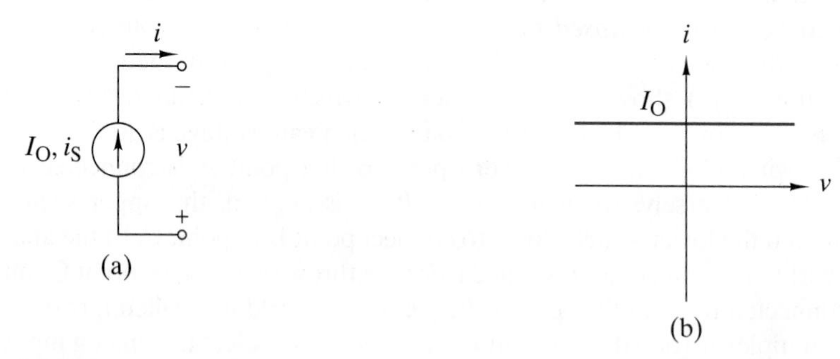

FIGURE 2–7 *Circuit symbols and i–v characteristic of an ideal independent current source: (a) Time-varying or constant source. (b) Constant source i–v characteristics.*

The voltage or current produced by these ideal sources is called a **forcing function** or a **driving function** because it represents an input that causes a circuit response.

EXAMPLE 2–3

Given an ideal voltage source with the time-varying voltage shown in Figure 2–8(a), sketch its i–v characteristic at the times $t = 0$, 1, and 2 ms.

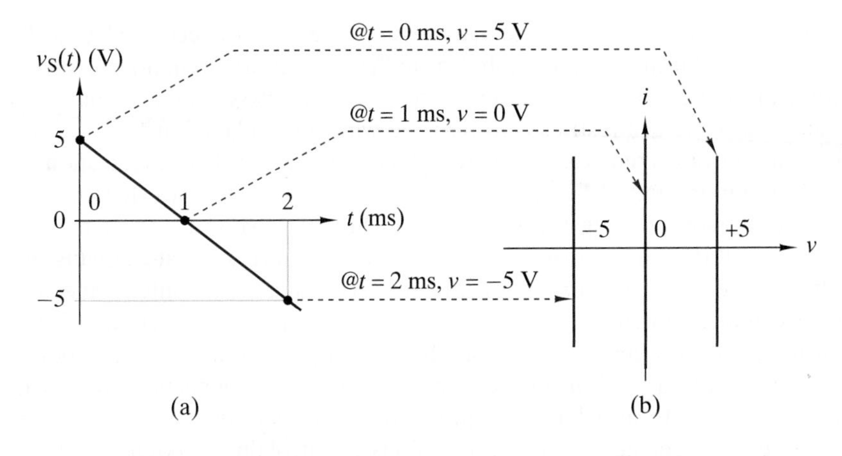

FIGURE 2–8

SOLUTION:
At any instant of time, the time-varying source voltage has only one value. We can treat the voltage and current at each instant of time as constants representing a snapshot of the source i–v characteristic. For example, at $t = 0$, the equations defining the i–v characteristic are $v_S = 5$ V and i = any value. Figure 2–8(b) shows the i–v relationship at the other instants of time. Curiously, the voltage source i–v characteristic at $t = 1$ ms ($v_S = 0$ and i = any value) is the same as that of a short circuit [see Eq. (2–5a) or Figure 2–3(b)]. ■

PRACTICAL SOURCES

The practical models for voltage and current sources in Figure 2–9 may be more appropriate in some situations than the ideal models used up to this point. These circuits are called practical models because they more accurately represent the characteristics of real-world sources than do the ideal models. It is important to remember that models are interconnections of elements, not devices. For example, the resistance in a model does not always represent an actual resistor. As a case in point, the resistances R_S in the practical source models in Figure 2–9 do not represent physical resistors but are circuit elements used to account for resistive effects within the devices being modeled.

The linear resistor, open circuit, short circuit, ideal switch, ideal voltage source, and ideal current source are the initial entries in our catalog of circuit elements. In Chapter 4 we will develop models for active devices like the transistor and OP AMP. Models for dynamic elements like capacitors and inductors are introduced in Chapter 6.

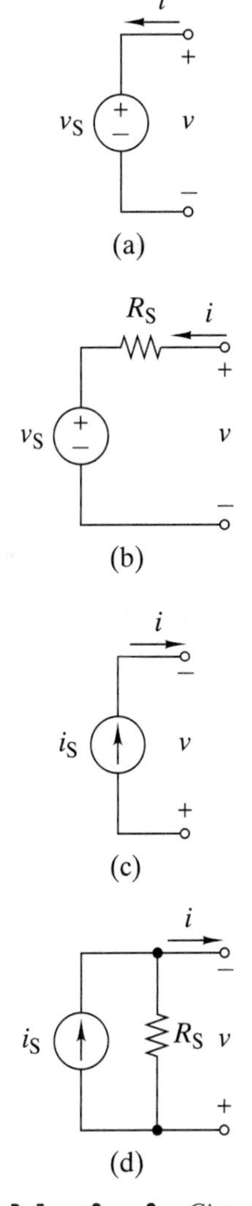

FIGURE 2–9 *Circuit symbols for ideal and practical independent sources: (a) Ideal voltage source. (b) Practical voltage source. (c) Ideal current source. (d) Practical current source.*

2–2 CONNECTION CONSTRAINTS

In the previous section, we considered individual devices and models. In this section, we turn our attention to the constraints introduced by interconnections of devices to form circuits. The laws governing circuit behavior are based on the meticulous work of the German scientist Gustav Kirchhoff (1824–1887). **Kirchhoff's laws** are derived from conservation laws as applied to circuits. They tell us that element voltages and currents are forced to behave in certain ways when the devices are interconnected to form a circuit. These conditions are called **connection constraints** because they are based only on the circuit connections, not on the specific devices in the circuit.

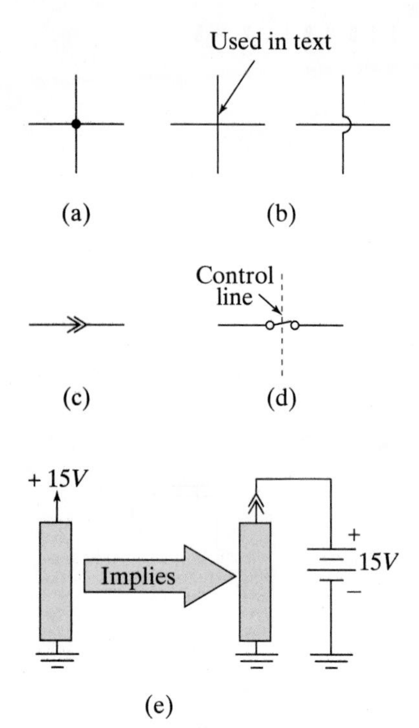

FIGURE 2–10 *Symbols used in circuit diagrams: (a) Electrical connection. (b) Crossover with no connection. (c) Jack connection. (d) Control line. (e) Power supply connection.*

In this book, we will indicate that crossing wires are connected (electrically tied together) using the dot symbol, as in Figure 2–10(a). Sometimes crossing wires are not connected (electrically insulated) but pass over or under each other. Since we are restricted to drawing wires on a planar surface, we will indicate unconnected crossovers by *not* placing a dot at their intersection, as indicated in the left of Figure 2–10(b). Other books sometimes show unconnected crossovers using the semicircular "hopover" shown on the right of Figure 2–10(b). In engineering systems two or more separate circuits are often tied together to form a larger circuit (for example, the interconnection of two integrated circuit packages). Interconnecting different circuits forms an interface between the circuits. The special jack or interface symbol in Figure 2–10(c) is used in this book because interface connections represent important points at which the interaction between two circuits can be observed or specified. On certain occasions a control line is required to show a mechanical or other nonelectrical dependency. Figure 2–10(d) shows how this dependency is indicated in this book. Figure 2–10(e) shows how power supply connections are often shown in electronic circuit diagrams. The implied power supply connection is indicated by an arrow pointing to the supply voltage, which may be given in numerical (+15 V) or symbolic form ($+V_{CC}$).

The treatment of Kirchhoff's laws uses the following definitions:

- A **circuit** is an interconnection of electrical devices.
- A **node** is an electrical juncture of two or more devices.
- A **loop** is a closed path formed by tracing through an ordered sequence of nodes without passing through any node more than once.

While it is customary to designate the juncture of two or more elements as a node, it is important to realize that a node is not confined to a point but includes all the zero-resistance wire from the point to each element. In the circuit of Figure 2–11, there are only three different nodes: A, B, and C. The points 2, 3, and 4, for example, are part of node B, while the points 5, 6, 7, and 8 are all part of node C.

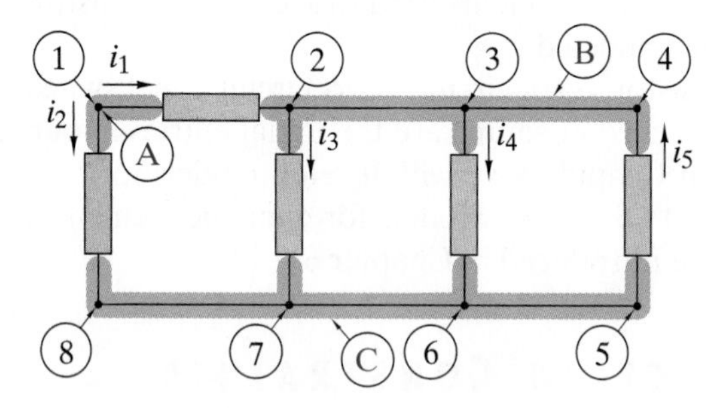

FIGURE 2–11 *Circuit for demonstrating Kirchhoff's current law.*

KIRCHHOFF'S CURRENT LAW

Kirchhoff's first law is based on the principle of conservation of charge. **Kirchhoff's current law (KCL)** states that

> ***the algebraic sum of the currents entering a node is zero at every instant.***

In forming the algebraic sum of currents, we must take into account the current reference direction associated with each device. If the current reference direction is into the node, then we assign a positive sign to the corresponding current in the algebraic sum. If the reference direction is away from the node, we assign a negative sign. Applying this convention to the nodes in Figure 2–11, we obtain the following set of KCL connection equations:

$$\begin{aligned} &\text{Node A: } -i_1 - i_2 = 0 \\ &\text{Node B: } i_1 - i_3 - i_4 + i_5 = 0 \\ &\text{Node C: } i_2 + i_3 + i_4 - i_5 = 0 \end{aligned} \qquad (2\text{–}8)$$

The KCL equation at node A does not mean that the currents i_1 and i_2 are both negative. The minus signs in this equation simply mean that the reference direction for each current is directed away from node A. Likewise, the equation at node B could be written as

$$i_3 + i_4 = i_1 + i_5 \qquad (2\text{–}9)$$

This form illustrates an alternate statement of KCL:

The sum of the currents entering a node equals the sum of the currents leaving the node.

There are two algebraic signs associated with each current in the application of KCL. First is the sign given to a current in writing a KCL connection equation. This sign is determined by the orientation of the current reference direction relative to a node. The second sign is determined by the actual direction of the current relative to the reference direction. The actual direction is found by solving the set of KCL equations, as illustrated in the following example.

EXAMPLE 2–4

Given $i_1 = +4$ A, $i_3 = +1$ A, $i_4 = +2$ A in the circuit shown in Figure 2–11, find i_2 and i_5.

SOLUTION:

Using the node A constraint in Eq. (2–8) yields

$$-i_1 - i_2 = -(+4) - i_2 = 0$$

The sign outside the parentheses comes from the node A KCL connection constraint in Eq. (2–8). The sign inside the parentheses comes from the actual direction of the current. Solving this equation for the unknown current, we find that $i_2 = -4$ A. In this case, the minus sign indicates that the actual direction of the current i_2 is directed upward in Figure 2–11, which is opposite to the reference direction assigned. Using the second KCL equation in Eq. (2–8), we can write

$$i_1 - i_3 - i_4 + i_5 = (+4) - (+1) - (+2) + i_5 = 0$$

which yields the result $i_5 = -1$ A.

Again, the signs inside the parentheses are associated with the actual direction of the current, and the signs outside come from the node B KCL connection constraint in Eq. (2–8). The minus sign in the final answer means that the current i_5 is directed in the opposite direction from its assigned reference direction. We can check our work by substituting the values found into the node C constraint in Eq. (2–8). These substitutions yield

$$+ i_2 + i_3 + i_4 - i_5 = (-4) + (+1) + (+2) - (-1) = 0$$

as required by KCL. Given three currents, we determined all the remaining currents in the circuit using only KCL without knowing the element constraints. ■

In Example 2–4, the unknown currents were found using only the KCL constraints at nodes A and B. The node C equation was shown to be valid, but it did not add any new information. If we look back at Eq. (2–8), we see that the node C equation is the negative of the sum of the node A and B equations. In other words, the KCL connection constraint at node C is not independent of the two previous equations. This example illustrates the following general principle:

In a circuit containing a total of N nodes there are only N – 1 independent KCL connection equations.

Current equations written at $N - 1$ nodes contain all the independent connection constraints that can be derived from KCL. To write these equations, we select one node as the reference or ground node and then write KCL equations at the remaining $N - 1$ nonreference nodes.

Exercise 2–1

Refer to Figure 2–12.

(a) Write KCL equations at nodes A, B, C, and D.
(b) Given $i_1 = -1$ mA, $i_3 = 0.5$ mA, $i_6 = 0.2$ mA, find i_2, i_4, and i_5.

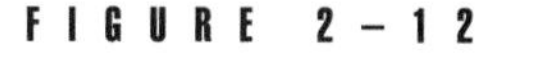

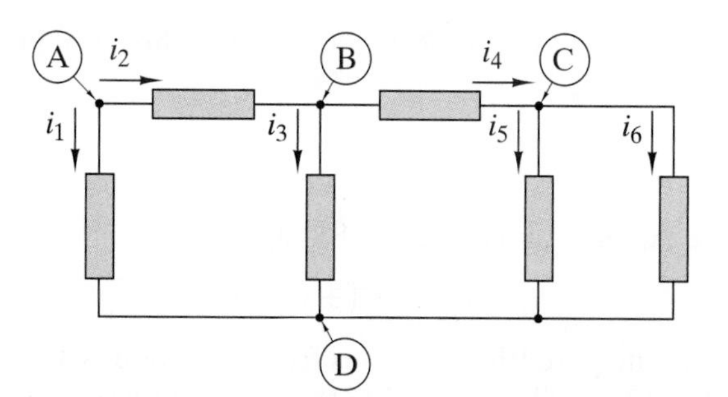

Answers:

(a) Node A: $-i_1 - i_2 = 0$; node B: $i_2 - i_3 - i_4 = 0$; node C: $i_4 - i_5 - i_6 = 0$; node D: $i_1 + i_3 + i_5 + i_6 = 0$
(b) $i_2 = 1$ mA, $i_4 = 0.5$ mA, $i_5 = 0.3$ mA

KIRCHHOFF'S VOLTAGE LAW

The second of Kirchhoff's circuit laws is based on the principle of conservation of energy. **Kirchhoff's voltage law** (abbreviated **KVL**) states that

the algebraic sum of all the voltages around a loop is zero at every instant.

For example, three loops are shown in the circuit of Figure 2–13. In writing the algebraic sum of voltages, we must account for the assigned reference marks. As a loop is traversed, a positive sign is assigned to a voltage when we go from a "+" to "−" reference mark. When we go from "−" to "+", we use a minus sign. Traversing the three loops in Figure 2–13 in the indicated clockwise direction yields the following set of KVL connection equations:

$$\begin{aligned} &\text{Loop 1: } -v_1 + v_2 + v_3 = 0 \\ &\text{Loop 2: } -v_3 + v_4 + v_5 = 0 \\ &\text{Loop 3: } -v_1 + v_2 + v_4 + v_5 = 0 \end{aligned} \quad (2\text{–}10)$$

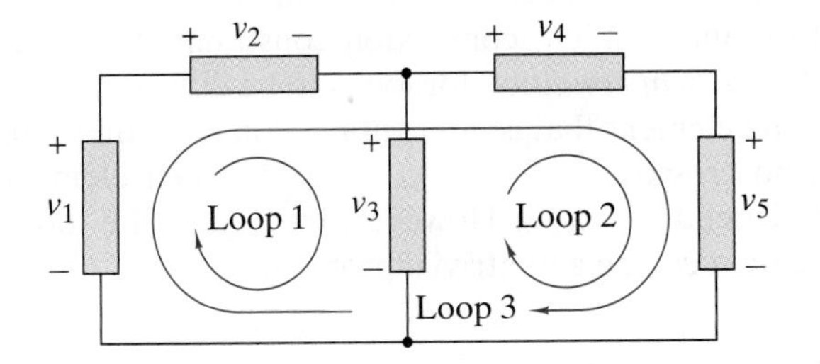

FIGURE 2–13 *Circuit for demonstrating Kirchhoff's voltage law.*

There are two signs associated with each voltage. The first is the sign given the voltage when writing the KVL connection equation. The second is the sign determined by the actual polarity of a voltage relative to its assigned reference polarity. The actual polarities are found by solving the set of KVL equations, as illustrated in the following example.

EXAMPLE 2–5

Given $v_1 = 5$ V, $v_2 = -3$ V, and $v_4 = 10$ V in the circuit shown in Figure 2–13, find v_3 and v_5.

SOLUTION:

Inserting the given numerical values into Eq. (2–10) yields the following KVL equation for loop 1:

$$-v_1 + v_2 + v_3 = -(+5) + (-3) + (v_3) = 0$$

The sign outside the parentheses comes from the loop 1 KVL constraint in Eq. (2–10). The sign inside comes from the actual polarity of the voltage. This equation yields $v_3 = +8$ V. Using this value in the loop 2, KVL constraint in Eq. (2–10) produces

$$-v_3 + v_4 + v_5 = -(+8) + (+10) + v_5 = 0$$

The result is $v_5 = -2$ V. The minus sign here means that the actual polarity of v_5 is the opposite of the assigned reference polarity indicated in Figure 2–13. The results can be checked by substituting all the aforementioned values into the loop 3 KVL constraint in Eq. (2–10). These substitutions yield

$$-(+5) + (-3) + (+10) + (-2) = 0$$

as required by KVL. ■

In Example 2–5, the unknown voltages were found using only the KVL constraints for loops 1 and 2. The loop 3 equation was shown to be valid, but it did not add any new information. If we look back at Eq. (2–10), we see that the loop 3 equation is equal to the sum of the loop 1 and 2 equations. In other words, the KVL connection constraint around loop 3 is not independent of the previous two equations. This example illustrates the following general principle:

> ***In a circuit containing a total of* E *two-terminal elements and* N *nodes, there are only* E − N + *1 independent KVL connection equations.***

Writing voltage summations around a total of $E - N + 1$ *different* loops produces all the independent connection constraints that can be derived from KVL. A *sufficient condition* for loops to be different is that each contains at least one element that is not contained in any other loop. In simple circuits with no crossovers, the open space between elements produces $E - N + 1$ independent loops. However, finding all the loops in a more complicated circuit can be a nontrivial problem.

Exercise 2–2

Find the voltages v_x and v_y in Figure 2–14.

FIGURE 2–14

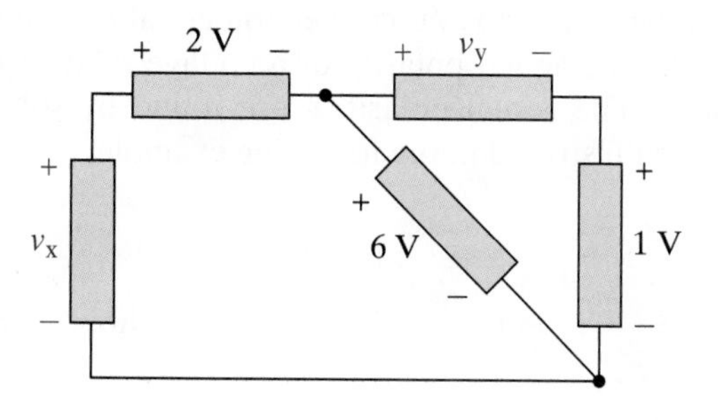

Answers: $v_x = +8$ V; $v_y = +5$ V

Exercise 2–3

Find the voltages v_x, v_y, and v_z in Figure 2–15.

FIGURE 2–15

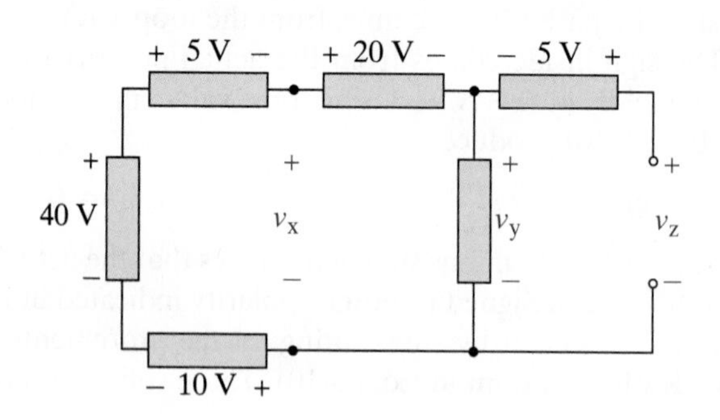

Answers: $v_x = +25$ V; $v_y = +5$ V; $v_z = +10$ V. *Note:* KVL yields the voltage v_z even though it appears across an open circuit.

PARALLEL AND SERIES CONNECTIONS

Two types of connections occur so frequently in circuit analysis that they deserve special attention. Elements are said to be connected in **parallel** when they form a loop containing no other elements. For example, loop A in Figure 2–16 contains only elements 1 and 2. As a result, the KVL connection constraint around loop A is

$$-v_1 + v_2 = 0 \tag{2–11}$$

which yields $v_1 = v_2$. In other words, in a parallel connection KVL requires equal voltages across the elements. The parallel connection is not restricted to two elements. For example, loop B in Figure 2–16 contains only elements 2 and 3; hence, by KVL $v_2 = v_3$. As a result, in this circuit we have $v_1 = v_2 = v_3$, and we say that elements 1, 2, and 3 are connected in parallel. In general, then, any number of elements connected between two common nodes are in parallel, and as a result, the same voltage appears across each of them. Existence of a parallel connection does not depend on the graphical position of the elements. For example, the position of elements 1 and 3 could be switched, and the three elements are still connected in parallel.

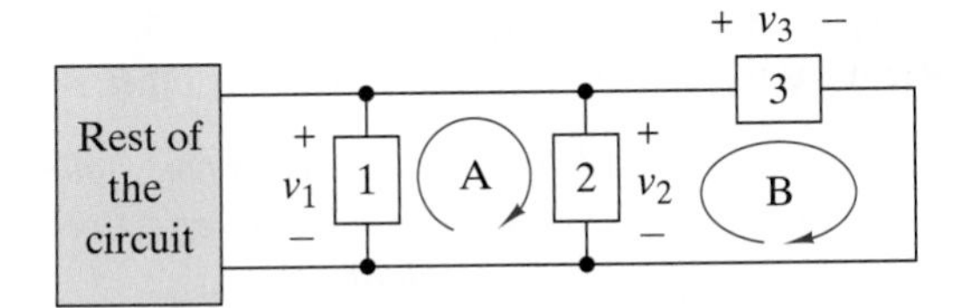

FIGURE 2–16 A parallel connection.

Two elements are said to be connected in **series** when they have one common node to which no other element is connected. In Figure 2–17 elements 1 and 2 are connected in series, since only these two elements are connected at node A. Applying KCL at node A yields

$$i_1 - i_2 = 0 \quad \text{or} \quad i_1 = i_2 \tag{2–12}$$

In a series connection, KCL requires equal current through each element. Any number of elements can be connected in series. For example, element 3 in Figure 2–17 is connected in series with element 2 at node B, and KCL requires $i_2 = i_3$. Therefore, in this circuit $i_1 = i_2 = i_3$, we say that elements 1, 2, and 3 are connected in series, and the same current exists in each of the elements. In general, elements are connected in series when they form a single path between two nodes such that only elements in the path are connected to the intermediate nodes along the path.

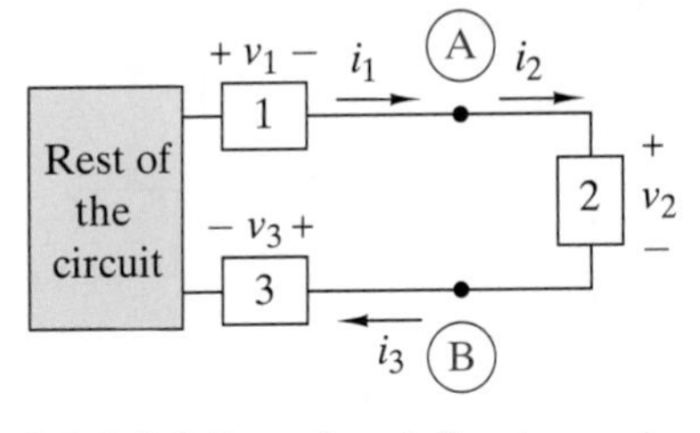

FIGURE 2–17 A series connection.

EXAMPLE 2–6

Identify the elements connected in parallel and in series in each of the circuits in Figure 2–18.

SOLUTION:

In Figure 2–18(a) elements 1 and 2 are connected in series at node A and elements 3 and 4 are connected in parallel between nodes B and C. In Figure 2–18(b) elements 1 and 2 are connected in series at node A, as are elements 4 and 5 at node D. There are no single elements connected in parallel in this circuit. In Figure 2–18(c) there are no elements connected

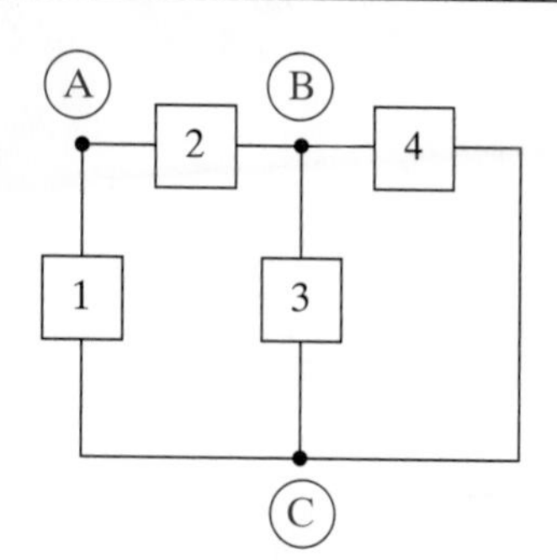

(a)

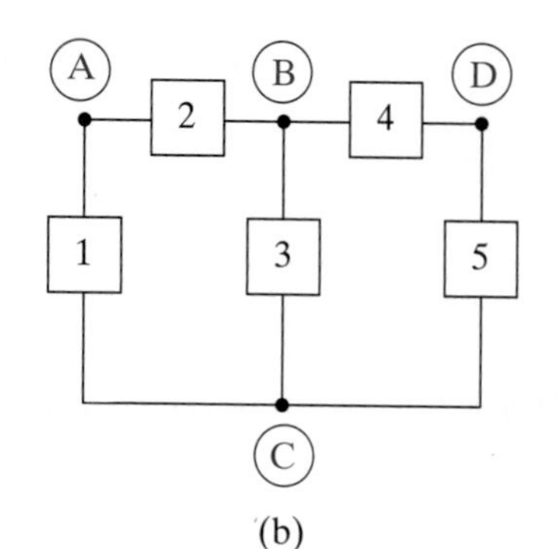

(b)

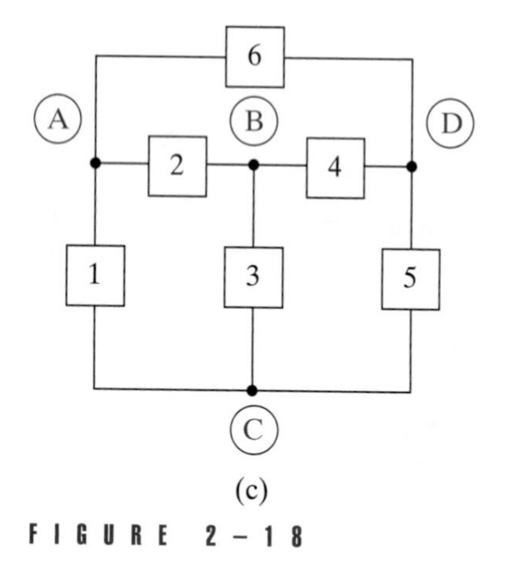

(c)

FIGURE 2–18

in either series or parallel. It is important to realize that in some circuits there are elements that are not connected in either series or in parallel. ■

Exercise 2–4

Identify the elements connected in series or parallel when a short circuit is connected between nodes A and B in each of the circuits of Figure 2–18.

Answers:

Circuit in Figure 2–18(a): Elements 1, 3, and 4 are all in parallel.
Circuit in Figure 2–18(b): Elements 1 and 3 are in parallel; elements 4 and 5 are in series.
Circuit in Figure 2–18(c): Elements 1 and 3 are in parallel; elements 4 and 6 are in parallel.

Exercise 2–5

Identify the elements in Figure 2–19 that are connected in (a) parallel, (b) series, or (c) neither.

Answers:

(a) The following elements are in parallel: 1, 8, and 11; 3, 4, and 5.
(b) The following elements are in series: 9 and 10; 6 and 7.
(c) Only element 2 is not in series or parallel with any other element.

DISCUSSION: *The ground symbol indicates the reference node. When ground symbols are shown at several nodes, the nodes are effectively connected together by a short circuit to form a single node.*

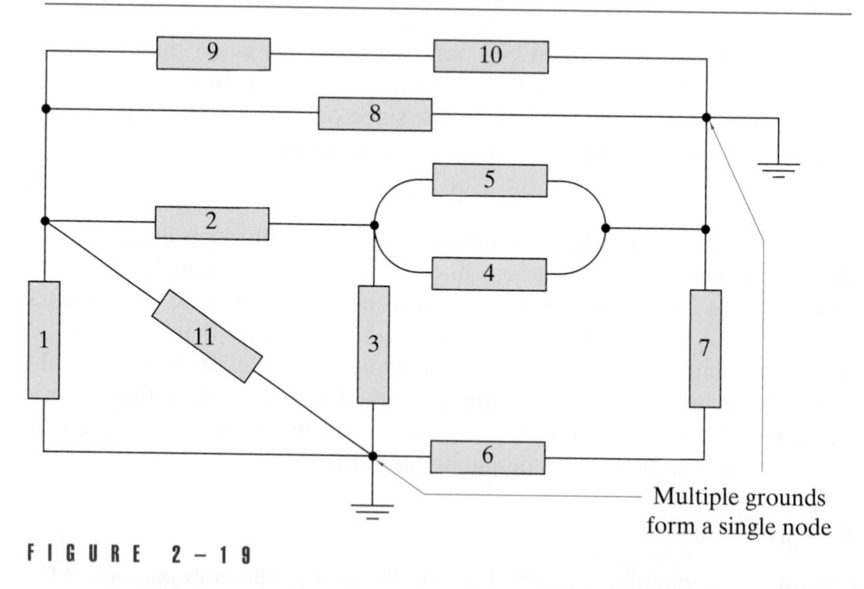

FIGURE 2–19

2–3 COMBINED CONSTRAINTS

The usual goal of circuit analysis is to determine the currents or voltages at various places in a circuit. This analysis is based on constraints of two distinctly different types. The element constraints are based on the models of the specific devices connected in the circuit. The connection constraints are based on Kirchhoff's laws and the circuit connections. The element equations are independent of the circuit connections. Likewise, the connection equations are independent of the devices in the circuit. Taken together, however, the combination of the element and connection constraints supply the equations needed to describe a circuit.

Our study of the combined constraints begins by considering the simple but important example in Figure 2–20. This circuit is driven by a current source i_S and the resulting responses are current/voltage pairs (i_x, v_x) and (i_O, v_O). The reference marks for the response pairs have been assigned using the passive sign convention.

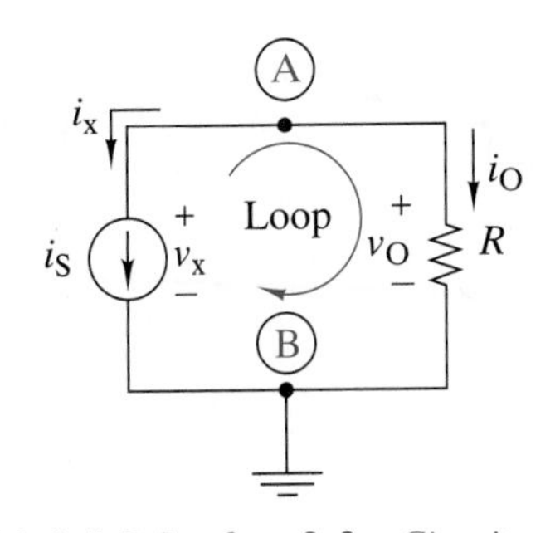

FIGURE 2–20 *Circuit used to demonstrate combined constraints.*

To solve for all four responses, we must write four equations. The first two are the element equations

$$i_x = i_S$$
$$v_O = Ri_O \tag{2–13}$$

The first element equation states that the response current i_x and the input driving force i_S are equal in magnitude and direction. The second element equation is Ohm's law relating v_O and i_O under the passive sign convention.

The connection equations are obtained by applying Kirchhoff's laws. The circuit in Figure 2–20 has two elements ($E = 2$) and two nodes ($N = 2$), so we need $E - N + 1 = 1$ KVL equation and $N - 1 = 1$ KCL equation. Selecting node B as the reference node, we apply KCL at node A and apply KVL around the loop to write

$$\text{KCL: } -i_x - i_O = 0$$
$$\text{KVL: } -v_x + v_O = 0 \tag{2–14}$$

We now have two element constraints in Eq. (2–13) and two connection constraints in Eq. (2–14), so we can solve for all four responses in terms of the input driving force i_S. Combining the KCL connection equation and the first element equation yields $i_O = -i_x = -i_S$. Substituting this result into the second element equation (Ohm's law) produces

$$v_O = -Ri_S \tag{2–15}$$

The minus sign in this equation does not mean that v_O is always negative. Nor does it mean the resistance is negative. It means that when the input driving force i_S is positive, then the response v_O is negative, and vice versa. This sign reversal is a result of the way we assigned reference marks at the beginning of our analysis. The reference marks defined the circuit input and outputs in such a way that i_S and v_O always have opposite algebraic signs. Put differently, Eq. (2–15) is an input-output relationship, not an element i–v relationship.

EXAMPLE 2–7

(a) Find the responses i_x, v_x, i_O, and v_O in the circuit in Figure 2–20 when $i_S = +2$ mA and $R = 2$ kΩ.
(b) Repeat for $i_S = -2$ mA.

SOLUTION:

(a) From Eq. (2–13) we have $i_x = i_S = +2$ mA and $v_O = 2000\ i_O$. From Eq. (2–14) we have $i_O = -i_x = -2$ mA and $v_x = v_O$. Combining these results, we obtain

$$v_x = v_O = 2000i_O = 2000(-0.002) = -4\text{ V}$$

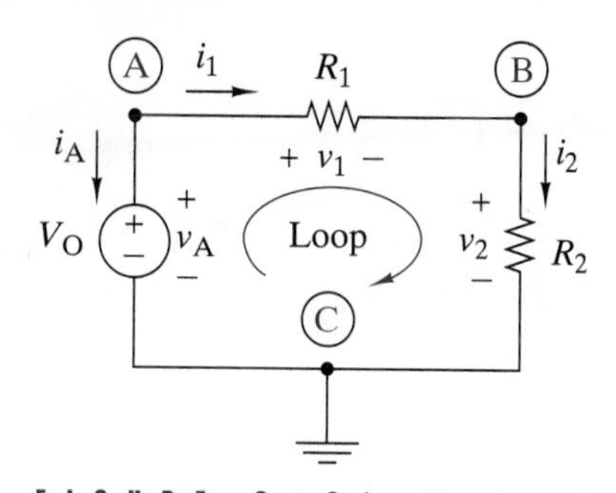

FIGURE 2–21 *Circuit used to demonstrate combined constraints.*

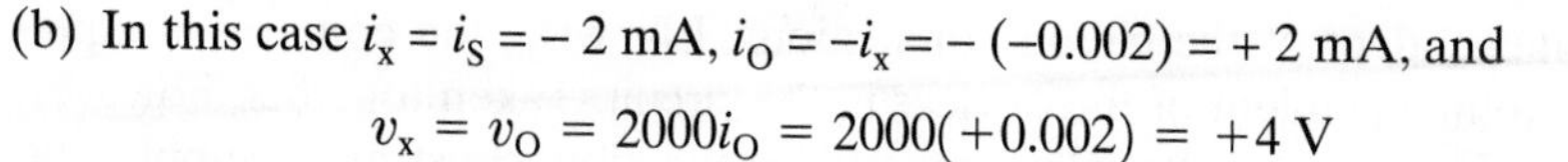

(b) In this case $i_x = i_S = -2$ mA, $i_O = -i_x = -(-0.002) = +2$ mA, and

$$v_x = v_O = 2000 i_O = 2000(+0.002) = +4 \text{ V}$$

This example confirms that the algebraic signs of the outputs v_x, v_O, and i_O are always the opposite sign from that of the input driving force i_S. ■

Analyzing the circuit in Figure 2–21 illustrates the formulation of combined constraints. We first assign reference marks for all the voltages and currents using the passive sign convention. Then, using these definitions we can write the element constraints as

$$\begin{aligned} v_A &= V_O \\ v_1 &= R_1 i_1 \\ v_2 &= R_2 i_2 \end{aligned} \tag{2–16}$$

These equations describe the three devices and do not depend on how the devices are connected in the circuit.

The connection equations are obtained from Kirchhoff's laws. To apply these laws, we must first label the different loops and nodes. The circuit contains $E = 3$ elements and $N = 3$ nodes, so there are $E - N + 1 = 1$ independent KVL constraints and $N - 1 = 2$ independent KCL constraints. There is only one loop, but there are three nodes in this circuit. We will select one node as the reference point and write KCL equations at the other two nodes. Any node can be chosen as the reference, so we select node C as the reference node and indicate this choice by drawing the ground symbol there. The connection constraints are

$$\begin{aligned} &\text{KCL: Node A} \quad -i_A - i_1 &&= 0 \\ &\text{KCL: Node B} \quad i_1 - i_2 &&= 0 \\ &\text{KVL: Loop} \quad -v_A + v_1 + v_2 &&= 0 \end{aligned} \tag{2–17}$$

These equations are independent of the specific devices in the circuit. They depend only on Kirchhoff's laws and the circuit connections.

This circuit has six unknowns: three element currents and three element voltages. Taken together, the element and connection equations give us six independent equations. For a network with (N) nodes and (E) two-terminal elements, we can write ($N - 1$) independent KCL connection equations, ($E - N + 1$) independent KVL connection equations, and (E) element equations. The total number of equations generated is

Element equations	E
KCL equations	$N - 1$
KVL equations	$\underline{E - N + 1}$
Total	$2E$

The grand total is then ($2E$) combined connection and element equations, which is exactly the number of equations needed to solve for the voltage across and current through every element—a total of ($2E$) unknowns.

EXAMPLE 2–8

Find all of the element currents and voltages in Figure 2–21 for $V_O = 10$ V, $R_1 = 2000\ \Omega$, and $R_2 = 3000\ \Omega$.

SOLUTION:

Substituting the element constraints from Eq. (2–16) into the KVL connection constraint in Eq. (2–17) produces

$$-V_O + R_1 i_1 + R_2 i_2 = 0$$

This equation can be used to solve for i_1 since the second KCL connection equation requires that $i_2 = i_1$.

$$i_1 = \frac{V_O}{R_1 + R_2} = \frac{10}{2000 + 3000} = 2 \text{ mA}$$

In effect, we have found all of the element currents since the elements are connected in series. Hence, collectively the KCL connection equations require that

$$-i_A = i_1 = i_2$$

Substituting all of the known values into the element equations gives

$$v_A = 10 \text{ V} \quad v_1 = R_1 i_1 = 4 \text{ V} \quad v_2 = R_2 i_2 = 6 \text{ V}$$

Every element voltage and current has been found. Note the analysis strategy used. We first found all the element currents and then used these values to find the element voltages. ■

EXAMPLE 2–9

Use element and connection equations to find the voltages across the resistors in Figure 2–22.

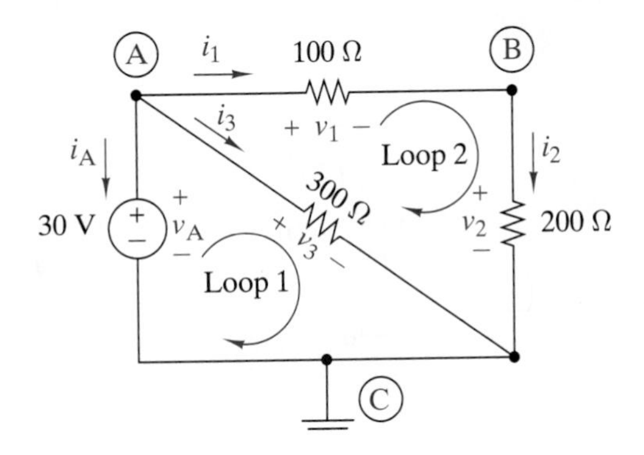

FIGURE 2–22

SOLUTION:

A complete description of this circuit involves four element equations and four connection equations. The element equations are

$$v_1 = 100\, i_1$$
$$v_2 = 200\, i_2$$
$$v_3 = 300\, i_3$$
$$v_A = 30 \text{ V}$$

The four connection equations are

$$\begin{aligned}
&\text{KCL: Node A} \ -i_A - i_1 - i_3 = 0\\
&\text{KCL: Node B} \ i_1 - i_2 = 0\\
&\text{KVL: Loop 1} \ -v_A + v_3 = 0\\
&\text{KVL: Loop 2} \ -v_3 + v_1 + v_2 = 0
\end{aligned}$$

Combining the last element equation and the KVL equation around loop 1 shows that

$$v_3 = v_A = 30 \text{ V}$$

which is nothing more than a statement that the voltage source and R_3 are connected in parallel. Using this result in the loop 2 equation yields $v_1 + v_2 = v_3 = 30$ V. Substituting the first two element equations into this equation produces

$$100i_1 + 200i_2 = 30$$

But the KCL equation at node B points out that $i_1 = i_2$, and this result reduces to $300i_2 = 30$ or $i_1 = i_2 = 0.1$ A. Finally, the first two element equations yield

$$v_1 = 100i_1 = 10 \text{ V and } v_2 = 200i_2 = 20 \text{ V}$$

In summary, the voltages across the three resistors are $v_1 = 10$ V, $v_2 = 20$ V, and $v_3 = 30$ V. ■

Exercise 2–6

In Figure 2–23 $i_1 = 200$ mA and $i_3 = -100$ mA. Find the voltage v_x.

FIGURE 2–23

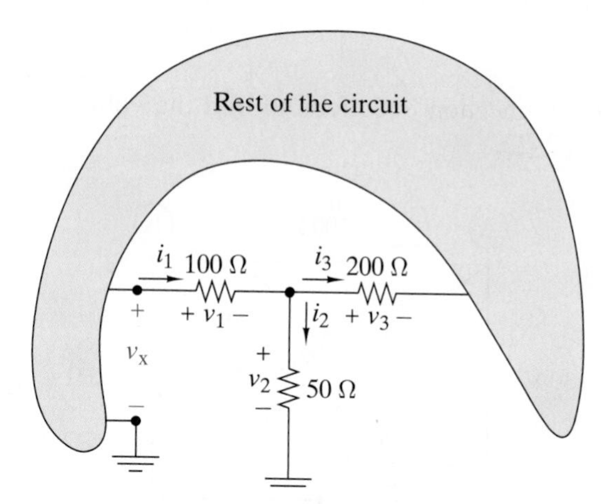

Answer: $v_x = 35$ V

ASSIGNING REFERENCE MARKS

In all of our previous examples and exercises, the reference marks for the element currents (arrows) and voltages (+ and −) were given. When reference marks are not shown on a circuit diagram, they must be assigned by the person solving the problem. Beginners sometimes wonder how to assign reference marks when the actual voltage polarities and current directions are unknown. It is important to remember that the reference marks do not indicate what is actually happening in the circuit. They are benchmarks assigned at the beginning of the analysis. If it turns out that the actual direction and reference direction agree, then the algebraic sign of the response will be positive. If they disagree, the algebraic sign will be negative. In other words, the sign of the answer together with assigned reference marks tell us the actual voltage polarity or current direction.

In this book the reference marks always follow the passive sign convention. This means that for any given two-terminal element we can arbitrarily assign either the + voltage reference mark or the current reference arrow, but not both. For example, we can arbitrarily assign the voltage reference marks to the terminals of a two-terminal device. Once the voltage reference is assigned, however, the passive sign convention requires that the current reference arrow be directed into the element at the terminal with the + mark. On the other hand, we could start by arbitrarily selecting the terminal at which the current reference arrow is directed into the device. Once the current reference is assigned, however, the passive sign convention requires that the + voltage reference be assigned to the selected terminal.

Following the passive sign convention avoids confusion about the direction of power flow in a device. In addition, the element constraints, such as Ohm's law, assume that the passive sign convention is used to assign the voltage and current reference marks to a device.

The next example illustrates the assignment of reference marks.

EXAMPLE 2–10

Find the voltages across the resistors and current sources in Fig. 2–24(a).

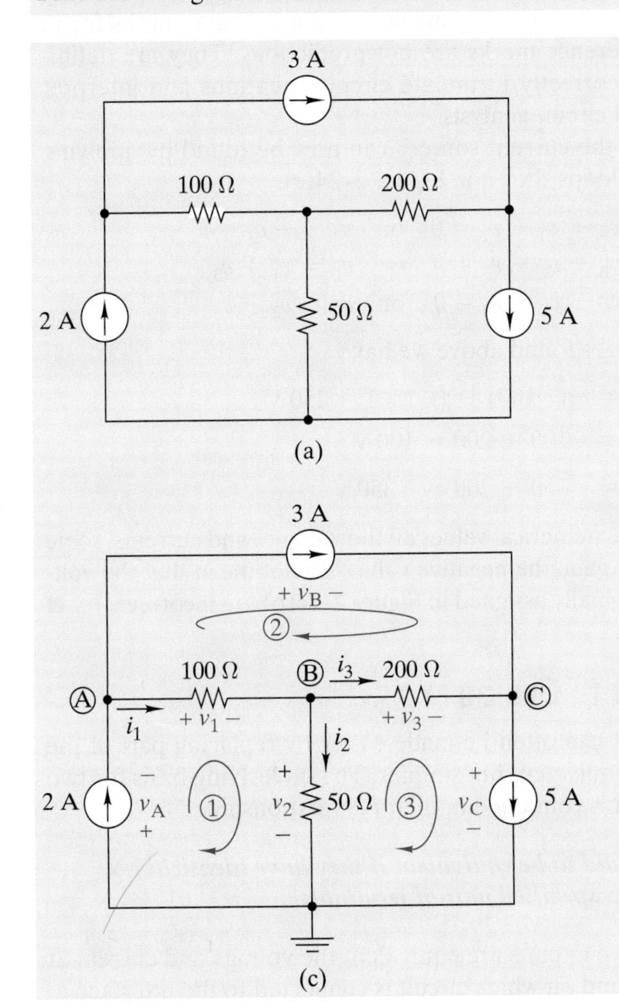

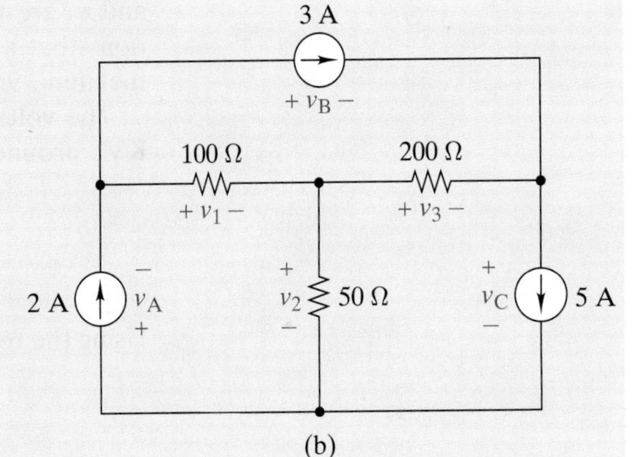

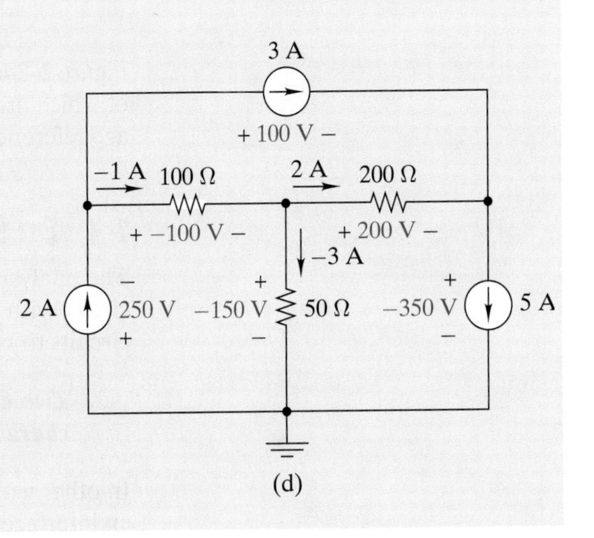

FIGURE 2–24

SOLUTION:
No voltage reference marks are given in Figure 2–24(a), so we assign those shown in Figure 2–24(b). Other choices are possible, of course, but once the voltage marks for v_1, v_2, and v_3 are assigned, the passive sign convention requires that the current reference directions for i_1, i_2, and i_3 be assigned as shown in Figure 2–24(c). KCL can be used to find the resistor currents directly. Using KCL at node A gives $2 - i_1 - 3 = 0$; hence $i_1 = -1$ A. KCL applied at node C yields $3 + i_3 - 5 = 0$; hence $i_3 = 2$ A. Finally, at node B KCL requires $i_1 - i_2 - i_3 = 0$; hence $i_2 = i_1 - i_3 = -1 - 2 = -3$ A. Given the three resistor currents, we use Ohm's law to find the three resistor voltages.

$$v_1 = 100\, i_1 = -100 \text{ V}$$

$$v_2 = 50\, i_2 = -150 \text{ V}$$

$$v_3 = 100\, i_3 = +200 \text{ V}$$

The plus on the numerical value of v_3 means that the assigned reference marks agree with the actual voltage polarity. The minus sign on the numerical values of v_1 and v_2 mean that the assigned marks and physical reality disagree. This disagreement does not mean that the assigned marks for v_1 and v_2 are wrong. Reference marks are not predictions. They are definitions that allow us to correctly formulate circuit equations and interpret the numerical results of circuit analysis.

The voltages across the current sources can now be found by applying KVL around the three loops shown in Figure 2–24(c).

$$\begin{array}{llll} \text{Loop 1} & v_A + v_1 + v_2 = 0 & \text{or} & v_A = -v_1 - v_2 \\ \text{Loop 2} & v_B - v_1 - v_3 = 0 & \text{or} & v_B = v_1 + v_3 \\ \text{Loop 3} & v_C - v_2 + v_3 = 0 & \text{or} & v_C = v_2 - v_3 \end{array}$$

Using the resistor voltages found above we have

$$v_A = -(-100) - (-150) = 250 \text{ V}$$

$$v_B = -100 + 200 = 100 \text{ V}$$

$$v_C = -150 - 200 = -350 \text{ V}$$

Figure 2–24(d) shows the numerical values all the voltages and currents, some of which are negative. Again, the negative values do not mean that the voltage reference marks originally assigned in Figure 2–24(b) are incorrect. ■

2–4 EQUIVALENT CIRCUITS

The analysis of a circuit can often be made easier by replacing part of the circuit with one that is equivalent but simpler. The underlying basis for two circuits to be equivalent is contained in their i–v relationships.

Two circuits are said to be equivalent if they have identical* i–v *characteristics at a specified pair of terminals.

In other words, when two circuits are equivalent the voltage and current at an interface do not depend on which circuit is connected to the interface.

EQUIVALENT RESISTANCE

The two resistors in Figure 2–25(a) are connected in series between a pair of terminals A and B. The objective is to simplify the circuit without altering the electrical behavior of the rest of the circuit.

The KVL equation around the loop from A to B is

$$v = v_1 + v_2 \tag{2–18}$$

Since the two resistors are connected in series, the same current i exists in both. Applying Ohm's law, we get $v_1 = R_1 i$ and $v_2 = R_2 i$. Substituting these relationships into Eq. (2–18) and then simplifying yields

$$v = R_1 i + R_2 i = i(R_1 + R_2)$$

We can write this equation in terms of an equivalent resistance R_{EQ} as

$$v = iR_{EQ} \quad \text{where} \quad R_{EQ} = R_1 + R_2 \tag{2–19}$$

This result means the circuits in Figs. 2–25(a) and 2–25(b) have the same i–v characteristic at terminals A and B. As a result, the response of the rest of the circuit is unchanged when the series connection of R_1 and R_2 is replaced by a resistance R_{EQ}.

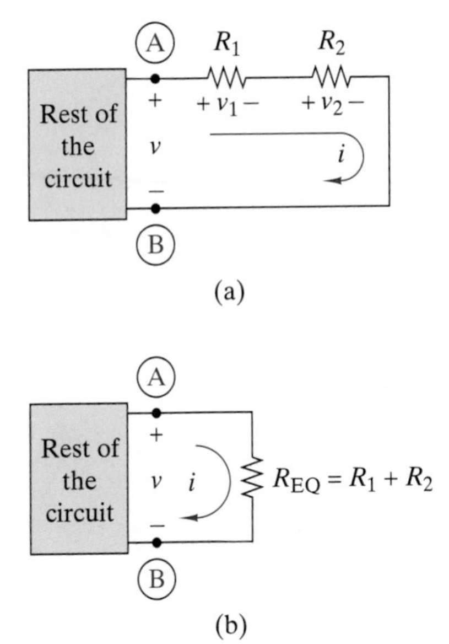

FIGURE 2–25 *A series resistance circuit: (a) Original circuit. (b) Equivalent circuit.*

The parallel connection of two conductances in Figure 2–26(a) is the dual[2] of the series circuit in Figure 2–25(a). Again the objective is to replace the parallel connection by a simpler equivalent circuit without altering the response of the rest of the circuit.

A KCL equation at node A produces

$$i = i_1 + i_2 \tag{2–20}$$

Since the conductances are connected in parallel, the voltage v appears across both. Applying Ohm's law, we obtain $i_1 = G_1 v$ and $i_2 = G_2 v$. Substituting these relationships into Eq. (2–20) and then simplifying yields

$$i = vG_1 + vG_2 = v(G_1 + G_2)$$

This result can be written in terms of an equivalent conductance G_{EQ} as follows:

$$i = vG_{EQ}, \quad \text{where} \quad G_{EQ} = G_1 + G_2 \tag{2–21}$$

This result means the circuits in Figures 2–26(a) and 2–26(b) have the same i–v characteristic at terminals A and B. As a result, the response of the rest of the circuit is unchanged when the parallel connection of G_1 and G_2 is replaced by a conductance G_{EQ}.

Since conductance is not normally used to describe a resistor, it is sometimes useful to rewrite Eq. (2–21) as an equivalent resistance $R_{EQ} = 1/G_{EQ}$. That is,

$$R_1 \| R_2 = R_{EQ} = \frac{1}{G_{EQ}} = \frac{1}{G_1 - G_2} = \frac{1}{\dfrac{1}{R_1} + \dfrac{1}{R_2}} = \frac{R_1 R_2}{R_1 + R_2} \tag{2–22}$$

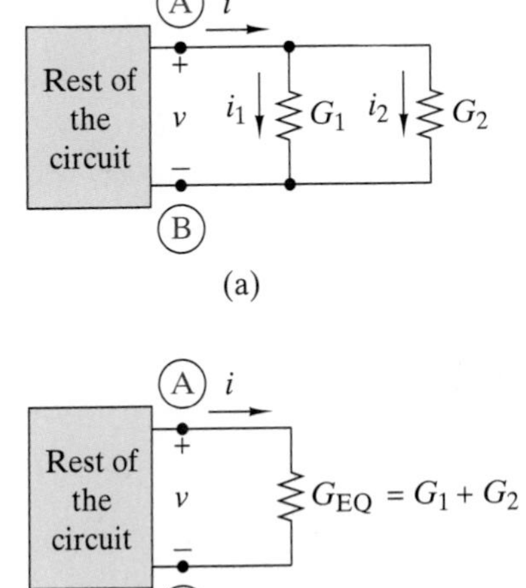

(b)

FIGURE 2–26 *A parallel resistance circuit: (a) Original circuit. (b) Equivalent circuit.*

2 Dual circuits have identical behavior patterns when we interchange the roles of the following parameters: (1) voltage and current, (2) series and parallel, and (3) resistance and conductance. In later chapters we will see duality exhibited by other circuit parameters as well.

where the symbol "||" is shorthand for "in parallel." The expression on the far right in Eq. (2–22) is called the product over the sum rule for two resistors in parallel. This rule is useful in derivations in which the resistances are left in symbolic form (see Example 2–11), but it is not an efficient algorithm for calculating numerical values of equivalent resistance.

Caution: The product over sum rule only applies to two resistors connected in parallel. When more than two resistors are in parallel, we must use the following general result to obtain the equivalent resistance:

$$R_{EQ} = \frac{1}{G_{EQ}} = \frac{1}{\dfrac{1}{R_1} + \dfrac{1}{R_2} + \dfrac{1}{R_3} + \cdots} \tag{2–23}$$

EXAMPLE 2–11

Given the circuit in Figure 2–27(a),

(a) Find the equivalent resistance R_{EQ1} connected between terminals A and B.

(b) Find the equivalent resistance R_{EQ2} connected between terminals C and D.

FIGURE 2–27

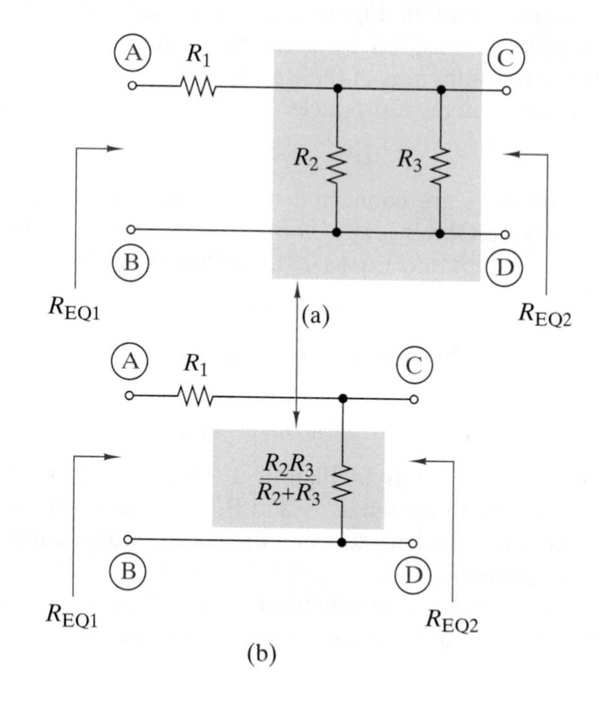

SOLUTION:

First we note that resistors R_2 and R_3 are connected in parallel. Applying the product over sum rule [Eq. (2–22)], we obtain

$$R_2 \| R_3 = \frac{R_2 R_3}{R_2 + R_3}$$

As an interim step, we redraw the circuit, as shown in Figure 2–27(b).

(a) To find the equivalent resistance between terminals A and B, we note that R_1 and the equivalent resistance $R_2 \| R_3$ are connected in series. The total equivalent resistance R_{EQ1} between terminals A and B is

$$R_{EQ1} = R_1 + (R_2 \| R_3)$$

$$R_{EQ1} = R_1 + \frac{R_2 R_3}{R_2 + R_3}$$

$$R_{EQ1} = \frac{R_1 R_2 + R_1 R_3 + R_2 R_3}{R_2 + R_3}$$

(b) Looking between terminals C and D yields a different result. In this case R_1 is not involved, since there is an open circuit (an infinite resistance) between terminals A and B. Therefore, only $R_2 \| R_3$ affect the resistance between terminals C and D. As a result, R_{EQ2} is simply

$$R_{EQ2} = R_2 \| R_3 = \frac{R_2 R_3}{R_2 + R_3}$$

This example shows that equivalent resistance depends on the pair of terminals involved. ■

One final note on checking numerical calculations of equivalent resistance. When several resistances are connected in parallel, the equivalent resistance must be smaller than the smallest resistance in the connection. Conversely, when several resistances are connected in series, the equivalent resistance must be larger than the largest resistance in the connection.

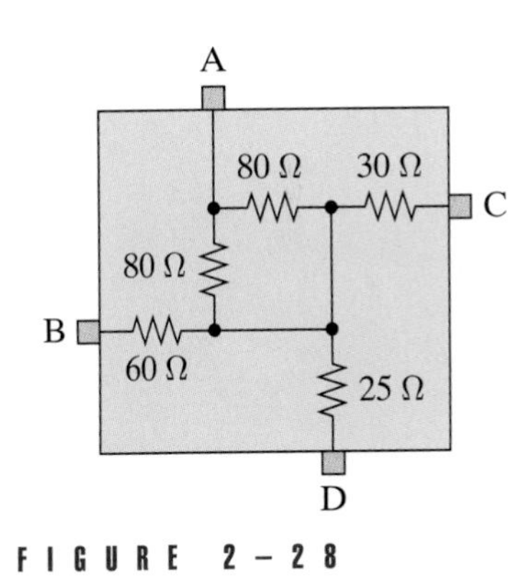

FIGURE 2–28

Exercise 2–7

Find the equivalent resistance between terminals A–C, B–D, A–D, and B–C in the circuit in Figure 2–27.

Answers: $R_{A\text{-}C} = R_1$; $R_{B\text{-}D} = 0\ \Omega$ (a short circuit); $R_{A\text{-}D} = R_1 + R_2 \| R_3$; $R_{B\text{-}C} = R_2 \| R_3$;

Exercise 2–8

Find the equivalent resistance between terminals A–B, A–C, A–D, B–C, B–D, and C–D in the circuit of Figure 2–28.

Answers: $R_{A\text{-}B} = 100\ \Omega$; $R_{A\text{-}C} = 70\ \Omega$; $R_{A\text{-}D} = 65\ \Omega$; $R_{B\text{-}C} = 90\ \Omega$; $R_{B\text{-}D} = 85\ \Omega$; $R_{C\text{-}D} = 55\ \Omega$;

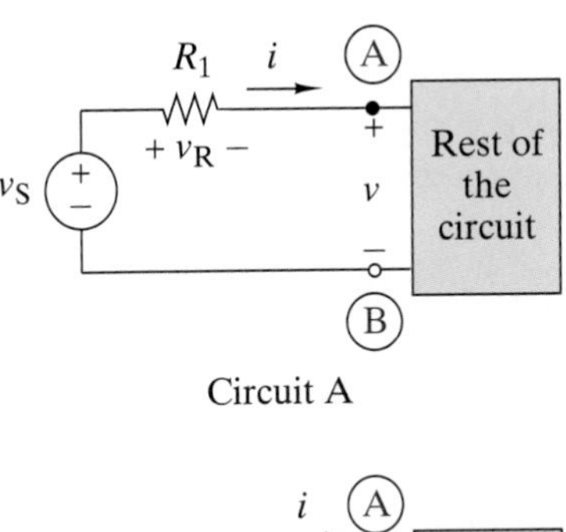

Circuit A

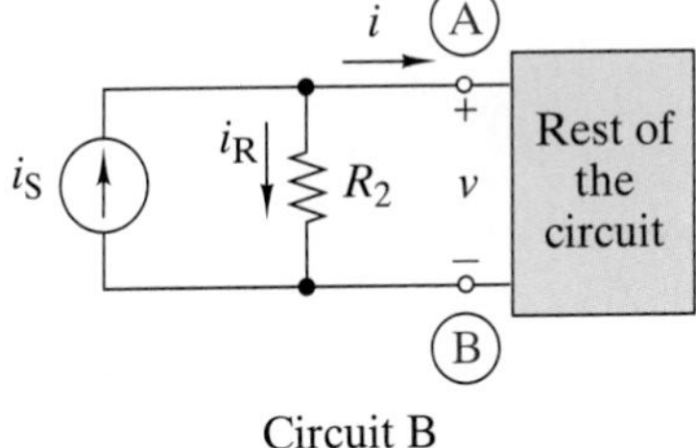

Circuit B

FIGURE 2–29 *Practical source models that are equivalent when Eq. (2–24) is satisfied.*

EQUIVALENT SOURCES

The practical source models introduced previously are shown in Figure 2–29. These models consist of an ideal voltage source in series with a resistance and an ideal current source in parallel with a resistance. We now determine the conditions under which the practical voltage source and the practical current sources are equivalent.

Figure 2–29 shows the two practical sources connected between terminals labeled A and B. A parallel analysis of these circuits yields the conditions for equivalency at terminals A and B. First, Kirchhoff's laws are applied as

Circuit A	Circuit B
KVL	KCL
$v_S = v_R + v$	$i_S = i_R + i$

Next, Ohm's law is used to obtain

Circuit A	Circuit B
$v_R = R_1 i$	$i_R = \dfrac{v}{R_2}$

Combining these results, we find that the i–v relationships of each of the circuits at terminals A and B are

Circuit A	Circuit B
$i = -\dfrac{v}{R_1} + \dfrac{v_S}{R_1}$	$i = -\dfrac{v}{R_2} + i_S$

These i–v characteristics take the form of the straight lines shown in Figure 2–30. The two lines are identical when the intercepts are equal. This requires that $v_S/R_1 = i_S$ and $v_S = i_S R_2$, which, in turn, requires that

$$R_1 = R_2 = R \quad \text{and} \quad v_S = i_S R \tag{2–24}$$

When conditions in Eq. (2–24) are met, the response of the rest of the circuit is unaffected when we replace a practical voltage source by an equivalent practical current source, or vice versa. Exchanging one practical source model for an equivalent model is called *source transformation.*

Source transformation means that either model will deliver the same voltage and current to the rest of the circuit. It does not mean the two

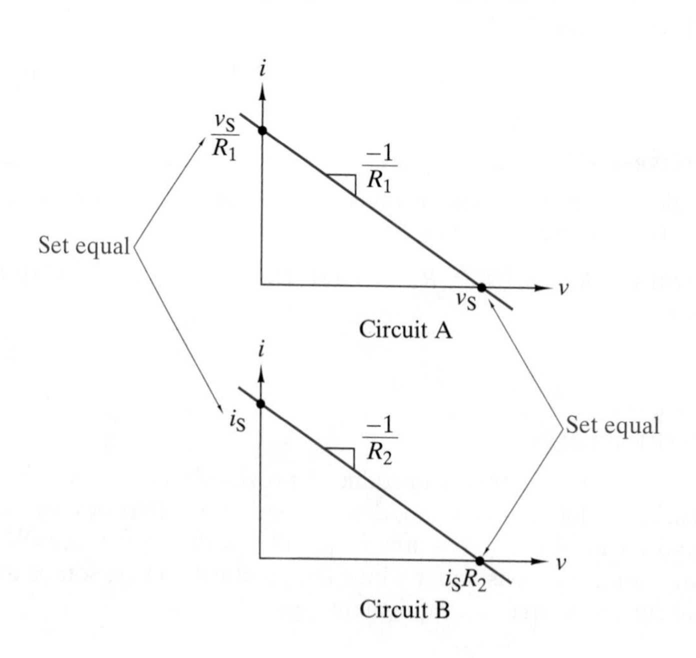

FIGURE 2–30 The i–v characteristics of practical sources in Fig. 2–29.

models are identical in every way. For example, when the practical voltage source drives an open circuit, there is no i^2R power loss since the current in the series resistance is zero. However, the current in the parallel resistance of a practical current source is not zero when the load is an open circuit. Thus, equivalent sources do not have the same internal power loss even though they deliver the same current and voltage to the rest of the circuit.

EXAMPLE 2–12

Convert the practical voltage source in Figure 2–31(a) into an equivalent current source.

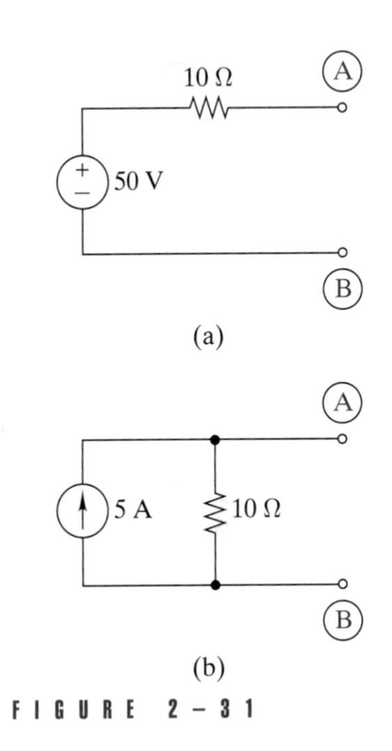

FIGURE 2–31

SOLUTION:
Using Eq. (2–24), we have

$$R_1 = R_2 = R = 10\ \Omega$$

$$i_S = v_S/R = 5\text{ A}$$

The equivalent practical current source is shown in Figure 2–31(b). ■

Exercise 2–9

A practical current source consists of a 2-mA ideal current source in parallel with a 0.002-S conductance. Find the equivalent practical voltage source.

Answer: The equivalent is a 1-V ideal voltage source in series with a 500-Ω resistance.

Figure 2–32 shows another source transformation in which a voltage source and resistor in parallel is replaced by a voltage source acting alone. The two circuits are equivalent because i–v constraint at the input to the rest of the circuit is $v = v_S$ in both circuits. In other words, the response of the rest of the circuit is unchanged if a resistor in parallel with a voltage source is removed from the circuit. However, removing the resistor does reduce the total current supplied by the voltage source by v_S/R. While the resistor does not affect the current and voltage delivered to the rest of the circuit, it does dissipate power that must be supplied by the source.

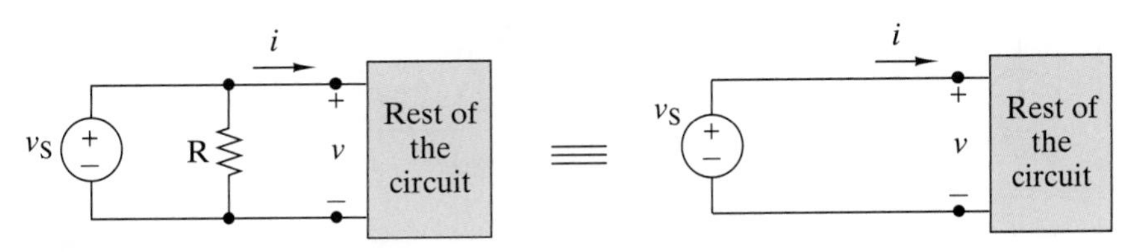

FIGURE 2–32 *Equivalent circuit of a voltage source and a resistor in parallel.*

The dual situation is shown in Figure 2–33. In this case a current source connected in series with a resistor can be replaced by a current source acting alone because the i–v constraint at the input to the rest of the circuit is $i = i_S$ for both circuits. In other words, the response of the rest of the circuit is unchanged if a resistor in series with a current source is removed from the circuit.

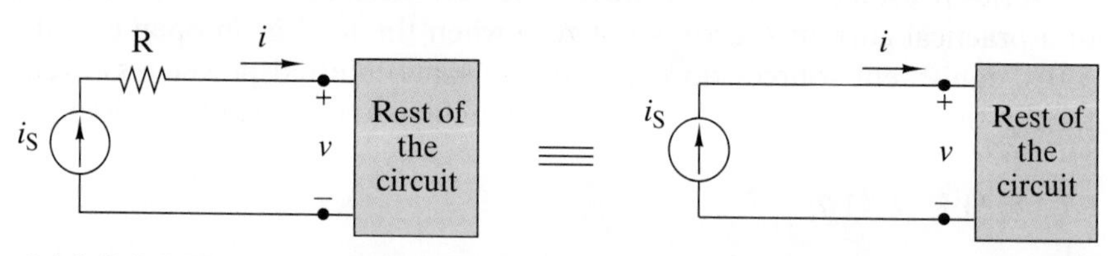

FIGURE 2–33 *Equivalent circuit of a current source and a resistor in series.*

SUMMARY OF EQUIVALENT CIRCUITS

Figure 2–34 is a summary of two-terminal equivalent circuits involving resistors and sources connected in series or parallel. The series and parallel equivalences in the first row and the source transformations in the second

FIGURE 2–34 *Summary of two-terminal equivalent circuits.*

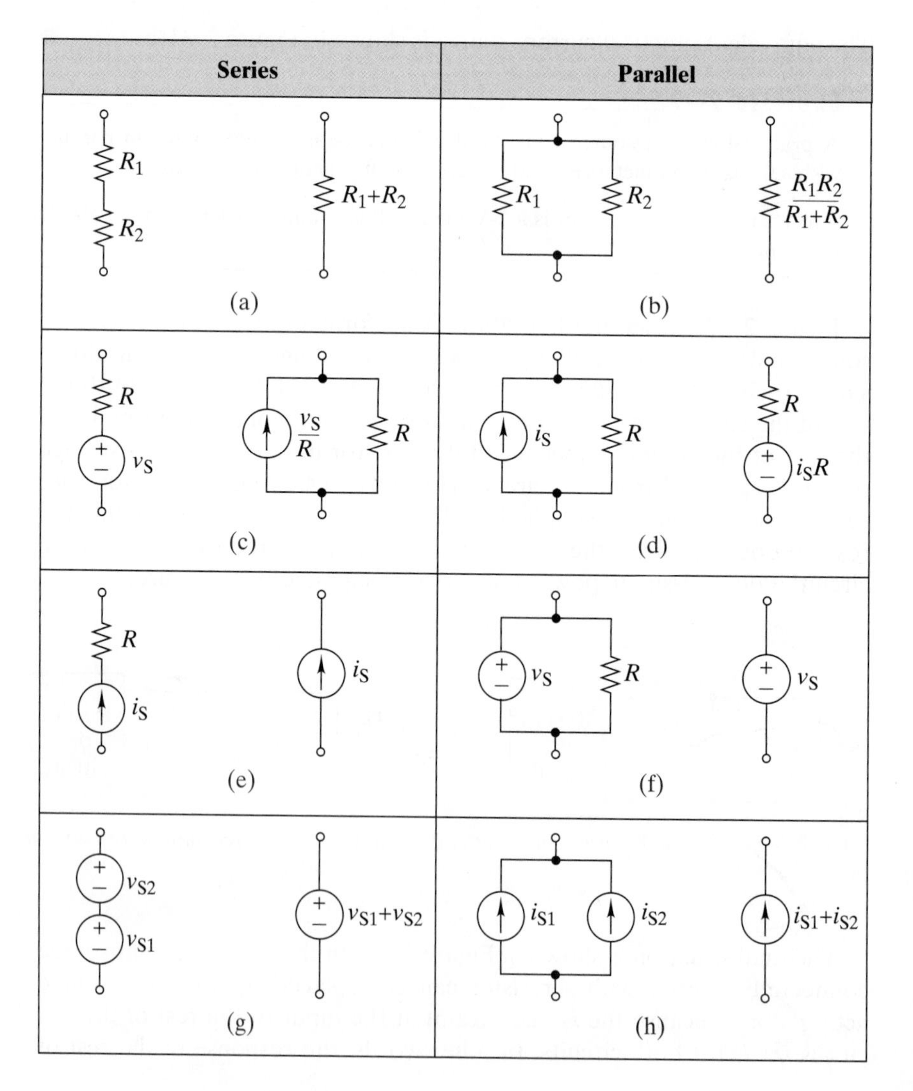

row are used regularly in subsequent discussions. The last row in Figure 2–34 presents additional source transformations that reduce series or parallel connections to a single ideal current or voltage source. Proof of these equivalences involves showing that the final single-source circuits have the same i–v characteristics as the original connections. The details of such a derivation are left as an exercise for the reader.

2–5 VOLTAGE AND CURRENT DIVISION

We complete our treatment of series and parallel circuits with a discussion of voltage and current division. These two analysis tools find wide application in circuit analysis and design.

VOLTAGE DIVISION

Voltage division provides a simple way to find the voltage across each element in a series circuit. Figure 2–35 shows a circuit that lends itself to solution by voltage division. Applying KVL around the loop in Figure 2–35 yields

$$v_S = v_1 + v_2 + v_3 \tag{2–25}$$

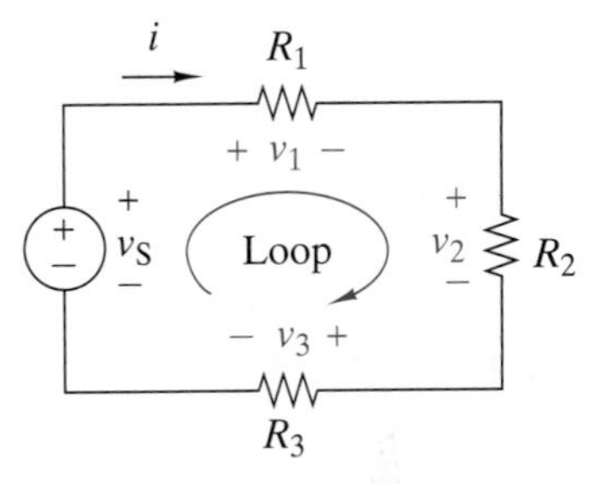

FIGURE 2–35 *A voltage divider circuit.*

The elements in Figure 2–35 are connected in series, so the same current i exists in each of the resistors. Using Ohm's law, we find that

$$v_S = R_1 i + R_2 i + R_3 i \tag{2–26}$$

Solving for i yields

$$i = \frac{v_S}{R_1 + R_2 + R_3} \tag{2–27}$$

Once the current in the series circuit is found, the voltage across each resistor is computed using Ohm's law:

$$v_1 = R_1 i = \left(\frac{R_1}{R_1 + R_2 + R_3}\right) v_S \tag{2–28}$$

$$v_2 = R_2 i = \left(\frac{R_2}{R_1 + R_2 + R_3}\right) v_S \tag{2–29}$$

$$v_3 = R_3 i = \left(\frac{R_3}{R_1 + R_2 + R_3}\right) v_S \tag{2–30}$$

Looking over these results, we see an interesting pattern. In a series connection, the voltage across each resistor is equal to its resistance divided by the equivalent series resistance of the connection times the voltage across the series circuit. Thus, the general expression of the *voltage division rule* is

$$v_k = \left(\frac{R_k}{R_{EQ}}\right) v_{TOTAL} \tag{2–31}$$

In other words, the total voltage divides among the series resistors in proportion to their resistance over the equivalent resistance of the series connection. The following examples show several applications of this rule.

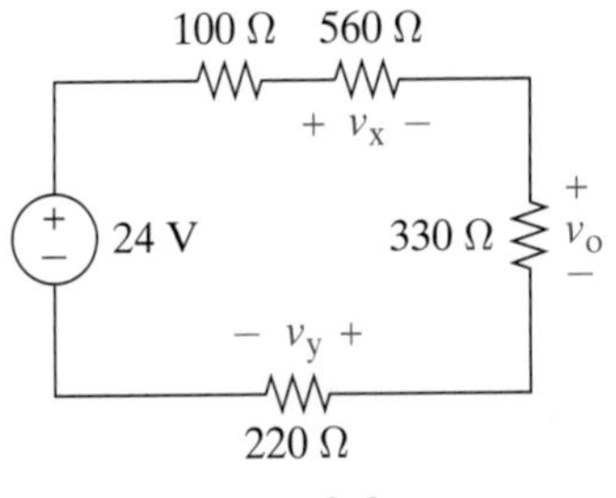

FIGURE 2–36

EXAMPLE 2–13

Find the voltage across the 330-Ω resistor in the circuit of Figure 2–36.

SOLUTION:

Applying the voltage division rule, we find that

$$v_O = \left(\frac{330}{100 + 560 + 330 + 220}\right)24 = 6.55 \text{ V}$$ ■

Exercise 2–10

Find the voltages v_x and v_y in Figure 2–36.

Answers: $v_x = 11.1$ V; $v_y = 4.36$ V

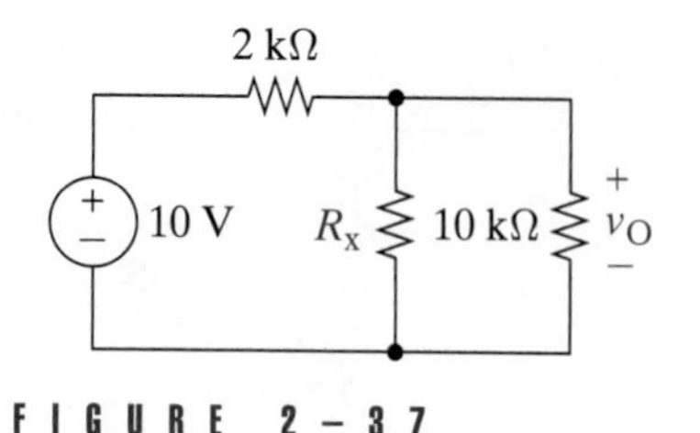

FIGURE 2–37

EXAMPLE 2–14

Select a value for the resistor R_x in Figure 2–37 so $v_O = 8$ V.

SOLUTION:

The unknown resistor is in parallel with the 10-kΩ resistor. Since voltages across parallel elements are equal, the voltage $v_O = 8$ V appears across both. We first define an equivalent resistance $R_{EQ} = R_x \parallel 10$ kΩ as

$$R_{EQ} = \frac{R_x \times 10000}{R_x + 10000}$$

We write the voltage division rule in terms of R_{EQ} as

$$v_O = 8 = \left(\frac{R_{EQ}}{R_{EQ} + 2000}\right)10$$

which yields $R_{EQ} = 8$ kΩ. Finally, we substitute this value into the equation defining R_{EQ} and solve for R_x to obtain $R_x = 40$ kΩ. ■

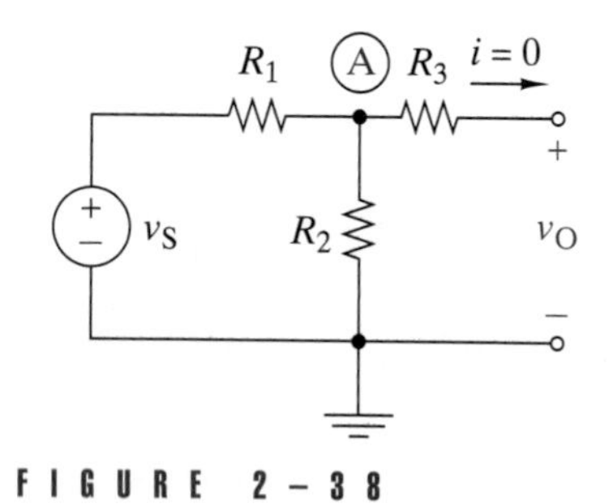

FIGURE 2–38

EXAMPLE 2–15

Use the voltage division rule to find the output voltage v_O of the circuit in Figure 2–38.

SOLUTION:

At first glance it appears that the voltage division rule does not apply, since the resistors are not connected in series. However, the current through R_3 is zero since the output of the circuit is an open circuit. Therefore, Ohm's law shows that $v_3 = R_3 i_3 = 0$. Applying KCL at node A shows that the same current exists in R_1 and R_2, since the current through R_3 is zero. Applying KVL around the output loop shows that the voltage across R_2 must be equal to v_O since the voltage across R_3 is zero. In essence, it is as if R_1 and R_2 were connected in series. Therefore, voltage division can be used and yields the output voltage as

$$v_O = \left(\frac{R_2}{R_1 + R_2}\right) v_S$$

The reader should carefully review the logic leading to this result because voltage division applications of this type occur frequently. ■

APPLICATION EXAMPLE 2–16

The operation of a potentiometer is based on the voltage division rule. The device is a three-terminal element that uses voltage (potential) division to meter out a fraction of the applied voltage. Simply stated, a **potentiometer** is an adjustable voltage divider. Figure 2–39 shows the circuit symbol of a potentiometer, sketches of three different types of actual potentiometers, and a typical application.

The voltage v_O in Figure 2–39(c) can be adjusted by turning the shaft on the potentiometer to move the wiper arm contact. Using the voltage division rule, the voltage v_O is found as

$$v_O = \left(\frac{R_{TOTAL} - R_1}{R_{TOTAL}}\right) v_S \tag{2-32}$$

Adjusting the movable wiper arm all the way to the top makes R_1 zero, and voltage division yields

$$v_O = \left(\frac{R_{TOTAL} - 0}{R_{TOTAL}}\right) v_S = v_S \tag{2-33}$$

In other words, 100% of the applied voltage is delivered to the rest of the circuit. Moving the wiper all the way to the bottom makes R_1 equal to R_{TOTAL}, and voltage division yields

$$v_O = \left(\frac{R_{TOTAL} - R_{TOTAL}}{R_{TOTAL}}\right) v_S = 0 \tag{2-34}$$

This opposite extreme delivers zero voltage. By adjusting the wiper arm position, we can obtain an output voltage anywhere between zero and the applied voltage v_S. When the wiper is positioned halfway between the top and bottom, we naturally expect to obtain half of the applied voltage. Setting $R_1 = \frac{1}{2}R_{TOTAL}$ yields

$$v_O = \left(\frac{R_{TOTAL} - \frac{1}{2}R_{TOTAL}}{R_{TOTAL}}\right) v_S = \frac{v_S}{2} \tag{2-35}$$

as expected. The many applications of the potentiometer include volume controls, voltage balancing, and fine-tuning adjustment.

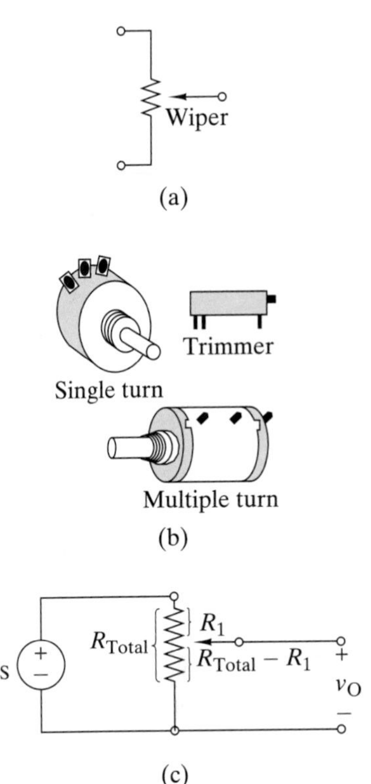

FIGURE 2–39 *The potentiometer: (a) Circuit symbol. (b) Actual device. (c) An application.*

EXAMPLE 2–17

Figure 2–40 shows a programmable voltage divider in which the digital signals D_1 and D_0 control the divider resistance R_1 and R_2 by opening and closing the analog switches shown. Determine the output v_O when $(D_1, D_0) = (0, 0), (0, 1), (1, 0)$, and $(1, 1)$. Assume that the analog switches are ideal switches.

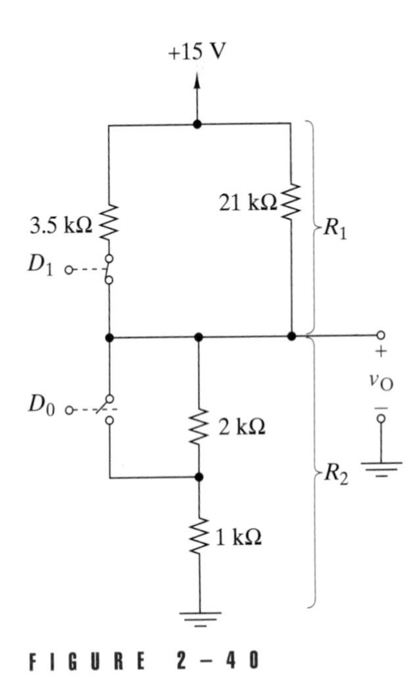

FIGURE 2–40

SOLUTION:
When $(D_1, D_0) = (0, 0)$ the upper switch is closed and the lower switch is open. In this configuration the divider resistances are

$$R_1 = 3.5\text{ k}\Omega \| 21\text{ k}\Omega = 3\text{ k}\Omega \qquad \text{and} \qquad R_2 = 2\text{ k}\Omega + 1\text{ k}\Omega = 3\text{ k}\Omega$$

and the output voltage is

$$v_O = \frac{R_2}{R_1 + R_2} \times 15 = \frac{1}{2} \times 15 = 7.5 \text{ V}$$

When $(D_1, D_0) = (0, 1)$ both switches are closed and the divider resistances are

$$R_1 = 3.5 \text{ k}\Omega \| 21 \text{ k}\Omega = 3 \text{ k}\Omega \qquad \text{and} \qquad R_2 = 1 \text{ k}\Omega$$

and the output voltage is

$$v_O = \frac{R_2}{R_1 + R_2} \times 15 = \frac{1}{4} \times 15 = 3.75 \text{ V}$$

When $(D_1, D_0) = (1, 0)$ both switches are open and the divider resistances are

$$R_1 = 21 \text{ k}\Omega \qquad \text{and} \qquad R_2 = 2 \text{ k}\Omega + 1 \text{ k}\Omega = 3 \text{ k}\Omega$$

and the output voltage is

$$v_O = \frac{R_2}{R_1 + R_2} \times 15 = \frac{1}{8} \times 15 = 1.875 \text{ V}$$

Finally, when $(D_1, D_0) = (1, 1)$ the upper switch is open, the lower switch is closed, and the divider resistances are

$$R_1 = 21 \text{ k}\Omega \qquad \text{and} \qquad R_2 = 1 \text{ k}\Omega$$

and the output voltage is

$$v_O = \frac{R_2}{R_1 + R_2} \times 15 = \frac{1}{22} \times 15 = 0.682 \text{ V}$$

This example illustrates using digital signals to control an analog signal processor (the divider). ■

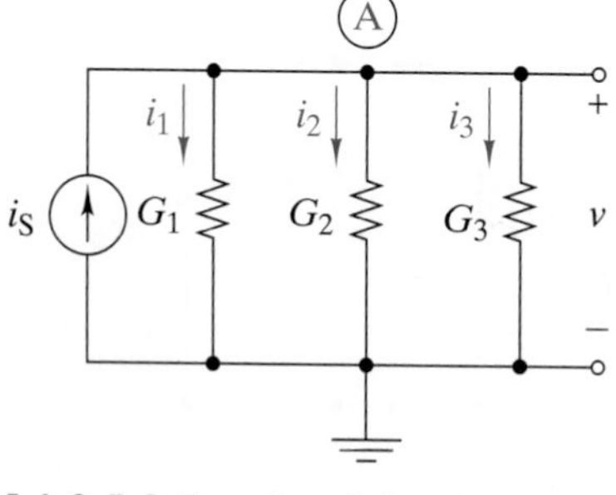

FIGURE 2–41 *A current divider circuit.*

CURRENT DIVISION

Current division provides a simple way to find the current through each element in a parallel circuit. Figure 2–41 shows a parallel circuit that lends itself to solution by current division. Applying KCL at node A yields

$$i_S = i_1 + i_2 + i_3$$

The voltage v appears across all three conductances since they are connected in parallel. Using Ohm's law, we can write

$$i_S = vG_1 + vG_2 + vG_3$$

Solving for v yields

$$v = \frac{i_S}{G_1 + G_2 + G_3}$$

Given the voltage v, the current through any element is found using Ohm's law as

$$i_1 = vG_1 = \left(\frac{G_1}{G_1 + G_2 + G_3}\right) i_S \tag{2–36}$$

$$i_2 = vG_2 = \left(\frac{G_2}{G_1 + G_2 + G_3}\right) i_S \tag{2–37}$$

$$i_3 = vG_3 = \left(\frac{G_3}{G_1 + G_2 + G_3}\right) i_S \tag{2–38}$$

These results show that the source current divides among the parallel resistors in proportion to their conductances divided by the equivalent conductances in the parallel connection. Thus, the general expression for the *current division rule* is

$$i_k = \left(\frac{G_k}{G_{EQ}}\right) i_{TOTAL} \tag{2–39}$$

Sometimes it is useful to express the current division rule in terms of resistance rather than conductance. For the two-resistor case in Figure 2–42, the current i_1 is found using current division as

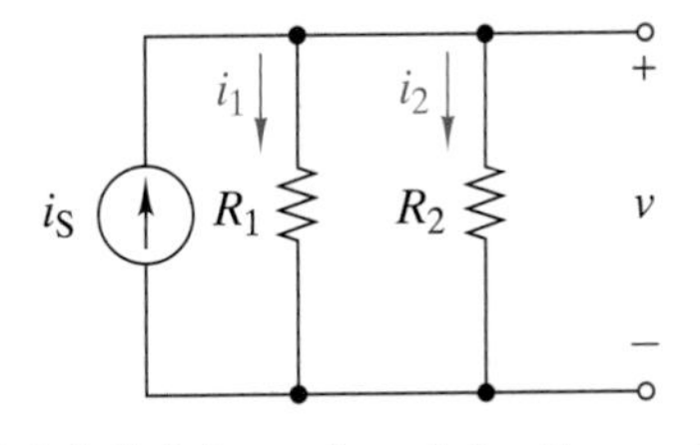

FIGURE 2–42 Two-path current divider circuit.

$$i_1 = \left(\frac{G_1}{G_1 + G_2}\right) i_S = \frac{\frac{1}{R_1}}{\frac{1}{R_1} + \frac{1}{R_2}} i_S = \left(\frac{R_2}{R_1 + R_2}\right) i_S \tag{2–40}$$

Similarly, the current i_2 in Figure 2–42 is found to be

$$i_2 = \left(\frac{G_2}{G_1 + G_2}\right) i_S = \frac{\frac{1}{R_2}}{\frac{1}{R_1} + \frac{1}{R_2}} i_S = \left(\frac{R_1}{R_1 + R_2}\right) i_S \tag{2–41}$$

These two results lead to the following *two-path current division rule:* When a circuit can be reduced to two equivalent resistances in parallel, the current through one resistance is equal to the other resistance divided by the sum of the two resistances times the total current entering the parallel combination.

Caution: Equations (2–40) and (2–41) only apply when the circuit is reduced to two parallel paths in which one path contains the desired current and the other path is the equivalent resistance of all other paths.

EXAMPLE 2–18

Find the current i_x in Figure 2–43(a).

SOLUTION:

To find i_x, we reduce the circuit to two paths, a path containing i_x and a path equivalent to all other paths, as shown in Figure 2–43(b). Now we can use the two-path current divider rule as

$$i_x = \frac{6.67}{20 + 6.67} \times 5 = 1.25 \text{ A}$$

■

FIGURE 2–43

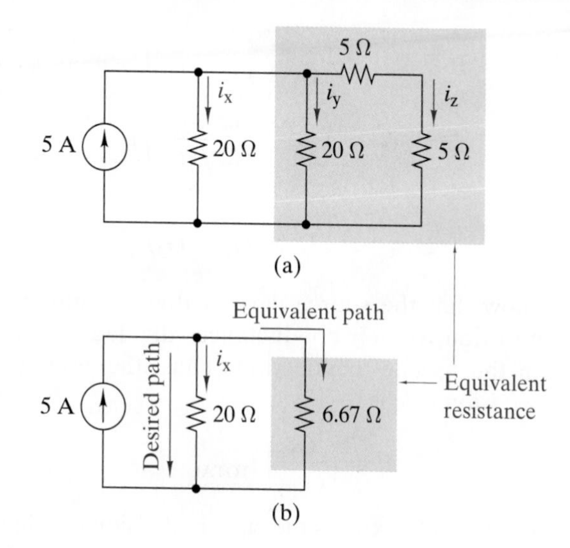

Exercise 2–11

(a) Find i_y and i_z in the circuit of Figure 2–43(a).
(b) Show that the sum of i_x, i_y, and i_z equals the source current.

Answers:

(a) $i_y = 1.25$ A, $i_z = 2.5$ A
(b) $i_x + i_y + i_z = 5$ A

Exercise 2–12

The circuit in Figure 2–44 shows a delicate device that is modeled by a 90-Ω equivalent resistance. The device requires a current of 1 mA to operate properly. A 1.5-mA fuse is inserted in series with the device to protect it from overheating. The resistance of the fuse is 10 Ω. Without the shunt resistance R_x, the source would deliver 5 mA to the device, causing the fuse to blow. Inserting a shunt resistor R_x diverts a portion of the available source current around the fuse and device. Select a value of R_x so only 1 mA is delivered to the device.

FIGURE 2–44

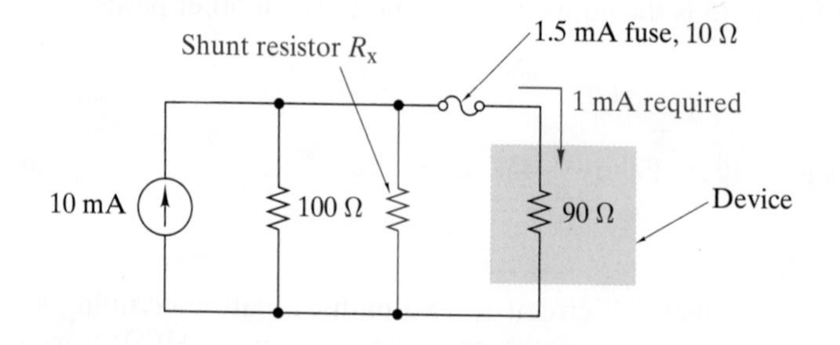

Answer: $R_x = 12.5$ Ω.

APPLICATION EXAMPLE 2–19

The *R*-2*R* ladder circuit in Figure 2–45 is a *binary current divider* that finds applications in digital-to-analog signal conversion. The operation of this circuit can be explained using current division together with series and parallel equivalent resistance. The equivalent resistance connected to ground at node 3 is $2R \| 2R = R$, which means that the equivalent resistance seen to the right of node 2 of $R + R = 2R$. This in turn means that the total equivalent resistance connected to ground at node 2 is $2R \| 2R = R$ and hence equivalent resistance seen to the right of node 1 of $R + R = 2R$. The net result is that the equivalent resistance seen to the right of each numbered node is $2R$.

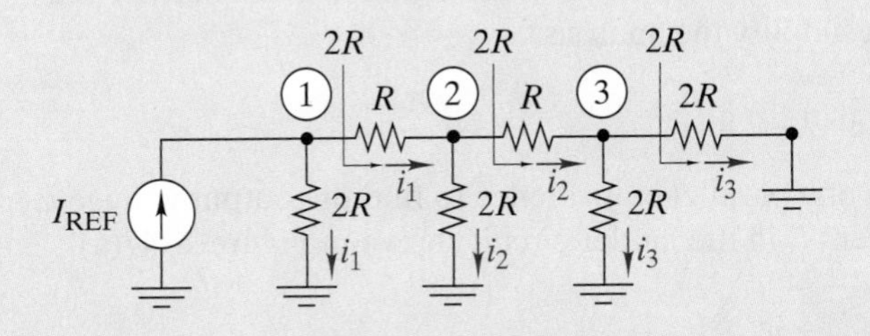

FIGURE 2–45 R-2R *binary circuit divider.*

The reference current I_{REF} entering node 1 divides equally between the two available $2R$ paths with the result that $i_1 = I_{REF}/2$, and the current into node 2 is also $i_1 = I_{REF}/2$. At node 2 this current again divides equally between the two $2R$ paths with the result that $i_2 = i_1/2 = I_{REF}/4$ and the current into node 3 is $i_2 = I_{REF}/4$. Finally, at node 3 this current divides equally once more so that $i_3 = i_2/2 = I_{REF}/8$. In sum, the currents in the $2R$ resistors connected to ground are all of the form $i_k = I_{REF}/2^k$, where k is the node number to which the resistor is connected. Thus, the *R*-2*R* ladder circuit produces signals (currents in this case) that decrease in a binary fashion as we proceed down the ladder.

Clearly the *R*-2*R* ladder can be extended to a larger number of nodes. Commercially available integrated circuit ladders have as many as 8 numbered nodes producing binary currents ranging from $I_{REF}/2$ to $I_{REF}/256$. The advantage of this circuit is that it produces this wide range of precisely related signals using only two values of resistance, namely R and $2R$. This greatly simplifies the fabrication of the *R*-2*R* ladder in integrated circuit form.

2–6 CIRCUIT REDUCTION

The concepts of series/parallel equivalence, voltage/current division, and source transformations can be used to analyze **ladder circuits** of the type shown in Figure 2–46. The basic analysis strategy is to reduce the circuit to a simpler equivalent in which the output is easily found by voltage or current division or Ohm's law. There is no fixed pattern to the reduction process, and much depends on the insight of the analyst. In any case, with circuit reduction we work directly with the circuit model, and so the process gives us insight into circuit behavior.

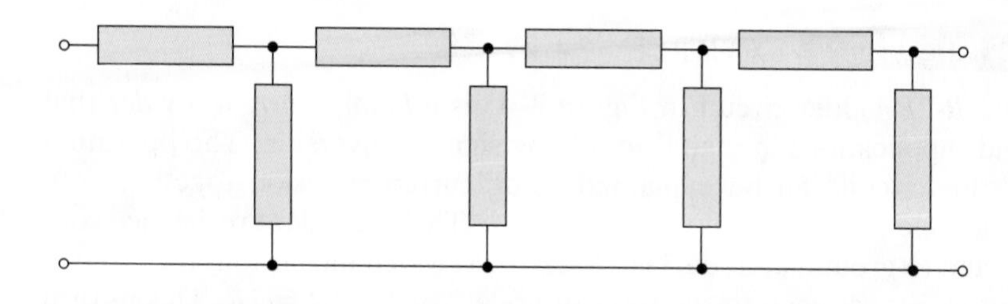

FIGURE 2–46 A ladder circuit.

With circuit reduction the desired unknowns are found by simplifying the circuit and, in the process, eliminating certain nodes and elements. However, we must be careful not to eliminate a node or element that includes the desired unknown voltage or current. The next three examples illustrate circuit reduction. The final example shows that rearranging the circuit can simplify the analysis.

EXAMPLE 2–20

Use series and parallel equivalence to find the output voltage v_O and the input current i_S in the ladder circuit shown in Figure 2–47(a).

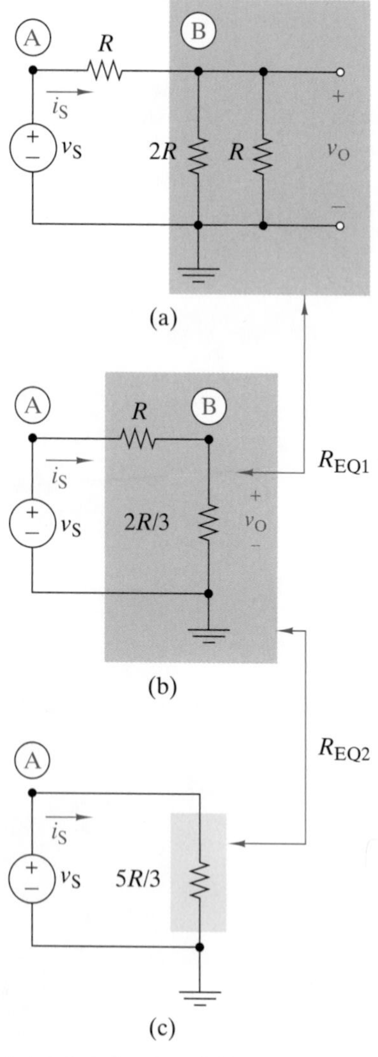

FIGURE 2–47

SOLUTION:

One approach is to combine parallel resistors and use voltage division to find v_O, and then combine all resistances into a single equivalent to find the input current i_S. Figure 2–47(b) shows the step required to determine the equivalent resistance between the terminals B and ground. The equivalent resistance of the parallel $2R$ and R resistors is

$$R_{EQ1} = \frac{R \times 2R}{R + 2R} = \frac{2}{3}R$$

The reduced circuit in Figure 2–47(b) is a voltage divider. Notice that the two nodes needed to find the voltage v_O have been retained. The unknown voltage is found in terms of the source voltage as

$$v_O = \frac{\frac{2}{3}R}{\frac{2}{3}R + R} v_S = \frac{2}{5} v_S$$

The input current is found by combining the equivalent resistance found previously with the remaining resistor R to obtain

$$R_{EQ2} = R + R_{EQ1}$$

$$= R + \frac{2}{3}R = \frac{5}{3}R$$

Application of series/parallel equivalence has reduced the ladder circuit to the single equivalent resistance shown in Figure 2–47(c). Using Ohm's law, the input current is

$$i_S = \frac{v_S}{R_{EQ2}} = \frac{3}{5}\frac{v_S}{R}$$

Notice that the reduction step between Figures 2–47(b) and 2–47(c) eliminates node B, so the output voltage v_O must be calculated before this reduction step is taken. ■

EXAMPLE 2–21

Use source transformations to find the output voltage v_O and the input current i_S in the ladder circuit shown in Figure 2–48(a).

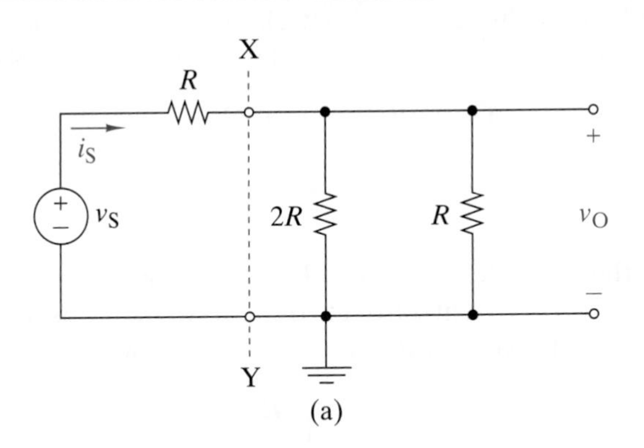

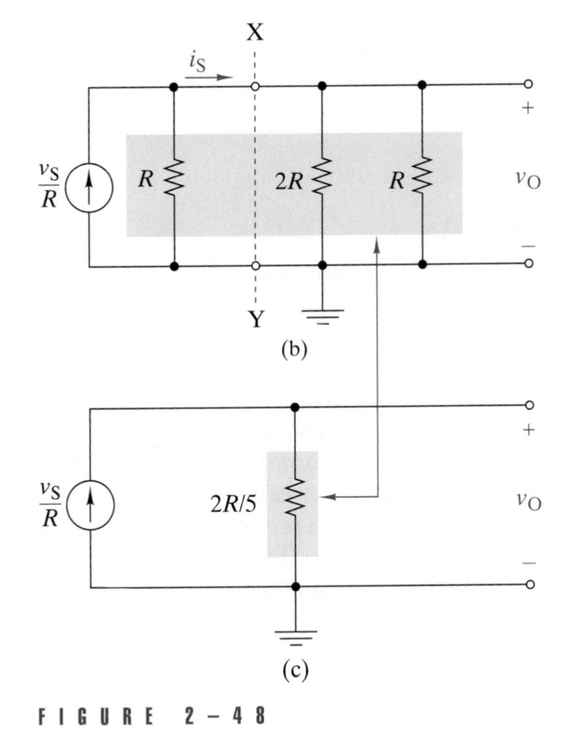

FIGURE 2–48

SOLUTION:

Figure 2–48 shows another way to reduce the circuit analyzed in Example 2–20. Breaking the circuit at points X and Y in Figure 2–48(a) produces a voltage source v_S in series with a resistor R. Using source transformation, this combination can be replaced by an equivalent current source in parallel with the same resistor, as shown in Figure 2–48(b).

Caution: The current source v_S/R is *not* the input current i_S, as is indicated in Figure 2–48(b). Applying the two-path current division rule to the circuit in Figure 2–48(b) yields the input current i_S as

$$i_S = \frac{R}{\frac{2}{3}R + R} \times \frac{v_S}{R} = \frac{v_S}{\frac{5}{3}R} = \frac{3}{5}\frac{v_S}{R}$$

The three parallel resistances in Figure 2–48(b) can be combined into a single equivalent conductance without eliminating the node pair used to define the output voltage v_O. Using parallel equivalence, we obtain

$$\begin{aligned} G_{EQ} &= G_1 + G_2 + G_3 \\ &= \frac{1}{R} + \frac{1}{2R} + \frac{1}{R} = \frac{5}{2R} \end{aligned}$$

which yields the equivalent circuit in Figure 2–48(c). The current source v_S/R determines the current through the equivalent resistance in Figure 2–48(c). The output voltage is found using Ohm's law.

$$v_O = \left(\frac{v_S}{R}\right) \times \left(\frac{2R}{5}\right) = \frac{2}{5}v_S$$

Of course, these results are the same as the result obtained in Example 2–20, except that here they were obtained using a different sequence of circuit reduction steps. ■

EXAMPLE 2–22

Find v_x in the circuit shown in Figure 2–49(a).

SOLUTION:

In the two previous examples the unknown responses were defined at the circuit input and output. In this example the unknown voltage appears across a 10-Ω resistor in the center of the network. The approach is to reduce the circuit at both ends while retaining the 10-Ω resistor defining v_x. Applying a source transformation to the left of terminals X–Y and a series reduction to the two 10-Ω resistors on the far right yields the reduced circuit shown in Figure 2–49(b). The two pairs of 20-Ω resistors connected in parallel can be combined to produce the circuit in Figure 2–49(c). At this point there are several ways to proceed. For example, a source transformation at the points W–Z in Figure 2–49(c) produces the circuit in Figure 2–49(d). Using voltage division in Figure 2–49(d) yields v_x.

$$v_x = \frac{10}{10 + 10 + 10} \times 7.5 = 2.5 \text{ V}$$

Yet another approach is to use the two-path current division rule in Figure 2–49(c) to find the current i_x.

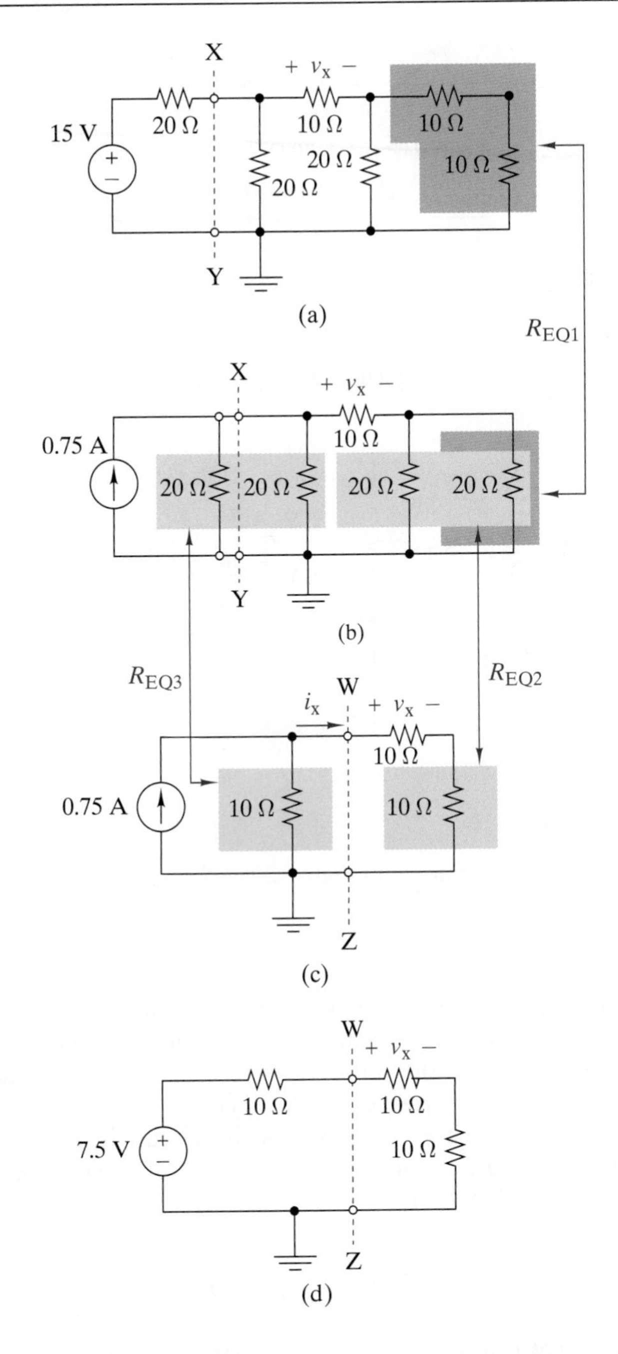

FIGURE 2–49

$$i_x = \frac{10}{10 + 10 + 10} \times \frac{3}{4} = \frac{1}{4} \text{ A}$$

Then, applying Ohm's law to obtain v_x,

$$v_x = 10 \times i_x = 2.5 \text{ V}$$ ■

Exercise 2–13

Find v_x and i_x using circuit reduction on the circuit in Figure 2–50.

FIGURE 2–50

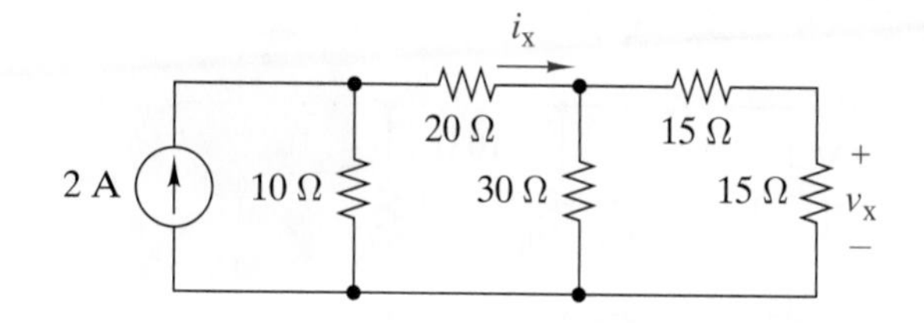

Answers: $v_x = 3.33$ V; $i_x = 0.444$ A

Exercise 2–14

Find v_x and v_y using circuit reduction on the circuit in Figure 2–51.

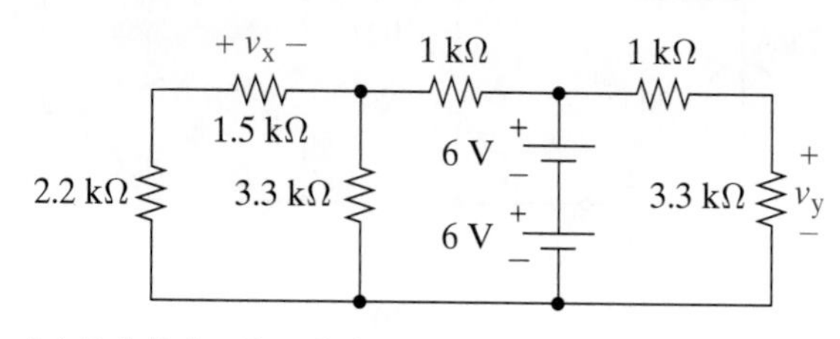

FIGURE 2–51

Answers: $v_x = -3.09$ V; $v_y = 9.21$ V

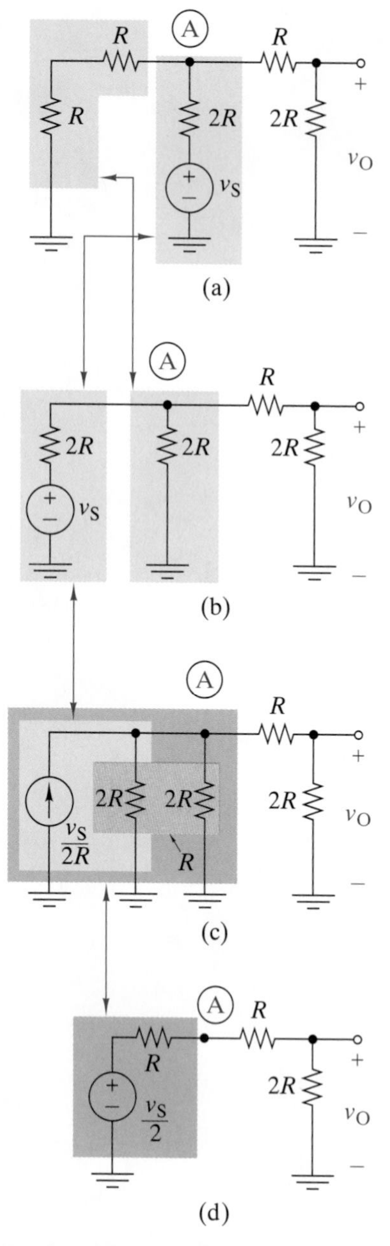

FIGURE 2–52

EXAMPLE 2–23

Using circuit reduction, find v_O in Figure 2–52(a).

SOLUTION:

One way to solve this problem is to notice that the source branch and the leftmost two-resistor branch are connected in parallel between node A and ground. Switching the order of these branches and replacing the two resistors by their series equivalent yields the circuit of Figure 2–52(b). A source transformation yields the circuit in Figure 2–52(c). This circuit contains a current source $v_S/2R$ in parallel with two $2R$ resistances whose equivalent resistance is

$$R_{EQ} = 2R \| 2R = \frac{2R \times 2R}{2R + 2R} = R$$

Applying a source transformation to the current source $v_S/2R$ in parallel with R_{EQ} results in the circuit of Figure 2–52(d), where

$$V_{EQ} = \left(\frac{v_S}{2R}\right) \times R_{EQ} = \left(\frac{v_S}{2R}\right) R = \frac{v_S}{2}$$

Finally, applying voltage division in Figure 2–52(d) yields

$$v_O = \left(\frac{2R}{R + R + 2R}\right)\frac{v_S}{2} = \frac{v_S}{4}$$

■

Exercise 2–15

Find the voltage across the current source in Figure 2–53.

Answer: $v_S = -0.225$ V

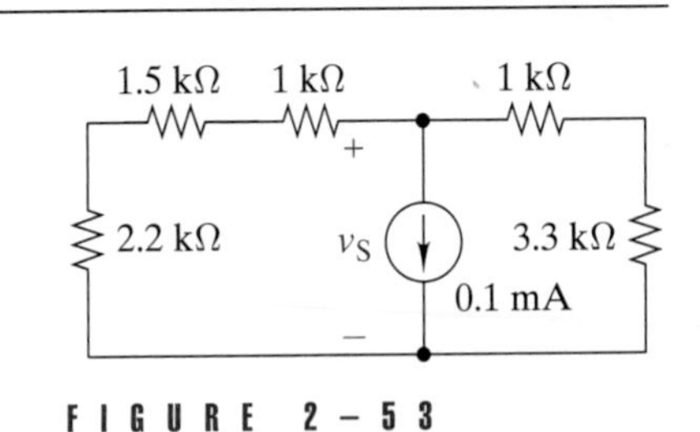

FIGURE 2–53

2–7 COMPUTER-AIDED CIRCUIT ANALYSIS

In this book we use three types of computer programs to illustrate computer-aided circuit analysis, namely spreadsheets, math solvers, and circuit simulators. Practicing engineers routinely use these tools to analyze and design circuits, and so it is important to learn how to use them effectively. The purpose of having computer examples in this book is to help you develop an analysis style that includes the intelligent use of computer tools. As you develop your style, always keep in mind that computer tools are not problem solvers. *You* are the problem solver. Computer tools can be very useful, even essential once the problem is defined. But they do not substitute for an understanding of the fundamentals needed to formulate the problem, identify a practical approach, and interpret analysis results.

There are 32 worked examples in the text that use computer tools. The spreadsheet examples use Microsoft Excel. The math solver examples use Mathcad by MathSoft and MATLAB by the MathWorks, Inc. The circuit simulation examples use Orcad Family Lite Edition by Cadence Design Systems, Inc, and Electronics WorkBench by Interactive Image Technologies, Inc.

However, keep in mind that our objective is to illustrate the effective use of computer tools rather than develop your ability to operate these specific software programs. Accordingly, this book does not emphasize details of how to operate any of these software tools. We assume that you learned how to operate computer tools in previous courses or have enough familiarity with the WINDOWS operating environment to learn how to do so using online tutorials or any of a number of commercially available paperback manuals.[3]

The following discussion gives a brief overview of circuit simulation, since you may be less familiar with this process than with the use of spreadsheets and math solvers.

CIRCUIT SIMULATION

Most circuit simulation programs are based on a circuit analysis package called SPICE, which is an acronym for **S**imulation **P**rogram with **I**ntegrated **C**ircuit **E**mphasis. The original SPICE program was developed in the 1970s at the University of California at Berkeley. Since then, various companies have added proprietary features to the basic SPICE program to produce an array of SPICE-based commercial products for personal computer and workstation platforms.

3 For example see, *PSpice for Linear Circuits*, by James A Svoboda, John Wiley & Sons, 2002, or *Simulation of Electric Circuits Using Electronic Workbench*, James L. Antonakos, Prentice Hall, 2001.

Figure 2–54 is a block diagram summarizing the major features of a SPICE-based circuit simulation program. The inputs are a circuit diagram and the type of analysis required. In contemporary programs the circuit diagram is drawn on the monitor screen using a graphical schematic editor. When the circuit diagram is complete, the input processor performs a *schematic capture,* a process that documents the circuit in what is called a *netlist.* To initiate circuit simulation, the input processor sends the netlist and analysis commands to the simulation processor. If the netlist file is not properly prepared, the simulation will not run or (worse) will return erroneous results. Hence, it is important to check the netlist to be sure that the circuit it defines is the one you want to analyze.

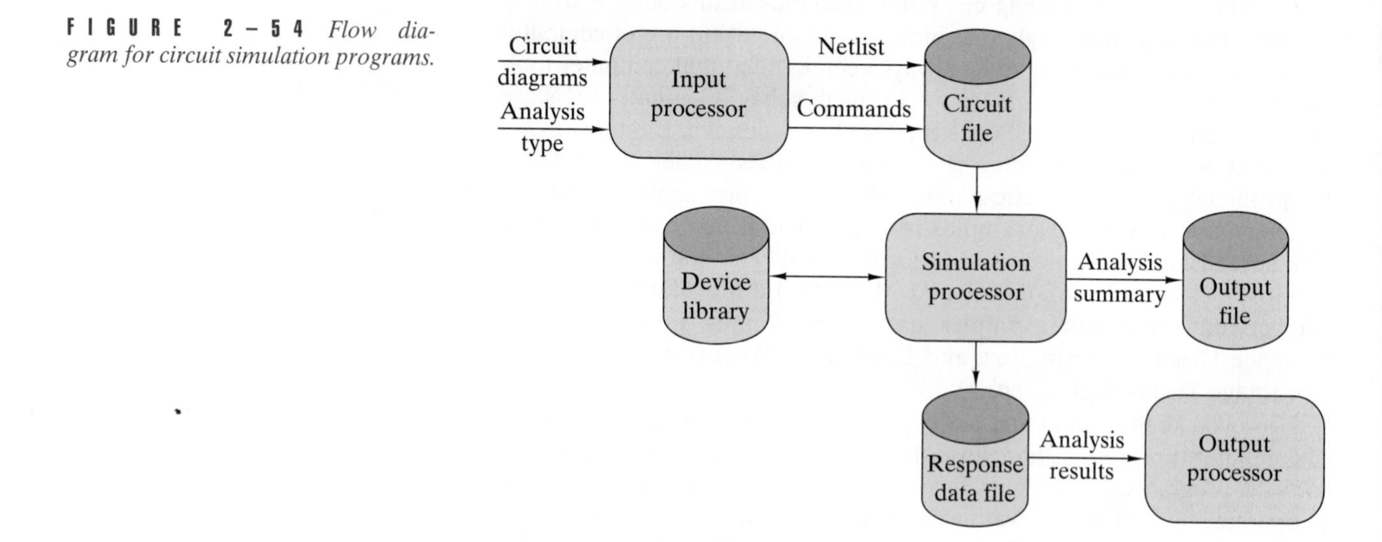

FIGURE 2–54 *Flow diagram for circuit simulation programs.*

The simulation processor uses the netlist together with data from the device library to formulate a set of equations that describes the circuit. The simulation processor then solves the equations, writes a dc analysis summary to a standard SPICE output file, and writes the other analysis results to a response data file. For simple dc analysis, the desired response data are accessible by examining the SPICE output file. For other types of analysis, the output processor can be used to generate graphical plots of the data in the response data file.

In the Orcad Family Lite Edition the input processor is called *Orcad Capture,* the simulation processor is a version of SPICE called *PSpice,* and the output processor is called *Probe*. These three programs allow the circuit diagram to be entered and analysis results viewed graphically on a computer monitor. Electronics WorkBench has three similar functions integrated into a single program called *MultiSIM.*

The following example illustrates a circuit simulation using Orcad.

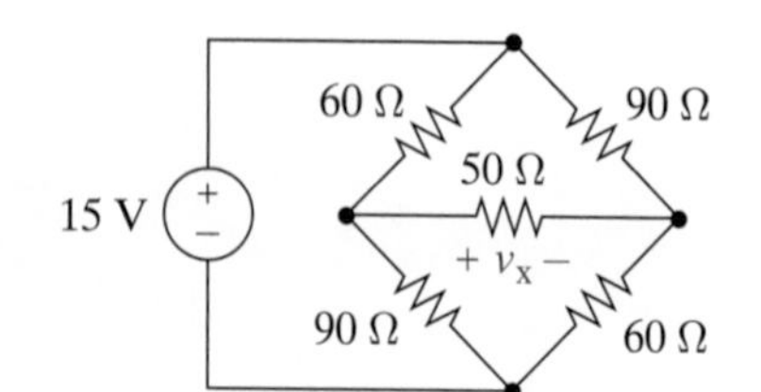

FIGURE 2–55

EXAMPLE 2–24

Use Orcad to find the voltage v_x across the 50-Ω resistor in the circuit of Fig. 2–55.

SOLUTION:

This example illustrates the steps involved in using Orcad to analyze a circuit. To begin, open Orcad Capture and the opening screen will come into view. Select **File\ New** from the main menu. Since this is an analog circuit, select **Analog or Mixed A/D**. Every new project must have a name, hence the circuit was named "Orcad Exercise." Drawing the circuit diagram on the monitor screen requires three actions:

1. Selecting the desired element and placing it on the Orcad Capture workspace.
2. Changing the element's default parameters to the desired values.
3. Connecting the elements together using virtual wires.

Selecting the **Place** button and **Part** causes the "Place Part" dialog box to appear. The various elements needed to build the circuit are found by scrolling down the "Part List." For example, Figure 2–56 shows the "Place Part" dialog box with a resistor selected. Clicking **OK** causes a resistor to appear on the screen with a default label of "R1" and a default value of "1k." These default assignments can be changed by double-clicking on the element designator (R1) and the element value (1k). After the resistors and voltage source are selected and arranged in the capture workspace, the circuit is wired using the **Place** button and selecting **Wire**.

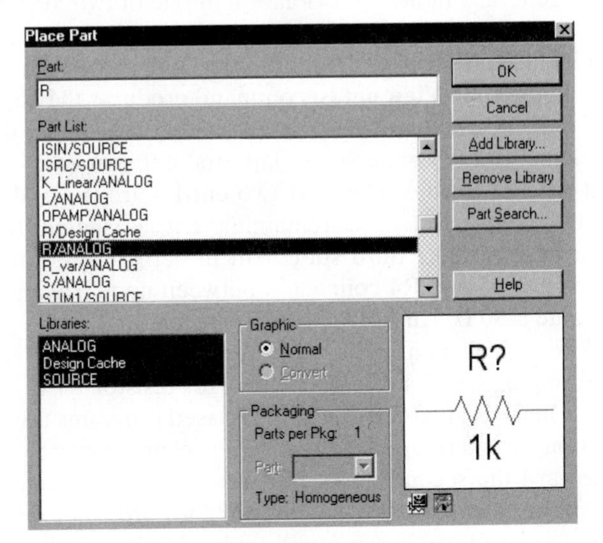

FIGURE 2–56 *Orcad Capture place part menu.*

Figure 2–57 shows the circuit diagram as it might appear in Orcad Capture. Note that in an Orcad circuit diagram, connection points are indicated by a small dot and are assigned a node number. Orcad (as well as Electronics WorkBench) requires that one of the nodes be chosen as ground. This is done by attaching the ground symbol found under the **Place\Ground** button to one of the nodes. The ground node is automatically assigned the number zero.

When the diagram is completed, the circuit is documented using the **PSpice** button and then by selecting the **Create Netlist** command. If there are mistakes, the program will issue error messages. Beginning users may receive an error message reporting that one or more nodes are "floating."

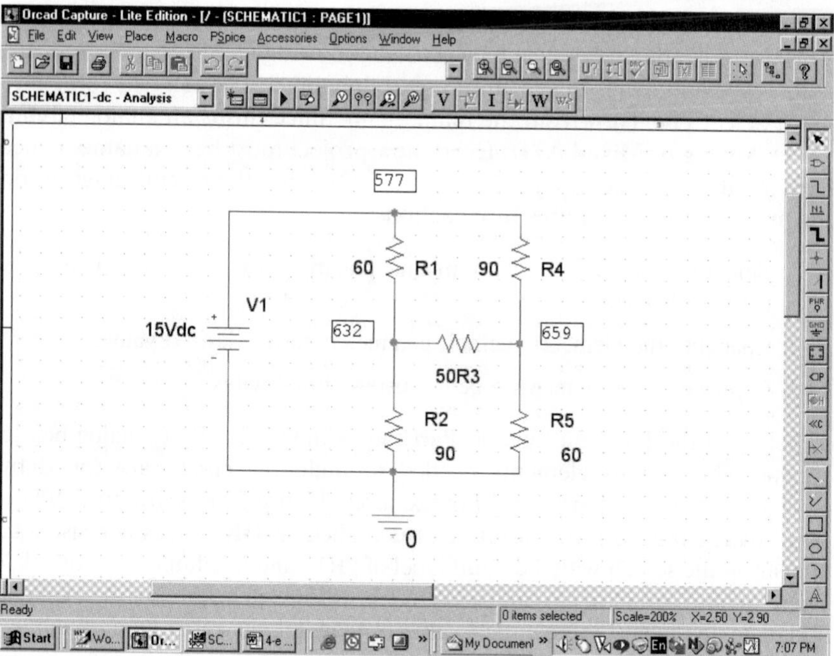

FIGURE 2–57 *Orcad Capture schematic.*

This message generally indicates a violation of one of two SPICE rules: (1) There must be at least two elements connected to every node; or (2) the circuit must include a reference (ground) node.

```
1: * source ORCAD EXERCISE
2: R_R3          N00632 N00659  50
3: R_R1          N00577 N00632  90
4: R_R4          N00577 N00659  90
5: R_R2          N00632 0  60
6: R_R5          N00659 0  60
7: V_V1          N00577 0  15Vdc
8:
```

FIGURE 2–58 *PSpice netlist.*

Using the **PSpice** and **View netlist** command produces the net list display shown in Figure 2–58. A netlist is a sequence of statements that defines the circuit elements and connections. The first entry in each statement is the element type and name. The next two entries are the nodes to which the element is connected. The remaining entries define the element value(s). For example, the third statement in netlist says that the circuit contains a resistor named R4 connected between node 659 and node 577, and whose value is 90 Ω. All SPICE-based programs assign the positive reference mark for the element voltage to the first node in the element statement. For example, the plus reference mark for resistor R1 is at Node 577 and the minus mark is at Node 632. SPICE-based programs use the passive sign convention, so the reference direction for element current is from the first node toward the second node. For example, the reference direction for the current in resistor R2 is from Node 632 toward Node 0 (Ground).

Orcad analysis commands are found under **PSpice** on the main menu. Selecting **New Simulation Profile** produces the dialog box shown in Figure 2–59. In Orcad each different simulation must be given a name—"dc Analysis" in this example. Once a name is provided, a new dialog box appears called "Simulations Settings" containing the various types of analysis that PSpice performs. In the present case we select "Bias Point."[4] This option produces a dc analysis of the circuit and writes the results to the output file. This analysis type is the default option so PSpice always performs a dc analysis even when another analysis type is selected. Also select **General Settings**. Conclude by selecting **OK**.

---

4 In later chapters we will use other analysis options on this menu such as Time Domain (transient), AC Sweep/Noise, and DC Sweep.

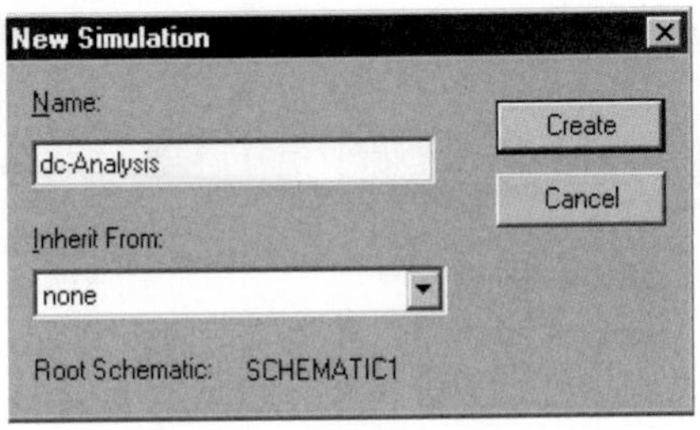

New Simulation

Name: dc-Analysis

Inherit From: none

Root Schematic: SCHEMATIC1

Create

Cancel

FIGURE 2–59 *New simulation menu.*

To perform the analysis select the **PSpice\Run** from the main menu. This causes PSpice to analyze the circuit defined by the netlist and to write the calculated dc responses in the output file. When the analysis is completed select **View\Output File** from the main menu. Paging down through the file we find the data shown in Figure 2–60. The output file lists the voltage between the numbered nodes and ground, namely $v_{577\text{-}0} = 15$ V, $v_{632\text{-}0} = 8.1148$ V, and $v_{639\text{-}0} = 6.8852$ V. The problem statement in this example asks for the voltage v_x across resistor R3 in Figure 2–57. To find this voltage we apply KVL to the loop formed by R2, R3, and R5.

$$-v_{632\text{-}0} + v_x + v_{659\text{-}0} = 0$$

Hence,

$$\begin{aligned} v_x &= v_{632\text{-}0} - v_{659\text{-}0} \\ &= 8.1148 - 6.8852 = 1.2296 \text{ V} \end{aligned}$$

A nice feature of Orcad is that the calculated voltages and/or currents can be displayed directly on the circuit diagram. To do this, return to the diagram on the Orcad Capture screen and select the **V** and **I** buttons on the analysis bar. Selecting these buttons produces Figure 2–61 which displays the calculated values of current in each branch and voltage between each node and ground. ■

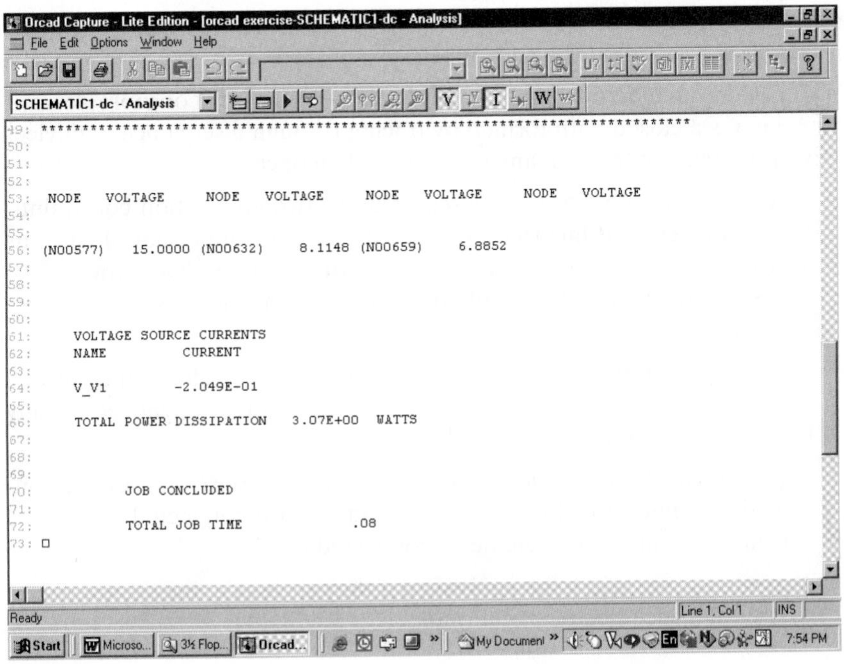

Orcad Capture - Lite Edition - [orcad exercise-SCHEMATIC1-dc - Analysis]

File Edit Options Window Help

SCHEMATIC1-dc - Analysis

```
49: ***************************************************************************
50:
51:
52:
53:  NODE   VOLTAGE     NODE   VOLTAGE     NODE   VOLTAGE     NODE   VOLTAGE
54:
55:
56: (N00577)   15.0000 (N00632)    8.1148 (N00659)    6.8852
57:
58:
59:
60:
61:     VOLTAGE SOURCE CURRENTS
62:     NAME           CURRENT
63:
64:     V_V1          -2.049E-01
65:
66:     TOTAL POWER DISSIPATION   3.07E+00  WATTS
67:
68:
69:
70:           JOB CONCLUDED
71:
72:           TOTAL JOB TIME            .08
73: □
```

Ready   Line 1, Col 1   INS

Start   Microso...   3½ Flop...   Orcad...   My Document   7:54 PM

FIGURE 2–60 *Simulation output file.*

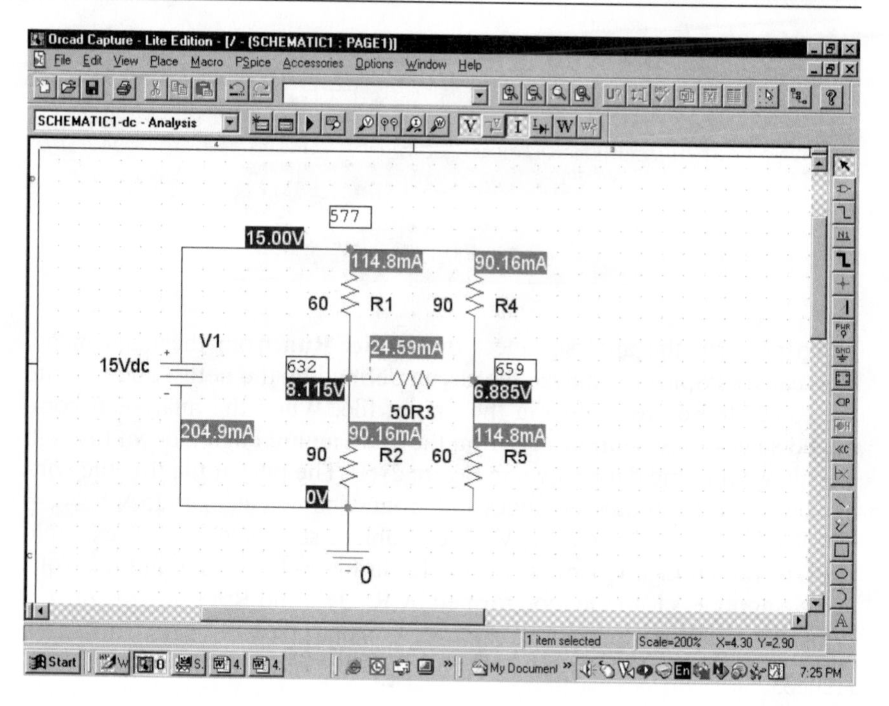

FIGURE 2–61 *Orcad Capture schematic with simulation results.*

SUMMARY

- An **electrical device** is a real physical entity, while a **circuit element** is a mathematical or graphical model that approximates major features of the device.
- Two-terminal circuit elements are represented by a circuit symbol and are characterized by a single constraint imposed on the associated current and voltage variables.
- An **electrical circuit** is an interconnection of electrical devices. The interconnections form nodes and loops.
- A **node** is an electrical juncture of the terminals of two or more devices. A **loop** is a closed path formed by tracing through a sequence of devices without passing through any node more than once.
- Device interconnections in a circuit lead to two connection constraints: **Kirchhoff's current law (KCL)** states that the algebraic sum of currents at a node is zero at every instant; and **Kirchhoff's voltage law (KVL)** states that the algebraic sum of voltages around any loop is zero at every instant.
- A pair of two-terminal elements are connected in **parallel** if they form a loop containing no other elements. The same voltage appears across any two elements connected in parallel.
- A pair of two-terminal elements are connected in **series** if they are connected at a node to which no other elements are connected. The same current exists in any two elements connected in series.
- Two circuits are said to be **equivalent** if they each have the same i–v constraints at a specified pair of terminals.

- Series and parallel equivalence and **voltage** and **current division** are important tools in circuit analysis and design.
- **Source transformation** changes a voltage source in series with a resistor into an equivalent current source in parallel with a resistor, or vice versa.
- **Circuit reduction** is a method of solving for selected signal variables in ladder circuits. The method involves sequential application of the series/parallel equivalence rules, source transformations, and the voltage/current division rules. The reduction sequence used depends on the variables to be determined and the structure of the circuit and is not unique.

PROBLEMS

OBJECTIVE 2–1 ELEMENT CONSTRAINTS (SECT. 2–1)

Given a two-terminal element with one or more electrical variables specified, use the element *i-v* constraint to find the magnitude and direction of the unknown variables.
See Example 2–1

2–1 The current through a 5-kΩ resistor is 12 mA. Find the voltage across the resistor.

2–2 A 6.2-kΩ resistor dissipates 12 mW. Find the current through the resistor.

2–3 The voltage across and current through a resistor are 10 V and 2.5 mA. Find the conductance of the resistor.

2–4 In Figure P2–4 the resistor dissipates 25 mW. Find R_x.

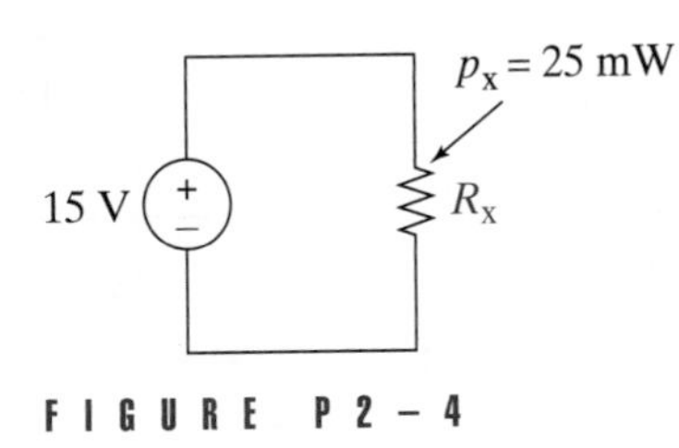

FIGURE P2–4

2–5 In Figure P2–5 find R_x and the power delivered to the resistor.

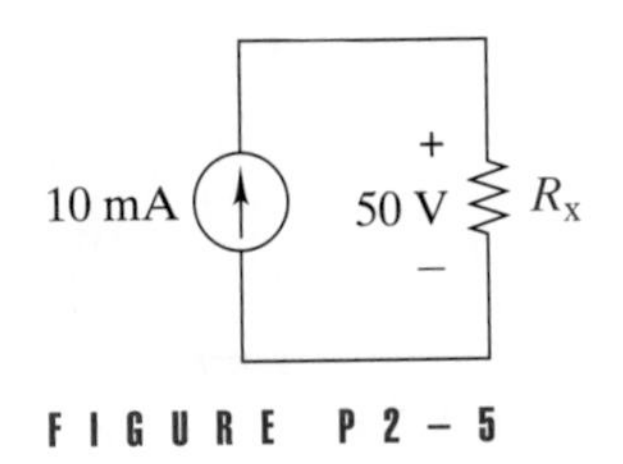

FIGURE P2–5

2–6 The *i–v* relationship of a nonlinear resistor is $v = 75i + 0.2i^3$.

(a) Calculate v and p for $i = \pm 0.5, \pm 1, \pm 2, \pm 5$, and ± 10 A.

(b) Find the maximum error in v when the device is treated as a 75-Ω linear resistance on the range $|i| < 0.5$ A.

2–7 A 10-kΩ resistor has a power rating of 0.25 W. Find the maximum voltage that can be applied to the resistor.

2–8 A certain type of film resistor is available with resistance values between 10 Ω and 100 MΩ. The maximum ratings for all resistors of this type are 500 V and ¼ W. Show that the voltage rating is the controlling limit for $R > 1$ MΩ, and that the power rating is the controlling limit when $R < 1$ MΩ.

2–9 Figure P2–9 shows the circuit symbol for a class of two-terminal devices called diodes. The $i-v$ relationship for a specific pn junction diode is

$$i = 2 \times 10^{-16}(e^{40v} - 1)$$

(a) Use this equation to find i and p for $v = 0, \pm 0.1, \pm 0.2, \pm 0.4$, and ± 0.8 V. Use these data to plot the $i-v$ characteristic of the element.

(b) Is the diode linear or nonlinear, bilateral or nonbilateral, and active or passive?

(c) Use the diode model to predict i and p for $v = 5$ V. Do you think the model applies to voltages in this range? Explain.

(d) Repeat part (c) for $v = -5$ V.

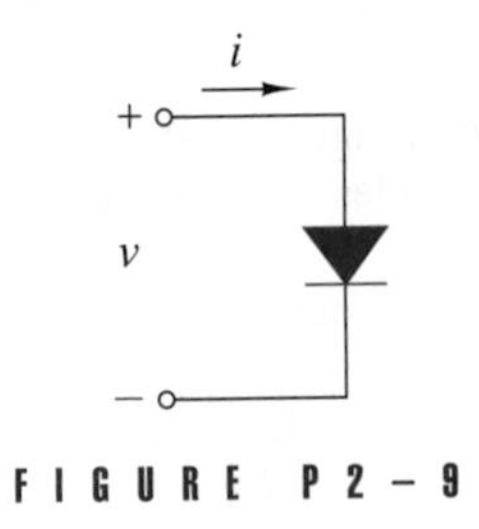

FIGURE P2–9

OBJECTIVE 2–2 CONNECTION CONSTRAINTS (SECT. 2–2)

Given a circuit composed of two-terminal elements:
(a) Identify nodes and loops in the circuit.
(b) Identify elements connected in series and in parallel.
(c) Use Kirchhoff's laws (KCL and KVL) to find selected signal variables.
See Examples 2–4, 2–5, 2–6 and Exercises 2–1, 2–2, 2–3, 2–4, 2–5

2–10 In Figure P2–10 $i_2 = 2$ A and $i_3 = -5$ A. Find i_1 and i_4.

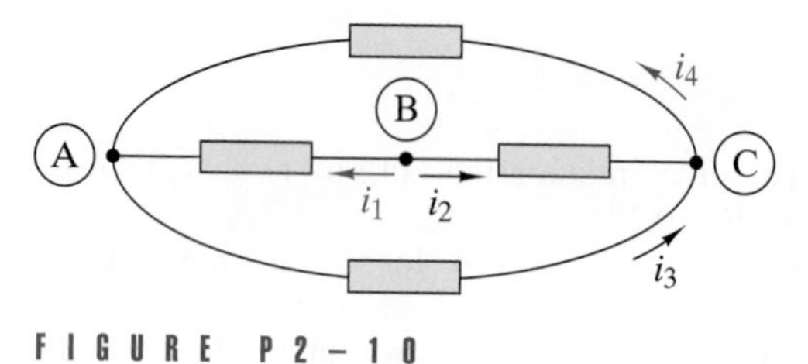

FIGURE P2–10

2–11 For the circuit in Figure P2–11:
(a) Identify the nodes and at least two loops.
(b) Identify any elements connected in series or in parallel.
(c) Write KCL and KVL connection equations for the circuit.

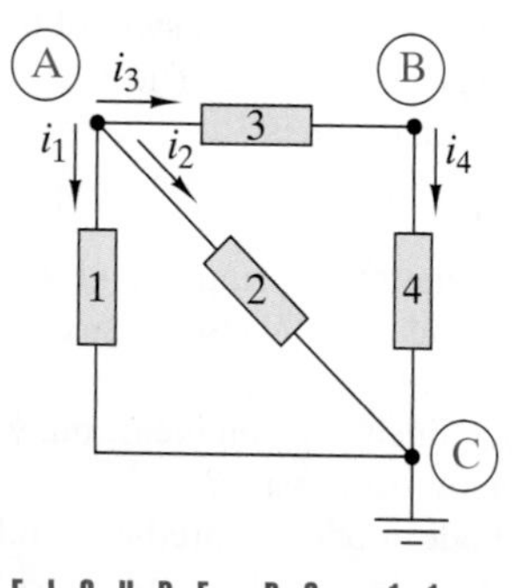

FIGURE P2–11

2–12 In Figure P2–11 $i_2 = 10$ mA and $i_4 = 20$ mA. Find i_1 and i_3.

2–13 For the circuit in Figure P2–13:
(a) Identify the nodes and at least three loops in the circuit.
(b) Identify any elements connected in series or in parallel.
(c) Write KCL and KVL connection equations for the circuit.

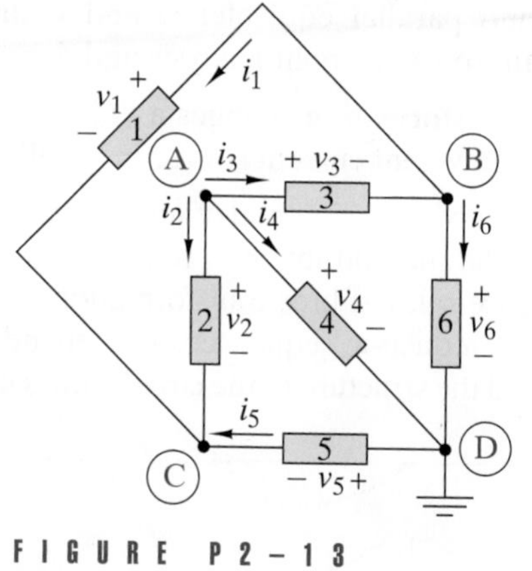

FIGURE P2–13

2–14 In Figure P2–13 $v_2 = 5$V, $v_3 = -8$V, and $v_4 = 3$V. Find v_1, v_5, and v_6.

2–15 The circuit in Figure P2–15 is organized around the three signal lines A, B, and C.
(a) Identify the nodes and at least three loops in the circuit
(b) Write KCL connection equations for the circuit.
(c) If $i_1 = -20$ mA, $i_2 = -12$ mA, and $i_3 = 50$ mA, find i_4, i_5, and i_6.

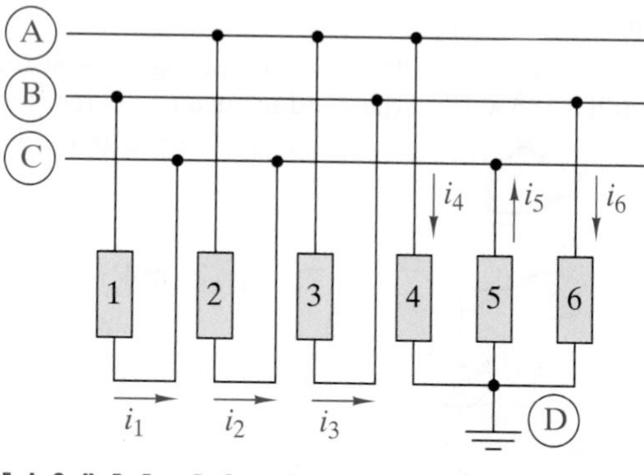

FIGURE P2–15

2–16 In Figure P2–16 $v_2 = 10$ V, $v_3 = 10$ V, and $v_4 = 10$ V. Find v_1 and v_5.

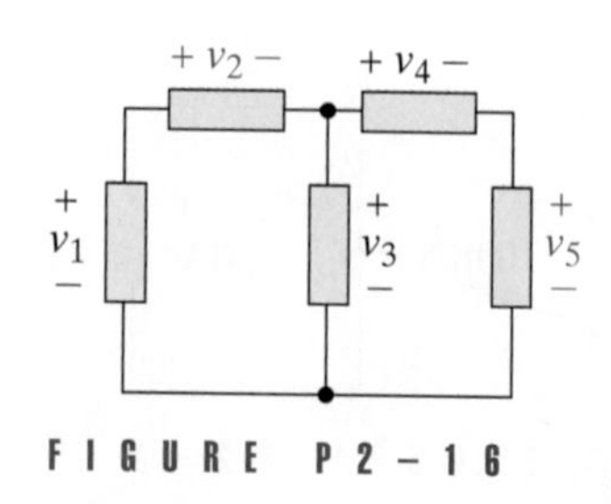

FIGURE P2–16

2–17 In Figure P2–17 $i_2 = 2$ A, $i_3 = -5$ A, and $i_4 = 4$ A. Find i_1 and i_5.

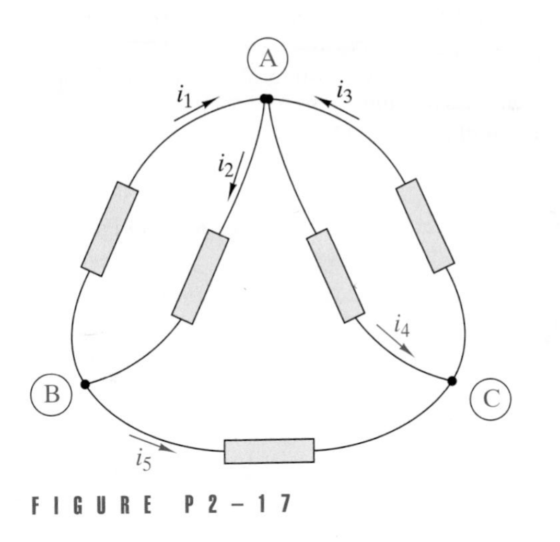

FIGURE P2–17

2–18 Use the passive sign convention to assign voltages variables consistent with the currents in Figure P2–17. Write three KVL connection equations using these voltage variables.

2–19 The KCL equations for a three-node circuit are:

$$\text{Node A: } -i_1 + i_2 - i_4 = 0$$

$$\text{Node B: } -i_2 - i_3 + i_5 = 0$$

$$\text{Node C: } i_1 + i_3 + i_4 - i_5 = 0$$

Draw the circuit diagram and indicate the reference directions for the element currents.

OBJECTIVE 2–3 COMBINED CONSTRAINTS (SECT. 2–3)

Given a circuit consisting of independent sources and linear resistors, use the element constraints and connection constraints to find selected signal variables.
See Examples 2–7, 2–8, 2–9, 2–10 and Exercise 2–6

2–20 Find v_x and i_x in Figure P2–20.

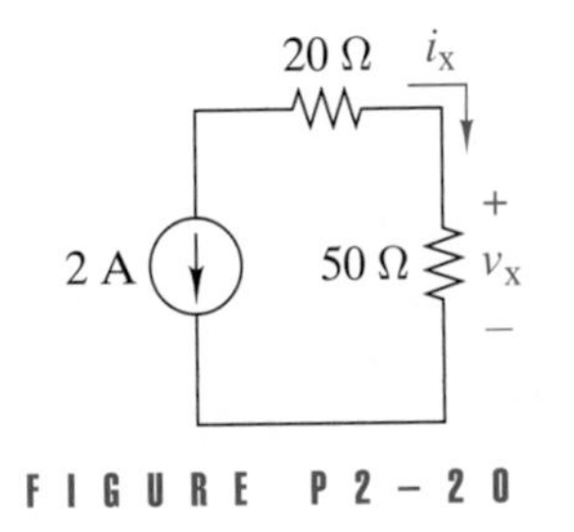

FIGURE P2–20

2–21 Find v_x and i_x in Figure P2–21.

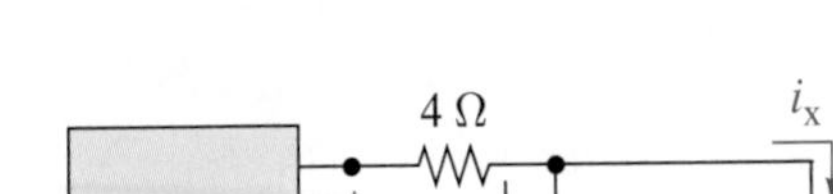
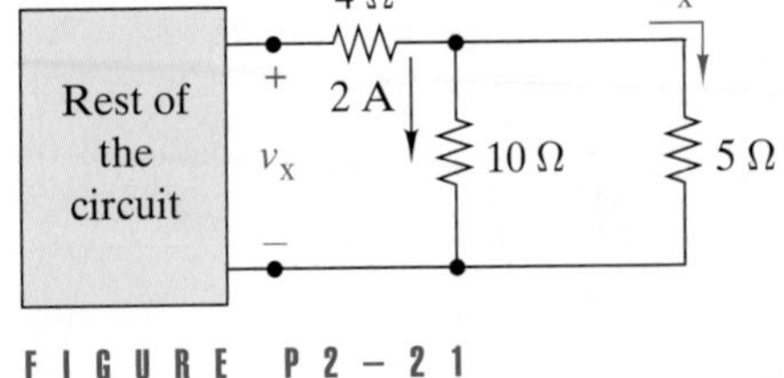

FIGURE P2–21

2–22 In Figure P2–22:

(a) Use the passive sign convention to assign a voltage and current variable to every element.

(b) Use KVL to find the voltage across each resistor.

(c) Use Ohm's law to find the current through each resistor.

(d) Use KCL to find the current through each voltage source.

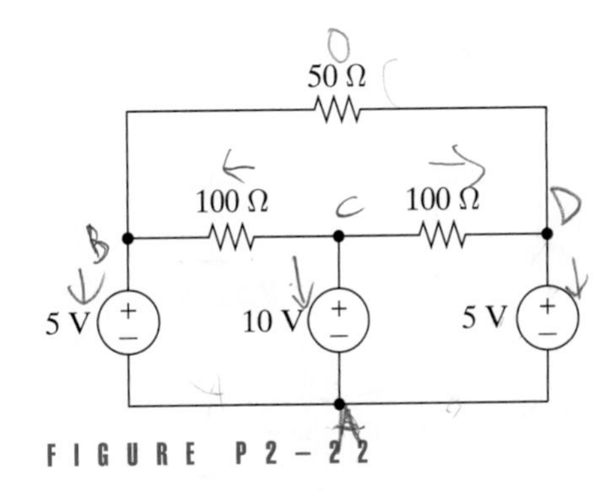

FIGURE P2–22

2–23 Find v_x in Figure P2–23.

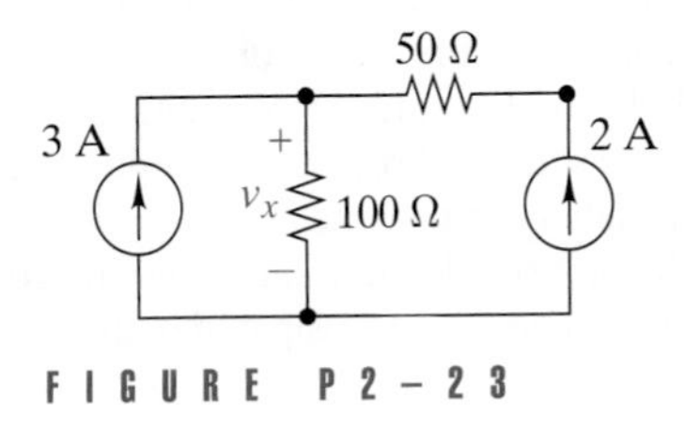

FIGURE P2–23

2–24 Figure P2–24 shows a subcircuit connected to the rest of the circuit at four points.

(a) Use element and connection constraints to find v_x and i_x.

(b) Show that the sum of the currents into the rest of the circuit is zero.

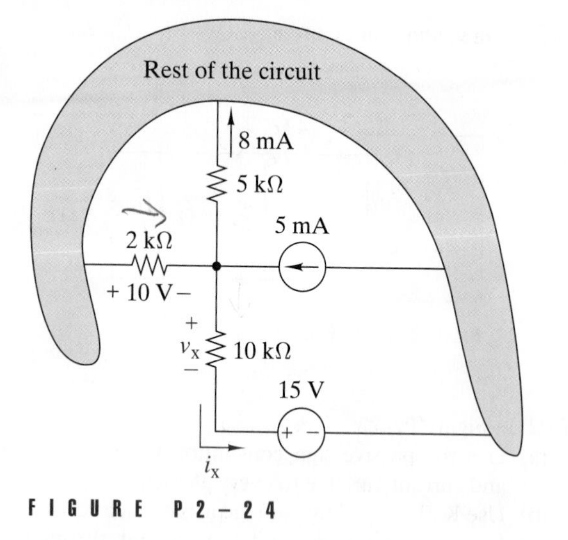

FIGURE P2–24

2–25 In Figure P2–25 $i_x = -0.5$ mA. Find the value of R.

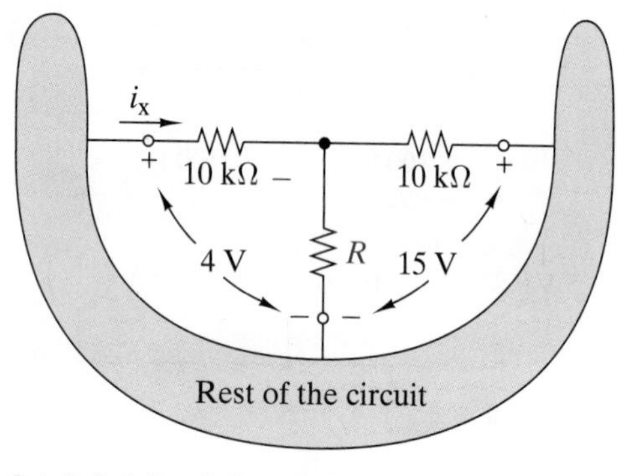

FIGURE P2–25

2–26 Figure P2–26 shows a resistor with one terminal connected to ground and the other connected to an arrow. The arrow symbol is used to indicate a connection to one terminal of a voltage source whose other terminal is connected to ground. The label next to the arrow indicates the source voltage at the ungrounded terminal. Find the voltage across, current through, and power dissipated in the resistor.

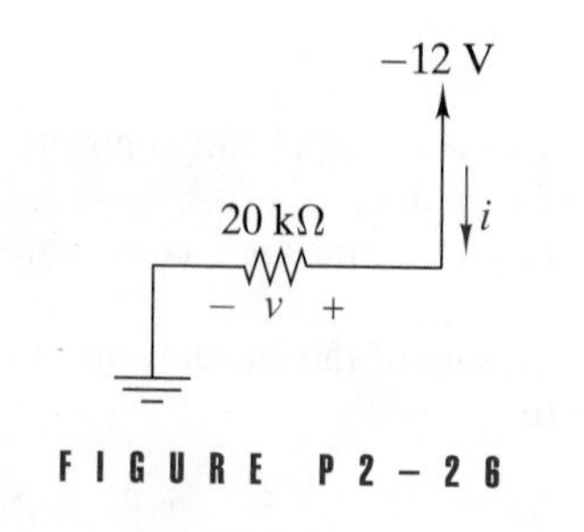

FIGURE P2–26

OBJECTIVE 2–4 EQUIVALENT CIRCUITS (SECT. 2–4)

(a) Given a circuit consisting of linear resistors, find the equivalent resistance between a specified pair of terminals.

(b) Given a circuit consisting of a source-resistor combination, find an equivalent source-resistor circuit.

See Example 2–11, 2–12 and Exercises 2–7, 2–8, 2–9

2–27 Find the equivalent resistance R_{EQ} in Figure P2–27.

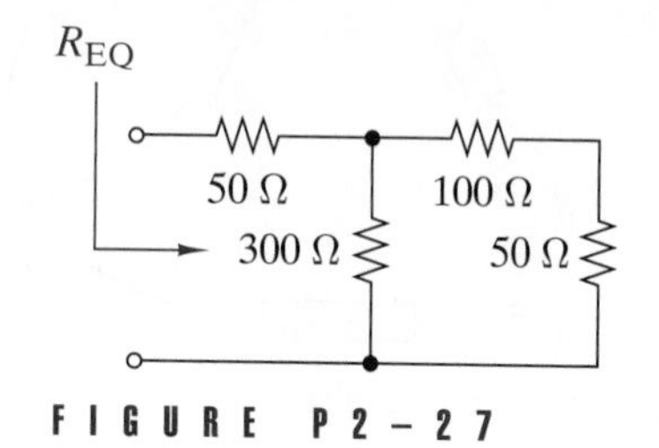

FIGURE P2–27

2–28 Find the equivalent resistance R_{EQ} in Figure P2–28.

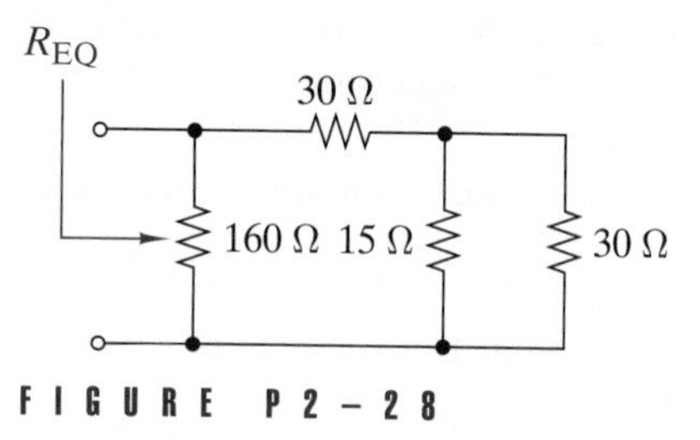

FIGURE P2–28

2–29 Find R_{EQ} in Figure P2–29 when the switch is open. Repeat when the switch is closed.

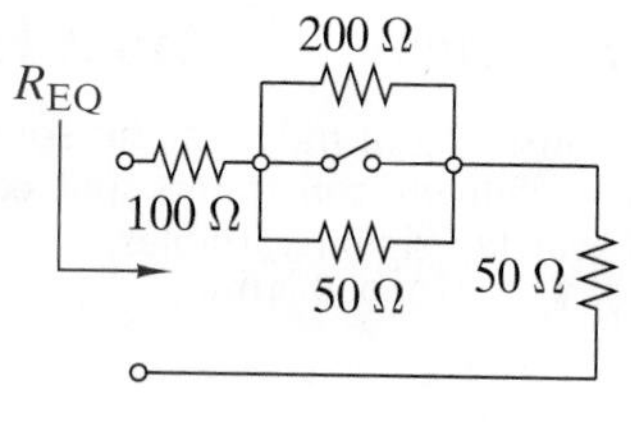

FIGURE P2–29

2–30 In Figure P2–30 find the equivalent resistance between terminals A–B, A–C, A–D, B–C, B–D, and C–D.

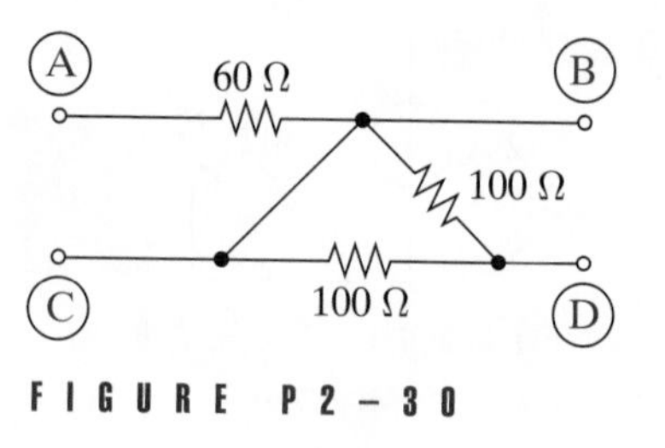

FIGURE P2–30

2–31 In Figure P2–31 find the equivalent resistance between terminals A–B, A–C, A–D, B–C, B–D, and C–D.

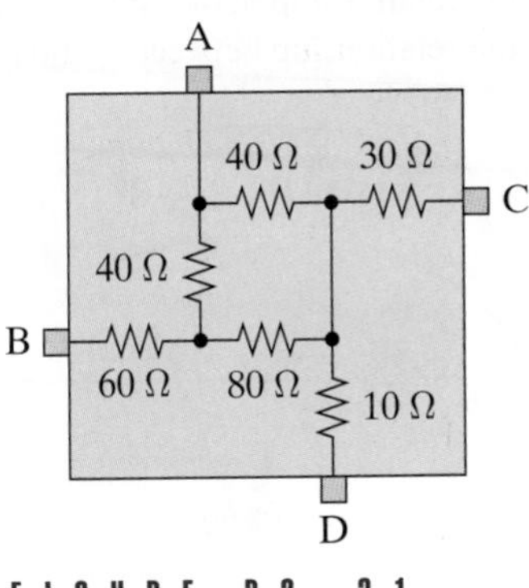

FIGURE P2–31

2–32 Select a value of R_L in Figure P2–32 so that $R_{EQ} = 25\ k\Omega$.

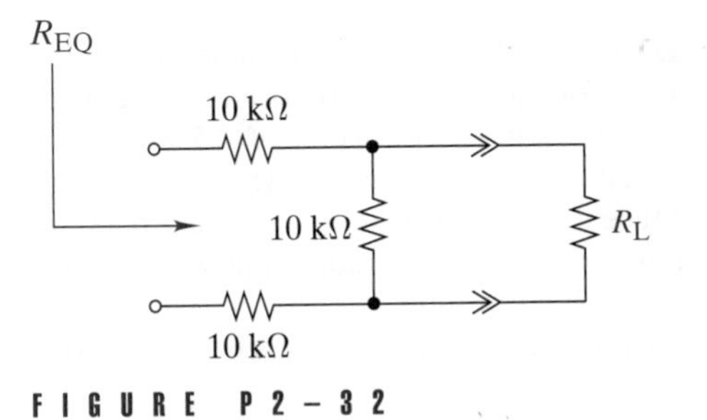

FIGURE P2–32

2–33 Repeat Problem 2–32 for $R_{EQ} = 15\ k\Omega$. *Caution:* R_L must be positive.

2–34 Find the equivalent practical voltage source at terminals A and B in Figure P2–34.

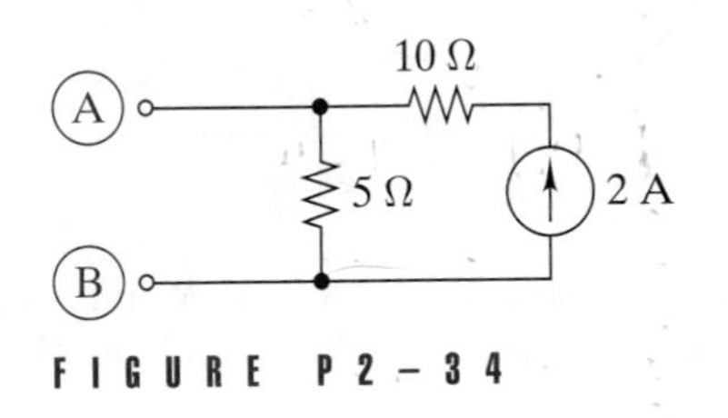

FIGURE P2–34

2–35 In Figure P2–35 the i–v characteristic network N is $v + 50i = 5$ V. Find the equivalent practical current source for the network.

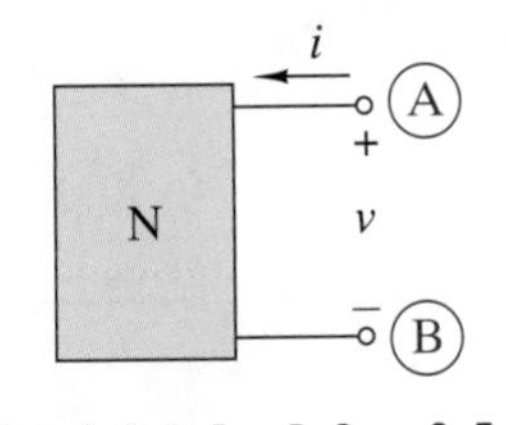

FIGURE P2–35

2–36 Select the value of R_x in Figure P2–36 so that $R_{EQ} = 60\ k\Omega$.

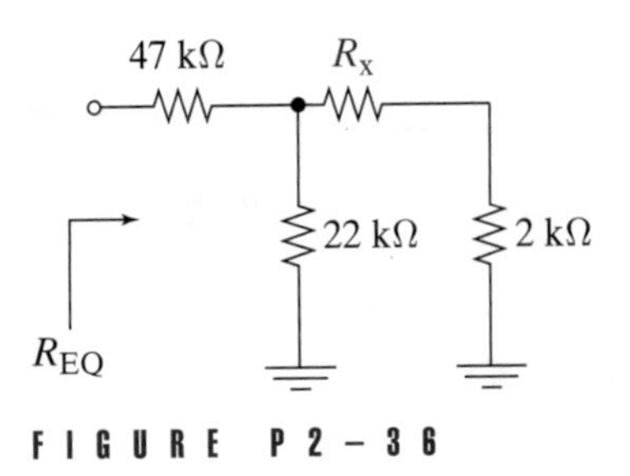

FIGURE P2–36

2–37 **D** Show how to interconnect standard 3.9-kΩ resistors to obtain equivalent resistances of $1\ k\Omega \pm 5\%$, $5\ k\Omega \pm 5\%$, and $10\ k\Omega \pm 5\%$.

2–38 Select the value of R in Figure P2–38 so that $R_{AB} = R_L$.

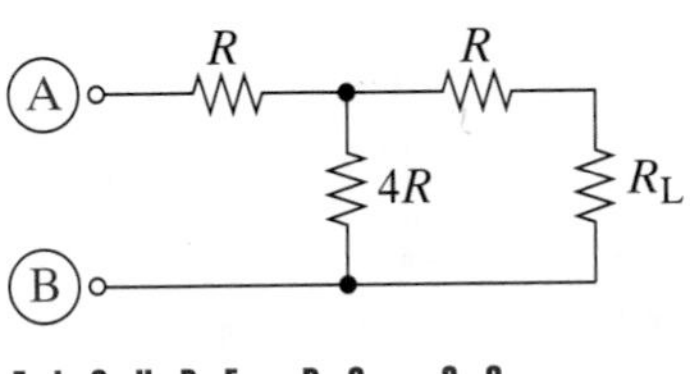

FIGURE P2–38

2–39 What is the range of R_{EQ} in Figure P2–39?

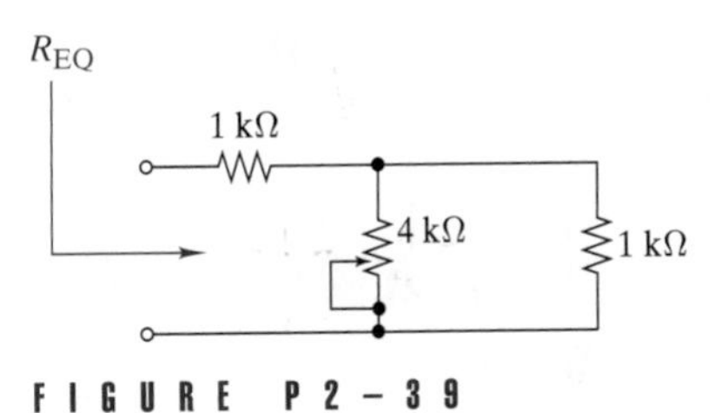

FIGURE P2–39

2–40 Find the equivalent resistance between terminals A and B in Figure P2–40.

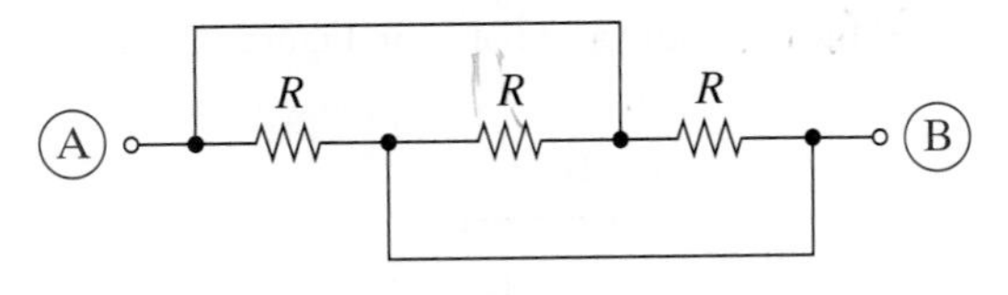

FIGURE P2–40

OBJECTIVE 2–5 VOLTAGE AND CURRENT DIVISION (SECT. 2–5)

(a) Given a circuit with elements connected in series or in parallel, use voltage or current division to find specified voltages or currents.

(b) Design a voltage or current divider that delivers specified output signals within stated constraints.

See Examples 2–13, 2–14, 2–16, 2–17, 2–18 and Exercises 2–10, 2–11, 2–12

2–41 Use voltage division in Figure P2–41 to obtain an expression for v_L in terms of R, R_L, and v_S.

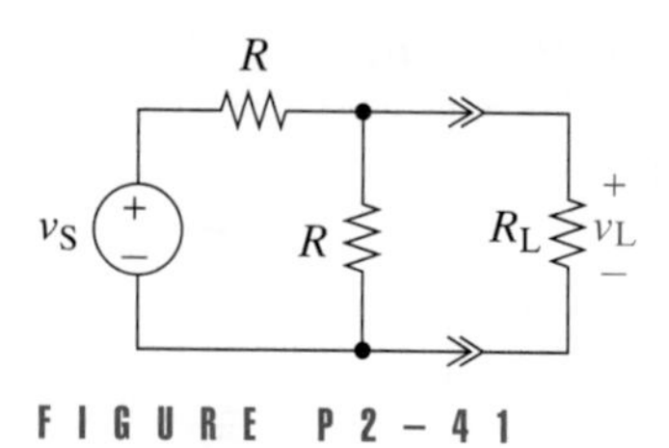

FIGURE P2–41

2–42 Use current division in Figure P2–42 to obtain an expression for v_L in terms of R, R_L, and i_S.

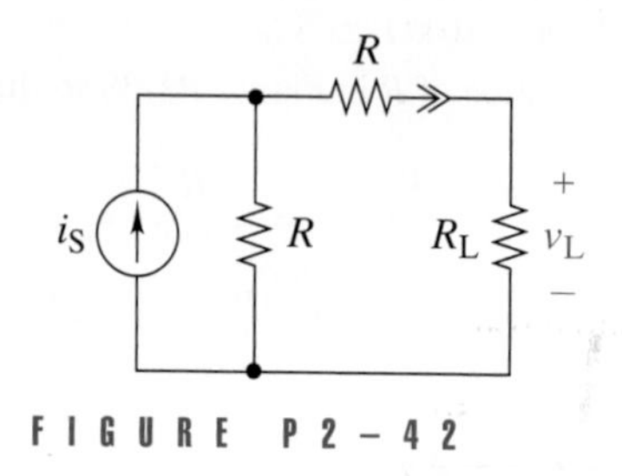

FIGURE P2–42

2–43 Find i_x in Figure P2–43.

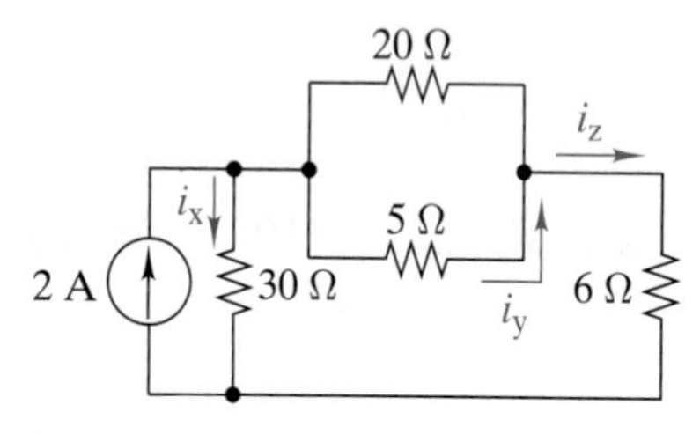

FIGURE P2–43

2–44 Find i_y and i_z in Figure P2–43.

2–45 Find the range of values of v_O in Figure P2–45.

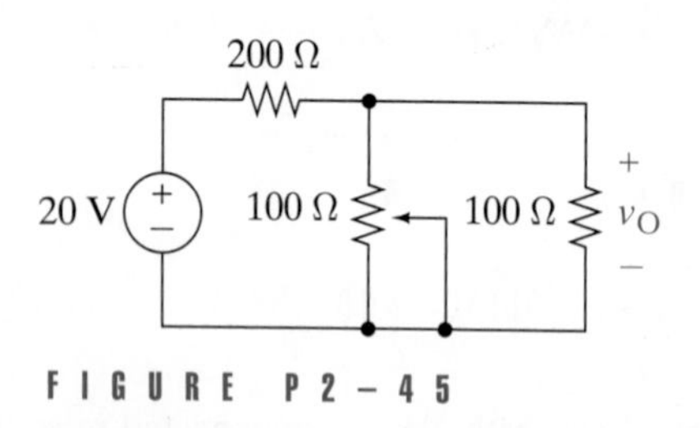

FIGURE P2–45

2–46 Figure P2–46 shows a resistance divider connected in a general circuit.

(a) What is the relationship between v_1 and v_2 when $i_1 = 0$?

(b) What is the relationship between v_1 and v_2 when $i_2 = 0$?

(c) What is the relationship between i_1 and i_2 when $v_1 = 0$?

(d) What is the relationship between i_1 and i_2 when $v_2 = 0$?

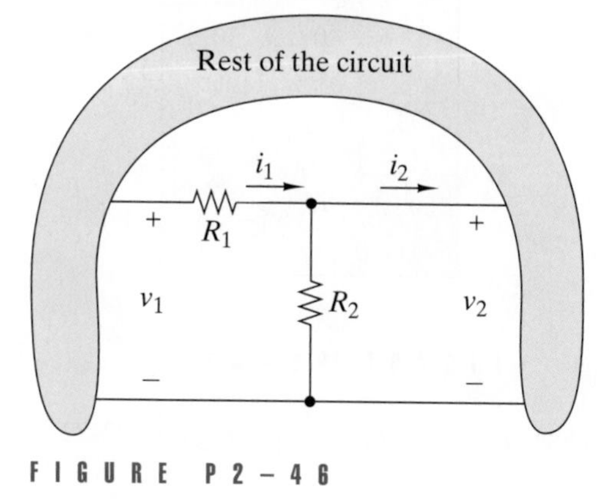

FIGURE P2–46

2–47 **D** Figure P2–47 shows an ammeter circuit consisting of a D'Arsonval meter, a two-position selector switch, and two shunt resistors. A current of 0.5 mA produces full-scale deflection of the D'Arsonval meter, whose internal resistance is $R_M = 50\ \Omega$. Select the shunt resistance R_1 and R_2 so that $i_x = 10$ mA produces full-scale deflection when the switch is in position A, and $i_x = 50$ mA produces full-scale deflection when the switch is in position B.

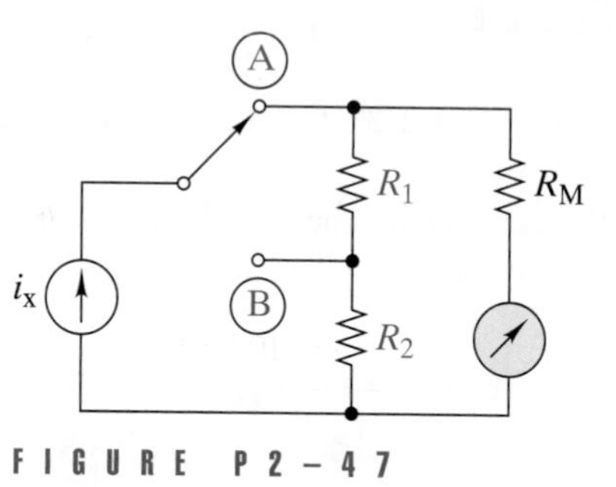

FIGURE P2–47

2–48 **D** Select values for R_1, R_2, and R_3 in Figure P 2–48 so the voltage divider produces the two output voltages shown.

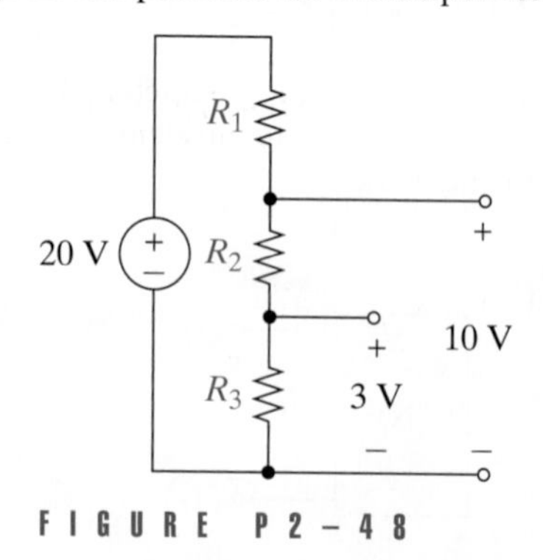

FIGURE P2–48

2–49 **D** Select a value of R_x in Figure P2–49 so that $v_L = 3$ V.

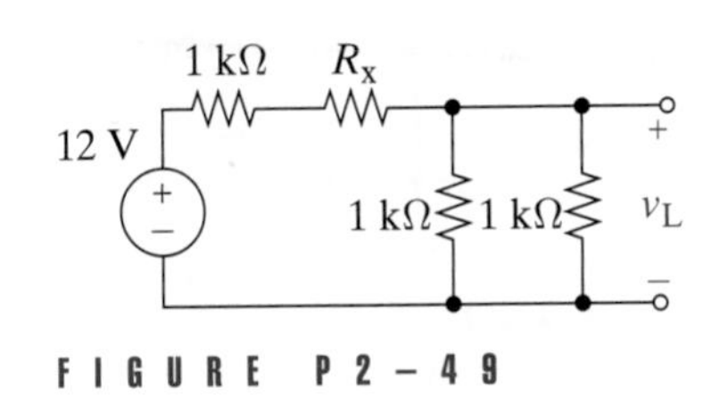

FIGURE P2–49

2–50 **D** Select a value of R_x in Figure P2–50 so that $v_L = 6$ V. *Caution:* R_x must be positive.

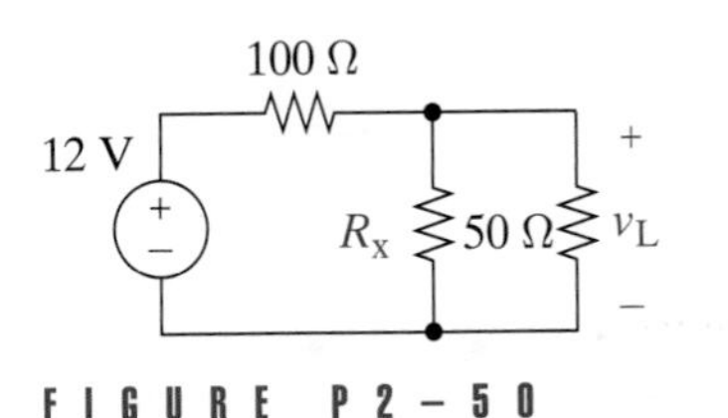

FIGURE P2–50

OBJECTIVE 2–6 CIRCUIT REDUCTION (SECT. 2–6)

Given a circuit consisting of linear resistors and an independent source, find selected signal variables using successive application of series/parallel equivalence, source transformations, and voltage/current division.
See Example 2–20, 2–21, 2–22, 2–23 and Exercises 2–13, 2–14, 2–15

2–51 Use circuit reduction to find v_x and i_x in Figure P2–51.

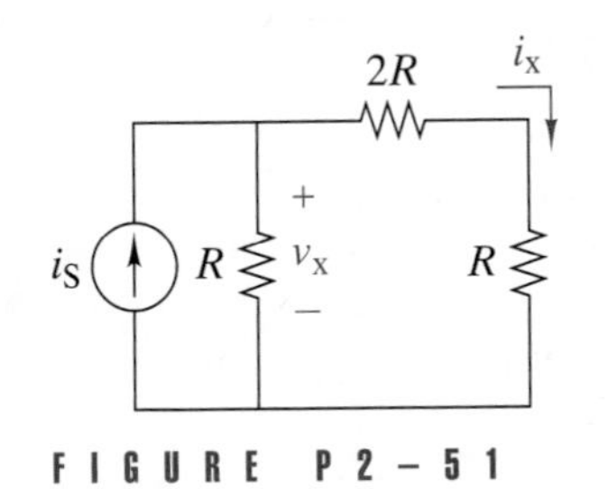

FIGURE P2–51

2–52 Use circuit reduction to find i_x in Figure P2–52.

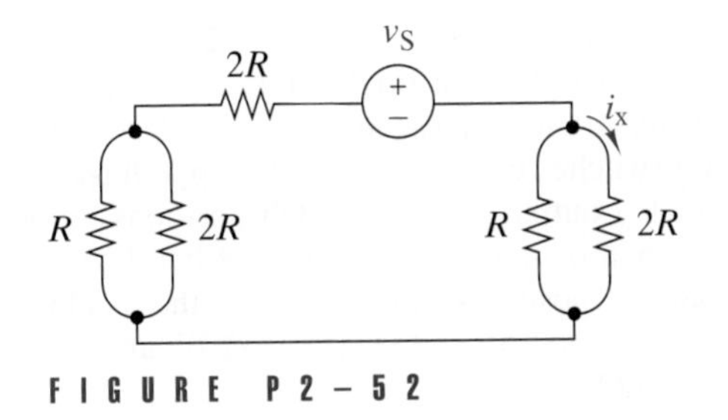

FIGURE P2–52

2–53 Use circuit reduction to find v_x in Figure P2–53.

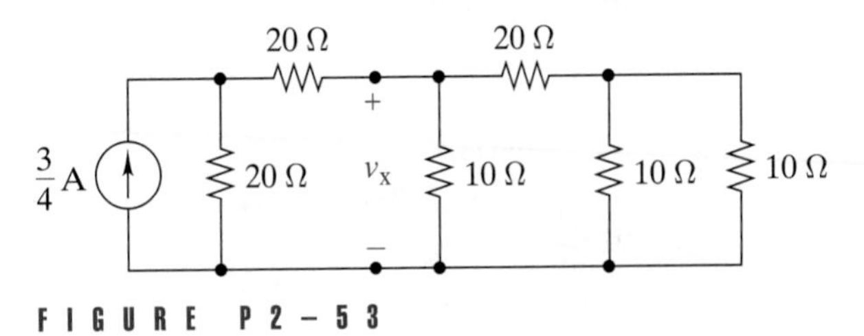

FIGURE P2–53

2–54 Use circuit reduction to find v_x and i_x in Figure P2–54.

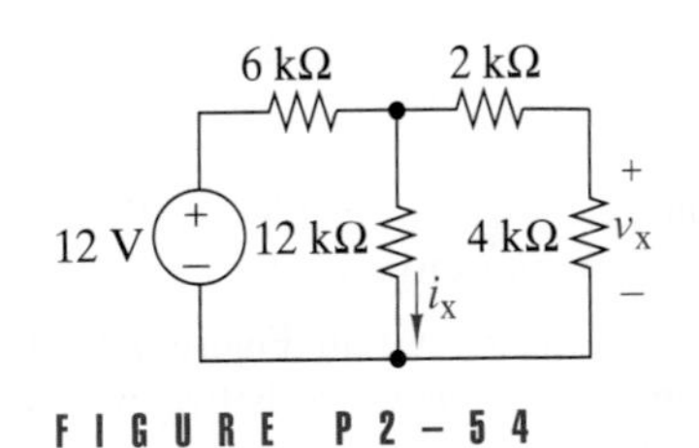

FIGURE P2–54

2–55 Use source transformation to find i_x in Figure P2–55.

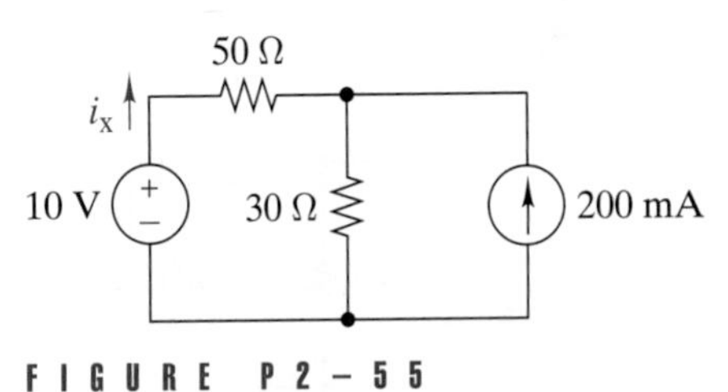

FIGURE P2–55

2–56 Use source transformation to find i_x in Figure P2–56.

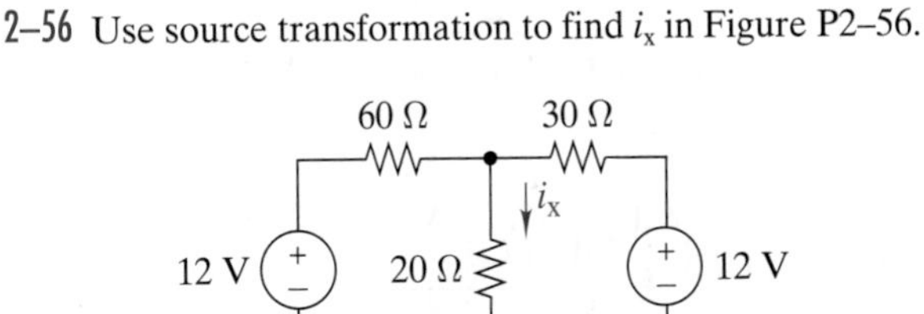

FIGURE P2–56

2–57 Use source transformations in Figure P2–57 to relate v_O to v_1, v_2, and v_3.

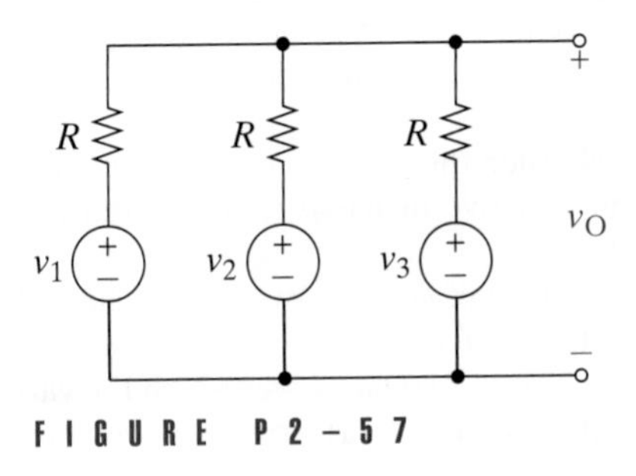

FIGURE P2–57

2–58 The current through R_L in Figure P2–58 is 40 mA. Use source transformations to find R_L.

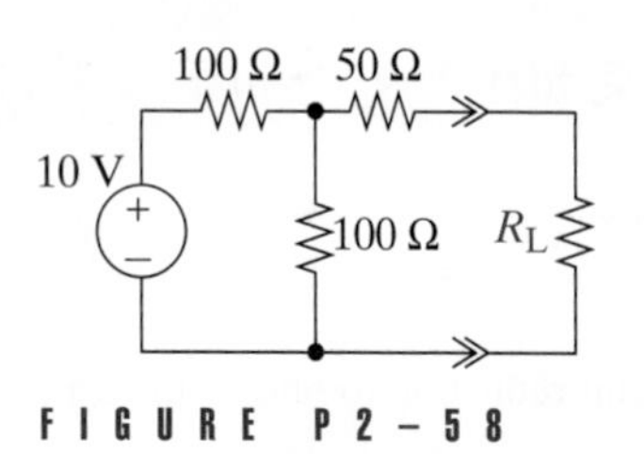

FIGURE P2–58

2–59 The voltage across R_L in Figure P2–58 is 2.5 V. Use source transformations to find R_L.

2–60 The box in the circuit in Figure P2–60 is a resistor whose value can be anywhere between 8 kΩ and 80 kΩ. Use circuit reduction to find the range of values of v_x.

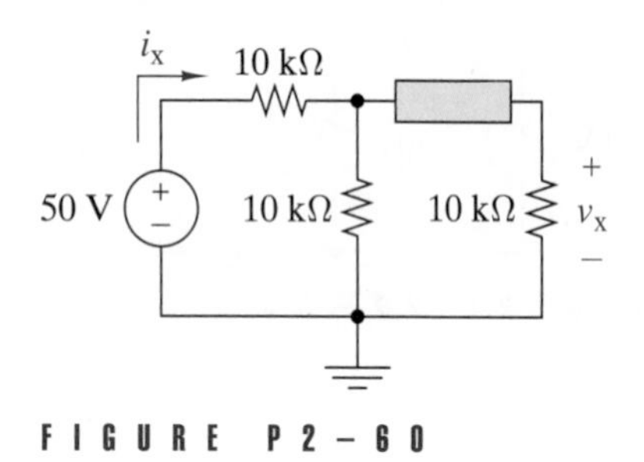

FIGURE P2–60

INTEGRATING PROBLEMS

2–61 Nonlinear Device Characteristics A

The circuit in Figure P2–61 is a parallel combination of a 50-Ω linear resistor and a varistor whose i–v characteristic is $i_V = 2.6 \times 10^{-5} v^3$. For small voltage the varistor current is quite small compared to the resistor current. For large voltages the varistor dominates because its current increases more rapidly with voltage.

(a) Plot the i–v characteristic of the parallel combination.

(b) State whether the parallel combination is linear or nonlinear, active or passive, and bilateral or nonbilateral.

(c) Find the range of voltages over which the resistor current is at least 10 times as large as the varistor current.

(d) Find the range of voltages over which the varistor current is at least 10 times as large as the resistor current.

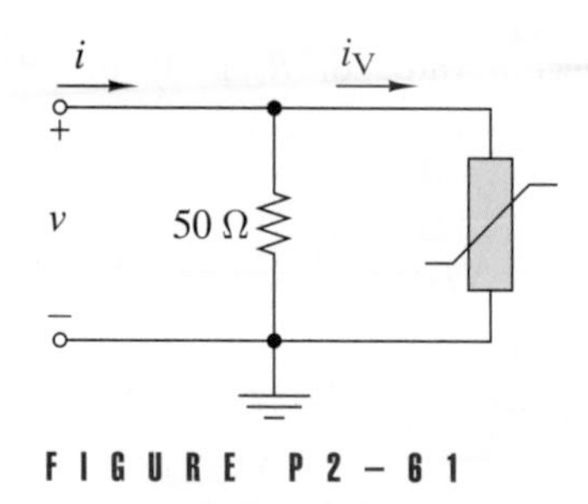

FIGURE P2–61

2–62 Center Tapped Voltage Divider A

Figure P2–62 shows a voltage divider with the center tap connected to ground. Derive equations relating v_A and v_B to v_S, R_1, and R_2.

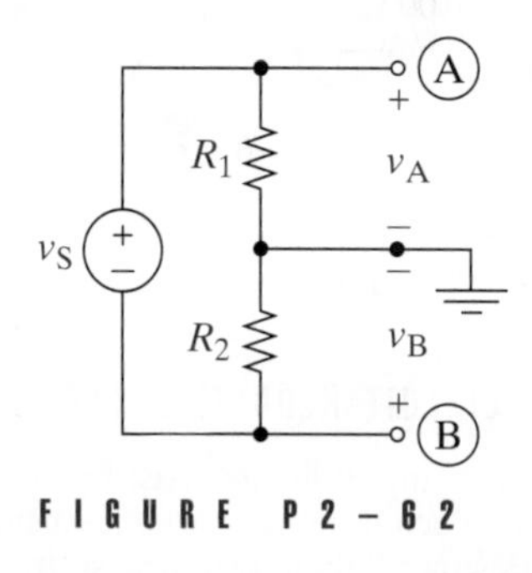

FIGURE P2–62

2–63 Combined i–v Characteristics A

Figure P2–63 shows an interconnection of practical voltage source and a two-terminal network N. The i–v relationship of the network N is $i_1 = (0.01v_1 + 3)$ A. Find the equivalent practical voltage source of the combination at terminals A and B.

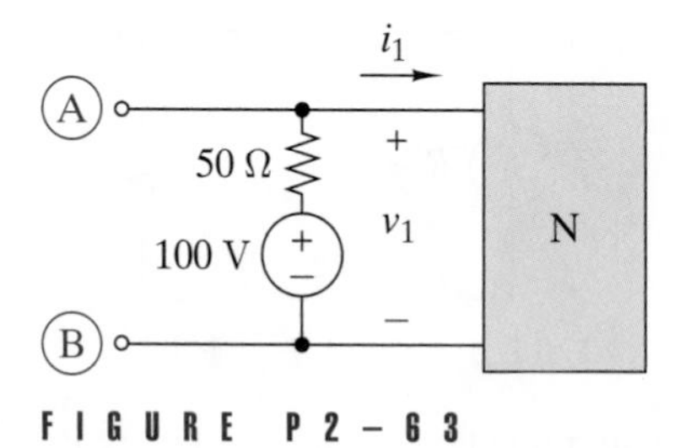

FIGURE P2–63

2–64 Programmable Voltage Divider A

Figure P2–64 shows a programmable voltage divider in which digital inputs b_0 and b_1 control complementary analog switches connecting a multitap voltage divider to the analog output v_O. The switch positions in the figure apply when digital inputs are low. When inputs go high, the switch positions reverse. Find the analog output voltage for $(b_1, b_0) = (0,0), (0,1), (1,0)$, and $(1,1)$ when $V_{REF} = 12$ V.

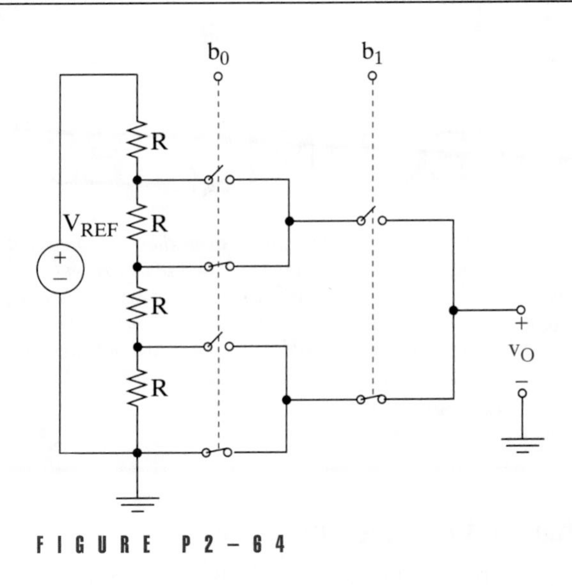

FIGURE P2–64

2–65 Analog Voltmeter Design A D E

Figure P2–65(a) shows a voltmeter circuit consisting of a D'Arsonval meter, two series resistors, and a two-position selector switch. A current of $I_{FS} = 400\ \mu A$ produces full-scale deflection of the D'Arsonval meter, whose internal resistance is $R_M = 25\ \Omega$.

(a) D Select the series resistance R_1 and R_2 so a voltage $v_x = 50$ V produces full-scale deflection when the switch is in position A, and voltage $v_x = 10$ V produces full-scale deflection when the switch is in position B.

(b) A What is the voltage across the 20-kΩ resistor in Figure P2–65(b)? What is the voltage when the voltmeter in part (a) is connected across the 20-kΩ resistor? What is the percentage error introduced connecting the voltmeter?

(c) E A different D'Arsonval meter is available with an internal resistance of 200 Ω and a full-scale deflection current of 100 μA. If the voltmeter in part (a) is redesigned using this D'Arsonval meter, would the error found in part (b) be smaller or larger? Explain.

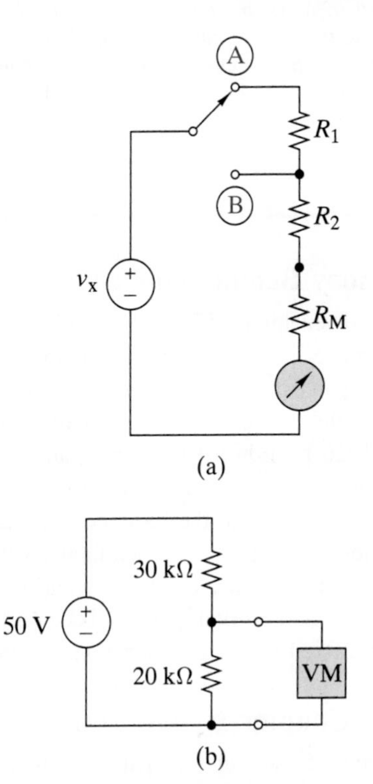

FIGURE P2–65

CHAPTER 3

CIRCUIT ANALYSIS TECHNIQUES

Assuming any system of linear conductors connected in such a manner that to the extremities of each one of them there is connected at least one other, a system having electromotive forces E_1, E_2.... E_3, no matter how distributed, we consider two points A and A′ belonging to the system and having potentials V and V′. If the points A and A′ are connected by a wire ABA′, which has a resistance r, with no electromotive forces, the potentials of points A and A′ assume different values from V and V′, but the current i flowing through this wire is given by $i = (V - V')/(r + R)$ in which R represents the resistance of the original wire, this resistance being measured between the points A and A′, which are considered to be electrodes.

Leon Charles Thévenin, 1883,
French Telegraph Engineer

Some History Behind This Chapter

Leon Charles Thévenin (1857–1926), a distinguished French telegraph engineer and teacher, was led to his theorem in 1883 following an extensive study of Kirchhoff''s laws. Norton's theorem, the dual of Thévenin's theorem, was not proposed until 1926 by Edward L. Norton, an American electrical engineer working on long-distance telephony. Curiously, it turns out that the basic concept had been discovered earlier by Herman von Helmholtz, while studying electricity in animal tissue. Electrical engineering tradition credits Thévenin and Norton, perhaps because they worked in areas that offered practical applications for their results.

Why This Chapter Is Important Today

In this chapter you advance to studying general methods of analyzing circuits and to the major theorems that describe linear circuits. These theorems are conceptual tools that give new insight into circuit behavior. Most importantly, you will be introduced to the design of interface circuits; your first exposure to devising a circuit to perform a predetermined function.

Chapter Sections

3–1 Node-Voltage Analysis

3–2 Mesh-Current Analysis

3–3 Linearity Properties

3–4 Thévenin and Norton Equivalent Circuits

3–5 Maximum Signal Transfer

3–6 Interface Circuit Design

Chapter Learning Objectives

3-1 General Circuit Analysis (Sects. 3-1 to 3-2)

Given a linear resistance circuit:

(a) (Formulation) Write node-voltage or mesh-current equations for the circuit.

(b) (Solution) Solve the equations from part (a) for selected signal variables or input-output relationships.

3-2 Linearity Properties (Sect. 3-3)

Given a linear resistance circuit:

(a) Use the proportionality principle to find selected signal variables.

(b) Use the superposition principle to find selected signal variables.

3-3 Thévenin and Norton Equivalent Circuits (Sect. 3-4)

Given a linear resistance circuit:

(a) Find the Thévenin or Norton equivalent at a specified pair of terminals.

(b) Use the Thévenin or Norton equivalent to find the signals delivered to linear or nonlinear loads.

3-4 Maximum Signal Transfer (Sect. 3-5)

Given a linear resistance circuit:

(a) Find the maximum voltage, current, and power available at a specified pair of terminals.

(b) Find the resistive loads required to obtain the maximum available signal levels.

3-5 Interface Circuit Design (Sect. 3-6)

Given the signal transfer goals at a source-load interface, design one or more two-port interface circuits to achieve the goals and evaluate the alternative design solutions.

3–1 NODE-VOLTAGE ANALYSIS

Before describing node-voltage analysis, we first review the foundation for every method of circuit analysis. As noted in Sec. 2–3, circuit behavior is based on constraints of two types: (1) connection constraints (Kirchhoff's laws) and (2) device constraints (element i–v relationships). As a practical matter, however, using element voltages and currents to express the circuit constraints produces a large number of equations that must be solved simultaneously to find the circuit responses. For example, a circuit with only six devices requires us to treat 12 equations with 12 unknowns. Although this is not an impossible task using software tools like Mathcad, it is highly desirable to reduce the number of equations that must be solved simultaneously.

You should not abandon the concept of element and connection constraints. This method is vital because it provides the foundation for all methods of circuit analysis. In subsequent chapters, we use element and connection constraints many times to develop important ideas in circuit analysis.

Using node voltages instead of element voltages as circuit variables can reduce the number of equations that must be treated simultaneously. To define a set of node voltages, we first select a reference node. The **node voltages** are then defined as the voltages between the remaining nodes and the selected reference node. Figure 3–1 shows the notation used to define node-voltage variables. In this figure the reference node indicated by the ground symbol and the node voltages are identified by a voltage symbol next to all the other nodes. This notation means that the positive reference mark for the node voltage is located at the node in question while the negative mark is at the reference node. Obviously, any circuit with N nodes involves $N-1$ node voltages.

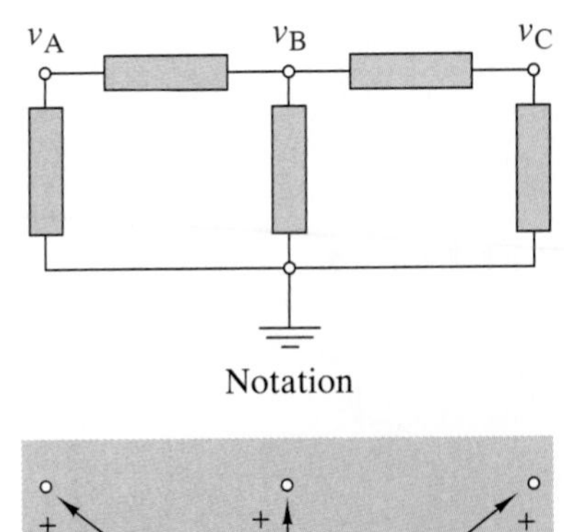

Notation

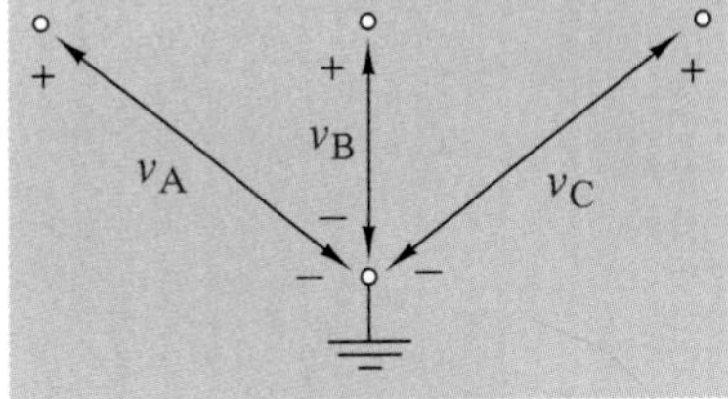

Interpretation

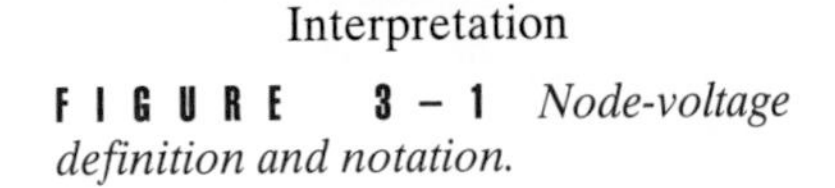

FIGURE 3–1 *Node-voltage definition and notation.*

A fundamental property of node voltages needs to be covered at the outset. Suppose we are given a two-terminal element whose element voltage is labeled v_1. Suppose further that the terminal with the plus reference mark is connected to a node, say node A. The two cases shown in Figure 3–2 are the only two possible ways the other element terminal can be connected. In case A, the other terminal is connected to the reference node, in which case KVL requires $v_1 = v_A$. In case B, the other terminal is connected to a nonreference node, say node B, in which case KVL requires $v_1 = v_A - v_B$. This example illustrates the following fundamental property of node voltages:

If the* K*th two-terminal element is connected between nodes* X *and* Y*, then the element voltage can be expressed in terms of the two node voltages as

$$v_K = v_X - v_Y \tag{3–1}$$

***where* X *is the node connected to the positive reference for element voltage* v$_K$.**

Equation (3–1) is a KVL constraint at the element level. If node Y is the reference node, then by definition $v_Y = 0$ and Eq. (3–1) reduces to $v_K = v_X$. On the other hand, if node X is the reference node, then $v_X = 0$ and therefore $v_K = -v_Y$. The minus sign occurs here because the positive reference for the element is connected to the reference node. In any case, the important fact is that the voltage across any two-terminal element can be expressed as the difference of two node voltages, one of which may be zero.

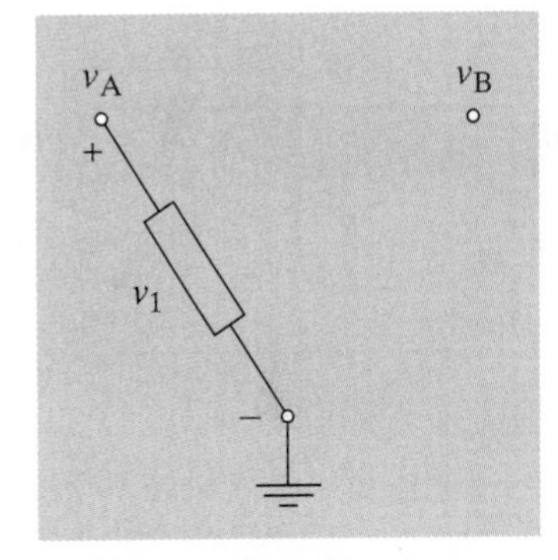

Case A

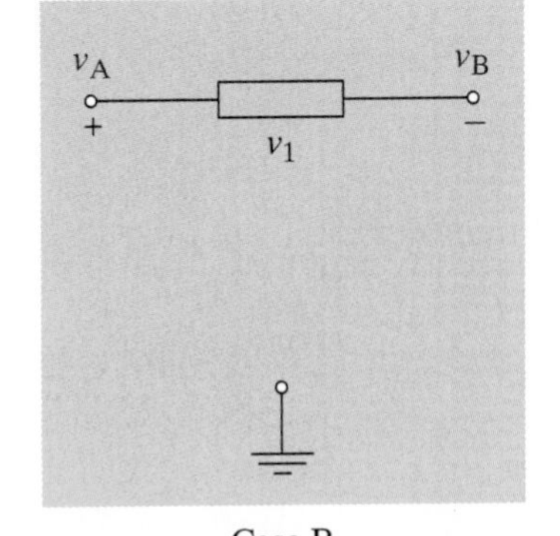

Case B

FIGURE 3–2 *Two possible connections of a two-terminal element.*

Exercise 3–1

The reference node and node voltages in the bridge circuit of Figure 3–3 are $v_A = 5$ V, $v_B = 10$ V, and $v_C = -3$ V. Find the element voltages.

FIGURE 3 – 3

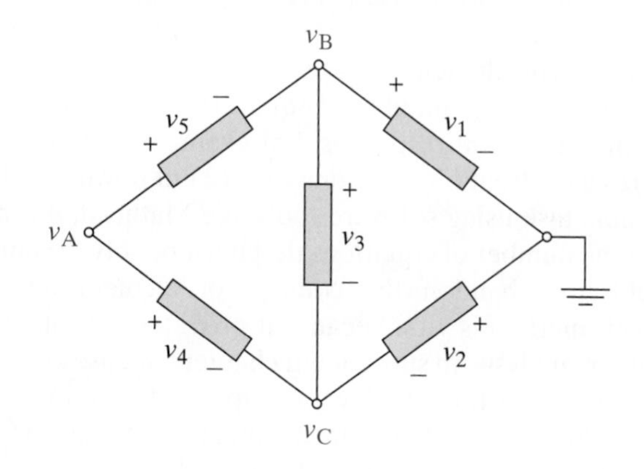

Answers: $v_1 = 10$ V; $v_2 = 3$ V; $v_3 = 13$ V; $v_4 = 8$ V; $v_5 = -5$ V

FORMULATING NODE-VOLTAGE EQUATIONS

To formulate a circuit description using node voltages, we use element and connection constraints, except that the KVL connection equations are not explicitly written. Instead we use the fundamental property of node analysis to express the element voltages in terms of the node voltages.

The circuit in Figure 3–4 will demonstrate the formulation of node-voltage equations. In Figure 3–4 we have identified a reference node (indicated by the ground symbol), four element currents (i_0, i_1, i_2, and i_3), and two node voltages (v_A and v_B).

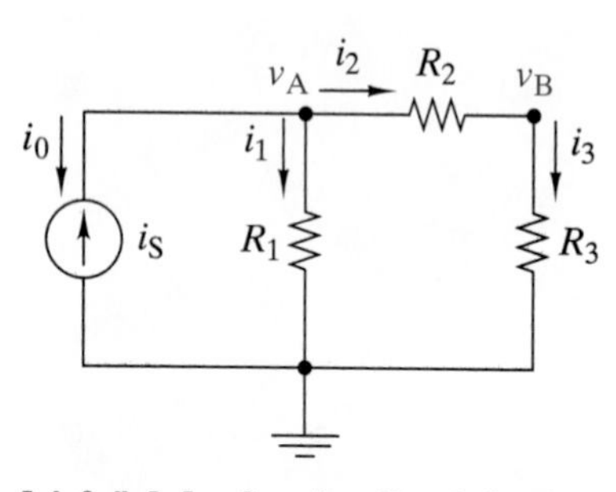

FIGURE 3 – 4 *Circuit for demonstrating node-voltage analysis.*

The KCL constraints at the two nonreference nodes are

$$\text{Node A:} \quad -i_0 - i_1 - i_2 = 0$$

$$\text{Node B:} \quad i_2 - i_3 = 0 \tag{3–2}$$

Using the fundamental property of node analysis, we use the element equations to relate the element currents to the node voltages.

$$\text{Resistor } R_1\text{:} \quad i_1 = G_1 v_A$$

$$\text{Resistor } R_2\text{:} \quad i_2 = G_2(v_A - v_B)$$

$$\text{Resistor } R_3\text{:} \quad i_3 = G_3 v_B$$

$$\text{Current Source:} \quad i_0 = -i_S \tag{3–3}$$

We have written six equations in six unknowns—four element currents and two node voltages. The right side of the element equations in Eq. (3–3) involves unknown node voltages and the input signal i_S. Substituting the element constraints in Eq. (3–3) into the KCL connection constraints in Eq. (3–2) yields

$$\text{Node A: } i_S - G_1 v_A - G_2(v_A - v_B) = 0$$

$$\text{Node B: } \quad G_2(v_A - v_B) - G_3 v_B = 0$$

which can be arranged in the following standard form:

$$\text{Node A: } \quad (G_1 + G_2)v_A - G_2 v_B = i_S$$

$$\text{Node B: } -G_2 v_A + (G_2 + G_3)v_B = 0 \tag{3–4}$$

In this standard form all of the unknown node voltages are grouped on one side and the independent sources on the other.

By systematically eliminating the element currents, we have reduced the circuit description to two linear equations in the two unknown node voltages. The coefficients in the equations on the left side ($G_1 + G_2$, G_2, $G_2 + G_3$) depend only on circuit parameters, while the right side contains the known input driving force i_S.

As noted previously, every method of circuit analysis must satisfy KVL, KCL, and the element i–v relationships. In developing the node-voltage equations in Eq. (3–4), it may appear that we have not used KVL. However, KVL is satisfied because the equations $v_1 = v_A$, $v_2 = v_A - v_B$, and $v_3 = v_B$ were used to write the right side of the element equations in Eq. (3–3). The KVL constraints do not appear explicitly in the formulation of node equations, but are implicitly included when the fundamental property of node analysis is used to write the element voltages in terms of the node voltages.

In summary, four steps are needed to develop node-voltage equations.

STEP 1 Select a reference node. Identify a node voltage at each of the remaining $N - 1$ nodes and a current with every element in the circuit.

STEP 2 Write KCL connection constraints in terms of the element currents at the $N - 1$ nonreference nodes.

STEP 3 Use the element i–v relationships and the fundamental property of node analysis to express the element currents in terms of the node voltages.

STEP 4 Substitute the element constraints from step 3 into the KCL connection constraints from step 2 and arrange the resulting $N - 1$ equations in a standard form.

Writing node-voltage equations leads to $N - 1$ equations that must be solved simultaneously. If we write the element and connection constraints in terms of element voltages and currents, we must solve $2E$ simultaneous equations. The reduction from $2E$ to $N - 1$ is particularly impressive in circuits with a large number of elements (large E) connected in parallel (small N).

EXAMPLE 3–1

Formulate node-voltage equations for the bridge circuit in Figure 3–5.

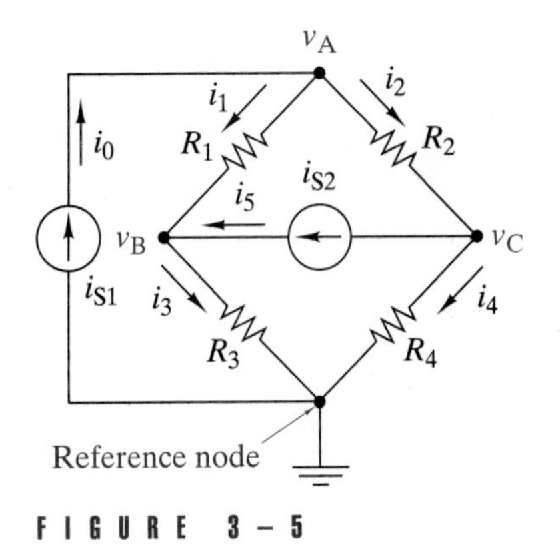

FIGURE 3–5

SOLUTION:

Step 1: The reference node, node voltages, and element currents are shown in Figure 3–5.

Step 2: The KCL constraints at the three nonreference nodes are:

$$\text{Node A: } i_0 - i_1 - i_2 = 0$$

$$\text{Node B: } i_1 - i_3 + i_5 = 0$$

$$\text{Node C: } i_2 - i_4 - i_5 = 0$$

Step 3: We write the element equations in terms of the node voltages and input signal sources.

$$i_0 = i_{S1} \qquad i_3 = G_3 v_B$$

$$i_1 = G_1(v_A - v_B) \quad i_4 = G_4 v_C$$

$$i_2 = G_2(v_A - v_C) \quad i_5 = i_{S2}$$

Step 4: Substituting the element equations into the KCL constraints and arranging the result in standard form yields three equations in the three unknown node voltages.

$$\text{Node A: } (G_1 + G_2)v_A - G_1 v_B - G_2 v_C = i_{S1}$$

$$\text{Node B: } \quad - G_1 v_A + (G_1 + G_3)v_B = i_{S2}$$

$$\text{Node C: } \quad - G_2 v_A + (G_2 + G_4)v_C = -i_{S2}$$

■

WRITING NODE-VOLTAGE EQUATIONS BY INSPECTION

The node-voltage equations derived in Example 3–1 have a symmetrical pattern. The coefficient of v_B in the node A equation and the coefficient of v_A in the node B equation are both the negative of the conductance connected between the nodes ($-G_1$). Likewise, the coefficients of v_A in the node C equation and v_C in the node A equation are both $-G_2$. Finally, coefficients of v_A in the node A equation', v_B in the node B equation, and v_C in the node C equation are the sum of the conductances connected to the node in question.

The symmetrical pattern always occurs in circuits containing only resistors and independent current sources. To understand why, consider any general two-terminal conductance G with one terminal connected to, say, node A. According to the fundamental property of node analysis, there are only two possibilities. Either the other terminal of G is connected to the reference node, in which case the current *leaving* node A via conductance G is

$$i = G(v_A - 0) = Gv_A$$

or else it is connected to another nonreference node, say, node B, in which case the current *leaving* node A via G is

$$i = G(v_A - v_B)$$

The pattern for node equations follows from these observations. The sum of the currents leaving any node A via conductances involves the following terms:

1. v_A times the sum of conductances connected to node A
2. Minus v_B times the sum of conductances connected between nodes A and B and similar terms for all other nodes connected to node A by conductances

Because of KCL, the sum of currents leaving node A via conductances plus the sum of currents directed away from node A by independent current sources must equal zero.

The aforementioned process allows us to write node-voltage equations by inspection without going through the intermediate steps involving the KCL constraints and the element equations. For example, the circuit in Figure 3–6 contains two independent current sources and four resistors. Starting with node A, the sum of conductances connected to node A is $G_1 + G_2$. The conductance between nodes A and B is G_2. The reference direction for the source current i_{S1} is into node A, while the reference direction for i_{S2} is directed away from node A. Pulling all of the observations together, we write the sum of currents directed out of node A as

$$\text{Node A: } (G_1 + G_2)v_A - G_2 v_B - i_{S1} + i_{S2} = 0 \tag{3–5}$$

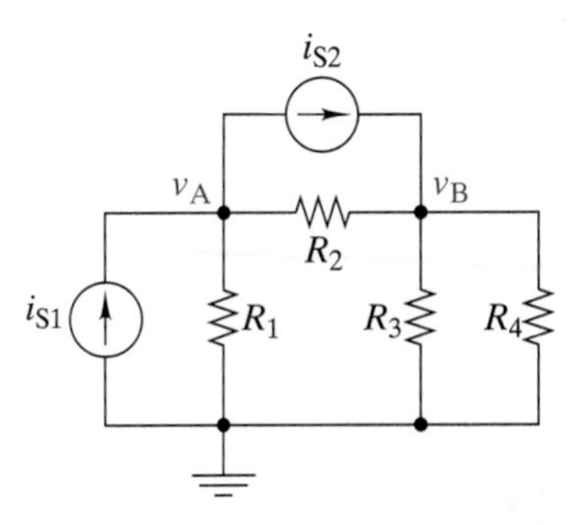

FIGURE 3–6 *Circuit for demonstrating writing node-voltage equations by inspection.*

Similarly, the sum of conductances connected to node B is $G_2 + G_3 + G_4$, the conductance connected between nodes B and A is again G_2, and the source current i_{S2} is directed toward node B. These observations yield the following node-voltage equation:

$$\text{Node B: } (G_2 + G_3 + G_4)v_B - G_2 v_A - i_{S2} = 0 \tag{3–6}$$

Rearranging Eqs. (3–5) and (3–6) in standard form yields

$$\begin{aligned} &\text{Node A:} && (G_1 + G_2)v_A - G_2 v_B = i_{S1} - i_{S2} \\ &\text{Node B:} && -G_2 v_A + (G_2 + G_3 + G_4)v_B = i_{S2} \end{aligned} \tag{3–7}$$

We have two symmetrical equations in the two unknown node voltages. The equations are symmetrical because the conductance G_2 connected between nodes A and B appears as the cross-coupling term in each equation.

EXAMPLE 3–2

Formulate node-voltage equations for the bridged-T circuit in Figure 3–7.

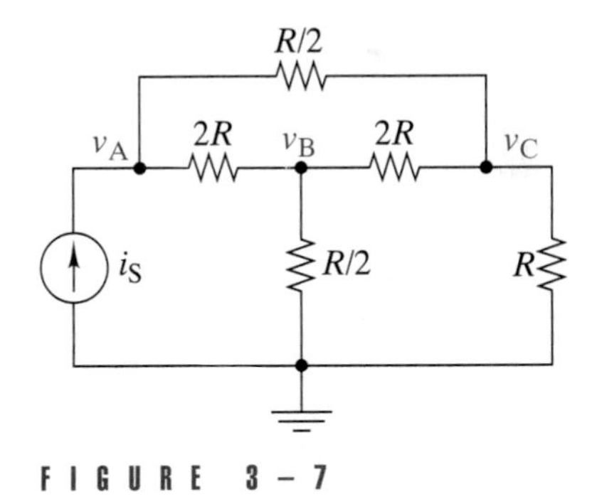

FIGURE 3–7

SOLUTION:

The total conductance connected to node A is $1/2R + 2/R = 2.5G$, to node B is $1/2R + 1/2R + 2/R = 3G$, and to node C is $1/R + 2/R + 1/2R = 3.5G$. The conductance connected between nodes A and B is $1/2R = 0.5G$, between nodes A and C is $2/R = 2G$, and between nodes B and C is $1/2R = 0.5G$. The independent current source is directed into node A. By inspection, the node-voltage equations are

$$\begin{aligned} &\text{Node A:} && 2.5Gv_A - 0.5Gv_B - 2Gv_C = i_S \\ &\text{Node B:} && -0.5Gv_A + 3Gv_B - 0.5Gv_C = 0 \\ &\text{Node C:} && -2Gv_A - 0.5Gv_B + 3.5Gv_C = 0 \end{aligned}$$

Written in matrix form,

$$\begin{bmatrix} 2.5G & -0.5G & -2G \\ -0.5G & 3G & -0.5G \\ -2G & -0.5G & 3.5G \end{bmatrix} \begin{bmatrix} v_A \\ v_B \\ v_C \end{bmatrix} = \begin{bmatrix} i_S \\ 0 \\ 0 \end{bmatrix}$$

This matrix equation is of the form $\mathbf{Ax} = \mathbf{B}$, where $\mathbf{A}$ is a 3×3 square matrix describing the circuit, $\mathbf{x}$ is a 3×1 column matrix of unknown node voltages, and $\mathbf{B}$ is a 3×1 column matrix of known inputs. Note that the $\mathbf{A}$ matrix is symmetrical about its main diagonal.[1] ■

Exercise 3–2

Formulate node-voltage equations for the circuit in Figure 3–8.

FIGURE 3–8

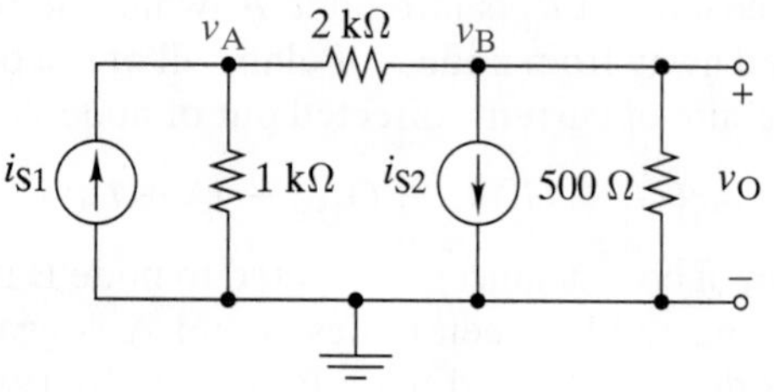

Answers:

$$(1.5 \times 10^{-3})v_A - (0.5 \times 10^{-3})v_B = i_{S1}$$
$$-(0.5 \times 10^{-3})v_A + (2.5 \times 10^{-3})v_B = -i_{S2}$$

SOLVING LINEAR ALGEBRAIC EQUATIONS

So far, we have only dealt with the problem of formulating node-voltage equations. To complete a circuit analysis problem, we must solve these linear equations for selected responses. Cramer's rule and Gaussian elimination are standard mathematical tools commonly used for hand solution of circuit equations. These tools are assumed to be part of the reader's mathematical background. Those needing a review of these matters are referred to Appendix B.

Cramer's rule and Gaussian elimination are suitable for hand calculations involving up to three or perhaps four simultaneous equations. Cramer's rule is useful when the circuit parameters are left in symbolic form, while the Gaussian method is more efficient for numerical examples involving four or more equations. However, any problem in which Gaussian methods offer significant advantages over Cramer's rule is probably best handled using computer tools. In other words, the ready availability of computer tools makes hand solution by Gaussian elimination an obsolete skill.

At about four or five simultaneous equations, numerical solutions are best obtained using computer-aided analysis. Many scientific hand-held calculators have a built-in capability to solve five or more linear equations. Virtually all PC-based mathematical software can solve systems of linear

1 See Appendix B for a discussion of matrix algebra, including the definition of a symmetrical matrix.

equations or, what is equivalent, perform matrix manipulations. In particular, Appendix B illustrates that both MATLAB and Mathcad can solve systems of linear equations written in matrix form.

In this book we often use Cramer's rule to solve simultaneous equations. This does not mean Cramer's rule is the optimum method, but only that it can easily handle and compactly document the solution of the class of problems treated in this book. We will also use MATLAB and Mathcad to demonstrate the use of computer tools to solve circuit equations in the matrix form. Readers needing a review of or an introduction to matrix methods are advised to study Appendix B.

Earlier in this section we formulated node-voltage equations for the circuit in Figure 3–4. [See Eq. (3–4).]

$$\text{Node A:} \quad (G_1 + G_2)v_A - G_2 v_B = i_S$$

$$\text{Node B:} \quad -G_2 v_A + (G_2 + G_3)v_B = 0$$

We use Cramer's rule to solve these equations because it easily handles the case in which the circuit parameters are left in symbolic form:

$$v_A = \frac{\Delta_A}{\Delta} = \frac{\begin{vmatrix} i_S & -G_2 \\ 0 & G_2 + G_3 \end{vmatrix}}{\begin{vmatrix} G_1 + G_2 & -G_2 \\ -G_2 & G_2 + G_3 \end{vmatrix}} = \left(\frac{G_2 + G_3}{G_1G_2 + G_1G_3 + G_2G_3}\right) i_S \tag{3–8}$$

$$v_B = \frac{\Delta_B}{\Delta} = \frac{\begin{vmatrix} G_1 + G_2 & i_S \\ -G_2 & 0 \end{vmatrix}}{\Delta} = \left(\frac{G_2}{G_1G_2 + G_1G_3 + G_2G_3}\right) i_S \tag{3–9}$$

These results express the two node voltages in terms of the circuit parameters and the input signal. Given the two node voltages v_A and v_B, we can now determine every element voltage and every current using Ohm's law and the fundamental property of node voltages.

$$v_1 = v_A \qquad v_2 = v_A - v_B \qquad v_3 = v_B$$

$$i_1 = G_1 v_A \qquad i_2 = G_2(v_A - v_B) \qquad i_3 = G_3 v_B$$

In solving the node equations, we left everything in symbolic form to emphasize that responses depend on the values of the circuit parameters (G_1, G_2, G_3) and the input signal (i_S). Even when numerical values are given, it is sometimes useful to leave some parameters in symbolic form to obtain input-output relationships or to reveal the effect of specific parameters on circuit response.

EXAMPLE 3–3

Given the circuit in Figure 3–9, find the input resistance R_{IN} seen by the current source and the output voltage v_O.

FIGURE 3–9

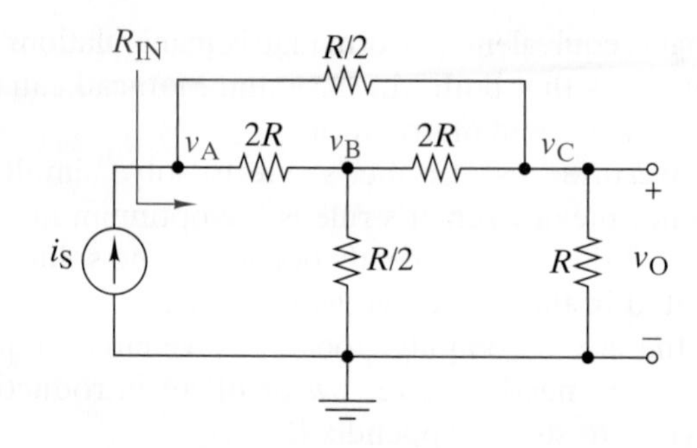

SOLUTION:

In Example 3–2 we formulated node-voltage equations for this circuit as follows:

$$\begin{aligned}
\text{Node A:}\quad & 2.5Gv_A - 0.5Gv_B - 2Gv_C = i_S \\
\text{Node B:}\quad & -0.5Gv_A + 3Gv_B - 0.5Gv_C = 0 \\
\text{Node C:}\quad & -2Gv_A - 0.5Gv_B + 3.5Gv_C = 0
\end{aligned}$$

The input resistance is the ratio v_A/i_S, so we first solve for v_A:

$$v_A = \frac{\Delta_A}{\Delta} = \frac{\begin{vmatrix} i_S & -0.5G & -2G \\ 0 & 3G & -0.5G \\ 0 & -0.5G & 3.5G \end{vmatrix}}{\begin{vmatrix} 2.5G & -0.5G & -2G \\ -0.5G & 3G & -0.5G \\ -2G & -0.5G & 3.5G \end{vmatrix}}$$

$$= \frac{i_S \times G^2 \begin{vmatrix} 3 & -0.5 \\ -0.5 & 3.5 \end{vmatrix}}{2.5G^3 \begin{vmatrix} 3 & -0.5 \\ -0.5 & 3.5 \end{vmatrix} + 0.5G^3 \begin{vmatrix} -0.5 & -2 \\ -0.5 & 3.5 \end{vmatrix} - 2G^3 \begin{vmatrix} -0.5 & -2 \\ 3 & -0.5 \end{vmatrix}}$$

$$= \frac{10.25 i_S}{11.75G}$$

Hence the input resistance is

$$R_{IN} = \frac{v_A}{i_S} = \frac{10.25}{11.75G} = 0.872R$$

To find the output voltage, we solve for v_C:

$$v_C = \frac{\Delta_C}{\Delta} = \frac{\begin{vmatrix} 2.5G & -0.5G & i_S \\ -0.5G & 3G & 0 \\ -2G & -0.5G & 0 \end{vmatrix}}{\Delta} = \frac{i_S \times G^2 \begin{vmatrix} -0.5 & 3 \\ -2 & -0.5 \end{vmatrix}}{\Delta}$$

$$= \frac{6.25G^2 i_S}{11.75G^3} = 0.532 i_S R$$

■

Exercise 3–3

Solve the node-voltage equations in Exercise 3–2 for v_O in Figure 3–8.

Answer: $v_O = 1000(i_{S1} - 3i_{S2})/7$

Exercise 3–4

Use node-voltage equations to solve for v_1, v_2, and i_3 in Figure 3–10.

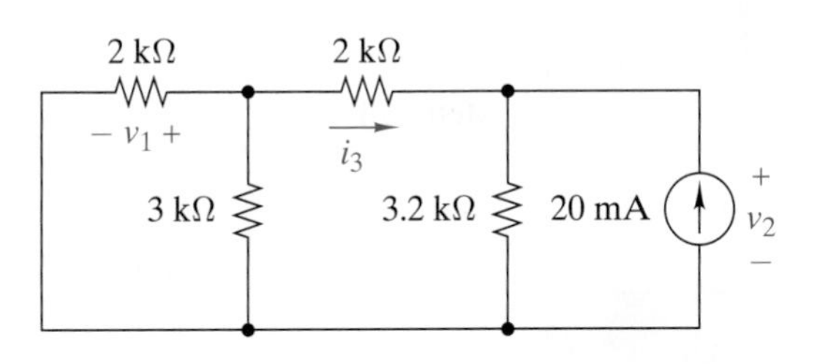

FIGURE 3–10

Answers: $v_1 = 12$ V; $v_2 = 32$ V; $i_3 = -10$ mA

NODE ANALYSIS WITH VOLTAGE SOURCES

Up to this point we have analyzed circuits containing only resistors and independent current sources. Applying KCL in such circuits is simplified because the sum of currents at a node only involves the output of current sources or resistor currents expressed in terms of the node voltages. Adding voltage sources to circuits modifies node analysis procedures because the current through a voltage source is not directly related to the voltage across it. While initially it may appear that voltage sources complicate the situation, they actually simplify node analysis by reducing the number of equations required.

Figure 3–11 shows three ways to deal with voltage sources in node analysis. Method 1 uses a source transformation to replace the voltage source and series resistance with an equivalent current source and parallel resistance. We can then formulate node equations at the remaining nonreference nodes in the usual way. The source transformation eliminates node C, so there are only $N - 2$ nonreference nodes left in the circuit. Obviously, method 1 only applies when there is a resistance in series with the voltage source.

Method 2 in Figure 3–11 can be used whether or not there is a resistance in series with the voltage source. When node B is selected as the reference node, then by definition $v_B = 0$ and the fundamental property of node voltages says that $v_A = v_S$. We do not need a node-voltage equation at node A because its voltage is known to be equal to the source voltage. We write the node equations at the remaining $N - 2$ nonreference nodes in the usual way. In the final step, we move all terms involving v_A to the right side, since it is a known input and not an unknown response. Method 2 reduces the number of node equations by 1 since no equation is needed at node A.

The third method in Figure 3–11 is needed when neither node A nor node B can be selected as the reference and the source is not connected in

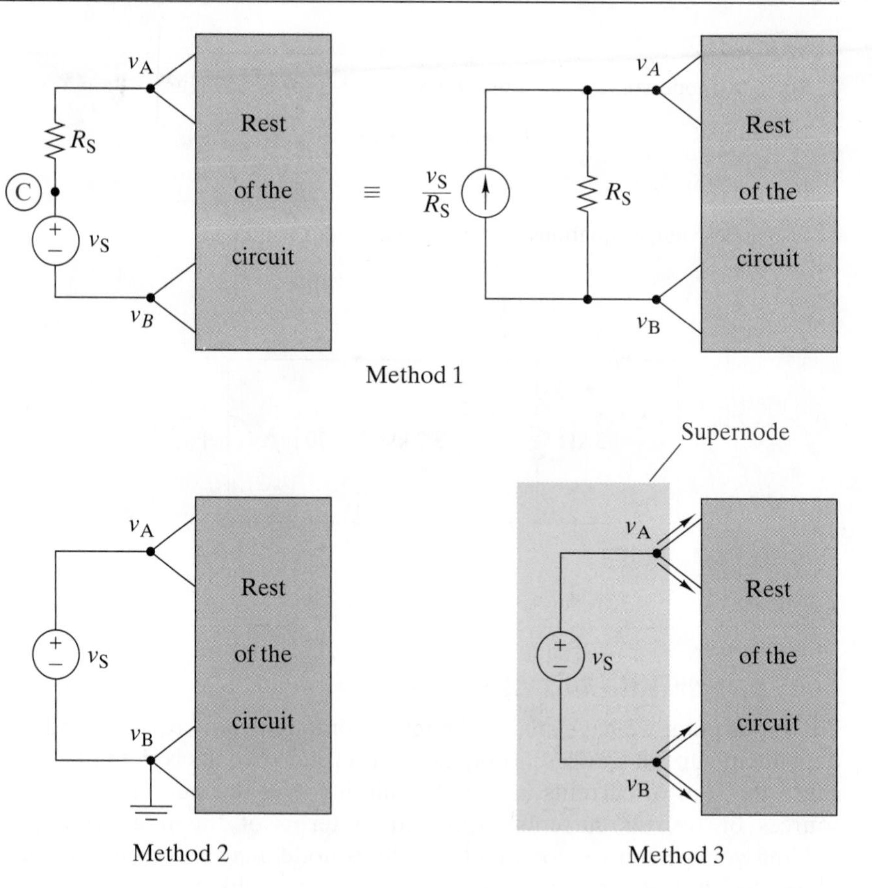

FIGURE 3–11 *Three methods of treating voltage sources in node analysis.*

series with a resistance. In this case we combine nodes A and B into a **supernode**, indicated by the boundary in Figure 3–11. We use the fact that KCL applies to the currents penetrating this boundary to write a node equation at the supernode. We then write node equations at the remaining $N-3$ nonreference nodes in the usual way. We now have $N-3$ node equations plus one supernode equation, leaving us one equation short of the $N-1$ required. Using the fundamental property of node voltages, we can write

$$v_A - v_B = v_S \tag{3–10}$$

The voltage source inside the supernode constrains the difference between the node voltages at nodes A and B. The voltage source constraint provides the additional relationship needed to write $N-1$ independent equations in $N-1$ node voltages.

For reference purposes we will call these modified node equations, since we either modify the circuit (method 1), use voltage source constraints to define node voltage at some nodes (method 2), or combine nodes to produce a supernode (method 3). The three methods are not mutually exclusive. We frequently use a combination of methods, as illustrated in the following examples.

EXAMPLE 3–4

Use node-voltage analysis to find v_O in the circuit in Figure 3–12(a).

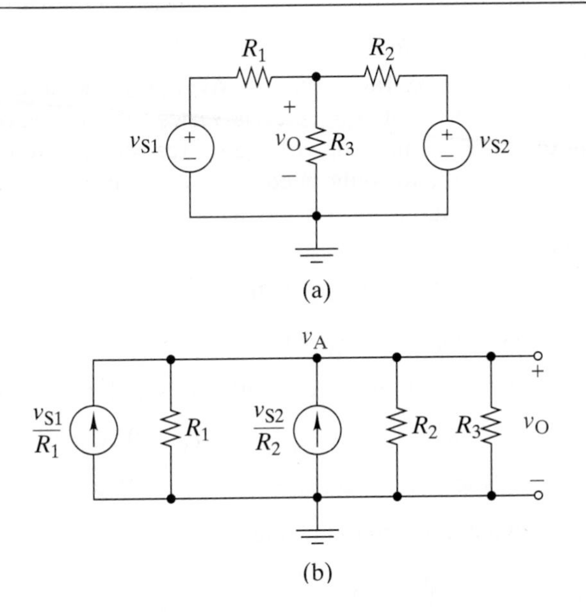

FIGURE 3–12

SOLUTION:
The given circuit in Figure 3–12(a) has four nodes, so we appear to need $N - 1 = 3$ node-voltage equations. However, applying source transformations to the two voltage sources (method 1) produces the two-node circuit in Figure 3–12(b). For the modified circuit, we need only one node equation.

$$(G_1 + G_2 + G_3)v_A = G_1 v_{S1} + G_2 v_{S2}$$

To find the output voltage, we solve for v_A:

$$v_O = v_A = \frac{G_1 v_{S1} + G_2 v_{S2}}{G_1 + G_2 + G_3}$$

Because of the two voltage sources, we need only one node equation in what appears to be a three-node circuit. The two voltage sources have a common node, so the number of unknown node voltages is reduced from three to one. The general principle illustrated is that the number of independent KCL constraints in a circuit containing N nodes and N_V voltage sources is $N - 1 - N_V$. ■

EXAMPLE 3–5

Find the input resistance of the circuit in Figure 3–13.

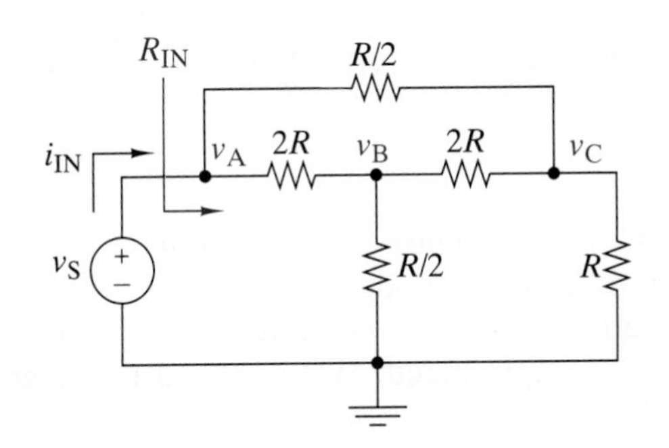

FIGURE 3–13

SOLUTION:

Method 1 of handling voltage sources will not work here because the source in Figure 3–13 is not connected in series with a resistor. Method 2 will work in this case because the voltage source is connected to the reference node. As a result, we only need node equations at nodes B and C since the node A voltage is $v_A = v_S$. By inspection, the two required node equations are

$$\text{Node B:} \quad -0.5Gv_A + 3Gv_B - 0.5Gv_C = 0$$

$$\text{Node C:} \quad -2Gv_A - 0.5Gv_B + 3.5Gv_C = 0$$

Since $v_A = v_S$, these equations can be written in standard form as follows:

$$\text{Node B:} \qquad 3Gv_B - 0.5Gv_C = 0.5Gv_S$$

$$\text{Node C:} \quad -0.5Gv_B + 3.5Gv_C = 2Gv_S$$

Solving for the two unknown node voltages yields

$$v_B = \frac{\Delta_B}{\Delta} = \frac{\begin{vmatrix} 0.5Gv_S & -0.5G \\ 2Gv_S & 3.5G \end{vmatrix}}{\begin{vmatrix} 3G & -0.5G \\ -0.5G & 3.5G \end{vmatrix}} = \frac{2.75G^2v_S}{10.25G^2} = \frac{2.75v_S}{10.25}$$

$$v_C = \frac{\Delta_C}{\Delta} = \frac{\begin{vmatrix} 3G & 0.5Gv_S \\ -0.5G & 2Gv_S \end{vmatrix}}{\Delta} = \frac{6.25G^2v_S}{10.25G^2} = \frac{6.25v_S}{10.25}$$

Given the two node voltages, we can now solve for the input current.

$$i_{IN} = \frac{v_S - v_B}{2R} + \frac{v_S - v_C}{R/2} = \frac{3.75v_S}{10.25R} + \frac{8v_S}{10.25R} = \frac{11.75v_S}{10.25R}$$

Hence, the input resistance is

$$R_{IN} = \frac{v_S}{i_{IN}} = \frac{10.25R}{11.75} = 0.872R$$

This is the same answer as in Example 3–3, where the same circuit was driven by a current source rather than a voltage source. Input resistance is an intrinsic property of a circuit that does not depend on how the circuit is driven. ■

EXAMPLE 3–6

For the circuit in Figure 3–14,

(a) Formulate node-voltage equations.
(b) Solve for the output voltage v_O using $R_1 = R_4 = 2$ kΩ and $R_2 = R_3 = 4$ kΩ.

SOLUTION:

(a) The voltage sources in Figure 3–14 do not have a common node, and we cannot select a reference node that includes both sources. Selecting node D as the reference forces the condition $v_B = v_{S2}$ (method 2) but leaves the other source v_{S1} ungrounded. We surround the ungrounded source,

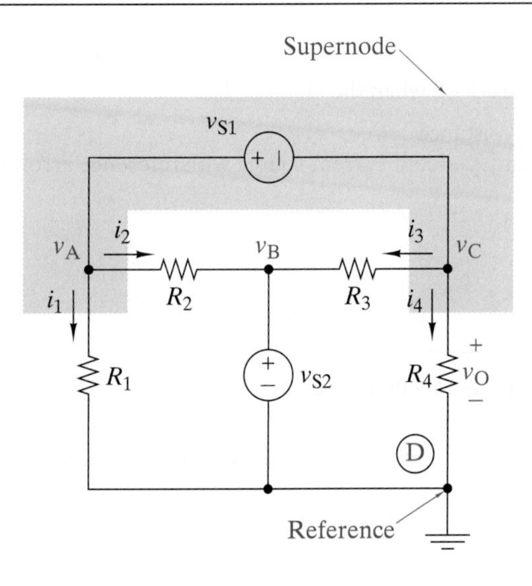

FIGURE 3–14

and all wires leading to it, by the supernode boundary shown in Figure 3–14 (method 3). KCL applies to the four element currents that penetrate the supernode boundary, and we can write

$$i_1 + i_2 + i_3 + i_4 = 0$$

These currents can easily be expressed in terms of the node voltages.

$$G_1 v_A + G_2(v_A - v_B) + G_3(v_C - v_B) + G_4 v_C = 0$$

Since $v_B = v_{S2}$, the standard form of this equation is

$$(G_1 + G_2)v_A + (G_3 + G_4)v_C = (G_2 + G_3)v_{S2}$$

We have one equation in the two unknown node voltages v_A and v_C. Applying the fundamental property of node voltages inside the supernode, we can write

$$v_A - v_C = v_{S1}$$

That is, the ungrounded voltage source constrains the difference between the two unknown node voltages inside the supernode. It thereby supplies the relationship needed to obtain two equations in two unknowns.

(b) Inserting the given numerical values yields

$$(7.5 \times 10^{-4})v_A + (7.5 \times 10^{-4})v_C = (5 \times 10^{-4})v_{S2}$$
$$v_A - v_C = v_{S1}$$

To find the output v_O, we need to solve these equations for v_C. The second equation yields $v_A = v_C + v_{S1}$, which, when substituted into the first equation, yields the required output:

$$v_O = v_C = \frac{v_{S2}}{3} - \frac{v_{S1}}{2}$$

■

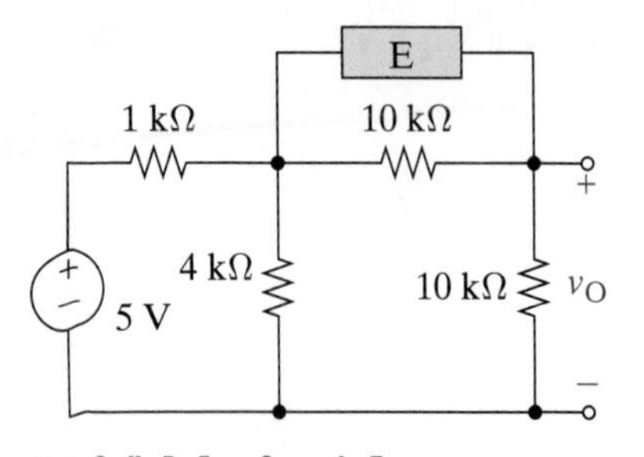

FIGURE 3–15

Exercise 3–5

Find v_O in Figure 3–15 when the element E is

(a) A 10-kΩ resistance
(b) A 4-mA independent current source with reference arrow pointing left

Answers:

(a) 2.53 V
(b) –17.3 V

Exercise 3–6

Find v_O in Figure 3–15 when the element E is

(a) An open circuit
(b) A 10-V independent voltage source with the plus reference on the right

Answers:

(a) 1.92 V
(b) 12.96 V

SUMMARY OF NODE-VOLTAGE ANALYSIS

We have seen that node-voltage equations are very useful in the analysis of a variety of circuits. These equations can always be formulated using KCL, the element constraints, and the fundamental property of node voltages. When in doubt, always fall back on these principles to formulate node equations in new situations. With practice and experience, however, we eventually develop an analysis approach that allows us to recognize shortcuts in the formulation process. The following guidelines summarize our approach and may help you develop your own analysis style:

1. Simplify the circuit by combining elements in series and parallel wherever possible.
2. If not specified, select a reference node so that as many voltage sources as possible are directly connected to the reference.
3. Node equations are required at supernodes and all other nonreference nodes except those that are directly connected to the reference by voltage sources.
4. Use KCL to write node equations at the nodes identified in step 3. Express element currents in terms of node voltages or the currents produced by independent current sources.
5. Write expressions relating the node voltages to the voltages produced by independent voltage sources.
6. Substitute the expressions from step 5 into the node equations from step 4 and arrange the resulting equations in standard form.
7. Solve the equations from step 6 for the node voltages of interest. Cramer's rule is often useful when circuit parameters are left in symbolic form. Computer tools are useful when there are four or more equations and numerical values are given.

3–2 MESH-CURRENT ANALYSIS

Mesh currents are analysis variables that are useful in circuits containing many elements connected in series. To review terminology, a loop is a closed path formed by passing through an ordered sequence of nodes without passing through any node more than once. A mesh is a special type of loop that does not enclose any elements. For example, loops A and B in Figure 3–16 are meshes, while the loop X is not a mesh because it encloses an element.

Mesh-current analysis is restricted to planar circuits. A **planar circuit** can be drawn on a flat surface without crossovers in the "window pane" fashion shown in Figure 3–16. To define a set of variables, we associate a **mesh current** (i_A, i_B, i_C, etc.) with each window pane and assign a reference direction. The reference directions for all mesh currents are customarily taken in a clockwise sense. There is no momentous reason for this, except perhaps tradition.

We think of these mesh currents as circulating through the elements in their respective meshes, as suggested by the reference directions shown in Figure 3–16. We should emphasize that this viewpoint is not based on the physics of circuit behavior. There are not red and blue electrons running around that somehow get assigned to mesh currents i_A or i_B. Mesh currents are variables used in circuit analysis. They are only somewhat abstractly related to the physical operation of a circuit and may be impossible to measure directly. For example, there is no way to cut the circuit in Figure 3–16 to insert an ammeter that only measures i_E.

Mesh currents have a unique feature that is the dual of the fundamental property of node voltages. If we examine Figure 3–16, we see the elements around the perimeter are contained in only one mesh, while those in the interior are in two meshes. In a planar circuit any given element is contained in at most two meshes. When an element is in two meshes, the two mesh currents circulate through the element in opposite directions. In such cases KCL declares that the net current through the element is the difference of the two mesh currents.

These observations lead us to the fundamental property of mesh currents:

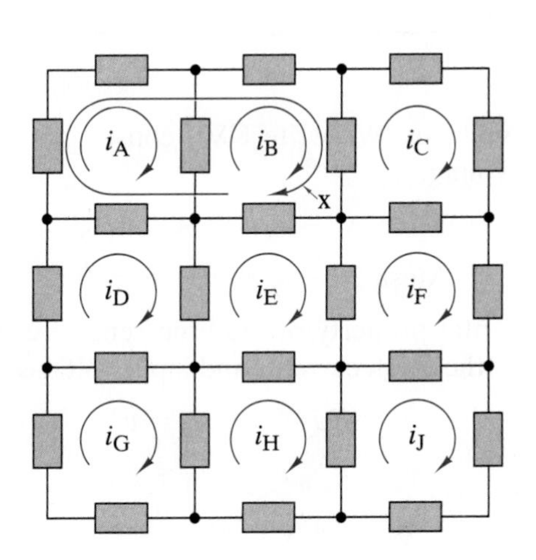

FIGURE 3–16 *Meshes in a planar circuit.*

If the Kth two-terminal element is contained in meshes X and Y, then the element current can be expressed in terms of the two mesh currents as

$$i_K = i_X - i_Y \tag{3–11}$$

where X is the mesh whose reference direction agrees with the reference direction of i_K.

Equation (3–11) is a KCL constraint at the element level. If the element is contained in only one mesh, then $i_K = i_X$ or $i_K = -i_Y$, depending on whether the reference direction for the element current agrees or disagrees with the reference direction of the mesh current. The key idea is that the current through every two-terminal element in a planar circuit can be expressed in terms of no more than two mesh currents.

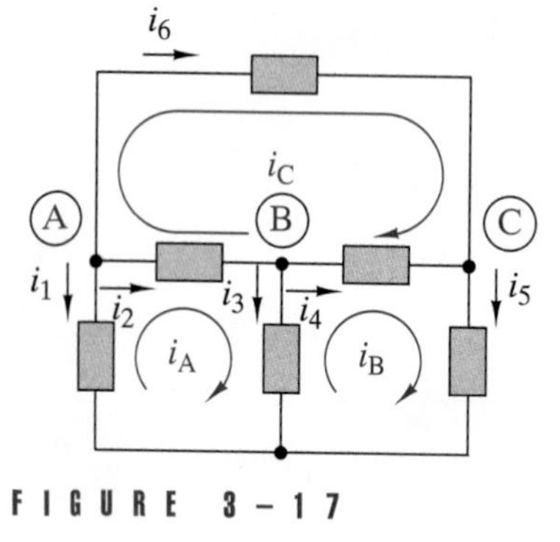

FIGURE 3–17

Exercise 3–7

In Figure 3–17 the mesh currents are $i_A = 10$ A, $i_B = 5$ A, and $i_C = -3$ A. Find the element currents i_1 through i_6 and show that KCL is satisfied at nodes A, B, and C.

Answers: $i_1 = -10$ A; $i_2 = 13$ A; $i_3 = 5$ A; $i_4 = 8$ A; $i_5 = 5$ A; $i_6 = -3$ A.

To use mesh currents to formulate circuit equations, we use elements and connection constraints, except that the KCL constraints are not explicitly written. Instead, we use the fundamental property of mesh currents to express the element voltages in terms of the mesh currents. By doing so we avoid using the element currents and work only with the element voltages and mesh currents.

For example, the planar circuit in Figure 3–18 can be analyzed using the mesh-current method. In the figure we have defined two mesh currents

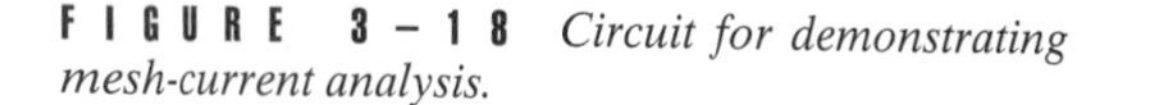
FIGURE 3–18 *Circuit for demonstrating mesh-current analysis.*

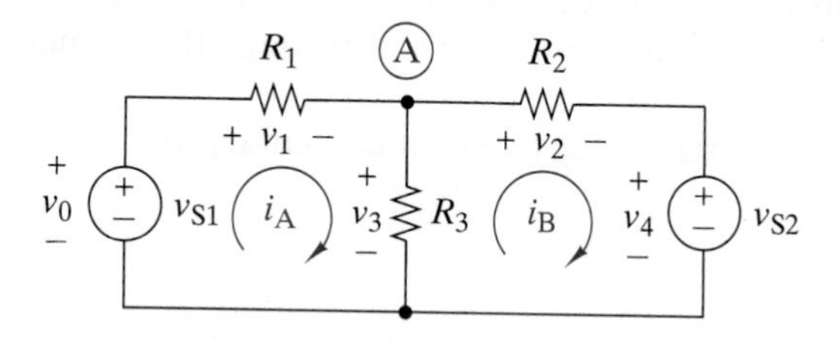

and five element voltages. We write KVL constraints around each mesh using the element voltages.

$$\text{Mesh A:} \quad -v_0 + v_1 + v_3 = 0$$

$$\text{Mesh B:} \quad -v_3 + v_2 + v_4 = 0 \tag{3–12}$$

Using the fundamental property of mesh currents, we write the element voltages in terms of the mesh currents and input voltages:

$$\begin{aligned} v_1 &= R_1 i_A & v_0 &= v_{S1} \\ v_2 &= R_2 i_B & v_4 &= v_{S2} \\ v_3 &= R_3(i_A - i_B) \end{aligned} \tag{3–13}$$

We substitute these element equations into the KVL connection equations and arrange the result in standard form.

$$\begin{aligned}(R_1 + R_3)i_A - R_3 i_B &= v_{S1} \\ -R_3 i_A + (R_2 + R_3)i_B &= -v_{S2}\end{aligned} \tag{3–14}$$

We have completed the formulation process with two equations in two unknown mesh currents.

As we have previously noted, every method of circuit analysis must satisfy KCL, KVL, and the element *i–v* relationships. When formulating mesh equations, it may appear that we have not used KCL. However, writing the element constraints in the form in Eq. (3–13) requires the KCL equations $i_1 = i_A$, $i_2 = i_B$, and $i_3 = i_A - i_B$. Mesh-current analysis implicitly satisfies KCL when the element constraints are expressed in terms of the mesh currents. In effect, the fundamental property of mesh currents ensures that the KCL constraints are satisfied.

We use Cramer's rule to solve for the mesh currents in Eq. (3–14):

$$i_A = \frac{\Delta_A}{\Delta} = \frac{\begin{vmatrix} v_{S1} & -R_3 \\ -v_{S2} & R_2 + R_3 \end{vmatrix}}{\begin{vmatrix} R_1 + R_3 & -R_3 \\ -R_3 & R_2 + R_3 \end{vmatrix}} = \frac{(R_2 + R_3)v_{S1} - R_3 v_{S2}}{R_1R_2 + R_1R_3 + R_2R_3} \tag{3–15}$$

and

$$i_B = \frac{\Delta_B}{\Delta} = \frac{\begin{vmatrix} R_1 + R_3 & v_{S1} \\ -R_3 & -v_{S2} \end{vmatrix}}{\begin{vmatrix} R_1 + R_3 & -R_3 \\ -R_3 & R_2 + R_3 \end{vmatrix}} = \frac{R_3 v_{S1} - (R_1 + R_3)v_{S2}}{R_1R_2 + R_1R_3 + R_2R_3} \tag{3–16}$$

Equations (3–15) and (3–16) can now be substituted into the element constraints in Eq. (3–13) to solve for every voltage in the circuit. For instance, the voltage across R_3 is

$$v_3 = R_3(i_A - i_B) = \frac{R_2R_3v_{S1} + R_1R_3v_{S2}}{R_1R_2 + R_1R_3 + R_2R_3} \tag{3–17}$$

You are invited to show that the result in Eq. (3–17) agrees with the node analysis result obtained in Example 3–4 for the same circuit.

The mesh-current analysis approach just illustrated can be summarized in four steps:

STEP 1 Identify a mesh current with every mesh and a voltage across every circuit element.

STEP 2 Write KVL connection constraints in terms of the element voltages around every mesh.

STEP 3 Use KCL and the *i–v* relationships of the elements to express the element voltages in terms of the mesh currents.

STEP 4 Substitute the element constraints from step 3 into the connection constraints from step 2 and arrange the resulting equations in standard form.

The number of mesh-current equations derived in this way equals the number of KVL connection constraints in step 2. When discussing combined constraints in Chapter 2, we noted that there are $E - N + 1$ independent KVL constraints in any circuit. Using the window panes in a planar circuit generates $E - N + 1$ independent mesh currents. Mesh analysis works best when the circuit has many elements (E large) connected in series (N also large).

WRITING MESH-CURRENT EQUATIONS BY INSPECTION

The mesh equations in Eq. (3–14) have a symmetrical pattern that is similar to the coefficient symmetry observed in node equations. The coefficients of i_B in the first equation and i_A in the second equation are the negative of the resistance common to meshes A and B. The coefficients of i_A in the first equation and i_B in the second equation are the sum of the resistances in meshes A and B, respectively.

This pattern will always occur in planar circuits containing resistors and independent voltage sources when the mesh currents are defined in the window panes of a planar circuit, as shown in Figure 3–16. To see why, consider a general resistance R that is contained in, say, mesh A. There are only two possibilities. Either R is not contained in any other mesh, in which case the voltage across it is

$$v = R(i_A - 0) = Ri_A$$

or else it is also contained in only one adjacent mesh, say mesh B, in which case the voltage across it is

$$v = R(i_A - i_B)$$

These observations lead to the following conclusions. The voltage across resistance in mesh A involves the following terms:

1. i_A times the sum of the resistances in mesh A
2. $-i_B$ times the sum of resistances common to mesh A and mesh B, and similar terms for any other mesh adjacent to mesh A

The sum of the voltages across resistors plus the sum of the independent voltage sources around mesh A must equal zero.

The aforementioned process makes it possible for us to write mesh-current equations by inspection without going through the intermediate steps involving the KVL connection constraints and the element constraints.

EXAMPLE 3–7

For the circuit of Figure 3–19,

(a) Formulate mesh-current equations.
(b) Find the output v_O using $R_1 = R_4 = 2\ \text{k}\Omega$ and $R_2 = R_3 = 4\ \text{k}\Omega$.

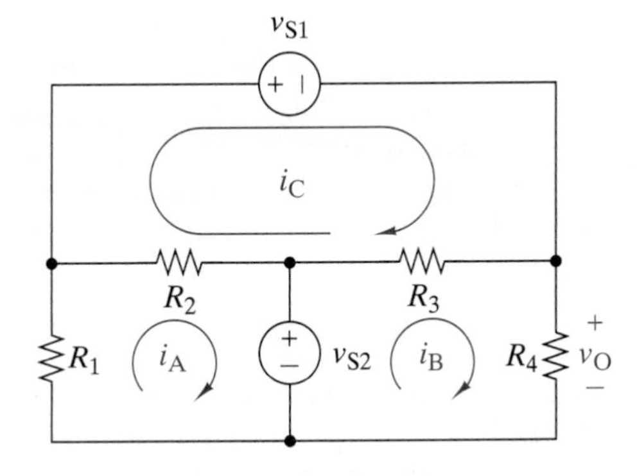

FIGURE 3 – 19

SOLUTION:

(a) To write mesh-current equations by inspection, we note that the total resistances in meshes A, B, and C are $R_1 + R_2$, $R_3 + R_4$, and $R_2 + R_3$, respectively. The resistance common to meshes A and C is R_2. The resistance common to meshes B and C is R_3. There is no resistance common to meshes A and B. Using these observations, we write the mesh equations as

$$\text{Mesh A: } (R_1 + R_2)i_A - 0\,i_B - R_2 i_C + v_{S2} = 0$$

$$\text{Mesh B: } (R_3 + R_4)i_B - 0\,i_A - R_3 i_C - v_{S2} = 0$$

$$\text{Mesh C: } (R_2 + R_3)i_C - R_2 i_A - R_3 i_B + v_{S1} = 0$$

The algebraic signs assigned to voltage source terms follow the passive convention for the mesh current in question. Arranged in standard form, these equations become

$$\begin{aligned} (R_1 + R_2)i_A - R_2 i_C &= -v_{S2} \\ +(R_3 + R_4)i_B - R_3 i_C &= +v_{S2} \\ -R_2 i_A - R_3 i_B + (R_2 + R_3)i_C &= -v_{S1} \end{aligned}$$

Coefficient symmetry greatly simplifies the formulation of these equations compared with the more fundamental, but time-consuming, process of writing element and connection constraints.

(b) Inserting the numerical values into these equations yields

$$\begin{aligned} 6000\,i_A \qquad\qquad - 4000\,i_C &= -v_{S2} \\ 6000\,i_B - 4000\,i_C &= v_{S2} \\ -4000\,i_A - 4000\,i_B + 8000\,i_C &= -v_{S1} \end{aligned}$$

Placing these three mesh equations in matrix form produces

$$\begin{bmatrix} 6000 & 0 & -4000 \\ 0 & 6000 & -4000 \\ -4000 & -4000 & +8000 \end{bmatrix} \begin{bmatrix} i_A \\ i_B \\ i_C \end{bmatrix} = \begin{bmatrix} -v_{S2} \\ v_{S2} \\ -v_{S1} \end{bmatrix}$$

This is a matrix equation of the form **AX** = **B**, where

$$\mathbf{A} = \begin{bmatrix} 6000 & 0 & -4000 \\ 0 & 6000 & -4000 \\ -4000 & -4000 & +8000 \end{bmatrix} \quad \mathbf{X} = \begin{bmatrix} i_A \\ i_B \\ i_C \end{bmatrix} \quad \mathbf{B} = \begin{bmatrix} -v_{S2} \\ v_{S2} \\ -v_{S1} \end{bmatrix}$$

The matrix **A** is a square matrix of the coefficients on the left side of the mesh equations, **X** is a column matrix of the unknown mesh currents, and **B** is a column matrix of the input voltages on the right side of the mesh equations. To solve this matrix equation, we multiply by the inverse of the coefficient matrix ($\mathbf{A}^{-1}$). On the left side the result is $\mathbf{A}^{-1}\,\mathbf{AX} = \mathbf{X}$, and the right side becomes $\mathbf{A}^{-1}\,\mathbf{B}$. In other words, multiplying by $\mathbf{A}^{-1}$ yields the solution for the unknown mesh currents as

$$\mathbf{X} = \mathbf{A}^{-1}\mathbf{B}$$

In sum, to solve a system of linear equations by matrix methods we form the matrix product $\mathbf{A}^{-1}\mathbf{B}$. It is at this point that MATLAB comes into play.

Using MATLAB to solve for the mesh currents, we first enter the **A** matrix by the statement

```
A=[6000 0 -4000;0 6000 -4000;-4000 -4000 8000];
```

The elements in **B** matrix are the symbolic variables v_{S1} and v_{S2}. These quantities are not unknowns, but symbols that represent all possible values of the input voltages. Thus, we must first write the statement

```
syms VS1 VS2
```

which declares these identifiers to be symbolic rather than numerical quantities. We can now enter the **B** matrix as

```
B=[-VS2;VS2;-VS1];
```

and solve for the unknown mesh currents using MATLAB statement

```
X=inv(A)*B
```

which yields

```
X =

[-1/6000*VS2-1/4000*VS1]
[1/6000*VS2-1/4000*VS1]
[             -3/8000*VS1]
```

The elements of the column matrix **X** are the three unknown mesh currents expressed in terms of the input voltages. The output voltage in Figure 3–19 is written in terms of the mesh currents as $v_O = R_4 i_B$. In MATLAB notation the required mesh current is

i_B = `X(2)=1/6000*VS2-1/4000*VS1`

Since $R_4 = 2000$ we conclude that

$$v_O = \frac{v_{S2}}{3} - \frac{v_{S1}}{2}$$

The mesh-current analysis result obtained here is the same as the node-voltage result obtained in Example 3–6. Either approach produces the same answer, but which method do you think is easier? ■

MESH EQUATIONS WITH CURRENT SOURCES

In developing mesh analysis, we assumed that circuits contain only voltage sources and resistors. This assumption simplifies the formulation process because the sum of voltages around a mesh is determined by voltage sources and the mesh currents through resistors. A current source complicates the picture because the voltage across it is not directly related to its current. We need to adapt mesh analysis to accommodate current sources just as we revised node analysis to deal with voltage sources.

There are three ways to handle current sources in mesh analysis:

1. If the current source is connected in parallel with a resistor, then it can be converted to an equivalent voltage source by source transformation. Each source conversion eliminates a mesh and reduces the number of equations required by 1. This method is the dual of method 1 for node analysis.
2. If a current source is contained in only one mesh, then that mesh current is determined by the source current and is no longer an unknown. We write mesh equations around the remaining meshes in the usual way and move the known mesh current to the source side of the equations in the final step. The number of equations obtained is one less than the number of meshes. This method is the dual of method 2 for node analysis.
3. Neither of the first two methods will work when a current source is contained in two meshes or is not connected in parallel with a resistance. In this case we create a **supermesh** by excluding the current source and any elements connected in series with it, as shown in Figure 3–20. We write one mesh equation around the supermesh

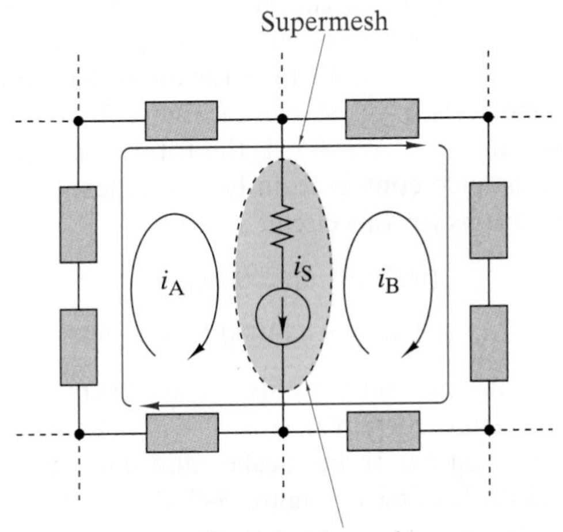

FIGURE 3–20 *Example of a supermesh.*

using the currents i_A and i_B. We then write mesh equations of the remaining meshes in the usual way. This leaves us one equation short because parts of meshes A and B are included in the supermesh. However, the fundamental property of mesh currents relates the currents i_S, i_A, and i_B as

$$i_A - i_B = i_S$$

This equation supplies the one additional relationship needed to get the requisite number of equations in the unknown mesh currents.

The aforementioned three methods are not mutually exclusive. We can use more than one method in a circuit, as the following examples illustrate.

EXAMPLE 3–8

Use mesh-current equations to find i_O in the circuit in Figure 3–21(a).

FIGURE 3–21

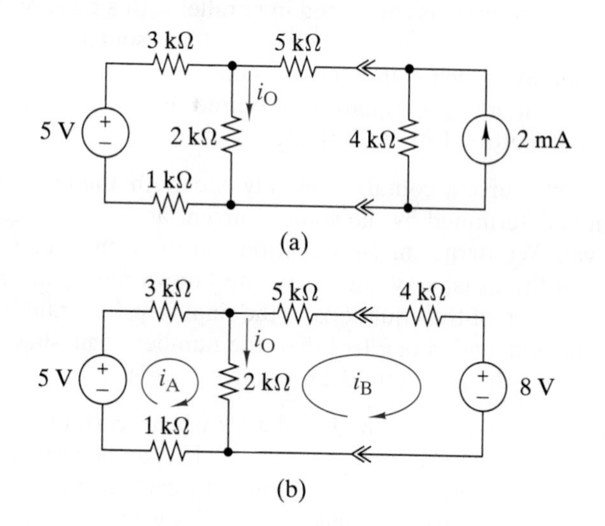

SOLUTION:

The current source in this circuit can be handled by a source transformation (method 1). The 2-mA source in parallel with the 4-kΩ resistor in Figure 3–21(a) can be replaced by an equivalent 8-V voltage source in series with the same resistor, as shown in Figure 3–21(b). In this circuit the total resistance in mesh A is 6 kΩ, the total resistance in mesh B is 11 kΩ, and the resistance contained in both meshes is 2 kΩ. By inspection, the mesh equations for this circuit are

$$(6000)i_A - (2000)i_B = 5$$

$$-(2000)i_A + (11000)i_B = -8$$

Solving for the two mesh currents yields $i_A = 0.6290$ mA and $i_B = -0.6129$ mA. By KCL the desired current is $i_O = i_A - i_B = 1.2419$ mA. The given circuit in Figure 3–21(a) has three meshes and one current source. The source transformation leading to Figure 3–21(b) produces a circuit with only two meshes. The general principle illustrated is that the number of independent mesh equations in a circuit containing E elements, N nodes, and N_I current sources is $E - N + 1 - N_I$. ■

EXAMPLE 3–9

Use mesh-current equations to find the v_O in Figure 3–22.

SOLUTION:

Source transformation (method 1) is not possible here since neither current source is connected in parallel with a resistor. The current source i_{S2} is in both mesh B and mesh C, so we exclude this element and create the

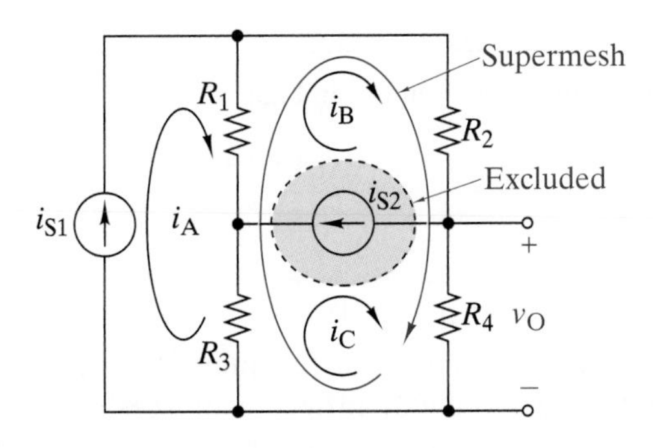

FIGURE 3–22

supermesh (method 3) shown in the figure. The sum of voltages around the supermesh is

$$R_1(i_B - i_A) + R_2(i_B) + R_4(i_C) + R_3(i_C - i_A) = 0$$

The supermesh voltage constraint yields one equation in the three unknown mesh currents. Applying KCL to each of the current sources yields

$$i_A = i_{S1}$$
$$i_B - i_C = i_{S2}$$

Because of KCL, the two current sources force constraints that supply two more equations. Using these two KCL constraints to eliminate i_A and i_B from the supermesh KVL constraint yields

$$(R_1 + R_2 + R_3 + R_4)i_C = (R_1 + R_3)i_{S1} - (R_1 + R_2)i_{S2}$$

Hence, the required output voltage is

$$v_O = R_4 i_C = R_4 \times \left[\frac{(R_1 + R_3)i_{S1} - (R_1 + R_2)i_{S2}}{R_1 + R_2 + R_3 + R_4}\right]$$

■

Exercise 3–8

Use mesh analysis to find the current i_O in Figure 3–23 when the element E is

(a) A 5-V voltage source with the positive reference at the top
(b) A 10-kΩ resistor

Answers:

(a) –0.136 mA
(b) –0.538 mA

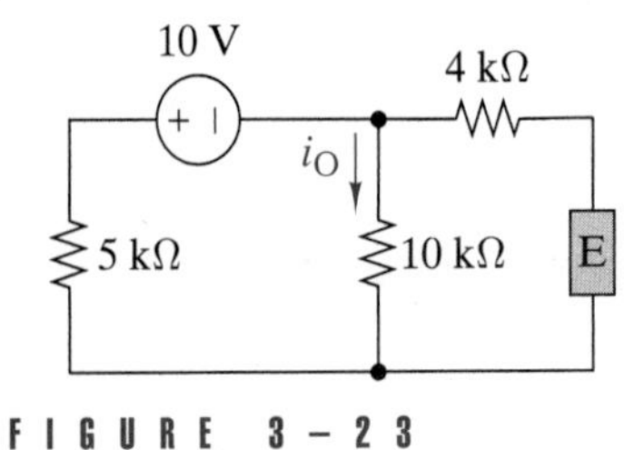

FIGURE 3–23

Exercise 3–9

Use mesh analysis to find the current i_O in Figure 3–23 when the element E is

(a) A 1-mA current source with the reference arrow directed down
(b) Two 20-kΩ resistors in parallel

Answers:

(a) –1 mA
(b) –0.538 mA

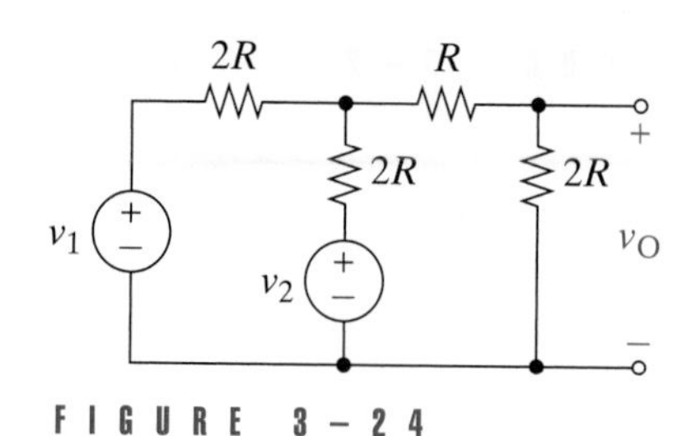

FIGURE 3–24

Exercise 3–10

Use mesh-current equations to find v_O in Figure 3–24.

Answer: $v_O = (v_1 + v_2)/4$

SUMMARY OF MESH-CURRENT ANALYSIS

Mesh-current equations can always be formulated from KVL, the element constraints, and the fundamental property of mesh currents. When in doubt, always fall back on these principles to formulate mesh equations in new situations. The following guidelines summarize an approach to formulating mesh equations for resistance circuits:

1. Simplify the circuit by combining elements in series or parallel wherever possible.
2. Mesh equations are required for supermeshes and all other meshes except those where current sources are contained in only one mesh.
3. Use KVL to write mesh equations for the meshes identified in step 2. Express element voltages in terms of mesh currents or the voltage produced by independent voltage sources.
4. Write expressions relating the mesh currents to the currents produced by independent current sources.
5. Substitute the expressions from step 4 into the mesh equations from step 3 and place the result in standard form.
6. Solve the equations from step 5 for the mesh currents of interest. Cramer's rule is often useful when circuit parameters are left in symbolic form. Computer tools are useful when there are four or more equations and numerical values are given.

3–3 LINEARITY PROPERTIES

This book treats the analysis and design of **linear circuits**. A circuit is said to be linear if it can be adequately modeled using only linear elements and independent sources. The hallmark feature of a linear circuit is that outputs are linear functions of the inputs. Circuit **inputs** are the signals produced by external sources, and **outputs** are any other designated signals. Mathematically, a function is said to be linear if it possesses two properties—homogeneity and additivity. In linear circuits, **homogeneity** means that the output is proportional to the input. **Additivity** means that the output due to two or more inputs can be found by adding the outputs obtained when each input is applied separately. Mathematically, these properties are written as follows:

$$f(Kx) = Kf(x) \quad \text{(homogeneity)} \tag{3–18}$$

and

$$f(x_1 + x_2) = f(x_1) + f(x_2) \quad \text{(additivity)} \tag{3–19}$$

where K is a scalar constant. In circuit analysis the homogeneity property is called **proportionality**, while the additivity property is called **superposition**.

THE PROPORTIONALITY PROPERTY

The **proportionality property** applies to linear circuits with one input. For linear resistive circuits, proportionality states that every input-output relationship can be written as

$$y = Kx \tag{3–20}$$

where x is the input current or voltage, y is an output current or voltage, and K is a constant. The block diagram in Figure 3–25 describes this linear input–output relationship. In a block diagram the lines headed by arrows indicate the direction of signal flow. The arrow directed into the block indicates the input, while the output is indicated by the arrow directed out of the block. The variable names written next to these lines identify the input and output signals. The scalar constant K written inside the block indicates that the input signal x is multiplied by K to produce the output signal as $y = Kx$.

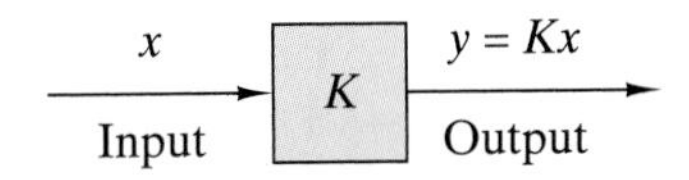

FIGURE 3–25 *Block diagram representation of the proportionality property.*

Caution: Proportionality only applies when the input and output are current or voltage. It does not apply to output power since power is equal to the product of current and voltage. In other words, output power is not linearly related to the input current or voltage.

We have already seen several examples of proportionality. For instance, using voltage division in Figure 3–26(a) produces

$$v_O = \left(\frac{R_2}{R_1 + R_2}\right)v_S$$

which means

$$x = v_S \qquad y = v_O$$

$$K = \frac{R_2}{R_1 + R_2}$$

Similarly, applying current division in Figure 3–26(b) yields

$$i_O = \left(\frac{G_2}{G_1 + G_2}\right)i_S$$

so that

$$x = i_S \qquad y = i_O$$

$$K = \frac{G_2}{G_1 + G_2}$$

In these two examples the proportionality constant K is dimensionless because the input and output have the same units. In other situations K could have the units of ohms or siemens when the input and output have different units.

The next example illustrates that the proportionality constant K can be positive, negative, or even zero.

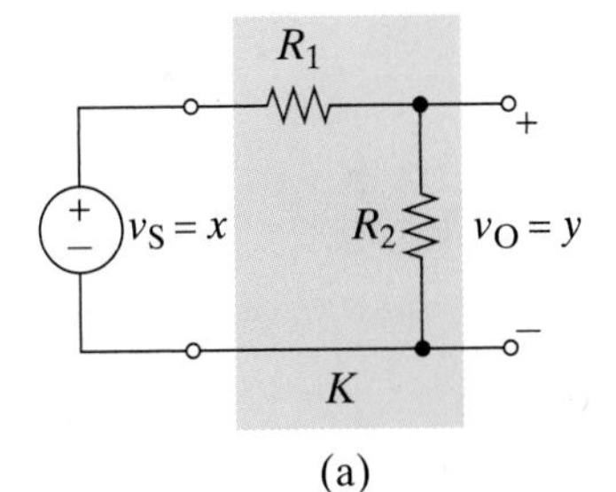

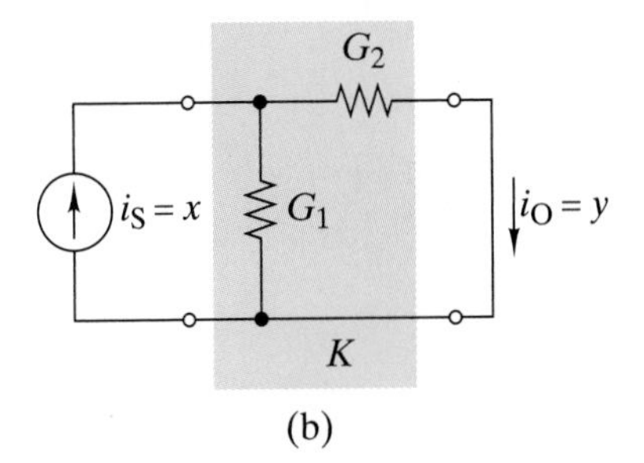

FIGURE 3–26 *Examples of circuit exhibiting proportionality: (a) Voltage divider. (b) Current divider.*

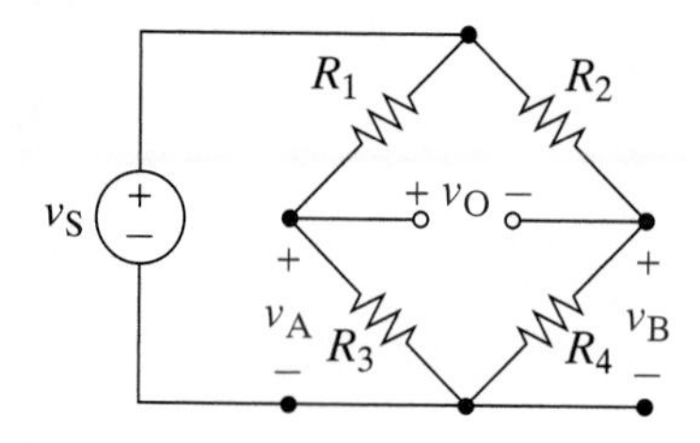

FIGURE 3–27

EXAMPLE 3–10

Given the bridge circuit of Figure 3–27,

(a) Find the proportionality constant K in the input-output relationship $v_O = Kv_S$.
(b) Find the sign of K when $R_2R_3 > R_1R_4$, $R_2R_3 = R_1R_4$, and $R_2R_3 < R_1R_4$.

SOLUTION:
(a) We observe that the circuit consists of two voltage dividers. Applying the voltage division rule to each side of the bridge circuit yields

$$v_A = \frac{R_3}{R_1 + R_3}v_S \quad \text{and} \quad v_B = \frac{R_4}{R_2 + R_4}v_S$$

The fundamental property of node voltages allows us to write

$$v_O = v_A - v_B$$

Substituting the equations for v_A and v_B into this KVL equation yields

$$\begin{aligned} v_O &= \left(\frac{R_3}{R_1 + R_3} - \frac{R_4}{R_2 + R_4}\right)v_S \\ &= \left(\frac{R_2R_3 - R_1R_4}{(R_1 + R_3)(R_2 + R_4)}\right)v_S \\ &= \qquad (K) \qquad v_S \end{aligned}$$

(b) The proportionality constant K can be positive, negative, or zero. Specifically,

$$\text{If } R_2R_3 > R_1R_4 \text{ then } K > 0$$
$$\text{If } R_2R_3 = R_1R_4 \text{ then } K = 0$$
$$\text{If } R_2R_3 < R_1R_4 \text{ then } K < 0$$

When the products of the resistances in opposite legs of the bridge are equal, then $K = 0$ and the bridge is said to be balanced. ■

UNIT OUTPUT METHOD

The **unit output method** is an analysis technique based on the proportionality property of linear circuits. The method involves finding the input-output proportionality constant K by assuming an output of one unit and determining the input required to produce that unit output. This technique is most useful when applied to ladder circuits, and it involves the following steps:

1. A unit output is assumed; that is, $v_O = 1$ V or $i_O = 1$ A.
2. The input required to produce the unit output is then found by successive application of KCL, KVL, and Ohm's law.

3. Because the circuit is linear, the proportionality constant relating input and output is

$$K = \frac{\text{Output}}{\text{Input}} = \frac{1}{\text{Input for unit output}}$$

Given the proportionality constant K, we can find the output for any input using Eq. (3–20).

In a way, the unit output method solves the circuit response problem backwards—that is, from output to input—as illustrated by the next example.

EXAMPLE 3–11

Use the unit output method to find v_O in the circuit shown in Figure 3–28(a).

FIGURE 3–28

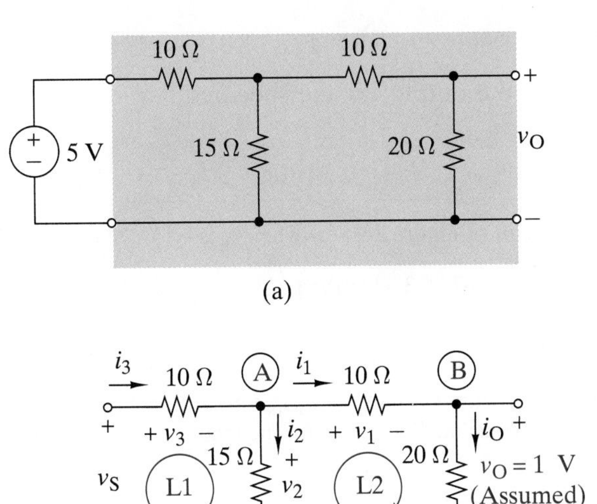

SOLUTION:

We start by assuming $v_O = 1$, as shown in Figure 3–28(b). Then, using Ohm's law, we find i_O.

$$i_O = \frac{v_O}{20} = 0.05 \text{ A}$$

Next, using KCL at node B, we find i_1.

$$i_1 = i_O = 0.05 \text{ A}$$

Again, using Ohm's law, we find v_1.

$$v_1 = 10i_1 = 0.5 \text{ V}$$

Then, writing a KVL equation around loop L2, we find v_2 as

$$v_2 = v_1 + v_O = 0.5 + 1.0 = 1.5 \text{ V}$$

Using Ohm's law once more produces

$$i_2 = \frac{v_2}{15} = \frac{1.5}{15} = 0.1 \text{ A}$$

Next, writing a KCL equation at node A yields

$$i_3 = i_1 + i_2 = 0.05 + 0.1 = 0.15 \text{ A}$$

Using Ohm's law one last time,

$$v_3 = 10 i_3 = 1.5 \text{ V}$$

We can now find the source voltage v_S by applying KVL around loop L1:

$$v_S\big|_{\text{for } v_O = 1\text{V}} = v_3 + v_2 = 1.5 + 1.5 = 3 \text{ V}$$

A 3-V source voltage is required to produce a 1-V output. From this result, we calculate the proportionality constant K to be

$$K = \frac{v_O}{v_S} = \frac{1}{3}$$

Once K is known, the output for the specified 5-V input is $v_O = (1/3)5 = 1.667$ V. ■

Exercise 3–11

Find v_O in the circuit of Figure 3–28(a) when v_S is –5 V, 10 mV, and 3 kV.

Answers: $v_O = -1.667$ V; 3.333 mV; 1 kV

Exercise 3–12

Use the unit output method to find $K = i_O/i_{IN}$ for the circuit in Figure 3–29. Then use the proportionality constant K to find i_O for the input current shown in the figure.

FIGURE 3–29

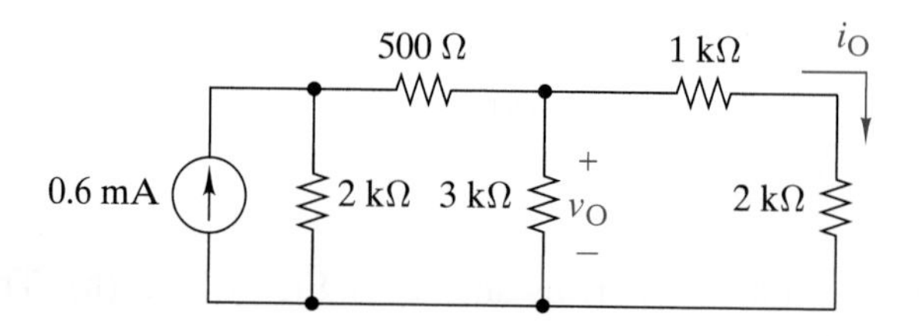

Answers: $K = 1/4$; $i_O = 0.15$ mA

Exercise 3–13

Use the unit output method to find $K = v_O/i_{IN}$ for the circuit in Figure 3–29. Then use K to find v_O for the input current shown in the figure.

Answers: $K = 750\ \Omega$; $v_O = 450$ mV

Note: In this exercise K has the dimensions of ohms because the input is a current and the output a voltage.

ADDITIVITY PROPERTY

The **additivity property** states that any output current or voltage of a linear resistive circuit with multiple inputs can be expressed as a linear combination of the several inputs:

$$y = K_1x_1 + K_2x_2 + K_3x_3 + \ldots \tag{3–21}$$

where x_1, x_2, x_3, . . . are current or voltage inputs, and K_1, K_2, K_3 . . . are constants that depend on the circuit parameters. Figure 3–30 shows how we represent this relationship in block diagram form. Again the arrows indicate the direction of signal flow and the K's within the blocks are scalar multipliers. The circle in Figure 3–30 is a new block diagram element called a summing point that implements the operation $y = \sum K_ix_i$. Although the block diagram in Figure 3–30 is nothing more than a pictorial representation of Eq. (3–21), the diagram often helps us gain a clearer picture of how signals interact in different parts of a circuit.

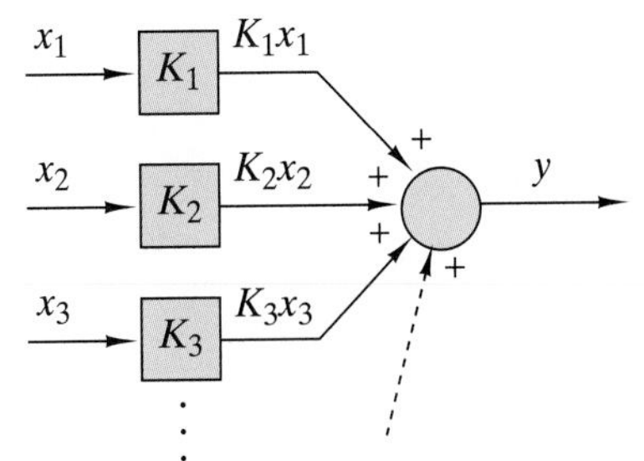

FIGURE 3–30 Block diagram representation of the additivity property.

To illustrate this property, we analyze the two-input circuit in Figure 3–31 using node-voltage analysis. Applying KCL at node A, we obtain

$$\frac{v_A - v_S}{R_1} - i_S + \frac{v_A}{R_2} = 0$$

Moving the inputs to the right side of this equation yields

$$\left[\frac{1}{R_1} + \frac{1}{R_2}\right]v_A = \frac{v_S}{R_1} + i_S$$

Since $v_O = v_A$, we obtain the input-output relationship in the form

$$v_O = \left[\frac{R_2}{R + R_2}\right]v_S + \left[\frac{R_1R_2}{R_1 + R_2}\right]i_S$$

$$y = [K_1]x_1 + [K_2]x_2 \tag{3–22}$$

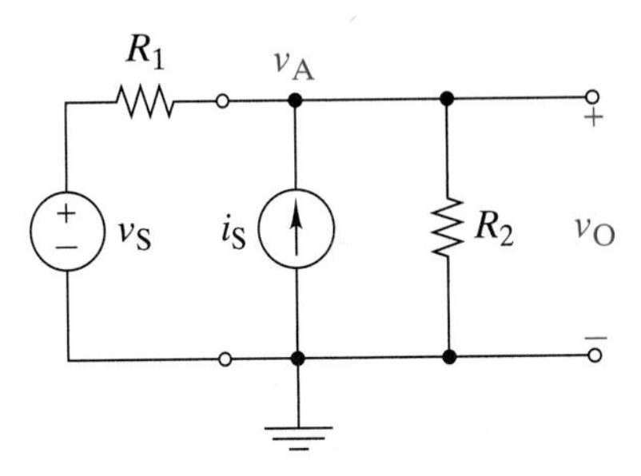

FIGURE 3–31 Circuit used to demonstrate superposition.

This result shows that the output is a linear combination of the two inputs. Note that K_1 is dimensionless since its input and output are voltages, and that K_2 has the units of ohms since its input is a current and its output is a voltage.

SUPERPOSITION PRINCIPLE

Since the output in Eq. (3–21) is a linear combination, the contribution of each input source is independent of all other inputs. This means that the output can be found by finding the contribution from each source acting alone and then adding the individual responses to obtain the total response. This suggests that the output of a multiple-input linear circuit can be found by the following steps:

STEP 1 "Turn off" all independent sources except one and find the output of the circuit due to that source acting alone.

STEP 2 Repeat the process in step 1 until each independent source has been turned on and the output due to that source found.

STEP 3 The total output with all independent sources turned on is the algebraic sum of the outputs caused by each source acting alone.

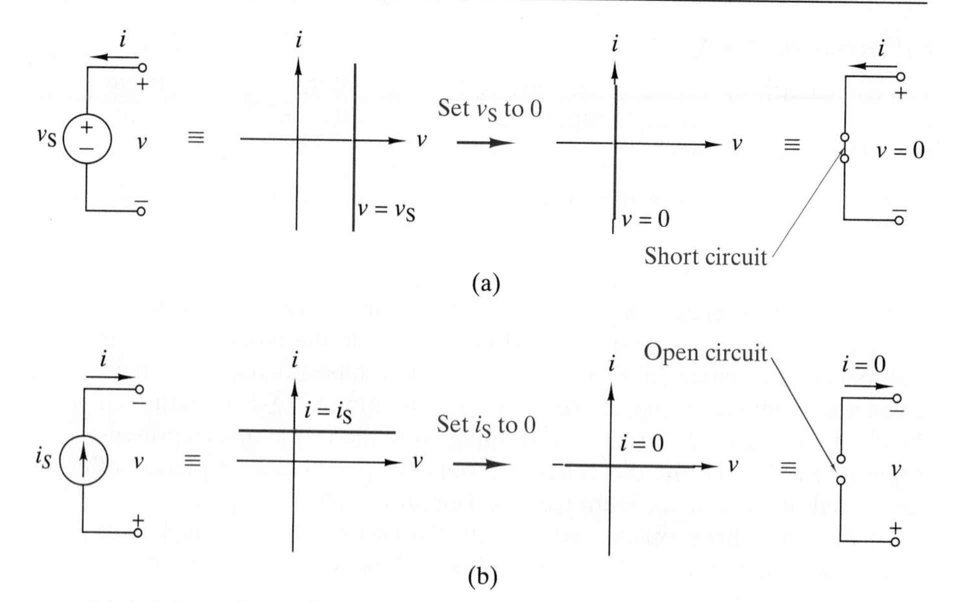

FIGURE 3–32 *Turning off an independent source: (a) Voltage source. (b) Current source.*

These steps describe a circuit analysis technique called the **superposition principle**. Before applying this method, we must discuss what happens when a voltage or current source is "turned off."

The i–v characteristics of voltage and current sources are shown in Figure 3–32. A voltage source is "turned off" by setting its voltage to zero ($v_S = 0$). This step translates the voltage source i–v characteristic to the i-axis, as shown in Figure 3–32(a). In Chapter 2 we found that a vertical line on the i-axis is the i–v characteristic of a short circuit. Similarly, "turning off" a current source ($i_S = 0$) in Figure 3–32(b) translates its i–v characteristic to the v-axis, which is the i–v characteristic of an open circuit. Therefore, when a voltage source is "turned off" we replace it by a short circuit, and when a current source is "turned off" we replace it by an open circuit.

The superposition principle is now applied to the circuit in Figure 3–31 to duplicate the response in Eq. (3–22), which was found by node analysis. Figure 3–33 shows the steps involved in applying superposition to the circuit in Figure 3–31. Figure 3–33(a) shows that the circuit has two input sources. We will first "turn off" i_S and replace it with an open circuit, as shown in Figure 3–33(b). The output of the circuit in Figure 3–33(b) is called v_{O1} and represents that part of the total output caused by the voltage source. Using voltage division in Figure 3–33(b) yields v_{O1} as

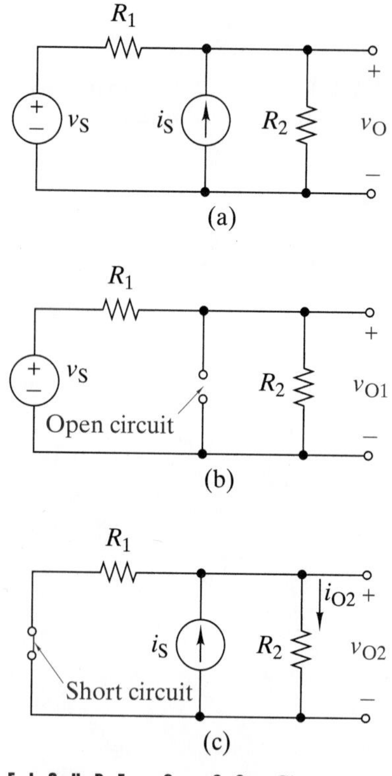

FIGURE 3–33 *Circuit analysis using superposition: (a) Current source off. (b) Voltage source off.*

$$v_{O1} = \frac{R_2}{R_1 + R_2} v_S$$

Next we "turn off" the voltage source and "turn on" the current source, as shown in Figure 3–33(c). Using Ohm's law, we get $v_{O2} = i_{O2}R_2$. We use current division to express i_{O2} in terms of i_S to obtain v_{O2}:

$$v_{O2} = i_{O2}R_2 = \left[\frac{R_1}{R_1 + R_2} i_S\right] R_2 = \frac{R_1 R_2}{R_1 + R_2} i_S$$

Applying the superposition principle, we find the response with both sources "turned on" by adding the two responses v_{O1} and v_{O2}.

$$v_O = v_{O1} + v_{O2}$$
$$= \left[\frac{R_2}{R_1 + R_2}\right]v_S + \left[\frac{R_1R_2}{R_1 + R_2}\right]i_S$$

This superposition result is the same as the circuit reduction result given in Eq. (3–22).

EXAMPLE 3–12

Figure 3–34(a) shows a resistance circuit used to implement a signal-summing function. Use superposition to show that the output v_O is a weighted sum of the inputs v_{S1}, v_{S2}, and v_{S3}.

SOLUTION:
To determine v_O using superposition, we first turn off sources 1 and 2 ($v_{S1} = 0$ and $v_{S2} = 0$) to obtain the circuit in Figure 3–34(b). This circuit is a voltage divider in which the output leg consists of two equal resistors in parallel. The equivalent resistance of the output leg is $R/2$, so the voltage division rule yields

$$v_{O3} = \frac{R/2}{R + R/2}v_{S3} = \frac{v_{S3}}{3}$$

Because of the symmetry of the circuit, it can be seen that the same technique applies to all three inputs; therefore

$$v_{O2} = \frac{v_{S2}}{3} \text{ and } v_{O1} = \frac{v_{S1}}{3}$$

Applying the superposition principle, the output with all sources "turned on" is

$$v_O = v_{O1} + v_{O2} + v_{O3}$$
$$= \frac{1}{3}[v_{S1} + v_{S2} + v_{S3}]$$

That is, the output is proportional to the sum of the three input signals with $K_1 = K_2 = K_3 = 1/3$. ■

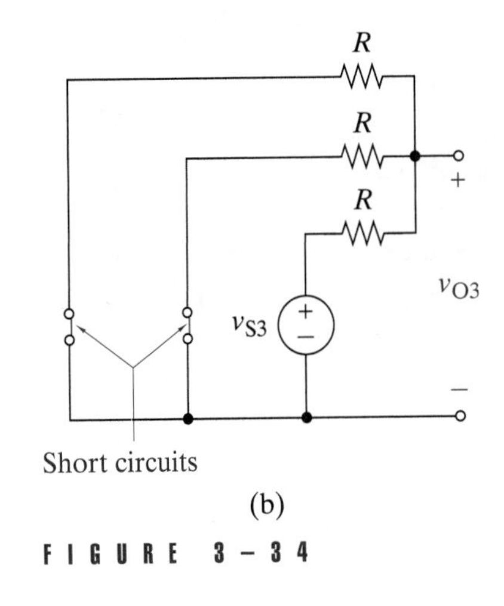

FIGURE 3–34

Exercise 3–14

The circuit of Figure 3–35 contains two R-$2R$ modules. Use superposition to find v_O.

Answer: $v_O = \frac{1}{2}v_{S1} + \frac{1}{4}v_{S2}$

FIGURE 3–35

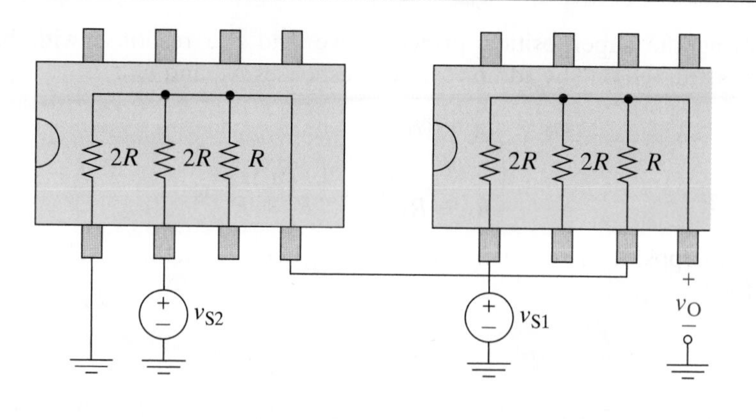

Exercise 3–15

Repeat Exercise 3–14 with the voltage source v_{S2} replaced by a current source i_{S2} with the current reference arrow directed toward ground.

Answer: $v_O = 3v_{S1}/5 - 4i_{S2}R/5$

The preceding examples and exercises illustrate the applications of the superposition principle. You should not conclude that superposition is used primarily to solve for the response of circuits with multiple independent sources. In fact, superposition is not a particularly attractive method of analysis since a circuit with N sources requires N different circuit analyses to obtain the final result. Unless the circuit is relatively simple, superposition does not reduce the analysis effort compared with, say, node-voltage analysis. Superposition is still an important property of linear circuits because it is often used as a conceptual tool to develop other circuit analysis techniques. For example, superposition is used in the next section to prove Thévenin's theorem.

3–4 THÉVENIN AND NORTON EQUIVALENT CIRCUITS

An *interface* is a connection between circuits. Circuit interfaces occur frequently in electrical and electronic systems, so special analysis methods are used to handle them. For the two-terminal interface shown in Figure 3–36, we normally think of one circuit as the source S and the other as the load L. We think of signals as being produced by the source circuit and delivered to the load circuit. The source-load interaction at an interface is one of the central problems of circuit analysis and design.

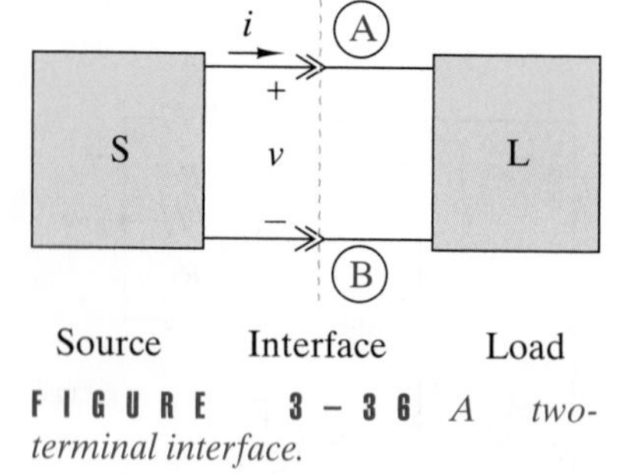

FIGURE 3–36 *A two-terminal interface.*

The Thévenin and Norton equivalent circuits shown in Figure 3–37 are valuable tools for dealing with circuit interfaces. The conditions under which these equivalent circuits exist can be stated as a theorem:

> ***If the source circuit in a two-terminal interface is linear, then the interface signals* v *and* i *do not change when the source circuit is replaced by its Thévenin or Norton equivalent circuit.***

The equivalence requires the source circuit to be linear but places no restriction on the linearity of the load circuit. Later in this section we

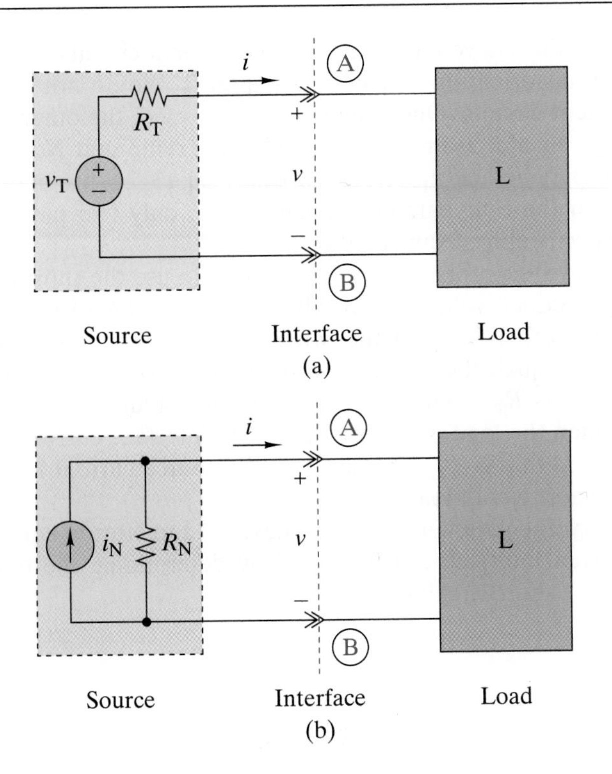

FIGURE 3–37 *Equivalent circuits for the source: (a) Thévenin equivalent. (b) Norton equivalent.*

consider cases in which the load is nonlinear. In subsequent chapters we will study circuits in which the loads are energy storage elements called capacitors and inductors.

The Thévenin equivalent circuit consists of a voltage source (v_T) in series with a resistance (R_T). The Norton equivalent circuit is a current source (i_N) in parallel with a resistance (R_N). Note that the Thévenin and Norton equivalent circuits are practical sources in the sense discussed in Chapter 2.

The two circuits have the same i–v characteristics, since replacing one by the other leaves the interface signals unchanged. To derive the equivalency conditions, we apply KVL and Ohm's law to the Thévenin equivalent in Figure 3–37(a) to obtain its i–v relationship at the terminals A and B:

$$v = v_T - iR_T \tag{3–23}$$

Next, applying KCL and Ohm's law to the Norton equivalent in Figure 3–37(b) yields its i–v relationship at terminals A–B:

$$i = i_N - \frac{v}{R_N} \tag{3–24}$$

Solving Eq. (3–24) for v yields

$$v = i_N R_N - iR_N \tag{3–25}$$

The Thévenin and Norton circuits have identical i–v relationships. Comparing Eqs. (3–23) and (3–25), we conclude that

$$R_N = R_T$$
$$i_N R_N = v_T \tag{3–26}$$

In essence, the Thévenin and Norton equivalent circuits are related by the source transformation studied in Chapter 2. We do not need to find both equivalent circuits. Once one of them is found, the other can be determined by a source transformation. The Thévenin and Norton circuits involve four parameters (v_T, R_T, i_N, R_N), and Eq. (3–26) provides two relations between the four parameters. Therefore, only two parameters are needed to specify either equivalent circuit.

In circuit analysis problems it is convenient to use the short-circuit current and open-circuit voltage to specify Thévenin and Norton circuits. The circuits in Figure 3–38(a) show that when the load is an open circuit the interface voltage equals the Thévenin voltage; that is, $v_{OC} = v_T$, since there is no voltage across R_T when $i = 0$. Similarly, the circuits in Figure 3–38(b) show that when the load is a short circuit the interface current equals the Norton current; that is, $i_{SC} = i_N$, since all the source current i_N is diverted through the short-circuit load.

In summary, the parameters of the Thévenin and Norton equivalent circuits at a given interface can be found by determining the open-circuit voltage and the short-circuit current.

$$
\begin{aligned}
v_T &= v_{OC} \\
i_N &= i_{SC} \\
R_N = R_T &= v_{OC}/i_{SC}
\end{aligned}
\tag{3–27}
$$

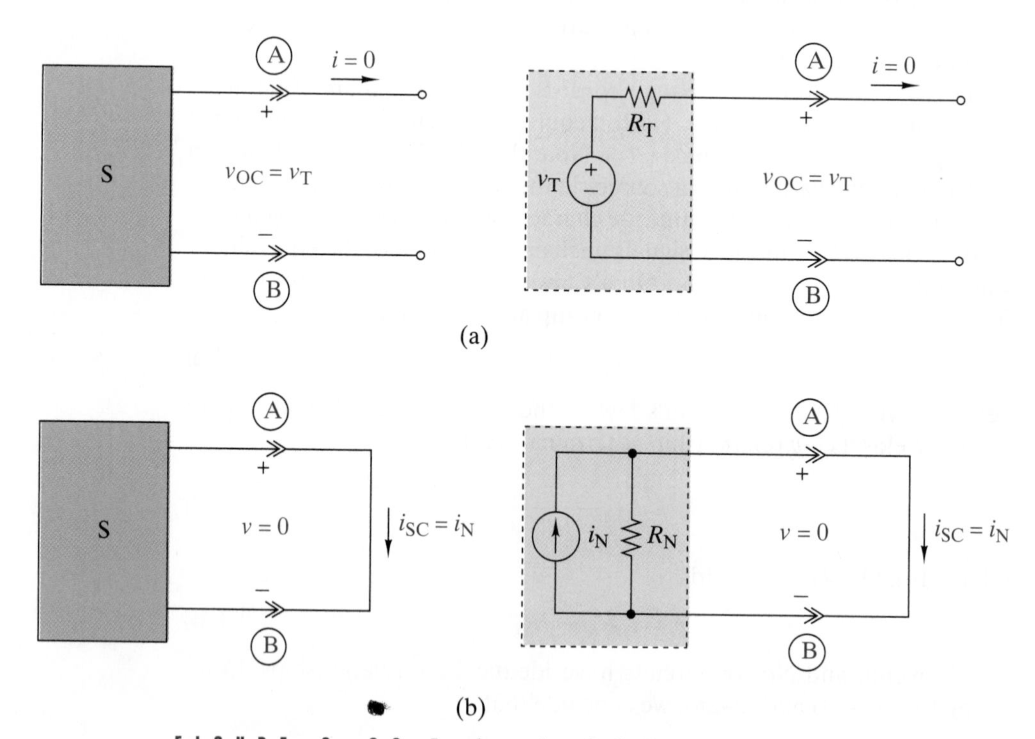

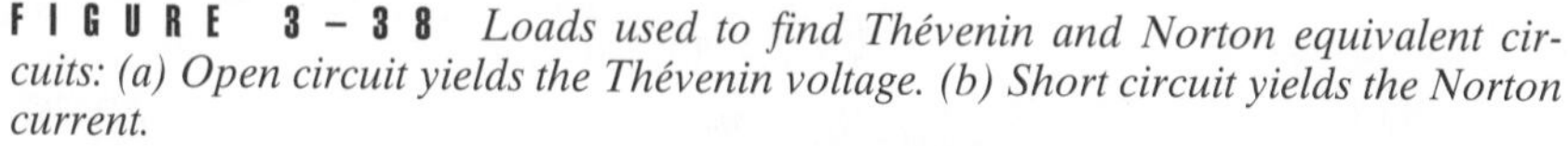

FIGURE 3–38 *Loads used to find Thévenin and Norton equivalent circuits: (a) Open circuit yields the Thévenin voltage. (b) Short circuit yields the Norton current.*

APPLICATIONS OF THÉVENIN AND NORTON EQUIVALENT CIRCUITS

Replacing a complex circuit by its Thévenin or Norton equivalent can greatly simplify the analysis and design of interface circuits. For example, suppose we need to select a load resistance in Figure 3–39(a) so the source circuit to the left of the interface A–B delivers 4 volts to the load. This task is easily handled once we have the Thévenin or Norton equivalent for the source circuit.

To obtain the Thévenin and Norton equivalents, we need v_{OC} and i_{SC}. The open-circuit voltage v_{OC} is found by disconnecting the load at the terminals A and B, as shown in Figure 3–39(b). The voltage across the 15-Ω resistor is zero because the open circuit causes the current through the resistor to be zero. The open-circuit voltage at the interface is the same as the voltage across the 10-Ω resistor. Using voltage division, this voltage is

$$v_T = v_{OC} = \frac{10}{10 + 5} \times 15 = 10 \text{ V}$$

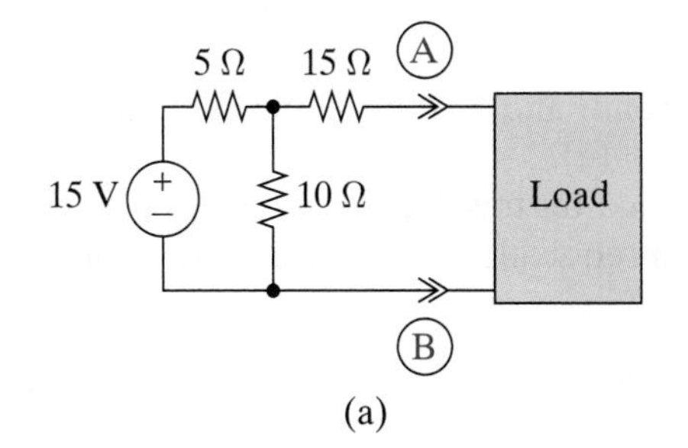

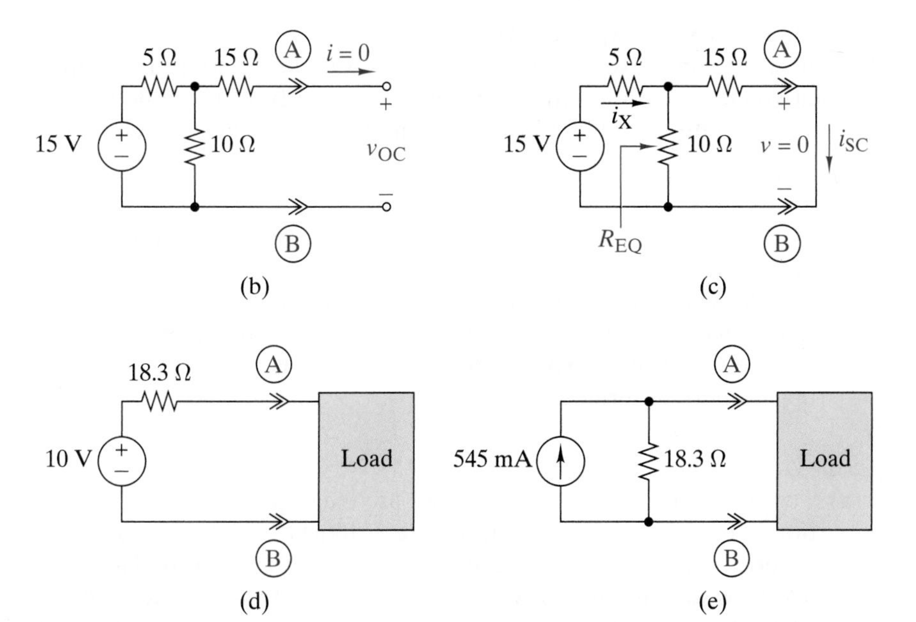

FIGURE 3–39 *Example of finding the Thévenin and Norton equivalent circuits: (a) The given circuit. (b) Open circuit yields the Thévenin voltage. (c) Short circuit yields the Norton current. (d) Thévenin equivalent circuit. (e) Norton equivalent circuit.*

Next we find the short-circuit current i_{SC} using the circuit in Figure 3–39(c). The total current i_X delivered by the 15-V voltage source is

$$i_X = 15/R_{EQ}$$

where R_{EQ} is the equivalent resistance seen by the voltage source with a short circuit at the interface.

$$R_{EQ} = 5 + \frac{1}{\frac{1}{10} + \frac{1}{15}} = 11\ \Omega$$

We find $i_X = 15/11 = 1.36$ A. Given i_X, we now use current division to obtain the short-circuit current:

$$i_N = i_{SC} = \frac{10}{10 + 15} \times i_X = 0.545\ \text{A}$$

Finally, we compute the Thévenin and Norton resistances:

$$R_T = R_N = \frac{v_{OC}}{i_{SC}} = 18.3\ \Omega$$

The resulting Thévenin and Norton equivalent circuits are shown in Figures 3–39(d) and 3–39(e).

It now is an easy matter to select a load R_L so 4 V is supplied to the load. Using the Thévenin equivalent circuit, the problem reduces to a voltage divider:

$$\frac{R_L}{R_L + R_T} \times v_T = \frac{R_L}{R_L + 18.3} \times 10 = 4\ \text{V}$$

Solving for R_L yields $R_L = 12.2\ \Omega$.

The Thévenin or Norton equivalent can always be found from the open-circuit voltage and short-circuit current at the interface. The following examples illustrate other methods of determining these equivalent circuits.

EXAMPLE 3–13

(a) Find the Thévenin equivalent circuit of the source circuit to the left of the interface in Figure 3–40(a).

(b) Use the Thévenin equivalent to find the power delivered to two different loads. The first load is a 10-kΩ resistor and the second is a 5-V voltage source whose positive terminal is connected to the upper interface terminal.

SOLUTION:

(a) To find the Thévenin equivalent, we use the sequence of circuit reductions shown in Figure 3–40. In Figure 3–40(a) the 15-V voltage source in series with the 3-kΩ to the left of terminals A and B is replaced by a 3-kΩ resistor in parallel with an equivalent current source with i_S = 15/3000 = 5 mA. In Figure 3–40(b), looking to the left at terminals C and D, we see two resistors in parallel whose equivalent resistance is (3 kΩ)||(6 kΩ) = 2 kΩ. We also see two current sources in parallel whose equivalent is i_S = 5 mA − 2 mA = 3 mA. This equivalent current source is shown in Figure 3–40(c) to the left of terminals C and D. Figure 3–40(d) shows this current source converted to an equivalent voltage source $v_S = 3\text{ mA} \times 2\text{ k}\Omega = 6\text{ V}$ in series with 2 kΩ. In Figure 3–40(d) the three resistors are connected in series and can be replaced by an equivalent resistance $R_{EQ} = 2\text{ k}\Omega + 3\text{ k}\Omega + 4\text{ k}\Omega = 9\text{ k}\Omega$. This step produces the Thévenin equivalent shown in Figure 3–40(e).

Note: The steps leading from Figure 3–40(a) to 3–40(e) involve circuit reduction techniques studied in Chapter 2, so we know that this approach only works on ladder circuits like the one in Figure 3–40(a).

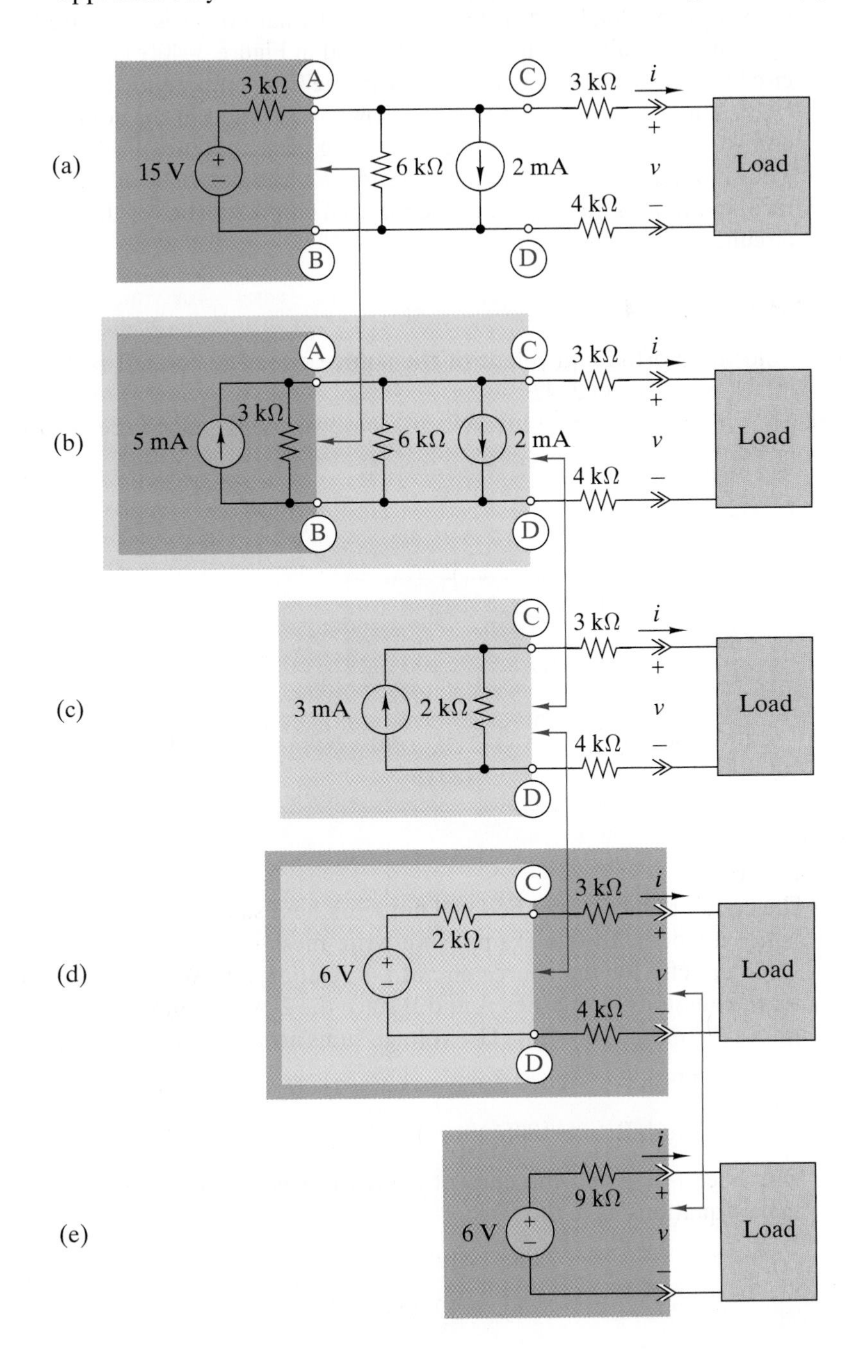

FIGURE 3–40

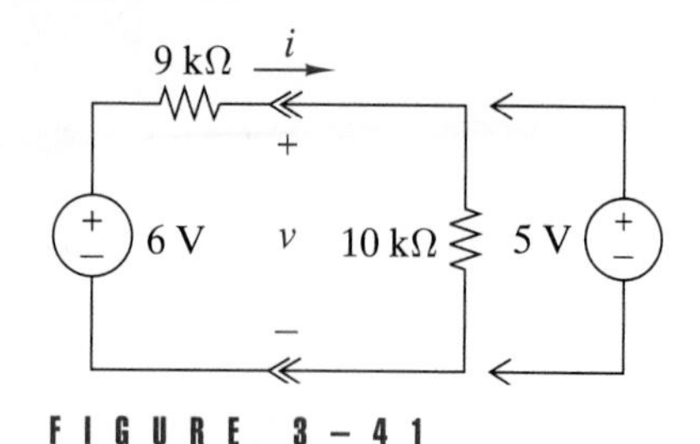

FIGURE 3–41

(b) Figure 3–41 shows the Thévenin equivalent found in (a) and the two loads. When the load is a 10-kΩ resistor, the interface current is $i = (6)/(9000 + 10{,}000) = 0.3158$ mA, and the power delivered to the load is $i^2R_L = 0.9973$ mW. When the load is a 5-V source, the interface voltage and current are $v = 5$ V and $i = (6 - 5)/9000 = 0.1111$ mA, and the power to the load is $vi = 0.5555$ mW. Since $p > 0$ in the latter case, we see that the voltage source load is absorbing rather than delivering power. A practical example of this situation is a battery charger.

Caution: The Thévenin equivalent allows us to calculate the power delivered to a load, but it does not tell us what power is dissipated in the source circuit. For instance, if the load in Figure 3–40(e) is an open circuit, then no power is dissipated in the Thévenin equivalent since $i = 0$. This does not mean that the power dissipated in the source circuit is zero, as we can easily see by looking back at Figure 3–40(a). The Thévenin equivalent circuit has the same i–v characteristic at the interface, but it does not duplicate the internal characteristics of the source circuit. ■

EXAMPLE 3–14

(a) Find the Norton equivalent of the source circuit to the left of the interface in Figure 3–42.

(b) Find the interface current i when the power delivered to the load is 5 W.

FIGURE 3–42

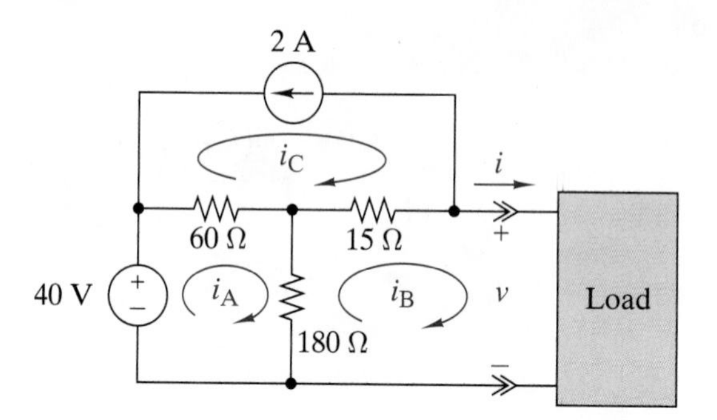

SOLUTION:

(a) The circuit reduction method will not work here since the source circuit is not a ladder. In this example we write mesh-current equations and solve directly for the source circuit i–v relationship. We only need to write equations for meshes A and B since the 2-A current source determines the mesh C current. The voltage sums around these meshes are

$$\text{Mesh A:}\quad -40 + 60(i_A - i_C) + 180(i_A - i_B) = 0$$

$$\text{Mesh B:}\quad -180(i_A - i_B) + 15(i_B - i_C) + v = 0$$

But since $i_B = i$ and the current source forces the condition $i_C = -2$, these equations have the form

$$240i_A - 180i = -80$$

$$-180i_A + 195i = -30 - v$$

Solving for i in terms of v yields

$$i = \frac{\begin{vmatrix} 240 & -80 \\ -180 & -30 - v \end{vmatrix}}{\begin{vmatrix} 240 & -180 \\ -180 & 195 \end{vmatrix}} = \frac{-21600 - 240v}{14400}$$

$$= -1.5 - \frac{v}{60}$$

At the interface the i–v relationship of the source circuit is $i = -1.5 - v/60$. Equation (3–24) gives the i–v relationship of the Norton circuit as $i = i_N - v/R_N$. By direct comparison, we conclude that $i_N = -1.5$ A and $R_N = 60\ \Omega$. This equivalent circuit is shown in Figure 3–43.

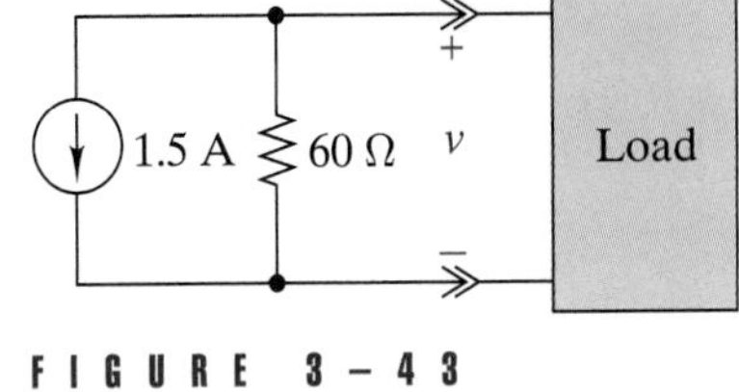

FIGURE 3–43

(b) When 5 W is delivered to the load, we have $vi = 5$ or $v = 5/i$. Substituting $v = 5/i$ into the source i–v relationship $i = -1.5 - v/60$ yields a quadratic equation

$$12i^2 + 18i + 1 = 0$$

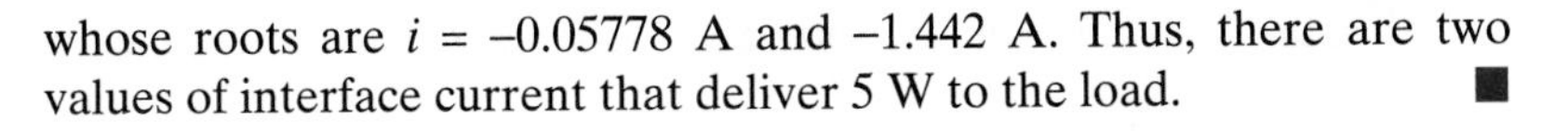

whose roots are $i = -0.05778$ A and -1.442 A. Thus, there are two values of interface current that deliver 5 W to the load. ■

DERIVATION OF THÉVENIN'S THEOREM

The derivation of Thévenin's theorem is based on the superposition principle. We begin with the circuit in Figure 3–44(a), where the source circuit S is linear. Our approach is to use superposition to show that the source circuit and the Thévenin circuit have the same i–v relationship at the interface. To find the source circuit i–v relationship, we first disconnect the load and apply a current source i_{TEST}, as shown in Figure 3–44(b). Using superposition to find v_{TEST}, we first turn i_{TEST} off and leave all the sources inside S on, as shown in Figure 3–44(c). Turning a current source off leaves an open circuit, so

$$v_{TEST1} = v_{OC}$$

Next we turn i_{TEST} back on and turn off all of the sources inside S. Since the source circuit S is linear, it reduces to the equivalent resistance shown in Figure 3–44(d) when all internal sources are turned off. Using Ohm's law, we write

$$v_{TEST2} = (R_{EQ})(-i_{TEST})$$

The minus sign in this equation results from the reference directions originally assigned to i_{TEST} and v_{TEST} in Figure 3–44(b). Using the superposition principle, we find the i–v relationship of the source circuit at the interface to be

$$v_{TEST} = v_{TEST1} + v_{TEST2}$$
$$= v_{OC} - R_{EQ}i_{TEST}$$

This equation has the same form as the i–v relationship of the Thévenin equivalent circuit in Eq. (3–23) when $v_{TEST} = v$, $i_{TEST} = i$, $v_{OC} = v_T$, and $R_T = R_{EQ}$.

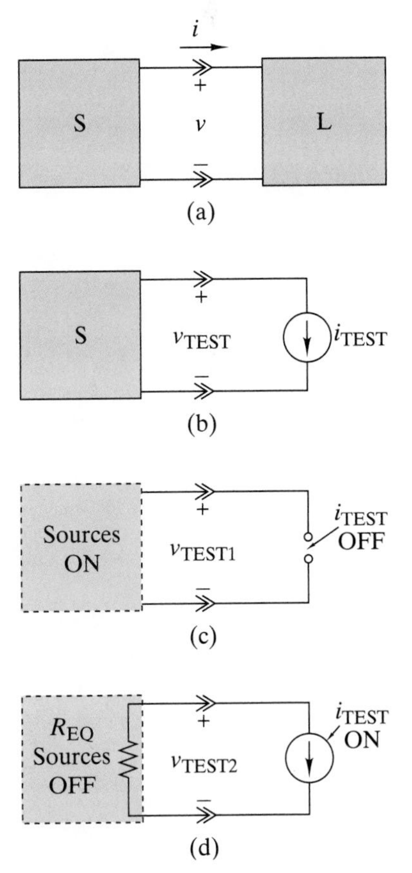

FIGURE 3–44 Using superposition to prove Thévenin's theorem.

The derivation points out another method of finding the Thévenin resistance. As indicated in Figure 3–44(d), when all the sources are turned off, the i–v relationship of the source circuit reduces to $v = -iR_{EQ}$. Similarly, the i–v relationship of a Thévenin equivalent circuit reduces to $v = -iR_T$ when $v_T = 0$. We conclude that

$$R_T = R_{EQ} \tag{3–28}$$

We can find the value of R_T by determining the resistance seen looking back into the source circuit with all sources turned off. For this reason the Thévenin resistance R_T is sometimes called the **lookback resistance**.

The next example shows how lookback resistance contributes to finding a Thévenin equivalent circuit.

EXAMPLE 3–15

(a) Find the Thévenin equivalent of the source circuit to the left of the interface in Figure 3–45.
(b) Use the Thévenin equivalent to find the voltage delivered to the load.

FIGURE 3–45

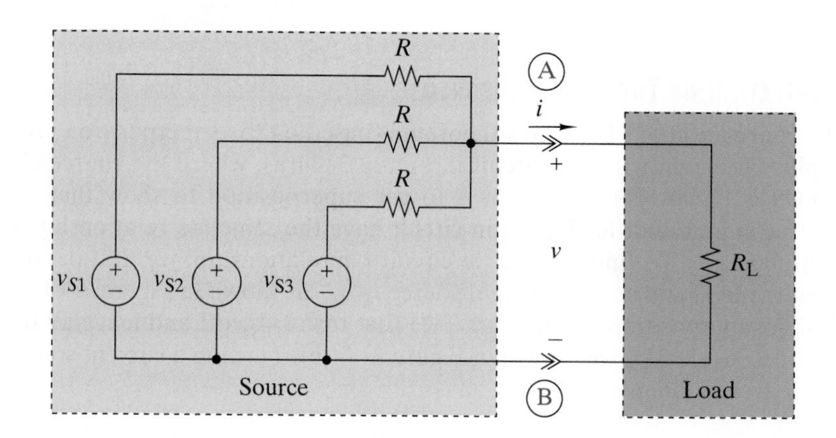

SOLUTION:

(a) The source circuit in Figure 3–45 is treated in Example 3–12 by using superposition to calculate the open-circuit voltage between terminals A and B. Using the results from Example 3–12, we have

$$v_T = v_{OC} = \frac{1}{3}(v_{S1} + v_{S2} + v_{S3})$$

Turning all sources off in Figure 3–45 leads to the resistance circuit in Figure 3–46. Looking back into the source circuit in Figure 3–46, we see three equal resistances connected in parallel whose equivalent resistance is $R/3$. Hence, the Thévenin resistance is

$$R_T = R_{EQ} = \frac{R}{3}$$

(b) Given the Thévenin circuit parameters v_T and R_T, we apply voltage division in Figure 3–45 to find the interface voltage.

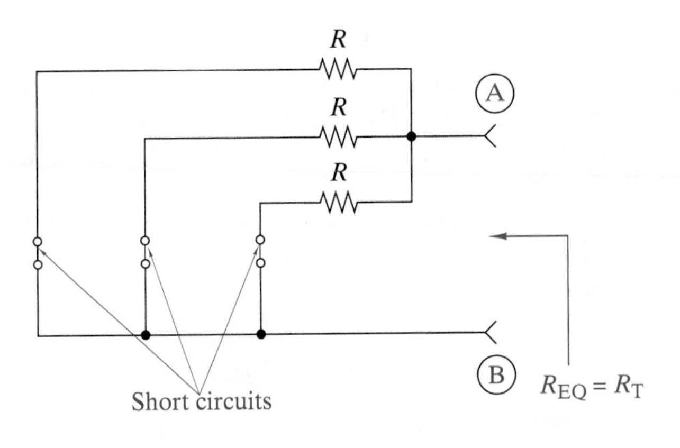

FIGURE 3–46

$$v = \frac{R_L}{R_L + R_T} v_T = \left(\frac{R_L}{R_L + R/3}\right)\left(\frac{v_{S1} + v_{S2} + v_{S3}}{3}\right)$$

$$= \left(\frac{R_L}{3R_L + R}\right)(v_{S1} + v_{S2} + v_{S3})$$

The interface voltage is proportional to the sum of the three source voltages. The proportionality constant $K = R_L/(3R_L + R)$ depends on both the source and the load since these two circuits are connected at the interface. ■

Exercise 3–16

(a) Find the Thévenin and Norton equivalent circuits seen by the load in Figure 3–47.
(b) Find the voltage, current, and power delivered to a 50-Ω load resistor.

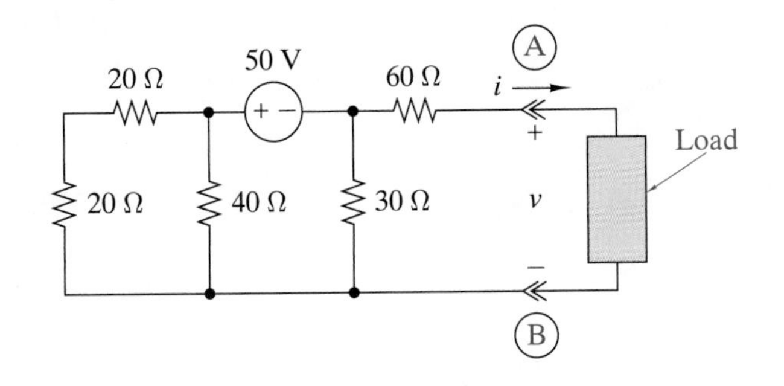

FIGURE 3–47

Answers:

(a) $v_T = -30$ V; $i_N = -417$ mA; $R_N = R_T = 72$ Ω
(b) $v = -12.3$ V; $i = -246$ mA; $p = 3.03$ W

Exercise 3–17

Find the current and power delivered to an unknown load in Figure 3–47 when $v = +6$ V.

Answers: $i = -½$A; $p = -3$ W

APPLICATION TO NONLINEAR LOADS

Thévenin and Norton equivalent circuits can be used to find the response of a two-terminal nonlinear element (NLE). The method of analysis is a straightforward application of device and interface i–v characteristics. An interface is defined at the terminals of the nonlinear element, and the linear part of the circuit is reduced to the Thévenin equivalent in Figure 3–48(a). The i–v relationship of the Thévenin equivalent can be written with interface current as the dependent variable:

$$i = \left(-\frac{1}{R_T}\right)v + \left(\frac{v_T}{R_T}\right) \quad (3\text{–}29)$$

This is the equation of a straight line in the i–v plane shown in Figure 3–48(b). The line intersects the i-axis ($v = 0$) at $i = v_T/R_T = i_{SC}$ and intersects

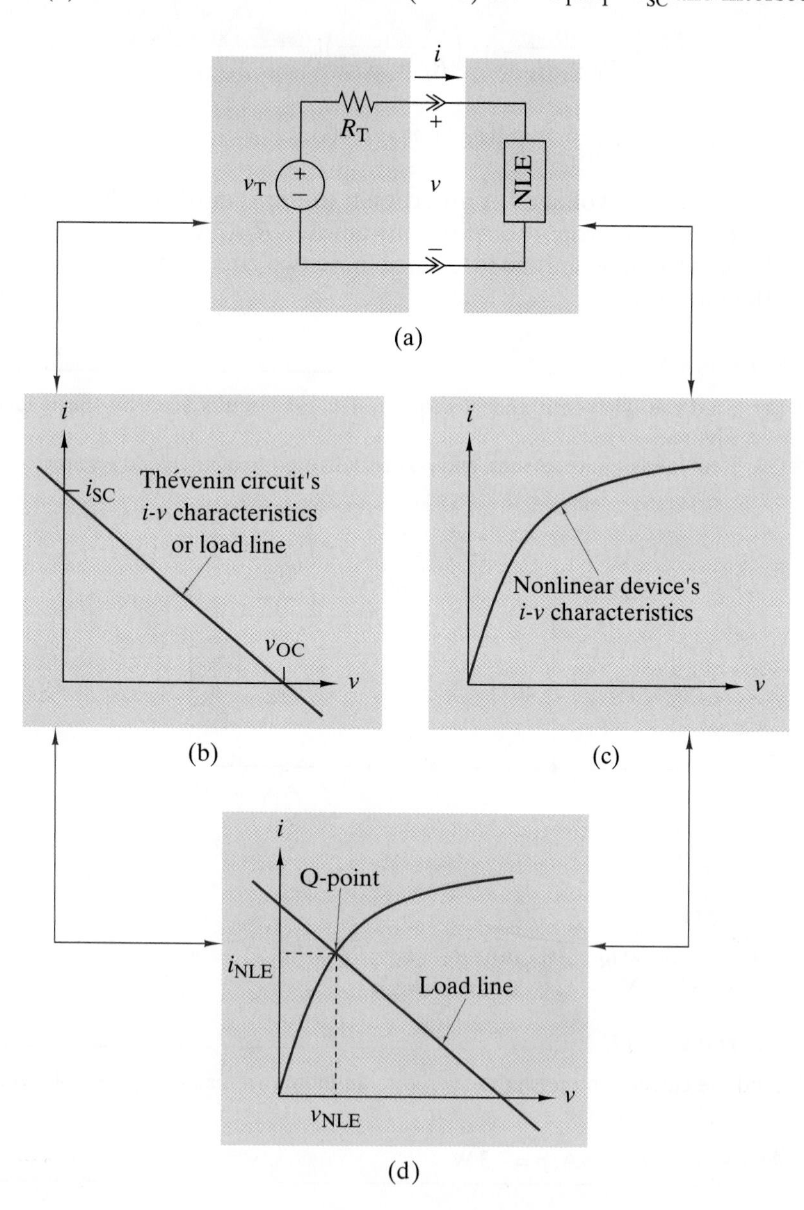

FIGURE 3–48 *Graphical analysis of a nonlinear circuit: (a) Given circuit. (b) Load line. (c) Nonlinear device i–v characteristics. (d) Q-point.*

the v-axis ($i = 0$) at $v = v_T = v_{OC}$. This line could logically be called the source line since it is determined by the Thévenin parameters of the source circuit. Logic notwithstanding, electrical engineers call this the **load line** for reasons that have blurred with the passage of time.

The nonlinear element has the i–v characteristic shown in Figure 3–48(c). Mathematically, this nonlinear characteristic has the form

$$i = f(v) \tag{3–30}$$

To find the circuit response, we must solve Eqs. (3–29) and (3–30) simultaneously. Computer software tools like Mathcad can easily solve this problem when a numerical expression for the function $f(v)$ is known explicitly. However, in practice an approximate graphical solution is often adequate, particularly when $f(v)$ is only given in graphical form.

In Figure 3–48(d) we superimpose the load line on the i–v characteristic curve of the nonlinear element. The two curves intersect at the point $i = i_{NLE}$ and $v = v_{NLE}$, which yields the values of interface variables that satisfy both the source constraints in Eq. (3–29) and the nonlinear element constraints in Eq. (3–30). In the terminology of electronics, the point of intersection is called the operating point or **Q-point**, where Q stands for *quiescent.*

EXAMPLE 3–16

Find the voltage, current, and power delivered to the diode in Figure 3–49(a).

SOLUTION:

We first find the Thévenin equivalent of the circuit to the left of the terminals A and B. By voltage division, the open-circuit voltage is

$$v_T = v_{OC} = \frac{100}{100 + 100} \times 5 = 2.5 \text{ V}$$

When the voltage source is turned off, the lookback equivalent resistance seen between terminals A and B is

$$R_T = 10 + 100\|100 = 60 \ \Omega$$

The source circuit load line is given by

$$i = -\frac{1}{60}v + \frac{1}{60} \times 2.5$$

This line intersects the i-axis ($v = 0$) at $i = i_{SC} = 2.5/60 = 41.7$ mA and intersects the v-axis ($i = 0$) at $v = v_{OC} = 2.5$ V. Figure 3–49(b) superimposes the source circuit load line on the diode's i–v curve. The intersection (Q-point) is at $i = i_D = 15$ mA and $v = v_D = 1.6$ V. This is the point (i_D, v_D) at which both the source and diode device constraints are satisfied. Finally, the power delivered to the diode is

$$p_D = i_D v_D = (15 \times 10^{-3})(1.6) = 24 \text{ mW}$$

Because of the nonlinear element, the proportionality and superposition properties do not apply to this circuit. For instance, if the source voltage in Figure 3–49 is decreased from 5 V to 2.5 V, the diode current and voltage do not decrease by one-half. Try it. ■

FIGURE 3–49

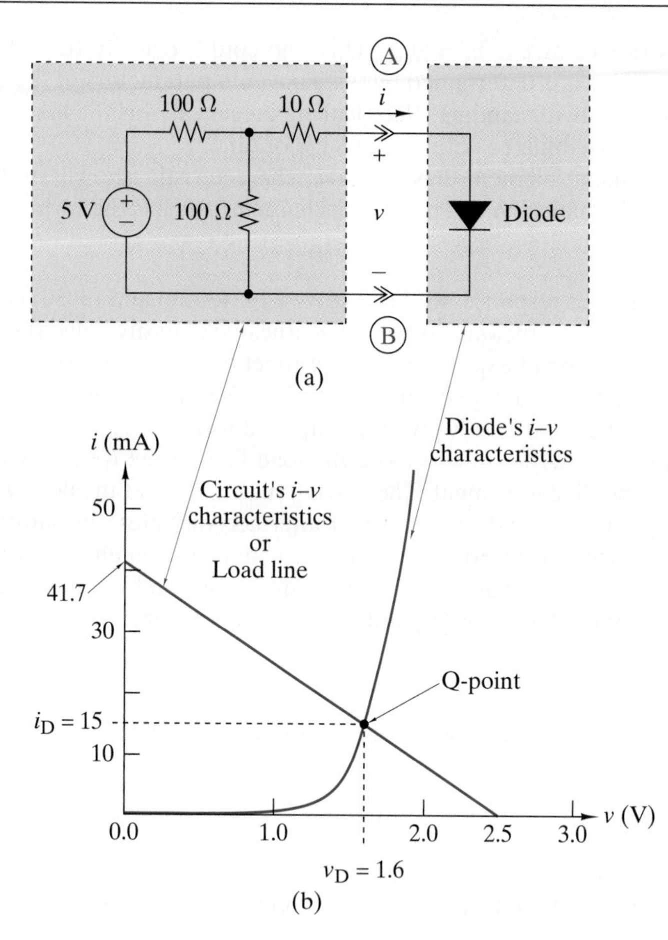

Exercise 3–18

Find the voltage, current, and power delivered to the diode in Figure 3–49(a) when the 10-Ω resistor is replaced by a short circuit.

Answer: $v_D = 1.7$ V; $i_D = 18$ mA; $p_D = 30.6$ mW

In summary, any two of the following parameters determine the Thévenin or Norton equivalent circuit at a specified interface:

- The open-circuit voltage at the interface
- The short-circuit current at the interface
- The source circuit lookback resistance

Alternatively, for ladder circuits the Thévenin or Norton equivalent circuit can be found by a sequence of circuit reductions (see Example 3–13). For general circuits they can always be found by directly solving for the i–v relationship of the source circuit using node-voltage or mesh-current equations that include the interface current and voltage as unknowns (see Example 3–14).

3–5 MAXIMUM SIGNAL TRANSFER

An interface is a connection between two circuits at which the signal levels may be observed or specified. In this regard an important consideration is the maximum signal levels that can be transferred across a given interface. This section defines the maximum voltage, current, and power available at an interface between a *fixed source* and an *adjustable load.*

For simplicity we will treat the case in which both the source and load are linear resistance circuits. The source can be represented by a Thévenin equivalent and the load by an equivalent resistance R_L, as shown in Figure 3–50. For a fixed source, the parameters v_T and R_T are given and the interface signal levels are functions of the load resistance R_L.

By voltage division, the interface voltage is

$$v = \frac{R_L}{R_L + R_T} v_T \tag{3–31}$$

For a fixed source and a variable load, the voltage will be a maximum if R_L is made very large compared with R_T. Ideally, R_L should be made infinite (an open circuit), in which case

$$v_{MAX} = v_T = v_{OC} \tag{3–32}$$

Therefore, the maximum voltage available at the interface is the source open-circuit voltage v_{OC}.

The current delivered at the interface is

$$i = \frac{v_T}{R_L + R_T} \tag{3–33}$$

For a fixed source and a variable load, the current will be a maximum if R_L is made very small compared with R_T. Ideally, R_L should be zero (a short circuit), in which case

$$i_{MAX} = \frac{v_T}{R_T} = i_N = i_{SC} \tag{3–34}$$

Therefore, the maximum current available at the interface is the source short-circuit current i_{SC}.

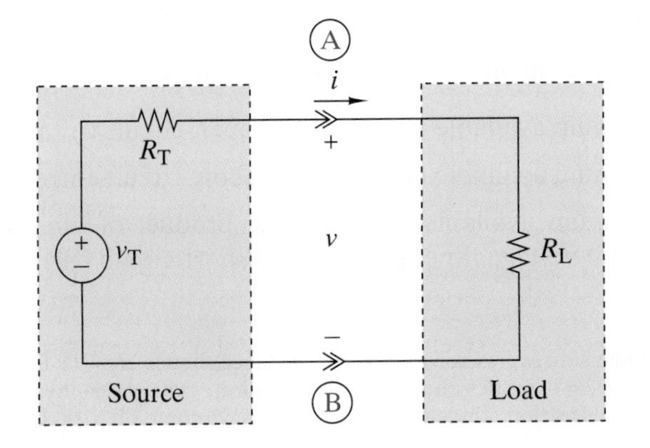

FIGURE 3–50 *Two-terminal interface for deriving the maximum signal transfer conditions.*

The power delivered at the interface is equal to the product $v \times i$. Using Eqs. (3–31) and (3–33), the power is

$$p = v \times i$$
$$= \frac{R_L v_T^2}{(R_L + R_T)^2} \tag{3–35}$$

For a given source, the parameters v_T and R_T are fixed and the delivered power is a function of a single variable R_L. The condition for maximum voltage ($R_L \to \infty$) and the condition for maximum current ($R_L = 0$) both produce zero power. The value of R_L that maximizes the power lies somewhere between these two extremes. To find this value, we differentiate Eq. (3–35) with respect to R_L and solve for the value of R_L for which $dp/dR_L = 0$.

$$\frac{dp}{dR_L} = \frac{[(R_L + R_T)^2 - 2R_L(R_L + R_T)]v_T^2}{(R_L + R_T)^4} = \frac{R_T - R_L}{(R_L + R_T)^3} v_T^2 = 0 \tag{3–36}$$

Clearly, the derivative is zero when $R_L = R_T$. Therefore, **maximum power transfer** occurs when the load resistance equals the Thévenin resistance of the source. When the condition $R_L = R_T$ exists, the source and load are said to be **matched**.

Substituting the condition $R_L = R_T$ back into Eq. (3–35) shows the maximum power to be

$$p_{MAX} = \frac{v_T^2}{4R_T} \tag{3–37}$$

Since $v_T = i_N R_T$, this result can also be written as

$$p_{MAX} = \frac{i_N^2 R_T}{4} \tag{3–38}$$

or

$$p_{MAX} = \frac{v_T i_N}{4} = \left[\frac{v_{OC}}{2}\right]\left[\frac{i_{SC}}{2}\right] \tag{3–39}$$

These equations are consequences of what is known as the **maximum power transfer theorem**:

> ***A source with a fixed Thévenin resistance* $\mathbf{R_T}$ *delivers maximum power to an adjustable load* $\mathbf{R_L}$ *when* $\mathbf{R_L = R_T}$.**[2]

To summarize, at an interface with a fixed source,

1. The maximum available voltage is the open-circuit voltage.
2. The maximum available current is the short-circuit current.
3. The maximum available power is the product of one-half the open-circuit voltage times one-half the short-circuit current.

---

2 An ideal voltage source has zero internal resistance, hence $R_T = 0$. Equation (3–37) points out that $R_T = 0$ implies an infinite p_{MAX}. Infinite power is a physical impossibility, which reminds us that all ideal circuit models have some physical limitations.

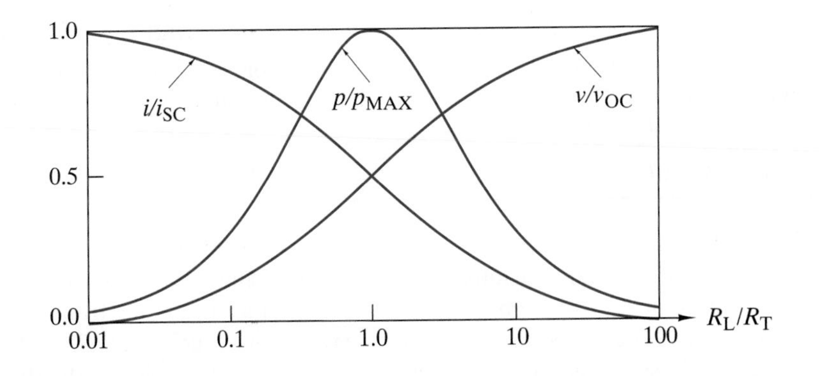

FIGURE 3–51 *Normalized plots of current, voltage, and power versus* R_L/R_T.

Figure 3–51 shows plots of the interface voltage, current, and power as functions of R_L/R_T. The plots of v/v_{OC}, i/i_{SC}, and p/p_{MAX} are normalized to the maximum available signal levels, so the ordinates in Figure 3–51 range from 0 to 1. The plot of the normalized power p/p_{MAX} in the neighborhood of the maximum is not a particularly strong function of R_L/R_T. Changing the ratio R_L/R_T by a factor of 2 in either direction from the maximum reduces p/p_{MAX} by less than 20%. The normalized voltage v/v_{OC} is within 20% of its maximum when $R_L/R_T = 4$. Similarly, the normalized current is within 20% of its maximum when $R_L/R_T = 1/4$. In other words, for engineering purposes we can get close to the maximum signal levels with load resistances that only approximate the theoretical requirements.

EXAMPLE 3–17

A source circuit with $v_T = 2.5$ V and $R_T = 60\ \Omega$ drives a load with $R_L = 30\ \Omega$.

(a) Determine the maximum signal levels available from the source circuit.
(b) Determine the actual signal levels delivered to the load.

SOLUTION:

(a) The maximum available voltage and current are

$$v_{MAX} = v_{OC} = v_T = 2.5 \text{ V } (R_L \to \infty)$$

$$i_{MAX} = i_{SC} = \frac{v_T}{R_T} = 41.7 \text{ mA } (R_L = 0)$$

The maximum available power is found using Eq. (3–39).

$$p_{MAX} = \left[\frac{v_{OC}}{2}\right]\left[\frac{i_{SC}}{2}\right] = 26.0 \text{ mW } (R_L = R_T = 60\ \Omega)$$

(b) The actual signal levels delivered to the 30-Ω load are

$$v_L = \frac{30}{30 + 60} 2.5 = 0.833 \text{ V}$$

$$i_L = \frac{2.5}{30 + 60} = 27.8 \text{ mA}$$

$$p_L = v_L i_L = 23.1 \text{ mW}$$

Although these levels are less than the maximum available values, the power delivered to the 30-Ω load is nearly 90% of the maximum. ■

Exercise 3–19

A source circuit delivers 4 V when a 50-Ω resistor is connected across its output and 5 V when a 75-Ω resistor is connected. Find the maximum voltage, current, and power available from the source.

Answers: 10 V, 133 mA; 333 mW

Remember that the maximum signal levels just derived are for a fixed source resistance and an adjustable load resistance. This situation often occurs in communication systems where devices such as antennas, transmitters, and signal generators have fixed source resistances such as 50, 75, 300, or 600 ohms. In such cases the load resistance is selected to achieve the desired interface conditions, which often involves matching.

Matching source and load applies when the load resistance R_L in Figure 3–50 is adjustable and the Thévenin source resistance R_T is fixed. When R_L is fixed and R_T is adjustable, then Eqs. (3–31), (3–33), and (3–35) point out that the maximum voltage, current, and power are delivered when the Thévenin source resistance is zero. If the source circuit at an interface is adjustable, then ideally the Thévenin source resistance should be zero. In the next chapter we will see that OP AMP circuits approach this ideal.

3–6 INTERFACE CIRCUIT DESIGN

The maximum signal levels discussed in the previous section place bounds on what is achievable at an interface. However, those bounds are based on a fixed source and an adjustable load. In practice, there are circumstances in which the source or the load, or both, can be adjusted to produce prescribed interface signal levels. Sometimes it is necessary to insert a circuit between the source and the load to achieve the desired results. Figure 3–52 shows the general situations and some examples of resistive interface circuits. By its nature, the inserted circuit has two terminal pairs, or interfaces, at which voltage and current can be observed or specified. These terminal pairs are also called *ports*, and the interface circuit is referred to as a **two-port network**. The port connected to the source is called the input, and the port connected to the load is called the output. The purpose of this two-port network is to make certain that the source and load interact in a prescribed way.

BASIC CIRCUIT DESIGN CONCEPTS

Before we treat examples of different interface situations, you should recognize that we are now discussing a limited form of circuit design, as contrasted with circuit analysis. Although we use circuit analysis tools in design, there are important differences. A linear circuit analysis problem generally has a unique solution. A circuit design problem may have many solutions or even no solution. The maximum available signal levels found in the preceding section provide bounds that help us test for the existence of a solution. Generally there will be several ways to meet the interface constraints, and it then becomes necessary to evaluate the alternatives using other factors, such as cost, power consumption, or reliability.

At this point in our study, resistors are the only elements we can use to design interface circuits. In subsequent chapters we will introduce other

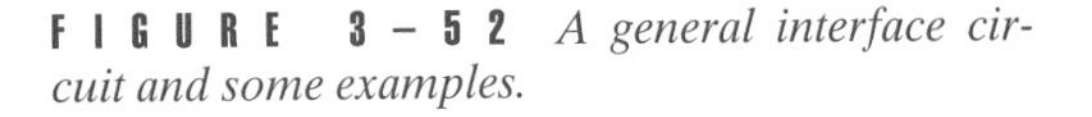

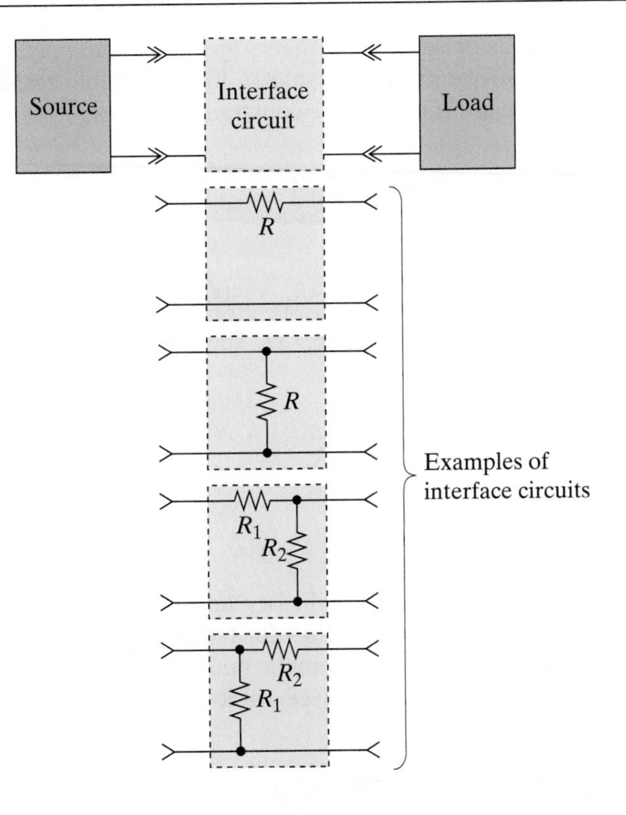

FIGURE 3–52 *A general interface circuit and some examples.*

devices, such as OP AMPs (Chapter 4) and capacitors and inductors (Chapter 6). In a design situation the engineer must choose the resistance values in a proposed circuit. This decision is influenced by a host of practical considerations, such as standard values and tolerances, power ratings, temperature sensitivity, cost, and fabrication methods. We will occasionally introduce some of these considerations into our design examples. Gaining a full understanding of these practical matters is not one of our objectives. Rather, our goal is simply to illustrate how different constraints can influence the design process.

DESIGN EXAMPLE 3–18

Select the load resistance in Figure 3–53 so that the interface signals are in the range defined by $v \geq 4$ V and $i \geq 30$ mA.

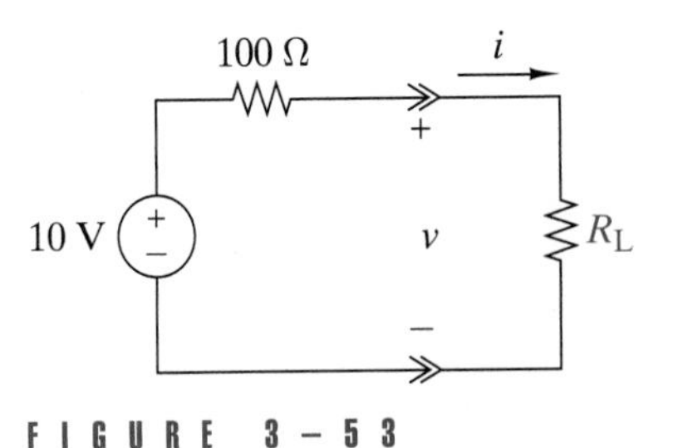

FIGURE 3–53

SOLUTION:

In this design problem, the source circuit is given and we are free to select the load. For a fixed source, the maximum signal levels available at the interface are

$$v_{\text{MAX}} = v_{\text{T}} = 10 \text{ V}$$

$$i_{\text{MAX}} = \frac{v_{\text{T}}}{R_{\text{T}}} = 100 \text{ mA}$$

The bounds given as design requirements are below the maximum available signal levels, so we should be able to find a suitable resistor. Using voltage division, the interface voltage constraint requires

$$\frac{R_L}{100 + R_L} \times 10 \geqslant 4$$

or

$$10R_L \geqslant 4R_L + 400$$

This condition yields $R_L \geq 400/6 = 66.7\ \Omega$. The interface current constraint can be written as

$$\frac{10}{100 + R_L} \geqslant 0.03$$

or

$$10 \geqslant 3 + 0.03R_L$$

which requires $R_L \leq 7/0.03 = 233\ \Omega$. In theory, any value of R_L between 66.7 Ω and 233 Ω will work. However, to allow for parameter variations we select $R_L = 150\ \Omega$ because it lies at the arithmetic midpoint of the allowable range and is a standard value of resistance (see Table A–1, Appendix A). ■

DESIGN EXAMPLE 3–19

A light-emitting diode (LED) converts electric current into an optical signal. LEDs operate at low signal levels with voltages from about 1 V to perhaps 3 V and at currents between about 10 mA and 40 mA. Voltages or currents above these levels may damage or destroy the device.

Figure 3–54 shows an LED operating at $v = 1.5$ V and connected to a 5-V source by an interface circuit. Design the interface circuit so that the LED current is $i = 15$ mA ± 10% using one or more of the following standard resistors: 110 Ω, 160 Ω, 240 Ω, 360 Ω, and 510 Ω. These resistors all have a tolerance of ±5%, which you must account for in your design.

FIGURE 3–54

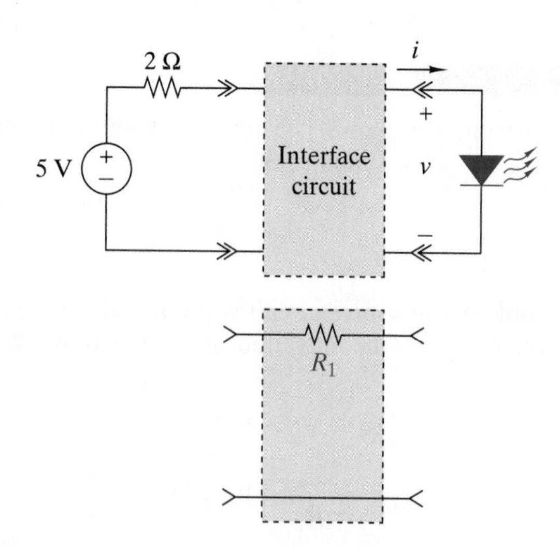

SOLUTION:

If the source is directly connected to the LED, the delivered current would be

$$i = \frac{5 - 1.5}{2} = 1.75 \text{ A}$$

This much current would destroy (vaporize?) the device. The series resistor R_1 in the interface is needed to limit the current to the prescribed level. Applying KVL around the series loop yields

$$-5 + (2 + R_1)i + 1.5 = 0$$

Setting $i = 15$ mA and solving for R_1 yields

$$R_1 = \frac{3.5}{0.015} - 2 = 231 \ \Omega$$

The nearest standard value listed is 240 Ω ± 5%, which means that R_1 would fall in the range $228 \leq R_1 \leq 252$ Ω. At the end points of this range, the LED current is

$$i = \frac{3.5}{252 + 2} = 13.8 \text{ mA} \quad \text{and} \quad i = \frac{3.5}{228 + 2} = 15.2 \text{ mA}$$

Both of these values are within the 15 mA ± 10% tolerance on the LED current. ■

DESIGN AND EVALUATION EXAMPLE 3–20

Design two versions of the interface circuit in Figure 3–55 that deliver $v_2 = 5$ V to the 200-Ω load. Evaluate the two designs in terms of power loss in the interface circuit.

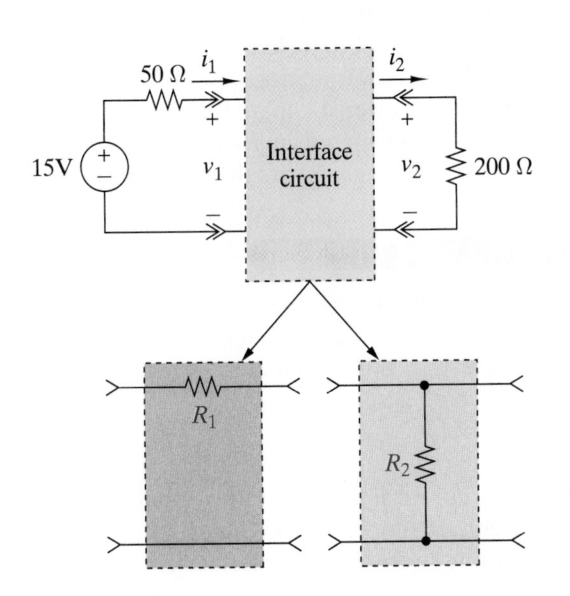

FIGURE 3–55

SOLUTION:

If the 15-V source is directly connected to the load, the delivered voltage would be

$$v_2 = \left(\frac{200}{50 + 200}\right)15 = 12 \text{ V}$$

An interface circuit is required to reduce this voltage to the prescribed 5-V level.

The figure shows two possible interface circuits. In either case

$$i_2 = \frac{5}{200} = 0.025 \text{ A}$$

In the series case $i_1 = i_2 = 0.025$, and the same current flows through all elements in the loop. Applying KVL to the series loop

$$-15 + 0.025 \times (50 + R_1) + 5 = 0$$

and solving for R_1

$$R_1 = \frac{10}{0.025} - 50 = 350 \ \Omega$$

In the parallel case $v_1 = v_2 = 5$ V. Applying KCL to the parallel resistor R_2

$$i_1 - \frac{v_2}{R_2} - i_2 = \frac{15 - 5}{50} - \frac{5}{R_2} - 0.025 = 0$$

and solving for R_2

$$R_2 = \frac{5}{0.2 - 0.025} = 28.57 \ \Omega$$

We have two alternative designs, both of which deliver $v_2 = 5$ V to the 200-Ω load.

In practice, engineers use additional factors to evaluate alternatives that meet the same design goal. The power dissipation in the interface circuit is an important factor for two reasons. First, less interface dissipation means less power demand on the source. Second, less dissipation in the interface resistors means they can have lower power ratings, which are generally less expensive.

In the series case the power dissipated in R_1 is $i_2^2 R_1 = 0.219$ W. In the parallel case the power dissipated in R_2 is $v_2^2/R_2 = 0.875$ W. Clearly the power dissipation factor strongly favors the series design. ■

D DESIGN EXAMPLE 3–21

Design the interface circuit in Figure 3–56 so that the 40-V source delivers $v_2 = 2$ V to the output load and the resistance seen at the input port is $R_{IN} = 300 \ \Omega$. Note that this means that input resistance of the two port matches the source resistance.

SOLUTION:

This example places constraints at both the output port and the input port of the interface circuit. In most cases two independent constraints cannot be satisfied using only one resistor in the interface circuit. To see why, suppose

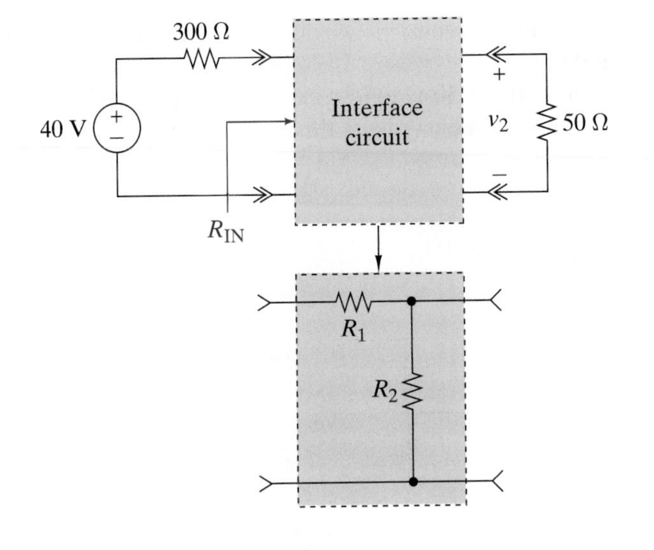

FIGURE 3–56

we use a single 650-Ω series resistor in the interface circuit. By voltage division, the output voltage would be

$$v_2 = \left(\frac{50}{300 + 650 + 50}\right) 40 = 2 \text{ V}$$

as required. However, the input resistance would be $R_{IN} = 650 + 50 = 700\ \Omega$, which does not meet the input port requirement of 300 Ω.

To meet both requirements, we need a two-resistor L-circuit such as the one shown in the figure. To design this circuit, we first define $R_{EQ} = R_2 \| 50$. Using this notation, the input port constraint is $R_{IN} = R_1 + R_{EQ} = 300\ \Omega$ and the output port constraint becomes

$$v_2 = \left(\frac{R_{EQ}}{300 + R_1 + R_{EQ}}\right) 40 = 2 \text{ V}$$

But $R_1 + R_{EQ} = 300$; hence, the output constraint reduces to $40R_{EQ} = 2 \times 600$, which means that $R_{EQ} = 30\ \Omega$. By definition

$$R_{EQ} = \frac{50R_2}{50 + R_2} = 30$$

which leads to $50R_2 = 1500 + 30R_2$, or $R_2 = 75\ \Omega$. Finally, since $R_{EQ} = 30\ \Omega$, the input port constraint then tells us that $R_1 = 300 - R_{EQ} = 270\ \Omega$. In sum, the L-circuit in the figure with $R_1 = 270\ \Omega$ and $R_2 = 75\ \Omega$ will meet both the input port and the output port constraints. ■

E EVALUATION EXAMPLE 3–22

In Example 3–21 we designed the interface circuit in Figure 3–56 to meet the requirements $v_2 = 2$ V and $R_{IN} = 300\ \Omega$. It is claimed that the interface circuit in Figure 3–57 meets the same requirements.

(a) Verify that the circuit in Figure 3–57 produces $v_2 = 2$ V and $R_{IN} = 300\ \Omega$.
(b) It is desired that the 50-Ω load "see" a low output resistance. Which of these two circuits best meets this requirement?

FIGURE 3–57

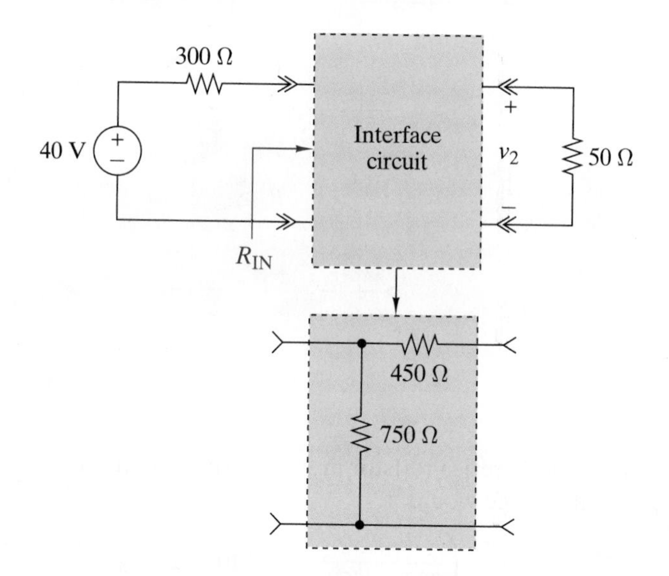

SOLUTION:

(a) The circuit in Figure 3–57 meets the input port constraint since

$$R_{IN} = 750 \| (450 + 50) = 750 \| 500 = 300\ \Omega$$

as required. Using this fact and voltage division, we find the voltage at the input port of the interface circuit to be

$$v_1 = \left(\frac{R_{IN}}{300 + R_{IN}}\right) 40 = \left(\frac{300}{600}\right) 40 = 20\ \text{V}$$

Using this voltage as the input to the voltage divider made up of the 450-Ω series resistor and the 50-Ω load gives us

$$v_2 = \left(\frac{50}{450 + 50}\right) v_1 = \left(\frac{50}{500}\right) 20 = 2\ \text{V}$$

This verifies that the circuit in Figure 3–57 produces $v_2 = 2$ V and $R_{IN} = 300\ \Omega$.

(b) To compare the output resistances, we turn the 40-V source off (replace it by a short) and find the lookback resistance seen at the output port. For the circuit in Figure 3–56

$$R_{OUT} = R_2 \| (R_1 + 300) = 75 \| (270 + 300) = 66.3\ \Omega$$

For the circuit in Figure 3–57

$$R_{OUT} = 450 + 750 \| 300 = 664\ \Omega$$

Clearly the circuit in Figure 3–56 has a much lower output resistance. ■

DESIGN EXAMPLE 3–23

Design the interface circuit in Figure 3–58 so the 50-Ω load "sees" a Thévenin resistance of 50 Ω between terminals C and D, while simultaneously the input voltage source "sees" an input resistance of 300 Ω between terminals A and B. Meeting these two constraints produces matched conditions at the input and output ports of the interface circuit.

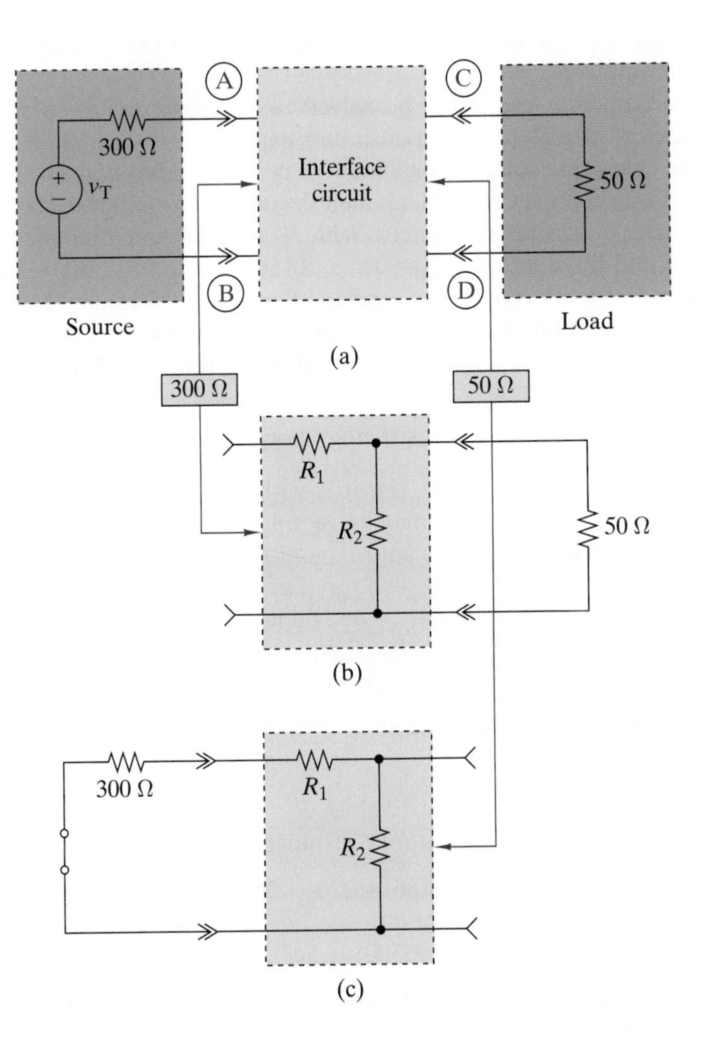

FIGURE 3–58

SOLUTION:

To meet the two constraints in this example, the interface circuit should be a two-resistors L-circuit. We have chosen the L-circuit configuration shown in Figure 3–58 for the following reasons. The source must see a larger resistance (300 Ω) at the input port than the load sees at the output port (50 Ω). This indicates that the source should "look" into a large series resistor R_1 while the load "looks" into a smaller parallel resistor R_2.

The design constraints in this example can be expressed in equation form. At the input port (terminals A and B) the equation is

$$R_1 + \frac{50R_2}{R_2 + 50} = 300\ \Omega$$

At the output port (terminals C and D) the equation is

$$\frac{(R_1 + 300)R_2}{R_1 + 300 + R_2} = 50\ \Omega$$

The design requirements reduce to two equations in two unknowns. What could be simpler?

These equations can easily be solved using programs like Mathcad or MATLAB. Solving them using pencil and paper is a bit of a chore. At this point we encourage you to think about the problem in physical terms. For instance, if we simply set $R_2 = 50\ \Omega$, then the conditions at terminals C and D will be met, at least approximately. With $R_2 = 50\ \Omega$ the requirement at terminals A and B reduces to $R_{AB} = R_1 + 50\|50 = R_1 + 25 = 300\ \Omega$. In other words, by physical reasoning we conclude that $R_1 = 275\ \Omega$ and $R_2 = 50\ \Omega$ is an approximate solution. How good is the approximation?

These values yield input and output resistances of $R_{AB} = 300\ \Omega$ as required, and

$$R_{CD} = 50\|(275 + 300) = 50\|575 = 46\ \Omega$$

This value is not exactly 50 Ω, but it is within 10% of the desired value. Since electrical components may have tolerances in the 10% range, a design based on our first-guess approximation might be adequate.

Our first-guess solution can also serve as the starting place for improving the design. The fact that $R_{CD} = 46\ \Omega$ tells us that $R_2 = 50\ \Omega$ is just a bit too low. Suppose we increase R_2 slightly to, say, $R_2 = 56\ \Omega$ (a standard value—see Appendix A, Table A-l). Then at the input port we have

$$R_{AB} = R_1 + 50\|56 = R_1 + 26.42 = 300\ \Omega$$

which would require $R_1 = 273.58\ \Omega$. The nearest standard value to this is $R_1 = 270\ \Omega$. Using the standard values $R_1 = 270\ \Omega$ and $R_2 = 56\ \Omega$ as our second guess, the input and output resistances are

$$R_{AB} = 270 + 50\|56 = 270 + 26.42 = 296.42\ \Omega$$

$$R_{CD} = 56\|(270 + 300) = 56\|570 = 50.99\ \Omega$$

both of which are within 2% of the desired values. Thus, finding an approximate solution can serve as the first step in the design process. Finding that first step is often the most creative and challenging part of circuit design. ■

APPLICATION EXAMPLE 3–24

The source-load interface in Figure 3–59 serves to introduce an important concept that we will encounter many times in subsequent chapters. By simple voltage division, the interface voltage is

$$v = \frac{R_L}{R_L + R_T} v_T$$

If the source is ideal ($R_T = 0$), then the interface voltage is $v = v_T$ regardless of the value of the load R_L. Conversely, if the load is an open circuit ($R_L = \infty$), then the interface voltage is $v = v_T$ regardless of the value of the source resistance R_T. Real-world applications typically fall between these two extremes with the result that $v < v_T$. Since $v_T = v_{OC} = v_{MAX}$, interface voltage is generally less than the maximum available voltage. The reduction in interface voltage is an example of an effect called *loading*. In general

Loading is the reduction in load voltage due to the effect of load resistance on the signal source driving it.

A loading problem is fundamentally different than the fixed-source, maximum power transfer problem. With loading, the source and load are both adjustable and the question is how they should be chosen to *minimize loading*. The undesirable effects of loading can be mitigated by making $R_T \ll R_L$, either by reducing the output resistance of the source or increasing the load resistance, or both. As a rule of thumb, the loading effect is less than 10% when $R_L \geq 10R_T$ and less than 1% when $R_L \geq 100R_T$.

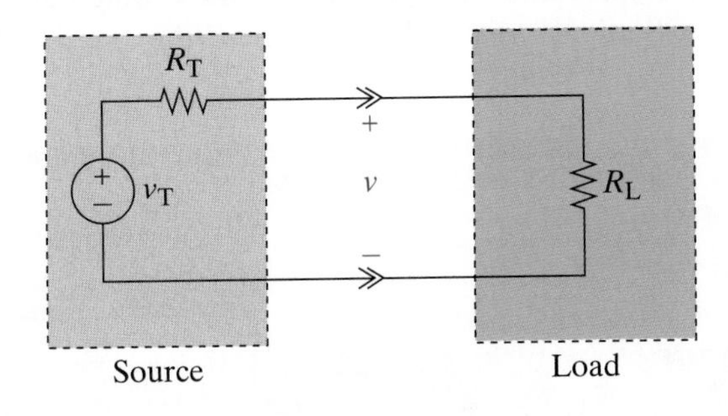

FIGURE 3–59

APPLICATION EXAMPLE 3–25

An **attenuation pad** is a two-port resistance circuit that provides a non-adjustable reduction in signal level while also providing resistance matching at the input and output ports. Figure 3–60 shows an example of an attenuation pad. The manufacturer's data sheet for this pad specifies the following characteristics at the input and output ports.

PORT	CHARACTERISTICS	CONDITION	VALUE	UNITS
Output	Thévenin voltage	600-Ω source connected at the input port	$v_S/4$	V
Output	Thévenin resistance	600-Ω source connected at the input port	600	Ω
Input	Input resistance	600-Ω load connected at the output port	600	Ω

Use Orcad to verify these characteristics.

FIGURE 3–60

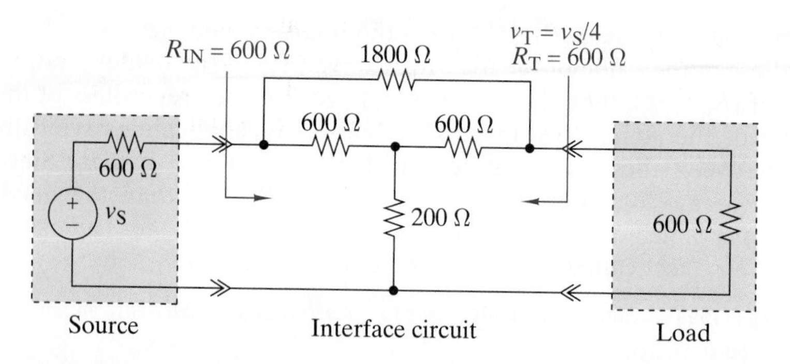

SOLUTION:

Orcad Capture can calculate the two-port characteristics of the pad in Figure 3–60. A circuit diagram created using Orcad is shown in the upper right of Figure 3–61. In this schematic a 600-Ω source is connected at the pad's input port. A 600-Ω load is *not* connected at the output port. The reason is that we need to find the open-circuit (Thévenin) voltage at the output port. We have labeled the pad's output as node D. Orcad has labeled our node D as N03172. This Orcad assignment is not under our control and will change with every new simulation.[3]

Selecting the **PSpice** and the **Edit Simulation Profile** commands causes the "Simulation Settings" dialog box shown in the upper left portion of

FIGURE 3–61

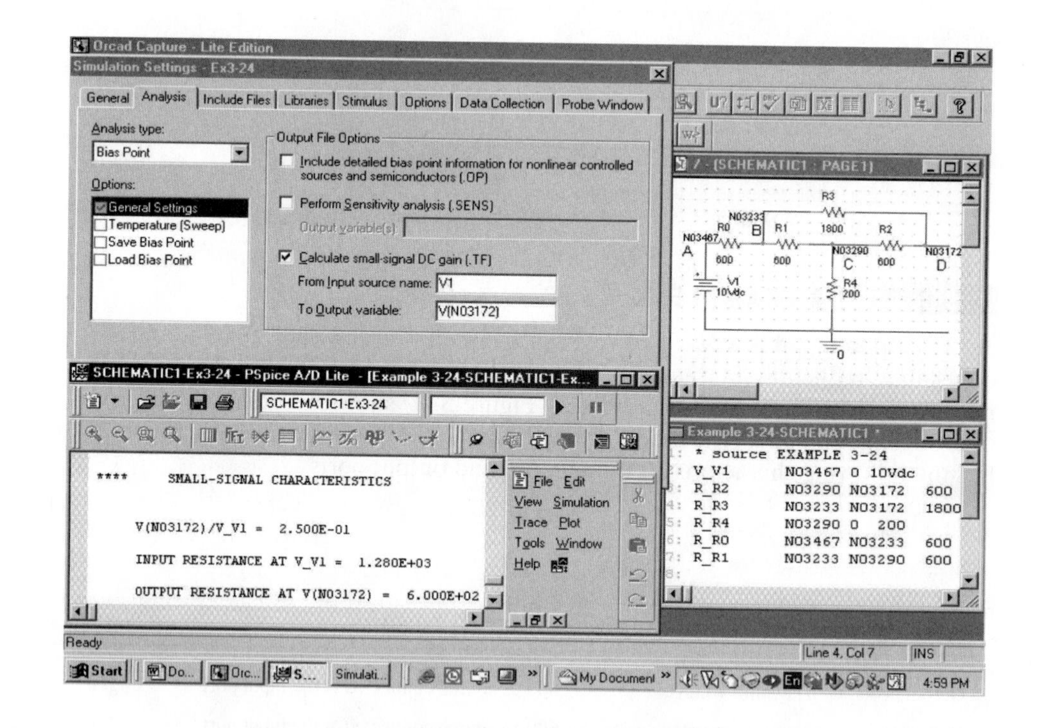

3 Every time a schematic is drawn (even the same schematic) Orcad generates a new random number for each node name. To display the Orcad assigned names, double-click on each node to open a new window that gives information about the node. Click on the node's name then select **Display**. This opens a second window. Select **Value Only**. Then click **OK**. Finally select **Apply** to place it on the drawing. Alternately, the Orcad assigned names can be viewed in the netlist file.

Figure 3-61 to appear. We first check the "Calculate small-signal DC gain (.TF)" box as shown in the figure. To calculate a gain, Orcad must know the input and output variables. We inform Orcad of these variables by supplying entries in the "From Input source name" and "to Output variable" boxes. In this example, voltage source V1 is the input source and the output variable is the node voltage V(N03172) (the voltage at our node D). Note that V(N03172) is the open-circuit voltage at the output port. With these entries Orcad PSpice will calculate three things: (1) the small-signal gain V(N03172)/V_V1, (2) the input resistance seen by V1, and (3) the output resistance.

Returning to the main menu, we select **PSpice/ Run**. Orcad PSpice performs a dc analysis of the circuit and writes the results to the output file. Included in the output file are the small-signal characteristics shown at the bottom left in Figure 3–61. PSpice reports that the small-signal gain V(N03172)/V_V1 = 2.500E-01, the input resistance seen by V1 is 1.280E+ 03, and the output resistance is 6.000E+02. This means that the open-circuit (Thévenin) voltage at the output port is one-fourth of the source voltage and the output (Thévenin) resistance is 600 Ω as stated in the data sheet. The input resistance (1280 Ω) given in the output file is the resistance seen by the source V1 when the output port is open-circuited. The manufacturer's data sheet specifies the input resistance when a 600-Ω load is connected at the output port. Thus, the simulation has not directly verified the input resistance characteristic.

However, a moment's reflection shows that the two-port pad circuit is symmetrical. The simulation results show that the output resistance is 600 Ω when a 600-Ω source is connected at the input. It follows from the circuit's symmetry that the input resistance is 600 Ω when a 600-Ω load is connected at the output. In summary, the simulation has verified all of the entries in the manufacturer's data sheet. ■

SUMMARY

- Node-voltage analysis involves identifying a reference node and the node to datum voltages at the remaining $N-1$ nodes. The KCL connection constraints at the $N-1$ nonreference nodes combined with the element constraints written in terms of the node voltages produce $N-1$ linear equations in the unknown node voltages.
- Mesh-current analysis involves identifying mesh currents that circulate around the perimeter of each mesh in a planar circuit. The KVL connection constraints around $E-N+1$ meshes combined with the element constraints written in terms of the mesh currents produce $E-N+1$ linear equations in the unknown mesh currents.
- Node and mesh analysis can be modified to handle both types of independent sources using a combination of three methods: (1) source transformations, (2) selecting circuit variables so independent sources specify the values of some of the unknowns, and (3) using supernodes or supermeshes.
- A circuit is linear if it contains only linear elements and independent sources. For single-input linear circuits, the proportionality property states that any output is proportional to the input. For multiple-input linear circuits, the superposition principle states that any output can be found by summing the output produced when each input acts alone.

- A Thévenin equivalent circuit consists of a voltage source in series with a resistance. A Norton equivalent circuit consists of a current source in parallel with a resistance. The Thévenin and Norton equivalent circuits are related by a source transformation.
- The parameters of the Thévenin and Norton equivalent circuits can be determined using any two of the following: (1) the open-circuit voltage at the interface, (2) the short-circuit current at the interface, and (3) the equivalent resistance of the source circuit with all sources turned off.
- The parameters of the Thévenin and Norton equivalent circuits can also be determined using circuit reduction methods or by directly solving for the source i–v relationship using node-voltage or mesh-current analysis.
- For a fixed source and an adjustable load, the maximum interface signal levels are $v_{MAX} = v_{OC}$ ($R_L = \infty$), $i_{MAX} = i_{SC}$ ($R_L = 0$), and $p_{MAX} = v_{OC}i_{SC}/4$ ($R_L = R_T$). When $R_L = R_T$, the source and load are said to be matched.
- Interface signal transfer conditions are specified in terms of the voltage, current, or power delivered to the load. The design constraints depend on the signal conditions specified and the circuit parameters that are adjustable. Some design requirements may require a two-port interface circuit. An interface design problem may have one, many, or no solutions.

PROBLEMS

OBJECTIVE 3–1 GENERAL CIRCUIT ANALYSIS (SECT. 3–1 TO 3–2)

Given a circuit consisting of linear resistors and independent sources:

(a) (Formulation) Write node-voltage or mesh-current equations for the circuit.

(b) (Solution) Solve the equations from part (a) for selected signal variables or input-output relationships.

Node-voltage method:
See Examples 3–1, 3–2, 3–3, 3–4, 3–5, 3–6 and Exercises 3–2, 3–3, 3–4, 3–5, 3–6

Mesh-current method:
See Examples 3–7, 3–8, 3–9 and Exercises 3–8, 3–9, 3–10

3–1 **(a)** Formulate node-voltage equations for the circuit in Figure P3–1.
(b) Use these equations to find v_x and i_x.

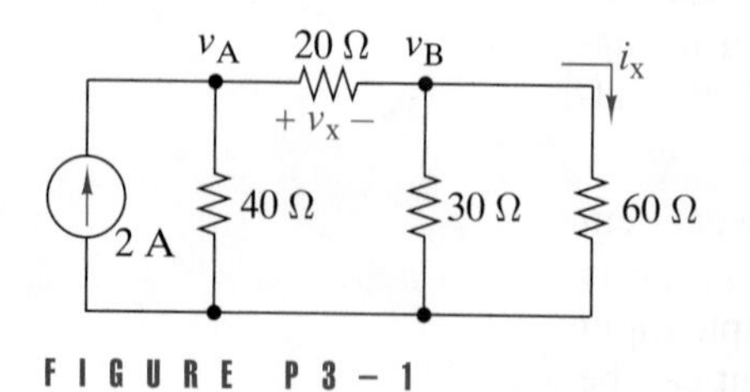

FIGURE P3–1

3–2 **(a)** Formulate node-voltage equations for the circuit in Figure P3–2.
(b) Use these equations to find v_x and i_x.

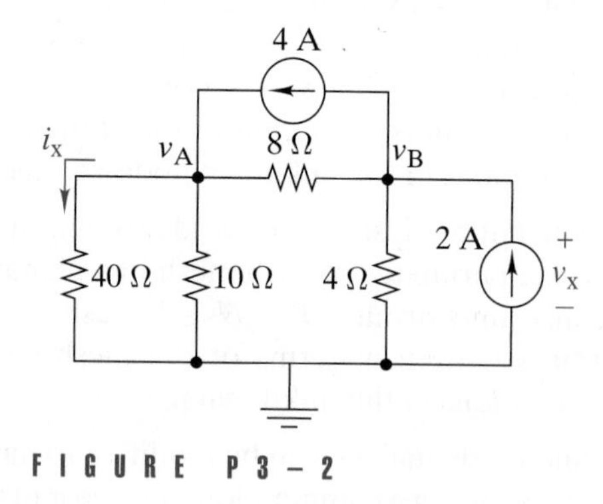

FIGURE P3–2

3–3 **(a)** Formulate node-voltage equations for the circuit in Figure P3–3.
(b) Use these equations to find v_x and i_x.

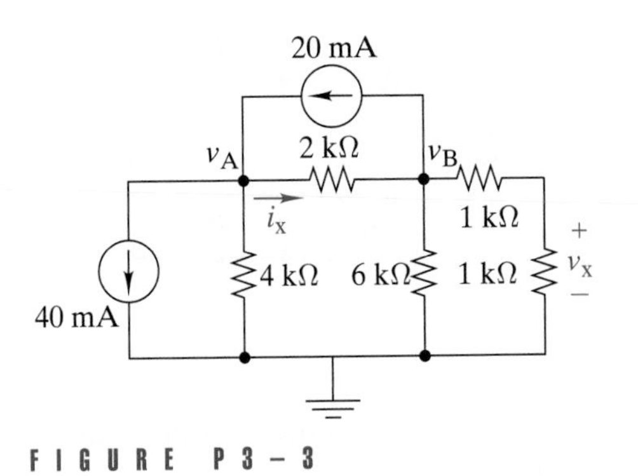

FIGURE P3–3

3–4 (a) Formulate node-voltage equations for the circuit in Figure P3–4.
(b) Use these equations to find v_x and i_x.

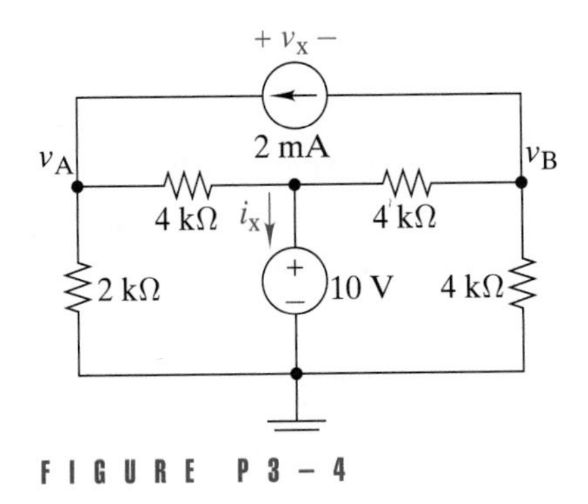

FIGURE P3–4

3–5 (a) Formulate node-voltage equations for the circuit in Figure P3–5.
(b) Solve for v_x and i_x when $R_1 = R_2 = R_3 = R_4 = 10\ \text{k}\Omega$, $v_S = 20$ V, and $i_S = 2$ mA.

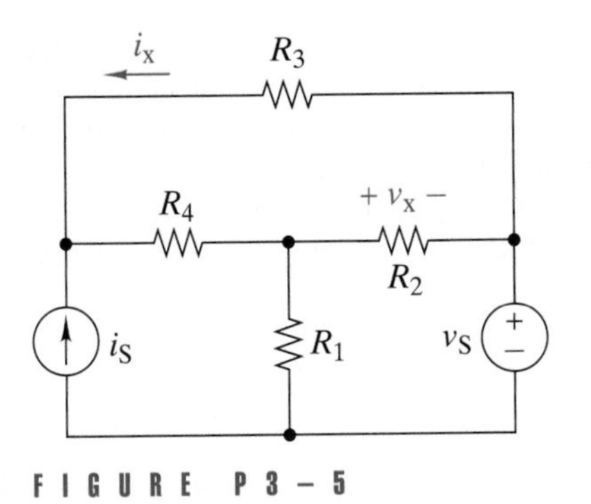

FIGURE P3–5

3–6 (a) Formulate node-voltage equations for the circuit in Figure P3–6.
(b) Solve for v_x and i_x when $R_1 = R_2 = R_3 = R_4 = R_5 = 10\ \text{k}\Omega$, and $v_S = 20$ V.

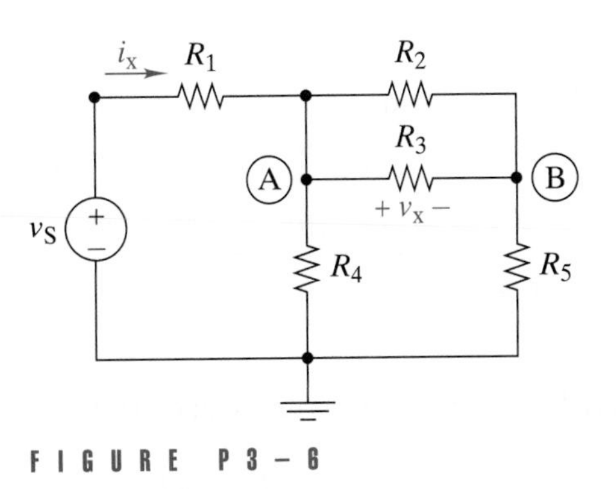

FIGURE P3–6

3–7 (a) Formulate node-voltage equations for the circuit in Figure P3–7.
(b) Solve for v_A, v_B, and v_C when $R_1 = 1\ \text{k}\Omega$, $R_2 = 2\ \text{k}\Omega$, $R_3 = 4\ \text{k}\Omega$, $R_4 = 2\ \text{k}\Omega$, and $i_{S1} = i_{S2} = 2$ mA.

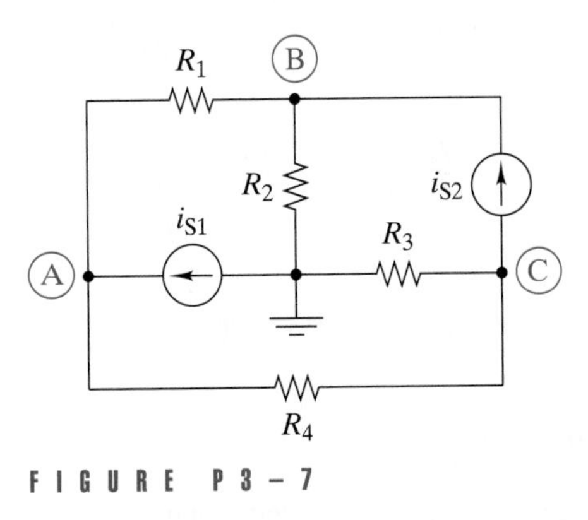

FIGURE P3–7

3–8 (a) Formulate node-voltage equations for the circuit in Figure P3–8.
(b) Solve for v_x and i_x when $R_1 = R_4 = 1\ \text{k}\Omega$, $R_2 = R_3 = 250\ \Omega$, $R_x = 500\ \Omega$, and $v_S = 15$ V.

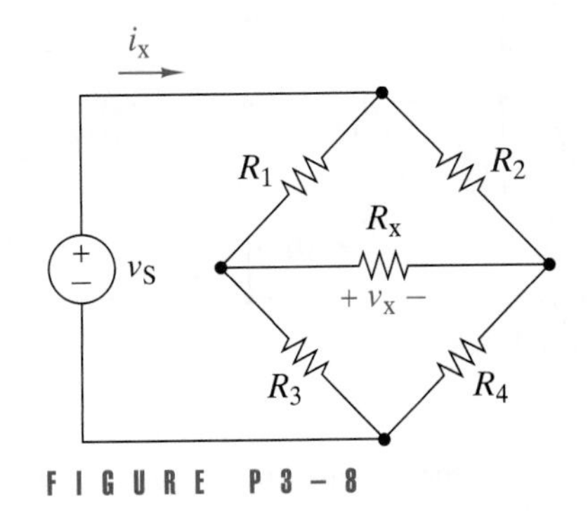

FIGURE P3–8

3–9 (a) Formulate mesh-current equations for the circuit in Figure P3–9.
(b) Use these equations to find v_x and i_x.

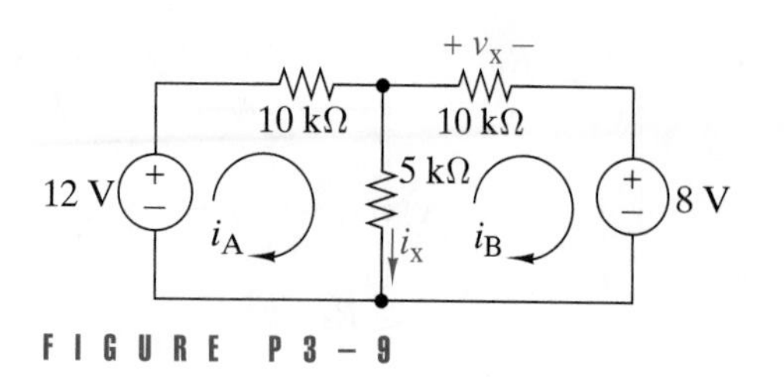

FIGURE P3–9

3–10 **(a)** Formulate mesh-current equations for the circuit in Figure P3–10.
(b) Use these equations to find v_x and i_x.

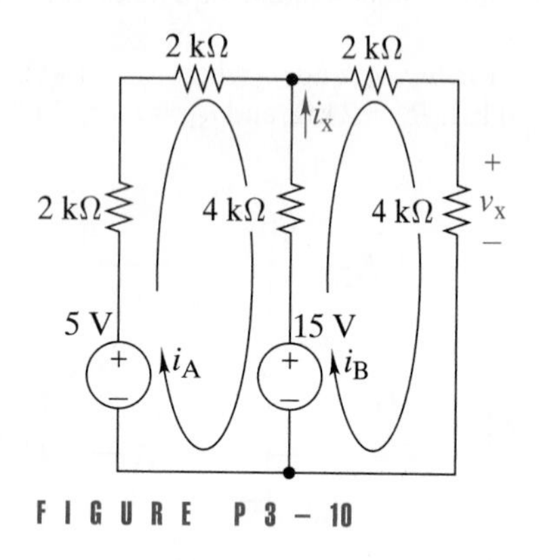

FIGURE P3–10

3–11 **(a)** Formulate mesh-current equations for the circuit in Figure P3–11.
(b) Use these equations to find v_x and i_x.

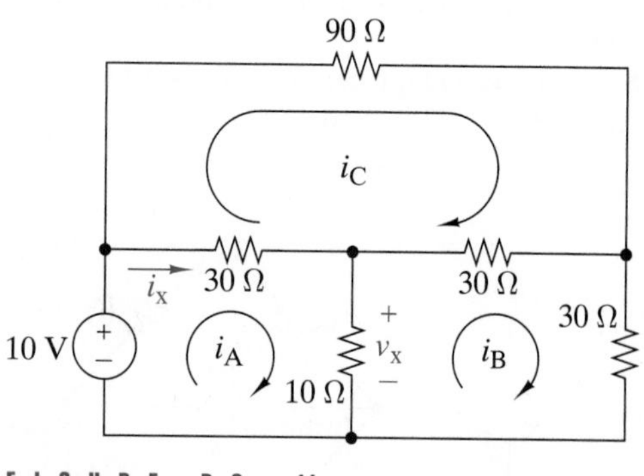

FIGURE P3–11

3–12 **(a)** Formulate mesh-current equations for the circuit in Figure P3–12.
(b) Solve for v_x and i_x when $R_1 = 200\ \Omega$, $R_2 = 300\ \Omega$, $R_3 = 50\ \Omega$, $R_4 = 250\ \Omega$, $R_5 = 200\ \Omega$, $i_S = 50$ mA, and $v_S = 15$ V.
(c) Find the total power dissipated in the circuit.

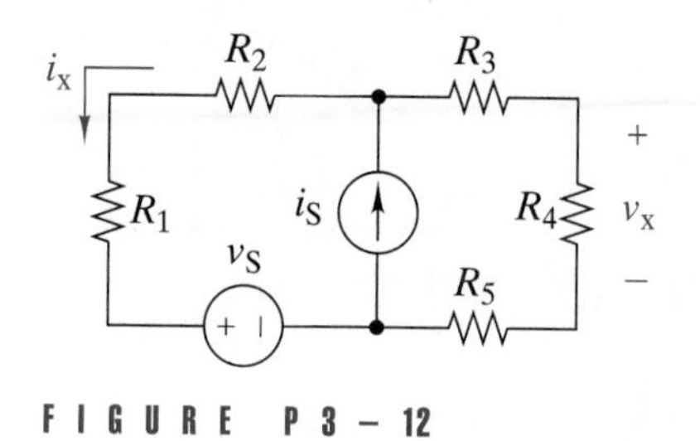

FIGURE P3–12

3–13 **(a)** Formulate mesh-current equations for the circuit in Figure P3–13.
(b) Solve for v_x and i_x when $R_1 = R_2 = 10$ kΩ, $R_3 = 2$ kΩ, $R_4 = 1$ kΩ, $i_S = 2.5$ mA, $v_{S1} = 12$ V, and $v_{S2} = 0.5$ V.
(c) Find the power supplied by v_{S1}.

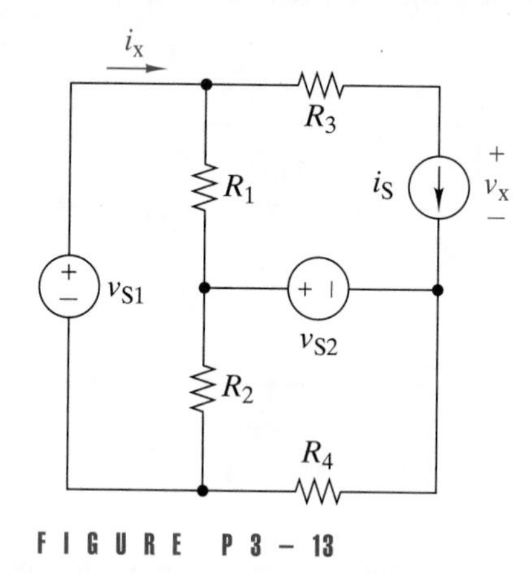

FIGURE P3–13

3–14 The circuit in Figure P3–14 seems to require two supermeshes since both current sources appear in two meshes. However, sometimes rearranging the circuit diagram will eliminate the need for a supermesh.
(a) Show that supermeshes can be avoided in Figure P3–14 by rearranging the connection of resistor R_6.
(b) Formulate mesh-current equations for the modified circuit as redrawn in part (a).
(c) Solve for v_x when $R_1 = R_2 = R_3 = R_4 = 2$ kΩ, $R_5 = R_6 = 1$ kΩ, $i_{S1} = 40$ mA, and $i_{S2} = 20$ mA.

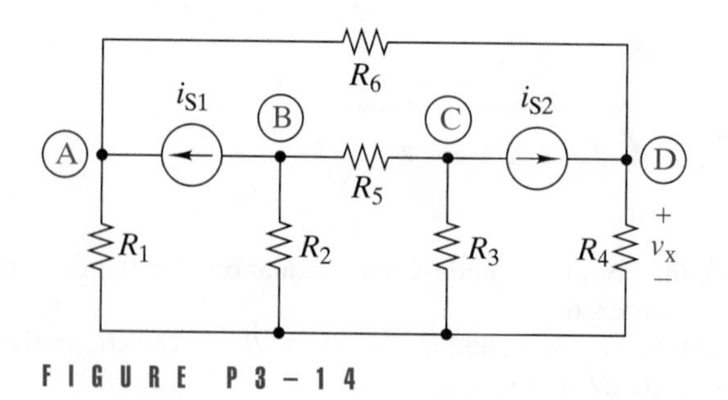

FIGURE P3–14

3–15 **(a)** Formulate mesh-current equations for the circuit in Figure P3–15.
(b) Use these equations to find v_x and i_x.
(c) Find the total power delivered to the resistors.

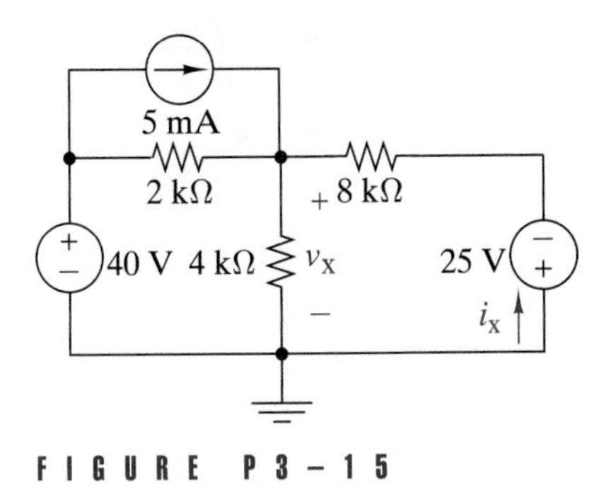

FIGURE P3–15

3–16 **(a)** Formulate mesh-current equations for the circuit in Figure P3–16.
(b) Use these equations to find the input resistance.

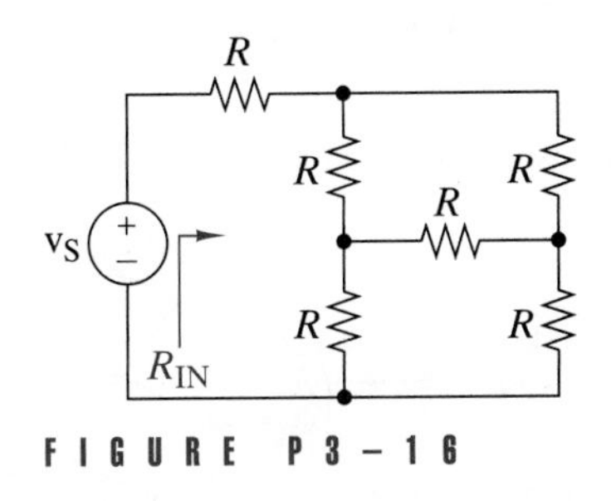

FIGURE P3–16

3–17 In Figure P3–17 all of the resistors are 1 kΩ. The voltage at node C is found to be $v_C = -2$ V when node B is connected to ground. Find the node voltages v_A and v_D, and the mesh currents i_A and i_B.

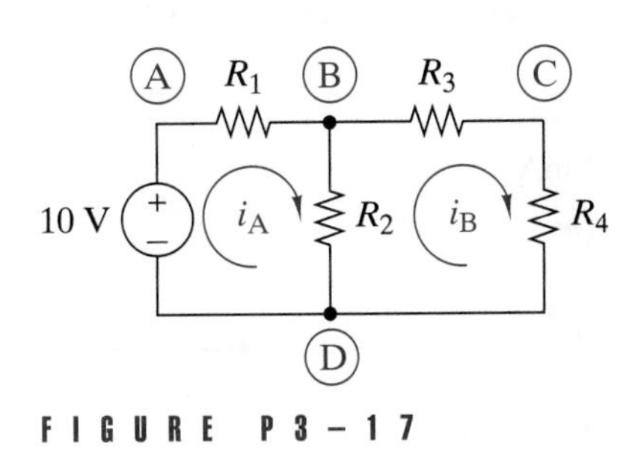

FIGURE P3–17

3–18 In Figure P3–17 all of the resistors are 1 kΩ. The voltage at node A is observed to be $v_A = 8$ V when node C is connected to ground. Find the node voltages v_B and v_D, and the mesh currents i_A and i_B.

3–19 Find the node voltages v_A and v_B in Figure P3–19.

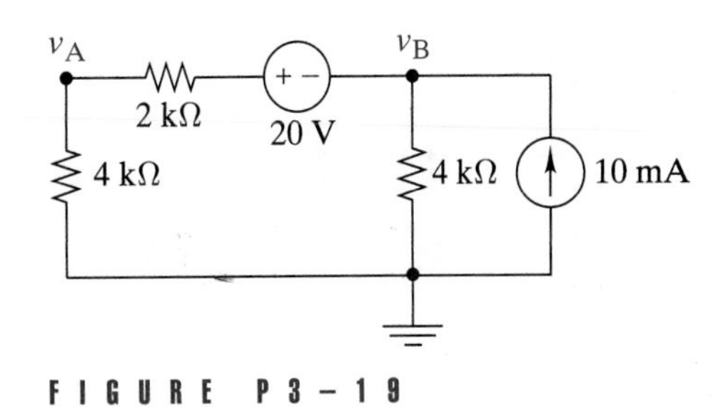

FIGURE P3–19

3–20 Find the mesh currents i_A, i_B, and i_C in Figure P3–20.

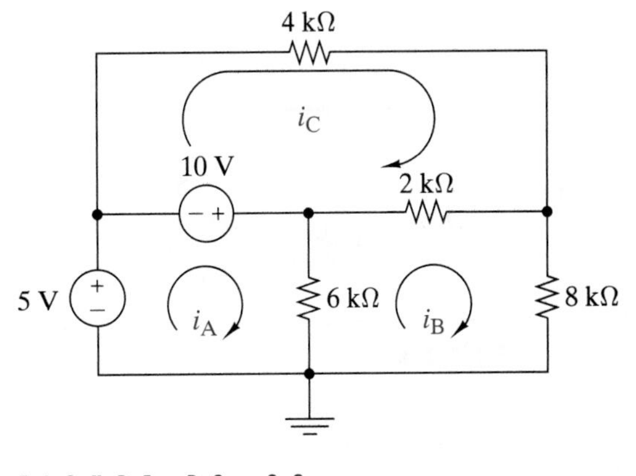

FIGURE P3–20

OBJECTIVE 3–2 LINEARITY PROPERTIES (SECT. 3–3)

(a) Given a circuit containing linear resistors and one independent source, use the proportionality principle to find selected signal variables.
(b) Given a circuit containing linear resistors and two or more independent sources, use the superposition principle to find selected signal variables.

See Examples 3–10, 3–11, 3–12 and Exercises 3–11, 3–12, 3–13, 3–14, 3–15

3–21 Find the proportionality constant $K = v_O/i_S$ for the circuit in Figure P3–21.

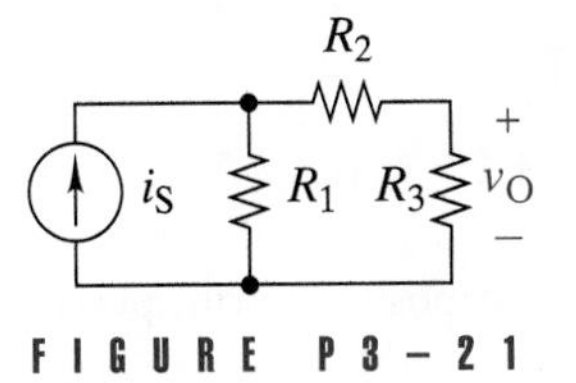

FIGURE P3–21

3–22 Find the proportionality constant $K = i_O/v_S$ for the circuit in Figure P3–22.

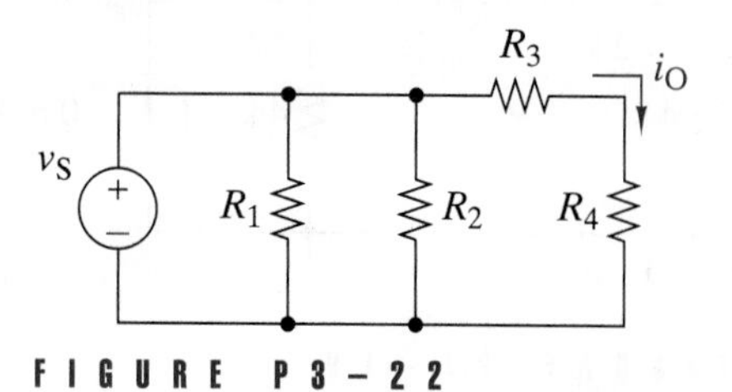

FIGURE P3–22

3–23 Find the proportionality constant $K = v_O/i_S$ for the circuit in Figure P3–23.

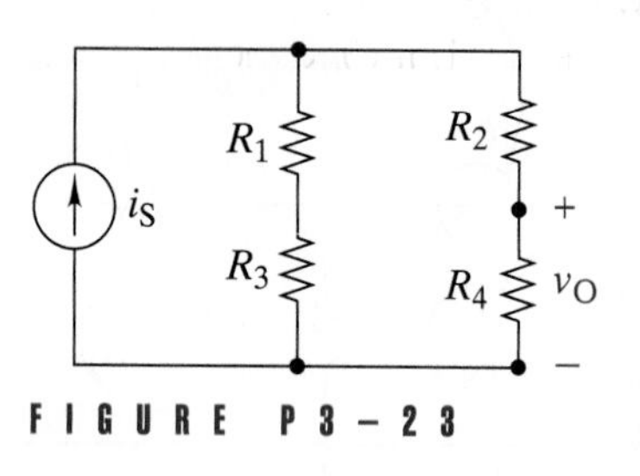

FIGURE P3–23

3–24 Use the unit output method to find v_O in Figure P3–24.

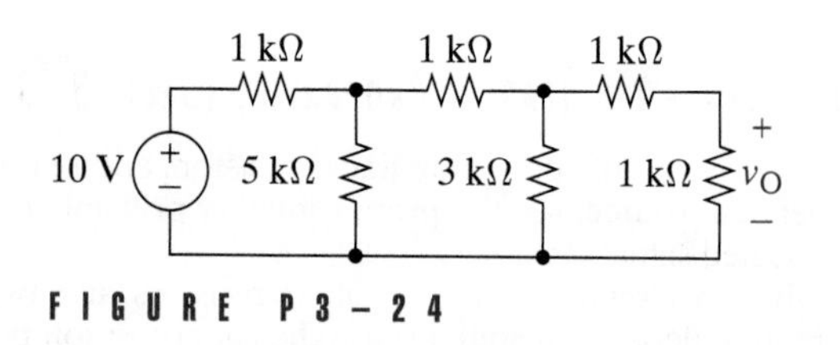

FIGURE P3–24

3–25 Use the unit output method to find i_O in Figure P3–25.

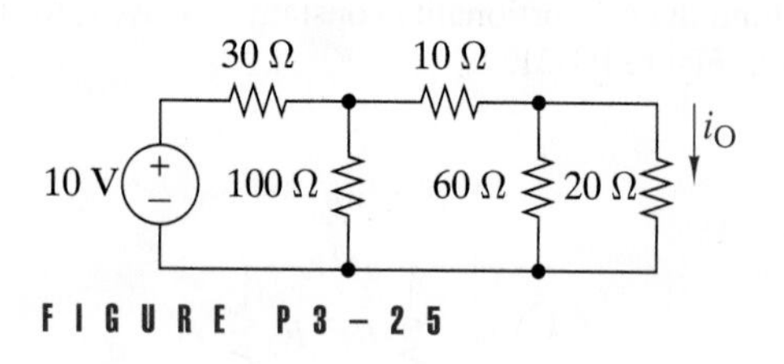

FIGURE P3–25

3–26 Use the superposition principle to find v_O in Figure P3–26.

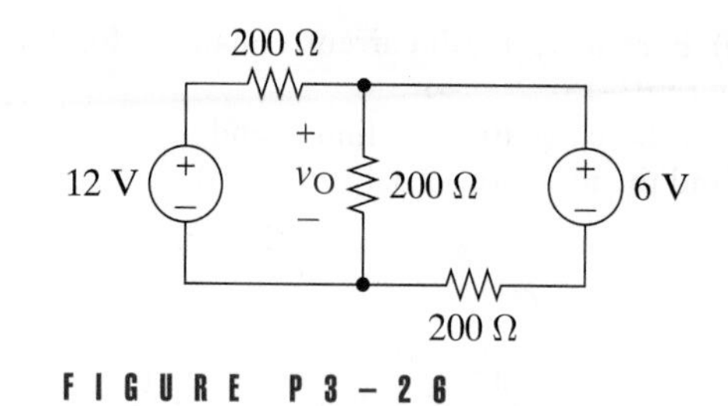

FIGURE P3–26

3–27 Use the superposition principle to find i_O in Figure P3–27.

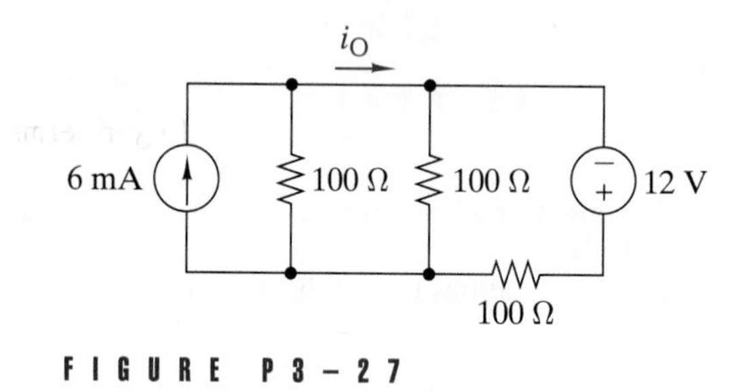

FIGURE P3–27

3–28 Use the superposition principle to find v_O in Figure P3–28.

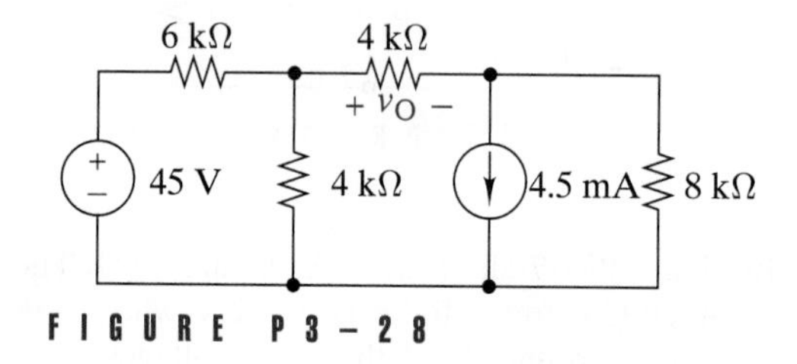

FIGURE P3–28

3–29 Use the superposition principle to find i_O in Figure P3–29.

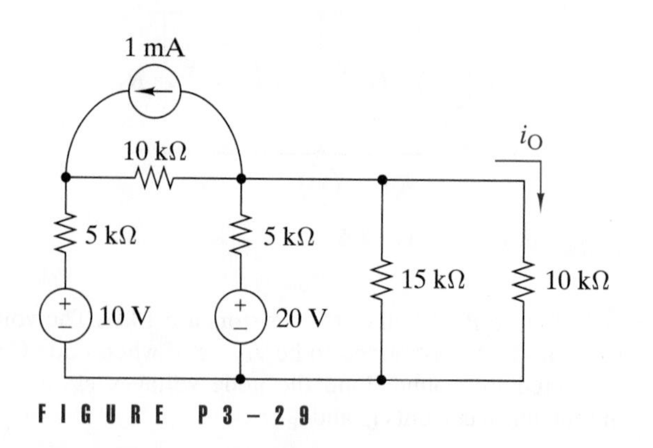

FIGURE P3–29

3–30 Use the superposition principle to find v_O in terms of v_1, v_2, and R in Figure P3–30.

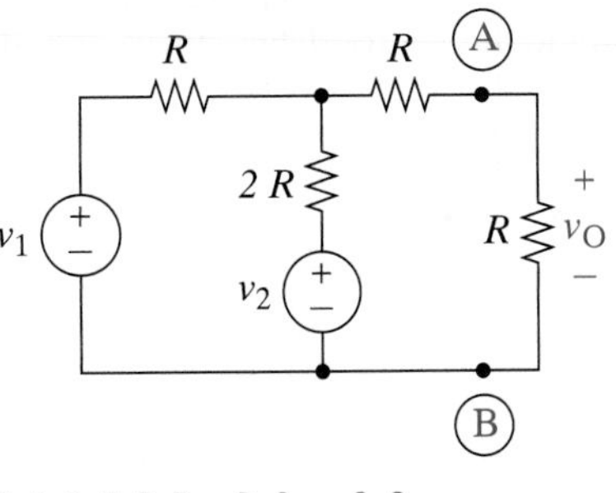

FIGURE P3–30

3–31 Use the superposition principle to find v_O in terms of i_1, i_2, and R in Figure P3–31.

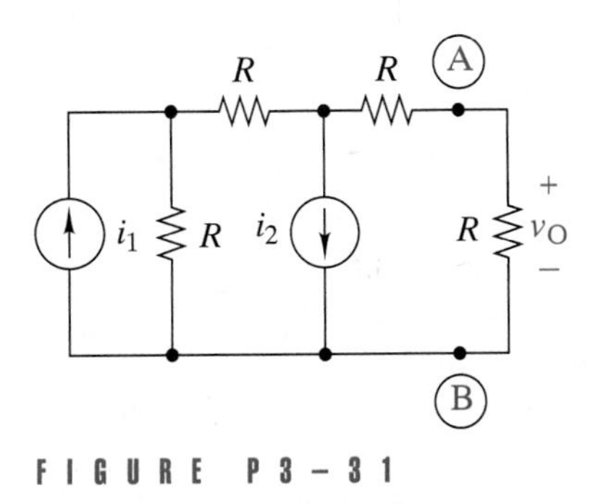

FIGURE P3–31

3–32 A linear circuit containing two sources drives a 100-Ω load resistor. Source number 1 delivers 250 mW to the load when source number 2 is off. Source number 2 delivers 4 W to the load when source number 1 is off. Find the power delivered to the load when both sources are on. *Hint*: The answer is not 4.25 W. Why?

3–33 A linear circuit is driven by an independent voltage source $v_S = 10$ V and an independent current source $i_S = 10$ mA. The output voltage is $v_O = 2$ V when the voltage source is on and the current source off. The output is $v_O = 1$ V when both sources are on. Find the output voltage when $v_S = 5$ V and $i_S = -10$ mA.

3–34 A certain linear circuit has three input voltages and one output voltage v_O. The following table lists the output for different values of the three inputs. Find the input-output relationship for the circuit.

v_{S1}(V)	v_{S2}(V)	v_{S3}(V)	v_O(V)
2	4	−4	1
1	2	2	1.5
1	4	2	2

3–35 **D** This problem involves designing a resistive circuit with two inputs v_{S1} and v_{S2} and a single output voltage v_O. Design the circuit so that $v_O = K(v_{S1} + 4v_{S2})$ is delivered across a 500-Ω load. The value of K is not specified but should be greater than 1/10.

OBJECTIVE 3–3 THÉVENIN AND NORTON EQUIVALENT CIRCUITS (SECT. 3–4)

Given a circuit containing linear resistors and independent sources:

(a) Find the Thévenin or Norton equivalent at a specified pair of terminals.

(b) Use the Thévenin or Norton equivalent to find the signals delivered to linear or nonlinear loads.

See Examples 3–13, 3–14, 3–15, 3–16 and Exercises 3–16, 3–17, 3–18

3–36 **(a)** Find the Thévenin or Norton equivalent circuit seen by R_L in Figure P3–36.

(b) Use the equivalent circuit found in part (a) to find i_L in terms of i_S, R_1, R_2, and R_L.

(c) Check your answer in part (b) using current division.

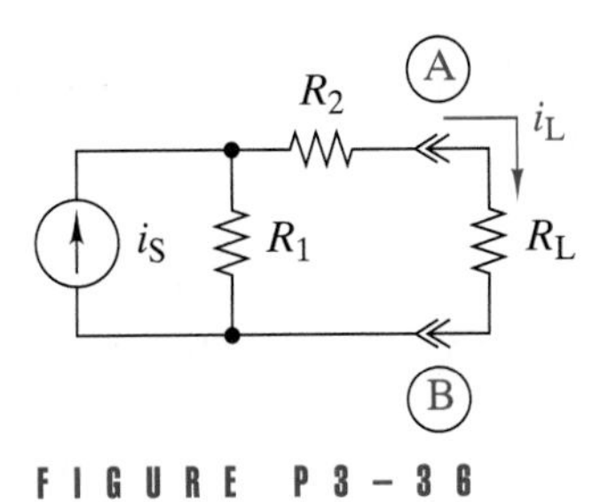

FIGURE P3–36

3–37 Find the Thévenin equivalent circuit seen by R_L in Figure P3–37. Find the voltage across the load when $R_L = 5$ Ω, 10 Ω, and 50 Ω.

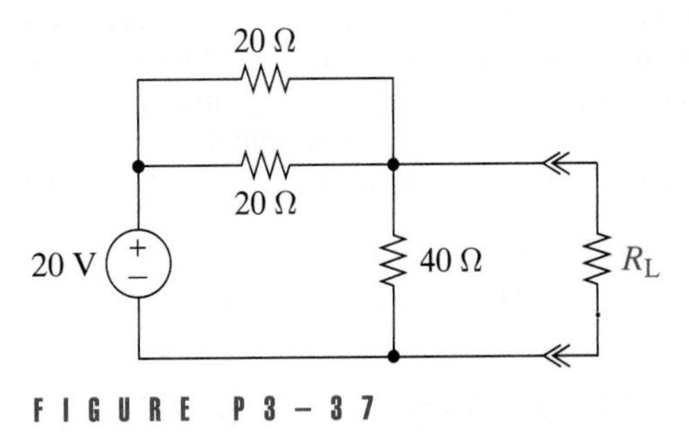

FIGURE P3–37

3–38 Find the Norton equivalent seen by R_L in Figure P3–38. Find the current through the load when $R_L = 6$ Ω, 12 Ω, and 60 Ω.

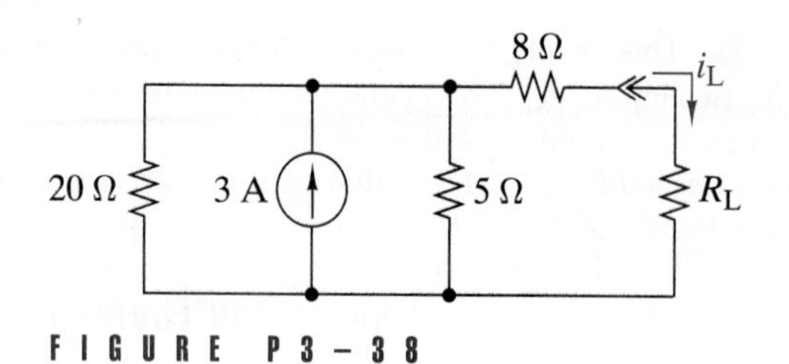

FIGURE P3–38

3–39 Find the Thévenin equivalent seen by R_L in Figure P3–39. Find the power delivered to the load when $R_L = 50$ kΩ and 200 kΩ.

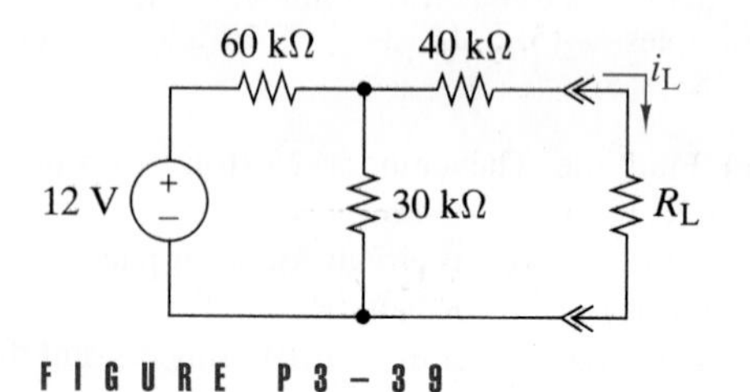

FIGURE P3–39

3–40 Find the Norton equivalent at terminals A and B in Figure P3–40.

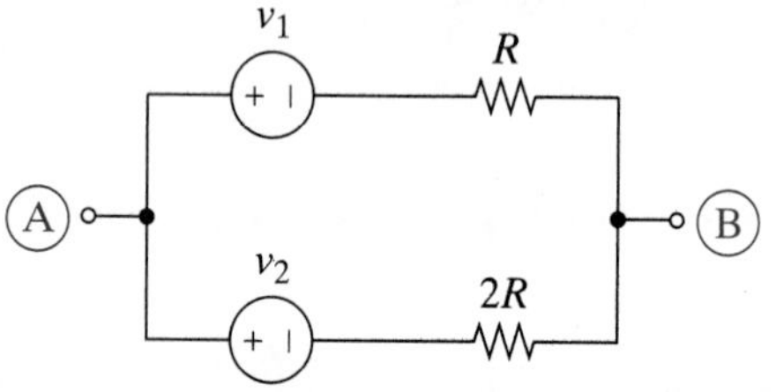

FIGURE P3–40

3–41 The purpose of this problem is to use Thévenin equivalent circuits to find the voltage v_X in Figure P3–41. Find the Thévenin equivalent circuit seen looking to the left of terminals A and B. Find the Thévenin equivalent circuit seen looking to the right of terminals A and B. Connect these equivalent circuits together and find the voltage v_X.

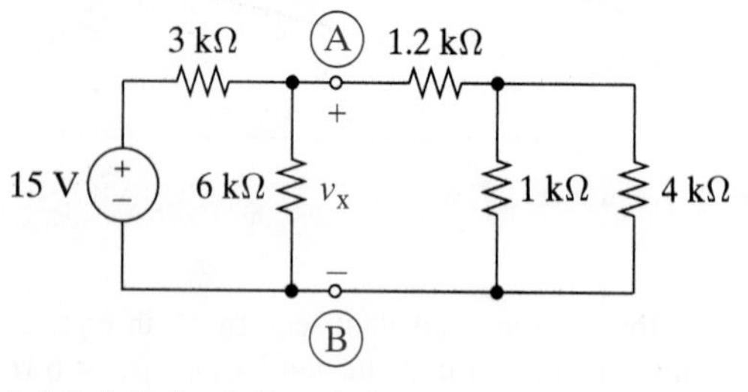

FIGURE P3–41

3–42 Figure P3–42 shows an active circuit with two accessible terminals. The output current is $i = 10$ mA when $v = 0$. The output voltage is $v = 6$ V when a 2.4-kΩ resistor is connected between the terminals. How much current would this source deliver to a 6-V battery?

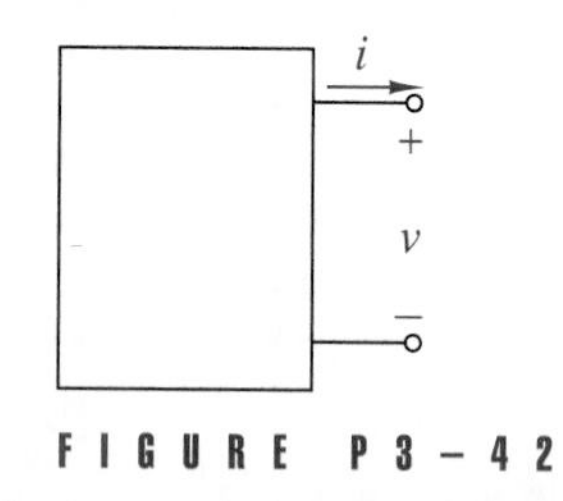

FIGURE P3–42

3–43 The $i-v$ characteristic of the active circuit in P3–42 is $5v + 500i = 60$. Find the output voltage when a 500-Ω resistive load is connected across the two accessible terminals.

3–44 Figure P3–42 shows a source circuit with two accessible terminals. Some voltage and current measurements at the accessible terminals are

v(V)	−10	−5	0	+5	+10	12	13	14
i(mA)	+5	+4	+3	+2	+1	0	−1	−2

(a) Use these data to plot the source $i-v$ characteristic.
(b) Develop a Thévenin equivalent circuit valid on the range $|v| < 10$ V.
(c) Use the equivalent circuit to predict the source v_{OC} and i_{SC}.
(d) Compare your results in part (c) with the given measurements and explain any differences.

3–45 The Thévenin equivalent parameters of a practical voltage source are $v_T = 25$ V and $R_T = 150$ Ω. Find the smallest load resistance for which the load voltage exceeds 15 V.

3–46 Use a sequence of source transformations to find the Thévenin equivalent at terminals A and B in Figure P3–46.

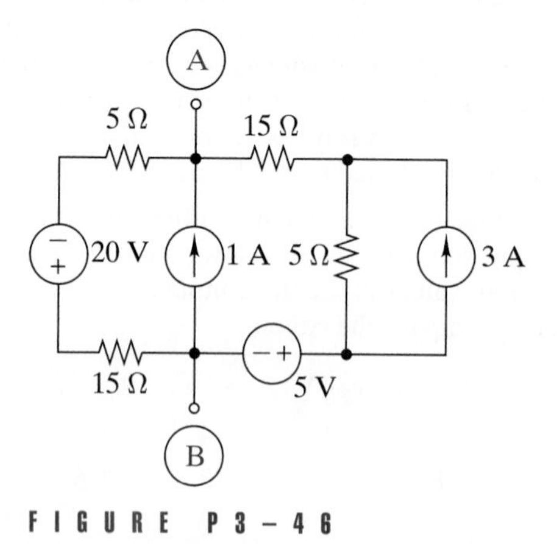

FIGURE P3–46

3–47 Select the value of R_L in Figure P3–47 so that $i_O = 80\ \mu A$.

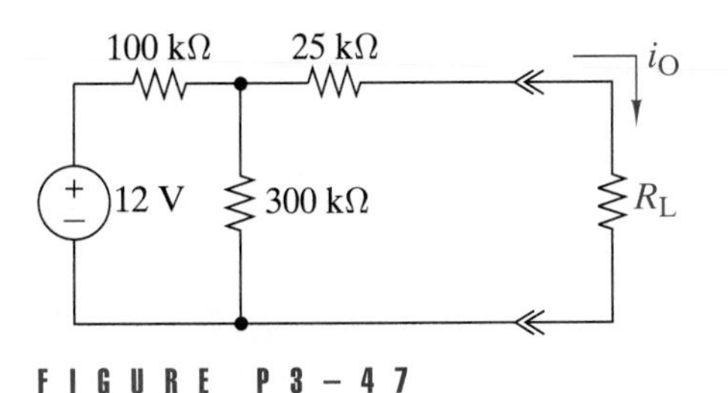

FIGURE P3–47

3–48 Find the Thévenin equivalent at terminals A and B in Figure P3–48.

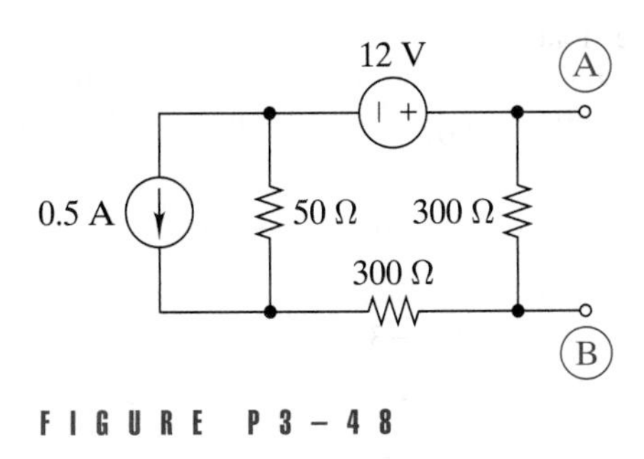

FIGURE P3–48

3–49 A nonlinear resistor is connected across a two-terminal source whose Thévenin equivalent is $v_T = 5$ V and $R_T = 500\ \Omega$. The i–v characteristic of the resistor is $i = 10^{-4}(v + 2\,v^{3.3})$. Plot the i–v characteristic of the source and the resistor and graphically determine the voltage across and current through the nonlinear resistor.

3–50 A nonlinear resistor is connected across a two-terminal source whose Thévenin equivalent is $v_T = 10$ V and $R_T = 200\ \Omega$. The i–v characteristic of the resistor is $v = 4000\,i^2$. Plot the i–v characteristic of the source and the resistor and graphically determine the voltage across and current through the nonlinear resistor.

3–51 Find the Norton equivalent seen by R_L in Figure P3–51. Select the value of R_L so that 100 mW is delivered to the load.

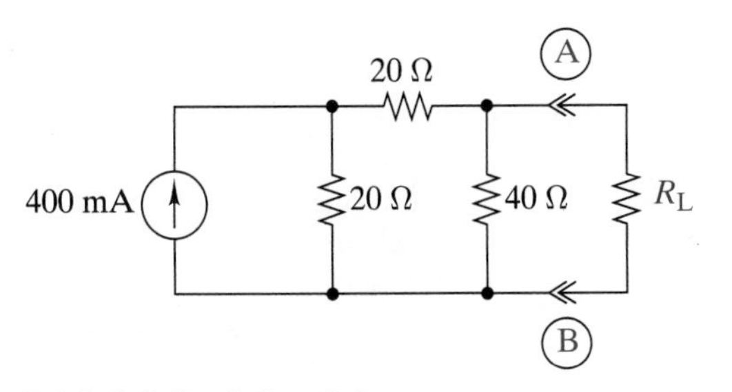

FIGURE P3–51

3–52 Find the Norton equivalent seen by R_L in Figure P3–52.

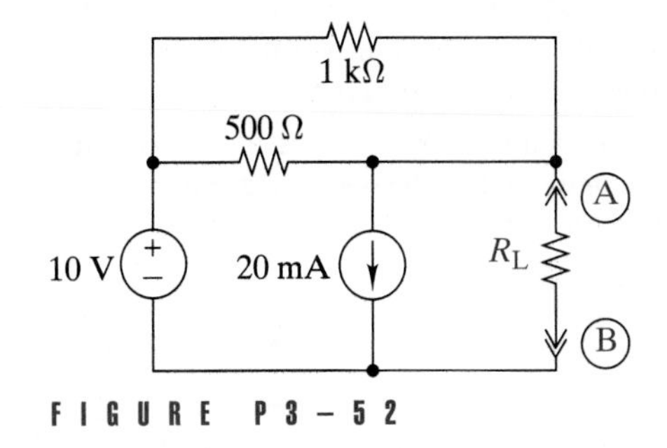

FIGURE P3–52

3–53 Find the Thévenin equivalent seen by R_L in Figure P3–53.

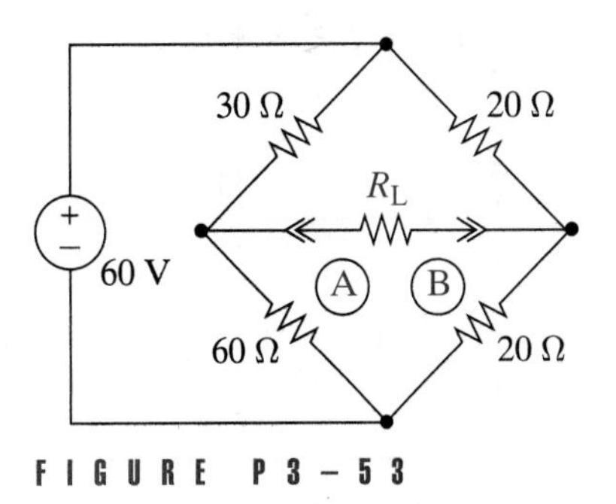

FIGURE P3–53

OBJECTIVE 3–4 MAXIMUM SIGNAL TRANSFER (SECT. 3–5)

Given a circuit containing linear resistors and independent sources:

(a) Find the maximum voltage, current, and power available at a specified pair of terminals.

(b) Find the resistive loads required to obtain the maximum available signal levels.

See Example 3–17 and Exercise 3–19

3–54 The resistance R in Figure P3–54 is adjusted until maximum power is delivered to the load consisting of R and the 6-kΩ resistor in parallel. Find the required value of R.

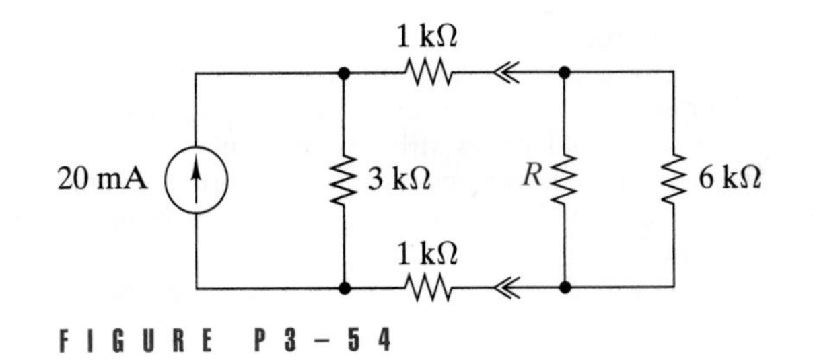

FIGURE P3–54

3–55 When a 5-kΩ resistor is connected across a two-terminal source, a current of 15 mA is delivered to the load. When a second 5-kΩ resistor is connected in parallel with the first, a total current of 20 mA is delivered. Find the maximum power available from the source.

3–56 Find the value of R in Figure P3–56 so that maximum power is delivered to the 4-kΩ load. Find the value of the maximum power.

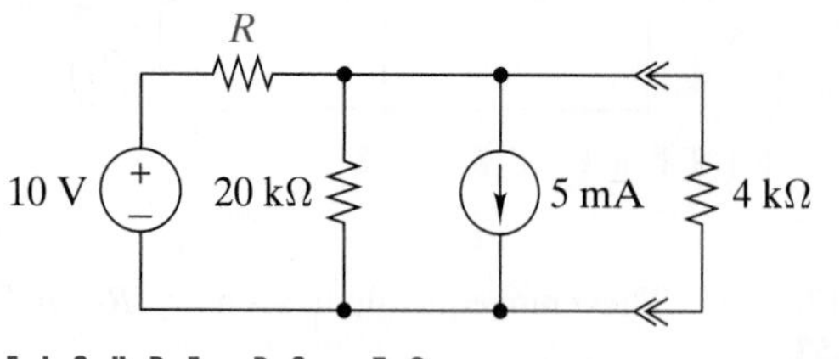

FIGURE P3–56

3–57 Find the value of R_L in Figure P3–57 such that $v_L = 80$ V.

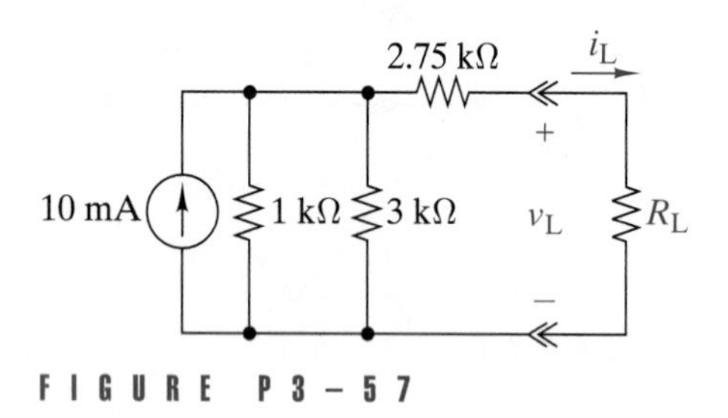

FIGURE P3–57

3–58 Find the value of R_L in Figure P3–57 such that maximum power is delivered to the load. Find the value of the maximum power.

3–59 A 15-V source with negligible internal resistance is connected in series with a resistor R_S. The purpose of R_S is to limit the signals delivered to any load to the ranges $i < 50$ mA and $p < 200$ mW. Find the range of R_S that will meet both constraints.

3–60 A practical source delivers 50 mA to a 300-Ω load. The source delivers 12 V to a 120-Ω load. Find the maximum power available from the source.

OBJECTIVE 3–5 INTERFACE CIRCUIT DESIGN AND EVALUATION (SECT. 3–6)

(a) **D** Given the signal transfer objectives at a source-load interface, adjust the circuit parameters or design one or more two-port interface circuits to achieve the specified objectives within stated constraints.

(b) **E** Given two or more circuits that perform the same interface function, rank-order the circuits using stated criteria.

See Examples 3–18, 3–19, 3–20, 3–21, 3–22, 3–23 and Exercise 3–20

3–61 **D** The output current of the voltage source in Figure P3–61 must be less then 100 mA. Design an interface circuit so that the load voltage is $v_2 = 4$ V and the source current is $i_1 < 100$ mA.

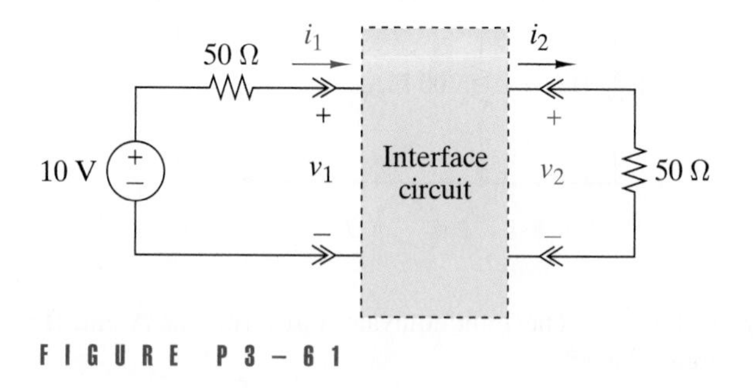

FIGURE P3–61

3–62 **D** Figure P3–62 shows an interface circuit connecting a 15-V source to a diode load. The $i-v$ characteristic of the diode is

$$i = 10^{-14}(e^{40v} - 1)$$

Design an interface circuit so that $v = 0.7$ V.

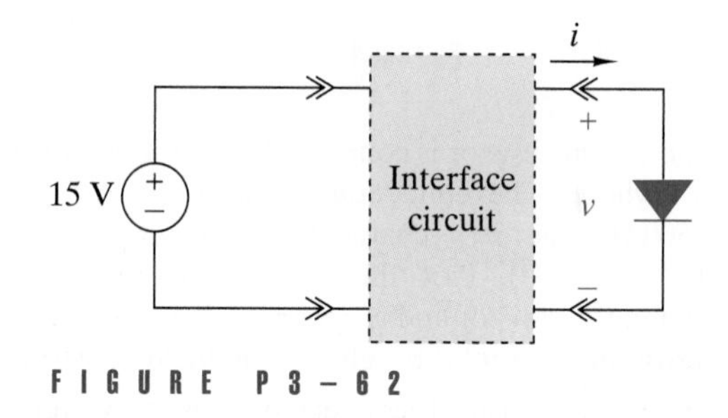

FIGURE P3–62

3–63 **D** Design the interface circuit in Figure P3–63 so that the voltage delivered to the load is $v = 10 \text{ V} \pm 10\%$. Use the following standard resistors in your design: 130 Ω, 200 Ω, 300 Ω, 430 Ω, 620 Ω, and 910 Ω. These resistors all have a tolerance of ±5%, which you must account for in your design.

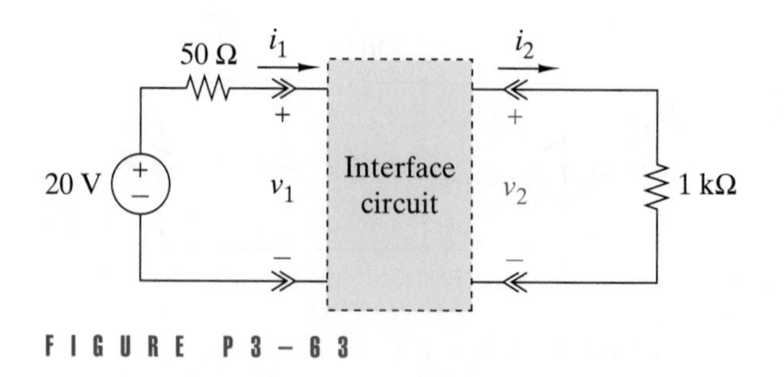

FIGURE P3–63

3–64 Design the interface circuit in Figure P3–63 so that 25 mW is delivered to the 1-kΩ load resistor.

3–65 **D** Use the resistor array in Figure P3–65 to design a voltage divider with $K = v_2/v_1 \geq 0.6$ and an input resistance close to 100 Ω.

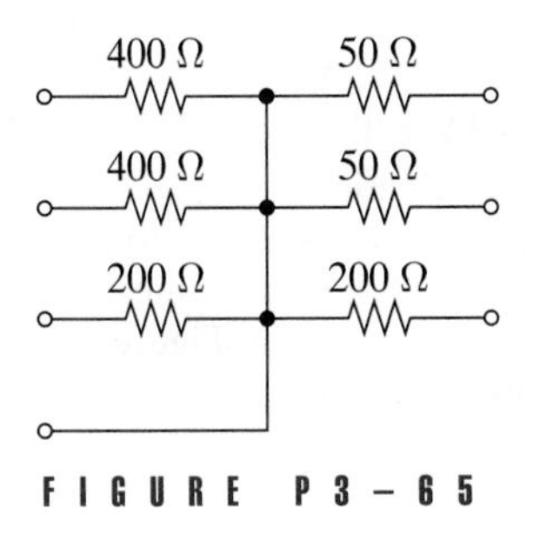

FIGURE P3–65

3–66 **D** Select R_L in Figure P3–66 so that the power delivered to the load is 400 mW.

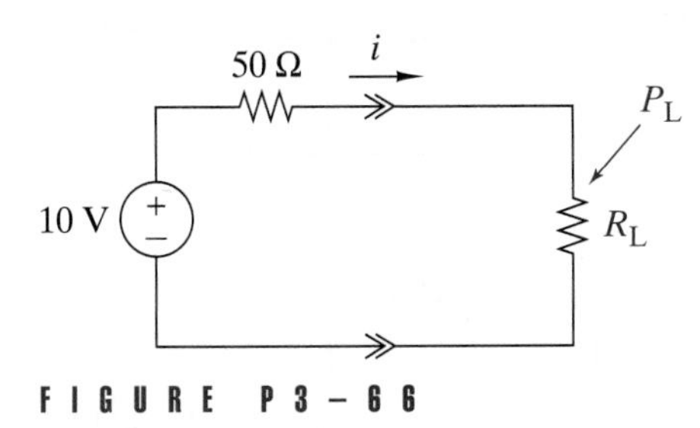

FIGURE P3–66

3–67 **D** Design the interface circuit in Figure P3–67 so that $R_{IN} = 100\ \Omega$ and the current delivered to the 50-Ω load is $i = 50$ mA.

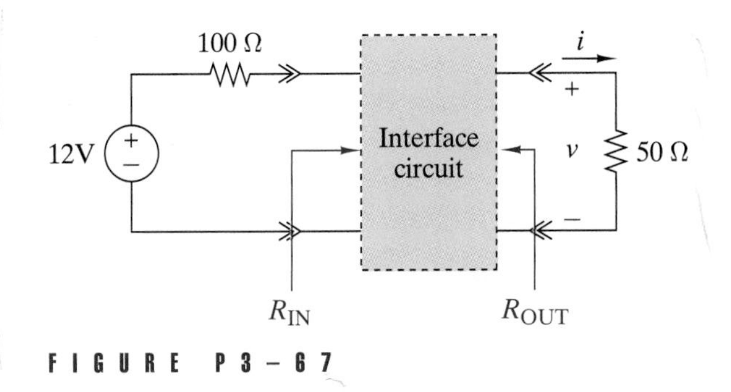

FIGURE P3–67

3–68 **D** Design the interface circuit in Figure P3–67 so that $R_{OUT} = 50\ \Omega$ and the voltage delivered to the 50-Ω load is $v = 2$ V.

3–69 **D** The circuit in Figure P3–69 has a source resistance of 75 Ω and a load resistance of 600 Ω. Design the interface circuit so that the input resistance is $R_{IN} = 75\ \Omega \pm 10\%$ and the output resistance is $R_{OUT} = 600\ \Omega \pm 10\%$.

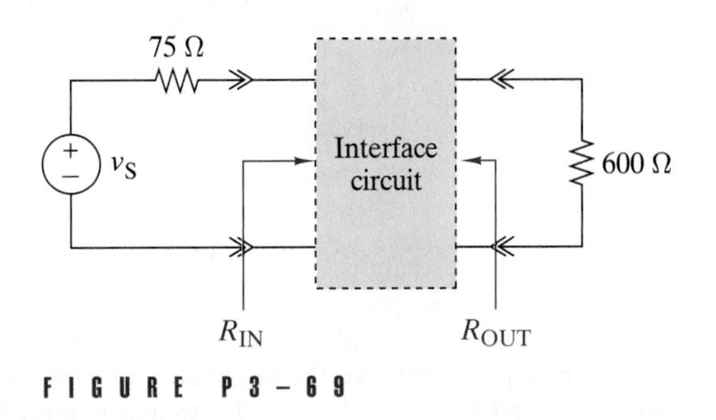

FIGURE P3–69

3–70 **E** It is claimed that both interface circuits in Figure 3–70 will deliver $v = 4$ V to the 75-Ω load. Verify this claim. Which interface circuit consumes the least power? Which has an output resistance that best matches the 75-Ω load?

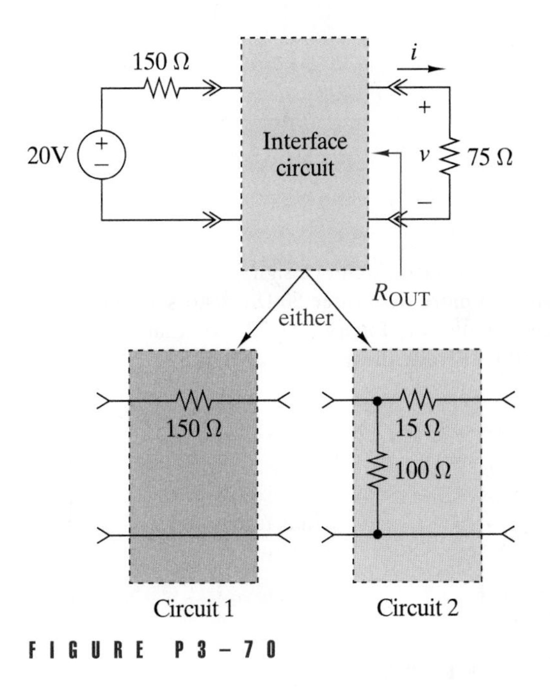

FIGURE P3–70

INTEGRATING PROBLEMS

3–71 **A** Composite Circuit Analysis
The i–v characteristic of network N in Figure P3–71 is $i = (5 \times 10^{-3}v - 0.5)$ A. Find the voltage at node A.

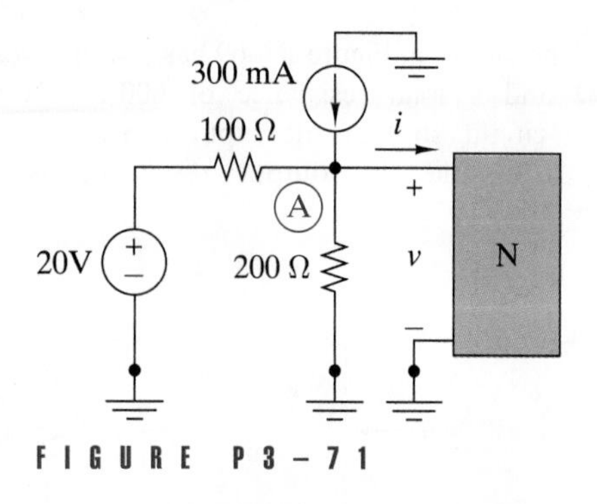

FIGURE P3–71

3–72 A Attenuation Analysis

In Figure P3–72 a two-port attenuator connects a 600-Ω source to a 600-Ω load. Find the power delivered to the load in terms of v_S. Remove the attenuator and find the power delivered to the load when the source is directly connected to the load. By what fraction does the attenuator reduce the power delivered to the 600-Ω load? Express the fraction in dB.

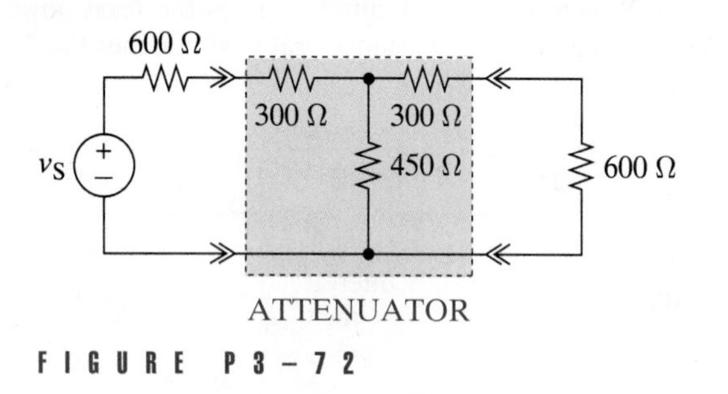

FIGURE P3–72

3–73 D Interface Circuit Design

Using no more than three 50-Ω resistors, design the interface circuit in Figure P3–73 so that $v \leq 4$ V and $i \leq 100$ mA regardless of the value of R_L.

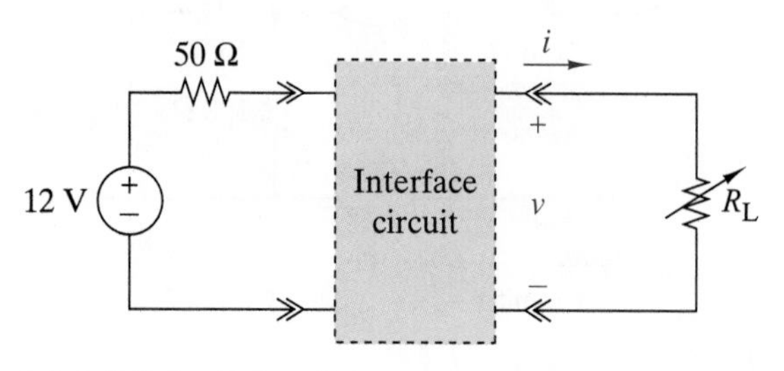

FIGURE P3–73

3–74 D Battery Design

A satellite requires a battery with an open-circuit voltage $v_{OC} = 36$ V and a Thévenin resistance $R_T \leq 10\ \Omega$. The battery is to be constructed using series and parallel combinations of one of two types of cells. The first type has $v_{OC} = 9$ V, $R_T = 4\ \Omega$, and a weight of 40 grams. The second type has $v_{OC} = 4$ V, $R_T = 0.5\ \Omega$, and a weight of 15 grams. Design a minimum-weight battery that meets the open-circuit voltage and Thévenin resistance requirements.

3–75 E Design Evaluation

A requirement exists for a circuit to deliver 0 to 5 V to a 100-Ω load from a 20-V source rated at 2.5 W. Two proposed circuits are shown in Figure P3–75. Which one would you choose and why?

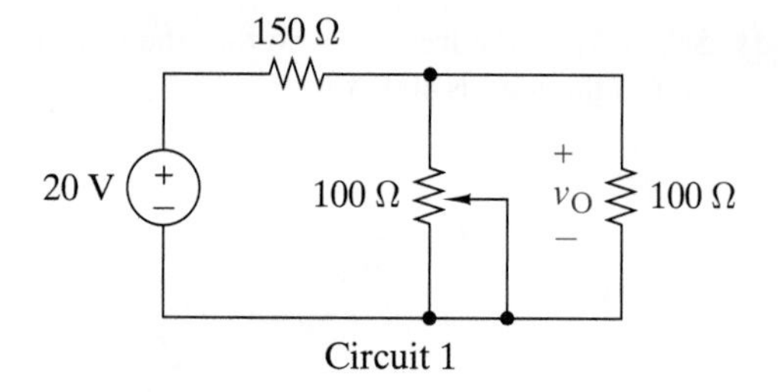

Circuit 1

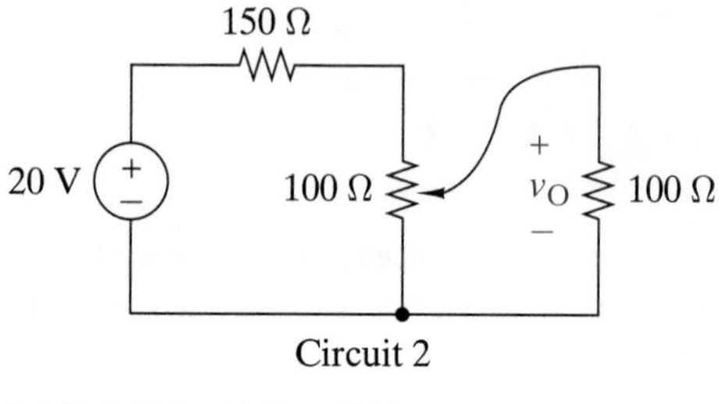

Circuit 2

FIGURE P3–75

CHAPTER 4

ACTIVE CIRCUITS

Then came the morning of Tuesday, August 2, 1927, when the concept of the negative feedback amplifier came to me in a flash while I was crossing the Hudson River on the Lacakawana Ferry, on my way to work.

Harold S. Black, 1927,
American Electrical Engineer

Some History Behind This Chapter

The negative feedback amplifier is one of the key inventions of all time. During the 1920's Harold S. Black (1898–1983) worked for several years without much success on the problem of improving the performance of vacuum tube amplifiers in telephone systems. The feedback amplifier solution came to him suddenly on his way to work. He documented his invention by sketching the key concepts on his morning copy of the *New York Times*. His invention paved the way for the development of whole new areas of technology such as feedback control systems and robotics.

Why This Chapter Is Important Today

This is an important chapter for all engineering disciplines. You will be introduced to modern electronic devices and how they can be modeled. The utility of these devices will be apparent when you design OP AMP circuits that provide signal conditioning in instrumentation systems. You will also be introduced to criteria used to evaluate alternative designs. That is, you will begin to function as an engineer making judgments about the best solution to a problem.

Chapter Sections

4–1 Linear Dependent Sources

4–2 Analysis of Circuits with Dependent Sources

4–3 The Transistor

4–4 The Operational Amplifier

4–5 OP AMP Circuit Analysis

4–6 OP AMP Circuit Design

4–7 OP AMP Circuit Applications

Chapter Learning Objectives

4-1 Linear Active Circuits (Sects. 4-1, 4-2)

Given a linear resistance circuit containing dependent sources, find selected output signals, input-output relationships, or input-output resistances.

4-2 Transistor Circuits (Sect. 4-3)

Given a linear resistance circuit with one transistor:

(a) Find the transistor operating mode and output signals.

(b) Select circuit parameters to obtain a specified operating mode or output signals.

4-3 OP AMP Circuit Analysis (Sects. 4-4, 4-5)

Given a linear resistance circuit containing OP AMPs, find selected output signals or input-output relationships.

4-4 OP AMP Circuit Design (Sect. 4-6)

Given an input-output relationship, design resistive OP AMP circuits that implement the relationship. Evaluate the alternative designs using stated criteria.

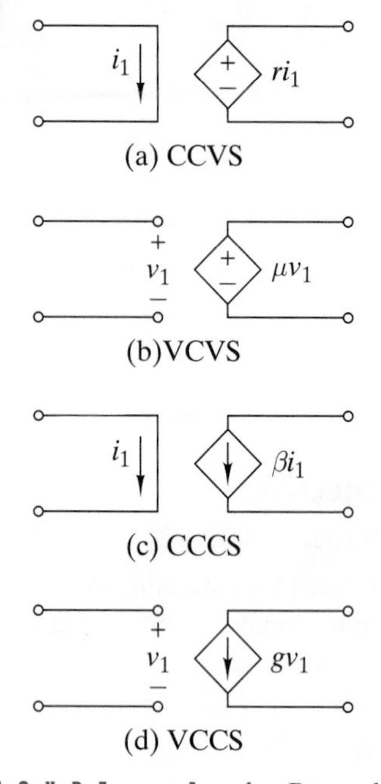

FIGURE 4–1 *Dependent source circuit symbols: (a) Current-controlled voltage source. (b) Voltage-controlled voltage source. (c) Current-controlled current source. (d) Voltage-controlled current source.*

4–1 LINEAR DEPENDENT SOURCES

This chapter treats the analysis and design of circuits containing active devices, such as transistors or operational amplifiers (OP AMPs). An **active device** is a component that requires an external power supply to operate correctly. An **active circuit** is one that contains one or more active devices. An important property of active circuits is that they are capable of providing signal amplification, one of the most important signal-processing functions in electrical engineering. Linear active circuits are governed by the proportionality property, so their input-output relationships are of the form $y = Kx$. The term **signal amplification** means the proportionality factor K is greater than 1 when the input x and output y have the same dimensions. Thus, active circuits can deliver more signal voltage, current, and power at their output than they receive from the input signal. The passive resistance circuits studied thus far cannot produce voltage, current, or power gains greater than unity.

Active devices operating in a linear mode are modeled using resistors and one or more of the dependent sources shown in Figure 4–1. A **dependent source** is a voltage or current source whose output is controlled by a voltage or current in a different part of the circuit. As a result, there are four possible types of dependent sources: a current-controlled voltage source (CCVS), a voltage-controlled voltage source (VCVS), a current-controlled current source (CCCS), and a voltage-controlled current source (VCCS). The properties of these dependent sources are very different from those of the independent sources described in Chapter 2. The output voltage (current) of an independent voltage (current) source is a specified value that does not depend on the circuit to which it is connected. To distinguish between the two types of sources, the dependent sources are represented by the diamond symbol in Figure 4–1, in contrast to the circle symbol used for independent sources.

Caution: This book uses the diamond symbol to indicate a dependent source and the circle to show an independent source. However, some books and circuit analysis programs use the circle symbol for both dependent and independent sources.

A **linear dependent source** is one whose output is proportional to the controlling voltage or current. The defining relationship for dependent sources in Figure 4–1 are all of the form $y = Kx$, where x is the controlling variable, y is the source output variable, and K is the proportionality factor. Each type of dependent source is characterized by a proportionality factor, either μ, β, r, or g. These parameters are often called simply the **gain** of the controlled source. Strictly speaking, the parameters μ and β are dimensionless quantities called the **voltage gain** and **current gain**, respectively. The parameter r has the dimensions of ohms and is called the **transresistance**, a contraction of transfer resistance. The parameter g is then called **transconductance** and has the dimensions of siemens.

Although dependent sources are elements used in circuit analysis, they are conceptually different from the other circuit elements we have studied. The linear resistor and ideal switch are models of actual devices called resistors and switches. However, you will not find dependent sources listed in electronic part catalogs. For this reason, dependent sources are more abstract, since they are not models of identifiable physical devices. Dependent sources are used in combination with other resistive elements to create models of active devices.

In Chapter 3 we found that a voltage source acts as a short circuit when it is turned off. Likewise, a current source behaves as an open circuit when it is turned off. The same results apply to dependent sources, with one important difference. Dependent sources cannot be turned on and off individually because they depend on excitation supplied by independent sources.

Some consequences of this dependency are illustrated in Figure 4–2. When the independent current source is turned on, KCL requires that $i_1 = i_S$. Through controlled source action, the current controlled voltage source is on and its output is

$$v_O = ri_1 = ri_S$$

When the independent current source is off ($i_S = 0$), it acts as an open circuit and KCL requires that $i_1 = 0$. The dependent source is now off and its output is

$$v_O = ri_1 = 0$$

When the independent current source is off, the dependent voltage source acts as a short circuit.

In other words, turning the independent source on and off turns the dependent source on and off as well. We must be careful when applying the superposition principle and Thévenin's theorem to active circuits, since the state of a dependent source depends on the excitation supplied by independent sources. To account for this possibility, we modify the superposition principle to state that the response due to all *independent* sources acting simultaneously is equal to the sum of the responses due to each *independent* source acting one at a time.

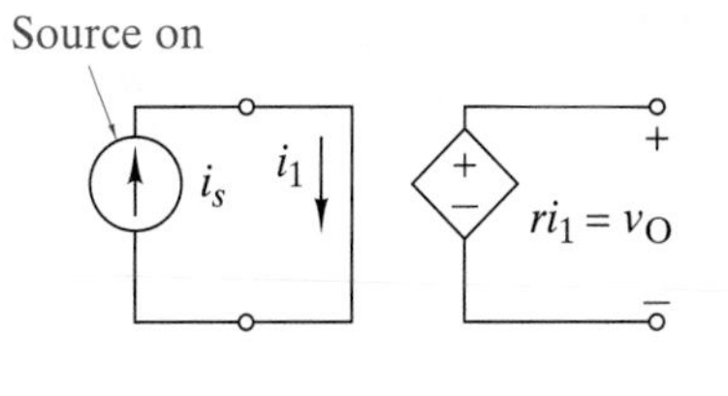

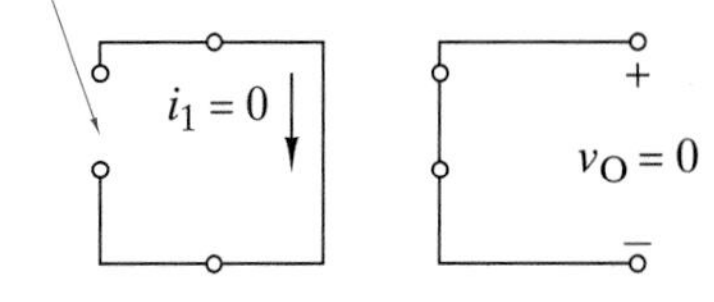

FIGURE 4–2 *Turning off the independent source affects the dependent source.*

EXAMPLE 4–1

Figure 4–3(a) shows the symbols used in Orcad Capture to represent the four dependent sources. These elements are found in the **Place Part/Analog** symbol library as part names starting with the letters:

- **E/Analog** for a voltage-controlled voltage source (VCVS)
- **F/Analog** for a current-controlled current source (CCCS)
- **G/Analog** for a voltage-controlled current source (VCCS)
- **H/Analog** for a current-controlled voltage source (CCVS)

The input ports are open circuits for the voltage-controlled sources and short circuits for the current-controlled sources. The output ports are voltage sources or current sources depending on the controlled variable. Note that Orcad uses circles rather than diamonds to represent dependent sources. All four dependent sources are characterized by a single parameter called "GAIN." This parameter is found in the "Property Editor" which appears when you double-click on the controlled source symbol. Once the desired gain is entered, click on **Apply** to update the value for subsequent simulations. The units of the gain depend on the units of the signals at the input and output ports.

Figure 4–3(b) shows the same symbols for Electronics WorkBench. These elements are available in the **Sources** group in the **Parts Bin** menu bar. The Electronics WorkBench screen symbols for dependent sources are diamonds.

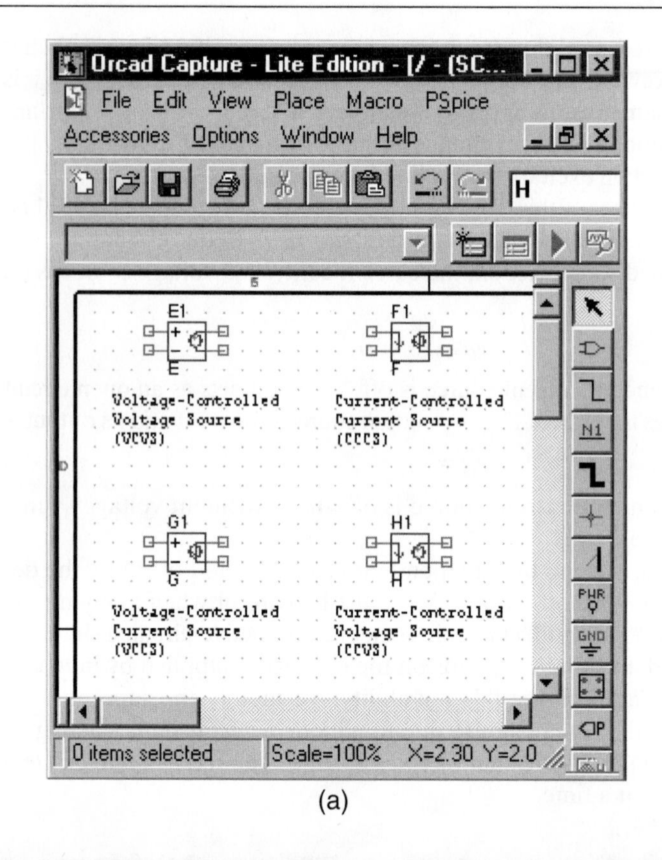

(a)

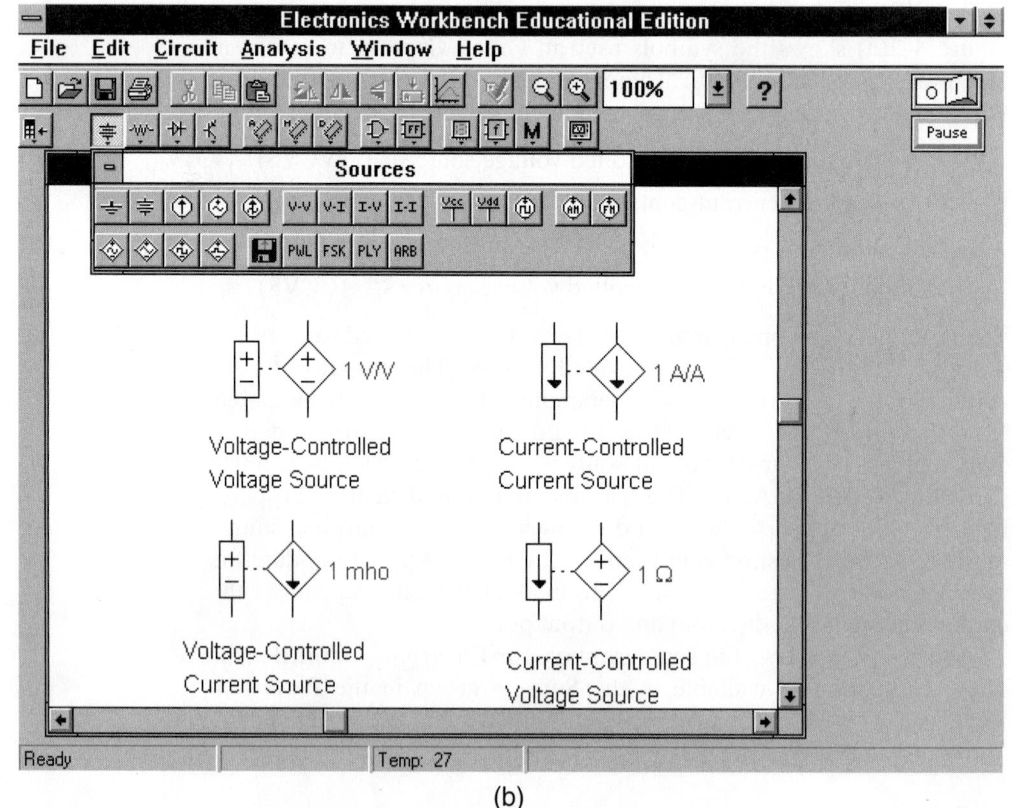

(b)

FIGURE 4–3 *Dependent source representations. (a) Orcad Capture. (b) Electronics Workbench.*

4–2 ANALYSIS OF CIRCUITS WITH DEPENDENT SOURCES

With certain modifications the analysis tools developed for passive circuits apply to active circuits as well. Circuit reduction applies to active circuits, but in so doing we must not eliminate the control variable for a dependent source. As noted previously, when applying the superposition principle or Thévenin's theorem, we must remember that dependent sources cannot be turned on and off independently since their states depend on excitation supplied by one or more independent sources. Applying a source transformation to a dependent source is sometimes helpful, but again we must not lose the identity of a controlling signal for a dependent source. Methods like node and mesh analysis can be adapted to include dependent sources as well.

However, the main difference is that the properties of active circuits can be significantly different from those of the passive circuits treated in Chapters 2 and 3. Our analysis examples are chosen to highlight these differences.

In our first example, the objective is to determine the current, voltage, and power delivered to the 500-V output load in Figure 4–4. The control current ix is found using current division in the input circuit:

$$i_x = \left(\frac{50}{50 + 25}\right) i_S = \frac{2}{3} i_S \tag{4–1}$$

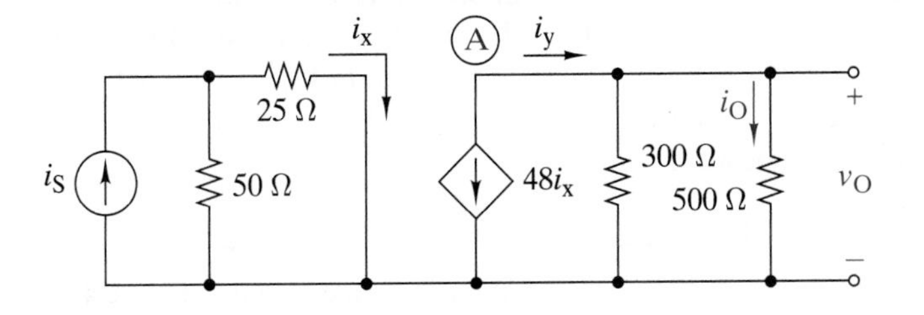

FIGURE 4–4 A circuit with a dependent source.

Similarly, the output current i_O is found using current division in the output circuit:

$$i_O = \left(\frac{300}{300 + 500}\right) i_y = \frac{3}{8} i_y \tag{4–2}$$

At node A, KCL requires that $i_y = -48i_x$. Combining this result with Eqs. (4–1) and (4–2) yields the output current:

$$i_O = \left(\frac{3}{8}\right)(-48)i_x = (-18)\left(\frac{2}{3}i_S\right)$$

$$= -12\, i_S \tag{4–3}$$

The output voltage v_O is found using Ohm's law:

$$v_O = i_O\, 500 = -6000\, i_S \tag{4–4}$$

The input-output relationships in Eqs. (4–3) and (4–4) are of the form $y = Kx$ with $K < 0$. The proportionality constants are negative because the reference direction for i_O in Figure 4–4 is the opposite of the orientation of the dependent source reference arrow. Active circuits often produce negative values of K, which means that the input and output signals have opposite algebraic signs. Circuits for which $K < 0$ are said to provide **signal inversion**. In the analysis and design of active circuits, it is important to keep track of signal inversions.

Using Eqs. (4–3) and (4–4), the power delivered to the 500-Ω load in Figure 4–4 is

$$p_O = v_O i_O = (-6000\, i_S)(-12\, i_S) = 72{,}000 i_S^2 \tag{4–5}$$

The independent source at the input delivers its power to the parallel combination of 50 Ω and 25 Ω. Hence, the input power supplied by the independent source is

$$p_S = (50 \,\|\, 25) i_S^2 = \left(\frac{50}{3}\right) i_S^2$$

Given the input power and output power, we find the power gain in the circuit:

$$\text{Power gain} = \frac{p_O}{p_S} = \frac{72{,}000 i_S^2}{(50/3) i_S^2} = 4320$$

A power gain greater than unity means that the circuit delivers more power at its output than it receives from the input source. At first glance this appears to be a violation of energy conservation, until we remember that dependent sources are models of active devices that require an external power supply to operate. Usually the external power supply is not shown in circuit diagrams. When using a dependent source to model an active circuit, we assume that the external supply and the active device itself can handle whatever power is required by the circuit. When designing the actual circuit, the engineer must make certain that the active device and its power supply operate within their power ratings.

Exercise 4–1

Find the output v_O in terms of the input v_S in the circuit in Figure 4–5.

FIGURE 4–5

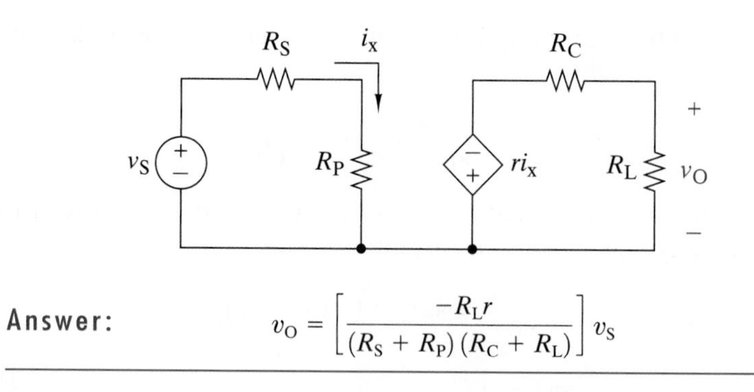

Answer:

$$v_O = \left[\frac{-R_L r}{(R_S + R_P)(R_C + R_L)}\right] v_S$$

NODE-VOLTAGE ANALYSIS WITH DEPENDENT SOURCES

Node analysis of active circuits is much the same as for passive circuits except that we must account for the additional constraints caused by the dependent sources. For example, the circuit in Figure 4–6 has five nodes. With node E as the reference, each independent voltage source has one terminal connected to ground. These connections force the conditions $v_A = v_{S1}$ and $v_B = v_{S2}$. Therefore, we only need to write node equations at nodes C and D because voltages at nodes A and B are already known.

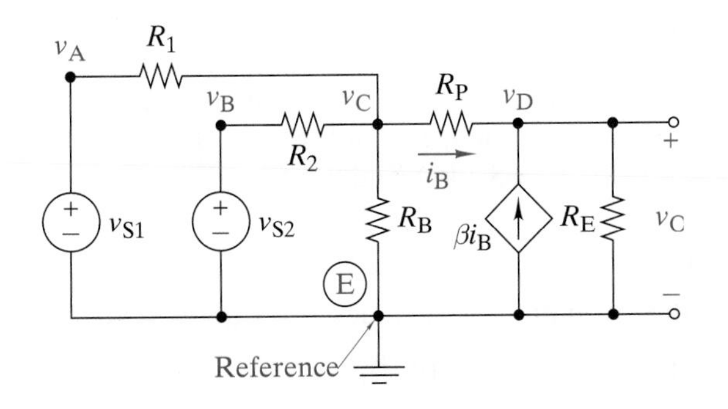

FIGURE 4–6 *Circuit used for node-voltage analysis with dependent sources.*

Node analysis involves expressing element currents in terms of the node voltages and applying KCL at each unknown node. The sum of the currents leaving node C is

$$G_1(v_C - v_{S1}) + G_2(v_C - v_{S2}) + G_B v_C + G_P(v_C - v_D) = 0$$

Similarly, the sum of currents leaving node D is

$$G_P(v_D - v_C) + G_E v_D - \beta i_B = 0$$

These two node equations can be rearranged into the form

$$\text{Node C: } (G_1 + G_2 + G_B + G_P)v_C - G_P v_D = G_1 v_{S1} + G_2 v_{S2}$$

$$\text{Node D: } \quad - G_P v_C + (G_P + G_E)v_D = \beta i_B \tag{4–6}$$

We could write these two symmetrical node equations by inspection if the dependent current source βi_B had been an independent source. Since it is not independent, we must express its controlling variable i_B in terms of the unknown node voltages. Using the fundamental property of node voltages and Ohm's law, we express the current i_B in terms of the node voltages as

$$i_B = G_P(v_C - v_D)$$

Substituting this expression for i_B into Eq. (4–6) and putting the results in standard form yields

$$\text{Node C: } \quad (G_1 + G_2 + G_B + G_P)v_C - G_P v_D = G_1 v_{S1} + G_2 v_{S2}$$

$$\text{Node D: } -(\beta + 1)G_P v_C + [(\beta + 1)G_P + G_E]v_D = 0 \tag{4–7}$$

The result in Eq. (4–7) involves two equations in the two unknown node voltages and includes the effect of the dependent source. Note that including the dependent source constraint destroys the coefficient symmetry in Eq. (4–6).

This example illustrates a general approach to writing node-voltage equations for circuits with dependent sources. We start out treating the dependent sources as if they are independent sources and write node equations for the resulting passive circuit using the inspection method developed in Chapter 3. This step produces a set of symmetrical node-voltage equations with the independent and dependent source terms on the right-hand side. Then we express the dependent source terms in terms of the unknown node voltages and move them to the left-hand side of the equations with the other terms involving the unknown node voltages. This step destroys the coefficient symmetry but leads to a set of node-voltage equations that describe the active circuit.

EXAMPLE 4–2

For the circuit in Figure 4–6, use the node-voltage equations in Eq. (4–7) to find the output voltage v_O when $R_1 = 1\ k\Omega$, $R_2 = 3\ k\Omega$, $R_B = 100\ k\Omega$, $R_P = 1.3\ k\Omega$, $R_E = 3.3\ k\Omega$, and $\beta = 50$.

SOLUTION:

Substituting the given numerical values into Eq. (4–7) yields

$$(2.11 \times 10^{-3})v_C - (7.69 \times 10^{-4})v_D = (10^{-3})v_{S1} + (3.33 \times 10^{-4})v_{S2}$$

$$-(3.92 \times 10^{-2})v_C + (3.95 \times 10^{-2})v_D = 0$$

We solve the second equation for $v_C = 1.008v_D$. When this equation is substituted into the first equation, we obtain

$$v_O = v_D = 0.736v_{S1} + 0.245v_{S2}$$

This circuit is a signal summer that does not involve a signal inversion. The fact that the output is a linear combination of the two inputs reminds us that the circuit is linear. ■

EXAMPLE 4–3

The circuit in Figure 4–7(a) is a model of an inverting OP AMP circuit.

FIGURE 4–7

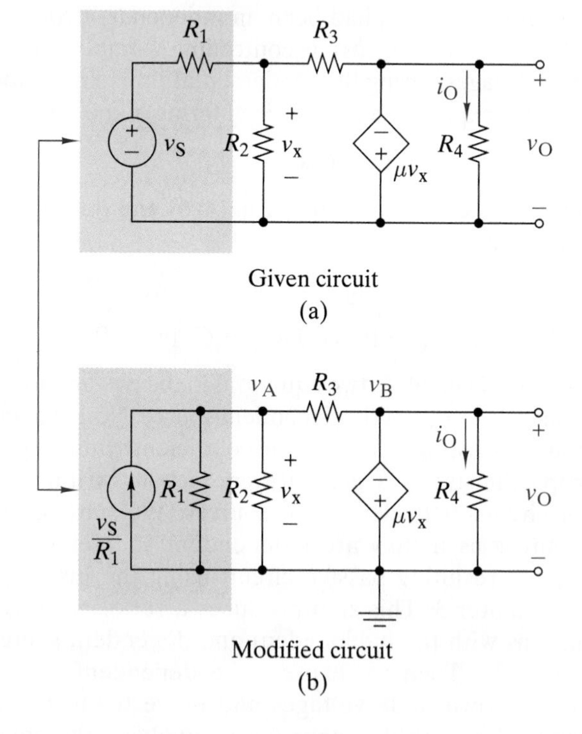

(a) Use node-voltage analysis to find the output v_O in terms of the input v_S.
(b) Evaluate the input-output relationship found in (a) as the gain μ becomes very large.

SOLUTION:

(a) Applying a source transformation to the independent source leads to the modified three-node circuit shown in Figure 4–7(b). With the indicated reference node the dependent voltage source constrains the voltage at node B. The control voltage is $v_x = v_A$, and the controlled source forces the node B voltage to be

$$v_B = -\mu v_x = -\mu v_A$$

Thus, node A is the only independent node in the circuit. We can write the node A equation by inspection as

$$(G_1 + G_2 + G_3)v_A - G_3 v_B = G_1 v_S$$

Substituting in the control source constraint yields the standard form for this equation:

$$[G_1 + G_2 + G_3(1 + \mu)]\, v_A = G_1 v_S$$

We end up with only one node equation even though at first glance the given circuit appears to need three node equations. The reason is that there are two voltage sources in the original circuit in Figure 4–7(a). Since the two sources share the reference node, the number of unknown node voltages is reduced from three to one. The general principle illustrated is that the number of independent KCL constraints in a circuit containing N nodes and N_V voltage sources (dependent or independent) is $N - 1 - N_V$.

The one-node equation can easily be solved for the output voltage $v_O = v_B$.

$$v_O = v_B = -\mu v_A = \left(\frac{-\mu G_1}{G_1 + G_2 + G_3(1 + \mu)}\right) v_S$$

The minus sign in the numerator means that the circuit provides signal inversion. The output voltage does not depend on the value of the load resistor R_4, since the load is connected across an ideal (though dependent) voltage source.

(b) For large gains μ we have $(1 + \mu)G_3 >> G_1 + G_2$ and the input-output relationship reduces to

$$v_O \approx \left[\frac{-\mu G_1}{(1 + \mu)\, G_3}\right] v_S \approx -\left[\frac{R_3}{R_1}\right] v_S$$

That is, when the active device gain is large, the voltage gain of the active circuit depends on the ratio of two resistances. We will see this situation again with OP AMP circuits. ■

Exercise 4–2

(a) Formulate node-voltage equations for the circuit in Figure 4–8.
(b) Solve the node-voltage equations for v_O and i_O in terms of i_S.

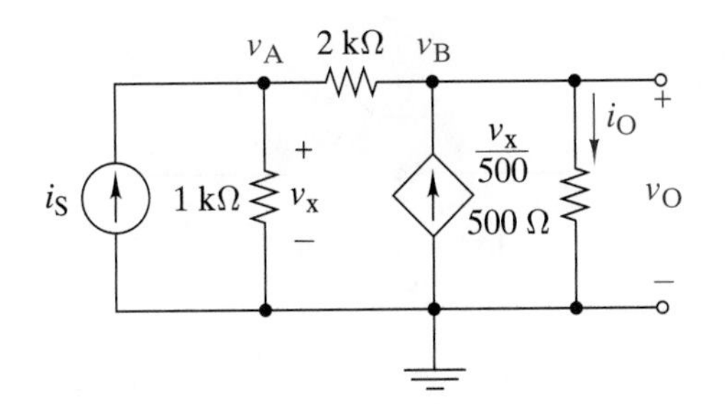

FIGURE 4–8

Answers:

(a)
$$(1.5 \times 10^{-3})\, v_A - (0.5 \times 10^{-3})\, v_B = i_S$$
$$-(2.5 \times 10^{-3})\, v_A + (2.5 \times 10^{-3})\, v_B = 0$$

(b) $v_O = 1000 i_S$; $i_O = 2 i_S$.

Exercise 4–3

Use node-voltage analysis to find v_O in Figure 4–9.

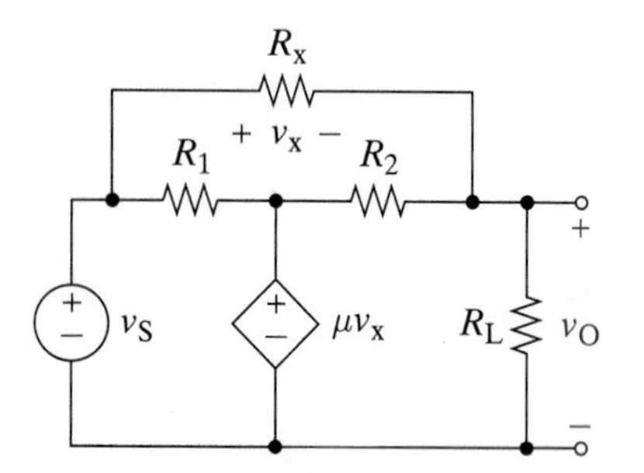

FIGURE 4–9

Answer:
$$v_O = \frac{G_x + \mu G_2}{G_x + G_L + (\mu + 1)G_2}\, v_S$$

MESH-CURRENT ANALYSIS WITH DEPENDENT SOURCES

Mesh-current analysis of active circuits follows the same pattern noted for node-voltage analysis. We initially treat the dependent sources as independent sources and write the mesh equations of the resulting passive circuit using the inspection method from Chapter 3. We then account for the dependent sources by expressing their constraints in terms of unknown mesh currents. The following example illustrates the method.

EXAMPLE 4–4

(a) Formulate mesh-current equations for the circuit in Figure 4–10.
(b) Use the mesh equations to find v_O and R_{IN} when $R_1 = 50\ \Omega$, $R_2 = 1\ \text{k}\Omega$, $R_3 = 100\ \Omega$, $R_4 = 5\ \text{k}\Omega$, and $g = 100$ mS.

FIGURE 4–10

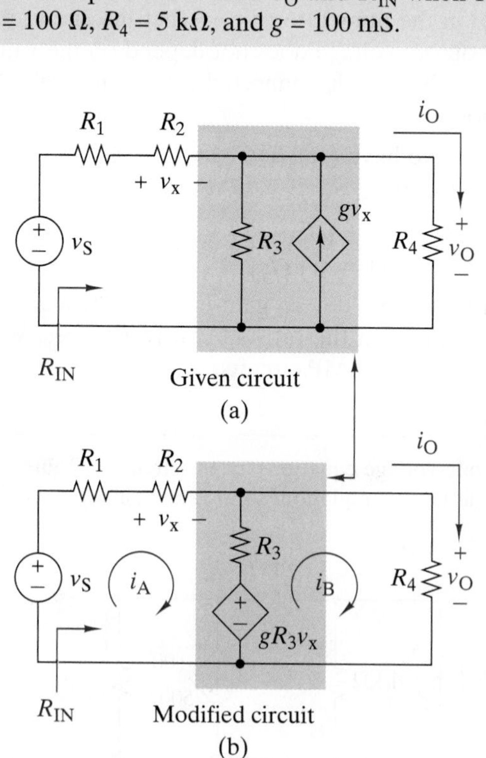

SOLUTION:

(a) Applying source transformation to the parallel combination of R_3 and gv_x in Figure 4–10(a) produces the dependent voltage source $R_3gv_x = \mu v_x$ in Figure 4–10(b). In the modified circuit we have identified two mesh currents. Initially treating the dependent source $(gR_3)v_x$ as an independent source leads to two symmetrical mesh equations.

$$\text{Mesh A: } (R_1 + R_2 + R_3)i_A - R_3i_B = v_S - (gR_3)v_x$$

$$\text{Mesh B: } \quad -R_3i_A + (R_3 + R_4)i_B = (gR_3)v_x$$

The control voltage v_x can be written in terms of mesh currents.

$$v_x = R_2i_A$$

Substituting this equation for v_x into the mesh equations and putting the equations in standard form yields

$$(R_1 + R_2 + R_3 + gR_2R_3)i_A - R_3i_B = v_S$$

$$-(R_3 + gR_2R_3)i_A + (R_3 + R_4)i_B = 0$$

The resulting mesh equations are not symmetrical because of the controlled source.

(b) Substituting the numerical values into the mesh equations gives

$$(1.115 \times 10^4)i_A - (10^2)i_B = v_S$$

$$-(1.01 \times 10^4)i_A + (5.1 \times 10^3)i_B = 0$$

Solving for the two mesh currents yields

$$i_A = \frac{\Delta_A}{\Delta} = \frac{\begin{vmatrix} v_S & -10^2 \\ 0 & 5.1 \times 10^3 \end{vmatrix}}{\begin{vmatrix} 1.115 \times 10^4 & -10^2 \\ -1.01 \times 10^4 & 5.1 \times 10^3 \end{vmatrix}} = \frac{5.1 \times 10^3 v_S}{5.5855 \times 10^7}$$

$$= 0.9131 \times 10^{-4} v_S$$

$$i_B = \frac{\Delta_B}{\Delta} = \frac{\begin{vmatrix} 1.115 \times 10^4 & v_S \\ -1.01 \times 10^4 & 0 \end{vmatrix}}{5.5885 \times 10^7} = 1.808 \times 10^{-4} v_S$$

The output voltage and input resistance are found using Ohm's law.

$$v_O = R_4i_B = 0.904v_S$$

$$R_{IN} = \frac{v_S}{i_A} = 10.95 \text{ k}\Omega$$

■

EXAMPLE 4–5

The circuit in Figure 4–11 is a model of a bipolar junction transistor operating in the active mode. Use mesh analysis to find the transistor base current i_B.

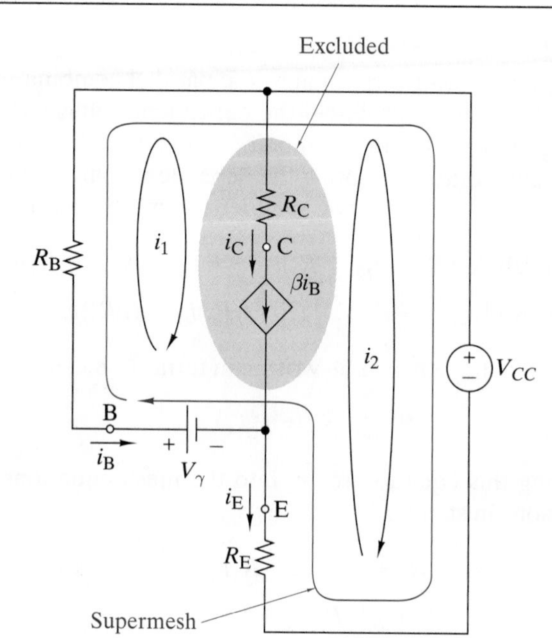

FIGURE 4–11 *Junction transistor circuit model.*

SOLUTION:

The two mesh currents in Figure 4–11 are labeled i_1 and i_2 to avoid possible confusion with the transistor base current i_B. As drawn, the circuit requires a supermesh since the dependent current source βi_B is included in both meshes and is not connected in parallel with a resistance. A supermesh is created by combining meshes 1 and 2 after excluding the series subcircuit consisting of βi_B and R_C. Beginning at the bottom of the circuit, we write a KVL mesh equation around the supermesh using unknowns i_1 and i_2:

$$i_2 R_E - V_\gamma + i_1 R_B + V_{CC} = 0$$

This KVL equation provides one equation in the two unknown mesh currents. Since the two mesh currents have opposite directions through the dependent current source βi_B, the currents i_1, i_2, and βi_B are related by KCL as

$$i_1 - i_2 = \beta i_B$$

This constraint supplies the additional relationship needed to obtain two equations in the two unknown mesh-current variables. Since $i_B = -i_1$, the preceding KCL constraint means that $i_2 = (\beta + 1)i_1$. Substituting $i_2 = (\beta + 1)i_1$ into the supermesh KVL equation and solving for i_B yields

$$i_B = -i_1 = \frac{V_{CC} - V_\gamma}{R_B + (\beta + 1)R_E}$$

■

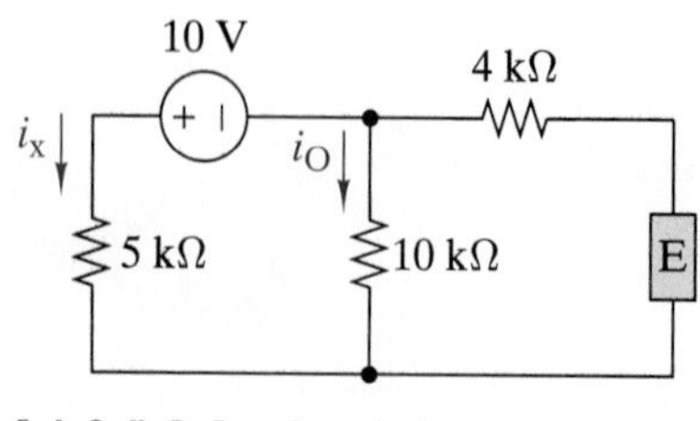

FIGURE 4–12

Exercise 4–4

Use mesh analysis to find the current i_O in Figure 4–12 when the element E is a dependent current source $2i_x$ with the reference arrow directed down.

Answer: –0.857 mA

Exercise 4–5

Use mesh analysis to find the current i_O in Figure 4–12 when the element E is a dependent voltage source $2000i_x$ with the plus reference at the top.

Answer: –0.222 mA

EXAMPLE 4–6

The circuit in Figure 4–13 is a small signal model of a field effect transistor (FET) amplifier with two inputs v_{S1} and v_{S2}. Use Orcad Capture to find the input-output relationship of the circuit.

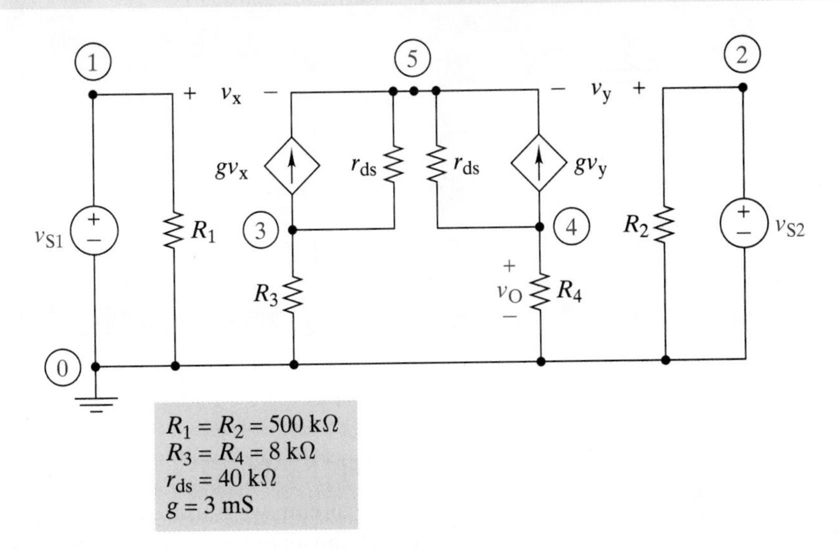

FIGURE 4–13 *Field effect transistor amplifier.*

SOLUTION:

Since the circuit is linear, the input-output relationship is of the form

$$v_O = K_1 v_{S1} + K_2 v_{S2}$$

The gain K_1 can be found by setting $v_{S1} = 1$ and $v_{S2} = 0$ and solving for v_O. The gain K_2 is then found by setting $v_{S1} = 0$ and $v_{S2} = 1$ and again solving for v_O. We now turn to Orcad to perform these operations.

Figure 4–14 shows an Orcad schematic circuit diagram for the case $v_{S1} =$ 1 V and $v_{S2} = 0$. In Figure 4–14 the node voltage V(N00521) corresponds to the output v_O in Figure 4–13. Figure 4–14 also shows the resulting netlist and a portion of the output file that includes the dc solution for the five node voltages. In particular, the output file reports that V(N00521) = 10, which means that $K_1 = \text{V(N00521)}/v_{S1} = 10$.

To find K_2 we change the source attributes to $v_{S1} = 0$ and $v_{S2} = 1$ and again simulate the circuit using Orcad Capture. For this case the output file reports V(N00521) = –10.000, which means that $K_2 = \text{V(N00521)}/v_{S2} = -10$. Taken together, the two Orcad simulations mean that the input-output relationship for the circuit is

$$v_O = 10(v_{S1} - v_{S2})$$

The circuit is a differential amplifier of a type often used as the input stage of an OP AMP. ■

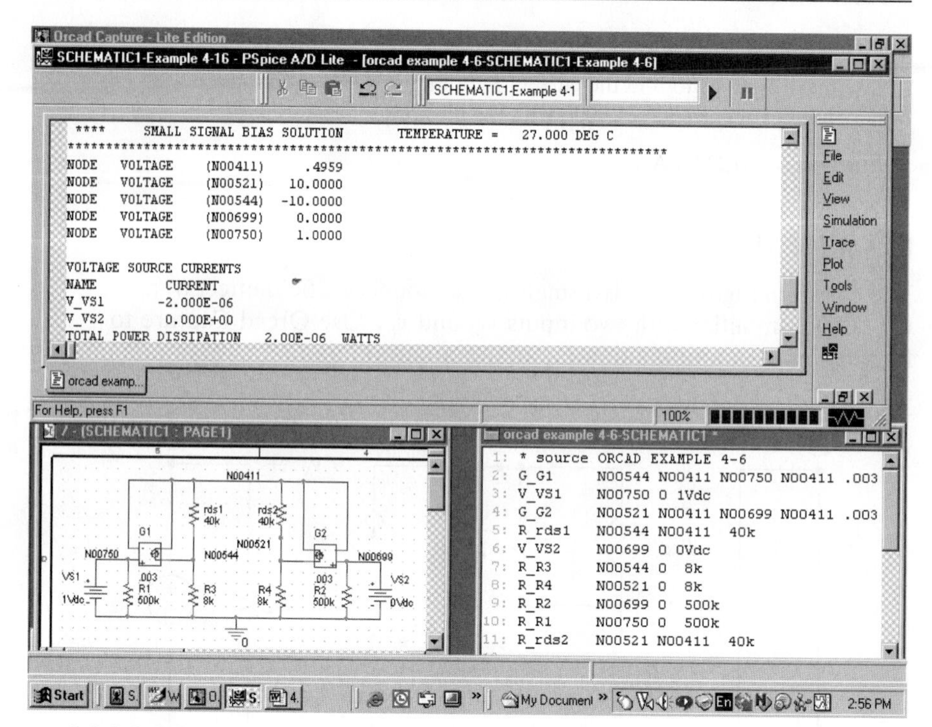

FIGURE 4–14 *Orcad Capture schematic diagram of the circuit in Figure 4–13.*

THÉVENIN EQUIVALENT CIRCUITS WITH DEPENDENT SOURCES

To find the Thévenin equivalent of an active circuit, we must leave the independent sources on or else supply excitation from an external test source. This means that the Thévenin resistance cannot be found by the lookback method because that method requires that all independent sources be turned off. Turning off the independent sources deactivates the dependent sources as well and can result in a profound change in the input and output characteristics of an active circuit. Thus, there are two ways of finding active circuit Thévenin equivalents. We can either find the open-circuit voltage and short-circuit current at the interface or directly solve for the interface i–v relationship.

EXAMPLE 4–7

Find the input resistance of the circuit in Figure 4–15.

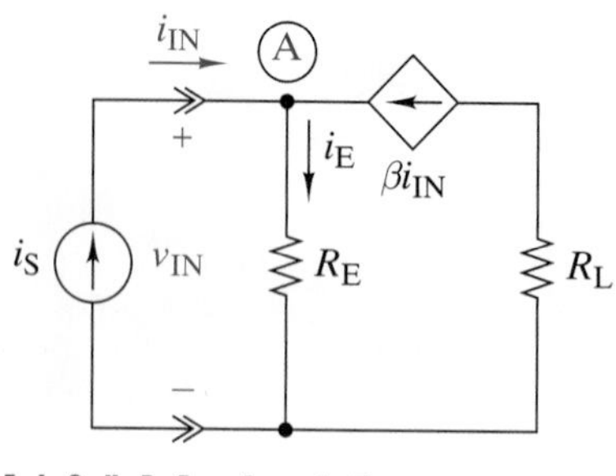

FIGURE 4–15

SOLUTION:

With the independent source turned off ($i_{IN} = i_S = 0$), the resistance seen at the input port is R_E since the dependent current source βi_{IN} is inactive and acts like an open circuit. Applying KCL at node A with the input source turned on yields

$$i_E = i_{IN} + \beta i_{IN} = (\beta + 1)i_{IN}$$

By Ohm's law, the input voltage is

$$v_{IN} = i_E R_E = (\beta + 1)i_{IN} R_E$$

Hence, the active input resistance is

$$R_{IN} = \frac{v_{IN}}{i_{IN}} = (\beta + 1)R_E$$

The circuit in Figure 4–15 is a model of a transistor circuit in which the gain parameter β typically lies between 50 and 100. The input resistance with external excitation is $(\beta + 1)R_E$, which is significantly different from the value of R_E without external excitation. ■

EXAMPLE 4–8

Find the Thévenin equivalent at the output interface of the circuit in Figure 4–16.

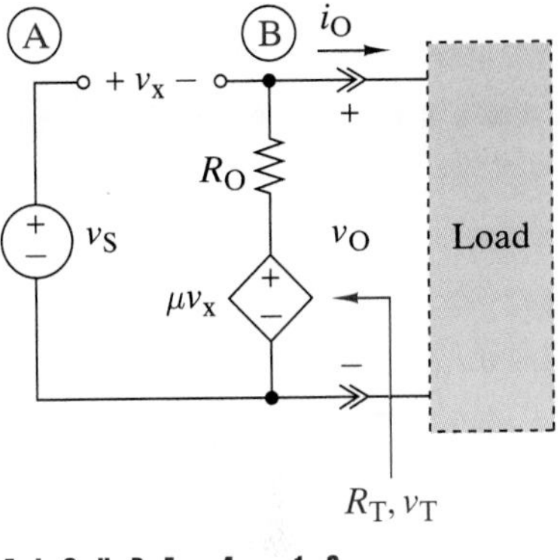

FIGURE 4–16

SOLUTION:

In this circuit the controlled voltage v_x appears across an open circuit between nodes A and B. By the fundamental property of node voltages, $v_x = v_S - v_O$. With the load disconnected and the input source turned off ($v_x = 0$), the dependent voltage source μv_x acts like a short circuit, and the Thévenin resistance looking back into the output port is R_O. With the load connected and the input source turned on, the sum of currents leaving node B is

$$\frac{v_O - \mu v_x}{R_O} + i_O = 0$$

Using the relationship $v_x = v_S - v_O$ to eliminate v_x and then solving for v_O produces the i–v characteristic at the output interface as

$$v_O = \frac{\mu v_S}{\mu + 1} - i_O\left[\frac{R_O}{\mu + 1}\right]$$

The i–v relationship of a Thévenin circuit is $v = v_T - iR_T$. By direct comparison, we find the Thévenin parameters of the active circuit to be

$$v_T = \frac{\mu v_S}{\mu + 1} \quad \text{and} \quad R_T = \frac{R_O}{\mu + 1}$$

The circuit in Figure 4–16 is a model of an OP AMP circuit called a voltage follower. The resistance R_O for a general-purpose OP AMP is around 100 Ω, while the gain μ is about 10^5. Thus, the active Thévenin resistance of the voltage follower is not 100 Ω, as the lookback method suggests, but is only a milliohm.

Exercise 4–6

Find the input resistance and output Thévenin equivalent circuit of the circuit in Figure 4–17.

Answers:

$$R_{IN} = (1 + \mu)R_F$$

$$v_T = \frac{\mu}{\mu + 1} v_S$$

$$R_T = R_O$$

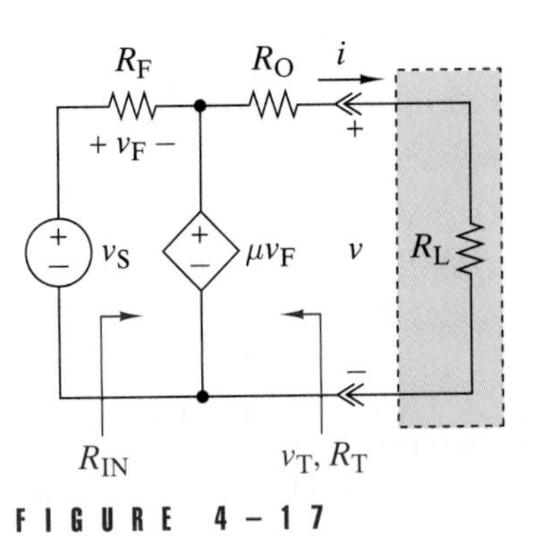

FIGURE 4–17

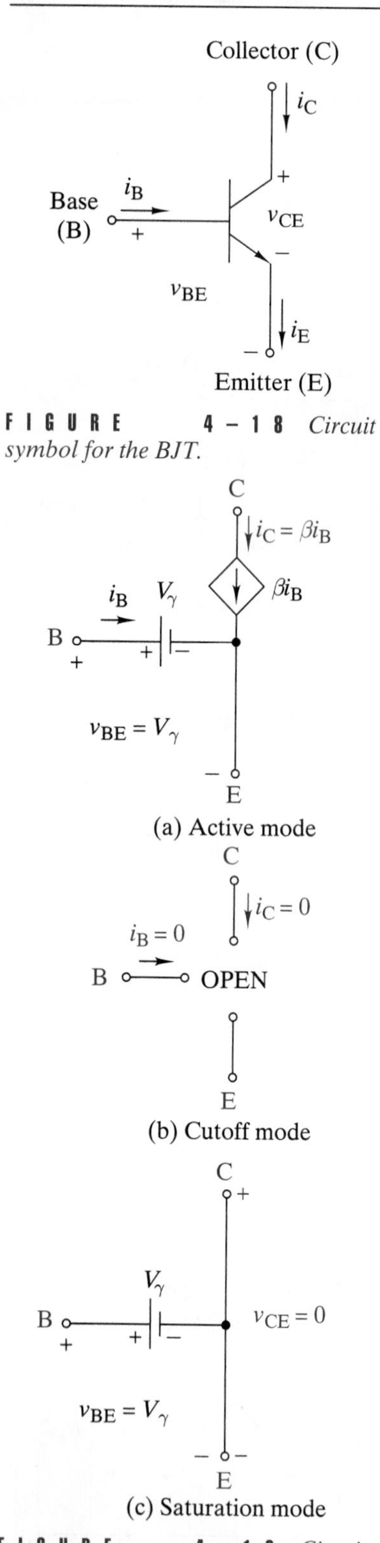

FIGURE 4–18 *Circuit symbol for the BJT.*

FIGURE 4–19 *Circuit models for BJT operating modes. (a) Active mode (b) Cutoff mode (c) Saturation mode.*

4–3 THE TRANSISTOR

We have defined four linear dependent sources and shown how to analyze circuits containing these active elements. In this and the next section we show how dependent sources are used to model semiconductor devices like transistors and OP AMPs. The transistor model used here describes the voltages and currents at its external terminals. The model does not describe the transistor's physical structure or internal charge flow. Those subjects are left to subsequent courses in semiconductor materials and devices.

The two basic transistor types are the *bipolar junction transistor* (BJT) and the *field effect transistor* (FET). Both types have several possible operating modes, each with a different set of *i–v* characteristics. This is something new in our study. Up to this point the characteristics of circuit elements have been fixed. With the transistor we encounter a device whose *i–v* characteristics can change. We concentrate on the BJT because its *i–v* characteristics are much easier to understand than the FET. Because it is easier to understand, the simpler BJT best serves as a prelude to our study of the OP AMP—an important semiconductor device that also has several possible operating modes.

The circuit symbol of the BJT is shown in Figure 4–18. The device has three terminals called the **emitter (E)**, the **base (B)**, and the **collector (C)**. The voltages v_{BE} and v_{CE} *are called the base-emitter and collector-emitter* voltage, respectively. The three currents i_E, i_B, and i_C are called the emitter, base, and collector currents.

Applying KCL to the BJT as a whole yields

$$i_E = i_B + i_C$$

which means that only two of the three currents can be independently specified. We normally work with i_B and i_C, and use KCL to find i_E when it is needed.

The BJT's large-signal model is defined in terms of input signals i_B and v_{BE}, and output signals i_C and v_{CE}. For the BJT shown in Figure 4–18, the model applies to a region in which these signals are never negative. Within this region there are three possible operating modes. The **active mode** is the dominant feature of a BJT. In this mode the collector current i_C is controlled by the base current i_B and v_{BE} is constant.

$$\text{Active Mode:} \quad i_C = \beta i_B \quad \text{and} \quad v_{BE} = V_\gamma \tag{4–8}$$

The proportionality factor β is called the *forward current gain* and typically ranges from about 50 to several hundred. The constant V_γ is called the *threshold voltage*, which is normally less than a volt. Figure 4–19(a) shows the circuit elements that model the active mode *i–v* characteristics as defined in Eq. (4–8). In the active mode i_B and v_{CE} are determined by the interaction of these *i–v* characteristics with the rest of the circuit.

Two additional operating modes exist at the boundary of the BJT's operating region. When $i_B = 0$ and $i_C = 0$, the transistor is in the **cutoff mode** and the device acts like an open circuit between the collector and emitter. When $v_{CE} = 0$ and $v_{BE} = V_\gamma$ the transistor is in the **saturation mode**, and the device acts like a short circuit between the collector and emitter. These two modes are summarized by writing

$$\begin{aligned} \text{Cutoff Mode:} &\quad i_B = 0 \quad \text{and} \quad i_C = 0 \\ \text{Saturation Mode:} &\quad v_{CE} = 0 \quad \text{and} \quad v_{BE} = V_\gamma \end{aligned} \tag{4–9}$$

Figures 4–19(b) and 4–19(c) show the circuit elements that model the i–v characteristics defined in Eq. (4–9).

The circuit in Figure 4–19(b) points out that in the cutoff mode, v_{CE} must equal the open-circuit voltage available from the external circuit. The circuit in Figure 4–19(c) points out that in the saturation mode, i_C must equal the short-circuit current available from the external circuit. The net result is that the BJT's output variables must fall within the following bounds:

$$\begin{array}{ccccc} \text{Cutoff Bounds} & & & & \text{Saturation Bounds} \\ 0 & \leq & i_C & \leq & i_{SC} \\ v_{OC} & \geq & v_{CE} & \geq & 0 \end{array} \qquad (4\text{–}10)$$

where v_{OC} and i_{SC} are the open-circuit voltage and short-circuit current available between the collector and emitter terminals. In the cutoff mode the transistor outputs i_C and v_{CE} are equal to their respective cutoff bounds. In saturation mode the outputs equal their saturation bounds. In the active mode the outputs fall between the cutoff and saturation bounds.

With this background we are prepared to analyze the transistor circuit in Figure 4–20.[1] Our analysis objective is to find the outputs i_C and v_{CE}. To do this we must know the transistor's operating mode. To find the operating mode we make use of two facts:

1. The lower bounds in Eq. (4–10) mean that i_C and v_{CE} cannot be negative.
2. The upper bounds in Eq. (4–10) depend on the rest of the circuit.

For the circuit in Figure 4–20 these upper bounds are $v_{OC} = V_{CC}$ and $i_{SC} = V_{CC}/R_C$.

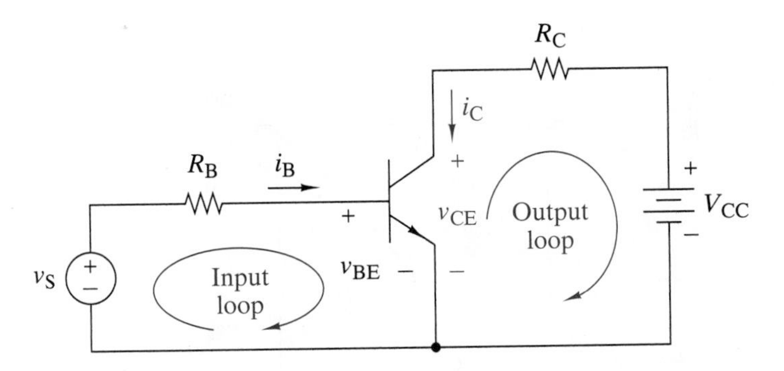

FIGURE 4–20 *BJT common-emitter circuit.*

Our analysis strategy *assumes* the device is in the active mode and uses the active mode device equations to find i_C. According to Eq. (4–8) the active mode element equations are $v_{BE} = V_\gamma$ and $i_C = \beta i_B$. Using these element constraints and applying KVL around the input loop in Figure 4–20 yields the collector current as

$$i_C = \beta\, i_B = \beta\left(\frac{v_S - V_\gamma}{R_B}\right) \qquad (4\text{–}11)$$

1 This circuit is called the common-emitter configuration because the emitter terminal is common to the input and output loops.

This equation indicates that if $v_S > V_\gamma$ then, $i_C > 0$. However, if $v_S < V_\gamma$, then $i_C < 0$, which would violate its cutoff bound. Thus, if the input voltage v_S is greater than the threshold voltage V_γ, then the BJT can be in the active mode. But if $v_S < V_\gamma$, the BJT is in the cutoff mode and the outputs equal their cutoff bounds in Eq. (4–10), namely $i_C = 0$ and $v_{CE} = v_{OC} = V_{CC}$.

When $v_S > V_\gamma$, Eq. (4–11) predicts a positive collector current that increases linearly with v_S. To find the collector-emitter voltage we apply KVL around the output loop in Figure 4–20 to obtain

$$v_{CE} = V_{CC} - i_C R_C \tag{4–12}$$

This equation predicts that $v_{CE} > 0$ as long as $i_C < V_{CC}/R_C$. But V_{CC}/R_C is the short-circuit current available from the external circuit. Thus, as long as $v_S > V_\gamma$ and $i_C < i_{SC}$, the BJT is in the active mode and Eqs. (4–11) and (4–12) correctly predict the outputs i_C and v_{CE}. However, if Eq. (4–11) predicts that $i_C > i_{SC}$, then Eq. (4–12) says that $v_{CE} < 0$. Both of these results violate the saturation bounds in Eq. (4–10). When this happens, the BJT is actually in the saturation mode and the outputs equal their saturation bounds in Eq. (4–10), namely $i_C = i_{SC} = V_{CC}/R_C$ and $v_{CE} = 0$.

Figure 4–21 summarizes this discussion using graphs of the outputs v_{CE} and i_C versus the input voltage v_S. When $v_S < V_\gamma$, the BJT is in the cutoff mode and the outputs are $i_C = 0$ and $v_{CE} = v_{OC} = V_{CC}$. When $v_S > V_\gamma$, the BJT enters the active mode and the outputs i_C and v_{CE} are governed by Eqs. (4–11) and (4–12). Under these equations, i_C increases linearly as v_S increases, with the result that v_{CE} decreases linearly. The collector current continues to increase as v_S increases until it reaches its saturation bound at $i_C = i_{SC}$. At that point the transistor switches into the saturation mode and thereafter the outputs remain constant at $i_C = i_{SC} = V_{CC}/R_C$ and $v_{CE} = 0$.

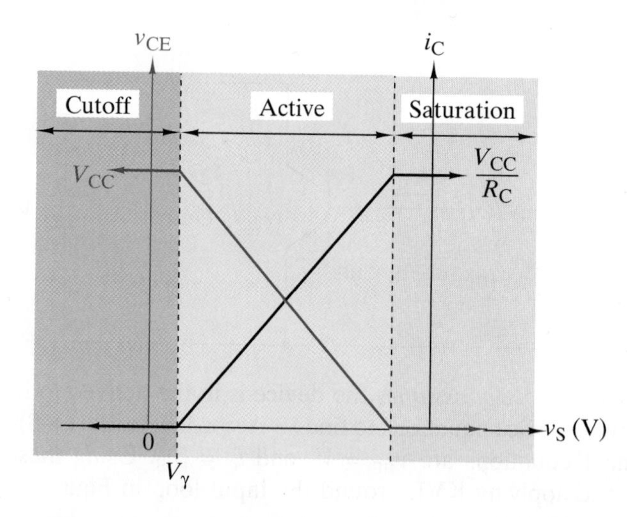

FIGURE 4–21 *Output responses of the BJT circuit in Figure 4–20.*

Before working some examples it is worth noting that the graph of v_{CE} versus v_S is called the *transfer characteristic* of the transistor circuit in Figure 4–20. In digital applications the input voltage drives the transistor between the cutoff and saturation modes passing through the active mode as quickly as possible. In analog circuit applications the transistor remains in the active mode where the slope of the transfer characteristic provides voltage amplification. In the next section we find that the OP AMP has similar transfer characteristics.

EXAMPLE 4–9

Suppose that the circuit parameters in Figure 4–20 are $\beta = 100$, $V_\gamma = 0.7$ V, $R_B = 100$ kΩ, $R_C = 1$ kΩ, and $V_{CC} = 5$ V. Find i_C and v_{CE} when $v_S = 2$ V. Repeat when $v_S = 6$ V.

SOLUTION:

Since $v_S = 2$ is greater than $V_\gamma = 0.7$, the transistor is *not* in the cutoff mode. We assume that it is in the active mode and use Eq. (4–11) to calculate i_C.

$$i_C = \beta\left(\frac{v_S - V_\gamma}{R_B}\right) = 100\left(\frac{2 - 0.7}{100 \times 10^3}\right) = 1.3 \text{ mA}$$

The available short-circuit current is $i_{SC} = V_{CC}/R_C = 5$ mA. Since the calculated i_C is less than i_{SC}, the transistor is in fact in the active mode and we use Eq. (4–12) to find v_{CE}.

$$v_{CE} = V_{CC} - i_C R_C = 5 - 1.3 \times 10^{-3} \times 1000 = 3.7 \text{ V}$$

For $v_S = 2$ V, the transistor is in the active mode and the outputs are $i_C = 1.3$ mA and $v_{CE} = 3.7$ V.

For $v_S = 6$ V, we again assume that the transistor is in the active mode and calculate the collector current from Eq. (4–11).

$$i_C = \beta\left(\frac{v_S - V_\gamma}{R_B}\right) = 100\left(\frac{6 - 0.7}{100 \times 10^3}\right) = 5.3 \text{ mA}$$

The calculated i_C is greater than the available i_{SC}. For this input the transistor is in the saturation mode and the outputs equal their saturation bounds, namely $i_C = i_{SC} = 5$ mA and $v_{CE} = 0$. ■

EXAMPLE 4–10

Suppose the circuit parameters in Figure 4–20 are $\beta = 50$, $V_\gamma = 0.7$ V, $R_B = 20$ kΩ, $R_C = 2.2$ kΩ, and $V_{CC} = 10$ V. Find the range of the input voltage v_S for which the transistor remains in the active mode.

SOLUTION:

To avoid the cutoff mode, the input voltage must exceed the transistor threshold voltage. Hence, the lower bound is $v_S > V_\gamma = 0.7$ V. To avoid saturation the collector current must be less than the available short-circuit current of $i_{SC} = V_{CC}/R_C = 4.545$ mA. To ensure active mode operation, we use Eq. (4–11) to bound the collector current as

$$i_C = \beta\left(\frac{v_S - V_g}{R_B}\right) = 50\left(\frac{v_S - 0.7}{20 \times 10^3}\right) < 4.545 \times 10^{-3} \text{ A}$$

Solving the inequality for v_S yields

$$v_S < 0.7 + \frac{4.545 \times 20}{50} = 2.518 \text{ V}$$

The transistor operates in the active mode when the input falls in the range $0.7 < v_S < 2.518$ V. ■

DESIGN EXAMPLE 4–11

The known parameters in Figure 4–22 are $\beta = 100$, $V_\gamma = 0.7$ V, $R_C = 1$ kΩ, and $V_{CC} = 5$ V. The circuit is to function as a digital inverter that meets two conditions:

1. An input of $v_S = 0$ V must produce an output $v_{CE} = 5$ V.
2. An input of $v_S = 5$ V must produce an output $v_{CE} = 0$ V.

Select a value of R_B so that the circuit meets these conditions.

FIGURE 4–22

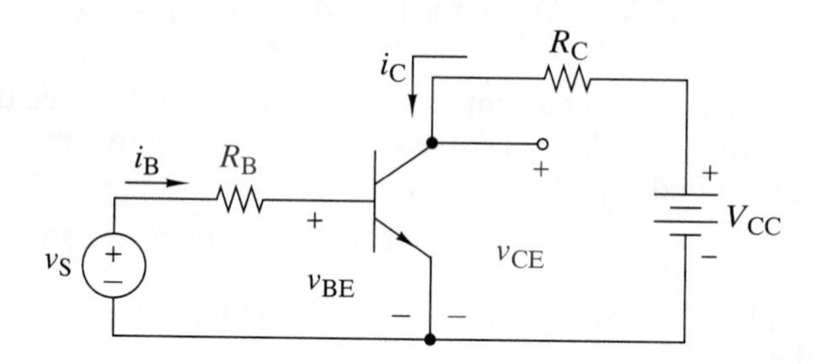

SOLUTION:
To meet the first condition, the transistor must be in the cutoff mode since the required output is $v_{CE} = 5\text{ V} = v_{OC} = V_{CC}$. For $v_S = 0$ the transistor will be in cutoff regardless of the value of R_B since the input is less than the threshold voltage $V_\gamma = 0.7$ V. To meet the second condition, the transistor must be in the saturation mode since the required output is $v_{CE} = 0$. In this case we know that the input is $v_S = 5$ V. We know that the outputs are $v_{CE} = 0$ and $i_C = i_{SC} = 5/1000 = 5$ mA. We also know that in the saturation mode $v_{BE} = V_\gamma = 0.7$ V. What we don't know is a value of R_B that will make all this happen.

To be in saturation the collector current *predicted* by the active mode i–v characteristics, namely, $i_C = \beta i_B$, must exceed the available short-circuit current. This does not mean that the collector current must exceed the short-circuit current. In fact, it can't because the transistor is in saturation with $i_C = i_{SC}$. What it does mean is that $\beta i_B > i_{SC} = 5$ mA. Using Eq. (4–11) to place a bound on the quantity βi_B yields

$$\beta i_B = \beta\left(\frac{v_S - V_\gamma}{R_B}\right) = 100\left(\frac{5 - 0.7}{R_B}\right) > 5 \times 10^{-3}\ \text{A}$$

Solving the inequality for R_B yields

$$R_B < \frac{100 \times 4.3}{5 \times 10^{-3}} = 86\ \text{k}\Omega$$

Any reasonable value less than 86 kΩ (say 56 kΩ, a standard value) will work. ■

Exercise 4–7

The circuit parameters in Figure 4–22 are $\beta = 100$, $V_\gamma = 0.7$ V, $R_C = 1$ kΩ, $R_B = 100$ kΩ, and $V_{CC} = 5$ V. Find i_C and v_{CE} when $v_S = 5$ V.

Answers: $i_C = 4.3$ mA, $v_{CE} = 0.7$ V

Exercise 4–8

The circuit parameters in Figure 4–22 are $\beta = 100$, $V_\gamma = 0.7$ V, $R_C = 1$ kΩ, and $V_{CC} = 5$ V. Find the value of R_B so that $v_{CE} = 2.5$ V when $v_S = 5$ V.

Answer: $R_B = 172$ kΩ

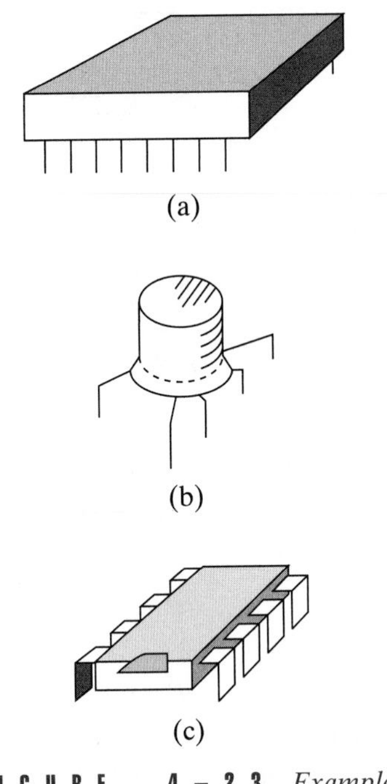

FIGURE 4–23 *Examples of integrated circuit OP AMP packages: (a) Encapsulated hybrid. (b) TO-5 can. (c) Dual in-line package.*

4–4 THE OPERATIONAL AMPLIFIER

The integrated circuit operational amplifier is the premier linear active device in present-day analog circuit applications. The term *operational amplifier* was apparently first used in a 1947 paper by John R. Ragazzini and his colleagues, who reported on work carried out for the National Defense Research Council during World War II. The paper described high-gain dc amplifier circuits that perform mathematical operations (addition, subtraction, multiplication, division, integration, etc.); hence the name *operational amplifier.* For more than a decade the most important applications were general- and special-purpose analog computers using vacuum tube amplifiers. In the early 1960s general-purpose, discrete-transistor operational amplifiers became readily available, and by the mid-1960s the first commercial integrated circuit OP AMPs entered the market. The transition from vacuum tubes to integrated circuits decreased the size, power consumption, and cost of OP AMPs by nearly three orders of magnitude. By the early 1970s the integrated circuit version became the dominant active device in analog circuits.

The device itself is a complex array of transistors, resistors, diodes, and capacitors, all fabricated and interconnected on a tiny silicon chip. Figure 4–23 shows examples of ways OP AMPs are packaged for use in circuits. In spite of its complexity, the device can be modeled by rather simple *i–v* characteristics. We do not need to concern ourselves with what is going on inside the package; rather, we treat the OP AMP using a behavioral model that constrains the voltages and currents at the external terminals of the device.

OP AMP NOTATION

Certain matters of notation and nomenclature must be discussed before developing a circuit model for the OP AMP. The OP AMP is a five-terminal device, as shown in Figure 4–24(a). The "+" and "–" symbols identify the input terminals and are a shorthand notation for the noninverting and inverting input terminals, respectively. These "+" and "–" symbols identify the two input terminals and have nothing to do with the polarity of the voltages applied. The other terminals are the output and the positive and negative supply voltages, usually labeled $+V_{CC}$ and $-V_{CC}$. While some OP AMPs have more than five terminals, these five are always present and are the only ones we will use in this text. Figure 4–24(b) shows how these terminals are arranged in a common eight-pin integrated circuit package.

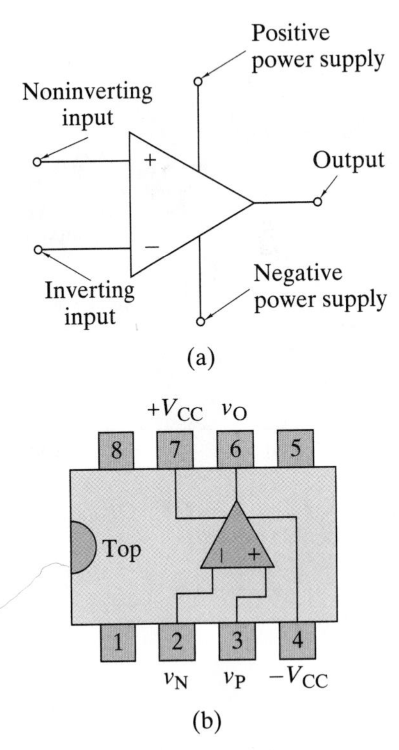

FIGURE 4–24 *The OP AMP: (a) Circuit symbol. (b) Pin out diagram for an eight-pin package.*

The two power supply terminals in Figure 4–24 are not usually shown in circuit diagrams. Be assured that they are always there because the external power supplies are required for the OP AMP to operate as an active device. The power required for signal amplification comes through these terminals from an external power source. The $+V_{CC}$ and $-V_{CC}$ voltages applied to these terminals also determine the upper and lower limits on the OP AMP output voltage.

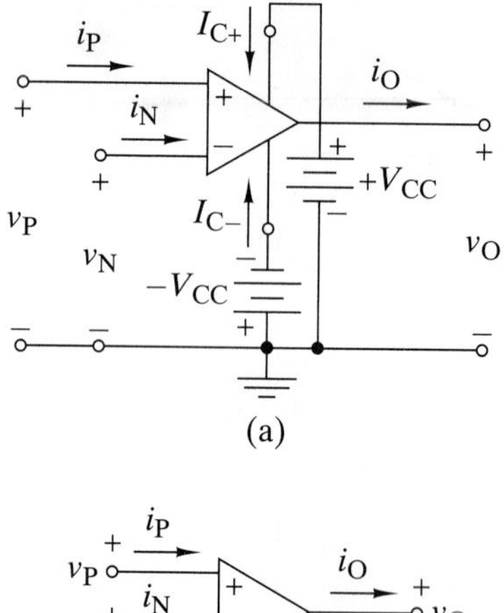

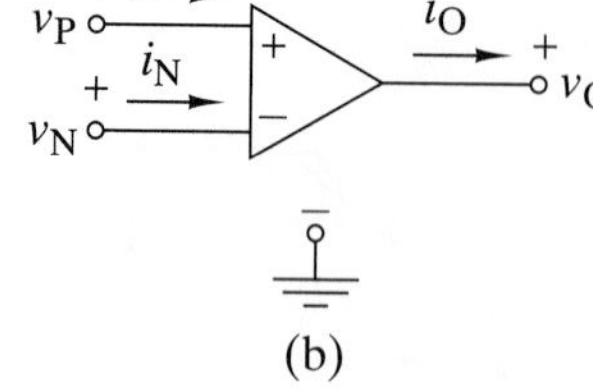

FIGURE 4–25 *OP AMP voltage and current definitions: (a) Complete set. (b) Shorthand set.*

Figure 4–25(a) shows a complete set of voltage and current variables for the OP AMP, while Figure 4–25(b) shows the abbreviated set of signal variables we will use. All voltages are defined with respect to a common reference node, usually ground. Voltage variables v_P, v_N, and v_O are defined by writing a voltage symbol beside the corresponding terminals. This notation means the "+" reference mark is at the terminal in question and the "–" reference mark is at the reference or ground terminal. In this book the reference directions for the currents are directed in at input terminals and out at the output. At times the abbreviated set of current variables may appear to violate KCL. For example, a global KCL equation for the complete set of variables in Figure 4–25(a) is

$$i_O = I_{C+} + I_{C-} + i_P + i_N \quad \text{(correct equation)} \tag{4–13}$$

A similar equation using the shorthand set of current variables in Figure 4–25(b) reads

$$i_O = i_N + i_P \quad \text{(incorrect equation)} \tag{4–14}$$

This equation is *not* correct, since it does not include all the currents. What is more important, it implies that the output current comes from the inputs. In fact, this is wrong. The input currents are very small, ideally zero. The output current comes from the supply voltages, as Eq. (4–13) points out, even though these terminals are not shown on the abbreviated circuit diagram.

TRANSFER CHARACTERISTICS

The dominant feature of the OP AMP is the transfer characteristic shown in Figure 4–26. This characteristic provides the relationships between the **noninverting input** v_P, the **inverting input** v_N, and the **output voltage** v_O. The transfer characteristic is divided into three regions or modes called **+saturation**, **–saturation**, and **linear**. In the linear region the OP AMP is a **differential amplifier** because the output is proportional to the difference between the two inputs. The slope of the line in the linear range is called the voltage gain. In this linear region the input-output relation is

$$v_O = A(v_P - v_N) \tag{4–15}$$

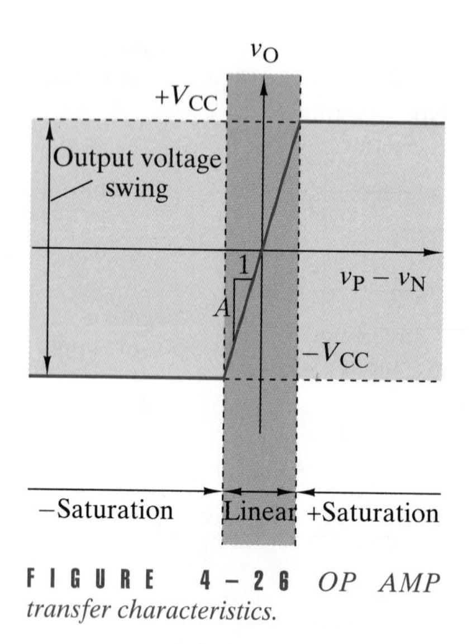

FIGURE 4–26 *OP AMP transfer characteristics.*

The voltage gain of an OP AMP is very large, usually greater than 10^5. As long as the net input $(v_P - v_N)$ is very small, the output will be proportional to the input. However, when $A|v_P - v_N| > V_{CC}$, the OP AMP is saturated and the output voltage is limited by the supply voltages (less some small internal losses).

In the previous section, we found that the transistor has three operating modes. The input-output characteristic in Figure 4–26 points out that the OP AMP also has three operating modes:

1. +Saturation mode when $A(v_P - v_N) > V_{CC}$ and $v_O = +V_{CC}$.
2. –Saturation mode when $A(v_P - v_N) < -V_{CC}$ and $v_O = -V_{CC}$.
3. Linear mode when $A|v_P - v_N| < V_{CC}$ and $v_O = A(v_P - v_N)$.

Usually we analyze and design OP AMP circuits using the model for the linear mode. When the operating mode is not given, we use a self-consistent approach similar to the one used for the transistor. That is, we assume that the OP AMP is in the linear mode and then calculate the output voltage v_O. If it turns out that $-V_{CC} < v_O < +V_{CC}$, then the assumption is correct and the OP AMP is indeed in the linear mode. If $v_O < -V_{CC}$, then the assumption is wrong and the OP AMP is in the –saturation mode with $v_O = -V_{CC}$. If $v_O > +V_{CC}$, then the assumption is wrong and the OP AMP is in the +saturation mode with $v_O = +V_{CC}$.

IDEAL OP AMP MODEL

A dependent-source model of an OP AMP operating in its linear range is shown in Figure 4–27. This model includes an input resistance (R_I), an output resistance (R_O), and a voltage-controlled voltage source whose gain is A. Numerical values of these OP AMP parameters typically fall in the following ranges:

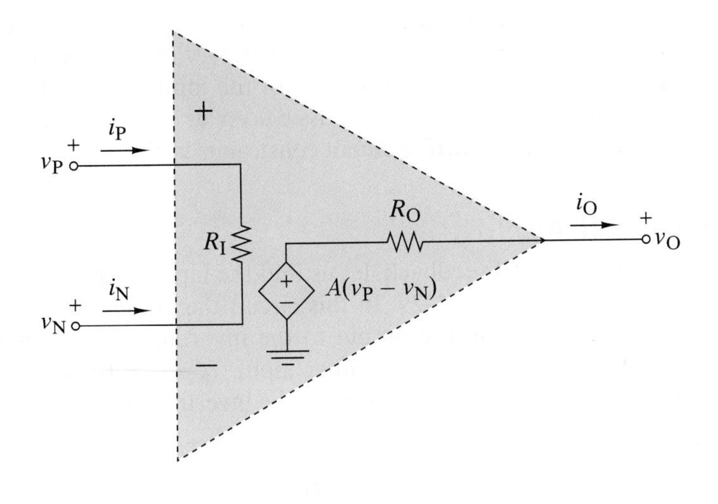

FIGURE 4–27 *Dependent-source model of an OP AMP operating in the linear mode.*

$$10^6 < R_I < 10^{12}\ \Omega$$
$$10 < R_O < 100\ \Omega$$
$$10^5 < A < 10^8$$

Clearly, high input resistance, low output resistances, and high voltage gain are the key attributes of an OP AMP.

The dependent-source model can be used to develop the i–v relationships of the ideal model. For the OP AMP to operate in its linear mode, the output voltage is bounded by

$$-V_{CC} \leqslant v_O \leqslant +V_{CC}$$

Using Eq. (4–15), we can write this bound as

$$-\frac{V_{CC}}{A} \leqslant (v_P - v_N) \leqslant +\frac{V_{CC}}{A}$$

The supply voltage V_{CC} is typically about 15 V, while A is a very large number, usually 10^5 or greater. Consequently, linear operation requires that $v_P \approx v_N$. In the ideal OP AMP model, the voltage gain is assumed to be infinite ($A \to \infty$), in which case linear operation requires $v_P = v_N$. The input resistance R_I of the ideal OP AMP is assumed to be infinite, so the currents

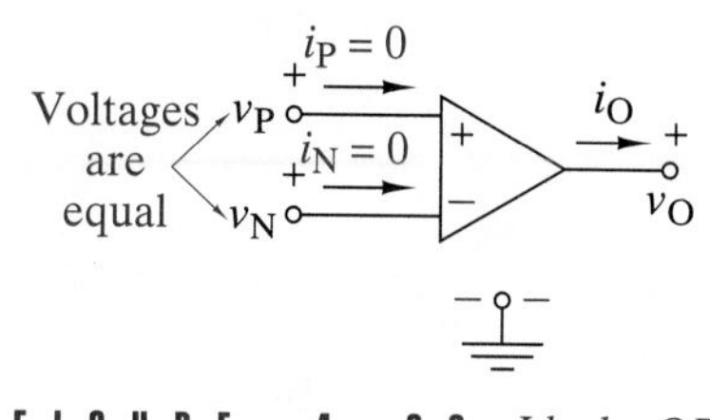

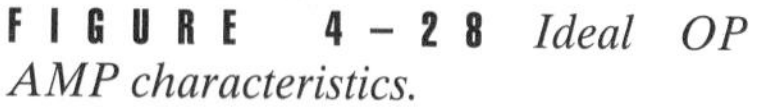

FIGURE 4–28 *Ideal OP AMP characteristics.*

entering both input terminals are both zero. In summary, the *i–v* relationships of the **ideal model** of the OP AMP are

$$v_P = v_N$$

$$i_P = i_N = 0 \tag{4–16}$$

The implications of these element equations are illustrated on the OP AMP circuit symbol in Figure 4–28.

At first glance the element constraints of the ideal OP AMP appear to be fairly useless. They look more like connection constraints and are totally silent about the output quantities (v_O and i_O), which are usually the signals of greatest interest. They seem to say that the OP AMP input terminals are simultaneously a short circuit ($v_P = v_N$) and an open circuit ($i_P = i_N = 0$). In practice, however, the ideal model of the OP AMP is very useful because in linear applications feedback is always present. That is, for the OP AMP to operate in a linear mode, it is necessary for there to be feedback paths from the output to one or both of the inputs. These feedback paths ensure that $v_P \approx v_N$ and make it possible for us to analyze OP AMP circuits using the ideal OP AMP element constraints in Eq. (4–16).

NONINVERTING OP AMP

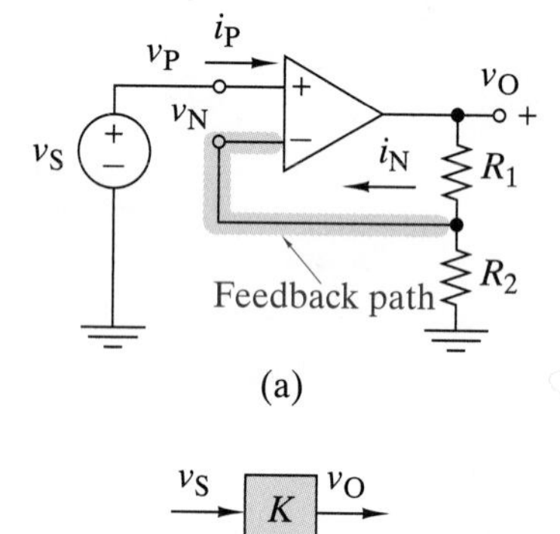

FIGURE 4–29 *The noninverting amplifier circuit.*

To illustrate the effects of feedback, let us find the input-output characteristics of the circuit in Figure 4–29. In this circuit the voltage divider provided a feedback path from the output to the inverting input.[2] Since the ideal OP AMP draws no current at either input ($i_P = i_N = 0$), we can use voltage division to determine the voltage at the inverting input:

$$v_N = \frac{R_2}{R_1 + R_2} v_O \tag{4–17}$$

The input source connection at the noninverting input requires the condition

$$v_P = v_S \tag{4–18}$$

The ideal OP AMP element constraints demand that $v_P = v_N$; therefore, we can equate the right sides of Eqs. (4–17) and (4–18) to obtain the input-output relationship of the overall circuit.

$$v_O = \frac{R_1 + R_2}{R_2} v_S \tag{4–19}$$

The preceding analysis illustrates a general strategy for analyzing OP AMP circuits. We use normal circuit analysis methods to express the OP AMP input voltages v_P and v_N in terms of circuit parameters. We then use the ideal OP AMP constraint $v_P = v_N$ to solve for the overall circuit input-output relationship.

The circuit in Figure 4–29(a) is called a **noninverting amplifier**. The input-output relationship is of the form $v_O = Kv_S$, which reminds us that

---

2 The feedback must always be to the inverting input; otherwise the circuit will be unstable for reasons that cannot be explained by what we have learned up to now.

the circuit is linear. Figure 4–29(b) shows the functional building block for this circuit, where the proportionality constant K is

$$K = \frac{R_1 + R_2}{R_2} \qquad (4\text{–}20)$$

In an OP AMP circuit the proportionality constant K is sometimes called the **closed-loop** gain, because it defines the input-output relationship when the feedback loop is connected (closed).

When discussing OP AMP circuits, it is necessary to distinguish between two types of gains. The first is the large voltage gain provided by the OP AMP device itself. The second is the voltage gain of the OP AMP circuit with a negative feedback path. Note that Eq. (4–20) indicates that the circuit gain is determined by the resistors in the feedback path, not by the value of the OP AMP gain. The gain in Eq. (4–20) is really the voltage division rule upside down. Variation of the value of K depends on the tolerance on the resistors in the feedback path, not the variation in the value of the OP AMP's gain. In effect, feedback converts the OP AMP's very large but variable gain into a much smaller but well-defined gain.

D DESIGN EXAMPLE 4–12

Design an amplifier with a gain of $K = 10$.

SOLUTION:

Using a noninverting OP AMP circuit, the design problem is to select the values of the resistors in the feedback path. From Eq. (4–20) the design constraint is

$$10 = \frac{(R_1 + R_2)}{R_2}$$

We have one constraint with two unknowns. Arbitrarily selecting $R_2 = 10\ \text{k}\Omega$, we find $R_1 = 90\ \text{k}\Omega$. These resistors would normally have low tolerances (± 1% or less) to produce a precisely controlled closed-loop gain.

Comment: The problem of choosing resistance values in OP AMP circuit design problems deserves some discussion. Although values of resistance from a few ohms to several hundred megohms are commercially available, we generally limit ourselves to the range from about 1 kΩ to perhaps 1 MΩ. The lower limit of 1 kΩ is imposed in part because of power dissipation in the resistors. Typically, we use resistors with ¼-W power ratings or less. The maximum voltage in OP AMP circuits is often around 15 V. The smallest ¼-W resistance we can use is $R_{\text{MIN}} \geq (15)^2/0.25 = 900\ \Omega$, or about 1 kΩ. The upper bound of 1 MΩ comes about because surface leakage makes it difficult to maintain the tolerance in a high-value resistance. High-value resistors are also noisy, which leads to problems when they are connected in the feedback path. The 1-kΩ to 1-MΩ range should be used as a guideline, not an inviolate design rule. Actual design choices are influenced by system-specific factors and changes in technology. ■

Exercise 4–9

The noninverting amplifier circuit in Figure 4–29(a) is operating with $R_1 = 2R_2$ and $V_{CC} = \pm 12$ V. Over what range of input voltages v_S is the OP AMP in the linear mode?

Answer: $-4\text{ V} < v_S < +4\text{ V}$

EFFECTS OF FINITE OP AMP GAIN

The ideal OP AMP model has an infinite gain. Actual OP AMP devices have very large, but finite voltage gains. We now address the effect of large but finite gain on the input-output relationships of OP AMP circuits.

The circuit in Figure 4–30 shows a finite gain OP AMP circuit model in which the input resistance R_I is infinite. The actual values of OP AMP input resistance range from 10^6 to 10^{12} Ω, so no important effect is left out by ignoring this resistance. Examining the circuit, we see that the noninverting input voltage is determined by the independent voltage source. The inverting input can be found by voltage division, since the current i_N is zero. In other words, Eqs. (4–17) and (4–18) apply to this circuit as well.

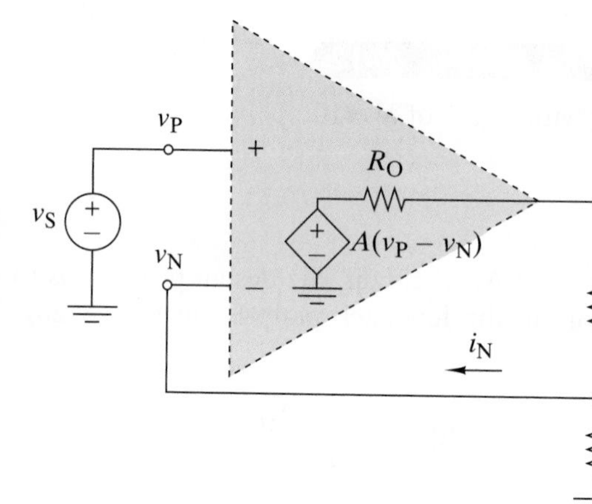

FIGURE 4–30 *The noninverting amplifier circuit with the dependent-source model.*

We next determine the output voltage in terms of the controlled-source voltage using voltage division on the series connection of the three resistors R_O, R_1, and R_2:

$$v_O = \frac{R_1 + R_2}{R_O + R_1 + R_2} A(v_P - v_N)$$

Substituting v_P and v_N from Eqs. (4–17) and (4–18) yields

$$v_O = \left[\frac{R_1 + R_2}{R_O + R_1 + R_2}\right] A \left[v_S - \frac{R_2}{R_1 + R_2} v_O\right] \tag{4–21}$$

The intermediate result in Eq. (4–21) shows that feedback is present since v_O appears on both sides of the equation. Solving for v_O yields

$$v_O = \frac{A(R_1 + R_2)}{R_O + R_1 + R_2(1 + A)} v_S \tag{4–22}$$

In the limit, as $A \to \infty$, Eq. (4–22) reduces to

$$v_O = \frac{R_1 + R_2}{R_2} v_S = K v_S$$

where K is the closed-loop gain we previously found using the ideal OP AMP model.

To see the effect of a finite A, we ignore R_O in Eq. (4–22) since it is generally quite small compared with $R_1 + R_2$. With this approximation Eq. (4–22) can be written in the following form:

$$v_O = \frac{K}{1 + (K/A)} v_S \tag{4–23}$$

When written in this form, we see that the closed-loop gain reduces to K as $A \to \infty$. Moreover, we see that the finite-gain model yields a good approximation to the ideal model results as long as $K << A$. In other words, the ideal model yields good results as long as the closed-loop gain is much less than the gain of the OP AMP device. One practical rule of thumb is to limit the closed-loop gain to less than 1% of the OP AMP gain (i.e., $K < A/100$).

The feedback path also affects the active output resistance. To see this, we construct a Thévenin equivalent circuit using the open-circuit voltage and the short-circuit current. Equation (4–23) is the open-circuit voltage, and we need only find the short-circuit current. Connecting a short-circuit at the output in Figure 4–30 forces $v_N = 0$ but leaves $v_P = v_S$. Therefore, the short-circuit current is

$$i_{SC} = A(v_S/R_O)$$

As a result, the Thévenin resistance is

$$R_T = \frac{v_{OC}}{i_{SC}} = \frac{K/A}{1 + K/A} R_O$$

When $K << A$, this expression reduces to

$$R_T = \frac{K}{A} R_O \approx 0\ \Omega$$

The OP AMP circuit with feedback has an output Thévenin resistance that is much smaller than the output Thévenin resistance of the OP AMP device itself. In fact, the Thévenin resistance is very small since R_O is typically less than 100 Ω and A is greater than 10^5.

At this point we can summarize our discussion. We introduced the OP AMP as an active five-terminal device including two supply terminals not normally shown on the circuit diagram. We then developed an ideal model of this device that is used to analyze and design circuits that have feedback. Feedback must be present for the device to operate in the linear mode. The most dramatic feature of the ideal model is the assumption of infinite gain. Using a finite-gain model, we found that the ideal model predicts the circuit input-output relationship quite closely as long as the circuit gain K is much smaller than the OP AMP gain A. We also discovered that the Thévenin output resistance of an OP AMP with feedback is essentially zero.

In the rest of this book we use the ideal i–v constraints in Eq. (4–16) to analyze OP AMP circuits. The OP AMP circuits have essentially zero output resistance, which means that the output voltage does not change with different loads. Unless otherwise stated, from now on the term *OP AMP* refers to the ideal model.

4–5 OP AMP CIRCUIT ANALYSIS

This section introduces OP AMP circuit analysis using examples that are building blocks for analog signal-processing systems. We have already introduced one of these circuits—the noninverting amplifier discussed in the preceding section. The other basic circuits are the voltage follower, the inverting amplifier, the summer, and the subtractor. The key to using the building block approach is to recognize the feedback pattern and to isolate the basic circuit as a building block. The first example illustrates this process.

EXAMPLE 4–13

Find the input-output relationship of the circuit in Figure 4–31(a).

FIGURE 4–31

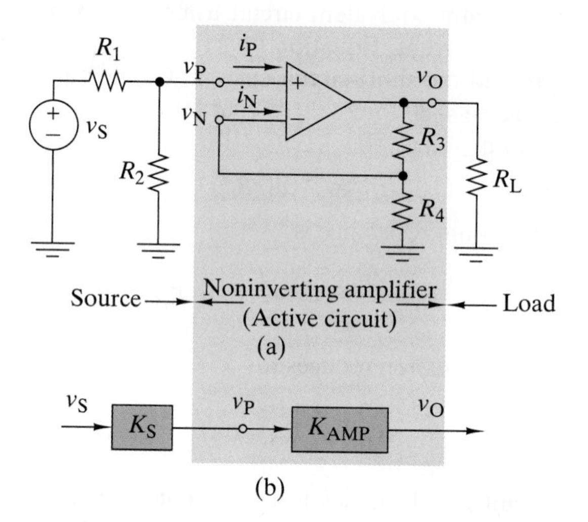

SOLUTION:
When the circuit is partitioned as shown in Figure 4–31(a), we recognize two building block gains: (1) K_S, the proportionality constant of the source circuit, and (2) K_{AMP}, the gain of the noninverting amplifier. The OP AMP circuit input current $i_P = 0$; hence, we can use voltage division to find K_S as

$$K_S = \frac{v_P}{v_S} = \frac{R_2}{R_1 + R_2}$$

Since the noninverting amplifier has zero output resistance, the load R_L has no effect on the output voltage v_O. Using Eq. (4–19), the gain of the noninverting amplifier circuit is

$$K_{AMP} = \frac{v_O}{v_P} = \frac{R_3 + R_4}{R_4}$$

The overall circuit gain is found as

$$K_{\text{CIRCUIT}} = \frac{v_O}{v_S} = \left[\frac{v_P}{v_S}\right]\left[\frac{v_O}{v_P}\right]$$
$$= K_S \times K_{\text{AMP}}$$
$$= \left[\frac{R_2}{R_1 + R_2}\right]\left[\frac{R_3 + R_4}{R_4}\right]$$

The gain K_{CIRCUIT} is the product of K_S times K_{AMP} because the amplifier circuit does not load the source circuit since $i_P = 0$. ■

VOLTAGE FOLLOWER

The OP AMP in Figure 4–32(a) is connected as a **voltage follower** or **buffer**. In this case, the feedback path is a direct connection from the output to the inverting input. The feedback connection forces the condition $v_N = v_O$. The input current $i_P = 0$, so there is no voltage across the source resistance R_S. Applying KVL, we have the input condition $v_P = v_S$. The ideal OP AMP model requires $v_P = v_N$, so we conclude that $v_O = v_S$. By inspection, the closed-loop gain is $K = 1$. Since the output exactly equals the input, we say that the output follows the input (hence the name *voltage follower*).

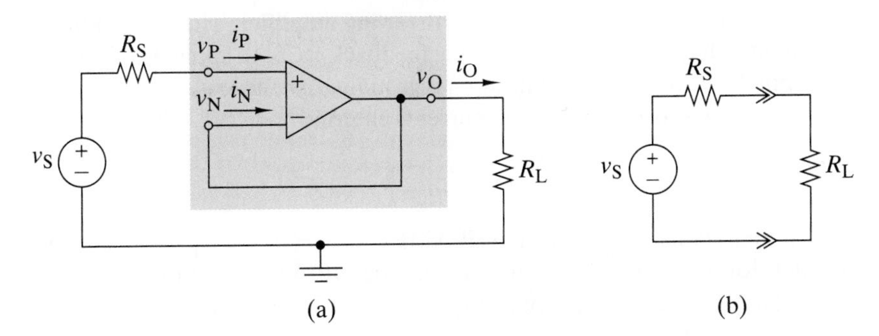

FIGURE 4–32 *(a) Source-load interface with a voltage follower. (b) Interface without the voltage follower.*

The voltage follower is used in interface circuits because it isolates the source from the load. Note that the input-output relationship $v_O = v_S$ does not depend on the source or load resistance. When the source is connected directly to the load, as in Figure 4–32(b), the voltage delivered to the load depends on R_S and R_L. The source and load interaction limits the signals that can be transferred across the interface, as discussed in Chapter 3. When the voltage follower is inserted between the source and load, the signal levels are limited by the capability of the OP AMP.

By Ohm's law, the current delivered to the load is $i_O = v_O/R_L$. But since $v_O = v_S$, the output current can be written in the form

$$i_O = v_S/R_L$$

Applying KCL at the reference node, we discover an apparent dilemma:

$$i_P = i_O$$

For the ideal model $i_P = 0$, but the preceding equations say that i_O cannot be zero unless v_S is zero. It appears that KCL is violated.

The dilemma is resolved by noting that the circuit diagram does not include the supply terminals. The output current comes from the power supply, not from the input. This dilemma arises only at the reference node (the ground terminal). In OP AMP circuits, as in all circuits, KCL must be satisfied. However, we must be alert to the fact that a KCL equation at the reference node could yield misleading results because the power supply terminals are not usually included in circuit diagrams.

Exercise 4–10

The circuits in Figure 4–32 have $v_S = 1.5$ V, $R_S = 2$ kΩ, and $R_L = 1$ kΩ. Compute the maximum power available from the source. Compute the power absorbed by the load resistor in the direct connection in Figure 4–32(b) and in the voltage follower circuit in Figure 4–32(a). Discuss any differences.

Answers: 281 μW; 250 μW; 2250 μW

DISCUSSION: *With the direct connection, the power delivered to the load is less than the maximum power available. With the voltage follower circuit, the power delivered to the load is greater than the maximum value specified by the maximum power transfer theorem. However, the maximum power transfer theorem does not apply to the voltage follower circuit since the load power comes from the OP AMP power supply rather than the signal source.*

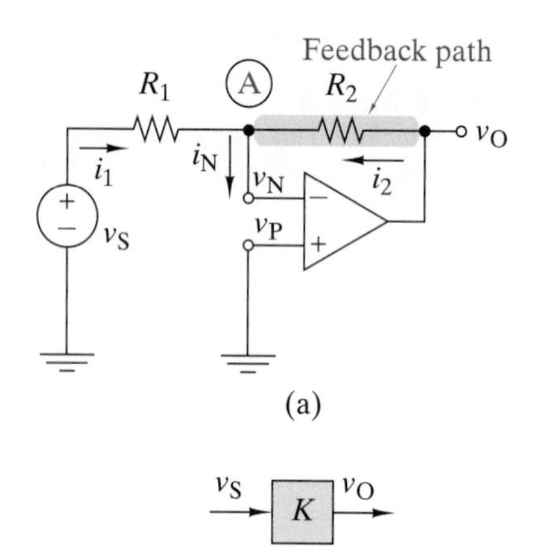

FIGURE 4–33 *The inverting amplifier circuit.*

THE INVERTING AMPLIFIER

The circuit in Figure 4–33 is called an **inverting amplifier**. The key feature of this circuit is that the input signal and the feedback are both applied at the inverting input. Since the noninverting input is grounded, we have $v_P = 0$, an observation we will use shortly. The sum of currents entering node A can be written as

$$\frac{v_S - v_N}{R_1} + \frac{v_O - v_N}{R_2} - i_N = 0 \tag{4–24}$$

The element constraints for the OP AMP are $v_P = v_N$ and $i_P = i_N = 0$. Since $v_P = 0$, it follows that $v_N = 0$. Substituting the OP AMP constraints into Eq. (4–24) and solving for the input-output relationship yields

$$v_O = -\left(\frac{R_2}{R_1}\right) v_S \tag{4–25}$$

This result is of the form $v_O = Kv_S$, where K is the closed-loop gain. However, in this case the voltage gain $K = -R_2/R_1$ is negative, indicating a signal inversion (hence the name *inverting amplifier*). We use the block diagram symbol in Figure 4–33(b) to indicate either the inverting or the noninverting OP AMP configuration.

The OP AMP constraints mean that the input current i_1 in Figure 4–33(a) is

$$i_1 = \frac{v_S - v_N}{R_1} = \frac{v_S}{R_1}$$

This, in turn, shows that the input resistance seen by the source v_S is

$$R_{IN} = \frac{v_S}{i_1} = R_1 \tag{4–26}$$

In other words, the inverting amplifier has a finite input resistance determined by the external resistor R_1.

The next example shows that the finite input resistance must be taken into account when analyzing circuits with OP AMPs in the inverting amplifier configuration.

EXAMPLE 4–14

Find the input-output relationship of the circuit in Figure 4–34(a).

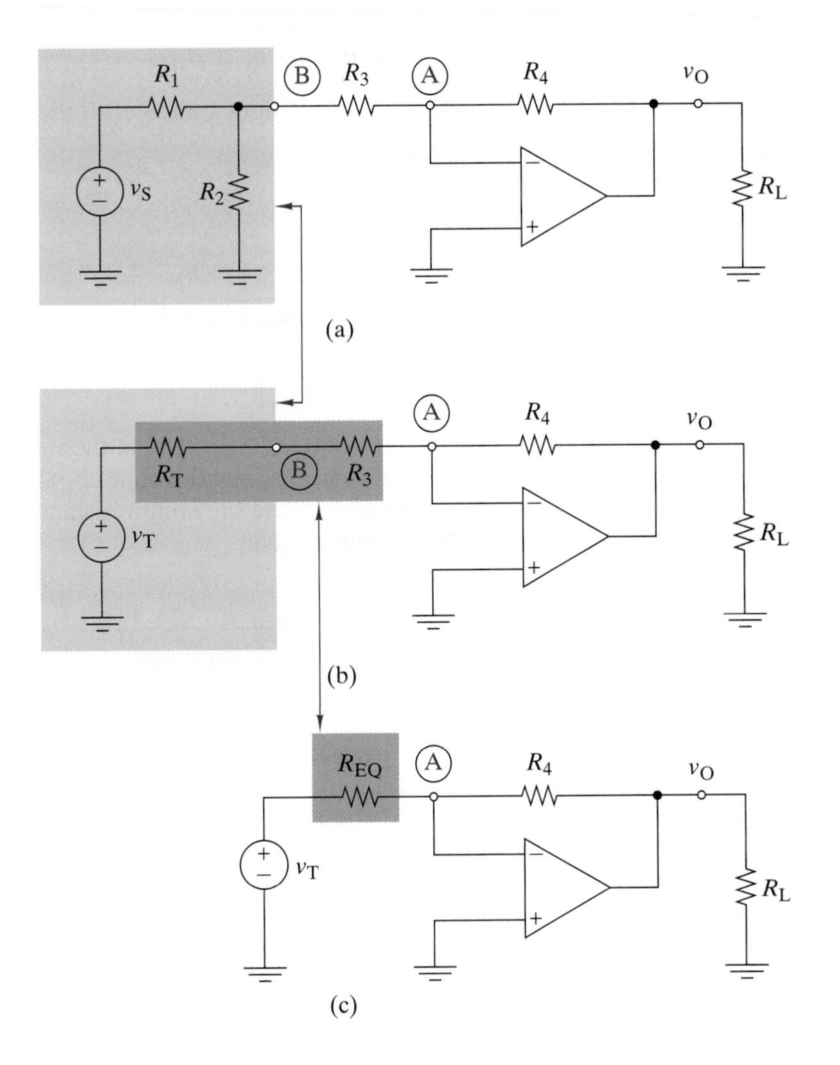

FIGURE 4–34

SOLUTION:

The circuit to the right of node B is an inverting amplifier. The load resistance R_L has no effect on the circuit transfer characteristics since the OP AMP has zero output resistance. However, the source circuit to the left of node B is influenced by the input resistance of the inverting amplifier circuit. The effect can be seen by constructing a Thévenin equivalent of the circuit to the left of node B, as shown in Figure 4–34(b). By inspection of Figure 4–34(a),

$$v_T = \frac{R_2}{R_1 + R_2} v_S$$

$$R_T = \frac{R_1 R_2}{R_1 + R_2}$$

In Figure 4–34(b) this Thévenin resistance is connected in series with the input resistor R_3, yielding the equivalence resistance $R_{EQ} = R_T + R_3$ shown in Figure 4–34(c). This reduced circuit is in the form of an inverting amplifier, so we can write the input-output relationship relating v_O and v_T as

$$K_1 = \frac{v_O}{v_T} = -\frac{R_4}{R_{EQ}} = -\frac{R_4(R_1 + R_2)}{R_1R_2 + R_1R_3 + R_2R_3}$$

The overall input-output relationship from the input source v_S to the OP AMP output v_O is obtained by writing

$$\begin{aligned} K_{CIRCUIT} = \frac{v_O}{v_S} &= \left[\frac{v_O}{v_T}\right]\left[\frac{v_T}{v_S}\right] \\ &= -\left[\frac{R_4(R_1 + R_2)}{R_1R_2 + R_1R_3 + R_2R_3}\right]\left[\frac{R_2}{R_1 + R_2}\right] \\ &= -\left[\frac{R_2R_4}{R_1R_2 + R_1R_3 + R_2R_3}\right] \end{aligned}$$

It is important to note that the overall gain is *not* the product of the source circuit voltage gain $R_2/(R_1 + R_2)$ and the inverting amplifier gain $-R_4/R_3$. In this circuit the two building blocks interact because the input resistance of the inverting amplifier circuit loads the source circuit. ■

Exercise 4–11

Find v_O in Figure 4–35 when $v_S = 2$ V. Repeat for $v_S = -4$ V and $v_S = 6$ V.

FIGURE 4–35

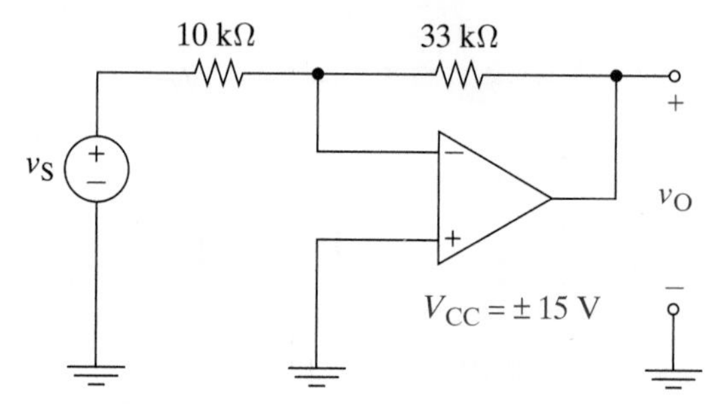

Answers: −6.6V; 13.2V; −15V

THE SUMMING AMPLIFIER

The **summing amplifier** or **adder** circuit is shown in Figure 4–36(a). This circuit has two inputs connected at node A, which is called the **summing point**. Since the noninverting input is grounded, we have the condition $v_P = 0$. This configuration is similar to the inverting amplifier, so we start by applying KCL to write the sum of currents entering the node A summing point.

$$\frac{v_1 - v_N}{R_1} + \frac{v_2 - v_N}{R_2} + \frac{v_O - v_N}{R_F} - i_N = 0 \qquad (4\text{–}27)$$

With the noninverting input grounded, the OP AMP element constraints are $v_N = v_P = 0$ and $i_N = 0$. Substituting these OP AMP constraints into Eq. (4–27), we can solve for the circuit input-output relationship.

$$v_O = \left(-\frac{R_F}{R_1}\right) v_1 + \left(-\frac{R_F}{R_2}\right) v_2 \quad (4\text{–}28)$$

$$= (K_1)\, v_1 + (K_2)\, v_2$$

The output is a weighted sum of the two inputs. The scale factors (or gains, as they are called) are determined by the ratio of the feedback resistor R_F to the input resistor for each input: that is, $K_1 = -R_F/R_1$ and $K_2 = -R_F/R_2$. In the special case $R_1 = R_2 = R$, Eq. (4–28) reduces to

$$v_O = -\frac{R_F}{R}(v_1 + v_2)$$

In this special case the output is proportional to the sum of the two inputs (hence the name *summing amplifier* or, more precisely, *inverting summer*). A block diagram representation of this circuit is shown in Figure 4–36(b).

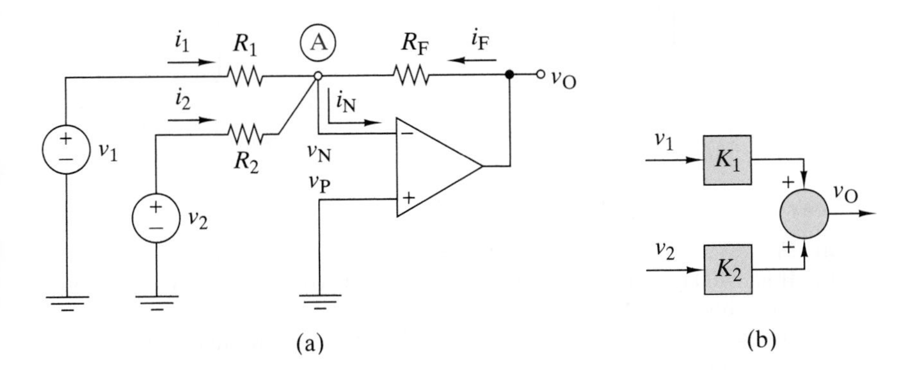

FIGURE 4–36 *The inverting summer.*

The summing amplifier in Figure 4–36 has two inputs, so there are two gains to contend with, one for each input. The input-output relationship in Eq. (4–28) is easily generalized to the case of n inputs as

$$v_O = \left(-\frac{R_F}{R_1}\right) v_1 + \left(-\frac{R_F}{R_2}\right) v_2 + \cdots + \left(-\frac{R_F}{R_n}\right) v_n$$

$$= K_1 v_1 + K_2 v_2 + \cdots + K_n v_n \quad (4\text{–}29)$$

where R_F is the feedback resistor and $R_1, R_2, \ldots R_n$ are the input resistors for the n input voltages $v_1, v_2, \ldots v_n$. You can easily verify this result by expanding the KCL sum in Eq. (4–27) to include n inputs, invoking the OP AMP constraints, and then solving for v_O.

D DESIGN EXAMPLE 4–15

Design an inverting summer that implements the input-output relationship

$$v_O = -(5\, v_1 + 13\, v_2)$$

SOLUTION:

The design problem involves selecting the input and feedback resistors so that

$$\frac{R_F}{R_1} = 5 \quad \text{and} \quad \frac{R_F}{R_2} = 13$$

One solution is arbitrarily to select $R_F = 65$ kΩ, which yields $R_1 = 13$ kΩ and $R_2 = 5$ kΩ. The resulting circuit is shown in Figure 4–37(a). The design can be modified to use standard resistance values for resistors with ±5% tolerance (see Appendix A, Table A-1). Selecting the standard value $R_F = 56$ kΩ requires $R_1 = 11.2$ kΩ and $R_2 = 4.31$ kΩ. The nearest standard values are 11 kΩ and 4.3 kΩ. The resulting circuit shown in Figure 4–37(b) incorporates standard value resistors and produces gains of $K_1 = 56/11 = 5.09$ and $K_2 = 56/4.3 = 13.02$. These nominal gains are within 2% of the values in the specified input-output relationship. ■

FIGURE 4–37

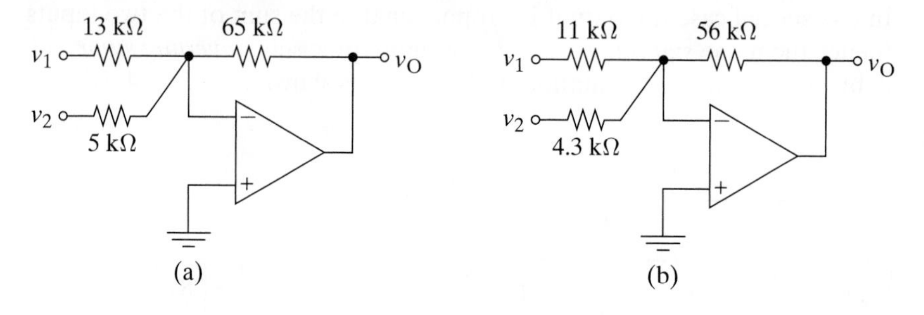

Exercise 4–12

(a) Find v_O in Figure 4–37(a) when $v_1 = 2$ V and $v_2 = -0.5$ V.

(b) If $v_1 = 400$ mV and $V_{CC} = \pm 15$ V, what is the maximum value of v_2 for linear mode operation?

(c) If $v_1 = 500$ mV and $V_{CC} = \pm 15$ V, what is the minimum value of v_2 for linear mode operation?

Answers: (a) −3.5 V; (b) 1 V; (c) −1.346 V

THE DIFFERENTIAL AMPLIFIER

The circuit in Figure 4–38(a) is called a **differential amplifier** or **subtractor**. Like the summer, this circuit has two inputs, one applied at the inverting

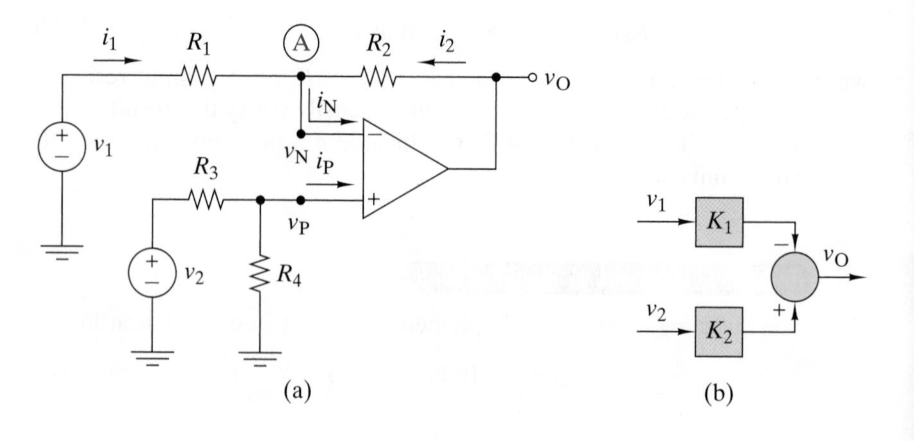

FIGURE 4–38 *The differential amplifier.*

input and one at the noninverting input of the OP AMP. The input-output relationship can be obtained using the superposition principle.

First, we turn off source v_2, in which case there is no excitation at the noninverting input and $v_P = 0$. In effect, the noninverting input is grounded and the circuit acts like an inverting amplifier with the result that

$$v_{O1} = -\frac{R_2}{R_1} v_1 \tag{4–30}$$

Next, turning v_2 back on and turning v_1 off, we see that the circuit looks like a noninverting amplifier with a voltage divider connected at its input. This case was treated in Example 4–13, so we can write

$$v_{O2} = \left[\frac{R_4}{R_3 + R_4}\right]\left[\frac{R_1 + R_2}{R_1}\right] v_2 \tag{4–31}$$

Using superposition, we add outputs in Eqs. (4–30) and (4–31) to obtain the output with both sources on:

$$\begin{aligned} v_O &= v_{O1} + v_{O2} \\ &= -\left[\frac{R_2}{R_1}\right] v_1 + \left[\frac{R_4}{R_3 + R_4}\right]\left[\frac{R_1 + R_2}{R_1}\right] v_2 \\ &= -[K_1]\, v_1 + [K_2] v_2 \end{aligned} \tag{4–32}$$

where K_1 and K_2 are the inverting and noninverting gains. Figure 4–38(b) shows how the differential amplifier is represented in a block diagram.

For the special case of $R_3/R_1 = R_4/R_2$, Eq. (4–32) reduces to

$$v_O = \frac{R_2}{R_1}(v_2 - v_1) \tag{4–33}$$

In this case the output is proportional to the difference between the two inputs (hence the name *differential amplifier* or *subtractor*).

Exercise 4–13

(a) Find the input-output relationship of the subtractor circuit in Figure 4–39.
(b) If $V_{CC} = \pm 15$ V and $v_1 = 3$ V, what is the allowable range of v_2 for linear operation of the OP AMP?

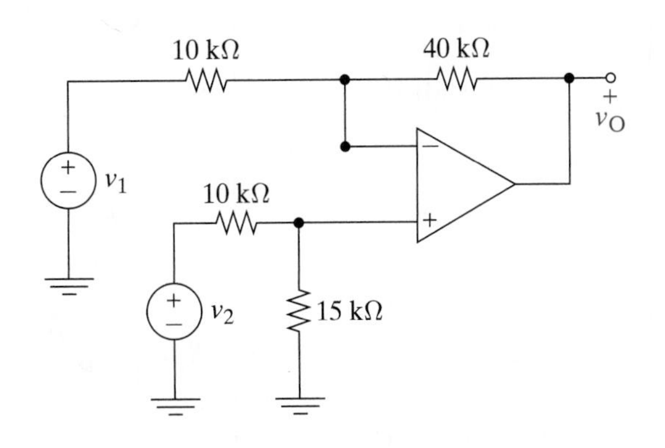

FIGURE 4–39

Answers: (a) $\omega_O = -4v_1 + 3v_2$; (b) $-1\text{ V} \leq v_2 \leq 9\text{ V}$

BASIC OP AMP BUILDING BLOCKS

The block diagram representations of the basic OP AMP circuit configurations are shown in Figure 4–40. The noninverting and inverting amplifiers are represented as gain blocks. The summing amplifier and differential amplifier require both gain blocks and the summing point symbol. Considerable care must be used when translating from a block diagram to a circuit, or vice versa, since some gain blocks involve negative gains. For example, the gains of the inverting summer are negative. The required minus sign is sometimes moved to the summing point and the value of K within the gain block changed to a positive number. Since there is no

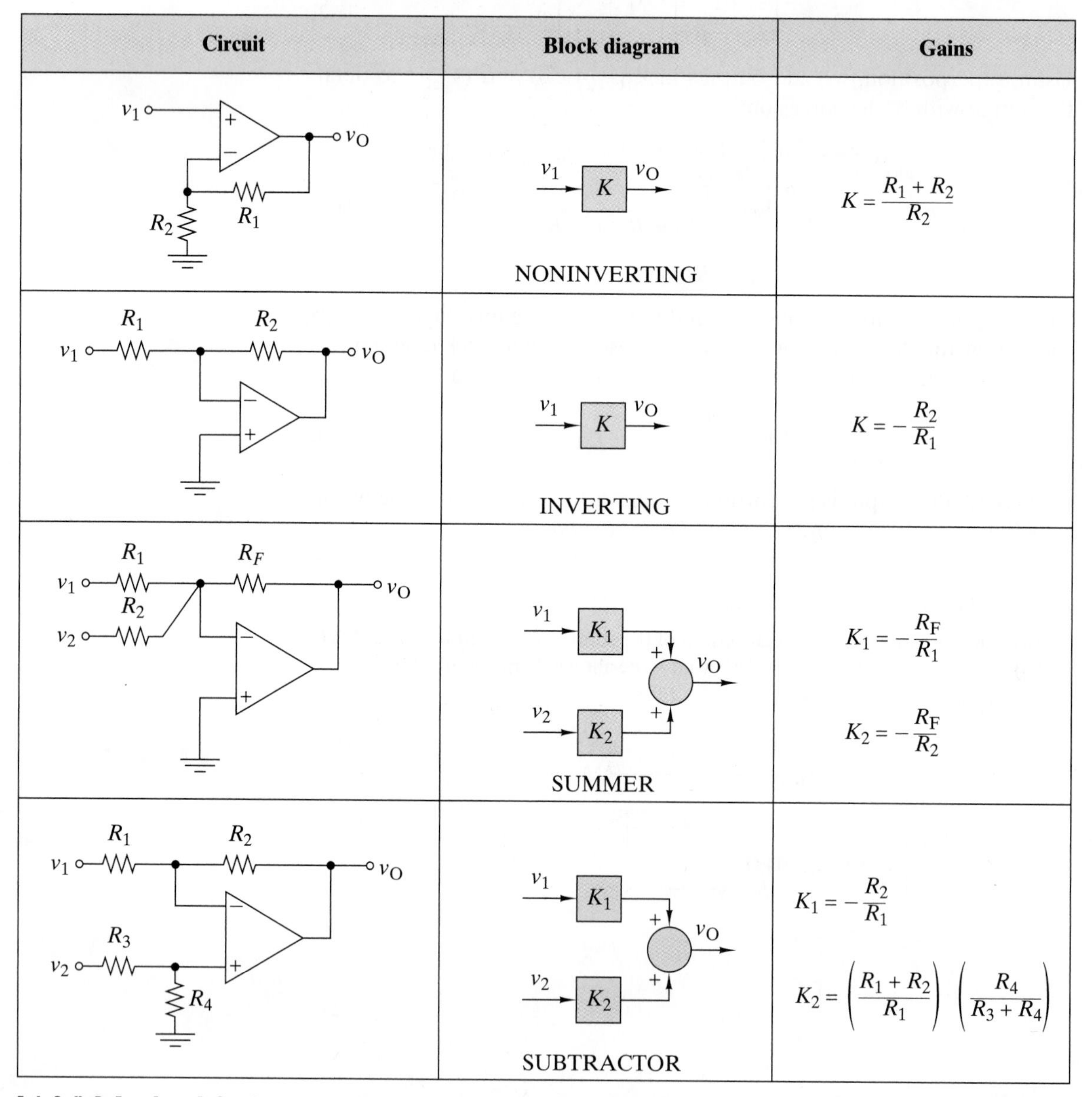

Circuit
Block diagram
Gains
NONINVERTING
$K = \frac{R_1 + R_2}{R_2}$
INVERTING
$K = -\frac{R_2}{R_1}$
SUMMER
$K_1 = -\frac{R_F}{R_1}$
$K_2 = -\frac{R_F}{R_2}$
SUBTRACTOR
$K_1 = -\frac{R_2}{R_1}$
$K_2 = \left(\frac{R_1 + R_2}{R_1}\right)\left(\frac{R_4}{R_3 + R_4}\right)$

FIGURE 4–40 *Summary of basic OP AMP signal-processing circuits.*

standard convention for doing this, it is important to keep track of the signs associated with gain blocks and summing points.

The OP AMP building blocks in Figure 4–40 can be interconnected to obtain complex signal-processing functions. These interconnects do not change the input-output relationships of each block, provided the connections are all between outputs and inputs. Each of the building blocks in Figure 4–40 is a feedback circuit with an OP AMP output. These feedback circuits have insignificant output resistances and can drive any load within the OP AMP's output current capacity.[3] In other words, the building block outputs act like ideal voltage sources just like the voltage sources connected to the block inputs. This observation leads to the following conclusion:

Connecting the output of one building block circuit to an input of another does not change the signal-processing function performed by either circuit.

This property allows us to find the function performed by an interconnection using the functions performed by the individual building blocks. Conversely, this property allows us to design an interconnection by breaking a required function down into separate building block functions.

EXAMPLE 4–16

Derive an expression for v_O in Figure 4–41 in terms of the two inputs.

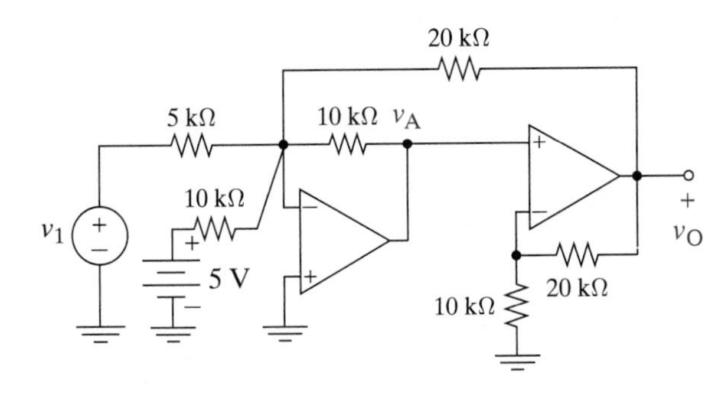

FIGURE 4–41

SOLUTION:

The circuit is an interconnection of two basic building blocks: a three-input summer and a noninverting amplifier. The circuit meets the connection requirement since building block outputs are connected to other building block inputs. The node voltage v_A in the figure is the output of the summer. The summer inputs are a fixed 5-V source, a signal source v_1, and the noninverting amplifier output v_O. Using the inverting summer input-output relationship in Eq. (4–29), we have

$$v_A = -5 - 2v_1 - \frac{1}{2}v_O$$

3 Maximum OP AMP output currents are typically around 20 mA and generally range from 1 to 100 mA.

The summer output v_A is the input to the noninverting amplifier whose voltage gain is $K = 3$. Using the input-output relationship of the noninverting amplifier, we have

$$v_O = 3v_A$$
$$= -15 - 6v_1 - \frac{3}{2}v_O$$

Solving for v_O yields

$$v_O = -6 - 2.4\,v_1$$

The signal-processing function on the interconnection was found using the input-output relationships of the individual building blocks. The method even works with a feedback path because the 20-kΩ resistor is connected from the noninverting amplifier output to an input of the inverting summer. ■

Exercise 4–14

Derive an expression for v_O in Figure 4–42 in terms of the inputs v_1 and v_2.

FIGURE 4–42

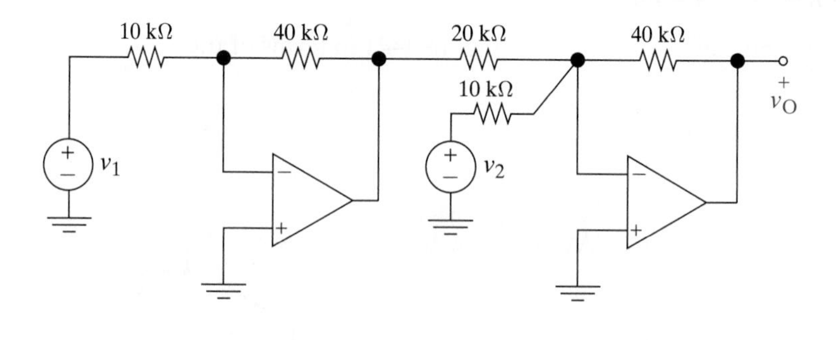

Answer: $v_O = 8\,v_1 - 4v_2$

EXAMPLE 4–17

Derive an expression for v_O in Figure 4–43 in terms of the two inputs v_1 and v_2.

FIGURE 4–43

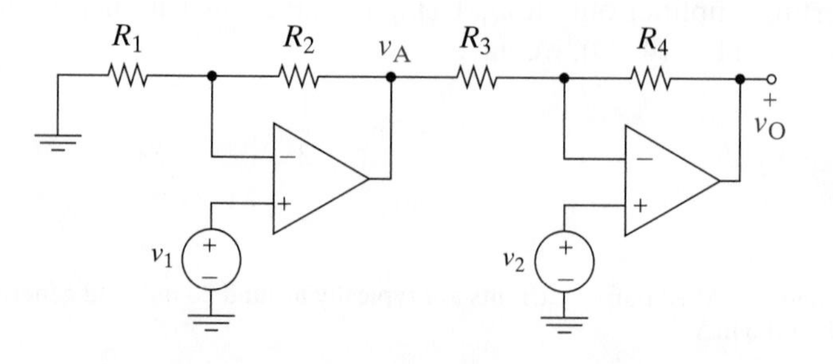

SOLUTION:

The circuit is an interconnection of a noninverting amplifier and an inverting amplifier with an additional signal v_2 applied at its noninverting input. The voltage v_A shown in the figure is the output of the noninverting amplifier:

$$v_A = \left[\frac{R_1 + R_2}{R_1}\right] v_1$$

We use superposition to find v_O. First, set $v_2 = 0$, which connects the noninverting input of the second stage to ground. In this case the second stage acts like an inverting amplifier whose input is v_A and whose output is the response due to v_1 acting alone:

$$\begin{aligned} v_{O1} &= \left[-\frac{R_4}{R_3}\right] v_A \\ &= \left[-\frac{R_4}{R_3}\right]\left[\frac{R_1 + R_2}{R_1}\right] v_1 \end{aligned}$$

Next, set $v_1 = 0$, which sets $v_A = 0$ in turn. In effect this connects resistor R_3 to ground. In this case the second stage acts like a noninverting amplifier whose input is v_2 and whose output is the response due to v_2 acting alone:

$$v_{O2} = \left[\frac{R_3 + R_4}{R_3}\right] v_2$$

Applying superposition, the total output is

$$\begin{aligned} v_O &= v_{O1} + v_{O2} \\ &= -\left[\frac{R_4}{R_3}\right]\left[\frac{R_1 + R_2}{R_1}\right] v_1 + \left[\frac{R_3 + R_4}{R_3}\right] v_2 \\ &= -[K_1] v_1 + [K_2] v_2 \end{aligned}$$

where K_1 and K_2 are the inverting and noninverting gains, respectively.

The dual OP AMP circuit in Figure 4–43 performs the same signal-processing function as the single OP AMP subtractor circuit in Fig. 4–40. Why use two OP AMPs to obtain a function that can be achieved using only one? The answer is that both input signals in Figure 4–43 are applied to noninverting OP AMP inputs that have very high input resistances. This means that the two–OP AMP subtractor does not load the input signal sources. The basic subtractor in Figure 4–40 has finite input resistances that may load the input signal sources. This difference could be important when the input signal sources have high Thévenin resistances. ■

NODE-VOLTAGE ANALYSIS WITH OP AMPS

There are many useful OP AMP circuits that are not simply interconnections of basic building blocks. In such cases we use a modified form of node-voltage analysis that is based on the OP AMP connections in Figure 4–44. The overall circuit contains N nodes, including the three associated with the OP AMP. Normally the objective is to find the OP AMP output voltage (v_O) relative to the reference node (ground). We assign node voltage variables to the $N - 1$ nonreference nodes, including a variable at the OP AMP output. However, an ideal

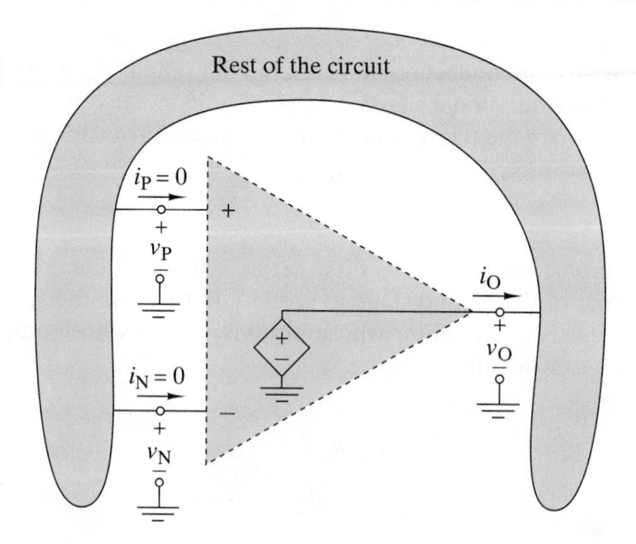

FIGURE 4–44 *General OP AMP circuit analysis.*

OP AMP acts like a dependent voltage source connected between the output terminal and ground. As a result, the OP AMP output voltage is determined by the other node voltages, so we do not need to write a node equation at the OP AMP output node.

We formulate node equations at the other $N - 2$ nonreference nodes in the usual way. Since there are $N - 1$ node voltages, we seem to have more unknowns than equations. However, the OP AMP forces the condition $v_P = v_N$ in Figure 4–44. This eliminates one unknown node voltage since these two nodes are forced to have identical voltages. Finally, remember that the ideal OP AMP draws no current at its inputs ($i_P = i_N = 0$) in Figure 4–44, so these currents can be ignored when formulating node equations.

The following steps outline an approach to the formulation of node equations for OP AMP circuits:

STEP 1 Identify a node voltage at all nonreference nodes, including OP AMP outputs, but do *not* formulate node equations at the OP AMP output nodes.

STEP 2 Formulate node equations at the remaining nonreference nodes and then use the ideal OP AMP voltage constraint $v_P = v_N$ to reduce the number of unknowns.

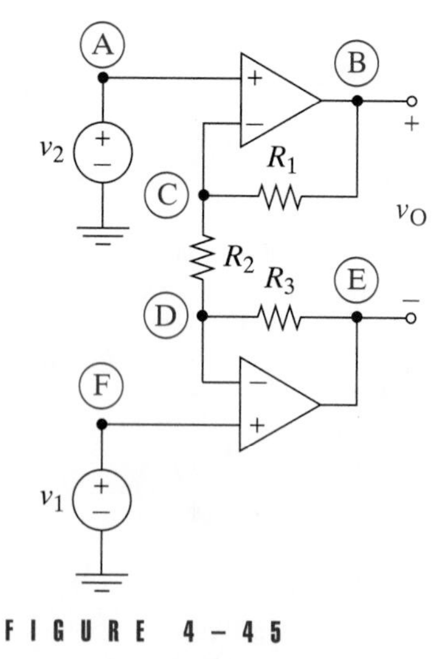

FIGURE 4–45

EXAMPLE 4–18

Derive an expression for v_O in Figure 4–45 in terms of the inputs $v_1$and v_2.

SOLUTION:

This circuit is not an interconnection of OP AMP building blocks because the resistor R_2 is connected between two inputs. We use the node-voltage method outlined above to find the required input-output relationship.

STEP 1 The circuit has a total of six nonreference nodes as shown in the figure. Nodes B and E are OP AMP outputs while nodes A and F are connected to ground by the input voltage sources v_2 and v_1, respectively. As a result, we only need node equations at C and D. Writing the sum of currents leaving these nodes

$$\text{Node C:} \quad \frac{v_C - v_B}{R_1} + \frac{v_C - v_D}{R_2} = 0$$

$$\text{Node D:} \quad \frac{v_D - v_C}{R_2} + \frac{v_D - v_E}{R_3} = 0$$

yields two equations in the four node voltages v_B, v_C, v_D, and v_E.

STEP 2 The noninverting OP AMP inputs are connected to independent voltages sources v_1 and v_2. The OP AMP voltage constraints ($v_P = v_N$) mean that $v_D = v_1$ and $v_C = v_2$. These constraints eliminate v_C and v_D as unknowns and reduce the two node equations to

$$\text{Node C:} \quad \frac{v_2 - v_B}{R_1} + \frac{v_2 - v_1}{R_2} = 0$$

$$\text{Node D:} \quad \frac{v_1 - v_2}{R_2} + \frac{v_1 - v_E}{R_3} = 0$$

The node-voltage formulation method outlined above leads to two equations in the two unknown node voltages v_B and v_E.

We solve the node C equation for v_B

$$v_B = v_2 + \frac{R_1}{R_2}(v_2 - v_1)$$

and the node D for v_E

$$v_E = v_1 + \frac{R_3}{R_2}(v_1 - v_2)$$

Using the fundamental property of node voltages yields the required output as

$$v_O = v_B - v_E = \left[\frac{R_1 + R_2 + R_3}{R_2}\right](v_2 - v_1)$$

The circuit in Figure 4–45 is a differential amplifier of the form $v_O = K(v_2 - v_1)$ in which the input voltages v_1 and v_2 are applied at noninverting OP AMP inputs that have very high input resistances. The circuit is the first stage of a commercially available integrated circuit called an instrumentation amplifier.

Exercise 4–15

Select values of R_1, R_2, and R_3 in Figure 4–45 so that $v_O = 50(v_2 - v_1)$.

Answer:

Selecting $R_1 = R_3 = 24.5\ \text{k}\Omega$ and $R_2 = 1\ \text{k}\Omega$ is one of many possible solutions.

EXAMPLE 4–19

Circuit simulation programs like PSpice or Electronics WorkBench do not support the theoretical characteristics of the ideal OP AMP model we use in circuit analysis and design. The default OP AMP in the Electronics WorkBench library lists its model as "ideal." The purpose of this example is to evaluate this default model as a stand-in for the theoretical ideal OP AMP.

(a) Determine the parameters of the default OP AMP model in Electronics WorkBench.
(b) Compare the theoretical voltage gain, input resistance, and output resistance of an inverting OP AMP circuit with $R_1 = 10$ kΩ and $R_2 = 50$ kΩ with the simulation results predicted by Electronics WorkBench using its default model.

SOLUTION:

(a) Figure 4–46 shows the circuit symbol for the three-terminal OP AMP component available in the Analog IC's group of the Parts Bin toolbar. Double-clicking on the circuit symbol brings up a menu of OP AMPs including the default OP AMP whose model is listed as "ideal." Clicking on edit brings up the parameter list shown in Figure 4–46. The voltage gain, input resistance, and output resistance of the "ideal" model are listed as $A = 10^6$, $R_I = 10^{10}$, and $R_O = 1$. The corresponding theoretical values for an ideal OP AMP are $A = \infty$, $R_I = \infty$, and $R_O = 0$. Thus, the default OP AMP approximates, but does not duplicate, the theoretical ideal. The question is, how much difference does this make?

FIGURE 4–46

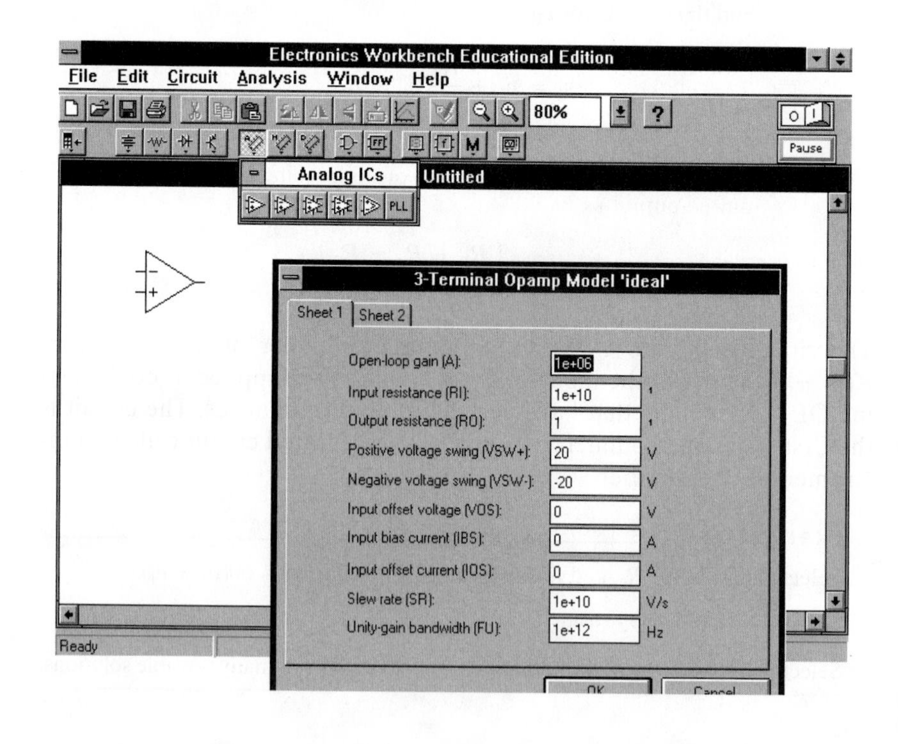

(b) Figure 4–47 shows different stages of constructing the circuit diagram for an inverting amplifier with $R_1 = 10\ \text{k}\Omega$ and $R_2 = 50\ \text{k}\Omega$. The OP AMP is the default model whose parameters were found in part (a). The left side of the figure shows the component placements prior to interconnection. The resistors are available in the Basic Group of the Parts Bin toolbar. The input voltage source and ground are found in the Sources Group. To wire components together, we point at a component terminal, click and drag a wire to the terminal of another component, and release the mouse button. The program then automatically routes the wire at right angles, without overlapping any of the components. The circuit with all components interconnected is shown in the right side of Figure 4–47.

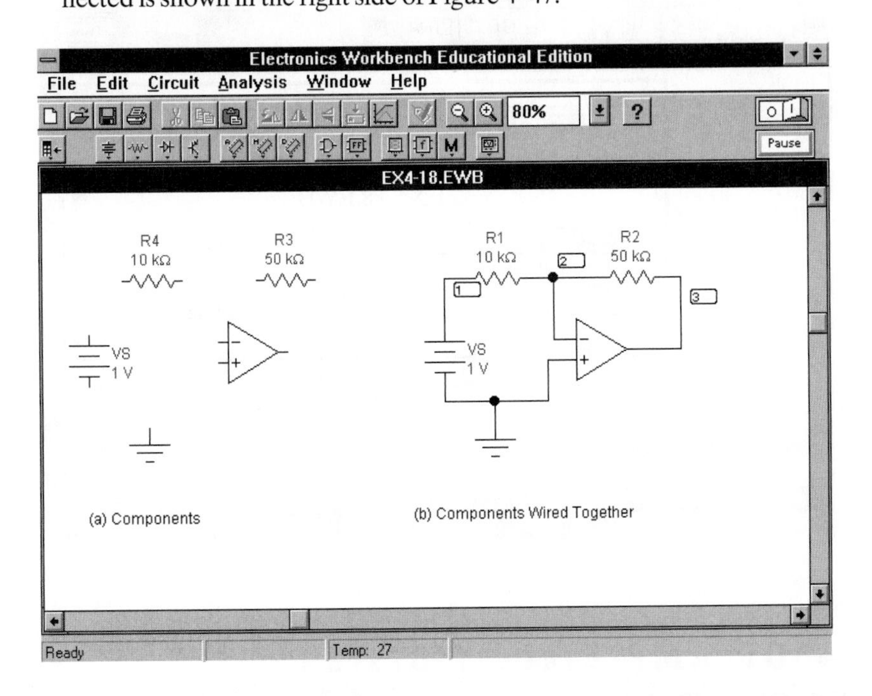

FIGURE 4–47

From the Analysis Menu we select the Transfer Function option, which brings up the window shown at the top of Figure 4–48. In this window we identify the output at the voltage between node 3 and node 0 (ground), and the input as the voltage source VS. Clicking on the simulate button causes the program to perform a dc analysis of the circuit and report the results shown at the bottom of Figure 4–48. Using the default model, the program predicts an output resistance of 5.999997 μΩ, a voltage gain of –4.99997, and an input resistance of 10.00005 kΩ. An ideal OP AMP in the same inverting configuration would have an output resistance of zero, a voltage gain of $-R_2/R_1 = -5$, and an input resistance of $R_{IN} = R_1 = 10\ \text{k}\Omega$. The results using the default model agree with those of the ideal model to within four significant figures for voltage gain and input resistance. The output resistance is not exactly zero, but the value 6 μΩ is negligible compared with other circuit resistances. In summary, the default model and the ideal model predict essentially identical results in typical OP AMP circuit applications. From this point on we will use the default model in circuit simulations as a stand-in for the ideal OP AMP. ■

FIGURE 4–48

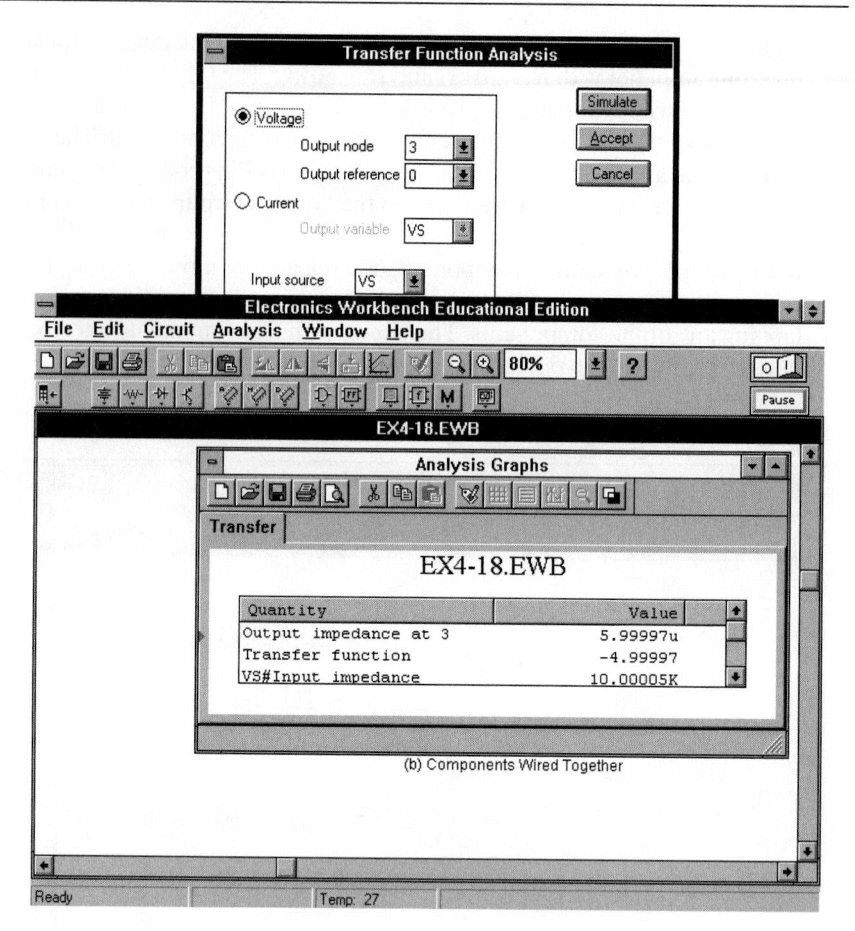

4–6 OP AMP CIRCUIT DESIGN

With OP AMP circuit analysis, we are asked to find the input-output relationship for a given circuit configuration. An OP AMP circuit analysis problem has a unique answer. In OP AMP circuit design, we are given an equation or block diagram representation of a signal-processing function and asked to devise a circuit configuration that implements the desired function. Circuit design can be accomplished by interconnecting the amplifier, summer, and subtractor building blocks shown in Figure 4–40. The design process is greatly simplified by the nearly one-to-one correspondence between the OP AMP circuits and the elements in a block diagram. However, a design problem may not have a unique answer since often there are several OP AMP circuits that meet the design objective. The following example illustrates the design process.

DESIGN EXAMPLE 4–20

Design the interface circuit in Figure 4–49 so that 200 mW is delivered to the 500-Ω load.

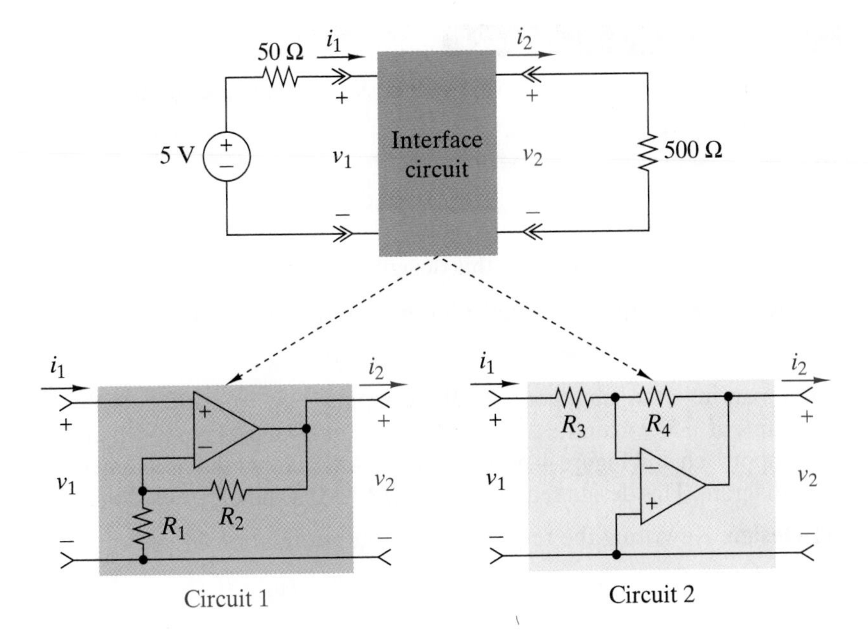

FIGURE 4–49

SOLUTION:

The maximum power available from the 5-V source is

$$p_{\mathrm{MAX}} = \frac{v_{\mathrm{T}}^2}{4R_{\mathrm{T}}} = \frac{5^2}{4 \times 50} = 125 \text{ mW}$$

which is less than the required 200-mW output. This means that the interface circuit must contain an active device that provides voltage gain. To determine the required voltage gain, we express the output power as

$$p_2 = \frac{v_2^2}{500} = 0.2 \text{ W}$$

which yields the required output voltage as $v_2 = \pm\sqrt{0.2 \times 500} = \pm 10$ V. Since the input is a 5-V source, we need overall voltage gains of $K = \pm 2$. We can get these gains using either the noninverting or the inverting amplifier shown in the figure.

1st Design: The noninverting amplifier shown in the figure has very high input resistance; hence, $i_1 = 0$ and $v_1 = 5$ V. To get a voltage gain of $K = +2$, we need $(1 + R_2/R_1) = 2$. Selecting $R_1 = R_2 = 10$ kΩ completes the design. Many other choices are possible.

2nd Design: The inverting amplifier shown in the figure has an input resistance of R_3; hence, the input current is not zero, and $v_1 = 5 - 50i_1 < 5$ V. In other words, the interface circuit loads the input source. However, if we specify that $R_3 >> 50$ Ω, the effect of loading can be ignored since $i_1 \approx 0$ and the input voltage is essentially 5 V. Neglecting the effect of loading, a voltage gain of $K = -2$ requires $R_4/R_3 = 2$. Selecting $R_3 = 25$ kΩ and $R_4 = 50$ kΩ completes the design. Many other choices are possible as long as $R_3 >> 50$ Ω.

DESIGN EXAMPLE 4–21

Design a circuit using OP AMPs that implement the input-output relationship

$$v_O = 5v_1 + 4v_2 - 2v_3$$

SOLUTION:

We will show two ways to solve this design problem.

1st Design: Rewriting the required input-output relationship as

$$v_O = -5(-v_1) - 4(-v_2) - 2v_3$$

suggests an inverting summer with three inputs $-v_1$, $-v_2$, and v_3 with summing gains of −5, −4, and −2, respectively. Figure 4–50(a) is a block diagram of this approach and Figure 4–50(b) is an OP AMP circuit that implements the block diagram. This design requires three OP AMPs and eight resistors.

2nd Design: Rewriting the required input-output relationship as

$$v_O = -2[-2.5v_1 - 2v_2] - 2v_3$$

FIGURE 4–50

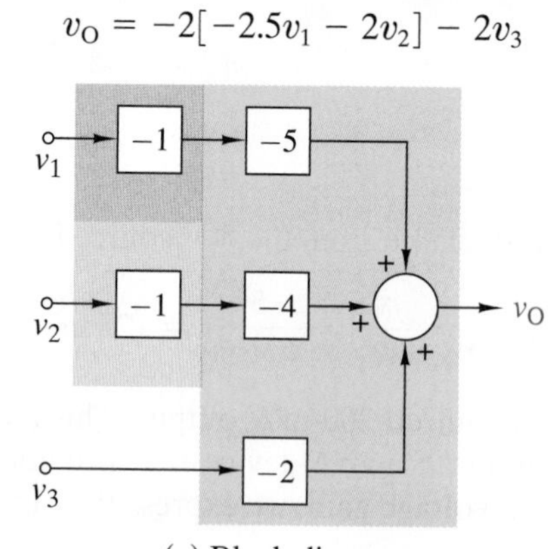

(a) Block diagram

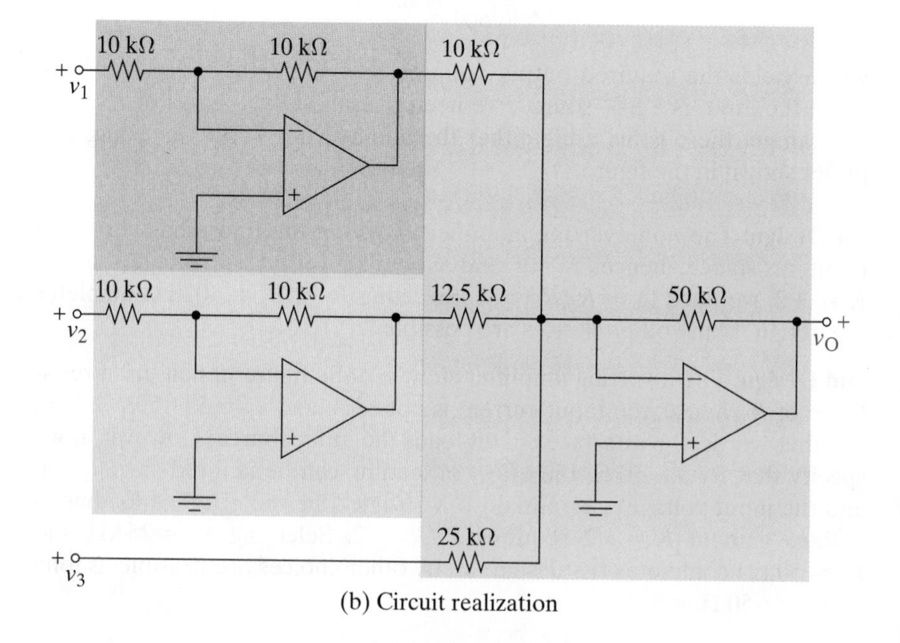

(b) Circuit realization

suggests an inverting summer with two inputs $[-2.5v_1 - 2v_2]$ and v_3, both with a summing gain of -2. The input $[-2.5v_1 - 2v_2]$ can be obtained using a second inverting summer whose inputs are v_1 and v_2. Figure 4–51(a) is a block diagram of this approach and Figure 4–51(b) is an OP AMP circuit that implements the block diagram. This design requires two OP AMPs and six resistors.

FIGURE 4–51

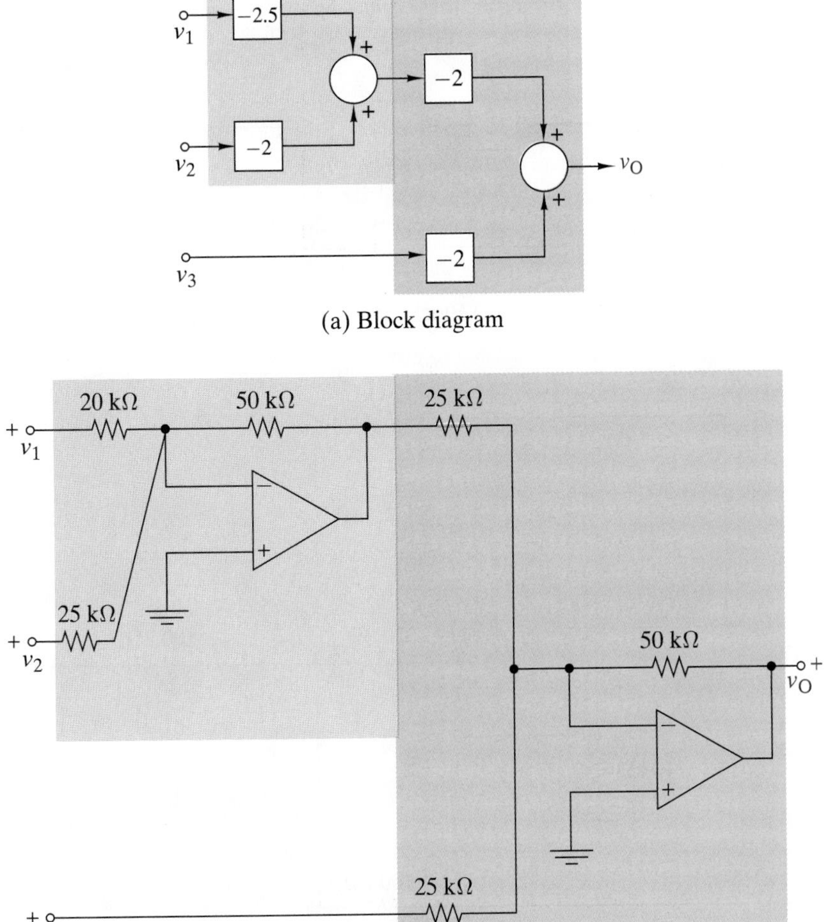

(a) Block diagram

(b) Circuit realization

Evaluation Discussion: This example illustrates that there are often several ways to solve a given design problem. This leads to the question of how to choose between different design solutions. Candidate designs are evaluated using criteria beyond the fact that they implement a given signal-processing function. One common measure is to choose the design that uses the fewest components. Applying this "smallest-part-count" criteria in this example leads us to choose the second design.

Using this criteria, it is important to understand the impact of part count on the cost of fabricating a circuit. Reducing part count does much more than simply reduce component costs. It also (and often more importantly) reduces printed circuit board space, assembly costs, packaging costs, testing, and logistics costs.

Exercise 4–16

A requirement exists for a circuit that implements the block diagram in Figure 4–52(a). The circuit in Figure 4–52(b) is a proposed solution. A breadboard prototype of this circuit failed to pass preliminary testing. Why? *Hint:* The circuit contains four errors. What are they?

FIGURE 4–52

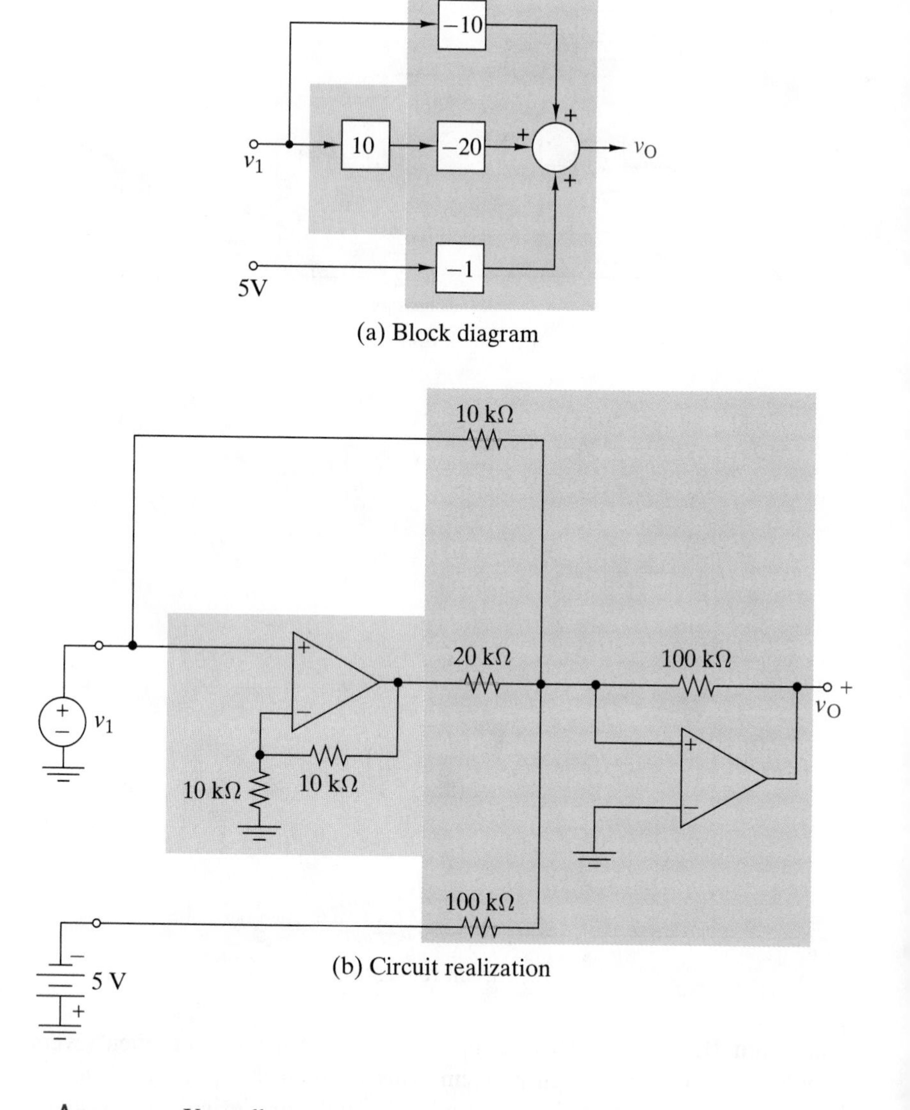

(a) Block diagram

(b) Circuit realization

Answer: You tell me.

4–7 OP AMP CIRCUIT APPLICATIONS

OP AMP circuits are fundamental building blocks in a wide range of signal-processing applications, especially instrumentation, status monitoring, process control, filtering, digital-to-analog conversion, and analog-to-digital conversion. This section provides a brief introduction to three of these applications.

DIGITAL-TO-ANALOG CONVERSION

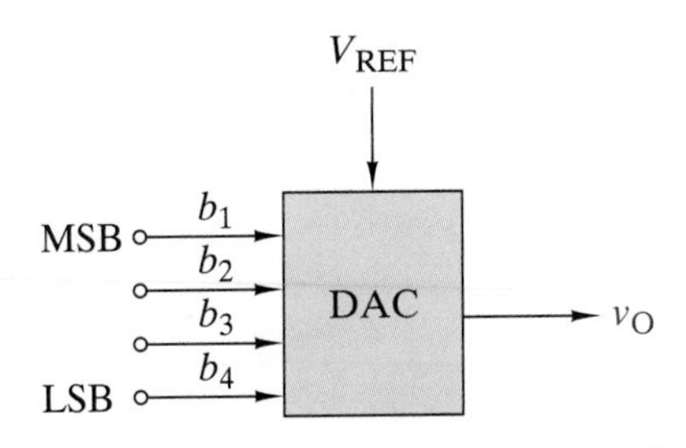

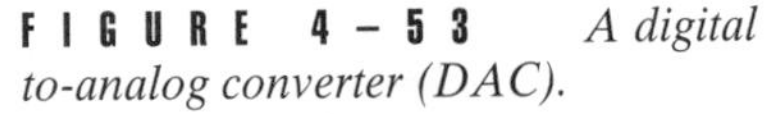
FIGURE 4–53 A digital-to-analog converter (DAC).

A digital-to-analog converter (DAC) is a mixed-signal device with a multibit binary input and an analog output proportional to the decimal equivalent of the binary input. Figure 4–53 is a block diagram of a DAC with a four-bit digital input (b_1, b_2, b_3, b_4), an analog output v_O, and a fixed reference voltage V_{REF}. The input bits can have only one to two values: a high (1) or low (0). The input-output relationship of a four-bit converter can be written as follows:

$$v_O = KV_{REF}\left(b_1 + \frac{b_2}{2} + \frac{b_3}{4} + \frac{b_4}{8}\right) \tag{4–34}$$

Bit b_1 is called the most significant bit (MSB) because it carries the largest weight in this sum. Conversely, bit b_4 is called the least significant bit (LSB) because it carries the smallest weight.

For example, a four-bit DAC with $K = 0.5$ and $V_{REF} = 10$ V and a digital input (0, 1, 0, 1) produces an analog output of

$$v_O = 0.5 \times 10\left(0 + \frac{1}{2} + \frac{0}{4} + \frac{1}{8}\right) = 3.125 \text{ V}$$

Similarly, inputs of (0, 1, 1, 1) and (1, 1, 0, 0) produce outputs of 4.375 V and 8.75 V, respectively. With a four-bit converter there are $2^4 = 16$ possible input codes and hence 16 possible analog output levels. The *full-scale output* of a DAC is defined as the output when the input bits are all one ($v_O = 9.375$ V in this example). The *resolution* of a DAC is defined as the change in the analog output caused by an input change of one LSB (0.625 V in this example). We can think of resolution as the voltage increment between adjacent output levels.

The N-bit converter generalization of four-bit input-output relationship in Eq. (4–34) is

$$v_O = KV_{REF}\sum_{k=1}^{N}\frac{b_k}{2^{k-1}} \tag{4–35}$$

Integrated circuit DACs are available with $N = 8$ to $N = 24$ bit inputs. Increasing the number of bits improves the conversion precision since the resolution is inversely proportional to 2^{N-1}. For this reason DAC resolution is often somewhat loosely quoted in terms of bits.

We now discuss two OP AMP circuits that implement Eq. (4–35) for $N = 4$. The first method uses the four-input OP AMP summer in Figure 4–54. The summer inputs v_1, v_2, v_3, and v_4 are applied to the binary-weighted input resistors R, $2R$, $4R$, and $8R$, respectively. The output of the summer is related to these inputs by the input-output relationship in Eq. (4–29).

$$v_O = -\frac{R_F}{R}v_1 - \frac{R_F}{2R}v_2 - \frac{R_F}{4R}v_3 - \frac{R_F}{8R}v_4$$

$$= -\frac{R_F}{R}\left(v_1 + \frac{v_2}{2} + \frac{v_3}{4} + \frac{v_4}{8}\right)$$

The input voltages are determined by switches controlled by the input bits b_1, b_2, b_3, and b_4. When a bit is low (0), the switch is in the lower position connecting the related input to ground. When a bit is high (1), the switch is in the upper position connecting the related input to the reference voltage V_{REF}. In

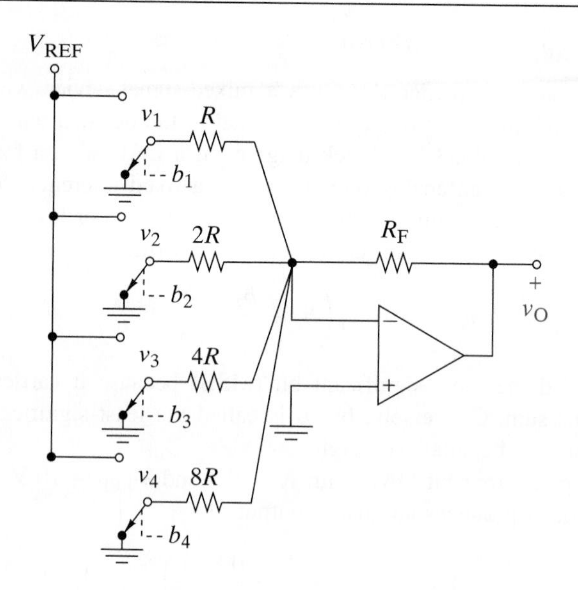

FIGURE 4–54 A binary-weighted summer DAC.

other words, an input voltage is zero when its control bit is 0 and equal to V_{REF} when its control bit is 1. In effect, the input voltages are related to the input bits as $v_k = b_k V_{REF}$. As a result, we can write the summer output as

$$v_O = -\frac{R_F}{R}\left(b_1 V_{REF} + \frac{b_2 V_{REF}}{2} + \frac{b_3 V_{REF}}{4} + \frac{b_4 V_{REF}}{8}\right)$$
$$= -\frac{R_F}{R} V_{REF}\left(b_1 + \frac{b_2}{2} + \frac{b_3}{4} + \frac{b_4}{8}\right)$$

This result is of the form of Eq. (4–35) with $N = 4$ and $K = -R_F/R$.

A four-bit DAC can also be realized using the *R*-2*R* ladder circuit in Figure 4–55. In this case the voltage inputs v_1, v_2, v_3, and v_4 are applied to

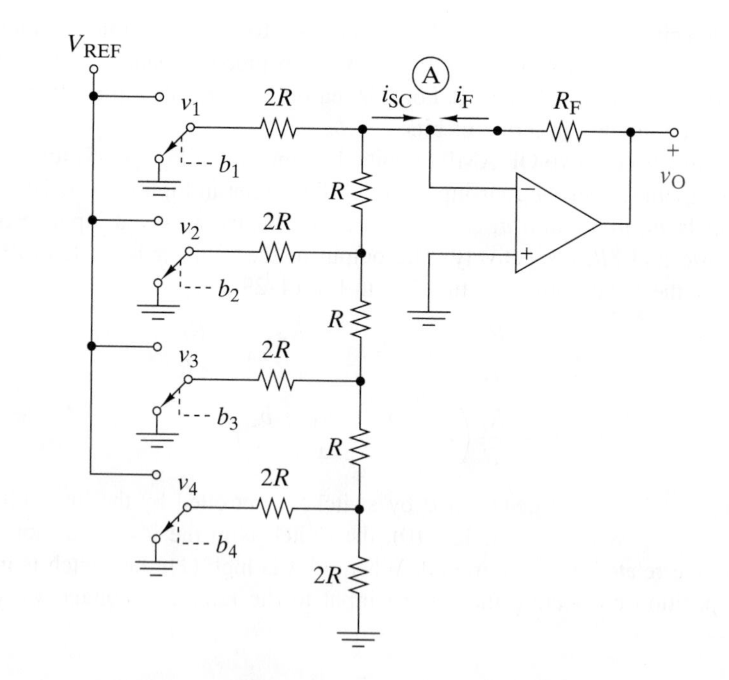

FIGURE 4–55 An *R*-2*R* ladder DAC.

the $2R$ legs of the ladder. The OP AMPs' noninverting input is connected to ground ($v_P = 0$). Since $v_N = v_P = 0$, the node A in Figure 4–55 is a "virtual" ground. In other words, the R-$2R$ ladder is effectively shorted to ground at node A. The short-circuit current the ladder delivers to this virtual short can be shown to be

$$i_{SC} = \frac{v_1}{2R} + \frac{v_2}{4R} + \frac{v_3}{8R} + \frac{v_4}{16R}$$

Since node A is a virtual ground, the current through the feedback resistor is $i_F = v_O/R_F$. For an ideal OP AMP, $i_N = 0$, so the KCL constraint at node A requires $i_F + i_{SC} = 0$, or

$$\frac{v_O}{R_F} + \frac{v_1}{2R} + \frac{v_2}{4R} + \frac{v_3}{8R} + \frac{v_4}{16R} = 0$$

Solving for v_O leads to the following results

$$\begin{aligned} v_O &= -\frac{R_F}{2R}v_1 - \frac{R_F}{4R}v_2 - \frac{R_F}{8R}v_3 - \frac{R_F}{16R}v_4 \\ &= -\frac{R_F}{2R}\left(v_1 + \frac{v_2}{2} + \frac{v_3}{4} + \frac{v_4}{8}\right) \end{aligned}$$

The input volltages v_1, v_2, v_3, and v_4 are defined by the same switching arrangement as the inverting summer DAC in Figure 4–54. Hence, they are related to the input bits as $v_k = b_k V_{REF}$ and the output voltage becomes

$$\begin{aligned} v_O &= -\frac{R_F}{2R}\left(b_1 V_{REF} + \frac{b_2 V_{REF}}{2} + \frac{b_3 V_{REF}}{4} + \frac{b_4 V_{REF}}{8}\right) \\ &= -\frac{R_F}{2R}V_{REF}\left(b_1 + \frac{b_2}{2} + \frac{b_3}{4} + \frac{b_4}{8}\right) \end{aligned}$$

This result is of the form of Eq. (4–35) with $N = 4$ and $K = -R_F/2R$.

In theory the four-bit inverting summer DAC in Figure 4–54 could be extended to a larger number of bits. However, the range of input resistances rapidly gets out of hand. For example, a 10-bit DAC would require input resistances ranging from R to $1024R$. With integrated circuit technology it is virtually impossible to maintain tight resistance tolerances over a wide range of resistance values.

The 4-bit R-$2R$ ladder DAC in Figure 4–55 can be extended to a larger number of bits using only two values of resistance, namely R and $2R$. The absolute value of R does not matter. What matters is that the added resistances stand in a two-to-one ratio. Controlling the ratio of two resistances is much easier to accomplish with integrated circuit technology than controlling the absolute value of a resistance. This fact accounts for the widespread use of R-$2R$ ladder architecture in integrated circuit DACs.

Exercise 4–17

The R-$2R$ ladder DAC in Figure 4–55 has $R_F = 40\ \text{k}\Omega$, $R = 10\ \text{k}\Omega$, and $V_{REF} = -3$ V. Find the full-scale output and resolution of the converter.

Answers: 11.25 V; 0.75 V

TRANSDUCER INTERFACE CIRCUITS

Transducers are key elements of instrumentation systems that monitor, measure, and control physical processes. A **transducer** is a device that converts a process variable such as temperature, pressure, motion, or light into an electrical signal.[4] Electrical signal processing then converts the transducer output into a form required by the rest of the instrumentation system. The key first step is the interface circuit shown in Figure 4–56. This circuit translates the transducer output signal range onto a range suitable for further signal processing.

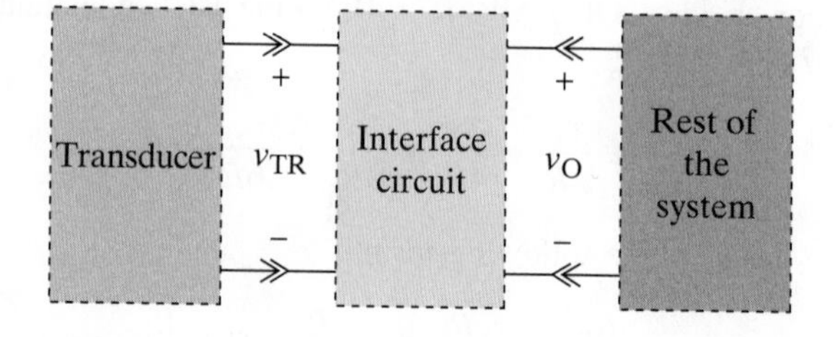

FIGURE 4–56 *Transducer interface circuit.*

The input to the interface circuit in Figure 4–56 is the transducer voltage v_{TR}. The output of the interface circuit must match the requirements of the rest of the system. To accomplish this goal, the interface circuit usually performs two functions, *amplification* and *level shifting*. Amplification is needed because transducer signals are often in the millivolt range. Level shifting is required because the amplified transducer signal may not fall in a range required by the rest of the system. Thus, the input-output relationship of the interface circuit is of the form

$$v_O = Kv_{TR} + V_{BIAS} \tag{4–36}$$

where K is the required voltage gain and V_{BIAS} provides the necessary level shifting.

Let us consider a specific example of transducer interfacing. Figure 4–57 shows the characteristics of a light transducer that converts illuminance in the range from 200 to 1000 lumens/m^2 into an electrical signal in the range from 4 to 20 mV. The output of the interface circuit is required to drive an analog-to-digital converter whose full-scale input range is 0 to 5 V. The interface circuit must convert its 4- to 20-mV input range into a 0- to 5-V output range.

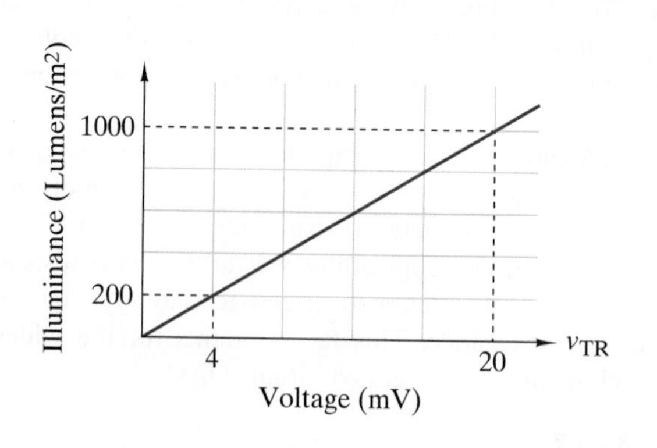

FIGURE 4–57 *Photocell transducer characteristics.*

4 The word *sensor* is also commonly used in place of *transducer*. The two words are used interchangeably, but both refer to the first element in an instrumentation system that detects a process variable and produces a corresponding electrical signal.

Inserting these values into Eq. (4–36) gives

$$0 = K \times 4 \times 10^{-3} + V_{BIAS} \quad \text{at 200 lumens/m}^2$$

$$5 = K \times 20 \times 10^{-3} + V_{BIAS} \quad \text{at 1000 lumens/m}^2$$

Subtracting the second equation from the first and solving for the gain K yields

$$K = \frac{5 - 0}{0.02 - 0.004} = 312.5$$

Inserting this value of K into the first equation and solving for the bias gives $V_{BIAS} = -0.004K = -1.25$ V. Thus, the interface circuit must implement an input-output relationship defined by

$$v_O = (312.5)\, v_{TR} - 1.25 \tag{4–37}$$

To design the interface circuit, we partition the required gain as $K = (-25)(-12.5) = 312.5$. This allows us to get the overall gain of $+312.5$ using two inverting amplifier stages with gains of -25 and -12.5. We also rewrite the bias voltage as $V_{BIAS} = (-0.25)5 = -1.25$ V. This allows us to get the required bias using an inverting gain of -0.25 and a standard 5-V power supply as the reference source. Inserting these results into Eq. (4–37) yields

$$v_{TR} = (-25)(-12.5)v_{TR} - (0.25)5 \tag{4–38}$$

Figure 4–58(a) shows a block diagram of Eq. (4–38) while Figure 4–58(b) shows an OP AMP circuit that implements the block diagram. Obviously this design is not unique since we can rearrange Eq. (4–37) in many different ways.

D DESIGN EXAMPLE 4–22

A commercially available temperature transducer has the characteristics shown in Figure 4–59. Design an OP AMP circuit to interface the transducer output for temperatures in the range from−20°C to 120°C with a panel meter whose full-scale input range is 0 to 3 V. Use a standard 1.5-V battery as the reference source for the required bias.

SOLUTION:

Figure 4–59 shows that the transducer output voltage ranges from 2.1 V to 0.42 V for the temperature range from −20°C to 120°C. The interface circuit must convert a 2.1-V to 0.42-V range into a 0-V to 3-V range. Inserting these values into Eq. (4–36) gives

$$0 = K \times 2.1 + V_{BIAS} \quad \text{at } -20^\circ\text{C}$$

$$3 = K \times 0.42 + V_{BIAS} \quad \text{at } 120^\circ\text{C}$$

Subtracting the second equation from the first and solving for the gain K yields

$$K = \frac{3 - 0}{0.42 - 2.1} = -\frac{1}{0.56}$$

Inserting this value of K into the first equation and solving for the bias gives $V_{BIAS} = -2.1K = 3.75$ V. The bias voltage can be rewritten as $V_{BIAS} = 3.75 = (2.5)1.5$ V, which means we need a gain of 2.5 to get

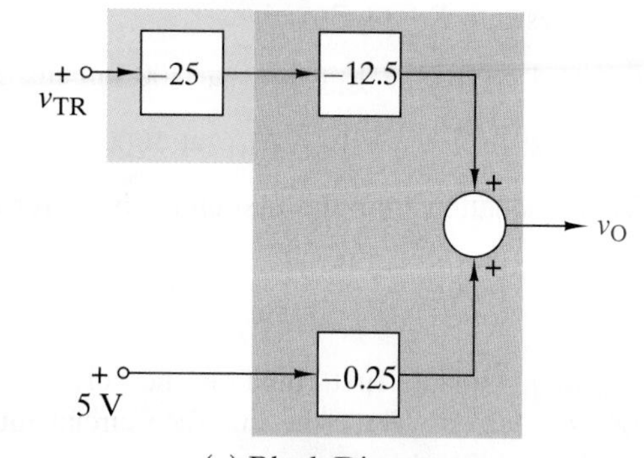

(a) Block Diagram

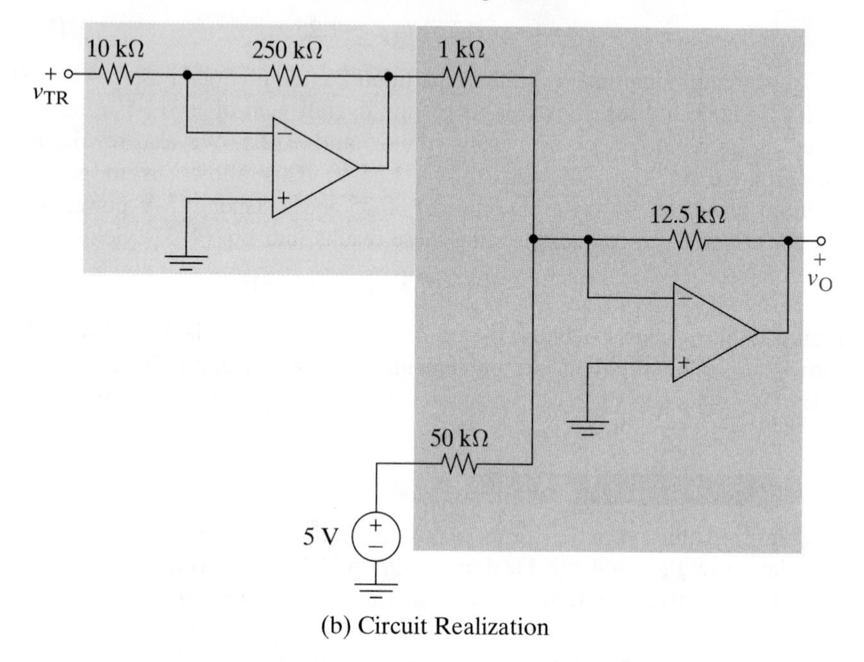

(b) Circuit Realization

FIGURE 4–58 *Photocell interface circuit realization.*

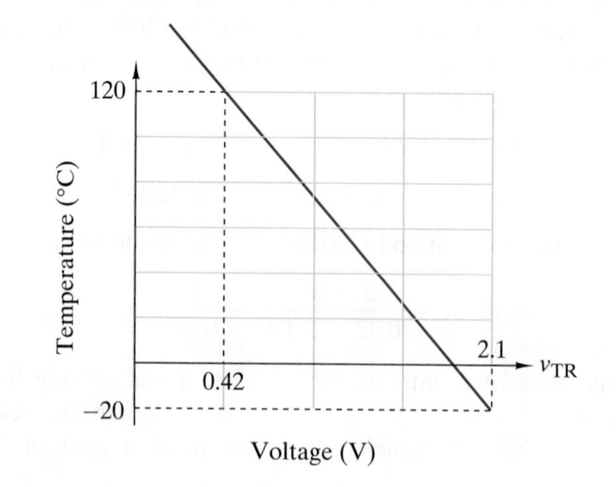

FIGURE 4–59

required bias from the specified 1.5-V battery reference source. Thus, the interface circuit must realize the following input-output relationship.

$$v_O = -\left(\frac{1}{0.56}\right)v_{TR} + (2.5)1.5$$

We now use each of the subtractor circuits in Figure 4–60 to realize this relationship.

FIGURE 4–60

(a) Basic Subtractor

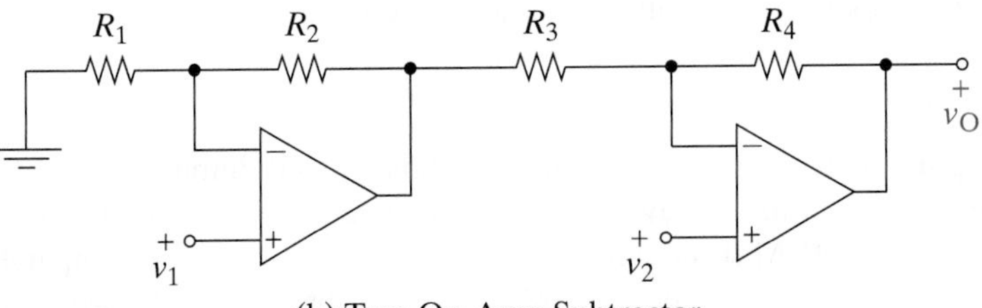

(b) Two-Op-Amp Subtractor

1st Design: The required function is of the form $v_O = -K_1v_1 + K_2v_2$, where $K_1 = 1/0.56$, $v_1 = v_{TR}$, $K_2 = 2.5$, and $v_2 = 1.5$ V. This function can be realized using the basic subtractor building block in Figure 4–60(a). Equation (4–32) relates the two gains to the subtractor circuit parameters as

$$K_1 = \frac{R_2}{R_1} \quad \text{and} \quad K_2 = \left(\frac{R_1 + R_2}{R_1}\right)\left(\frac{R_4}{R_3 + R_4}\right)$$

The gain $K_1 = 1/0.56$ can be realized by selecting $R_2 = 10$ kΩ and $R_1 = 0.56R_2 = 5.6$ kΩ. Using these values in the K_2 equation produces

$$K_2 = \left(1 + \frac{10}{5.6}\right)\left(\frac{R_4}{R_3 + R_4}\right) = 2.5$$

Solving for R_4 yields $R_4 = 8.75\,R_3$. Selecting $R_3 = 1$ kΩ requires that $R_4 = 8.75$ kΩ. This completes the first design. Obviously many other choices are possible

2nd Design: The required function $v_O = -K_1v_1 + K_2v_2$ can also be realized using the two-OP-AMP subtractor shown in Figure 4–60(b). This circuit is analyzed in Example 4–17, where the two gains are related to circuit parameters as

$$K_1 = \frac{R_4}{R_3}\left(1 + \frac{R_2}{R_1}\right) \quad \text{and} \quad K_2 = 1 + \frac{R_4}{R_3}$$

The gain $K_2 = 2.5$ can be realized by selecting $R_3 = 10\text{ k}\Omega$, which requires $R_4 = 1.5\ R_3 = 15\text{ k}\Omega$. Using these values in the K_1 equation produces

$$K_1 = \frac{15}{10}\left(1 + \frac{R_2}{R_1}\right) = \frac{1}{0.56}$$

Solving for R_1 yields $R_1 = 5.25\ R_2$. Selecting $R_2 = 10\text{ k}\Omega$ requires that $R_1 = 52.5\text{ k}\Omega$. This completes the second design. Many other choices are possible.

Evaluation Discussion: In terms of part counts, both designs use four resistors. The first design uses a one-OP-AMP subtractor while the second design uses the two-OP-AMP version. This difference makes the first design the best choice under a "smallest-part-count" criteria. However, with the two-OP-AMP subtractor, both input signals are applied directly to noninverting OP AMP inputs. These inputs have very high input resistances, which means that the second design does not load the transducer or drain energy from the 1.5-V battery. This difference could be an important advantage of the second design if the transducer has a high Thévenin resistance or the interface circuit must operate for extended periods of time without servicing the battery.

COMPARATOR CIRCUITS

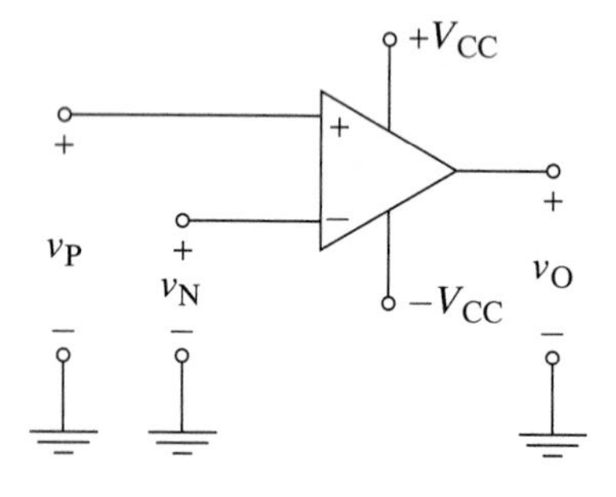

FIGURE 4–61 *An OP AMP comparator.*

A **comparator** is a mixed-signal device whose digital output is either high or low depending on the relative amplitudes of two analog inputs. For example, the open-loop OP AMP circuit in Figure 4–61 functions as a comparator. In the absence of feedback, the OP AMP is driven into one of its two saturation modes by the analog inputs v_P and v_N. Specifically, if $v_P > v_N$, the OP AMP is in +saturation with $v_O = +V_{CC}$. Conversely, if $v_P < v_N$, the OP AMP is in –saturation with $v_O = -V_{CC}$.

In digital logic terminology, the comparator output is said to be high (1) when $v_P > v_N$ and low (0) when $v_P > v_N$. The output voltage levels associated with the high and low states are usually denoted V_{OH} and V_{OL}, respectively. These levels are determined by the positive and negative power supply voltages, which can be different from the ± 15 V commonly used in linear applications of OP AMPs.

Exercise 4–18

Find the comparator output voltage in Figure 4–62 for the following:

(a) $v_1 = 2\text{ V}, v_2 = 3\text{ V}, V_{CC} = 5\text{ V}$
(b) $v_1 = 0, v_2 = -3\text{ V}, V_{CC} = 10\text{ V}$
(c) $v_1 = -2\text{ V}, v_2 = -3\text{ V}, V_{CC} = 3\text{ V}$

Answers: (a) $v_O = 0$ V; (b) $v_O = 10$ V; (c) $v_O = 3$ V

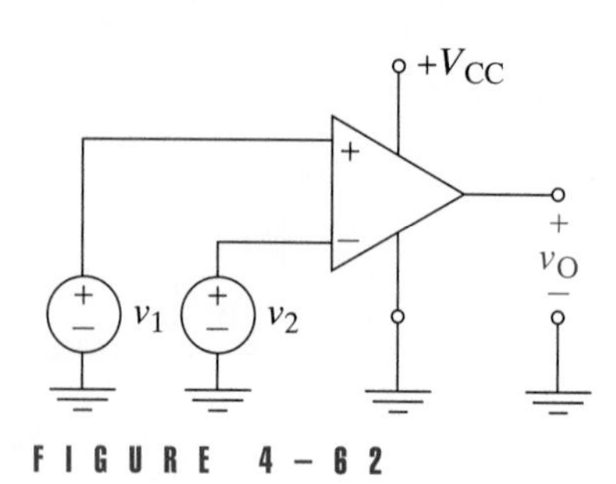

FIGURE 4–62

Figure 4–63(a) shows a comparator circuit often called a *zero-crossing detector.* A time-varying analog signal is applied to the noninverting input $[v_P = v_S(t)]$, and the inverting input is connected to ground $[v_N = 0]$. In this

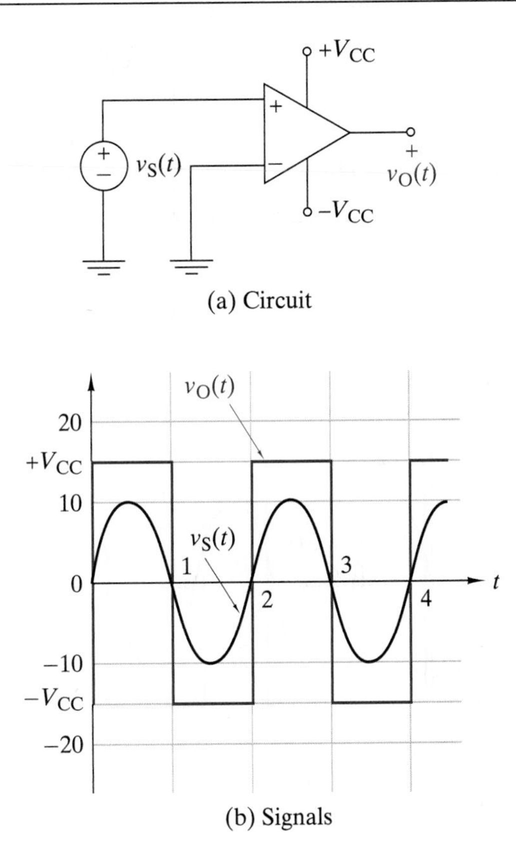

FIGURE 4–63 *Zero-crossing detector.*

configuration the comparator output is $v_O = +V_{CC}$ when $v_S(t) > 0$ and $v_O = -V_{CC}$ when $v_S(t) < 0$. Figure 4–63(b) shows plots of the input and output signals versus time for $v_S(t) = 10\sin(\pi t)$ V and $V_{CC} = 15$ V. In digital logic terms, the comparator output changes state (toggles) whenever the analog input passes through zero; hence the name *zero-crossing detector.* This circuit is also called a *polarity detector* because the digital output is high when the polarity of the analog input is positive and low when the polarity is negative.

A modified version of the zero-crossing detector is shown in Figure 4–64(a). A time-varying analog signal is still applied to the noninverting input $[v_P = v_S(t)]$ as before, and the modification connects a fixed reference voltage to the inverting input $[v_N = V_{REF}]$. In this configuration the comparator output is $v_O = +V_{CC}$ when $v_S(t) > V_{REF}$ and $v_O = -V_{CC}$ when $v_S(t) < V_{REF}$. In effect the applied reference voltage shifts the switching threshold from zero to V_{REF}. Figure 4–64(b) shows how the threshold shift changes the output signal for $v_S(t) = 10\sin(\pi t)$ V, $V_{CC} = 15$ V, and $V_{REF} = 8$ V.

An analog-to-digital converter (ADC) is a mixed-signal device that converts an analog input into a multibit digital output. Figure 4–65 is a block diagram of an ADC with an analog input $v_S(t)$, a four-bit digital output (b_1, b_2, b_3, b_4), and a fixed reference voltage V_{REF}. The output bits can have only one to two values: a high (1) or low (0). The code used by these bits to represent the analog input depends on the architecture of the ADC.

The circuit in Figure 4–66 is a four-comparator ADC that converts the analog input $v_S(t)$ into a four-bit digital output. The analog input is applied to the noninverting input of each of the comparators. A fixed reference voltage is applied to a voltage divider string. Successive taps on the voltage divider supply a reference

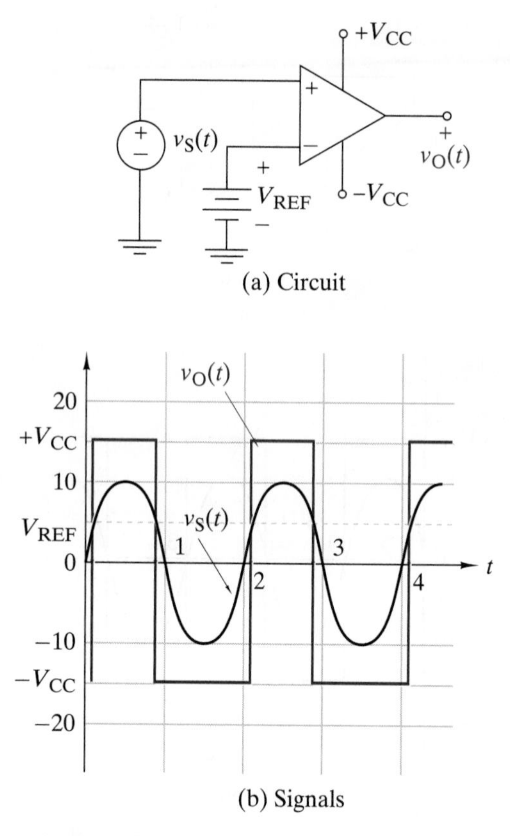

FIGURE 4–64 *Modified zero-crossing detector.*

FIGURE 4–65 *An analog-to-digital converter (ADC).*

voltage to the inverting input of the comparators. These reference voltages are all different, and each is larger than the reference voltage of the comparator immediately below it.

The output of any one comparator is high (1) when the analog input exceeds its applied reference voltage; otherwise it is low (0). When $v_S(t) < 0.2V_{REF}$, the input is smaller than all of the reference voltages, so the digital output is (0, 0, 0, 0). When $0.2V_{REF} < v_S(t) < 0.4V_{REF}$, the digital output is (0, 0, 0, 1). In this range the input exceeds the reference voltage of the lower comparator, so its output switches to $b_4 = 1$. The outputs of the other comparators remain at zero since the input is smaller than their reference voltages. When $0.4V_{REF} < v_S(t) < 0.6V_{REF}$, the output is (0, 0, 1, 1) because the input exceeds the reference voltages of the bottom two comparators but not the top two. From these observations reveal a pattern that is summarized in Table 4–1.

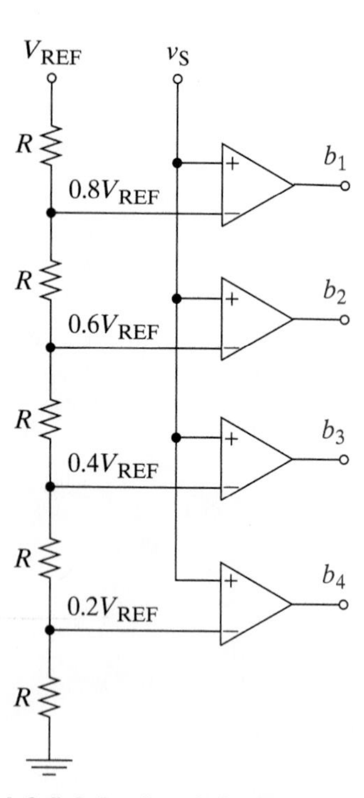

FIGURE 4–66 *Flash ADC.*

TABLE 4–1

ANALOG INPUT RANGE	DIGITAL OUTPUTS B_1	B_2	B_3	B_4
$0 < v_S < 0.2V_{REF}$	0	0	0	0
$0.2V_{REF} < v_S < 0.4V_{REF}$	0	0	0	1
$0.4V_{REF} < v_S < 0.6V_{REF}$	0	0	1	1
$0.6V_{REF} < v_S < 0.8V_{REF}$	0	1	1	1
$0.8V_{REF} < v_S < V_{REF}$	1	1	1	1

The circuit in Figure 4–66 is called a *flash converter* because the comparators operate in parallel and the conversion takes place almost instantaneously. The circuit divides the input amplitude range into five bins and converts each bin into a unique four-bit code. The *full-scale input range* is the voltage range over which the input amplitude falls within one of the bins (0 to V_{REF} in this example). This range can obviously be increased by increasing the reference voltage. The price of doing so is reduced *resolution*, defined as the largest input voltage change that falls entirely within one bin ($0.2V_{REF}$ in this example). Resolution can be improved by expanding the voltage divider string and adding more comparators. Some integrated circuit flash converters have as many as 256 comparators.

The pattern of 1's and 0's in Table 4–1 is called a *thermometer code* because the number of 1's increases monotonically as v_S increases, like the way the mercury column in a thermometer increases with temperature. Integrated circuit flash converters have built-in decoders that convert the thermometer code into a standard binary code. The flash converter is by far the fastest ADC architecture.

Exercise 4–19

The reference voltage in Figure 4–66 is $V_{REF} = 15$ V. What are the output codes corresponding to $v_S = 1, 2, 5, 10$, and 15 V?

Answers: (0, 0, 0, 0); (0, 0, 0, 0); (0, 0, 0, 1); (0, 1, 1, 1); (1, 1, 1, 1)

SUMMARY

- A linear dependent source generates a voltage or a current whose value is proportional to a voltage or current at another point in a circuit. There are four such sources: the current-controlled voltage source, the voltage-controlled voltage source, the current-controlled current source, and the voltage-controlled current source.
- Circuits containing dependent sources can be analyzed by node-voltage or mesh-current methods. Such circuits can have input-output relationships that produce voltage, current, or power gain. The presence of feedback can dramatically influence the input and output resistances of active circuits.
- The bipolar junction transistor (BJT) is a three-terminal active device with three possible operating modes: active, cutoff, and saturation. To determine the operating mode of a transistor, we compare the outputs v_{CE} and i_C predicted by the active mode model with known bounds.
- The OP AMP is an active device with five terminals called the inverting input, the noninverting input, the output, and two power supply terminals. The device is a high-gain differential amplifier with three possible operating modes: +saturation, –saturation, and linear. The output predicted by the linear mode circuit model is compared with known bounds to determine the actual operating mode.
- The ideal OP AMP model has an infinite voltage gain, an infinite input resistance, and zero output resistance. The i–v characteristics of an ideal OP AMP are $i_P = i_N = 0$ and $v_P = v_N$. The ideal model is a good working approximation in linear applications.
- The four basic OP AMP circuit building blocks are the inverting amplifier, the noninverting amplifier, the inverting summer, and the subtractor. The

analysis or design of complex OP AMP circuits can be based on these four building blocks provided the interconnections are made between the output of one to the input of another.

- Important applications of OP AMPs include digital-to-digital converters, transducer interface circuits, and comparator circuits.

PROBLEMS

OBJECTIVE 4–1 LINEAR ACTIVE CIRCUITS (SECTS. 4–1, 4–2)

Given a circuit containing linear resistors, dependent sources, and independent sources, find selected output signal variables, input-output relationships, or input-output resistances.
See Examples 4–2, 4–3, 4–4, 4–5, 4–6, 4–7, 4–8 and Exercises 4–1, 4–2, 4–3, 4–5, 4–6

4–1 Find the voltage gain v_O/v_S and current gain i_O/i_x in Figure P4–1 for $r = 4$ kΩ.

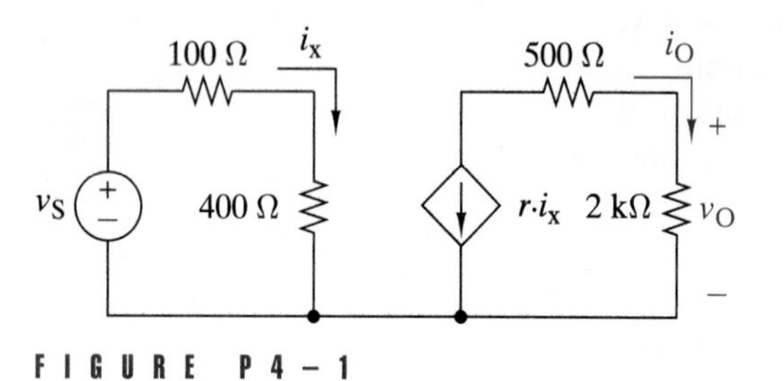

FIGURE P4–1

4–2 Find the voltage gain v_O/v_1 and the current gain i_O/i_S in Figure P4–2. For $i_S = 2$ mA, find the power supplied by the input current source and the power delivered to the 2-kΩ load resistor.

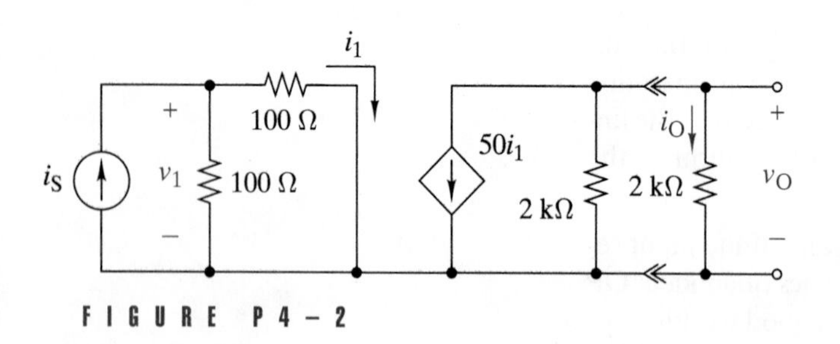

FIGURE P4–2

4–3 Find the voltage gain v_O/v_S and current gain i_O/i_x in Figure P4–3 for $g = 10^{-2}$ S.

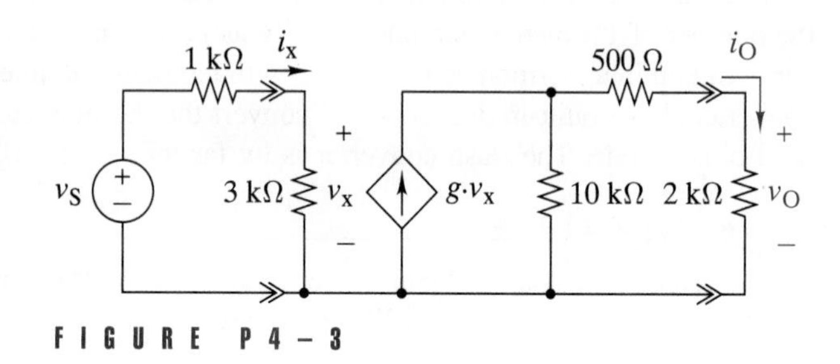

FIGURE P4–3

4–4 Find the voltage gain v_O/v_S and current gain i_O/i_x in Figure P4–4.

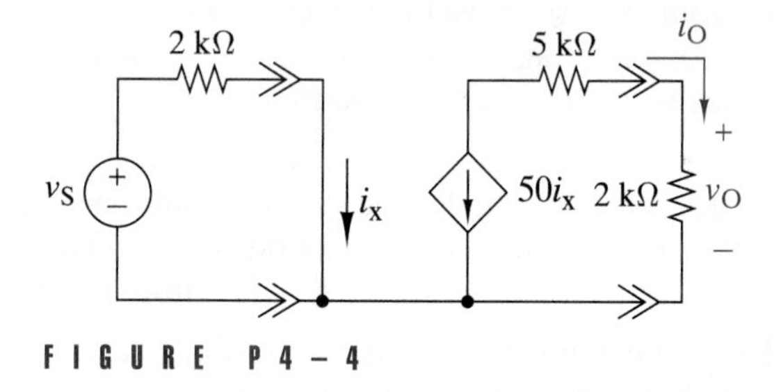

FIGURE P4–4

4–5 Find the current gain i_O/i_S in Figure P4–5.

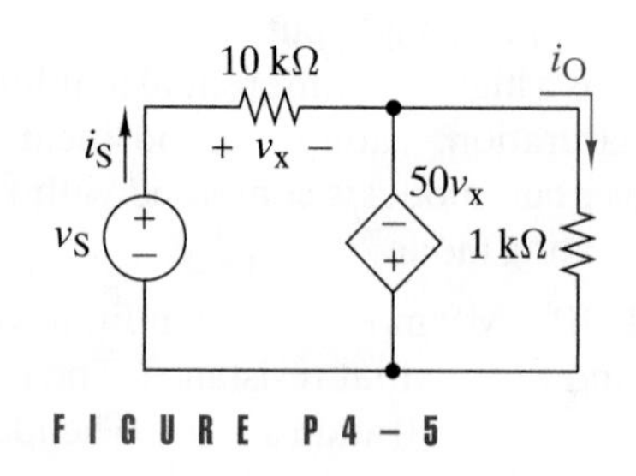

FIGURE P4–5

4–6 Find an expression for the current gain i_O/i_S in Figure P4–6. *Hint:* Apply KCL at node A.

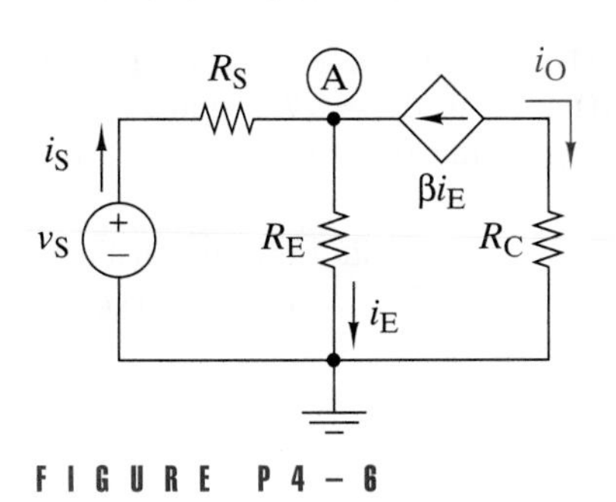

FIGURE P4–6

4–7 Find the voltage v_O in Figure P4–7.

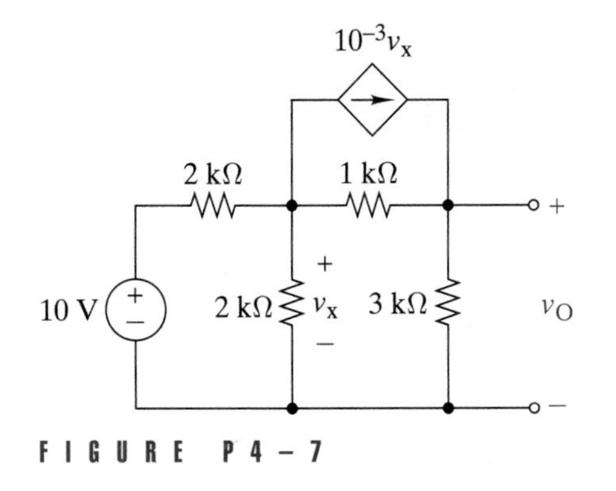

FIGURE P4–7

4–8 Find an expression for the current gain i_O/i_S in Figure P4–8.

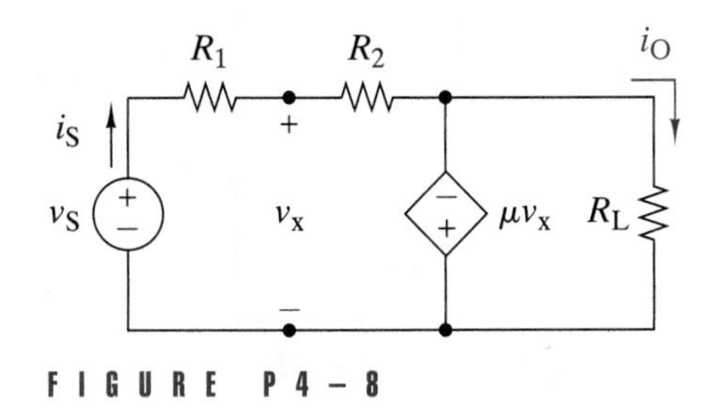

FIGURE P4–8

4–9 Find an expression for the voltage gain v_O/v_S in Figure P4–9.

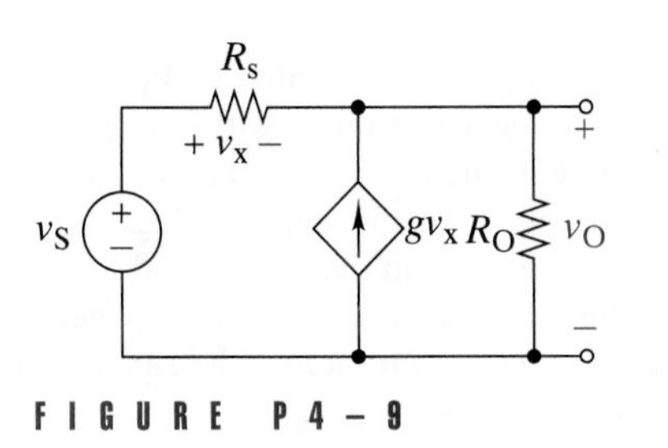

FIGURE P4–9

4–10 Find an expression for the voltage gain v_O/v_S in Figure P4–10.

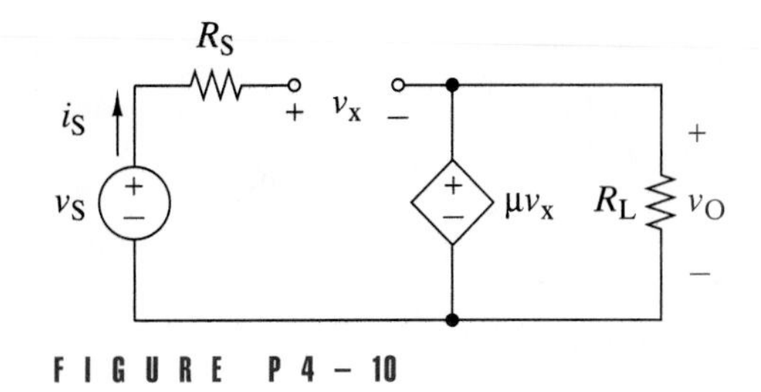

FIGURE P4–10

4–11 Find an expression for the voltage gain v_O/v_S in Figure P4–11.

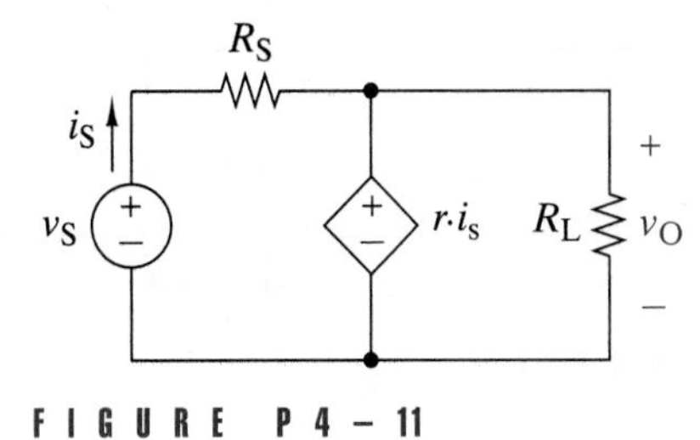

FIGURE P4–11

4–12 Find R_{IN} in Figure P4–12.

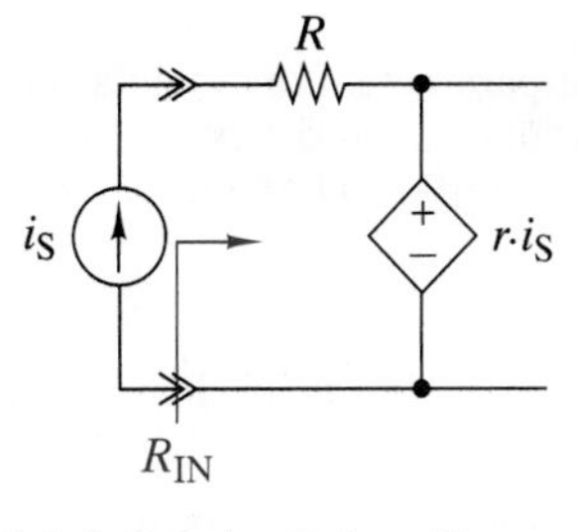

FIGURE P4–12

4–13 Find R_{IN} in Figure P4–13.

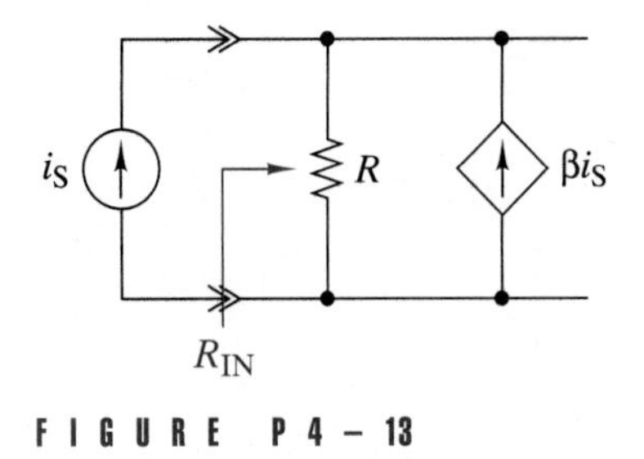

FIGURE P4–13

4–14 Find the Thévenin equivalent circuit seen by the load in Figure P4–14.

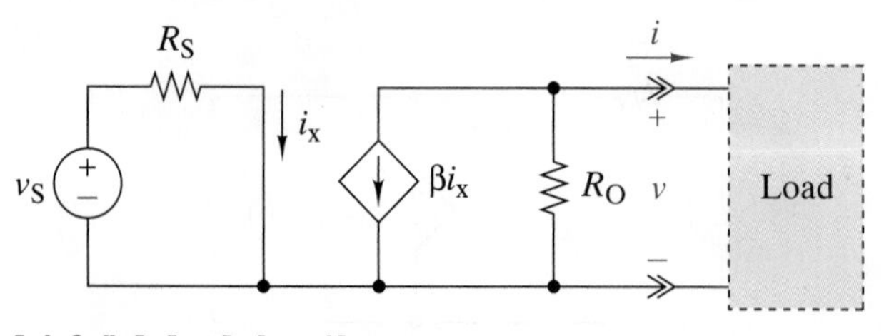

FIGURE P4–14

4–15 Find the Norton equivalent circuit seen by the load in Figure P4–15.

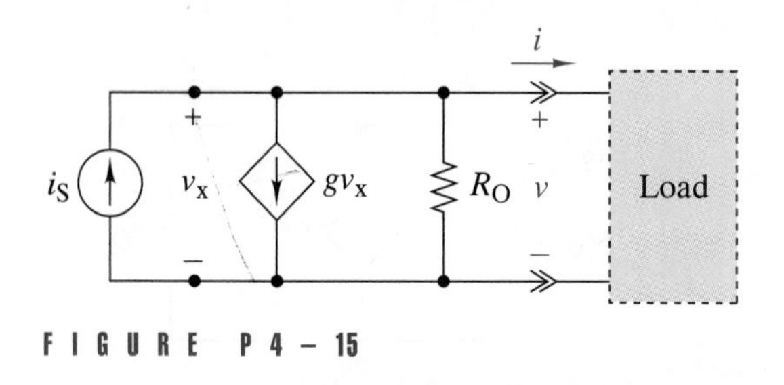

FIGURE P4–15

OBJECTIVE 4–2 TRANSISTOR CIRCUITS (SECT. 4–3)

Given a linear resistive circuit with one transistor:
(a) Find the transistor operating mode and outputs i_C and v_{CE}.
(b) Select circuit parameters to obtain a specified operating mode or output characteristics.
See Examples 4–9, 4–10, 4–11 and Exercises 4–7, 4–8

4–16 The circuit parameters in Figure P4–16 are R_B = 50 kΩ, R_C = 3 kΩ, β = 100, V_γ = 0.7 V, and V_{CC} = 15 V. Find i_C and v_{CE} for v_S = 2 V. Repeat for v_S = 5 V.

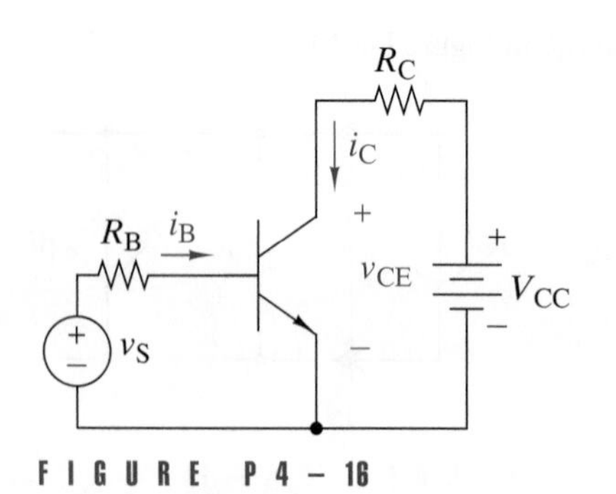

FIGURE P4–16

4–17 **D** The circuit parameters in Figure P4–16 are R_C = 3 kΩ, β = 50, V_γ = 0.7 V, and V_{CC} = 5 V. Select a value of R_B so the transistor is in the saturation mode when $v_S \geq 2$ V.

4–18 The parameters of the transistor in Figure P4–18 are β = 50 and V_γ = 0.7 V. Find i_C and v_{CE} for v_S = 0.8 V. Repeat for v_S = 2 V.

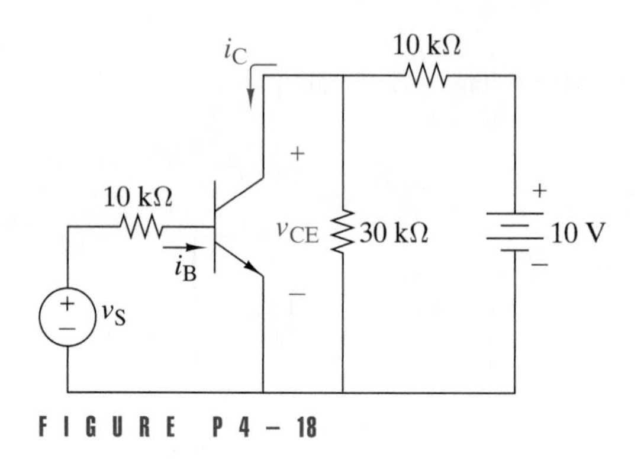

FIGURE P4–18

4–19 In Figure P4–19 the transistor parameters are β = 150 and V_γ = 0.7 V. Find i_C and v_{CE}.

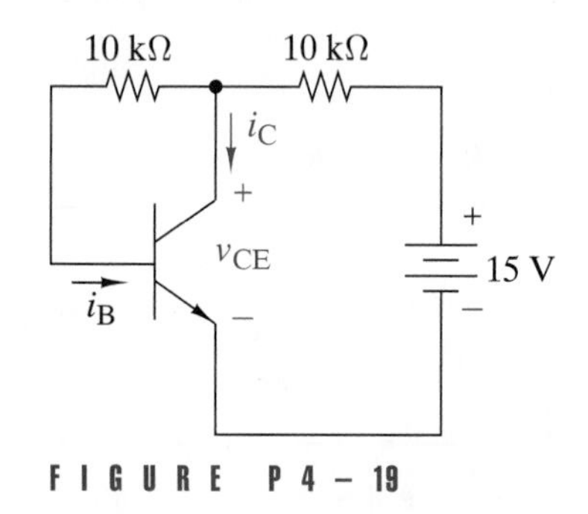

FIGURE P4–19

4–20 The input in Figure P4–20 is a series connection of a dc source V_{BB} and a signal source v_S. The circuit parameters are R_B = 500 kΩ, R_C = 5 kΩ, β = 100, V_γ = 0.7 V, and V_{CC} = 15 V.

(a) With v_S = 0, select the value of V_{BB} so that the transistor is in the active mode with $v_{CE} = V_{CC}/2$.

(b) Using the value of V_{BB} found in part (a), find the range of values of the signal voltage v_S for which the transistor remains in the active mode.

(c) Plot the transfer characteristic v_{CE} versus v_S as the signal voltage sweeps across the range from −10 V to +10 V.

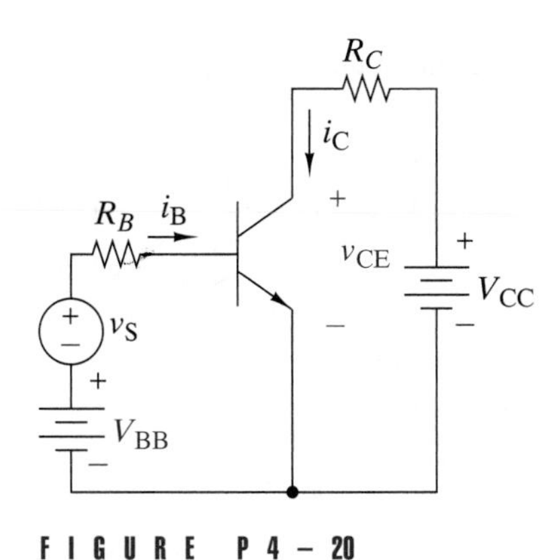

FIGURE P4–20

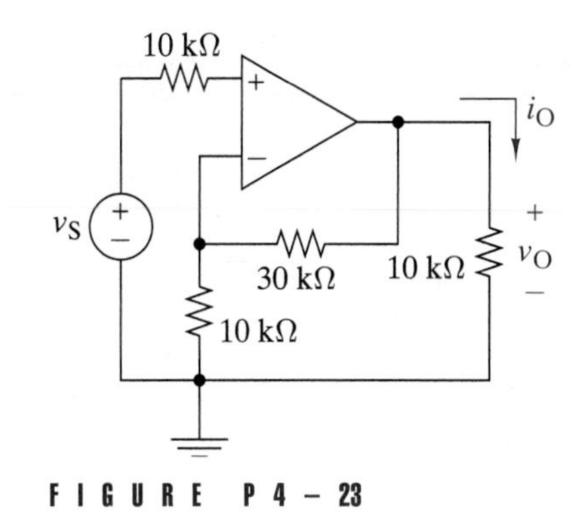

FIGURE P4–23

OBJECTIVE 4–3 OP AMP CIRCUIT ANALYSIS (SECTS. 4–4, 4–5)

Given a circuit consisting of linear resistors, OP AMPs, and independent sources, find selected output signals or input-output relationships.
See Examples 4–13, 4–14, 4–16, 4–17, 4–18, 4–19 and Exercises 4–10, 4–11, 4–12, 4–13, 4–15

4–21 Find v_O in terms of v_S in Figure P4–21.

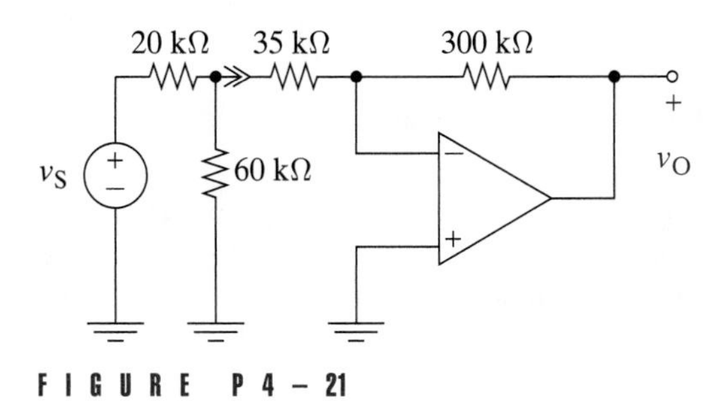

FIGURE P4–21

4–22 What is the range of the gain v_O/v_S in Figure P4–22?

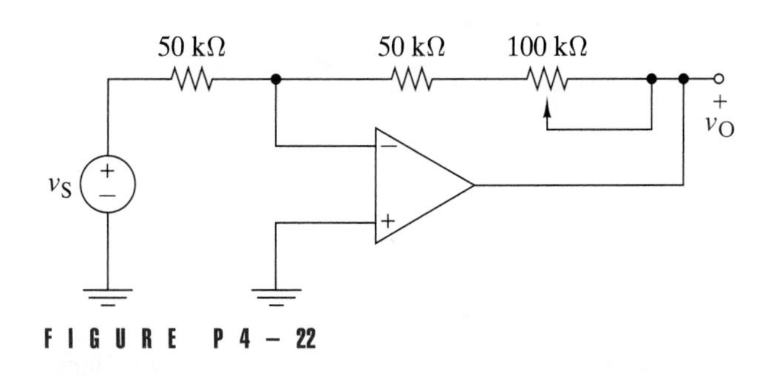

FIGURE P4–22

4–23 **(a)** Find v_O in terms of v_S in Figure P4–23.
(b) Find i_O for v_S = 1.5 V.

4–24 **(a)** Find v_O in terms of v_S in Figure P4–24.
(b) Find i_O for v_S = 2 V.

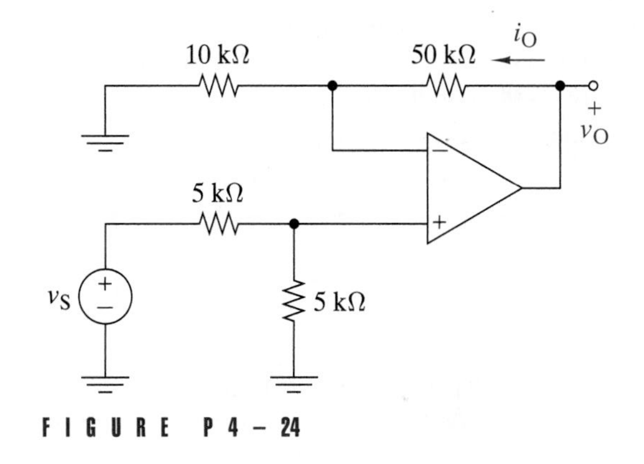

FIGURE P4–24

4–25 What is the gain v_O/v_S in Figure P4–25 when the switch is in position 1? Repeat for positions 2 and 3.

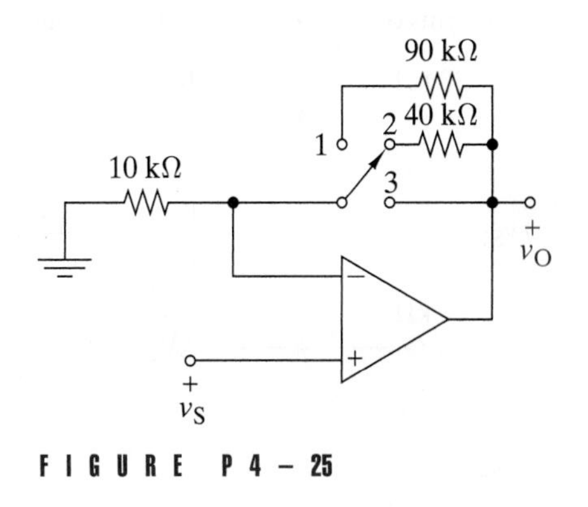

FIGURE P4–25

4–26 Find v_O in terms of the inputs v_1 and v_2 in Figure P4–26.

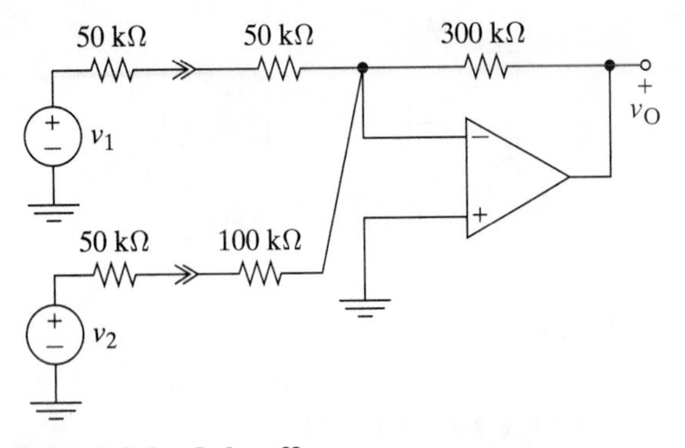

FIGURE P4–26

4–27 The input-output relationship for a three-input inverting summer is

$$v_O = -[v_1 + 2v_2 + 5v_3]$$

The resistance of the feedback resistor is 50 kΩ and the supply voltages are $V_{CC} = \pm 15$ V.

(a) Find the values of the input resistors R_1, R_2, and R_3.

(b) For $v_2 = 0.5$ V and $v_3 = -1$ V, find the allowable range of v_1 for linear operation.

4–28 Find v_O in terms of the inputs v_1 and v_2 in Figure P4–28.

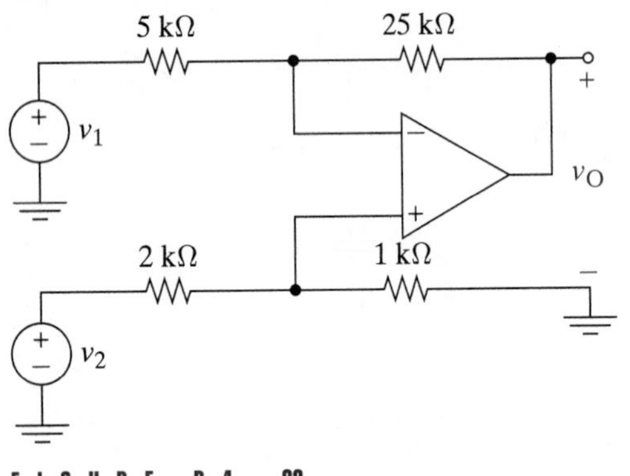

FIGURE P4–28

4–29 Find v_O in terms of the inputs v_{S1} and v_{S2} in Figure P4–29.

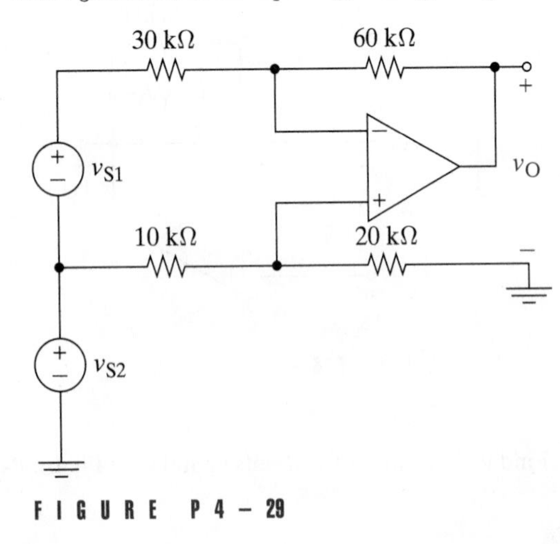

FIGURE P4–29

4–30 Find v_O in terms of the inputs v_{S1} and v_{S2} in Figure P4–30.

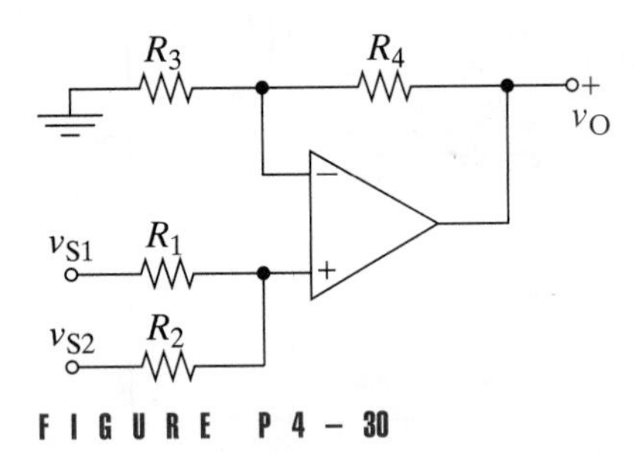

FIGURE P4–30

4–31 E It is claimed that $v_O = v_S$ when the switch is closed in Figure P4–31 and that $v_O = -v_S$ when the switch is open. Prove or disprove this claim.

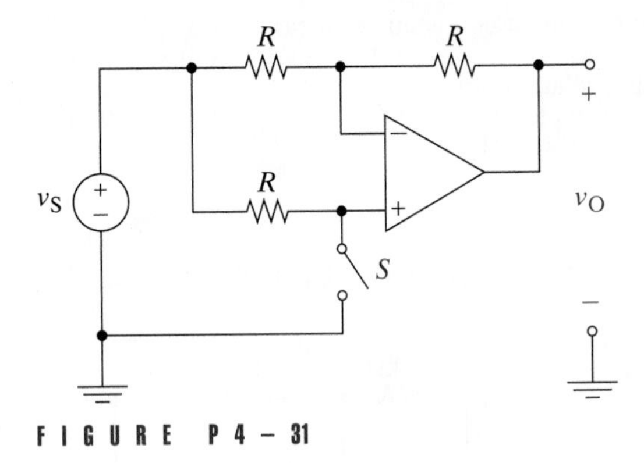

FIGURE P4–31

4–32 The inputs to the circuit in Figure P4–32 are a current source i_S and a voltage source v_S. When the OP AMP is in its linear range, the output voltage has the form $v_O = K_1v_S + K_2i_S$. Find the constants K_1 and K_2.

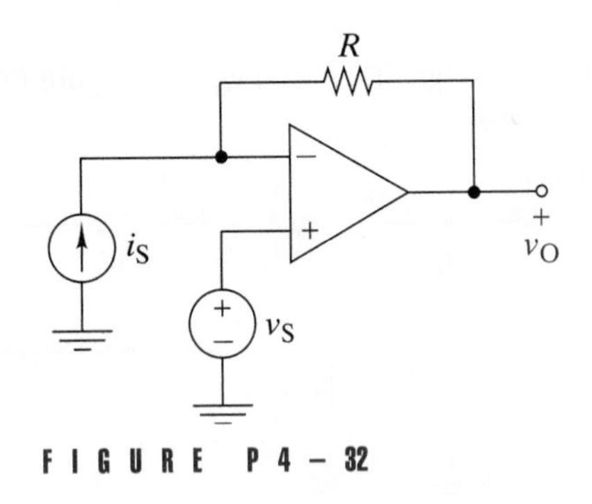

FIGURE P4–32

4–33 Use node-voltage analysis to show that the input-output relationship of the circuit in Figure P4–33 is

$$v_O = -\left(\frac{R_1R_2 + R_1R_3 + R_2R_3}{RR_2}\right)v_S$$

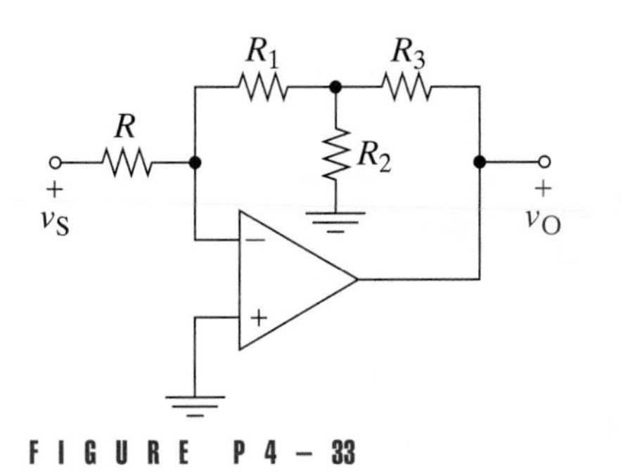

FIGURE P4–33

4–34 Use node-voltage analysis in Figure P4–34 to show that $i_O = -v_S/2R$ regardless of the load. That is, show that the circuit is a voltage-controlled current source.

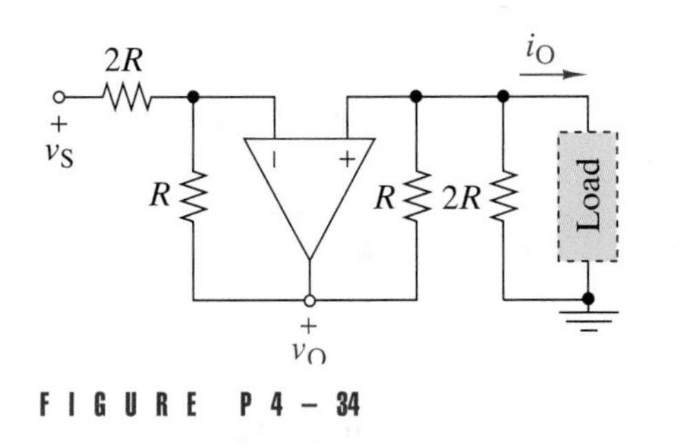

FIGURE P4–34

4–35 Find v_O in terms of the inputs v_{S1}, v_{S2}, and v_{S3}, in Figure P4–35.

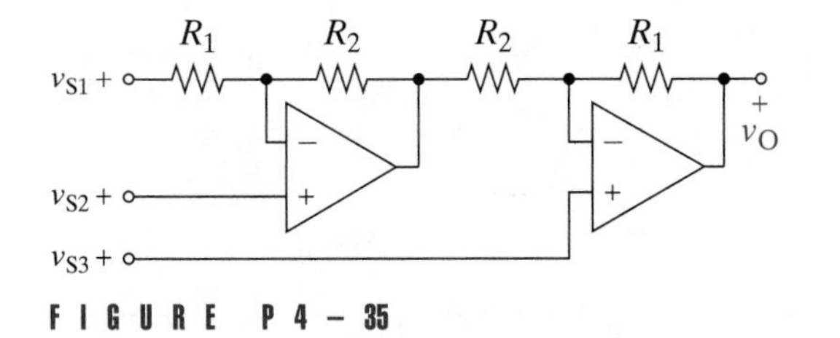

FIGURE P4–35

4–36 Find v_O in terms of the inputs v_{S1} and v_{S2} in Figure P4–36.

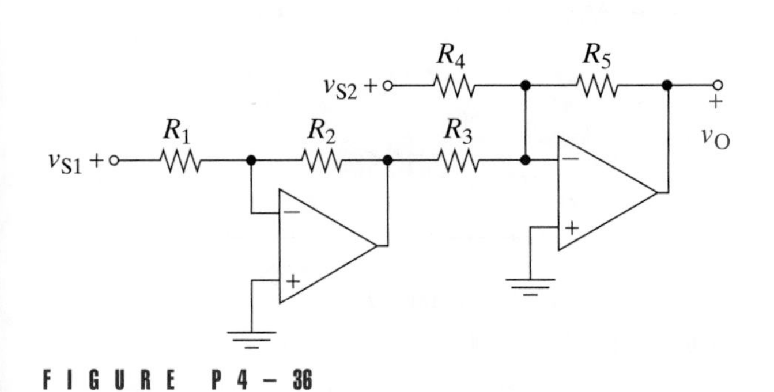

FIGURE P4–36

4–37 Find the output v_2 in terms of the input v_1 in Figure P4–37.

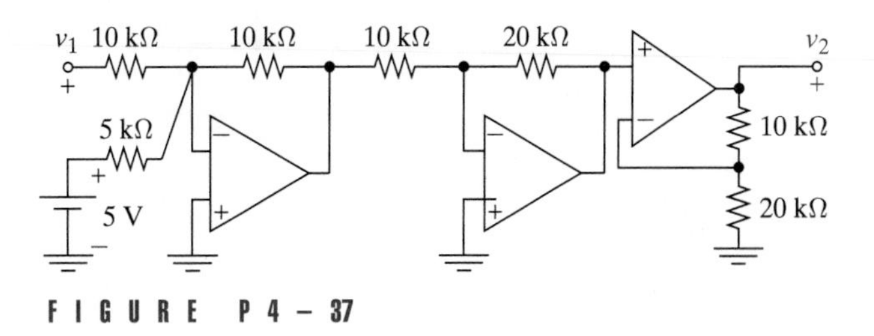

FIGURE P4–37

4–38 Find the output v_2 in terms of the input v_1 in Figure P4–38.

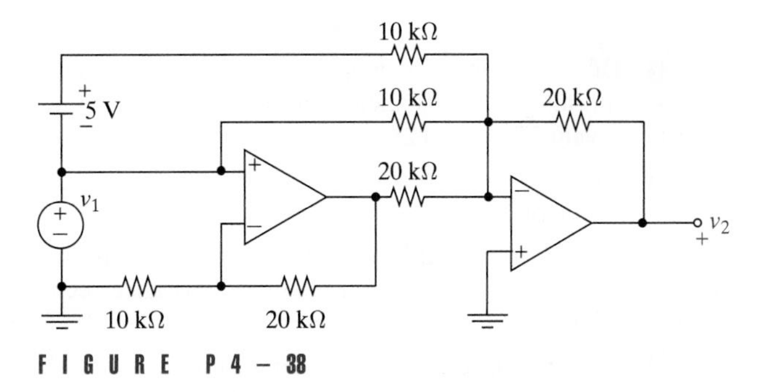

FIGURE P4–38

4–39 Find the output v_O in terms of the input v_S in Figure P4–39.

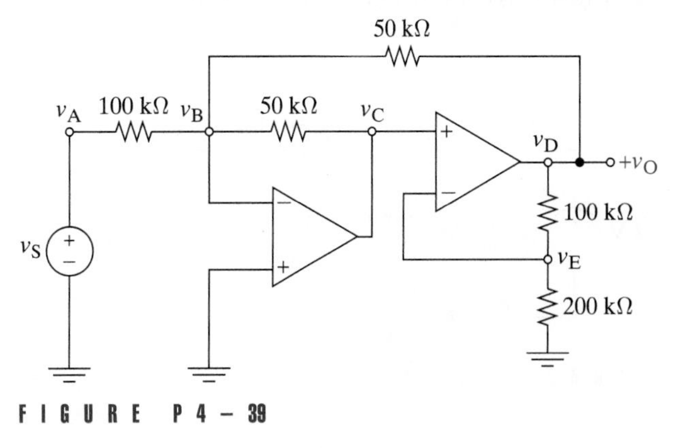

FIGURE P4–39

4–40 Find the output v_O in terms of the inputs v_1 and v_2 in Figure P4–40.

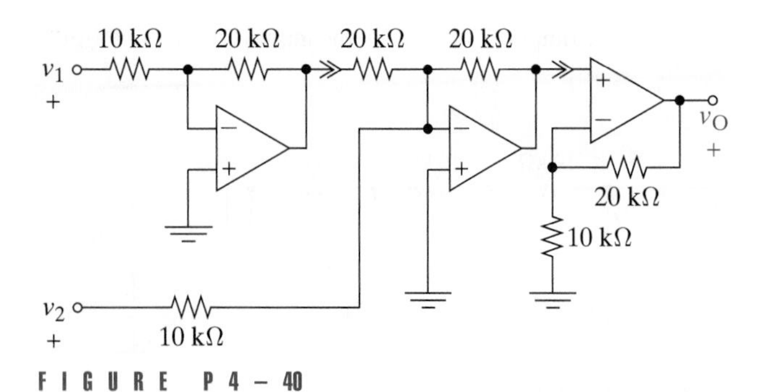

FIGURE P4–40

OBJECTIVE 4–4 OP AMP CIRCUIT DESIGN (SECT. 4–6)

Given an input-output relationship, use resistors and OP AMPs to design one or more circuits that implement the relationship within stated constraints.
See Examples 4–15, 4–20, 4–21, 4–22, 4–23 and Exercise 4–16

4–41 **D** Design an inverting amplifier with a voltage gain of −40 and an input resistance greater than 10 kΩ using resistance values less than 250 kΩ.

4–42 **D** Design a noninverting amplifier with a voltage gain of 5 using only 10-kΩ resistors and one OP AMP.

4–43 **D** Design an OP AMP circuit with inputs v_1 and v_2 and an output $v_O = 4v_2 - 2v_1$. The input resistance seen by both inputs should be greater than 5 kΩ.

4–44 **D** Design a differential amplifier with inputs v_1 and v_2 and an output $v_O = 20(v_2 - v_1)$ using only one OP AMP. All resistances must be between 5 kΩ and 200 kΩ.

4–45 **D** Using no more than two OP AMPs, design an OP AMP circuit with inputs v_1, v_2, and v_3 and an output $v_O = 3v_1 - 2v_2 - 5v_3$.

4–46 **D** Design the interface circuit in Figure P4–46 so that 10 mW is delivered to the 100-kΩ load.

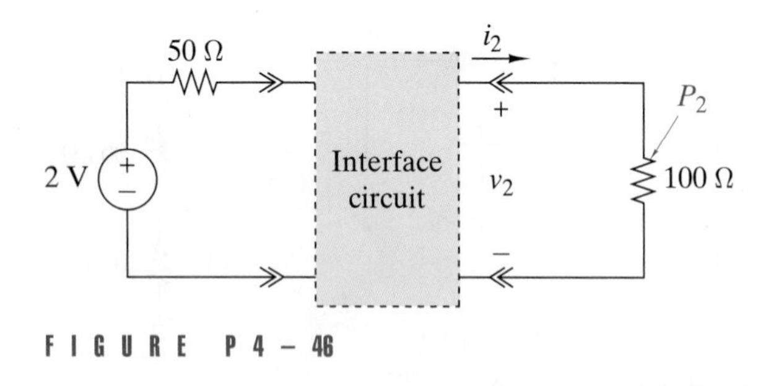

FIGURE P4–46

4–47 **D** Design the interface circuit in Figure P4–47 so that the output is $v_2 = (50v_1 + 3)$ V.

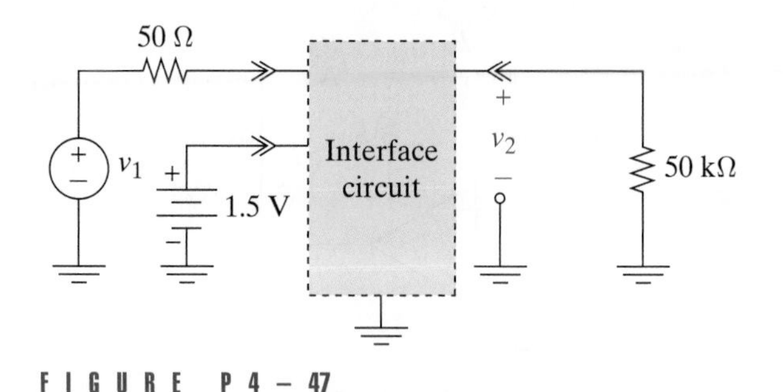

FIGURE P4–47

4–48 **D** Design the interface circuit in Figure P4–48 so that $v_2 = 2000i_S$.

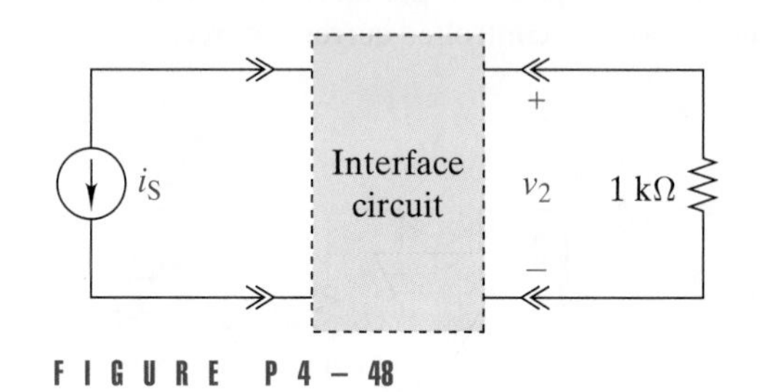

FIGURE P4–48

4–49 **E** A requirement exists for an OP AMP circuit with the input-output relationship

$$v_O = 5v_{S1} - 2v_{S2}$$

Two proposed designs are shown in Figure P4–49. As project engineer you must recommend one of these circuits for

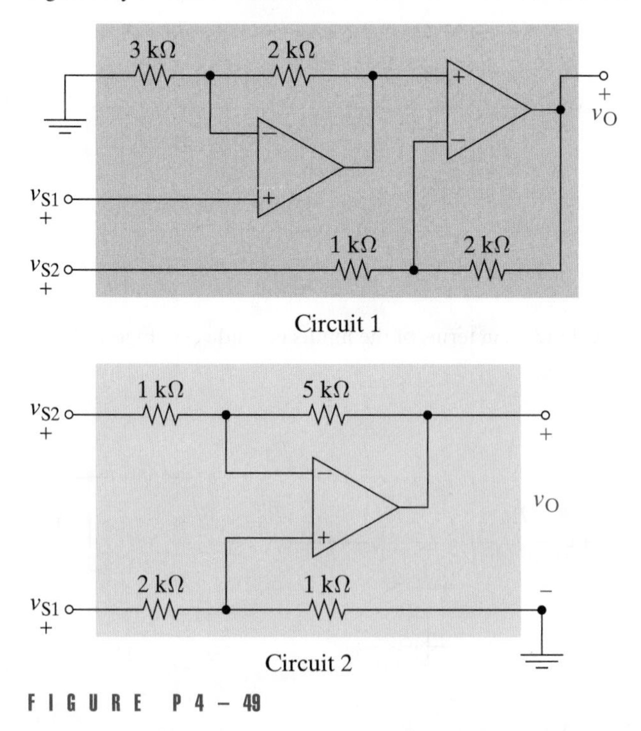

FIGURE P4–49

production. Which of these circuits would you recommend for production and why? *Hint:* First verify that the circuits perform the required function.

4–50 **E** A requirement exists for an OP AMP circuit to deliver 12 V to a 1-kΩ load using a 4-V source as an input voltage. Two proposed designs are shown in Figure P4–50. The OP AMP to be used in the design has the following ratings:

CHARACTERISTIC	MIN	TYPICAL	MAX	UNITS
Open-loop gain	100	200	—	V/mV
Input resistance	10^{10}	10^{11}	—	Ω
Output voltage	−12	—	+15	V
Output current	—	—	25	mA

Which of these circuits would you recommend for production and why? *Hint:* First verify that the circuits perform the required function.

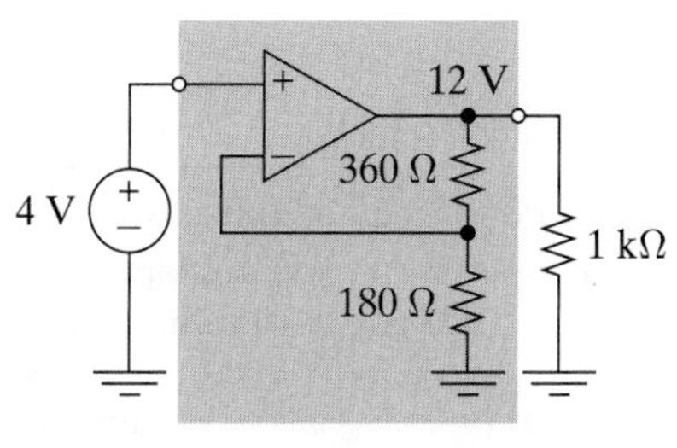

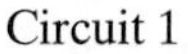
Circuit 1

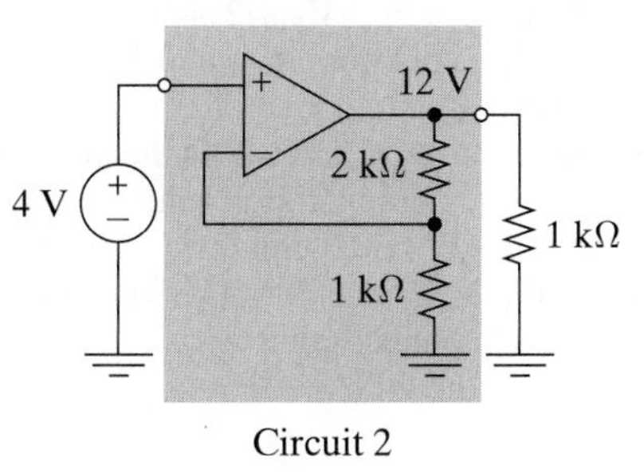

Circuit 2

FIGURE P4–50

APPLICATION PROBLEMS

4–51 The analog output of a five-bit DAC is 5.5 V when the input code is (1, 0, 1, 1, 0). What is the full-scale output of the DAC? How much does the analog output change when the input LSB changes?

4–52 The full-scale output of a six-bit DAC is 10.5 V. What is the analog output when the input code is (0, 1, 0, 1, 0, 1)? How much does the analog output change when the input LSB changes?

4–53 Show that the short-circuit current at the output interface in Figure P4–53 is

$$i_{SC} = \frac{1}{2R}\left(v_1 + \frac{v_2}{2} + \frac{v_3}{4} + \frac{v_4}{8}\right)$$

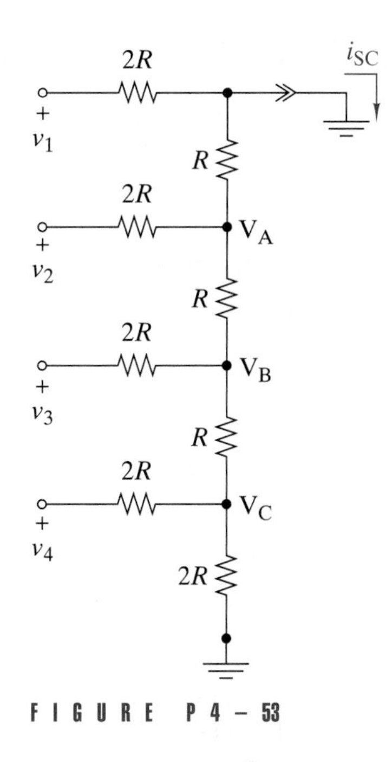

FIGURE P4–53

4–54 **D** The output voltage of a semiconductor temperature sensor is $v_S = (10T + 500)$ mV, where T is the temperature in °C. The sensor is to be used to measure temperatures in the range −40°C to 110°C. Design an interface circuit that translates this temperature range to a 0-V to 3-V output voltage range. Use a standard 5-V reference source to obtain any required bias voltage.

4–55 **D** A medical-grade pressure transducer has been developed for use in invasive blood pressure monitoring. The output voltage of the transducer is $v_{TR} = (0.05P - 0.75)$ mV, where P is pressure in mmHg. The output resistance of the transducer is 300Ω. The blood pressure measurement is to be an input to an existing multisensor monitoring system. This system treats a 1-V input as a blood pressure of 20 mmHg and a 10-V input as 200 mmHg. Design an OP AMP circuit to interface the new pressure transducer with the existing monitoring system.

4–56 **D** The acid/alkaline balance of a fluid is measured by the pH scale. The scale runs from 0 (extremely acid) to 14 (extremely alkaline), with pH 7 being neutral. A pH electrode is a sensor that produces a small voltage that is directly proportional to the pH of the fluid in a test chamber. For a certain pH electrode the proportionality factor is 60 mV/pH. A preamplifier is needed to interface this sensor with a variety of laboratory instruments. The output of the preamp must be 1 V when the sensor is immersed in a test solution with pH = 4 and 1.75 V when it is immersed in a solution with pH = 7. Design an amplifier to meet these requirements.

4–57 The OP AMP in Figure P4–57 operates as a comparator. Find the output voltage when $v_S = 3$ V. Repeat for $v_S = -3$ V and $v_S = 6$ V.

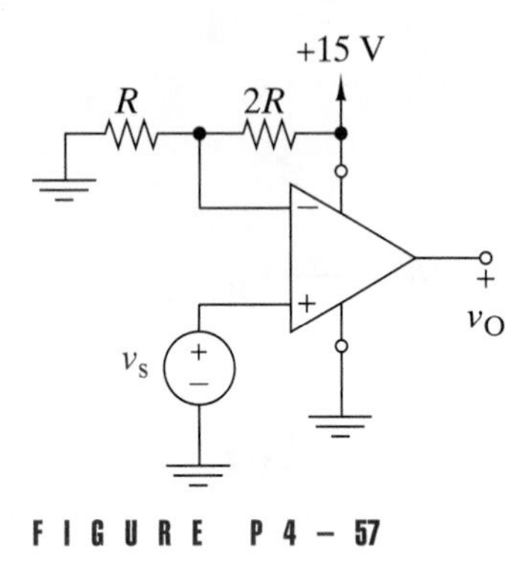

FIGURE P4 – 57

4–58 The circuit in Figure P4–58 has $V_{CC} = 10$ V and $V_{BB} = -2$ V. Sketch the output voltage v_O in the range $0 \leq t \leq 2$ s for $v_S = 10 \sin(2\pi t)$ V.

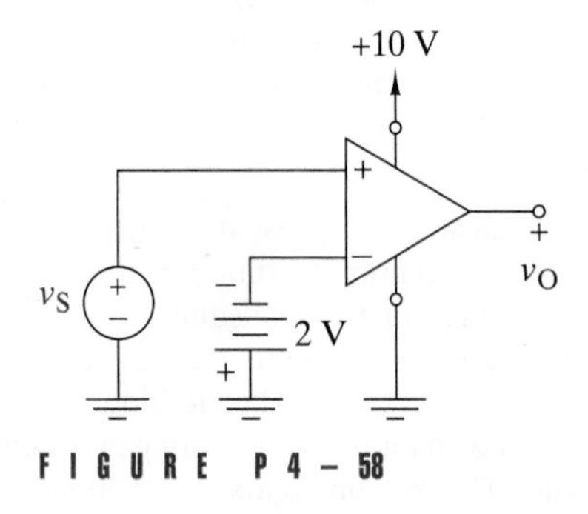

FIGURE P4 – 58

4–59 Repeat Problem 4–58 with $v_S(t) = -2t$ V.

4–60 A five-bit flash ADC in Figure P4–60 uses a reference voltage of 6 V. Find the output code for the analog inputs $v_S = 3.5$ V, 2.3 V, and 5.3 V. If the reference voltage is changed to 8 V, which of these codes would change?

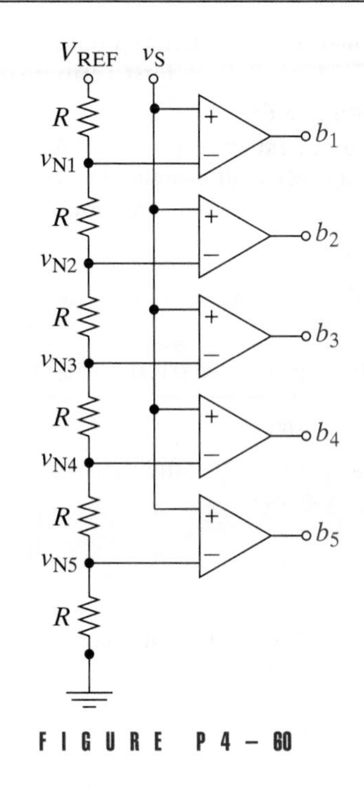

FIGURE P4 – 60

INTEGRATING PROBLEMS

4–61 **A** Bipolar Power Supply Voltages

The circuit in Figure P4–61 produces bipolar power supply voltages $V_{POS} > 0$ and $V_{NEG} < 0$ from a floating unipolar voltage source $V_{REF} > 0$. Note that the OP AMP output is grounded and that its $+V_{CC}$ and $-V_{CC}$ terminals are connected to V_{POS} and V_{NEG}, respectively.

(a) Show that $V_{POS} = +V_{REF}/2$ and $V_{NEG} = -V_{REF}/2$ even if the load resistors R_{POS} and R_{NEG} are not equal.

(b) If R_{POS} and R_{NEG} are not equal, a current i_G must flow into or out of ground. How does the ungrounded voltage source V_{REF} supply this ground current?

(c) In effect the OP AMP creates a "virtual ground" at point A ($V_A = 0$) but draws no current in doing so. Why not just connect point A to a "real ground" and do away with the OP AMP?

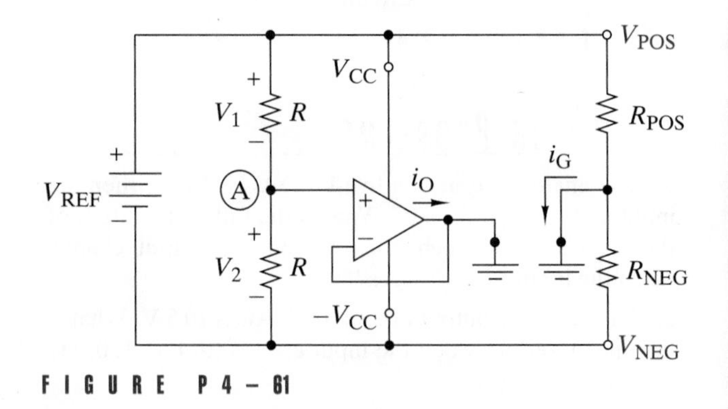

FIGURE P4 – 61

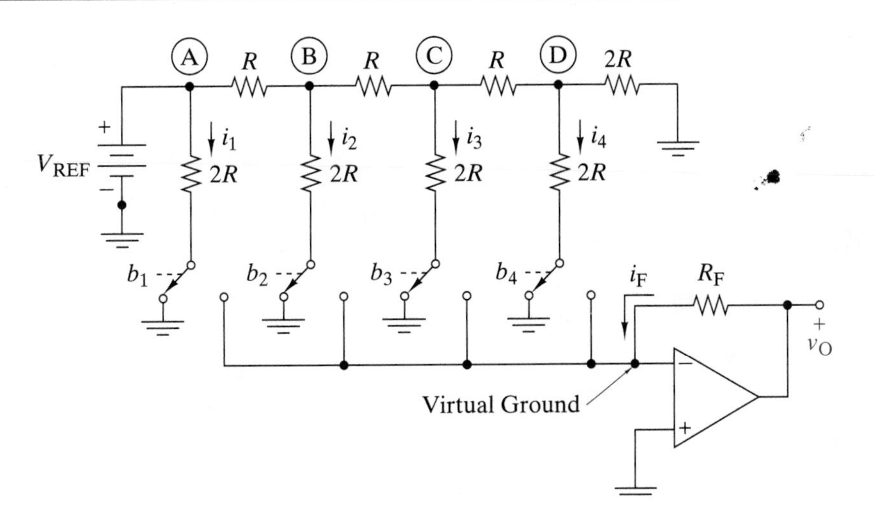

FIGURE P4–62

4–62 **A** Current Switching DAC

The circuit in Figure P4–62 is a four-bit digital-to-analog converter (DAC). The DAC output is the voltage v_O and the input is the binary code represented by bits b_1, b_2, b_3, and b_4. The input bits are either 0 (low) or 1 (high), and each controls one of the four switches in the figure. When bits are low, their switches are in the left position, directing the $2R$ leg currents to ground. When bits are high, their switches move to the right position, directing the $2R$ leg currents to the OP AMP's inverting input. The $2R$ leg currents do not change when switching from left to right because the inverting input is a virtual ground ($v_N = v_P = 0$). The purpose of this problem is to show that this constant-current switching produces the following input-output relationship:

$$v_O = -\frac{R_F}{2R} V_{REF}\left(b_1 + \frac{b_2}{2} + \frac{b_3}{4} + \frac{b_4}{8}\right)$$

(a) Since the inverting input is a virtual ground, show that the currents in the $2R$ legs are $i_1 = V_{REF}/2R$, $i_2 = V_{REF}/4R$, $i_3 = V_{REF}/8R$, and $i_4 = V_{REF}/16R$ regardless of switch positions.

(b) Show that the sum of currents at the inverting input is

$$b_1 i_1 + b_2 i_2 + b_3 i_3 + b_4 i_4 + i_F = 0$$

where bits b_k ($k = 1, 2, 3, 4$) is either 0 or 1.

(c) Use the results in part (a) and (b) to show that the OP AMP output voltage is

$$v_O = -\frac{R_F}{2R} V_{REF}\left(b_1 + \frac{b_2}{2} + \frac{b_3}{4} + \frac{b_4}{8}\right)$$

as stated.

4–63 **A D** OP AMP Circuit Analysis and Design

(a) **A** Find the input-output relationship of the circuit in Figure P4–63.

(b) **D** Design a circuit that realizes the relationship found in part (a) using only 10–kΩ resistors and one OP AMP.

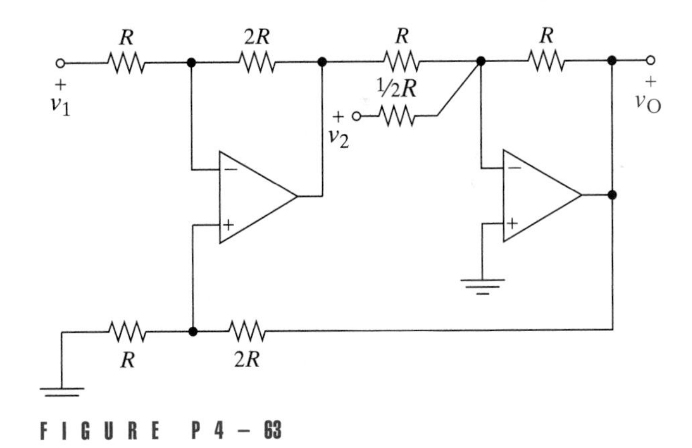

FIGURE P4–63

4–64 **E** Resistance Temperature Transducer

A resistive transducer uses a sensing element whose resistance varies with temperature. For a particular transducer the resistance varies as $R_{TR} = 0.375T + 100\ \Omega$, where T is temperature in °C. This transducer is to be included in a circuit to measure temperatures in the range from −200°C to 800°C. The circuit must convert the transducer resistance variation over this temperature range into an output voltage in the range from 0 V to 5 V. Two proposed circuit designs are shown in Figure P4–64. Which of these circuits would you recommend for production and why? *Hint:* First verify that the circuits perform the required function.

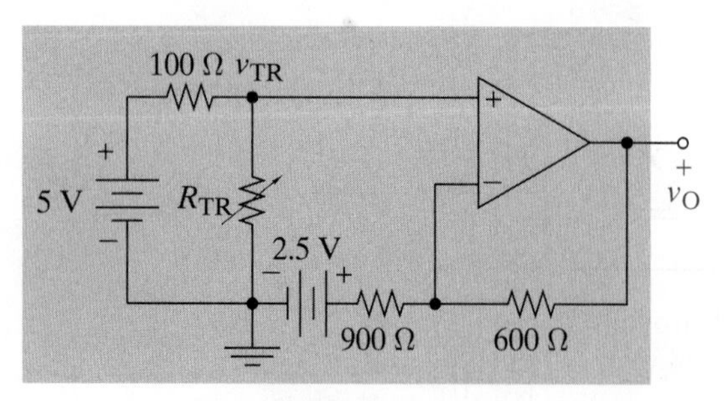

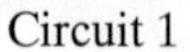

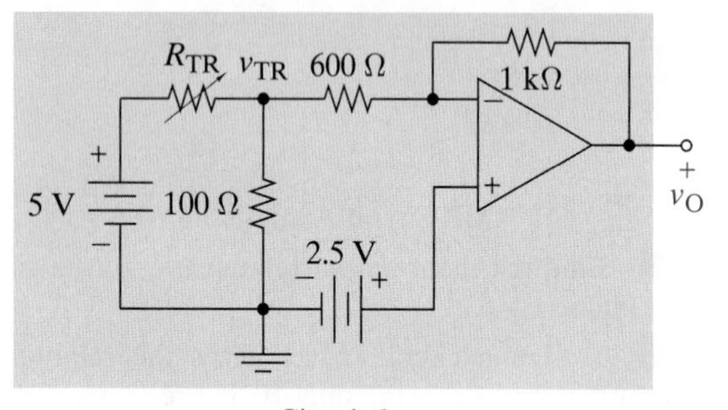

Circuit 2

FIGURE P4–64

4–65 D Computer-Aided Circuit Design

Use computer-aided circuit analysis to find the value of R_F in Figure P4–65 that causes the input resistance seen by i_S to be 50 Ω. Find the current gain i_O/i_S for this value of R_F. Use $\beta = 100$, $r_\pi = 1.1\ k\Omega$, $R_C = 10\ k\Omega$, $R_E = 100\ \Omega$, and $R_L = 100\ \Omega$.

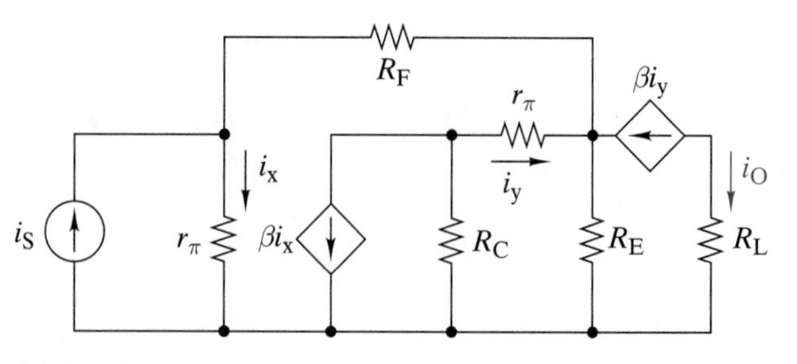

FIGURE P4–65

CHAPTER 5

SIGNAL WAVEFORMS

Under the sea, under the sea mark how the telegraph motions to me. Under the sea, under the sea signals are coming along.

James Clerk Maxwell, 1873,
Scottish Physicist and
Occasional Humorous Poet.

Some History Behind This Chapter

James Clerk Maxwell (1831–1879) is considered the unifying founder of the mathematical theory of electromagnetics. This genial Scotsman often communicated his thoughts to friends and colleagues via whimsical poetry. In the short excerpt given above, Maxwell reminds us that the purpose of a communication system (the submarine cable telegraph in this case) is to transmit signals and that those signals must be changing, or *in motion* as he put it.

Why This Chapter Is Important Today

Up to this point we have only treated dc signals that are time invariant. These constant signals are a logical place to begin the study of circuit analysis and design. However, to carry information, signals must change, otherwise they keep telling us the same thing over and over again. This chapter introduces the three basic, time-varying signals used in the analysis and design of linear circuits.

Chapter Sections

5–1 Introduction
5–2 The Step Waveform
5–3 The Exponential Waveform
5–4 The Sinusoidal Waveform
5–5 Composite Waveforms
5–6 Waveform Partial Descriptors

Chapter Learning Objectives

5-1 Basic Waveforms (Sects. 5-2, 5-3, 5-4)

Given an equation, graph, or word description of step, ramp, exponential, or sinusoid waveforms:

(a) Construct an alternative description of the waveform.
(b) Find the parameters or properties of the waveform.
(c) Construct new waveforms by integrating or differentiating the given waveform.

5-2 Composite Waveforms (Sect. 5-5)

Given an equation, graph, or word description of a composite waveform:

(a) Construct an alternative description of the waveform.
(b) Find the parameters or properties of the waveform.
(c) Find new waveforms by integrating or differentiating the given waveform.

5-3 Waveform Partial Descriptors (Sect. 5-6)

Given a complete description of a basic or composite waveform:

(a) Classify the waveform as periodic or aperiodic and causal or noncausal.
(b) Find the applicable partial waveform descriptors.

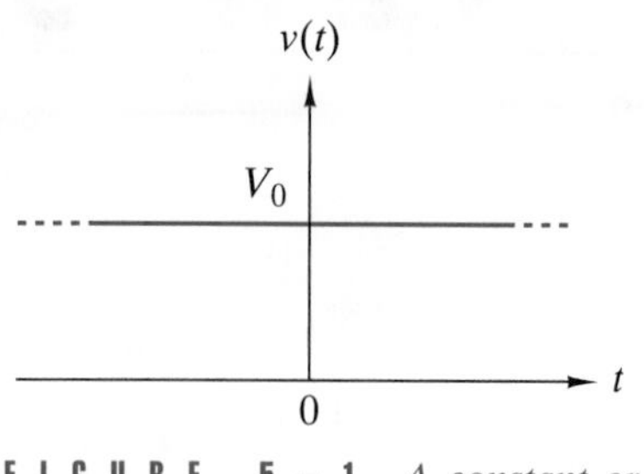

FIGURE 5–1 *A constant or dc waveform.*

5–1 INTRODUCTION

We normally think of a signal as an electrical current $i(t)$ or voltage $v(t)$. The time variation of the signal is called a waveform. More formally,

> ***A waveform** is an equation or graph that defines the signal as a function of time.*

Up to this point our study has been limited to the type of waveform shown in Figure 5–1. Waveforms that are constant for all time are called **dc signals**. The abbreviation *dc* stands for direct current, but it applies to either voltage or current. Mathematical expressions for a dc voltage $v(t)$ or current $i(t)$ take the form

$$\left.\begin{aligned} v(t) &= V_0 \\ i(t) &= I_0 \end{aligned}\right\} \quad \text{for} \quad -\infty < t < \infty \tag{5–1}$$

This equation is only a model. No physical signal can remain constant forever. It is a useful model, however, because it approximates the signals produced by physical devices such as batteries.

There are two matters of notation and convention that must be discussed before continuing. First, quantities that are constant (non-time-varying) are usually represented by uppercase letters (V_A, I, T_O) or lowercase letters in the early part of the alphabet (a, b_7, f_0). Time-varying electrical quantities are represented by the lowercase letters i, v, p, q, and w. The time variation is expressly indicated when we write these quantities as $v_1(t)$, $i_A(t)$, or $w_C(t)$. Time variation is implicit when they are written as v_1, i_A, or w_C.

Second, in a circuit diagram signal variables are normally accompanied by the reference marks (+, –) for voltage and (→) for current. It is important to remember that these reference marks *do not* indicate the polarity of a voltage or the direction of current. The marks provide a baseline for determining the sign of the numerical value of the actual waveform. When the actual voltage polarity or current direction coincides with the reference directions, the signal has a positive value. When the opposite occurs, the value is negative. Figure 5–2 shows examples of voltage waveforms, including some that assume both positive and negative values. The bipolar waveforms indicate that the actual voltage polarity is changing as a function of time.

The waveforms in Figure 5–2 are examples of signals used in electrical engineering. Since there are many such signals, it may seem that the study of signals involves the uninviting task of compiling a lengthy catalog of waveforms. However, it turns out that a long list is not needed. In fact, we can derive most of the waveforms of interest using just three basic signal models: the step, exponential, and sinusoidal functions. The small number of basic signals illustrates why models are so useful to engineers. In reality, waveforms are very complex, but their time variation can be approximated adequately using only a few basic building blocks.

Finally, in this chapter we will generally use voltage $v(t)$ to represent a signal waveform. Remember, however, that a signal can be either a voltage $v(t)$ or current $i(t)$.

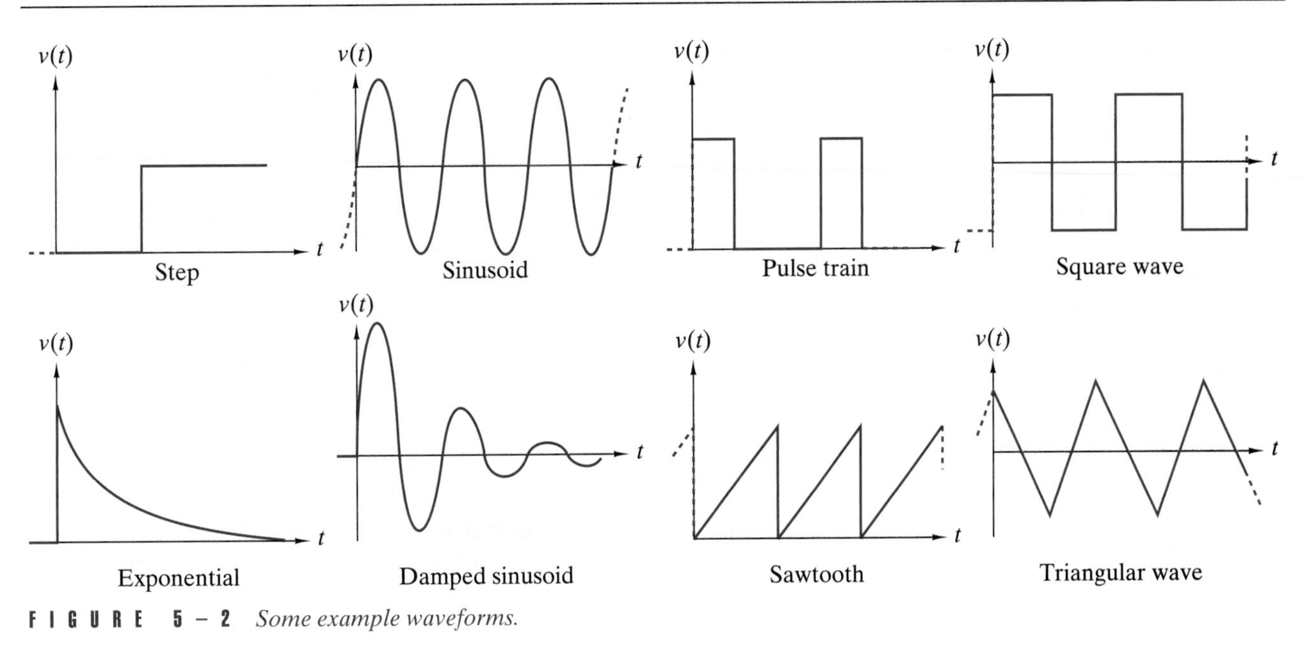

FIGURE 5–2 *Some example waveforms.*

5–2 THE STEP WAVEFORM

The first basic signal in our catalog is the step waveform. The general step function is based on the **unit step function** defined as

$$u(t) = \begin{cases} 0 & \text{for } t < 0 \\ 1 & \text{for } t > 0 \end{cases} \tag{5–2}$$

The step function waveform is equal to zero when its argument t is negative, and is equal to unity when its argument is positive. Mathematically, the function $u(t)$ has a jump discontinuity at $t = 0$.

Strictly speaking, it is impossible to generate a true step function since signal variables like current and voltage cannot jump from one value to another in zero time. Practically speaking, we can generate very good approximations to the step function. What is required is that the transition time be short compared with other response times in the circuit. Actually, the generation of approximate step functions is an everyday occurrence since people frequently turn things like TVs, stereos, and lights on and off.

On the surface, it may appear that the step function is not a very exciting waveform or, at best, only a source of temporary excitement. However, the step waveform is a versatile signal used to construct a wide range of useful waveforms. Multiplying $u(t)$ by a constant V_A produces the waveform

$$V_A u(t) = \begin{cases} 0 & \text{for } t < 0 \\ V_A & \text{for } t \geqslant 0 \end{cases} \tag{5–3}$$

Replacing t by $(t - T_S)$ produces a waveform $V_A u(t - T_S)$, which takes on the values

$$V_A u(t - T_S) = \begin{cases} 0 & \text{for } t < T_S \\ V_A & \text{for } t \geqslant T_S \end{cases} \tag{5–4}$$

The **amplitude** V_A scales the size of the step discontinuity, and the **time-shift** parameter T_S advances or delays the time at which the step occurs, as shown in Figure 5–3.

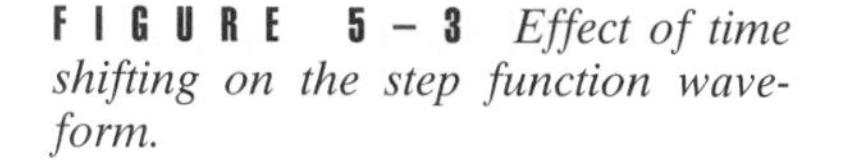

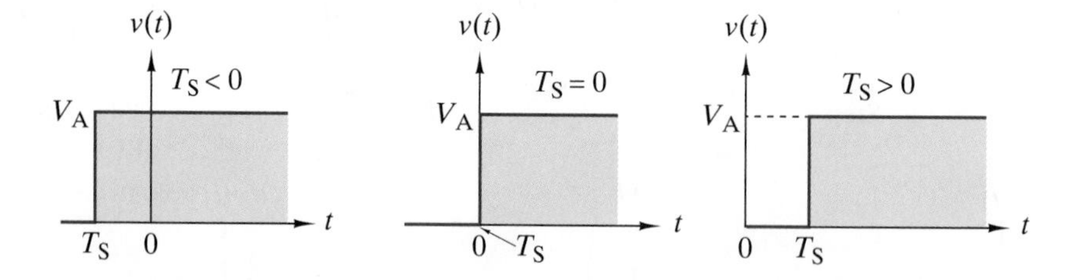

FIGURE 5–3 *Effect of time shifting on the step function waveform.*

Amplitude and time-shift parameters are required to define the general step function. The amplitude V_A carries the units of volts. The amplitude of step function in electric current is I_A and carries the units of amperes. The constant T_S carries the units of time, usually seconds. The parameters V_A (or I_A) and T_S can be positive, negative, or zero. By combining several step functions, we can represent a number of important waveforms. One possibility is illustrated in the following example:

EXAMPLE 5–1

Express the waveform in Figure 5–4(a) in terms of step functions.

SOLUTION:

The amplitude of the pulse jumps to a value of 3 V at $t = 1$ s; therefore, $3\,u(t - 1)$ is part of the equation for the waveform. The pulse returns to zero at $t = 3$ s, so an equal and opposite step must occur at $t = 3$ s. Putting these observations together, we express the rectangular pulse as

$$v(t) = 3u\,(t - 1) - 3u\,(t - 3)$$

Figure 5–4(b) shows how the two step functions combine to produce the given rectangular pulse. ■

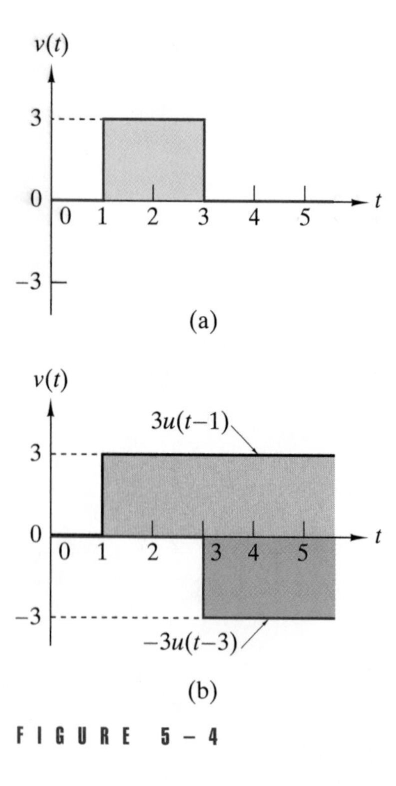

FIGURE 5–4

THE IMPULSE FUNCTION

The generalization of Example 5–1 is the waveform

$$v(t) = V_A[u(t - T_1) - u(t - T_2)]$$

This waveform is a rectangular pulse of amplitude V_A that turns on at $t = T_1$ and off at $t = T_2$. The pulse train and square wave signals in Figure 5–2 can be generated by a series of these pulses. Pulses that turn on at some time T_1 and off at some later time T_2 are sometimes called **gating functions** because they are used in conjunction with electronic switches to enable or inhibit the passage of another signal.

A unit-area pulse centered on $t = 0$ is written in terms of step functions as

$$v(t) = \frac{1}{T}\left[u\left(t + \frac{T}{2}\right) - u\left(t - \frac{T}{2}\right)\right] \tag{5–5}$$

The pulse in Eq. (5–5) is zero everywhere except in the range $-T/2 \leq t \leq T/2$, where its value is $1/T$. The area under the pulse is 1 because its scale factor is inversely proportional to its duration. As shown in Figure 5–5(a), the pulse becomes narrower and higher as T decreases but maintains its unit area. In the limit as $T \rightarrow 0$ the scale factor approaches infinity but the area remains 1. The function obtained in the limit is called a **unit impulse**, symbolized as $\delta(t)$. The graphical representation of $\delta(t)$ is shown in Figure 5–5(b). The impulse is an idealized model of a large-amplitude, short-duration pulse.

A formal definition of the unit impulse is

$$\delta(t) = 0 \text{ for } t \neq 0 \quad \text{and} \quad \int_{-\infty}^{t} \delta(x)dx = u(t) \tag{5–6}$$

The first condition says the impulse is zero everywhere except at $t = 0$. The second condition suggests that the unit impulse is the derivative of a unit step function:

$$\delta(t) = \frac{du(t)}{dt} \tag{5–7}$$

The conclusion in Eq. (5–7) cannot be justified using elementary mathematics since the function $u(t)$ has a discontinuity at $t = 0$ and its derivative at that point does not exist in the usual sense. However, the concept can be justified using limiting conditions on continuous functions, as discussed in texts on signals and systems.[1] Accordingly, we defer the question of mathematical rigor to later courses and think of the unit impulse as the derivative of a unit step function. Note that this means that the unit impulse $\delta(t)$ has units of reciprocal time, or s^{-1}.

An impulse of strength K is denoted $v(t) = K\delta(t)$. Consequently, the scale factor K has the units of V-s and is the area under the impulse $K\delta(t)$. In the graphical representation of the impulse the value of K is written in parentheses beside the arrow, as shown in Figure 5–5(b).

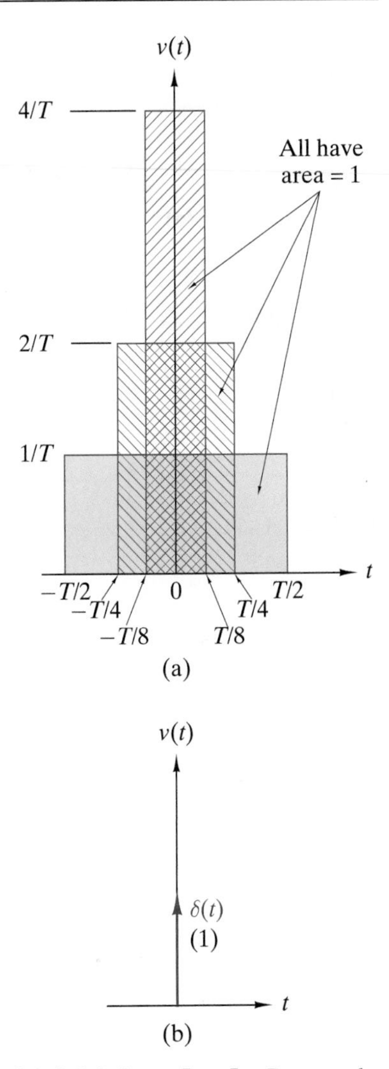

FIGURE 5–5 *Rectangular pulse waveforms and the impulse.*

EXAMPLE 5–2

Calculate and sketch the derivative of the pulse in Figure 5–6(a).

SOLUTION:

In Example 5–1 the pulse waveform was written as

$$v(t) = 3u(t-1) - 3u(t-3) \text{ V}$$

Using the derivative property of the step function, we write

$$\frac{dv(t)}{dt} = 3\delta(t-1) - 3\delta(t-3)$$

---

1 For example, see Alan V. Oppenheim and Allan S. Willsky, *Signals and Systems Analysis* (Englewood Cliffs, N.J.: Prentice Hall, 1983), pp. 22–23.

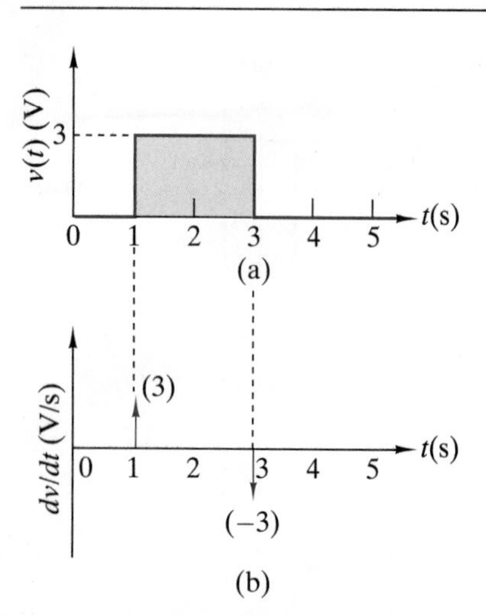

FIGURE 5–6

The derivative waveform consists of a positive-going impulse at $t = 1$ s and a negative-going impulse at $t = 3$ s. Figure 5–6(b) shows how the impulse train is represented graphically. The waveform $v(t)$ has the units of volts (V), so its derivative $dv(t)/dt$ has the units of V/s. ■

THE RAMP FUNCTION

The **unit ramp** is defined as the integral of a step function:

$$r(t) = \int_{-\infty}^{t} u(x)dx = tu(t) \tag{5–8}$$

The unit ramp waveform $r(t)$ in Figure 5–7(a) is zero for $t < 0$ and is equal to t for $t > 0$. Notice that the slope of $r(t)$ is 1 and has the units of time, or s. A ramp of strength K is denoted $v(t) = Kr(t)$, where the scale factor K has the units of V/s and is the slope of the ramp. The general ramp waveform shown in Figure 5–7(b) written as $v(t) = Kr(t - T_S)$ is zero for $t < T_S$ and equal to $K(t - T_S)$ for $t \geq T_S$. By adding a sequence of ramps, we can create the triangular and sawtooth waveforms shown in Figure 5–2.

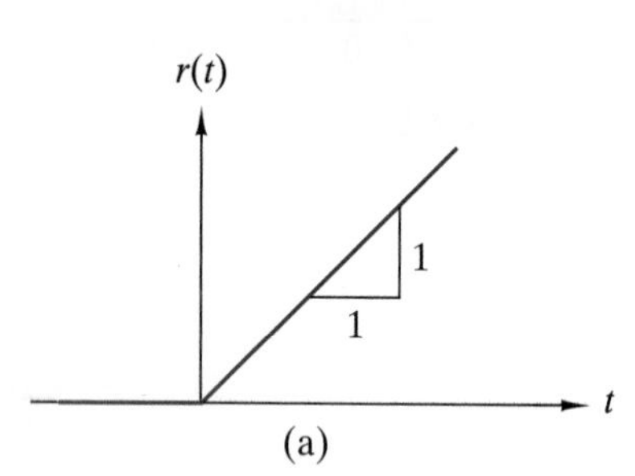

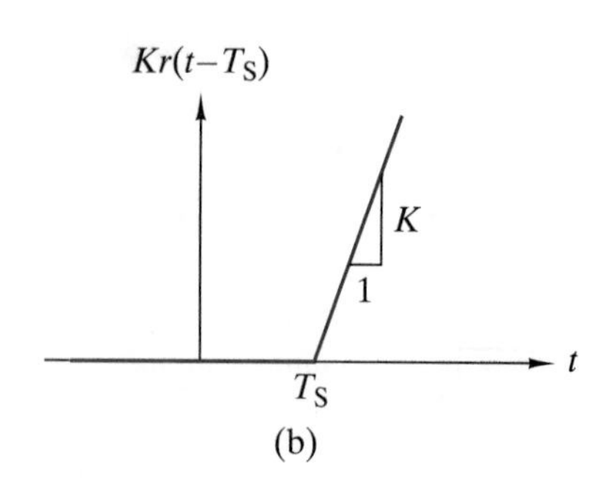

FIGURE 5–7 *(a) Unit ramp waveform. (b) General ramp waveform.*

SINGULARITY FUNCTIONS

The unit impulse, unit step, and unit ramp form a triad of related signals that are referred to as **singularity functions**. They are related by integration as

$$\begin{aligned} u(t) &= \int_{-\infty}^{t} \delta(x)dx \\ r(t) &= \int_{-\infty}^{t} u(x)dx \end{aligned} \tag{5–9}$$

or by differentiation as

$$\begin{aligned} \delta(t) &= \frac{du(t)}{dt} \\ u(t) &= \frac{dr(t)}{dt} \end{aligned} \tag{5–10}$$

These signals are used to generate other waveforms and as test inputs to linear systems to characterize their responses. When applying the singularity functions in circuit analysis, it is important to remember that $u(t)$ is a dimensionless function. But Eqs. (5–9) and (5–10) point out that $\delta(t)$ carries the units of s^{-1} and $r(t)$ carries units of seconds.

EXAMPLE 5–3

Derive an expression for the waveform for the integral of the pulse in Figure 5–8(a).

SOLUTION:

In Example 5–1 the pulse waveform was written as

$$v(t) = 3u(t - 1) - 3u(t - 3)\ \text{V}$$

Using the integration property of the step function, we write

$$\int_{-\infty}^{t} v(x)dx = 3r(t-1) - 3r(t-3)$$

The integral is zero for $t < 1$ s. For $1 < t < 3$ the waveform is $3(t-1)$. For $t > 3$ it is $3(t-1) - 3(t-3) = 6$. These two ramps produce the pulse integral shown in Figure 5–8(b). The waveform $v(t)$ has the units of volts (V), so the units of its integral are V-s. ■

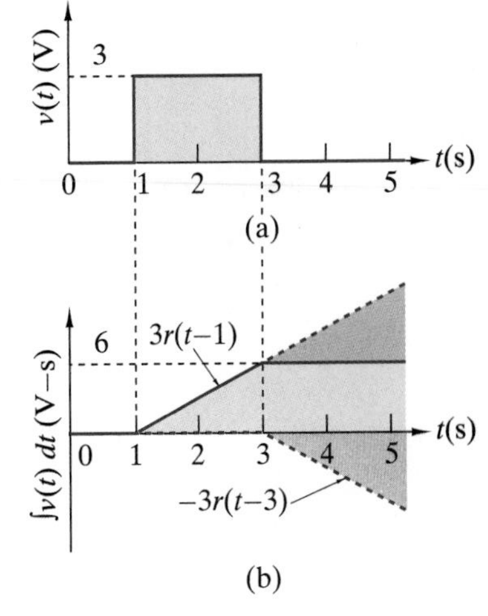

FIGURE 5–8

EXAMPLE 5–4

Figure 5–9(a) shows an ideal electronic switch whose input is a ramp $2r(t)$, where the scale factor $K = 2$ carries the units of V/s. Find the switch output $v_O(t)$ when the gate function in Example 5–1 is applied to the control terminal (G) of the switch.

SOLUTION:

In Example 5–1 the gate function was written as

$$v_G(t) = 3u(t-1) - 3u(t-3) \text{ V}$$

The gate function turns the switch on at $t = 1$ s and off at $t = 3$ s. The output voltage of the switch is

$$v_O(t) = \begin{cases} 0 & t < 1 \\ 2t & 1 < t < 3 \\ 0 & 3 < t \end{cases}$$

Only the portion of the input waveform within the gate interval appears at the output. Figures 5–9(b), 5–9(c), and 5–9(d) show how the gate function $v_G(t)$ controls the passage of the input signal through the electronic switch.

This waveform can be written as a sum of singularity functions as follows. First we write $v_O(t)$ in terms of a gate function

$$v_O(t) = 2t\,\underbrace{[u(t-1) - u(t-3)]}_{\text{Gate Function}}$$

We then manipulate this equation as follows:

$$\begin{aligned} v_O(t) &= 2tu(t-1) - 2tu(t-3) \\ &= 2(t-1+1)u(t-1) - 2(t-3+3)u(t-3) \\ &= \underbrace{2(t-1)u(t-1)}_{r(t-1)} + 2u(t-1) - \underbrace{2(t-3)u(t-3)}_{r(t-3)} - 6u(t-3) \end{aligned}$$

So finally

$$v_O(t) = 2r(t-1) + 2u(t-1) - 2r(t-3) - 6u(t-3)$$

which describes the gated ramp in terms of step and ramp waveforms. ■

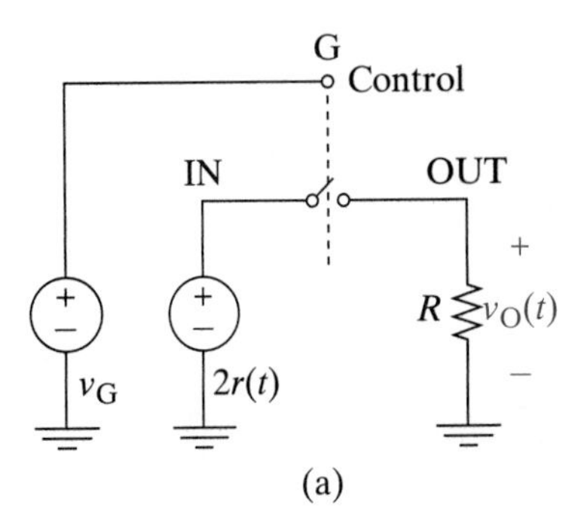

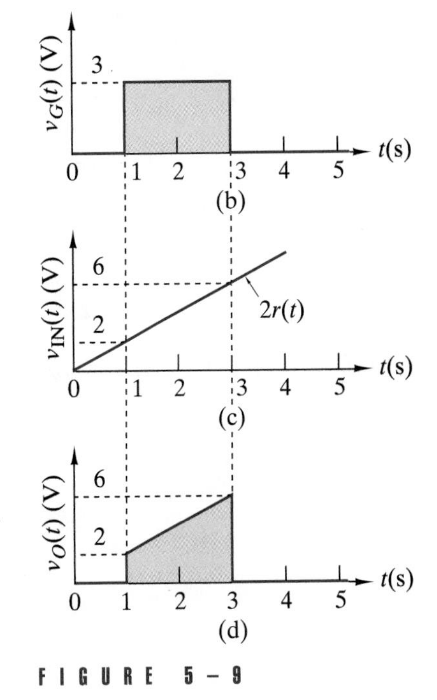

FIGURE 5–9

Exercise 5–1

Express the following signals in terms of singularity functions:

$$\text{(a) } v_1(t) = \begin{cases} 0 & t < 2 \\ 4 & 2 < t < 4 \\ -4 & 4 < t \end{cases} \quad \text{(b) } v_2(t) = \begin{cases} 0 & t < 2 \\ 4 & 2 < t < 4 \\ -2t + 12 & 4 < t \end{cases}$$

$$\text{(c) } v_3(t) = \int_{-\infty}^{t} v_1(x)dx \qquad \text{(d) } v_4(t) = \frac{dv_2(t)}{dt}$$

Answers:

(a) $v_1(t) = 4\,u(t-2) - 8\,u(t-4)$ (b) $v_2(t) = 4\,u(t-2) - 2\,r(t-4) + 8\,u(t-4)$
(c) $v_3(t) = 4\,r(t-2) - 8\,r(t-4)$ (d) $v_4(t) = 4\,\delta(t-2) - 2\,u(t-4)$

Exercise 5–2

(a) Write an expression for a rectangular pulse with an amplitude of 15 V that begins at $t = -5$ s and ends at $t = 10$ s.
(b) Write an expression for the derivative of the pulse defined in (a).
(c) Write an expression for the integral of the pulse in (a).

Answers:

(a) $15[u(t+5) - u(t-10)]$
(b) $15[\delta(t+5) - \delta(t-10)]$
(c) $15(t+5)u(t+5) - 15(t-10)u(t-10) = 15[r(t+5) - r(t-10)]$

5–3 THE EXPONENTIAL WAVEFORM

The **exponential waveform** is a step function whose amplitude factor gradually decays to zero. The equation for this waveform is

$$v(t) = [V_A e^{-t/T_C}]\, u(t) \tag{5–11}$$

A graph of $v(t)$ versus t/T_C is shown in Figure 5–10. The exponential starts out like a step function. It is zero for $t < 0$ and jumps to a maximum amplitude of V_A at $t = 0$. Thereafter it monotonically decays toward zero as time marches on. The two parameters that define the waveform are the **amplitude**

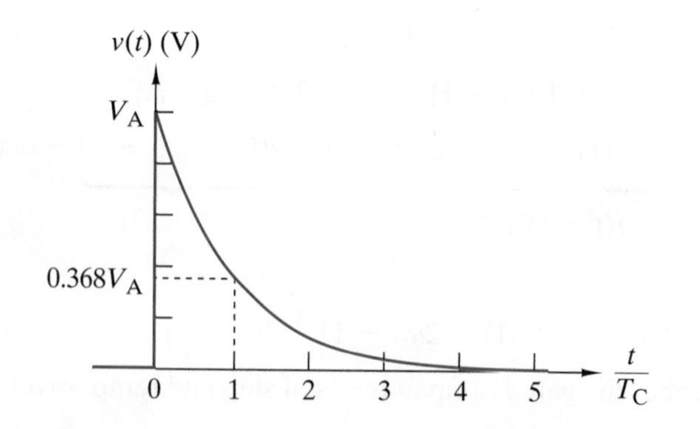

FIGURE 5–10 *The exponential waveform.*

V_A (in volts) and the **time constant** T_C (in seconds). The amplitude of a current exponential would be written I_A and carry the units of amperes.

The time constant is of special interest, since it determines the rate at which the waveform decays to zero. An exponential decays to about 37% of its initial amplitude $v(0) = V_A$ in one time constant, because at $t = T_C$, $v(T_C) = V_A e^{-1}$, or approximately $0.368 \times V_A$. At $t = 5T_C$, the value of the waveform is $V_A e^{-5}$, or approximately 0.00674 V_A. An exponential signal decays to less than 1% of its initial amplitude in a time span of five time constants. In theory, an exponential endures forever, but practically speaking after about $5T_C$ the waveform amplitude becomes negligibly small. We define the **duration** of a waveform to be the interval of time outside of which the waveform is everywhere less than a stated value. Using this concept, we say the duration of an exponential waveform is $5T_C$.

EXAMPLE 5–5

Plot the waveform $v(t) = [-17e^{-100t}]u(t)$ V.

SOLUTION:

From the form of $v(t)$, we recognize that $V_A = -17$ V and $T_C = 1/100$ s or 10 ms. The minimum value of $v(t)$ is $v(0) = -17$ V, and the maximum value is approximately 0 V as t approaches $5T_C = 50$ ms. These observations define appropriate scales for plotting the waveform. Spreadsheet programs are especially useful for the repetitive calculations and graphical functions involved in waveform plotting. Figure 5–11 shows how this example can be handled using Excel. We begin by filling in the "A" column with time values ranging from $t = 0$ to $t = 50$ ms. The data in the "B" column are obtained by

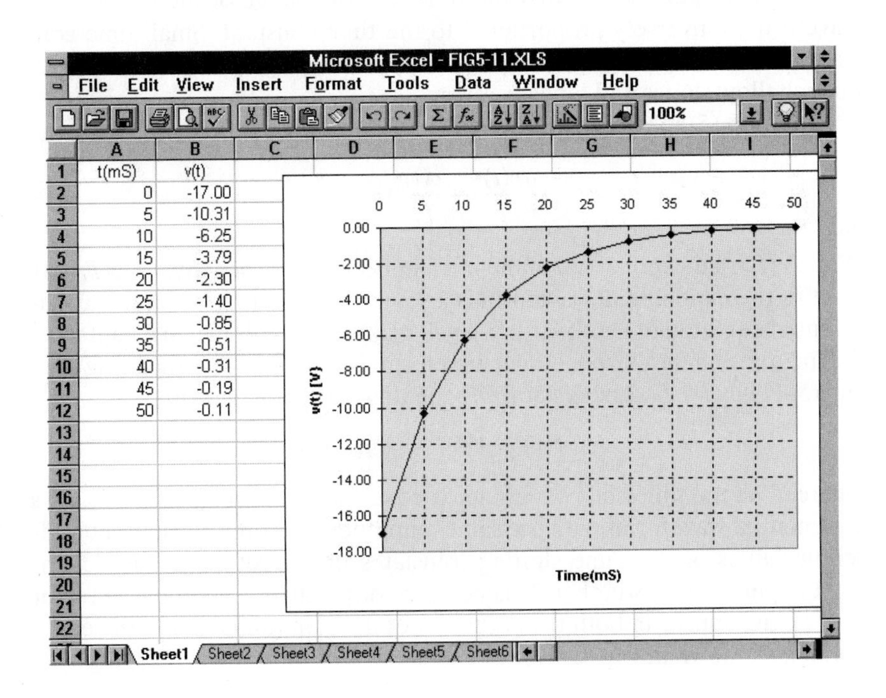

FIGURE 5–11

calculating the value of $-17e^{-1001t}$ for each of the values of t (ms) in the "A" column. Then using the graphing tool, we create a plot showing that the waveform starts out at $v(t) = -17$ V at $t = 0$ and then increases toward $v(t) = 0$ as $t \to 50$ ms. ■

PROPERTIES OF EXPONENTIAL WAVEFORMS

The **decrement property** describes the decay rate of an exponential signal. For $t > 0$ the exponential waveform is given by

$$v(t) = V_A e^{-t/T_C} \tag{5–12}$$

The step function can be omitted since it is unity for $t > 0$. At time $t + \Delta t$ the amplitude is

$$v(t + \Delta t) = V_A e^{-(t+\Delta t)/T_C} = V_A e^{-t/T_C} e^{-\Delta t/T_C} \tag{5–13}$$

The ratio of these two amplitudes is

$$\frac{v(t + \Delta t)}{v(t)} = \frac{V_A e^{-t/T_C} e^{-\Delta t/T_C}}{V_A e^{-t/T_C}} = e^{-\Delta t/T_C} \tag{5–14}$$

The decrement ratio is independent of amplitude and time. In any fixed time period Δt, the fractional decrease depends only on the time constant. The decrement property states that the same percentage decay occurs in equal time intervals.

The slope of the exponential waveform (for $t > 0$) is found by differentiating Eq. (5–12) with respect to time:

$$\frac{dv(t)}{dt} = -\frac{V_A}{T_C} e^{-t/T_C} = -\frac{v(t)}{T_C} \tag{5–15}$$

The **slope property** states that the time rate of change of the exponential waveform is inversely proportional to the time constant. Small time constants lead to large slopes or rapid decays, while large time constants produce shallow slopes and long decay times.

Equation (5–15) can be rearranged as

$$\frac{dv(t)}{dt} + \frac{v(t)}{T_C} = 0 \tag{5–16}$$

When $v(t)$ is an exponential of the form in Eq. (5–12), then $dv/dt + v/T_C = 0$. That is, the exponential waveform is a solution of the first-order linear differential equation in Eq. (5–16). We will make use of this fact in Chapter 7.

The time-shifted exponential waveform is obtained by replacing t in Eq. (5–11) by $t - T_S$. The general exponential waveform is written as

$$v(t) = [V_A e^{-(t-T_S)/T_C}]\, u(t - T_S) \tag{5–17}$$

where T_S is the time-shift parameter for the waveform. Figure 5–12 shows exponential waveforms with the same amplitude and time constant but different values of T_S. Time shifting translates the waveform to the left or right depending on whether T_S is negative or positive. *Caution:* The factor $t - T_S$ must appear in both the argument of the step function and the exponential, as shown in Eq. (5–17).

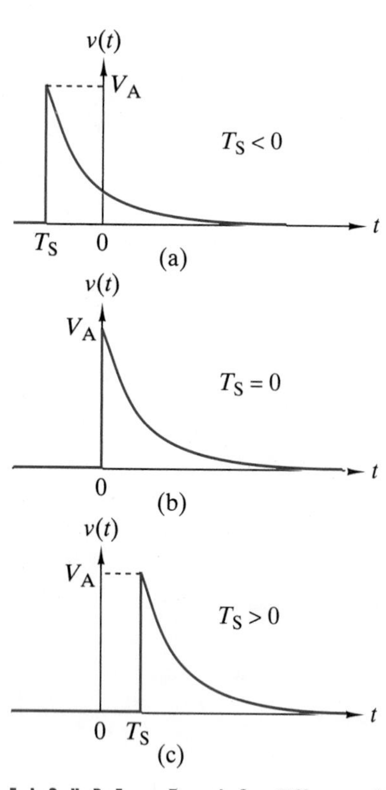

FIGURE 5–12 *Effect of time shifting on the exponential waveform.*

EXAMPLE 5–6

An oscilloscope is a laboratory instrument that displays the instantaneous value of waveform versus time. Figure 5–13 shows an oscilloscope display of a portion of an exponential waveform. In the figure the vertical (amplitude) axis is calibrated at 2 V per division, and the horizontal (time) axis is calibrated at 1 ms per division. Find the time constant of the exponential.

SOLUTION:

For $t > 0$ the general expression for an exponential in Eq. (5–17) becomes

$$v(t) = V_A e^{-(t-T_S)/T_C}$$

We have only a portion of the waveform, so we do not know the location of the $t = 0$ time origin; hence, we cannot find the amplitude V_A or the time shift T_S from the display. But, according to the decrement property, we should be able to determine the time constant since the decrement ratio is independent of amplitude and time. Specifically, Eq. (5–14) points out that

$$\frac{v(t + \Delta t)}{v(t)} = e^{-\Delta t/T_C}$$

Solving for the time constant T_C yields

$$T_C = \frac{\Delta t}{\ln\left[\dfrac{v(t)}{v(t + \Delta t)}\right]}$$

Taking the starting point at the left edge of the oscilloscope display yields

$$v(t) = (3.6 \text{ div})(2 \text{ V/div}) = 7.2 \text{ V}$$

Next, defining Δt to be the full width of the display produces

$$\Delta t = (8 \text{ div})(1 \text{ ms/div}) = 8 \text{ ms}$$

and

$$v(t + \Delta t) = (0.5 \text{ div})(2 \text{ V/div}) = 1 \text{ V}$$

FIGURE 5–13

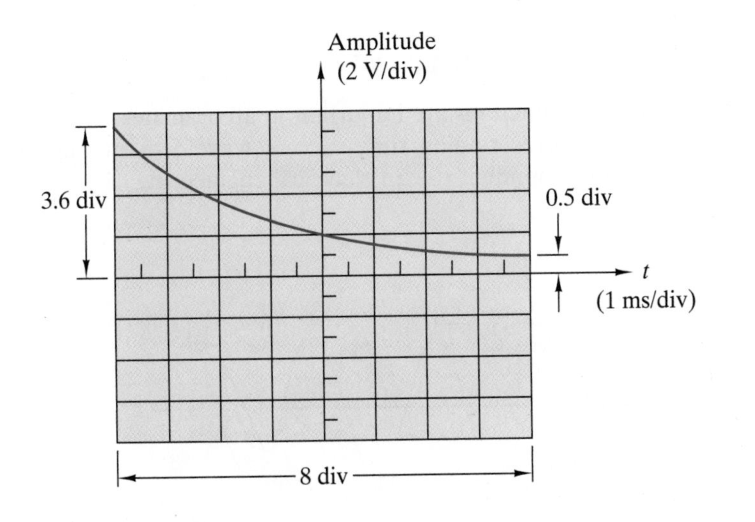

As a result, the time constant of the waveform is found to be

$$T_C = \frac{\Delta t}{\ln\left[\dfrac{v(t)}{v(t+\Delta t)}\right]} = \frac{8 \times 10^{-3}}{\ln(7.2/1)} = 4.05 \text{ ms}$$ ■

Exercise 5–3

(a) An exponential waveform has $v(0) = 1.2$ V and $v(3) = 0.5$ V. What are V_A and T_C for this waveform?
(b) An exponential waveform has $v(0) = 5$ V and $v(2) = 1.25$ V. What are values of $v(t)$ at $t = 1$ and $t = 4$?
(c) An exponential waveform has $v(0) = 5$ and an initial ($t = 0$) slope of -25 V/s. What are V_A and T_C for this waveform?
(d) An exponential waveform decays to 10% of its initial value in 3 ms. What is T_C for this waveform?
(e) A waveform has $v(2) = 4$ V, $v(6) = 1$ V, and $v(10) = 0.5$ V. Is it an exponential waveform?

Answers:

(a) $V_A = 1.2$ V, $T_C = 3.43$ s
(b) $v(1) = 2.5$ V, $v(4) = 0.3125$ V
(c) $V_A = 5$ V, $T_C = 200$ ms
(d) $T_C = 1.303$ ms
(e) No, it violates the decrement property.

Exercise 5–4

Find the amplitude and time constant of the following exponential signals:

(a) $v_1(t) = [-15e^{-1000t}]u(t)$ V
(b) $v_2(t) = [+12e^{-t/10}]u(t)$ mV
(c) $i_3(t) = [15e^{-500t}]u(-t)$ mA
(d) $i_4(t) = [4e^{-200(t-100)}]u(t-100)$ A

Answers:

(a) $V_A = -15$ V, $T_C = 1$ ms (b) $V_A = 12$ mV, $T_C = 10$ s
(c) $I_A = 15$ mA, $T_C = 2$ ms (d) $I_A = 4$ A, $T_C = 5$ ms

5–4 THE SINUSOIDAL WAVEFORM

The cosine and sine functions are important in all branches of science and engineering. The corresponding time-varying waveform in Figure 5–14 plays an especially prominent role in electrical engineering.

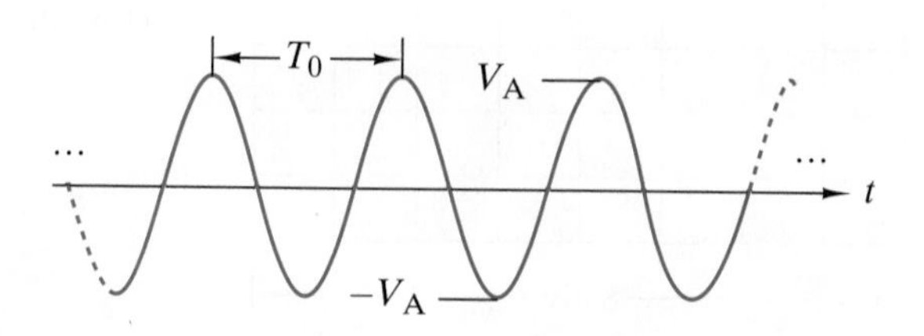

FIGURE 5–14 *The eternal sinusoid.*

In contrast with the step and exponential waveforms studied earlier, the sinusoid, like the dc waveform in Figure 5–1, extends indefinitely in time in both the positive and negative directions. The sinusoid has neither a beginning nor an end. Of course, real signals have finite durations. They were turned on at some finite time in the past and will be turned off at some time in the future. While it may seem unrealistic to have a signal model that lasts forever, it turns out that the eternal sinewave is a very good approximation in many practical applications.

The sinusoid in Figure 5–14 is an endless repetition of identical oscillations between positive and negative peaks. The **amplitude** V_A (in volts) defines the maximum and minimum values of the oscillations. The **period** T_0 (usually seconds) is the time required to complete one cycle of the oscillation. The sinusoid can be expressed mathematically using either the sine or the cosine function. The choice between the two depends on where we choose to define $t = 0$. If we choose $t = 0$ at a point where the sinusoid is zero, then it can be written as

$$v(t) = V_A \sin(2\pi t/T_0) \tag{5–18a}$$

On the other hand, if we choose $t = 0$ at a point where the sinusoid is at a positive peak, we can write an equation for it in terms of a cosine function:

$$v(t) = V_A \cos(2\pi t/T_0) \tag{5–18b}$$

Although either choice will work, it is common practice to choose $t = 0$ at a positive peak; hence Eq. (5–18b) applies. Thus, we will continue to call the waveform a sinusoid even though we use a cosine function to describe it.

As in the case of the step and exponential functions, the general sinusoid is obtained by replacing t by $(t - T_S)$. Inserting this change in Eq. (5–18) yields a general expression for the sinusoid as

$$v(t) = V_A \cos[2\pi(t - T_S)/T_0] \tag{5–19}$$

where the constant T_S is the time-shift parameter. Figure 5–15 shows that the sinusoid shifts to the right when $T_S > 0$ and to the left when $T_S < 0$. In effect, time shifting causes the positive peak nearest the origin to occur at $t = T_S$.

The time-shifting parameter can also be represented by an angle:

$$v(t) = V_A \cos[2\pi t/T_0 + \phi] \tag{5–20}$$

The parameter ϕ is called the **phase angle**. The term *phase angle* is based on the circular interpretation of the cosine function. We think of the period as being divided into 2π radians. In this sense the phase angle is the angle between $t = 0$ and the nearest positive peak. Comparing Eqs. (5–19) and (5–20), we find the relation between T_S and ϕ to be

$$\phi = -2\pi \frac{T_S}{T_0} \tag{5–21}$$

Changing the phase angle moves the waveform to the left or right, revealing different phases of the oscillating waveform (hence the name *phase angle*).

The phase angle should be expressed in radians, but is often reported in degrees. Care must be used when numerically evaluating the argument of

FIGURE 5–15 *Effect of time shifting on the sinusoidal waveform.*

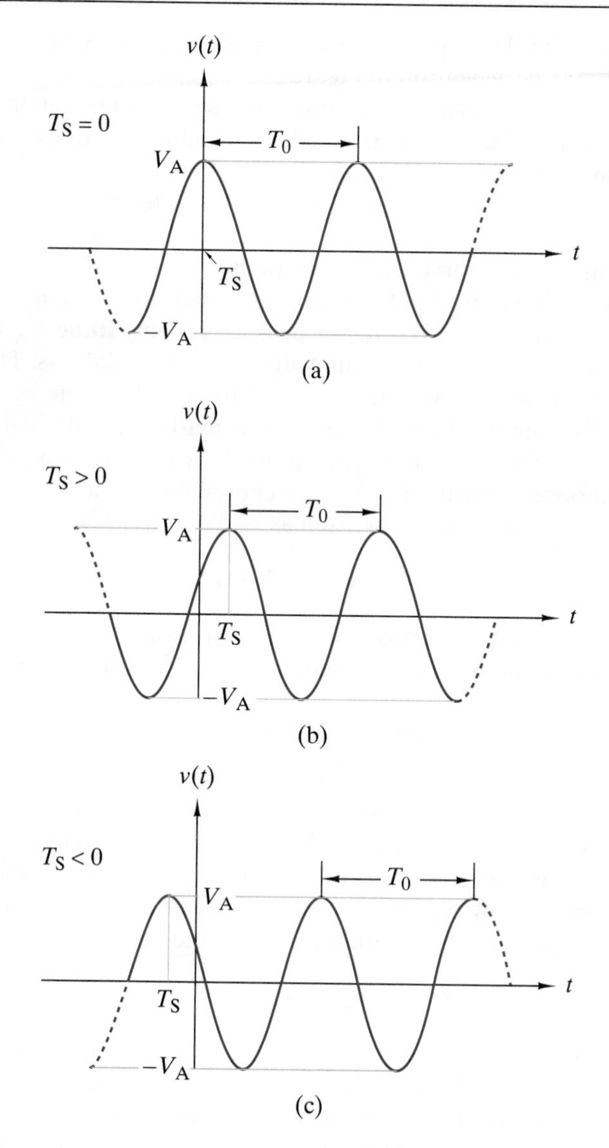

the cosine $(2\pi t/T_0 + \phi)$ to ensure that both terms have the same units. The term $2\pi t/T_0$ has the units of radians, so it is necessary to convert ϕ to radians when it is given in degrees.

An alternative form of the general sinusoid is obtained by expanding Eq. (5–20) using the identity $\cos(x + y) = \cos(x)\cos(y) - \sin(x)\sin(y)$,

$$v(t) = [V_A \cos \phi] \cos (2\pi t/T_0) + [-V_A \sin \phi] \sin (2\pi t/T_0)$$

The quantities inside the brackets in this equation are constants; therefore, we can write the general sinusoid in the form

$$v(t) = a \cos(2\pi t/T_0) + b \sin(2\pi t/T_0) \qquad (5\text{–}22)$$

The two amplitudelike parameters a and b have the same units as the waveform (volts in this case) and are called Fourier coefficients. By definition, the Fourier coefficients are related to the amplitude and phase parameters by the equations

$$a = V_A \cos \phi$$
$$b = -V_A \sin \phi \tag{5–23}$$

The inverse relationships are obtained by squaring and adding the expressions in Eq. (5–23):

$$V_A = \sqrt{a^2 + b^2} \tag{5–24}$$

and by dividing the second expression in Eq. (5–23) by the first:

$$\phi = \tan^{-1} \frac{-b}{a} \tag{5–25}$$

Caution: The inverse tangent function on a calculator has a ±180° ambiguity that can be resolved by considering the signs of the Fourier coefficients a and b.

It is customary to describe the time variation of the sinusoid in terms of a frequency parameter. **Cyclic frequency** f_0 is defined as the number of periods per unit time. By definition, the period T_0 is the number of seconds per cycle; consequently, the number of cycles per second is

$$f_0 = \frac{1}{T_0} \tag{5–26}$$

where f_0 is the cyclic frequency or simply the frequency. The unit of frequency (cycles per second) is the **hertz** (Hz). The **angular frequency** ω_0 in radians per second is related to the cyclic frequency by the relationship

$$\omega_0 = 2\pi f_0 = \frac{2\pi}{T_0} \tag{5–27}$$

because there are 2π radians per cycle.

There are two ways to express the concept of sinusoidal frequency: cyclic frequency (f_0, hertz) and angular frequency (ω_0, radians per second). When working with signals, we tend to use the former. For example, radio stations transmit carrier signals at frequencies specified as 690 kHz (AM band) or 101 MHz (FM band). Radian frequency is more convenient when describing the characteristics of circuits driven by sinusoidal inputs.

In summary, there are several equivalent ways to describe the general sinusoid:

$$v(t) = V_A \cos\left[\frac{2\pi(t - T_S)}{T_0}\right] = V_A \cos\left(\frac{2\pi t}{T_0} + \phi\right) = a \cos\left(\frac{2\pi t}{T_0}\right) + b \sin\left(\frac{2\pi t}{T_0}\right)$$
$$= V_A \cos[2\pi f_0(t - T_S)] = V_A \cos(2\pi f_0 t + \phi) = a \cos(2\pi f_0 t) + b \sin(2\pi f_0 t)$$
$$= V_A \cos[\omega_0(t - T_S)] = V_A \cos(\omega_0 t + \phi) = a \cos(\omega_0 t) + b \sin(\omega_0 t)$$

To use any one of these expressions, we need three types of parameters:

1. *Amplitude:* either V_A or the Fourier coefficients a and b
2. *Time shift:* either T_S or the phase angle ϕ
3. *Time/frequency:* either T_0, f_0, or ω_0.

In different parts of this book we use different forms to represent a sinusoid. Therefore, it is important for you to understand thoroughly the relationships among the various parameters in Eqs. (5–21) through (5–27).

EXAMPLE 5–7

Figure 5–16 shows an oscilloscope display of a sinusoid. The vertical axis (amplitude) is calibrated at 5 V per division, and the horizontal axis (time) is calibrated at 0.1 ms per division. Derive an expression for the sinusoid displayed in Figure 5–16.

FIGURE 5–16

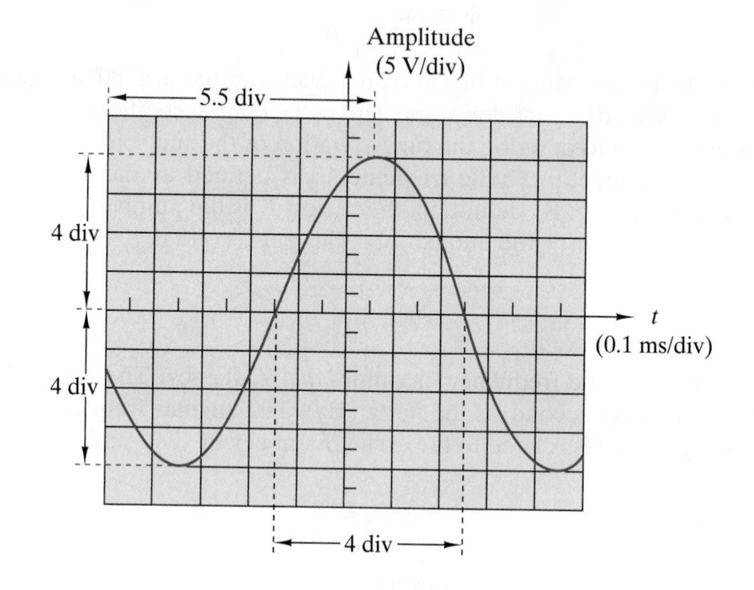

SOLUTION:

The maximum amplitude of the waveform is seen to be four vertical divisions; therefore,

$$V_A = (4 \text{ div})(5 \text{ V/div}) = 20 \text{ V}$$

There are four horizontal divisions between successive zero crossings, which means there are a total of eight divisions in one cycle. The period of the waveform is

$$T_0 = (8 \text{ div})(0.1 \text{ ms/div}) = 0.8 \text{ ms}$$

The two frequency parameters are $f_0 = 1/T_0 = 1.25$ kHz and $\omega_0 = 2\pi f_0 =$ 7854 rad/s. The parameters V_A, T_0, f_0, and ω_0 do not depend on the location of the $t = 0$ axis.

To determine the time shift T_S, we need to define a time origin. The $t =$ 0 axis is arbitrarily taken at the left edge of the display in Figure 5–16. The positive peak shown in the display is 5.5 divisions to the right of $t = 0$, which is more than half a cycle (four divisions). The positive peak closest to $t = 0$ is not shown in Figure 5–16 because it must lie beyond the left edge of the display. However, the positive peak shown in the display is located at $t = T_S + T_0$ since it is one cycle after $t = T_S$. We can write

$$T_S + T_0 = (5.5 \text{ div})(0.1 \text{ ms/div}) = 0.55 \text{ ms}$$

which yields $T_S = 0.55 - T_0 = -0.25$ ms. As expected, T_S is negative because the nearest positive peak is to the left of $t = 0$.

Given T_S, we can calculate the remaining parameters of the sinusoid as follows:

$$\phi = -\frac{2\pi T_S}{T_0} = 1.96 \text{ rad or } 112.5°$$

$$a = V_A \cos\phi = -7.65 \text{ V}$$

$$b = -V_A \sin\phi = -18.5 \text{ V}$$

Finally, the three alternative expressions for the displayed sinusoid are

$$v(t) = 20\cos[(7854\,t + 0.25 \times 10^{-3})]$$
$$= 20\cos(7854\,t + 112.5°)$$
$$= -7.65\cos 7854t - 18.5\sin 7854t$$

■

Exercise 5–5

Derive an expression for the sinusoid displayed in Figure 5–16 when $t = 0$ is placed in the middle of the display.

Answer: $v(t) = 20\cos(7854t - 22.5°)$

PROPERTIES OF SINUSOIDS

In general, a waveform is said to be **periodic** if

$$v(t + T_0) = v(t)$$

for all values of t. The constant T_0 is called the period of the waveform if it is the smallest nonzero interval for which $v(t + T_0) = v(t)$. Since this equality must be valid for all values of t, it follows that periodic signals must have eternal waveforms that extend indefinitely in time in both directions. Signals that are not periodic are called **aperiodic**.

The sinusoid is a periodic signal since

$$v(t + T_0) = V_A \cos[2\pi\,(t + T_0)/T_0 + \phi]$$
$$= V_A \cos[2\pi\,(t)/T_0 + \phi + 2\,\pi]$$

But $\cos(x + 2\pi) = \cos(x)$. Consequently,

$$v(t + T_0) = V_A \cos(2\pi t/T_0 + \phi) = v(t)$$

for all t.

The **additive property** of sinusoids states that summing two or more sinusoids with the same frequency yields a sinusoid with different amplitude and phase parameters but the same frequency. To illustrate, consider two sinusoids

$$v_1(t) = a_1 \cos(2\pi f_0 t) + b_1 \sin(2\pi f_0 t)$$

$$v_2(t) = a_2 \cos(2\pi f_0 t) + b_2 \sin(2\pi f_0 t)$$

The waveform $v_3(t) = v_1(t) + v_2(t)$ can be written as

$$v_3(t) = (a_1 + a_2) \cos(2\pi f_0 t) + (b_1 + b_2) \sin(2\pi f_0 t)$$

because cosine and sine are linearly independent functions. We obtain the Fourier coefficients of the sum of two sinusoids by adding their Fourier coefficients, provided the two have the same frequency. *Caution:* The summation must take place with the sinusoids in Fourier coefficient form. Sums of sinusoids *cannot* be found by adding amplitudes and phase angles.

The **derivative** and **integral** properties state that when we differentiate or integrate a sinusoid, the result is another sinusoid with the same frequency:

$$\frac{d(V_A \cos \omega t)}{dt} = -\omega V_A \sin \omega t = \omega V_A \cos(\omega t + \pi/2)$$

$$\int V_A \cos(\omega t)\, dt = \frac{V_A}{\omega} \sin \omega t = \frac{V_A}{\omega} \cos(\omega t - \pi/2)$$

These operations change the amplitude and phase angle but do not change the frequency. The fact that differentiation and integration preserve the underlying waveform is a key property of the sinusoid. No other periodic waveform has this shape-preserving property.

EXAMPLE 5–8

(a) Find the period's cyclic and radian frequencies of the sinusoids
$v_1(t) = 17 \cos(2000t - 30°)$
$v_2(t) = 12 \cos(2000t + 30°)$.
(b) Find the waveform of $v_3(t) = v_1(t) + v_2(t)$.

SOLUTION:

(a) The two sinusoids have the same frequency $\omega_0 = 2000$ rad/s since a term $2000t$ appears in the arguments of $v_1(t)$ and $v_2(t)$. Therefore, $f_0 = \omega_0/2\pi = 318.3$ Hz and $T_0 = 1/f_0 = 3.14$ ms.

(b) We use the additive property, since the two sinusoids have the same frequency. Beyond this checkpoint, the frequency plays no further role in the calculation. The two sinusoids must be converted to the Fourier coefficient form using Eq. (5–23).

$$a_1 = 17 \cos(-30°) = +14.7 \text{ V}$$

$$b_1 = -17 \sin(-30°) = +8.50 \text{ V}$$

$$a_2 = 12 \cos(30°) = +10.4 \text{ V}$$

$$b_2 = -12 \sin(30°) = -6.00 \text{ V}$$

The Fourier coefficients of the signal $v_3 = v_1 + v_2$ are found as

$$a_3 = a_1 + a_2 = 25.1 \text{ V}$$

$$b_3 = b_1 + b_2 = 2.50 \text{ V}$$

The amplitude and phase angle of $v_3(t)$ are found using Eqs. (5–24) and (5–25):

$$V_A = \sqrt{a_3^2 + b_3^2} = 25.2 \text{ V}$$

$$\phi = \tan^{-1}(-2.5/25.1) = -5.69°$$

Two equivalent representations of $v_3(t)$ are

$$v_3(t) = 25.1\cos(2000t) + 2.5\sin(2000t) \text{ V}$$

and

$$v_3(t) = 25.2\cos(2000t - 5.69°) \text{ V}$$

■

EXAMPLE 5–9

The balanced three-phase voltages used in electrical power systems can be written as

$$v_A(t) = V_m \cos(2\pi f_0 t)$$

$$v_B(t) = V_m \cos(2\pi f_0 t + 120°)$$

$$v_C(t) = V_m \cos(2\pi f_0 t + 240°)$$

Show that the sum of these voltages is zero.

SOLUTION:

The three voltages are given in amplitude/phase angle form. They can be converted to the Fourier coefficient form using Eq. (5–23):

$$v_A(t) = +V_m \cos(2\pi f_0 t)$$

$$v_B(t) = -\frac{V_m}{2}\cos(2\pi f_0 t) + \frac{\sqrt{3}V_m}{2}\sin(2\pi f_0 t)$$

$$v_C(t) = -\frac{V_m}{2}\cos(2\pi f_0 t) - \frac{\sqrt{3}V_m}{2}\sin(2\pi f_0 t)$$

The sum of the cosine and sine terms is zero; consequently, $v_A(t) + v_B(t) + v_C(t) = 0$. The zero sum occurs because the three sinusoids have equal amplitudes and are disposed at 120° intervals. If the amplitudes are not equal or the phase angles do not differ by 120°, then the voltages are said to be unbalanced. ■

Exercise 5–6

Write an equation for the waveform obtained by integrating and differentiating the following signals:

(a) $v_1(t) = 30\cos(10t - 60°)$

(b) $v_2(t) = 3\cos(4000\pi t) - 4\sin(4000\pi t)$

Answers:

(a) $\dfrac{dv_1}{dt} = 300\cos(10t + 30°)$

$$\int v_1(t)\,dt = 3\cos(10t - 150°)$$

$$\text{(b)} \frac{dv_2}{dt} = 2\pi \times 10^4 \cos(4000\pi t + 143.1°)$$

$$\int v_2\, dt = \frac{1}{800\pi} \cos(4000\pi t - 36.87°)$$

Exercise 5–7

A sinusoid has a period of 5 μs. At $t = 0$ the amplitude is 12 V. The waveform reaches its first positive peak after $t = 0$ at $t = 4$ μs. Find its amplitude, frequency, and phase angle.

Answers: $V_A = 38.8$ V; $f_0 = 200$ kHz; $\phi = +72°$

5–5 COMPOSITE WAVEFORMS

In the previous sections we introduced the step, exponential, and sinusoidal waveforms. These waveforms are basic signals because they can be combined to synthesize all other signals used in this book. Signals generated by combining the three basic waveforms are called **composite signals**. This section provides examples of composite waveforms.

EXAMPLE 5–10

Characterize the composite waveform generated by

$$v(t) = V_A u(t) - V_A u(-t)$$

SOLUTION:

The first term in this waveform is simply a step function of amplitude V_A that occurs at $t = 0$. The second term involves the function $u(-t)$, whose waveform requires some discussion. Strictly speaking, the general step function $u(x)$ is unity when $x > 0$ and zero when $x < 0$. That is, $u(x)$ is unity when its argument is positive and zero when it is negative. Under this rule the function $u(-t)$ is unity when $-t > 0$ and zero when $-t < 0$, that is,

$$u(-t) = \begin{cases} 1 & \text{for } t < 0 \\ 0 & \text{for } t > 0 \end{cases}$$

which is the reverse of the step function $u(t)$. Figure 5–17 shows how the two components combine to produce a composite waveform that extends indefinitely in both directions and has a jump discontinuity of $2V_A$ at $t = 0$. This composite waveform is called a **signum** function.

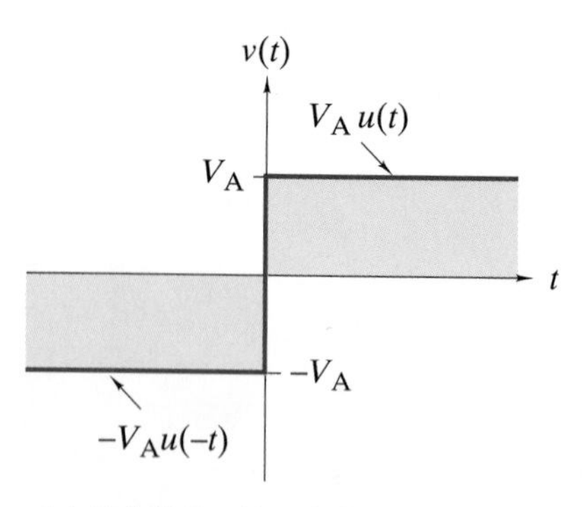

FIGURE 5–17 *The signum waveform.*

EXAMPLE 5–11

Characterize the composite waveform generated by subtracting an exponential from a step function with the same amplitude.

SOLUTION:

The equation for this composite waveform is

$$v(t) = V_A u(t) - [V_A e^{-t/T_C}]u(t)$$
$$= V_A[1 - e^{-t/T_C}]u(t)$$

For $t < 0$ the waveform is zero because of the step function. At $t = 0$ the waveform is still zero since the step and exponential cancel:

$$v(0) = V_A[1 - e^0](1) = 0$$

For $t >> T_C$ the waveform approaches a constant value V_A because the exponential term decays to zero. For practical purposes $v(t)$ is within less than 1% of its final value V_A when $t = 5T_C$. At $t = T_C$, $v(T_C) = V_A(1 - e^{-1}) = 0.632V_A$. The waveform rises to about 63% of its final value in one time constant. All of the observations are summarized in the plot shown in Figure 5–18. This waveform is called an **exponential rise**. It is also sometimes referred to as a "charging exponential," since it represents the behavior of signals that occur during the buildup of voltage in resistor-capacitor circuits studied in Chapter 7. ■

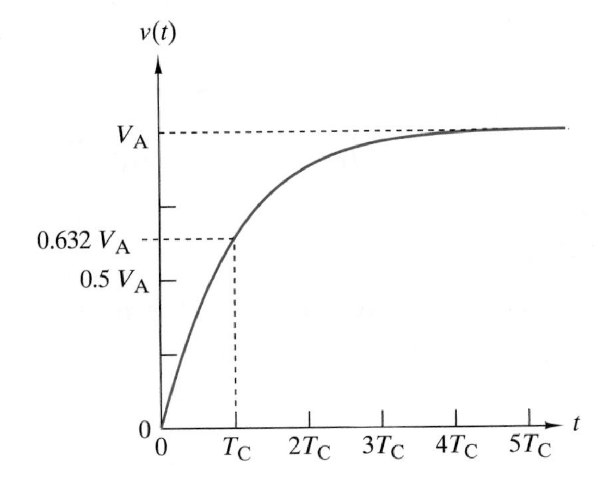

FIGURE 5–18 *The exponential rise waveform.*

EXAMPLE 5–12

Characterize the composite waveform obtained by multiplying the ramp $r(t)/T_C$ times an exponential.

SOLUTION:

The equation for this composite waveform is

$$v(t) = \frac{r(t)}{T_C}[V_A e^{-t/T_C}]u(t)$$
$$= V_A[(t/T_C)\, e^{-t/T_C}]u(t)$$

For $t < 0$ the waveform is zero because of the step function. At $t = 0$ the waveform is zero because $r(0) = 0$. For $t > 0$ there is a competition between two effects—the ramp increases linearly with time while the exponential decays to zero. Since the composite waveform is the product of these terms, it is important to determine which effect dominates. In the limit, as $t \to \infty$, the product of the ramp and exponential takes on the indeterminate form of infinity times zero. A single application of *l'Hôpital's* rule, then, shows that the exponential dominates, forcing the $v(t)$ to zero as t becomes large. That is, the exponential decay overpowers the linearly increasing ramp, as shown by the graph in Figure 5–19. The waveform obtained by multiplying a ramp by a decaying exponential is called a **damped ramp**. ■

FIGURE 5–19 *The damped ramp waveform.*

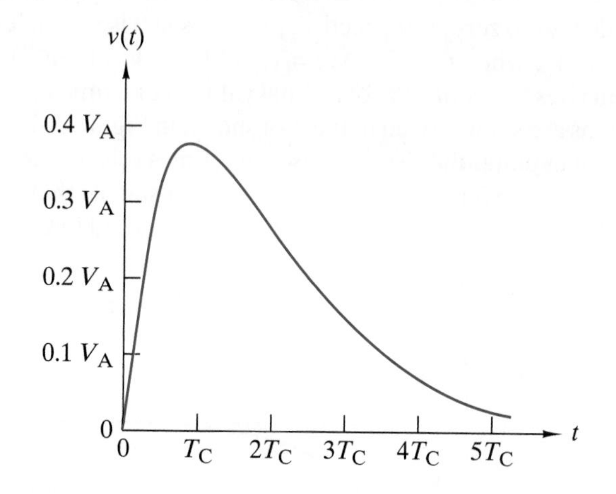

EXAMPLE 5–13

Characterize the composite waveform obtained by multiplying $\sin \omega_0 t$ by an exponential.

SOLUTION:
In this case the composite waveform is expressed as

$$\begin{aligned} v(t) &= \sin \omega_0 t[V_A e^{-t/T_C}]\, u(t) \\ &= V_A[e^{-t/T_C} \sin \omega_0 t]\, u(t) \end{aligned}$$

Figure 5–20 shows a graph of this waveform for $T_0 = 2T_C$. For $t < 0$ the step function forces the waveform to be zero. At $t = 0$, and periodically thereafter, the waveform passes through zero because $\sin(n\pi) = 0$. The waveform is not periodic, however, because the decaying exponential gradually reduces the amplitude of the oscillation. For all practical purposes the oscillations become negligibly small for $t > 5T_C$. The waveform obtained by multiplying a sinusoid by a decaying exponential is called a **damped sine**. ■

EXAMPLE 5–14

Characterize the composite waveform obtained as the difference of two exponentials with the same amplitude.

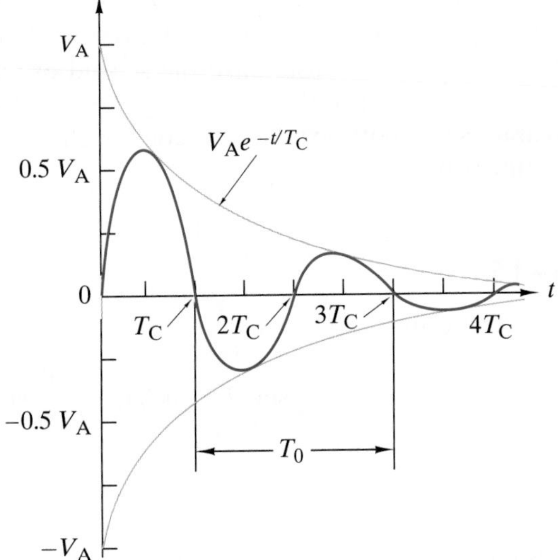

FIGURE 5–20 *The damped sine waveform.*

SOLUTION:

The equation for this composite waveform is

$$v(t) = [V_A\, e^{-t/T_1}]\, u(t) - [V_A\, e^{-t/T_2}]\, u(t)$$
$$= V_A\, (e^{-t/T_1} - e^{-t/T_2})\, u(t)$$

For $T_1 > T_2$ the resulting waveform is illustrated in Figure 5–21 (plotted for $T_1 = 2T_2$). For $t < 0$ the waveform is zero. At $t = 0$ the waveform is still zero, since

$$v(0) = V_A\, (e^{-0} - e^{-0})$$
$$= V_A\, (1 - 1) = 0$$

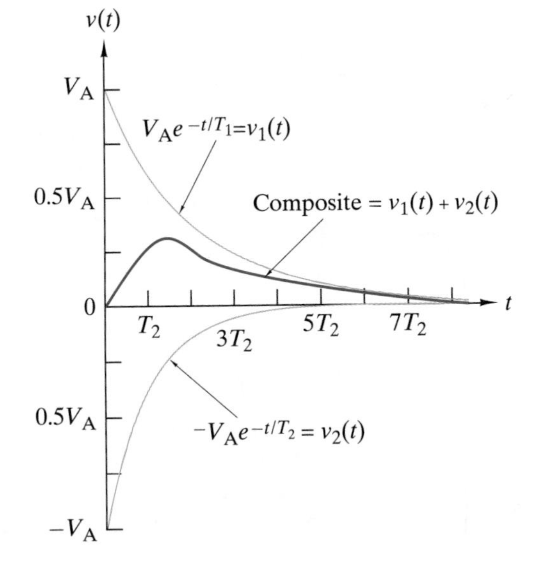

FIGURE 5–21 *The double exponential waveform.*

For $t >> T_1$ the waveform returns to zero because both exponentials decay to zero. For $5T_1 > t > 5T_2$ the second exponential is negligible and the waveform essentially reduces to the first exponential. Conversely, for $t << T_1$ the first exponential is essentially constant, so the second exponential determines the early time variation of the waveform. The waveform is called a **double exponential**, since both exponential components make important contributions to the waveform. ■

EXAMPLE 5–15

Characterize the composite waveform defined by

$$v(t) = 5 - \frac{10}{\pi}\sin(2\pi 500\, t) - \frac{10}{2\pi}\sin(2\pi 1000\, t) - \frac{10}{3\pi}\sin(2\pi 1500\, t)$$

SOLUTION:

The waveform is the sum of a constant (dc) term and three sinusoids at different frequencies. The first sinusoidal component is called the **fundamental** because it has the lowest frequency. As a result the frequency $f_0 = 500$ Hz is called the **fundamental frequency**. The other sinusoidal terms are said to be harmonics because their frequencies are integer multiples of f_0. Specifically, the second sinusoidal term is called the **second harmonic** ($2f_0 = 1000$ Hz) while the third term is the **third harmonic** ($3f_0 = 1500$ Hz). Figure 5–22 shows a plot of this waveform. Note that the waveform is periodic with a period equal to that of the fundamental component, namely, $T_0 = 1/f_0 = 2$ ms. The decomposition of a periodic waveform into a sum of harmonic sinusoids is called a **Fourier series**, a topic we will study in detail in Chapter 13. In fact, the waveform in this example is the first four terms in the Fourier series for a 10-V sawtooth wave of the type shown in Figure 5–1. ■

FIGURE 5–22

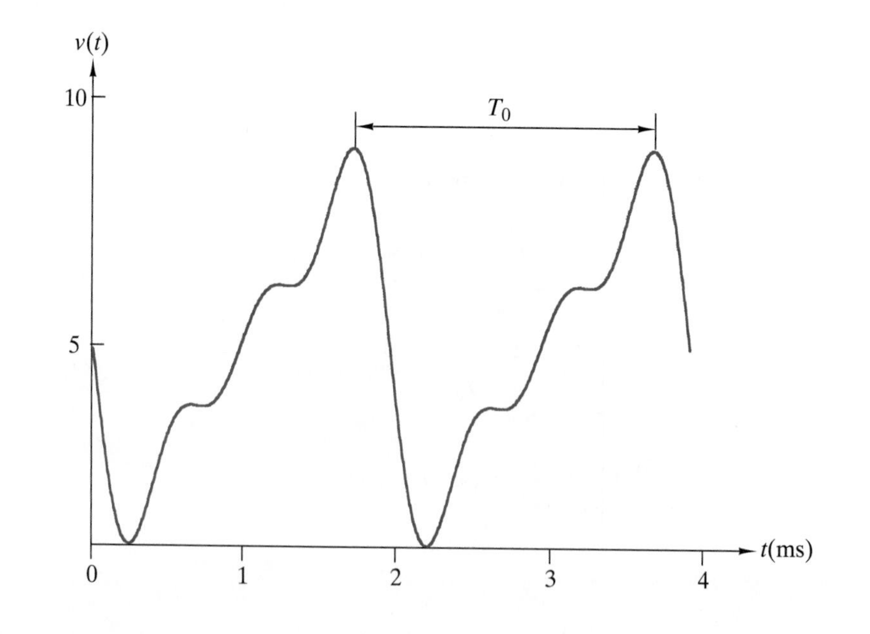

Exercise 5–8

Write an expression for the pulse in Figure 5–23 using step functions.

Answer:

$$v(t) = V_A[u(t) - u(t-T) - u(-t) + u(-t-T)]$$

or

$$v(t) = V_A[-u(t+T) + 2u(t) - u(t-T)]$$

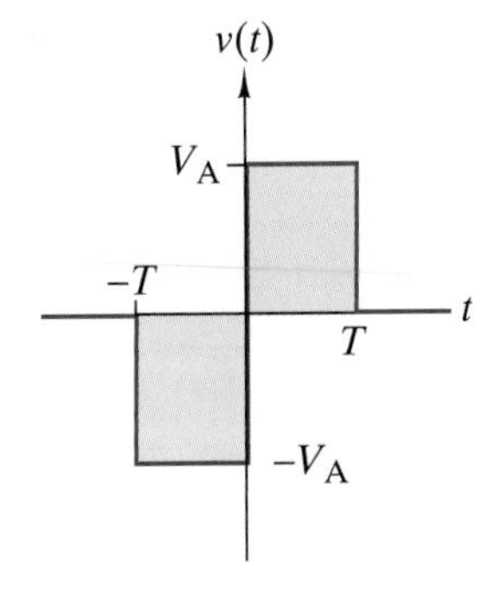

FIGURE 5–23

Exercise 5–9

Find the maximum amplitude and the approximate duration of the following composite waveforms:

(a) $v_1(t) = [25 \sin 1000t][u(t) - u(t-10)]$ V
(b) $v_2(t) = [50 \cos 1000t][e^{-200t}]u(t)$ V
(c) $i_3(t) = [3000te^{-1000t}]u(t)$ mA
(d) $i_4(t) = [10e^{5000t}]u(-t) + [10e^{-5000t}]u(t)$ A

Answers:

(a) 25 V, 10 s
(b) 50 V, 25 ms
(c) 1.10 mA, 5 ms
(d) 10 A, 2 ms

5–6 WAVEFORM PARTIAL DESCRIPTORS

An equation or graph defines a waveform for all time. The value of a waveform $v(t)$ or $i(t)$ at time t is called the **instantaneous value** of the waveform. We often use parameters called **partial descriptors** that characterize important features of a waveform but do not give a complete description. These partial descriptors fall into two categories: (1) those that describe temporal features and (2) those that describe amplitude features.

TEMPORAL DESCRIPTORS

Temporal descriptors identify waveform attributes relative to the time axis. For example, waveforms that repeat themselves at fixed time intervals are said to be **periodic**. Stated formally,

> ***A signal v(t) is periodic if v(t + T$_0$) = v(t) for all t, where the period T$_0$ is the smallest value that meets this condition. Signals that are not periodic are called aperiodic.***

The fact that a waveform is periodic provides important information about the signal but does not specify all of its characteristics. Thus, the fact that a signal is periodic is itself a partial description, as is the value of the period. The eternal sinewave is the premier example of a periodic signal. The square wave and triangular wave in Figure 5–2 are also periodic. Examples of aperiodic waveforms are the step function, exponential, and damped sine.

Waveforms that are identically zero prior to some specified time are said to be **causal**. Stated formally,

> ***A signal v(t) is causal if there exists a value of T such that v(t) ≡ 0 for all t < T; otherwise it is noncausal.***

It is usually assumed that a causal signal is zero for $t < 0$, since we can always use time shifting to make the starting point of a waveform at $t = 0$. Examples of causal waveforms are the step function, exponential, and damped sine. The eternal sinewave is, of course, noncausal.

Causal waveforms play a central role in circuit analysis. When the input driving force $x(t)$ is causal, the circuit response $y(t)$ must also be causal. That is, a physically realizable circuit cannot anticipate and respond to an input before it is applied. Causality is an important temporal feature, but only a partial description of the waveform.

AMPLITUDE DESCRIPTORS

Amplitude descriptors are positive scalars that describe signal strength. Generally, a waveform varies between two extreme values denoted as V_{MAX} and V_{MIN}. The **peak-to-peak value** (V_{pp}) describes the total excursion of $v(t)$ and is defined as

$$V_{pp} = V_{MAX} - V_{MIN} \tag{5–28}$$

Under this definition V_{pp} is always positive even if V_{MAX} and V_{MIN} are both negative. The **peak value** (V_p) is the maximum of the absolute value of the waveform. That is,

$$V_p = \text{Max}\{|V_{MAX}|, |V_{MIN}|\} \tag{5–29}$$

The peak value is a positive number that indicates the maximum absolute excursion of the waveform from zero. Figure 5–24 shows examples of these two amplitude descriptors.

The peak and peak-to-peak values describe waveform variation using the extreme values. The average value smooths things out to reveal the

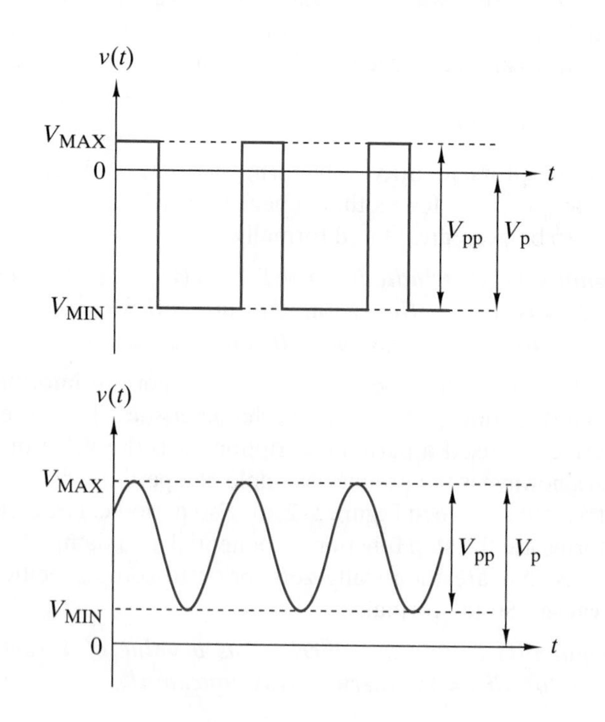

FIGURE 5–24 *Peak value (V_p) and peak-to-peak value (V_{pp}).*

underlying waveform baseline. Average value is the area under the waveform over some period of time T, divided by that time period. Mathematically, we define **average value** (V_{avg}) over the time interval T as

$$V_{\text{avg}} = \frac{1}{T}\int_{t}^{t+T} v(x)dx \tag{5–30}$$

For periodic signals the period T_0 is used as the averaging interval T.

For some periodic waveforms the integral in Eq. (5–30) can be estimated graphically. The net area under the waveform is the area above the time axis minus the area below the time axis. For example, the two waveforms in Figure 5–24 obviously have nonzero average values. The top waveform has a negative average value because the negative area below the time axis more than cancels the area above the axis. Similarly, the bottom waveform clearly has a positive average value.

The average value indicates whether the waveform contains a constant, non-time-varying component. The average value is also called the **dc component** because dc signals are constant for all t. On the other hand, the **ac components** have zero average value and are periodic. For example, the waveform in Example 5–15

$$v(t) = 5 - \frac{10}{\pi}\sin(2\,\pi\,500\,t) - \frac{10}{2\,\pi}\sin(2\,\pi\,1000\,t) - \frac{10}{3\,\pi}\sin(2\,\pi\,1500\,t)$$

has a 5-V average value due to its dc component. The three sinusoids are ac components because they are periodic and have zero average value. Sinusoids have zero average value because over any given cycle the positive area above the time axis is exactly canceled by the negative area below.

EXAMPLE 5–16

Find the peak, peak-to-peak, and average values of the periodic input and output waveforms in Figure 5–25.

SOLUTION:

The input waveform is a sinusoid whose amplitude descriptors are

$$V_{\text{pp}} = 2V_{\text{A}} \quad V_{\text{p}} = V_{\text{A}} \quad V_{\text{avg}} = 0$$

The output waveform is obtained by clipping off the negative half-cycle of the input sinusoid. The amplitude descriptors of the output waveform are

$$V_{\text{pp}} = V_{\text{p}} = V_{\text{A}}$$

The output has a nonzero average value, since there is a net positive area under the waveform. The upper limit in Eq. (5–30) can be taken as $T_0/2$, since the waveform is zero from $T_0/2$ to T_0.

$$V_{\text{avg}} = \frac{1}{T_0}\int_0^{T_0/2} V_{\text{A}}\sin(2\pi t/T_0)dt = -\frac{V_{\text{A}}}{2\pi}\cos(2\pi\, t/T_0)\Big|_0^{T_0/2}$$
$$= \frac{V_{\text{A}}}{\pi}$$

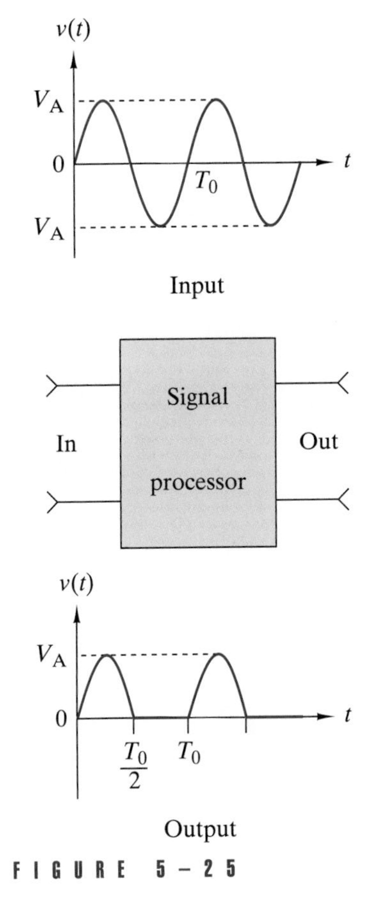

FIGURE 5–25

The signal processor produces an output with a dc value from an input with no dc component. Rectifying circuits described in electronics courses produce waveforms like the output in Figure 5–25. ■

ROOT-MEAN-SQUARE VALUE

The **root-mean-square value** (V_{rms}) is a measure of the average power carried by the signal. The instantaneous power delivered to a resistor R by a voltage $v(t)$ is

$$p(t) = \frac{1}{R}[v(t)]^2 \tag{5–31}$$

The average power delivered to the resistor in time span T is defined as

$$P_{avg} = \frac{1}{T}\int_{t}^{t+T} p(t)dt \tag{5–32}$$

Combining Eqs. (5–31) and (5–32) yields

$$P_{avg} = \frac{1}{R}\left[\frac{1}{T}\int_{t}^{t+T} [v(t)]^2\, dt\right] \tag{5–33}$$

The quantity inside the large brackets in Eq. (5–33) is the average value of the square of the waveform. The units of the bracketed term are volts squared. The square root of this term defines the amplitude partial descriptor V_{rms}:

$$V_{rms} = \sqrt{\frac{1}{T}\int_{t}^{t+T} [v(t)]^2\, dt} \tag{5–34}$$

The amplitude descriptor V_{rms} is called the root-mean-square (rms) value because it is obtained by taking the square root of the average (mean) of the square of the waveform. For periodic signals the averaging interval is one cycle since such a waveform repeats itself every T_0 seconds.

We can express the average power delivered to a resistor in terms of V_{rms} as

$$P_{avg} = \frac{1}{R}V_{rms}^2 \tag{5–35}$$

The equation for average power in terms of V_{rms} has the same form as the power delivered by a dc signal. For this reason the rms value was originally called the **effective value**, although this term is no longer common. If the waveform amplitude is doubled, its rms value is doubled, and the average power is quadrupled. Commercial electrical power systems use transmission voltages in the range of several hundred kilovolts (rms).

EXAMPLE 5–17

Find the average and rms values of the sinusoid and sawtooth in Figure 5–26.

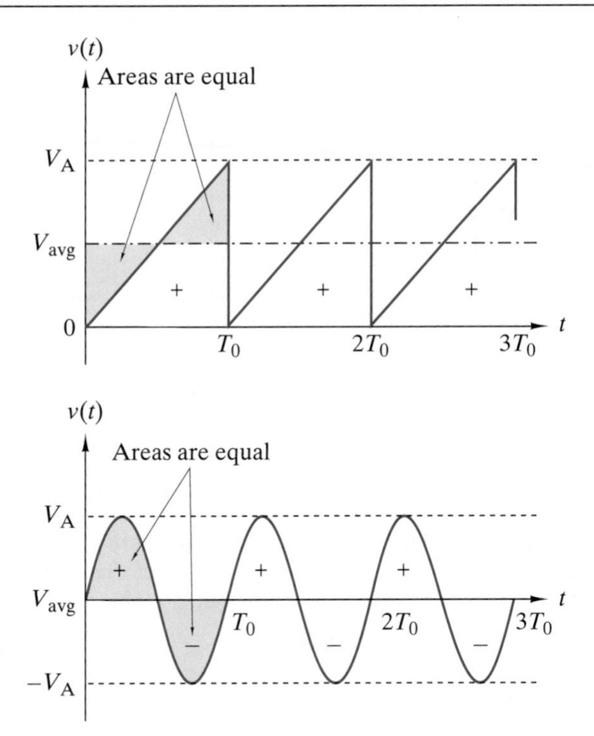

FIGURE 5–26

SOLUTION:

As noted previously, the sinusoid has an average value of zero. The sawtooth clearly has a positive average value. By geometry, the net area under one cycle of the sawtooth waveform is $V_A T_0/2$, so its average value is $(1/T_0)(V_A T_0/2) = V_A/2$. To obtain the rms value of the sinusoid we apply Eq. (5–34) as

$$V_{rms} = \sqrt{\frac{(V_A)^2}{T_0}\int_0^{T_0} \sin^2(2\pi t/T_0)dt}$$

$$= \sqrt{\frac{(V_A)^2}{T_0}\left[\frac{t}{2} - \frac{\sin(4\pi t/T_0)}{8\pi/T_0}\right]_0^{T_0}} = \frac{V_A}{\sqrt{2}}$$

For the sawtooth the rms value is found as:

$$V_{rms} = \sqrt{\frac{1}{T_0}\int_0^{T_0} (V_A t/T_0)^2\, dt} = \sqrt{\frac{(V_A)^2}{T_0^3}\left[\frac{t^3}{3}\right]_0^{T_0}} = \frac{V_A}{\sqrt{3}}$$ ■

Exercise 5–10

Find the peak, peak-to-peak, average, and rms values of the periodic waveform in Figure 5–27.

Answers:

$$V_p = 2V_A;\quad V_{pp} = 3V_A;\quad V_{avg} = \frac{V_A}{4};\quad V_{rms} = \frac{\sqrt{5}}{2}V_A$$

FIGURE 5–27

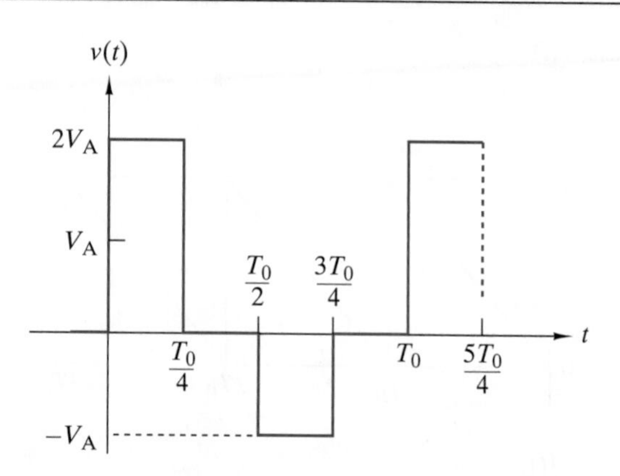

Exercise 5–11

Classify each of the following signals as periodic or aperiodic and causal or noncausal. Then calculate the average and rms values of the periodic waveforms, and the peak and peak-to-peak values of the other waveforms.

(a) $v_1(t) = 99 \cos 3000t - 132 \sin 3000t$ V
(b) $v_2(t) = 34\,[\sin 800\pi t][u(t) - u(t - 0.03)]$ V
(c) $i_3(t) = 120[u(t+5) - u(t-5)]$ mA
(d) $t_4(t) = 50$ A

Answers:

(a) Periodic, noncausal, $V_{avg} = 0$ and $V_{rms} = 117$ V
(b) Aperiodic, causal, $V_p = 34$ V and $V_{pp} = 68$ V
(c) Aperiodic, causal, $V_p = V_{pp} = 120$ mA
(d) Aperiodic, noncausal, $V_p = 50$ A and $V_{pp} = 0$

Exercise 5–12

Construct waveforms that have the following characteristics:

(a) Aperiodic and causal with $V_p = 8$ V and $V_{pp} = 15$ V
(b) Periodic and noncausal with $V_{avg} = 10$ V and $V_{pp} = 50$ V
(c) Periodic and noncausal with $V_{avg} = V_p/2$
(d) Aperiodic and causal with $V_p = V_{pp} = 10$ V

Answers: There are many possible correct answers since the given parameters are only partial descriptions of the required waveforms. Some examples that meet the requirements are as follows:

(a) $v(t) = 8\,u(t) - 15\,u(t-1) + 7\,u(t-2)$ V
(b) $v(t) = 10 + 25 \sin 1000t$ V
(c) A sawtooth wave
(d) $v(t) = 10[e^{-100t}]u(t)$ V

APPLICATION EXAMPLE 5–18

The operation of a digital system is coordinated and controlled by a periodic waveform called a clock. The *clock waveform* provides a standard timing reference to maintain synchronization between signal-processing results that become valid at different times during the clock cycle. Because of differences in digital circuit delays, there must be agreed-upon instants of time when circuit outputs can be treated as valid. The clock defers further signal processing until slower and faster outputs settle down when the clock signals the start of the next signal-processing cycle.

Figure 5–28 shows an idealized clock waveform as a periodic sequence of rectangular pulses. While we could easily write an exact expression for the clock waveform, we are interested here in discussing its partial descriptors. The first descriptor is the period T_0 or equivalently the **clock frequency** $f_0 = 1/T_0$. Clock frequency is a common measure of signal-processing speed and can range up into the hundreds of MHz. The pulse duration T is the time interval in each cycle when the pulse amplitude is high (not zero). In waveform terminology the ratio of the time in the high state to the period, that is, T/T_0, is called the **duty cycle**, usually expressed as a percentage. The **pulse edges** are the transition points at which the pulse changes states. There is a **rising edge** at the low-to-high transition and a **falling edge** at the high-to-low transition.

The pulse edges define the agreed-upon time instants at which the circuit outputs can be treated as valid inputs to other circuits. This means that circuit outputs must settle down during the time period between successive edges. Some synchronous operations are triggered by the rising edge and others by the falling edge. To provide equal settling times for both cases requires equal time between edges. In other words, it is desirable for the clock duty cycle to be 50%. As a result, the clock waveform is essentially a raised square wave whose dc offset equals one half of the peak-to-peak value.

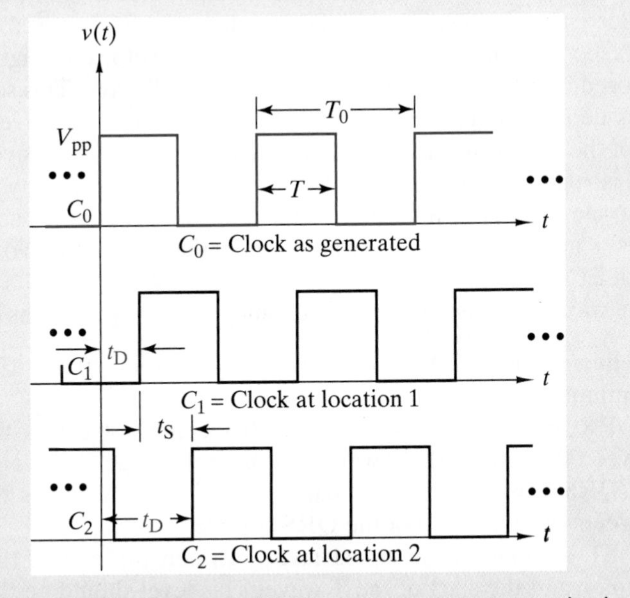

FIGURE 5–28

The system clock C_0 in Figure 5–28 is generated at some point in a circuit and then distributed to other locations. The clock distribution network almost invariably introduces delays, as illustrated by C_1 and C_2 in Figure 5–28. **Clock delay (t_D)** is defined as the time difference between a clock edge at a given location and the corresponding edge in the system clock at the point where it was generated. Delay is not necessarily a bad thing unless unequal delays cause the edges to be skewed, as indicated by the offset between C_1 and C_2 in Figure 5–28. When delays are significantly different, there is uncertainty as to instants of time at which further signal processing can safely proceed. This delay dispersion is called **clock skew** (t_S), defined as the time difference between a clock edge at a given location and the corresponding edge at another location. Controlling clock skew is an important consideration in the design of the clock distribution network in high-speed VLSI circuits.

Thus, partial descriptors of clock waveforms include *frequency*, *duty cycle*, *edges*, *delay*, and *skew*. The coming chapters treat dynamic circuits that modify input waveforms to produce outputs with different partial descriptors. In particular, dynamic circuit elements cause changes in a clock waveform, especially the partial descriptors of edges, delay, and skew.

APPLICATION EXAMPLE 5–19

An electrocardiogram (ECG) is a valuable diagnostic tool used in cardiovascular medicine. The ECG is based on the fact that the heart emits measurable bioelectric signals that can be recorded to evaluate the functioning of the heart as a mechanical pump. These signals were first observed in the late 19th century and subsequent signal processing developments have led to the advanced technology of present day ECG equipment.

The bioelectric signals of the heart muscle are measured and recorded through the placement of skin electrodes at various sites on the surface of the body. The site selection as well as discussion of the functions of the cardiac muscle are beyond the scope of this example. Rather, our purpose is to introduce some of the useful partial descriptors of ECG waveforms.

In bioelectric terminology the normal ECG waveform in Figure 5–29 is composed of a P wave, a QRS complex, and a T wave. This sequence of pulses depicts the electrical activity that stimulates the correct functioning of the cardiac muscle. The flat baseline between successive events is called isoelectric, which means there is no bioelectric activity and the heart muscle returns to a resting state. The body's natural pacemaker produces a nominally periodic waveform under the resting conditions used with ECG tests.

Partial waveform descriptors used to analyze ECG waveforms include:

1. The **heart rate** ($1/T_0$), which is normally between 60 and 100 beats per minute.
2. The **PR interval** (normally 0.12–0.20 seconds), which is the time between the start of the P wave and the start of the QRS complex.
3. The **QRS interval** (normally 0.06–0.10 seconds), which is the time between onset and end of the QRS complex.
4. The **ST segment** is the signal level between the end of the QRS complex and the start of the T wave. This level should be the same as the isoelectric baseline between successive pulses.

Departures from these normal conditions serve as diagnostic tools in cardiovascular medicine. Some of the abnormal waveform features of concern include an irregular heart rate, a missing P wave, a prolonged QRS interval, or an elevated ST segment. Departures from nominal conditions allow the trained clinician to diagnose the situation, especially when abnormal features occur in certain combinations. However, it is not our purpose to discuss the medical interpretation of ECG waveform abnormalities. Rather, this examples illustrates that bioelectric signals carry information and that the information is decoded by analyzing the signal's partial waveform descriptors.

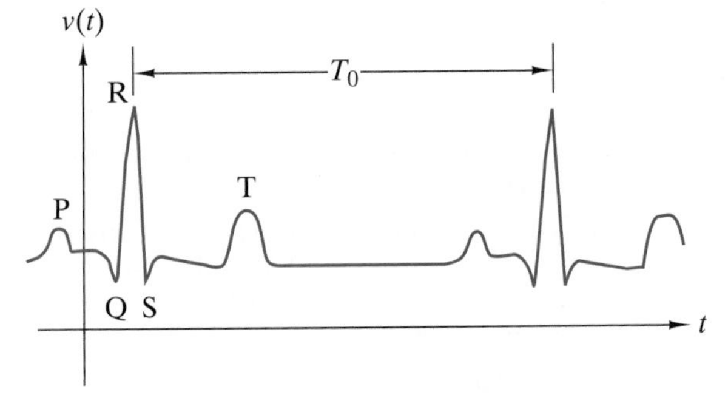

FIGURE 5–29

SUMMARY

- A waveform is an equation or graph that describes a voltage or current as a function of time. Most signals of interest in electrical engineering can be derived using three basic waveforms: the step, exponential, and sinusoid.
- The step function is defined by its amplitude and time-shift parameters. The impulse, step, and ramp are called singularity functions and are often used as test inputs for circuit analysis purposes.
- The exponential waveform is defined by its amplitude, time constant, and time-shift parameter. For practical purposes, the duration of the exponential waveform is five time constants.
- A sinusoid can be defined in terms of three types of parameters: *amplitude* (either V_A or the Fourier coefficients a and b), *time shift* (either T_S or the phase angle ϕ), and *time/frequency* (either T_0 or f_0 or ω_0).
- Many composite waveforms can be derived using the three basic waveforms. Some examples are the impulse, ramp, damped ramp, damped sinusoid, exponential rise, and double exponential.
- Partial descriptors are used to classify or describe important signal attributes. Two important temporal attributes are periodicity and causality. Periodic waveforms repeat themselves every T_0 seconds. Causal signals are zero for $t < 0$. Some important amplitude descriptors are peak value V_p, peak-to-peak value V_{pp}, average value V_{avg}, and root-mean-square value V_{rms}.
- A spectrum is an equation or graph that defines the amplitudes and phase angles of sinusoidal components contained in a signal. The signal bandwidth (B) is a partial descriptor that defines the range of frequencies outside of which the component amplitudes are less than a specified value. Periodic signals can be resolved into a dc component and a sum of ac components at harmonic frequencies.

PROBLEMS

OBJECTIVE 5–1 BASIC WAVEFORMS (SECTS. 5–2, 5–3, 5–4)

Given an equation, graph, or word description of a linear combination of step, ramp, exponential, or sinusoid waveforms:
(a) Construct an alternative description of the waveform.
(b) Find the parameters or properties of the waveform.
(c) Construct new waveforms by summing, integrating, or differentiating the given waveform.
See Examples 5–1, 5–2, 5–3, 5–5, 5–6, 5–7, 5–8, 5–9 and Exercises 5–1, 5–2, 5–3, 5–4, 5–5, 5–6, 5–7

5–1 Sketch the following waveforms:

(a) $v_1(t) = -2u(t+1) + 4u(t-1)$ V

(b) $v_2(t) = u(t+2) - 2u(t)$ V

(c) $v_3(t) = \int_{-\infty}^{t} v_1(x)\,dx$

(d) $v_4(t) = \int_{-\infty}^{t} v_2(x)\,dx$

5–2 Sketch the following waveforms:

(a) $v_1(t) = 1 - u(t) - 2u(t-1)$ V

(b) $v_2(t) = -u(t+3) + 3u(t+1) - 2u(t)$ V

5–3 Sketch the following waveforms:

(a) $v_1(t) = -2r(t+1) + 2r(t-1)$ V

(b) $v_2(t) = r(t+1) - 2r(t-1) + r(t-3)$ V

(c) $v_3(t) = \dfrac{dv_1(t)}{dt}$

(d) $v_4(t) = \dfrac{dv_2(t)}{dt}$

5–4 Express the following signals as a sum of singularity functions.

(a) $v_1(t) = \begin{cases} 2 & t < 1 \\ -5 & 1 \le t < 2 \\ 0 & 2 \le t \end{cases}$

(b) $v_2(t) = \begin{cases} 0 & t < 0 \\ -4t & 0 \le t < 2 \\ -12 + 2t & 2 \le t < 6 \\ 0 & 6 \le t \end{cases}$

5–5 Express the waveforms in Figure P5–5 as a sum of singularity functions.

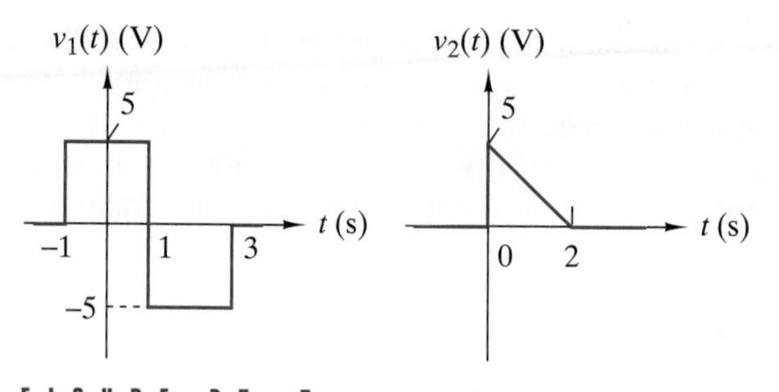

FIGURE P5–5

5–6 Write expressions for the derivatives of the waveforms in Figure P5–5.

5–7 A waveform $v(t)$ is zero for $t < 0$, rises linearly to 10 V at $t = 2$ s, remains at 10 V until $t = 4$ s, and abruptly drops to zero thereafter. Express the waveform as a sum of singularity functions.

5–8 Sketch the following exponential waveforms. Find the amplitude and time constant of each waveform.
(a) $v_1(t) = [2\,e^{-10t}]u(t)$ V
(b) $v_2(t) = [10\;e^{-t/2}]u(t-1)$ V
(c) $v_3(t) = [5e^{-2(t-1)}]u(t-1)$ V
(d) $v_4(t) = [-10\,e^{-t/20}]u(t)$ V

5–9 Write expressions for the derivative ($t > 0$) and integral (from 0 to t) of the exponential waveform $v(t) = [0.5e^{-20t}]u(t)$ V.

5–10 An exponential waveform decays to 50% of its initial ($t = 0$) amplitude in 5 ms. Find the time constant of the waveform.

5–11 The amplitude of an exponential waveform is 5 V at $t = 0$ and 3.5 V at $t = 3$ ms. What is its amplitude at $t = 6$ ms?

5–12 Construct an exponential waveform that fits entirely within the unshaded region in Figure P5–12.

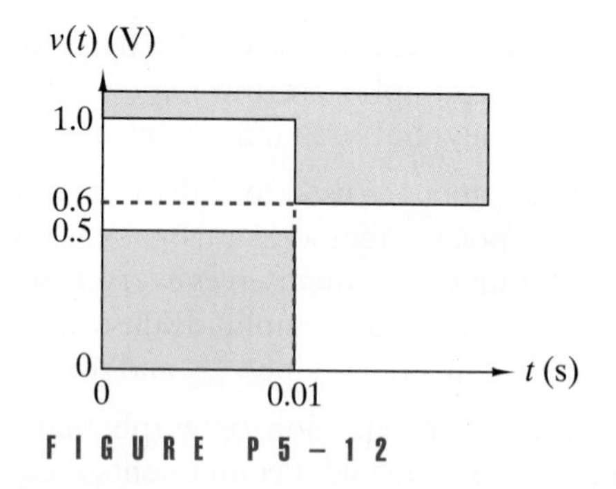

FIGURE P5–12

5–13 By direct substitution, show that the exponential function $v(t) = V_A e^{-\alpha t}$ satisfies the following first-order differential equation.

$$\frac{dv(t)}{dt} + \alpha v(t) = 0$$

5–14 Find the period, frequency, amplitude, time shift, and phase angle of the following sinusoids:
(a) $v_1(t) = 10\cos(500\pi t) - 10\sin(500\pi t)$ V
(b) $v_2(t) = -30\cos(2000\pi t) + 20\sin(2000\pi t)$ V

5–15 Find the amplitude and phase angle of the derivative of each sinusoid in Problem 5–14.

5–16 Write an expression for the sinusoid in Figure P5–16. What are the phase angle and time shift of the waveform?

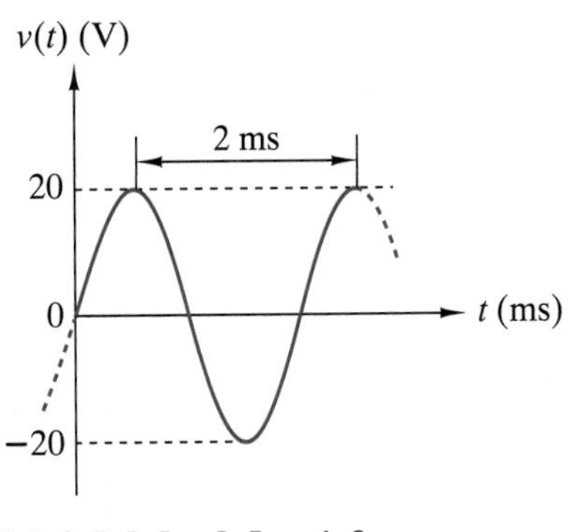

FIGURE P5–16

5–17 Find the Fourier coefficients, cyclic frequency, and radian frequency of the following sinusoids:
(a) $v_1(t) = 20\cos(500\pi t + 180°)$ V
(b) $v_2(t) = 20\cos(500\pi t - 90°)$ V

5–18 Find the sum of the two sinusoids in Problem 5–17.

5–19 Find the Fourier coefficients, cyclic frequency, and radian frequency of the following sinusoids:
(a) $v_1(t) = 30\cos(2\pi 400t + 30°)$ V
(b) $v_2(t) = 20\cos(2000\pi t - 60°)$ V

5–20 Find the time shift of each sinusoid in Problem 5–19.

OBJECTIVE 5–2 COMPOSITE WAVEFORMS (SECT. 5–5)

Given an equation, graph, or word description of a composite waveform:
(a) Construct an alternative description of the waveform.
(b) Find the parameters or properties of the waveform.
(c) Find new waveforms by integrating, or differentiating the given waveform.
See Examples 5–10, 5–11, 5–12, 5–13, 5–14, 5–15 and Exercises 5–8, 5–9

5–21 Sketch the following composite waveforms. What are the maximum and minimum values of each waveform?
(a) $v_1(t) = 10[1 - e^{-5t}]u(t)$ V
(b) $v_2(t) = 20[e^{-5t} - e^{-10t}]u(t)$ V

5–22 Sketch the following composite waveforms. What are the maximum and minimum values of each waveform?
(a) $v_1(t) = 20 - 10\sin(10\pi t)$ V
(b) $v_2(t) = 10\,[e^{-5t} + \sin(10\pi t)]u(t)$ V

5–23 Sketch the damped ramp $v(t) = 20\,te^{-5t}u(t)$. Find the maximum value of the waveform and the time at which it occurs.

5–24 The value of the waveform $v(t) = (V_A - V_B\,e^{-\alpha t})u(t)$ is 5 V at $t = 0$, 8 V at $t = 5$ ms, and approaches 12 V as $t \to \infty$. Find V_A, V_B, and α then sketch the waveform.

5–25 Write an expression for the composite sinusoidal waveform in Figure P5–25.

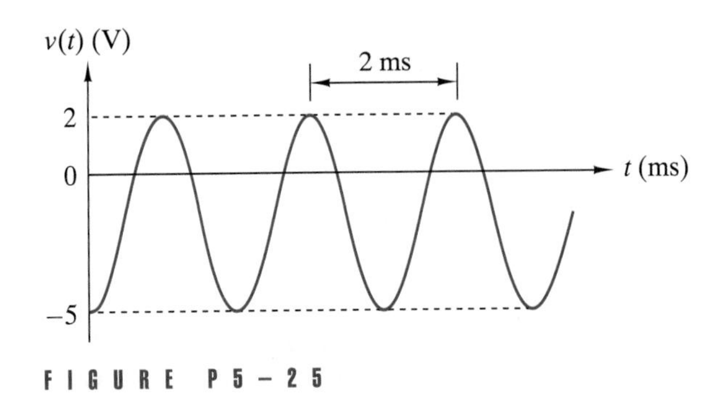

FIGURE P5–25

5–26 A waveform of the form $v(t) = 6 - 10\sin(\beta t)$ periodically reaches a minimum every 4 ms. Find the values of V_{max}, V_{min}, and β; then sketch the waveform.

5–27 Write an expression for the composite exponential waveform in Figure P5–27.

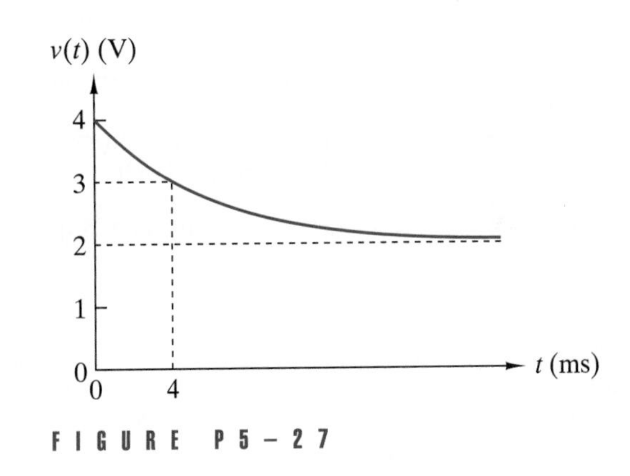

FIGURE P5–27

5–28 Sketch the double exponential $v(t) = 30(e^{-20t} - e^{-100t})$ $u(t)$. Find the maximum value of the waveform and the time at which it occurs.

5–29 Write an expression for the damped sine waveform in Figure P5–29.

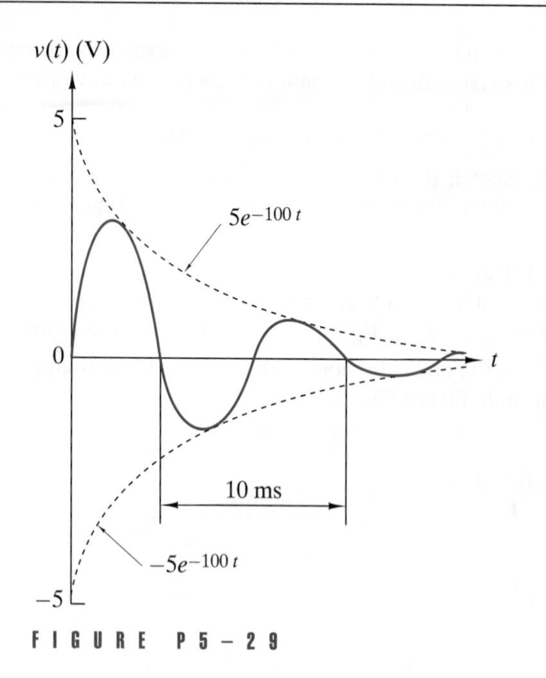

FIGURE P5–29

5–30 The basic waveform $u(x)$ is zero for all $x < 0$. The function $f(t) = u(\sin \pi t)$ defines a periodic waveform. Sketch the waveform of $\sin \pi t$ and then sketch $f(t)$. Is the $f(t)$ zero for all $t < 0$? If not, why not? How would you describe $f(t)$? Is it periodic? If so, what is the period of $f(t)$?

OBJECTIVE 5–3 WAVEFORM PARTIAL DESCRIPTORS (SECT. 5–6)

Given a complete description of a basic or composite waveform:

(a) Classify the waveform as periodic or aperiodic and causal or noncausal.
(b) Find the applicable partial waveform descriptors.
(c) Find the parameters of a waveform given its partial descriptors.

See Examples 5–16, 5–17 and Exercises 5–10, 5–11, 5–12

5–31 Find V_p, V_{pp}, V_{avg}, and V_{rms} for each of the following sinusoids.

(a) $v_1(t) = 10\cos(2000\pi t) + 10\sin(2000\pi t)$ V
(b) $v_2(t) = -30\cos(2000\pi t) - 20\sin(2000\pi t)$ V

5–32 Find V_p, V_{pp}, V_{avg}, and V_{rms} for each of the following sinusoids.

(a) $v_1(t) = 20\cos(4000\pi t - 90°)$ V
(b) $v_2(t) = 30\cos(2\pi t/400 - 45°)$ V

5–33 Find V_p, V_{pp}, V_{avg}, and V_{rms} for the periodic waveform in Figure P5–33.

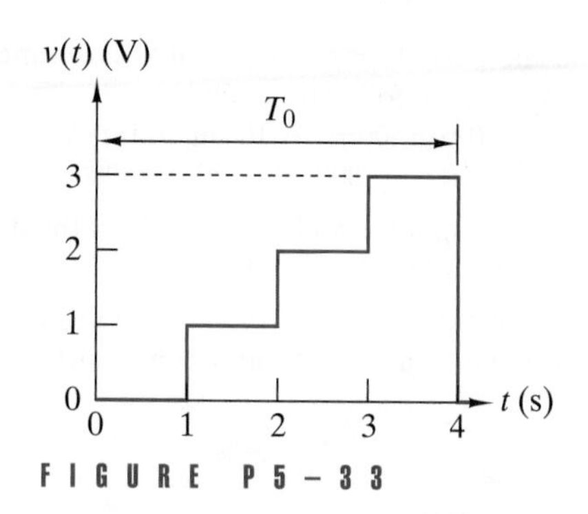

FIGURE P5–33

5–34 Find V_p, V_{pp}, V_{avg}, and V_{rms} for the periodic waveform in Figure P5–34.

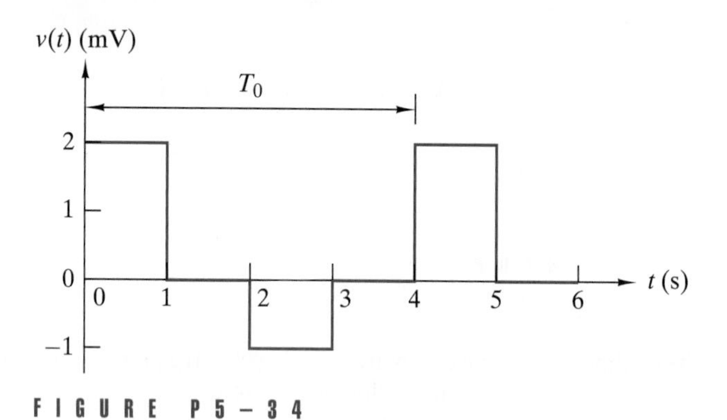

FIGURE P5–34

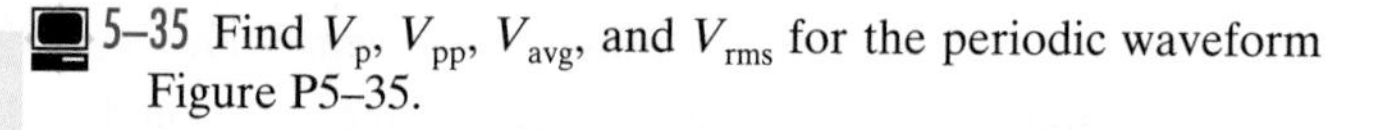

5–35 Find V_p, V_{pp}, V_{avg}, and V_{rms} for the periodic waveform Figure P5–35.

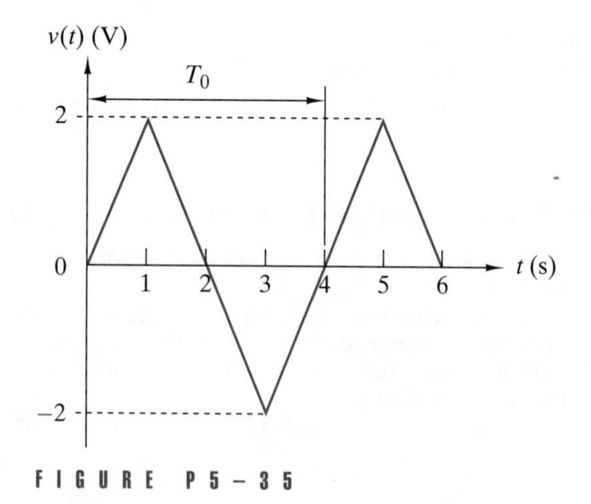

FIGURE P5–35

5–36 Find V_{avg} and V_{rms} of the offset sinewave $v(t) = V_0 + V_A \sin(2\pi t/T_0)$ in terms of V_0 and V_A.

5–37 Find V_{avg} and V_{rms} of the full-wave rectified sinewave $v(t) = V_A|\sin(2\pi t/T_0)|$ in terms of V_A.

5–38 The first cycle ($t > 0$) of a periodic waveform with $T_0 = 500$ ms can be expressed as

$$v(t) = 2u(t) - 3u(t - 0.1) + 3u(t - 0.4) \text{ V}$$

Sketch the waveform and find V_{max}, V_{min}, V_p, V_{pp}, and V_{avg}.

5–39 A periodic waveform can be expressed as

$$v(t) = 20 - 10\cos(2\pi\, 1000t) - 5\sin(2\pi\, 2000t) + 3\cos(2\pi\, 4000t) \text{ V}$$

What is the period of the waveform? What is the average value of the waveform? What is the amplitude of the fundamental component? What is the highest frequency in the waveform?

5–40 The first cycle ($0 \le t < T_0$) of a periodic clock pulse train is

$$v(t) = V_A[u(t) - u(t - T)]$$

where $T < T_0$. The duty cycle of the clock pulse is defined as the fraction of the period during which the pulse is not zero. Derive expressions for V_{avg} and V_{rms} in terms of V_A and $D = T/T_0$.

INTEGRATING PROBLEMS

5–41 **A** Exponential Signal Descriptors

Several of the time descriptors used in digital data communication systems are based on exponential signals. In this problem we explore three of these descriptors.

(a) The *time constant of fall* is defined as the time required for a pulse to fall from 70.7% to 26.0% of its maximum value. Assuming that the pulse decreases as e^{-t/T_C}, find the relationship between the time constant of fall and the time constant of the exponential decay.

(b) The *rise time* of a pulse is the time required for a pulse to rise from 10% to 90% of its maximum value. Assuming the pulse increases as $1 - e^{-t/T_C}$, find the relationship between rise time and the time constant of the exponential rise.

(c) The *leading-edge pulse time* is defined as the time at which a pulse rises to 50% of its maximum value. Assuming the pulse increases as $1 - e^{-t/T_C}$, find the relationship between leading-edge pulse time and the time constant of the exponential rise.

5–42 **A** Defibrillation Waveforms

Ventricular fibrillation is a life-threatening loss of synchronous activity in the heart. To restore normal activity, a defibrillator delivers a brief but intense pulse of electrical current through the patient's chest. The pulse waveform is of interest because different waveforms may lead to different outcomes. Figure P5–42 shows a waveform known as a *biphasic truncated exponential* used in implantable defibrillators. The waveform is an exponential current whose direction of flow reverses after 4 ms and terminates after 8 ms. Write an expression for this waveform using the basic signals discussed in this chapter.

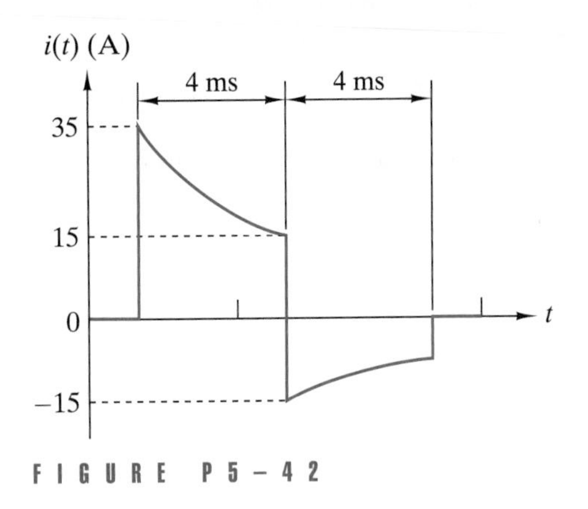

FIGURE P5–42

5–43 **A** Timing Diagrams

Timing diagrams are separate plots of digital waveforms that are stacked on top of each other to show how signals change over time and to reveal temporal relationships between signals. The waveforms in a timing diagram can take on only two states: 0 (low) and 1 (high). A three-bit binary counter has outputs ($b_3\, b_2\, b_1$), where b_3 is the most significant bit (MSB). At $t = 0$ the outputs are in the zero state, that is, ($b_3\, b_2\, b_1$) = (0 0 0). Thereafter the outputs undergo the following state transitions:

$$000 \rightarrow 001 \rightarrow 010 \rightarrow 011 \rightarrow 100 \rightarrow 101 \rightarrow 110 \rightarrow 111 \rightarrow 000$$

The first transition occurs at $t = 50$ ns and subsequent transitions occur every 50 ns.

(a) Sketch a timing diagram for these signals with b_1 at the top and b_3 at the bottom.

(b) A digital signal b_4 is 1 (high) when b_3 AND b_1 are high and is 0 (low) otherwise. Add the waveform of b_4 at the bottom of the timing diagram in part (a).

(c) The counter outputs ($b_3\, b_2\, b_1$) are inputs to a digital-to-analog converter whose output is

$$v_O(t) = 4\left(b_3 + \frac{b_2}{2} + \frac{b_1}{4}\right)$$

Add the analog waveform $v_O(t)$ at the bottom of the timing diagram in part (b).

5–44 **A** Fourier Series

Reasonably well-behaved periodic waveforms can be expressed as a sum of sinusoidal components called a Fourier series. For example, the sum

$$v(t) = \frac{V_A}{2} - \sum_{n=1}^{n=\infty} \frac{V_A}{n\pi} \sin(2\pi n f_0 t)$$

is the Fourier series of one of the periodic waveforms in Figure 5–2. Use a spreadsheet or math solver to plot the sum of the first 10 terms of this series for $V_A = 10$ V and $f_0 = 50$ kHz. Use a time interval $0 \leq t \leq 2T_0$. Can you identify this waveform in Figure 5–2?

5–45 **A** Voltmeter Calibration

Most dc voltmeters measure the average value of the applied signal. A dc meter that measures the average value can be adapted to indicate the rms value of an ac signal. The input is passed through a rectifier circuit. The rectifier output is the absolute value of the input and is applied to a dc meter whose deflection is proportional to the average value of the rectified signal. The meter scale is calibrated to indicate the rms value of the input signal. A calibration factor is needed to convert the average absolute value into the rms value of the ac signal. What is the required calibration factor for a sinusoid? Would the same calibration factor apply to a square wave?

CHAPTER 6

CAPACITANCE AND INDUCTANCE

From the foregoing facts, it appears that a current of electricity is produced, for an instant, in a helix of copper wire surrounding a piece of soft iron whenever magnetism is induced in the iron; also that an instantaneous current in one or the other direction accompanies every change in the magnetic intensity of the iron.

Joseph Henry, 1831,
American Physicist

Some History Behind This Chapter

Joseph Henry (1797–1878) and the British physicist Michael Faraday (1791–1867) independently discovered magnetic induction almost simultaneously. The quotation above is Henry's summary of the experiments leading to his discovery of magnetic induction. Although Henry and Faraday used similar apparatus and observed almost the same results, Henry was the first to fully recognize the importance of the discovery. The units of circuit inductance (henrys) honors Henry, while the mathematical generalization of magnetic induction is called Faraday's law.

Why This Chapter Is Important Today

Electric circuits owe much of their utility to devices that store energy, even if only for a short period of time. In this chapter you will be introduced to two new circuit elements: the capacitor and the inductor. These energy storing elements lead to circuits that perform various mathematical operations like integration and differentiation. The energy storing capability makes possible the signal processing operations in modern communications systems and audio equipment.

Chapter Sections

6–1 The Capacitor

6–2 The Inductor

6–3 Dynamic OP AMP Circuits

6–4 Equivalent Capacitance and Inductance

Chapter Learning Objectives

6-1 Capacitor and Inductor Responses (Sects. 6-1, 6-2)

(a) Given the current through a capacitor or an inductor, find the voltage across the element.

(b) Given the voltage across a capacitor or an inductor, find the current through the element.

(c) Find the power and energy associated with a capacitor or inductor.

6-2 Dynamic OP AMP Circuits (Sect. 6-3)

(a) Given an OP AMP integrator or differentiator, determine the output for specified inputs.

(b) Given an *RC* circuit containing OP AMPs, find the input-output relationship and construct a block diagram.

(c) Design an *RC* circuit containing OP AMPs that implements a given input-output relationship.

6-3 Equivalent Inductance and Capacitance (Sect. 6-4)

(a) Derive equivalence properties of inductors and capacitors or use equivalence properties to simplify *LC* circuits.

(b) Given an *RLC* circuit with dc inputs, find the dc currents and voltage responses.

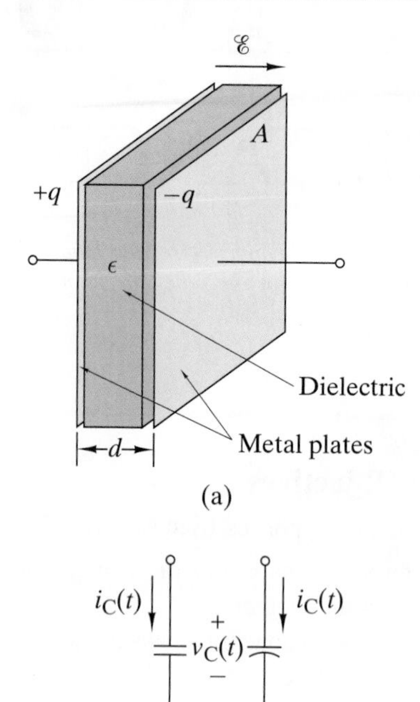

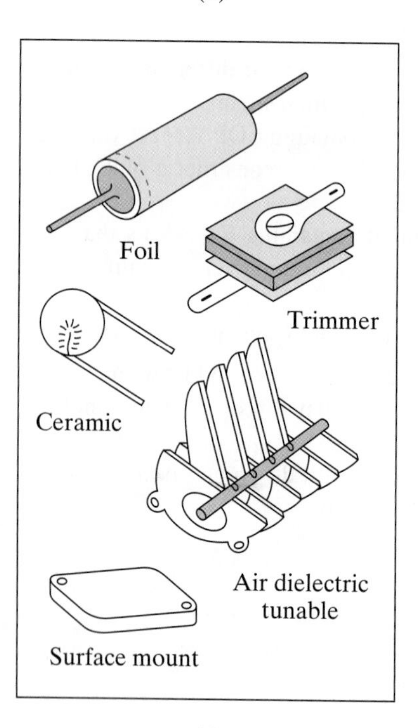

FIGURE 6–1 *The capacitor: (a) Parallel plate device. (b) Circuit symbol. (c) Example devices.*

6–1 THE CAPACITOR

A capacitor is a dynamic element involving the time variation of an electric field produced by a voltage. Figure 6–1(a) shows the parallel plate capacitor, which is the simplest physical form of a capacitive device. Figure 6–1 also shows two alternative circuit symbols and sketches of actual devices. Some of the physical features of commercially available devices are given in Appendix A.

Electrostatics shows that a uniform electric field $\mathscr{E}(t)$ exists between the metal plates in Figure 6–1(a) when a voltage exists across the capacitor.[1] The electric field produces charge separation with equal and opposite charges appearing on the capacitor plates. When the separation d is small compared with the dimension of the plates, the electric field between the plates is

$$\mathscr{E}(t) = \frac{q(t)}{\varepsilon A} \tag{6–1}$$

where ε is the permittivity of the dielectric, A is the area of the plates, and $q(t)$ is the magnitude of the electric charge on each plate. The relationship between the electric field and the voltage across the capacitor $v_C(t)$ is given by

$$\mathscr{E}(t) = \frac{v_C(t)}{d} \tag{6–2}$$

Substituting Eq. (6–2) into Eq. (6–1) and solving for the charge $q(t)$ yields

$$q(t) = \left[\frac{\varepsilon A}{d}\right] v_C(t) \tag{6–3}$$

The proportionality constant inside the brackets in this equation is the **capacitance** C of the capacitor. That is, by definition,

$$C = \frac{\varepsilon A}{d} \tag{6–4}$$

The unit of capacitance is the **farad** (F), a term that honors the British physicist Michael Faraday. Values of capacitance range from a few pF (10^{-12} F) in semiconductor devices to tens of mF (10^{-3} F) in industrial capacitor banks. Using Eq. (6–4), the defining relationship for the capacitor becomes

$$q(t) = Cv_C(t) \tag{6–5}$$

Figure 6–2 graphically displays the element constraint in Eq. (6–5). The graph points out that the capacitor is a linear element since the defining relationship between voltage and charge is a straight line through the origin.

I–V RELATIONSHIP

To express the element constraint in terms of voltage and current, we differentiate Eq. (6–5) with respect to time t:

$$\frac{dq(t)}{dt} = \frac{d[Cv_C(t)]}{dt}$$

Since C is constant and $i_C(t)$ is the time derivative of $q(t)$, we obtain a capacitor i–v relationship in the form

$$i_C(t) = C\frac{dv_C(t)}{dt} \tag{6–6}$$

1 An electric field is a vector quantity. In Figure 6–1(a) the field is confined to the space between the two plates and is perpendicular to the plates.

The relationship assumes that the reference marks for the current and voltage follow the passive sign convention shown in Figure 6–3.

The time derivative in Eq. (6–6) means the current is zero when the voltage across the capacitor is constant, and vice versa. In other words, the capacitor acts like an open circuit ($i_C = 0$) when dc excitations are applied. The capacitor is a dynamic element because the current is zero unless the voltage is changing. However, a discontinuous change in voltage would require an infinite current, which is physically impossible. Therefore, the capacitor voltage must be a continuous function of time.

Equation (6–6) relates the capacitor current to the rate of change of the capacitor voltage. To express the voltage in terms of the current, we multiply both sides of Eq. (6–6) by dt, solve for the differential dv_C, and integrate:

$$\int dv_C = \frac{1}{C}\int i_C(t)dt$$

Selecting the integration limits requires some discussion. We assume that at some time t_0 the voltage across the capacitor $v_C(t_0)$ is known and we want to determine the voltage at some later time $t > t_0$. Therefore, the integration limits are

$$\int_{v_C(t_0)}^{v_C(t)} dv_C = \int_{t_0}^{t} i_C(x)dx$$

where x is a dummy integration variable. Integrating the left side of this equation yields

$$v_C(t) = v_C(t_0) + \frac{1}{C}\int_{t_0}^{t} i_C(x)dx \tag{6–7}$$

In practice, the time t_0 is established by a physical event such as closing a switch or the start of a particular clock pulse. Nothing is lost in the integration in Eq. (6–7) if we arbitrarily define t_0 to be zero. Using $t_0 = 0$ in Eq. (6–7) yields

$$v_C(t) = v_C(0) + \frac{1}{C}\int_{0}^{t} i_C(x)dx \tag{6–8}$$

Equation (6–8) is the integral form of the capacitor i–v constraint. Both the integral form and the derivative form in Eq. (6–6) assume that the reference marks for current and voltage follow the passive sign convention in Figure 6–3.

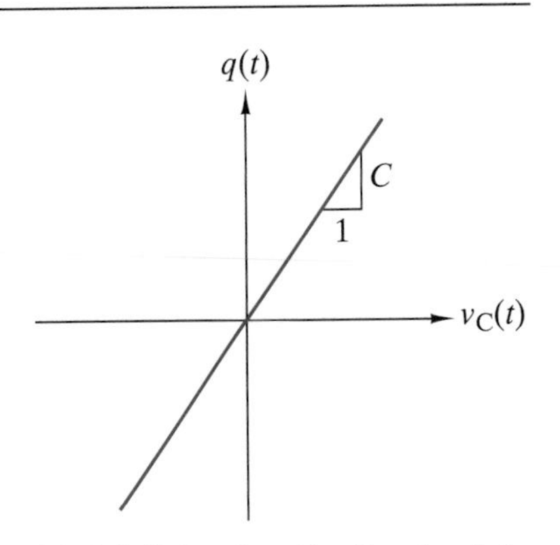

FIGURE 6–2 Graph of the defining relationship of a linear capacitor.

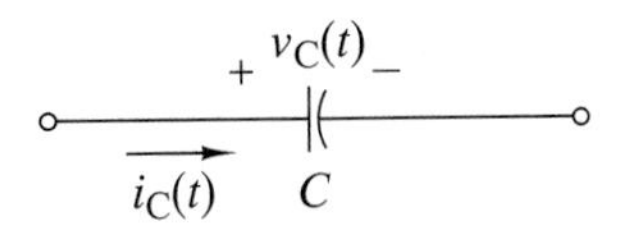

FIGURE 6–3 Capacitor current and voltage.

POWER AND ENERGY

With the passive sign convention the capacitor power is

$$p_C(t) = i_C(t)v_C(t) \tag{6–9}$$

Using Eq. (6–6) to eliminate $i_C(t)$ from Eq. (6–9) yields the capacitor power in the form

$$p_C(t) = Cv_C(t)\frac{dv_C(t)}{dt} = \frac{d}{dt}[{}^1\!/_2 Cv_C^2(t)] \tag{6–10}$$

This equation shows that the power can be either positive or negative because the capacitor voltage and its time rate of change can have opposite signs. With the passive sign convention, a positive sign means the element absorbs power, while a negative sign means the element delivers power. The ability to deliver power implies that the capacitor can store energy.

To determine the stored energy, we note that the expression for power in Eq. (6–10) is a perfect derivative. Since power is the time rate of change of energy, the quantity inside the brackets must be the energy stored in the capacitor. Mathematically, we can infer from Eq. (6–10) that the energy at time t is

$$w_C(t) = {}^1/_2 C v_C^2(t) + \text{constant}$$

The constant in this equation is the value of stored energy at some instant t when $v_C(t) = 0$. At such an instant the electric field is zero; hence the stored energy is also zero. As a result, the constant is zero and we write the capacitor energy as

$$w_C(t) = \frac{1}{2} C v_C^2(t) \tag{6–11}$$

The stored energy is never negative, since it is proportional to the square of the voltage. The capacitor absorbs power from the circuit when storing energy and returns previously stored energy when delivering power to the circuit.

The relationship in Eq. (6–11) also implies that voltage is a continuous function of time, since an abrupt change in the voltage implies a discontinuous change in energy. Since power is the time derivative of energy, a discontinuous change in energy implies infinite power, which is physically impossible. The capacitor voltage is called a **state variable** because it determines the energy state of the element.

To summarize, the capacitor is a dynamic circuit element with the following properties:

1. *The current through the capacitor is zero unless the voltage is changing. The capacitor acts like an open circuit to dc excitations.*
2. *The voltage across the capacitor is a continuous function of time. A discontinuous change in capacitor voltage would require infinite current and power, which is physically impossible.*
3. *The capacitor absorbs power from the circuit when storing energy and returns previously stored energy when delivering power. The net energy transfer is nonnegative, indicating that the capacitor is a passive element.*

The following examples illustrate these properties.

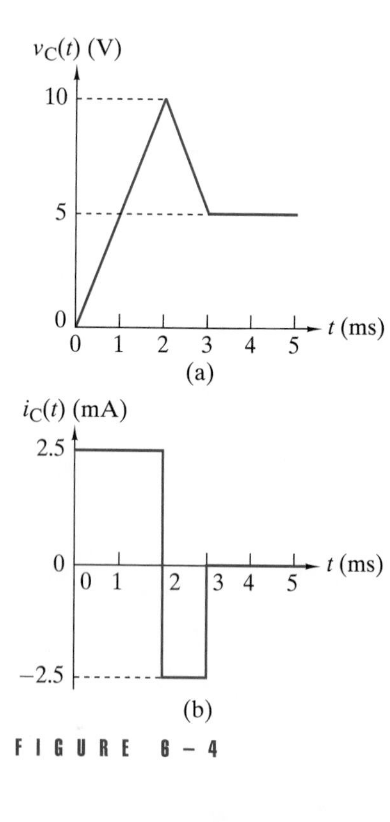

FIGURE 6–4

EXAMPLE 6–1

The voltage in Figure 6–4(a) appears across a ½-μF capacitor. Find the current through the capacitor.

SOLUTION:
The capacitor current is proportional to the time rate of change of the voltage. For $0 < t < 2$ ms the slope of the voltage waveform has a constant value

$$\frac{dv_C}{dt} = \frac{10}{2 \times 10^{-3}} = 5000 \text{ V/s}$$

The capacitor current during this interval is

$$i_C(t) = C\frac{dv_C}{dt} = (0.5 \times 10^{-6}) \times (5 \times 10^3) = 2.5 \text{ mA}$$

For $2 < t < 3$ ms the rate of change of the voltage is –5000 V/s. Since the rate of change of voltage is negative, the current changes direction and takes on the value $i_C(t) = -2.5$ mA. For $t > 3$ ms, the voltage is constant, so its slope is zero; hence the current is zero. The resulting current waveform is shown in Figure 6–4(b). Note that the voltage across the capacitor (the state variable) is continuous, but the capacitor current can be, and in this case is, discontinuous. ■

EXAMPLE 6–2

The $i_C(t)$ in Figure 6–5(a) is given by

$$i_C(t) = I_0(e^{-t/T_C})u(t)$$

Find the voltage across the capacitor if $v_C(0) = 0$ V.

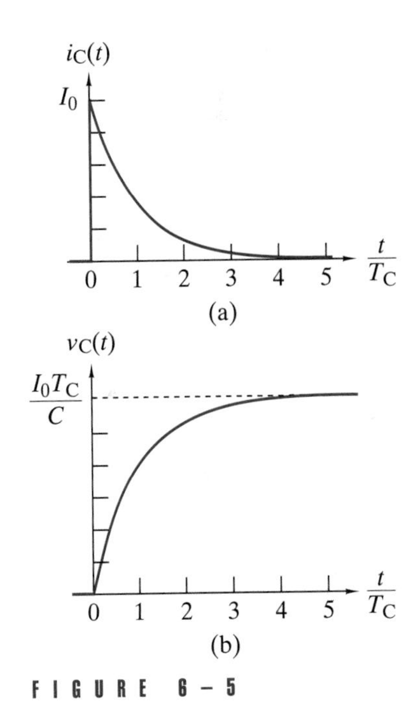

FIGURE 6–5

SOLUTION:
Using the capacitor i–v relationship in integral form,

$$v_C(t) = v_C(0) + \frac{1}{C}\int_0^t i_C(x)dx$$

$$= 0 + \frac{1}{C}\int_0^t I_0 e^{-x/T_C}\,dx = \frac{I_0 T_C}{C}(-e^{-x/T_C})\Big|_0^t$$

$$= \frac{I_0 T_C}{C}(1 - e^{-t/T_C})$$

The graphs in Figure 6–5(b) show that the voltage is continuous while the current is discontinuous. ■

Exercise 6–1

(a) The voltage across a 10-μF capacitor is $25[\sin 2000t]u(t)$. Derive an expression for the current through the capacitor.
(b) At $t = 0$ the voltage across a 100-pF capacitor is –5 V. The current through the capacitor is $10[u(t) - u(t - 10^{-4})]$ μA. What is the voltage across the capacitor for $t > 0$?

Answers:

(a) $i_C(t) = 0.5\,[\cos 2000t]u(t)$ A
(b) $v_C(t) = -5 + 10^5 t$ V for $0 < t < 0.1$ ms and $v_C(t) = 5$ V for $t > 0.1$ ms

Exercise 6–2

For $t \geq 0$ the voltage across a 200-pF capacitor is $5e^{-4000t}$ V.

(a) What is the charge on the capacitor at $t = 0$ and $t = +\infty$?
(b) Derive an expression for the current through the capacitor for $t \geq 0$.
(c) For $t > 0$ is the device absorbing or delivering power?

Answers:

(a) 1 nC and 0 C
(b) $i_C(t) = -4e^{-4000t}$ μA
(c) Delivering

EXAMPLE 6–3

Figure 6–6(a) shows the voltage across a 0.5-μF capacitor. Find the capacitor's energy and power.

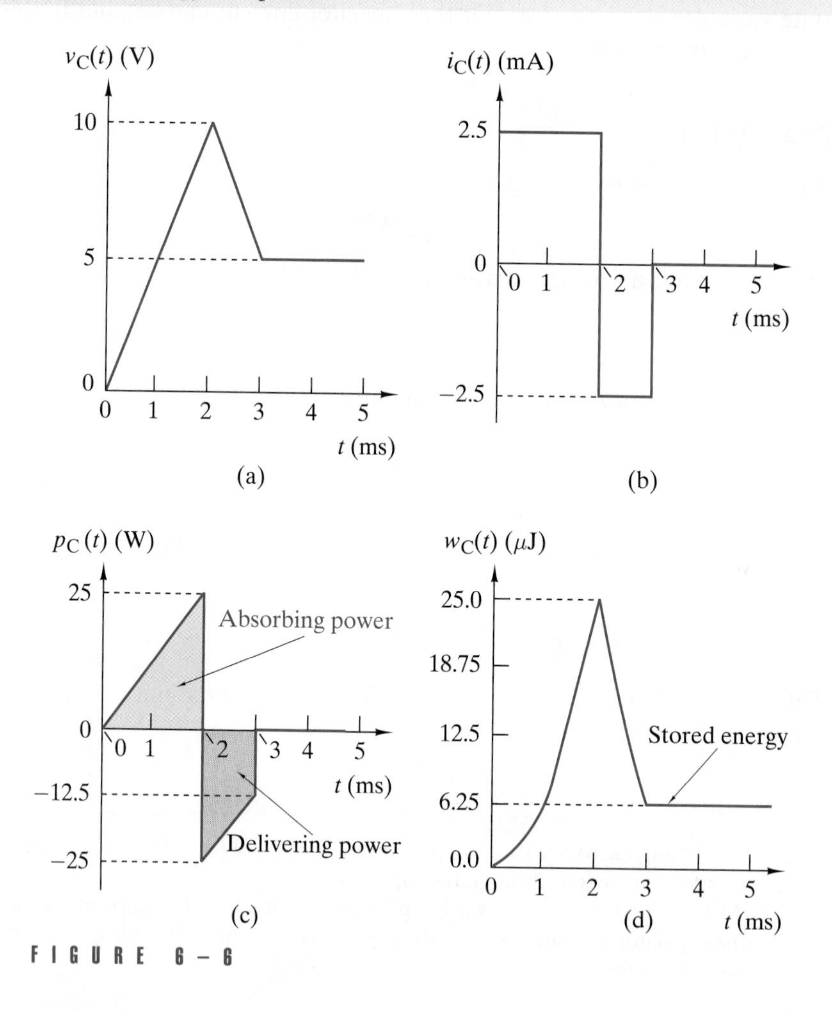

FIGURE 6–6

SOLUTION:

The current through the capacitor was found in Example 6–1. The power waveform is the point-by-point product of the voltage and current waveforms.

The energy is found by either integrating the power waveform or by calculating $\frac{1}{2}C[v_C(t)]^2$ point by point. The current, power, and energy are shown in Figures 6–6(b), 6–6(c), and 6–6(d). Note that the capacitor energy increases when it is absorbing power $[p_C(t) > 0]$ and decreases when delivering power $[p_C(t) < 0]$. ■

EXAMPLE 6–4

The current through a capacitor is given by

$$i_C(t) = I_0[e^{-t/T_C}]u(t)$$

Find the capacitor's energy and power.

SOLUTION:

The current and voltage were found in Example 6–2 and are shown in Figures 6–7(a) and 6–7(b). The power waveform is found as the product of current and voltage:

$$\begin{aligned} p_C(t) &= i_C(t)v_C(t) \\ &= [I_0 e^{-t/T_C}]\left[\frac{I_0 T_C}{C}(1 - e^{-t/T_C})\right] \\ &= \frac{I_0^2 T_C}{C}(e^{-t/T_C} - e^{-2t/T_C}) \end{aligned}$$

The waveform of the power is shown in Figure 6–7(c). The energy is

$$w_C(t) = \frac{1}{2}Cv_C^2(t) = \frac{(I_0 T_C)^2}{2C}(1 - e^{-t/T_C})^2$$

The time history of the energy is shown in Figure 6–7(d). In this example, both power and energy are always positive. ■

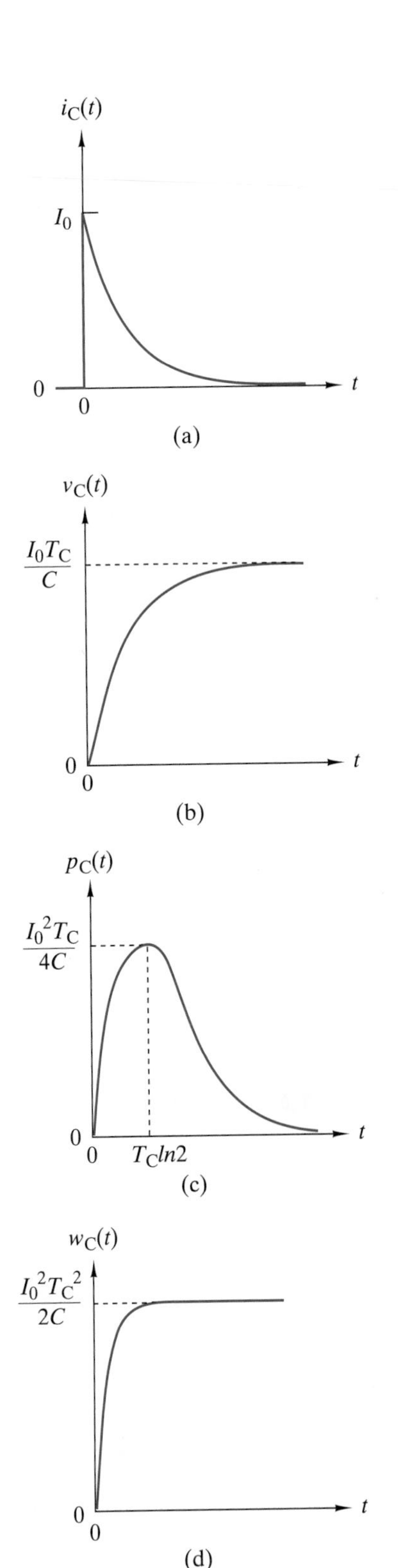

FIGURE 6–7

Exercise 6–3

Find the power and energy for the capacitors in Exercise 6–1.

Answers:

(a) $p_C(t) = 6.25[\sin 4000t]u(t)$ W
$w_C(t) = 3.125 \sin^2 2000t$ mJ

(b) $p_C(t) = -0.05 + 10^3 t$ mW for $0 < t < 0.1$ ms
$p_C(t) = 0$ for $t > 0.1$ ms
$w_C(t) = 1.25 - 5 \times 10^4 t + 5 \times 10^8 t^2$ nJ for $0 < t < 0.1$ ms
$w_C(t) = 1.25$ nJ for $t > 0.1$ ms

Exercise 6–4

Find the power and energy for the capacitor in Exercise 6–2.

Answers:

$p_C(t) = -20e^{-8000t}$ μW

$w_C(t) = 2.5e^{-8000t}$ nJ

APPLICATION EXAMPLE 6–5

A **sample-and-hold circuit** is usually found at the input to an analog-to-digital converter (ADC). The purpose of the circuit is to sample a time-varying input waveform at a specified instant and then hold that value constant until conversion to digital form is complete. This example discusses the role of a capacitor in such a circuit.

The basic sample-and-hold circuit in Figure 6–8(a) includes an input buffer, a digitally controlled electronic switch, a holding capacitor, and an output buffer. The input buffer is a voltage follower whose output replicates the analog input $v_S(t)$ and supplies charging current to the capacitor. The output buffer is also a voltage follower whose output replicates the capacitor voltage.

To see how the circuit operates, we describe one cycle of the sample-and-hold process. At time t_1 shown in Figure 6–8(b), the digital control $v_G(t)$ goes high, which causes the switch to close. Thereafter, the input buffer supplies a charging current $i_C(t)$ to drive the capacitor voltage to the level of the analog input. At time t_2 shown in Figure 6–8(b) the digital control goes low, the switch opens, and thereafter the capacitor current $i_C(t) = 0$. Zero current means that capacitor voltage is constant since dv_C/dt is zero. In sum, closing the switch causes the capacitor voltage to track the input and opening the switch causes the capacitor voltage to hold a sample of the input.

Figure 6–8(b) shows several more cycles of the sample-and-hold process. Samples of the input waveform are acquired during the time intervals labelled t_C. During these intervals the control signal is high, the switch is closed, and the capacitor charges or discharges in order to track the analog input voltage. Analog-to-digital conversion of the circuit output voltage takes place during the time intervals t_{ADC}. During these intervals the control signal is low, the switch is open, and the capacitor holds the output voltage constant.

Sample-and-hold circuits are available as monolithic integrated circuits that include the two buffers, the electronic switch, but not the holding

FIGURE 6–8

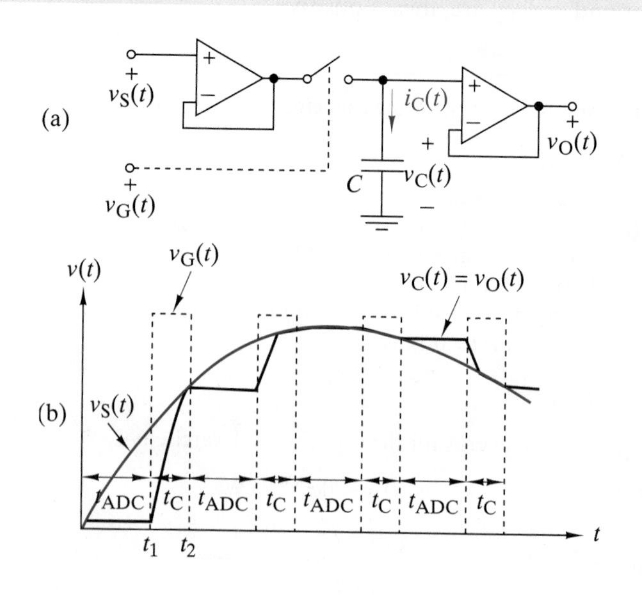

capacitor. The capacitor is supplied externally, and its selection involves a trade-off. In an ideal sample-and-hold circuit, the capacitor voltage tracks the input when the switch is closed (sample mode) and holds the value indefinitely when the switch is open (hold mode). In real circuits the input buffer has a maximum output current, which means that some time is needed to charge the capacitor in the sample mode. Minimizing this sample acquisition time argues for a small capacitor. On the other hand, in the hold mode the output buffer draws a small current which gradually discharges the capacitor, causing the output voltage to slowly decrease. Minimizing this output droop calls for a large capacitor. Thus, selecting the capacitance of the holding capacitor involves a compromise between the sample acquisition time and the output voltage droop in the hold mode.

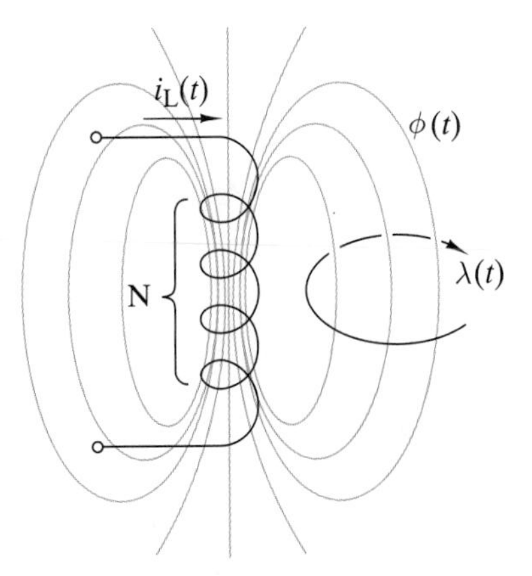

FIGURE 6–9 *Magnetic flux surrounding a current-carrying coil.*

6–2 THE INDUCTOR

The inductor is a dynamic circuit element involving the time variation of the magnetic field produced by a current. Magnetostatics shows that a magnetic flux ϕ surrounds a wire carrying an electric current. When the wire is wound into a coil, the lines of flux concentrate along the axis of the coil, as shown in Figure 6–9. In a linear magnetic medium, the flux is proportional to both the current and the number of turns in the coil. Therefore, the total flux is

$$\phi(t) = k_1 N i_L(t) \tag{6–12}$$

where k_1 is a constant of proportionality.

The magnetic flux intercepts or links the turns of the coil. The flux linkage in a coil is represented by the symbol λ, with units of webers (Wb), named after the German scientist Wilhelm Weber (1804–1891). The flux linkage is proportional to the number of turns in the coil and to the total magnetic flux, so $\lambda(t)$ is

$$\lambda(t) = N\phi(t) \tag{6–13}$$

Substituting Eq. (6–12) into Eq. (6–13) gives

$$\lambda(t) = [k_1 N^2] i_L(t) \tag{6–14}$$

The proportionality constant inside the brackets in this equation is the **inductance** L of the coil. That is, by definition

$$L = k_1 N^2 \tag{6–15}$$

The unit of inductance is the henry (H) (plural: henrys), a name that honors American scientist Joseph Henry. Figure 6–10 shows the circuit symbol for an inductor and some examples of actual devices.

Using Eq. (6–15), the defining relationship for the inductor becomes

$$\lambda(t) = L i_L(t) \tag{6–16}$$

Figure 6–11 graphically displays the inductor's element constraint in Eq. (6–16). The graph points out that the inductor is a linear element since the defining relationship is a straight line through the origin.

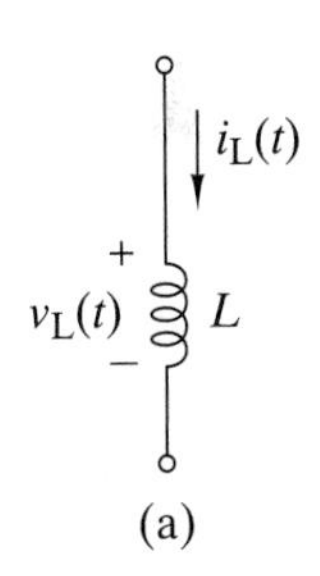

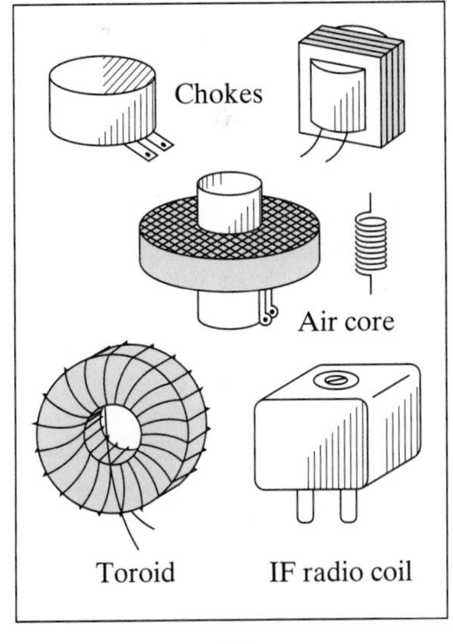

FIGURE 6–10 *The inductor: (a) Circuit symbol. (b) Example devices.*

I–V RELATIONSHIP

Equation (6–16) is the inductor element constraint in terms of current and flux linkage. To obtain the element characteristic in terms of voltage and

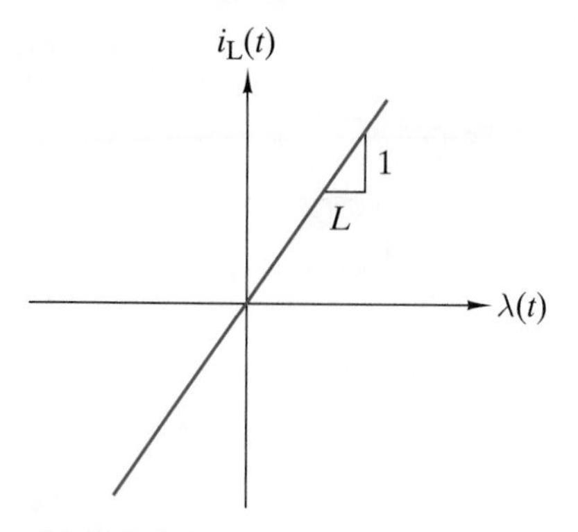

FIGURE 6–11 *Graph of the defining relationship of a linear inductor.*

current, we differentiate Eq. (6–16) with respect to time:

$$\frac{d[\lambda(t)]}{dt} = \frac{d[Li_L(t)]}{dt} \tag{6–17}$$

The inductance L is a constant. According to Faraday's law, the voltage across the inductor is equal to the time rate of change of flux linkage. Therefore, we obtain an inductor i–v relationship in the form

$$v_L(t) = L\frac{di_L(t)}{dt} \tag{6–18}$$

The time derivative in Eq. (6–18) means that the voltage across the inductor is zero unless the current is time varying. Under dc excitation the current is constant and $v_L = 0$, so the inductor acts like a short circuit. The inductor is a dynamic element because only a changing current produces a nonzero voltage. However, a discontinuous change in current would produce an infinite voltage, which is physically impossible. Therefore, the current $i_L(t)$ must be a continuous function of time t.

Equation (6–18) relates the inductor voltage to the rate of change of the inductor current. To express the inductor current in terms of the voltage, we multiply both sides of Eq. (6–18) by dt, solve for the differential di_L, and integrate:

$$\int di_L = \frac{1}{L}\int v_L(t)dt \tag{6–19}$$

To set the limits of integration, we assume that the inductor current $i_L(t_0)$ is known at some time t_0. Under this assumption the integration limits are

$$\int_{i_L(t_0)}^{i_L(t)} di_L = \frac{1}{L}\int_{t_0}^{t} v_L(x)dx \tag{6–20}$$

where x is a dummy integration variable. The left side of Eq. (6–20) integrates to produce

$$i_L(t) = i_L(t_0) + \frac{1}{L}\int_{t_0}^{t} v_L(x)dx \tag{6–21}$$

The reference time t_0 is established by some physical event, such as closing or opening a switch. Without losing any generality, we can assume $t_0 = 0$ and write Eq. (6–21) in the form

$$i_L(t) = i_L(0) + \frac{1}{L}\int_{0}^{t} v_L(x)dx \tag{6–22}$$

Equation (6–22) is the integral form of the inductor i–v characteristic. Both the integral form and the derivative form in Eq. (6–18) assume that the reference marks for the inductor voltage and current follow the passive sign convention shown in Figure 6–10.

POWER AND ENERGY

With the passive sign convention the inductor power is

$$p_L(t) = i_L(t)v_L(t) \tag{6–23}$$

Using Eq. (6–18) to eliminate $v_L(t)$ from this equation puts the inductor power in the form

$$p_L(t) = [i_L(t)]\left[L\frac{di_L(t)}{dt}\right] = \frac{d}{dt}[{}^1/_2 L i_L^2(t)] \tag{6–24}$$

This expression shows that power can be positive or negative because the inductor current and its time derivative can have opposite signs. With the passive sign convention a positive sign means the element absorbs power, while a negative sign means the element delivers power. The ability to deliver power indicates that the inductor can store energy.

To find the stored energy, we note that the power relation in Eq. (6–24) is a perfect derivative. Since power is the time rate of change of energy, the quantity inside the brackets must represent the energy stored in the magnetic field of the inductor. From Eq. (6–24), we infer that the energy at time t is

$$w_L(t) = \frac{1}{2}Li_L^2(t) + \text{constant}$$

As is the case with capacitor energy, the constant in this expression is zero since it is the energy stored at an instant t at which $i_L(t) = 0$. As a result, the energy stored in the inductor is

$$w_L(t) = \frac{1}{2}Li_L^2(t) \tag{6–25}$$

The energy stored in an inductor is never negative because it is proportional to the square of the current. The inductor stores energy when absorbing power and returns previously stored energy when delivering power, so that the net energy transfer is never negative.

Equation (6–25) implies that inductor current is a continuous function of time because an abrupt change in current causes a discontinuity in the energy. Since power is the time derivative of energy, an energy discontinuity implies infinite power, which is physically impossible. Current is called the **state variable** of the inductor because it determines the energy state of the element.

In summary, the inductor is a dynamic circuit element with the following properties:

1. *The voltage across the inductor is zero unless the current through the inductor is changing. The inductor acts like a short circuit for dc excitations.*
2. *The current through the inductor is a continuous function of time. A discontinuous change in inductor current would require infinite voltage and power, which is physically impossible.*
3. *The inductor absorbs power from the circuit when storing energy and delivers power to the circuit when returning previously stored energy. The net energy is nonnegative, indicating that the inductor is a passive element.*

EXAMPLE 6–6

The current through a 2-mH inductor is $i_L(t) = 4 \sin 1000t + 1 \sin 3000t$ A. Find the resulting inductor voltage.

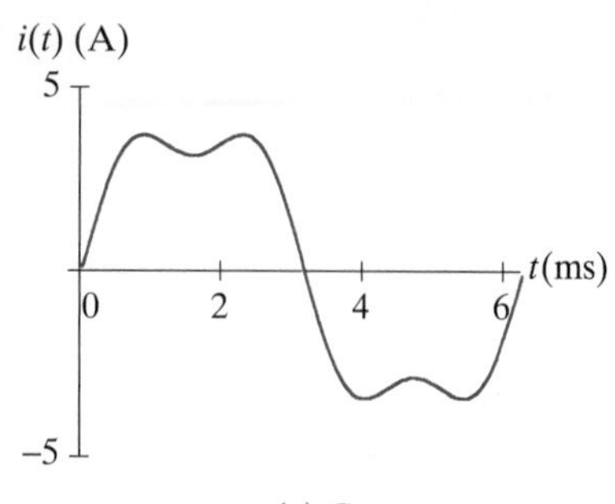

(a) Current

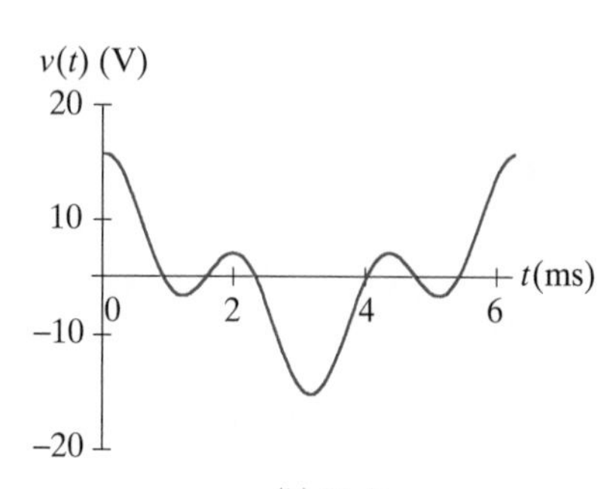

(b) Voltage

FIGURE 6–12 *(a) Current (b) Voltage*

SOLUTION:

The voltage is found from the derivative form of the i–v relationship:

$$v_L(t) = L\frac{di_L(t)}{dt} = 0.002\,[4 \times 1000 \cos 1000\,t + 1 \times 3000 \cos 3000\,t]$$

$$= 8 \cos 1000\,t + 6 \cos 3000\,t \text{ V}$$

The current and voltage waveforms are shown in Figure 6–12. Note that the current and voltage each contain two sinusoids at different frequencies. However, the relative amplitudes of the two sinusoids are different. In $i_L(t)$ the ratio of the amplitude of the ac component at $\omega = 3$ krad/s to the ac component at $\omega = 1$ krad/s is 1-to-3, whereas in $v_L(t)$ this ratio is 3-to-4. The fact that the ac responses of energy storage elements depend on frequency allows us to create frequency selective signal processors called filters. ■

EXAMPLE 6–7

Figure 6–13 shows the current through and voltage across an unknown energy storage element.

(a) What is the element and what is its numerical value?

(b) If the energy stored in the element at $t = 0$ is zero, how much energy is stored in the element at $t = 1$ s?

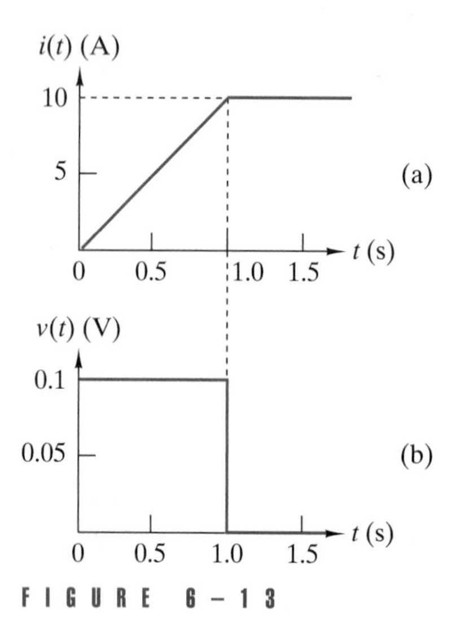

FIGURE 6–13

SOLUTION:

(a) By inspection, the voltage across the device is proportional to the derivative of the current, so the element is a linear inductor. During the interval $0 < t < 1$ s, the slope of the current waveform is 10 A/s. During the same interval the voltage is a constant 100 mV. Therefore, the inductance is

$$L = \frac{v}{di/dt} = \frac{0.1 \text{ V}}{10 \text{ A/s}} = 10 \text{ mH}$$

(b) The energy stored at $t = 1$ s is

$$w_L(1) = \frac{1}{2}Li_L^2(1) = 0.5(0.01)(10)^2 = 0.5 \text{ J}$$

■

EXAMPLE 6–8

The current through a 2.5-mH inductor is a damped sine $i(t) = 10e^{-500t}$ sin $2000t$. Plot the waveforms of the element current, voltage, power, and energy.

SOLUTION:

Figure 6–14 shows a Mathcad file that produces the required graphs. The first two lines in the file define the inductance and the current. The inductor voltage is found by differentiation, the power as the product of voltage and current, and the energy by integrating the power. In the plots shown in Figure 6–14, note that the current, voltage, and power alternate signs, whereas the total energy is always positive.

In Mathcad the time scale statement shown in the figure uses the following format:

$$t = T_{\text{Start}}, T_{\text{Start}} + \Delta t \ldots T_{\text{Stop}}$$

Numerical values for T_{Start}, Δt, and T_{Stop} are chosen by considering the damped sine waveform. The period of the sinusoid is

$$T_0 = \frac{2 \times \pi}{2000} = 0.00314 \text{ s}$$

The plots start at $T_{Start} = 0$. To include at least one cycle requires an end time of $T_{End} > T_0 = 0.00314$ s; hence we select $T_{Stop} = 0.004$ s. To include at least 20 points per cycle requires a time increment ($t < T_O/20 = 0.000158$); hence we select $\Delta t = 0.0001$ s. ■

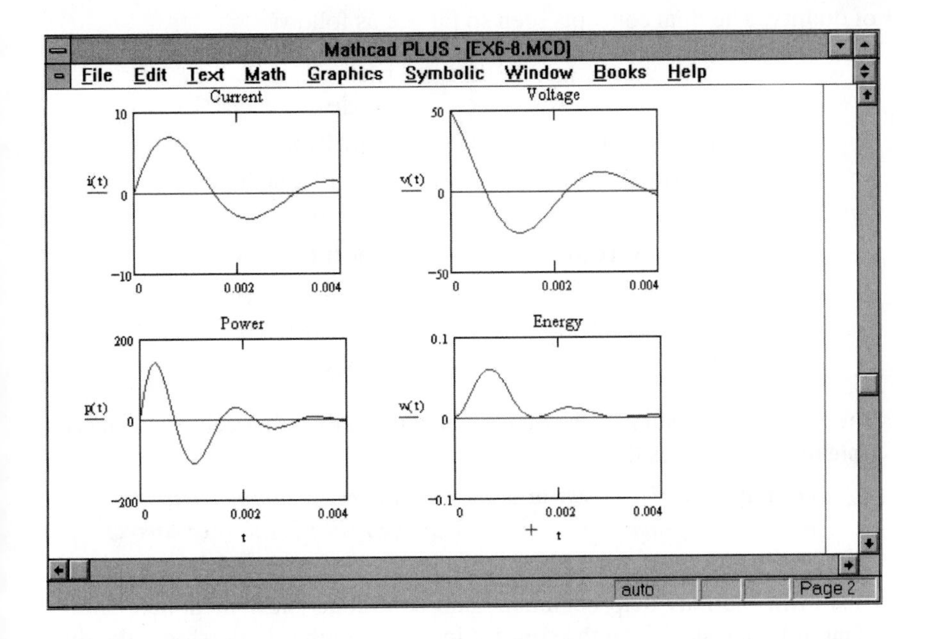

FIGURE 6–14

Exercise 6–5

For $t > 0$, the voltage across a 4-mH inductor is $v_L(t) = 20e^{-2000t}$ V. The initial current is $i_L(0) = 0$.

(a) What is the current through the inductor for $t > 0$?
(b) What is the power and (c) what is the energy for $t > 0$?

Answers:

(a) $i_L(t) = 2.5(1 - e^{-2000t})$ A
(b) $p_L(t) = 50(e^{-2000t} - e^{-4000t})$ W
(c) $w_L(t) = 12.5(1 - 2e^{-2000t} + e^{-4000t})$ mJ

Exercise 6–6

For $t < 0$, the current through a 100-mH inductor is zero. For $t \geq 0$, the current is $i_L(t) = 20e^{-2000t} - 20e^{-4000t}$ mA.

(a) Derive an expression for the voltage across the inductor for $t > 0$.
(b) Find the time $t > 0$ at which the inductor voltage passes through zero.
(c) Derive an expression for the inductor power for $t > 0$.
(d) Find the time interval over which the inductor absorbs power and the interval over which it delivers power.

Answers:

(a) $v_L(t) = -4e^{-2000t} + 8e^{-4000t}$ V
(b) $t = 0.347$ ms
(c) $p_L(t) = -80e^{-4000t} + 240e^{-6000t} - 160e^{-8000t}$ mW
(d) Absorbing for $0 < t < 0.347$ ms, delivering for $t > 0.347$ ms

MORE ABOUT DUALITY

The capacitor and inductor characteristics are quite similar. Interchanging C and L, and i and v converts the capacitor equations into the inductor equations, and vice versa. This interchangeability illustrates the principle of duality. The dual concepts seen so far are as follows:

KVL	↔	KCL
Loop	↔	Node
Resistance	↔	Conductance
Voltage source	↔	Current source
Thévenin	↔	Norton
Short circuit	↔	Open circuit
Series	↔	Parallel
Capacitance	↔	Inductance
Flux linkage	↔	Charge

The term in one column is the dual of the term in the other column. The **principle of duality** states that

> ***If every electrical term in a correct statement about circuit behavior is replaced by its dual, then the result is another correct statement.***

This principle may help beginners gain confidence in their understanding of circuit analysis. When the concept in one column is understood, the dual concept in the other column becomes easier to remember and apply.

6–3 DYNAMIC OP AMP CIRCUITS

The dynamic characteristics of capacitors and inductors produce signal processing functions that cannot be obtained using resistors. The OP AMP circuit in Figure 6–15 is similar to the inverting amplifier circuit except for the capacitor in the feedback path. To determine the signal-processing function of the circuit, we need to find its input-output relationship.

We begin by writing a KCL equation at node A.

$$i_R(t) + i_C(t) = i_N(t)$$

The resistor and capacitor device equations are written using their i–v relationships and the fundamental property of node voltages:

$$i_C(t) = C\frac{d[v_O(t) - v_A(t)]}{dt}$$

$$i_R(t) = \frac{1}{R}[v_S(t) - v_A(t)]$$

The ideal OP AMP device equations are $i_N(t) = 0$ and $v_A(t) = 0$. Substituting all of the element constraints into the KCL connection constraint produces

$$\frac{v_S(t)}{R} + C\frac{dv_O(t)}{dt} = 0$$

To solve for the output $v_O(t)$, we multiply this equation by dt, solve for the differential dv_O, and integrate:

$$\int dv_O = -\frac{1}{RC}\int v_S(t)dt$$

Assuming the output voltage is known at time $t_0 = 0$, the integration limits are

$$\int_{v_O(0)}^{v_O(t)} dv_O = -\frac{1}{RC}\int_0^t v_S(x)dx$$

which yields

$$v_O(t) = v_O(0) - \frac{1}{RC}\int_0^t v_S(x)dx$$

The initial condition $v_O(0)$ is actually the voltage on the capacitor at $t = 0$, since by KVL, we have $v_C(t) = v_O(t) - v_A(t)$. But $v_A = 0$ for the OP AMP, so in general $v_O(t) = v_C(t)$. When the voltage on the capacitor is zero at $t = 0$, the circuit input-output relationship reduces to

$$v_O(t) = -\frac{1}{RC}\int_0^t v_S(x)dx \qquad (6\text{–}26)$$

The output voltage is proportional to the integral of the input voltage when the initial capacitor voltage is zero. The circuit in Figure 6–15 is an **inverting integrator** since the proportionality constant is negative. The constant $1/RC$ has the units of reciprocal seconds (s^{-1}) so that both sides of Eq. (6–26) have the units of volts.

Interchanging the resistor and capacitor in Figure 6–15 produces the OP AMP differentiator in Figure 6–16. To find the input-output relationship of this circuit, we start by writing the element and connection equations. The KCL connection constraint at node A is

$$i_R(t) + i_C(t) = i_N(t)$$

FIGURE 6–15 *The inverting OP AMP integrator.*

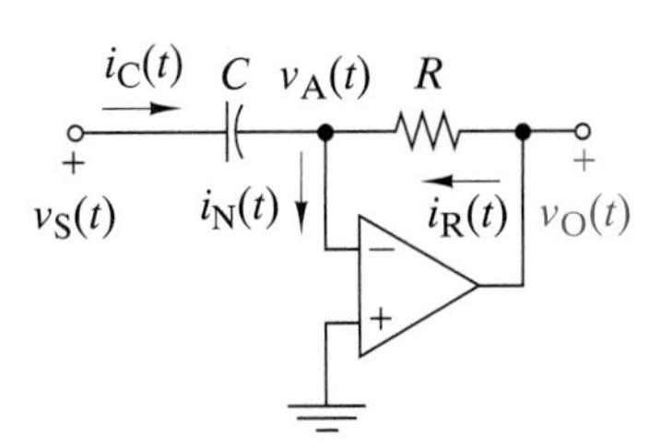

FIGURE 6–16 *The inverting OP AMP differentiator.*

The device equations for the input capacitor and feedback resistor are

$$i_C(t) = C\frac{d[v_S(t) - v_A(t)]}{dt}$$

$$i_R(t) = \frac{1}{R}[v_O(t) - v_A(t)]$$

The device equations for the OP AMP are $i_N(t) = 0$ and $v_A(t) = 0$. Substituting all of these element constraints into the KCL connection constraint produces

$$\frac{v_O(t)}{R} + C\frac{dv_S(t)}{dt} = 0$$

Solving this equation for $v_O(t)$ produces the circuit input-output relationship:

$$v_O(t) = -RC\frac{dv_S(t)}{dt} \tag{6–27}$$

The output voltage is proportional to the derivative of the input voltage. The circuit in Figure 6–16 is an **inverting differentiator** since the proportionality constant ($-RC$) is negative. The units of the constant RC are seconds so that both sides of Eq. (6–27) have the units of volts.

There are OP AMP inductor circuits that produce the inverting integrator and differentiator functions; however, they are of little practical interest because of the physical size and resistive losses in real inductor devices.

Figure 6–17 shows OP AMP circuits and block diagrams for the inverting integrator and differentiator, together with signal-processing functions studied in Chapter 4. The term *operational amplifier* results from the various mathematical operations implemented by these circuits. The following examples illustrate using the collection of circuits in Figure 6–17 in the analysis and design of signal-processing functions.

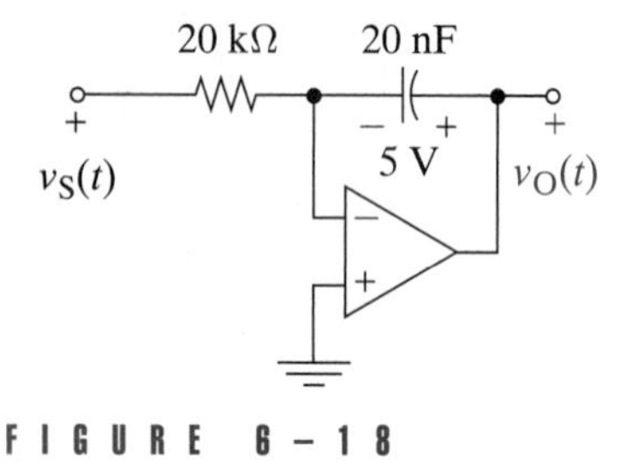

FIGURE 6–18

EXAMPLE 6–9

The input to the circuit in Figure 6–18 is $v_S(t) = 10u(t)$. Derive an expression for the output voltage. The OP AMP saturates when $v_O(t) = \pm 15$ V.

SOLUTION:

The circuit is the inverting integrator with an initial voltage of 5 V across the capacitor. For the reference marks shown in Figure 6–18, this means that $v_O(0) = +5$ V. Assuming the OP AMP is operating in the linear mode, the output voltage is

$$\begin{aligned} v_O(t) &= v_O(0) - \frac{1}{RC}\int_0^t v_S\,dt \\ &= 5 - 2500\int_0^t 10\,dt \\ &= 5 - 25{,}000t \quad t > 0 \end{aligned}$$

The output contains a negative-going ramp because the circuit is an inverting integrator. The ramp output response is valid only as long as the OP AMP

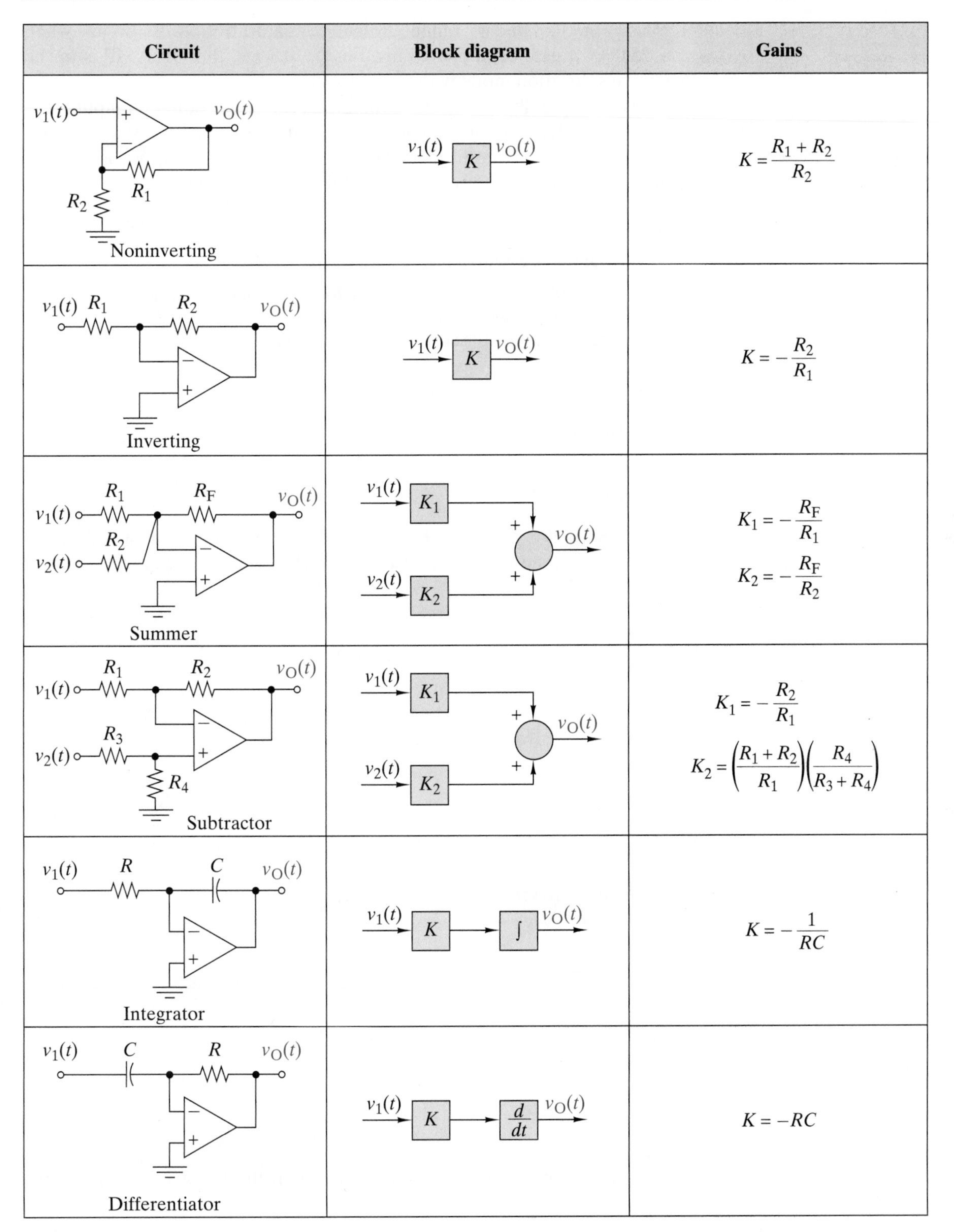

FIGURE 6–17 *Summary of basic OP AMP signal-processing circuits.*

remains in its linear range. Negative saturation will occur when $5 - 25{,}000t = -15$, or at $t = 0.8$ ms. For $t > 0.8$ ms, the OP AMP is in the negative saturation mode with $v_O = -15$ V.

This example illustrates that dynamic circuits with bounded inputs may have unbounded responses. The circuit input here is a 10-V step function that has a bounded amplitude. The circuit output is a ramp whose output would be unbounded except that the OP AMP saturates. ■

Exercise 6–7

The input to the circuit in Figure 6–18 is $v_S(t) = 10\,[e^{-5000t}]u(t)$ V.

(a) For $v_C(0) = 0$, derive an expression for the output voltage, assuming the OP AMP is in its linear range.
(b) Does the OP AMP saturate with the given input?

Answers:

(a) $v_O(t) = 5(e^{-5000t} - 1)u(t)$
(b) Does not saturate

EXAMPLE 6–10

The input to the circuit in Figure 6–19(a) is a trapezoidal waveform shown in (b). Find the output waveform. The OP AMP saturates when $v_O(t) = \pm 15$ V.

FIGURE 6–19

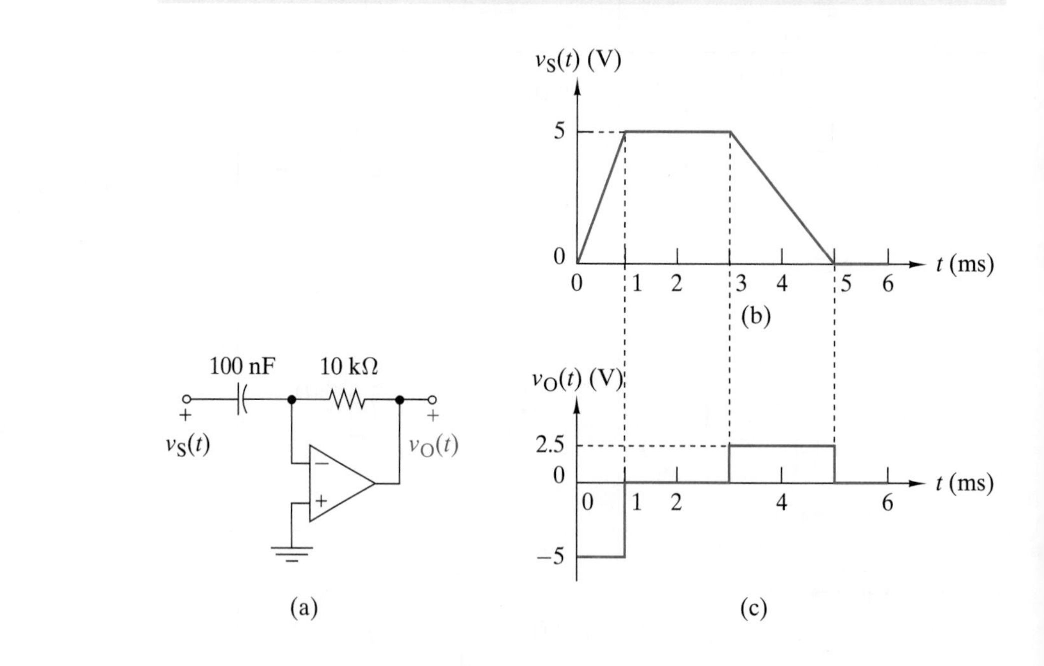

SOLUTION:
The circuit is the inverting differentiator with the following input-output relationship:

$$v_O(t) = -RC\frac{dv_S(t)}{dt} = -\frac{1}{1000}\frac{dv_S(t)}{dt}$$

The output voltage is constant over each of the following three time intervals:

1. For $0 < t < 1$ ms, the input slope is 5000 V/s and the output is $v_O = -5$ V.
2. For $1 < t < 3$ ms, the input slope is zero, so the output is zero as well.
3. For $3 < t < 5$ ms, the input slope is –2500 V/s and the output is +2.5 V.

The resulting output waveform is shown in Figure 6–19(c).

The output voltage remains within ±15-V limits, so the OP AMP operates in the linear mode. ■

Exercise 6–8

The input to the circuit in Figure 6–19 is $v_S(t) = V_A \cos 2000t$. The OP AMP saturates when $v_O = \pm 15$ V.

(a) Derive an expression for the output, assuming that the OP AMP is in the linear mode.

(b) What is the maximum value of V_A for linear operation?

Answers:

(a) $v_O(t) = 2V_A \sin 2000t$

(b) $|V_A| \leq 7.5$ V

EXAMPLE 6–11

Determine the input-output relationship of the circuit in Figure 6–20(a).

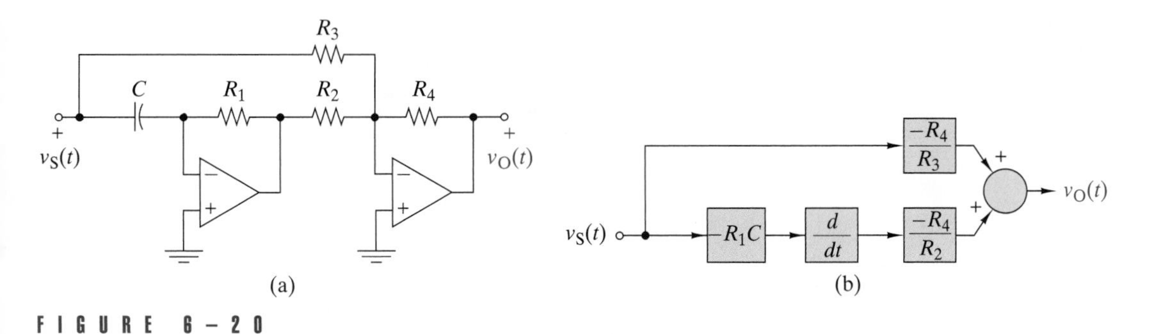

FIGURE 6–20

SOLUTION:

The circuit contains an inverting differentiator and an inverting summer. To find the input-output relationship, it is helpful to develop a block diagram for the circuit. Figure 6–20(b) shows a block diagram using the functional blocks in Figure 6–17. The product of the gains along the lower path yields its contribution to the output as

$$(-R_1C)\left[\frac{d}{dt}\left(-\frac{R_4}{R_2}\right)v_S(t)\right]$$

The upper path contributes $(-R_4/R_3)v_S(t)$ to the output. The total output is the sum of the contributions from each path:

$$v_O(t) = \left(\frac{R_1CR_4}{R_2}\right)\frac{dv_S(t)}{dt} - \left(\frac{R_4}{R_3}\right)v_S(t)$$

This equation assumes that both OP AMPs remain within their linear range. ■

Exercise 6–9

Find the input-output relationship of the circuit in Figure 6–21.

Answer:
$$v_O(t) = v_O(0) + \frac{1}{RC}\int_0^t (v_{S1} - v_{S2})dt$$

FIGURE 6–21

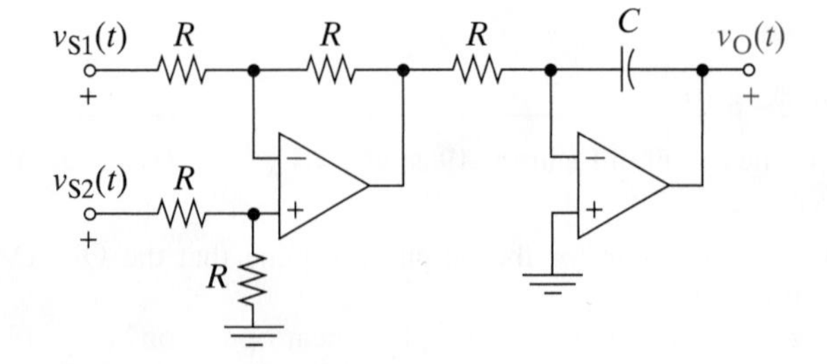

DESIGN EXAMPLE 6–12

Use the functional blocks in Figure 6–17 to design an OP AMP circuit to implement the following input-output relationship:

$$v_O(t) = 10v_S(t) + \int_0^t v_S\, dt$$

FIGURE 6–22

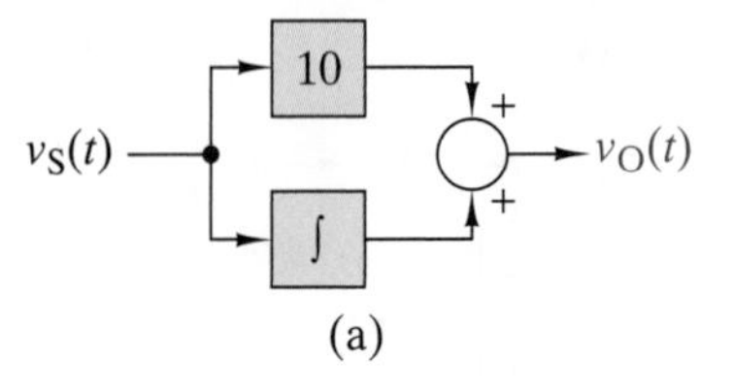

(a)

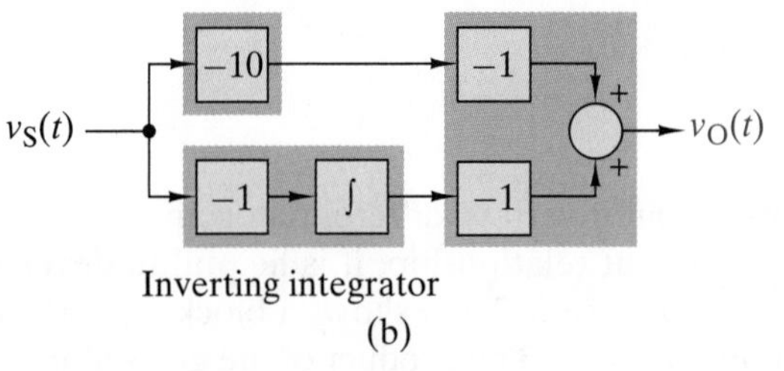

(b)

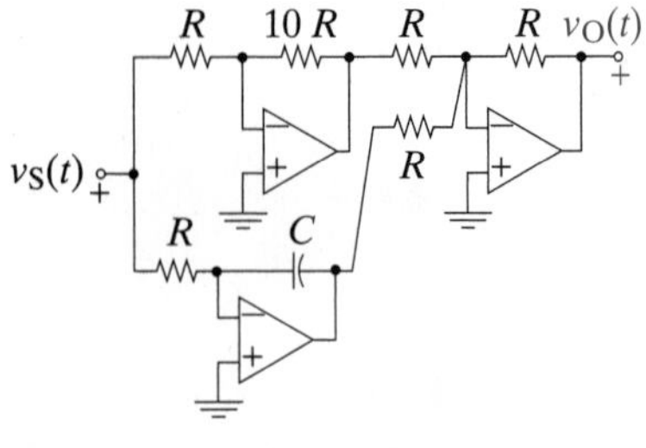

FIGURE 6–23

SOLUTION:

There is no unique solution to this design problem. We begin by drawing the block diagram in Figure 6–22(a), which shows that we need a gain block, an integrator, and a summer. However, the integrator and summer in Figure 6–17 are inverting circuits. Figure 6–22(b) shows how to overcome this problem by including an even number of signal inversions in each path. The inverting building blocks are realizable using OP AMP circuits in Figure 6–17, and the overall transfer characteristic is noninverting as required. One (of many) possible circuit realization of these processors is shown in Figure 6–23. The element parameter constraint on this circuit is $RC = 1$. Selecting the OP AMPs and the values of R and C depends on many additional factors, such as accuracy, internal resistance of the input source, and output load. ■

APPLICATION EXAMPLE 6–13

Dynamic OP AMP circuits can function as integrators or differentiators. Given an equation or graph of an input waveform, we can predict output waveforms using the mathematical operations of integration or differentiation. Integrators and differentiators work flawlessly on ideal signal models. In practice there are some limitations caused by relatively small imperfections in real-world signals.

First consider an inverting integrator with an input $v_S(t) + V_0$, where $v_S(t)$ is the desired signal and V_0 is a relatively small dc offset. For this input an ideal inverting integrator has an output of

$$v_O(t) = -\frac{1}{RC}\Bigg[\underbrace{\int_0^t v_S(x)dx}_{\text{Desired Output}} + \underbrace{V_0 t}_{\text{Ramp}}\Bigg]$$

The desired output is accompanied by a ramp waveform caused by a small dc offset. Even if V_0 is very small, the ramp $V_0 t$ will eventually overwhelm the desired signal and saturate the OP AMP.

The dc offset problem is dealt with using the reset switch in Figure 6–24(a) to limit the time interval during which the circuit performs integration. When the reset signal is high, the switch is closed, any voltage on the capacitor is removed, and the integrator output voltage is forced to zero. When the reset signal goes low, the switch opens and the circuit operates as an integrator. At the end of a fixed time interval, the reset signal goes high again and the switch closes, forcing the output to zero. Practical integrator applications use *time-limited integration* in which the integrator output is periodically reset to zero before any offset driven ramp becomes important.

Next, consider an inverting differentiator with an input $v_S(t) + V_A \sin(\omega t)$, where $v_S(t)$ is the desired signal and $V_A \sin(\omega t)$ represents relatively small high-frequency noise. For this input an ideal inverting differentiator has an output of

$$v_O(t) = -RC\Bigg[\underbrace{\frac{dv_S(t)}{dt}}_{\text{Desired Output}} + \underbrace{\omega V_A \cos(\omega t)}_{\text{Noise}}\Bigg]$$

The desired output is accompanied by an amplified noise component. Even if the input noise amplitude V_A is very small, the term ωV_A can be large for high-frequency noise. The basic problem is that differentiation amplifies high-frequency noise to a degree that can overwhelm the desired signal.

The high-frequency noise problem is dealt with by adding the series resistor shown in Figure 6–24(b). This addition limits the frequency range over which the circuit actually performs differentiation. At low frequencies

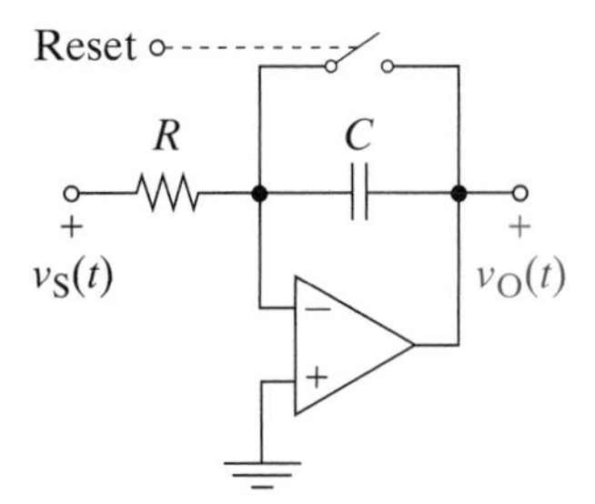

(a) Time-Limited Integrator

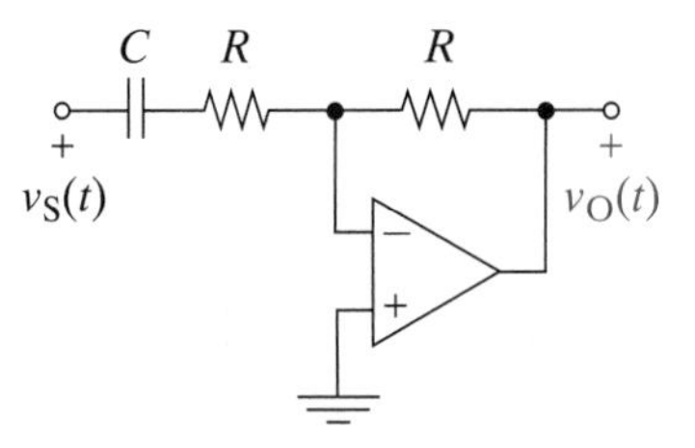

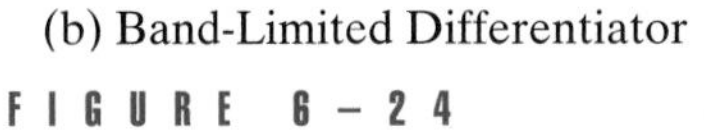

(b) Band-Limited Differentiator

FIGURE 6–24

the capacitor is the dominant element, the series resistor plays no role, and the circuit performs differentiation. At high frequencies the roles reverse. The added resistor dominates, the capacitor can be ignored, and the circuit functions as an inverter. As a result the modified circuit only differentiates signals in a low-frequency band.[2] Practical applications use *band-limited differentiation* to avoid high-frequency noise problems.

6–4 EQUIVALENT CAPACITANCE AND INDUCTANCE

In Chapter 2 we found that resistors connected in series or parallel can be replaced by equivalent resistances. The same principle applies to connections of capacitors and inductors—for example, to the parallel connection of capacitors in Figure 6–25(a). Applying KCL at node A yields

$$i(t) = i_1(t) + i_2(t) + \ldots + i_N(t)$$

Since the elements are connected in parallel, KVL requires

$$v_1(t) = v_2(t) = \ldots = v_N(t) = v(t)$$

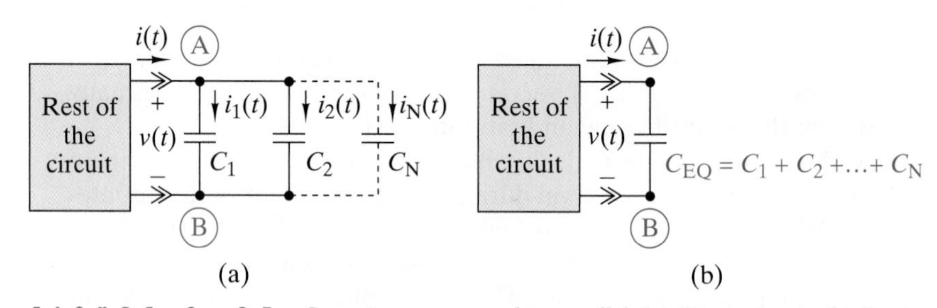

FIGURE 6–25 *Capacitors connected in parallel. (a) Given circuit. (b) Equivalent circuit.*

Because the capacitors all have the same voltage, their *i–v* relationships are all of the form $i_k(t) = C_k dv(t)/dt$. Substituting the *i–v* relationships into the KCL equation yields

$$i(t) = C_1\frac{dv(t)}{dt} + C_2\frac{dv(t)}{dt} + \ldots + C_N\frac{dv(t)}{dt}$$

Factoring the derivative out of each term produces

$$i(t) = (C_1 + C_2 + \ldots + C_N)\frac{dv(t)}{dt}$$

This equation states that the responses $v(t)$ and $i(t)$ in Figure 6–25(a) do not change when the N parallel capacitors are replaced by an equivalent capacitance:

$$C_{EQ} = C_1 + C_2 + \ldots + C_N \quad \text{(parallel connection)} \qquad (6\text{–}28)$$

2 In Chapter 12 we find that this low-frequency band falls below $\omega_C = 1/RC$.

The equivalent capacitance simplification is shown in Figure 6–25(b). The initial voltage, if any, on the equivalent capacitance is $v(0)$, the common voltage across all of the original N capacitors at $t = 0$.

Next consider the series connection of N capacitors in Figure 6–26(a). Applying KVL around loop 1 in Figure 6–26(a) yields the equation

$$v(t) = v_1(t) + v_2(t) + \ldots + v_N(t)$$

Since the elements are connected in series, KCL requires

$$i_1(t) = i_2(t) = \ldots = i_N(t) = i(t)$$

Since the same current exists in all capacitors, their i–v relationships are all of the form

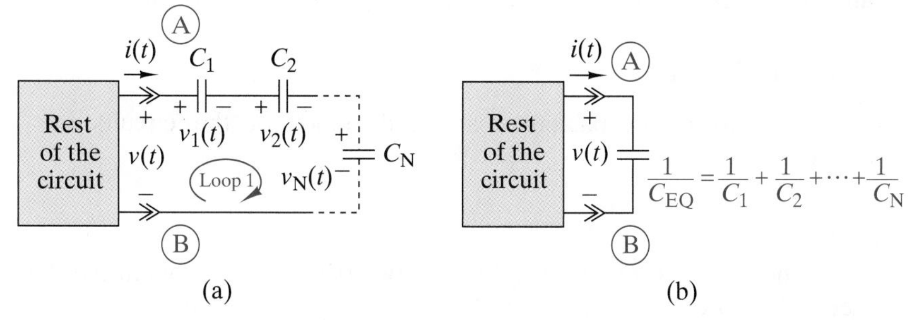

FIGURE 6–26 *Capacitors connected in series. (a) Given circuit. (b) Equivalent circuit.*

$$v_k(t) = v_k(0) + \frac{1}{C_k}\int_0^t i(x)dx$$

Substituting these i–v relationships into the loop 1 KVL equation yields

$$v(t) = v_1(0) + \frac{1}{C_1}\int_0^t i(x)dx + v_2(0) + \frac{1}{C_2}\int_0^t i(x)dx$$
$$+ \quad + v_N(0) + \frac{1}{C_N}\int_0^t i(x)dx$$

We can factor the integral out of each term to obtain

$$v(t) = [v_1(0) + v_2(0) + \ldots + v_N(0)] + \left(\frac{1}{C_1} + \frac{1}{C_2} + \ldots + \frac{1}{C_N}\right)\int_0^t i(x)dx$$

This equation indicates that the responses $v(t)$ and $i(t)$ in Figure 6–26(a) do not change when the N series capacitors are replaced by an equivalent capacitance:

$$\frac{1}{C_{EQ}} = \frac{1}{C_1} + \frac{1}{C_2} + \ldots + \frac{1}{C_N} \quad \text{(series connection)} \qquad (6\text{–}29)$$

The equivalent capacitance is shown in Figure 6–26(b). The initial voltage on the equivalent capacitance is the sum of the initial voltages on each of the original N capacitors.

The equivalent capacitance of a parallel connection is the sum of the individual capacitances. The reciprocal of the equivalent capacitance of a series connection is the sum of the reciprocals of the individual capacitances.

Since the capacitor and inductor are dual elements, the corresponding results for inductors are found by interchanging the series and parallel equivalence rules for the capacitor. That is, in a series connection the equivalent inductance is the sum of the individual inductances:

$$L_{EQ} = L_1 + L_2 + \ldots + L_N \qquad \text{(series connection)} \tag{6–30}$$

For the parallel connection, the reciprocals add to produce the reciprocal of the equivalent inductance:

$$\frac{1}{L_{EQ}} = \frac{1}{L_1} + \frac{1}{L_2} + \ldots + \frac{1}{L_N} \qquad \text{(parallel connection)} \tag{6–31}$$

Derivation of Eqs. (6–30) and (6–31) uses the approach given previously for the capacitor except that the roles of voltage and current are interchanged. Completion of the derivation is left as a problem for the reader.

EXAMPLE 6–14

Find the equivalent capacitance and inductance of the circuits in Figure 6–27.

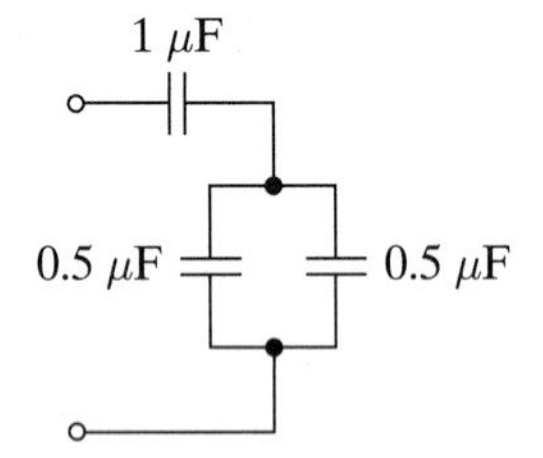

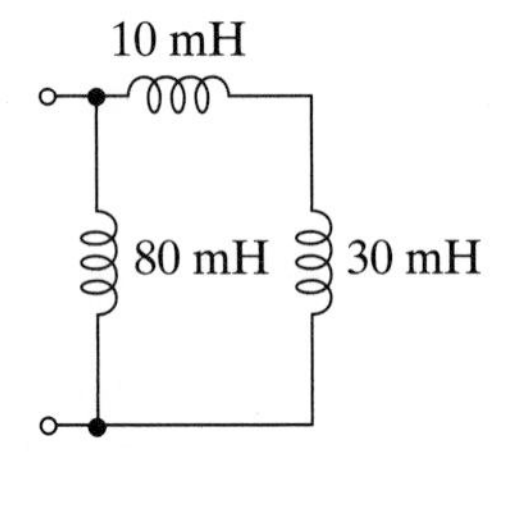

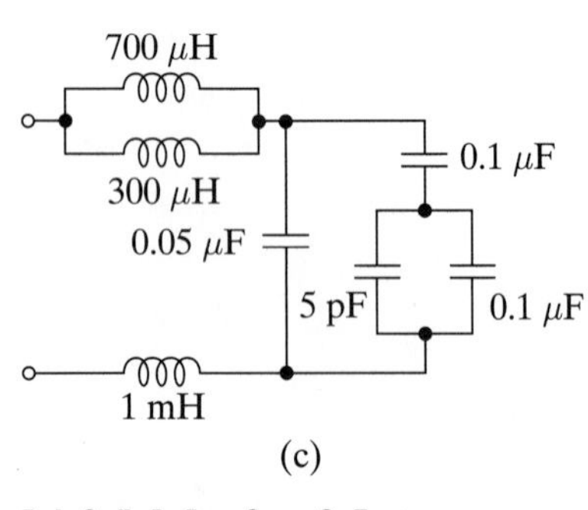

(c)

FIGURE 6–27

SOLUTION:

(a) For the circuit in Figure 6–27(a), the two 0.5-μF capacitors in parallel combine to yield an equivalent 0.5 + 0.5 = 1-μF capacitance. This 1-μF equivalent capacitance is in series with a 1-μF capacitor, yielding an overall equivalent of $C_{EQ} = 1/(1/1 + 1/1) = 0.5$ μF.

(b) For the circuit of Figure 6–27(b), the 10-mH and the 30-mH inductors are in series and add to produce an equivalent inductance of 40 mH. This 40-mH equivalent inductance is in parallel with the 80-mH inductor. The equivalent inductance of the parallel combination is $L_{EQ} = 1/(1/40 + 1/80) = 26.67$ mH.

(c) The circuit of Figure 6–27(c) contains both inductors and capacitors. In later chapters, we will learn how to combine all of these into a single equivalent element. For now, we combine the inductors and the capacitors separately. The 5-pF capacitor in parallel with the 0.1-μF capacitor yields an equivalent capacitance of 0.100005 μF. For all practical purposes, the 5-pF capacitor can be ignored, leaving two 0.1-μF capacitors in series with equivalent capacitance of 0.05 μF. Combining this equivalent capacitance in parallel with the remaining 0.05-μF capacitor yields an overall equivalent capacitance of 0.1 μF. The parallel 700-μH and 300-μH inductors yield an equivalent inductance of 1/(1/700 + 1/300) = 210 μH. This equivalent inductance is effectively in series with the 1-mH inductor at the bottom, yielding 1000 + 210 = 1210 μH as the overall equivalent inductance.

Figure 6–28 shows the simplified equivalent circuits for each of the circuits of Figure 6–27. ■

Exercise 6–10

The current through a series connection of two 1-μF capacitors is a rectangular pulse with an amplitude of 2 mA and a duration of 10 ms. At $t = 0$ the voltage across the first capacitor is +10 V and across the second is zero.

(a) What is the voltage across the series combination at $t = 10$ ms?
(b) What is the maximum instantaneous power delivered to the series combination?
(c) What is the energy stored on the first capacitor at $t = 0$ and $t = 10$ ms?

Answers:

(a) 50 V
(b) 100 mW at $t = 10$ ms
(c) 50 μJ and 450 μJ

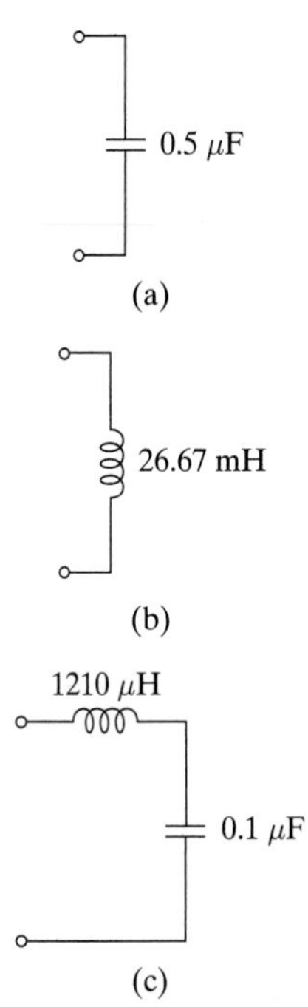

FIGURE 6–28

DC EQUIVALENT CIRCUITS

Sometimes we need to find the dc response of circuits containing capacitors and inductors. In the first two sections of this chapter, we found that under dc conditions a capacitor acts like an open circuit and an inductor acts like a short circuit. In other words, under dc conditions, an equivalent circuit for a capacitor is an open circuit and an equivalent circuit of an inductor is a short circuit.

To determine dc responses, we replace capacitors by open circuits and inductors by short circuits and analyze the resulting resistance circuit using any of the methods in Chapters 2 through 4. The circuit analysis involves only resistance circuits and yields capacitor voltages and inductor currents along with any other variables of interest. Computer programs like SPICE use this type of dc analysis to find the initial operating point of a circuit to be analyzed. The dc capacitor voltages and inductor currents become initial conditions for a transient response that begins at $t = 0$ when something in the circuit changes, such as the position of a switch.

EXAMPLE 6–15

Determine the voltage across the capacitors and current through the inductors in Figure 6–29(a).

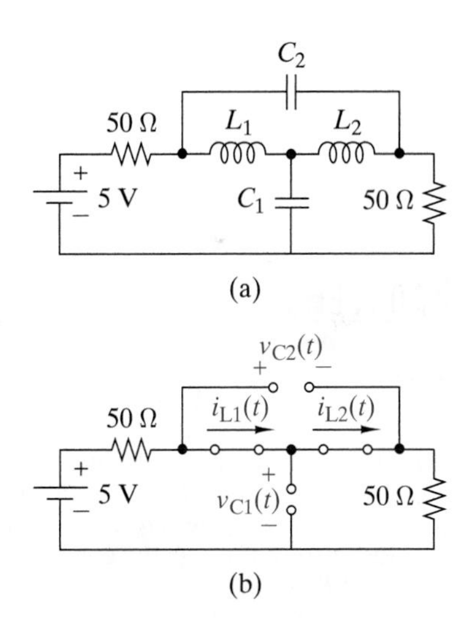

FIGURE 6–29

SOLUTION:
The circuit is driven by a 5-V dc source. Figure 6–29(b) shows the equivalent circuit under dc conditions. The current in the resulting series circuit is 5/(50 + 50) = 50 mA. This dc current exists in both inductors, so $i_{L1} = i_{L2} = 50$ mA. By voltage division the voltage across the 50-Ω output resistor is $v = 5 \times 50/(50 + 50) = 2.5$ V; therefore, $v_{C1}(0) = 2.5$ V. The voltage across C_2 is zero because of the short circuits produced by the two inductors. ■

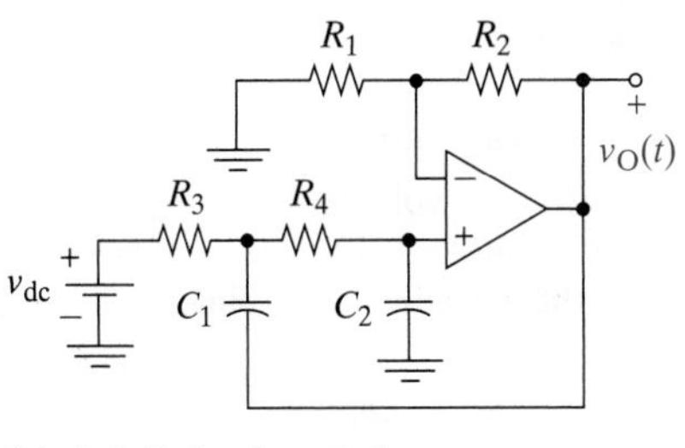

FIGURE 6–30

Exercise 6–11

Find the OP AMP output voltage in Figure 6–30.

Answer:

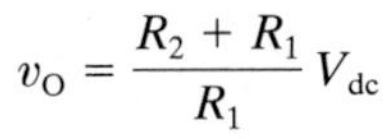

$$v_O = \frac{R_2 + R_1}{R_1} V_{dc}$$

SUMMARY

- The linear capacitor and inductor are dynamic circuit elements that can store energy. The instantaneous element power is positive when they are storing energy and negative when they are delivering previously stored energy. The net energy transfer is never negative because inductors and capacitors are passive elements.
- The current through a capacitor is zero unless the voltage is changing. A capacitor acts like an open circuit to dc excitations.
- The voltage across an inductor is zero unless the current is changing. An inductor acts like a short circuit to dc excitations.
- Capacitor voltage and inductor current are called state variables because they define the energy state of a circuit. Circuit state variables are continuous functions of time as long as the circuit driving forces are finite.
- OP AMP capacitor circuits perform signal integration or differentiation. These operations, together with the summer and gain functions, provide the building blocks for designing dynamic input-output characteristics.
- Capacitors or inductors in series or parallel can be replaced with an equivalent element found by adding the individual capacitances or inductances or their reciprocals. The dc response of a dynamic circuit can be found by replacing all capacitors with open circuits and all inductors with short circuits.

PROBLEMS

OBJECTIVE 6–1 CAPACITOR AND INDUCTOR RESPONSES (SECTS. 6–1, 6–2)

(a) Given the current through a capacitor or an inductor, find the voltage across the element.
(b) Given the voltage across a capacitor or an inductor, find the current through the element.
(c) Find the power and energy associated with a capacitor or inductor.

See Examples 6–1, 6–2, 6–3, 6–4, 6–6, 6–7, 6–8 and Exercises 6–1, 6–2, 6–3, 6–4, 6–5, 6–6

6–1 For $t \geq 0$ the voltage across a 4-μF capacitor is $v_C(t) = 3\,e^{-2000t}$ V. Derive expressions for $i_C(t)$ and $p_C(t)$. Is the capacitor absorbing power, delivering power, or both?

6–2 The voltage across a 3-mF capacitor is $v_C(t) = 20\sin(2\pi 10t)$ V. Derive expressions for $i_C(t)$ and $p_C(t)$. Is the capacitor absorbing power, delivering power, or both?

6–3 For $t \geq 0$ the current through a 0.25-μF capacitor is $i_C(t) = 10\sin(2000t)$ mA. Derive expressions for $v_C(t)$ when $v_C(0) = -10$ V.

6–4 The current through a 2-μF capacitor is a rectangular pulse with an amplitude of 3 mA and a duration of 5 ms. Find the capacitor voltage at the end of the pulse when the voltage at the beginning is −6 V.

6–5 For $t \geq 0$ the current through a capacitor is $i_C(t) = 10\,r(t)$ mA. At $t = 0$ the capacitor voltage is 3 V. At $t =$ 1ms the voltage is 8 V. Find the capacitance of the device.

6–6 The voltage across a 0.5-μF capacitor is shown in Figure P6–6. Prepare sketches of $i_C(t)$ and $p_C(t)$. Is the capacitor absorbing power, delivering power, or both?

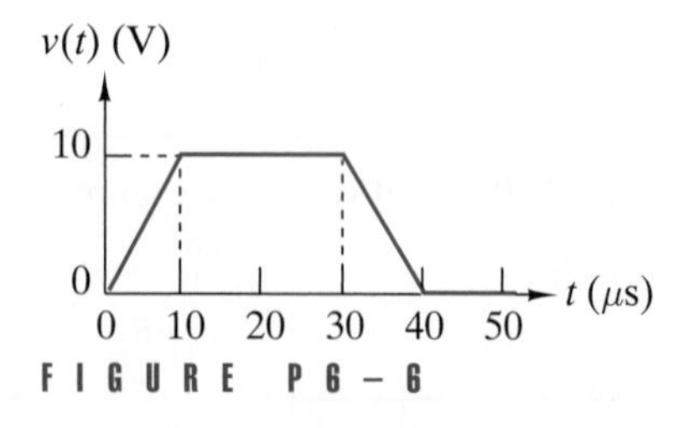

FIGURE P6–6

6–7 The current through a 10-nF capacitor is shown in Figure P6–7. Given that $v_C(0) = -5$ V, find the value of $v_C(t)$ at $t = 5$, 10, and 20 μs.

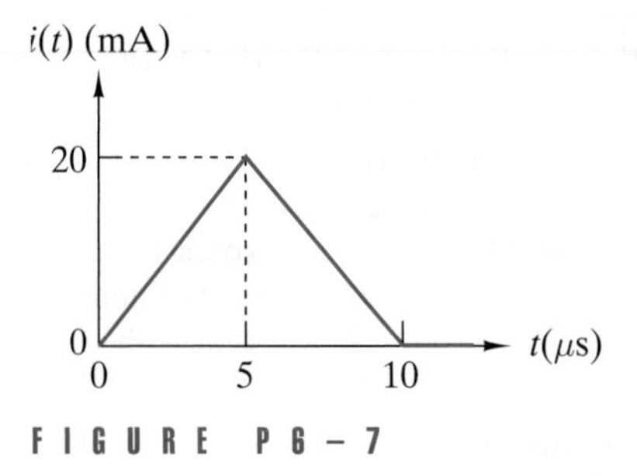

FIGURE P6–7

6–8 The current through a 5-mH inductor is shown in Figure P6–7. Prepare sketches of $v_L(t)$, $p_L(t)$, and $w_L(t)$.

6–9 For $t \geq 0$ the current through a 100-mH inductor is $i_L(t) = (30t - 60)$ A. Find $p_L(t)$ and $w_L(t)$ for $t \geq 0$. Is the inductor absorbing power, delivering power, or both?

6–10 For $t \geq 0$ the voltage across a 25-mH inductor is $v_L(t) = 20\,e^{-400t}$ V. Find $i_L(t)$ for $t \geq 0$ when $i_L(0) = 3$ A.

6–11 A voltage $v_L(t) = 5\cos(1000t) - 2\sin(3000t)$ V appears across a 50-mH inductor. Derive an expression for $i_L(t)$. Assume $i_L(0) = 0$. Discuss the effect of frequency on the relative amplitudes of the sinusoidal components in $v_L(t)$ and $i_L(t)$.

6–12 For $t \geq 0$ the voltage across a 0.1-H inductor is $v_L(t) = 1000r(t)$ V. At $t = 5$ ms the inductor current is observed to be zero. Find the value of $i_L(0)$.

6–13 For $t > 0$ the current through a 2-mH inductor is $i_L(t) = 100te^{-1000t}$ A. Derive an expression for $v_L(t)$. Is the inductor absorbing power, delivering power, or both?

6–14 The capacitor in Figure P6–14 carries an initial voltage $v_C(0) = 50$ V. At $t = 0$, the switch is closed, and thereafter the voltage across the capacitor is $v_C(t) = 50\,e^{-2000t}$ V. Derive expressions for $i_C(t)$ and $p_C(t)$ for $t > 0$. Is the capacitor absorbing power, delivering power, or both?

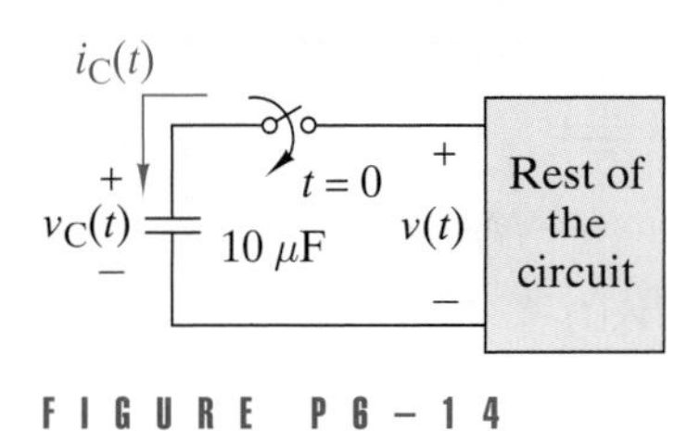

FIGURE P6–14

6–15 The capacitor in Figure P6–14 carries an initial voltage $v_C(0) = 10$ V. At $t = 0$ the switch is closed, and thereafter the voltage across the capacitor is $v_C(t) = (20\,e^{-4000t} - 10)$ V. Derive expressions for $i_C(t)$ and $p_C(t)$ for $t > 0$. Is the capacitor absorbing power, delivering power, or both?

6–16 The inductor in Figure P6–16 carries an initial current of $i_L(0) = 0.5$ A. At $t = 0$ the switch opens, and thereafter the current into the rest of the circuit is $i(t) = -0.5\,e^{-2000t}$ A.Derive expressions for $v_L(t)$ and $p_L(t)$ for $t > 0$. Is the inductor absorbing or delivering power?

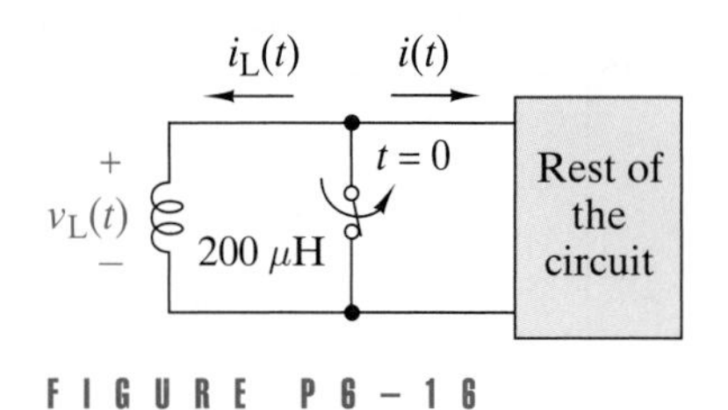

FIGURE P6–16

6–17 The inductor in Figure P6–16 carries an initial current of $i_L(0) = 20$ mA. AT $t = 0$ the switch opens, and thereafter the voltage across the inductor is $v_L(t) = -8\,e^{-1000t}$ mV. Derive expressions for $i_L(t)$ and $p_L(t)$ for $t > 0$. Is the inductor absorbing or delivering power?

6–18 A 2.2-μF capacitor is connected in series with a 200-Ω resistor. The voltage across the capacitor is $v_C(t) = 10\cos(2000t)$ V. What is the voltage across the resistor?

6–19 A 300-mH inductor is connected in parallel with a 10-kΩ resistor. The current through the inductor is $i_L(t) = 10\,e^{-1000t}$ mA. What is the current through the resistor?

6–20 For $t > 0$ the voltage across an energy storage element is $v(t) = (5 - 20\,e^{-500t})$ V and the current through the element is $i(t) = (2000t + 16\,e^{-500t})$ mA. What is the element and the element value?

OBJECTIVE 6–2 DYNAMIC OP AMP CIRCUITS (SECT. 6–3)

(a) Given an *RC* OP AMP integrator or differentiator, determine the output for specified inputs.
(b) Given a general *RC* OP AMP circuit, determine its input-output relationship and construct a block diagram.
(c) Design an *RC* OP AMP circuit to implement a given input-output relationship or a block diagram.

See Examples 6–9, 6–10, 6–11, 6–12 and Exercises 6–7, 6–8, 6–9

6–21 The OP AMP integrator in Figure P6–21 has $R = 40$ kΩ, $C = 50$ nF, and $v_O(0) = 10$ V. The input is $v_S(t) = 10\,e^{-500t}u(t)$ V. Find $v_O(t)$ for $t > 0$.

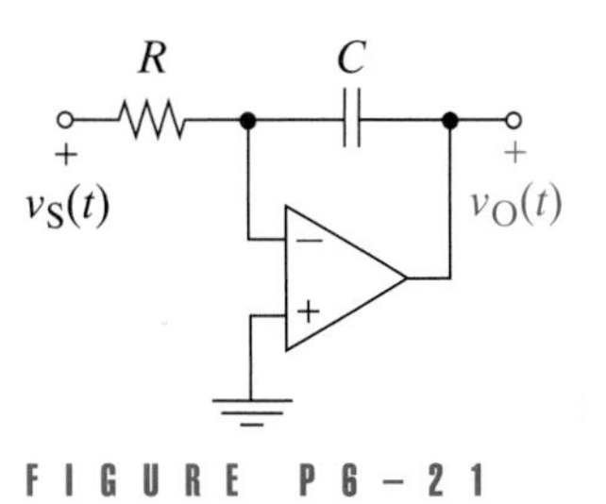

FIGURE P6–21

6–22 The OP AMP integrator in Figure P6–21 has $R = 20\ \text{k}\Omega$, $C = 1\ \mu\text{F}$, and $v_O(0) = 0$ V. The input waveform is shown in Figure P6–22. Sketch $v_O(t)$ for $t > 0$.

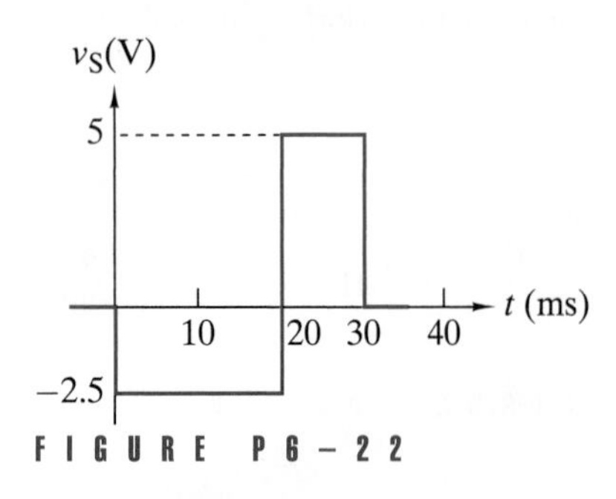

FIGURE P6–22

6–23 The OP AMP integrator in Figure P6–21 has $R = 50\ \text{k}\Omega$, $C = 100\ \mu\text{F}$, and $v_O(0) = 12$ V. The input is $v_S(t) = 10\,u(t)$ V. How long does it take for the OP AMP to saturate when $V_{CC} = \pm 15$ V.

6–24 The OP AMP integrator in Figure P6–21 has $R = 20\ \text{k}\Omega$, $C = 100$ nF, and $v_O(0) = 0$ V. The input is $v_S(t) = 5\sin(\omega t)\,u(t)$ V. Derive an expression for $v_O(t)$ and find the smallest allowable value of ω for linear operation of the OP AMP. Assume $V_{CC} = \pm 15$ V.

6–25 The OP AMP differentiator in Figure P6–25 with $R = 100\ \text{k}\Omega$ and $C = 50$ nF has an input $v_S(t) = 10(1 - e^{-50t})u(t)$ V. Find $v_O(t)$ for $t > 0$.

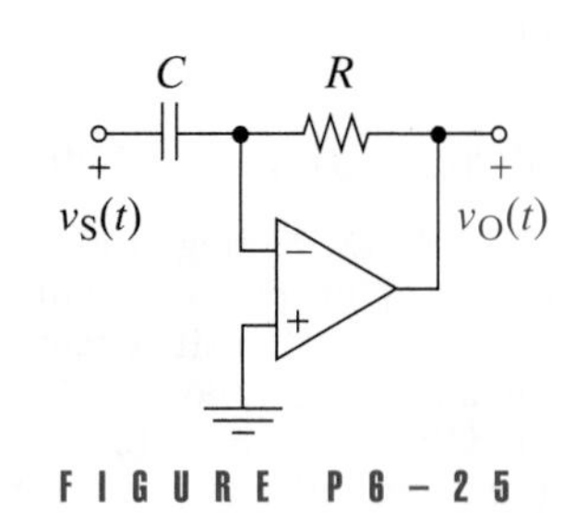

FIGURE P6–25

6–26 The OP AMP differentiator in Figure P6–25 with $R = 50\ \text{k}\Omega$ and $C = 20$ nF has an input waveform shown in Figure P6–26. Sketch $v_O(t)$ for $t > 0$.

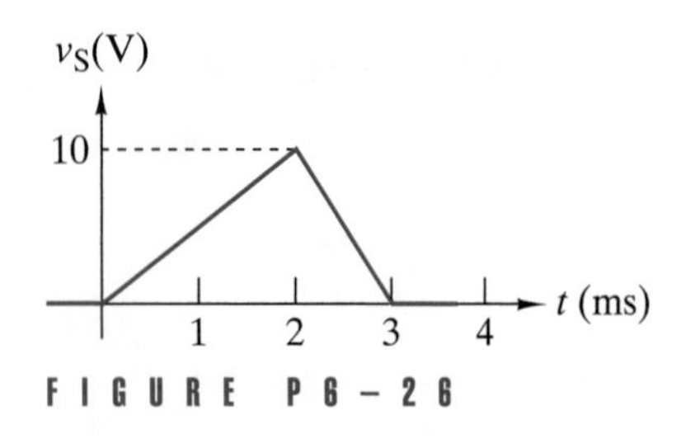

FIGURE P6–26

6–27 The OP AMP differentiator in Figure P6–25 with $R = 100\ \text{k}\Omega$ and $C = 10$ pF has an input $v_S(t) = V_A[\sin(10^6 t)]\,u(t)$ V. Derive an expression for the output voltage when the OP AMP is in its linear mode. If the OP AMP saturates at ± 15 V, what is the maximum value of V_A for linear operation?

6–28 The OP AMP differentiator in Figure P6–25 with $R = 100\ \text{k}\Omega$ and $C = 10$ pF has an input $v_S(t) = 5[\sin(\omega t)]u(t)$ V. Derive an expression for the output voltage when the OP AMP is in its linear mode. If the OP AMP saturates at ± 15 V, what is the maximum value of ω for linear operation?

6–29 Find the input-output relationship of the *RC* OP AMP circuit in Figure P6–29.

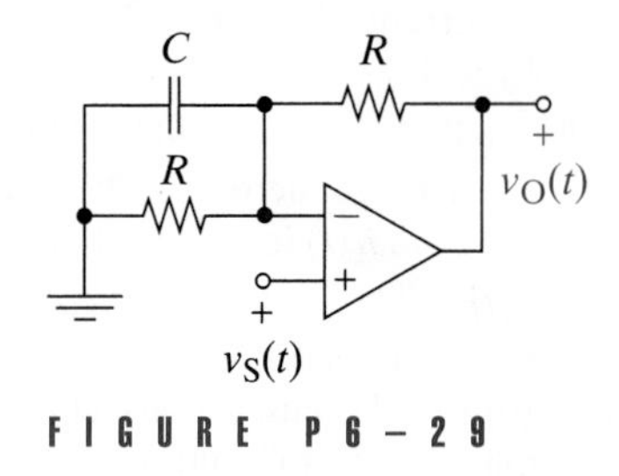

FIGURE P6–29

6–30 Show that the *RC* OP AMP circuit in Figure P6–30 is a noninverting integrator whose input-output relationship is

$$v_O(t) = \frac{1}{RC}\int_0^t v_S(x)dx + v_O(0)$$

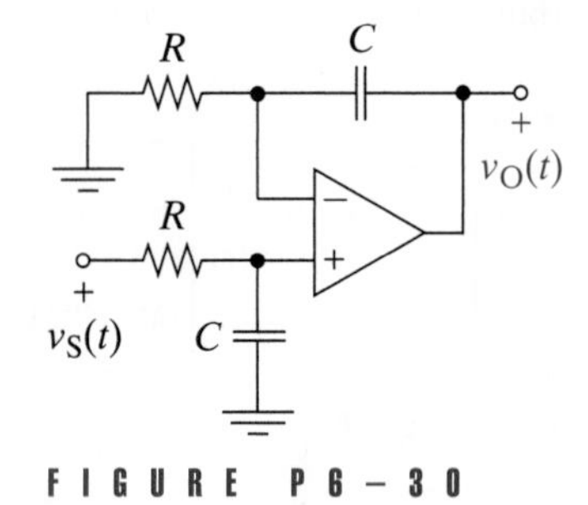

FIGURE P6–30

6–31 Find the input-output relationship of the *RC* OP AMP circuit in Figure P6–31.

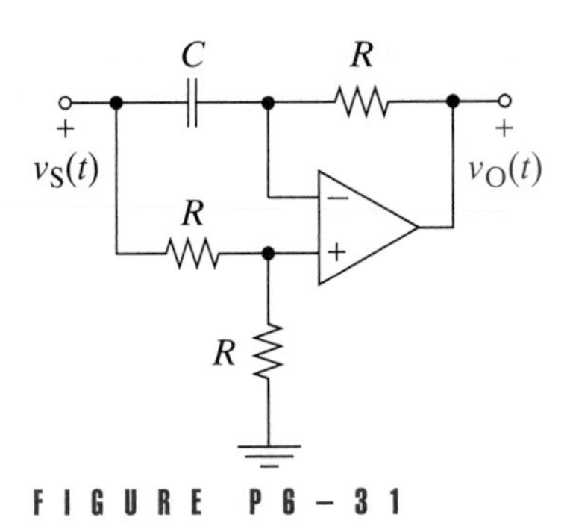

FIGURE P6–31

6–32 **D** Design an *RC* OP AMP circuit to implement the block diagram in Figure P6–32.

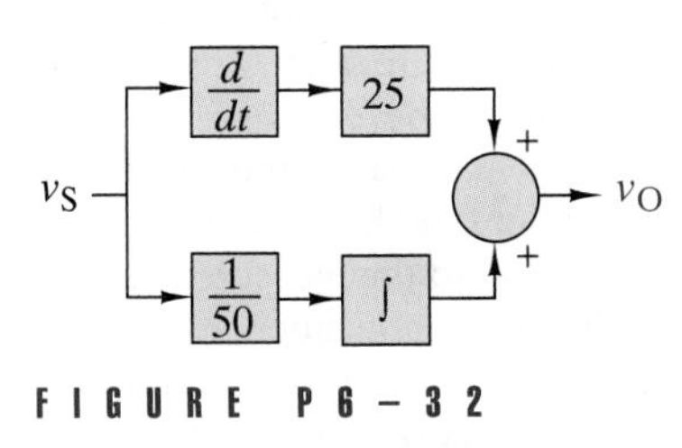

FIGURE P6–32

6–33 **D** Design an *RC* OP AMP circuit to implement the input-output relationship

$$v_O(t) = -5\, v_S(t) + 20\int_0^t v_S(x)dx$$

6–34 **D** Design an *RC* OP AMP circuit to implement the input-output relationship

$$v_O(t) = 5v_S(t) + \frac{1}{20}\frac{dv_S(t)}{dt}$$

6–35 **D** Design a circuit using one OP AMP and one capacitor that implements the input-output relationship

$$v_O(t) = -5\int_0^t v_{S1}(x)dx - 10\int_0^t v_{S2}(x)dx$$

OBJECTIVE 6–3 EQUIVALENT INDUCTANCE AND CAPACITANCE (SECT. 6–4)

(a) Derive equivalence properties of inductors and capacitors or use equivalence properties to simplify *LC* circuits.
(b) Solve for currents and voltages in *RLC* circuits with dc input signals.
See Examples 6–14, 6–15 and Exercises 6–10, 6–11

6–36 Find a single equivalent element for each circuit in Figure P6–36.

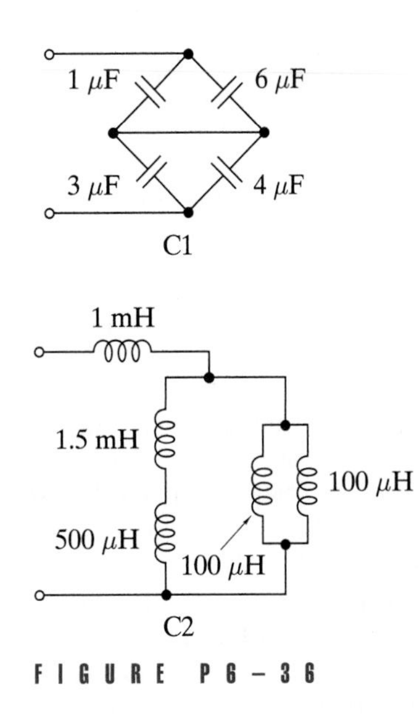

FIGURE P6–36

6–37 Verify Eqs. (6–30) and (6–31).

6–38 A 2-mH inductor is connected in series with a 1-mH inductor and the combination is connected in parallel with a 12-mH inductor. Find the equivalent inductance of the connection.

6–39 A series connection of a 1.8-μF capacitor and a 3.6-μF capacitor is connected in parallel with a series connection of two 1.6-μF capacitors. Find the equivalent capacitance of the connection.

6–40 For the circuit in Figure P6–40, find an equivalent circuit consisting of one inductor and one capacitor.

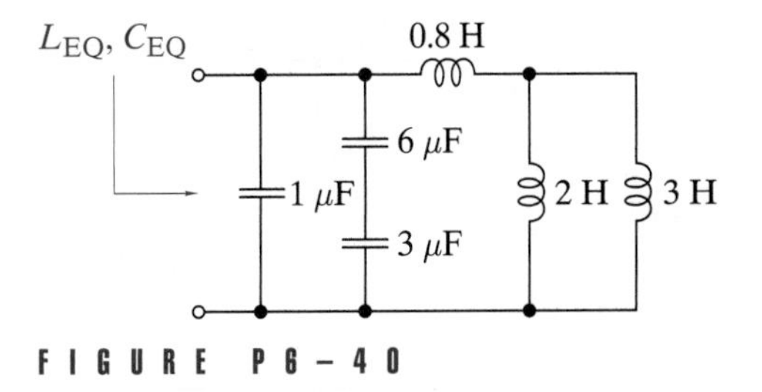

FIGURE P6–40

6–41 Figure P6–41 is the equivalent circuit of a two-wire feed through capacitor.

(a) What is the capacitance between terminal 1 and ground when terminal 2 is open?

(b) What is the capacitance between terminal 1 and ground when terminal 2 is grounded?

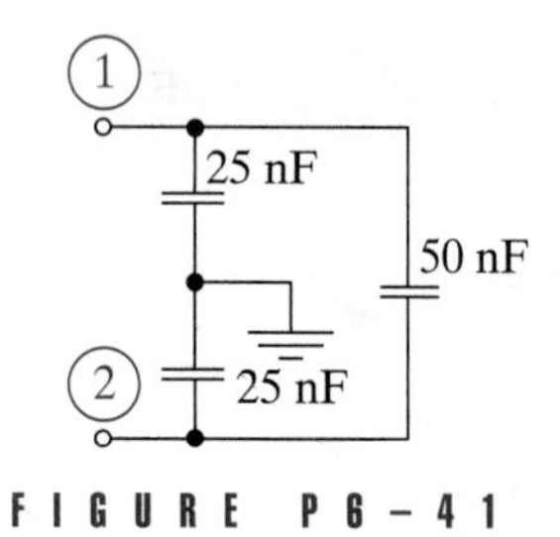

FIGURE P6–41

6–42 **D** A capacitor bank is required that can be charged to 4 kV and store at least 200 J of energy. Design a series/parallel combination that meets the voltage and energy requirements using 20-μF capacitors, each rated at 1.5 kV max.

6–43 **E** A switching power supply requires an inductor that can store at least 1 mJ of energy. A list of available inductors is shown below. Select the inductor that best meets the requirement.

L (μH)	I_{max} (A)
10	9.3
20	7.2
50	5.5
100	4.5
150	3.5
250	2.6
500	1.8

6–44 The circuits in Figure P6–44 are driven by dc sources. Find the current through the 100-Ω resistor under dc conditions.

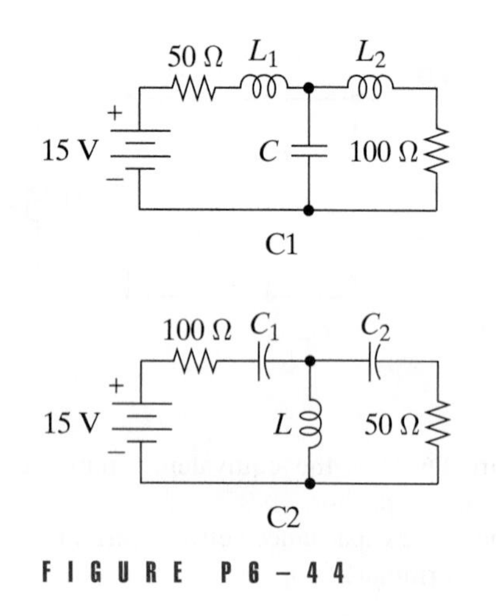

FIGURE P6–44

6–45 The circuit in Figure P6–45 is driven by 15-V dc source. Find the energy stored in the capacitor and inductor under dc conditions.

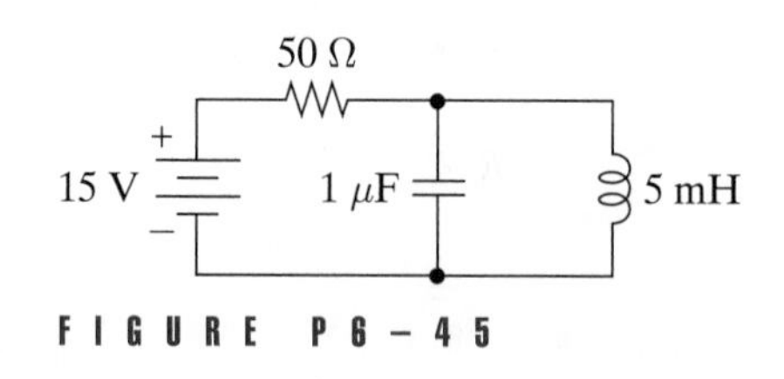

FIGURE P6–45

INTEGRATING PROBLEMS

6–46 **A** Piezoelectric Transducer

Piezoelectric transducers (sensors) measure dynamic phenomena such as pressure and force. These phenomena cause stresses that "squeeze" a quantity of electric charge from piezoelectric material in transducer (the term *piezo* means "squeeze" in Greek). The amount of charge $q(t)$ is directly proportional to the measured variable $x(t)$, that is, $q(t) = \alpha x(t)$. Signal amplification is needed because the amount of charge produced is of the order of pC. Figure P6–46 shows an OP AMP charge amplifier that provides the necessary gain. First show that the OP AMP output is $v_O(t) = -Kq(t)$. Then select a value of C so that the charge amplifier gain is $K = 5$ mV/pC.

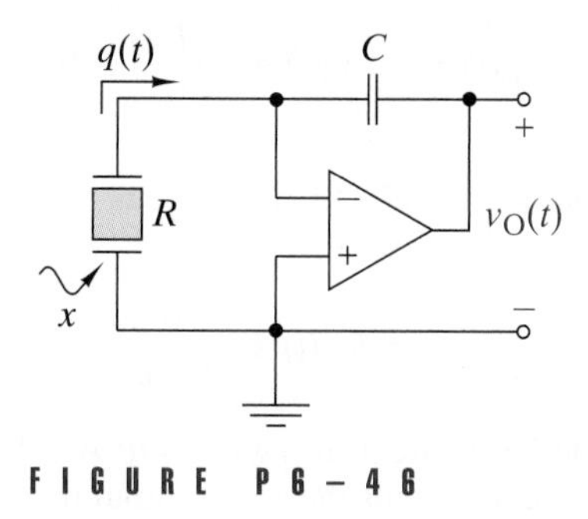

FIGURE P6–46

6–47 **A** LC Circuit Response

At $t = 0$ the switch in Figure P6–47 is closed and thereafter the voltage across the capacitor is

$$v_C(t) = (10 + 40000\,t)\,e^{-4000t}\ \text{V}$$

(a) Use the capacitor's i–v characteristic to find the current $i(t)$ for $t \geq 0$.

(b) Use the inductor's i–v characteristic and $i(t)$ to find $v_L(t)$ for $t \geq 0$.

(c) Use $v_C(t)$, $v_L(t)$, and KVL to find the voltage $v(t)$ delivered to the rest of the circuit.

(d) The $v(t)$ found in part (c) should be proportional to the $i(t)$ found in part (a). If so, what is the equivalent resistance looking into the rest of the circuit?

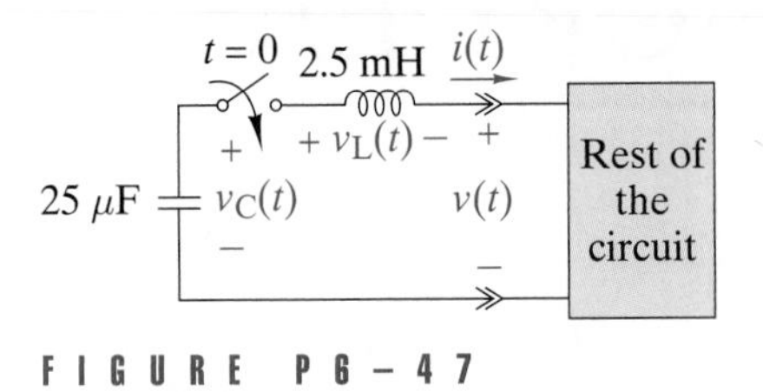

FIGURE P6–47

6–48 A Supercapacitor

Supercapacitors have very large capacitances (typically from 0.1 to 50 F), very long charge holding times, and small sizes, making them useful in nonbattery backup power applications. To measure its capacitance, a supercapacitor is charged to a initial voltage $V_0 = 5.5$ V. At $t = 0$ the device undergoes a constant current discharge of $i_D = 2$ mA. At $t = 2500$ s the voltage remaining on the capacitor is 3 V. Find the device capacitance.

6–49 D Differentiator Design

The input to the *RC* OP AMP differentiator in Figure P6–25 is a sinusoid with a peak-to-peak amplitude of 20 V. Select the values of *R* and *C* so that the OP AMP operates in its linear mode for all input frequencies less than 5 kHz. Assume the OP AMP saturates at ±15 V.

6–50 A E *RC* OP AMP Circuit Design

An upgrade to one of your company's robotics products requires a proportional plus integral compensator that implements the input-output relationship

$$v_O(t) = v_S(t) + 50\int_0^t v_S(x)dx$$

The input voltage $v_S(t)$ comes from an OP AMP, and the output voltage $v_O(t)$ drives a 10-kΩ resistive load. Two competing designs are shown in Figure P6–50. As the project engineer you are responsible for recommending one of these designs for production. Which design would you recommend and why? Your mentor, a wise senior engineer, suggests that you first check that both designs implement the required signal-processing function.

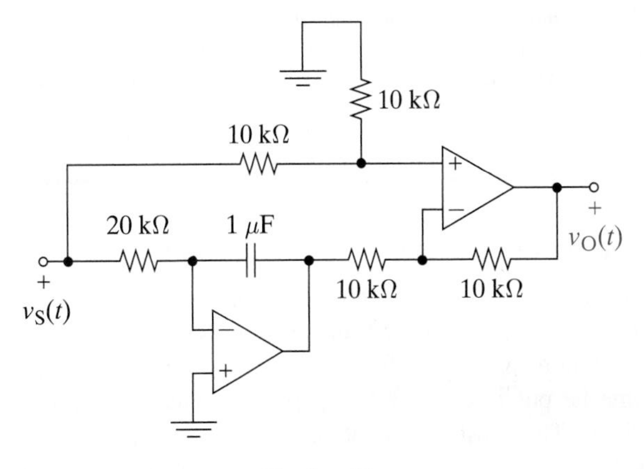

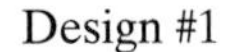

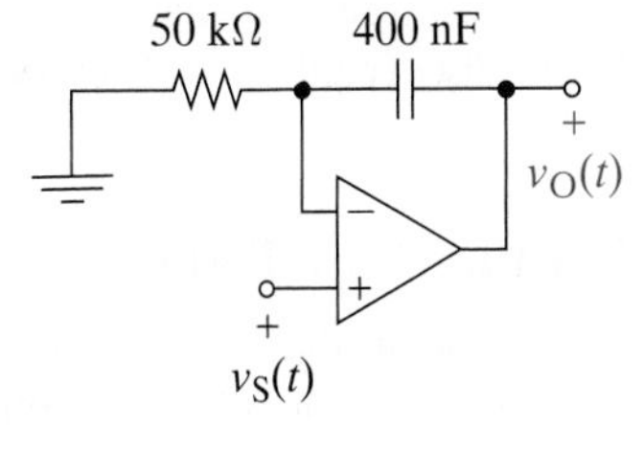

FIGURE P6–50

CHAPTER 7

FIRST- AND SECOND-ORDER CIRCUITS

When a mathematician engaged in investigating physical actions and results has arrived at his own conclusions, may they not be expressed in common language as fully, clearly and definitely as in mathematical formula? If so, would it not be a great boon to such as we to express them so—translating them out of their hieroglyphics that we also might work upon them by experiment.

Michael Faraday, 1857,
British Physicist

Some History Behind This Chapter

Michael Faraday (1791–1867) was appointed a Fellow in the Royal Society at age 32 and was a lecturer at the Royal Institution in London for more than 50 years. During this time he published over 150 papers on chemistry and electricity. The most important of these papers was the series *Experimental Researches in Electricity,* which included a description of his discovery of magnetic induction. A gifted experimentalist, Faraday apparently felt that mathematics obscured the physical truths he discovered through experimentation.

Why This Chapter Is Important Today

OK, this is a tough chapter. It looks at the classical methods of finding the transient response of circuits containing resistors, capacitors, and inductors. Mathematically this requires us to solve first- and second-order differential equations. These solutions help us understand applications such as timing circuits and digital gate delays. It is important to understand first- and second-order transients because we will revisit these concepts frequently in subsequent chapters.

Chapter Sections

7–1 *RC* and *RL* Circuits

7–2 First-Order Circuit Step Response

7–3 Initial and Final Conditions

7–4 First-Order Circuit Sinusoidal Response

7–5 The Series *RLC* Circuit

7–6 The Parallel *RLC* Circuit

7–7 Second-Order Circuit Step Response

7–8 Other Second-Order Circuits

Chapter Learning Objectives

7-1 First-order Circuit Analysis (Sects. 7-1, 7-2, 7-3, 7-4)

Given a first-order *RC* or *RL* circuit:

(a) Find the circuit differential equation, the circuit time constant, and the initial conditions (if not given).
(b) Find the zero-input response.
(c) Find the complete response for step function and sinusoidal inputs.

7-2 First-order Circuit Responses (Sects. 7-1, 7-2, 7-3)

Given responses in a first-order *RC* or *RL* circuit:

(a) Find the circuit parameters or other responses.
(b) Design a circuit to produce the given responses.

7-3 Second-order Circuit Analysis (Sects. 7-5, 7-6, 7-7, 7-8)

Given a second-order circuit:

(a) Find the circuit differential equation.
(b) Find the circuit natural frequencies and the initial conditions (if not given).
(c) Find the zero-input response.
(d) Find the complete response for a step function input.

7-4 Second-order Circuit Responses (Sects. 7-5, 7-6, 7-7, 7-8)

Given responses in a second-order *RLC* circuit:

(a) Find the circuit parameters or other responses.
(b) Design a circuit to produce the given responses.

7–1 *RC* AND *RL* CIRCUITS

The flow diagram in Figure 7–1 shows the two major steps in the analysis of a dynamic circuit. In the first step we use device and connection equations to formulate a differential equation describing the circuit. In the second step we solve the differential equation to find the circuit response. In this chapter we examine basic methods of formulating circuit differential equations and the time-honored, classical methods of solving for responses. Solving for the responses of simple dynamic circuits gives us insight into the physical behavior of the basic modules of the complex networks in subsequent chapters. This insight will help us correlate circuit behavior with the results obtained by other methods of dynamic circuit analysis. The treatment of other methods includes phasor circuit analysis (Chapter 8) and the Laplace transform methods (beginning in Chapter 9).

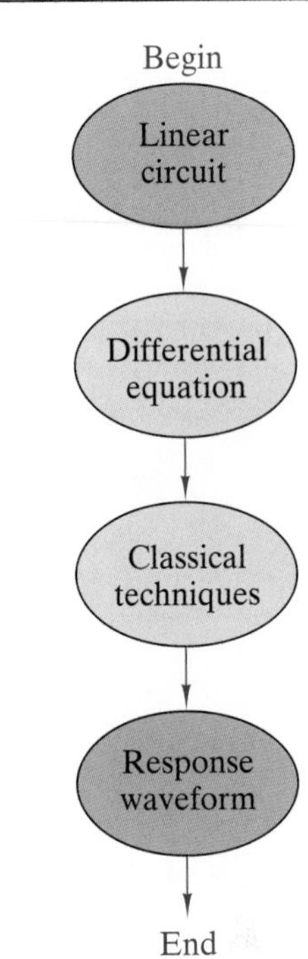

FIGURE 7–1 *Flow diagram for dynamic circuit analysis.*

FORMULATING *RC* AND *RL* CIRCUIT EQUATIONS

RC and *RL* circuits contain linear resistors and a single capacitor or a single inductor. Figure 7–2 shows how we can divide *RC* and *RL* circuits into two parts: (1) the dynamic element and (2) the rest of the circuit, containing only linear resistors and sources. To formulate the equation governing either of these circuits, we replace the resistors and sources by their Thévenin and Norton equivalents shown in Figure 7–2.

Dealing first with the *RC* circuit in Figure 7–2(a), we note that the Thévenin equivalent source is governed by the constraint

$$R_T i(t) + v(t) = v_T(t) \tag{7–1}$$

The capacitor *i–v* constraint is

$$i(t) = C\frac{dv(t)}{dt} \tag{7–2}$$

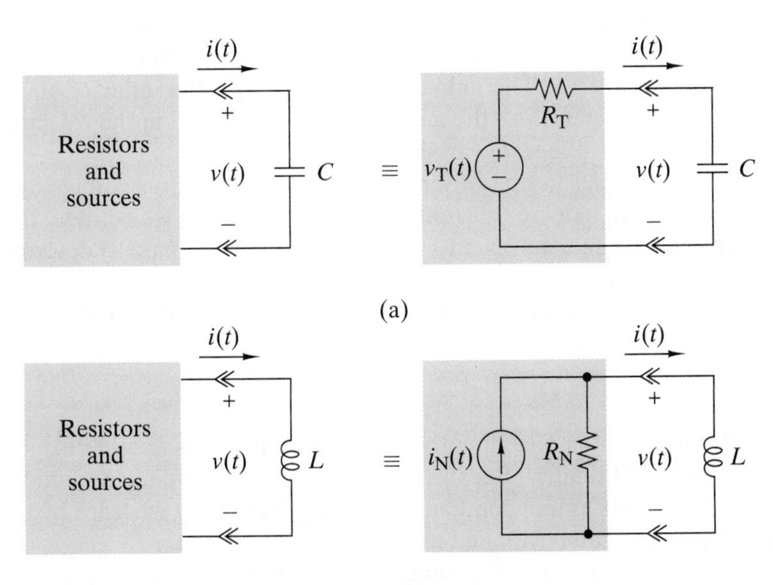

FIGURE 7–2 *First-order circuits: (a)* RC *circuit. (b)* RL *circuit.*

Substituting the i–v constraint into the source constraint yields

$$R_T C \frac{dv(t)}{dt} + v(t) = v_T(t) \tag{7–3}$$

Combining the source and element constraints produces the equation governing the *RC* series circuit. The unknown in Eq. (7–3) is the capacitor voltage $v(t)$ that determines the amount of energy stored in the *RC* circuit and is referred to as the **state variable**.

Mathematically, Eq. (7–3) is a first-order linear differential equation with constant coefficients. The equation is first order because the first derivative of the dependent variable is the highest-order derivative in the equation. The product $R_T C$ is a constant coefficient because it depends on fixed circuit parameters. The signal $v_T(t)$ is the Thévenin equivalent of the independent sources driving the circuit. The voltage $v_T(t)$ is the input, and the capacitor voltage $v(t)$ is the circuit response.

The Norton equivalent source in the *RL* circuit in Figure 7–2(b) is governed by the constraint

$$G_N v(t) + i(t) = i_N(t) \tag{7–4}$$

The element constraint for the inductor can be written

$$v(t) = L \frac{di(t)}{dt} \tag{7–5}$$

Combining the element and source constraints produces the differential equation for the *RL* circuit:

$$G_N L \frac{di(t)}{dt} + i(t) = i_N(t) \tag{7–6}$$

The response of the *RL* circuit is also governed by a first-order linear differential equation with constant coefficients. The dependent variable in Eq. (7–6) is the inductor current. The circuit parameters enter as the constant product $G_N L$, and the driving forces are represented by a Norton equivalent current $i_N(t)$. The unknown in Eq. (7–6) is the inductor current $i(t)$. This current determines the amount of energy stored in the *RL* circuit and is referred to as the **state variable**.

The state variables in first-order circuits are the capacitor voltage in the *RC* circuit and the inductor current in the *RL* circuit. As we will see, these state variables contain sufficient information about the past to determine future circuit responses.

We observe that Eqs. (7–3) and (7–6) have the same form. In fact, interchanging the quantities

$$R_T \leftrightarrow G_N \qquad C \leftrightarrow L \qquad v \leftrightarrow i \qquad v_T \leftrightarrow i_N$$

converts one equation into the other. This interchange is another example of the principle of duality. Because of duality we do not need to study the *RC* and *RL* circuits as independent problems. Everything we learn by solving the *RC* circuit can be applied to the *RL* circuit as well.

We refer to the *RC* and *RL* circuits as **first-order circuits** because they are described by a first-order differential equation. The first-order differential equations in Eqs. (7–3) and (7–6) describe general *RC* and *RL* circuits shown in Figure 7–2. Any circuit containing a single capacitor or inductor and linear resistors and sources is a first-order circuit.

ZERO-INPUT RESPONSE OF FIRST-ORDER CIRCUITS

The response of a first-order circuit is found by solving the circuit differential equation. For the *RC* circuit the response $v(t)$ must satisfy the differential equation in Eq. (7–3) and the initial condition $v(0)$. By examining Eq. (7–3) we see that the response depends on three factors:

1. The inputs driving the circuit $v_T(t)$
2. The values of the circuit parameters R_T and C
3. The value of $v(t)$ at $t = 0$ (i.e., the initial condition)

The first two factors apply to any linear circuit, including resistance circuits. The third factor relates to the initial energy stored in the circuit. The initial energy can cause the circuit to have a nonzero response even when the input $v_T(t) = 0$ for $t \geq 0$. The existence of a response with no input is something new in our study of linear circuits.

To explore this discovery we find the **zero-input response**. Setting all independent sources in Figure 7–2 to zero makes $v_T = 0$ in Eq. (7–3):

$$R_T C \frac{dv}{dt} + v = 0 \tag{7–7}$$

Mathematically, Eq. (7–7) is a **homogeneous equation** because the right side is zero. The classical approach to solving a linear homogeneous differential equation is to try a solution in the form of an exponential

$$v(t) = Ke^{st} \tag{7–8}$$

where K and s are constants to be determined.

The form of the homogenous equation suggests an exponential solution for the following reasons. Equation (7–7) requires that $v(t)$ plus $R_T C$ times its derivative must add to zero for all time $t \geq 0$. This can only occur if $v(t)$ and its derivative have the same form. In Chapter 5 we saw that an exponential signal and its derivative are both of the form $e^{-t/Tc}$. Therefore, the exponential is a logical starting place.

If Eq. (7–8) is indeed a solution, then it must satisfy the differential equation in Eq. (7–7). Substituting the trial solution into Eq. (7–7) yields

$$R_T CKse^{st} + Ke^{st} = 0$$

or

$$Ke^{st}(R_T Cs + 1) = 0$$

The exponential function e^{st} cannot be zero for all t. The condition $K = 0$ is a trivial solution because it implies that $v(t)$ is zero for all time t. The only nontrivial way to satisfy the equation involves the condition

$$R_T Cs + 1 = 0 \tag{7–9}$$

Equation (7–9) is the circuit **characteristic equation** because its root determines the attributes of $v(t)$. The characteristic equation has a single root at $s = -1/R_T C$, so the zero-input response of the *RC* circuit has the form

$$v(t) = Ke^{-t/R_T C} \qquad t \geqslant 0$$

The constant K can be evaluated using the value of $v(t)$ at $t = 0$. Using the notation $v(0) = V_0$ yields

$$v(0) = Ke^0 = K = V_0$$

The final form of the zero-input response is

$$v(t) = V_0 e^{-t/R_T C} \qquad t \geqslant 0 \tag{7–10}$$

The zero-input response of the RC circuit is the familiar exponential waveform shown in Figure 7–3. At $t = 0$ the exponential response starts out at $v(0) = V_0$ and then decays to zero at $t \to \infty$. The time constant $T_C = R_T C$ depends only on fixed circuit parameters. From our study of the exponential signals in Chapter 5, we know that the $v(t)$ decays to about 37% of its initial amplitude in one time constant and to essentially zero after about five time constants. The zero-input response of the RC circuit is determined by two quantities: (1) the circuit time constant and (2) the value of the capacitor voltage at $t = 0$.

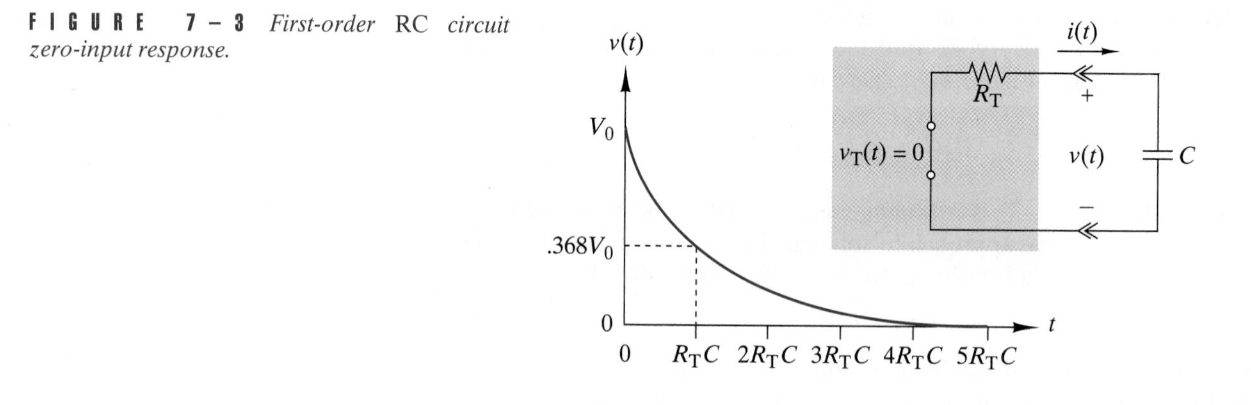

FIGURE 7–3 *First-order* RC *circuit zero-input response.*

The zero-input response of the RL circuit in Figure 7–2(b) is found by setting the Norton current $i_N(t) = 0$ in Eq. (7–6).

$$G_N L \frac{di}{dt} + i = 0 \tag{7–11}$$

The unknown in this homogeneous differential equation is the inductor current $i(t)$. Equation (7–11) has the same form as the homogeneous equation for the RC circuit, which suggests a trial solution of the form

$$i(t) = Ke^{st}$$

where K and s are constants to be determined. Substituting the trial solution into Eq. (7–11) yields the RL circuit characteristic equation.

$$G_N Ls + 1 = 0 \tag{7–12}$$

The root of this equation is $s = -1/G_N L$. Denoting the initial value of the inductor current by I_0, we evaluate the constant K:

$$i(0) = I_0 = Ke^0 = K$$

The final form of the zero-input response of the RL circuit is

$$i(t) = I_0 e^{-t/G_N L} \qquad t \geqslant 0 \tag{7–13}$$

For the RL circuit the zero-input response of the state variable $i(t)$ is an exponential function with a time constant of $T_C = G_N L = L/R_T$. This response connects the initial state $i(0) = I_0$ with the final state $i(\infty) = 0$.

The zero-input responses in Eqs. (7–10) and (7–13) show the duality between first-order RC and RL circuits. These results point out that the zero-input response in a first-order circuit depends on two quantities: (1) the circuit time constant and (2) the value of the state variable at $t = 0$. Capacitor voltage and inductor current are called state variables because they determine the amount of energy stored in the circuit at any time t. The following examples show that the zero-input response of the state variable provides enough information to determine the zero-input response of every other voltage and current in the circuit.

EXAMPLE 7–1

The switch in Figure 7–4 is closed at $t = 0$, connecting a capacitor with an initial voltage of 30 V to the resistances shown. Find the responses $v_C(t)$, $i(t)$, $i_1(t)$, and $i_2(t)$ for $t \geq 0$.

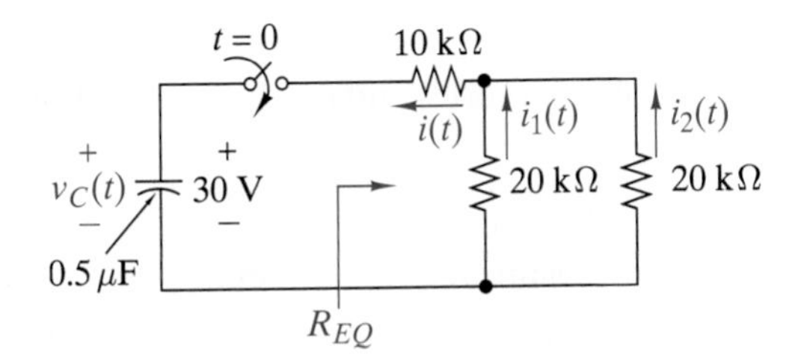

FIGURE 7–4

SOLUTION:

This problem involves the zero-input response of an RC circuit since there is no independent source in the circuit. To find the required responses, we first determine the circuit time constant with the switch closed ($t \geq 0$). The equivalent resistance seen by the capacitor is

$$R_{EQ} = 10 + (20\|20) = 20 \text{ k}\Omega$$

For $t \geq 0$ the circuit time constant is

$$T_C = R_T C = 20 \times 10^3 \times 0.5 \times 10^{-6} = 10 \text{ ms}$$

The initial capacitor voltage is given by $V_0 = 30$ V. Using Eq. (7–10), the zero-input response of the capacitor voltage is

$$v_C(t) = 30e^{-100t} \text{ V} \qquad t \geqslant 0$$

The capacitor voltage provides the information needed to solve for all other zero-input responses. The current $i(t)$ through the capacitor is

$$\begin{aligned} i(t) &= C\frac{dv_C}{dt} = (0.5 \times 10^{-6})(30)(-100)\, e^{-100t} \\ &= -1.5 \times 10^{-3}\, e^{-100t} \text{ A} \qquad t \geqslant 0 \end{aligned}$$

The minus sign means the actual current direction is opposite of the reference direction shown in Figure 7–4. The minus sign makes physical sense because the initial voltage on the capacitor is positive, which forces current into the resistances to the right of the switch. The other current responses are found by current division.

$$i_1(t) = i_2(t) = \frac{20}{20 + 20} i(t) = -0.75 \times 10^{-3} e^{-100t} \text{ A} \qquad t \geq 0$$

Notice the analysis pattern. We first determine the zero-input response of the capacitor voltage. The state variable response together with resistance circuit analysis techniques were then used to find other voltages and currents. The circuit time constant and the value of the state variable at $t = 0$ provide enough information to determine the zero-input response of every voltage or current in the circuit. ■

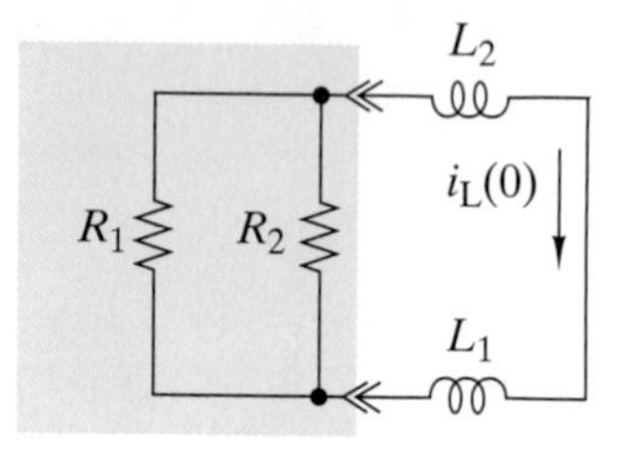

Given circuit

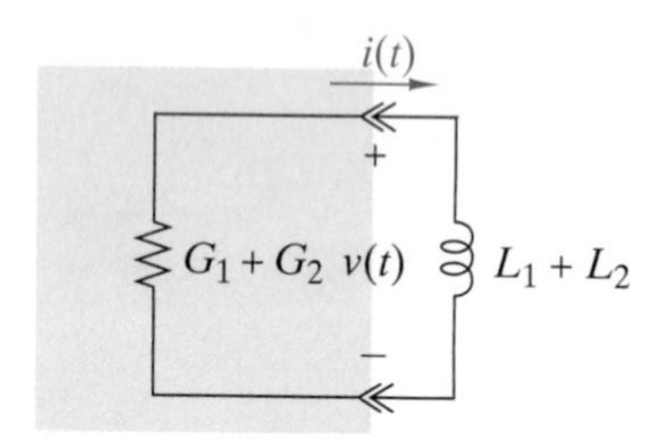

Equivalent circuit

FIGURE 7–5

EXAMPLE 7–2

Find the response of the state variable of the *RL* circuit in Figure 7–5 using $L_1 = 10$ mH, $L_2 = 30$ mH, $R_1 = 2$ kΩ, $R_2 = 6$ kΩ, and $i_L(0) = 100$ mA.

SOLUTION:

The inductors are connected in series and can be replaced by an equivalent inductor

$$L_{EQ} = L_1 + L_2 = 10 + 30 = 40 \text{ mH}$$

Likewise, the resistors are connected in parallel and the conductance seen by L_{EQ} is

$$G_{EQ} = G_1 + G_2 = 10^{-3}/2 + 10^{-3}/6 = 2 \times 10^{-3}/3 \text{ S}$$

Figure 7–5 shows the resulting equivalent circuit. The interface signals $v(t)$ and $i(t)$ are the voltage across and current through $L_{EQ} = L_1 + L_2$. The time constant of the equivalent *RL* circuit is

$$T_C = G_{EQ} L_{EQ} = 8 \times 10^{-5}/3 \text{ s} = 1/37500 \text{ s}$$

The initial current through L_{EQ} is $i_L(0) = 0.1$ A. Using Eq. (7–13) with $I_0 = 0.1$ yields the zero-state response of the inductor current.

$$i(t) = 0.1 e^{-37500t} \text{ A} \qquad t \geq 0$$

Given the state variable response, we can find every other response in the original circuit. For example, by KCL and current division the currents through R_1 and R_2 are

$$i_{R_1}(t) = \frac{R_2}{R_1 + R_2} i(t) = 0.075\, e^{-37500t} \text{ A} \qquad t \geq 0$$

$$i_{R_2}(t) = \frac{R_1}{R_1 + R_2} i(t) = 0.025\, e^{-37500t} \text{ A} \qquad t \geq 0$$

■

Example 7–2 illustrates an important point. The *RL* circuit in Figure 7–5 is a first-order circuit even though it contains two inductors. The two inductors are connected in series and can be replaced by a single equivalent inductor. In general, capacitors or inductors in series and parallel can be replaced by a single equivalent element. Thus, any circuit containing the *equivalent* of a single inductor or a single capacitor is a first-order circuit.

Sometimes it may be difficult to determine the Thévenin or Norton equivalent seen by the dynamic element in a first-order circuit. In such cases we use other circuit analysis techniques to derive the differential equation in terms of a more convenient signal variable. For example, the OP AMP *RC* circuit in Figure 7–6 is a first-order circuit because it contains a single capacitor.

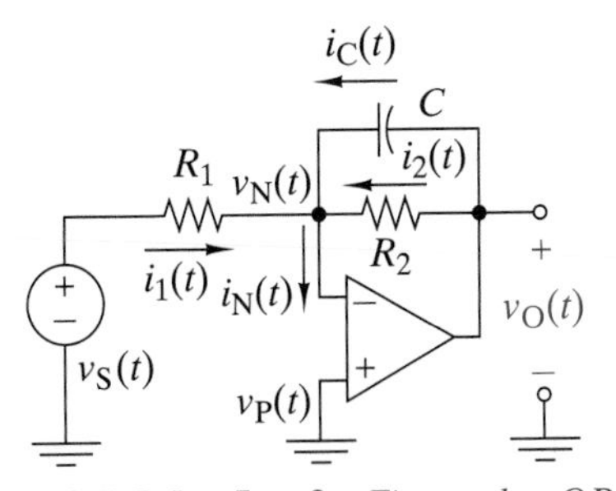

FIGURE 7–6 *First-order OP AMP RC circuit.*

From previous experience we know that the key to analyzing an inverting OP AMP circuit is to write a KCL equation at the inverting input. The sum of currents entering the inverting input is

$$\underbrace{G_1(v_S - v_N)}_{i_1(t)} + \underbrace{G_2(v_O - v_N)}_{i_2(t)} + \underbrace{C\frac{d(v_O - v_N)}{dt}}_{i_C(t)} - i_N(t) = 0$$

The element equations for the OP AMP are $i_N(t) = 0$ and $v_N(t) = v_P(t)$. However, the noninverting input is grounded; hence $v_N(t) = v_P(t) = 0$. Substituting the OP AMP element constraints into the KCL constraint yields

$$G_1 v_S + G_2 v_O + C\frac{dv_O}{dt} = 0$$

which can be rearranged in standard form as

$$R_2 C\frac{dv_O}{dt} + v_O = -\frac{R_2}{R_1}v_S(t) \qquad (7\text{–}14)$$

The unknown Eq. (7–14) is the OP AMP output voltage rather than the capacitor voltage. The form of the differential equation indicates that the circuit time constant is $T_C = R_2C$.

EXAMPLE 7–3

Use Orcad Capture to calculate the response $v_C(t)$ in the circuit in Figure 7–4 for $t \geq 0$.

SOLUTION:

An analytical solution for this problem was given in Example 7–1. The problem is repeated here to introduce computer simulation of dynamic circuits. The simulation option used is called *Time Domain (Transient)* analysis, which predicts the variation of currents, voltages, and powers as a function of time. Figure 7–7 shows the circuit diagram as drawn in Orcad Capture *Schematics*. This diagram does not include the switch in Figure 7–4 because for $t \geq 0$ the switch is closed. The netlist in Figure 7–7 shows that the 0.5 μF is connected between Nodes N00102 and Node 0 (ground) and has an initial condition of IC = 30 V. The **Property Editor** dialog box in Figure 7–7 shows how the capacitor's initial condition attribute is given a numerical value. The **Property Editor** is accessed by double-clicking on the desired part—C1 in this case (highlighted by a "dashed square" in the figure). The 30-V initial condition is entered in the **IC** box, and then the

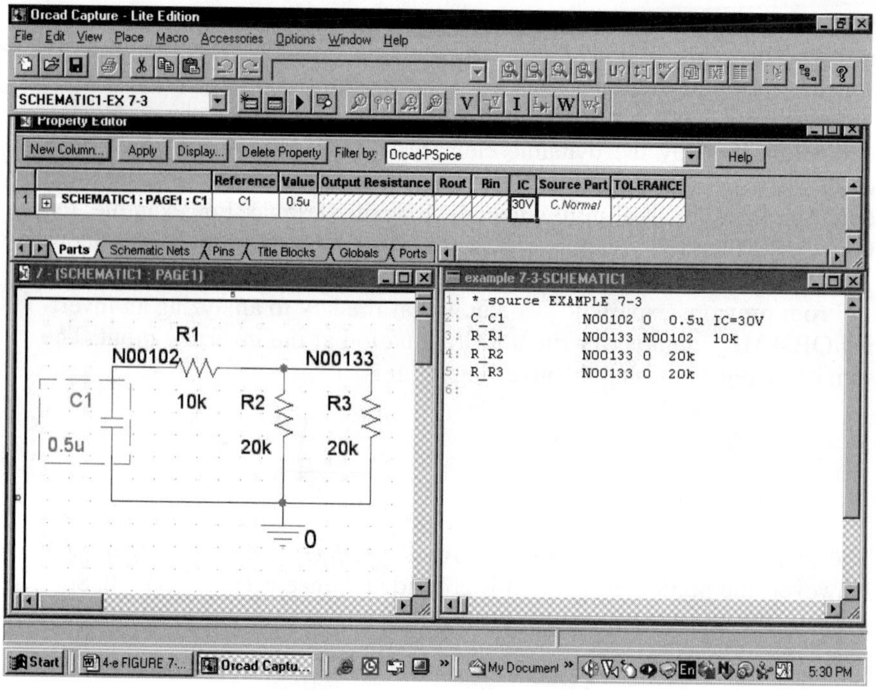

Orcad Capture - Lite Edition

File Edit View Place Macro Accessories Options Window Help

SCHEMATIC1-EX 7-3

Property Editor

New Column... Apply Display... Delete Property Filter by: Orcad-PSpice Help

		Reference	Value	Output Resistance	Rout	Rin	IC	Source Part	TOLERANCE
1	SCHEMATIC1 : PAGE1 : C1	C1	0.5u				30V	C.Normal	

Parts Schematic Nets Pins Title Blocks Globals Ports

/ - (SCHEMATIC1 : PAGE1)

R1 N00102 N00133 C1 10k R2 R3 0.5u 20k 20k 0

example 7-3-SCHEMATIC1

```
1: * source EXAMPLE 7-3
2: C_C1      N00102 0  0.5u IC=30V
3: R_R1      N00133 N00102  10k
4: R_R2      N00133 0  20k
5: R_R3      N00133 0  20k
6:
```

Start 4-e FIGURE 7-... Orcad Captu... My Document 5:30 PM

FIGURE 7–7 *Orcad Capture schematic and netlist..*

Apply button is pressed. With capacitors and inductors, simulation programs require the user to specify both the element value and the initial condition attributes.

Once the circuit diagram is complete, we must set up a transient analysis run. Selecting **PSpice/New Simulation/Time Domain (Transient)** from the schematics menu bar brings up the dialog box in Figure 7–8. The key entry in this box is the **Run to time** which specifies the time duration of

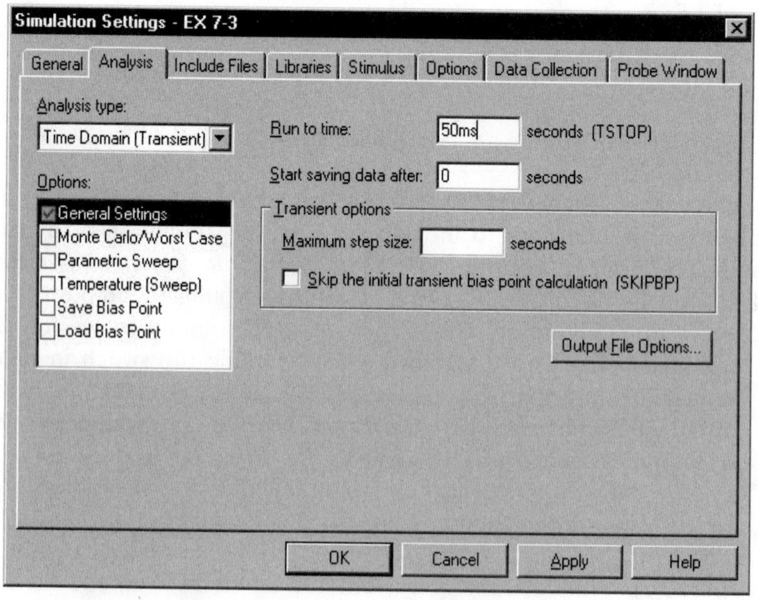

FIGURE 7–8 *Time domain analysis setup.*

Simulation Settings - EX 7-3

General Analysis Include Files Libraries Stimulus Options Data Collection Probe Window

Analysis type: Time Domain (Transient)

Run to time: 50ms seconds (TSTOP)

Start saving data after: 0 seconds

Options:
General Settings
Monte Carlo/Worst Case
Parametric Sweep
Temperature (Sweep)
Save Bias Point
Load Bias Point

Transient options
Maximum step size: seconds
Skip the initial transient bias point calculation (SKIPBP)

Output File Options...

OK Cancel Apply Help

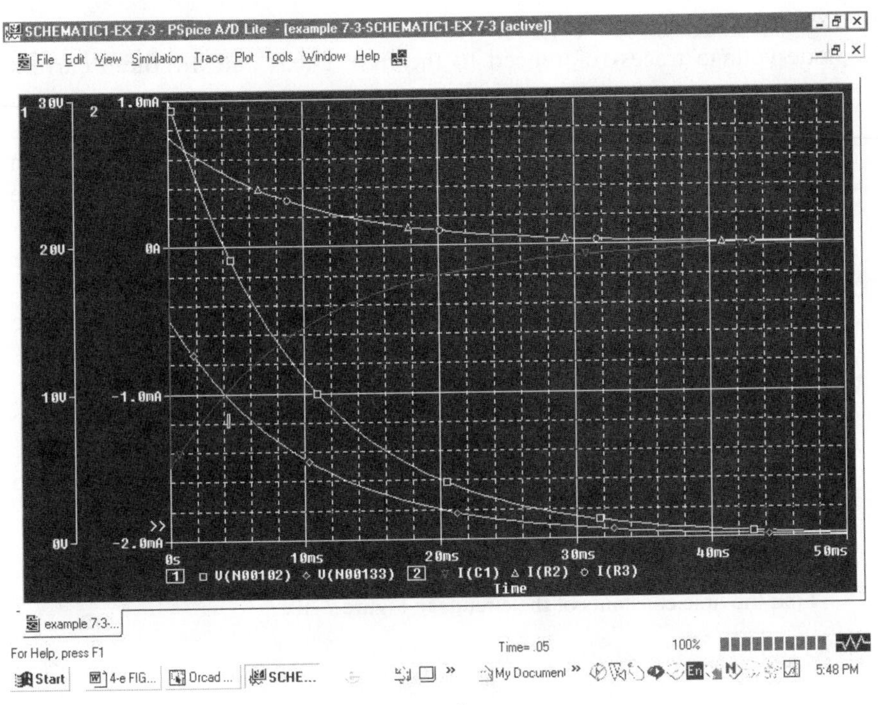

FIGURE 7–9 *Simulation response plots.*

simulation. A PSpice transient analysis always starts at $t = 0$ and runs to $t =$ **(TSTOP)**. In the present case we know that the circuit time constant is 10 ms, so we can safely specify **Run to time** $= 5T_C = 50$ ms.[1]

The **Start saving data after [ ] seconds** and the **Maximum step size:[ ] seconds** parameters in Figure 7–8 are not important to us here because we observe the calculated response using Orcad **Probe**.

With the circuit defined and the transient analysis run set up, we can start Orcad Capture PSpice by selecting **PSpice/Run** from the schematics menu bar. Orcad Capture **Probe** starts automatically once PSpice successfully completes the transient simulation. The results shown in Figure 7–9 were created using the probe graphical interface. **Probe** first displays a blank plot with only the time axis labeled. To add circuit responses to the plot, select **Trace/Add Trace**. In this example this brings up a menu of 28 possible responses to choose from. Selecting node voltages V(N00102) and V(N00123) produces the two exponential voltage traces in Figure 7–9. We can also display current responses on the same plot. To display voltages and currents together, we first select **Plot** and **Add Y Axis**; this adds a second axis that is automatically scaled to the values of the currents. Adding an axis for currents must be done after selecting the voltage responses but before selecting the current responses. The legend at the bottom of

---

1 Specifying a transient analysis **Run to time** presents a dilemma. We must know something about the response to specify an appropriate simulation time. If we know everything about the response, there is no need to run the simulation. If we know absolutely nothing, we may specify a simulation time that is too short or too long to observe the response. Using circuit simulation tools effectively requires that we operate in a wide gray area between no knowledge and complete knowledge.

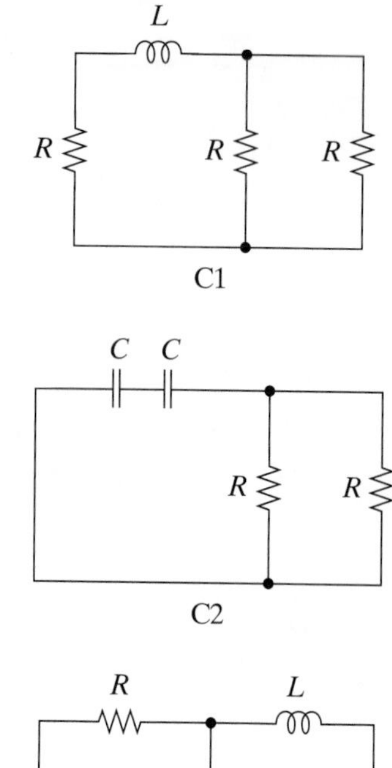

FIGURE 7–10

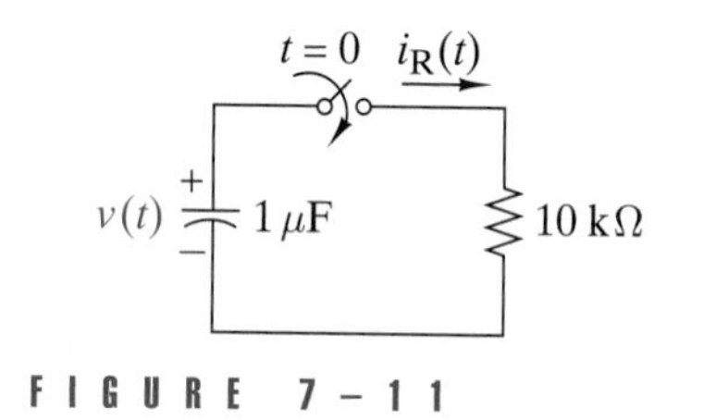

FIGURE 7–11

Figure 7–9 indicates that we have chosen a total of five responses—the two node-voltage traces referenced to the first Y-axis and the three current traces referenced to the second Y-axis.

The node voltage V(N00102) is the voltage across the capacitor. This trace begins at 30 V and decays to zero in about 50 ms. This agrees with our analytical solution $v_C(t) = 30\ e^{-100t}$ V, $t \geq 0$. The node voltage V(N00133) is the voltage across the two 20-kΩ resistors in parallel. This exponentially decaying trace begins at 15 V and is exactly half of the capacitor voltage because of the voltage divider formed by the three resistors. The capacitor current trace I(C1) is negative, indicating that the capacitor is delivering power to the circuit. The two resistor currents I(R2) and I(R3) are positive, indicating that they are absorbing power. The traces for I(R2) and I(R3) fall on top of each other because the two resistors have the same value. Although not shown in the figure, the waveform of the power provided or dissipated by each element can also be displayed.

Exercise 7–1

Find the time constants of the circuits in Figure 7–10.

Answers: C1: $\dfrac{2L}{3R}$ C2: $\dfrac{RC}{4}$ C3: $\dfrac{L}{4R}$

Exercise 7–2

The switch in Figure 7–11 closes at $t = 0$. For $t \geq 0$ the current through the resistor is $i_R(t) = e^{-100t}$ mA.

(a) What is the capacitor voltage at $t = 0$?
(b) Write an equation for $v(t)$ for $t \geq 0$.
(c) Write an equation for the power absorbed by the resistor for $t \geq 0$.
(d) How much energy does the resistor dissipate for $t \geq 0$?
(e) How much energy is stored in the capacitor at $t = 0$?

Answers:

(a) 10 V
(b) $v(t) = 10e^{-100t}$ V
(c) $p_R(t) = 10e^{-200t}$ mW
(d) 50 μJ
(e) 50 μJ

Exercise 7–3

For $t > 0$ the current through the 40-mH inductor in a first-order circuit is $20e^{-500t}$ mA.

(a) What is the circuit time constant?
(b) How much energy is stored in the inductor at $t = 0$, $t = T_C$, and $t = 5T_C$?
(c) Write an equation for the voltage across the inductor.
(d) What is the equivalent resistance seen by the inductor?

Answers:

(a) 2 ms
(b) 8, 1.08, 0.000363 μJ
(c) $-400e^{-500t}$ mV
(d) 20 Ω

7–2 FIRST-ORDER CIRCUIT STEP RESPONSE

Linear circuits are often characterized by applying step function and sinusoid inputs. This section introduces the step response of first-order circuits. Later in this chapter we treat the sinusoidal response of first-order circuits and step response of second-order circuits. The step response analysis introduces the concepts of forced, natural, and zero-state responses that appear extensively in later chapters.

Our development of first-order step response treats the *RC* circuit in detail and then summarizes the corresponding results for its dual, the *RL* circuit. When the input to the *RC* circuit in Figure 7–2 is a step function, we can write the Thévenin source as $v_T(t) = V_A u(t)$. The circuit differential equation in Eq. (7–3) becomes

$$R_T C \frac{dv}{dt} + v = V_A u(t) \tag{7–15}$$

The step response is a function $v(t)$ that satisfies this differential equation for $t \geq 0$ and meets the initial condition $v(0)$. Since $u(t) = 1$ for $t \geq 0$ we can write Eq. (7–15) as

$$R_T C \frac{dv(t)}{dt} + v(t) = V_A \qquad \text{for } t \geqslant 0 \tag{7–16}$$

Mathematics provides a number of approaches to solving this equation, including separation of variables and integrating factors. However, because the circuit is linear, we chose a method that uses superposition to divide solution $v(t)$ into two components:

$$v(t) = v_N(t) + v_F(t) \tag{7–17}$$

The first component $v_N(t)$ is the **natural response** and is the general solution of Eq. (7–16) when the input is set to zero. The natural response has its origin in the physical characteristic of the circuit and does not depend on the form of the input. The component $v_F(t)$ is the **forced response** and is a particular solution of Eq. (7–16) when the input is the step function. We call this the forced response because it represents what the circuit is compelled to do by the form of the input.

Finding the natural response requires the general solution of Eq. (7–16) with the input set to zero:

$$R_T C \frac{dv_N(t)}{dt} + v_N(t) = 0 \qquad t \geqslant 0$$

But this is the homogeneous equation that produces the zero-input response in Eq. (7–8). Therefore, we know that the natural response takes the form

$$v_N(t) = Ke^{-t/R_T C} \qquad t \geqslant 0 \tag{7–18}$$

This is a general solution of the homogeneous equation because it contains an arbitrary constant *K*. At this point we cannot evaluate *K* from the initial condition, as we did for the zero-input response. The initial condition applies to the total response (natural plus forced), and we have yet to find the forced response.

Turning now to the forced response, we seek a particular solution of the equation

$$R_T C \frac{dv_F(t)}{dt} + v_F(t) = V_A \qquad t \geq 0 \tag{7–19}$$

The equation requires that a linear combination of $v_F(t)$ and its derivative equal a constant V_A for $t \geq 0$. Setting $v_F(t) = V_A$ meets this condition since $dv_F/dt = dV_A/dt = 0$. Substituting $v_F = V_A$ into Eq. (7–19) reduces it to the identity $V_A = V_A$.

Now combining the forced and natural responses, we obtain

$$\begin{aligned} v(t) &= v_N(t) + v_F(t) \\ &= Ke^{-t/R_T C} + V_A \qquad t \geq 0 \end{aligned}$$

This equation is the general solution for the step response because it satisfies Eq. (7–16) and contains an arbitrary constant K. This constant can now be evaluated using the initial condition:

$$v(0) = V_0 = Ke^0 + V_A = K + V_A$$

The initial condition requires that $K = (V_0 - V_A)$. Substituting this conclusion into the general solution yields the step response of the *RC* circuit.

$$v(t) = (V_0 - V_A)e^{-t/R_T C} + V_A \qquad t \geq 0 \tag{7–20}$$

A typical plot of $v(t)$ is shown in Figure 7–12.

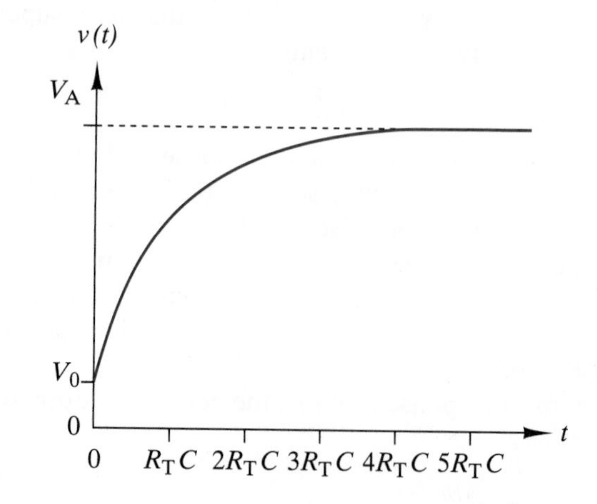

FIGURE 7 – 12 *Step response of a first-order* RC *circuit.*

The *RC* circuit step response in Eq. (7–20) starts out at the initial condition V_0 and is driven to a final condition V_A, which is determined by the amplitude of the step function input. That is, the initial and final values of the response are

$$\lim_{t \to 0+} v(t) = (V_0 - V_A)e^{-0} + V_A = V_0$$
$$\lim_{t \to \infty} v(t) = (V_0 - V_A)e^{-\infty} + V_A = V_A$$

The path between the two end points is an exponential waveform whose time constant is the circuit time constant. We know from our study of exponential signals that the step response will reach its final value after about five time constants. In other words, after about five time constants the natural response decays to zero and we are left with a constant forced response caused by the step function input.

The *RL* circuit in Figure 7–2 is the dual of the *RC* circuit, so the development of its step responses follows the same pattern discussed previously. Briefly sketching the main steps, the Norton equivalent input is a step function $I_A u(t)$, and for $t \geq 0$ the *RL* circuit differential equation Eq. (7–6) becomes

$$G_N L \frac{di(t)}{dt} + i(t) = I_A \qquad t \geq 0 \tag{7–21}$$

The solution of this equation is found by superimposing the natural and forced components. The natural response is the solution of the homogeneous equation [right side of Eq. (7–21) set to zero] and takes the same form as the zero-input response found in the previous section.

$$i_N(t) = Ke^{-t/G_N L} \qquad t \geq 0$$

where K is a constant to be evaluated from the initial condition once the complete response is known. The forced response is a particular solution of the equation

$$G_N L \frac{di_F(t)}{dt} + i_F(t) = I_A \qquad t \geq 0$$

Setting $i_F = I_A$ satisfies this equation since $dI_A/dt = 0$.

Combining the forced and natural responses, we obtain the general solution of Eq. (7–21) in the form

$$\begin{aligned} i(t) &= i_N(t) + i_F(t) \\ &= Ke^{-t/G_N L} + I_A \qquad t \geq 0 \end{aligned}$$

The constant K is now evaluated from the initial condition:

$$i(0) = I_0 = Ke^{-0} + I_A = K + I_A$$

The initial condition requires that $K = I_0 - I_A$, so the step response of the *RL* circuit is

$$i(t) = (I_0 - I_A)e^{-t/G_N L} + I_A \qquad t \geq 0 \tag{7–22}$$

The *RL* circuit step response has the same form as the *RC* circuit step response in Eq. (7–20). At $t = 0$ the starting value of the response is $i(0) = I_0$, as required by the initial condition. The final value is the forced response $i(\infty) = i_F = I_A$, since the natural response decays to zero as time increases.

A step function input to the *RC* or *RL* circuit drives the state variable from an initial value determined by what happened prior to $t = 0$ to a final value determined by amplitude of the step function applied at $t = 0$. The time needed to transition from the initial to the final value is about $5T_C$, where T_C is the circuit time constant. We conclude that the step response of a first-order circuit depends on three quantities:

1. The amplitude of the step input (V_A or I_A)
2. The circuit time constant ($R_T C$ or $G_N L$)
3. The value of the state variable at $t = 0$ (V_0 or I_0).

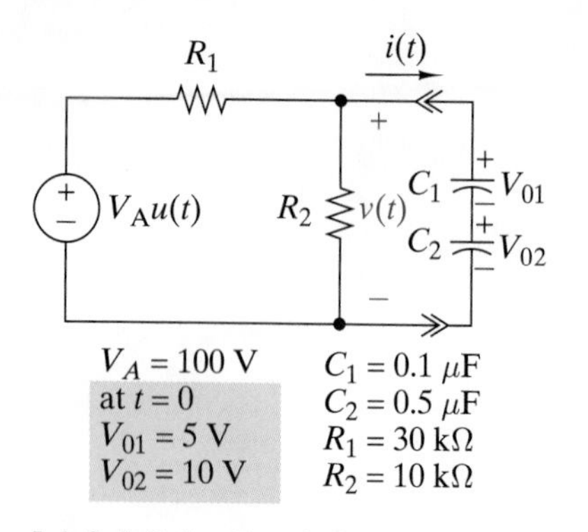

FIGURE 7–13

EXAMPLE 7–4

Find the response of the *RC* circuit in Figure 7–13.

SOLUTION:

The circuit is first order, since the two capacitors in series can be replaced by a single equivalent capacitor

$$C_{EQ} = \frac{1}{\dfrac{1}{C_1} + \dfrac{1}{C_2}} = 0.0833\ \mu\text{F}$$

The initial voltage on C_{EQ} is the sum of the initial voltages on the original capacitors.

$$V_0 = V_{01} + V_{02} = 5 + 10 = 15\ \text{V}$$

To find the Thévenin equivalent seen by C_{EQ}, we first find the open-circuit voltage. Disconnecting the capacitors in Figure 7–13 and using voltage division at the interface yields

$$v_T = v_{OC} = \frac{R_2}{R_1 + R_2} V_A u(t) = \frac{10}{40} 100u(t) = 25u(t)\ \text{V}$$

Replacing the voltage source by a short circuit and looking to the left at the interface, we see R_1 in parallel with R_2. The Thévenin resistance of this combination is

$$R_T = \frac{1}{\dfrac{1}{R_1} + \dfrac{1}{R_2}} = 7.5\ \text{k}\Omega$$

The circuit time constant is

$$T_C = R_T C_{EQ} = (7.5 \times 10^3)(8.33 \times 10^{-8}) = \frac{1}{1600}\ \text{s}$$

For the Thévenin equivalent circuit, the initial capacitor voltage is $V_0 =$ 15 V, the step input is $25u(t)$, and the time constant is 1/1600 s. Using the *RC* circuit step response in Eq. (7–20) yields

$$v(t) = (15 - 25)\, e^{-1600t} + 25$$
$$= 25 - 10\, e^{-1600t}\ \text{V} \qquad t \geqslant 0$$

The initial ($t = 0$) value of $v(t)$ is 25 – 10 = 15 V, as required. The equivalent capacitor voltage is driven to a final value of 25 V by the step input in the Thévenin equivalent circuit. For practical purposes, $v(t)$ reaches 25 V after about $5T_C = 3.125$ ms. ■

EXAMPLE 7–5

Find the step response of the *RL* circuit in Figure 7–14(a). The initial condition is $i(0) = I_0$.

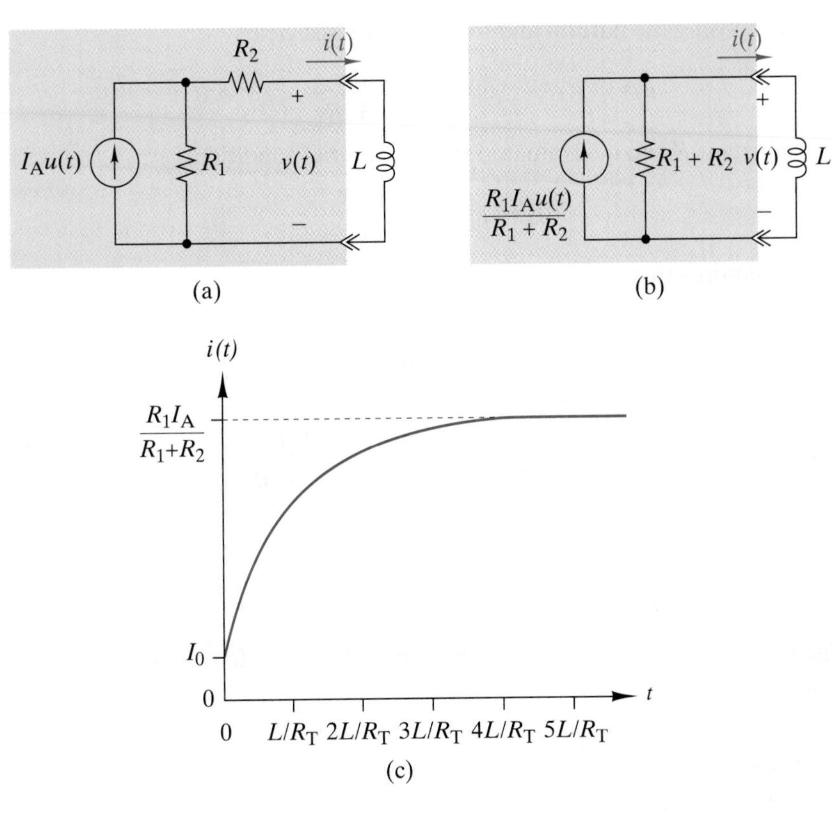

FIGURE 7–14

SOLUTION:

We first find the Norton equivalent to the left of the interface. By current division, the short-circuit current at the interface is

$$i_{SC}(t) = \frac{R_1}{R_1 + R_2} I_A u(t)$$

Looking to the left at the interface with the current source off (replaced by an open circuit), we see R_1 and R_2 in series producing a Thévenin resistance

$$R_T = \frac{1}{G_N} = R_1 + R_2$$

The time constant of the Norton equivalent circuit in Figure 7–14(b) is

$$T_C = G_N L = \frac{L}{(R_1 + R_2)}$$

The natural response of the Norton equivalent circuit is

$$i_N(t) = Ke^{-(R_1+R_2)t/L} \qquad t \geq 0$$

The short-circuit current $i_{SC}(t)$ is the step function input in the Norton circuit. Therefore, the forced response is

$$i_F(t) = i_{SC}(t) = \frac{R_1}{(R_1 + R_2)} I_A u(t)$$

Superimposing the natural and forced responses yields

$$i(t) = Ke^{-(R_1+R_2)t/L} + \frac{R_1 I_A}{R_1 + R_2} \qquad t \geqslant 0$$

The constant K can be evaluated from the initial condition:

$$i(0) = I_0 = K + \frac{R_1 I_A}{R_1 + R_2}$$

which requires that

$$K = I_0 - \frac{R_1 I_A}{R_1 + R_2}$$

So circuit step response is

$$i(t) = \left[I_0 - \frac{R_1 I_A}{R_1 + R_2}\right] e^{-(R_1+R_2)t/L} + \frac{R_1 I_A}{R_1 + R_2} \qquad t \geqslant 0$$

An example of this response is shown in Figure 7–14(c). ■

EXAMPLE 7–6

The state variable response of a first-order *RC* circuit for a step function input is

$$v_C(t) = 20e^{-200t} - 10 \text{ V} \qquad t \geqslant 0$$

(a) What is the circuit time constant?
(b) What is the initial voltage across the capacitor?
(c) What is the amplitude of the forced response?
(d) At what time is $v_C(t) = 0$?

SOLUTION:

(a) The natural response of a first-order circuit is of the form Ke^{-t/T_C}. Therefore, the time constant of the given responses is $T_C = 1/200 = 5$ ms.
(b) The initial ($t = 0$) voltage across the capacitor is

$$v_C(0) = 20e^{-0} - 10 = 20 - 10 = 10 \text{ V}$$

(c) The natural response decays to zero, so the forced response is the final value $v_C(t)$.

$$v_C(\infty) = 20e^{-\infty} - 10 = 0 - 10 = -10 \text{ V}$$

(d) The capacitor voltage must pass through zero at some intermediate time, since the initial value is positive and the final value negative. This time is found by setting the step response equal to zero:

$$20e^{-200t} - 10 = 0$$

which yields the condition $e^{200t} = 2$ or $t = \ln 2/200 = 3.47$ ms. ■

Exercise 7–4

Given the following first-order circuit step response

$$v_C(t) = 20 - 20e^{-1000t} \text{ V} \quad t \geq 0$$

(a) What is the amplitude of the step input?
(b) What is the circuit time constant?
(c) What is the initial value of the state variable?
(d) What is the circuit differential equation?

Answers:

(a) 20 V
(b) 1 ms
(c) 0 V
(d) $10^{-3}\, dv_C/dt + v_C = 20u(t)$

Exercise 7–5

Find the solution of the following first-order differential equations:

(a) $10^{-4}\dfrac{dv_C}{dt} + v_C = -5u(t) \qquad v_C(0) = 5\text{ V}$

(b) $5 \times 10^{-2}\dfrac{di_L}{dt} + 2000\, i_L = 10u(t) \qquad i_L(0) = -5\text{ mA}$

Answers:

(a) $v_C(t) = -5 + 10e^{-10000t}\text{ V} \qquad t \geq 0$
(b) $i_L(t) = 5 - 10e^{-40000t}\text{ mA} \qquad t \geq 0$

ZERO-STATE RESPONSE

Additional properties of dynamic circuit responses are revealed by rearranging the *RC* and *RL* circuit step responses in Eqs. (7–20) and (7–22) in the following way:

$$RC\text{ circuit: } v(t) = \underbrace{V_0 e^{-t/R_T C}}_{\text{zero-input response}} + \underbrace{V_A(1 - e^{-t/R_T C})}_{\text{zero-state response}} \qquad t \geqslant 0$$

$$RL\text{ circuit: } i(t) = \overbrace{I_0 e^{-t/G_N L}}^{\text{zero-input response}} + \overbrace{I_A(1 - e^{-t/G_N L})}^{\text{zero-state response}} \qquad t \geqslant 0$$

We recognize the first term on the right in each equation as the zero-input response discussed in Sect. 7–1. By definition, the **zero-input response** occurs when the input is zero ($V_A = 0$ or $I_A = 0$). The second term on the right in each equation is called the **zero-state response** because this part occurs when the initial state of the circuit is zero ($V_0 = 0$ or $I_0 = 0$).

The zero-state response is proportional to the amplitude of the input step function (V_A or I_A). However, the total response (zero input plus zero state) is not directly proportional to the input amplitude. When the initial state is not zero, the circuit appears to violate the proportionality property of linear circuits. However, bear in mind that the proportionality property applies to linear circuits with only one input.

The *RC* and *RL* circuits can store energy and have memory. In effect, they have two inputs: (1) the input that occurred before $t = 0$, and (2) the step function applied at $t = 0$. The first input produces the initial energy state of the circuit at $t = 0$, and the second causes the zero-state response for $t \geq 0$. In general, for $t \geq 0$, the total response of a dynamic circuit is the sum of two responses: (1) the zero-input response caused by the initial conditions produced by inputs applied before $t = 0$, and (2) the zero-state response caused by inputs applied after $t = 0$.

APPLICATION EXAMPLE 7–7

The operation of a digital system is controlled by a clock waveform that provides a standard timing reference. At its source a clock waveform can be described by a rectangular pulse of the form

$$v_S(t) = V_A\,[u(t) - u(t - T)]$$

In this example the pulse amplitude is $V_A = 5$ V and the pulse duration is $T = 10$ ns. This clock pulse drives a digital device that can be modeled by the circuit in Figure 7–15(a). In this model $v_S(t)$ is the rectangular clock pulse defined above and $v(t)$ is the clock waveform as received at the input to the digital device. The presence of a clock pulse at the device input will be detected only if $v(t)$ exceeds a specified logic "1" threshold level.

Find the zero-state response of the voltage $v(t)$ when $RC = 10$ ns. Will the clock pulse be detected if the logic "1" threshold level is 3.7 V?

FIGURE 7–15

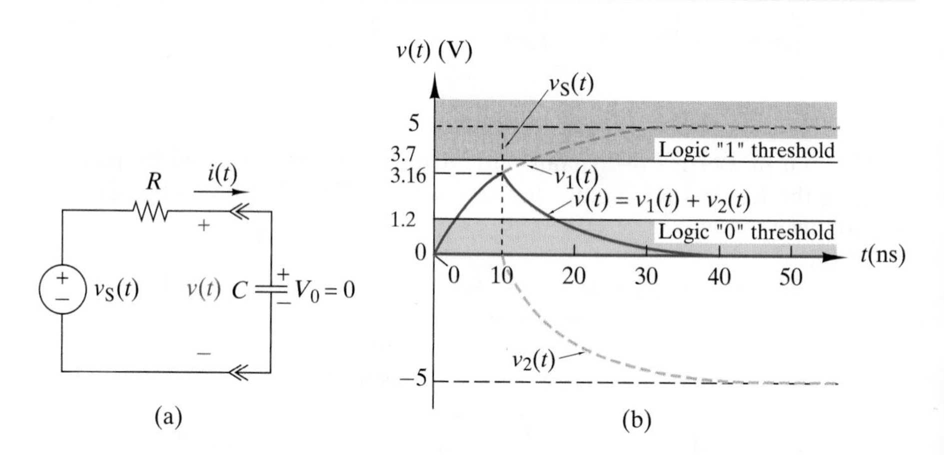

SOLUTION:

The rectangular pulse input $v_S(t)$ is indicated by dashed lines in Figure 7–15(b). The initial capacitor voltage is zero because we seek the zero-state response. The total response can be found as the sum of the zero-state responses caused by two inputs:

1. A positive 5-V step function applied at $t = 0$
2. A negative 5-V step function applied at $t = 10$ ns

The first input causes a zero-state response of

$$\begin{aligned} v_1(t) &= V_A(1 - e^{-t/RC})u(t) \\ &= 5(1 - e^{-10^8 t})u(t) \end{aligned}$$

The second input causes a zero-state response of

$$\begin{aligned} v_2(t) &= -V_A(1 - e^{-(t-T)/RC})u(t - T) \\ &= -5[1 - e^{-10^8(t-10^{-8})}]u(t - 10^{-8}) \end{aligned}$$

Notice that $v_2(t) = -v_1(t - T)$, that is, $v_2(t)$ is obtained by inverting and delaying $v_1(t)$ by $T = 10$ ns. The total response is the superposition of these two responses.

$$v(t) = v_1(t) + v_2(t)$$

Figure 7–15(b) shows how the two responses combine to produce the overall pulse response of the circuit. The response $v_1(t)$ begins at zero and eventually reaches a final value of +5 V. At $t = T = 10$ ns the first response reaches $v_1(T) = 5(1 - e^{-1}) = 3.16$ V. The second response $v_2(t)$ begins at $t = T = 10$ ns, and thereafter is equal and opposite to $v_1(t)$ except that it is delayed by $T = 10$ ns. The net result is that the total response reaches a maximum of $v(T) = 3.16$ V. In this example the clock pulse will not be detected because the logic "1" threshold level is 3.7 V. Clock pulse detection would be made possible by increasing the pulse duration so that $v(T) > 3.7$ V. This requires that

$$5(1 - e^{-10^8 T}) > 3.7 \quad \text{V}$$

or $1.3 > 5\,e^{-10^8 T}$, which yields $T > 1.347 \times 10^{-8}$. For the digital device in this example, the minimum detectable clock pulse duration is about 13.5 ns. ■

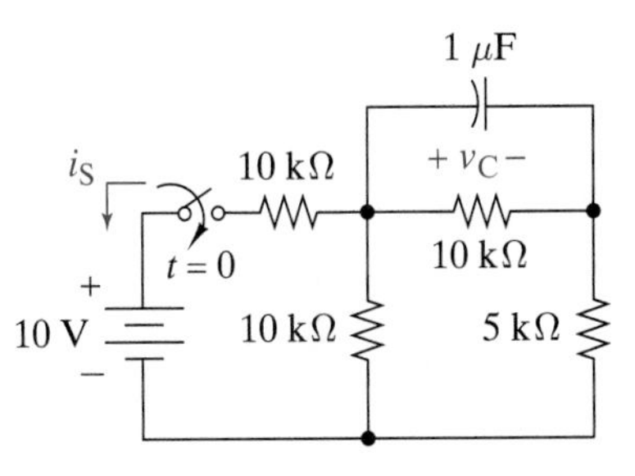

FIGURE 7–16

Exercise 7–6

The switch in Figure 7–16 closes at $t = 0$. Find the zero-state response of the capacitor voltage for $t \geq 0$.

Answer: $v_C(t) = 2.5(1 - e^{-200t})$ V

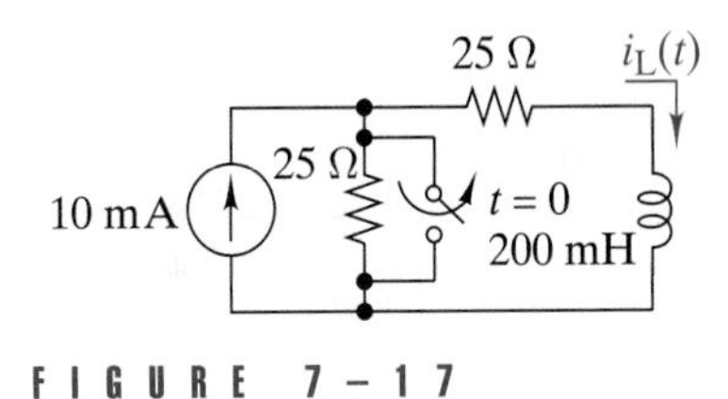

FIGURE 7–17

Exercise 7–7

The switch in Figure 7–17 opens at $t = 0$. Find the zero-state response of the inductor current for $t \geq 0$.

Answer: $i_L(t) = 5(1 - e^{-250t})$ mA

7–3 INITIAL AND FINAL CONDITIONS

Reviewing the first-order step responses of the last section shows that for $t \geq 0$ the state variable responses can be written in the form

$$\begin{aligned} &RC \text{ circuit: } v_C(t) = [v_C(0) - v_C(\infty)]e^{-t/T_C} + v_C(\infty) \qquad t \geqslant 0 \\ &RL \text{ circuit: } i_L(t) = [i_L(0) - i_L(\infty)]e^{-t/T_C} + i_L(\infty) \qquad t \geqslant 0 \end{aligned} \tag{7–23}$$

In both circuits the step response is of the general form

$$\begin{matrix}\text{The state}\\ \text{variable}\\ \text{response}\end{matrix} = \left[\begin{matrix}\text{The initial}\\ \text{value of the}\\ \text{state variable}\end{matrix} - \begin{matrix}\text{The final}\\ \text{value of the}\\ \text{state variable}\end{matrix}\right] \times e^{-t/T_C} + \begin{matrix}\text{The final}\\ \text{value of the}\\ \text{state variable}\end{matrix}$$

To determine the step response of a first-order circuit, we need three quantities: the initial value of the state variable, the final value of the state variable, and the time constant. Since we know how to get the time constant directly from the circuit, it would be useful to have a direct way to determine the initial and final values by inspecting the circuit itself.

The final value can be calculated directly from the circuit by observing that for $t > 5T_C$ the step responses approach a constant value or dc value. Under dc conditions a capacitor acts like an open circuit and an inductor acts like a short circuit. As a result, the final value of the state variable is found by applying dc analysis methods to the circuit configuration for $t > 0$, with capacitors replaced by open circuits and inductors replaced by short circuits.

We can also use dc analysis to determine the initial value in many practical situations. A common situation is a circuit containing dc sources and a switch that is in one position for a period of time much greater than the circuit time constant, and then is moved to a new position at $t = 0$. For example, if the switch is closed for a long period of time, then the dc sources drive the state variable to a final value. If the switch is now opened at $t = 0$, then the dc sources drive the state variable to a new final condition appropriate to the new circuit configuration for $t > 0$.

Note: The initial condition at $t = 0$ is the dc value of the state variable for the circuit configuration that existed before the switch changed positions at $t = 0$. The switching action cannot cause an instantaneous change in the initial condition because capacitor voltage and inductor current are continuous functions of time. In other words, opening a switch at $t = 0$ marks the boundary between two eras. The final condition of the state variable for the $t < 0$ era is the initial condition for the $t > 0$ era that follows.

The usual way to state a switched circuit problem is to say that a switch has been closed (open) for a long time and then is opened (closed) at $t = 0$. In this context, a long time means at least five time constants. Time constants rarely exceed a few hundred milliseconds in electrical circuits, so a long time passes rather quickly.

The state variable response in switched dynamic circuits is found using the following steps:

STEP 1 Find the initial value by applying dc analysis to the circuit configuration for $t < 0$.

STEP 2 Find the final value by applying dc analysis to the circuit configuration for $t > 0$.

STEP 3 Find the time constant T_C of the circuit in the configuration for $t > 0$.

STEP 4 Write the step response directly using Eq. (7–23) without formulating and solving the circuit differential equation.

For example, the switch in Figure 7–18(a) has been closed for a long time and is opened at $t = 0$. We want to find the capacitor voltage $v(t)$ for $t \geq 0$.

STEP 1 The initial condition is found by dc analysis of the circuit configuration in Figure 7–18(b), where the switch is closed. Using voltage division, the initial capacitor voltage is found to be

$$v(0) = \frac{R_2 V_A}{R_1 + R_2}$$

STEP 2 The final condition is found by dc analysis of the circuit configuration in Figure 7–18(c), where the switch is open. When the switch is open the circuit has no dc excitation, so the final value of the capacitor voltage is zero.

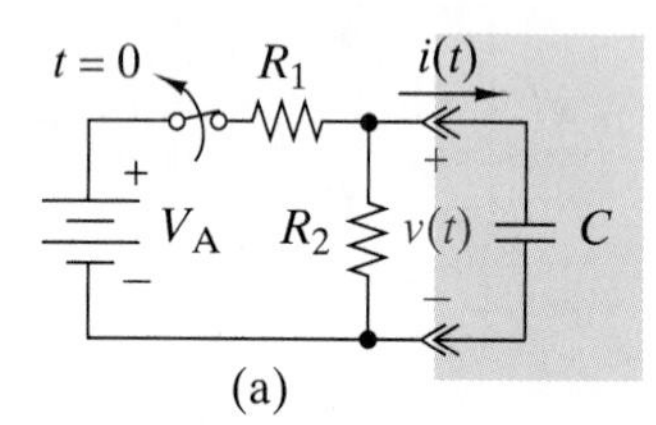

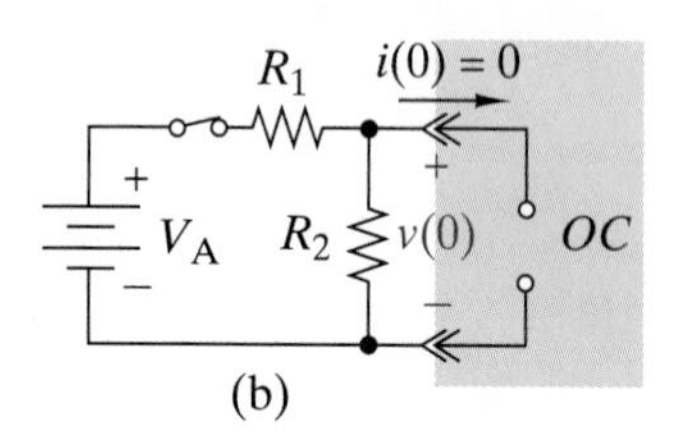

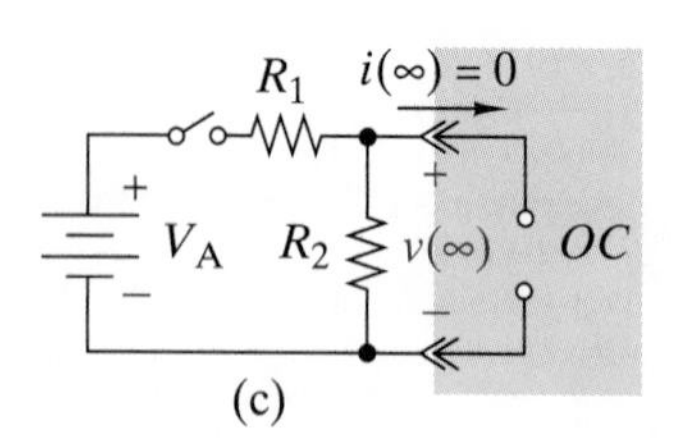

FIGURE 7–18 *Solving a switched dynamic circuit using the initial and final conditions.*

STEP 3 The circuit in Figure 7–18(c) also gives us the time constant. Looking back at the interface, we see an equivalent resistance of R_2, since R_1 is connected in series with an open switch. For $t \geq 0$ the time constant is R_2C. Using Eq. (7–23), the capacitor voltage for $t \geq 0$ is

$$v(t) = [v(0) - v(\infty)]e^{-t/T_C} + v_C(\infty)$$

$$= \frac{R_2 V_A}{R_1 + R_2} e^{-t/R_2C} \qquad t \geq 0$$

The result is a zero-input response, since there is no excitation for $t \geq 0$. But now we see how the initial condition for the zero-input response could be produced physically by opening a switch that has been closed for a long time.

To continue the analysis, we find the capacitor current using its element constraint:

$$i(t) = C\frac{dv}{dt} = -\frac{V_A}{R_1 + R_2} e^{-t/R_2C} \qquad t \geq 0$$

This is the capacitor current for $t \geq 0$. For $t < 0$ the circuit in Figure 7–18(b) points out that the capacitor current is zero since the capacitor acts like an open circuit.

The capacitor voltage and current responses are plotted in Figure 7–19. The capacitor voltage is continuous at $t = 0$, but the capacitor current has a jump discontinuity at $t = 0$. In other words, state variables are continuous, but nonstate variables can have discontinuities at $t = 0$. Since the state variable is continuous, we first find the circuit state variable and then solve for other circuit variables using the element and connection constraints.

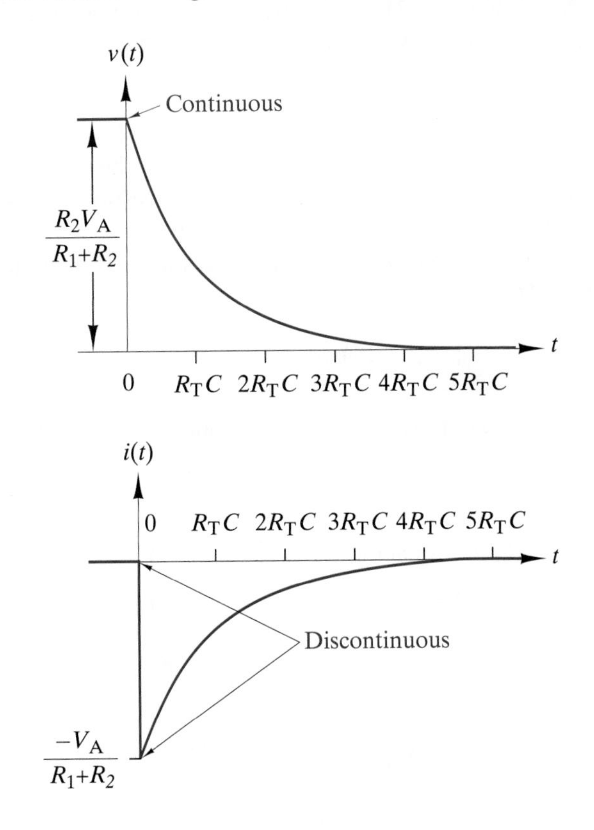

FIGURE 7–19 *Two responses in the* RC *circuit of Figure 7–18.*

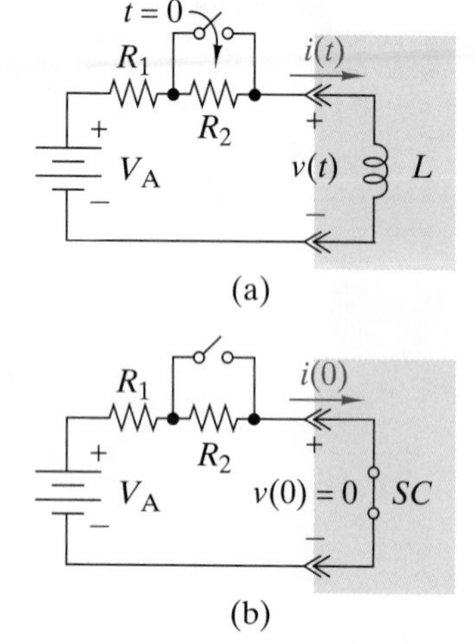

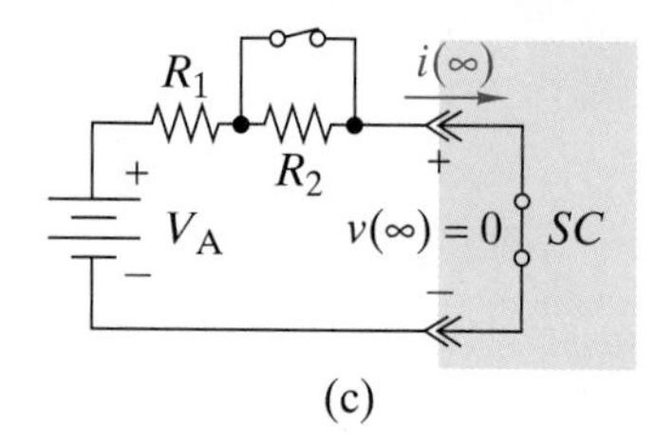

FIGURE 7–20

EXAMPLE 7–8

The switch in Figure 7–20(a) has been open for a long time and is closed at $t = 0$. Find the inductor current for $t > 0$.

SOLUTION:

We first find the initial condition using the circuit in Figure 7–20(b). By series equivalence the initial current is

$$i(0) = \frac{V_A}{R_1 + R_2}$$

The final condition and the time constant are determined from the circuit in Figure 7–17(c). Closing the switch shorts out R_2, and the final condition and time constant for $t > 0$ are

$$i(\infty) = \frac{V_A}{R_1} \quad \text{and} \quad T_C = G_N L = \frac{L}{R_1}$$

Using Eq. (7–23), the inductor current for $t \geq 0$ is

$$\begin{aligned} i(t) &= [i(0) - i(\infty)]e^{-t/T_C} + i(\infty) \\ &= \left[\frac{V_A}{R_1 + R_2} - \frac{V_A}{R_1}\right]e^{-R_1 t/L} + \frac{V_A}{R_1} \qquad t \geqslant 0 \end{aligned}$$

■

EXAMPLE 7–9

The switch in Figure 7–21(a) has been closed for a long time and is opened at $t = 0$. Find the voltage $v_O(t)$.

SOLUTION:

The problem asks for voltage $v_O(t)$, which is not the circuit state variable. Our approach is first to find the state variable response and then use this response to solve for the required nonstate variable.

For $t < 0$ the circuit in Figure 7–21(b) applies. By voltage division, the initial capacitor voltage is

$$v(0) = \frac{R_1 V_A}{R_1 + R_2}$$

The final value and time constant are found from the circuit in Figure 7–18(c).

$$v(\infty) = V_A \quad \text{and} \quad T_C = R_T C = R_2 C$$

Using Eq. (7–23), the capacitor voltage for $t \geq 0$ is

$$\begin{aligned} v(t) &= [v(0) - v(\infty)]e^{-t/T_C} + v(\infty) \\ &= \left[\frac{R_1 V_A}{R_1 + R_2} - V_A\right]e^{-t/R_2 C} + V_A \\ &= V_A + \frac{R_2 V_A}{R_1 + R_2} e^{-t/R_2 C} \qquad t \geqslant 0 \end{aligned}$$

Given the state variable, we can find the voltage $v_O(t)$ by writing a KVL equation around the perimeter of the circuit in Figure 7–21(a):

$$-V_A + v(t) + v_O(t) = 0$$

or

$$v_O(t) = V_A - v(t) = -\frac{R_2 V_A}{R_1 + R_2} e^{-t/R_2 C} \qquad t \geq 0$$

The output voltage response looks like a zero-input response even though the circuit input is not zero for $t \geq 0$. However, $v_O(t)$ is not the state variable but the voltage across the resistor R_2. The voltage across R_2 is proportional to the capacitor current, which eventually decays to zero in the final circuit configuration in Figure 7–21(c) because the capacitor acts like an open circuit. ■

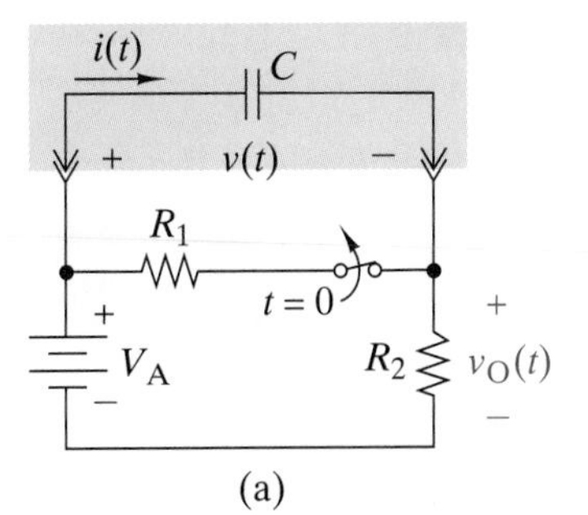

(a)

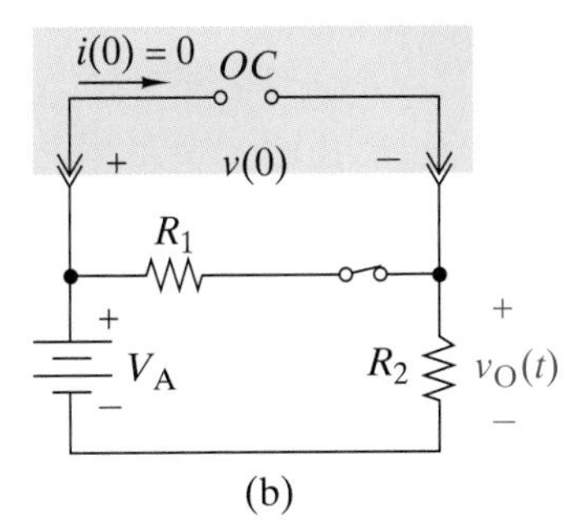

(b)

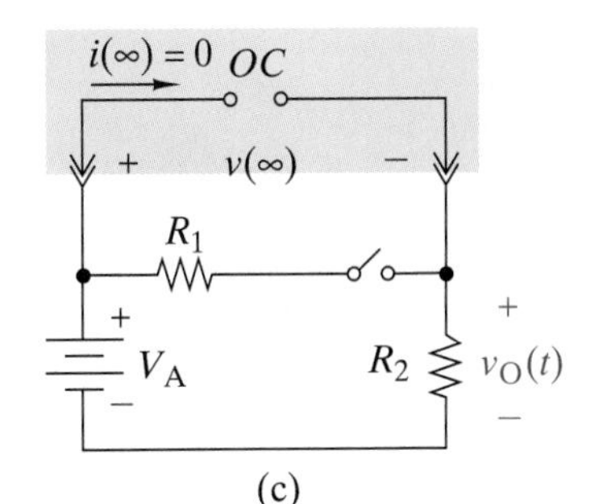

(c)

FIGURE 7–21

EXAMPLE 7–10

For $t \geq 0$ the state variable response of the RL circuit in Figure 7–22 is observed to be

$$i_L(t) = 50 + 100\, e^{-5000t} \quad \text{mA}$$

(a) Identify the forced and natural components of the response.
(b) Find the circuit time constant.
(c) Find the Thévenin equivalent circuit seen by the inductor.

SOLUTION:

(a) The natural component is the exponential term $100e^{-5000t}$ mA. The forced component is what remains after the natural component dies out as $t \to \infty$, namely $i_L(\infty) = 50$ mA. The forced response is a constant 50 mA, which means that the Thévenin equivalent is a dc source.
(b) The time constant is the reciprocal of the coefficient of t in e^{-5000t}, i. e., $T_C = 5000^{-1} = 0.2$ ms.
(c) Expressed in terms of circuit parameters the time constant is $T_C = L/R_T$, which yields the Thévenin resistance as $R_T = L/T_C = 100\ \Omega$. For dc excitation the inductor acts like a short circuit at $t = \infty$. Hence, $i_L(\infty) = v_T/R_T$ and the Thévenin voltage is

$$v_T = R_T i_L(\infty) = 100 \times 0.05 = 5 \quad \text{V}$$

■

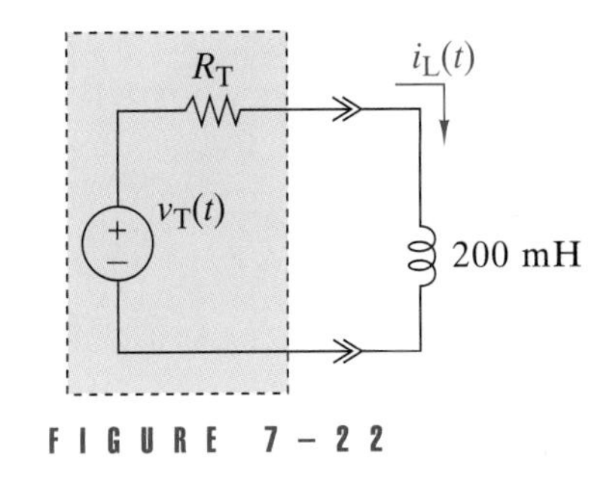

FIGURE 7–22

EVALUATION EXAMPLE 7–11

The switch in Figure 7–23 moves from position A to position B at $t = 0$. The first-order RC circuit in the figure must be designed to produce an output of

$$v_O(t) = 5(1 - e^{-1000t})\ \text{V} \qquad t > 0$$

Evaluate the two proposed circuit designs shown in the figure using the following criteria.
(a) A design must produce the required output.
(b) If both produce the desired output, then compare part counts and use of standard values to identify the best design.

FIGURE 7–23

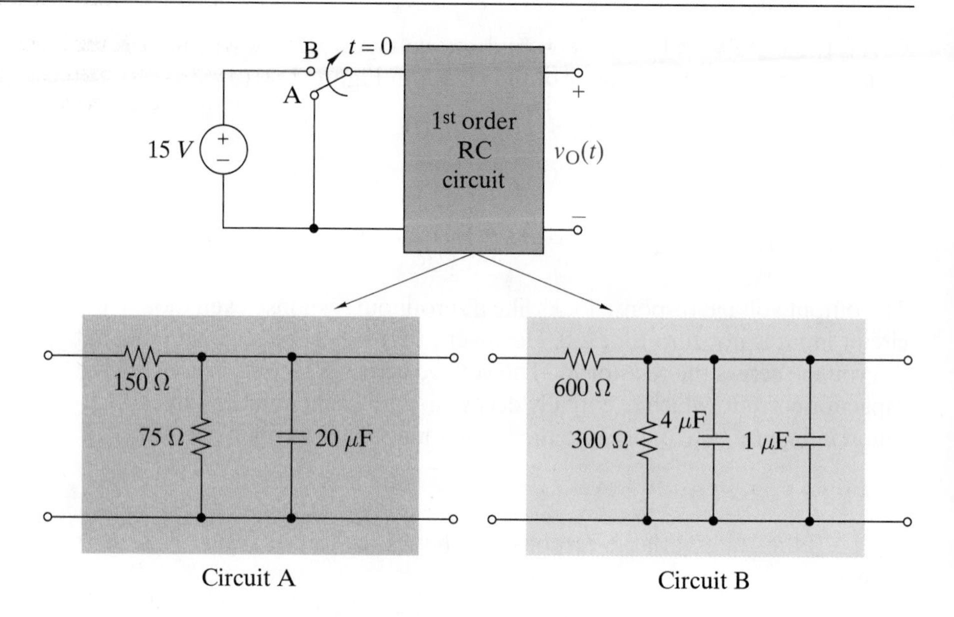

SOLUTION:

(a) The desired output is a first-order step response with $v_O(0) = 0$, $v_O(\infty) = 5$ V, and $T_C = 1$ ms. For $t < 0$ the switch is in position B and there is zero input; hence $v_O(0) = 0$ for both circuits. For $t > 0$ the switch is in position A. The final condition $v_O(\infty)$ is found using voltage division on the circuit with the capacitors replaced by open circuits. The circuit time constant is found using the Thévenin resistance seen by the capacitors. The final value and time constant of circuit A are

$$\text{Circuit A: } v_O(\infty) = \frac{75}{150 + 75} 15 = 5 \text{ V}$$

$$T_C = \frac{150 \times 75}{150 + 75} 20 \times 10^{-6} = 10^{-3} \text{ s}$$

The equivalent capacitance in circuit B is $C_{EQ} = 4 + 1 = 5$ μF. The final value and time constant of circuit B are

$$\text{Circuit B: } v_O(\infty) = \frac{300}{600 + 300} 15 = 5 \text{ V}$$

$$T_C = \frac{600 \times 300}{600 + 300} 5 \times 10^{-6} = 10^{-3} \text{ s}$$

Both circuits produce the desired output.

(b) Circuit A uses three components: a standard 75-Ω resistor, a standard 150-Ω resistor, and a standard 20-μF capacitor (see Appendix A for standard values). Circuit B uses four components: a standard 300-Ω resistor, a nonstandard 600-Ω resistor, a standard 1-μF capacitor, and a nonstandard 4-μF capacitor. Circuit A is a better design than circuit B in terms of both the number of parts and the use of standard values. ■

Exercise 7–8

In each circuit shown in Figure 7–24 the switch has been in position A for a long time and is moved to position B at $t = 0$. Find the circuit state variable for each circuit for $t \geq 0$.

Answers:

(a) $v_C(t) = V_A e^{-t/(R_1+R_2)C}$

(b) $i_L(t) = \dfrac{V_A}{R_2} e^{-(R_1-R_2)t/L}$

Exercise 7–9

In each circuit shown in Figure 7–24 the switch has been in position B for a long time and is moved to position A at $t = 0$. Find the circuit state variable for each circuit for $t \geq 0$.

Answers:

(a) $v_C(t) = V_A(1 - e^{-t/R_2C})$

(b) $i_L(t) = \dfrac{V_A}{R_2}(1 - e^{-R_2t/L})$

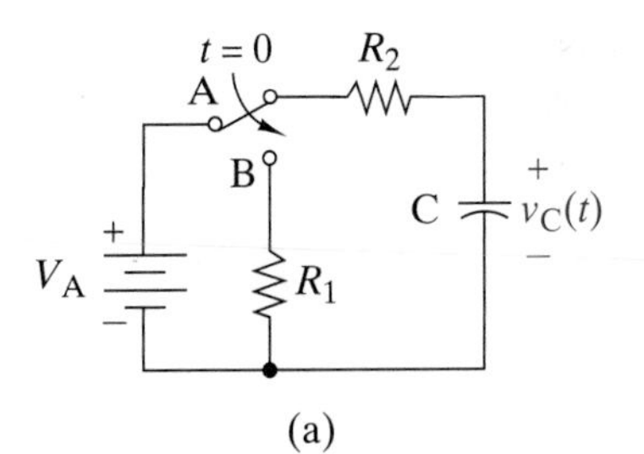

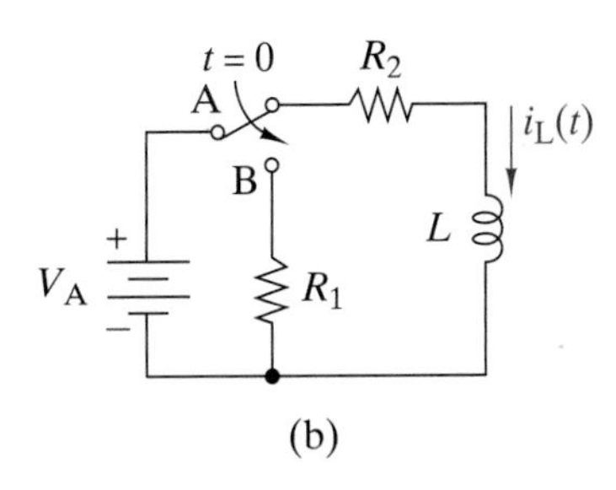

FIGURE 7–24

Exercise 7–10

In the circuit in Figure 7–25 the switch has been in position A for a long time and is moved to position B at $t = 0$. For $t \geq 0$ find the output voltage $v_O(t)$,

Answer: $v_O(t) = -4e^{-200t}$ V

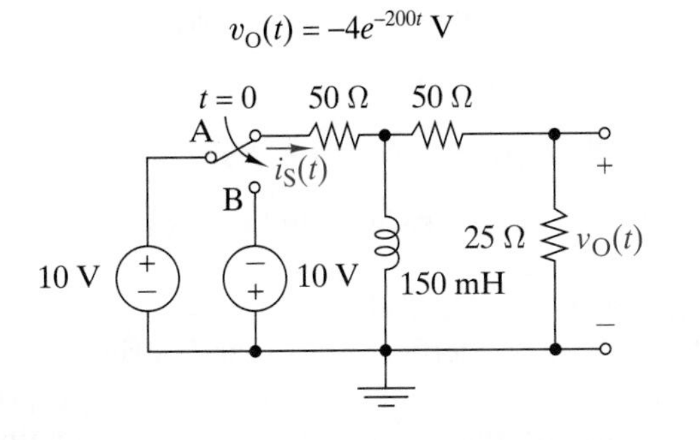

FIGURE 7–25

7–4 FIRST-ORDER CIRCUIT SINUSOIDAL RESPONSE

The response of linear circuits to sinusoidal inputs is one of the central themes of electrical engineering. In this introduction to the concept we treat the sinusoidal response of first-order circuits using differential equations. In later chapters we see that sinusoidal response can be found using other techniques, such as phasors and Laplace transforms. But for the moment, we concentrate on the classical method of finding the forced response from the circuit differential equation.

If the input to the *RC* circuit in Figure 7–2 is a causal sinusoid, then the circuit differential equation in Eq. (7–3) is written as

$$R_T C \frac{dv(t)}{dt} + v(t) = V_A[\cos \omega t]u(t) \quad (7\text{–}24)$$

The input on the right side of Eq. (7–24) is *not* an eternal sinewave but a causal sinusoid that starts at $t = 0$, through some action such as closing a switch. We seek a solution function $v(t)$ that satisfies Eq. (7–24) for $t \geq 0$ and that meets the prescribed initial condition $v(0) = V_0$.

As with the step response, we find the solution in two parts: natural response and forced response. The natural response is of the form

$$v_N(t) = K\,e^{-t/R_TC} \qquad t \geq 0$$

The natural response of a first-order circuit always has this form because it is a general solution of the homogeneous equation with input set to zero. The form of the natural response depends on the physical characteristics of the circuit and is independent of the input.

The forced response depends on both the circuit and the nature of the forcing function. The forced response is a particular solution of the equation

$$R_TC\frac{dv_F(t)}{dt} + v_F(t) = V_A \cos \omega t \qquad t \geq 0$$

This equation requires that $v_F(t)$ plus R_TC times its first derivative add to produce a cosine function for $t \geq 0$. The only way this can happen is for $v_F(t)$ and its derivative to be sinusoids of the same frequency. This requirement brings to mind the derivative property of the sinusoid. So we try a solution in the form of a general sinusoid. As noted in Chapter 5, a general sinusoid can be written in amplitude and phase angle form as

$$v_F(t) = V_F \cos(\omega t + \phi) \tag{7–25a}$$

or in terms of Fourier coefficients as

$$v_F(t) = a \cos \omega t + b \sin \omega t \tag{7–25b}$$

While either form will work, it is somewhat easier to work with the Fourier coefficient format.

The approach we are using is called the method of undetermined coefficients, where the unknown coefficients are the Fourier coefficients a and b in Eq. (7–25b). To find these unknowns we insert the proposed forced response in Eq. (7–25b) into the differential equation to obtain

$$R_TC\frac{d}{dt}(a \cos \omega t + b \sin \omega t) + (a \cos \omega t + b \sin \omega t) = V_A \cos \omega t \quad t \geq 0$$

Performing the differentiation gives

$$R_TC(-\omega a \sin \omega t + \omega b \cos \omega t) + (a \cos \omega t + b \sin \omega t) = V_A \cos \omega t$$

We next gather all sine and cosine terms on one side of the equation.

$$[R_TC\omega b + a - V_A]\cos \omega t + [-R_TC\omega a + b]\sin \omega t = 0$$

The left side of this equation is zero for all $t \geq 0$ only when the coefficients of the cosine and sine terms are identically zero. This requirement yields two linear equations in the unknown coefficients a and b:

$$a + (R_TC\omega)b = V_A$$
$$-(R_TC\omega)a + b = 0$$

The solutions of these linear equations are

$$a = \frac{V_A}{1 + (\omega R_T C)^2} \qquad b = \frac{\omega R_T C V_A}{1 + (\omega R_T C)^2}$$

These equations express the unknowns a and b in terms of known circuit parameters ($R_T C$) and known input signal parameters (ω and V_A).

We combine the forced and natural responses as

$$v(t) = Ke^{-t/R_T C} + \frac{V_A}{1 + (\omega R_T C)^2}(\cos \omega t + \omega R_T C \sin \omega t) \qquad t \geq 0 \tag{7–26}$$

The initial condition requires

$$v(0) = V_0 = K + \frac{V_A}{1 + (\omega R_T C)^2}$$

which means K is

$$K = V_0 - \frac{V_A}{1 + (\omega R_T C)^2}$$

We substitute this value of K into Eq. (7–26) to obtain the function $v(t)$ that satisfies the differential equation and the initial conditions.

$$v(t) = \underbrace{\left[V_0 - \frac{V_A}{1 + (\omega R_T C)^2}\right]e^{-t/R_T C}}_{\text{natural response}} + \underbrace{\frac{V_A}{1 + (\omega R_T C)^2}(\cos \omega t + \omega R_T C \sin \omega t)}_{\text{forced response}} \qquad t \geq 0$$

This expression seems somewhat less formidable when we convert the forced response to an amplitude and phase angle format

$$v(t) = \underbrace{\left[V_0 - \frac{V_A}{1 + (\omega R_T C)^2}\right]e^{-t/R_T C}}_{\text{natural response}} + \underbrace{\frac{V_A}{\sqrt{1 + (\omega R_T C)^2}} \cos(\omega t + \theta)}_{\text{forced response}} \qquad t \geq 0 \tag{7–27}$$

where

$$\theta = \tan^{-1}(-b/a) = \tan^{-1}(-\omega R_T C)$$

Equation (7–27) is the complete response of the RC circuit for an initial condition V_0 and a sinusoidal input $[V_A \cos \omega t]u(t)$. Several aspects of the response deserve comment:

1. After roughly five time constants the natural response decays to zero but the sinusoidal forced response persists.
2. The forced response is a sinusoid with the same frequency (ω) as the input but with a different amplitude and phase angle.

3. The forced response is proportional to V_A. This means that the amplitude of the forced component has the proportionality property because the circuit is linear.

In the terminology of electrical engineering, the forced component is called the **sinusoidal steady-state response**. The words *steady state* may be misleading since together they seem to imply a constant or "steady" value, whereas the forced response is a sustained oscillation. To electrical engineers *steady state* means the conditions reached after the natural response has died out. The sinusoidal steady-state response is also called the **ac steady-state response**. Often the words *steady state* are dropped and it is called simply the **ac response**. Hereafter, ac response, sinusoidal steady-state response, and the forced response for a sinusoidal input will be used interchangeably.

Finally, the forced response due to a step function input is called the **zero-frequency** or **dc steady-state response**. The zero-frequency terminology means that we think of a step function as a cosine $V_A[\cos \omega t]u(t)$ with $\omega = 0$. The reader can easily show that inserting $\omega = 0$ reduces Eq. (7–27) to the *RC* circuit step response in Eq. (7–20).

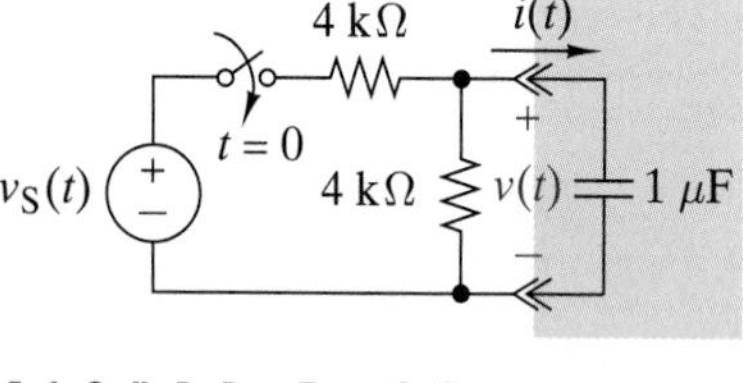

FIGURE 7–26

EXAMPLE 7–12

The switch in Figure 7–26 has been open for a long time and is closed at $t = 0$. Find the voltage $v(t)$ for $t \geq 0$ when $v_S(t) = [20 \sin 1000t]u(t)$ V.

SOLUTION:

We first derive the circuit differential equation. By voltage division, the Thévenin voltage seen by the capacitor is

$$v_T(t) = \frac{4}{4+4} v_S(t) = 10 \sin 1000t \quad \text{V}$$

The Thévenin resistance (switch closed and source off) looking back into the interface is two 4-kΩ resistors in parallel, so $R_T = 2$ kΩ. The circuit time constant is

$$T_C = R_T C = (2 \times 10^3)(1 \times 10^{-6}) = 2 \times 10^{-3} = 1/500 \text{ s}$$

Given the Thévenin equivalent seen by the capacitor and the circuit time constant, the circuit differential equation is

$$2 \times 10^{-3} \frac{dv(t)}{dt} + v(t) = 10 \sin 1000t \qquad t \geqslant 0$$

Note that the right side of the circuit differential equation is the Thévenin voltage $v_T(t)$, not the original source input $v_S(t)$. The natural response is of the form

$$v_N(t) = Ke^{-500t} \qquad t \geqslant 0$$

The forced response with undetermined Fourier coefficients is

$$v_F(t) = a \cos 1000t + b \sin 1000t$$

Substituting the forced response into the differential equation produces

$$2 \times 10^{-3}(-1000\, a \sin 1000t + 1000b \cos 1000t) +$$
$$a \cos 1000t + b \sin 1000t = 10 \sin 1000t$$

Collecting all sine and cosine terms on one side of this equation yields

$$(a + 2b)\cos 1000t + (-2a + b - 10)\sin 1000t = 0$$

The left side of this equation is zero for all $t \geq 0$ only when the coefficient of the sine and cosine terms vanish:

$$a + 2b = 0$$

$$-2a + b = 10$$

The solutions of these two linear equations are $a = -4$ and $b = 2$. We combine the forced and natural responses

$$v(t) = Ke^{-500t} - 4\cos 1000t + 2\sin 1000t \qquad t \geq 0$$

The constant K is found from the initial conditions

$$v(0) = V_0 = K - 4$$

The initial condition is $V_0 = 0$ because with the switch open the capacitor had no input for a long time prior to $t = 0$. The initial condition $v(0) = 0$ requires $K = 4$, so we can now write the complete response in the form

$$v(t) = 4e^{-500t} - 4\cos 1000t + 2\sin 1000t \text{ V} \qquad t \geq 0$$

or, in an amplitude, phase angle format as

$$v(t) = 4e^{-500t} + 4.47\cos(1000t + 153°) \text{ V} \qquad t \geq 0$$

Figure 7–27 shows an Excel worksheet that generates plots of the natural response, forced response, and total response. Column A is the time at

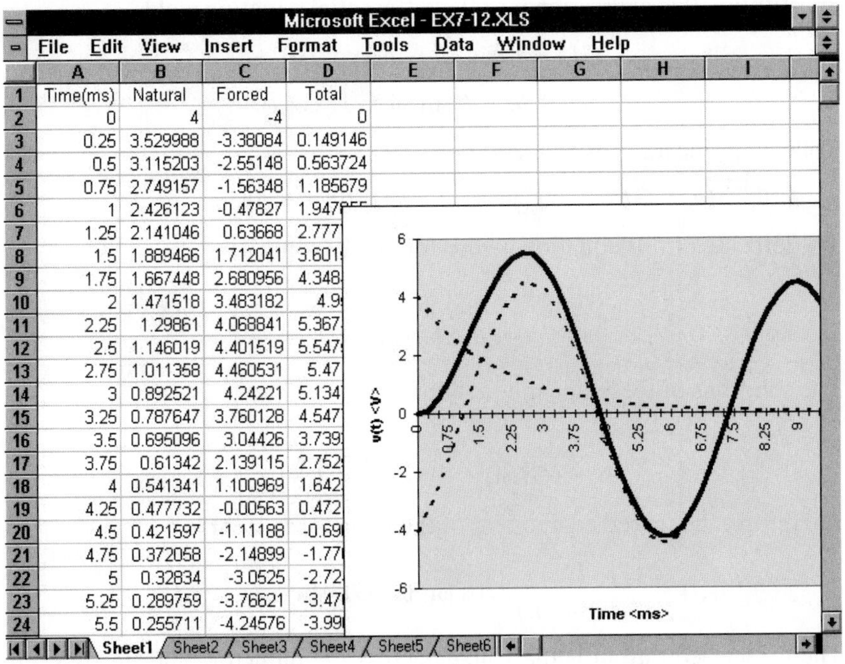

Microsoft Excel - EX7-12.XLS

	A	B	C	D
1	Time(ms)	Natural	Forced	Total
2	0	4	-4	0
3	0.25	3.529988	-3.38084	0.149146
4	0.5	3.115203	-2.55148	0.563724
5	0.75	2.749157	-1.56348	1.185679
6	1	2.426123	-0.47827	1.947
7	1.25	2.141046	0.63668	2.777
8	1.5	1.889466	1.712041	3.601
9	1.75	1.667448	2.680956	4.348
10	2	1.471518	3.483182	4.9
11	2.25	1.29861	4.068841	5.367
12	2.5	1.146019	4.401519	5.547
13	2.75	1.011358	4.460531	5.47
14	3	0.892521	4.24221	5.134
15	3.25	0.787647	3.760128	4.547
16	3.5	0.695096	3.04426	3.739
17	3.75	0.61342	2.139115	2.752
18	4	0.541341	1.100969	1.642
19	4.25	0.477732	-0.00563	0.472
20	4.5	0.421597	-1.11188	-0.69
21	4.75	0.372058	-2.14899	-1.77
22	5	0.32834	-3.0525	-2.72
23	5.25	0.289759	-3.76621	-3.47
24	5.5	0.255711	-4.24576	-3.99

v(t) <V>

Time <ms>

FIGURE 7–27

0.25-ms intervals. Columns B and C calculate the natural response ($4e^{-500t}$) and the forced response ($-4 \cos 1000t + 2 \sin 1000t$) at each of the times given in column A. The total response in column D is the sum of the entries in columns B and C. The plots show that the total response merges into the sinusoidal forced response since the natural response decays to zero after about $5T_C = 10$ ms. That is, after about 10 ms or so the circuit settles down to an ac steady-state condition. ■

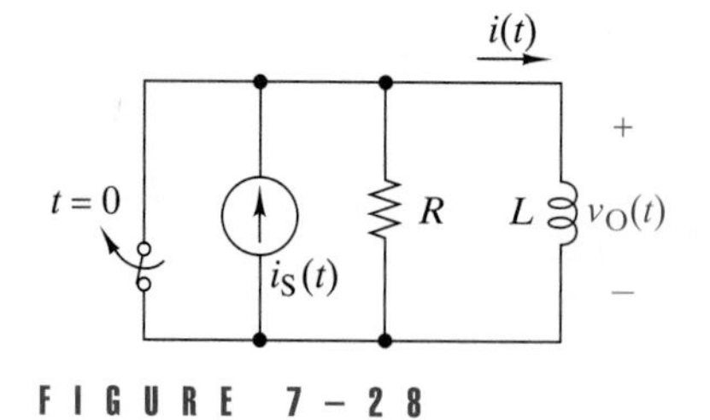

FIGURE 7 – 28

EXAMPLE 7–13

Find the sinusoidal steady-state response of the output voltage $v_O(t)$ in Figure 7–28 when the input current is $i_S(t) = [I_A \cos \omega t]u(t)$.

SOLUTION:

In keeping with our general analysis approach, we first find the steady-state response of the state variable $i(t)$ and use it to determine the required output voltage. The differential equation of the circuit in terms of the inductor current is

$$GL\frac{di}{dt} + i = I_A \cos \omega t \qquad t \geqslant 0$$

To find the steady-state response, we need to find the unknown Fourier coefficients a and b in the forced component:

$$i_F(t) = a \cos \omega t + b \sin \omega t \qquad t \geqslant 0$$

Substituting this expression into the differential equation yields

$$GL(-a\omega \sin \omega t + b\omega \cos \omega t) +$$
$$a \cos \omega t + b \sin \omega t = I_A \cos \omega t$$

Collecting sine and cosine terms produces

$$(a + GLb\omega - I_A)\cos \omega t + (-GLa\omega + b)\sin \omega t = 0$$

The left side of this equation is zero for all $t \geq 0$ only if

$$a + (GL\omega)b = I_A$$
$$-(GL\omega)a + b = 0$$

The solutions of these linear equations are

$$a = \frac{I_A}{1 + (GL\omega)^2} \qquad b = \frac{\omega GLI_A}{1 + (GL\omega)^2}$$

Therefore, the forced component of the inductor current is

$$i_F(t) = \frac{I_A}{1 + (\omega GL)^2}(\cos \omega t + \omega GL \sin \omega t) \qquad t \geqslant 0$$

The prescribed output is the voltage across the inductor. The steady-state output voltage is found using the inductor element equation:

$$v_O = L\frac{di_F}{dt} = \left[\frac{I_A L}{1 + (\omega GL)^2}\right]\frac{d}{dt}[\cos \omega t + \omega GL \sin \omega t]$$

$$= \left[\frac{I_A L}{1 + (\omega GL)^2}\right][-\omega \sin \omega t + \omega^2 GL \cos \omega t]$$

$$= \frac{I_A \omega L}{\sqrt{1 + (\omega GL)^2}} \cos (\omega t + \theta) \qquad t \geq 0$$

where $\theta = \tan^{-1}(1/\omega GL)$. The output voltage is a sinusoid with the same frequency as the input signal, but with a different amplitude and phase angle. In fact, in the sinusoidal steady state every voltage and current in a linear circuit is sinusoidal with the same frequency.

Notice that the amplitude of the steady-state output takes the form

$$\frac{I_A \omega L}{\sqrt{1 + (\omega GL)^2}}$$

Thus, the amplitude changes with the frequency of the sinusoidal current input. At $\omega = 0$ the input $i_S(t) = I_A \cos(0) = I_A$ is a constant dc waveform and the steady-state output is $v_O = 0$. This makes sense because at dc the inductor acts like a short circuit that forces the steady-state output in Figure 7–28 to be zero. At very high frequencies ($\omega GL >> 1$) the steady-state output approaches $v_O = I_A R$. This also makes sense because at very high frequency the inductor acts like an open circuit that forces all of the input current to pass through the resistor in Figure 7–28. In between these two extremes the input current divides between the two paths in a manner that depends on the frequency. We will study frequency dependent responses in detail in later chapters. ■

Exercise 7–11

Find the forced component solution of the differential equation

$$10^{-3}\frac{dv}{dt} + v = 10 \cos \omega t$$

for the following frequencies:

(a) $\omega = 500$ rad/s
(b) $\omega = 1000$ rad/s
(c) $\omega = 2000$ rad/s

Answers:

(a) $v_F(t) = 8 \cos 500t + 4 \sin 500t \qquad t \geq 0$
(b) $v_F(t) = 5 \cos 1000t + 5 \sin 1000t \qquad t \geq 0$
(c) $v_F(t) = 2 \cos 2000t + 4 \sin 2000t \qquad t \geq 0$

DISCUSSION: *Converting these answers to an amplitude and phase angle as*

(a) $v_F(t) = 8.94 \cos (500t - 26.6°) \qquad t \geq 0$
(b) $v_F(t) = 7.07 \cos (1000t - 45°) \qquad t \geq 0$
(c) $v_F(t) = 4.47 \cos (2000t - 63.4°) \qquad t \geq 0$

we see that increasing the frequency of the input sinusoid decreases the amplitude and phase angle of the sinusoidal steady-state output of the circuit.

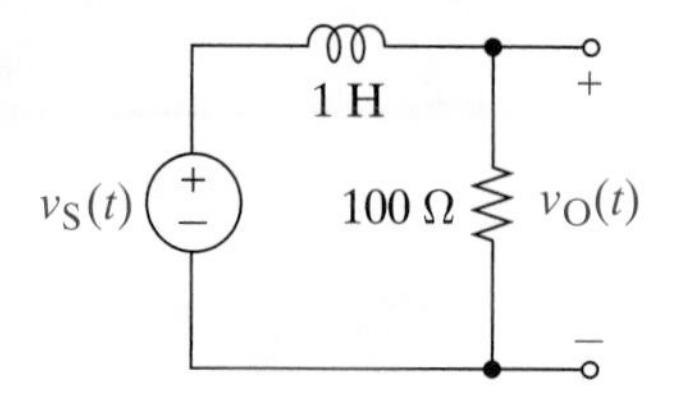

FIGURE 7–29

Exercise 7–12

The circuit in Figure 7–29 is operating in the sinusoidal steady state with

$$v_O(t) = 10\cos(100t - 45°)\ \text{V}$$

Find the source voltage $v_S(t)$.

Answer: $v_S(t) = 10\sqrt{2}\cos 100t$ V

7–5 THE SERIES *RLC* CIRCUIT

Second-order circuits contain two energy storage elements that cannot be replaced by a single equivalent element. They are called **second-order circuits** because the circuit differential equation involves the second derivative of the dependent variable. Although there is an endless number of such circuits, in this chapter we will concentrate on two classical forms: (1) the series *RLC* circuit and (2) the parallel *RLC* circuit. These two circuits illustrate almost all of the basic concepts of second-order circuits and serve as vehicles for studying the solution of second-order differential equations. In subsequent chapters we use Laplace transform techniques to analyze any second-order circuit.

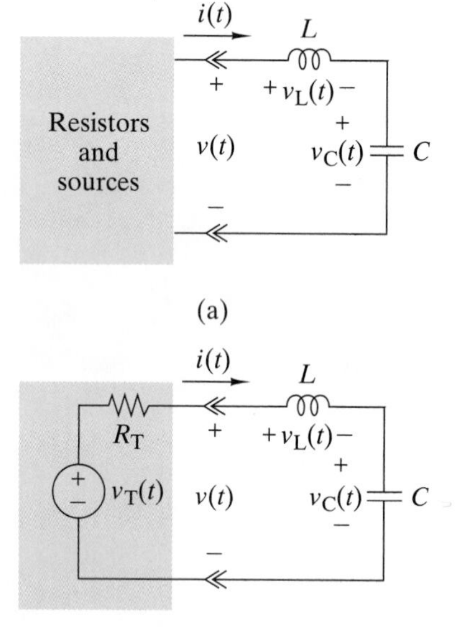

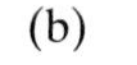

FIGURE 7–30 The series RLC circuit.

FORMULATING SERIES *RLC* CIRCUIT EQUATIONS

We begin with the circuit in Figure 7–30(a), where the inductor and capacitor are connected in series. The source-resistor circuit can be reduced to the Thévenin equivalent shown in Figure 7–25(b). The result is a circuit in which a voltage source, resistor, inductor, and capacitor are connected in series (hence the name **series *RLC* circuit**).

The first task is to develop the equations that describe the series *RLC* circuit. The Thévenin equivalent to the left of the interface in Figure 7–30(b) produces the KVL constraint

$$v + R_T i = v_T \tag{7–28}$$

Applying KVL around the loop on the right side of the interface yields

$$v = v_L + v_C \tag{7–29}$$

Finally, the *i–v* characteristics of the inductor and capacitor are

$$v_L = L\frac{di}{dt} \tag{7–30}$$

$$i = C\frac{dv_C}{dt} \tag{7–31}$$

Equations (7–28) through (7–31) are four independent equations in four unknowns (i, v, v_L, v_C). Collectively, this set of equations provides a complete description of the dynamics of the series *RLC* circuit. To find the circuit response using classical methods, we must derive a circuit equation containing only one of these unknowns.

We use circuit state variables as solution variables because they are continuous functions of time. In the series *RLC* circuit in Figure 7–25(b), there are two state variables: (1) the capacitor voltage $v_C(t)$ and (2) the inductor current $i(t)$. We first show how to describe the circuit using the capacitor voltage as the solution variable:

To derive a single equation in $v_C(t)$, we substitute Eqs. (7–29) and (7–31) into Eq. (7–28).

$$v_L + v_C + R_T C\frac{dv_C}{dt} = v_T \tag{7–32}$$

These substitutions eliminate the unknowns except v_C and v_L. To eliminate the inductor voltage, we substitute Eq. (7–31) into Eq. (7–30) to obtain

$$v_L = LC\frac{d^2v_C}{dt^2}$$

Substituting this result into Eq. (7–32) produces

$$\underbrace{LC\frac{d^2v_C}{dt^2}}_{v_L} + \underbrace{R_T C\frac{dv_C}{dt}}_{v_R} + v_C = v_T \tag{7–33}$$

In effect, this is a KVL equation around the loop in Figure 7–30(b), where the inductor and resistor voltages have been expressed in terms of the capacitor voltage.

Equation (7–33) is a second-order linear differential equation with constant coefficients. It is a second-order equation because the highest-order derivative is the second derivative of the dependent variable $v_C(t)$. The coefficients are constant because the circuit parameters L, C, and R_T do not change. The Thévenin voltage $v_T(t)$ is a known driving force. The initial conditions

$$v_C(0) = V_0 \text{ and } \frac{dv_C}{dt}(0) = \frac{1}{C}i(0) = \frac{I_0}{C} \tag{7–34}$$

are determined by the values of the capacitor voltage and inductor current at $t = 0$, V_0 and I_0.

In summary, the second-order differential equation in Eq. (7–33) characterizes the response of the series *RLC* circuit in terms of the capacitor voltage $v_C(t)$. Once the solution $v_C(t)$ is found, we can solve for every other voltage or current, including the inductor current, using the element and connection constraints in Eqs. (7–28) to (7–31).

Alternatively, we can characterize the series *RLC* circuit using the inductor current. We first write the capacitor *i–v* characteristics in integral form:

$$v_C(t) = \frac{1}{C}\int_0^t i(x)dx + v_C(0) \tag{7–35}$$

Equations (7–35), (7–30), and (7–29) are inserted into the interface constraint of Eq. (7–28) to obtain a single equation in the inductor current $i(t)$:

$$\underbrace{L\frac{di}{dt}}_{v_L} + \underbrace{\frac{1}{C}\int_0^t i(x)dx + v_C(0)}_{v_C} + \underbrace{R_T i}_{v_R} = v_T \tag{7–36}$$

In effect, this is a KVL equation around the loop in Figure 7–25(b), where the capacitor and resistor voltages have been expressed in terms of the inductor current.

Equation (7–36) is a second-order linear integrodifferential equation with constant coefficients. It is second order because it involves the first derivative and the first integral of the dependent variable $i(t)$. The coefficients are constant because the circuit parameters L, C, and R_T do not change.

The Thévenin equivalent voltage $v_T(t)$ is a known driving force, and the initial conditions are $v_C(0) = V_0$ and $i(0) = I_0$.

Equations (7–33) and (7–36) involve the same basic ingredients: (1) an unknown state variable, (2) three circuit parameters (R_T, L, C), (3) a known input $v_T(t)$, and (4) two initial conditions (V_0 and I_0). The only difference is that one expresses the sum of voltages around the loop in terms of the capacitor voltage, while the other uses the inductor current. Either equation characterizes the dynamics of the series *RLC* circuit because once a state variable is found, every other voltage or current can be found using the element and connection constraints.

ZERO-INPUT RESPONSE OF THE SERIES *RLC* CIRCUIT

The circuit dynamic response for $t \geq 0$ can be divided into two components: (1) the zero-input response caused by the initial conditions and (2) the zero-state response caused by driving forces applied after $t = 0$. Because the circuit is linear, we can solve for these responses separately and superimpose them to get the total response. We first deal with the zero-input response.

With $v_T = 0$ (zero-input) Eq. (7–33) becomes

$$LC\frac{d^2v_C}{dt^2} + R_TC\frac{dv_C}{dt} + v_C = 0 \tag{7–37}$$

This result is a second-order homogeneous differential equation in the capacitor voltage. Alternatively, we set $v_T = 0$ in Eq. (7–36) and differentiate once to obtain the following homogeneous differential equation in the inductor current:

$$LC\frac{d^2i}{dt^2} + R_TC\frac{di}{dt} + i = 0 \tag{7–38}$$

We observe that Eqs. (7–37) and (7–38) have exactly the same form except that the dependent variables are different. The zero-input response of the capacitor voltage and inductor current have the same general form. We do not need to study both to understand the dynamics of the series *RLC* circuit. In other words, in the series *RLC* circuit we can use either state variable to describe the zero-input response.

In the following discussion we will concentrate on the capacitor voltage response. Equation (7–37) requires the capacitor voltage, plus *RC* times its first derivative, plus *LC* times its second derivative to add to zero for all $t \geq 0$. The only way this can happen is for $v_C(t)$, its first derivative, and its second derivative to have the same waveform. No matter how many times we differentiate an exponential of the form e^{st}, we are left with a signal with the same waveform. This observation, plus our experience with first-order circuits, suggests that we try a solution of the form

$$v_C(t) = Ke^{st}$$

where the parameters K and s are to be evaluated. When the trial solution is inserted in Eq. (7–37), we obtain the condition

$$Ke^{st}(LCs^2 + R_TCs + 1) = 0$$

The function e^{st} cannot be zero for all $t \geq 0$. The condition $K = 0$ is not allowed because it is a trivial solution declaring that $v_C(t)$ is zero for all t. The only useful way to meet the condition is to require

$$LCs^2 + R_TCs + 1 = 0 \tag{7–39}$$

Equation (7–39) is the **characteristic equation** of the series *RLC* circuit. The characteristic equation is a quadratic because the circuit contains two energy storage elements. Inserting Ke^{st} into the homogeneous equation of the inductor current in Eq. (7–38) produces the same characteristic equation. Thus, Eq. (7–39) relates the zero-input response to circuit parameters for both state variables (hence the name *characteristic equation*).

In general, a quadratic characteristic equation has two roots:

$$s_1, s_2 = \frac{-R_T C \pm \sqrt{(R_T C)^2 - 4LC}}{2LC} \tag{7–40}$$

From the form of the expression under the radical in Eq. (7–40), we see that there are three distinct possibilities:

Case A: If $(R_T C)^2 - 4LC > 0$, there are two real, unequal roots ($s_1 = -\alpha_1 \neq s_2 = -\alpha_2$).

Case B: If $(R_T C)^2 - 4LC = 0$, there are two real, equal roots ($s_1 = s_2 = -\alpha$).

Case C: If $(R_T C)^2 - 4LC < 0$, there are two complex conjugate roots ($s_1 = -\alpha - j\beta$ and $s_2 = -\alpha + j\beta$).

The symbol j represents the imaginary number $\sqrt{-1}$.[2] Before dealing with the form of the zero-input response for each case, we consider an example.

EXAMPLE 7–14

A series *RLC* circuit has $C = 0.25\ \mu\text{F}$ and $L = 1$ H. Find the roots of the characteristic equation for $R_T = 8.5\ \text{k}\Omega$, $4\ \text{k}\Omega$, and $1\ \text{k}\Omega$.

SOLUTION:

For $R_T = 8.5\ \text{k}\Omega$, the characteristic equation is

$$0.25 \times 10^{-6} s^2 + 2.125 \times 10^{-3} s + 1 = 0$$

whose roots are

$$s_1, s_2 = -4250 \pm \sqrt{(3750)^2} = -500, \quad -8000$$

These roots illustrate case A. The quantity under the radical is positive, and there are two real, unequal roots at $s_1 = -500$ and $s_2 = -8000$.

For $R_T = 4\ \text{k}\Omega$, the characteristic equation is

$$0.25 \times 10^{-6} s^2 + 10^{-3} s + 1 = 0$$

whose roots are

$$s_1, s_2 = -2000 \pm \sqrt{4 \times 10^6 - 4 \times 10^6} = -2000$$

This is an example of case B. The quantity under the radical is zero, and there are two real, equal roots at $s_1 = s_2 = -2000$.

For $R_T = 1\ \text{k}\Omega$ the characteristic equation is

$$0.25 \times 10^{-6} s^2 + 0.25 \times 10^{-3}\, s + 1 = 0$$

---

2 Mathematicians use the letter i to represent $\sqrt{-1}$. Electrical engineers use j, since the letter i represents electric current.

whose roots are

$$s_1, s_2 = -500 \pm 500\sqrt{-15}$$

The quantity under the radical is negative, illustrating case C.

$$s_1, s_2 = -500 \pm j500\sqrt{15}$$

In case C the two roots are complex conjugates. ■

Exercise 7–13

For a series *RLC* circuit:

(a) Find the roots of the characteristic equation when $R_T = 2$ kΩ, $L = 100$ mH, and $C = 0.4$ μF.
(b) For $L = 100$ mH, select the values of R_T and C so the roots of the characteristic equation are $s_1, s_2 = -1000 \pm j2000$.
(c) Select the values of R_T, L, and C so $s_1 = s_2 = -10^4$.

Answers:

(a) $s_1 = -1340, s_2 = -18{,}660$
(b) $R_T = 200$ Ω, $C = 2$ μF
(c) There is no unique answer to part (c) since the requirement

$$\begin{aligned}(10^{-4}s + 1)^2 &= LCs^2 + R_TCs + 1\\ &= 10^{-8}s^2 + 10^{-4}s + 1\end{aligned}$$

gives two equations $R_TC = 10^{-4}$ and $LC = 10^{-8}$ in three unknowns. One solution is to select $C = 1$ μF, which yields $L = 10$ mH and $R_T = 200$ Ω.

We have not introduced complex numbers simply to make things complex. Complex numbers arise quite naturally in practical physical situations involving nothing more than factoring a quadratic equation. The ability to deal with complex numbers is essential to our study. For those who need a review of such matters, there is a concise discussion in Appendix C.

FORM OF THE ZERO-INPUT RESPONSE

Since the characteristic equation has two roots, there are two solutions to the homogeneous differential equation:

$$\begin{aligned}v_{C1}(t) &= K_1e^{s_1t}\\ v_{C2}(t) &= K_2e^{s_2t}\end{aligned}$$

That is,

$$LC\frac{d^2}{dt^2}(K_1e^{s_1t}) + R_TC\frac{d}{dt}(K_1e^{s_1t}) + K_1e^{s_1t} = 0$$

and

$$LC\frac{d^2}{dt^2}(K_2e^{s_2t}) + R_TC\frac{d}{dt}(K_2e^{s_2t}) + K_2e^{s_2t} = 0$$

The sum of these two solutions is also a solution since

$$LC\frac{d^2}{dt^2}(K_1e^{s_1t} + K_2e^{s_2t}) + R_TC\frac{d}{dt}(K_1e^{s_1t} + K_2e^{s_2t}) + K_1e^{s_1t} + K_2e^{s_2t} = 0$$

Therefore, the general solution for the zero-input response is of the form

$$v_C(t) = K_1e^{s_1t} + K_2e^{s_2t} \tag{7–41}$$

The constants K_1 and K_2 can be found using the initial conditions given in Eq. (7–34). At $t = 0$ the condition on the capacitor voltage yields

$$v_C(0) = V_0 = K_1 + K_2 \tag{7–42}$$

To use the initial condition on the inductor current, we differentiate Eq. (7–41).

$$\frac{dv_C}{dt} = K_1s_1e^{s_1t} + K_2s_2e^{s_2t}$$

Using Eq. (7–34) to relate the initial value of the derivative of the capacitor voltage to the initial inductor current $i(0)$ yields

$$\frac{dv_C(0)}{dt} = \frac{I_0}{C} = K_1s_1 + K_2s_2 \tag{7–43}$$

Equations (7–42) and (7–43) provide two equations in the two unknown constants K_1 and K_2:

$$K_1 + K_2 = V_0$$
$$s_1K_1 + s_2K_2 = I_0/C$$

The solutions of these equations are

$$K_1 = \frac{s_2V_0 - I_0/C}{s_2 - s_1} \text{ and } K_2 = \frac{-s_1V_0 + I_0/C}{s_2 - s_1}$$

Inserting these solutions back into Eq. (7–41) yields

$$v_C(t) = \frac{s_2V_0 - I_0/C}{s_2 - s_1}e^{s_1t} + \frac{-s_1V_0 + I_0/C}{s_2 - s_1}e^{s_2t} \qquad t \geqslant 0 \tag{7–44}$$

Equation (7–44) is the general zero-input response of the series RLC circuit. The response depends on two initial conditions V_0 and I_0, and the circuit parameters R_T, L, and C since s_1 and s_2 are the roots of the characteristic equation $LCs^2 + R_TCs + 1 = 0$. The response takes on different forms depending on whether the roots s_1 and s_2 fall under case A, B, or C.

For case A the two roots are real and distinct. Using the notation $s_1 = -\alpha_1$ and $s_2 = -\alpha_2$, the form zero-input response for $t \geq 0$ is

$$v_C(t) = \left[\frac{\alpha_2V_0 + I_0/C}{\alpha_2 - \alpha_1}\right]e^{-\alpha_1t} + \left[\frac{-\alpha_1V_0 - I_0/C}{\alpha_2 - \alpha_1}\right]e^{-\alpha_2t} \tag{7–45}$$

For case A the response is the sum of two exponential functions similar to the double exponential signal treated in Example 5–14. The function has two time constants $1/\alpha_1$ and $1/\alpha_2$. The time constants can be greatly different, or nearly equal, but they cannot be equal because we would have case B.

With case B the roots are real and equal. Using the notation $s_1 = s_2 = -\alpha$, the general form in Eq. (7–44) becomes

$$v_C(t) = \frac{(\alpha V_0 + I_0/C)e^{-\alpha t} + (-\alpha V_0 - I_0/C)e^{-\alpha t}}{\alpha - \alpha}$$

We immediately see a problem here because the denominator vanishes. However, a closer examination reveals that the numerator vanishes as well, so the solution reduces to the indeterminate form 0/0. To investigate the indeterminacy, we let $s_1 = -\alpha$ and $s_2 = -\alpha + x$, and we explore the situation as x approaches zero. Inserting s_1 and s_2 in this notation in Eq. (7–44) produces

$$v_C(t) = V_0 e^{-\alpha t} + \left[\frac{-\alpha V_0 - I_0/C}{x}\right] e^{-\alpha t} + \left[\frac{\alpha V_0 + I_0/C}{x}\right] e^{-\alpha t} e^{xt}$$

which can be arranged in the form

$$v_C(t) = e^{-\alpha t}\left[V_0 - (\alpha V_0 + I_0/C)\frac{1 - e^{xt}}{x}\right]$$

We see that the indeterminacy comes from the term $(1 - e^{xt})/x$, which reduces to 0/0 as x approaches zero. Application of l'Hôpital's rule reveals

$$\lim_{x \to 0} \frac{1 - e^{xt}}{x} = \lim_{x \to 0} \frac{-te^{xt}}{1} = -t$$

This result removes the indeterminacy, and as x approaches zero the zero-input response reduces to

$$v_C(t) = V_0 e^{-\alpha t} + (\alpha V_0 + I_0/C)\, te^{-\alpha t} \qquad t \geq 0 \tag{7–46}$$

For case B the response includes an exponential and the damped ramp studied in Example 5–12. The damped ramp is required, rather than two exponentials, because in case B the two equal roots produce the same exponential function.

Case C produces complex conjugate roots of the form

$$s_1 = -\alpha - j\beta \text{ and } s_2 = -\alpha + j\beta$$

Inserting these roots into Eq. (7–44) yields

$$v_C(t) = \left[\frac{(-\alpha + j\beta)\, V_0 - I_0/C}{j2\beta}\right] e^{-\alpha t} e^{-j\beta t} + \left[\frac{(\alpha + j\beta)\, V_0 + I_0/C}{j2\beta}\right] e^{-\alpha t} e^{j\beta t}$$

which can be arranged in the form

$$v_C(t) = V_0 e^{-\alpha t}\left[\frac{e^{j\beta t} + e^{-j\beta t}}{2}\right] + \frac{\alpha V_0 + I_0/C}{\beta} e^{-\alpha t}\left[\frac{e^{j\beta t} - e^{-j\beta t}}{j2}\right] \tag{7–47}$$

The expressions within the brackets have been arranged in a special way for the following reasons. Euler's relationships for an imaginary exponential are written as

$$e^{j\theta} = \cos\theta + j\sin\theta$$

and

$$e^{-j\theta} = \cos\theta - j\sin\theta$$

When we add and subtract these equations, we obtain

$$\cos\theta = \frac{e^{j\theta} + e^{-j\theta}}{2} \quad \text{and} \quad \sin\theta = \frac{e^{j\theta} - e^{-j\theta}}{2j}$$

Comparing these expressions for sin θ and cos θ with the complex terms in Eq. (7–47) reveals that we can write $v_C(t)$ in the form

$$v_C(t) = V_0 e^{-\alpha t} \cos \beta t + \frac{\alpha V_0 + I_0/C}{\beta} e^{-\alpha t} \sin \beta t \qquad t \geqslant 0$$

For case C the response contains the damped sinusoid studied in Example 5–13. The real part of the roots (α) provides the exponent coefficient in the exponential function, while the imaginary part (β) defines the frequency of the sinusoidal oscillation.

In summary, the roots of the characteristic equation affect the form of the zero-input response in the following ways. In case A the two roots are real and unequal ($s_1 = -\alpha_1 \neq s_2 = -\alpha_2$) and the zero-input response is the sum of two exponentials of the form

$$v_C(t) = K_1 e^{-\alpha_1 t} + K_2 e^{-\alpha_2 t} \tag{7–48a}$$

In case B the two roots are real and equal ($s_1 = s_2 = -\alpha$) and the zero-input response is the sum of an exponential and a damped ramp.

$$v_C(t) = K_1 e^{-\alpha t} + K_2 t e^{-\alpha t} \tag{7–48b}$$

In case C the two roots are complex conjugates ($s_1 = -\alpha - j\beta$, $s_2 = -\alpha + j\beta$) and the zero-input response is the sum of a damped cosine and a damped sine.

$$v_C(t) = K_1 e^{-\alpha t} \cos \beta t + K_2 e^{-\alpha t} \sin \beta t \tag{7–48c}$$

In determining the zero-input response we use the parameters s, α, and β. At various points in the development, these parameters appear in expressions such as e^{st}, $e^{-\alpha t}$, and $e^{j\beta t}$. Since the exponent of e must be dimensionless, the parameters s, α, and β all have the dimensions of the reciprocal of time, or equivalently, frequency. Collectively, we say that s, α, and β define the **natural frequencies** of the circuit. When it is necessary to distinguish between these three parameters we say that s is the **complex frequency**, α is the **neper frequency**, and β is the **radian frequency**. The importance of this notation will become clear as we proceed through subsequent chapters of this book. To be consistent with expressions such as $s = -\alpha + j\beta$, we specify numerical values of s, α, and β in units of radians per second (rad/s).[3]

The constants K_1 and K_2 in Eqs. (7–48a), (7–48b), and (7–48c) are determined by the initial conditions on two state variables, as illustrated in the following example.

EXAMPLE 7–15

The circuit of Figure 7–31 has $C = 0.25$ μF and $L = 1$ H. The switch has been open for a long time and is closed at $t = 0$. Find the capacitor voltage for $t \geq 0$ for (a) $R = 8.5$ kΩ, (b) $R = 4$ kΩ, and (c) $R = 1$ kΩ. The initial conditions are $I_0 = 0$ and $V_0 = 15$ V.

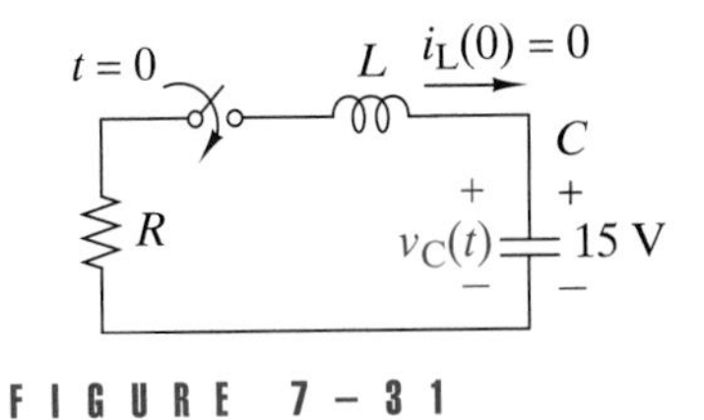

FIGURE 7–31

3 The term *neper frequency* honors the sixteenth-century mathematician John Napier, who invented the base *e* or natural logarithms. The term *complex frequency* was apparently first used about 1900 by the British engineer Oliver Heaviside.

SOLUTION:

The roots of the characteristic equation for these three values of resistance are found in Example 7–14. We are now in a position to use those results to find the corresponding zero-input responses.

(a) In Example 7–14 the value R = 8.5 kΩ yields case A with roots s_1 = –500 and s_2 = –8000. The corresponding zero-input solution takes the form in Eq. (7–48a).

$$v_C(t) = K_1 e^{-500t} + K_2 e^{-8000t}$$

The initial conditions yield two equations in the constants K_1 and K_2:

$$v_C(0) = V_0 = 15 = K_1 + K_2$$

$$\frac{dv_C(0)}{dt} = \frac{I_0}{C} = 0 = -500K_1 - 8000K_2$$

Solving these equations yields K_1 = 16 and K_2 = –1, so that the zero-input response is

$$v_C(t) = 16e^{-500t} - e^{-8000t} \text{ V} \qquad t \geq 0$$

(b) In Example 7–14 the value R = 4 kΩ yields case B with roots $s_1 = s_2$ = –2000. The zero-input response takes the form in Eq. (7–48b):

$$v_C(t) = K_1 e^{-2000t} + K_2 t e^{-2000t}$$

The initial conditions yield two equations in the constants K_1 and K_2:

$$v_C(0) = V_0 = 15 = K_1$$

$$\frac{dv_C(0)}{dt} = \frac{I_0}{C} = 0 = -2000\,K_1 + K_2$$

Solving these equations yields K_1 = 15 and K_2 = 2000 × 15, so the zero-input response is

$$v_C(t) = 15e^{-2000t} + 15(2000t)e^{-2000t} \text{ V} \qquad t \geq 0$$

(c) In Example 7–14 the value R = 1 kΩ yields case C with roots s_1, s_2 = $-500 \pm j500\sqrt{15}$. The zero-input response takes the form in Eq. (7–48c):

$$v_C(t) = K_1 e^{-500t} \cos(500\sqrt{15})t + K_2 e^{-500t} \sin(500\sqrt{15})t$$

The initial conditions yield two equations in the constants K_1 and K_2:

$$v_C(0) = V_0 = 15 = K_1$$

$$\frac{dv_C(0)}{dt} = \frac{I_0}{C} = 0 = -500K_1 + 500\sqrt{15}K_2$$

which yield K_1 = 15 and $K_2 = \sqrt{15}$, so the zero-input response is

$$v_C(t) = 15e^{-500t} \cos(500\sqrt{15})t + \sqrt{15}\, e^{-500t} \sin(500\sqrt{15})t \text{ V} \qquad t \geq 0$$

Figure 7–32 shows plots of these responses. All three responses start out at 15 V (the initial condition) and all eventually decay to zero. The temporal decay of the responses is caused by energy loss in the circuit and is called **damping**. The case A response does not change sign and is called the **overdamped** response. The case C response undershoots and then oscillates about the final value. This response is said to be **underdamped** because there is not enough damping to prevent these oscillations. The case B response is said to be **critically damped** since it is a special case at the boundary between overdamping and underdamping.

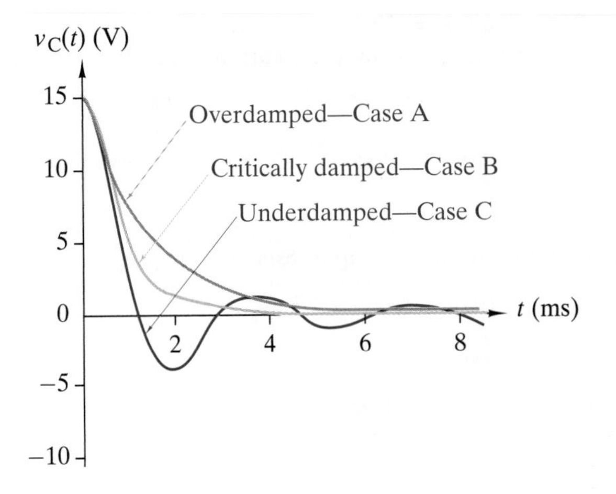

FIGURE 7–32

EXAMPLE 7–16

In a series *RLC* circuit the zero-input voltage across the 1 μF capacitor is

$$v_C(t) = 10\, e^{-1000t} \sin 2000\, t \quad \text{V} \quad t \geq 0$$

(a) Find the circuit characteristic equation.
(b) Find the *R* and *L*.
(c) Find $i_L(t)$ for $t \geq 0$.
(d) Find the initial values of the state variables.

SOLUTION:

(a) The circuit is underdamped because the zero-input response is a damped sine with $\alpha = 1000$ and $\beta = 2000$ rad/s. The characteristic equation is

$$(s + 1000 - j2000)(s + 1000 + j2000) = s^2 + 2000\, s + 5 \times 10^6 = 0$$

(b) The characteristic equation of a series *RLC* circuit [Eq. (7–39)] can be written as

$$s^2 + \frac{R}{L} s + \frac{1}{LC} = 0$$

Comparing this term by term to the result in (a) yields the constraints

$$\frac{R}{L} = 2000 \quad \text{and} \quad \frac{1}{LC} = 5 \times 10^6$$

Since $C = 1$ μF, we find that $L = 0.2$ H and $R = 400\ \Omega$.

(c) In a series circuit KCL requires $i_L(t) = i_C(t)$. Hence, the inductor current is

$$i_L(t) = C\frac{dv_C(t)}{dt} = 10^{-6}\frac{d}{dt}[10\, e^{-1000t} \sin 2000\, t]$$
$$= -10\, e^{-1000t} \sin 2000\, t + 10\, e^{-1000t} \cos 2000\, t \quad \text{mA} \quad t \geq 0$$

(d) By inspection, the initial values of the state variables are $v_C(0) = 0$ and $i_L(0) = 20$ mA. ■

Exercise 7–14

In a series *RLC* circuit, $R = 250\ \Omega$, $L = 10$ mH, $C = 1\mu$F, $V_0 = 0$, and $I_0 = 30$ mA. Find the capacitor voltage and inductor current for $t \geq 0$.

Answers:

$$v_C(t) = 2\,e^{-5000t} - 2\,e^{-20000t} \quad \text{V}$$

$$i_L(t) = -10\,e^{-5000t} + 40\,e^{-20000t} \quad \text{mA}$$

Exercise 7–15

In a series *RLC* circuit the zero-input responses are

$$v_C(t) = 2000\,t\,e^{-500t} \quad \text{V}$$

$$i_L(t) = 3.2\,e^{-500t} - 1600\,t\,e^{-500t} \quad \text{mA}$$

(a) Find the circuit characteristic equation.
(b) Find the initial values of the state variables.
(c) Find *R*, *L*, and *C*.

Answers:

(a) $s^2 + 1000s + 25 \times 10^4 = 0$
(b) $V_0 = 0$, $I_0 = 3.2$ mA
(c) $R = 2.5$ kΩ, $L = 2.5$ H, $C = 1.6\ \mu$F

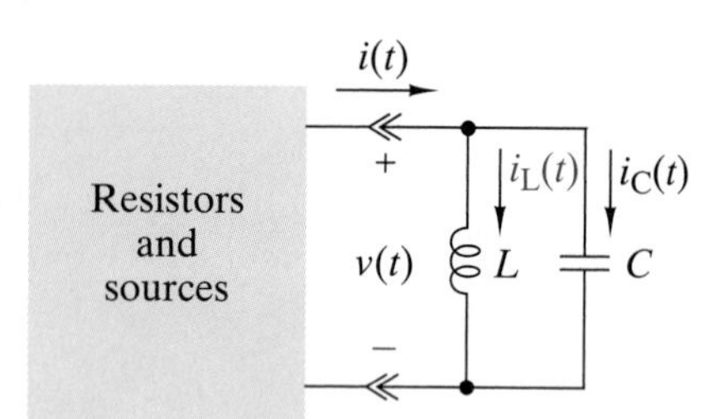

(a)

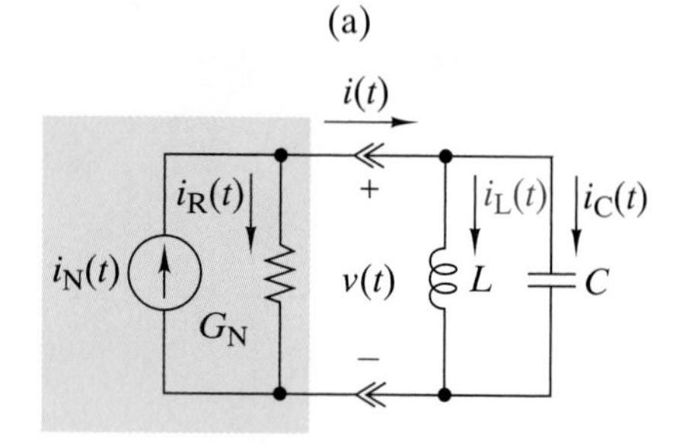

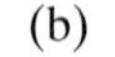

(b)

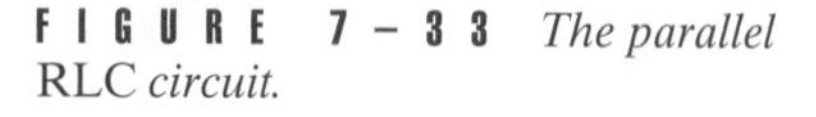

FIGURE 7–33 *The parallel* RLC *circuit.*

7–6 THE PARALLEL *RLC* CIRCUIT

The inductor and capacitor in Figure 7–33(a) are connected in parallel. The source-resistor circuit can be reduced to the Norton equivalent shown in Figure 7–33(b). The result is a **parallel *RLC* circuit** consisting of a current source, resistor, inductor, and capacitor. Our first task is to develop a differential equation for this circuit. We expect to find a second-order differential equation because there are two energy storage elements.

The Norton equivalent to the left of the interface introduces the constraint

$$i + G_N v = i_N \tag{7–49}$$

Writing a KCL equation at the interface yields

$$i = i_L + i_C \tag{7–50}$$

The *i–v* characteristics of the inductor and capacitor are

$$i_C = C\frac{dv}{dt} \tag{7–51}$$

$$v = L\frac{di_L}{dt} \tag{7–52}$$

Equations (7–49) through (7–52) provide four independent equations in four unknowns (i, v, i_L, i_C). Collectively these equations describe the dynamics of the parallel *RLC* circuit. To solve for the circuit response using classical methods, we must derive a circuit equation containing only one of these four variables.

We prefer using state variables because they are continuous. To obtain a single equation in the inductor current, we substitute Eqs. (7–50) and (7–52) into Eq. (7–49):

$$i_L + i_C + G_N L \frac{di_L}{dt} = i_N \tag{7–53}$$

The capacitor current can be eliminated from this result by substituting Eq. (7–52) into Eq. (7–51) to obtain

$$i_C = LC \frac{d^2 i_L}{dt^2} \tag{7–54}$$

Inserting this equation into Eq. (7–53) produces

$$LC \frac{d^2 i_L}{dt^2} + G_N L \frac{di_L}{dt} + i_L = i_N \tag{7–55}$$

$$i_C \quad + \quad i_R \quad + i_L = i_N$$

This result is a KCL equation in which the resistor and capacitor currents are expressed in terms of the inductor current.

Equation (7–55) is a second-order linear differential equation of the same form as the series *RLC* circuit equation in Eq. (7–33). In fact, if we interchange the following quantities:

$$v_C \leftrightarrow i_L \qquad L \leftrightarrow C \qquad R_T \leftrightarrow G_N \qquad v_T \leftrightarrow i_N$$

we change one equation into the other. The two circuits are duals, which means that the results developed for the series case apply to the parallel circuit with the preceding duality interchanges.

However, it is still helpful to outline the major features of the analysis of the parallel *RLC* circuit. The initial conditions in the parallel circuit are the initial inductor current I_0 and capacitor voltage V_0. The initial inductor current provides the condition $i_L(0) = I_0$ for the differential equation in Eq. (7–55). By using Eq. (7–52), the initial capacitor voltage specifies the initial rate of change of the inductor current as

$$\frac{di_L(0)}{dt} = \frac{1}{L} v_C(0) = \frac{1}{L} V_0$$

These initial conditions are the dual of those obtained for the series *RLC* circuit in Eq. (7–34).

To solve for the zero-input response, we set $i_N = 0$ in Eq. (7–55) and obtain a homogeneous equation in the inductor current:

$$LC \frac{d^2 i_L}{dt^2} + G_N L \frac{di_L}{dt} + i_L = 0$$

A trial solution of the form $i_L = Ke^{st}$ leads to the characteristic equation

$$LCs^2 + G_N Ls + 1 = 0 \tag{7–56}$$

The characteristic equation is quadratic because there are two energy storage elements in the parallel *RLC* circuit. The characteristic equation has two roots:

$$s_1, s_2 = \frac{-G_N L \pm \sqrt{(G_N L)^2 - 4LC}}{2LC}$$

and, as in the series case, there are three distinct cases:

Case A: If $(G_N L)^2 - 4LC > 0$, there are two unequal real roots $s_1 = -\alpha_1$ and $s_2 = -\alpha_2$ and the zero-input response is the overdamped form

$$i_L(t) = K_1 e^{-\alpha_1 t} + K_2 e^{-\alpha_2 t} \qquad t \geq 0 \tag{7–57}$$

Case B: If $(G_N L)^2 - 4LC = 0$, there are two real equal roots $s_1 = s_2 = -\alpha$ and the zero-input response is the critically damped form

$$i_L(t) = K_1 e^{-\alpha t} + K_2 t e^{-\alpha t} \qquad t \geq 0 \tag{7–58}$$

Case C: If $(G_N L)^2 - 4LC < 0$, there are two complex, conjugate roots $s_1, s_2 = -\alpha \pm j\beta$ and the zero-input response is the underdamped form

$$i_L(t) = K_1 e^{-\alpha t} \cos \beta t + K_2 e^{-\alpha t} \sin \beta t \qquad t \geq 0 \tag{7–59}$$

The analysis results for the series *RLC* circuit apply to the parallel *RLC* case with the appropriate duality replacements. In particular, the form of overdamped, critically damped, and underdamped response applies to both circuits. The forms of the responses in Eqs. (7–57), (7–58), and (7–59) have been written with two arbitrary constants K_1 and K_2. The following example shows how to evaluate these constants using the initial conditions for the two state variables.

EXAMPLE 7–17

In a parallel *RLC* circuit $R_T = 1/G_N = 500\ \Omega$, $C = 1\ \mu F$, $L = 0.2$ H. The initial conditions are $I_0 = 50$ mA and $V_0 = 0$. Find the zero-input response of inductor current, resistor current, and capacitor voltage.

SOLUTION:

From Eq. (7–56) the circuit characteristic equation is

$$LCs^2 + G_N Ls + 1 = 2 \times 10^{-7} s^2 + 4 \times 10^{-4} s + 1 = 0$$

The roots of the characteristic equation are

$$s_1, s_2 = \frac{-4 \times 10^{-4} \pm \sqrt{16 \times 10^{-8} - 8 \times 10^{-7}}}{4 \times 10^{-7}} = -1000 \pm j2000$$

Since roots are complex conjugates, we have the underdamped case. The zero-input response of the inductor current takes the form of Eq. (7–59).

$$i_L(t) = K_1 e^{-1000t} \cos 2000t + K_2 e^{-1000t} \sin 2000t \qquad t \geq 0$$

The constants K_1 and K_2 are evaluated from the initial conditions. At $t = 0$ the inductor current reduces to

$$i_L(0) = I_0 = K_1 e^0 \cos 0 + K_2 e^0 \sin 0 = K_1$$

We conclude that $K_1 = I_0 = 50$ mA. To find K_2 we use the initial capacitor voltage. In a parallel *RLC* circuit the capacitor and inductor voltages are equal.

$$L\frac{di_L}{dt} = v_C(t)$$

In this example the initial capacitor voltage is zero, so the initial rate of change of inductor current is zero at $t = 0$. Differentiating the zero-input response produces

$$\begin{aligned}\frac{di_L}{dt} &= -2000K_1e^{-1000t}\sin 2000t - 1000K_1e^{-1000t}\cos 2000t \\ &\quad - 1000K_2e^{-1000t}\sin 2000t + 2000K_2e^{-1000t}\cos 2000t\end{aligned}$$

Evaluating this expression at $t = 0$ yields

$$\begin{aligned}\frac{di_L}{dt}(0) &= -2000K_1e^0\sin 0 - 1000K_1e^0\cos 0 \\ &\quad - 1000K_2e^0\sin 0 + 2000K_2e^0\cos 0 \\ &= -1000K_1 + 2000K_2 = 0\end{aligned}$$

The derivative initial condition gives condition $K_2 = K_1/2 = 25$ mA. Given the values of K_1 and K_2, the zero-input response of the inductor current is

$$i_L(t) = 50e^{-1000t}\cos 2000t + 25e^{-1000t}\sin 2000t \text{ mA} \qquad t \geqslant 0$$

The zero-input response of the inductor current allows us to solve for every voltage and current in the parallel *RLC* circuit. For example, using the *i*–*v* characteristic of the inductor, we obtain the inductor voltage:

$$v_L(t) = L\frac{di_L}{dt} = -25e^{-1000t}\sin 2000t \text{ V} \qquad t \geqslant 0$$

Since the elements are connected in parallel, we obtain the capacitor voltage and resistor current as

$$v_C(t) = v_L(t) = L\frac{di_L}{dt} = -25e^{-1000t}\sin 2000t \text{ V} \qquad t \geqslant 0$$

$$i_R(t) = \frac{v_L(t)}{R} = -50e^{-1000t}\sin 2000t \text{ mA} \qquad t \geqslant 0$$

■

EXAMPLE 7–18

The switch in Figure 7–34 has been open for a long time and is closed at $t = 0$.

(a) Find the initial conditions at $t = 0$.
(b) Find the inductor current for $t \geq 0$.
(c) Find the capacitor voltage and current through the switch for $t \geq 0$.

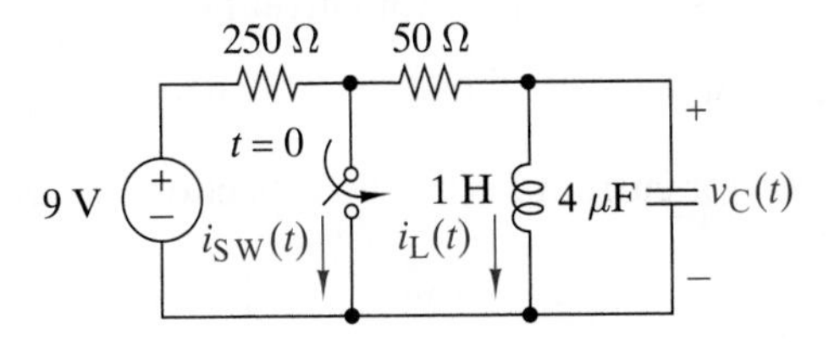

FIGURE 7–34

SOLUTION:

(a) For $t < 0$ the circuit is in the dc steady state, so the inductor acts like a short circuit and the capacitor like an open circuit. Since the inductor shorts out the capacitor, the initial conditions just prior to closing the switch at $t = 0$ are

$$v_C(0) = 0 \quad . \; i_L(0) = \frac{9}{250 + 50} = 30 \text{ mA}$$

(b) For $t \geq 0$ the circuit is a zero-input parallel RLC circuit with initial conditions found in (a). The circuit characteristic equation is

$$LCs^2 + GLs + 1 = 4 \times 10^{-6}s^2 + 2 \times 10^{-2}s + 1 = 0$$

The roots of this equation are

$$s_1 = -50.51 \quad \text{and} \quad s_2 = -4950$$

The circuit is overdamped (case A), since the roots are real and unequal. The general form of the inductor current zero-input response is

$$i_L(t) = K_1 e^{-50.51t} + K_2 e^{-4950t} \qquad t \geqslant 0$$

The constants K_1 and K_2 are found using the initial conditions. At $t = 0$ the zero-input response is

$$i_L(0) = K_1 e^0 + K_2 e^0 = K_1 + K_2 = 30 \times 10^{-3}$$

The initial capacitor voltage establishes an initial condition on the derivative of the inductor current since

$$L\frac{di_L}{dt}(0) = v_C(0) = 0$$

The derivative of the inductor response at $t = 0$ is

$$\frac{di_L}{dt}(0) = (-50.51K_1 e^{-50.51t} - 4950K_2 e^{-4950t})\big|_{t=0}$$

$$= -50.51K_1 - 4950K_2 = 0$$

The initial conditions on inductor current and capacitor voltage produce two equations in the unknown constants K_1 and K_2:

$$K_1 + K_2 = 30 \times 10^{-3}$$

$$-50.51K_1 - 4950K_2 = 0$$

Solving these equations yields $K_1 = 30.3$ mA and $K_2 = -0.309$ mA. The zero-input response of the inductor current is

$$i_L(t) = 30.3e^{-50.51t} - 0.309e^{-4950t} \text{ mA} \qquad t \geqslant 0$$

(c) Given the inductor current in (b), the capacitor voltage is

$$v_C(t) = L\frac{di_L}{dt} = -1.53e^{-50.51t} + 1.53e^{-4950t} \text{ V} \qquad t \geqslant 0$$

For $t \geq 0$ the current $i_{SW}(t)$ is the current through the 50-Ω resistor plus the current through the 250-Ω resistor.

$$i_{SW}(t) = i_{250} + i_{50} = \frac{9}{250} + \frac{v_C(t)}{50}$$

$$= 36 - 30.6e^{-50.51t} + 30.6e^{-4950t} \text{ mA} \qquad t \geq 0$$

■

Exercise 7–16

The zero-input responses of a parallel *RLC* circuit are observed to be

$$i_L(t) = 10te^{-2000t} \text{ A}$$

$$v_C(t) = 10e^{-2000t} - 20000te^{-2000t} \text{ V} \qquad t \geq 0$$

(a) What is the circuit characteristic equation?
(b) What are the initial values of the state variables?
(c) What are the values of *R*, *L*, and *C*?
(d) Write an expression for the current through the resistor.

Answers:

(a) $s^2 + 4000s + 4 \times 10^6 = 0$
(b) $i_L(0) = 0$, $v_C(0) = 10$ V
(c) $L = 1$ H, $C = 0.25$ μF, $R = 1$ kΩ
(d) $i_R(t) = 10e^{-2000t} - 20000te^{-2000t}$ mA $\quad t \geq 0$

7–7 SECOND-ORDER CIRCUIT STEP RESPONSE

The step response provides important insights into the response of dynamic circuits in general. So it is natural that we investigate the step response of second-order circuits. In Chapter 11 we will develop general techniques for determining the step response of any linear circuit. However, in this introduction we use classical methods of solving differential equations to find the step response of second-order circuits.

The general second-order linear differential equation with a step function input has the form

$$a_2 \frac{d^2y(t)}{dt^2} + a_1 \frac{dy(t)}{dt} + a_0 y(t) = Au(t) \tag{7–60}$$

where $y(t)$ is a voltage or current response, $Au(t)$ is the step function input, and a_2, a_1, and a_0 are constant coefficients. The step response is the general solution of this differential equation for $t \geq 0$. The step response can be found by partitioning $y(t)$ into forced and natural components:

$$y(t) = y_N(t) + y_F(t) \tag{7–61}$$

The natural response $y_N(t)$ is the general solution of the homogeneous equation (input set to zero), while the forced response $y_F(t)$ is a particular solution of the equation

$$a_2 \frac{d^2y_F}{dt^2} + a_1 \frac{dy_F}{dt} + a_0 y_F = A \qquad t \geq 0$$

Since A is a constant, it follows that dA/dt and d^2A/dt^2 are both zero, so it is readily apparent that $y_F = A/a_0$ is a particular solution of this differential equation. So much for the forced response.

Turning now to the natural response, we seek a general solution of the homogeneous equation. The natural response has the same form as the zero-state response studied in the previous section. In a second-order circuit the zero-state and natural responses take one of the three possible forms: overdamped, critically damped, or underdamped. To describe the three possible forms, we introduce two new parameters: ω_0 (omega zero) and ζ (zeta). These parameters are defined in terms of the coefficients of the general second-order equation in Eq. (7–60):

$$\omega_0^2 = \frac{a_0}{a_2} \quad \text{and} \quad 2\zeta\omega_0 = \frac{a_1}{a_2} \tag{7–62}$$

The parameter ω_0 is called the **undamped natural frequency** and ζ is called the **damping ratio**. Using these two parameters, the general homogeneous equation is written in the form

$$\frac{d^2y_N(t)}{dt^2} + 2\zeta\omega_0\frac{dy_N(t)}{dt} + \omega_0^2 y_N(t) = 0 \tag{7–63}$$

The left side of Eq. (7–63) is called the **standard form** of the second-order linear differential equation. When a second-order equation is arranged in this format, we can determine its damping ratio and undamped natural frequency by equating its coefficients with those in the standard form. For example, in standard form the homogeneous equation for the series *RLC* circuit in Eq. (7–37) is

$$\frac{d^2v_C}{dt^2} + \frac{R_T}{L}\frac{dv_C}{dt} + \frac{1}{LC}v_C = 0$$

Equating like terms yields

$$\omega_0^2 = \frac{1}{LC} \quad \text{and} \quad 2\zeta\omega_0 = \frac{R_T}{L}$$

for the series *RLC* circuit. In an analysis situation the circuit element values determine the value of the parameters ω_0 and ζ. In a design situation we select L and C to obtain a specified ω_0 and then select R_T to obtain a specified ζ.

To determine the form of the natural response using ω_0 and ζ, we insert a trial solution $y_N(t) = Ke^{st}$ into the standard form in Eq. (7–63). The trial function Ke^{st} is a solution provided that

$$Ke^{st}[s^2 + 2\zeta\omega_0 s + \omega_0^2] = 0$$

Since $K = 0$ is the trivial solution and $e^{st} \neq 0$ for all $t \geq 0$, the only useful way for the right side of this equation to be zero for all t is for the quadratic expression within the brackets to vanish. The quadratic expression is the characteristic equation for the general second-order differential equation:

$$s^2 + 2\zeta\omega_0 s + \omega_0^2 = 0$$

The roots of the characteristic equation are

$$s_1, s_2 = \omega_0(-\zeta \pm \sqrt{\zeta^2 - 1})$$

We begin to see the advantage of using the parameters ω_0 and ζ. The constant ω_0 is a scale factor that designates the size of the roots. The expression under the radical defines the form of the roots and depends only on the damping ratio ζ. As a result, we can express the three possible forms of the natural response in terms of the damping ratio.

Case A: For $\zeta > 1$ the discriminant is positive, there are two unequal, real roots

$$s_1, s_2 = -\alpha_1, -\alpha_2 = \omega_0(-\zeta \pm \sqrt{\zeta^2 - 1}) \tag{7–64a}$$

and the natural response has the overdamped form

$$y_N(t) = K_1 e^{-\alpha_1 t} + K_2 e^{-\alpha_2 t} \qquad t \geqslant 0 \tag{7–64b}$$

Case B: For $\zeta = 1$ the discriminant vanishes, there are two real, equal roots

$$s_1 = s_2 = -\alpha = -\zeta\omega_0 \tag{7–65a}$$

and the natural response has the critically damped form

$$y_N(t) = K_1 e^{-\alpha t} + K_2 t e^{-\alpha t} \qquad t \geqslant 0 \tag{7–65b}$$

Case C: For $\zeta < 1$ the discriminant is negative, leading to two complex, conjugate roots $s_1, s_2 = -\alpha \pm j\beta$, where

$$\alpha = \zeta\omega_0 \quad \text{and} \quad \beta = \omega_0\sqrt{1 - \zeta^2} \tag{7–66a}$$

and the natural response has the underdamped form

$$y_N(t) = K_1 e^{-\alpha t} \cos \beta t + K_2 e^{-\alpha t} \sin \beta t \qquad t \geqslant 0 \tag{7–66b}$$

Equations (7–64a), (7–65a), and (7–66a) provide relationships between the natural frequency parameters α and β and the new parameters ζ and ω_0. The reasons for using two equivalent sets of parameters to describe the natural frequencies of a second-order circuit will become clear as we continue our study of dynamic circuits. Since the units of complex frequency s are radians per second, the standard form of the characteristic equation $s^2 + 2\zeta\omega_0 s + \omega_0^2$ shows that ω_0 is specified in radians per second and ζ is dimensionless.

Combining the forced and natural responses yields the step response of the general second-order differential equation in the form

$$y(t) = y_N(t) + A/a_0 \qquad t \geqslant 0 \tag{7–67}$$

The factor A/a_0 is the forced response. The natural response $y_N(t)$ takes one of the forms in Eqs. (7–64b), (7–65b), or (7–66b), depending on the value of the damping ratio. The constants K_1 and K_2 in natural response can be evaluated from the initial conditions.

In summary, the step response of a second-order circuit is determined by

1. The amplitude of the step function input $Au(t)$
2. The damping ratio ζ and natural frequency ω_0
3. The initial conditions $y(0)$ and $dy/dt\ (0)$

In this regard the damping ratio and natural frequency play the same role for second-order circuits that the time constant plays for first-order circuits. That is, these circuit parameters determine the basic form of the natural response, just as the time constant defines the form of the natural response in a first-order circuit. It is not surprising that a second-order circuit takes two parameters, since it contains two energy storage elements.

EXAMPLE 7–19

The series RLC circuit in Figure 7–35 is driven by a step function and is in the zero state at $t = 0$. Find the capacitor voltage for $t \geq 0$.

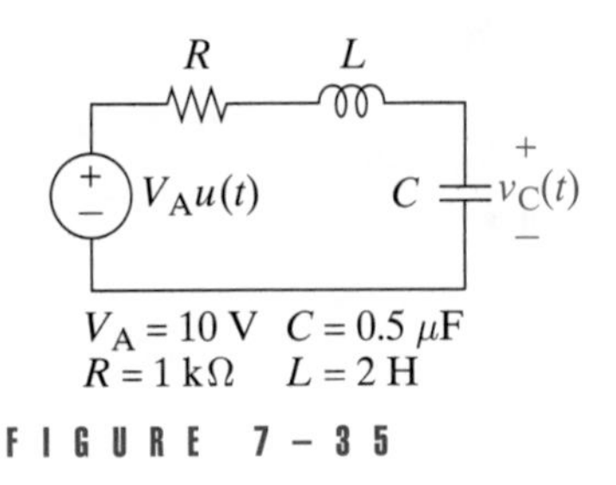

FIGURE 7–35

SOLUTION:

This is a series *RLC* circuit, so the differential equation for the capacitor voltage is

$$10^{-6}\frac{d^2v_C}{dt^2} + 0.5 \times 10^{-3}\frac{dv_C}{dt} + v_C = 10 \qquad t \geq 0$$

By inspection, the forced response is $v_{CF} = 10$ V. In standard format the homogeneous equation is

$$\frac{d^2v_{CN}}{dt^2} + 500\frac{dv_{CN}}{dt} + 10^6\, v_{CN} = 0 \qquad t \geq 0$$

Comparing this format to the standard form in Eq. (7–63) yields

$$\omega_0^2 = 10^6 \quad \text{and} \quad 2\zeta\omega_0 = 500$$

so $\omega_0 = 1000$ and $\zeta = 0.25$. Since $\zeta < 1$, the natural response is underdamped (case C). Using Eqs. (7–66a) and (7–66b), we have

$$\alpha = \zeta\omega_0 = 250$$

$$\beta = \omega_0\sqrt{1 - \zeta^2} = 968$$

$$v_{CN}(t) = K_1 e^{-250t}\cos 968t + K_2 e^{-250t}\sin 968t$$

The general solution of the circuit differential equation is the sum of the forced and natural responses:

$$v_C(t) = 10 + K_1 e^{-250t}\cos 968t + K_2 e^{-250t}\sin 968t \qquad t \geq 0$$

The constants K_1 and K_2 are determined by the initial conditions. The circuit is in the zero state at $t = 0$, so the initial conditions are $v_C(0) = 0$ and $i_L(0) = 0$. Applying the initial condition constraints to the general solution yields two equations in the constants K_1 and K_2:

$$v_C(0) = 10 + K_1 = 0$$

$$\frac{dv_C}{dt}(0) = -250\,K_1 + 968\,K_2 = 0$$

These equations yield $K_1 = -10$ and $K_2 = -2.58$. The step response of the capacitor voltage step response is

$$v_C(t) = 10 - 10e^{-250t}\cos 968t - 2.58e^{-250t}\sin 968t \text{ V} \qquad t \geq 0$$

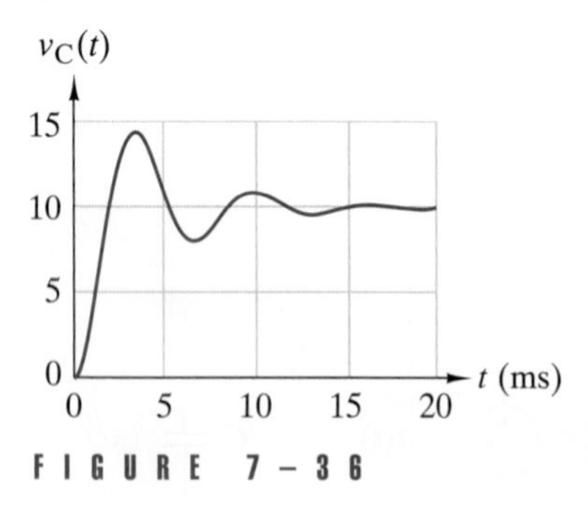

FIGURE 7–36

A plot of $v_C(t)$ versus time is shown in Figure 7–36. The response and its first derivative at $t = 0$ satisfy the initial conditions. The natural response decays to zero, so the forced response determines the final value of $v_C(\infty) = 10$ V. Beginning at $t = 0$ the response climbs rapidly but overshoots the mark several times before eventually settling down to the final value. The damped sinusoidal behavior results from the fact that $\zeta < 1$, producing an underdamped natural response. ■

EXAMPLE 7–20

The step response of the series *RLC* circuit in Figure 7–35 is found analytically in Example 7–19. In this example we explore the effect of varying the resistance on the step response of this series *RLC* circuit. Use Electronics WorkBench to calculate the zero-state step response of the capacitor voltage as R sweeps across the range from 1 kΩ to 8 kΩ with $L = 2$ H, $C = 0.5$ μF, and $V_A = 10$ V.

SOLUTION:

The Electronics WorkBench circuit diagram in Figure 7–37 includes numerical values for the inductor, capacitor, voltage source, and the starting value of the resistance. The 10-V dc source simulates the step function input for $t \geq 0$. To get the zero-state response the initial inductor current and capacitor voltage must be set to zero. Note that the node voltage at node 3 is the voltage across the capacitor.

To obtain step responses as the resistance varies, we select the **Parameter Sweep** option from the **Analysis** menu. This brings up the parameter sweep dialog box shown at the top of Figure 7–38. Here we select resistance **R1** as the component to be varied, specify an analysis **Start value** of 1 kΩ and an **End value** of 8 kΩ. Specifying an octave **Sweep type** means that the parameter sweep values are obtained by doubling the start value until the end value is reached. That is, separate analyses will be run for R1 = 1, 2, 4, and 8 kΩ. Finally, the **Output node** to be monitored is set to node 3, which is the voltage across the capacitor in the series *RLC* circuit in Figure 7–37.

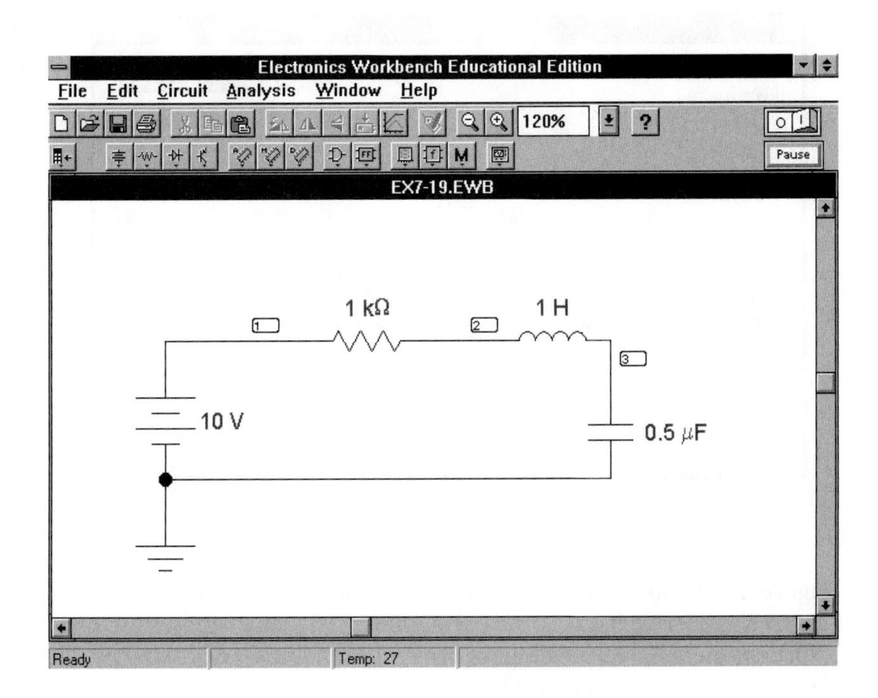

FIGURE 7–37

With the parameter sweep conditions set, we next press the **Set transient analysis** button, which brings up the transient analysis dialog box shown at the bottom of Figure 7–38. In this box the **Initial conditions** are set to zero (to get the zero-state response), the **Start time** is set to zero, the **End time** set to 0.02 s = 20 ms, and the **Minimum number of time points** set to 100. Pressing the **Accept** button returns us to the parameter sweep dialog box, where pressing the **Simulate** button launches a sequence of transient analysis runs as R1 sweeps across the specified range.

FIGURE 7–38

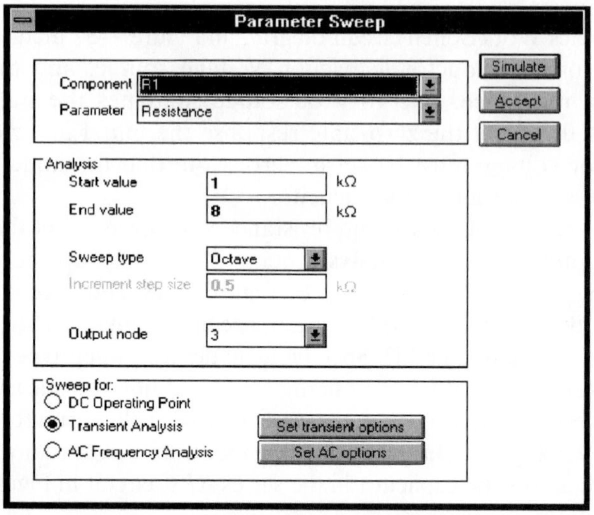

Parameter Sweep

Component R1
Parameter Resistance

Simulate
Accept
Cancel

Analysis
Start value 1 kΩ
End value 8 kΩ
Sweep type Octave
Increment step size 0.5 kΩ
Output node 3

Sweep for:
DC Operating Point
Transient Analysis
AC Frequency Analysis
Set transient options
Set AC options

(a)

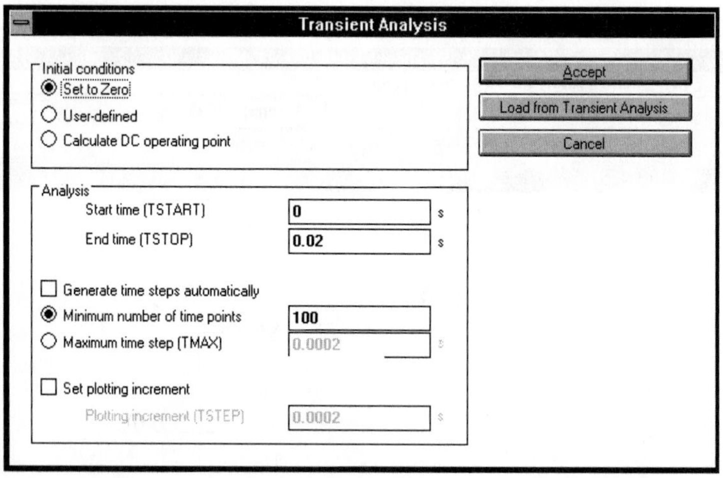

Transient Analysis

Initial conditions
Set to Zero
User-defined
Calculate DC operating point

Accept
Load from Transient Analysis
Cancel

Analysis
Start time (TSTART) 0 s
End time (TSTOP) 0.02 s

Generate time steps automatically
Minimum number of time points 100
Maximum time step (TMAX) 0.0002 s

Set plotting increment
Plotting increment (TSTEP) 0.0002 s

(b)

Figure 7–39 shows the family of step responses for R1 = 1, 2, 4, and 8 kΩ. As should be expected, increasing the resistance increases the damping so that the step response goes from the underdamped case at R1 = 1 kΩ to being overdamped at R1 = 8 kΩ.

We can qualitatively check these results by calculating ω_0 and ζ.

$$\omega_0 = \frac{1}{\sqrt{LC}} = \frac{1}{\sqrt{2 \times 0.5 \times 10^{-6}}} = 1000 \text{ rad/s}$$

and

$$\zeta = \frac{R_1}{2\omega_0 L} = \frac{R_1}{4000}$$

Hence, $R_1 = 1$ kΩ yields $\zeta = 0.25$, $R_1 = 2$ kΩ yields $\zeta = 0.5$, $R_1 = 4$ kΩ yields $\zeta = 1$, and $R_1 = 8$ kΩ yields $\zeta = 2$. These calculations confirm that sweeping the resistance from 1 kΩ to 8 kΩ takes the step response from an underdamped case to an overdamped case, passing through the critically damped case ($\zeta = 1$) at $R_1 = 4$ kΩ. ■

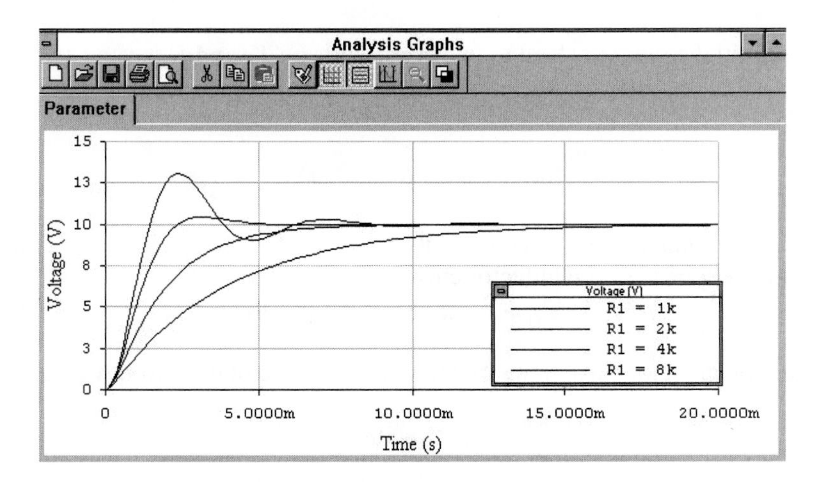

FIGURE 7–39

EXAMPLE 7–21

Find the zero-state response of $v_O(t)$ in Figure 7–40 for $v_S(t) = 60u(t)$ V.

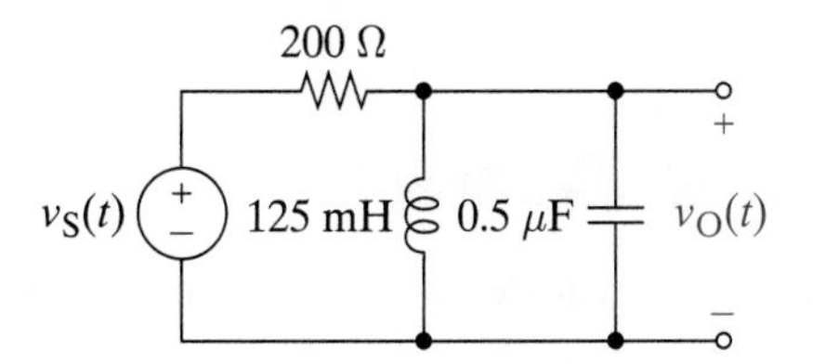

FIGURE 7–40

SOLUTION:

This is a parallel *LC* circuit in which the Norton equivalent seen by the inductor and capacitor is defined by $i_N = 60/200 = 0.3$ A and $G_N = G_N = 1/200 = 0.005$ S. To find the desired output response, we first find the inductor current. Since $LC = 6.25 \times 10^{-8}$ and $G_N L = 6.25 \times 10^{-4}$, Eq. (7–55) gives the differential equation for the inductor current as

$$6.25 \times 10^{-8}\frac{d^2 i_L}{dt^2} + 6.25 \times 10^{-4}\frac{di_L}{dt} + i_L = 0.3$$

from which we obtain the forced response as $i_{LF} = 0.3$ A. In standard form the homogeneous equation is

$$\frac{d^2 i_L}{dt^2} + 10^4\frac{di_L}{dt} + 16 \times 10^6 i_L = 0$$

The corresponding characteristic equation is $s^2 + 10^4 s + 16 \times 10^6 = 0$, whose roots are $s_1 = -2000$ rad/s and $s_2 = -8000$ rad/s. The general form of the natural response is $i_{LN} = K_1 e^{-2000t} + K_2 e^{-8000t}$. Adding the natural and forced responses gives the complete response as

$$i_L(t) = 0.3 + K_1 e^{-2000t} + K_2 e^{-8000t}$$

At $t = 0$ the circuit is in the zero state, so the initial conditions are

$$i_L(0) = 0 \quad \text{and} \quad \frac{di_L}{dt}(0) = \frac{1}{L} v_C(0) = 0$$

These initial conditions lead to two constraints on K_1 and K_2:

$$i_L(0) = 0.3 + K_1 + K_2 = 0$$

$$\frac{di_L}{dt}(0) = -2000\, K_1 - 8000\, K_2 = 0$$

Taken together these two constraints yield $K_1 = -0.4$ and $K_2 = 0.1$, and the step response of the inductor current is found to be

$$i_L(t) = 0.3 - 0.4 e^{-2000t} + 0.1 e^{-8000t} \text{ A}$$

Given the inductor current, we find the step response of the output voltage as

$$v_O(t) = L\frac{di_L}{dt} = 0.125(800 e^{-2000t} - 800 e^{-8000t})$$

$$= 100(e^{-2000t} - e^{-8000t}) \text{ V}$$

The analysis approach used here is one we have used before. First find a state variable response (i_L in this case) and then use this result to find the nonstate variable response (v_O in this case). ■

DESIGN EXAMPLE 7–22

Design a series *RLC* circuit whose zero-state step response is

$$v_C(t) = V_A - \frac{5}{4} V_A e^{-400t} + \frac{1}{4} V_A e^{-2000t} \quad t > 0$$

where V_A is the amplitude of the step function input.

SOLUTION:

To obtain the required response, the numerical characteristic equation must be

$$(s + 400)(s + 2000) = s^2 + 2400s + 8 \times 10^5 = 0$$

Using circuit parameters, the symbolic characteristic equation of a series *RLC* circuit is

$$s^2 + \frac{R}{L}s + \frac{1}{LC} = 0$$

Equating coefficients in these two equations leads to two design constraints, namely $R/L = 2400$ and $1/LC = 8 \times 10^5$. Let $R = 3\ \text{k}\Omega$; then $L = R/2400 = 1.25$ H and $C = 1/(L \times 8 \times 10^5) = 1 \times 10^{-6}$ F. Many other choices are possible. ■

Exercise 7–17

Find the zero-state solution of the following differential equations:

(a) $10^{-4}\dfrac{d^2v}{dt^2} + 2 \times 10^{-2}\dfrac{dv}{dt} + v = 100\,u(t)$

(b) $\dfrac{d^2i}{dt^2} + 2400\dfrac{di}{dt} + 4 \times 10^6 i = 8 \times 10^4 u(t)$

Answers:

(a) $v(t) = 100 - 100\,e^{-100t} - 10^4 te^{-100t}$ V $t \geq 0$
(b) $i(t) = 20 - 20\,e^{-1200t}\cos 1600\,t - 15\,e^{-1200t}\sin 1600t$ mA $t \geq 0$

Exercise 7–18

The step response of a series RLC circuit is observed to be

$$v_C(t) = 15 - 15e^{-1000t}\cos 1000t \text{ V} \qquad t \geqslant 0$$

$$i_L(t) = 45e^{-1000t}\cos 1000t + 45e^{-1000t}\sin 1000t \text{ mA} \qquad t \geqslant 0$$

(a) What is the circuit characteristic equation?
(b) What are the initial values of the state variables?
(c) What is the amplitude of the step input?
(d) What are the values of R, L, and C?
(e) What is the voltage across the resistor?

Answers:

(a) $s^2 + 2000s + 2 \times 10^6 = 0$
(b) $v_C(0) = 0$, $i_L(0) = 45$ mA
(c) $V_A = 15$ V
(d) $R = 333\ \Omega$, $L = 167$ mH, $C = 3\ \mu$F
(e) $v_R(t) = 15e^{-1000t}\cos 1000t + 15e^{-1000t}\sin 1000t$ V

7–8 OTHER SECOND-ORDER CIRCUITS

In general a second-order circuit contains two independent energy storage elements. Up to this point the series and parallel RLC circuits have been the only second-order circuits considered. These two time-honored examples introduce the role of second-order differential equations in the analysis of dynamic circuit. Using classical methods shows that the solution of a second-order equation has three possible forms: overdamped, critically damped, and underdamped. An important concept here is the connection between the characteristic equation and the natural response of a second-order circuit. To highlight this relationship, we introduced a **standard form** for the second-order characteristic equation, namely

$$s^2 + 2\zeta\omega_0 s + \omega_0^2 = 0$$

where ζ is the damping ratio and ω_0 the undamped natural frequency. Briefly stated, the natural response is overdamped when $\zeta > 1$, critically damped when $\zeta = 1$, and underdamped when $\zeta < 1$.

In subsequent chapters we will encounter a number of second-order circuits, some of which are shown in Figure 7–41. All of these circuits contain two energy storage elements and could be analyzed using the classical differential equation methods used in this chapter. However, in these later chapters we use Laplace transforms because this method readily handles all types of second-order circuits and higher-order circuits as well. Given that our ultimate goal is to use the more powerful Laplace transforms, why bother with the classical methods at all? The answer is that classical methods introduce two concepts that we use repeatedly in subsequent chapters.

1. The roots of a characteristic equation determine the form of the natural response.
2. The parameters ζ and ω_0 determine the roots of a second-order characteristic equation.

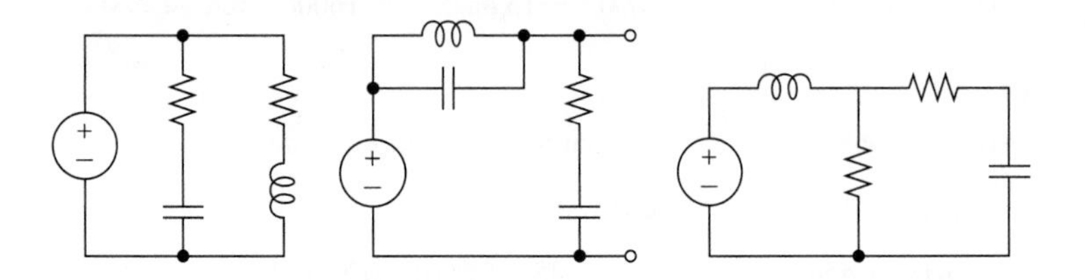

FIGURE 7–41 *Some second-order circuits.*

The following examples illustrate the use of these concepts in the analysis and design of second order circuits.

EXAMPLE 7–23

What range of source resistance will produce an underdamped natural response in a parallel *RLC* circuit with $L = 200$ mH and $C = 32$ nF?

SOLUTION:

According to Eq. (7–56) the characteristic equation of a parallel *RLC* circuit is

$$LCs^2 + G_N Ls + 1 = 0$$

Dividing the *LC* puts this equation in the form

$$s^2 + \frac{1}{R_N C}s + \frac{1}{LC} = 0$$

Comparing this with the standard form $s^2 + 2\zeta\omega_0 s + \omega_0^2$ leads to

$$\omega_0 = \frac{1}{\sqrt{LC}} = \frac{1}{\sqrt{200 \times 10^{-3} \times 32 \times 10^{-9}}} = 12.5 \times 10^3 \text{ rad/s}$$

and $2\zeta\omega_0 = 1/R_N C$. Underdamped response requires that $\zeta < 1$; hence

$$R_N > \frac{1}{2\omega_0 C} = \frac{1}{25 \times 10^3 \times 32 \times 10^{-9}} = 1250\ \Omega$$

■

D DESIGN EXAMPLE 7–24

Design a parallel RLC circuit with $\zeta = 0.5$ and $\omega_0 = 25$ krad/s.

SOLUTION:
The characteristic equation of a parallel RLC circuit can be written as

$$s^2 + \frac{1}{R_N C}s + \frac{1}{LC} = 0$$

Comparing this with the standard form $s^2 + 2\zeta\omega_0 s + \omega_0^2$ leads to two design constraints:

$$\frac{1}{R_N C} = 2\zeta\omega_0 = 25 \times 10^3 \quad \text{and} \quad \frac{1}{LC} = \omega_0^2 = 6.25 \times 10^8$$

Let the resistance $R_N = 10$ kΩ; then $C = 1/25 \times 10^7 = 4 \times 10^{-9}$ F and $L = 25 \times 10^7/6.25 \times 10^8 = 0.4$ H. Many other choices are possible since there are three circuit parameters and only two constraints. ■

SUMMARY

- Circuits containing linear resistors and the equivalent of one capacitor or one inductor are described by first-order differential equations in which the unknown is the circuit state variable.
- The zero-input response in a first-order circuit is an exponential whose time constant depends on circuit parameters. The amplitude of the exponential is equal to the initial value of the state variable.
- For linear circuits the total response is the sum of the forced and natural responses. The natural response is the general solution of the homogeneous differential equation obtained by setting the input to zero. The forced response is a particular solution of the differential equation for the given input.
- For linear circuits the total response is the sum of the zero-input and zero-state responses. The zero-input response is caused by the initial energy stored in capacitors or inductors. The zero-state response results from the input driving forces.
- The initial and final values of the step response of a first- and second-order circuit can be found by replacing capacitors by open circuits and inductors by short circuits and then using resistance circuit analysis methods.
- For a sinusoidal input the forced response is called the sinusoidal steady-state response, or the ac response. The ac response is a sinusoid with the same frequency as the input but with a different amplitude and phase angle. The ac response can be found from the circuit differential equation using the method of undetermined coefficients.

- Circuits containing linear resistors and the equivalent of two energy storage elements are described by second-order differential equations in which the dependent variable is one of the state variables. The initial conditions are the values of the two state variables at $t = 0$.
- The zero-input response of a second-order circuit takes different forms depending on the roots of the characteristic equation. Unequal real roots produce the overdamped response, equal real roots produce the critically damped response, and complex conjugate roots produce underdamped responses.
- The circuit damping ratio ζ and undamped natural frequency ω_0 determine the form of the zero-input and natural responses of any second-order circuit. The response is overdamped if $\zeta > 1$, critically damped if $\zeta = 1$, and underdamped if $\zeta < 1$. Active circuits can produce undamped ($\zeta = 0$) and unstable ($\zeta < 0$) responses.
- Computer-aided circuit analysis programs can generate numerical solutions for circuit transient responses. Some knowledge of analytical methods and an estimate of the general form of the expected response are necessary to use these analysis tools.

PROBLEMS

OBJECTIVE 7–1 FIRST-ORDER CIRCUIT ANALYSIS (SECTS. 7–1, 7–2, 7–3, 7–4)

Given a first-order RC or RL circuit:

(a) Find the circuit differential equation, the circuit characteristic equation, the circuit time constant, and the initial conditions (if not given).

(b) Find the zero-input response.

(c) Find the complete response for step function and sinusoidal inputs.

See Examples 7–1, 7–2, 7–3, 7–4, 7–5, 7–7, 7–8, 7–9, 7–12, 7–13 and Exercises 7–1, 7–2, 7–5, 7–6, 7–7

7–1 Find the function $y(t)$ that satisfies the following differential equation and the initial condition:

$$50\frac{dy(t)}{dt} + 250y(t) = 0 \quad y(0) = 10$$

7–2 Find the function $y(t)$ that satisfies the following differential equation and initial condition:

$$10^{-3}\frac{dy(t)}{dt} + y(t) = 0 \quad y(0) = 50$$

7–3 Find the time constants of the circuits in Figure P7–3.

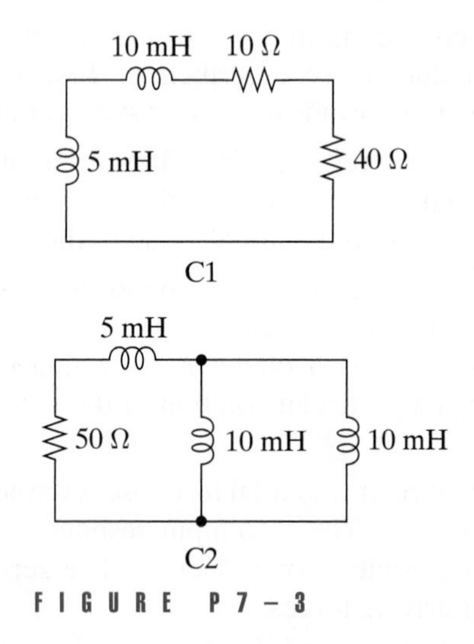

FIGURE P7–3

7–4 Find the time constants of the circuits in Figure P7–4.

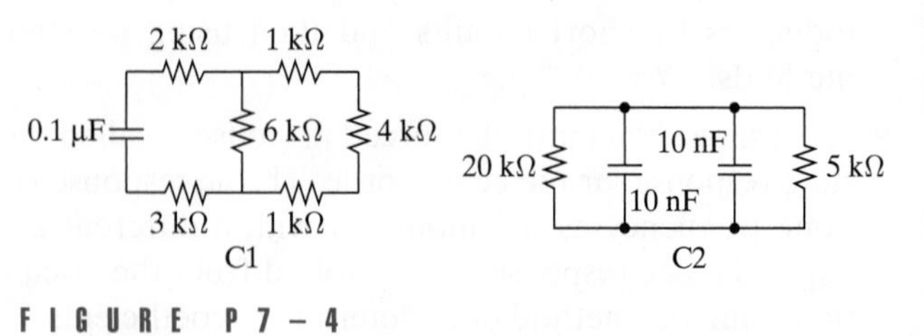

FIGURE P7–4

7–5 The switch in Figure P7–5 is closed at $t = 0$. The initial voltage on the capacitor is $v_C(0) = 30$ V. Find $v_C(t)$ and $i_O(t)$ for $t \geq 0$.

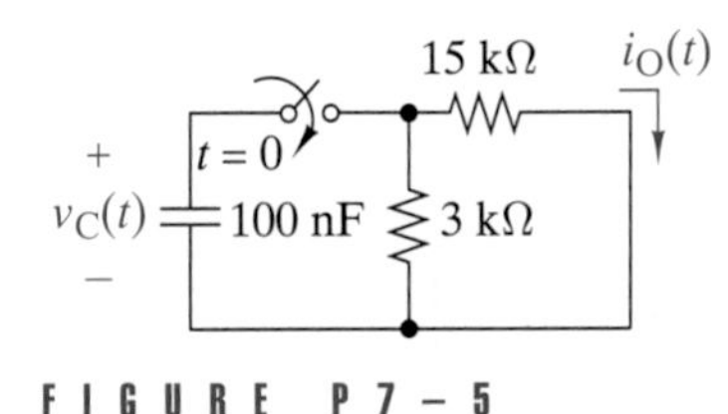

FIGURE P7–5

7–6 In Figure P7–6 the initial current through the inductor is $i_L(0) = 5$ mA. Find $i_L(t)$ and $v_O(t)$ for $t \geq 0$.

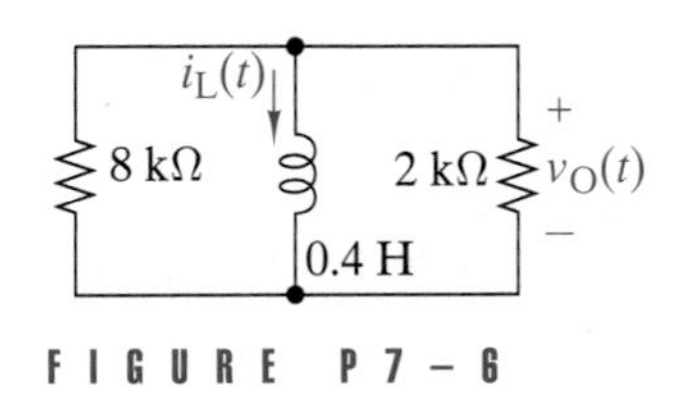

FIGURE P7–6

7–7 The switch in Figure P7–7 has been in position A for a long time and is moved to position B at $t = 0$. Find $v_C(t)$ for $t \geq 0$.

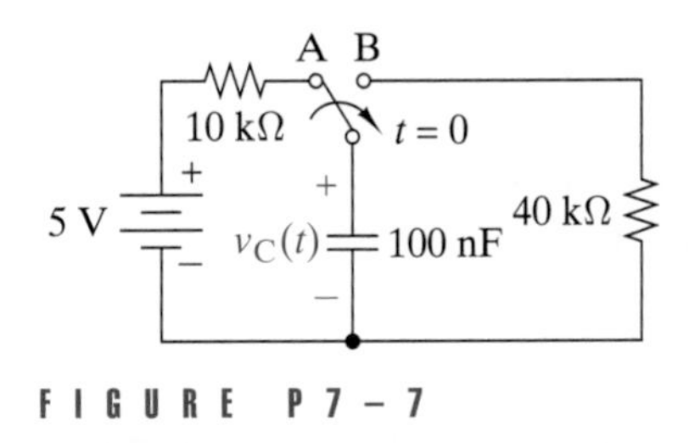

FIGURE P7–7

7–8 The switch in Figure P7–8 has been open for a long time and is closed at $t = 0$. Find $i_L(t)$ for $t \geq 0$.

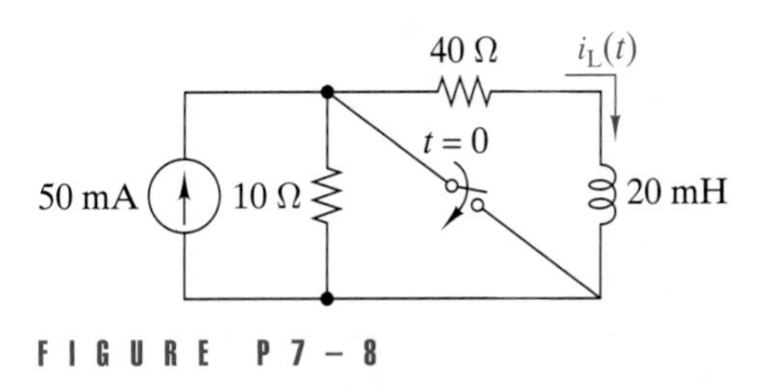

FIGURE P7–8

7–9 The circuit in Figure P7–9 is in the zero state when the input $i_S(t) = I_A u(t)$ is applied. Find the voltage $v_O(t)$ for $t \geq 0$. Identify the forced and natural components in the output.

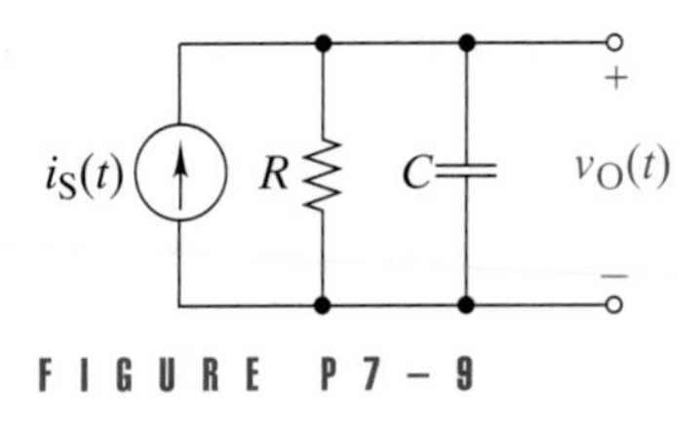

FIGURE P7–9

7–10 The circuit in Figure P7–10 is in the zero state when the input $v_S(t) = V_A u(t)$ is applied. Find $v_O(t)$ for $t \geq 0$. Identify the forced and natural components in the output.

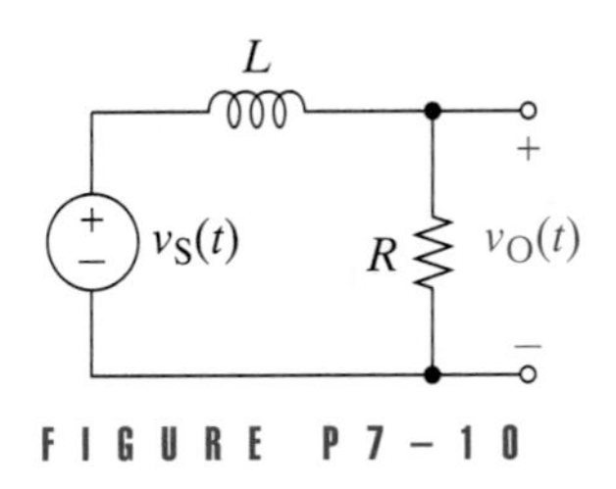

FIGURE P7–10

7–11 The circuit in Figure P7–11 is in the zero state when the input $v_S(t) = V_A u(t)$ is applied. Find $v_O(t)$ for $t \geq 0$. Identify the forced and natural components in the output.

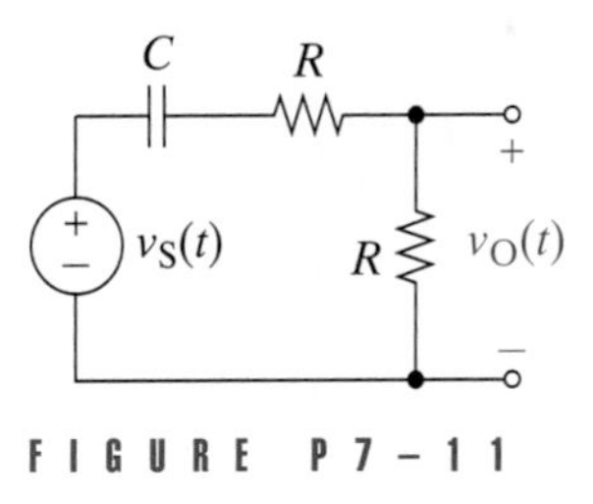

FIGURE P7–11

7–12 The circuit in Figure P7–12 is in the zero state when the input $v_S(t) = V_A u(t)$ is applied. Find $v_O(t)$ for $t \geq 0$. Identify the forced and natural components in the output.

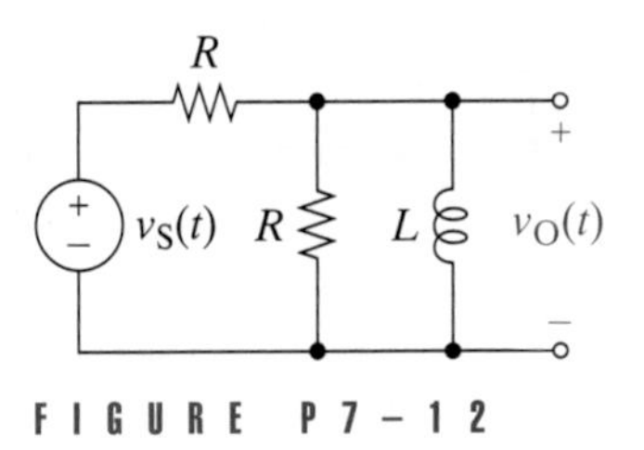

FIGURE P7–12

7–13 The switch in Figure P7–13 has been in position A for a long time and is moved to position B at $t = 0$. Find $v_C(t)$ for $t \geq 0$. Identify the forced and natural components in the response.

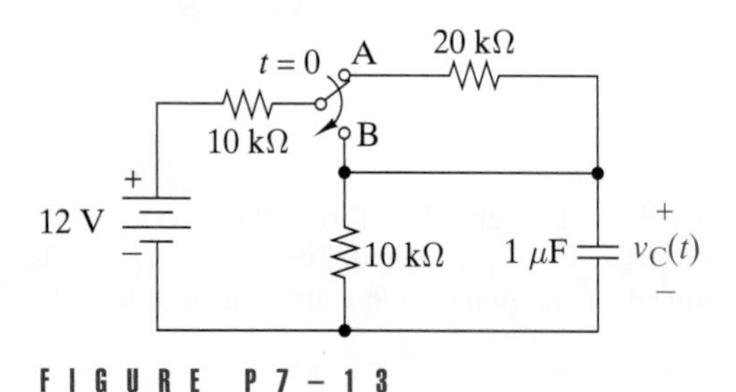

FIGURE P7–13

7–14 Repeat Problem 7–13 when the switch has been in position B for a long time and is moved to position A at $t = 0$.

7–15 Find the function that satisfies the following differential equation and the initial condition for an input $v_S(t) = 25\cos(20t)$ V:

$$\frac{dv(t)}{dt} + 10v(t) = v_S(t) \quad v(0) = 0\text{ V}$$

7–16 Repeat Problem 7–15 for $v_S(t) = 25\sin(20t)$ V.

7–17 The input in Figure P7–17 is $v_S(t) = 30\cos(5t)$V. The switch has been open for a long time and is closed at $t = 0$. Find $i_L(t)$ for $t \geq 0$.

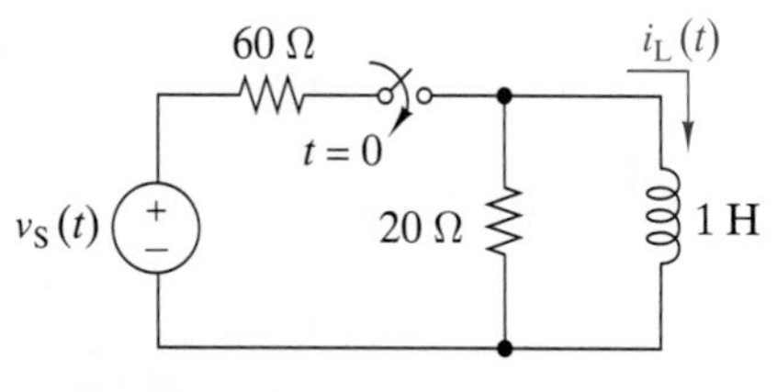

FIGURE P7–17

7–18 The switch in Figure P7–18 has been in position A for a long time and is moved to position B at $t = 0$. Find $i_L(t)$ for $t \geq 0$ and sketch its waveform.

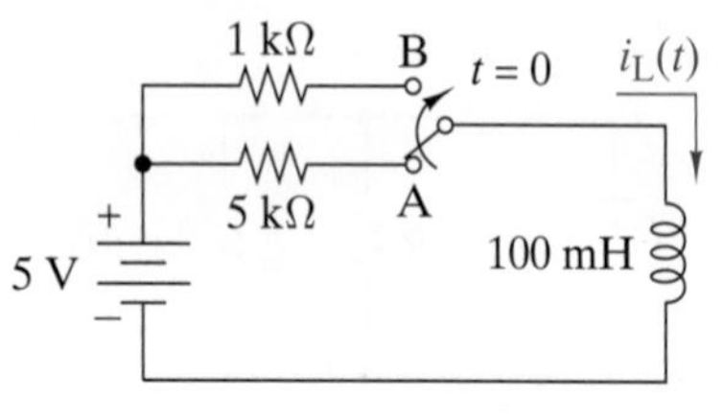

FIGURE P7–18

7–19 The switch in Figure P7–19 has been in position A for a long time and is moved to position B at $t = 0$. Find $v_C(t)$ for $t \geq 0$ and sketch its waveform.

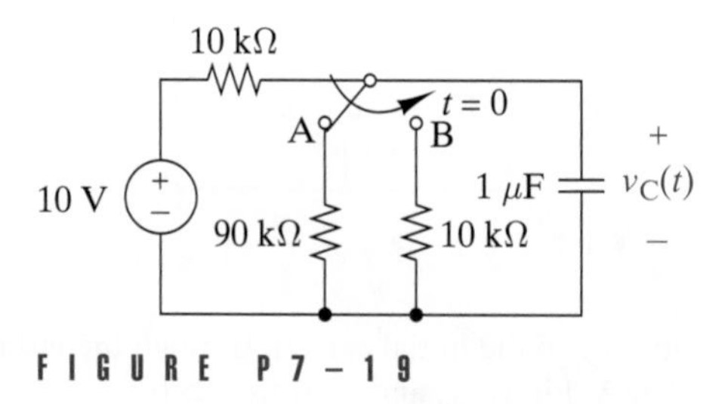

FIGURE P7–19

7–20 Switches 1 and 2 in Figure P7–20 have both been in position A for a long time. Switch 1 is moved to position B at $t = 0$ and switch 2 is moved to position B at $t = 20$ ms. Find the voltage across the 0.1-μF capacitor for $t > 0$ and sketch its waveform.

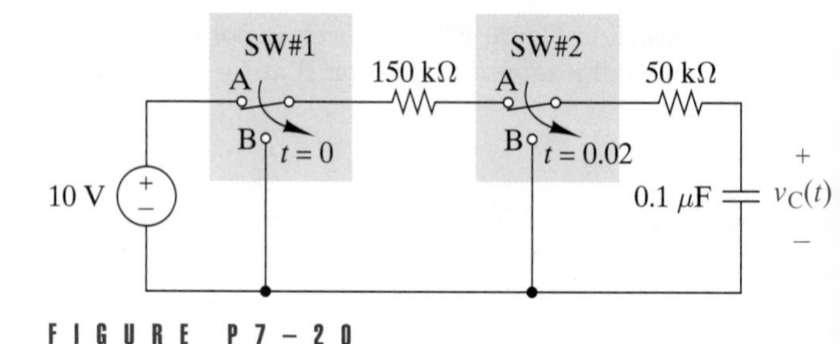

FIGURE P7–20

7–21 The switch in Figure P7–21 has been open for a long time and is closed at $t = 0$. The switch is reopened at $t = 2$ ms. Find $v_C(t)$ for $t \geq 2$ ms.

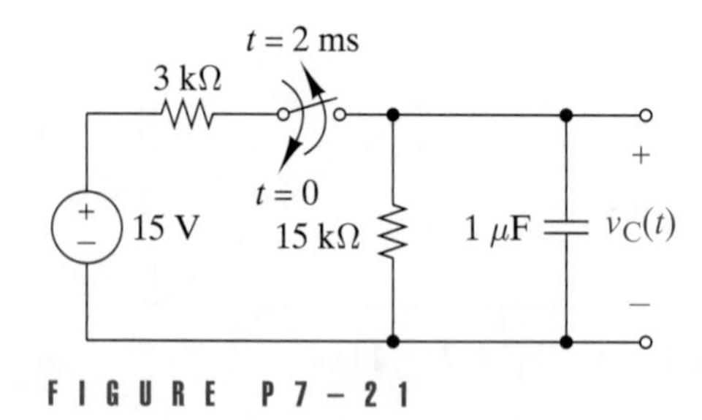

FIGURE P7–21

7–22 Find the sinusoidal steady-state response of $v_C(t)$ in Figure P7–22 when the input voltage is $v_S(t) = V_A[\sin(\omega t)]u(t)$ V.

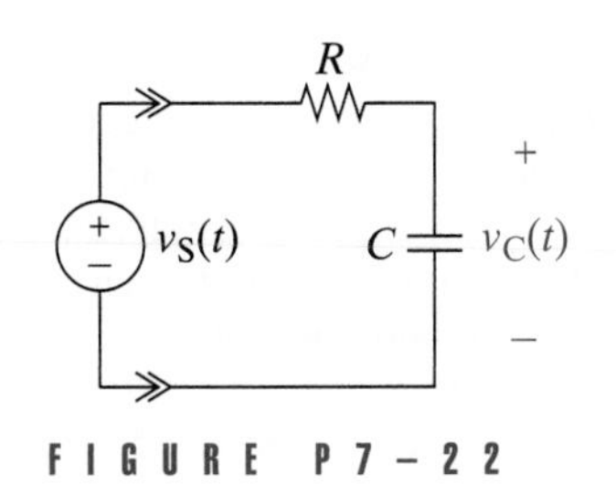

FIGURE P7–22

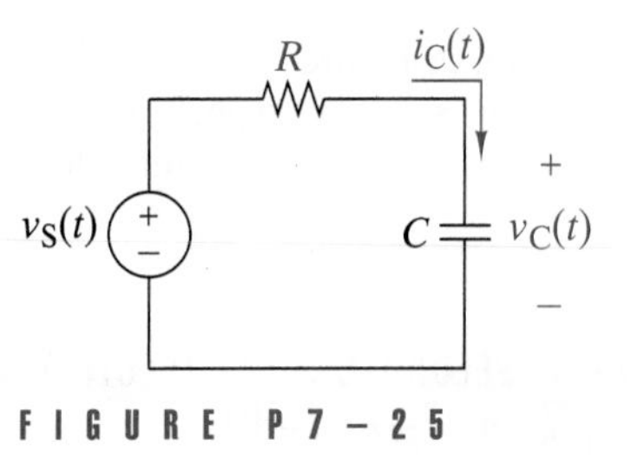

FIGURE P7–25

OBJECTIVE 7–2 FIRST-ORDER CIRCUIT RESPONSES (SECTS. 7–1, 7–2, 7–3)

Given responses in a first-order *RC* or *RL* circuit:
(a) Find the circuit parameters or other responses.
(b) Select element values to produce a given response.
See Examples 7–6, 7–10 and Exercises 7–2, 7–3, 7–4, 7–12

7–23 For $t \geq 0$ the zero-input response of the circuit in Figure P7–23 is $v_C(t) = 10e^{-1000t}$ V.

(a) Find C and $i_C(t)$ when $R = 10$ kΩ.
(b) Find the energy stored in the capacitor at $t = 2$ ms.

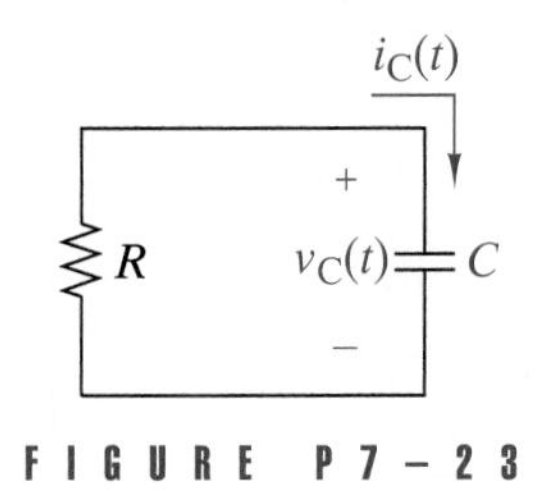

FIGURE P7–23

7–24 For $t \geq 0$ the zero-input response of the circuit in Figure P7–24 is $i_L(t) = 5e^{-5000t}$ mA.

(a) Find L and $v_L(t)$ when $R = 500$ Ω.
(b) Find the energy stored in the inductor at $t = 0.2$ ms.

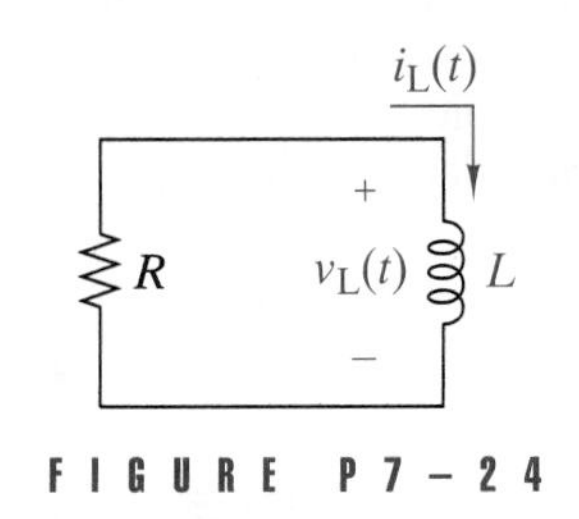

FIGURE P7–24

7–25 For $t \geq 0$ the step response of the voltage across the capacitor in Figure P7–25 is $v_C(t) = 5 - 10e^{-2000t}$ V.

(a) Find v_S, R, and $i_C(t)$ when $C = 1\mu$F.
(b) Find the energy stored in the capacitor at $t = \ln(2)/2$ ms.

7–26 For $t \geq 0$ the step response of the current through the capacitor in Figure P7–25 is $i_C(t) = 20e^{-2000t}$ mA. Find $v_C(t)$ for $t \geq 0$ when $C = 1\mu$F and $v_C(0) = 5$ V.

7–27 For $t \geq 0$ the step responses of the current through and voltage across the inductor in Figure P7–27 are

$$i_L(t) = 5 - 10e^{-2000t} \text{ mA} \quad \text{and} \quad v_L(t) = e^{-2000t} \text{ V}$$

(a) Find v_S, R, and L.
(b) Find the energy stored in the inductor at $t = \ln(2)/2$ ms.

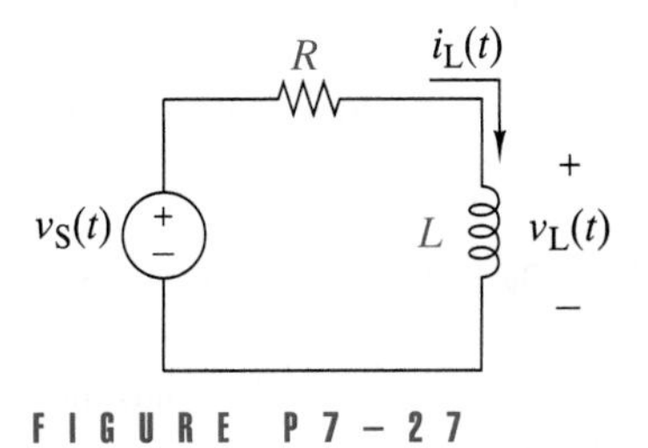

FIGURE P7–27

7–28 For $t \geq 0$ the step response of the current through the inductor in Figure P7–27 is $i_L(t) = 20 - 10e^{-500t}$ mA. For $L = 100$ mH:

(a) Find v_S, R, and $v_L(t)$.
(b) Find the energy stored in the inductor at $t = 1$ ms.

7–29 **D** The switch in Figure P7–29 has been in position B for a long time and is moved to position A at $t = 0$. Design the first-order *RC* interface circuit such that $v_O(t) = 10 - 10e^{-5000t}$ V.

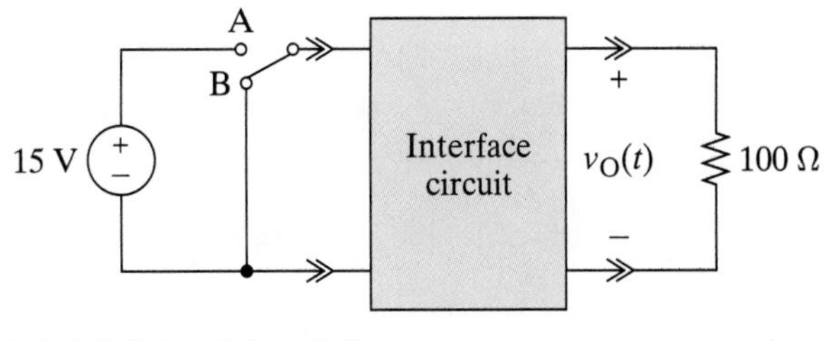

FIGURE P7–29

7–30 **D** The switch in Figure 7–29 has been in position A for a long time and is moved to position B at $t = 0$. Design the first-order RC interface circuit such that $v_O(t) = 5e^{-3000t}$ V.

OBJECTIVE 7–3 SECOND-ORDER CIRCUIT ANALYSIS (SECTS. 7–5, 7–6, 7–7, 7–8)

Given a second-order circuit:
(a) Find the circuit differential equation.
(b) Find the characteristic equation and the initial conditions (if not given).
(c) Find the zero-input response.
(d) Find the complete response for a step function input.
See Examples 7–14, 7–15, 7–17, 7–18, 7–19, 7–20, 7–21 and Exercises 7–14, 7–17, 7–19, 7–20

7–31 Find the $v(t)$ that satisfies the following differential equation and initial conditions.

$$\frac{d^2v}{dt^2} + 10\frac{dv}{dt} + 25v = 0, \quad v(0) = 0\text{ V}, \quad \frac{dv}{dt}(0) = 10\text{ V/s}$$

7–32 Find the $v(t)$ that satisfies the following differential equation and initial conditions.

$$\frac{d^2v}{dt^2} + 15\frac{dv}{dt} + 50v = 0, \quad v(0) = 5\text{ V}, \quad \frac{dv}{dt}(0) = 10\text{ V/s}$$

7–33 Find the $v(t)$ that satisfies the following differential equation and initial conditions.

$$\frac{d^2v}{dt^2} + 10\frac{dv}{dt} + 125v = 250u(t), \ v(0) = 5\text{ V}, \ \frac{dv}{dt}(0) = 25\text{ V/s}$$

7–34 Find the $i(t)$ that satisfies the following differential equation and initial conditions.

$$\frac{d^2i}{dt^2} + 4\frac{di}{dt} + 8i = 24u(t), \quad i(0) = 0, \quad \frac{dv}{dt}(0) = 0$$

7–35 The switch in Figure P7–35 has been open for a long time and is closed at $t = 0$. The circuit parameters are $L = 1$ H, $C = 0.5$ μF, $R = 3$ kΩ, and $v_C(0) = 10$ V. Find $v_C(t)$ and $i_L(t)$ for $t \geq 0$. Is the circuit overdamped, critically damped, or underdamped?

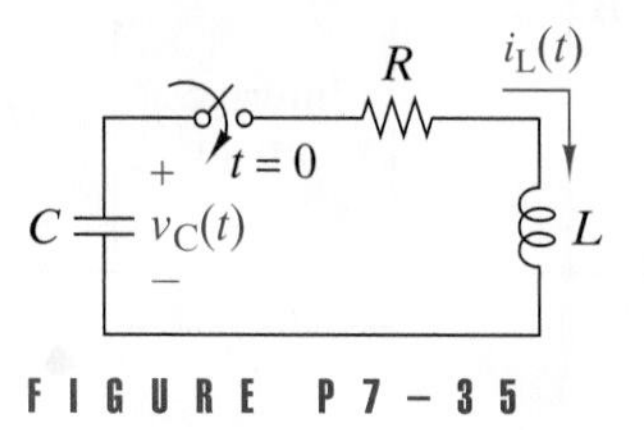

FIGURE P7–35

7–36 Repeat problem 7–35 with $R = 2$ kΩ.

7–37 The switch in Figure P7–37 has been open for a long time and is closed at $t = 0$. The circuit parameters are $L = 0.4$ H, $C = 0.25$ μF, $R = 2$ kΩ, and $v_C(0) = 60$ V. Find $i_L(t)$ and $v_C(t)$ for $t \geq 0$. Is the circuit overdamped, critically damped, or underdamped?

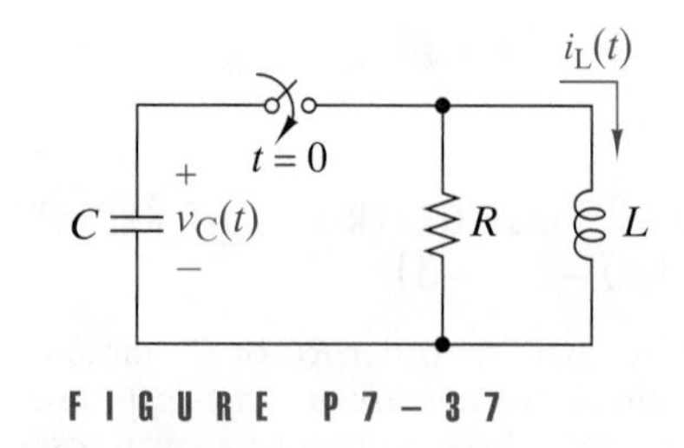

FIGURE P7–37

7–38 Repeat problem 7–37 with $C = 25$ nF.

7–39 The switch in Figure P7–39 has been open for a long time and is closed at $t = 0$. The circuit parameters are $L = 0.8$ H, $C = 1.25$ μF, $R_1 = 3$ kΩ, $R_2 = 2$ kΩ, and $V_A = 15$ V. Find $v_C(t)$ and $i_L(t)$ for $t \geq 0$. Is the circuit overdamped, critically damped, or underdamped?

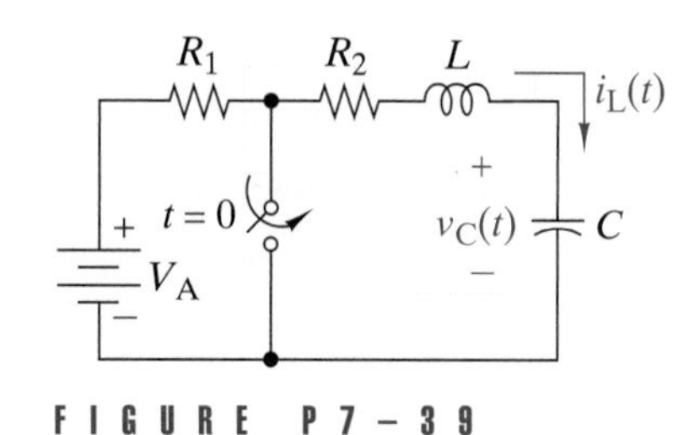

FIGURE P7–39

7–40 Repeat problem 7–39 with $L = 4$ H, $C = 0.1$ μF, $R_1 = 30$ kΩ, $R_2 = 22$ kΩ, and $V_A = 9$ V.

7–41 The switch in Figure P7–41 has been open for a long time and is closed at $t = 0$. The circuit parameters are $L = 0.8$ H, $C = 50$ nF, $R_1 = 4$ kΩ, $R_2 = 4$ kΩ, and $V_A = 20$ V. Find $v_C(t)$ and $i_L(t)$ for $t \geq 0$. Is the circuit overdamped, critically damped, or underdamped?

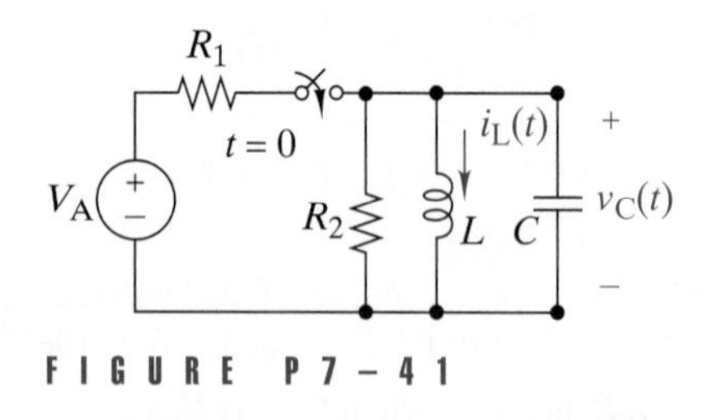

FIGURE P7–41

7–42 Repeat Problem 7–41 with $L = 1.25$ H.

7–43 The switch in Figure P7–43 has been in position A for a long time. At $t = 0$ it is moved to position B. The circuit parameters are $R_1 = 20$ kΩ, $R_2 = 4$ kΩ, $L = 1.6$ H, $C = 1.25$ μF, and $V_A = 24$ V. Find $v_C(t)$ and $i_L(t)$ for $t > 0$. Is the circuit overdamped, critically damped, or underdamped?

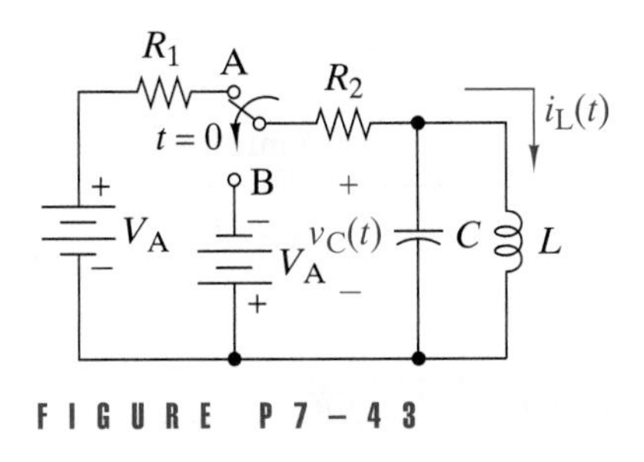

FIGURE P7–43

7–44 The switch in Figure P7–44 has been in position A for a long time and is moved to position B at $t = 0$. The circuit parameters are $R_1 = 500$ Ω, $R_2 = 500$ Ω, $L = 250$ mH, $C = 3.2$ μF, and $V_A = 5$ V. Find $v_C(t)$ and $i_L(t)$ for $t > 0$. Is the circuit overdamped, critically damped, or underdamped?

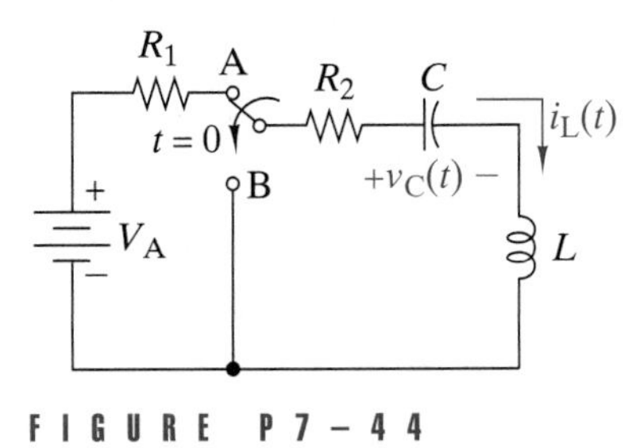

FIGURE P7–44

7–45 The switch in Figure P7–44 has been in position B for a long time and is moved to position A at $t = 0$. The circuit parameters are $R_1 = 500$ Ω, $R_2 = 500$ Ω, $L = 250$ mH, $C = 1$ μF, and $V_A = 5$ V. Find $v_C(t)$ and $i_L(t)$ for $t > 0$. Is the circuit overdamped, critically damped, or underdamped?

7–46 The circuit in Figure P7–46 is in the zero state when the step function input is applied. The circuit parameters are $L = 0.4$ H, $C = 1$ μF, $R = 2.2$ kΩ, and $V_A = 90$ V. Find $v_O(t)$ for $t > 0$.
Hint: First find the capacitor voltage.

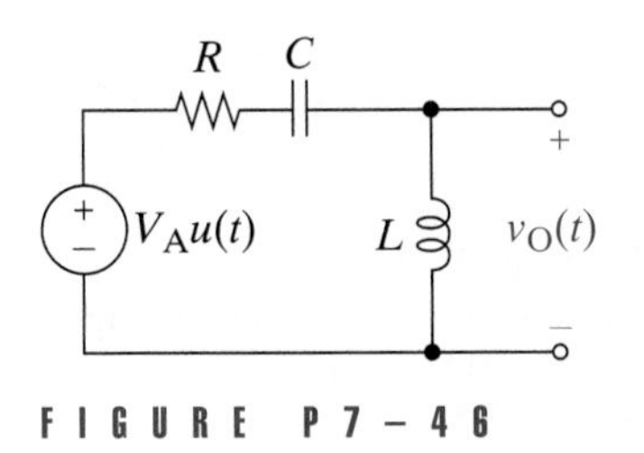

FIGURE P7–46

7–47 The circuit in Figure P7–47 is in the zero state when the step function input is applied. The circuit parameters are $R = 5$ kΩ, $L = 0.5$ H, $C = 250$ nF, and $V_A = 70$ V. Find $v_O(t)$ for $t > 0$.
Hint: First find the inductor current.

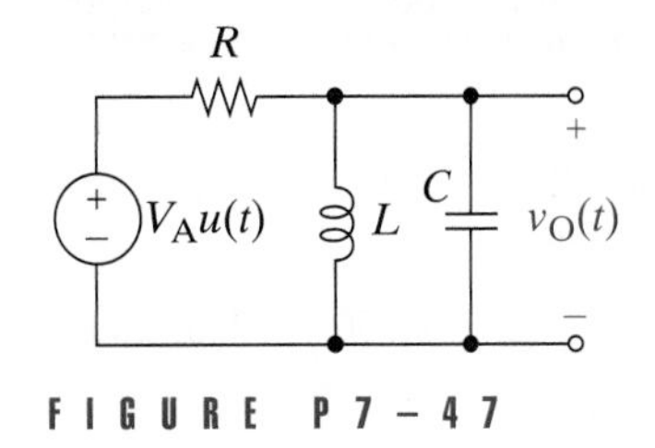

FIGURE P7–47

7–48 The circuit in Figure P7–48 is in the zero state when the step function input is applied. The circuit parameters are $R = 500$ Ω, $L = 2.5$ H, $C = 2.5$ μF, and $V_A = 100$ V. Find $v_O(t)$ for $t > 0$.
Hint: First find the inductor current.

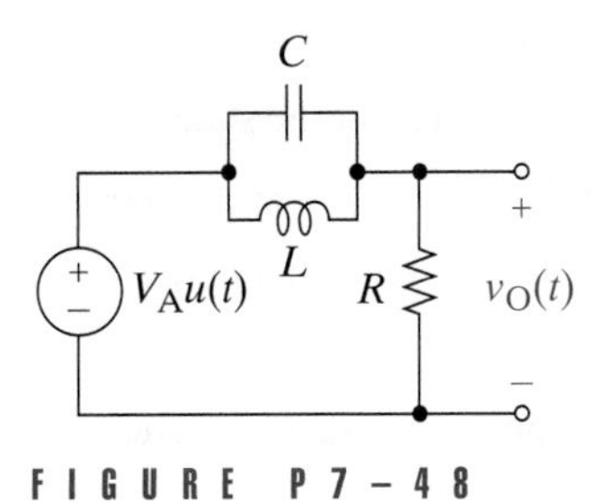

FIGURE P7–48

7–49 Derive expressions for the damping ratio and undamped natural frequency of the circuit in Figure P7–49 in terms of the circuit parameter R, L, and C.

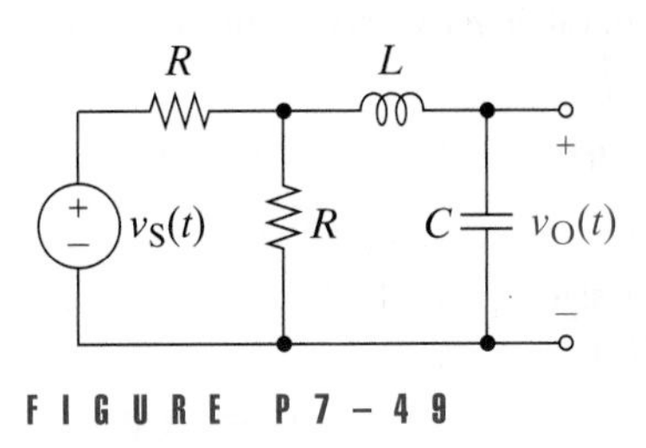

FIGURE P7–49

7–50 Repeat problem 7–49 for the circuit in Figure P7–50.

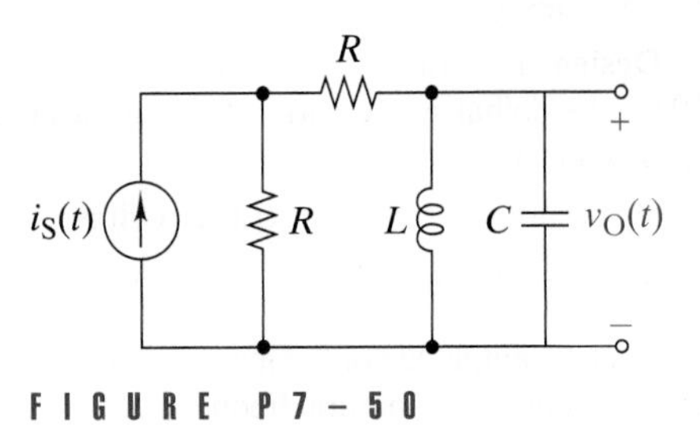

FIGURE P7–50

OBJECTIVE 7–4 SECOND-ORDER CIRCUIT RESPONSES (SECTS. 7–5, 7–6, 7–7, 7–8)

Given one or more responses of a second-order *RLC* circuit:
(a) Find the circuit parameters or other responses.
(b) Design a circuit to produce a given response.
See Examples 7–16, 7–22 and Exercises 7–15, 7–16, 7–18

7–51 In a series *RLC* circuit the step response across the 4-μF capacitor is

$$v_C(t) = 14 - e^{-50t}[14\cos(350t) + 2\sin(350t)]\text{ V} \qquad t \geq 0$$

(a) Find *R* and *L*.
(b) Find $i_L(t)$ for $t \geq 0$.

7–52 In a parallel *RLC* circuit the zero-input response in the 100-mH inductor is

$$i_L(t) = 50e^{-4000t} - 40e^{-5000t}\text{ mA} \qquad t \geq 0$$

(a) Find *R* and *C*.
(b) Find $v_C(t)$ for $t \geq 0$.

7–53 In a parallel *RLC* circuit the state variable responses are

$$v_C(t) = e^{-100t}[5\cos(300t) + 15\sin(300t)]\text{ V} \qquad t \geq 0$$

$$i_L(t) = 20 - 25e^{-100t}\cos(300t)\text{ mA} \qquad t \geq 0$$

Find *R*, *L*, and *C*.

7–54 The zero-input response of a series *RLC* circuit with $R = 80\ \Omega$ is

$$v_C(t) = 2e^{-2000t}\cos(1000t) - 4e^{-2000t}\sin(1000t)\text{ V} \qquad t \geq 0$$

If the initial conditions remain the same, what is the zero-input response when $R = 40\ \Omega$?

7–55 In a parallel *RLC* circuit the inductor current is observed to be

$$i_L(t) = 10e^{-10t}\cos(20t)\text{ mA} \qquad t \geq 0$$

Find $v_C(t)$ when $v_C(0) = -10$ V.

7–56 D Design a parallel *RLC* circuit whose natural response has the form

$$v_L(t) = K_1e^{-2000t} + K_2e^{-3000t}\text{ V} \qquad t \geq 0$$

7–57 D Design a series *RLC* circuit with $\zeta = 0.6$ and $\omega_0 = 4$ krad/s. What is the form of the natural response of $v_C(t)$ for your design?

7–58 D Design a series *RLC* circuit with $\zeta = 1$ and $\omega_0 = 10$ krad/s. What is the form of the natural response of $v_C(t)$ for your design?

7–59 The step response of a series *RLC* circuit is

$$v_C(t) = V_A - 2V_Ae^{-200t} + V_Ae^{-800t}\text{ V} \qquad t > 0$$

where V_A is the amplitude of the input. What are the damping ratio and undamped natural frequency of the circuit?

7–60 What range of damping ratios is available in the circuit in Figure P7–60?

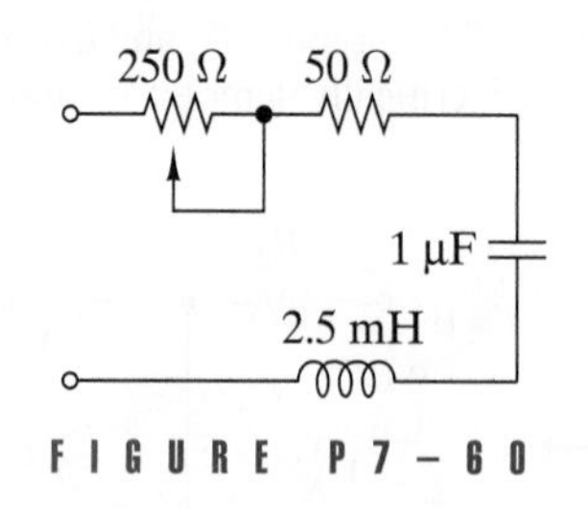

FIGURE P7–60

INTEGRATING PROBLEMS

7–61 A Reverse Step Response
The first-order *RC* circuit in Figure P7–61 is driven by a reverse step function input $v_S(t) = V_Au(-t)$. Derive an expression for $i_C(t)$ that is valid for $t \geq 0$.

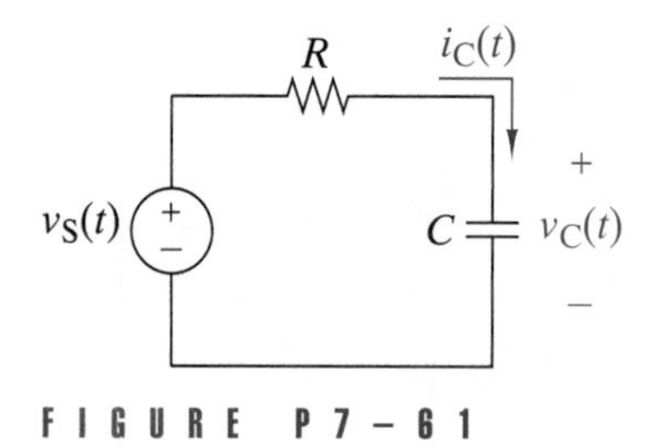

FIGURE P7–61

7–62 A First-Order OP AMP Circuit Step Response
Find the zero-state response of the OP AMP output voltage in Figure P7–62 when the input is $v_S(t) = V_Au(t)$.

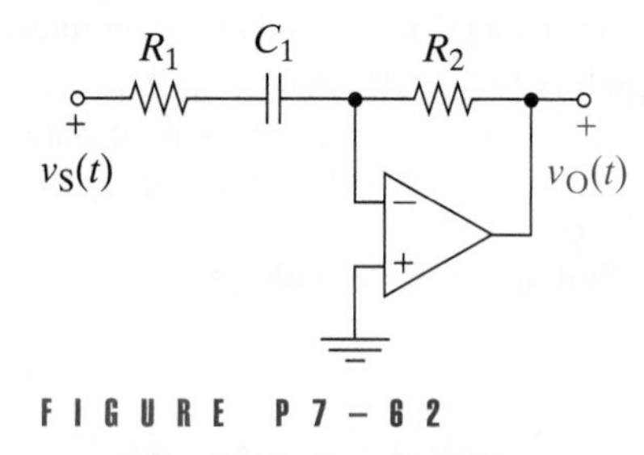

FIGURE P7–62

7–63 D *RC* Circuit Design
Design the first-order *RC* circuit in Figure P7–63 so an input $v_S(t) = 10u(t)$ V produces a zero-state response $v_O(t) = 10 - 5e^{-500t}$ V.

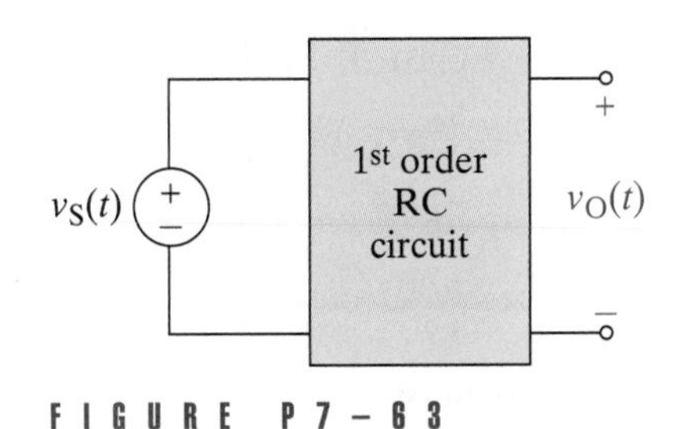

FIGURE P7–63

7–64 **D** Sample Hold Circuit
Figure P7–64 is a simplified diagram of a sample hold circuit. When the switch is in position A, the circuit is in the sample mode and the capacitor voltage must charge to at least 99% of the source voltage V_A in less than 1 μs. When the switch is moved to position B, the circuit is in the hold mode and the capacitor must retain at least 99% of V_A for at least 1 ms. Select a capacitance that meets these constraints.

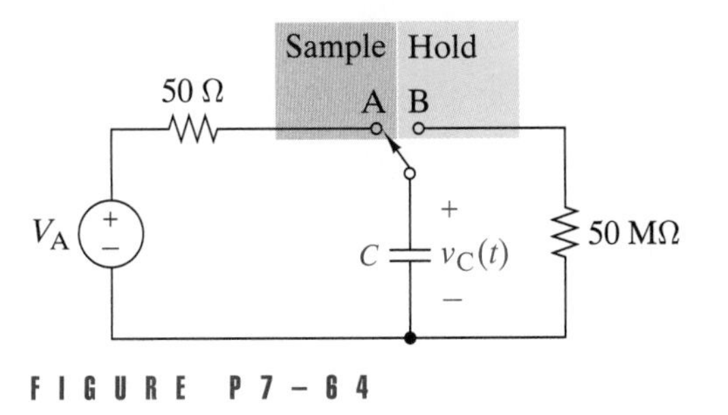

FIGURE P7–64

7–65 **A** Supercapacitor
Supercapacitors have very large capacitance (typically from 0.1 to 50 F), small sizes, and very long charge holding times, making them useful in nonbattery backup power applications. The charge holding quality of a supercapacitor is measured using the circuit in Figure P7–65. The switch is closed for a long time (say, 24 hours) and the capacitor is charged to 5 V. The switch is then opened and the capacitor allowed to self-discharge through any leakage resistance for 24 hours. Suppose that after 24 hours the voltage across a 0.47-F supercapacitor is 4.5 V. What is the equivalent leakage resistance in parallel with the capacitor?

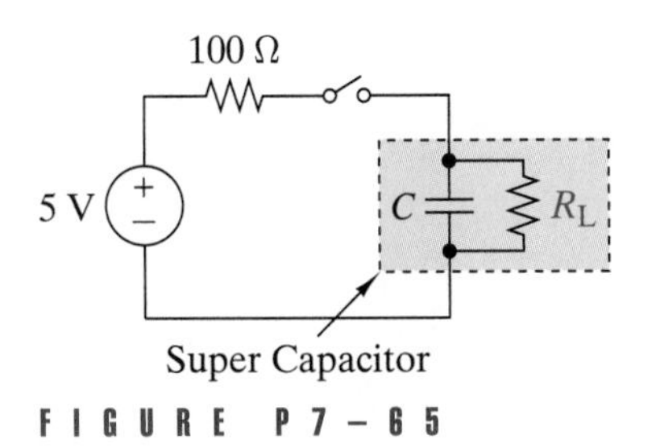

FIGURE P7–65

7–66 **A** Obtaining a Critically Damped Response
The voltage across a 500-Ω resistor in a series *RLC* circuit is $v_R(t) = 10e^{-500t}\sin(3500t)$ V. How much additional resistance is needed for the response to be critically damped?

7–67 **A** Combined First- and Second-Order Response
The switch in Figure P7–67 has been in position A for a long time and is moved to position B at $t = 0$ and then to position C when $t = 10$ ms. For $0 < t < 10$ ms the capacitor voltage is a charging exponential $v_C(t) = 10(1 - e^{-100t})$ V. For $t > 10$ ms the capacitor voltage is a sinusoid $v_C(t) = 6.321\cos[1000(t - 0.1)]$ V.

(a) Suppose the resistance is reduced to 1 kΩ and the switching sequence repeated. Will the amplitude of the sinusoid increase, decrease, or stay the same? Will the frequency of the sinusoid increase, decrease, or stay the same?

(b) Suppose the inductance is reduced to 100 mH and the switching sequence repeated. Will the amplitude of the sinusoid increase, decrease, or stay the same? Will the frequency of the sinusoid increase, decrease, or stay the same?

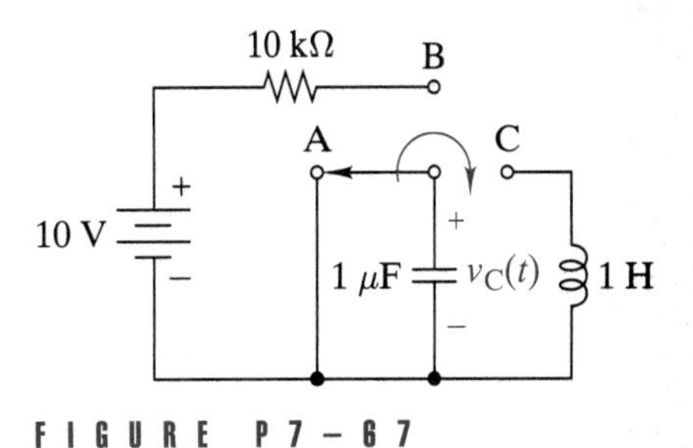

FIGURE P7–67

7–68 **A** Second-Order *RC* Circuit
Find the second-order differential equation relating v_O and i_S in Figure P7–68.
Hint: Write node-voltage equations.

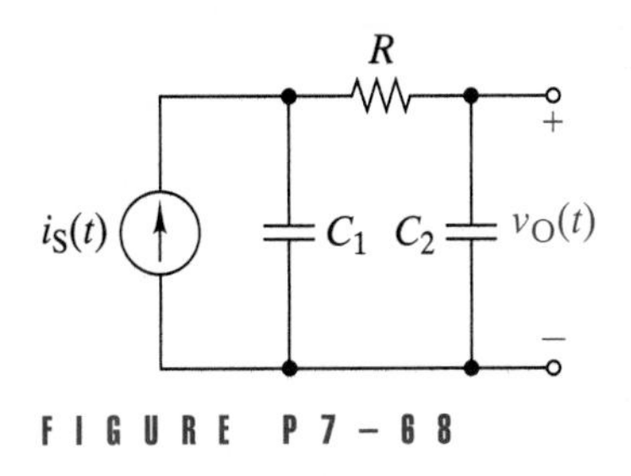

FIGURE P7–68

7–69 **D** Lightning Pulser Design
The circuit in Figure P7–69 is a simplified diagram of a pulser that delivers simulated lightning transients to the test article at the output interface. Closing the switch must produce a short-circuit current of the form $i_{SC}(t) = I_A e^{-\alpha t}\cos(\beta t)$, with $\alpha = 100$ krad/s, $\beta = 200$ krad/s, and $I_A = 2$ kA. Select the values of L, C, and V_0.

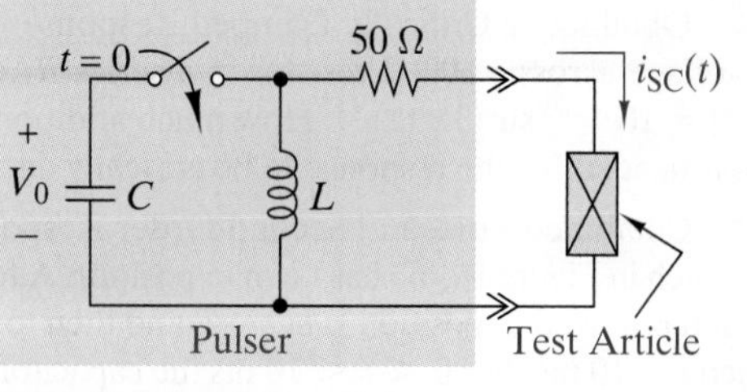

FIGURE P7–69

7–70 E *RLC* Circuit Design

Losses in real inductors can be modeled by the series resistor shown in Figure P7–70. In this problem we include the effect of this resistor on the design of the series *RLC* circuit shown in the figure. The design requirements include a source resistance of 50 Ω, an undamped natural frequency of 50 kHz, and a damping ratio less than 0.1. The characteristics of the available inductors are:

L (mH)	R_L (Ω)
8.2	433
6.8	371
5.6	306
4.7	266
3.9	299

Which inductor would you use in your design and why?

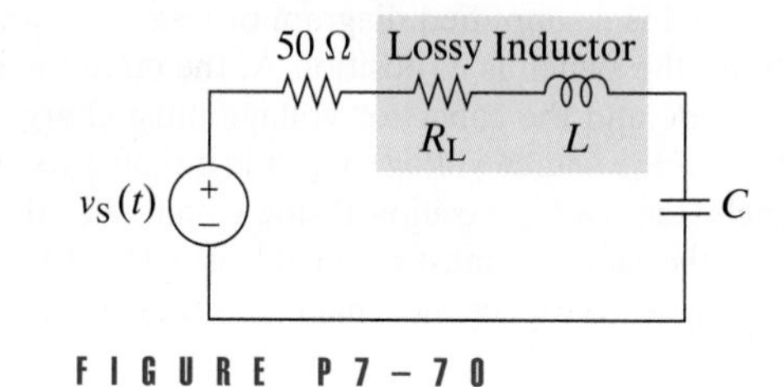

FIGURE P7–70

CHAPTER 8

SINUSOIDAL STEADY-STATE RESPONSE

The vector diagram of sine waves gives the best insight into the mutual relationships of alternating currents and emf's.

Charles P. Steinmetz, 1893,
American Engineer

Some History Behind This Chapter

The vector description of sinusoids was first discussed in detail by Charles Steinmetz (1865–1923) at the International Electric Congress of 1893. Although Oliver Heaviside may have used the vector concept earlier, Steinmetz is credited with popularizing the approach by demonstrating its many applications. By the turn of the century the vector method was well established in engineering practice and education. In the 1950s Steinmetz's vectors came to be called phasors to avoid possible confusion with the space vectors used to describe electromagnetic fields.

Why This Chapter Is Important Today

In this chapter we learn how to analyze ac circuits driven by a single-frequency sinusoid. We do this using a neat technique called phasor analysis that allows us to deal with ac circuits using the same tools we used on dc circuits. Phasor analysis is the key to understanding the electrical power systems that supply our homes and businesses. Using complex numbers is the price we pay for the simplicity of phasor analysis. Yet complex numbers are easy to master, after all, you were first introduced to them in high school.

Chapter Sections

8–1 Sinusoids and Phasors

8–2 Phasor Circuit Analysis

8–3 Basic Circuit Analysis with Phasors

8–4 Circuit Theorems with Phasors

8–5 General Circuit Analysis with Phasors

8–6 Energy and Power

Chapter Learning Objectives

8-1 Sinusoids and Phasors (Sect. 8-1)

Use the additive and derivative properties of phasors to convert sinusoids into phasors and vice versa.

8-2 Equivalent Impedance (Sects. 8-2, 8-3)

Given a linear circuit in the sinusoidal steady state, use series and parallel equivalence to find the equivalent impedance at a specified pair of terminals.

8-3 Basic Phasor Circuit Analysis (Sects. 8-3, 8-4)

Given a linear circuit in the sinusoidal steady state, find phasor responses using series and parallel equivalence, voltage and current division, circuit reduction, Thévenin or Norton equivalent circuits, and proportionality or superposition.

8-4 General Circuit Analysis (Sect. 8-5)

Given a linear circuit in the sinusoidal steady state, find equivalent impedances and phasor responses using node-voltage or mesh-current analysis.

8-5 Average Power and Maximum Power Transfer (Sect. 8-6)

Given a linear circuit in the sinusoidal steady state:

(a) Find the average power delivered at a specified interface.

(b) Find the maximum average power available at a specified interface.

(c) Find the load impedance required to draw the maximum available power.

8–1 SINUSOIDS AND PHASORS

The phasor concept is the foundation for the analysis of linear circuits in the sinusoidal steady state. Simply put, a **phasor** is a complex number representing the amplitude and phase angle of a sinusoidal voltage or current. The connection between sinewaves and complex numbers is provided by Euler's relationship:

$$e^{j\theta} = \cos\theta + j\sin\theta \tag{8–1}$$

Equation (8–1) relates the sine and cosine functions to the complex exponential $e^{j\theta}$. To develop the phasor concept, it is necessary to adopt the point of view that the cosine and sine functions can be written in the form

$$\cos\theta = \text{Re}\{e^{j\theta}\} \tag{8–2}$$

and

$$\sin\theta = \text{Im}\{e^{j\theta}\} \tag{8–3}$$

where Re stands for the "real part of" and Im for the "imaginary part of." Development of the phasor concept can begin with either Eq. (8–2) or (8–3). The choice between the two involves deciding whether to describe the eternal sinewave using a sine or cosine function. In Chapter 5 we chose the cosine, so we will reference phasors to the cosine function.

When Eq. (8–2) is applied to the general sinusoid defined in Chapter 5, we obtain

$$\begin{aligned} v(t) &= V_A \cos(\omega t + \phi) \\ &= V_A \text{Re}\{e^{j(\omega t+\phi)}\} = V_A \text{Re}\{e^{j\omega t}e^{j\phi}\} \\ &= \text{Re}\{(V_A e^{j\phi})e^{j\omega t}\} \end{aligned} \tag{8–4}$$

In the last line of Eq. (8–4), moving the amplitude V_A inside the real part operation does not change the final result because it is a real constant.

By definition, the quantity $V_A e^{j\phi}$ in the last line of Eq. (8–4) is the **phasor representation** of the sinusoid $v(t)$. The phasor **V** is written as

$$\mathbf{V} = V_A e^{j\phi} = V_A \cos\phi + jV_A \sin\phi \tag{8–5}$$

Note that **V** is a complex number determined by the amplitude and phase angle of the sinusoid. Figure 8–1 shows a graphical representation commonly called a phasor diagram.

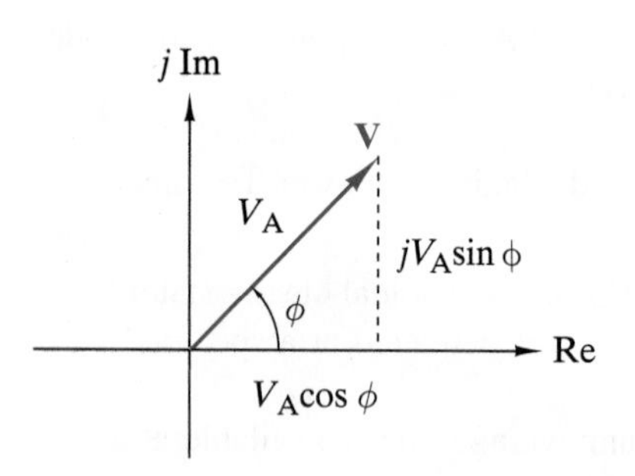

FIGURE 8–1 *Phasor diagram.*

The phasor is a complex number that can be written in either polar or rectangular form. An alternative way to write the polar form is to replace the exponential $e^{j\phi}$ by the shorthand notation $\angle\phi$. In subsequent discussions, we will often express phasors as $\mathbf{V} = V_A\angle\phi$, which is equivalent to the polar form in Eq. (8–5).

Two features of the phasor concept need emphasis:

1. Phasors are written in boldface type like **V** or $\mathbf{I}_1$ to distinguish them from signal waveforms such as $v(t)$ and $i_1(t)$.
2. A phasor is determined by amplitude and phase angle and does not contain any information about the frequency of the sinusoid.

The first feature points out that signals can be described in different ways. Although the phasor **V** and waveform $v(t)$ are related concepts, they have different physical interpretations and our notation must clearly distinguish between them. The absence of frequency information in the phasors results from the fact that in the sinusoidal steady state, all currents and voltages are sinusoids with the same frequency. Carrying frequency information in the phasor would be redundant, since it is the same for all phasors in any given steady-state circuit problem.

In summary, given a sinusoidal signal $v(t) = V_A \cos(\omega t + \phi)$, the corresponding phasor representation is $\mathbf{V} = V_A e^{j\phi}$. Conversely, given the phasor $\mathbf{V} = V_A e^{j\phi}$, the corresponding sinusoid is found by multiplying the phasor by $e^{j\omega t}$ and reversing the steps in Eq. (8–4) as follows:

$$\begin{aligned} v(t) &= \text{Re}\{\mathbf{V}e^{j\omega t}\} = \text{Re}\{(V_A e^{j\phi})e^{j\omega t}\} \\ &= V_A \text{Re}\{e^{j(\omega t+\phi)}\} = V_A \text{Re}\{\cos(\omega t + \phi) + j\sin(\omega t + \phi)\} \\ &= V_A \cos(\omega t + \phi) \end{aligned} \tag{8–6}$$

The frequency ω in the complex exponential $\mathbf{V}e^{j\omega t}$ in Eq. (8–6) must be expressed or implied in a problem statement, since by definition it is not contained in the phasor. Figure 8–2 shows a geometric interpretation of the complex exponential $\mathbf{V}e^{j\omega t}$ as a vector in the complex plane of length V_A, which rotates counterclockwise with a constant angular velocity of ω. The real part operation projects the rotating vector onto the horizontal (real) axis and thereby generates $v(t) = V_A \cos(\omega t + \phi)$. The complex exponential is sometimes called a **rotating phasor**, and the phasor **V** is viewed as a snapshot of the situation at $t = 0$.

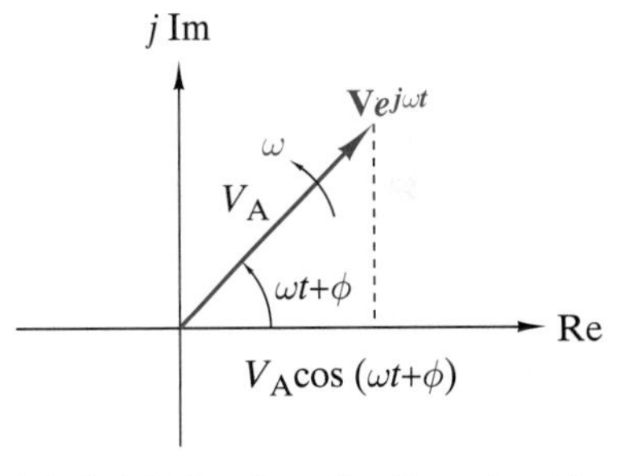

FIGURE 8–2 *Complex Exponential* $Ve^{j\omega t}$.

PROPERTIES OF PHASORS

Two important properties of phasors play key roles in circuit analysis. First, the **additive property** states that the phasor representing a sum of sinusoids of the same frequency is obtained by adding the phasor representations of the component sinusoids. To establish this property we write the expression

$$\begin{aligned} v(t) &= v_1(t) + v_2(t) + \ldots + v_N(t) \\ &= \text{Re}\{\mathbf{V}_1 e^{j\omega t}\} + \text{Re}\{\mathbf{V}_2 e^{j\omega t}\} + \ldots + \text{Re}\{\mathbf{V}_N e^{j\omega t}\} \end{aligned} \tag{8–7}$$

where $v_1(t)$, $v_2(t)$, ... and $v_N(t)$ are sinusoids of the same frequency whose phasor representations are $\mathbf{V}_1$, $\mathbf{V}_2$, ... and $\mathbf{V}_N$. The real part operation is additive, so the sum of real parts equals the real part of the sum. Consequently, Eq. (8–7) can be written in the form

$$\begin{aligned} v(t) &= \text{Re}\{\mathbf{V}_1 e^{j\omega t} + \mathbf{V}_2 e^{j\omega t} + \ldots + \mathbf{V}_N e^{j\omega t}\} \\ &= \text{Re}\{(\mathbf{V}_1 + \mathbf{V}_2 + \ldots + \mathbf{V}_N)e^{j\omega t}\} \end{aligned} \tag{8–8}$$

Comparing the last line in Eq. (8–8) with the definition of a phasor, we conclude that the phasor **V** representing $v(t)$ is

$$\mathbf{V} = \mathbf{V}_1 + \mathbf{V}_2 + \ldots + \mathbf{V}_N \tag{8–9}$$

The result in Eq. (8–9) applies only if the component sinusoids all have the same frequency so that $e^{j\omega t}$ can be factored out as shown in the last line in Eq. (8–8).

In Chapter 5 we found that the time derivative of a sinusoid is another sinusoid with the same frequency. Since they have the same frequency, the signal and its derivative can be represented by phasors. The **derivative property** of phasors allows us easily to relate the phasor representing a sinusoid to the phasor representing its derivative.

Equation (8–6) relates a sinusoid function and its phasor representation as

$$v(t) = \text{Re}\{\mathbf{V}e^{j\omega t}\}$$

Differentiating this equation with respect to time t yields

$$\begin{aligned} \frac{dv(t)}{dt} &= \frac{d}{dt}\text{Re}\{\mathbf{V}e^{j\omega t}\} = \text{Re}\left\{\mathbf{V}\frac{d}{dt}e^{j\omega t}\right\} \\ &= \text{Re}\{(j\omega\mathbf{V})e^{j\omega t}\} \end{aligned} \quad (8\text{–}10)$$

From the definition of a phasor, we see that the quantity $(j\omega\mathbf{V})$ on the right side of this equation is the phasor representation of the time derivative of the sinusoidal waveform. This phasor can be written in the form

$$\begin{aligned} j\omega\mathbf{V} &= (\omega e^{j90^\circ})(V_A e^{j\theta}) \\ &= \omega V_A e^{j(\theta+90^\circ)} \end{aligned} \quad (8\text{–}11)$$

which points out that differentiating a sinusoid changes its amplitude by a multiplicative factor ω and shifts the phase angle by 90°.

In summary, the additive property states that adding phasors is equivalent to adding sinusoids of the same frequency. The derivative property states that multiplying a phasor by $j\omega$ is equivalent to differentiating the corresponding sinusoid. The following examples show applications of these two properties of phasors.

EXAMPLE 8–1

(a) Construct the phasors for the following signals:

$$v_1(t) = 10\cos(1000t - 45^\circ) \text{ V}$$

$$v_2(t) = 5\cos(1000t + 30^\circ) \text{ V}$$

(b) Use the additive property of phasors and the phasors found in (a) to find $v(t) = v_1(t) + v_2(t)$.

SOLUTION:

(a) The phasor representations of $v_1(t)$ and $v_2(t)$ are

$$\begin{aligned} \mathbf{V}_1 &= 10e^{-j45^\circ} = 10\cos(-45^\circ) + j10\sin(-45^\circ) \\ &= 7.07 - j7.07 \end{aligned}$$

$$\begin{aligned} \mathbf{V}_2 &= 5\,e^{+j30^\circ} = 5\cos(30^\circ) + j5\sin(30^\circ) \\ &= 4.33 + j2.5 \end{aligned}$$

(b) The two sinusoids have the same frequency, so the additive property of phasors can be used to obtain their sum:

$$\mathbf{V} = \mathbf{V}_1 + \mathbf{V}_2 = 11.4 - j4.57 = 12.3e^{-j21.8^\circ}$$

The waveform corresponding to this phasor sum is

$$v(t) = \mathrm{Re}\{(12.3e^{-j21.8°})e^{j1000t}\}$$
$$= 12.3\cos(1000t - 21.8°) \text{ V}$$

The phasor diagram in Figure 8–3 shows that summing sinusoids can be viewed geometrically in terms of phasors. ■

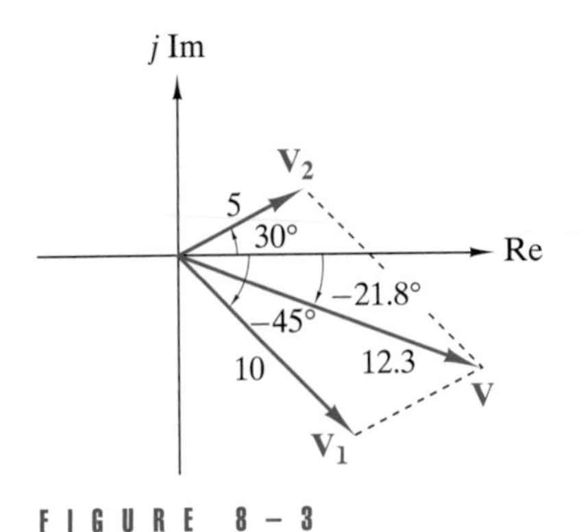

FIGURE 8–3

EXAMPLE 8–2

(a) Construct the phasors representing the following signals:

$$i_A(t) = 5\cos(377t + 50°) \text{ A}$$
$$i_B(t) = 5\cos(377t + 170°) \text{ A}$$
$$i_C(t) = 5\cos(377t - 70°) \text{ A}$$

(b) Use the additive property of phasors and the phasors found in (a) to find the sum of these waveforms.

SOLUTION:

(a) The phasor representation of the three sinusoidal currents are

$$\mathbf{I}_A = 5e^{j50°} = 5\cos(50°) + j5\sin(50°) = 3.21 + j3.83 \text{ A}$$
$$\mathbf{I}_B = 5e^{j170°} = 5\cos(170°) + j5\sin(170°) = -4.92 + j0.87 \text{ A}$$
$$\mathbf{I}_C = 5e^{-j70} = 5\cos(-70°) + j5\sin(-70°) = 1.71 - j4.70 \text{ A}$$

(b) The currents have the same frequency, so the additive property of phasors applies. The phasor representing the sum of these currents is

$$\mathbf{I}_A + \mathbf{I}_B + \mathbf{I}_C = (3.21 - 4.92 + 1.71) + j(3.83 + 0.87 - 4.70)$$
$$= 0 + j0 \text{ A}$$

It is not obvious by examining the waveforms that these three currents add to zero. However, the phasor diagram in Figure 8–4 makes this fact clear, since the sum of any two phasors is equal and opposite to the third. Phasors of this type occur in balanced three-phase power systems. The balanced condition occurs when three equal-amplitude phasors are displaced in phase by exactly 120°. ■

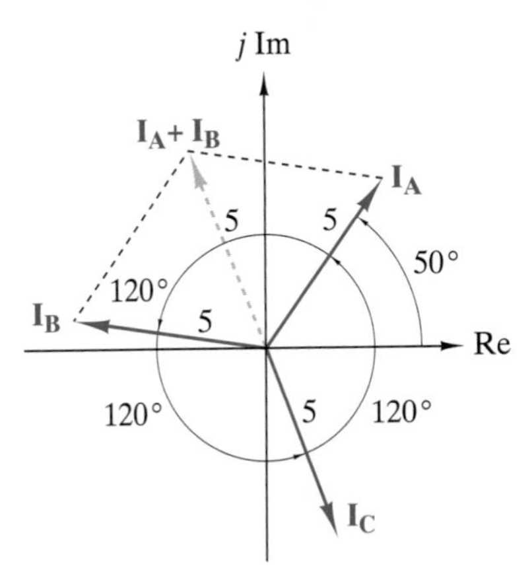

FIGURE 8–4

EXAMPLE 8–3

Use the derivative property of phasors to find the time derivative of $v(t) = 15\cos(200t - 30°)$ V.

SOLUTION:

The phasor for the sinusoid is $\mathbf{V} = 15\angle{-30°}$. According to the derivative property, the phasor representing the dv/dt is found by multiplying $\mathbf{V}$ by $j\omega$.

$$(j200) \times (15|-30°) = (200e^{j90°}) \times (15e^{-j30°}) = 3000e^{j60°}$$

The sinusoid corresponding to the phasor $j\omega\mathbf{V}$ is

$$\frac{dv(t)}{dt} = \text{Re}\{(3000e^{j60^\circ})e^{j200t}\} = 3000\ \text{Re}\{e^{j(200t+60^\circ)}\}$$

$$= 3000\cos(200t + 60^\circ)\ \text{V/s}$$

Finding the derivative of a sinusoid is easily carried out in phasor form, since it only involves manipulating complex numbers. ■

EXAMPLE 8–4

(a) Convert the following phasors into sinusoidal waveforms:

$$\mathbf{V}_1 = 20 + j20\quad \text{V},\ \omega = 500\quad \text{rad/s}$$

$$\mathbf{V}_2 = 10\sqrt{2}\,e^{-j45^\circ}\quad \text{V},\ \omega = 500\quad \text{rad/s}$$

(b) Use phasors addition to find the sinusoidal waveform $v_3(t) = v_1(t) + v_2(t)$.

SOLUTION:

(a) Since $\mathbf{V}_1 = 20 + j20 = 20\sqrt{2}\,e^{j45^\circ}$, the waveforms corresponding to the phasors $\mathbf{V}_1$ and $\mathbf{V}_2$ are

$$v_1(t) = \text{Re}\,\{20\sqrt{2}\,e^{j45^\circ}\,e^{j500t}\} = 20\sqrt{2}\cos(500\,t + 45^\circ)\quad \text{V}$$

$$v_2(t) = \text{Re}\,\{10\sqrt{2}\,e^{-j45^\circ}\,e^{j500t}\} = 10\sqrt{2}\cos(500\,t - 45^\circ)\quad \text{V}$$

(b) Since $\mathbf{V}_2 = 10\sqrt{2}\,e^{-j45^\circ} = 10 - j10$, the additive property of phasors yields

$$\mathbf{V}_3 = \mathbf{V}_1 + \mathbf{V}_2 = 20 + j20 + 10 - j10$$

$$= 30 + j10 = 31.6\,e^{j18.4^\circ}$$

Hence,

$$v_3(t) = \text{Re}\,\{31.6\,e^{j18.4^\circ}\,e^{j500t}\} = 31.6\cos(500\,t + 18.4^\circ)\quad \text{V}$$ ■

Exercise 8–1

Convert the following sinusoids to phasors in polar and rectangular form:

(a) $v(t) = 20\cos(150t - 60^\circ)$ V
(b) $v(t) = 10\cos(1000t + 180^\circ)$ V
(c) $i(t) = -4\cos 3t + 3\cos(3t - 90^\circ)$ A

Answers:

(a) $\mathbf{V} = 20\angle -60^\circ = 10 - j17.3$ V
(b) $\mathbf{V} = 10\angle 180^\circ = -10 + j0$ V
(c) $\mathbf{I} = 5\angle -143^\circ = -4 - j3$ A

Exercise 8–2

Convert the following phasors to sinusoids:

(a) $\mathbf{V} = 169\angle -45^\circ$ V at $f = 60$ Hz
(b) $\mathbf{V} = 10\angle 90^\circ + 66 - j10$ V at $\omega = 10$ krad/s
(c) $\mathbf{I} = 15 + j5 + 10\angle 180^\circ$ mA at $\omega = 1000$ rad/s

Answers:

(a) $v(t) = 169 \cos(377t - 45°)$ V
(b) $v(t) = 66 \cos 10^4 t$ V
(c) $i(t) = 7.07 \cos(1000t + 45°)$ mA

Exercise 8–3

Find the phasor corresponding to the waveform $v(t) = V_A \cos(\omega t) + 2V_A \sin(\omega t)$.

Answer: $\mathbf{V} = V_A - j2V_A$

Exercise 8–4

Find the phasor corresponding to the time derivative of the waveform $v(t) = V_A \cos(\omega t) + 2V_A \sin(\omega t)$.

Answer: $\mathbf{V} = 2\omega V_A + j\omega V_A$

8–2 PHASOR CIRCUIT ANALYSIS

Phasor circuit analysis is a method of finding sinusoidal steady-state responses directly from the circuit without using differential equations. How do we perform phasor circuit analysis? At several points in our study we have seen that circuit analysis is based on two kinds of constraints: (1) connection constraints (Kirchhoff's laws) and (2) device constraints (element equations). To analyze phasor circuits, we must see how these constraints are expressed in phasor form.

CONNECTION CONSTRAINTS IN PHASOR FORM

The sinusoidal steady-state condition is reached after the circuit's natural response decays to zero. In the steady state all of the voltages and currents are sinusoids with the same frequency as the driving force. Under these conditions, the application of KVL around a loop could take the form

$$V_1 \cos(\omega t + \phi_1) + V_2 \cos(\omega t + \phi_2) \ldots + V_N \cos(\omega t + \phi_N) = 0$$

These sinusoids have the same frequency but have different amplitudes and phase angles. The additive property of phasors discussed in the preceding section shows that there is a one-to-one correspondence between waveform sums and phasor sums. Therefore, if the sum of the waveforms is zero, then the corresponding phasors must also sum to zero.

$$\mathbf{V}_1 + \mathbf{V}_2 \ldots + \mathbf{V}_N = 0$$

Clearly the same result applies to phasor currents and KCL. In other words, we can state Kirchhoff's laws in phasor form as follows:

KVL:* *The algebraic sum of phasor voltages around a loop is zero.

KCL:* *The algebraic sum of phasor currents at a node is zero.

DEVICE CONSTRAINTS IN PHASOR FORM

Turning now to the device constraints, we note that the i–v characteristics of the three passive elements are

$$\begin{aligned} \text{Resistor: } v_R(t) &= Ri_R(t) \\ \text{Inductor: } v_L(t) &= L\frac{di_L(t)}{dt} \\ \text{Capacitor: } i_C(t) &= C\frac{dv_C(t)}{dt} \end{aligned} \tag{8–12}$$

In the sinusoidal steady state, all of these currents and voltages are sinusoids. Given that the signals are sinusoid, how do these i–v relationships constrain the corresponding phasors?

In the sinusoidal steady state, the voltage and current of the resistor can be written in terms of phasors as $v_R(t) = \text{Re}\{\mathbf{V}_R e^{j\omega t}\}$ and $i_R(t) = \text{Re}\{\mathbf{I}_R e^{j\omega t}\}$. Consequently, the resistor i–v relationship in Eq. (8–12) can be expressed in terms of phasors as follows:

$$\text{Re}\{\mathbf{V}_R e^{j\omega t}\} = R \times \text{Re}\{\mathbf{I}_R e^{j\omega t}\}$$

Since R is a real constant, moving it inside the real part operation on the right side of this equation does not change things:

$$\text{Re}\{\mathbf{V}_R e^{j\omega t}\} = \text{Re}\{R\mathbf{I}_R e^{j\omega t}\}$$

This relationship holds only if the phasor voltage and current for a resistor are related as

$$\mathbf{V}_R = R\mathbf{I}_R \tag{8–13}$$

To explore this relationship, we assume that the current through a resistor is $i_R(t) = I_A \cos(\omega t + \phi)$. Then the phasor current is $\mathbf{I}_R = I_A e^{j\phi}$, and according to Eq. (8–13), the phasor voltage across the resistor is

$$\mathbf{V}_R = RI_A e^{j\phi}$$

This result shows that the voltage has the same phase angle (ϕ) as the current. Phasors with the same phase angle are said to be **in phase**; otherwise they are said to be **out of phase**. Figure 8–5 shows the phasor diagram of the resistor current and voltage. Two scale factors are needed to construct a phasor diagram showing both voltage and current, since the two phasors do not have the same dimensions.

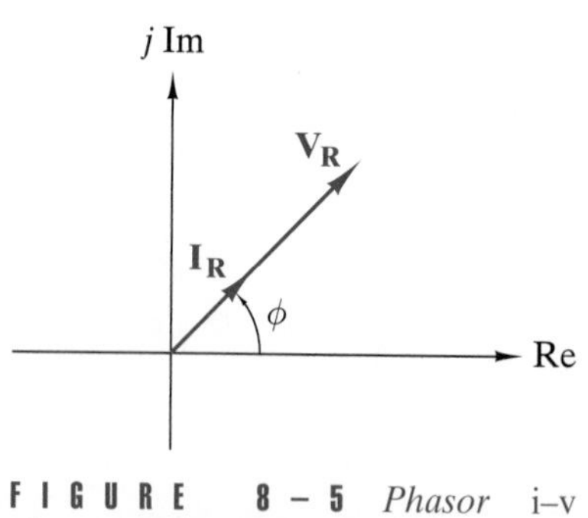

FIGURE 8–5 *Phasor* i–v *characteristics of the resistor.*

In the sinusoidal steady state, the voltage and phasor current for the inductor can be written in terms of phasors as $v_L(t) = \text{Re}\{\mathbf{V}_L e^{j\omega t}\}$ and $i_L(t) = \text{Re}\{\mathbf{I}_L e^{j\omega t}\}$. Using the derivative property of phasors, the inductor i–v relationship in Eq. (8–12) can be expressed as follows:

$$\begin{aligned} \text{Re}\{\mathbf{V}_L e^{j\omega t}\} &= L \times \text{Re}\{j\omega\mathbf{I}_L e^{j\omega t}\} \\ &= \text{Re}\{j\omega L\mathbf{I}_L e^{j\omega t}\} \end{aligned}$$

Since L is a real constant, moving it inside the real part operation does not change things. Written this way, we see that the phasor voltage and current for an inductor are related as

$$\mathbf{V}_L = j\omega L\mathbf{I}_L \tag{8–14}$$

When the current is $i_L(t) = I_A \cos(\omega t + \phi)$, the corresponding phasor is $\mathbf{I}_L = I_A e^{j\phi}$ and the i–v constraint in Eq. (8–14) yields

$$\mathbf{V}_L = j\omega L \mathbf{I}_L = (\omega L e^{j90°})(I_A e^{j\phi})$$
$$= \omega L I_A e^{j(\phi + 90°)}$$

The resulting phasor diagram in Figure 8–6 shows that the inductor voltage and current are 90° out of phase. The voltage phasor is advanced by 90° counterclockwise, which is in the direction of rotation of the complex exponential $e^{j\omega t}$. When the voltage phasor is advanced counterclockwise (that is, ahead of the rotating current phasor), we say that the voltage phasor *leads* the current phasor by 90° or, equivalently, the current *lags* the voltage by 90°.

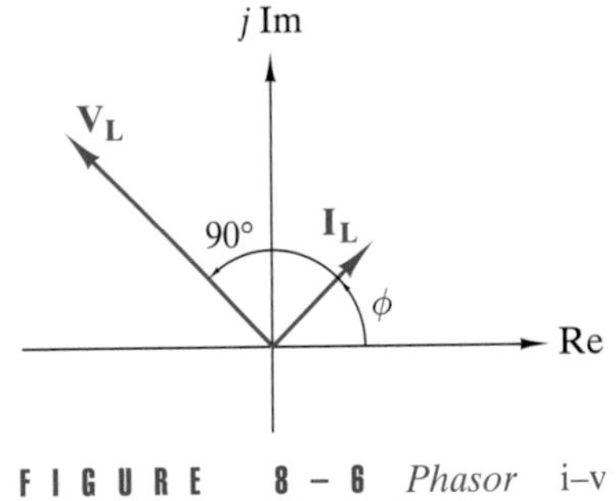

FIGURE 8–6 *Phasor i–v characteristics of the inductor.*

Finally, the capacitor voltage and current in the sinusoidal steady state can be written in terms of phasors as $v_C(t) = \text{Re}\{\mathbf{V}_C e^{j\omega t}\}$ and $i_C(t) = \text{Re}\{\mathbf{I}_C e^{j\omega t}\}$. Using the derivative property of phasors, the i–v relationship of the capacitor in Eq. (8–12) becomes

$$\text{Re}\{\mathbf{I}_C e^{j\omega t}\} = C \times \text{Re}\{j\omega \mathbf{V}_C e^{j\omega t}\}$$
$$= \text{Re}\{j\omega C \mathbf{V}_C e^{j\omega t}\}$$

Moving the real constant C inside the real part operation does not change the final result, so we conclude that the phasor voltage and current for a capacitor are related as

$$\mathbf{I}_C = j\omega C \mathbf{V}_C$$

Solving for $\mathbf{V}_C$ yields

$$\mathbf{V}_C = \frac{1}{j\omega C}\mathbf{I}_C \qquad (8\text{–}15)$$

When $i_C(t) = I_A \cos(\omega t + \phi)$, then according to Eq. (8–15), the phasor voltage across the capacitor is

$$\mathbf{V}_C = \frac{1}{j\omega C}\mathbf{I}_C = \left(\frac{1}{\omega C}e^{-j90°}\right)(I_A e^{j\phi})$$
$$= \frac{I_A}{\omega C}e^{j(\phi - 90°)}$$

The resulting phasor diagram in Figure 8–7 shows that voltage and current are 90° out of phase. In this case, the voltage phasor is retarded by 90° clockwise, which is in a direction opposite to rotation of the complex exponential $e^{j\omega t}$. When the voltage is retarded clockwise (that is, behind the rotating current phasor), we say that the voltage phasor *lags* the current phasor by 90° or, equivalently, the current *leads* the voltage by 90°.

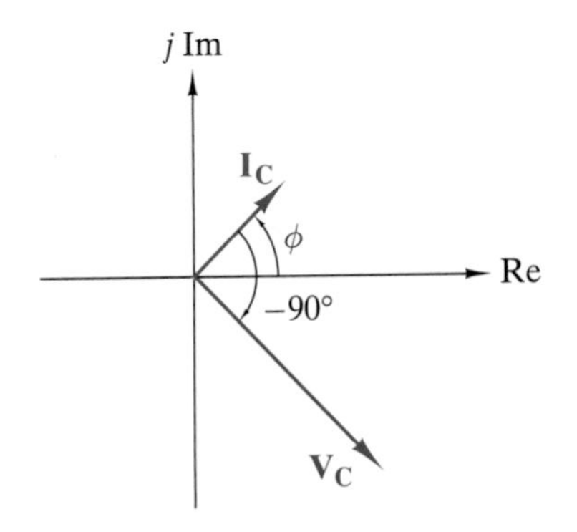

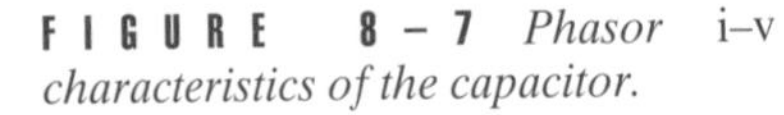
FIGURE 8–7 *Phasor i–v characteristics of the capacitor.*

THE IMPEDANCE CONCEPT

The phasor **I–V** constraints in Eqs. (8–13), (8–14), and (8–15) are all of the form

$$\mathbf{V} = Z\mathbf{I} \qquad (8\text{–}16)$$

where Z is called the impedance of the element. Equation (8–16) is analogous to Ohm's law in resistive circuits. **Impedance** is the proportionality constant relating phasor voltage and phasor current in linear, two-terminal elements. The impedances of the three passive elements are

$$\begin{aligned} &\text{Resistor: } Z_R = R \\ &\text{Inductor: } Z_L = j\omega L \\ &\text{Capacitor: } Z_C = \frac{1}{j\omega C} = -\frac{j}{\omega C} \end{aligned} \tag{8–17}$$

Since impedance relates phasor voltage to phasor current, it is a complex quantity whose units are ohms. Although impedance can be a complex number, it is not a phasor. Phasors represent sinusoidal signals, while impedances characterize circuit elements in the sinusoidal steady state. Finally, it is important to remember that the generalized two-terminal device constraint in Eq. (8–16) assumes that the passive sign convention is used to assign the reference marks to the voltage and current.

EXAMPLE 8–5

The circuit in Figure 8–8 is operating in the sinusoidal steady state with $i(t) = 4\cos(5000t)$ A. Find the steady-state voltage $v(t)$.

FIGURE 8–8

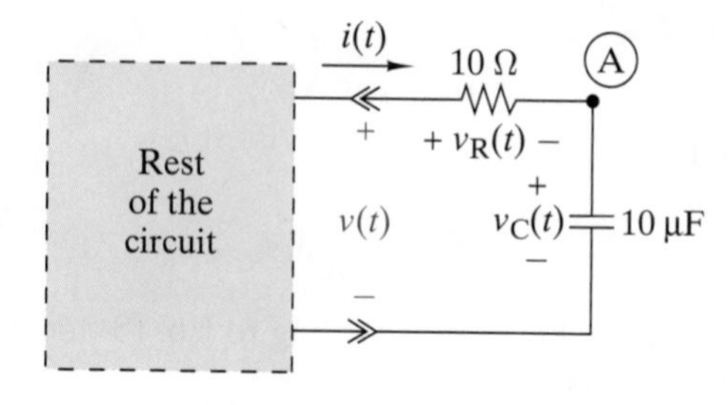

SOLUTION:

The resistor and the capacitor have impedances

$$Z_R = R = 10 + j0 \quad \text{and} \quad Z_C = \frac{1}{j\,\omega\,C} = \frac{1}{j5000 \times 10^{-5}} = 0 - j20$$

Applying KCL at node A shows that $i(t) = 4\cos(5000t)$ is the current through both the resistor and the capacitor. The corresponding phasor current is $\mathbf{I} = 4 + j0$. Using the element impedances, the phasor voltages across the resistor and the capacitor are

$$\mathbf{V_R} = Z_R\mathbf{I} = 10 \times (4 + j0) = 40 + j0$$

$$\mathbf{V_C} = Z_C\mathbf{I} = (-j20) \times (4 + j0) = 0 - j80$$

Applying KVL around the loop yields $\mathbf{V} = \mathbf{V}_R + \mathbf{V}_C$; hence,

$$\mathbf{V} = 40 - j80 = 89.4 \angle -63.4°$$

and the steady-state voltage waveform is

$$\begin{aligned} v(t) &= \text{Re}\,\{89.4\, e^{-j63.4°}\, e^{j5000t}\} \\ &= 89.4\cos(5000\,t - 63.4°) \quad \text{V} \end{aligned}$$

■

Exercise 8–5

The circuit in Figure 8–9 is operating in the sinusoidal steady state with $v(t) = 50 \cos(500t)$ V and $i(t) = 4 \cos(500t - 60°)$ A. Find the impedance of the elements in the box.

Answer: $Z = 6.25 + j10.8\ \Omega$

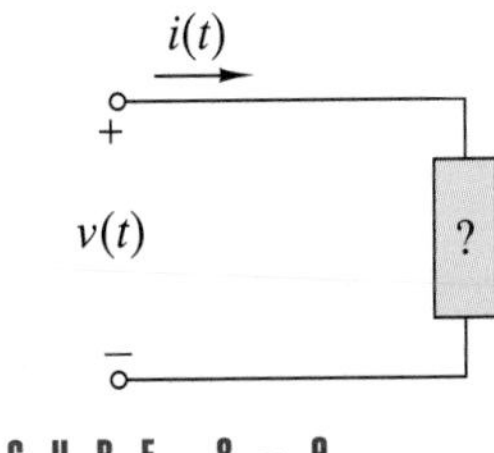

FIGURE 8–9

Exercise 8–6

The current through a 12-mH inductor is $i_L(t) = 20 \cos(10^6 t)$ mA. Determine:

(a) The impedance of the inductor
(b) The phasor voltage across the inductor
(c) The waveform of the voltage across the inductor.

Answers:

(a) $Z_L = j12\ \text{k}\Omega$
(b) $\mathbf{V}_L = 240\angle 90°$ V
(c) $v_L = 240 \cos(10^6 t + 90°)$ V

Exercise 8–7

The current through a 20-pF capacitor is $i_C(t) = 0.3 \cos(10^6 t)$ mA.

(a) Find the impedance of the capacitor.
(b) Find the phasor voltage across the capacitor.
(c) Find the waveform of the voltage across the capacitor.

Answers:

(a) $Z_C = -j50\ \text{k}\Omega$
(b) $\mathbf{V}_C = 15\angle -90°$ V
(c) $v_C = 15 \cos(10^6 t - 90°)$ V

8–3 BASIC CIRCUIT ANALYSIS WITH PHASORS

Functions of time like $v(t) = V_A \cos(\omega t + \phi_V)$ and $i(t) = I_A \cos(\omega t + \phi_I)$ are time-domain representations of sinusoidal signals. Producing the corresponding phasors can be thought of as a transformation that carries $v(t)$ and $i(t)$ into a complex-number domain where signals are represented as phasors **V** and **I**. We call this complex-number domain the **phasor domain**. When we analyze circuits in this phasor domain, we obtain sinusoidal steady-state responses in terms of phasors like **V** and **I**. Performing the inverse phasor transformation as $v(t) = \text{Re}\{\mathbf{V}e^{j\omega t}\}$ and $i(t) = \text{Re}\{\mathbf{I}e^{j\omega t}\}$ carries the responses back into the time domain. To perform ac circuit analysis in this way, we obviously need to develop methods of analyzing circuits in the phasor domain.

In the preceding section, we showed that KVL and KCL apply in the phasor domain and that the phasor element constraints all have the form $\mathbf{V} = Z\mathbf{I}$. These element and connection constraints have the same format as the underlying constraints for resistance circuit analysis as developed in Chapters 2, 3, and 4. Therefore, familiar algebraic circuit analysis tools, such as series and parallel equivalence, voltage and current division, proportionality and superposition, and Thévenin and Norton equivalent circuits, are applicable in the phasor domain. In other words, we do not need new analysis techniques to handle circuits in the phasor domain. The only difference is that circuit responses are phasors (complex numbers) rather than dc signals (real numbers).

We can think of phasor-domain circuit analysis in terms of the flow diagram in Figure 8–10. The analysis begins in the time domain with a linear circuit operating in the sinusoidal steady state and involves three major steps:

STEP 1 The circuit is transformed into the phasor domain by representing the input and response sinusoids as phasors and the passive circuit elements by their impedances.

STEP 2 Standard algebraic circuit analysis techniques are applied to solve the phasor-domain circuit for the desired unknown phasor responses.

STEP 3 The phasor responses are inverse transformed back into time-domain sinusoids to obtain the response waveforms.

The third step assumes that the required end product is a time-domain waveform. However, a phasor is just another representation of a sinusoid. With some experience, we learn to think of the response as a phasor without converting it back into a time-domain waveform.

Figure 8–10 points out that there is another route to time-domain response using the classical differential equation method from Chapter 7. However, the phasor-domain method works directly with the circuit model and is far simpler. More important, phasor-domain analysis provides insights into ac

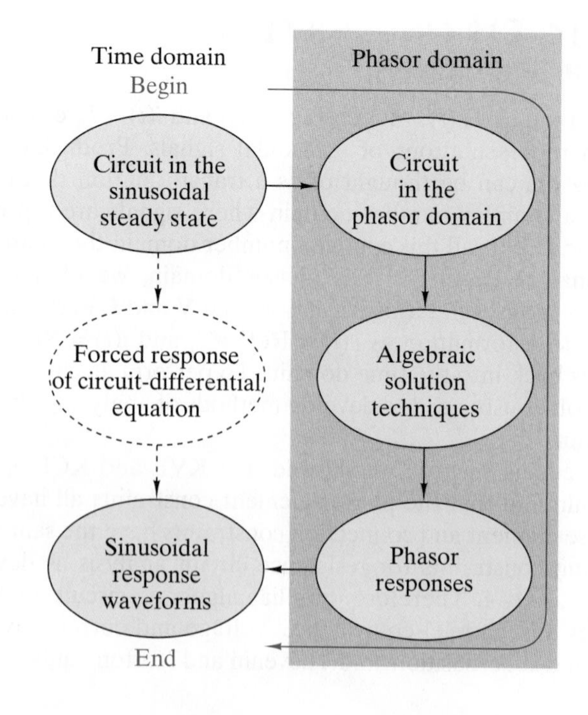

FIGURE 8–10 *Flow diagram for phasor circuit analysis.*

circuit analysis that are essential to understanding much of the terminology and viewpoint of electrical engineering.

SERIES EQUIVALENCE AND VOLTAGE DIVISION

We begin the study of phasor-domain analysis with two basic analysis tools—series equivalence and voltage division. In Figure 8–11 the two-terminal elements are connected in series, so by KCL, the same phasor current **I** exists in impedances $Z_1, Z_2, \ldots Z_N$. Using KVL and the element constraints, the voltage across the series connection can be written as

$$\begin{aligned} \mathbf{V} &= \mathbf{V}_1 + \mathbf{V}_2 + \ldots + \mathbf{V}_N \\ &= Z_1\mathbf{I} + Z_2\mathbf{I} + \ldots + Z_N\mathbf{I} \\ &= (Z_1 + Z_2 + \ldots + Z_N)\mathbf{I} \end{aligned} \quad (8\text{–}18)$$

The last line in this equation points out that the phasor responses **V** and **I** do not change when the series connected elements are replaced by an equivalent impedance:

$$Z_{EQ} = Z_1 + Z_2 + \ldots + Z_N \quad (8\text{–}19)$$

In general, the equivalent impedance Z_{EQ} is a complex quantity of the form

$$Z_{EQ} = R + jX$$

where R is the real part and X is the imaginary part. The real part of Z is called **resistance** and the imaginary part (X, not jX) is called **reactance**. Both resistance and reactance are expressed in ohms (Ω), and both can be functions of frequency (ω). For passive circuits, resistance is always positive, while reactance X can be either positive or negative. A positive X is called an **inductive** reactance because the reactance of an inductor is ωL, which is always positive. A negative X is called a **capacitive** reactance because the reactance of a capacitor is $-1/\omega C$, which is always negative.

Combining Eqs. (8–18) and (8–19), we can write the phasor voltage across the kth element in the series connection as

$$\mathbf{V}_k = Z_k\mathbf{I} = \frac{Z_k}{Z_{EQ}}\mathbf{V} \quad (8\text{–}20)$$

Equation (8–20) is the phasor version of the voltage division principle. The phasor voltage across any element in a series connection is equal to the ratio of its impedance to the equivalent impedance of the connection times the total phasor voltage across the connection.

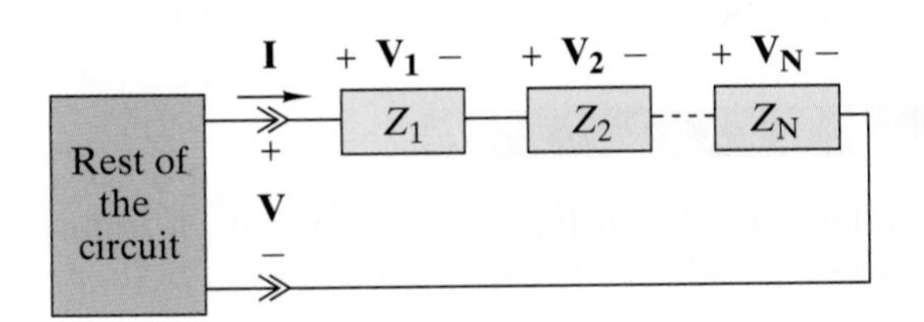

FIGURE 8–11 *A series connection of impedances.*

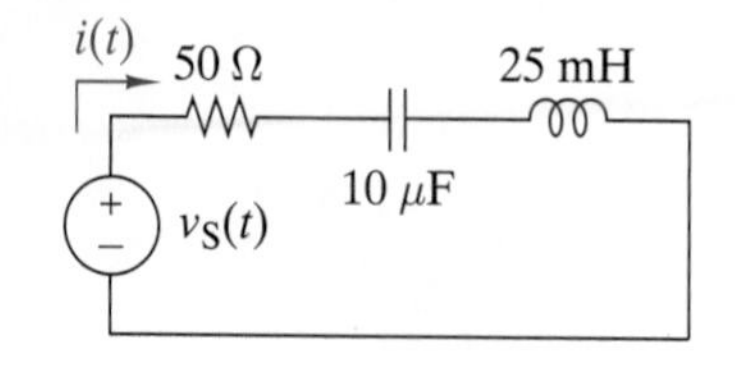

FIGURE 8–12

EXAMPLE 8–6

The circuit in Figure 8–12 is operating in the sinusoidal steady state with $v_S(t) = 35 \cos 1000t$ V.

(a) Transform the circuit into the phasor domain.
(b) Solve for the phasor current **I**.
(c) Solve for the phasor voltage across each element.
(d) Construct the waveforms corresponding to the phasors found in (b) and (c).

SOLUTION:

(a) The phasor representing the input source voltage is $\mathbf{V}_S = 35\angle 0°$. The impedances of the three passive elements are

$$Z_R = R = 50\ \Omega$$

$$Z_L = j\omega L = j1000 \times 25 \times 10^{-3} = j25\ \Omega$$

$$Z_C = \frac{1}{j\omega C} = \frac{1}{j1000 \times 10^{-5}} = -j100\ \Omega$$

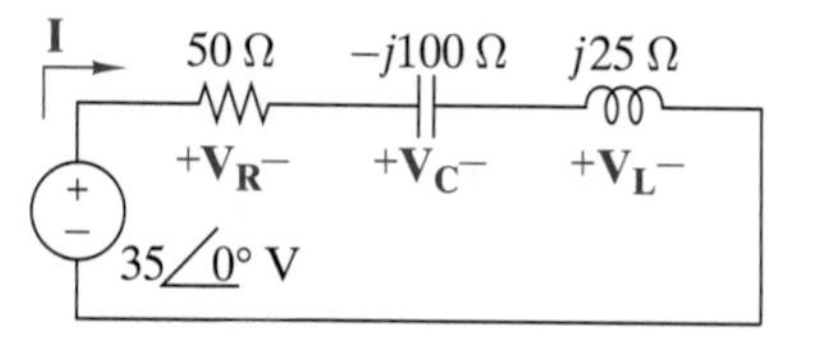

FIGURE 8–13

Using these results, we obtain the phasor-domain circuit in Figure 8–13.

(b) The equivalent impedance of the series connection is

$$Z_{EQ} = 50 + j25 - j100 = 50 - j75 = 90.1\angle -56.3°\ \Omega$$

The current in the series circuit is

$$\mathbf{I} = \frac{\mathbf{V}_S}{Z_{EQ}} = \frac{35\angle 0°}{90.1\angle -56.3°} = 0.388\angle 56.3°\ \text{A}$$

(c) The current **I** exists in all three series elements, so the voltage across each passive element is

$$\mathbf{V}_R = Z_R\mathbf{I} = 50 \times 0.388\angle 56.3° = 19.4\angle 56.3°\ \text{V}$$

$$\mathbf{V}_L = Z_L\mathbf{I} = j25 \times 0.388\angle 56.3° = 9.70\angle 146.3°\ \text{V}$$

$$\mathbf{V}_C = Z_C\mathbf{I} = -j100 \times 0.388\angle 56.3° = 38.8\angle -33.7°\ \text{V}$$

Note that the voltage across the resistor is in phase with the current, the voltage across the inductor leads the current by 90°, and the voltage across the capacitor lags the current by 90°.

(d) The sinusoidal steady-state waveforms corresponding to the phasors in (b) and (c) are

$$i(t) = \text{Re}\{0.388e^{j56.3°}e^{j1000t}\} = 0.388 \cos(1000t + 56.3°)\ \text{A}$$

$$v_R(t) = \text{Re}\{19.4e^{j56.3°}e^{j1000t}\} = 19.4 \cos(1000t + 56.3°)\ \text{V}$$

$$v_L(t) = \text{Re}\{9.70e^{j146.3°}e^{j1000t}\} = 9.70 \cos(1000t + 146.3°)\ \text{V}$$

$$v_C(t) = \text{Re}\{38.8e^{-j33.7°}e^{j1000t}\} = 38.8 \cos(1000t - 33.7°)\ \text{V}$$

■

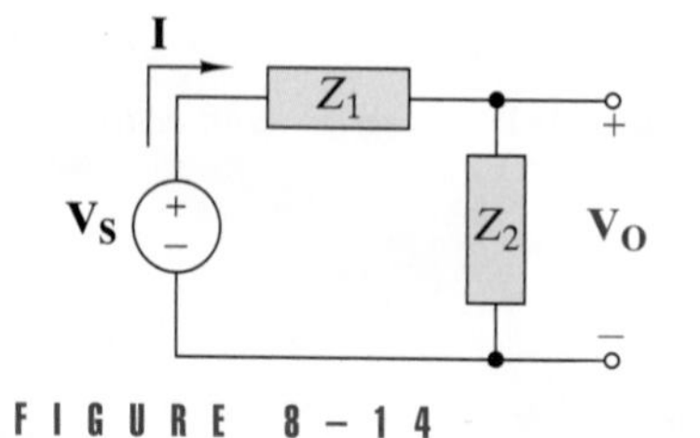

FIGURE 8–14

D DESIGN EXAMPLE 8–7

Design the voltage divider in Figure 8–14 so that an input $v_S = 15 \cos 2000t$ V produces a steady-state output $v_O(t) = 2 \sin 2000t$ V.

SOLUTION:

Using voltage division, we can relate the input and output phasor as follows:

$$\mathbf{V}_O = \frac{Z_2}{Z_1 + Z_2}\mathbf{V}_S$$

The phasor representation of the input voltage is $\mathbf{V}_S = 15\angle 0 = 15 + j0$. Using the identity $\cos(x - 90°) = \sin x$, we write the required output phasor as $\mathbf{V}_O = 2\angle{-90°} = 0 - j2$. The design problem is to select the impedances Z_1 and Z_2 so that

$$0 - j2 = \frac{Z_2}{Z_1 + Z_2}(15 + j0)$$

Solving this design constraint for Z_1 yields

$$Z_1 = \frac{15 + j2}{-j2}Z_2$$

Evidently, we can choose Z_2 and then solve for Z_1. In making this choice, we must keep some physical realizability conditions in mind. In general, an impedance has the form $Z = R + jX$. The reactance X can be either positive (an inductor) or negative (a capacitor), but the resistance R must be positive. With these constraints in mind, we select $Z_2 = -j1000$ (a capacitor) and solve for $Z_1 = 7500 + j1000$ (a resistor in series with an inductor). Figure 8–15 shows the resulting phasor circuit. To find the values of L and C, we note that the input is $v_S = 15 \cos 2000t$; hence, the frequency is $\omega = 2000$. The inductive reactance $\omega L = 1000$ requires $L = 0.5$ H, while the capacitive reactance requires $-(\omega C)^{-1} = -1000$ or $C = 0.5$ μF. Other possible designs are obtained by selecting different values of Z_2. To be physically realizable, the selected value of Z_2 must produce $R \geq 0$ for Z_1 and Z_2. ■

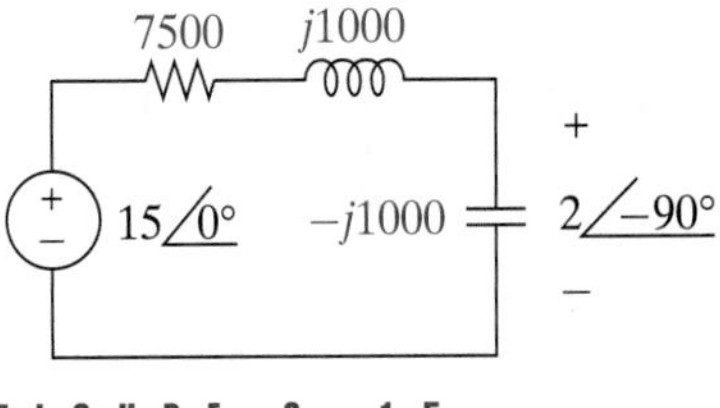

FIGURE 8–15

APPLICATION EXAMPLE 8–8

The purpose of the impedance bridge in Figure 8–16 is to measure the unknown impedance Z_X by adjusting known impedances Z_1, Z_2, and Z_3 until the detector voltage $\mathbf{V}_{DET}$ is zero. The circuit consists of a sinusoidal source $\mathbf{V}_S$ driving two voltage dividers connected in parallel. Using the voltage division principle, we find that the detector voltage is

$$\mathbf{V}_{DET} = \mathbf{V}_A - \mathbf{V}_B = \frac{Z_2}{Z_1 + Z_2}\mathbf{V}_S - \frac{Z_X}{Z_3 + Z_X}\mathbf{V}_S$$

$$= \left[\frac{Z_2 Z_3 - Z_1 Z_X}{(Z_1 + Z_2)(Z_3 + Z_X)}\right]\mathbf{V}_S$$

This equation shows that the detector voltage will be zero when $Z_2 Z_3 = Z_1 Z_X$. When the branch impedances are adjusted so that the detector voltage is zero, the unknown impedance can be written in terms of the known impedances as follows:

$$Z_X = R_X + jX_X = \frac{Z_2 Z_3}{Z_1}$$

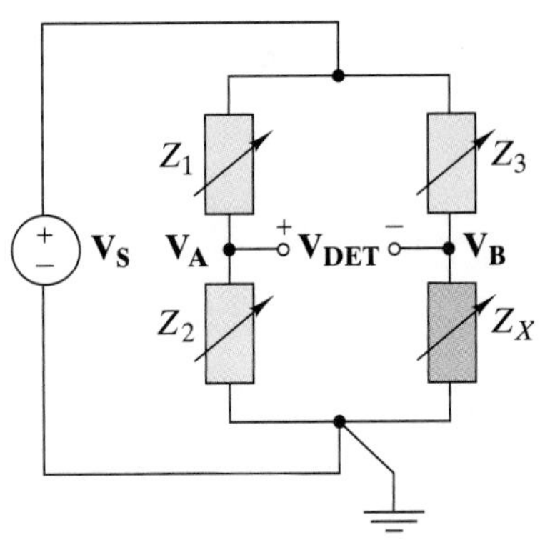

FIGURE 8–16 *Impedance bridge.*

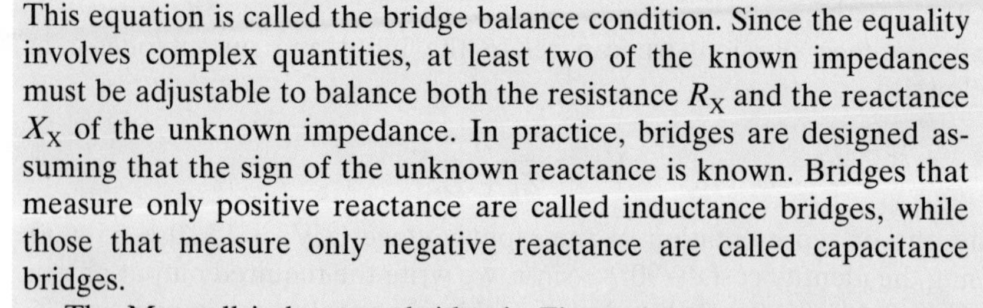

This equation is called the bridge balance condition. Since the equality involves complex quantities, at least two of the known impedances must be adjustable to balance both the resistance R_X and the reactance X_X of the unknown impedance. In practice, bridges are designed assuming that the sign of the unknown reactance is known. Bridges that measure only positive reactance are called inductance bridges, while those that measure only negative reactance are called capacitance bridges.

The Maxwell inductance bridge in Figure 8–17 is used to measure the resistance R_X and inductance L_X of an inductive device by alternately adjusting resistances R_1 and R_2 to balance the bridge circuit. The impedances of the legs of this bridge are

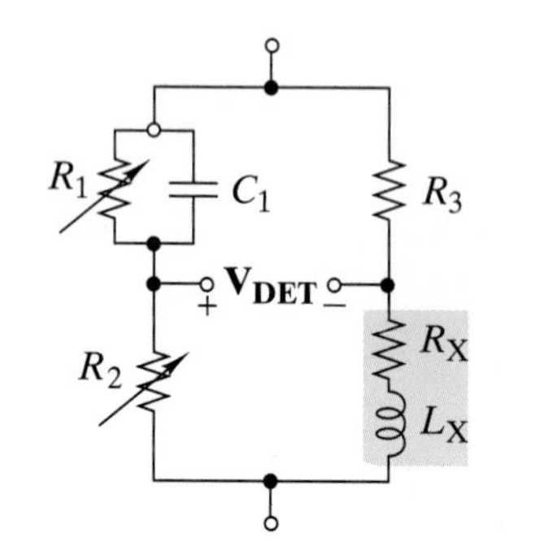

FIGURE 8–17 *Maxwell bridge.*

$$Z_1 = \frac{1}{j\omega C_1 + \dfrac{1}{R_1}}$$

$$Z_2 = R_2 \quad Z_3 = R_3$$

For the Maxwell bridge, the balance condition $Z_X = Z_2 Z_3 / Z_1$ yields

$$R_X + j\omega L_X = \frac{R_2 R_3}{R_1} + j\omega C_1 R_2 R_3$$

Equating the real and imaginary parts on each side of this equation yields the parameters of the unknown impedance in terms of the known impedances:

$$R_X = \frac{R_2 R_3}{R_1} \quad \text{and} \quad L_X = R_2 R_3 C_1$$

Note that adjusting R_1 only affects R_X. The Maxwell bridge measures inductance by balancing the positive reactance of an unknown inductive device with a calibrated fraction of negative reactance of the known capacitor C_1. If the reactance of the unknown device is actually capacitive (negative), then the Maxwell bridge cannot be balanced.

PARALLEL EQUIVALENCE AND CURRENT DIVISION

In Figure 8–18 the two-terminal elements are connected in parallel, so the same phasor voltage $\mathbf{V}$ appears across the impedances $Z_1, Z_2, \ldots Z_N$. Using the phasor element constraints, the current through each impedance is $\mathbf{I}_k = \mathbf{V}/Z_k$. Next, using KCL, the total current entering the parallel connection is

$$\begin{aligned}\mathbf{I} &= \mathbf{I}_1 + \mathbf{I}_2 + \ldots + \mathbf{I}_N \\ &= \frac{\mathbf{V}}{Z_1} + \frac{\mathbf{V}}{Z_2} + \ldots + \frac{\mathbf{V}}{Z_N} \\ &= \left(\frac{1}{Z_1} + \frac{1}{Z_2} + \ldots + \frac{1}{Z_N}\right)\mathbf{V}\end{aligned} \qquad (8\text{–}21)$$

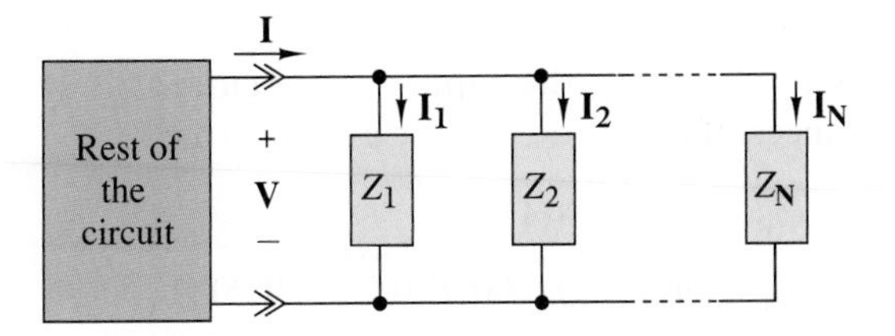

FIGURE 8–18 *Parallel connection of impedances.*

The same phasor responses **V** and **I** exist when the parallel connected elements are replaced by an equivalent impedance.

$$\frac{1}{Z_{EQ}} = \frac{\mathbf{I}}{\mathbf{V}} = \frac{1}{Z_1} + \frac{1}{Z_2} + \ldots + \frac{1}{Z_N} \tag{8–22}$$

These results can also be written in terms of **admittance** Y, which is defined as the reciprocal of impedance:

$$Y = \frac{1}{Z} = G + jB$$

The real part of Y is called **conductance** G and the imaginary part is called **susceptance**, both of which are expressed in units of siemens (S).

Using admittances to rewrite Eq. (8–21) yields

$$\begin{aligned}\mathbf{I} &= \mathbf{I}_1 + \mathbf{I}_2 + \ldots + \mathbf{I}_N \\ &= Y_1\mathbf{V} + Y_2\mathbf{V} + \ldots + Y_N\mathbf{V} \\ &= (Y_1 + Y_2 + \ldots + Y_N)\mathbf{V}\end{aligned} \tag{8–23}$$

Hence, the equivalent admittance of the parallel connection is

$$Y_{EQ} = \frac{\mathbf{I}}{\mathbf{V}} = Y_1 + Y_2 + \ldots + Y_N \tag{8–24}$$

Combining Eqs. (8–23) and (8–24), we find that the phasor current through the kth element in the parallel connection is

$$\mathbf{I}_k = Y_k\mathbf{V} = \frac{Y_k}{Y_{EQ}}\mathbf{I} \tag{8–25}$$

Equation (8–25) is the phasor version of the current division principle. The phasor current through any element in a parallel connection is equal to the ratio of its admittance to the equivalent admittance of the connection times the total phasor current entering the connection.

EXAMPLE 8–9

The circuit in Figure 8–19 is operating in the sinusoidal steady state with $i_S(t) = 50 \cos 2000t$ mA.

(a) Transform the circuit into the phasor domain.
(b) Solve for the phasor voltage **V**.
(c) Solve for the phasor current through each element.
(d) Construct the waveforms corresponding to the phasors found in (b) and (c).

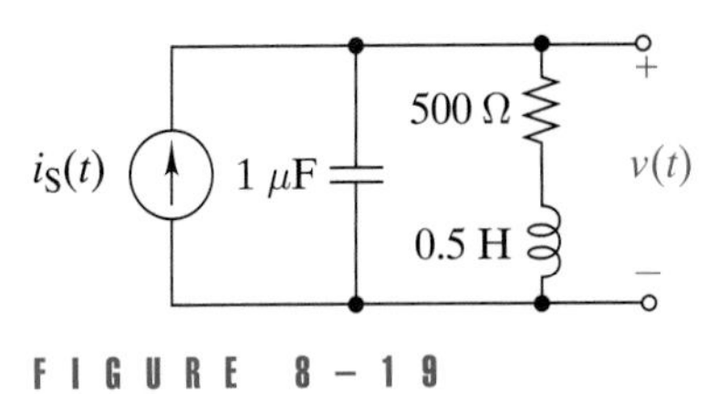

FIGURE 8–19

SOLUTION:

(a) The phasor representing the input source current is $\mathbf{I}_S = 0.05\angle 0°$ A. The impedances of the three passive elements are

$$Z_R = R = 500\ \Omega$$

$$Z_L = j\omega L = j2000 \times 0.5 = j1000\ \Omega$$

$$Z_C = \frac{1}{j\omega C} = \frac{1}{j2000 \times 10^{-6}} = -j500\ \Omega$$

Using these results, we obtain the phasor-domain circuit in Figure 8–20.

FIGURE 8 – 20

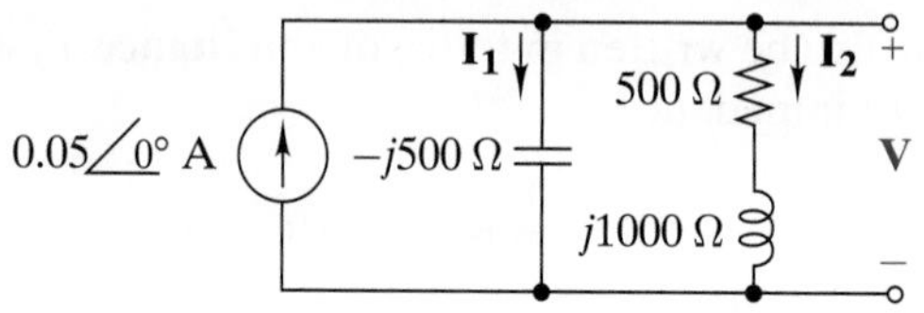

(b) The admittances of the two parallel branches are

$$Y_1 = \frac{1}{-j500} = j2 \times 10^{-3}\ \text{S}$$

$$Y_2 = \frac{1}{500 + j1000} = 4 \times 10^{-4} - j8 \times 10^{-4}\ \text{S}$$

The equivalent admittance of the parallel connection is

$$Y_{EQ} = Y_1 + Y_2 = 4 \times 10^{-4} + j12 \times 10^{-4}$$
$$= 12.6 \times 10^{-4}\angle 71.6°\ \text{S}$$

and the voltage across the parallel circuit is

$$\mathbf{V} = \frac{\mathbf{I}_S}{Y_{EQ}} = \frac{0.05\angle 0°}{12.6 \times 10^{-4}\angle 71.6°}$$
$$= 39.7\angle -71.6°\ \text{V}$$

(c) The current through each parallel branch is

$$\mathbf{I}_1 = Y_1\mathbf{V} = j2 \times 10^{-3} \times 39.7\angle -71.6° = 79.4\angle 18.4°\ \text{mA}$$

$$\mathbf{I}_2 = Y_2\mathbf{V} = (4 \times 10^{-4} - j8 \times 10^{-4}) \times 39.7\angle -71.6°$$
$$= 35.5\ \angle -135°\ \text{mA}$$

(d) The sinusoidal steady-state waveforms corresponding to the phasors in (b) and (c) are

$$v(t) = \text{Re}\{39.7e^{-j71.6°}e^{j2000t}\} = 39.7\cos(2000t - 71.6°)\ \text{V}$$
$$i_1(t) = \text{Re}\{79.4e^{j18.4°}e^{j2000t}\} = 79.4\cos(2000t + 18.4°)\ \text{mA}$$
$$i_2(t) = \text{Re}\{35.5e^{-j135°}e^{j2000t}\} = 35.5\cos(2000t - 135°)\ \text{mA}$$

■

EXAMPLE 8–10

Find the steady-state currents $i(t)$, $i_C(t)$, and $i_R(t)$ in the circuit of Figure 8–21 for $v_S = 100 \cos 2000t$ V, $L = 250$ mH, $C = 0.5$ μF, and $R = 3$ kΩ.

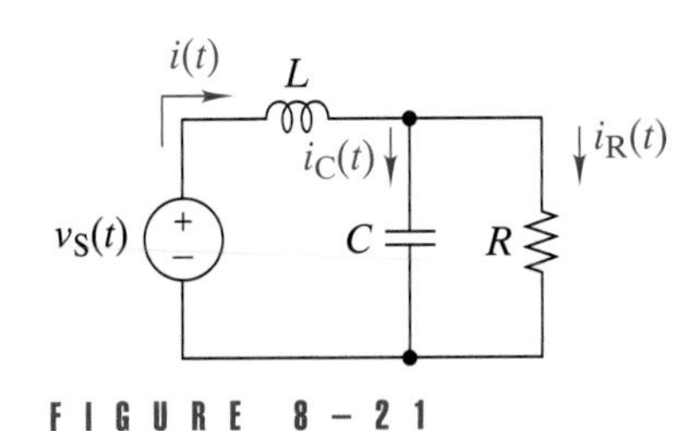

FIGURE 8–21

SOLUTION:

The phasor representation of the input voltage is $100\angle 0°$. The impedances of the passive elements are

$$Z_L = j500\ \Omega \quad Z_C = -j1000\ \Omega \quad Z_R = 3000\ \Omega$$

Figure 8–22(a) shows the phasor-domain circuit.

To solve for the required phasor responses, we reduce the circuit using a combination of series and parallel equivalence. Using parallel equivalence, we find that the capacitor and resistor can be replaced by an equivalent impedance

$$Z_{EQ1} = \frac{1}{Y_{EQ1}} = \frac{1}{\dfrac{1}{-j1000} + \dfrac{1}{3000}}$$
$$= 300 - j900\ \Omega$$

The resulting circuit reduction is shown in Figure 8–22(b). The equivalent impedance Z_{EQ1} is connected in series with the impedance $Z_L = j500$. This series combination can be replaced by an equivalent impedance

$$Z_{EQ2} = j500 + Z_{EQ1} = 300 - j400\ \Omega$$

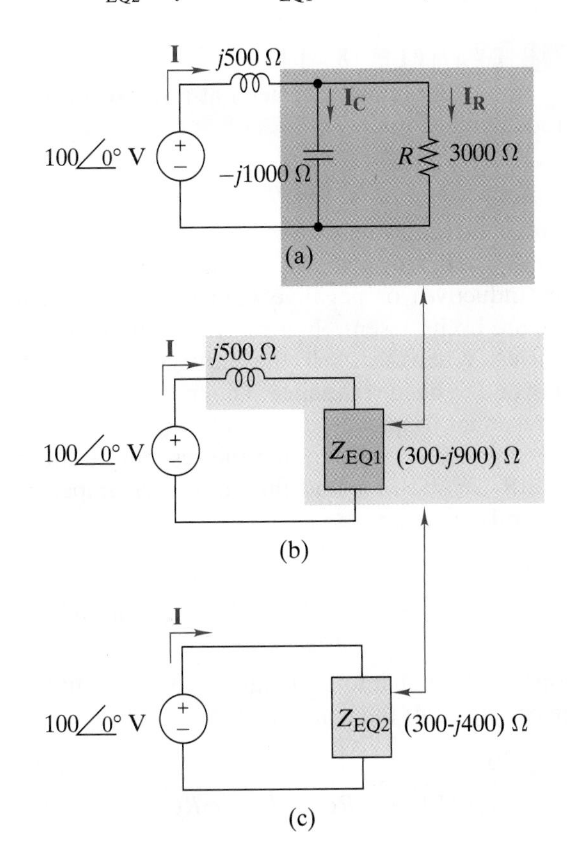

FIGURE 8–22

This step reduces the circuit to the equivalent input impedance shown in Figure 8–22(c). The phasor input current in Figure 8–22(c) is

$$\mathbf{I} = \frac{100\angle 0^\circ}{Z_{EQ2}} = \frac{100\angle 0^\circ}{300 - j400} = 0.12 + j0.16 = 0.2\angle 53.1^\circ \text{ A}$$

Given the phasor current $\mathbf{I}$, we use current division to find $\mathbf{I}_C$:

$$\mathbf{I}_C = \frac{Y_C}{Y_C + Y_R}\mathbf{I} = \frac{\dfrac{1}{-j1000}}{\dfrac{1}{-j1000} + \dfrac{1}{3000}} 0.2\angle 53.1^\circ$$
$$= 0.06 + j0.18 = 0.19\angle 71.6^\circ \text{ A}$$

By KCL, $\mathbf{I} = \mathbf{I}_C + \mathbf{I}_R$, so the remaining unknown current is

$$\mathbf{I}_R = \mathbf{I} - \mathbf{I}_C = 0.06 - j0.02 = 0.0632\angle -18.4^\circ \text{ A}$$

The waveforms corresponding to the phasor currents are

$$i(t) = \text{Re}\{\mathbf{I}e^{j2000t}\} = 0.2\cos(2000t + 53.1^\circ) \text{ A}$$
$$i_C(t) = \text{Re}\{\mathbf{I}_C e^{j2000t}\} = 0.19\cos(2000t + 71.6^\circ) \text{ A}$$
$$i_R(t) = \text{Re}\{\mathbf{I}_R e^{j2000t}\} = 0.0632\cos(2000t - 18.4^\circ) \text{ A}$$

■

APPLICATION EXAMPLE 8–11

In general, the equivalent impedance seen at any pair of terminals can be written in rectangular form as

$$Z_{EQ} = R_{EQ} + jX_{EQ} \tag{8–26}$$

In a passive circuit the equivalent resistance R_{EQ} must always be non-negative, that is, $R_{EQ} \geq 0$. However, the equivalent reactance X_{EQ} can be either positive (inductive) or negative (capacitive). When inductance and capacitance are both present, their reactances may exactly cancel at certain frequencies. When $X_{EQ} = 0$, the impedance is purely resistive and the circuit is said to be in **resonance**. The frequency at which this occurs is called a **resonant frequency**, denoted by ω_0.

For example, suppose we want to find the resonant frequency of the circuit in Figure 8–23. We first find the equivalent impedance of the parallel resistor and capacitor:

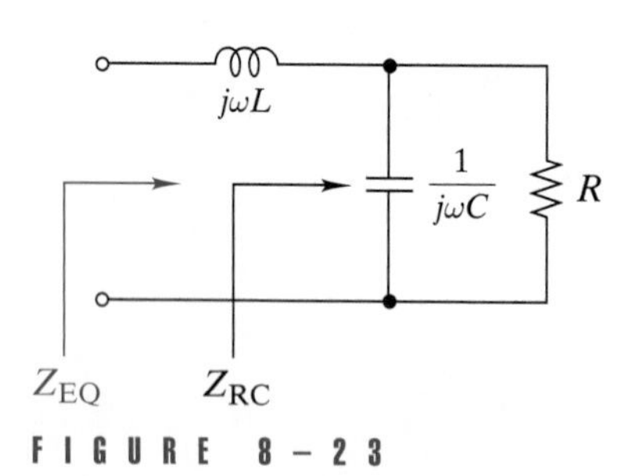

FIGURE 8–23

$$Z_{RC} = \frac{1}{Y_R + Y_C} = \frac{1}{\dfrac{1}{R} + j\omega C} = \frac{R}{1 + j\omega RC}$$

This expression can be put into rectangular form by multiplying and dividing by the conjugate of the denominator:

$$Z_{RC} = \frac{R}{1 + j\omega RC}\frac{1 - j\omega RC}{1 - j\omega RC} = \frac{R}{1 + (\omega RC)^2} - j\frac{\omega R^2 C}{1 + (\omega RC)^2}$$

The impedance Z_{RC} is connected in series with the inductor. Therefore, the overall equivalent impedance Z_{EQ} is

$$\begin{aligned} Z_{EQ} &= Z_L + Z_{RC} \\ &= \underbrace{\frac{R}{1+(\omega RC)^2}}_{} + j\underbrace{\left[\omega L - \frac{\omega R^2 C}{1+(\omega RC)^2}\right]}_{} \\ &= \quad R_{EQ} \quad + \quad jX_{EQ} \end{aligned}$$

Note that the equivalent resistance R_{EQ} is positive for all ω. However, the equivalent reactance X_{EQ} can be positive or negative. The resonant frequency is found by setting the reactance to zero

$$X_{EQ}(\omega_0) = \omega_0 L - \frac{\omega_0 R^2 C}{1+(\omega_0 RC)^2} = 0$$

and solving for the resonant frequency:

$$\omega_0 = \sqrt{\frac{1}{LC} - \frac{1}{(RC)^2}} \tag{8–27}$$

Note the reactance X_{EQ} is inductive (positive) when $\omega > \omega_0$ and capacitive (negative) with $\omega < \omega_0$.

EXAMPLE 8–12

The circuit in Figure 8–24 is operating in the sinusoidal steady state with ω = 5 krad/s.

(a) Find the value of capacitance C that causes the input impedance Z to be purely resistive.
(b) Find the real part of the input impedance for this value of C.

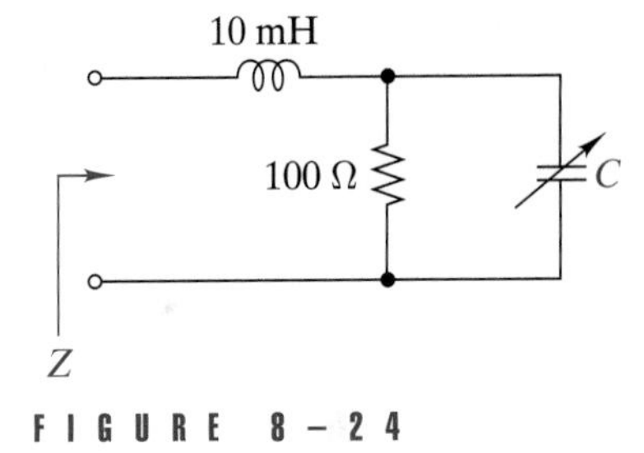

FIGURE 8–24

SOLUTION:

(a) At the specified frequency the impedance is

$$Z = j5000 \times 0.01 + \frac{1}{\dfrac{1}{100} + j5000\,C}$$

which can be written as

$$\begin{aligned} Z &= j50 + \frac{100}{1 + j5\times 10^5\,C} \times \frac{1 - j5\times 10^5\,C}{1 - j5\times 10^5\,C} \\ &= j50 + \frac{100}{1+(5\times 10^5\,C)^2} - j\frac{5\times 10^7\,C}{1+(5\times 10^5\,C)^2} \end{aligned}$$

A purely resistive input impedance means that the imaginary part of Z is zero, which requires that

$$50 - \frac{5\times 10^7\,C}{1+(5\times 10^5\,C)^2} = 0 \quad \text{or} \quad 25\times 10^{10}\,C^2 - 10^6\,C + 1 = 0$$

This quadratic has a double root at $C = 2 \times 10^{-6}$ F.

(b) For this value of capacitance the real part of Z is

$$Re\{Z\} = R_{EQ} = \frac{100}{1 + (5 \times 10^5\, C)^2} = 50\ \Omega$$

Exercise 8–8

(a) Find the equivalent impedance Z in Figure 8–25 at $\omega = 1$ krad/s.
(b) Repeat (a) with $\omega = 4$ krad/s.

FIGURE 8–25

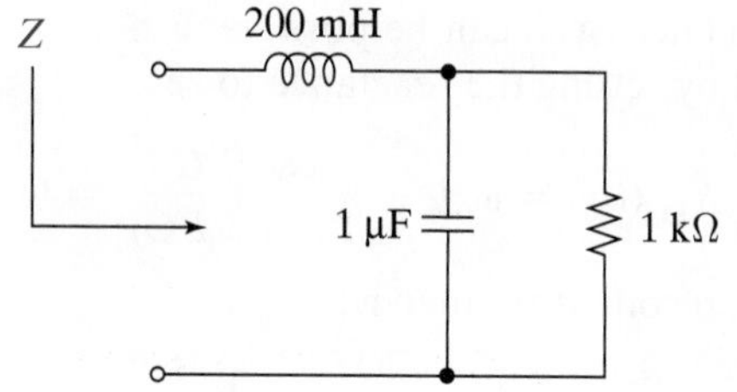

Answers:

(a) $Z = 500 - j300\ \Omega$
(b) $Z = 58.8 + j565\ \Omega$

Exercise 8–9

A voltage source $v_S(t) = 15 \cos 2000t$ V is applied at the input in Figure 8–25.

(a) Find the steady-state current through the inductor.
(b) Find the steady-state voltage across the 1-kΩ resistor.

Answers:

(a) $75 \cos 2000t$ mA
(b) $33.5 \cos (2000t - 63.4°)$ V

Exercise 8–10

The circuit in Figure 8–26 is operating at $\omega = 10$ krad/s.

(a) Find the equivalent impedance Z.
(b) What element should be connected in series with Z to make the total reactance zero?

FIGURE 8–26

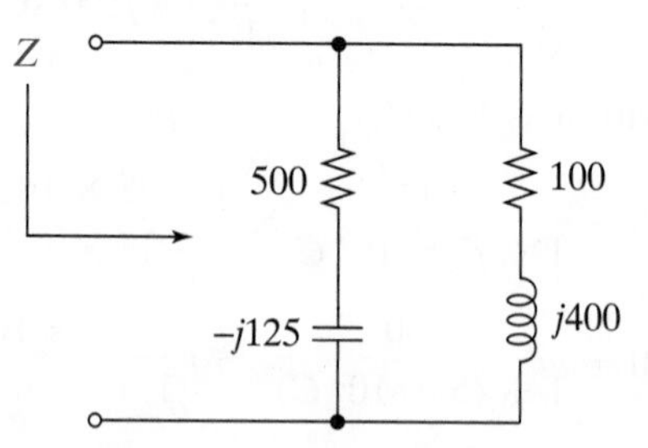

Answers:

(a) $Z = 256 + j195 = 322\angle 37.3°\ \Omega$
(b) A capacitor with $C = 513$ nF

Exercise 8–11

In Figure 8–27 $v_S(t) = 12.5 \cos 1000t$ V and $i_S(t) = 0.2 \cos(1000t - 36.9°)$ A. What is the impedance seen by the voltage source and what element is in the box?

Answer: $Z = 50 + j37.5\ \Omega$ and the element is a 37.5-mH inductor.

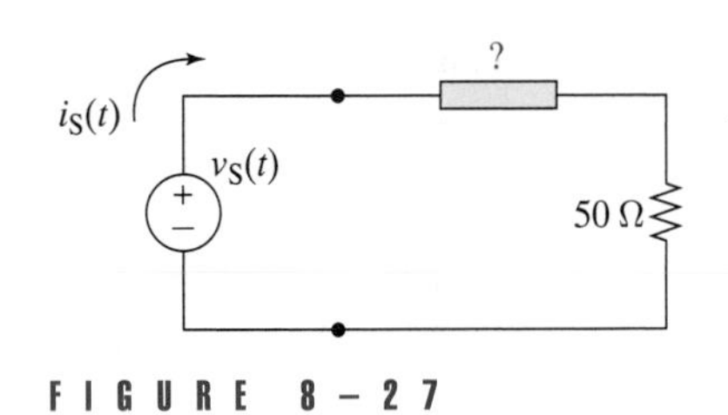

FIGURE 8–27

8–4 CIRCUIT THEOREMS WITH PHASORS

In this section we treat basic properties of phasor circuits that parallel the resistance circuit theorems developed in Chapter 3. Circuit linearity is the foundation for all of these properties. The proportionality and superposition properties are two fundamental consequences of linearity.

PROPORTIONALITY

The **proportionality** property states that phasor output responses are proportional to the input phasor. Mathematically, proportionality means that

$$\mathbf{Y} = K\mathbf{X} \tag{8–28}$$

where $\mathbf{X}$ is the input phasor, $\mathbf{Y}$ is the output phasor, and K is the proportionality constant. In phasor circuit analysis, the proportionality constant is generally a complex number.

The unit output method discussed in Chapter 3 is based on the proportionality property and is applicable to phasors. To apply the unit output method in the phasor domain, we assume that the output is a unit phasor $\mathbf{Y} = 1\angle 0°$. By successive application of KCL, KVL, and the element impedances, we solve for the input phasor required to produce the unit output. Because the circuit is linear, the proportionality constant relating input and output is

$$K = \frac{\text{Output}}{\text{Input}} = \frac{1\angle 0°}{\text{Input phasor for unit output}}$$

Once we have the constant K, we can find the output for any input or the input required to produce any specified output.

The next example illustrates the unit output method for phasor circuits.

EXAMPLE 8–13

Use the unit output method to find the input impedance, current $\mathbf{I}_1$, output voltage $\mathbf{V}_C$, and current $\mathbf{I}_3$ of the circuit in Figure 8–28 for $\mathbf{V}_S = 10\angle 0°$ V.

SOLUTION:

The following steps implement the unit output method for the circuit in Figure 8–28:

FIGURE 8 – 28

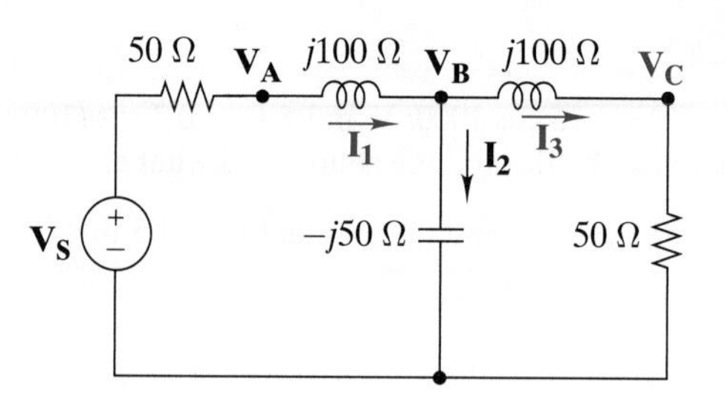

1. Assume a unit output voltage $\mathbf{V}_C = 1 + j0$ V.
2. By Ohm's law, $\mathbf{I}_3 = \mathbf{V}_C/50 = 0.02 + j0$ A.
3. By KVL, $\mathbf{V}_B = \mathbf{V}_C + (j100)\mathbf{I}_3 = 1 + j2$ V.
4. By Ohm's law, $\mathbf{I}_2 = \mathbf{V}_B/(-j50) = -0.04 + j0.02$ A.
5. By KCL, $\mathbf{I}_1 = \mathbf{I}_2 + \mathbf{I}_3 = -0.02 + j0.02$ A.
6. By KVL, $\mathbf{V}_S = (50 + j100)\mathbf{I}_1 + \mathbf{V}_B = -2 + j1$ V.

Given $\mathbf{V}_S$ and $\mathbf{I}_1$, the input impedance is

$$Z_{IN} = \frac{\mathbf{V}_S}{\mathbf{I}_1} = \frac{-2 + j1}{-0.02 + j0.02} = 75 + j25\ \Omega$$

The proportionality factor between the input $\mathbf{V}_S$ and output voltage $\mathbf{V}_C$ is

$$K = \frac{1}{\mathbf{V}_S} = \frac{1}{-2 + j} = -0.4 - j0.2$$

Given K and Z_{IN}, we can now calculate the required responses for an input $\mathbf{V}_S = 10\angle 0°$:

$$\mathbf{V}_C = K\mathbf{V}_S = -4 - j2 = 4.47\angle -153°\ \text{V}$$

$$\mathbf{I}_1 = \frac{\mathbf{V}_S}{Z_{IN}} = 0.12 - j0.04 = 0.126\angle -18.4°\ \text{A}$$

$$\mathbf{I}_3 = \frac{\mathbf{V}_C}{50} = -0.08 - j0.04 = 0.0894\angle -153°\ \text{A}$$

■

SUPERPOSITION

The superposition principle applies to phasor responses only if all of the independent sources driving the circuit have the *same frequency.* That is, when the input sources have the same frequency, we can find the phasor response due to each source acting alone and obtain the total response by adding the individual phasors. If the sources have different frequencies, then superposition can still be used but its application is different. With different frequency sources, each source must be treated in a separate steady-state analysis because the element impedances change with frequency. The phasor response for each source must be changed into waveforms and then superposition applied in the time domain. In other words, the superposition principle always applies in the time domain. It also applies in the phasor domain when all independent sources have the same frequency. The following examples illustrate both cases.

EXAMPLE 8–14

Use superposition to find the steady-state voltage $v_R(t)$ in Figure 8–29 for $R = 20\ \Omega$, $L_1 = 2$ mH, $L_2 = 6$ mH, $C = 20\ \mu$F, $v_{S1} = 100 \cos 5000t$ V, and $v_{S2} = 120 \cos(5000t + 30°)$ V.

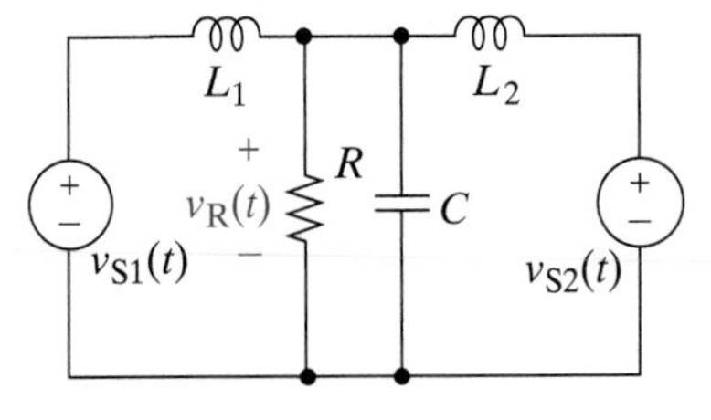

FIGURE 8–29

SOLUTION:

In this example, the two sources operate at the same frequency. Figure 8–30(a) shows the phasor-domain circuit with source 2 turned off and replaced by a short circuit. The three elements in parallel in Figure 8–30(a) produce an equivalent impedance of

$$Z_{EQ} = \frac{1}{\dfrac{1}{20} + \dfrac{1}{-j10} + \dfrac{1}{j30}} = 7.20 - j9.60\ \Omega$$

By voltage division, the phasor response $\mathbf{V}_{R1}$ is

$$\mathbf{V}_{R1} = \frac{Z_{EQ1}}{j10 + Z_{EQ1}} 100\angle 0°$$
$$= 92.3 - j138 = 166\angle -56.3°\ \text{V}$$

Figure 8–30(b) shows the phasor-domain circuit with source 1 turned off and source 2 on. The three elements in parallel in Figure 8–30(b) produce an equivalent impedance of

$$Z_{EQ2} = \frac{1}{\dfrac{1}{20} + \dfrac{1}{-j10} + \dfrac{1}{j10}} = 20 - j0\ \Omega$$

By voltage division, the response $\mathbf{V}_{R2}$ is

$$\mathbf{V}_{R2} = \frac{Z_{EQ2}}{j30 + Z_{EQ2}} 120\angle 30°$$
$$= 59.7 - j29.5 = 66.6\angle -26.3°\ \text{V}$$

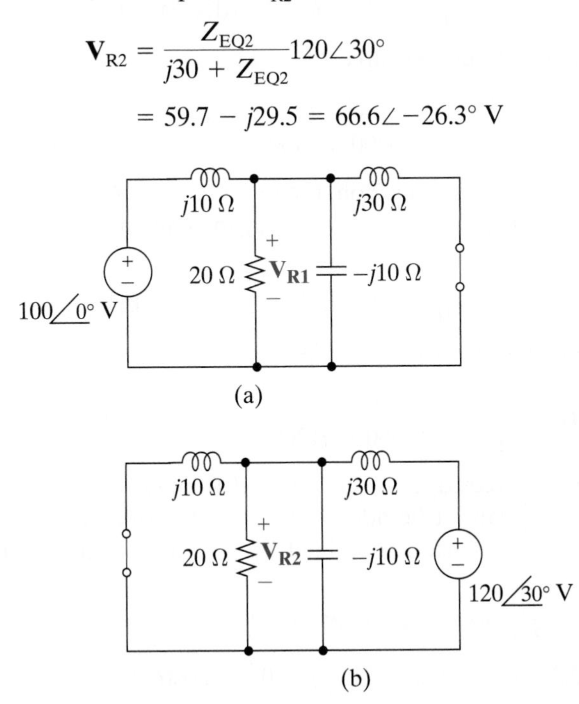

FIGURE 8–30

Since the sources have the same frequency, the total response can be found by adding the individual phasor responses $\mathbf{V}_{R1}$ and $\mathbf{V}_{R2}$:

$$\mathbf{V}_R = \mathbf{V}_{R1} + \mathbf{V}_{R2} = 152 - j167 = 226\angle -47.8^\circ \text{ V}$$

The time-domain function corresponding to the phasor sum is

$$v_R(t) = \text{Re}\{\mathbf{V}_R e^{j5000t}\} = 226\cos(5000t - 47.8^\circ) \text{ V}$$

The overall response can also be obtained by adding the time-domain functions corresponding to the individual phasor responses $\mathbf{V}_{R1}$ and $\mathbf{V}_{R2}$:

$$v_R(t) = \text{Re}\{\mathbf{V}_{R1} e^{j5000t}\} + \text{Re}\{\mathbf{V}_{R2} e^{j5000t}\}$$
$$= 166\cos(5000t - 56.3^\circ) + 66.6\cos(5000t - 26.3^\circ) \text{ V}$$

You are encouraged to show that the two expressions for $v_R(t)$ are equivalent using the additive property of sinusoids. ■

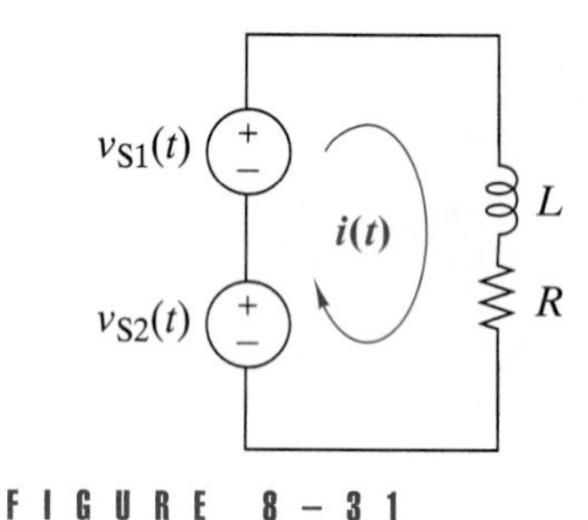

FIGURE 8–31

EXAMPLE 8–15

Use superposition to find the steady-state current $i(t)$ in Figure 8–31 for $R = 10$ kΩ, $L = 200$ mH, $v_{S1} = 24\cos 20000t$ V, and $v_{S2} = 8\cos(60000t + 30^\circ)$ V.

SOLUTION:

In this example the two sources operate at different frequencies. With source 2 off, the input phasor is $\mathbf{V}_{S1} = 24\angle 0^\circ$ V at a frequency of $\omega = 20$ krad/s. At this frequency the equivalent impedance of the inductor and resistor is

$$Z_{EQ1} = R + j\omega L = 10 + j4 \text{ k}\Omega$$

The phasor current due to source 1 is

$$\mathbf{I}_1 = \frac{\mathbf{V}_{S1}}{Z_{EQ1}} = \frac{24\angle 0^\circ}{10000 + j4000} = 2.23\angle -21.8^\circ \text{ mA}$$

With source 1 off and source 2 on, the input phasor $\mathbf{V}_{S2} = 8\angle 30^\circ$ V at a frequency of $\omega = 60$ krad/s. At this frequency the equivalent impedance of the inductor and resistor is

$$Z_{EQ2} = R + j\omega L = 10 + j12 \quad \text{k}\Omega$$

The phasor current due to source 2 is

$$\mathbf{I}_2 = \frac{\mathbf{V}_{S2}}{Z_{EQ2}} = \frac{8\angle 30^\circ}{10000 + j12000} = 0.512\angle -20.2^\circ \text{ mA}$$

The two input sources operate at different frequencies, so the phasors responses $\mathbf{I}_1$ and $\mathbf{I}_2$ cannot be added to obtain the overall response. In this case the overall response is obtained by adding the corresponding time-domain functions.

$$i(t) = \text{Re}\{\mathbf{I}_1 e^{j20000t}\} + \text{Re}\{\mathbf{I}_2 e^{j60000t}\}$$
$$= 2.23\cos(20000t - 21.8^\circ) + 0.512\cos(60000t - 20.2^\circ) \text{ mA}$$ ■

EXAMPLE 8–16

The voltage source in Figure 8–32 produces a 60-Hz sinusoid with a peak amplitude of 200 V plus a 180-Hz third harmonic with a peak amplitude of 10 V. The purpose of the *LC* circuit is to reduce the relative size of the third harmonic component delivered to the 100-Ω load resistor. Use Orcad Capture to calculate the amplitude of the 60-Hz fundamental and 180-Hz third harmonic in the voltage across the load.

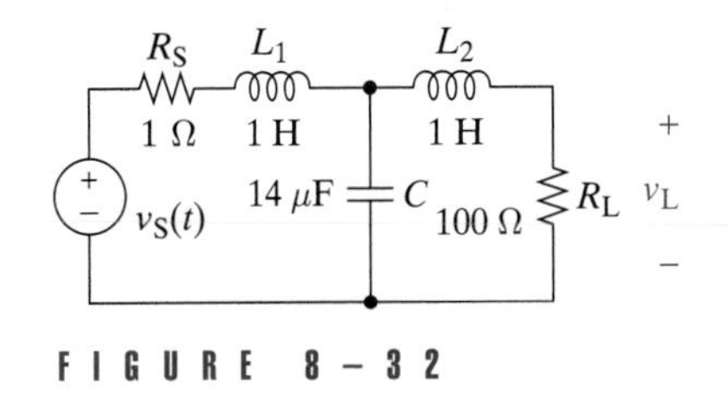

FIGURE 8–32

SOLUTION:

Figure 8–33 shows the circuit diagram as constructed in Orcad Capture. The values ACMAG = 1V and ACPHASE = 0 mean that $\mathbf{V}_S$ is an ac voltage source with an amplitude of 1 V and a phase angle of 0%. The printer symbol[1] attached to node N00576 causes Orcad PSpice to write the magnitude and phase of the node voltage **V(N00576)** in the output file. This example involves ac circuit analysis with input sinusoids at 60 Hz and 180 Hz. Selecting **PSpice/Simulations Settings/AC Sweep/Noise** from the Schematics menu bar brings up the AC sweep menu also shown in Figure 8–33. We select a two-point linear sweep that starts at 60 Hz and ends at 180 Hz. Returning to the main menu and selecting **PSpice/Run** cause Orcad PSpice to make two AC analysis runs, first at $f = 60$ Hz and then again at $f = 180$ Hz. The input phasor for each run is $\mathbf{V}_S = 1\angle 0°$. The relevant portion of the output file is shown at the bottom in Figure 8–33. The symbol **VM(N00576)** is the magnitude of the output voltage phasor in volts and **VP(N00576)** is the phase angle in degrees. These values are for the normalized input $\mathbf{V}_S = 1\angle 0°$. To obtain the actual output, we must scale the normalized results by the actual input amplitudes at each frequency.

Component	Freq (Hz)	Input(V)	Output(V)
Fundamental	60	200	$200 \times 0.9994 = 199.9$
Third Harmonic	180	10	$10 \times 0.005534 = 0.05534$

The amplitude of the 60-Hz component is virtually unchanged while the amplitude of the 180-Hz component is reduced from 10 V to 55.3 mV. This frequency selectivity or filtering occurs because the impedances of the inductor and capacitor change with frequency. ■

---

1 The printer is found under the "Place Part" menu. Choose **Add Library** and select **Special**. Scroll down and select **VPRINT1**. Connect the printer to the desired node. In order to set the parameters that the printer will record, it is necessary to double-click on the printer symbol. This brings up the Property Editor. Click on the **AC** box and place a "Y" for "Yes." Repeat this for the boxes labeled "MAG" and "PHASE." Before closing the Property Editor, click on the **Apply** button. After running the simulation, the Printer records the desired results in the Output File.

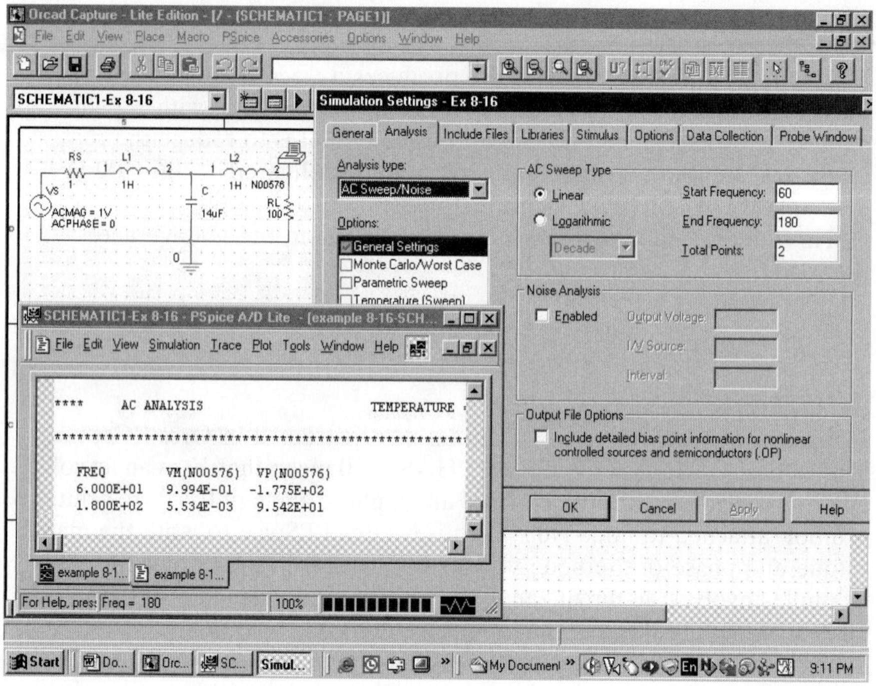

FIGURE 8–33 *Orcad Capture circuit diagram, ac sweep analysis setup, and analysis output.*

Exercise 8–12

The two sources in Figure 8–34 have the same frequency. Use superposition to find the phasor current $\mathbf{I}_X$.

FIGURE 8–34

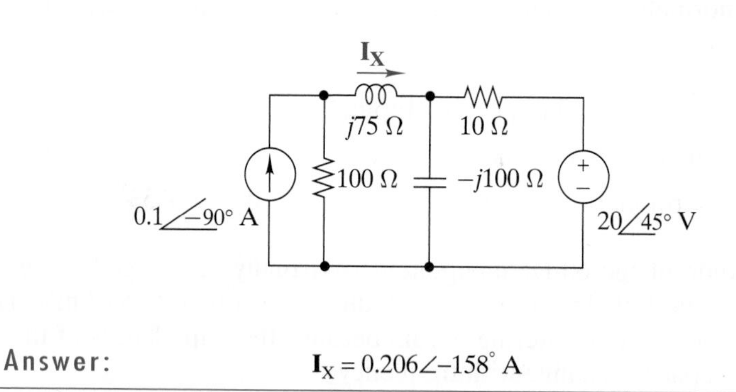

Answer: $\mathbf{I}_X = 0.206\angle{-158^\circ}$ A

THÉVENIN AND NORTON EQUIVALENT CIRCUITS

In the phasor domain, a two-terminal circuit containing linear elements and sources can be replaced by the Thévenin or Norton equivalent circuits shown in Figure 8–35. The general concept of Thévenin's and Norton's theorems and their restrictions are the same as in the resistive circuit studied in Chapter 3. The important difference here is that the signals $\mathbf{V}_T$, $\mathbf{I}_N$, $\mathbf{V}$, and $\mathbf{I}$ are phasors, and $Z_T = 1/Y_N$ and Z_L are complex numbers representing the source and load impedances.

Finding the Thévenin or Norton equivalent of a phasor circuit involves the same process as for resistance circuits, except that now we must manipulate complex numbers. The Thévenin and Norton circuits are equivalent to each other, so their circuit parameters are related as follows:

$$\mathbf{V}_{OC} = \mathbf{V}_T = \mathbf{I}_N Z_T$$
$$\mathbf{I}_{SC} = \frac{\mathbf{V}_T}{Z_T} = \mathbf{I}_N \qquad (8\text{–}29)$$
$$Z_T = \frac{1}{Y_N} = \frac{\mathbf{V}_{OC}}{\mathbf{I}_{SC}}$$

Algebraically, the results in Eq. (8–29) are identical to the corresponding equations for resistance circuits. The important difference is that these equations involve phasors and impedances rather than waveforms and resistances. These equations point out that we can determine a Thévenin or Norton equivalent by finding any two of the following quantities: (1) the open-circuit voltage $\mathbf{V}_{OC}$, (2) the short-circuit current $\mathbf{I}_{SC}$, and (when there are no dependent sources) (3) the impedance Z_T looking back into the source circuit with all independent sources turned off.

The relationships in Eq. (8–29) define source transformations that allow us to convert a voltage source in series with an impedance into a current source in parallel with the same impedance, or vice versa. Phasor-domain **source transformations** simplify circuits and are useful in formulating general node-voltage or mesh-current equations, discussed in the next section.

The next two examples illustrate applications of source transformation and Thévenin equivalent circuits.

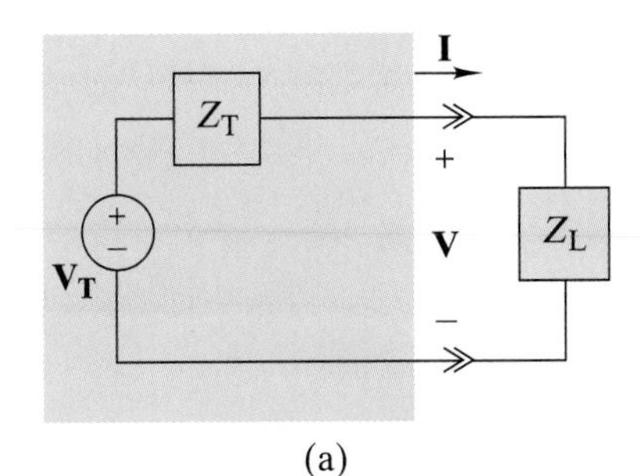

(a)

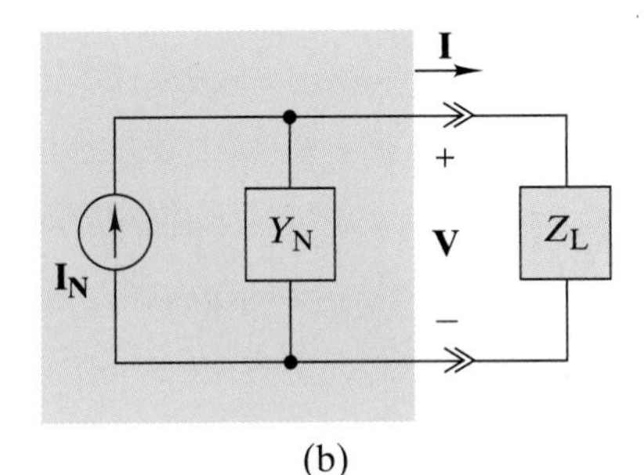

(b)

FIGURE 8–35 *Thévenin and Norton equivalent circuits in the phasor analysis.*

EXAMPLE 8–17

Both sources in Figure 8–36(a) operate at a frequency of ω = 5000 rad/s. Find the steady-state voltage $v_R(t)$ using source transformations.

SOLUTION:

Example 8–14 solves this problem using superposition. In this example we use source transformations. We observe that the voltage sources in Figure 8–36(a) are connected in series with an impedance and can be converted into the following equivalent current sources:

$$\mathbf{I}_{EQ1} = \frac{100\angle 0^\circ}{j10} = 0 - j10 \text{ A}$$

$$\mathbf{I}_{EQ2} = \frac{120\angle 30^\circ}{j30} = 2 - j3.46 \text{ A}$$

Figure 8–36(b) shows the circuit after these two source transformations. The two current sources are connected in parallel and can be replaced by a single equivalent current source:

$$\mathbf{I}_{EQ} = \mathbf{I}_{EQ1} + \mathbf{I}_{EQ2} = 2 - j13.46 = 13.6\angle -81.5^\circ \text{ A}$$

FIGURE 8–36

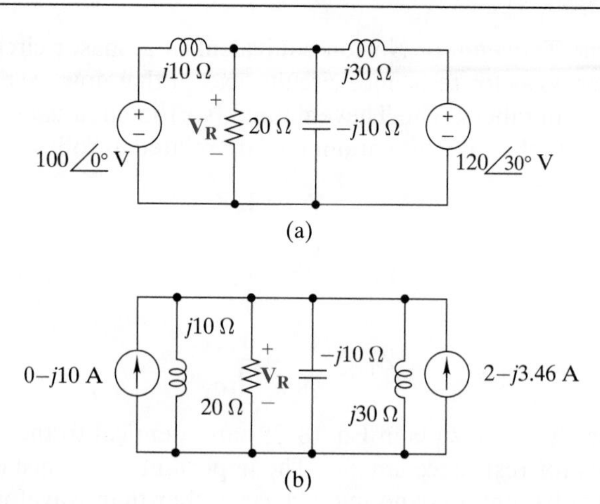

The four passive elements are connected in parallel and can be replaced by an equivalent impedance:

$$Z_{EQ} = \frac{1}{\dfrac{1}{20} + \dfrac{1}{-j10} + \dfrac{1}{j10} + \dfrac{1}{j30}} = 16.6\angle 33.7^\circ\ \Omega$$

The voltage across this equivalent impedance equals $\mathbf{V}_R$, since one of the parallel elements is the resistor R. Therefore, the unknown phasor voltage is

$$\mathbf{V}_R = \mathbf{I}_{EQ} Z_{EQ} = (13.6\angle -81.5^\circ) \times (16.6\angle 33.7^\circ) = 226\angle -47.8^\circ\ \text{V}$$

The value of $\mathbf{V}_R$ is the same as found in Example 8–14 using superposition. The corresponding time-domain function is

$$v_R(t) = \text{Re}\{\mathbf{V}_R e^{j5000t}\} = 226\cos(5000t - 47.8^\circ)\ \text{V}$$

■

EXAMPLE 8–18

Use Thévenin's theorem to find the current $\mathbf{I}_X$ in the bridge circuit shown in Figure 8–37.

FIGURE 8–37

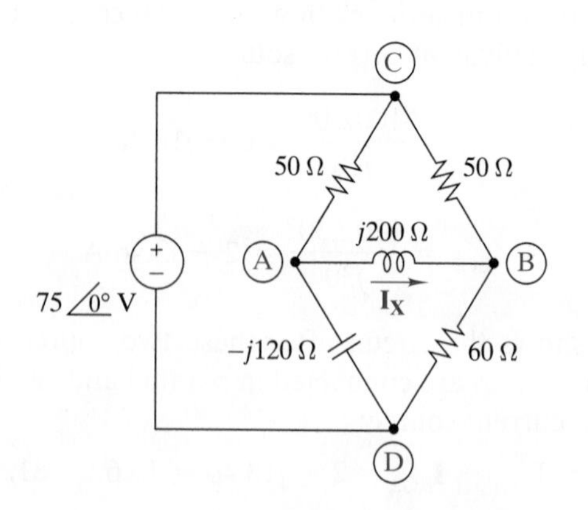

SOLUTION:

Disconnecting the impedance $j200$ from the circuit in Figure 8–37 produces the circuit shown in Figure 8–38(a). The voltage between nodes A and B is the Thévenin voltage since removing the impedance $j200$ leaves an open circuit. The voltages at nodes A and B can each be found by voltage division. Since the open-circuit voltage is the difference between these node voltages, we have

$$\begin{aligned}\mathbf{V}_\mathrm{T} &= \mathbf{V}_\mathrm{A} - \mathbf{V}_\mathrm{B}\\ &= \frac{-j120}{50 - j120}75\angle 0^\circ - \frac{60}{60+50}75\angle 0^\circ\\ &= 23.0 - j26.6 \text{ V}\end{aligned}$$

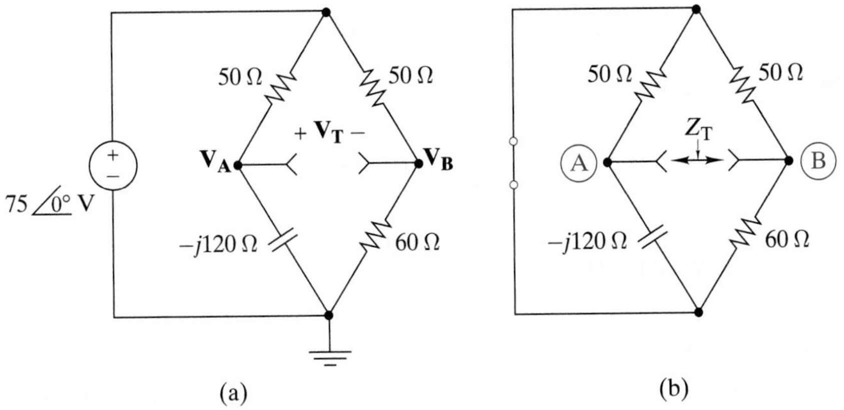

FIGURE 8–38

(a) (b)

Turning off the voltage source in Figure 8–38(a) and replacing it by a short circuit produces the situation shown in Figure 8–38(b). The lookback impedance seen at the interface is a series connection of two pairs of elements connected in parallel. The equivalent impedance of the series/parallel combination is

$$Z_\mathrm{T} = \frac{1}{\dfrac{1}{50} + \dfrac{1}{-j120}} + \frac{1}{\dfrac{1}{50} + \dfrac{1}{60}} = 69.9 - j17.8\ \Omega$$

Given the Thévenin equivalent circuit, we treat the impedance $j200$ as a load connected at the interface and calculate the resulting load current $\mathbf{I}_\mathrm{X}$ as

$$\mathbf{I}_\mathrm{X} = \frac{\mathbf{V}_\mathrm{T}}{Z_\mathrm{T} + j200} = \frac{23.0 - j26.6}{69.9 + j182} = 0.180\angle -118^\circ \text{ A}$$ ■

EXAMPLE 8–19

In the steady state, the open-circuit voltage at an interface is observed to be

$$v_\mathrm{OC}(t) = 12 \cos 2000t \quad \text{V}$$

When a 50-mH inductor is connected across the interface, the interface voltage is observed to be

$$v(t) = 17 \cos(2000t + 45^\circ) \quad \text{V}$$

Find the Thévenin equivalent circuit at the interface.

SOLUTION:

The phasors for $v_{OC}(t)$ and $v(t)$ are $\mathbf{V}_{OC} = 12\angle 0°$ and $\mathbf{V} = 17\angle 45°$. The phasor Thévenin voltage at the interface is $\mathbf{V}_T = \mathbf{V}_{OC} = 12\angle 0°$. The impedance of the inductor is $Z_L = j\omega L = j2000 \times 0.050 = j100\ \Omega$. When the inductor load is connected across the interface, we use voltage division to express the interface voltage as

$$\mathbf{V} = \frac{Z_L}{Z_T + Z_L}\mathbf{V}_T$$

Inserting the known numerical values yields

$$17\angle 45° = \frac{j100}{Z_T + j100} 12\angle 0°$$

Solving for Z_T, we have

$$Z_T = j100 \times \frac{12\angle 0°}{17\angle 45°} - j100 = 49.9 - j50.1\ \Omega$$

The Thévenin equivalent circuit at the interface is defined by $\mathbf{V}_T = 12\angle 0°$ and $Z_T = 49.9 - j50.1\ \Omega$. ■

Exercise 8–13

(a) Find the Thévenin equivalent circuit seen by the inductor in Figure 8–34.
(b) Use the Thévenin equivalent to calculate the current $\mathbf{I}_X$.

Answers:

(a) $\mathbf{V}_T = -15.4 - j22.6$ V, $Z_T = 109.9 - j0.990\ \Omega$
(b) $\mathbf{I}_X = 0.206\angle -158°$ A

Exercise 8–14

By inspection, determine the Thévenin equivalent circuit seen by the capacitor in Figure 8–28 for $\mathbf{V}_S = 10\angle 0°$ V.

Answer: $\mathbf{V}_T = 5\angle 0°$ V, $Z_T = 25 + j50\ \Omega$

Exercise 8–15

In the steady state the short-circuit current at an interface is observed to be

$$i_{SC}(t) = 0.75 \sin \omega t \text{ A}$$

When a 150-Ω resistor is connected across the interface, the interface current is observed to be

$$i(t) = 0.6 \cos(\omega t - 53.1°) \text{ A}$$

Find the Norton equivalent phasor circuit at the interface.

Answer: $\mathbf{I}_N = 0 - j0.75$ A, $Z_N = 0 + j200\ \Omega$

8–5 GENERAL CIRCUIT ANALYSIS WITH PHASORS

The previous sections discuss basic analysis methods based on equivalence, reduction, and circuit theorems. These methods are valuable because we work directly with element impedances and thereby gain insight into steady-state circuit behavior. We also need general methods, such as node and mesh analysis, to deal with more complicated circuits than the basic methods can easily handle. These general methods use node-voltage or mesh-current variables to reduce the number of equations that must be solved simultaneously.

Node-voltage equations involve selecting a reference node and assigning a node-to-datum voltage to each of the remaining nonreference nodes. Because of KVL, the voltage between any two nodes equals the difference of the two node voltages. This fundamental property of node voltages plus the element impedances allow us to write KCL constraints at each of the nonreference nodes.

For example, consider node A in Figure 8–39. The sum of currents leaving this node can be written as

$$\mathbf{I}_{S2} - \mathbf{I}_{S1} + \frac{\mathbf{V}_A}{Z_1} + \frac{\mathbf{V}_A - \mathbf{V}_B}{Z_2} + \frac{\mathbf{V}_A - \mathbf{V}_C}{Z_3} = 0$$

Rewriting this equation with unknowns grouped on the left and known inputs on the right yields

$$\left[\frac{1}{Z_1} + \frac{1}{Z_2} + \frac{1}{Z_3}\right]\mathbf{V}_A - \frac{1}{Z_2}\mathbf{V}_B - \frac{1}{Z_3}\mathbf{V}_C = \mathbf{I}_{S1} - \mathbf{I}_{S2}$$

Expressing this result in terms of admittances produces the following equation.

$$[Y_1 + Y_2 + Y_3]\mathbf{V}_A - [Y_2]\mathbf{V}_B - [Y_3]\mathbf{V}_C = \mathbf{I}_{S1} - \mathbf{I}_{S2}$$

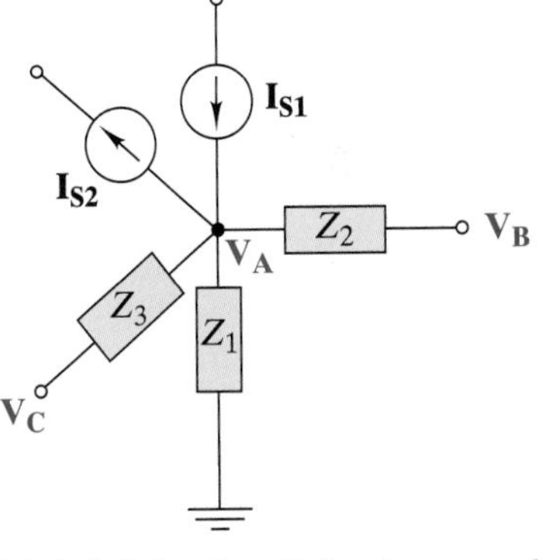

FIGURE 8–39 *An example node.*

This equation has a familiar pattern. The unknowns $\mathbf{V}_A$, $\mathbf{V}_B$, and $\mathbf{V}_C$ are the node-voltage phasors. The coefficient $[Y_1 + Y_2 + Y_3]$ of $\mathbf{V}_A$ is the sum of the admittances of all of the elements connected to node A. The coefficient $[Y_2]$ of $\mathbf{V}_B$ is admittance of the elements connected between nodes A and B, while $[Y_3]$ is the admittance of the elements connected between nodes A and C. Finally, $\mathbf{I}_{S1}$ and $\mathbf{I}_{S2}$ are the phasor current sources connected to node A, with $\mathbf{I}_{S1}$ directed into and $\mathbf{I}_{S2}$ directed away from the node. These observations suggest that we can write node-voltage equations for phasor circuits by inspection, just as we did with resistive circuits.

Circuits that can be drawn on a flat surface with no crossovers are called **planar** circuits. The mesh-current variables are the loop currents assigned to each mesh in a planar circuit. Because of KCL, the current through any two-terminal element is equal to the difference of the two adjacent meshes. This fundamental property of mesh currents together with the element impedances allow us to write KVL constraints around each of the meshes.

For example, the sum of voltages around mesh A in Figure 8–40 is

$$Z_1\mathbf{I}_A + Z_2[\mathbf{I}_A - \mathbf{I}_B] + Z_3[\mathbf{I}_A - \mathbf{I}_C] - \mathbf{V}_{S1} + \mathbf{V}_{S2} = 0$$

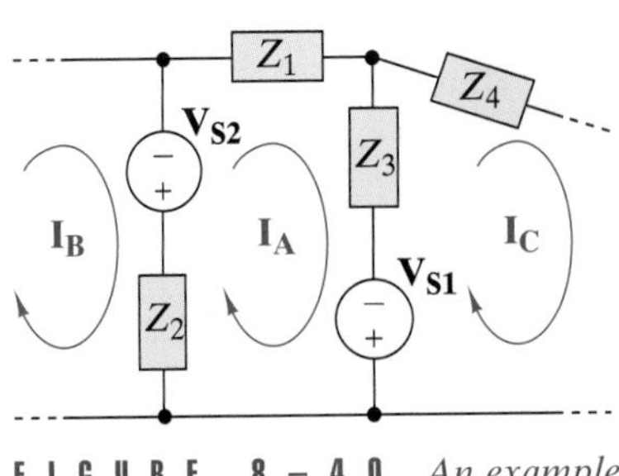

FIGURE 8–40 *An example mesh.*

The mesh A equation is obtained by equating this sum to the sum of the source voltages produced in mesh A. Arranging this equation in standard form yields

$$[Z_1 + Z_2 + Z_3]\mathbf{I}_A - [Z_2]\mathbf{I}_B - [Z_3]\mathbf{I}_C = \mathbf{V}_{S1} - \mathbf{V}_{S2}$$

This equation also displays a familiar pattern. The unknowns $\mathbf{I}_A$, $\mathbf{I}_B$, and $\mathbf{I}_C$ are mesh-current phasors. The coefficient $[Z_1 + Z_2 + Z_3]$ of $\mathbf{I}_A$ is the sum of the impedances in mesh A. The coefficient $[Z_2]$ of $\mathbf{I}_B$ is the impedance in both mesh A and mesh B, while $[Z_3]$ is the impedance common to meshes A and C. Finally, $\mathbf{V}_{S1}$ and $\mathbf{V}_{S2}$ are the phasor voltage sources in mesh A. These observations allow us to write mesh-current equations for phasor circuits by inspection.

The preceding discussion assumes that the circuit contains only current sources in the case of node analysis and voltage sources in mesh analysis. If there is a mixture of sources, we may be able to use the source transformations discussed in Sect. 8–4 to convert from voltage to current sources, or vice versa. A source transformation is possible only when there is an impedance connected in series with a voltage source or an admittance in parallel with a current source. When a source transformation is not possible, we use the phasor version of the modified node- and mesh-analysis methods described in Chapter 3.

Formulating a set of equilibrium equations in phasor form is a straightforward process involving concepts that we have used before in Chapters 3 and 4. Once formulated, we use Cramer's rule or Gaussian reduction to solve these equations for phasor responses, although this requires manipulating linear equations with complex coefficients. In principle, the solution process can be done by hand, but as a practical matter circuits with more than three nodes or meshes are best handled using computer tools. Modern hand-held scientific calculators and math analysis programs like Mathcad can deal with sets of linear equations with complex coefficients. Circuit analysis programs such as SPICE have ac analysis options that handle steady-state circuit analysis problems.

If computer tools are required for all but the simplest circuits, why bother with the hand solution at all? Why not always use SPICE or Mathcad? The answer is that hand analysis and computer-aided analysis are complementary rather than competitive. Computer-aided analysis excels at generating numerical responses when numerical values of circuit parameters are given. With hand analysis we can generate responses in symbolic form for all possible element values and input frequencies. Solving simple circuits in symbolic form gives insight as to how various circuit elements affect the steady-state response, which helps us intelligently use the numerical profusion generated by computer-based "what if" analysis of large-scale circuits.

EXAMPLE 8–20

Use node analysis to find the node voltages $\mathbf{V}_A$ and $\mathbf{V}_B$ in Figure 8–41(a).

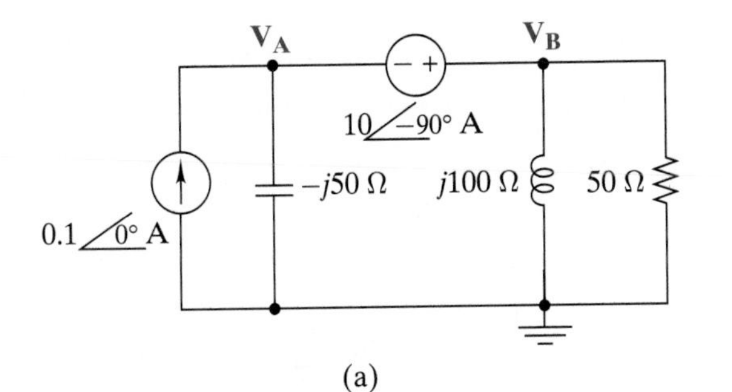

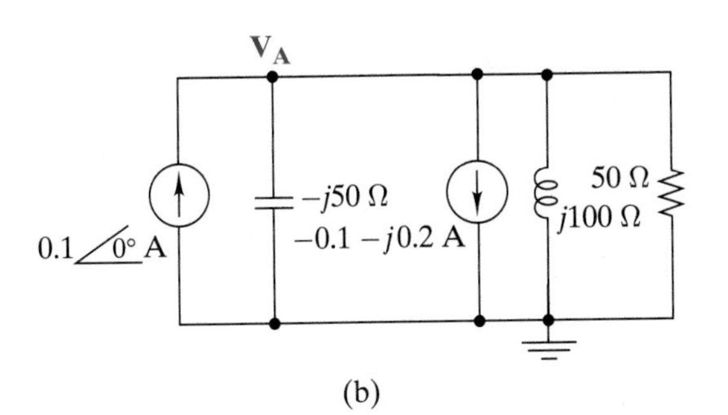

FIGURE 8–41

SOLUTION:

The voltage source in Figure 8–41(a) is connected in series with an impedance consisting of a resistor and inductor connected in parallel. The equivalent impedance of this parallel combination is

$$Z_{EQ} = \frac{1}{\dfrac{1}{50} + \dfrac{1}{j100}} = 40 + j20\ \Omega$$

Applying a source transformation produces an equivalent current source of

$$\mathbf{I}_{EQ} = \frac{10\angle -90°}{40 + j20} = -0.1 - j0.2 \text{ A}$$

Figure 8–41(b) shows the circuit produced by the source transformation. Note that the transformation eliminates node B. The node-voltage equation at the remaining nonreference node in Figure 8–41(b) is

$$\left(\frac{1}{-j50} + \frac{1}{j100} + \frac{1}{50}\right)\mathbf{V}_A = 0.1\angle 0° - (-0.1 - j0.2)$$

Solving for $\mathbf{V}_A$ yields

$$\mathbf{V}_A = \frac{0.2 + j0.2}{0.02 + j0.01} = 12 + j4 = 12.6\angle 18.4° \text{ V}$$

Referring to Figure 8–41(a), we see that KVL requires $\mathbf{V}_B = \mathbf{V}_A + 10\angle{-90°}$. Therefore, $\mathbf{V}_B$ is found to be

$$\mathbf{V}_B = (12 + j4) + 10\angle -90° = 12 - j6 = 13.4\angle -26.6° \text{ V}$$ ■

EXAMPLE 8–21

Use node analysis to find the current $\mathbf{I}_X$ in Figure 8–42.

FIGURE 8–42

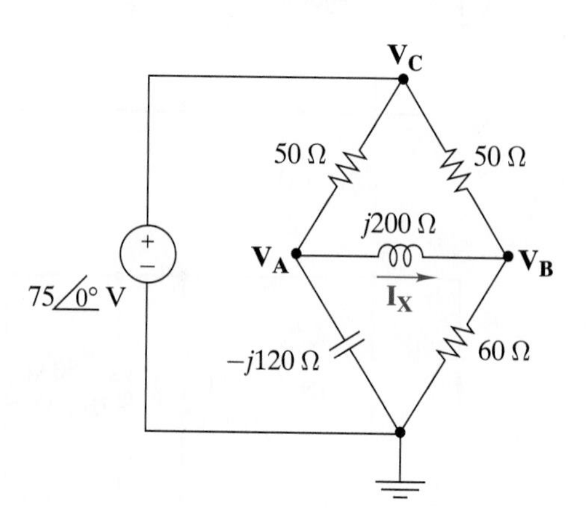

SOLUTION:

In this example we use node analysis on a problem solved in Example 8–18 using a Thévenin equivalent circuit. The voltage source cannot be replaced by source transformation because it is not connected in series with an impedance. By inspection, the node equations at nodes A and B are

$$\text{Node A: } \frac{\mathbf{V}_A}{-j120} + \frac{\mathbf{V}_A - \mathbf{V}_B}{j200} + \frac{\mathbf{V}_A - \mathbf{V}_C}{50} = 0$$

$$\text{Node B: } \frac{\mathbf{V}_B}{60} + \frac{\mathbf{V}_B - \mathbf{V}_A}{j200} + \frac{\mathbf{V}_B - \mathbf{V}_C}{50} = 0$$

A node equation at node C is not required because the voltage source forces the condition $\mathbf{V}_C = 75\angle 0°$. Substituting this constraint into the equations of nodes A and B and arranging the equations in standard form yields two equations in two unknowns:

$$\text{Node A: } \left(\frac{1}{50} + \frac{1}{-j120} + \frac{1}{j200}\right)\mathbf{V}_A - \left(\frac{1}{j200}\right)\mathbf{V}_B = \left(\frac{75|0°}{50}\right)$$

$$\text{Node B: } -\left(\frac{1}{j200}\right)\mathbf{V}_A + \left(\frac{1}{50} + \frac{1}{60} + \frac{1}{j200}\right)\mathbf{V}_B = \left(\frac{75\angle 0°}{50}\right)$$

Solving these equations for $\mathbf{V}_A$ and $\mathbf{V}_B$ yields

$$\mathbf{V}_A = 70.4 - j21.4 \text{ V}$$

$$\mathbf{V}_B = 38.6 - j4.33 \text{ V}$$

Using these values for $\mathbf{V}_A$ and $\mathbf{V}_B$, the unknown current is found to be

$$\mathbf{I}_X = \frac{\mathbf{V}_A - \mathbf{V}_B}{j200} = \frac{31.8 - j17.1}{j200} = 0.180\angle -118° \text{ A}$$

This value of $\mathbf{I}_X$ is the same as the answer obtained in Example 8–18. ■

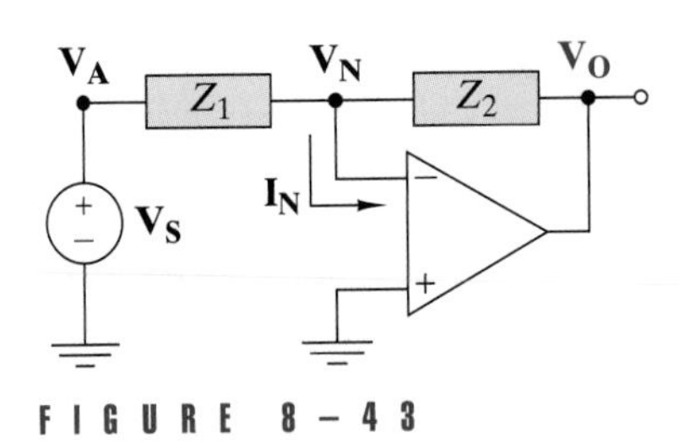

FIGURE 8–43

EXAMPLE 8–22

Use node-voltage analysis to determine the phasor input-output relationship of the OP AMP circuit in Figure 8–43.

SOLUTION:
In the sinusoidal steady state the sum of currents leaving the inverting input node is

$$\frac{\mathbf{V}_N - \mathbf{V}_S}{Z_1} + \frac{\mathbf{V}_N - \mathbf{V}_O}{Z_2} + \mathbf{I}_N = 0$$

This is the only required node equation, since the input source forces the condition $\mathbf{V}_A = \mathbf{V}_S$ and no node equation is ever required at an OP AMP output. In the time domain the i–v relationships of an ideal OP AMP are $v_P(t) = v_N(t)$ and $i_P(t) = i_N(t) = 0$. In the sinusoidal steady state these equations are written in phasor form as $\mathbf{V}_P = \mathbf{V}_N$ and $\mathbf{I}_P = \mathbf{I}_N = 0$. In the present case this means $\mathbf{V}_N = 0$ since the noninverting input is grounded. When the ideal OP AMP constraints are inserted in the node equation, we can solve for the OP AMP input-output relationship as

$$\mathbf{V}_O = -\frac{Z_2}{Z_1}\mathbf{V}_S$$

This result is the phasor-domain version of the inverting amplifier configuration. In the phasor domain, the "gain" $K = -Z_2/Z_1$ is determined by a ratio of impedances rather than resistances. Thus, the gain affects both the amplitude and the phase angle of the steady-state output. ■

EXAMPLE 8–23

Use node-voltage analysis to determine the phasor input–output relationship of the OP AMP circuit in Figure 8–44.

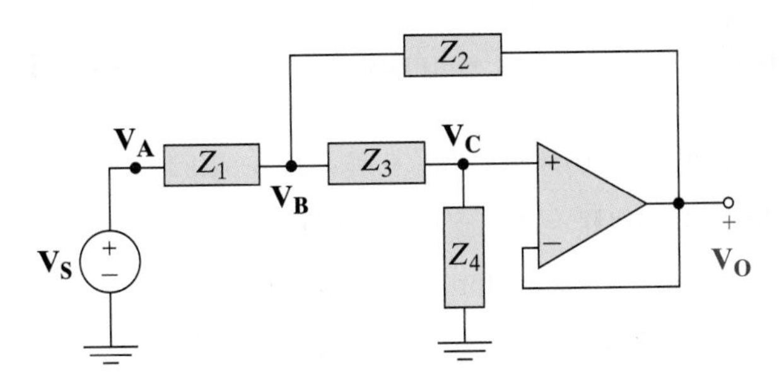

FIGURE 8–44

SOLUTION:
The input source forces the condition $\mathbf{V}_A = \mathbf{V}_S$. The OP AMP is connected as a voltage follower, so the inverting input voltage is $\mathbf{V}_N = \mathbf{V}_O$. The voltage at the noninverting input is $\mathbf{V}_P = \mathbf{V}_C = \mathbf{V}_O$ because OP AMP voltage

constraint requires $\mathbf{V}_P = \mathbf{V}_N$. Taking all these constraints into account, we can write the sum of currents leaving nodes B and C as

$$\text{Node B:} \quad \frac{\mathbf{V}_B - \mathbf{V}_S}{Z_1} + \frac{\mathbf{V}_B - \mathbf{V}_O}{Z_2} + \frac{\mathbf{V}_B - \mathbf{V}_O}{Z_3} = 0$$

$$\text{Node C:} \quad \frac{\mathbf{V}_O - \mathbf{V}_B}{Z_3} + \frac{\mathbf{V}_O}{Z_4} = 0$$

Solving the node C equation for $\mathbf{V}_B$ yields

$$\mathbf{V}_B = \frac{Z_3 + Z_4}{Z_4}\mathbf{V}_O$$

Using this result to eliminate $\mathbf{V}_B$ from the node B equation and then solving for the output $\mathbf{V}_O$ produces

$$\mathbf{V}_O = \frac{Z_2 Z_4}{Z_1 Z_2 + Z_1 Z_3 + Z_2 Z_3 + Z_2 Z_4} V_S$$

■

EXAMPLE 8–24

The circuit in Figure 8–45 is an equivalent circuit of an ac induction motor. The current $\mathbf{I}_S$ is called the stator current, $\mathbf{I}_R$ the rotor current, and $\mathbf{I}_M$ the magnetizing current. Use the mesh-current method to solve for the branch currents $\mathbf{I}_S$, $\mathbf{I}_R$, and $\mathbf{I}_M$.

FIGURE 8–45

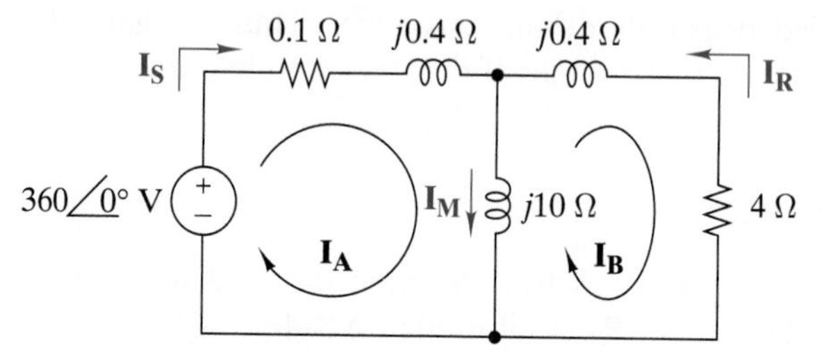

SOLUTION:

Applying KVL to the sum of voltages around each mesh in Figure 8–45 yields

$$\text{Mesh A:} \quad -360\angle 0° + (0.1 + j0.4)\mathbf{I}_A + j10(\mathbf{I}_A - \mathbf{I}_B) = 0$$

$$\text{Mesh B:} \quad j10(\mathbf{I}_B - \mathbf{I}_A) + (4 + j0.4)\mathbf{I}_B = 0$$

Arranging these equations in standard form yields

$$(0.1 + j10.4)\mathbf{I}_A - (j10)\mathbf{I}_B = 360|0°$$

$$-(j10)\mathbf{I}_A + (4 + j10.4)\mathbf{I}_B = 0$$

Solving these equations for $\mathbf{I}_A$ and $\mathbf{I}_B$ produces

$$\mathbf{I}_A = 79.0 - j48.2 \text{ A}$$

$$\mathbf{I}_B = 81.7 - j14.9 \text{ A}$$

The required stator, rotor, and magnetizing currents are related to these mesh currents, as follows:

$$\mathbf{I}_S = \mathbf{I}_A = 92.5\angle{-31.4°}\text{ A}$$

$$\mathbf{I}_R = -\mathbf{I}_B = -81.8 + j14.7 = 83.0\angle 170°\text{ A}$$

$$\mathbf{I}_M = \mathbf{I}_A - \mathbf{I}_B = -2.68 - j33.3 = 33.4\angle{-94.6°}\text{ A}$$

■

EXAMPLE 8–25

Use the mesh-current method to solve for output voltage $\mathbf{V}_2$ and input impedance Z_{IN} of the circuit in Figure 8–46.

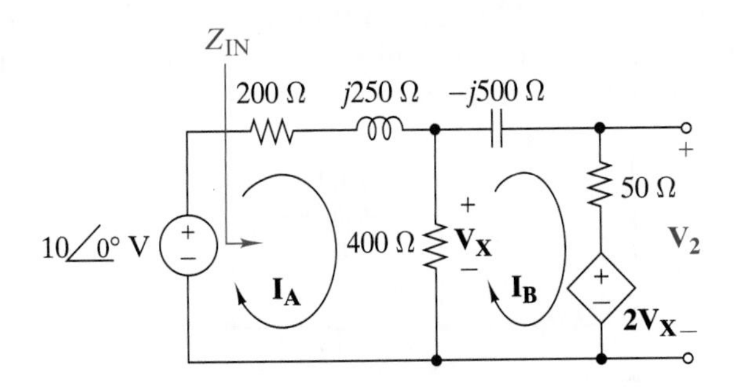

FIGURE 8–46

SOLUTION:

The circuit contains a voltage-controlled voltage source. We initially treat the dependent source as an independent source and use KCL to write the sum of voltages around each mesh:

$$\text{Mesh A:}\quad -10\angle 0° + (200 + j250)\mathbf{I}_A + 400(\mathbf{I}_A - \mathbf{I}_B) = 0$$

$$\text{Mesh B:}\quad 400(\mathbf{I}_B - \mathbf{I}_A) + (50 - j500)\mathbf{I}_B + 2\mathbf{V}_X = 0$$

Arranging these equations in standard form produces

$$\text{Mesh A:}\ (600 + j250)\mathbf{I}_A - 400\mathbf{I}_B = 10\angle 0°$$

$$\text{Mesh B:}\ -400\mathbf{I}_A + (450 - j500)\mathbf{I}_B = -2\mathbf{V}_X$$

Using Ohm's law, the control voltage $\mathbf{V}_X$ is

$$\mathbf{V}_X = 400(\mathbf{I}_A - \mathbf{I}_B)$$

Eliminating $\mathbf{V}_X$ from the mesh equations yields

$$\text{Mesh A:}\ (600 + j250)\mathbf{I}_A - 400\mathbf{I}_B = 10\angle 0°$$

$$\text{Mesh B:}\ 400\mathbf{I}_A + (-350 - j500)\mathbf{I}_B = 0$$

Solving for the two mesh currents produces

$$\mathbf{I}_A = 10.8 - j11.1\text{ mA}$$

$$\mathbf{I}_B = -1.93 - j9.95\text{ mA}$$

Using these values of the mesh currents, the output voltage and input impedance are

$$\begin{aligned}\mathbf{V}_2 &= 2\mathbf{V}_X + 50\mathbf{I}_B = 800(\mathbf{I}_A - \mathbf{I}_B) + 50\mathbf{I}_B \\ &= 800\mathbf{I}_A - 750\mathbf{I}_B = 10.1 - j1.42 \\ &= 10.2\angle -8.00° \text{ V}\end{aligned}$$

$$Z_{IN} = \frac{10\angle 0°}{\mathbf{I}_A} = \frac{10\angle 0°}{0.0108 - j0.0111} = 450 + j463\ \Omega$$

■

EXAMPLE 8–26

In the circuit in Figure 8–47 the input voltage is $v_S(t) = 10 \cos 10^5 t$ V. Use mesh equations and MATLAB to find the input impedance at the input interface and the proportionality constant relating the input voltage phasor to the phasor voltage across the 50-Ω load resistor.

FIGURE 8–47

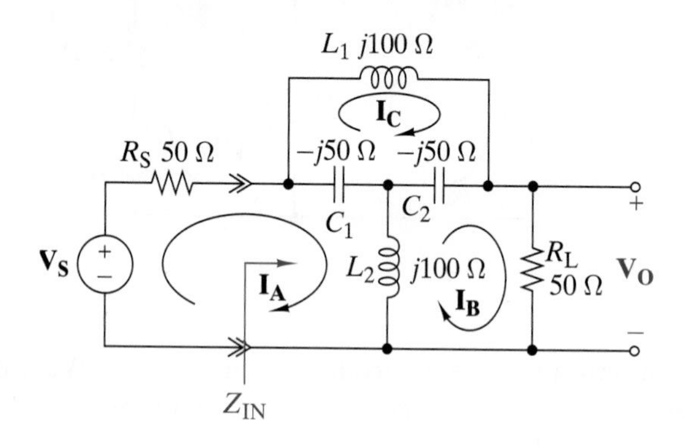

SOLUTION:

In matrix form, the mesh-current equations for this circuit are

$$\begin{bmatrix} R_S + Z_{C1} + Z_{L2} & -Z_{L2} & -Z_{C1} \\ -Z_{L2} & Z_{L2} + Z_{C2} + R_L & -Z_{C2} \\ -Z_{C1} & -Z_{C2} & Z_{C1} + Z_{C2} + Z_{L1} \end{bmatrix} \begin{bmatrix} \mathbf{I}_A \\ \mathbf{I}_B \\ \mathbf{I}_C \end{bmatrix} = \begin{bmatrix} \mathbf{V}_S \\ 0 \\ 0 \end{bmatrix}$$

This matrix equation is of the form $\mathbf{ZI} = \mathbf{V}$, where $\mathbf{Z}$ is a 3×3 matrix of element impedance, $\mathbf{I}$ is the 3×1 mesh current vector, and $\mathbf{V}$ is the 3×1 input voltage vector. The entries on the main diagonal of the $\mathbf{Z}$ matrix are the total impedances in each mesh. The $\mathbf{Z}$ matrix is symmetrical because the off diagonal entries are the negatives of the impedances common to two meshes. For example, the entry in the 2nd row and 3rd column ($-Z_{C2}$) is equal to the entry in the 3rd row and 2nd column.

With the circuit equations formulated, we can use MATLAB to solve for the required circuit descriptors. To solve for the mesh current vector, we need the matrix product $\mathbf{Z}^{-1}\mathbf{V}$. We have used MATLAB for this purpose in previous examples. The only thing new here is that the matrices contain complex numbers.

One of the advantages of MATLAB is the ease with which it handles complex numbers. Complex inputs are identified using either the mathematical convention ($i = \sqrt{-1}$) or the electrical engineering convention ($j = \sqrt{-1}$). That is, MATLAB accepts complex number inputs written as `z=x+yi` or `z=x+yj`. Accordingly, in the MATLAB command window, we enter the element parameters as

```
VS=10; RS=50; RL=50;
ZC1=-50j; ZC2=-50j;
ZL1=100j; ZL2=100j;
```

the **Z** and **V** matrices as

```
Z=[RS+ZC1+ZL2     -ZL2        -ZC1
     -ZL2     ZL2+ZC2+RL     -ZC2
     -ZC1          -ZC2    ZC1+ZC2+ZL1];
V=[VS ; 0 ; 0];
```

and solve for the mesh currents as

```
I=inv(Z)*V
I=
       0.0100 - 0.0300i
      -0.0100 + 0.0300i
            0 - 0.1000i
```

The form of the output points out that MATLAB accepts inputs in either convention, but it reports complex number outputs using the mathematical convention. In MATLAB matrix notation, the mesh currents are $\mathbf{I}_A$ = I(1), $\mathbf{I}_B$ = I(2), and $\mathbf{I}_C$ = I(3). We can now calculate the required parameters as

```
ZIN=(VS-I(1)*RS)/I(1)
ZIN =
      5.0000e+001 + 3.0000e+002i
K=I(2)*RL/VS
K=
      -0.05000 + 0.15000i
```

That is, in electrical engineering notation, $Z_{IN} = 50 + j300\ \Omega$ and $K = -0.05 + j0.15$. ■

Exercise 8–16

Use the mesh-current or node-voltage method to find the branch currents $\mathbf{I}_1$, $\mathbf{I}_2$, and $\mathbf{I}_3$ in Figure 8–48.

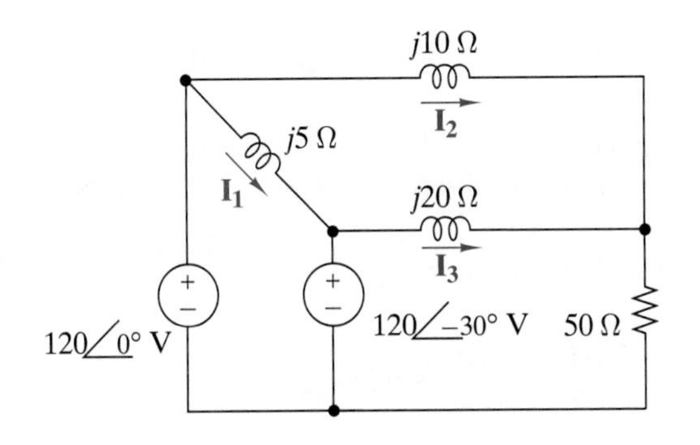

FIGURE 8–48

Answer: $\mathbf{I}_1 = 12.4\angle{-15°}$ A, $\mathbf{I}_2 = 3.61\angle{-16.1°}$ A, $\mathbf{I}_3 = 1.31\angle{166°}$ A

Exercise 8–17

Use the mesh-current or node-voltage method to find the output voltage $\mathbf{V}_2$ and input impedance Z_{IN} in Figure 8–49.

FIGURE 8–49

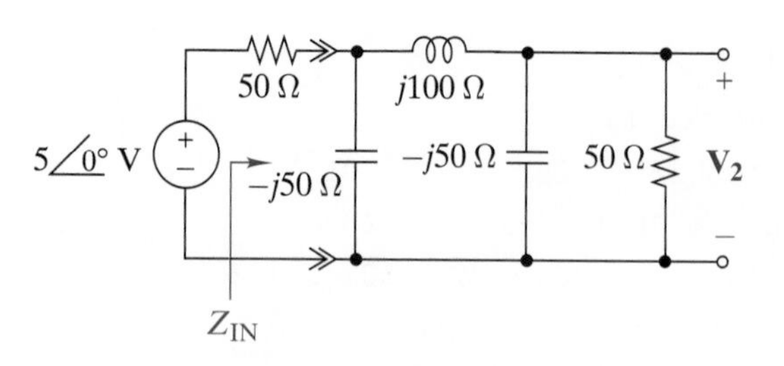

Answer: $\mathbf{V}_2 = 1.77\angle{-135°}$ V, $Z_{IN} = 100 - j100\ \Omega$

Exercise 8–18

Use the mesh-current or node-voltage method to find the current $\mathbf{I}_X$ in Figure 8–50.

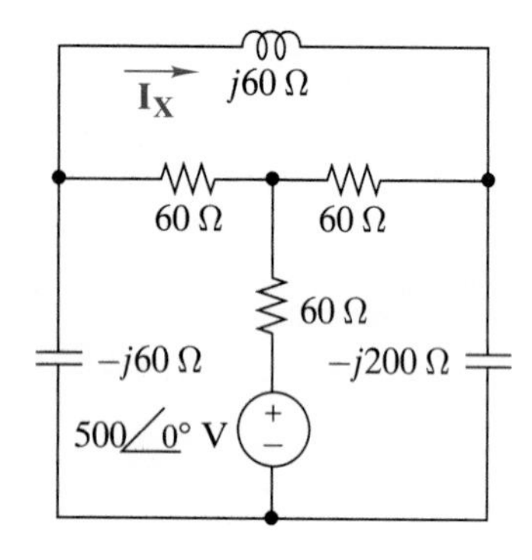

FIGURE 8–50

Answer: $\mathbf{I}_X = 1.44\angle{171°}$ mA

8–6 ENERGY AND POWER

In the sinusoidal steady state, ac power is transferred from sources to various loads. To study the transfer process, we must calculate the power delivered in the sinusoidal steady state to any specified load. It turns out that there is an upper bound on the available load power; hence, we need to understand how to adjust the load to extract the maximum power from the rest of the circuit. In this section the load is assumed to be made up of passive resistance, inductance, and capacitance. To reach our objectives, we must first study the power and energy delivered to these passive elements in the sinusoidal steady state.

In the sinusoidal steady state the current through a resistor can be expressed as $i_R(t) = I_A \cos(\omega t)$. The instantaneous power delivered to the resistor is

$$
\begin{aligned}
p_R(t) &= Ri_R^2(t) = RI_A^2 \cos^2(\omega t) \\
&= \frac{RI_A^2}{2}[1 + \cos(2\omega t)]
\end{aligned}
\tag{8–30}
$$

where the identity $\cos^2(x) = \frac{1}{2}[1 + \cos(2x)]$ is used to obtain the last line in Eq. (8–30). The energy delivered for $t \geq 0$ is found to be

$$
\begin{aligned}
w_R(t) &= \int_0^t p_R(x)dx = \frac{RI_A^2}{2}\int_0^t dx + \frac{RI_A^2}{2}\int_0^t \cos 2\omega x\, dx \\
&= \frac{RI_A^2}{2}t + \frac{RI_A^2}{4\omega}\sin 2\omega t
\end{aligned}
$$

Figure 8–51 shows the time variation of $p_R(t)$ and $w_R(t)$. Note that the power is a periodic function with twice the frequency of the current, that both $p_R(t)$ and $w_R(t)$ are always positive, and that $w_R(t)$ increases without bound. These observations remind us that a resistor is a passive element that dissipates energy.

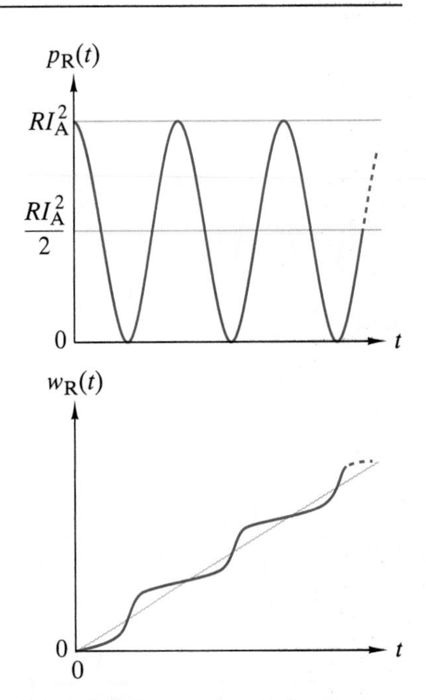

FIGURE 8–51 Resistor power and energy in the sinusoidal steady state.

In the sinusoidal steady state an inductor operates with a current $i_L(t) = I_A \cos(\omega t)$. The corresponding energy stored in the element is

$$w_L(t) = \frac{1}{2} Li_L^2(t) = \frac{1}{2} LI_A^2 \cos^2 \omega t$$

$$= \frac{1}{4} LI_A^2(1 + \cos 2\omega t)$$

where the identity $\cos^2(x) = ½ [1 + \cos(2x)]$ is again used to produce the last line. The instantaneous power delivered to the inductor is

$$p_L(t) = \frac{dw_L}{dt} = -\frac{\omega LI_A^2}{2} \sin(2\omega t) \tag{8–31}$$

Figure 8–52 shows the time variation of $p_L(t)$ and $w_L(t)$. Observe that both $p_L(t)$ and $w_L(t)$ are periodic functions at twice the frequency of the ac current, that $p_L(t)$ is alternately positive and negative, and that $w_L(t)$ is never negative. Since $w_L(t) \geq 0$, the inductor does not deliver net energy to the rest of the circuit. Unlike the resistor's energy in Figure 8–51, the energy in the inductor is bounded by $½ LI_A^2 \geq w_L(t)$, which means that the inductor does not dissipate energy. Finally, since $p_L(t)$ alternates signs, we see that the inductor stores energy during a positive half cycle and then returns the energy undiminished during the next negative half cycle. Thus, in the sinusoidal steady state there is a lossless interchange of energy between an inductor and the rest of the circuit.

In the sinusoidal steady state the voltage across a capacitor is $v_C(t) = V_A \cos(\omega t)$. The energy stored in the element is

$$w_C(t) = \frac{1}{2}Cv_C^2(t) = \frac{1}{2}CV_A^2 \cos^2 \omega t$$

$$= \frac{1}{4}CV_A^2(1 + \cos 2\omega t)$$

The instantaneous power delivered to the capacitor is

$$p_C(t) = \frac{dw_C}{dt} = -\frac{\omega CV_A^2}{2} \sin(2\omega t) \tag{8–32}$$

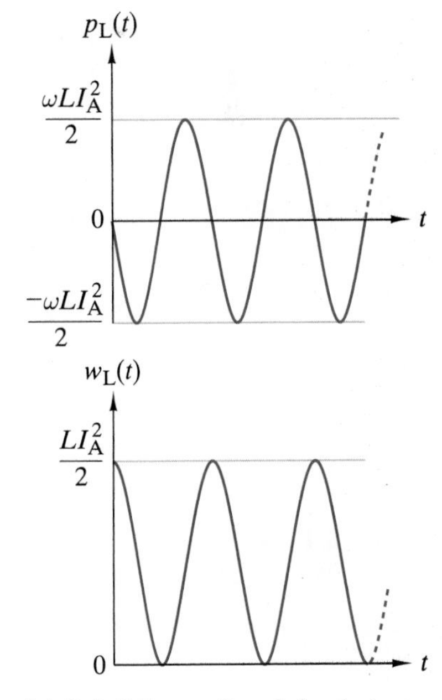

FIGURE 8–52 Inductor power and energy in the sinusoidal steady state.

Figure 8–53 shows the time variation of $p_C(t)$ and $w_C(t)$. Observe that these relationships are the duals of those found for the inductor. Thus, in the sinusoidal steady state the element power is sinusoidal and there is a lossless interchange of energy between the capacitor and the rest of the circuit.

AVERAGE POWER

We are now in a position to calculate the average power delivered to various loads. The instantaneous power delivered to any of the three passive elements is a periodic function that can be described by an average value. The **average power** in the sinusoidal steady state is defined as

$$P = \frac{1}{T_0}\int_0^{T_0} p(t)dt$$

The power variation of the inductor in Eq. (8–31) and capacitor in Eq. (8–32) have the same sinusoidal form. The average value of any sinusoid is zero since the areas under alternate cycles cancel. Hence, the average power delivered to an inductor or capacitor is zero:

$$\text{Inductor: } P_L = 0$$

$$\text{Capacitor: } P_C = 0$$

The resistor power in Eq. (8–30) has both a sinusoidal ac component and a constant dc component ½ RI_A^2. The average value of the ac component is zero, but the dc component yields

$$\text{Resistor: } P_R = \frac{1}{2}RI_A^2$$

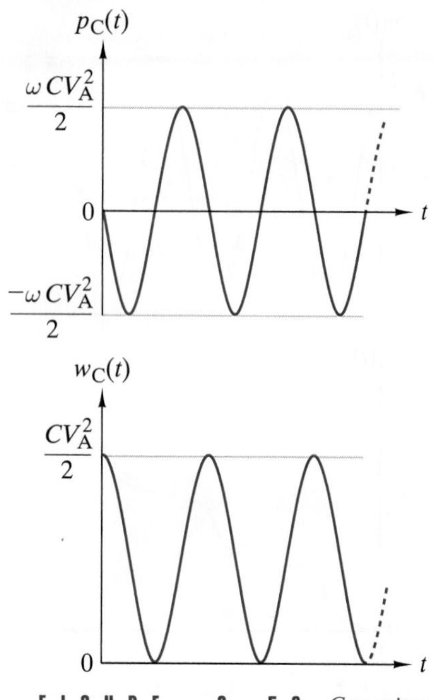

FIGURE 8–53 *Capacitor power and energy in the sinusoidal steady state.*

To calculate the average power delivered to an arbitrary load $Z_L = R_L + jX_L$, we use phasor circuit analysis to find the phasor current $\mathbf{I}_L$ through Z_L. The average power delivered to the load is dissipated in R_L, since the reactance X_L represents the net inductance or capacitance of the load. Hence the average power to the load is

$$P = \frac{1}{2}R_L|\mathbf{I}_L|^2 \tag{8–33}$$

Caution: When a circuit contains two or more sources, superposition applies only to the total load current and not to the total load power. You cannot find the total power to the load by summing the power delivered by each source acting alone.

The following example illustrates a power transfer calculation.

EXAMPLE 8–27

Find the average power delivered to the load to the right of the interface in Figure 8–54.

FIGURE 8–54

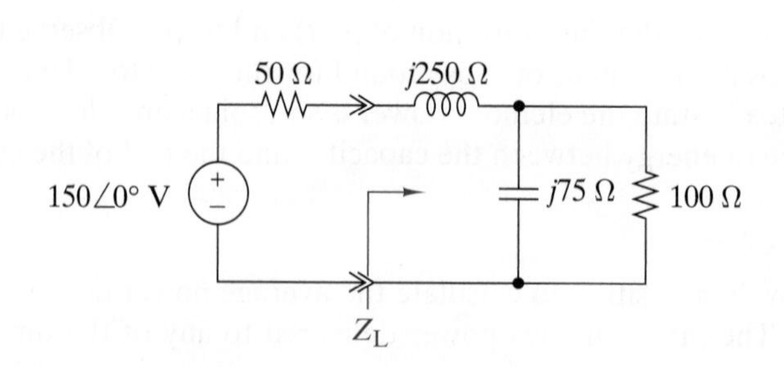

SOLUTION:

The equivalent impedance to the right of the interface is

$$Z_L = j250 + \cfrac{1}{\cfrac{1}{-j75} + \cfrac{1}{100}} = 36 + j202 \quad \Omega$$

The current delivered to the load is

$$\mathbf{I}_L = \frac{150\angle 0^\circ}{50 + Z_L} = 0.683\angle -66.9^\circ \quad \text{A}$$

Hence, the average power delivered across the interface is

$$P = \frac{1}{2} R_L |\mathbf{I}_L|^2 = \frac{36}{2} |0.683|^2 = 8.40 \quad \text{W}$$

Note: All of this power goes into the 100-Ω resistor since the inductor and capacitor do not absorb average power. ■

Exercise 8–19

Find the average power delivered to the 25-Ω load resistor in Figure 8–55.

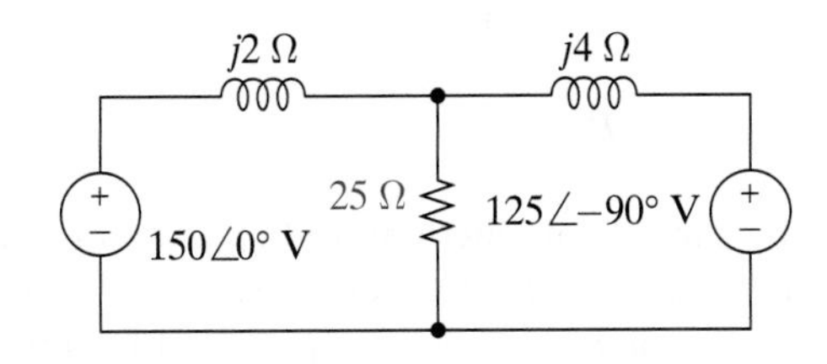

FIGURE 8–55

Answer: 234 W

MAXIMUM POWER

To address the maximum power transfer problem, we model the source-load interface as shown in Figure 8–56. The source circuit is represented by a Thévenin equivalent circuit with source voltage $\mathbf{V}_T$ and source impedance $Z_T = R_T + jX_T$. The load circuit is represented by an equivalent impedance $Z_L = R_L + jX_L$. In the maximum power transfer problem the source parameters $\mathbf{V}_T$, R_T, and X_T are given, and the objective is to adjust the load impedance R_L and X_L so that average power to the load is a maximum.

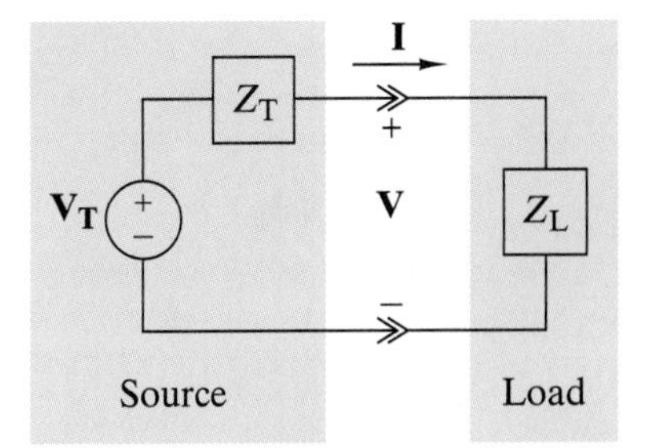

FIGURE 8–56 *A source-load interface in the sinusoidal steady state.*

The average power to the load is expressed in terms of the phasor current and load resistance:

$$P = \frac{1}{2} R_L |\mathbf{I}|^2$$

Then, using series equivalence, we express the magnitude of the interface current as

$$|\mathbf{I}| = \left|\frac{\mathbf{V}_T}{Z_T + Z_L}\right| = \frac{|\mathbf{V}_T|}{|(R_T + R_L) + j(X_T + X_L)|}$$
$$= \frac{|\mathbf{V}_T|}{\sqrt{(R_T + R_L)^2 + (X_T + X_L)^2}}$$

Combining the last two equations yields the average power delivered across the interface:

$$P = \frac{1}{2}\frac{R_L|\mathbf{V}_T|^2}{(R_T + R_L)^2 + (X_T + X_L)^2} \tag{8–34}$$

The quantities $|\mathbf{V}_T|$, R_T, and X_T in Eq. (8–34) are fixed. Our problem is to select R_L and X_L to maximize P.

Clearly, for every value of R_L the denominator in Eq. (8–34) is minimized and P maximized when $X_L = -X_T$. This choice of X_L is possible because a reactance can be positive or negative. When the source Thévenin equivalent has an inductive reactance ($X_T > 0$), we modify the load to have a capacitive reactance of the same magnitude, and vice versa. This step reduces the net reactance of the series connection in Figure 8–56 to zero, creating a condition in which the net impedance seen by the Thévenin voltage source is purely resistive.

When the source and load reactances cancel out, the expression for average power in Eq. (8–34) reduces to

$$P = \frac{1}{2}\frac{R_L|\mathbf{V}_T|^2}{(R_T + R_L)^2} \tag{8–35}$$

This equation has the same form encountered in Chapter 3 in dealing with maximum power transfer in resistive circuits. From the derivation in Sect. 3–5, we know P is maximized when $R_L = R_T$. In summary, to obtain maximum power transfer in the sinusoidal steady state, we select the load resistance and reactance so that

$$R_L = R_T \quad \text{and} \quad X_L = -X_T \tag{8–36}$$

These conditions can be compactly expressed in the following way:

$$Z_L = Z_T^* \tag{8–37}$$

The condition for maximum power transfer is called a **conjugate match**, since the load impedance is the conjugate of the source impedance. When the conjugate-match conditions are inserted into Eq. (8–34), we find that the maximum average power available from the source circuit is

$$P_{MAX} = \frac{|\mathbf{V}_T|^2}{8R_T} \tag{8–38}$$

where $|\mathbf{V}_T|$ is the peak amplitude of the Thévenin equivalent voltage.

It is important to remember that conjugate matching applies when the source is fixed and the load is adjustable. These conditions arise frequently in power-limited communication systems. However, as we will see in Chapter 16, conjugate matching does not apply to electrical power systems because the power transfer constraints are different.

EXAMPLE 8–28

(a) Calculate the average power delivered to the load in the circuit shown in Figure 8–57 for $v_S(t) = 5 \cos 10^6 t$ V, $R = 200\ \Omega$, $R_L = 200\ \Omega$, and $C = 10$ nF.

(b) Calculate the maximum average power available at the interface and specify the load required to draw the maximum power.

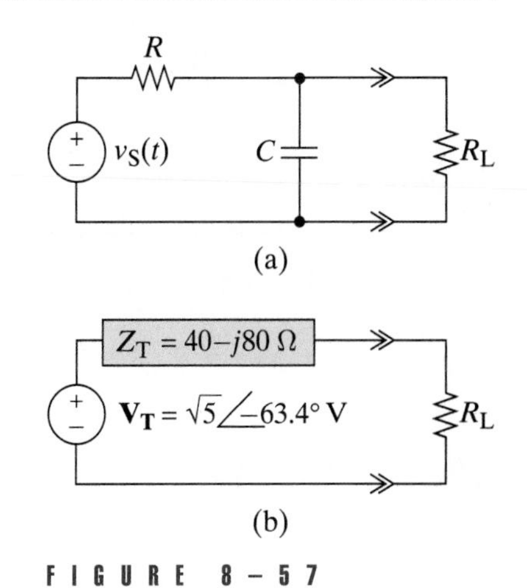

FIGURE 8–57

SOLUTION:

(a) To find the power delivered to the 200-Ω load resistor, we use a Thévenin equivalent circuit. By voltage division, the open-circuit voltage at the interface is

$$\mathbf{V}_T = \frac{Z_C}{Z_R + Z_C}\mathbf{V}_S = \frac{-j100}{200 - j100}5\angle 0^\circ$$
$$= 1 - j2 = \sqrt{5}\angle -63.4^\circ \text{ V}$$

By inspection, the short-circuit current at the interface is

$$\mathbf{I}_N = \frac{5\angle 0^\circ}{200} = 0.025 + j0 \text{ A}$$

Given $\mathbf{V}_T$ and $\mathbf{I}_N$, we calculate the Thévenin source impedance:

$$Z_T = \frac{\mathbf{V}_T}{\mathbf{I}_N} = \frac{1 - j2}{0.025} = 40 - j80\ \ \Omega$$

Using the Thévenin equivalent shown in Figure 8–57(b), we find that the current through the 200-Ω resistor is

$$\mathbf{I} = \frac{\mathbf{V}_T}{Z_T + Z_L} = \frac{\sqrt{5}\angle -63.4^\circ}{40 - j80 + 200} = 8.84\angle -45^\circ \text{ mA}$$

and the average power delivered to the load resistor is

$$P = \frac{1}{2}R_L|\mathbf{I}|^2 = 100(8.84 \times 10^{-3})^2 = 7.81 \text{ mW}$$

(b) Using Eq. (8–38), the maximum average power available at the interface is

$$P_{MAX} = \frac{|\mathbf{V}_T|^2}{8R_T} = \frac{(\sqrt{5})^2}{(8)(40)} = 15.6 \text{ mW}$$

The 200-Ω load resistor in (a) draws about half of the maximum available power. To extract maximum power, the load impedance must be

$$Z_L = Z_T^* = 40 + j80\ \ \Omega$$

This impedance can be obtained using a 40-Ω resistor in series with a reactance of +80 Ω. The required reactance is inductive (positive) and can be produced by an inductance of

$$L = \frac{|X_T|}{\omega} = \frac{80}{10^6} = 80\ \mu\text{H}$$

■

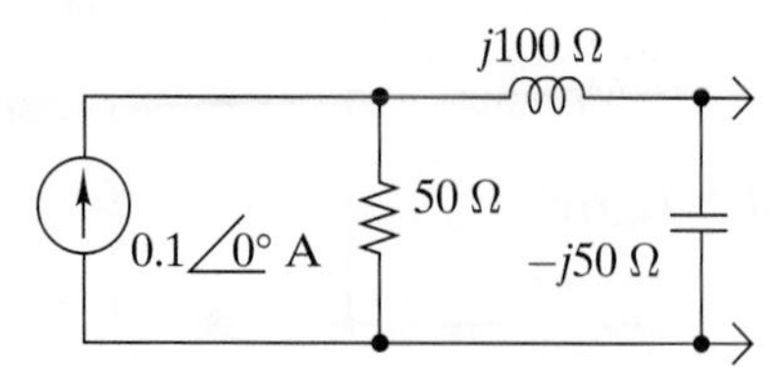

FIGURE 8–58

Exercise 8–20

Calculate the maximum average power available at the interface in Figure 8–58.

Answer: 125 mW.

SUMMARY

- A phasor is a complex number representing a sinusoidal waveform. The magnitude and angle of the phasor correspond to the amplitude and phase angle of the sinusoid. The phasor does not provide frequency information.
- The additive property states that adding phasors is equivalent to adding sinusoids of the same frequency. The derivative property states that multiplying a phasor by $j\omega$ is equivalent to differentiating the corresponding sinusoid.
- In the sinusoidal steady state, phasor currents and voltages obey Kirchhoff's laws and the element i–v relationships are written in terms of impedances. Impedance can be defined as the ratio of phasor voltage over phasor current. The device and connection constraints for phasor circuit analysis have the same form as resistance circuits.
- Phasor circuit analysis techniques include series equivalence, parallel equivalence, circuit reduction, Thévenin's and Norton's theorems, unit output method, superposition, node-voltage analysis, and mesh-current analysis.
- In the sinusoidal steady state the equivalent impedance at a pair of terminals is $Z(j\omega) = R(\omega) + jX(\omega)$, where $R(\omega)$ is called resistance and $X(\omega)$ is called reactance. A frequency at which an equivalent impedance is purely real is called a resonant frequency. Admittance is the reciprocal of impedance.
- In the sinusoidal steady state the instantaneous power to a passive element is a periodic function at twice the frequency of the driving force. The average power delivered to an inductor or capacitor is zero. The average power delivered to a resistor is ½ $R|\mathbf{I}_R|^2$. The maximum average power is delivered by a fixed source to an adjustable load when the source and load impedances are conjugates.

PROBLEMS

OBJECTIVE 8–1 SINUSOIDS AND PHASORS (SECT. 8–1)

Use the additive and derivative properties of phasors to convert sinusoidal waveforms into phasors and vice versa.
See Examples 8–1, 8–2, 8–3, 8–4 and Exercises 8–1, 8–2, 8–3, 8–4

8–1 Transform the following sinusoids into phasor form and draw a phasor diagram. Use the additive property of phasors to find $v_1(t) + v_2(t)$.

(a) $v_1(t) = 250\cos(\omega t + 60°)$ V

(b) $v_2(t) = 100\cos(\omega t) + 150\sin(\omega t)$ V

8–2 Transform the following sinusoids into phasor form and draw a phasor diagram. Use the additive property of phasors to find $i_1(t) + i_2(t)$.

(a) $i_1(t) = 4\cos(\omega t)$ A

(b) $i_2(t) = 3\cos(\omega t - 90°)$ A

8–3 Convert the following phasors into sinusoidal waveforms.

(a) $\mathbf{V}_1 = 10\,e^{-j30°}$ V, $\omega = 10^4$ rad/s

(b) $\mathbf{V}_2 = 60\,e^{-j220°}$ V, $\omega = 10^4$ rad/s

(c) $\mathbf{I}_1 = 5\,e^{j90°}$ A, $\omega = 200$ rad/s

(d) $\mathbf{I}_2 = 2\,e^{j270°}$ A, $\omega = 200$ rad/s

8–4 Use the phasors in Problem 8–3 and the additive property to find the sinusoidal waveforms $2v_1(t) + v_2(t)$ and $i_1(t) + 3i_2(t)$.

8–5 The phasor representation of a sinusoid with $\omega = 25$ rad/s is $\mathbf{V} = 15 + j10$ V. Use the phasor derivative property to find the time derivative of the sinusoid.

8–6 Convert the following phasors into sinusoids:

(a) $\mathbf{V}_1 = 20 + j25$ V, $\omega = 10$ rad/s

(b) $\mathbf{V}_2 = (8 - j3)5$ V, $\omega = 20$ rad/s

(c) $\mathbf{I}_1 = 12 - j5 + \dfrac{4}{j}$ A, $\omega = 300$ rad/s

(d) $\mathbf{I}_2 = \dfrac{3 + j8}{2 - j6}$ A, $\omega = 50$ rad/s

8–7 Use phasors to find the sinusoid $v_2(t)$, where

$$v_2(t) = \frac{1}{10}\frac{dv_1(t)}{dt} + 20\,v_1(t) \text{ and } v_1(t) = 10\cos(250t + 90^\circ)$$

8–8 Given the sinusoids

$$v_1(t) = 50\cos(\omega t - 45^\circ) \text{ and } v_2(t) = 25\sin(\omega t)$$

use the additive property of phasors to find $v_3(t)$ such that $v_1 + v_2 + v_3 = 0$.

8–9 Use phasors to find the sinusoid $v_1(t)$, where

$$\frac{dv_1(t)}{dt} + 250v_1(t) = 50v_2(t) \quad \text{and} \quad v_2(t) = 5\cos(100t)$$

8–10 Given a sinusoid $v_1(t)$ whose phasor is $\mathbf{V}_1 = -3 + j4$ V, use phasor methods to find the voltage $v_2(t)$ that leads $v_1(t)$ by 90° and has an amplitude of 10 V.

OBJECTIVE 8–2 EQUIVALENT IMPEDANCE (SECTS. 8–2, 8–3)

Given a linear circuit, use series and parallel equivalence to find the equivalent impedance at a specified pair of terminals. See Examples 8–5, 8–6, 8–9, 8–10, 8–12 and Exercises 8–8, 8–10, 8–11

8–11 Find the equivalent impedance Z in Figure P8–11 when $\omega = 1000$ rad/s. Express the result in both polar and rectangular form.

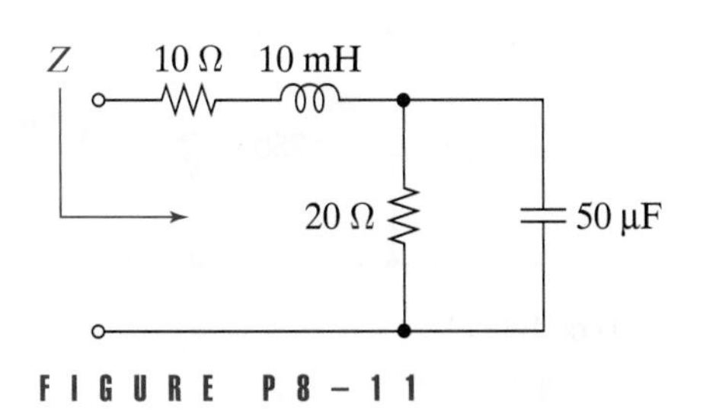

FIGURE P8–11

8–12 Find the equivalent impedance Z in Figure P8–12. Express the result in both polar and rectangular form.

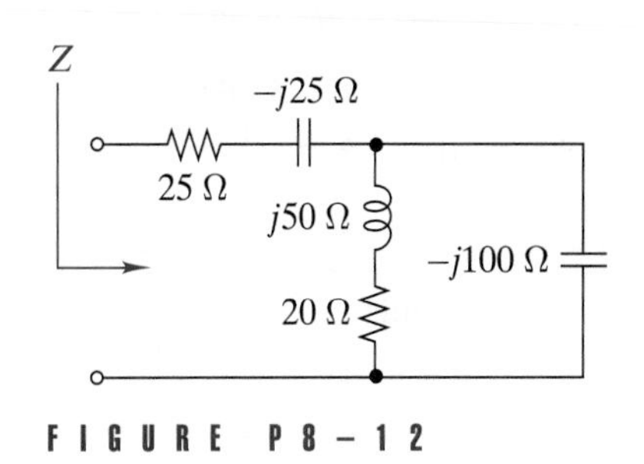

FIGURE P8–12

8–13 Find the equivalent impedance Z in Figure P8–13 when $\omega = 10$ krad/s. Express the result in both polar and rectangular form.

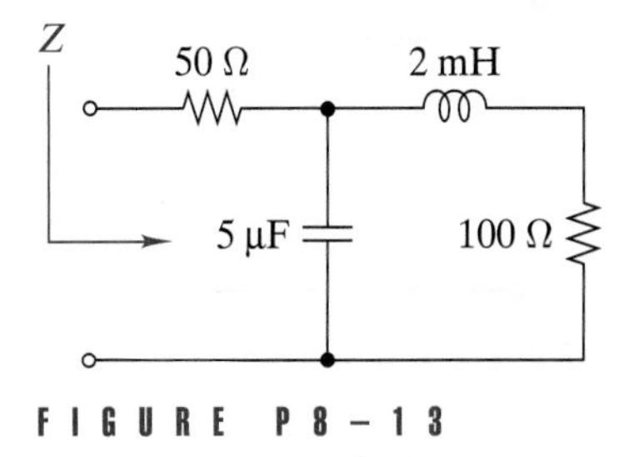

FIGURE P8–13

8–14 Find the equivalent impedance Z in Figure P8–14. Express the result in both polar and rectangular form.

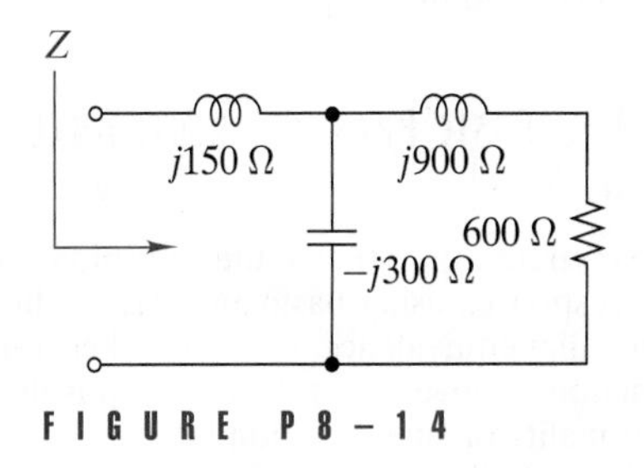

FIGURE P8–14

8–15 The circuit in Figure P8–14 is operating in the sinusoidal steady state with $\omega = 3$ krad/s. How would the element impedances change if the steady-state frequency is reduced to 1 krad/s? What is the equivalent impedance Z at this new frequency?

8–16 The circuit in Figure P8–16 is operating in the sinusoidal steady state with $\omega = 20$ krad/s. Find the equivalent impedance Z.

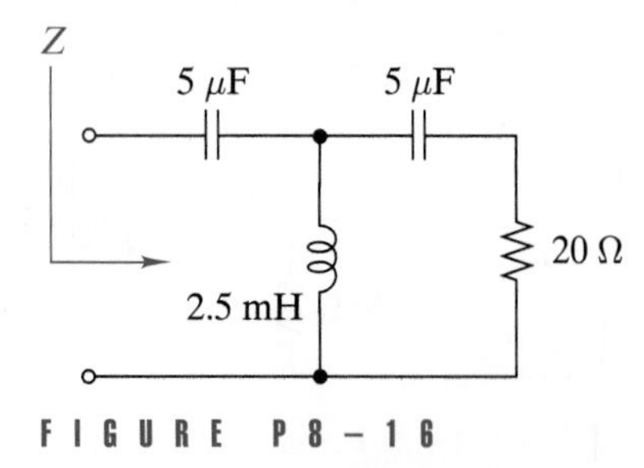

FIGURE P8–16

8–17 Repeat Problem 8–16 with $\omega = 10$ krad/s.

8–18 The equivalent impedance in Figure P8–18 is known to be $Z = 60 + j180\ \Omega$. Find the impedance of the inductor.

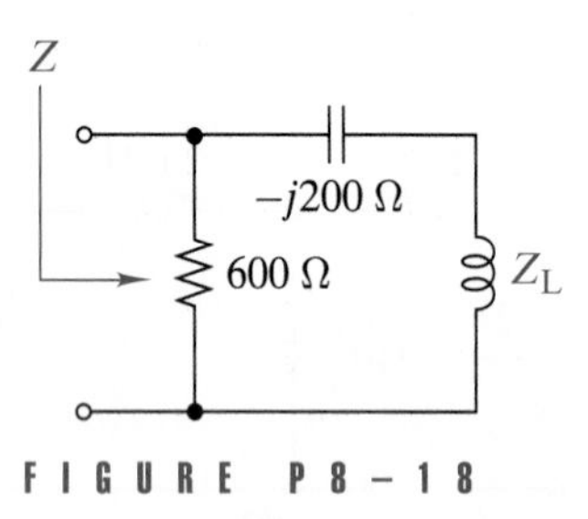

FIGURE P8–18

8–19 D A capacitor C is connected in parallel with a resistor R. Select values of R and C so that the equivalent impedance of the parallel combination is $300 - j400\ \Omega$ at $\omega = 1$ Mrad/s.

8–20 An 800-Ω resistor is connected in parallel with a 1-nF capacitor. The impedance of the parallel combination is $400 - j400\ \Omega$. Find the frequency.

OBJECTIVE 8–3 BASIC PHASOR CIRCUIT ANALYSIS (SECTS. 8–3, 8–4)

Given a linear circuit operating in the sinusoidal steady state, find phasor responses using basic analysis methods such as series and parallel equivalence, voltage and current division, circuit reduction, Thévenin or Norton equivalent circuits, and proportionality or superposition.
See Examples 8–6, 8–8, 8–9, 8–10, 8–13, 8–14, 8–15, 8–17, 8–18, 8–19 and Exercises 8–9, 8–13, 8–14, 8–15

8–21 A voltage $v(t) = 50\cos(2000t)$ V is applied across a series connection of a 100-Ω resistor and 50-mH inductor. Find the steady-state current $i(t)$ through the series connection.

8–22 The circuit in Figure P8–22 is operating in the sinusoidal steady state with $v_S(t) = V_A\cos(\omega t)$. Derive a general expression for the phasor response $\mathbf{I}_L$.

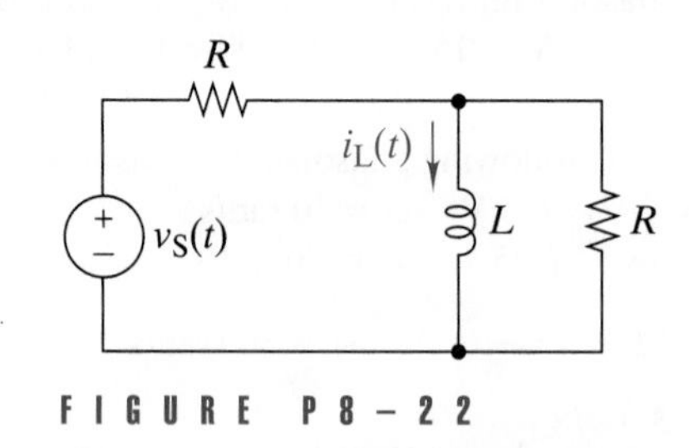

FIGURE P8–22

8–23 A current source delivering $i(t) = 300\cos(2500t)$ mA is connected across a parallel combination of a 1-kΩ resistor and a 400-nF capacitor. Find the steady-state current $i_R(t)$ through the resistor and the steady-state current $i_C(t)$ through the capacitor. Draw a phasor diagram showing $\mathbf{I}$, $\mathbf{I}_C$, and $\mathbf{I}_R$.

8–24 The circuit in Figure P8–24 is operating in the sinusoidal steady state with $i_S(t) = I_A\cos(\omega t)$. Derive a general expression for the steady-state response $\mathbf{V}_R$.

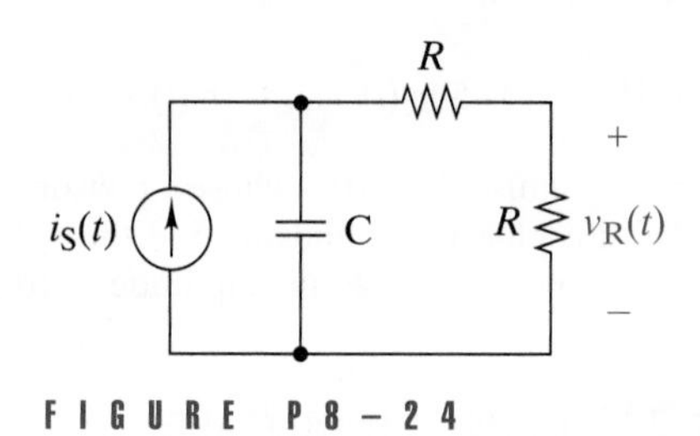

FIGURE P8–24

8–25 A voltage $v(t) = 50\sin(2000t)$ V is applied across a parallel connection of a 5-kΩ resistor and a 100-nF capacitor. Find the steady-state current $i_C(t)$ through the capacitor and the steady-state current $i_R(t)$ through the resistor. Draw a phasor diagram showing $\mathbf{V}$, $\mathbf{I}_C$, and $\mathbf{I}_R$.

8–26 The circuit in Figure P8–26 is operating in the sinusoidal steady state. Find the steady-state response $v_x(t)$.

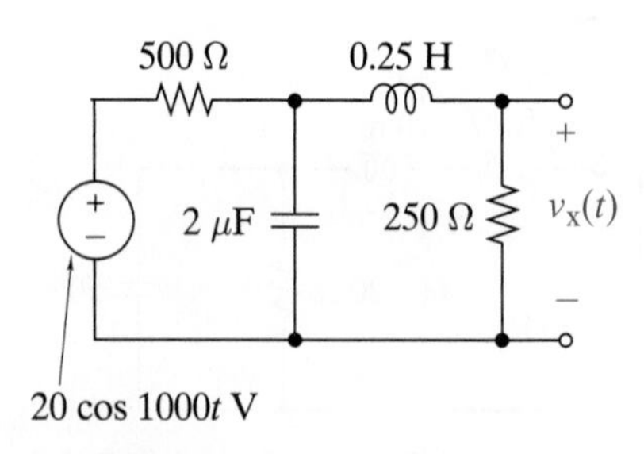

FIGURE P8–26

8–27 The circuit in Figure P8–27 is operating in the sinusoidal steady state. Find the steady-state response $v_x(t)$.

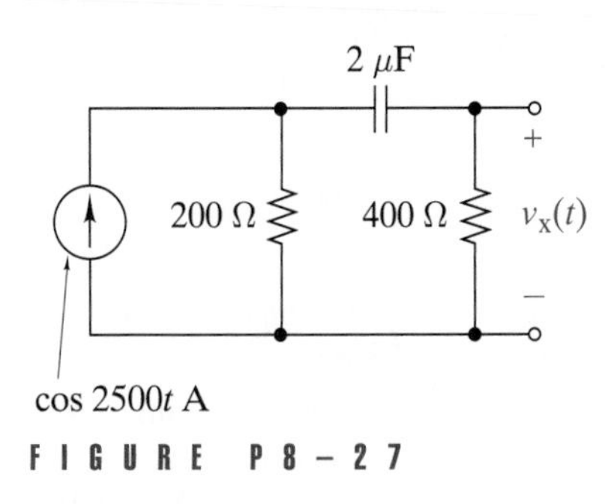

FIGURE P8–27

8–28 The circuit in Figure P8–28 is operating in the sinusoidal steady state. Find the steady-state phasor response $\mathbf{V}_x$.

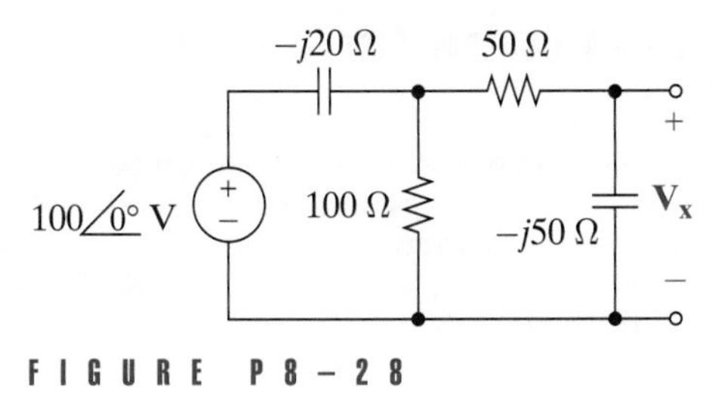

FIGURE P8–28

8–29 The circuit in Figure P8–29 is operating in the sinusoidal steady state. Use superposition to find the phasor response $\mathbf{I}_x$.

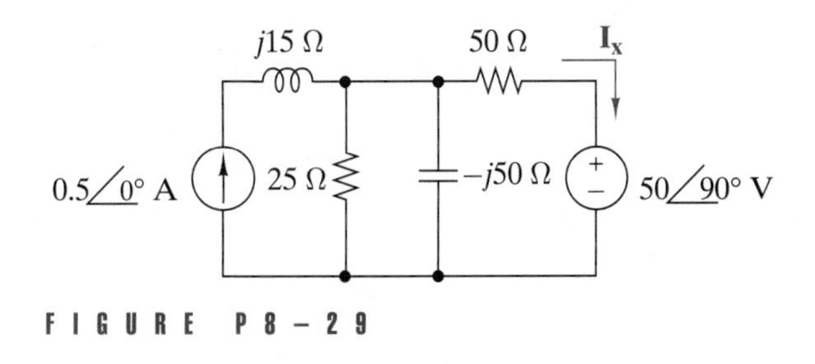

FIGURE P8–29

8–30 The circuit in Figure P8–30 is operating in the sinusoidal steady state. Use superposition to find the response $v_x(t)$. *Note:* The sources do not have the same frequency.

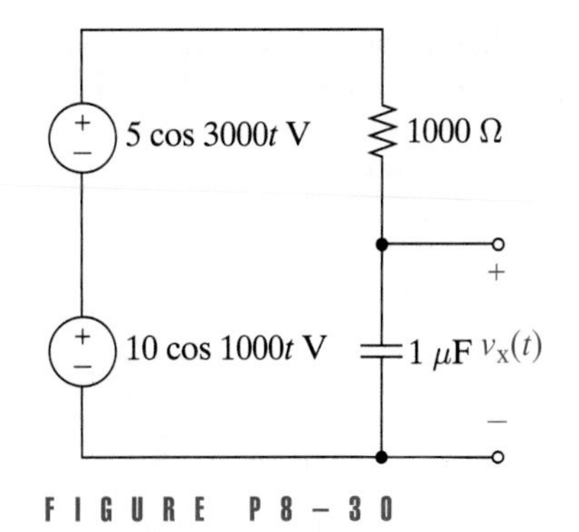

FIGURE P8–30

8–31 The circuit in Figure P8–31 is operating in the sinusoidal steady state. Use superposition to find the response $v_x(t)$. *Note:* The sources do not have the same frequency.

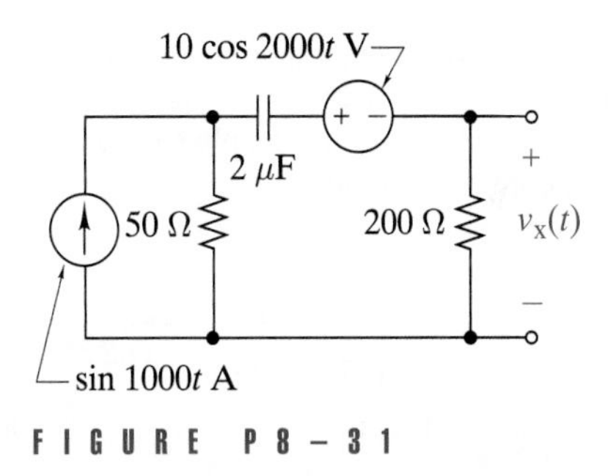

FIGURE P8–31

8–32 The circuit in Figure P8–32 is operating in the sinusoidal steady state. Find the phasor response $\mathbf{V}_x$.

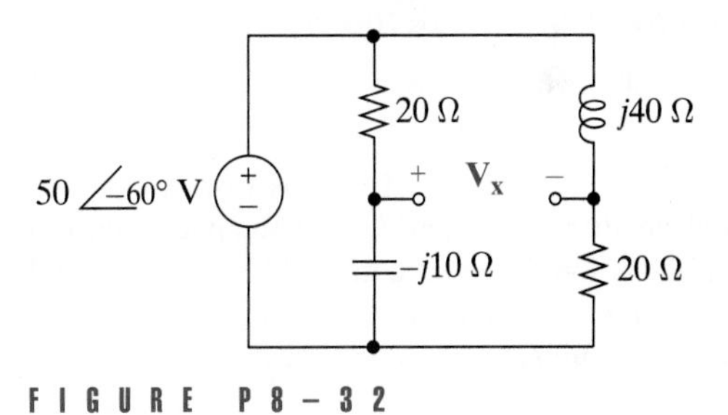

FIGURE P8–32

8–33 The circuit in Figure P8–33 is operating in the sinusoidal steady state. Use the unit output method to find the phasor response $\mathbf{V}_x$.

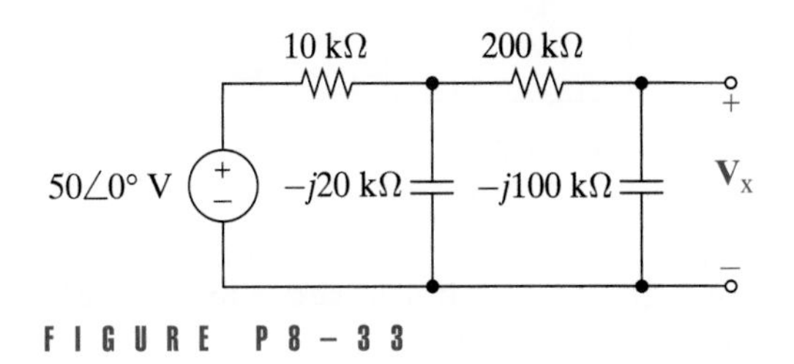

FIGURE P8–33

8–34 Find the phasor Thévenin equivalent of the source circuit to the left of the interface in Figure P8–34. Then use the equivalent circuit to find the steady-state voltage $v(t)$ and current $i(t)$ delivered to the load.

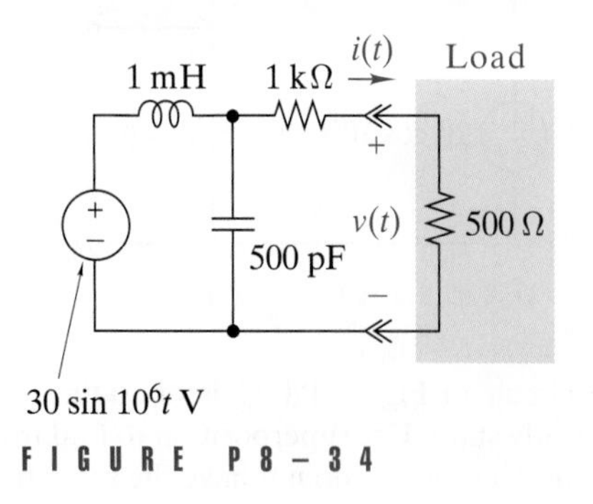

FIGURE P8–34

8–35 Find the phasor Thévenin equivalent of the source circuit to the left of the interface in Figure P8–35. Then use the equivalent circuit to find the phasor voltage **V** and current **I** delivered to the load.

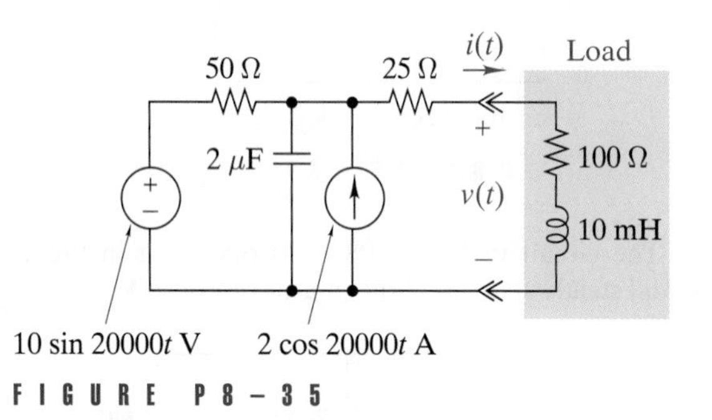

FIGURE P8–35

8–36 The circuit in Figure P8–36 is operating in the sinusoidal steady state. When $Z_L = 0$, the phasor current at the interface is $\mathbf{I} = 4.8 - j3.6$ mA. When $Z_L = -j20$ kΩ, the phasor interface current is $\mathbf{I} = 10 + j0$ mA. Find the Thévenin equivalent of the source circuit.

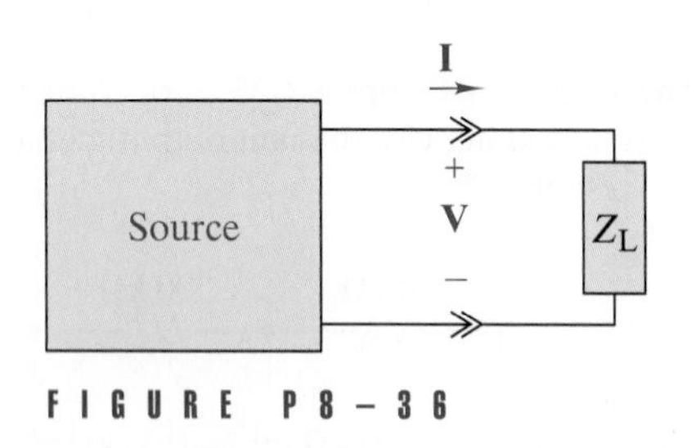

FIGURE P8–36

8–37 Use a Thévenin equivalent circuit to find the phasor response $\mathbf{V}_x$ in Figure P8–37.

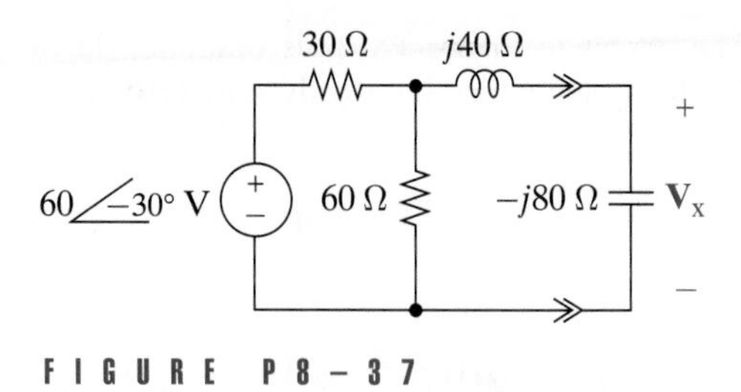

FIGURE P8–37

8–38 Use a Thévenin equivalent circuit to find the phasor response $\mathbf{V}_x$ in Figure P8–33.

8–39 **D** Design a two-port circuit so that an input voltage $v_S(t) = 100\cos(10^4 t)$ V delivers a steady-state output current of $i_O(t) = 10\cos(10^4 t - 30°)$ mA to a 100-Ω resistive load.

8–40 **D** Design a two-port circuit so that an input voltage $v_S(t) = 100\sin(1000t)$ V delivers a steady-state output voltage of $v_O(t) = 50\cos(1000t - 45°)$ V.

OBJECTIVE 8–4 GENERAL CIRCUIT ANALYSIS (SECT. 8–5)

Given a linear circuit operating in the sinusoidal steady state, find equivalent impedances and phasor responses using node-voltage or mesh-current analysis.
See Examples 8–20, 8–21, 8–22, 8–23, 8–24, 8–25, 8–26 and Exercises 8–16, 8–17, 8–18

8–41 The circuit in Figure P8–41 is operating in the sinusoidal steady state with ω = 4 krad/s. Use node-voltage analysis to find the steady-state response $v_x(t)$.

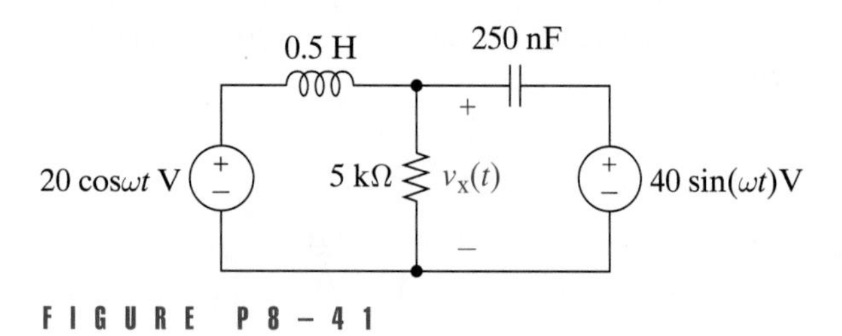

FIGURE P8–41

8–42 Use node-voltage analysis to find the steady-state phasor response $\mathbf{V}_O$ in Figure P8–42.

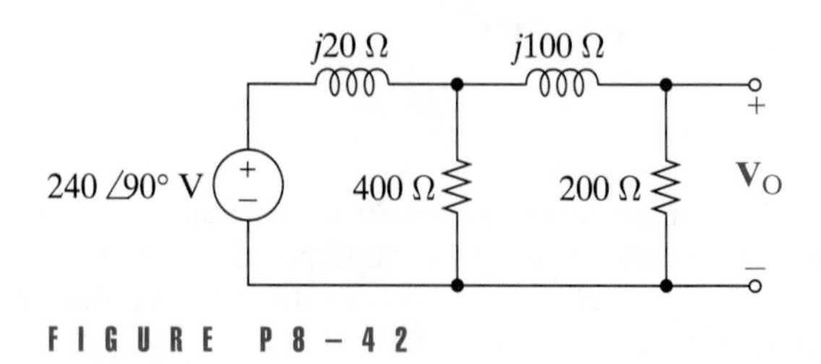

FIGURE P8–42

8–43 Find the node voltage phasors $\mathbf{V}_A$ and $\mathbf{V}_B$ in Figure P8–43.

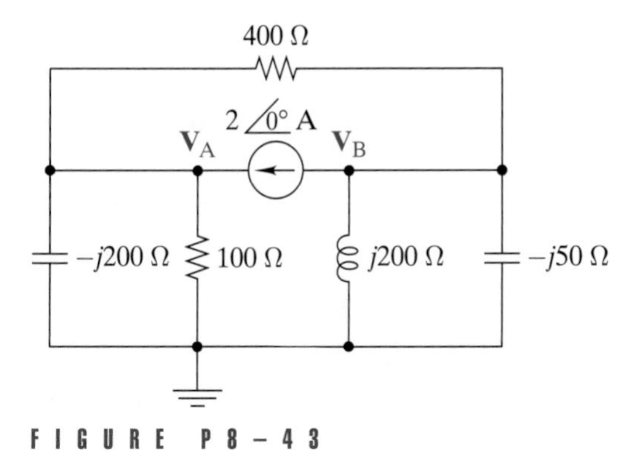

FIGURE P8–43

8–44 Use mesh-current analysis to find the phasor branch currents $\mathbf{I}_1$, $\mathbf{I}_2$, and $\mathbf{I}_3$ in the circuit shown in Figure P8–44.

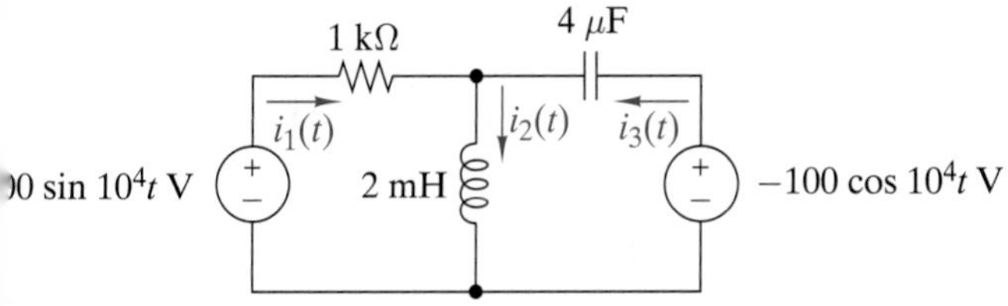

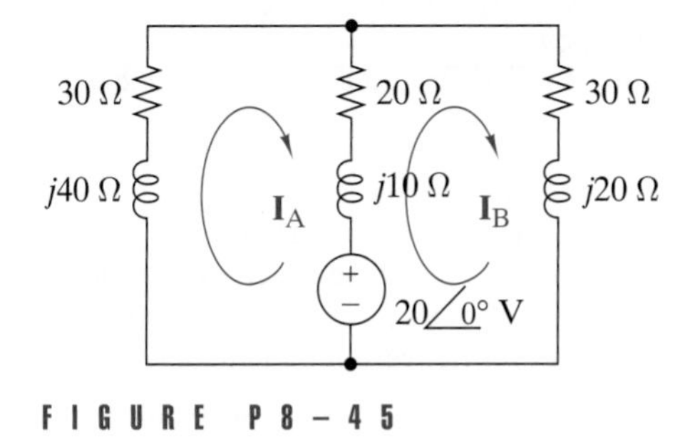

FIGURE P8–44

8–45 Use mesh-current to find the phasor currents $\mathbf{I}_A$ and $\mathbf{I}_B$ in Figure P8–45.

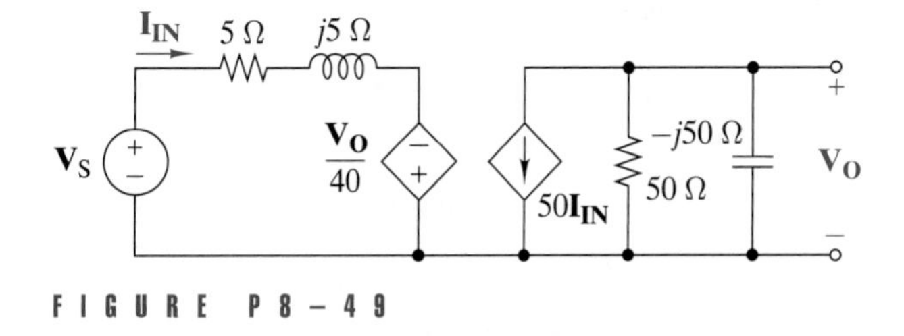

FIGURE P8–45

8–46 Use mesh-current analysis to solve Problem 8–42.

8–47 Find the phasor current $\mathbf{I}_O$ in Figure P8–47.

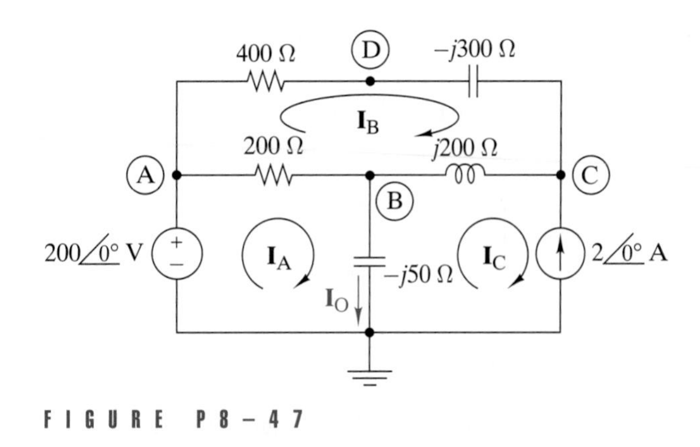

FIGURE P8–47

8–48 Find the phasor outputs $\mathbf{V}_O$ and $\mathbf{I}_O$ in Figure P8–48 when $\mu = 10$ and the phasor input is $\mathbf{I}_S = 1 + j0$ mA.

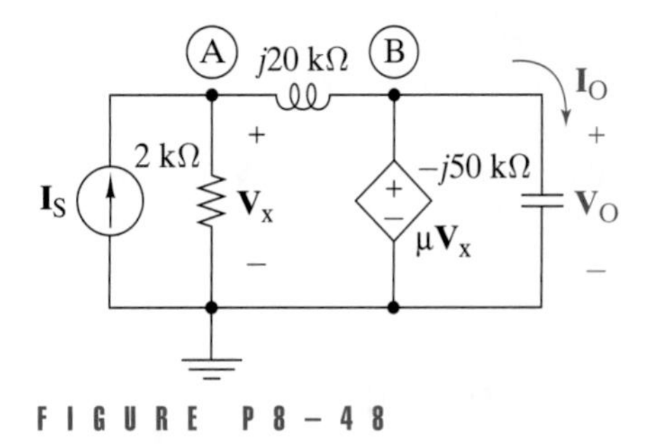

FIGURE P8–48

8–49 Find the phasor responses $\mathbf{I}_{IN}$ and $\mathbf{V}_O$ in Figure P8–49 when $\mathbf{V}_S = 1 + j0$ V.

FIGURE P8–49

8–50 The OP AMP circuit in Figure P8–50 is operating in the sinusoidal steady state with $\omega = 100$ krad/s. Find the phasor input $\mathbf{V}_S$ when the phasor output is $\mathbf{V}_O = 10 + j0$ V.

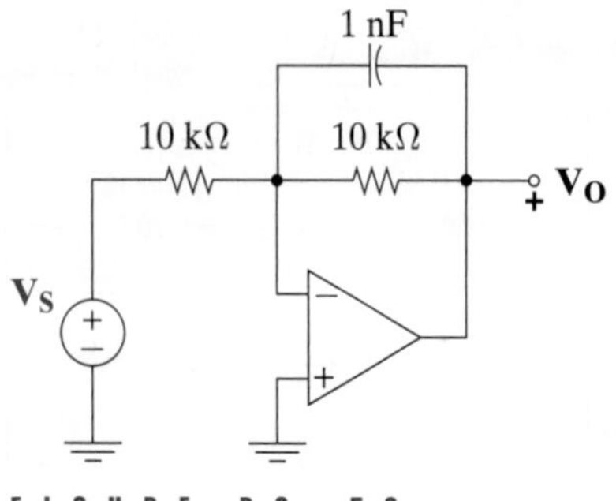

FIGURE P8–50

8–51 Find the phasor input $\mathbf{V}_S$ in Figure P8–51 when the phasor output is $\mathbf{V}_O = 200 + j100$ V.

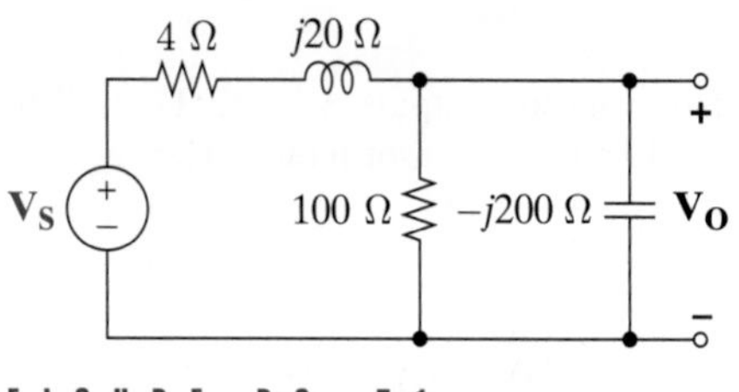

FIGURE P8–51

8–52 Find the phasor gain $K = \mathbf{V}_O/\mathbf{V}_S$ and input impedance Z_{IN} of the circuit in Figure P8–52 with $\mu = 100$.

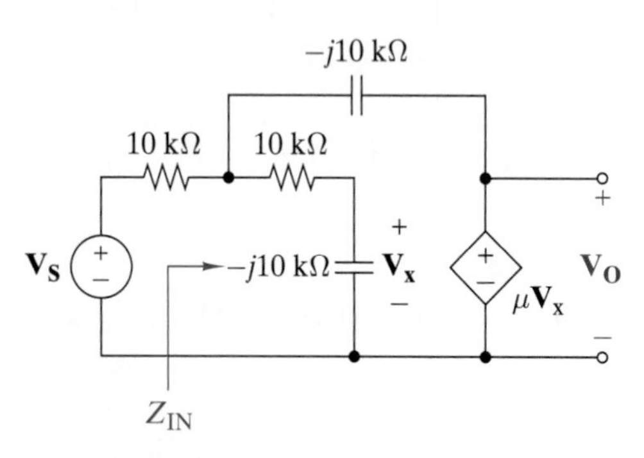

FIGURE P8–52

8–53 Find the phasor gain $K = \mathbf{V}_O/\mathbf{V}_S$ and input impedance Z_{IN} of the circuit in Figure P8–53.

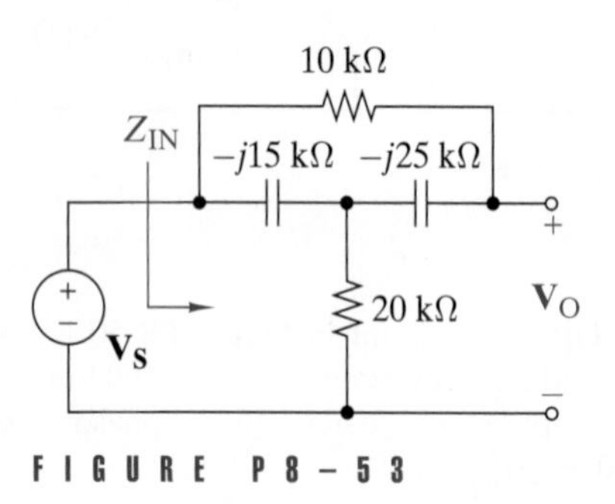

FIGURE P8–53

8–54 Find the phasor gain $K = \mathbf{V}_O/\mathbf{V}_S$ and input impedance Z_{IN} of the circuit in Figure P8–54.

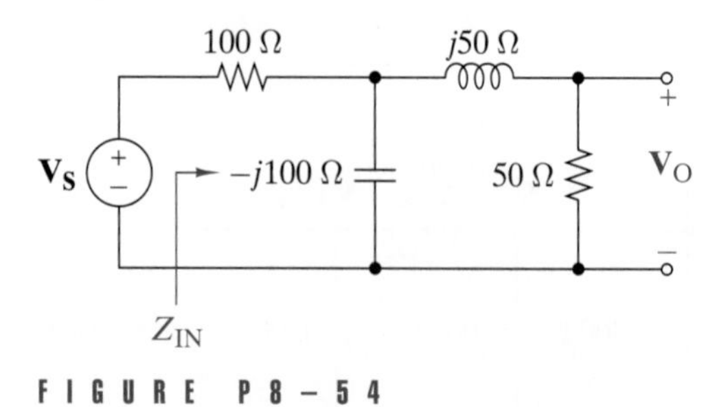

FIGURE P8–54

8–55 Find the phasor gain $K = \mathbf{V}_O/\mathbf{V}_S$ and input impedance Z_{IN} of the circuit in Figure P8–55.

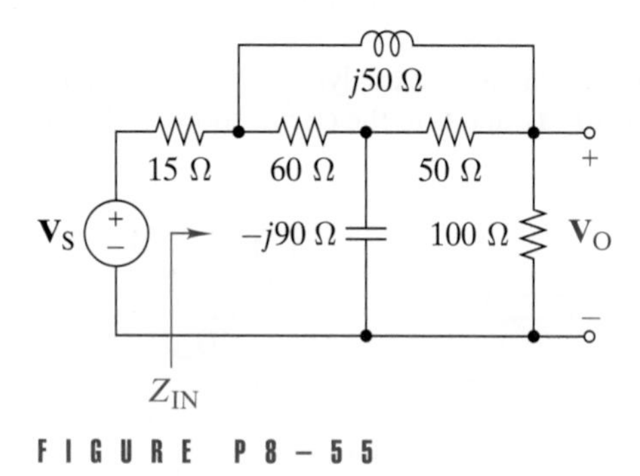

FIGURE P8–55

OBJECTIVE 8–5 AVERAGE POWER AND MAXIMUM POWER TRANSFER (SECT. 8–6)

Given a linear circuit operating in the sinusoidal steady state:
(a) Find the average power delivered at a specified interface.
(b) Find the maximum average power available at a specified interface.
(c) Find the load impedance required to draw the maximum available power.
See Examples 8–27, 8–28 and Exercises 8–19, 8–20

8–56 A load consisting of a 500-Ω resistor in series with a 2-μF capacitor is connected across a voltage source $v_S(t) = 150\cos(500t)$ V. Find the phasor voltage, current, and average power delivered to the load.

8–57 A load consisting of a 50-Ω resistor in parallel with a 8-μF capacitor is connected across a current source delivering $i_S(t) = 50\cos(2500t)$ mA. Find the average power delivered to the load.

8–58 Find the average power delivered to the 50-Ω resistor in Figure P8–58.

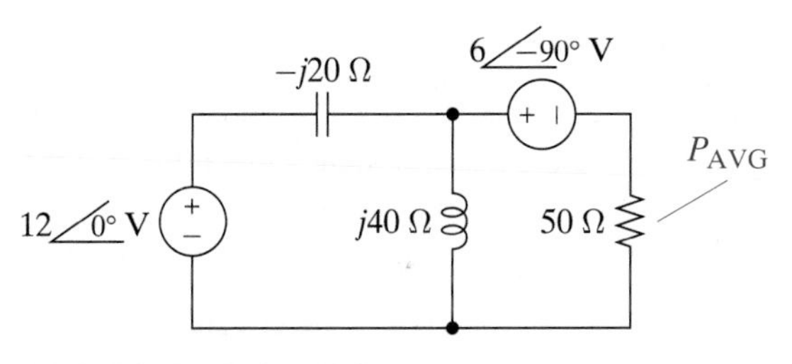

FIGURE P8–58

8–59 **(a)** Find the average power delivered to the load in Figure P8–59.
(b) Find the maximum available average power at the interface shown in the figure.
(c) Specify the load required to extract the maximum average power.

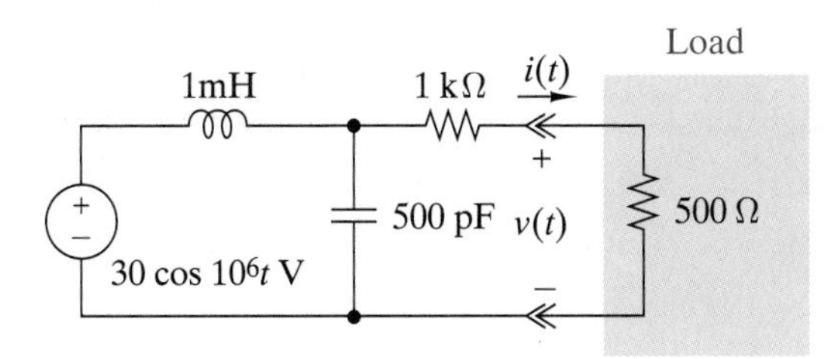

FIGURE P8–59

8–60 **(a)** Find the maximum average power available at the interface in Figure P8–60.
(b) Specify the values of R and C that will extract the maximum power from the source circuit.

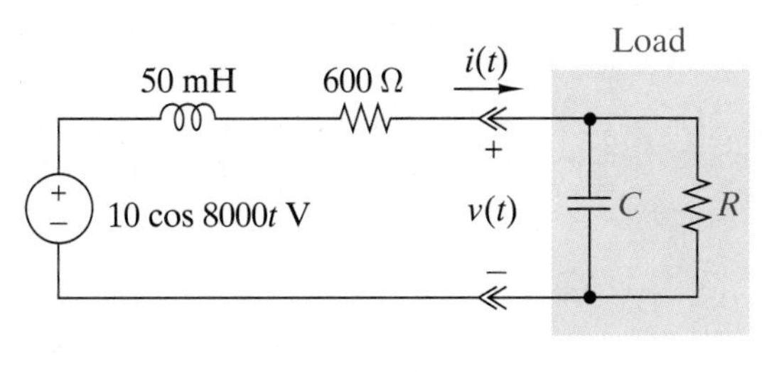

FIGURE P8–60

INTEGRATING PROBLEMS

8–61 **A** AC Voltage Measurement
An ac voltmeter measurement indicates the amplitude of a sinusoid, not its phase angle. The magnitude and phase can be inferred by making several measurements and using KVL. For example, Figure P8–61 shows a relay coil of unknown resistance and inductance. The following ac voltmeter readings are taken with the circuit operating in the sinusoidal steady state at $f = 1$ kHz: $|\mathbf{V}_S| = 12$ V, $|\mathbf{V}_1| = 5$ V, and $|\mathbf{V}_2| = 9$ V. Find R and L.

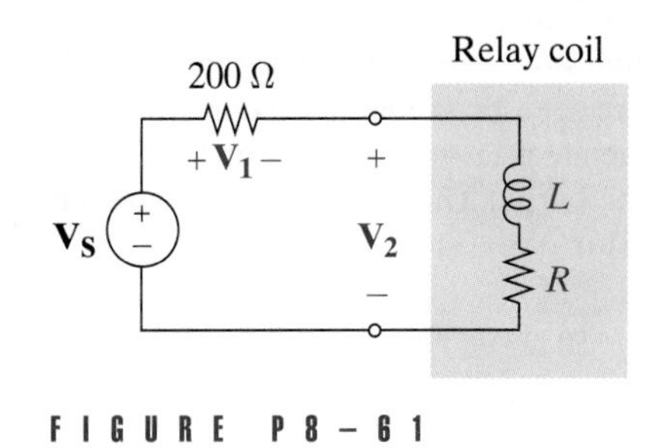

FIGURE P8–61

8–62 **A** OP AMP Circuits
The analysis methods used with resistive OP AMP circuits can be extended to dynamic OP AMP circuits using the concept of element impedances. For example, the circuit in Figure P8–62 is operating in the sinusoidal steady state with $v_S(t) = 5\cos(10^3 t)$ V. Find $v_O(t)$ using an extension of the methods developed in Chapter 4.

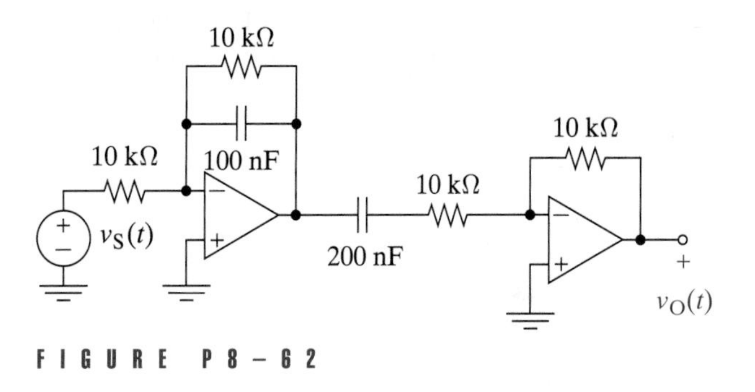

FIGURE P8–62

8–63 **A** Power Transmission Efficiency
A power transmission circuit with a source voltage of $\mathbf{V}_S = 440 + j0$ V can be modeled as shown in Figure P8–63. Find the average power produced by the source, lost in the wires, and delivered to the load. What is the transmission efficiency?

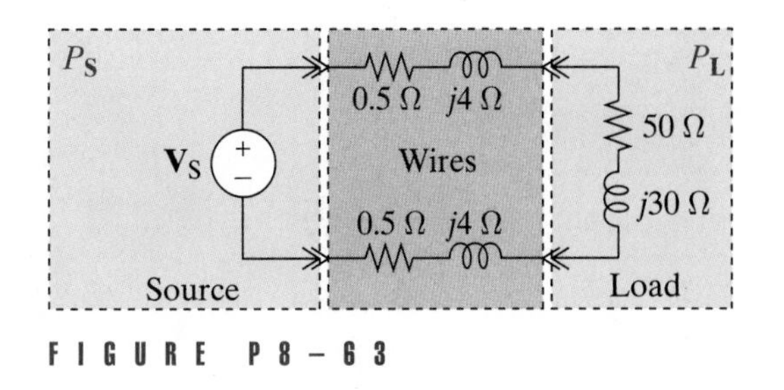

FIGURE P8–63

8–64 A Average Stored Energy
In the sinusoidal steady state the phasor voltage across a capacitor is $\mathbf{V}_C = V_A + j0$. Derive an expression for the average energy stored in the capacitor.

8–65 D AC Circuit Design
Select values of L and C in Figure P8–65 so that the input impedance seen by the voltage source is $50 + j0\ \Omega$ when the frequency is $\omega = 10^6$ rad/s. For these values of L and C, find the output Thévenin impedance seen by the 600-Ω load resistor.

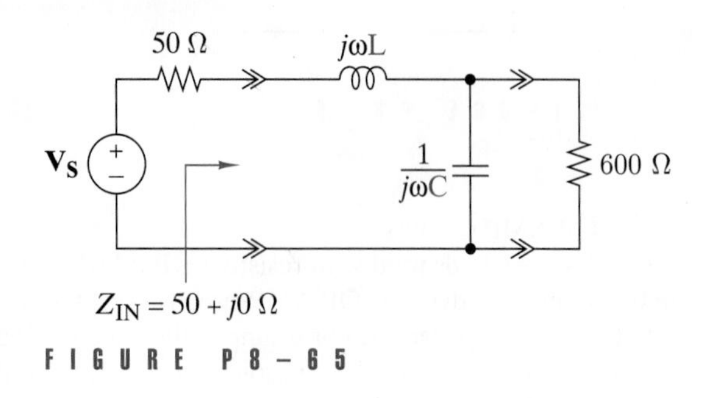

FIGURE P8–65

8–66 E AC Circuit Analysis
Ten years after graduating with a BSEE, you decide to go to graduate school for a master's degree. In desperate need of income, you agree to sign on as a grader in the basic circuit analysis course. One of the problems asks the students to find $v(t)$ in Figure P8–66 when the circuit operates in the sinusoidal steady state. One of the students offers the following solution:

$$\begin{aligned} v(t) &= (R + j\omega L) \times i(t) \\ &= (20 + j20) \times 0.5 \cos 200\,t \\ &= 10 \cos 200t + j10 \cos 200\,t \\ &= 10\sqrt{2} \cos(200\,t + 45°) \end{aligned}$$

Is the answer correct? If not, what grade would you give the student? If correct, what comments would you give the student about the method of solution?

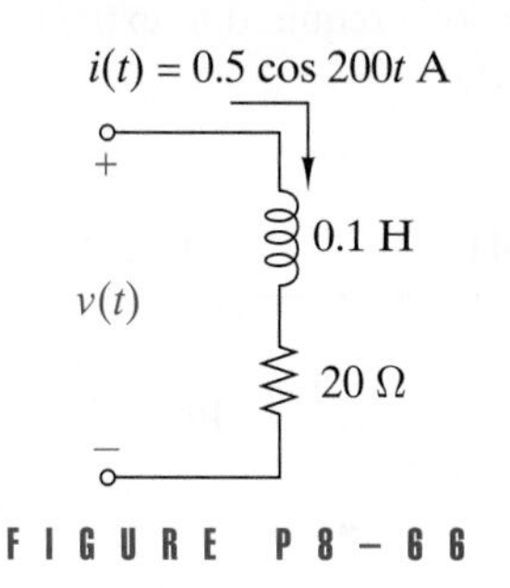

FIGURE P8–66

CHAPTER 9

LAPLACE TRANSFORMS

My method starts with a complex integral; I fear this sounds rather formidable; but it is really quite simple . . . I am afraid that no physical people will ever try to make out my method: but I am hoping that it may give them confidence to try your methods.

Thomas John Bromwich, 1915,
British Mathematician

Some History Behind This Chapter

Laplace transforms have their roots in the pioneering work of the eccentric British engineer Oliver Heaviside (1850–1925). His operational calculus was essentially a collection of intuitive rules that allowed him to formulate and solve important technical problems of his day. His intuitive approach drew bitter criticism from the scientists of his day. Eventually individuals like John Bromwich recognized the importance of Heaviside's methods and began to develop the necessary mathematical foundations. A complete development was eventually discovered in the 1780 writings of the French mathematician Pierre Simon Laplace.

Why This Chapter Is Important Today

The difficulties of finding transient responses using classical differential equations are avoided when we apply the techniques of Laplace transforms. These techniques not only simplify the solution of circuit differential equations but also give us a deeper insight into circuit behavior. Laplace transforms make the analysis and design of circuits easy and fun.

Chapter Sections

9–1 Signal Waveforms and Transforms
9–2 Basic Properties and Pairs
9–3 Pole-Zero Diagrams
9–4 Inverse Laplace Transforms
9–5 Some Special Cases
9–6 Circuit Response Using Laplace Transforms
9–7 The Initial and Final Value Properties

Chapter Learning Objectives

9-1 Laplace Transform (Sects. 9-1, 9-2, 9-3)

Find the Laplace transform of a given signal waveform using transform properties and pairs, or using the integral definition of the Laplace transformation. Locate the poles and zeros of the transform and construct a pole-zero diagram.

9-2 Inverse Transforms (Sects. 9-4, 9-5)

Find the signal waveform corresponding to a given Laplace transform using partial-fraction expansion or basic transform properties and pairs.

9-3 Circuit Response Using Laplace Transforms (Sect. 9-6)

Given a first- or second-order circuit:

(a) Determine the circuit differential equation and the initial conditions (if not given).
(b) Transform the differential equation into the s domain and solve for the response transform.
(c) Use the inverse transformation to find the response waveform.

9-4 Initial and Final Value Properties (Sect. 9-7)

Given the Laplace transform of a signal, find the initial and final values of the signal waveform.

9–1 SIGNAL WAVEFORMS AND TRANSFORMS

A mathematical transformation employs rules to change the form of data without altering its meaning. An example of a transformation is the conversion of numerical data from decimal to binary form. In engineering circuit analysis, transformations are used to obtain alternative representations of circuits and signals. These alternate forms provide a different perspective that can be quite useful or even essential. Examples of the transformations used in circuit analysis are the Fourier transformation, the Z-transformation, and the Laplace transformation. These methods all involve specific transformation rules, make certain analysis techniques more manageable, and provide a useful viewpoint for circuit and system design.

This chapter deals with the Laplace transformation. The discussion of the Laplace transformation follows the path shown in Figure 9–1 by the solid arrow. The process begins with a linear circuit. We derive a differential equation describing the circuit response and then transform this equation into the frequency domain, where it becomes an algebraic equation. Algebraic techniques are then used to solve the transformed equation for the circuit response. The inverse Laplace transformation then changes the frequency-domain response into the response waveform in the time domain. The dashed arrow in Figure 9–1 shows that there is another route to the time-domain response using the classical techniques discussed in Chapter 7. The classical approach appears to be more direct, but the advantage of the Laplace transformation is that solving a differential equation becomes an algebraic process.

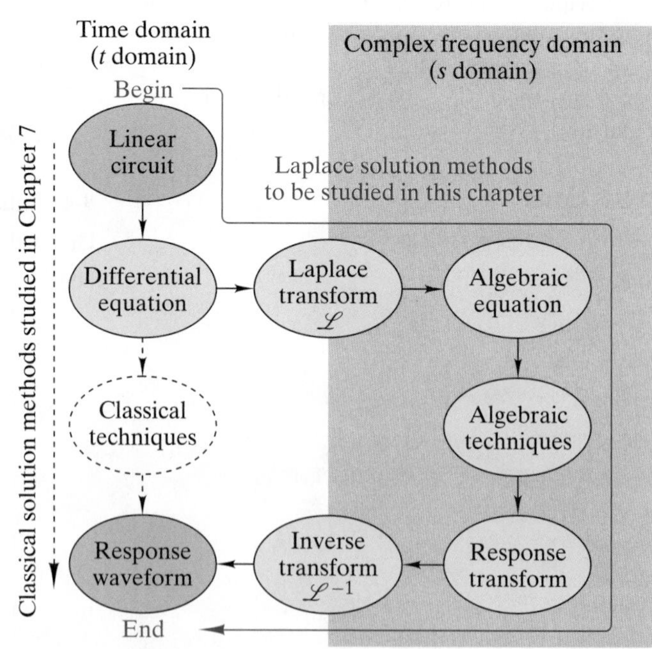

FIGURE 9–1 *Flow diagram dynamic circuit analysis with Laplace transforms.*

Symbolically, we represent the Laplace transformation as

$$\mathscr{L}\{f(t)\} = F(s) \tag{9–1}$$

This expression states that $F(s)$ is the Laplace transform of the waveform $f(t)$. The transformation operation involves two domains: (1) the time domain, in which the signal is characterized by its **waveform** $f(t)$, and (2) the complex frequency domain, in which the signal is represented by its **transform** $F(s)$.

The symbol s stands for the complex frequency variable, a notation we first introduced in Chapter 7 in connection with the zero-state response of linear circuits. The variable s has the dimensions of reciprocal time, or frequency, and is expressed in units of radians per second. In this chapter the complex frequency variable is written as $s = \sigma + j\omega$, where $\sigma = \text{Re}\{s\}$ is the real part and $\omega = \text{Im}\{s\}$ is the imaginary part. This variable is the independent variable in the s domain, just as t is the independent variable in the time domain. Although we cannot physically measure complex frequency in the same sense that we measure time, it is an extremely useful concept that pervades the analysis and design of linear systems.

A signal can be expressed as a waveform or a transform. Collectively, $f(t)$ and $F(s)$ are called a **transform pair**, where the pair involves two representations of the signal. To distinguish between the two forms, a lowercase letter denotes a waveform and an uppercase a transform. For electrical waveforms such as current $i(t)$ or voltage $v(t)$, the corresponding transforms are denoted $I(s)$ and $V(s)$. In this chapter we will use $f(t)$ and $F(s)$ to stand for signal waveforms and transforms in general.

The **Laplace transformation** is defined by the integral

$$F(s) = \int_{0-}^{\infty} f(t)e^{-st}dt \tag{9–2}$$

Since the definition involves an improper integral (the upper limit is infinite), we must discuss the conditions under which the integral exists (converges). The integral exists if the waveform $f(t)$ is piecewise continuous and of exponential order. **Piecewise continuous** means that $f(t)$ has a finite number of steplike discontinuities in any finite interval. **Exponential order** means that constants K and b exist such that $|f(t)| < Ke^{bt}$ for all $t > 0$. As a practical matter, the signals encountered in engineering applications meet these conditions.

From the integral definition of the Laplace transformation, we see that when a voltage waveform $v(t)$ has units of volts (V) the corresponding voltage transform $V(s)$ has units of volt-seconds (V-s). Similarly, when a current waveform $i(t)$ has units of amperes (A), the corresponding current transform $I(s)$ has units of ampere-seconds (A-s). Thus, waveforms and transforms do not have the same units. Even so, we often refer to both $V(s)$ and $v(t)$ as voltages and both $I(s)$ and $i(t)$ as currents despite the fact that they have different units. The reason is simply that it is awkward to keep adding the words *waveform* and *transform* to statements when the distinction is clear from the context.

Equation (9–2) uses a lower limit denoted $t = 0-$ to indicate a time just a whisker before $t = 0$. We use $t = 0-$ because in circuit analysis $t = 0$ is defined by a discrete event, such as closing a switch. Such an event may cause a discontinuity in $f(t)$ at $t = 0$. To capture this discontinuity, we set the lower limit at $t = 0-$, just prior to the event. Fortunately, in many situations there is no discontinuity so we will not distinguish between $t = 0-$ and $t = 0$ unless it is crucial.

Equally fortunate is the fact that the number of different waveforms encountered in linear circuits is relatively small. The list includes the three basic waveforms from Chapter 5 (the step, exponential, and sinusoid), as well as composite waveforms such as the impulse, ramp, damped ramp, and damped sinusoid. Since the number of waveforms of interest is relatively small, we do not often use the integral definition in Eq. (9–2) to find Laplace transforms. Once a transform pair has been found, it can be cataloged in a table for future reference and use. Table 9–2 in this chapter is sufficient for our purposes.

EXAMPLE 9–1

Show that the Laplace transform of the unit step function $f(t) = u(t)$ is $F(s) = 1/s$.

SOLUTION:

Applying Eq. (9–2) yields

$$F(s) = \int_0^\infty u(t)e^{-st}dt$$

Since $u(t) = 1$ throughout the range of integration this integral becomes

$$F(s) = \int_0^\infty e^{-st}dt = -\frac{e^{-st}}{s}\bigg|_0^\infty = -\frac{e^{-(\sigma+j\omega)t}}{\sigma + j\omega}\bigg|_0^\infty$$

The last expression on the right side vanishes at the upper limit since e^{-st} goes to zero as t approaches infinity provided that $\sigma > 0$. At the lower limit the expression reduces to $1/s$. The integral used to calculate $F(s)$ is only valid in the region for which $\sigma > 0$. However, once evaluated, the result $F(s) = 1/s$ can be extended to neighboring regions provided that we avoid the point at $s = 0$ where the function blows up. ■

EXAMPLE 9–2

Show that the Laplace transform $f(t) = [e^{-\alpha t}]u(t)$ is $F(s) = 1/(s + \alpha)$.

SOLUTION:

Applying Eq. (9–2) yields

$$F(s) = \int_0^\infty e^{-\alpha t}e^{-st}\,dt = \int_0^\infty e^{-(s+\alpha)t}dt = \frac{e^{-(s+\alpha)t}}{-(s+\alpha)}\bigg|_0^\infty$$

The last term on the right side vanishes at the upper limit since $e^{-(s+\alpha)t}$ vanishes as t approaches infinity provided that $\sigma > \alpha$. At the lower limit the last term reduces to $1/(s + \alpha)$. Again, the integral is only valid for a limited region, but the result $F(s) = 1/(s + \alpha)$ can be extended outside this region if we avoid the point at $s = -\alpha$. ■

EXAMPLE 9–3

Show that the Laplace transform of the impulse function $f(t) = \delta(t)$ is $F(s) = 1$.

SOLUTION:
Applying Eq. (9–2) yields

$$F(s) = \int_{0-}^{\infty} \delta(t)e^{-st}\,dt = \int_{0-}^{0+} \delta(t)e^{-st}\,dt = \int_{0-}^{0+} \delta(t)dt = 1$$

The difference between $t = 0-$ and $t = 0$ is important since the impulse is zero everywhere except at $t = 0$. To capture the impulse in the integration, we take a lower limit at $t = 0-$ and an upper limit at $t = 0+$. Since $e^{-st} = 1$ and $\int \delta(t)dt = 1$ on this integration interval, we find that $F(s) = 1$. ■

INVERSE TRANSFORMATION

So far we have used the direct transformation to convert waveforms into transforms. But Figure 9–1 points out the need to perform the inverse transformation to convert transforms into waveforms. Symbolically, we represent the inverse process as

$$\mathcal{L}^{-1}\{F(s)\} = f(t) \tag{9–3}$$

This equation states that $f(t)$ is the inverse Laplace transform of $F(s)$. The **inverse Laplace transformation** is defined by the complex inversion integral

$$f(t) = \frac{1}{2\pi j}\int_{\alpha - j\infty}^{\alpha + j\infty} F(s)e^{st}\,ds \tag{9–4}$$

The Laplace transformation is an integral transformation since both the direct process in Eq. (9–2) and the inverse process in Eq. (9–4) involve integrations.

Happily, formal evaluation of the complex inversion integral is not necessary because of the uniqueness property of the Laplace transformation. A symbolical statement of the **uniqueness property** is

$$\text{IF } \mathcal{L}\{f(t)\} = F(s) \text{ THEN } \mathcal{L}^{-1}\{F(s)\}(=)u(t)f(t)$$

The mathematical justification for this statement is beyond the scope of our treatment.[1] However, the notation (=) means "equal almost everywhere." The only points where equality may not hold is at the discontinuities of $f(t)$.

If we just look at the definition of the direct transformation in Eq. (9–2), we could conclude that $F(s)$ is not affected by the values of $f(t)$ for $t < 0$. However, when we use Eq. (9–2) we are not just looking for the Laplace transform of $f(t)$, but a Laplace transform pair such that $\mathcal{L}\{f(t)\} = F(s)$ and $\mathcal{L}^{-1}\{F(s)\} = f(t)$. The inverse Laplace transformation in Eq. (9–4) always produces a causal waveform, one that is zero for $t < 0$. Hence a transform pair $[f(t) \leftrightarrow F(s)]$ is unique if and only if $f(t)$ is causal. For instance, in Example 9–1 we show that $\mathcal{L}\{u(t)\} = 1/s$; hence by the uniqueness property we know that $\mathcal{L}^{-1}\{1/s\}\,(=)\,u(t)$.

---

1 See Wilber R. LePage, *Complex Variables and Laplace Transform for Engineering*, Dover Publishing Co., New York, 1980, p. 318.

For this reason Laplace transform–related waveforms are written as $[f(t)]u(t)$ to make their causality visible. For example, in the next section we find the Laplace transform of the sinusoid waveform cos βt. In the context of Laplace transforms this signal is not an eternal sinusoid but a causal waveform $f(t) = [\cos \beta t]u(t)$. It is important to remember that causality and Laplace transforms go hand in hand when interpreting the results of circuit analysis.

9–2 BASIC PROPERTIES AND PAIRS

The previous section gave the definition of the Laplace transformation and showed that the transforms of some basic signals can be found using the integral definition. In this section we develop the basic properties of the Laplace transformation and show how these properties can be used to obtain additional transform pairs.

The **linearity** property of the Laplace transformation states that

$$\mathcal{L}\{Af_1(t) + Bf_2(t)\} = AF_1(s) + BF_2(s) \tag{9–5}$$

where A and B are constants. This property is easily established using the integral definition in Eq. (9–2):

$$\begin{aligned}\mathcal{L}\{Af_1(t) + Bf_2(t)\} &= \int_0^\infty [Af_1(t) + Bf_2(t)]e^{-st}dt \\ &= A\int_0^\infty f_1(t)e^{-st}dt + B\int_0^\infty f_2(t)e^{-st}dt \\ &= AF_1(s) + BF_2(s)\end{aligned}$$

The integral definition of the inverse transformation in Eq. (9–4) is also a linear operation, so it follows that

$$\mathcal{L}^{-1}\{AF_1(s) + BF_2(s)\} = Af_1(t) + Bf_2(t) \tag{9–6}$$

An important consequence of linearity is that for any constant K

$$\mathcal{L}\{Kf(t)\} = KF(s) \text{ and } \mathcal{L}^{-1}\{KF(s)\} = Kf(t) \tag{9–7}$$

The linearity property is an extremely important feature that we will use many times in this and subsequent chapters. The next two examples show how this property can be used to obtain the transform of the exponential rise waveform and a sinusoidal waveform.

EXAMPLE 9–4

Show that the Laplace transform of $f(t) = A(1 - e^{-\alpha t})u(t)$ is

$$F(s) = \frac{A\alpha}{s(s+\alpha)}$$

SOLUTION:

This waveform is the difference between a step function and an exponential. We can use the linearity property of Laplace transforms to write

$$\mathcal{L}\{A(1 - e^{-\alpha t})u(t)\} = A\mathcal{L}\{u(t)\} - A\mathcal{L}\{e^{-\alpha t}u(t)\}$$

The transforms of the step and exponential functions were found in Examples 9–1 and 9–2. Using linearity, we find that the transform of the exponential rise is

$$F(s) = \frac{A}{s} - \frac{A}{s+\alpha} = \frac{A\alpha}{s(s+\alpha)}$$ ■

EXAMPLE 9–5

Show that the Laplace transform of the sinusoid $f(t) = A[\sin(\beta t)]u(t)$ is $F(s) = A\beta/(s^2 + \beta^2)$.

SOLUTION:

Using Euler's relationship, we can express the sinusoid as a sum of exponentials.

$$e^{+j\beta t} = \cos \beta t + j \sin \beta t$$

$$e^{-j\beta t} = \cos \beta t - j \sin \beta t$$

Subtracting the second equation from the first yields

$$f(t) = A \sin \beta t = \frac{A(e^{j\beta t} - e^{-j\beta t})}{2j} = \frac{A}{2j}e^{j\beta t} - \frac{A}{2j}e^{-j\beta t}$$

The transform pair $\mathscr{L}\{e^{-\alpha t}\} = 1/(s + \alpha)$ in Example 9–2 is valid even if the exponent α is complex. Using this fact and the linearity property, we obtain the transform of the sinusoid as

$$\mathscr{L}\{A \sin \beta t\} = \frac{A}{2j}\mathscr{L}\{e^{j\beta t}\} - \frac{A}{2j}\mathscr{L}\{e^{-j\beta t}\}$$

$$= \frac{A}{2j}\left[\frac{1}{s - j\beta} - \frac{1}{s + j\beta}\right]$$

$$= \frac{A\beta}{s^2 + \beta^2}$$ ■

INTEGRATION PROPERTY

In the time domain the i–v relationships for capacitors and inductors involve integration and differentiation. Since we will be working in the s domain, it is important to establish the s-domain equivalents of these mathematical operations. Applying the integral definition of the Laplace transformation to a time-domain integration yields

$$\mathscr{L}\left[\int_0^t f(\tau)d\tau\right] = \int_0^\infty \left[\int_0^t f(\tau)d\tau\right]e^{-st}\,dt \qquad (9\text{–}8)$$

The right side of this expression can be integrated by parts using

$$y = \int_0^t f(\tau)d\tau \quad \text{and} \quad dx = e^{-st}\,dt$$

These definitions result in

$$dy = f(t)dt \quad \text{and} \quad x = \frac{-e^{-st}}{s}$$

Using these factors reduces the right side of Eq. (9–8) to

$$\mathcal{L}\left[\int_0^t f(\tau)d\tau\right] = \left[\frac{-e^{-st}}{s}\int_0^t f(\tau)d\tau\right]_0^\infty + \frac{1}{s}\int_0^\infty f(t)e^{-st}\,dt \tag{9–9}$$

The first term on the right in Eq. (9–9) vanishes at the lower limit because the integral over a zero-length interval is zero provided that $f(t)$ is finite at $t = 0$. It vanishes at the upper limit because e^{-st} approaches zero as t goes to infinity for $\sigma > 0$. By the definition of the Laplace transformation, the second term on the right is $F(s)/s$. We conclude that

$$\mathcal{L}\left[\int_0^t f(\tau)d\tau\right] = \frac{F(s)}{s} \tag{9–10}$$

The **integration property** states that time-domain integration of a waveform $f(t)$ can be accomplished in the s domain by the algebraic process of dividing its transform $F(s)$ by s. The next example applies the integration property to obtain the transform of the ramp function.

EXAMPLE 9–6

Show that the Laplace transform of the ramp function $r(t) = tu(t)$ is $1/s^2$.

SOLUTION:

From our study of signals, we know that the ramp waveform can be obtained from $u(t)$ by integration.

$$r(t) = \int_0^t u(\tau)\,d\tau$$

In Example 9–1 we found $\mathcal{L}\{u(t)\} = 1/s$. Using these facts and the integration property of Laplace transforms, we obtain

$$\mathcal{L}\{r(t)\} = \mathcal{L}\left[\int_0^t u(\tau)\,d\tau\right] = \frac{1}{s}\mathcal{L}\{u(t)\} = \frac{1}{s^2}$$ ■

DIFFERENTIATION PROPERTY

The time-domain differentiation operation transforms into the s domain as follows:

$$\mathcal{L}\left[\frac{df(t)}{dt}\right] = \int_0^\infty \left[\frac{df(t)}{dt}\right]e^{-st}\,dt \tag{9–11}$$

The right side of this equation can be integrated by parts using

$$y = e^{-st} \quad \text{and} \quad dx = \frac{df(t)}{dt}\,dt$$

These definitions result in

$$dy = -se^{-st}\,dt \quad \text{and} \quad x = f(t)$$

Inserting these factors reduces the right side of Eq. (9–11) to

$$\mathcal{L}\left[\frac{df(t)}{dt}\right] = f(t)e^{-st}\Big|_{0-}^{\infty} + s\int_{0-}^{\infty} f(t)e^{-st}\,dt \tag{9–12}$$

For $\sigma > 0$ the first term on the right side of Eq. (9–12) is zero at the upper limit because e^{-st} approaches zero as t goes to infinity. At the lower limit it reduces to $-f(0-)$. By the definition of the Laplace transform, the second term on the right side is $sF(s)$. We conclude that

$$\mathcal{L}\left[\frac{df(t)}{dt}\right] = sF(s) - f(0-) \tag{9–13}$$

The **differentiation property** states that time-domain differentiation of a waveform $f(t)$ is accomplished in the s domain by the algebraic process of multiplying the transform $F(s)$ by s and subtracting the constant $f(0-)$. Note that the constant $f(0-)$ is the value of $f(t)$ at $t = 0-$ just prior to $t = 0$.

The s-domain equivalent of a second derivative is obtained by repeated application of Eq. (9–13). We first define a waveform $g(t)$ as

$$g(t) = \frac{df(t)}{dt} \quad \text{hence} \quad \frac{d^2f(t)}{dt^2} = \frac{dg(t)}{dt}$$

Applying the differentiation rule to these two equations yields

$$G(s) = sF(s) - f(0-) \quad \text{and} \quad \mathcal{L}\left[\frac{d^2f(t)}{dt^2}\right] = sG(s) - g(0-)$$

Substituting the first of these equations into the second results in

$$\mathcal{L}\left[\frac{d^2f(t)}{dt^2}\right] = s^2F(s) - sf(0-) - f'(0-)$$

where

$$f'(0-) = \frac{df}{dt}\bigg|_{t=0-}$$

Repeated application of this procedure produces the nth derivative:

$$\mathcal{L}\left[\frac{d^nf(t)}{dt^n}\right] = s^nF(s) - s^{n-1}f(0-) - s^{n-2}f'(0-)\ldots - f^{(n)}(0-) \tag{9–14}$$

where $f^{(n)}(0-)$ is the nth derivative of $f(t)$ evaluated at $t = 0-$.

A hallmark feature of the Laplace transformation is the fact that time integration and differentiation change into algebraic operations in the s domain. This observation gives us our first hint as to why it is often easier to work with circuits and signals in the s domain. The next example shows how the differentiation rule can be used to obtain additional transform pairs.

EXAMPLE 9–7

Show that the Laplace transform of $f(t) = [\cos \beta t]u(t)$ is $F(s) = s/(s^2 + \beta^2)$.

SOLUTION:
We can express $\cos \beta t$ in terms of the derivative of $\sin \beta t$ as

$$\cos \beta t = \frac{1}{\beta}\frac{d}{dt}\sin \beta t$$

In Example 9–5 we found $\mathcal{L}\{\sin \beta t\} = \beta/(s^2 + \beta^2)$. Using these facts and the differentiation rule, we can find the Laplace transform of $\cos \beta t$ as follows:

$$\mathcal{L}\{\cos \beta t\} = \frac{1}{\beta}\mathcal{L}\left\{\frac{d}{dt}\sin \beta t\right\} = \frac{1}{\beta}\left[s\left(\frac{\beta}{s^2 + \beta^2}\right) - \sin(0-)\right]$$
$$= \frac{s}{s^2 + \beta^2}$$

■

TRANSLATION PROPERTIES

The **s-domain translation property** of the Laplace transformation is

$$\text{IF } \mathcal{L}\{f(t)\} = F(s) \text{ THEN } \mathcal{L}\{e^{-\alpha t}f(t)\} = F(s + \alpha)$$

This theorem states that multiplying $f(t)$ by $e^{-\alpha t}$ is equivalent to replacing s by $s + \alpha$ (that is, translating the origin in the s plane by an amount α). In engineering applications the parameter α is always a real number, but it can be either positive or negative so the origin in the s domain can be translated to the left or right. Proof of the theorem follows almost immediately from the definition of the Laplace transformation.

$$\mathcal{L}\{e^{-\alpha t}f(t)\} = \int_0^\infty e^{-\alpha t}f(t)e^{-st}dt$$
$$= \int_0^\infty f(t)e^{-(s+\alpha)t}\,dt$$
$$= F(s + \alpha)$$

The s-domain translation property can be used to derive transforms of damped waveforms from undamped prototypes. For instance, the Laplace transform of the ramp, cosine, and sine functions are

$$\mathcal{L}\{tu(t)\} = \frac{1}{s^2}$$

$$\mathcal{L}\{[\cos \beta t]u(t)\} = \frac{s}{s^2 + \beta^2}$$

$$\mathcal{L}\{[\sin \beta t]u(t)\} = \frac{\beta}{s^2 + \beta^2}$$

To obtain the damped ramp, damped cosine, and damped sine functions, we multiply each undamped waveform by $e^{-\alpha t}$. Using the s-domain translation property, we replace s by $s + \alpha$ to obtain transforms of the corresponding damped waveforms.

$$\mathscr{L}\{te^{-\alpha t}u(t)\} = \frac{1}{(s+\alpha)^2}$$

$$\mathscr{L}\{[e^{-\alpha t}\cos\beta t]u(t)\} = \frac{s+\alpha}{(s+\alpha)^2+\beta^2}$$

$$\mathscr{L}\{[e^{-\alpha t}\sin\beta t]u(t)\} = \frac{\beta}{(s+\alpha)^2+\beta^2}$$

This completes the derivation of a basic set of transform pairs.

The **time-domain translation property** of the Laplace transformation is

$$\text{IF } \mathscr{L}\{f(t)\} = F(s) \text{ THEN for } a > 0 \; \mathscr{L}\{f(t-a)u(t-a)\} = e^{-as}F(s)$$

The theorem states that multiplying $F(s)$ by e^{-as} is equivalent to shifting $f(t)$ to the right in the time domain by an amount $a > 0$. In other words, it is equivalent to delaying $f(t)$ in time by an amount $a > 0$. Proof of this property follows from the definition of the Laplace transformation.

$$\mathscr{L}\{f(t-a)u(t-a)\} = \int_{0-}^{\infty} f(t-a)u(t-a)e^{-st}dt = \int_{a}^{\infty} f(t-a)e^{-st}dt$$

In this equation we have used the fact that $u(t-a)$ is zero for $t < a$ and is unity for $t \geq a$. We now change the integration variable from t to $\tau = t - a$. With this change of variable the last integral in this equation takes the form

$$\mathscr{L}\{f(t-a)u(t-a)\} = \int_{0}^{\infty} f(\tau)e^{-s\tau}e^{-as}d\tau$$

$$= e^{-as}\int_{0}^{\infty} f(\tau)e^{-s\tau}\,d\tau$$

$$= e^{-as}F(s)$$

which confirms the statement of the time-domain translation property. A simple application of this property is finding the Laplace transform of the delayed step function.

$$\mathscr{L}\{u(t-T)\} = e^{-sT}\mathscr{L}\{u(t)\} = \frac{e^{-sT}}{s}$$

In this section we derived the basic transform properties listed in Table 9–1. The Laplace transformation has other properties that are useful in signal-processing applications. We treat two of these properties in the last section of this chapter. However, the basic properties in Table 9–1 are used frequently in circuit analysis and are sufficient for nearly all of the applications in this book.

Similarly, Table 9–2 lists a basic set of Laplace transform pairs that is sufficient for most of the applications in this book. All of these pairs were derived in the preceding two sections.

All of the waveforms in Table 9–2 are causal. As a result, the Laplace transform pairs are unique and we can use the table in either direction. That is, given an $f(t)$ in the waveform column, we find its Laplace transform in the right column, or given an $F(s)$ in the right column, we find its inverse transform in the waveform column.

TABLE 9–1 BASIC LAPLACE TRANSFORMATION PROPERTIES

PROPERTIES	TIME DOMAIN	FREQUENCY DOMAIN
Independent variable	t	s
Signal representation	$f(t)$	$F(s)$
Uniqueness	$\mathscr{L}^{-1}\{F(s)\} (=) [f(t)]u(t)$	$\mathscr{L}\{f(t)\} = F(s)$
Linearity	$Af_1(t) + Bf_2(t)$	$AF_1(s) + BF_2(s)$
Integration	$\int_0^t f(\tau)\, d\tau$	$\frac{F(s)}{s}$
Differentiation	$\frac{df(t)}{dt}$	$sF(s) - f(0-)$
	$\frac{d^2f(t)}{dt^2}$	$s^2F(s) - sf(0-) - f'(0-)$
	$\frac{d^3f(t)}{dt^3}$	$s^3F(s) - s^2f(0-) - sf'(0-) - f''(0-)$
s-Domain translation	$e^{-\alpha t}f(t)$	$F(s + \alpha)$
t-Domain translation	$f(t - a)u(t - a)$	$e^{-as}F(s)$

TABLE 9–2 BASIC LAPLACE TRANSFORM PAIRS

SIGNAL	WAVEFORM $f(t)$	TRANSFORM $F(s)$
Impulse	$\delta(t)$	1
Step function	$u(t)$	$\frac{1}{s}$
Ramp	$tu(t)$	$\frac{1}{s^2}$
Exponential	$[e^{-\alpha t}]u(t)$	$\frac{1}{s + \alpha}$
Damped ramp	$[te^{-\alpha t}]u(t)$	$\frac{1}{(s + \alpha)^2}$
Sine	$[\sin \beta t]u(t)$	$\frac{\beta}{s^2 + \beta^2}$
Cosine	$[\cos \beta t]u(t)$	$\frac{s}{s^2 + \beta^2}$
Damped sine	$[e^{-\alpha t} \sin \beta t]u(t)$	$\frac{\beta}{(s + \alpha)^2 + \beta^2}$
Damped cosine	$[e^{-\alpha t} \cos \beta t]u(t)$	$\frac{(s + \alpha)}{(s + \alpha)^2 + \beta^2}$

The last example in this section shows how to use the properties and pairs in Tables 9–1 and 9–2 to obtain the transform of a waveform not listed in the tables.

EXAMPLE 9–8

Find the Laplace transform of the waveform

$$f(t) = 2u(t) - 5[e^{-2t}]u(t) + 3[\cos 2t]u(t) + 3[\sin 2t]u(t)$$

SOLUTION:

Using the linearity property, we write the transform of $f(t)$ in the form

$$\mathscr{L}\{f(t)\} = 2\mathscr{L}\{u(t)\} - 5\mathscr{L}\{e^{-2t}u(t)\} + 3\mathscr{L}\{[\cos 2t]u(t)\} + 3\mathscr{L}\{[\sin 2t]u(t)\}$$

The transforms of each term in this sum are listed in Table 9–2:

$$F(s) = \frac{2}{s} - \frac{5}{s+2} + \frac{3s}{s^2+4} + \frac{6}{s^2+4}$$

Normally, a Laplace transform is written as a quotient of polynomials rather than as a sum of terms. Rationalizing the preceding sum yields

$$F(s) = \frac{16(s^2+1)}{s(s+2)(s^2+4)}$$ ■

Exercise 9–1

Find the Laplace transforms of the following waveforms:

(a) $f(t) = [e^{-2t}]u(t) + 4tu(t) - u(t)$

(b) $f(t) = [2 + 2\sin 2t - 2\cos 2t]u(t)$

Answers:

(a) $F(s) = \dfrac{2(s+4)}{s^2(s+2)}$

(b) $F(s) = \dfrac{4(s+2)}{s(s^2+4)}$

Exercise 9–2

Find the Laplace transforms of the following waveforms:

(a) $f(t) = [e^{-4t}]u(t) + 5\int_0^t \sin 4x \, dx$

(b) $f(t) = 5[e^{-40t}]u(t) + \dfrac{d[5te^{-40t}]u(t)}{dt}$

Answers:

(a) $F(s) = \dfrac{s^3 + 36s + 80}{s(s+4)(s^2+16)}$

(b) $F(s) = \dfrac{10s + 200}{(s+40)^2}$

Exercise 9–3

Find the Laplace transforms of the following waveforms:

(a) $f(t) = A[\cos(\beta t - \phi)]u(t)$

(b) $f(t) = A[e^{-\alpha t} \cos(\beta t - \phi)]u(t)$

Answers:

(a) $F(s) = A \cos \phi \left[\dfrac{s + \beta \tan \phi}{s^2 + \beta^2}\right]$

(b) $F(s) = A \cos \phi \left[\dfrac{s + \alpha + \beta \tan \phi}{(s + \alpha)^2 + \beta^2}\right]$

Exercise 9–4

Find the Laplace transforms of the following waveforms:

(a) $f(t) = Au(t) - 2Au(t - T) + Au(t - 2T)$

(b) $f(t) = Ae^{-\alpha(t-T)}u(t - T)$

Answers:

(a) $F(s) = \dfrac{A(1 - e^{-Ts})^2}{s}$

(b) $F(s) = \dfrac{Ae^{-Ts}}{s + \alpha}$

9–3 POLE-ZERO DIAGRAMS

The transforms for signals in Table 9–2 are ratios of polynomials in the complex frequency variable s. Likewise, the transform found in Example 9–8 takes the form of a ratio of two polynomials in s. These results illustrate that the signal transforms of greatest interest to us usually have the form

$$F(s) = \frac{b_m s^m + b_{m-1} s^{m-1} + \ldots + b_1 s + b_0}{a_n s^n + a_{n-1} s^{n-1} + \ldots + a_1 s + a_0} \tag{9–15}$$

If numerator and denominator polynomials are expressed in factored form, then $F(s)$ is written as

$$F(s) = K\frac{(s - z_1)(s - z_2)\ldots(s - z_m)}{(s - p_1)(s - p_2)\ldots(s - p_n)} \tag{9–16}$$

where the constant $K = b_m/a_n$ is called the **scale factor**.

The roots of the numerator and denominator polynomials, together with the scale factor K, uniquely define a transform $F(s)$. The denominator roots are called **poles** because for $s = p_i$ $(i = 1, 2, \ldots n)$ the denominator vanishes and $F(s)$ becomes infinite. The roots of the numerator polynomial are called **zeros** because the transform $F(s)$ vanishes for $s = z_i$ $(i = 1, 2, \ldots m)$. Collectively the poles and zeros are called **critical frequencies** because they are values of s at which $F(s)$ does dramatic things, like vanish or blow up.

In the s domain we can specify a signal transform by listing the location of its critical frequencies together with the scale factor K. That is, in the frequency domain we describe signals in terms of poles and zeros. The description takes the form of a **pole-zero diagram**, which shows the location of poles and zeros in the complex s plane. The pole locations in such plots are indicated by an X and the zeros by an O. The independent variable in the frequency domain is the complex frequency variable s, so the poles or zeros can be complex as well. In the s plane we use a horizontal axis to plot the value of the real part of s and a vertical j-axis to plot the imaginary part. The j-axis is an important boundary in the frequency domain because it divides the s plane into two distinct half planes. The real part of s is negative in the left half plane and positive in the right half plane. As we will soon see, the sign of the real part of a pole has a profound effect on the form of the corresponding waveform.

For example, Table 9–2 shows that the transform of the exponential waveform $f(t) = e^{-\alpha t}u(t)$ is $F(s) = 1/(s + \alpha)$. The exponential signal has a single pole at $s = -\alpha$ and no finite zeros. The pole-zero diagram in Figure 9–2(a) is the s-domain portrayal of the exponential signal. In this diagram the X identifies the pole located at $s = -\alpha + j0$, a point on the negative real axis in the left half plane.

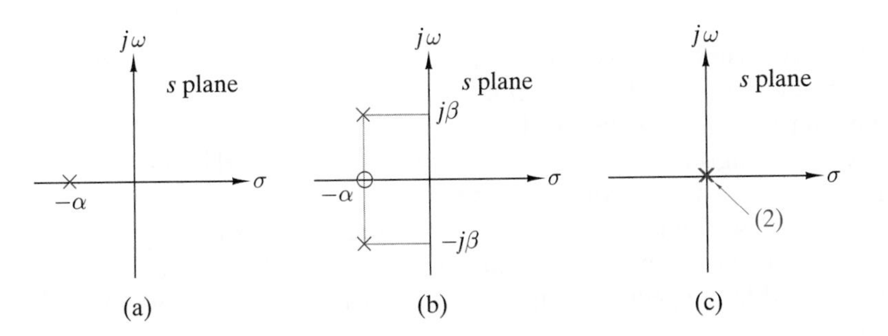

FIGURE 9–2 Pole-zero diagrams in the s plane.

The damped sinusoid $f(t) = [A\, e^{-\alpha t} \cos \beta t]u(t)$ is an example of a signal with complex poles. From Table 9–2 the corresponding transform is

$$F(s) = \frac{A(s + \alpha)}{(s + \alpha)^2 + \beta^2}$$

The transform $F(s)$ has a finite zero on the real axis at $s = -\alpha$. The roots of the denominator polynomial are $s = -\alpha \pm j\beta$. The resulting pole-zero diagram is shown in Figure 9–2(b). The poles of the damped cosine do not lie on either axis in the s plane because neither the real nor imaginary parts are zero.

Finally, the transform of a unit ramp $f(t) = tu(t)$ is $F(s) = 1/s^2$. This transform has no finite zeros and two poles at the origin ($s = 0 + j0$) in the s plane as shown in Figure 9–2(c). The poles in all of the diagrams of Figure 9–2 lie in the left half plane or on the j-axis boundary.

The diagrams in Figure 9–2 show the poles and zeros in the finite part of the s plane. Signal transforms may have poles or zeros at infinity as well. For example, the step function has a zero at infinity since $F(s) = 1/s$ approaches zero as $s \rightarrow \infty$. In general, a transform $F(s)$ given by Eq. (9–16) has a zero of order $n - m$ at infinity if $n > m$ and a pole of order $m - n$ at infinity if $n < m$. Thus, the number of zeros equals the number of poles if we include those at infinity.

The pole-zero diagram is the s-domain portrayal of the signal, just as a plot of the waveform versus time depicts the signal in the t domain. The utility of a pole-zero diagram as a description of circuits and signals will become clearer as we develop additional s-domain analysis and design concepts.

EXAMPLE 9–9

Find the poles and zeros of the waveform

$$f(t) = [e^{-2t} + \cos 2t - \sin 2t]u(t)$$

SOLUTION:

Using the linearity property and the basic pairs in Table 9–2, we write the transform in the form

$$F(s) = \frac{1}{s+2} + \frac{s}{s^2+4} - \frac{2}{s^2+4}$$

Rationalizing this expression yields $F(s)$.

$$F(s) = \frac{2s^2}{(s+2)(s^2+4)} = \frac{2s^2}{(s+2)(s+j2)(s-j2)}$$

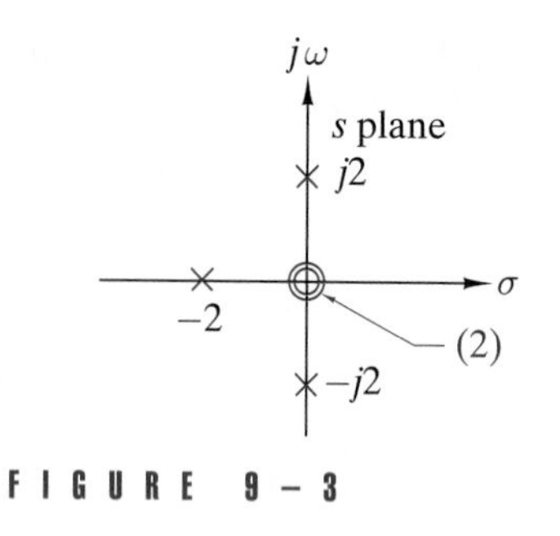

FIGURE 9–3

This transform has three zeros and three poles. There are two zeros at $s = 0$ and one at $s = \infty$. There is a pole on the negative real axis at $s = -2 + j0$, and there are two poles on the imaginary axis at $s = \pm j2$. The resulting pole-zero diagram is shown in Figure 9–3. Reviewing the analysis, we can trace the poles to the components of $f(t)$. The pole on the real axis at $s = -2$ came from the exponential e^{-2t}, while the complex conjugate poles on the j-axis came from the sinusoid $\cos 2t - \sin 2t$. The zeros, however, are not traceable to specific components. Their locations depend on all three components. ■

EXAMPLE 9–10

Use a math analysis program to find the Laplace transform of the waveform

$$f(t) = [200te^{-25t} + 10e^{-50t} \sin (40t)]u(t)$$

and construct a pole-zero map of $F(s)$.

SOLUTION:

In this example we will illustrate the Laplace transform capabilities of MATLAB. Operating in the MATLAB command window, we first write the statement

```
syms t s
```

This defines the symbols t (for time) and s (for complex frequency) as symbolic variables rather than numerical quantities. We can then write the waveform $f(t)$ as

```
f=200*t*exp(-25*t)+10*exp(-50*t)*sin(40*t);
```

The Laplace transformation is performed by the statement

```
F=laplace (f, t, s);
```

This statement causes MATLAB to transform the previously defined waveform f from a domain where the independent variable is t to a domain in which the independent variable is s. The result of the transformation yields

```
F=
200/(s+25)^2+400/((s+50)^2+1600)
```

This result is clearly a term-by-term transformation of the waveform $f(t)$. The first term comes from the damped ramp and has a double pole at $s = -25$. The second term comes from the damped sine and has a pair of conjugate complex poles at $s = -50 \pm j40$. This set of poles can be predicted in advance from the form of the terms in the input waveform. To find the zeros of $F(s)$ we must express this result as a quotient of polynomials. To accomplish this we use a powerful MATLAB function called `simplify` that attempts to simplify an expression using many different mathematical tools. In the present case the result is

```
simplify(F)
ans=
200*(3*s^2+200*s+5350)/(s+25)^2/(s^2+100*s+4100)
```

Interpreting this result requires careful application of the MATLAB rules of precedence. Fortunately, MATLAB offers another way out. The function `pretty` attempts to place an expression in a form that more closely resembles type set mathematics. In the present case the result is

```
pretty(ans)
                      2
                   3 s  + 200 s + 5350
          200 ---------------------------
                      2   2
               (s + 25)  (s + 100 s + 4100)
```

The zeros of $F(s)$ are roots of the numerator polynomial in this expression and are located at $s = -33.33 \pm j25.93$. Figure 9–4 shows the resulting pole-zero map. ■

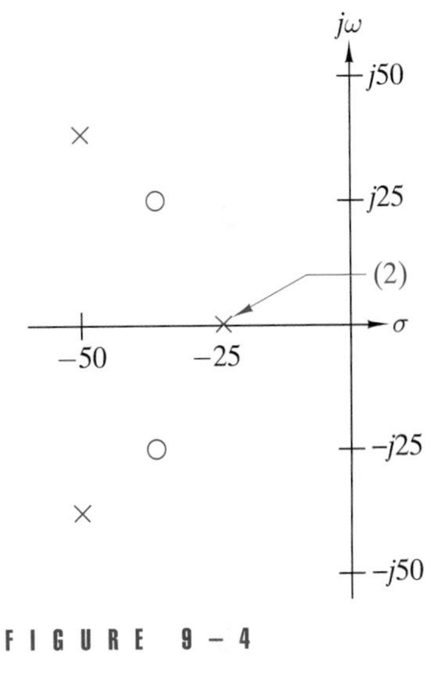

FIGURE 9 – 4

Exercise 9–5

Find the poles and zeros of transforms of each of the following waveforms:

(a) $f(t) = [-2e^{-t} - t + 2]u(t)$

(b) $f(t) = [4 - 3\cos\beta t]u(t)$

(c) $f(t) = [e^{-\alpha t}\cos\beta t + (\alpha/\beta)\, e^{-\alpha t}\sin\beta t]u(t)$

Answers:

(a) Zeros: $s = 1$, $s = \infty$ (2); poles: $s = 0$ (2), $s = -1$
(b) Zeros: $s = \pm j2\beta$, $s = \infty$; poles: $s = 0$, $s = \pm j\beta$
(c) Zeros: $s = -2\alpha$, $s = \infty$; poles: $s = -\alpha \pm j\beta$

Exercise 9–6

A transform has poles at $s = -3 \pm j6$ and $s = -2$ and finite zeros at $s = 0$ and $s = -1$. Write $F(s)$ as a quotient of polynomials in s.

Answer:

$$F(s) = K\frac{s^2 + s}{s^3 + 8s^2 + 57s + 90}$$

9–4 INVERSE LAPLACE TRANSFORMS

The inverse transformation converts a transform $F(s)$ into the corresponding waveform $f(t)$. Applying the inverse transformation in Eq. (9–4) requires knowledge of a branch of mathematics called the complex variable. Fortunately, we do not need Eq. (9–4) because the uniqueness of the Laplace transform pairs in Table 9–2 allows us to go from a transform to a waveform. This may not seem like much help since it does not take a very complicated circuit or signal before we exceed the listing in Table 9–2, or even the more extensive tables that are available. However, there is a general method of expanding $F(s)$ into a sum of terms that are listed in Table 9–2.

For linear circuits the transforms of interest are ratios of polynomials in s. In mathematics such functions are called **rational functions**. To perform the inverse transformation, we must find the waveform corresponding to rational functions of the form

$$F(s) = K\frac{(s - z_1)(s - z_2)\ldots(s - z_m)}{(s - p_1)(s - p_2)\ldots(s - p_n)} \tag{9–17}$$

where the K is the scale factor, z_i $(i = 1, 2, \ldots m)$ are the zeros, and p_i $(i = 1, 2, \ldots n)$ are the poles of $F(s)$.

If there are more finite poles than finite zeros $(n > m)$, then $F(s)$ is called a **proper rational function**. If the denominator in Eq. (9–17) has no repeated roots ($p_i \neq p_j$ for $i \neq j$), then $F(s)$ is said to have **simple poles**. In this section we treat the problem of finding the inverse transform of proper rational functions with simple poles. The problem of improper rational functions and multiple poles is covered in the next section.

If a proper rational function has only simple poles, then it can be decomposed into a partial fraction expansion of the form

$$F(s) = \frac{k_1}{s - p_1} + \frac{k_2}{s - p_2} + \ldots + \frac{k_n}{s - p_n} \tag{9–18}$$

In this case, $F(s)$ can be expressed as a linear combination of terms with one term for each of its n simple poles. The k's associated with each term are called **residues**.

Each term in the partial fraction decomposition has the form of the transform of an exponential signal. That is, we recognize that $\mathcal{L}^{-1}\{k/(s + \alpha)\} = [ke^{-\alpha t}]u(t)$. We can now write the corresponding waveform using the linearity property

$$f(t) = [k_1 e^{p_1 t} + k_2 e^{p_2 t} + \ldots + k_n e^{p_n t}]u(t) \tag{9–19}$$

In the time domain, the s-domain poles appear in the exponents of exponential waveforms and the residues at the poles become the amplitudes.

Given the poles of $F(s)$, finding the inverse transform $f(t)$ reduces to finding the residues. To illustrate the procedure, consider a case in which $F(s)$ has three simple poles and one finite zero.

$$F(s) = K\frac{(s - z_1)}{(s - p_1)(s - p_2)(s - p_3)} = \frac{k_1}{s - p_1} + \frac{k_2}{s - p_2} + \frac{k_3}{s - p_3}$$

We find the residue k_1 by first multiplying this equation through by the factor $(s - p_1)$:

$$(s - p_1)F(s) = K\frac{(s - z_1)}{(s - p_2)(s - p_3)} = k_1 + \frac{k_2(s - p_1)}{s - p_2} + \frac{k_3(s - p_1)}{s - p_3}$$

If we now set $s = p_1$, the last two terms on the right vanish, leaving

$$k_1 = (s - p_1)F(s)\big|_{s=p_1} = K\frac{(s - z_1)}{(s - p_2)(s - p_3)}\bigg|_{s=p_1}$$

Using the same approach for k_2 yields

$$k_2 = (s - p_2)F(s)\big|_{s=p_2} = K\frac{(s - z_1)}{(s - p_1)(s - p_3)}\bigg|_{s=p_2}$$

The technique generalizes so that the residue at any simple pole p_i is

$$k_i = (s - p_i)F(s)\big|_{s=p_i} \tag{9–20}$$

The process of determining the residue at any simple pole is sometimes called the **cover-up algorithm** because we temporarily remove (cover up) the factor $(s - p_i)$ in $F(s)$ and then evaluate the remainder at $s = p_i$.

EXAMPLE 9–11

Find the waveform corresponding to the transform

$$F(s) = 2\frac{(s + 3)}{s(s + 1)(s + 2)}$$

SOLUTION:

$F(s)$ is a proper rational function and has simple poles at $s = 0$, $s = -1$, $s = -2$. Its partial fraction expansion is

$$F(s) = \frac{k_1}{s} + \frac{k_2}{s + 1} + \frac{k_3}{s + 2}$$

The cover-up algorithm yields the residues as

$$k_1 = sF(s)\big|_{s=0} = \frac{2(s + 3)}{(s + 1)(s + 2)}\bigg|_{s=0} = 3$$

$$k_2 = (s + 1)F(s)\big|_{s=-1} = \frac{2(s + 3)}{s(s + 2)}\bigg|_{s=-1} = -4$$

$$k_3 = (s + 2)F(s)\big|_{s=-2} = \frac{2(s + 3)}{s(s + 1)}\bigg|_{s=-2} = 1$$

The inverse transform $f(t)$ is

$$f(t) = [3 - 4e^{-t} + e^{-2t}]u(t)$$

■

Exercise 9–7

Find the waveforms corresponding to the following transforms:

(a) $F_1(s) = \dfrac{4}{(s+1)(s+3)}$

(b) $F_2(s) = e^{-5s}\left[\dfrac{2s}{(s+1)(s+3)}\right]$

(c) $F_3(s) = \dfrac{4(s+2)}{(s+1)(s+3)}$

Answers:

(a) $f_1(t) = [2e^{-t} - 2e^{-3t}]u(t)$

(b) $f_2(t) = [-e^{-(t-5)} + 3e^{-3(t-5)}]u(t-5)$

(c) $f_3(t) = [2e^{-t} + 2e^{-3t}]u(t)$

Comment: Note that

$$F_2(s) = e^{-5s}\left[\frac{2s}{(s+1)(s+3)}\right] = e^{-5s}\left[\frac{-1}{s+1} + \frac{3}{s+3}\right]$$

so that the delay factor e^{-5s} is not involved in the partial fraction expansion but simply flags the amount by which the resulting waveform $[-e^{-t} + 3e^{-3t}]u(t)$ is delayed to produce $f_2(t)$.

Exercise 9–8

Find the waveforms corresponding to the following transforms:

(a) $F(s) = \dfrac{6(s+2)}{s(s+1)(s+4)}$

(b) $F(s) = \dfrac{4(s+1)}{s(s+1)(s+4)}$

Answers:

(a) $f(t) = [3 - 2e^{-t} - e^{-4t}]u(t)$

(b) $f(t) = [1 - e^{-4t}]u(t)$

COMPLEX POLES

Special treatment is necessary when $F(s)$ has a complex pole. In physical situations the function $F(s)$ is a ratio of polynomials with real coefficients. If $F(s)$ has a complex pole $p = -\alpha + j\beta$, then it must also have a pole $p^* = -\alpha - j\beta$; otherwise the coefficients of the denominator polynomial would not be real. In other words, for physical signals the complex poles of $F(s)$ must occur in conjugate pairs. As a consequence, the partial fraction decomposition of $F(s)$ will contain two terms of the form

$$F(s) = \ldots + \frac{k}{s+\alpha-j\beta} + \frac{k^*}{s+\alpha+j\beta} + \ldots \tag{9–21}$$

The residues k and k^* at the conjugate poles are themselves conjugates because $F(s)$ is a rational function with real coefficients. These residues can be calculated using the cover-up algorithm and, in general, they turn out to be complex numbers. If the complex residues are written in polar form as

$$k = |k|e^{j\theta} \text{ and } k^* = |k|e^{-j\theta}$$

then the waveform corresponding to the two terms in Eq. (9–21) is

$$f(t) = [\ldots + |k|e^{j\theta}e^{(-\alpha + j\beta)t} + |k|e^{-j\theta}e^{(-\alpha - j\beta)t} + \ldots]u(t)$$

This equation can be rearranged in the form

$$f(t) = \left[\ldots + 2|k|e^{-\alpha t}\left\{\frac{e^{+j(\beta t+\theta)} + e^{-j(\beta t+\theta)}}{2}\right\} + \ldots\right]u(t) \tag{9–22}$$

The expression inside the brackets is of the form

$$\cos x = \left\{\frac{e^{+jx} + e^{-jx}}{2}\right\}$$

Consequently, we combine terms inside the braces as a cosine function with a phase angle:

$$f(t) = [\ldots + 2|k|e^{-\alpha t}\cos(\beta t + \theta) + \ldots]u(t) \tag{9–23}$$

In summary, if $F(s)$ has a complex pole, then in physical applications there must be an accompanying conjugate complex pole. The inverse transformation combines the two poles to produce a damped cosine waveform. We only need to compute the residue at one of these poles because the residues at conjugate poles must be conjugates. Normally, we calculate the residue for the pole at $s = -\alpha + j\beta$ because its angle equals the phase angle of the damped cosine. Note that the imaginary part of this pole is positive, which means that the pole lies in the upper half of the s plane.

The inverse transform of a proper rational function with simple poles can be found by the partial fraction expansion method. The residues k at the simple poles can be found using the cover-up algorithm. The resulting waveform is a sum of terms of the form $[ke^{-\alpha t}]u(t)$ for real poles and $[2|k|e^{-\alpha t}\cos(\beta t + \theta)]u(t)$ for a pair of complex conjugate poles. The partial fraction expansion of the transform contains all of the data needed to construct the corresponding waveform.

EXAMPLE 9–12

Find the inverse transform of

$$F(s) = \frac{20(s + 3)}{(s + 1)(s^2 + 2s + 5)}$$

SOLUTION:

$F(s)$ has a simple pole at $s = -1$ and a pair of conjugate complex poles located at the roots of the quadratic factor

$$(s^2 + 2s + 5) = (s + 1 - j2)(s + 1 + j2)$$

The partial fraction expansion of $F(s)$ is

$$F(s) = \frac{k_1}{s+1} + \frac{k_2}{s+1-j2} + \frac{k_2^*}{s+1+j2}$$

The residues at the poles are found from the cover-up algorithm.

$$k_1 = \left. \frac{20(s+3)}{s^2+2s+5} \right|_{s=-1} = 10$$

$$k_2 = \left. \frac{20(s+3)}{(s+1)(s+1+j2)} \right|_{s=-1+j2} = -5 - j5 = 5\sqrt{2}e^{+j5\pi/4}$$

We now have all of the data needed to construct the inverse transform.

$$f(t) = [10e^{-t} + 10\sqrt{2}e^{-t}\cos(2t + 5\pi/4)]u(t)$$

In this example we used k_2 to obtain the amplitude and phase angle of the damped cosine term. The residue k_2^* is not needed, but to illustrate a point we note that its value is

$$k_2^* = (-5 - j5)^* = -5 + j5 = 5\sqrt{2}e^{-j5\pi/4}$$

If k_2^* is used instead, we get the same amplitude for the damped sine but the wrong phase angle. *Caution:* Remember that Eq. (9–23) uses the residue at the complex pole with a positive imaginary part. In this example this is the pole at $s = -1 + j2$, not the pole at $s = -1 - j2$.

Exercise 9–9

Find the inverse transforms of the following rational functions:

(a) $F(s) = \dfrac{16}{(s+2)(s^2+4)}$

(b) $F(s) = \dfrac{2(s+2)}{s(s^2+4)}$

Answers:

(a) $f(t) = [2e^{-2t} + 2\sqrt{2}\cos(2t - 3\pi/4)]u(t)$

(b) $f(t) = [1 + \sqrt{2}\cos(2t - 3\pi/4)]u(t)$

Exercise 9–10

Find the inverse transforms of the following rational functions:

(a) $F(s) = \dfrac{8}{s(s^2+4s+8)}$

(b) $F(s) = \dfrac{4s}{s^2+4s+8}$

Answers:

(a) $f(t) = [1 + \sqrt{2}e^{-2t}\cos(2t + 3\pi/4)]u(t)$

(b) $f(t) = [4\sqrt{2}e^{-2t}\cos(2t + \pi/4)]u(t)$

SUMS OF RESIDUES

The sums of the residues of a proper rational function are subject to certain conditions that are useful for checking the calculations in a partial fraction expansion. To derive these conditions, we multiply Eqs. (9–17) and (9–18) by s and take the limit as $s \to \infty$. These operations yield

$$\lim_{s\to\infty} sF(s) = \lim_{s\to\infty} \frac{Ks^{m+1}}{s^n} = \lim_{s\to\infty}\left(\frac{k_1 s}{s+p_1} + \ldots + \frac{k_n s}{s+p_n}\right)$$

In the limit this equation reduces to

$$K\left[\lim_{s\to\infty} \frac{s^{m+1}}{s^n}\right] = k_1 + k_2 + \ldots + k_n$$

Since $F(s)$ is a proper rational function with $n > m$, the limit process in this equation yields to following conditions:

$$k_1 + k_2 + \ldots + k_n = \begin{cases} 0 \text{ if } n > m+1 \\ K \text{ if } n = m+1 \end{cases} \tag{9–24}$$

For a proper rational function with simple poles, the sum of residues is either zero or else equal to the transform scale factor K.

Exercise 9–11

Use the sum of residues to find the unknown residue in the following expansions:

(a) $\dfrac{21\,(s+5)}{(s+3)(s+10)} = \dfrac{6}{s+3} + \dfrac{k}{s+10}$

(b) $\dfrac{58s}{(s+2)(s^2+25)} = \dfrac{k}{s+2} + \dfrac{2+j5}{s+j5} + \dfrac{2-j5}{s-j5}$

Answers:

(a) $k = 15$
(b) $k = -4$

9–5 SOME SPECIAL CASES

Most of the transforms encountered in physical applications are proper rational functions with simple poles. The inverse transforms of such functions can be handled by the partial fraction expansion method developed in the previous section. This section covers the problem of finding the inverse transform when $F(s)$ is an improper rational function or has multiple poles. These matters are treated as special cases because they only occur for certain discrete values of circuit or signal parameters. However, some of these special cases are important, so we need to learn how to handle improper rational functions and multiple poles.

$F(s)$ is an **improper rational function** when the order of the numerator polynomial equals or exceeds the order of the denominator ($m \geq n$). For example, the transform

$$F(s) = \frac{s^3 + 6s^2 + 12s + 11}{s^2 + 4s + 3} \tag{9–25}$$

is improper because $m = 3$ and $n = 2$. Using long division this improper rational function can be changed into the sum of a quotient plus a remainder which is a proper rational function. We proceed as follows:

$$\begin{array}{r}
s + 2 \\
s^2 + 4s + 3 \,\overline{)\, s^3 + 6s^2 + 12s + 11} \\
\underline{s^3 + 4s^2 + 3s} \\
2s^2 + 9s + 8 \\
\underline{2s^2 + 8s + 6} \\
s + 2
\end{array}$$

which yields

$$F(s) = s + 2 + \frac{s + 5}{s^2 + 4s + 3}$$

$$= \text{Quotient} + \text{Remainder}$$

The remainder is a proper rational function, which can be expanded by partial fractions to produce

$$F(s) = s + 2 + \frac{2}{s + 1} - \frac{1}{s + 3}$$

All of the terms in this expansion are listed in Table 9–2 except the first term. The inverse transform of the first term is found using the transform of an impulse and the differentiation property. The Laplace transform of the derivative of an impulse is

$$\mathcal{L}\left[\frac{d\delta(t)}{dt}\right] = s\mathcal{L}[\delta(t)] - \delta(0-) = s$$

since $\mathcal{L}\{\delta(t)\} = 1$ and $\delta(0-) = 0$. By the uniqueness property of the Laplace transformation, we have $\mathcal{L}^{-1}\{s\} = d\delta(t)/dt$. The first derivative of an impulse is called a doublet. The inverse transform of the improper rational function in Eq. (9–25) is

$$f(t) = \frac{d\delta(t)}{dt} + 2\delta(t) + [2e^{-t} - 1e^{-3t}]u(t)$$

The method illustrated by this example generalizes in the following way. When $m = n$, long division produces a quotient K plus a proper rational function remainder. The constant K corresponds to an impulse $K\delta(t)$, and the remainder can be expanded by partial fractions to find the corresponding waveform. If $m > n$, then long division yields a quotient with terms like $s, s^2, \ldots s^{m-n}$ before a proper remainder function is obtained. These higher powers of s correspond to derivatives of the impulse. These pathological waveforms are theoretically interesting, but they do not actually occur in real circuits.

Improper rational functions can arise during mathematical manipulation of signal transforms. When $F(s)$ is improper, it is essential to reduce it by long division prior to expansion; otherwise the resulting partial fraction expansion will be incomplete.

Exercise 9–12

Find the inverse transforms of the following functions:

(a) $F(s) = \dfrac{s^2 + 4s + 5}{s^2 + 4s + 3}$

(b) $F(s) = \dfrac{s^2 - 4}{s^2 + 4}$

Answers:

(a) $f(t) = \delta(t) + [e^{-t} - e^{-3t}]u(t)$

(b) $f(t) = \delta(t) - [4\sin(2t)]u(t)$

Exercise 9–13

Find the inverse transforms of the following functions:

(a) $F(s) = \dfrac{2s^2 + 3s + 5}{s}$

(b) $F(s) = \dfrac{s^3 + 2s^2 + s + 3}{s + 2}$

Answers:

(a) $f(t) = 2\dfrac{d\delta(t)}{dt} + 3\delta(t) + 5u(t)$

(b) $f(t) = \dfrac{d^2\delta(t)}{dt^2} + \delta(t) + [e^{-2t}]u(t)$

MULTIPLE POLES

Under certain special conditions, transforms can have multiple poles. For example, the transform

$$F(s) = \frac{K(s - z_1)}{(s - p_1)(s - p_2)^2} \tag{9–26}$$

has a simple pole at $s = p_1$ and a pole of order 2 at $s = p_2$. Finding the inverse transform of this function requires special treatment of the multiple pole. We first factor out one of the two multiple poles.

$$F(s) = \frac{1}{s - p_2}\left[\frac{K(s - z_1)}{(s - p_1)(s - p_2)}\right] \tag{9–27}$$

The quantity inside the brackets is a proper rational function with only simple poles and can be expanded by partial fractions using the method of the previous section.

$$F(s) = \frac{1}{s - p_2}\left[\frac{c_1}{s - p_1} + \frac{k_{22}}{s - p_2}\right]$$

We now multiply through by the pole factored out in the first step to obtain

$$F(s) = \frac{c_1}{(s - p_1)(s - p_2)} + \frac{k_{22}}{(s - p_2)^2}$$

The first term on the right is a proper rational function with only simple poles, so it too can be expanded by partial fractions as

$$F(s) = \frac{k_1}{s - p_1} + \frac{k_{21}}{s - p_2} + \frac{k_{22}}{(s - p_2)^2}$$

After two partial fraction expansions, we have an expression in which every term is available in Table 9–2. The first two terms are simple poles that lead to exponential waveforms. The third term is of the form $k/(s + \alpha)^2$, which is the transform of a damped ramp waveform $[kt\ e^{-\alpha t}]u(t)$. Therefore, the inverse transform of $F(s)$ in Eq. (9–26) is

$$f(t) = [k_1 e^{p_1 t} + k_{12} e^{p_2 t} + k_{22} t e^{p_2 t}]u(t) \tag{9–28}$$

Caution: If $F(s)$ in Eq. (9–26) had another finite zero, then the term in the brackets in Eq. (9–27) would be an improper rational function. When this occurs, long division must be used to reduce the improper rational function before proceeding to the partial-fraction expansion in the next step.

As with simple poles, the *s*-domain location of multiple poles determines the exponents of the exponential waveforms. The residues at the poles are the amplitudes of the waveforms. The only difference here is that the double pole leads to two terms rather than a single waveform. The first term is an exponential of the form e^{pt}, and the second term is a damped ramp of the form te^{pt}.

EXAMPLE 9–13

Find the inverse transform of

$$F(s) = \frac{4(s + 3)}{s(s + 2)^2}$$

SOLUTION:

The given transform has a simple pole at $s = 0$ and a double pole at $s = -2$. Factoring out one of the multiple poles and expanding the remainder by partial fractions yields

$$F(s) = \frac{1}{s + 2}\left[\frac{4(s + 3)}{s(s + 2)}\right] = \frac{1}{s + 2}\left[\frac{6}{s} - \frac{2}{s + 2}\right]$$

Multiplying through by the removed factor and expanding again by partial fractions produces

$$F(s) = \frac{6}{s(s + 2)} - \frac{2}{(s + 2)^2} = \frac{3}{s} - \frac{3}{s + 2} - \frac{2}{(s + 2)^2}$$

The last expansion on the right yields the inverse transform as

$$f(t) = [3 - 3e^{-2t} - 2te^{-2t}]u(t)$$

■

In principle, the procedure illustrated in this example can be applied to higher-order poles, although the process rapidly becomes quite tedious. For example, an *n*th-order pole would require *n* partial-fraction expansions, which is not an idea with irresistible appeal. Mathematics offers other methods of determining multiple pole residues that reduce the computational burden somewhat. However, these advanced mathematical tools are probably not worth learning because computer tools like Mathcad and MATLAB readily handle multiple pole situations (see Example 9–14). Although we do encounter functions with high-order multiple poles in later chapters, we rarely need to find their inverse transforms.

Thus, for practical reasons our interest in multiple pole transforms is limited to two possibilities. First, a double pole on the negative real axis leads to the damped ramp:

$$\mathcal{L}^{-1}\left[\frac{k}{(s+\alpha)^2}\right] = [kte^{-\alpha t}]u(t) \tag{9–29}$$

Second, a pair of double, complex poles leads to the damped cosine ramp:

$$\mathcal{L}^{-1}\left[\frac{k}{(s+\alpha-j\beta)^2} + \frac{k^*}{(s+\alpha+j\beta)^2}\right] = [2\,|k|\,te^{-\alpha t}\cos(\beta t + \angle k)]u(t) \tag{9–30}$$

These two cases illustrate a general principle: When a simple pole leads to a waveform $f(t)$, then a double pole at the same location leads to a waveform $tf(t)$. Multiplying a waveform by t tends to cause the waveform to increase without bound unless exponential damping is present. The following example illustrates a case in which the damping is not present.

EXAMPLE 9–14

Use a math analysis program to find the waveform corresponding to a transform with a zero at $s = -400$, a simple pole at $s = -1000$, a double pole at $s = j400$, a double pole at $s = -j400$, and a value at $s = 0$ of $F(0) = 2 \times 10^{-4}$. Plot the resulting waveform $f(t)$.

SOLUTION:

The symbolic analysis capability of Mathcad includes inverse Laplace transforms. Figure 9–5 is a worksheet demonstrating this capability. The specified critical frequencies define the $F(s)$ shown in the first line. The specified value at $s = 0$ allows us to evaluate the unknown scale factor K as shown in the second line. This leads directly to the transform in the third line. After highlighting (clicking on) an "s" in $F(s)$ to identify the independent variable, we select **Symbolic, Transforms, Inverse Laplace Transform** from the menu bar. Mathcad responds by producing the $f(t)$ shown in the figure. This waveform contains five terms: The exponential e^{-1000t} came from the simple pole at $s = -1000$, and the four sinusoidal terms $\cos(400t)$, $\sin(400t)$, $t \times \cos(400t)$, and $t \times \sin(400t)$ all came from the double poles at $s = \pm j400$. As time increases, the terms $t \times \cos(400t)$ and $t \times \sin(400t)$ increase without bound, leading to the linearly increasing amplitude shown in the waveform plot. A key point to remember is that poles on the j-axis lead to sustained waveforms, like the step function and sinusoid, that do not decay. Consequently, double poles on the j-axis lead to $tf(t)$ waveforms whose amplitudes increase without bound, producing an unstable response.

FIGURE 9–5

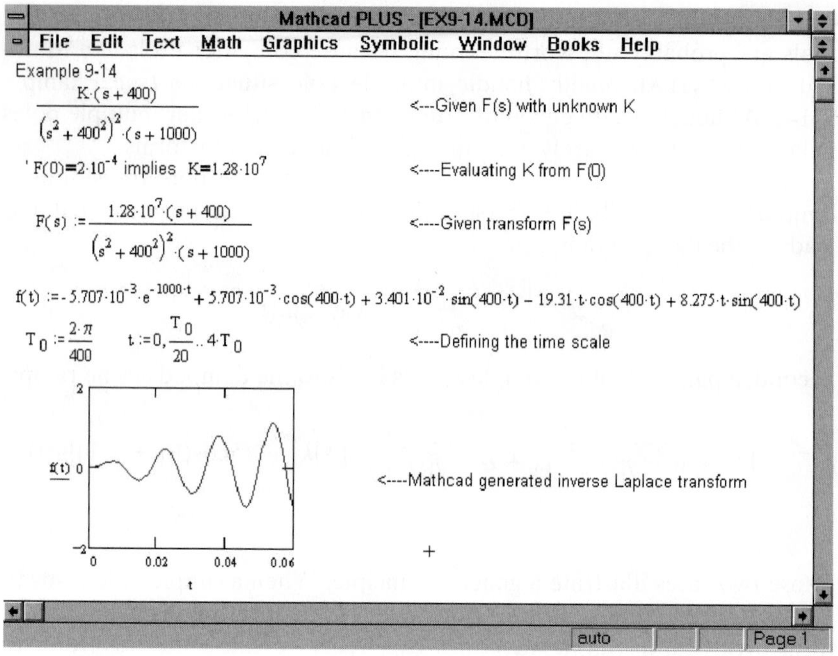

Mathcad PLUS - [EX9-14.MCD]

File Edit Text Math Graphics Symbolic Window Books Help

Example 9-14

$$\frac{K\cdot(s+400)}{\left(s^2+400^2\right)^2\cdot(s+1000)}$$ <---Given F(s) with unknown K

$F(0)=2\cdot10^{-4}$ implies $K=1.28\cdot10^{7}$ <----Evaluating K from F(0)

$$F(s):=\frac{1.28\cdot10^{7}\cdot(s+400)}{\left(s^2+400^2\right)^2\cdot(s+1000)}$$ <----Given transform F(s)

$$f(t):=-5.707\cdot10^{-3}\cdot e^{-1000\cdot t}+5.707\cdot10^{-3}\cdot\cos(400\cdot t)+3.401\cdot10^{-2}\cdot\sin(400\cdot t)-19.31\cdot t\cdot\cos(400\cdot t)+8.275\cdot t\cdot\sin(400\cdot t)$$

$T_0:=\frac{2\cdot\pi}{400}$ $\quad t:=0,\frac{T_0}{20}..4\cdot T_0$ <----Defining the time scale

f(t), t; 2, 0, −2; 0, 0.02, 0.04, 0.06 <----Mathcad generated inverse Laplace transform

auto Page 1

Exercise 9–14

Find the inverse transforms of the following functions:

(a) $F(s) = \dfrac{s}{(s+1)(s+2)^2}$

(b) $F(s) = \dfrac{16}{s^2(s+4)}$

(c) $F(s) = \dfrac{800\,s(s+1)}{(s+2)(s+10)^2}$

Answers:

(a) $f(t) = [-e^{-t} + e^{-2t} + 2te^{-2t}]u(t)$

(b) $f(t) = [4t - 1 + e^{-4t}]u(t)$

(c) $f(t) = [25e^{-2t} + 775e^{-10t} - 9000te^{-10t}]u(t)$

9–6 CIRCUIT RESPONSE USING LAPLACE TRANSFORMS

The payoff for learning about the Laplace transformation comes when we use it to find the response of dynamic circuits. The pattern for circuit analysis is shown by the solid line in Figure 9–1. The basic analysis steps are as follows:

STEP 1 Develop the circuit differential equation in the time domain.

STEP 2 Transform this equation into the *s* domain and algebraically solve for the response transform.

STEP 3 Apply the inverse transformation to this transform to produce the response waveform.

The first-order RC circuit in Figure 9–6 will be used to illustrate these steps.

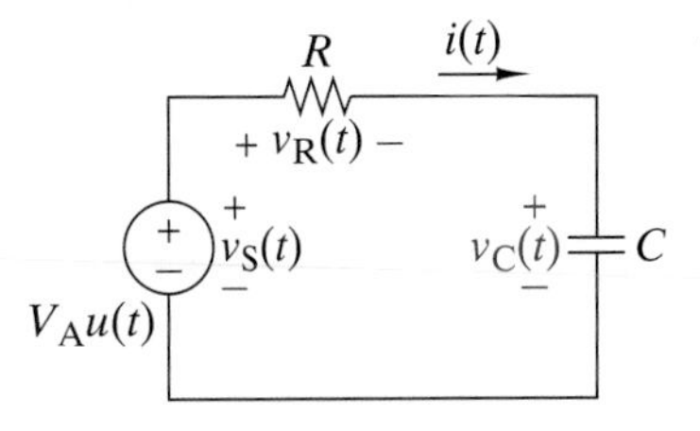

FIGURE 9–6 First-order RC circuit.

STEP 1

The KVL equation around the loop and the element i–v relationship or element equations are

$$\text{KVL:} \quad -v_S(t) + v_R(t) + v_C(t) = 0$$

$$\text{Source:} \ v_S(t) = V_A u(t)$$

$$\text{Resistor:} \ v_R(t) = i(t)R$$

$$\text{Capacitor:} \ i(t) = C\frac{dv_C(t)}{dt}$$

Substituting the i–v relationships into the KVL equation and rearranging terms produces a first-order differential equation

$$RC\frac{dv_C(t)}{dt} + v_C(t) = V_A u(t) \tag{9–31}$$

with an initial condition $v_C(0-) = V_0$ V.

STEP 2

The analysis objective is to use Laplace transforms to find the waveform $v_C(t)$ that satisfies this differential equation and the initial condition. We first apply the Laplace transformation to both sides of Eq. (9–31):

$$\mathcal{L}\left[RC\frac{dv_C(t)}{dt} + v_C(t)\right] = \mathcal{L}[V_A u(t)]$$

Using the linearity property leads to

$$RC\mathcal{L}\left[\frac{dv_C(t)}{dt}\right] + \mathcal{L}[v_C(t)] = V_A\mathcal{L}[u(t)]$$

Using the differentiation property and the transform of a unit step function produces

$$RC[sV_C(s) - V_0] + V_C(s) = V_A\frac{1}{s} \tag{9–32}$$

This result is an algebraic equation in $V_C(s)$, which is the transform of the response we seek. We rearrange Eq. (9–32) in the form

$$(s + 1/RC)V_C(s) = \frac{V_A/RC}{s} + V_0$$

and algebraically solve for $V_C(s)$.

$$V_C(s) = \frac{V_A/RC}{s(s + 1/RC)} + \frac{V_0}{s + 1/RC} \ \text{V-s} \tag{9–33}$$

The function $V_C(s)$ is the transform of the waveform $v_C(t)$ that satisfies the differential equation and the initial condition. The initial condition appears explicitly in this equation as a result of applying the differentiation rule to obtain Eq. (9–32).

STEP 3

To obtain the waveform $v_C(t)$, we find the inverse transform of the right side of Eq. (9–33). The first term on the right is a proper rational function with two simple poles on the real axis in the s plane. The pole at the origin was introduced by the step function input. The pole at $s = -1/RC$ came from the circuit. The partial fraction expansion of the first term in Eq. (9–33) is

$$\frac{V_A/RC}{s(s + 1/RC)} = \frac{k_1}{s} + \frac{k_2}{s + 1/RC}$$

The residues k_1 and k_2 are found using the cover-up algorithm.

$$k_1 = \left.\frac{V_A/RC}{s + 1/RC}\right|_{s=0} = V_A \quad \text{and} \quad k_2 = \left.\frac{V_A/RC}{s}\right|_{s=-1/RC} = -V_A$$

Using these residues, we expand Eq. (9–33) by partial fractions as

$$V_C(s) = \frac{V_A}{s} - \frac{V_A}{s + 1/RC} + \frac{V_0}{s + 1/RC} \tag{9–34}$$

Each term in this expansion is recognizable: The first is a step function and the next two are exponentials. Taking the inverse transform of Eq. (9–34) gives

$$\begin{aligned} v_C(t) &= [V_A - V_A e^{-t/RC} + V_0 e^{-t/RC}]u(t) \\ &= [V_A + (V_0 - V_A)e^{-t/RC}]u(t) \quad \text{V} \end{aligned} \tag{9–35}$$

The waveform $v_C(t)$ satisfies the differential equation in Eq. (9–31) and the initial condition $v_C(0-) = V_0$. The term $V_A u(t)$ is the forced response due to the step function input, and the term $[(V_0 - V_A)e^{t/RC}]u(t)$ is the natural response. The complete response depends on three parameters: the input amplitude V_A, the circuit time constant RC, and the initial condition V_0.

These results are identical to those found using the classical methods in Chapter 7. The outcome is the same, but the method is quite different. The Laplace transformation yields the complete response (forced and natural) by an algebraic process that inherently accounts for the initial conditions. The solid arrow in Figure 9–1 shows the overall procedure. Begin with Eq. (9–31) and relate each step leading to Eq. (9–35) to steps in Figure 9–1.

EXAMPLE 9–15

The switch in Figure 9–7 has been in position A for a long time. At $t = 0$ it is moved to position B. Find $i_L(t)$ for $t \geq 0$.

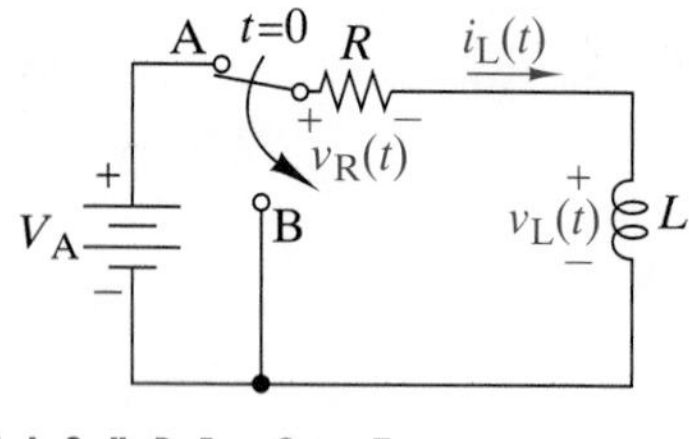

FIGURE 9–7

SOLUTION:

Step 1: The circuit differential equation is found by combining the KVL equation and element equations with the switch in position B.

$$\text{KVL: } v_R(t) + v_L(t) = 0$$

$$\text{Resistor: } v_R(t) = i_L(t)R$$

$$\text{Inductor: } v_L(t) = L\frac{di_L(t)}{dt}$$

Substituting the element equations into the KVL equation yields

$$L\frac{di_L(t)}{dt} + Ri_L(t) = 0$$

Prior to $t = 0$, the circuit was in a dc steady-state condition with the switch in position A. Under dc conditions the inductor acts like a short circuit, and the inductor current just prior to moving the switch is $i_L(0-) = I_0 = V_A/R$.

Step 2: Using the linearity and differentiation properties, we transform the circuit differential equation into the s domain as

$$L[sI_L(s) - I_0] + RI_L(s) = 0$$

Solving algebraically for $I_L(s)$ yields

$$I_L(s) = \frac{I_0}{s + R/L} \quad \text{A-s}$$

Step 3: The inverse transform of $I_L(s)$ is an exponential waveform:

$$i_L(t) = [I_0 e^{-Rt/L}]u(t) \quad \text{A}$$

where $I_0 = V_A/R$. Substituting $i_L(t)$ back into the differential equation yields

$$L\frac{di_L(t)}{dt} + Ri_L(t) = -RI_0 e^{-Rt/L} + RI_0 e^{-Rt/L} = 0$$

The waveform found using Laplace transforms does indeed satisfy the circuit differential equation and the initial condition. ■

EXAMPLE 9–16

The switch in Figure 9–8 has been open for a long time. At $t = 0$ the switch is closed. Find $i(t)$ for $t \geq 0$.

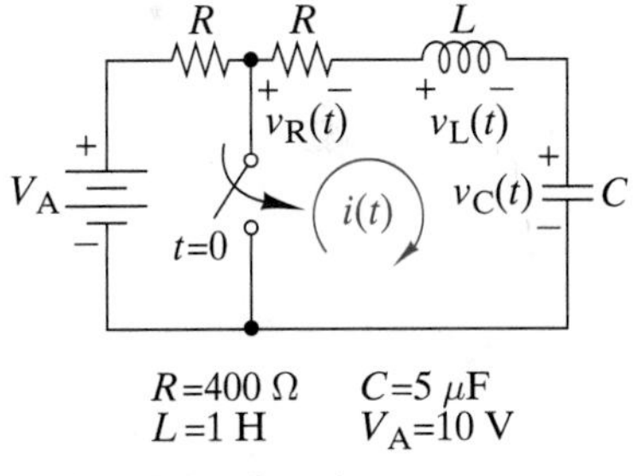

FIGURE 9–8

SOLUTION:

The governing equation for the second-order circuit in Figure 9–8 is found by combining the element equations and a KVL equation around the loop with the switch closed:

$$\text{KVL: } v_R(t) + v_L(t) + v_C(t) = 0$$

$$\text{Resistor: } v_R(t) = Ri(t)$$

$$\text{Inductor: } v_L(t) = L\frac{di(t)}{dt}$$

$$\text{Capacitor: } v_C(t) = \frac{1}{C}\int_0^t i(\tau)d\tau + v_C(0)$$

Substituting the element equations into the KVL equation yields

$$L\frac{di(t)}{dt} + Ri(t) + \frac{1}{C}\int_0^t i(\tau)d\tau + v_C(0) = 0$$

Using the linearity property, the differentiation property, and the integration property, we transform this second-order integrodifferential equation into the s domain as

$$L[sI(s) - i_L(0)] + RI(s) + \frac{1}{C}\frac{I(s)}{s} + v_C(0)\frac{1}{s} = 0$$

Solving for $I(s)$ results in

$$I(s) = \frac{si_L(0) - v_C(0)/L}{s^2 + \frac{R}{L}s + \frac{1}{LC}} \quad \text{A-s}$$

Prior to $t = 0$, the circuit was in a dc steady-state condition with the switch open. In dc steady state, the inductor acts like a short circuit and the capacitor like an open circuit, so the initial conditions are $i_L(0-) = 0$ A and $v_C(0-) = V_A = 10$ V. Inserting the initial conditions and the numerical values of the circuit parameters into the equation for $I(s)$ gives

$$I(s) = -\frac{10}{s^2 + 400s + 2 \times 10^5} \quad \text{A-s}$$

The denominator quadratic can be factored as $(s + 200)^2 + 400^2$ and $I(s)$ written in the following form:

$$I(s) = -\frac{10}{400}\left[\frac{400}{(s + 200)^2 + (400)^2}\right] \quad \text{A-s}$$

Comparing the quantity inside the brackets with the entries in the $F(s)$ column of Table 9–2, we find that $I(s)$ is a damped sine with $\alpha = 200$ and $\beta = 400$. By linearity, the quantity outside the brackets is the amplitude of the damped sine. The inverse transform is

$$i(t) = [-0.025\, e^{-200t} \sin 400t]u(t) \ \text{A}$$

Substituting this result back into the circuit integrodifferential equation yields the following term-by-term tabulation:

$$L\frac{di(t)}{dt} = +5e^{-200t} \sin 400t - 10e^{-200t} \cos 400t$$

$$Ri(t) = -10e^{-200t} \sin 400t$$

$$\frac{1}{C}\int_0^t i(\tau)d\tau = +5e^{-200t} \sin 400t + 10e^{-200t} \cos 400t - 10$$

$$v_C(0) = +10$$

The sum of the right-hand sides of these equations is zero. This result shows that the waveform $i(t)$ found using Laplace transforms does indeed satisfy the circuit integrodifferential equation and the initial conditions. ■

Exercise 9–15

Find the transform $V(s)$ that satisfies the following differential equations and the initial conditions:

(a) $\frac{dv(t)}{dt} + 6v(t) = 4u(t) \quad v(0-) = -3$

(b) $4\frac{dv(t)}{dt} + 12v(t) = 16\cos 3t, \quad v(0-) = 2$

Answers:

(a) $V(s) = \frac{4}{s(s+6)} - \frac{3}{s+6}$ V-s

(b) $V(s) = \frac{4s}{(s^2+9)(s+3)} + \frac{2}{s+3}$ V-s

Exercise 9–16

Find the $V(s)$ that satisfies the following equations:

(a) $\int_0^t v(\tau)d\tau + 10v(t) = 10u(t)$

(b) $\frac{d^2v(t)}{dt^2} + 4\frac{dv(t)}{dt} + 3v(t) = 5e^{-2t} \quad v'(0-) = 2 \quad v(0-) = -2$

Answers:

(a) $V(s) = \frac{1}{s+0.1}$ V-s

(b) $V(s) = \frac{5}{(s+1)(s+2)(s+3)} - \frac{2}{s+1}$ V-s

CIRCUIT RESPONSE WITH TIME-VARYING INPUTS

It is encouraging to find that the Laplace transformation yields results that agree with those obtained by classical methods. The transform method reduces solving circuit differential equations to an algebraic process that includes the initial conditions. However, before being overcome with euphoria we must remember that the Laplace transform method begins with the circuit differential equation and the initial conditions. It does not provide these quantities to us. The transform method simplifies the solution process, but it does not substitute for understanding how to formulate circuit equations.

The Laplace transform method is especially usefully when the circuit is driven by time-varying inputs. To illustrate, we return to the RC circuit in Figure 9–6 and replace the step function input by a general input signal denoted $v_S(t)$. The right side of the circuit differential equation in Eq. (9–31) changes to accommodate the new input by taking the form

$$RC\frac{dv_C(t)}{dt} + v_C(t) = v_S(t) \tag{9–36}$$

with an initial condition $v_C(0) = V_0$ V.

The only change here is that the driving force on the right side of the differential equation is a general time-varying waveform $v_S(t)$. The objective is to find the capacitor voltage $v_C(t)$ that satisfies the differential equation and the initial conditions. The classical methods of solving for the forced response depend on the form of $v_S(t)$. However, with the Laplace transform method we can proceed without actually specifying the form of the input signal.

We first transform Eq. (9–36) into the s domain:

$$RC[sV_C(s) - V_0] + V_C(s) = V_S(s)$$

The only assumption here is that the input waveform is Laplace transformable, a condition met by all causal signals of engineering interest. We now algebraically solve for the response $V_C(s)$:

$$V_C(s) = \frac{V_S(s)/RC}{s + 1/RC} + \frac{V_0}{s + 1/RC} \quad \text{V-s} \tag{9–37}$$

The function $V_C(s)$ is the transform of the response of the RC circuit in Figure 9–6 due to a general input signal $v_S(t)$. We have gotten this far without specifying the form of the input signal. In a sense, we have found the general solution in the s domain of the differential equation in Eq. (9–36) for any casual input signal.

All of the necessary ingredients are present in Eq. (9–37):

1. The transform $V_S(s)$ represents the applied input signal.
2. The pole at $s = -1/RC$ defines the circuit time constant.
3. The initial value $v_C(0-) = V_0$ summarizes all events prior to $t = 0$.

However, we must have a particular input in mind to solve for the waveform $v_C(t)$. The following examples illustrate the procedure for different input driving forces.

EXAMPLE 9–17

Find $v_C(t)$ in the RC circuit in Figure 9–6 when the input is the waveform $v_S(t) = [V_A e^{-\alpha t}]u(t)$.

SOLUTION:

The transform of the input is $V_S(s) = V_A/(s + \alpha)$. For the exponential input the response transform in Eq. (9–37) becomes

$$V_C(s) = \frac{V_A/RC}{(s + \alpha)(s + 1/RC)} + \frac{V_0}{s + 1/RC} \quad \text{V-s} \tag{9–38}$$

If $\alpha \neq 1/RC$, then the first term on the right is a proper rational function with two simple poles. The pole at $s = -\alpha$ came from the input and the pole at $s = -1/RC$ from the circuit. A partial fraction expansion of the first term has the form

$$\frac{V_A/RC}{(s + \alpha)(s + 1/RC)} = \frac{k_1}{s + \alpha} + \frac{k_2}{s + 1/RC}$$

The residues in this expansion are

$$k_1 = \left.\frac{V_A/RC}{s + 1/RC}\right|_{s=-\alpha} = \frac{V_A}{1 - \alpha RC}$$

$$k_2 = \left.\frac{V_A/RC}{s + \alpha}\right|_{s=-1/RC} = \frac{V_A}{\alpha RC - 1}$$

The expansion of the response transform $V_C(s)$ is

$$V_C(s) = \frac{V_A/(1 - \alpha RC)}{s + \alpha} + \frac{V_A/(\alpha RC - 1)}{s + 1/RC} + \frac{V_0}{s + 1/RC} \quad \text{V-s}$$

The inverse transform of $V_C(s)$ is

$$v_C(t) = \left[\frac{V_A}{1 - \alpha RC}e^{-\alpha t} + \frac{V_A}{\alpha RC - 1}e^{-t/RC} + V_0 e^{-t/RC}\right]u(t) \quad \text{V}$$

The first term is the forced response, and the last two terms are the natural response. The forced response is an exponential because the input introduced a pole at $s = -\alpha$. The natural response is also an exponential, but its time constant depends on the circuit's pole at $s = -1/RC$. In this case the forced and natural responses are both exponential signals with poles on the real axis. However, the forced response comes from the pole introduced by the input, while the natural response depends on the circuit's pole.

If $\alpha = 1/RC$, then the response just given is no longer valid (k_1 and k_2 become infinite). To find the response for this condition, we return to Eq. (9–38) and replace α by $1/RC$:

$$V_C(s) = \frac{V_A/RC}{(s + 1/RC)^2} + \frac{V_0}{s + 1/RC} \quad \text{V-s}$$

We now have a double pole at $s = -1/RC = -\alpha$. The double pole term is the transform of a damped ramp, so the inverse transform is

$$v_C(t) = \left[V_A \frac{t}{RC}e^{-t/RC} + V_0 e^{-t/RC}\right]u(t) \quad \text{W}$$

When $\alpha = 1/RC$, the s-domain poles of the input and the circuit coincide and the zero-state ($V_0 = 0$) response has the form $\alpha t e^{-\alpha t}$. We cannot separate this response into forced and natural components since the input and circuit poles coincide. ■

EXAMPLE 9–18

Find $v_C(t)$ when the input to the RC circuit in Figure 9–6 is $v_S(t) = [V_A \cos \beta t]u(t)$.

SOLUTION:

The transform of the input is $V_S(s) = V_A s/(s^2 + \beta^2)$. For a cosine input the response transform in Eq. (9–37) becomes

$$V_C(s) = \frac{sV_A/RC}{(s^2 + \beta^2)(s + 1/RC)} + \frac{V_0}{s + 1/RC} \quad \text{V-s}$$

The sinusoidal input introduces a pair of poles located at $s = \pm j\beta$. The first term on the right is a proper rational function with three simple poles. The partial fraction expansion of the first term is

$$\frac{sV_A/RC}{(s - j\beta)(s + j\beta)(s + 1/RC)} = \frac{k_1}{s - j\beta} + \frac{k_1^*}{s + j\beta} + \frac{k_2}{s + 1/RC}$$

To find the response, we need to find the residues k_1 and k_2:

$$k_1 = \left.\frac{sV_A/RC}{(s + j\beta)(s + 1/RC)}\right|_{s=j\beta} = \frac{V_A/2}{1 + j\beta RC} = |k_1|e^{j\theta}$$

where

$$|k_1| = \frac{V_A/2}{\sqrt{1 + (\beta RC)^2}} \quad \text{and} \quad \theta = -\tan^{-1}(\beta RC)$$

The residue k_2 at the circuit pole is

$$k_2 = \left.\frac{sV_A/RC}{s^2 + \beta^2}\right|_{s=-1/RC} = -\frac{V_A}{1 + (\beta RC)^2}$$

We now perform the inverse transform to obtain the response waveform:

$$v_C(t) = [2|k_1|\cos(\beta t + \theta) + k_2 e^{-t/RC} + V_0 e^{-t/RC}]u(t)$$

$$= \left[\frac{V_A}{\sqrt{1 + (\beta RC)^2}}\cos(\beta t + \theta) - \frac{V_A}{1 + (\beta RC)^2}e^{-t/RC} + V_0 e^{-t/RC}\right]u(t) \quad \text{V}$$

The first term is the forced response, and the remaining two are the natural response. The forced response is sinusoidal because the input introduces poles at $s = \pm j\beta$. The natural response is an exponential with a time constant determined by the location of the circuit's pole at $s = -1/RC$. ■

9–7 INITIAL VALUE AND FINAL VALUE PROPERTIES

The **initial value** and **final value properties** can be stated as follows:

$$\text{Initial value: } \lim_{t \to 0+} f(t) = \lim_{s \to \infty} sF(s) \qquad (9\text{–}39)$$
$$\text{Final value: } \lim_{t \to \infty} f(t) = \lim_{s \to 0} sF(s)$$

These properties display the relationship between the origin and infinity in the time and frequency domains. The value of $f(t)$ at $t = 0+$ in the time domain (initial value) is the same as the value of $sF(s)$ at infinity in the s plane. Conversely, the value of $f(t)$ as $t \to \infty$ (final value) is the same as the value of $sF(s)$ at the origin in the s plane.

Proof of both the initial value and final value properties starts with the differentiation property:

$$sF(s) - f(0-) = \int_{0-}^{\infty} \frac{df}{dt} e^{-st}\, dt \qquad (9\text{–}40)$$

To establish the initial value property, we rewrite the integral on the right side of this equation and take the limit of both sides as $s \to \infty$.

$$\lim_{s\to\infty}[sF(s) - f(0-)] = \lim_{s\to\infty}\int_{0-}^{0+}\frac{df}{dt}e^{-st}dt + \lim_{s\to\infty}\int_{0+}^{\infty}\frac{df}{dt}e^{-st}\,dt \quad (9\text{–}41)$$

The first integral on the right side reduces to $f(0+) - f(0-)$ since e^{-st} is unity on the interval from $t = 0-$ to $t = 0+$. The second integral vanishes because e^{-st} goes to zero as $s \to \infty$. In addition, on the left side of Eq. (9–41) the $f(0-)$ is independent of s and can be taken outside the limiting process. Inserting all of these considerations reduces Eq. (9–41) to

$$\lim_{s\to\infty} sF(s) = \lim_{t\to 0+} f(t) \quad (9\text{–}42)$$

which completes the proof of the initial value property.

Proof of the final value theorem begins by taking the limit of both sides of Eq. (9–40) as $s \to 0$:

$$\lim_{s\to 0}[sF(s) - f(0-)] = \lim_{s\to 0}\int_{0-}^{\infty}\frac{df}{dt}e^{-st}\,dt \quad (9\text{–}43)$$

The integral on the right side of this equation reduces to $f(\infty) - f(0-)$ because e^{-st} becomes unity as $s \to 0$. Again, the $f(0-)$ on the left side is independent of s and can be taken outside of the limiting process. Inserting all of these considerations reduces Eq. (9–43) to

$$\lim_{s\to 0} sF(s) = \lim_{t\to\infty} f(t) \quad (9\text{–}44)$$

which completes the proof of the final value property.

A damped cosine waveform provides an illustration of the application of these properties. The transform of the damped cosine is

$$\mathcal{L}\{[Ae^{-\alpha t}\cos\beta t]u(t)\} = \frac{A(s+\alpha)}{(s+\alpha)^2+\beta^2}$$

Applying the initial and final value limits, we obtain

$$\text{Initial value: } \lim_{t\to 0} f(t) = \lim_{t\to 0} Ae^{-\alpha t}\cos\beta t = A$$

$$\lim_{s\to\infty} sF(s) = \lim_{s\to\infty}\frac{sA\,(s+\alpha)}{(s+\alpha)^2+\beta^2} = A$$

$$\text{Final value: } \lim_{t\to\infty} f(t) = \lim_{t\to\infty} Ae^{-\alpha t}\cos\beta t = 0$$

$$\lim_{s\to 0} sF(s) = \lim_{s\to 0}\frac{sA\,(s+\alpha)}{(s+\alpha)^2+\beta^2} = 0$$

Note the agreement between the t-domain and s-domain limits in both cases.

There are restrictions on the initial and final value properties. The initial value property is valid when $F(s)$ is a proper rational function or, equivalently, when $f(t)$ does not have an impulse at $t = 0$. The final value property is valid when the poles of $sF(s)$ are in the left half plane or, equivalently, when $f(t)$ is a waveform that approaches a final value at $t \to \infty$. Note that the final value restriction allows $F(s)$ to have a simple pole at the origin since the limitation is on the poles of $sF(s)$.

Caution: The initial and final value properties will appear to work when the aforementioned restrictions are not met. In other words, these properties do not tell you they are giving nonsense answers when you violate their limitations. You must always check the restrictions on $F(s)$ before applying either of these properties.

For example, applying the final value property to a cosine waveform yields

$$\lim_{t \to \infty} \cos \beta t = \lim_{s \to 0} s \left[\frac{s}{s^2 + \beta^2} \right] = 0$$

The final value property appears to say that $\cos \beta t$ approaches zero as $t \to \infty$. This conclusion is incorrect since the waveform oscillates between ± 1. The problem is that the final value property does not apply to cosine waveform because $sF(s)$ has poles on the j-axis at $s = \pm j\beta$.

EXAMPLE 9–19

Use the initial and final value properties to find the initial and final values of the waveform whose transform is

$$F(s) = 2 \frac{(s + 3)}{s(s + 1)(s + 2)}$$

SOLUTION:

The given $F(s)$ is a proper rational function, so the initial value property can be applied as

$$f(0) = \lim_{s \to \infty} sF(s) = \lim_{s \to \infty} \left[2 \frac{(s + 3)}{(s + 1)(s + 2)} \right] = 0$$

The poles of $sF(s)$ are located in the left half plane at $s = -1$ and $s = -2$; hence, the final value property can be applied as

$$f(\infty) = \lim_{s \to 0} sF(s) = \lim_{s \to 0} \left[2 \frac{(s + 3)}{(s + 1)(s + 2)} \right] = 3$$

In Example 9–11 the waveform corresponding to this transform was found to be

$$f(t) = [3 - 4e^{-t} + e^{-2t}]u(t)$$

from which we find

$$f(0) = 3 - 4e^{-0} + e^{-0} = 0$$

$$f(\infty) = 3 - 4e^{-\infty} + e^{-\infty} = 3$$

which confirms the results found directly from $F(s)$. ■

Exercise 9–17

Find the initial and final values of the waveforms corresponding to the following transforms:

(a) $F_1(s) = 100 \dfrac{s + 3}{s(s + 5)(s + 20)}$

(b) $F_2(s) = 80 \dfrac{s(s + 5)}{(s + 4)(s + 20)}$

Answers:

(a) Initial value = 0, final value = 3.
(b) $F_2(s)$ is not a proper rational function, final value = 0.

APPLICATION EXAMPLE 9–20

Biological processes such as cellular osmosis or gluclose uptake involve the diffusion of a liquid (the solvent) containing a dissolved substance (the solute) from one volume to another. Figure 9–9 shows a schematic representation of two such volumes separated by a semipermeable membrane (γ). In this context V_1 and V_2 are the solvent volumes while m_1 and m_2 are the solute masses in each chamber.

The primary variables of interest are called the concentrations, defined as

$$c_1 = \frac{m_1}{V_1} \qquad \text{and} \qquad c_2 = \frac{m_2}{V_2}$$

Units used to quantify concentration include such things as milligrams per deciliter. The diffusion through the membrane is governed by two coupled first-order differential equations.

$$\frac{dc_1(t)}{dt} = \alpha_1[c_2(t) - c_1(t)]$$

$$\frac{dc_2(t)}{dt} = \alpha_2[c_1(t) - c_2(t)]$$

where $\alpha_1 = \gamma/V_1$, $\alpha_2 = \gamma/V_2$, and γ is the permeability coefficient of the membrane. These equations point out that when the two concentrations are equal ($c_1 = c_2$) the diffusion process is in equilibrium since both time derivatives are zero. Note that

$$\text{If } c_1 > c_2, \text{ then } \frac{dv_1}{dt} < 0 \text{ and } \frac{dv_2}{dt} > 0$$

and conversely

$$\text{If } c_1 < c_2, \text{ then } \frac{dv_1}{dt} > 0 \text{ and } \frac{dv_2}{dt} < 0$$

That is, when the two concentrations are unequal, the two derivatives have opposite signs and diffusion proceeds in such a direction as to bring the process back into equilibrium.

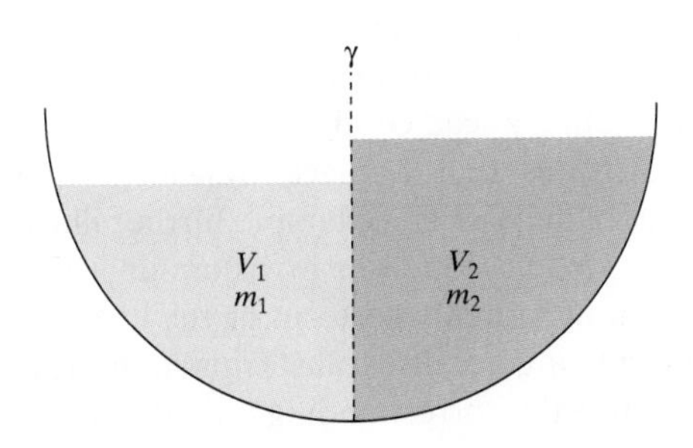

FIGURE 9–9

To describe the diffusion process we must solve these two differential equations simultaneously. One way to do this is to transform them into the s domain, where they become algebraic equations. Using the differentiation and linearity properties of the Laplace transformation, the two differential equations become

$$sC_1(s) - c_1(0) = \alpha_1[C_1(s) - C_2(s)]$$
$$sC_2(s) - c_2(0) = \alpha_2[C_2(s) - C_1(s)]$$

where $c_1(0)$ and $c_2(0)$ are the concentrations at $t = 0$. To solve these s-domain equations, we rearrange them as two linear algebraic equations in the transforms $C_1(s)$ and $C_2(s)$.

$$(s + \alpha_1)C_1(s) \quad - \alpha_1 C_2(s) \; = c_1(0)$$
$$-\alpha_2 C_1(s) + (s + \alpha_2)C_2(s) = c_2(0)$$

Applying Cramer's rule yields

$$C_1(s) = \frac{\Delta_1}{\Delta} = \frac{\begin{vmatrix} c_1(0) & -\alpha_1 \\ c_2(0) & s + \alpha_2 \end{vmatrix}}{\begin{vmatrix} s + \alpha_1 & -\alpha_1 \\ -\alpha_2 & s + \alpha_2 \end{vmatrix}} = \frac{(s + \alpha_2)c_1(0) + \alpha_1 c_2(0)}{s(s + \alpha_1 + \alpha_2)}$$

$$C_2(s) = \frac{\Delta_2}{\Delta} = \frac{\begin{vmatrix} s + \alpha_1 & c_1(0) \\ -\alpha_2 & c_2(0) \end{vmatrix}}{s(s + \alpha_1 + \alpha_2)} = \frac{(s + \alpha_1)c_2(0) + \alpha_2 c_1(0)}{s(s + \alpha_1 + \alpha_2)}$$

The corresponding time-domain waveforms can be found using the inverse Laplace transformation. However, let us see what we can infer by examining the response transforms. First, the two transforms have the same poles located at $s = 0$ and $s = -(\alpha_1 + \alpha_2)$. The first pole leads to a step function and the second pole to an exponential of the form $\exp[-(\alpha_1 + \alpha_2)t]$. This tells us that in the time domain the final state ($t \to \infty$) is a constant (the step function). The decaying exponential waveform connecting the initial ($t = 0$) to the final state has a time constant of $1/(\alpha_1 + \alpha_2)$. Thus, the time needed to transition from the initial to the final state is about $5/(\alpha_1 + \alpha_2)$.

Since the pole at $s = -(\alpha_1 + \alpha_2)$ is in the left half plane, we can apply the final value property to $sC_1(s)$ and $sC_2(s)$ to find the final state.

$$c_1(\infty) = \lim_{s \to 0} sC_1(s) = \lim_{s \to 0} \frac{(s + \alpha_2)c_1(0) + \alpha_1 c_2(0)}{(s + \alpha_1 + \alpha_2)} = \frac{\alpha_2 c_1(0) + \alpha_1 c_2(0)}{\alpha_1 + \alpha_2}$$

$$c_2(\infty) = \lim_{s \to 0} sC_2(s) = \lim_{s \to 0} \frac{(s + \alpha_1)c_2(0) + \alpha_2 c_1(0)}{(s + \alpha_1 + \alpha_2)} = \frac{\alpha_1 c_2(0) + \alpha_2 c_1(0)}{\alpha_1 + \alpha_2}$$

The two final values are equal, which means that the diffusion process proceeds from an initial state of $c_1(0) \neq c_2(0)$ to a final state with $c_1(\infty) = c_2(\infty)$, that is, from an unbalanced condition to an equilibrium condition.

This example illustrates that Laplace transforms have applications outside of linear circuit analysis. The example further illustrates that much can be learned about system responses by examining the response transforms themselves without formally performing the inverse transformation. This should not be surprising because transforms contain all of the information needed to determine the corresponding time-domain waveforms.

SUMMARY

- The Laplace transformation converts waveforms in the time domain to transforms in the s domain. The inverse transformation converts transforms into causal waveforms. A transform pair is unique if and only if $f(t)$ is causal.
- The Laplace transforms of basic signals like the step function, exponential, and sinusoid are easily derived from the integral definition. Other transform pairs can be derived using basic signal transforms and the uniqueness, linearity, time integration, time differentiation, and translation properties of the Laplace transformation.
- Proper rational functions with simple poles can be expanded by partial fraction to obtain inverse Laplace transforms. Simple real poles lead to exponential waveforms and simple complex poles to damped sinusoids. Partial-fraction expansions of improper rational functions and functions with multiple poles require special treatment.
- Using Laplace transforms to find the response of a linear circuit involves transforming the circuit differential equation into the s domain, algebraically solving for the response transform, and performing the inverse transformation to obtain the response waveform.
- The initial and final value properties determine the initial and final values of a waveform $f(t)$ from the value of $sF(s)$ at $s \to \infty$ and $s = 0$, respectively. The initial value property applies if $F(s)$ is a proper rational function. The final value property applies if all of the poles of $sF(s)$ are in the left half plane.

PROBLEMS

OBJECTIVE 9–1 LAPLACE TRANSFORM (SECTS. 9–1, 9–2, 9–3)

Find the Laplace transform of a given signal waveform using transform properties and pairs, or using the integral definition of the Laplace transformation. Locate the poles and zeros of the transform and construct a pole-zero diagram.
See Examples 9–1, 9–2, 9–3, 9–4, 9–5, 9–6, 9–7, 9–8, 9–9, 9–10

9–1 Find the Laplace transform of $f(t) = A[e^{-\alpha t} - 2e^{-\gamma t}]u(t)$. Locate the poles and zeros of $F(s)$.

9–2 Find the Laplace transform of $f(t) = A[(2 - \alpha t)e^{-\alpha t}]u(t)$. Locate the poles and zeros of $F(s)$.

9–3 Find the Laplace transform of $f(t) = A[1 - 2\cos(\beta t)]u(t)$. Locate the poles and zeros of $F(s)$.

9–4 Find the Laplace transform of $f(t) = A[\cos(\beta t) - \sin(\beta t)]u(t)$. Locate the poles and zeros of $F(s)$.

9–5 Find the Laplace transform of $f(t) = A\delta(t) - 2A\beta e^{-\beta t}\cos(\beta t)u(t)$. Locate the poles and zeros of $F(s)$.

9–6 Find the Laplace transform of $f(t) = A[2 - \alpha t - 2e^{-\alpha t}]u(t)$ for $\alpha > 0$. Locate the poles and zeros of $F(s)$.

9–7 Find the Laplace transforms of the following waveforms and plot their pole-zero diagrams:

(a) $f_1(t) = [5e^{-5t} - 10e^{-20t}]u(t)$
(b) $f_2(t) = [5\cos(10t) + 7\cos(20t)]u(t)$

9–8 Find the Laplace transforms of the following waveforms and plot their pole-zero diagrams.

(a) $f_1(t) = 3\delta(t) + [10e^{-10t} - 40e^{-40t}]u(t)$
(b) $f_2(t) = [20 - 15\cos(500t)]u(t)$

9–9 Find the Laplace transform of the following waveform. Locate the poles and zeros of $F(s)$.

(a) $f_1(t) = \delta(t) - (625te^{-50t})u(t)$
(b) $f_2(t) = [5 + e^{-20t} - 6\cos(10t) + 2\sin(10t)]u(t)$

9–10 Find the Laplace transforms of the following waveforms:

(a) $f_1(t) = 2\delta(t - 2)$
(b) $f_2(t) = e^{-50(t-1)}u(t - 1)$
(c) $f_3(t) = e^{-50(t-2)}u(t - 2)$

9–11 Find the Laplace transform of the following waveform:

$$f(t) = +[5 - 2e^{-5t}]u(t) - 3\cos[10(t - 0.5)]u(t - 0.5)$$

9–12 Find the Laplace transform of the following waveforms:

(a) $f_1(t) = \dfrac{d}{dt}(10e^{-5t}\cos 20t)$

(b) $f_2(t) = \displaystyle\int_0^t e^{-10x}dx + 10\,te^{-10t}u(t)$

9–13 **(a)** Write an expression for the waveform $f(t)$ in Figure P9–13 using step and ramp functions.

(b) Use the time-domain translation property to find the Laplace transform of the waveform $f(t)$ found in part (a).

(c) Verify the Laplace transform found in part (b) by applying the definition of the Laplace transformation in Eq. (9–2) to the waveform $f(t)$ found in part (a).

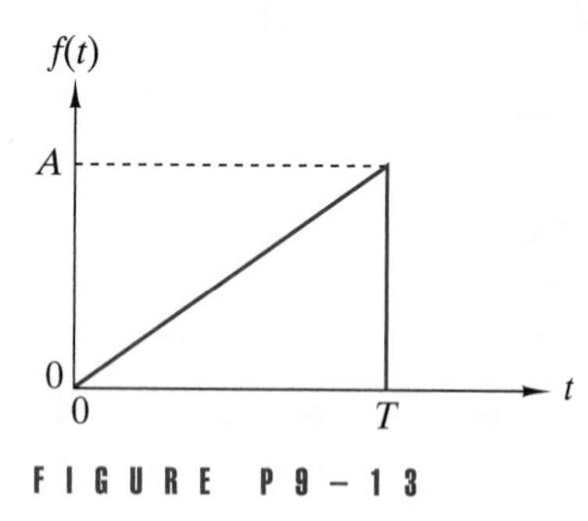

FIGURE P9–13

9–14 **(a)** Write an expression for the waveform $f(t)$ in Figure P9–14 using step functions.

(b) Use the time-domain translation property to find the Laplace transform of the waveform $f(t)$ found in part (a).

(c) Verify the Laplace transform found in part (b) by applying the definition of the Laplace transformation in Eq. (9–2) to the waveform $f(t)$ found in part (a).

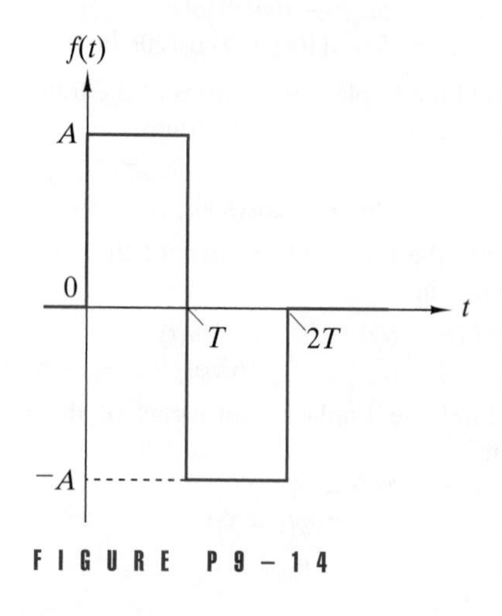

FIGURE P9–14

9–15 Find the Laplace transform of the following function. Locate the poles and zeros of $F(s)$.

$$f(t) = [1600t + 8 + 75e^{-10t} + 17e^{-50t}]u(t)$$

OBJECTIVE 9–2 INVERSE TRANSFORMS (SECTS. 9–4, 9–5)

Find the inverse transform of a given Laplace transform using partial-fraction expansion or basic transform properties and pairs.

See Examples 9–11, 9–12, 9–13, 9–14 and Exercises 9–7, 9–8, 9–9, 9–10, 9–11, 9–12, 9–13, 9–14

9–16 Find the inverse Laplace transforms of the following functions:

(a) $F_1(s) = \dfrac{s + 20}{s(s + 10)}$

(b) $F_2(s) = \dfrac{s^2 + 10s + 10}{s(s + 10)}$

9–17 Find the inverse Laplace transforms of the following functions:

(a) $F_1(s) = \dfrac{2(s + 5)}{s(s + 10)}$

(b) $F_2(s) = \dfrac{s^2}{(s + 5)(s + 10)}$

9–18 Find the inverse Laplace transforms of the following functions:

(a) $F_1(s) = \dfrac{20(s + 20)}{(s + 10)^2 + 400}$

(b) $F_2(s) = \dfrac{20(s - 20)}{(s + 10)^2 + 400}$

9–19 Find the inverse Laplace transforms of the following functions and sketch their waveforms for $\beta > 0$:

(a) $F_1(s) = \dfrac{\beta(s + \beta)}{s(s^2 + \beta^2)}$

(b) $F_2(s) = \dfrac{s(s + \beta)}{s^2 + \beta^2}$

9–20 Find the inverse Laplace transforms of the following functions:

(a) $F_1(s) = \dfrac{\alpha^2}{s^2(s + \alpha)}$

(b) $F_2(s) = \dfrac{\alpha^2}{s(s + \alpha)^2}$

9–21 Find the inverse Laplace transforms of the following functions:

(a) $F_1(s) = \dfrac{6\alpha^2}{(s + \alpha)(s + 2\alpha)(s + 4\alpha)}$

(b) $F_2(s) = \dfrac{6(s + 2\alpha)}{(s + \alpha)(s + 4\alpha)}$

9–22 Find the inverse transforms of the following functions:

(a) $F_1(s) = \dfrac{16}{(s+2)(s^2+12s+36)}$

(b) $F_2(s) = \dfrac{2(s^2+2)}{s(s^2+4)}$

9–23 Find the inverse transforms of the following functions:

(a) $F_1(s) = \dfrac{(s+4)(s+8)}{s(s+2)(s+6)}$

(b) $F_2(s) = \dfrac{3s^4+10s^2+4}{s(s^2+1)(s^2+4)}$

9–24 Find the inverse transforms for the following functions:

(a) $F_1(s) = \dfrac{30(s+2)}{s(s^2+4s+5)}$

(b) $F_2(s) = \dfrac{2s}{(s+4)(s^2+4s+8)}$

9–25 Find the inverse transforms for the following functions:

(a) $F_1(s) = \dfrac{2(s^2+16)}{s(s^2+8s+32)}$

(b) $F_2(s) = \dfrac{s^2+30s+800}{s(s^2+50s+400)}$

9–26 Find the inverse transforms of the following functions:

(a) $F_1(s) = \dfrac{(s+40)^2}{(s+10)^2(s+100)}$

(b) $F_2(s) = \dfrac{(s+10)^2}{(s+40)^2(s+100)}$

9–27 A certain transform has a simple pole at $s = -20$, a simple zero at $s = -\gamma$, and a scale factor of $K = 1$. Select values for γ so the inverse transform is

(a) $f(t) = \delta(t) - 5e^{-20t}$
(b) $f(t) = \delta(t)$
(c) $f(t) = \delta(t) + 5e^{-20t}$

9–28 Find the inverse transforms of the following functions:

(a) $F_1(s) = \dfrac{(s+40)e^{-2s}}{(s+10)(s+20)}$

(b) $F_2(s) = \dfrac{se^{-2s}+20}{(s+10)(s+40)}$

9–29 Use Mathcad or MATLAB to find the inverse transform of the following function:

$$F(s) = \frac{s(s^2+3s+4)}{(s+2)(s^3+6s^2+16s+16)}$$

9–30 Use Mathcad or MATLAB to find the inverse transform of the following function:

$$F(s) = \frac{40(s^3+2s^2+s+2)}{s(s^3+4s^2+4s+16)}$$

OBJECTIVE 9–3 CIRCUIT RESPONSE USING LAPLACE TRANSFORMS (SECT. 9–6)

Given a first- or second-order circuit:
(a) Determine the circuit differential equation and the initial conditions (if not given).
(b) Transform the differential equation into the s domain and solve for the response transform.
(c) Use the inverse transformation to find the response waveform.
See Examples 9–15, 9–16, 9–17, 9–18, 9–19

9–31 Use the Laplace transformation to find the $y(t)$ that satisfies the following first-order differential equations:

(a) $50\dfrac{dy}{dt} + 250y = 0$ with $y(0-) = 10$

(b) $\dfrac{dy}{dt} + 20y = 40u(t)$ with $y(0-) = -10$

9–32 Use the Laplace transformation to find the $y(t)$ that satisfies the following first-order differential equation:

$$\frac{dy}{dt} + 500y = [2500e^{-250t}]u(t) \quad \text{with} \quad y(0-) = 0$$

9–33 The switch in Figure P9–33 has been open for a long time and is closed at $t = 0$. The circuit parameters are $R = 200\ \Omega$, $L = 0.2$ H, and $V_A = 10$ V.
(a) Find the differential equation for the inductor current $i_L(t)$ and initial condition $i_L(0)$.
(b) Solve for $i_L(t)$ using the Laplace transformation.

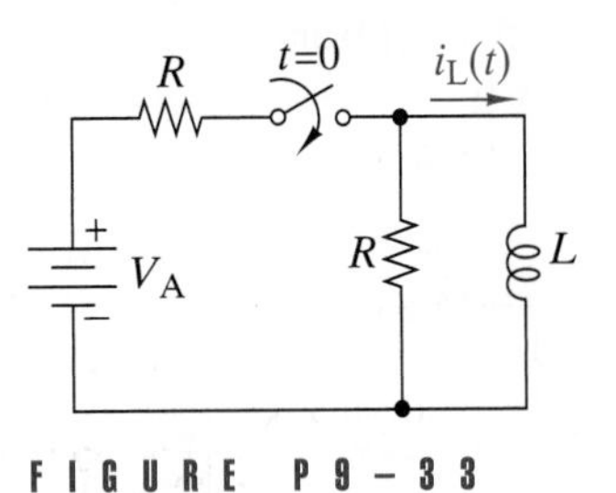

FIGURE P9–33

9–34 The switch in Figure P9–33 has been closed for a long time and is opened at $t = 0$. The circuit parameters are $R = 200\ \Omega$, $L = 200$ mH, and $V_A = 50$ V.
(a) Find the differential equation for the inductor current $i_L(t)$ and initial condition $i_L(0)$.
(b) Solve for $i_L(t)$ using the Laplace transformation.

9–35 The switch in Figure P9–35 has been open for a long time. At $t = 0$ the switch is closed.
(a) Find the differential equation for the capacitor voltage and initial condition.
(b) Find $v_O(t)$ using the Laplace transformation when $v_S(t) = 15[e^{-2500t}]u(t)$.

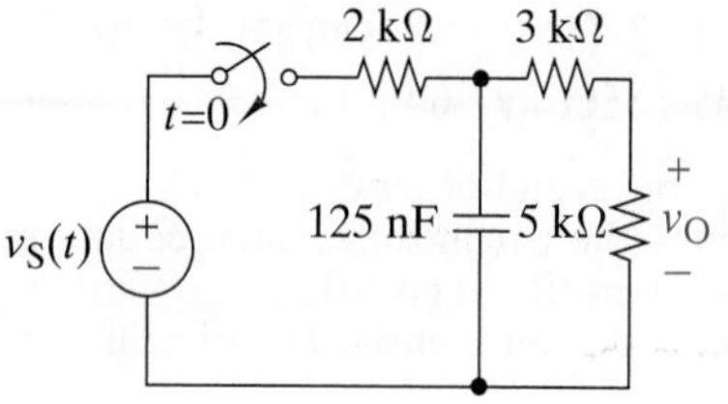

FIGURE P9–35

9–36 Repeat Problem 9–35 for the input waveform $v_S(t) = 10[\sin 1000t]u(t)$.

9–37 Use the Laplace transformation to find the $y(t)$ that satisfies the following second-order differential equation:

$$\frac{d^2y}{dt^2} + 15\frac{dy}{dt} + 50y = 0$$

with $y(0-) = 10$ and $y'(0-) = 0$.

9–38 Use the Laplace transformation to find the $y(t)$ that satisfies the following second-order differential equation:

$$\frac{d^2y}{dt^2} + 10\frac{dy}{dt} + 50y = 0$$

with $y(0-) = 0$ and $y'(0-) = 10$.

9–39 The switch in Figure P9–39 has been open for a long time and is closed at $t = 0$. The circuit parameters are $R = 500\ \Omega$, $L = 2.5$ H, $C = 2.5\ \mu$F, and $V_A = 20$ V.

(a) Find the circuit differential equation in $i_L(t)$ and the initial conditions $i_L(0)$ and $v_C(0)$.

(b) Use Laplace transforms to solve for the $i_L(t)$ for $t \geq 0$.

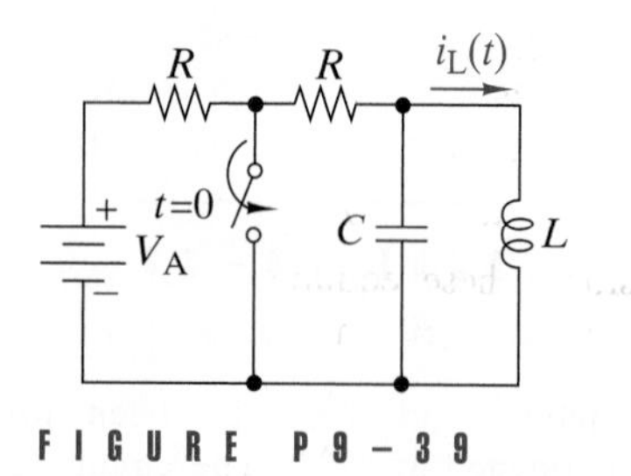

FIGURE P9–39

9–40 Repeat Problem 9–39 with $L = 1.25$ H.

9–41 The switch in Figure P9–41 has been closed for a long time and is opened at $t = 0$. The circuit parameters are $L = 5$ H, $C = 2.5\ \mu$F, $R_1 = 1$ kΩ, $R_2 = 2$ kΩ, and $v_S = 20u(t)$ V.

(a) Find the circuit differential equation in $v_C(t)$ and the initial conditions. $i_L(0)$ and $v_C(0)$.

(b) Use Laplace transforms to solve for the $v_C(t)$ for $t \geq 0$.

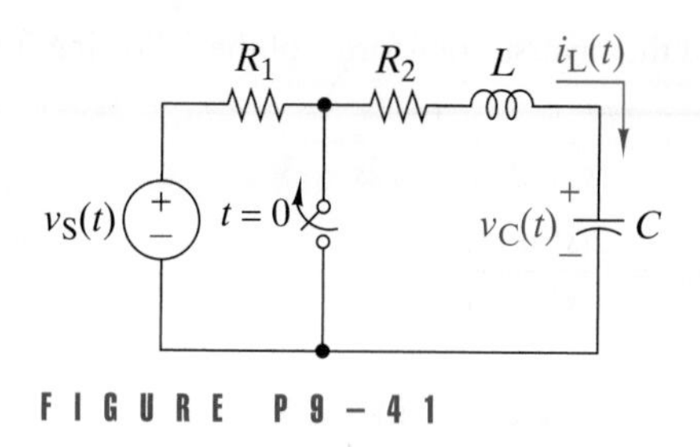

FIGURE P9–41

9–42 Repeat Problem 9–41 with $R_2 = 1$ kΩ.

9–43 In Figure P9–43 the input is $v_S(t) = V_A u(t)$ and the circuit initial conditions are $v_C(0) = 0$ and $i_L(0) = 0$. In the time domain the KVL equation around the loop can be written as

$$\underbrace{L\frac{di(t)}{dt}}_{v_L(t)} + \underbrace{Ri(t)}_{v_R(t)} + \underbrace{\frac{1}{C}\int_0^t i(x)\,dx + v_C(0)}_{v_C(t)} - v_S(t) = 0$$

Transform the KVL equation into the s domain and solve for $V_L(s)$, $V_C(s)$, and $V_R(s)$.

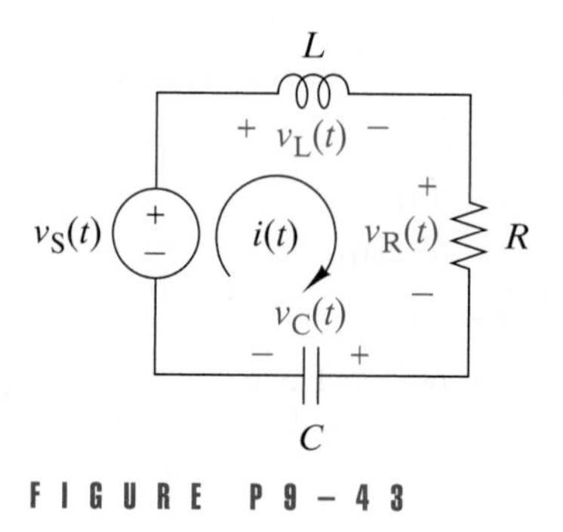

FIGURE P9–43

9–44 Repeat problem 9–43 with $v_S(t) = 0$, $v_C(0) = V_0$, and $i_L(0) = 0$.

OBJECTIVE 9–4 INITIAL AND FINAL VALUE PROPERTIES (SECT. 9–7)

Find the initial and final values of waveforms using the initial and final value properties of Laplace transforms.
See Example 9–19 and Exercise 9–17

9–45 Use the initial and final value properties to find the initial and final values of the waveform corresponding to the transforms in Problem 9–22. If either property is not applicable, explain why.

9–46 Use the initial and final value properties to find the initial and final values of the waveform corresponding to the transforms in Problem 9–23. If either property is not applicable, explain why.

9–47 Use the initial and final value properties to find the initial and final values of the waveform corresponding to the following transforms. If either property is not applicable, explain why.

(a) $F_1(s) = \dfrac{2(s^2 + 5s + 6)}{(s + 2)(s + 6)(s + 12)}$

(b) $F_2(s) = \dfrac{10(s^2 + 10s + 40)}{s(s^2 - 100)}$

9–48 Use the initial and final value properties to find the initial and final values of the waveform corresponding to the following transforms. If either property is not applicable, explain why.

(a) $F_1(s) = \dfrac{s(s + 5)}{s^2 + 6s + 9}$

(b) $F_2(s) = \dfrac{10(s^2 + 10s - 20)}{s(s^2 + 100)}$

9–49 Use the initial and final value properties to find the initial and final values of the waveform corresponding to the transforms in Problem 9–25. If either property is not applicable, explain why.

9–50 Use the initial and final value properties to find the initial and final values of the waveform corresponding to the transforms in Problem 9–26. If either property is not applicable, explain why.

INTEGRATING PROBLEMS

9–51 **A** The Dominant Pole Approximation

When a transform $F(s)$ has widely separated poles, then those closest to the j-axis tend to dominate the response because they have less damping. An approximation to the waveform can be obtained by ignoring the contributions of all except the dominant poles. We can ignore the other poles by simply discarding their terms in the partial-fraction expansion of $F(s)$. The purpose of this example is to examine a dominant pole approximation of the transform

$$F(s) = 10^6 \frac{s + 1000}{(s + 4000)[(s + 25)^2 + 100^2]}$$

(a) Construct a partial-fraction expansion of $F(s)$ and find $f(t)$.

(b) Construct a pole-zero diagram of $F(s)$ and identify the dominant poles.

(c) Construct a dominant pole approximation $g(t)$ by discarding the other poles in the partial fraction expansion in part (a).

(d) Plot $f(t)$ and $g(t)$ and comment on the accuracy of the approximation.

9–52 **A** First-Order Circuit Step Response

In Chapter 7 we found that the step response of a first-order circuit can be written as

$$f(t) = f(\infty) + [f(0) - f(\infty)]e^{-t/T_C}$$

where $f(0)$ is the initial value, $f(\infty)$ is the final value, and T_C is the time constant. Show that the corresponding transform has the form

$$F(s) = K\left[\frac{s + \gamma}{s(s + \alpha)}\right]$$

and relate the time-domain parameters $f(0)$, $f(\infty)$, and T_C to the s-domain parameters K, γ, and α.

9–53 **A** Inverse Transform for Complex Poles

In Section 9–4 we learned that complex poles occur in conjugate pairs and that for simple poles the partial-fraction expansion of $F(s)$ will contain two terms of the form

$$F(s) = \cdots \frac{k}{s + \alpha - j\beta} + \frac{k^*}{s + \alpha + j\beta} + \cdots$$

Show that when the complex conjugate residues are written in rectangular form as

$$k = a + jb \quad \text{and} \quad k^* = a - jb$$

the corresponding term in the waveform $f(t)$ is

$$f(t) = \cdots + 2e^{-\alpha t}[a\cos(\beta t) - b\sin(\beta t)] + \cdots$$

9–54 **A** Solving State Variable Equations

For zero input a series RLC circuit can be described by the following coupled first-order equations in the inductor current $i_L(t)$ and capacitor voltage $v_C(t)$.

$$\frac{dv_C(t)}{dt} = \frac{1}{C} i_L(t)$$

$$\frac{di_L(t)}{dt} = -\frac{1}{L} v_C(t) - \frac{R}{L} i_L(t)$$

(a) Transform these equations into the s domain and solve for the transforms $I_L(s)$ and $V_C(s)$ in terms of the initial conditions $i_L(0) = I_0$ and $v_C(0) = V_0$.

(b) Find $i_L(t)$ and $v_C(t)$ for $R = 3\text{ k}\Omega$, $L = 10$ H, $C = 5\ \mu$F, $I_0 = 5$ mA, and $V_0 = 5$ V.

9–55 **A** Complex Differentiation Property

The complex differentiation property of the Laplace transformation states that

$$\text{If } \mathcal{L}\{f(t)\} = F(s), \text{ then } \mathcal{L}\{tf(t)\} = -\frac{d}{ds}F(s)$$

Use this property to find the Laplace transforms of $f(t) = \{tg(t)\}u(t)$ when $g(t) = e^{-\alpha t}$. Repeat when $g(t) = \sin \beta t$ and $\cos \beta t$.

CHAPTER 10

s-DOMAIN CIRCUIT ANALYSIS

The resistance operator Z is a function of the electrical constants of the circuit components and of d/dt, the operator of time-differentiation, which will in the following be denoted by p simply.

Oliver Heaviside, 1887,
British Engineer

Some History Behind This Chapter

The use of operational methods to study electric circuits was pioneered by Oliver Heaviside (1850–1925). The quotation given here was taken from his book *Electrical Papers* originally published in 1887. His resistance operator Z, which he later called impedance, is a central theme for much of electrical engineering. Heaviside does not often receive the recognition he deserves, in part because his intuitive approach was not accepted by most Victorian scientists of his day. Mathematical justification for his methods was eventually supplied by John Bromwich and others. However, no important errors were found in Heaviside's results.

Why This Chapter Is Important Today

In this chapter we use Laplace transforms to make analyzing dynamic circuits no more difficult than dc circuits. You will see that all the circuit analysis tools learned in Chapters 1 through 4 can be extended to the study of the transient response of linear circuits. A new tool called pole-zero diagrams will give you a new way to visualize and predict circuit behavior.

Chapter Sections

10–1 Transformed Circuits
10–2 Basic Circuit Analysis in the s Domain
10–3 Circuit Theorems in the s Domain
10–4 Node-Voltage Analysis in the s Domain
10–5 Mesh-Current Analysis in the s Domain
10–6 Summary of s-Domain Circuit Analysis

Chapter Learning Objectives

10-1 Equivalent Impedance (Sects. 10-1, 10-2)

Given a linear circuit, use series and parallel equivalence to find the equivalent impedance at specified terminal pairs. Select element values to obtain specified pole locations.

10-2 Basic Circuit Analysis Techniques (Sects. 10-2, 10-3)

Given a linear circuit:

(a) Determine the initial conditions (if not given) and transform the circuit into the s domain.
(b) Solve for zero-state and zero-input responses using circuit reduction, the unit output method, Thévenin or Norton equivalent circuits, or superposition.
(c) Identify the forced and natural poles in the responses, or select circuit parameters to place the natural poles at specified locations.

10-3 General Circuit Analysis (Sects. 10-4, 10-5)

Given a linear circuit:

(a) Determine the initial conditions (if not given) and transform the circuit into the s domain.
(b) Solve for zero-state and zero-input response transforms and waveforms using node-voltage or mesh-current methods.
(c) Identify the forced and natural poles in the responses, or select circuit parameters to place the natural poles at specified locations.

10–1 TRANSFORMED CIRCUITS

So far we have used the Laplace transformation to change waveforms into transforms and convert circuit differential equations into algebraic equations. These operations provide a useful introduction to the s domain. However, the real power of the Laplace transformation emerges when we transform the circuit itself and study its behavior directly in the s domain.

The solid arrow in Figure 10–1 indicates the analysis path we will be following in this chapter. The process begins with a linear circuit in the time domain. We transform the circuit into the s domain, write the circuit equations directly in that domain, and then solve these algebraic equations for response transform. The inverse Laplace transformation then produces the response waveform. However, the s-domain approach is not just another way to derive response waveforms. This approach allows us to work directly with the circuit model using analysis tools such as voltage division and equivalence. By working directly with the circuit model, we gain insights into the interaction between circuits and signals that cannot be obtained using the classical approach indicated by the dotted path in Figure 10–1.

How are we to transform a circuit? We have seen several times that circuit analysis is based on device and connection constraints. The connection constraints are derived from Kirchhoff's laws and the device constraints from the i–v relationships used to model the physical devices in the circuit. To transform circuits, we must see how these two types of constraints are altered by the Laplace transformation.

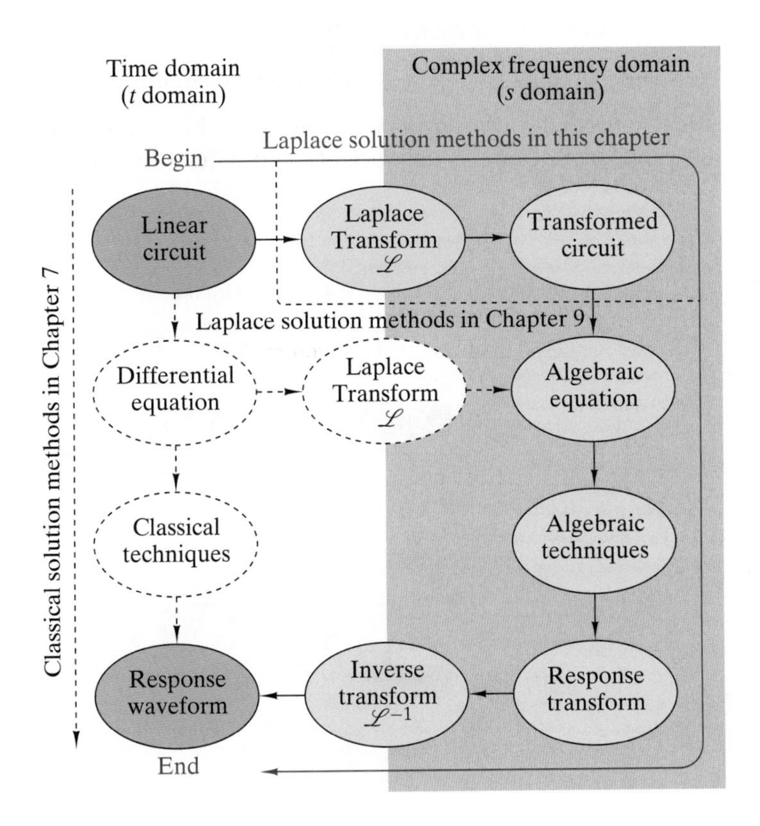

FIGURE 10–1 *Flow diagram for s-domain circuit analysis.*

CONNECTION CONSTRAINTS IN THE s DOMAIN

A typical KCL connection constraint could be written as

$$i_1(t) + i_2(t) - i_3(t) + i_4(t) = 0$$

This connection constraint requires that the sum of the current waveforms at a node be zero for all times t. Using the linearity property, the Laplace transformation of this equation is

$$I_1(s) + I_2(s) - I_3(s) + I_4(s) = 0$$

In the s domain the KCL connection constraint requires that the sum of the current transforms be zero for all values of s. This idea generalizes to any number of currents at a node and any number of nodes. In addition, this idea obviously applies to Kirchhoff's voltage law as well. The form of the connection constraints do not change because they are linear equations and the Laplace transformation is a linear operation. In summary, KCL and KVL apply to waveforms in the t domain and to transforms in the s domain.

ELEMENT CONSTRAINTS IN THE s DOMAIN

Turning now to the element constraints, we first deal with the independent signal sources shown in Figure 10–2. The i–v relationships for these elements are

$$\begin{aligned} \text{Voltage source: } v(t) &= v_S(t) \\ i(t) &= \text{Depends on circuit} \\ \text{Current source: } i(t) &= i_S(t) \\ v(t) &= \text{Depends on circuit} \end{aligned} \tag{10–1}$$

Independent sources are two-terminal elements. In the t domain they constrain the waveform of one signal variable and adjust the unconstrained

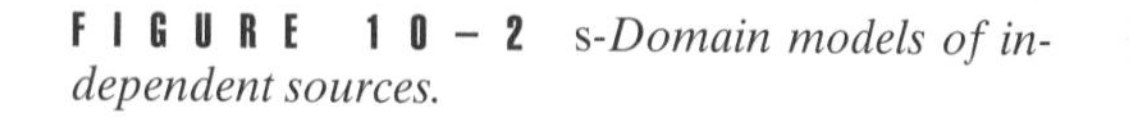

FIGURE 10 – 2 s-Domain models of independent sources.

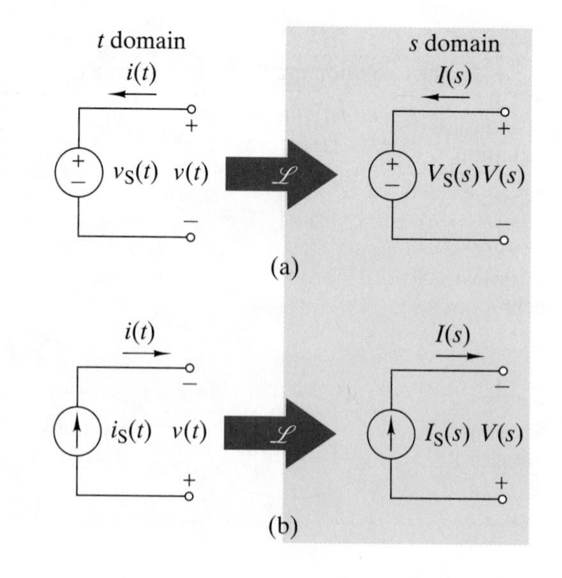

variable to meet the demands of the external circuit. We think of an independent source as a generator of a specified voltage or current waveform. The Laplace transformation of the expressions in Eq. (10–1) yields

$$\begin{aligned} \text{Voltage source: } V(s) &= V_S(s) \\ I(s) &= \text{Depends on circuit} \\ \text{Current source: } I(s) &= I_S(s) \\ V(s) &= \text{Depends on circuit} \end{aligned} \tag{10–2}$$

In the s domain, independent sources function the same way as in the t domain, except that we think of them as generating voltage or current transforms rather than waveforms.

Next we consider the active elements in Figure 10–3. In the time domain, the element constraints for linear dependent sources are linear algebraic equations. Because of the linearity property of the Laplace transformation,

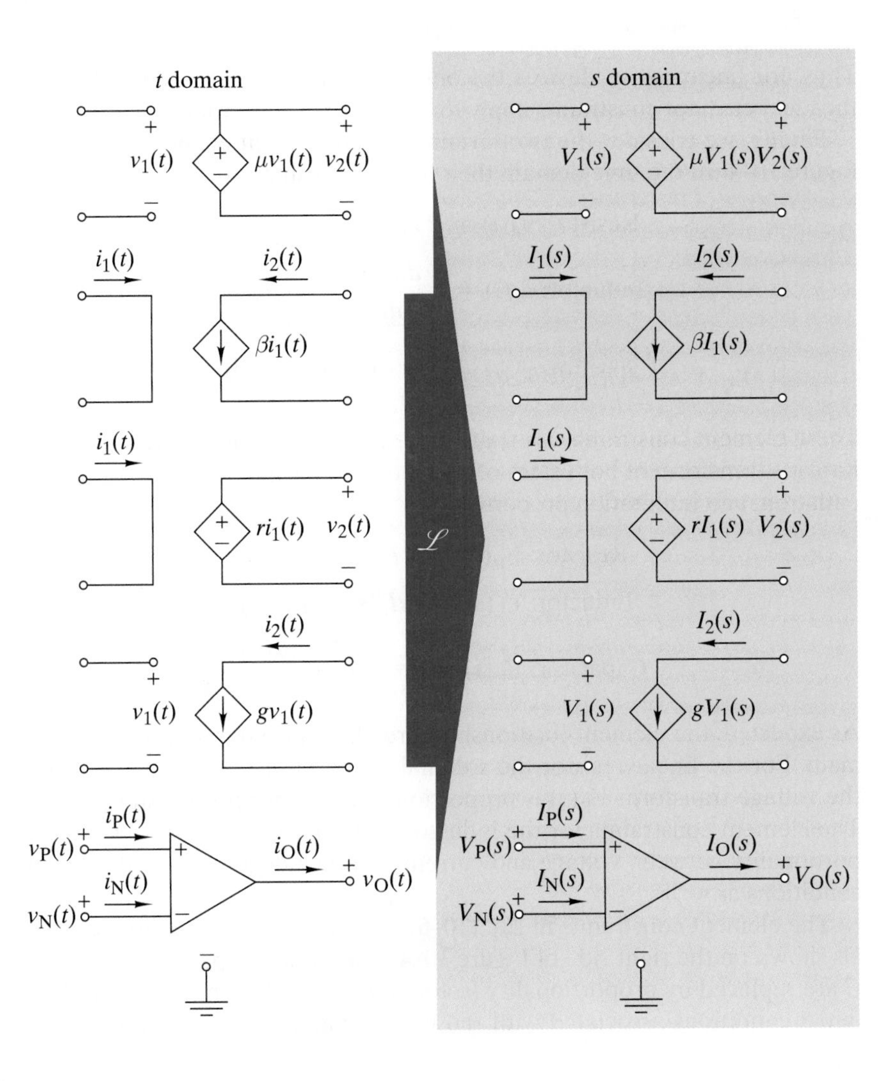

FIGURE 10–3 s-*Domain models of dependent sources and OP AMPs.*

the forms of these constraints are unchanged when they are transformed into the s domain:

	t domain	s domain
Voltage-controlled voltage source	$v_2(t) = \mu v_1(t)$	$V_2(s) = \mu V_1(s)$
Current-controlled current source	$i_2(t) = \beta i_1(t)$	$I_2(s) = \beta I_1(s)$
Current-controlled voltage source	$v_2(t) = r i_1(t)$	$V_2(s) = r I_1(s)$
Voltage-controlled current source	$i_2(t) = g v_1(t)$	$I_2(s) = g V_1(s)$

(10–3)

Similarly, the element constraints of the ideal OP AMP are linear algebraic equations that are unchanged in form by the Laplace transformation:

t domain	s domain
$v_P(t) = v_N(t)$	$V_P(s) = V_N(s)$
$i_N(t) = 0$	$I_N(s) = 0$
$i_P(t) = 0$	$I_P(s) = 0$

(10–4)

Thus, for linear active devices the only difference is that in the s domain the ideal element constraints apply to transforms rather than waveforms.

Finally, we consider the two-terminal passive circuit elements shown in Figure 10–4. In the time domain their i–v relationships are

$$\begin{aligned} \text{Resistor: } v_R(t) &= R\, i_R(t) \\ \text{Inductor: } v_L(t) &= L\frac{di_L(t)}{dt} \\ \text{Capacitor: } v_C(t) &= \frac{1}{C}\int_0^t i_C(\tau)d\tau + v_C(0) \end{aligned} \tag{10–5}$$

These element constraints are transformed into the s domain by taking the Laplace transform of both sides of each equation using the linearity, differentiation, and integration properties.

$$\begin{aligned} \text{Resistor: } V_R(s) &= R\, I_R(s) \\ \text{Inductor: } V_L(s) &= LsI_L(s) - Li_L(0) \\ \text{Capacitor: } V_C(s) &= \frac{1}{Cs}I_C(s) + \frac{v_C(0)}{s} \end{aligned} \tag{10–6}$$

As expected, the element relationships are algebraic equations in the s domain. For the linear resistor the s domain version of Ohm's law says that the voltage transform $V_R(s)$ is proportional to the current transform $I_R(s)$. The element constraints for the inductor and capacitor also involve a proportionality between voltage and current, but include a term for the initial conditions as well.

The element constraints in Eq. (10–6) lead to the s-domain circuit models shown on the right side of Figure 10–4. The t-domain parameters L and C are replaced by proportionality factors Ls and $1/Cs$ in the s domain. The initial conditions associated with the inductor and capacitor are modeled

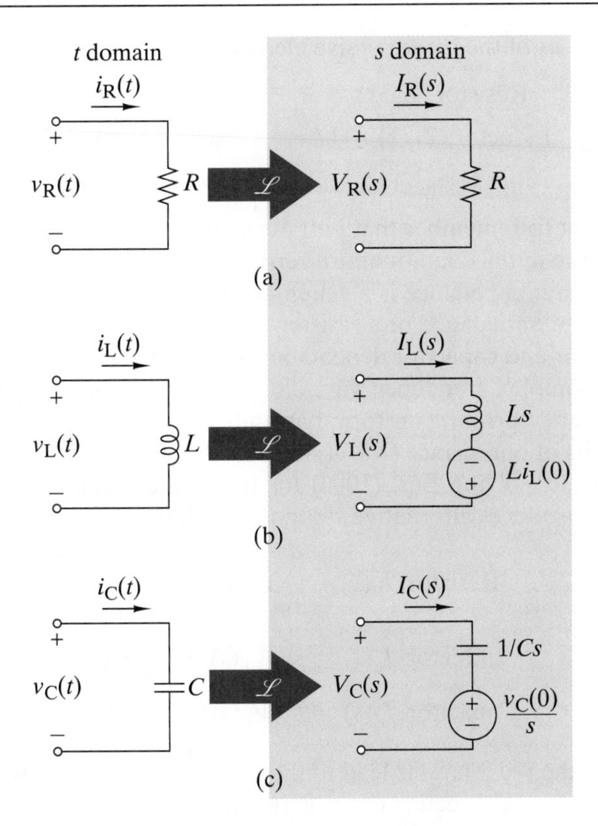

FIGURE 10–4 *s-Domain models of passive elements using voltage sources for initial conditions.*

as voltage sources in series with these elements. The polarities of these sources are determined by the sign of the corresponding initial condition terms in Eq. (10–6). These initial condition voltage sources must be included when using these models to calculate the voltage transforms $V_L(s)$ or $V_C(s)$.

IMPEDANCE AND ADMITTANCE

The concept of impedance is a basic feature of s-domain circuit analysis. For zero initial conditions the element constraints in Eq. (10–6) reduce to

$$\begin{aligned} \text{Resistor: } V_R(s) &= (R)\, I_R(s) \\ \text{Inductor: } V_L(s) &= (Ls) I_L(s) \\ \text{Capacitor: } V_C(s) &= (1/Cs)\, I_C(s) \end{aligned} \tag{10–7}$$

In each case the element constraints are all of the form $V(s) = Z(s)I(s)$, which means that in the s domain the voltage across the element is proportional to the current through it. The proportionality factor is called the element **impedance** $Z(s)$. Stated formally,

> **Impedance** ***is the proportionality factor relating the transform of the voltage across a two-terminal element to the transform of the current through the element with all initial conditions set to zero.***

The impedances of the three passive elements are

$$\begin{aligned} &\text{Resistor: } Z_R(s) = R \\ &\text{Inductor: } Z_L(s) = (Ls) \qquad \text{with } i_L(0) = 0 \\ &\text{Capacitor: } Z_C(s) = (1/Cs) \quad \text{with } v_C(0) = 0 \end{aligned} \tag{10–8}$$

It is important to remember that part of the definition of *s*-domain impedance is that the initial conditions are zero.

The *s*-domain impedance is a generalization of the *t*-domain concept of resistance. The impedance of a resistor is its resistance R. The impedance of the inductor and capacitor depend on the inductance L and capacitance C and the complex frequency variable s. Since a voltage transform has units of V-s and current transform has units of A-s, it follows that impedance has units of ohms since (V-s)/(A-s) = V/A = Ω.

Algebraically solving Eqs. (10–6) for the element currents in terms of the voltages produces alternative *s*-domain models.

$$\begin{aligned} &\text{Resistor: } I_R(s) = \frac{1}{R} V_R(s) \\ &\text{Inductor: } I_L(s) = \frac{1}{Ls} V_L(s) + \frac{i_L(0)}{s} \\ &\text{Capacitor: } I_C(s) = Cs\, V_C(s) - C\, v_C(0) \end{aligned} \tag{10–9}$$

In this form, the *i–v* relations lead to the *s*-domain models shown in Figure 10–5. The reference directions for the initial condition current sources are

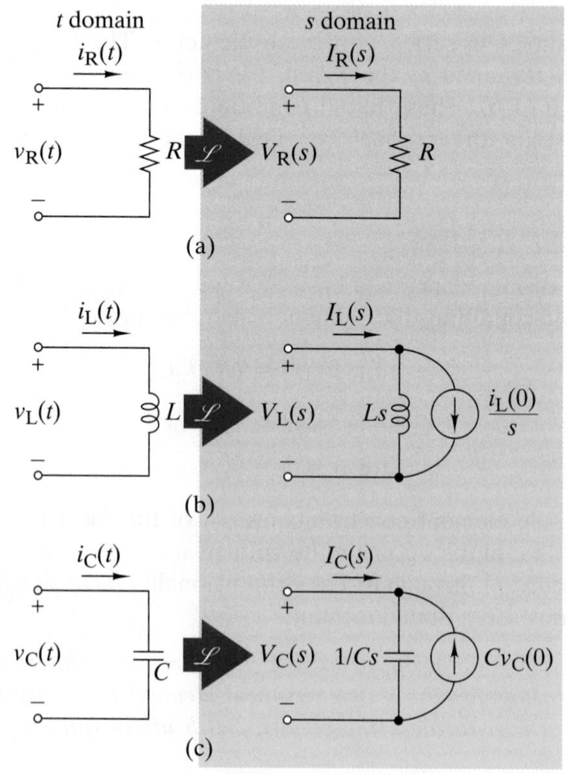

FIGURE 10–5 *s-Domain models of passive elements using current sources for initial conditions.*

determined by the sign of the corresponding terms in Eqs. (10–9). The initial condition sources are in parallel with the element impedance.

Admittance $Y(s)$ is the s-domain generalization of the t-domain concept of conductance and can be defined as the reciprocal of impedance.

$$Y(s) = \frac{1}{Z(s)} \tag{10–10}$$

Using this definition, the admittances of the three passive elements are

$$\begin{aligned}
&\text{Resistor: } Y_R(s) = \frac{1}{Z_R(s)} = \frac{1}{R} = G \\
&\text{Inductor: } Y_L(s) = \frac{1}{Z_L(s)} = \frac{1}{Ls} \quad \text{with } i_L(0) = 0 \\
&\text{Capacitor: } Y_C(s) = \frac{1}{Z_C(s)} = Cs \quad \text{with } v_C(0) = 0
\end{aligned} \tag{10–11}$$

Since $Y(s)$ is the reciprocal of impedance, its units are siemens since $\Omega^{-1} = \text{A/V} = \text{S}$.

In summary, to transform a circuit into the s domain we replace each element by an s-domain model. For independent sources, dependent sources, OP AMPs, and resistors, the only change is that these elements now constrain transforms rather than waveforms. For inductors and capacitors, we can use either the model with a series initial condition voltage source (Figure 10–4) or the model with a parallel initial condition current source (Figure 10–5). However, to avoid possible confusion we always write the inductor impedance Ls and capacitor impedance $1/Cs$ beside the transformed element regardless of which initial condition source is used.

To analyze the transformed circuit, we can use the tools developed for resistance circuits in Chapters 2 through 4. These tools are applicable because KVL and KCL apply to transforms, and the s-domain element constraints are linear equations similar to those for resistance circuits. These features make s-domain analysis of dynamic circuits an algebraic process that is akin to resistance circuit analysis.

EXAMPLE 10–1

The switch in Figure 10–6 has been in position 1 for a long time and is moved to position 2 at $t = 0$. For $t > 0$, transform the circuit into the s domain and use Laplace transforms to solve for the voltage $v_C(t)$.

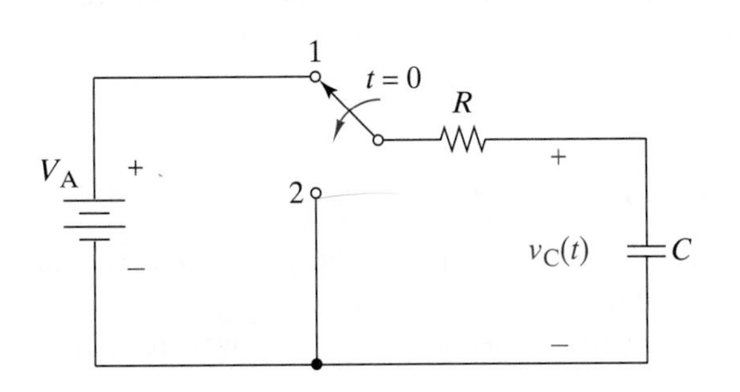

FIGURE 10–6

SOLUTION:

We have solved this type of problem before using various methods. In this example, we find the response by first transforming the circuit itself. For $t > 0$ the transformed circuit takes the form in Figure 10–7, where we have used a parallel current source $Cv_C(0)$ to account for the initial condition.

FIGURE 10–7

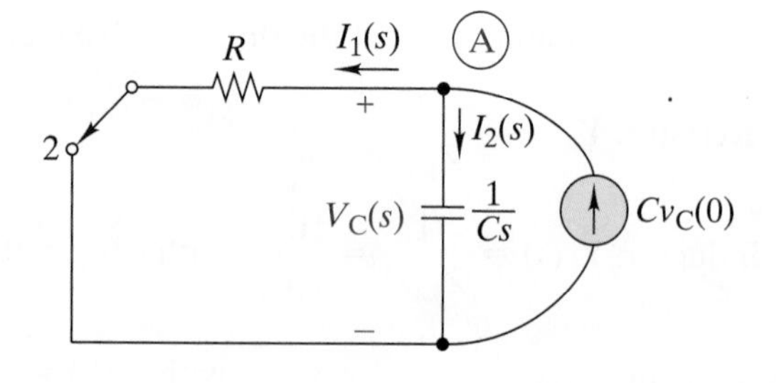

Applying KCL to the current transforms at node A produces

$$-I_1(s) - I_2(s) + Cv_C(0) = 0 \qquad \text{A-s}$$

For $t < 0$ the switch in Figure 10–6 is in position 1 and the circuit is in a dc steady-state condition. As a result, we have $v_C(0) = V_A$. In the s-domain circuit in Figure 10–7, the two branch current transforms can be written in terms of the capacitor voltage and element impedances as

$$\text{Resistor:} \quad I_1(s) = \frac{V_C(s)}{R}$$

$$\text{Capacitor:} \quad I_2(s) = \frac{V_C(s)}{1/Cs} = Cs\,V_C(s)$$

Substituting these observations into the KCL equation and solving for $V_C(s)$ yields

$$V_C(s) = \frac{CV_A}{Cs + \dfrac{1}{R}} = \frac{V_A}{s + \dfrac{1}{RC}} \qquad \text{V-s}$$

Performing the inverse Laplace transformation leads to

$$v_C(t) = [V_A e^{-t/RC}]u(t) \qquad \text{V}$$

The form of the response should be no great surprise. We could easily predict this response using the classical differential equation methods studied in Chapter 7. What is important in this example is that we obtained the response using only basic circuit concepts applied in the s domain. ■

EXAMPLE 10–2

(a) Transform the circuit in Figure 10–8(a) into the s domain.
(b) Solve for the current transform $I(s)$.
(c) Perform the inverse transformation to the waveform $i(t)$.

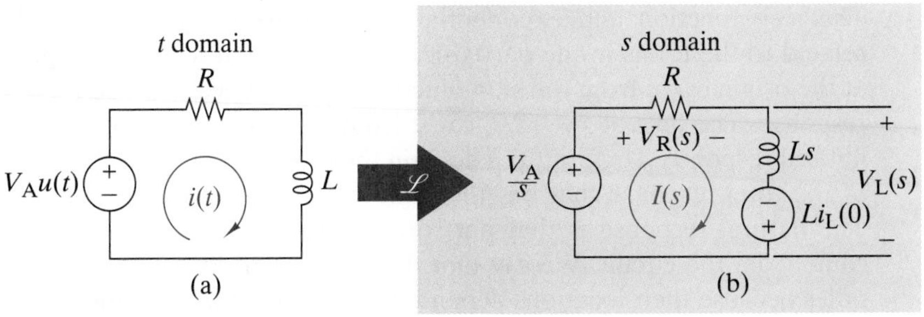

FIGURE 10–8

t domain
R
$V_A u(t)$
$i(t)$
L
(a)

s domain
R
$+ V_R(s) -$
$\frac{V_A}{s}$
$I(s)$
Ls
$V_L(s)$
$Li_L(0)$
(b)

SOLUTION:

(a) Figure 10–8(b) shows the transformed circuit using a series voltage source $Li_L(0)$ to represent the inductor initial condition. The impedances of the two passive elements are R and Ls. The independent source voltage $V_Au(t)$ transforms as V_A/s.

(b) By KVL, the sum of voltage transforms around the loop is

$$-\frac{V_A}{s} + V_R(s) + V_L(s) = 0$$

Using the impedance models, the s-domain element constraints are

$$\text{Resistor: } V_R(s) = RI(s)$$
$$\text{Inductor: } V_L(s) = LsI(s) - Li_L(0)$$

Substituting the element constraints into the KVL constraint and collecting terms yields

$$-\frac{V_A}{s} + (R + Ls)I(s) - Li_L(0) = 0$$

Solving for $I(s)$ produces

$$I(s) = \frac{V_A/L}{s(s + R/L)} + \frac{i_L(0)}{s + R/L} \quad \text{A-s}$$

The current $I(s)$ is the transform of the circuit response for a step function input. $I(s)$ is a rational function with simple poles at $s = 0$ and $s = -R/L$.

(c) To perform the inverse transformation, we expand $I(s)$ by partial fractions:

$$I(s) = \overbrace{\frac{V_A/R}{s}}^{\text{forced}} \overbrace{- \frac{V_A/R}{s + R/L} + \frac{i_L(0)}{s + R/L}}^{\text{natural}} \quad \text{A-s}$$

Taking the inverse transform of each term in this expansion gives

$$i(t) = \left[\overbrace{\frac{V_A}{R}}^{\text{forced}} \overbrace{- \frac{V_A}{R} e^{-Rt/L} + i_L(0)\, e^{-Rt/L}}^{\text{natural}}\right] u(t) \quad \text{A}$$

The forced response is caused by the step function input. The exponential terms in the natural response depend on the circuit time constant L/R.

The step function and exponential components in $i(t)$ are directly related to the terms in the partial-fraction expansion of $I(s)$. The pole at the origin came from the step function input and leads to the forced response. The pole at $s = -R/L$ came from the circuit and leads to the natural response. Thus, in the s domain the forced response is that part of the total response that has the same poles as the input excitation. The natural response is that part of the total response whose poles came from the circuit. We say that the circuit contributes the natural poles because their locations depend on circuit parameters, not on the input. In other words, poles in the response do not occur by accident. They are present because the physical response depends on two things—(1) the input and (2) the circuit. ■

10–2 BASIC CIRCUIT ANALYSIS IN THE s DOMAIN

In this section we develop the s-domain versions of series and parallel equivalence, and voltage and current division. These analysis techniques are the basic tools in s-domain circuit analysis, just as they are for resistance circuit analysis. These methods apply to circuits with elements connected in series or parallel. General analysis methods using node-voltage or mesh-current equations are covered later in Sects. 10–4 and 10–5.

SERIES EQUIVALENCE AND VOLTAGE DIVISION

The concept of a series connection applies in the s domain because Kirchhoff's laws do not change under the Laplace transformation. In Figure 10–9 the two-terminal elements are connected in series; hence by KCL the same current $I(s)$ exists in impedances $Z_1(s)$, $Z_2(s)$, . . . $Z_N(s)$. Using KVL and

FIGURE 10–9 Series equivalence in the s domain.

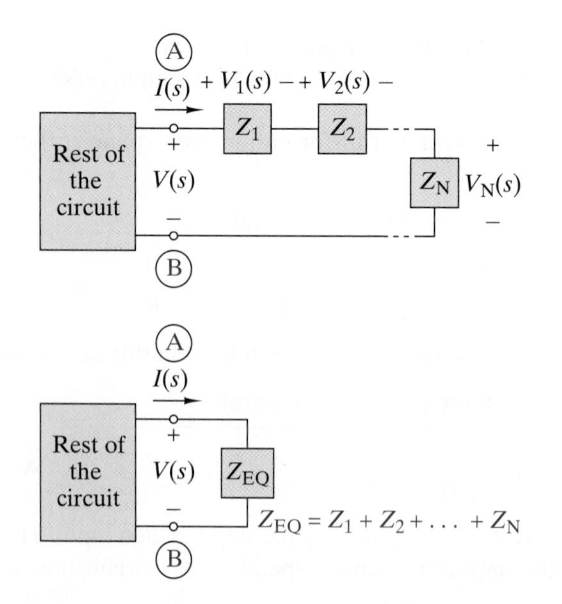

the element constraints, the voltage across the series connection can be written as

$$\begin{aligned} V(s) &= V_1(s) + V_2(s) + \cdots + V_N(s) \\ &= Z_1(s)I(s) + Z_2(s)I(s) + \cdots + Z_N(s)I(s) \\ &= [Z_1(s) + Z_2(s) + \cdots + Z_N(s)]\, I(s) \end{aligned} \quad (10\text{–}12)$$

The last line in this equation points out that the responses $V(s)$ and $I(s)$ do not change when the series-connected elements are replaced by an **equivalent impedance**:

$$Z_{EQ}(s) = Z_1(s) + Z_2(s) + \cdots + Z_N(s) \quad (10\text{–}13)$$

In general, the equivalent impedance $Z_{EQ}(s)$ is a quotient of polynomials in the complex frequency variable of the form

$$Z_{EQ}(s) = \frac{b_m s^m + b_{m-1}s^{m-1} + \cdots + b_1 s + b_0}{a_n s^n + a_{n-1}s^{n-1} + \cdots + a_1 s + a_0} \quad (10\text{–}14)$$

The roots of the numerator polynomial are the zeros of $Z_{EQ}(s)$, while the roots of the denominator are the poles.

Combining Eqs. (10–12) and (10–13), we can write the element voltages in the form

$$V_1(s) = \frac{Z_1(s)}{Z_{EQ}(s)}V(s) \quad V_2(s) = \frac{Z_2(s)}{Z_{EQ}(s)}V(s) \cdots V_N(s) = \frac{Z_N(s)}{Z_{EQ}(s)}V(s) \quad (10\text{–}15)$$

These equations are the *s*-domain **voltage division principle**:

Every element voltage in a series connection is equal to its impedance divided by the equivalent impedance of the connection times the voltage across the series circuit.

This statement parallels the corresponding rule for resistance circuits given in Chapter 4.

PARALLEL EQUIVALENCE AND CURRENT DIVISION

The parallel circuit in Figure 10–10 is the dual of the series circuit discussed previously. In this circuit the two-terminal elements are connected in parallel; hence by KVL the same voltage $V(s)$ appears across admittances $Y_1(s)$, $Y_2(s)$, . . . $Y_N(s)$. Using KCL and the element constraints, the current into the parallel connection can be written as

$$\begin{aligned} I(s) &= I_1(s) + I_2(s) + \cdots + I_N(s) \\ &= Y_1(s)V(s) + Y_2(s)V(s) + \cdots + Y_N(s)V(s) \\ &= [Y_1(s) + Y_2(s) + \cdots + Y_N(s)]V(s) \end{aligned} \quad (10\text{–}16)$$

The last line in this equation points out that the responses $V(s)$ and $I(s)$ do not change when the parallel connected elements are replaced by an **equivalent admittance**:

$$Y_{EQ}(s) = Y_1(s) + Y_2(s) + \cdots + Y_N(s) \quad (10\text{–}17)$$

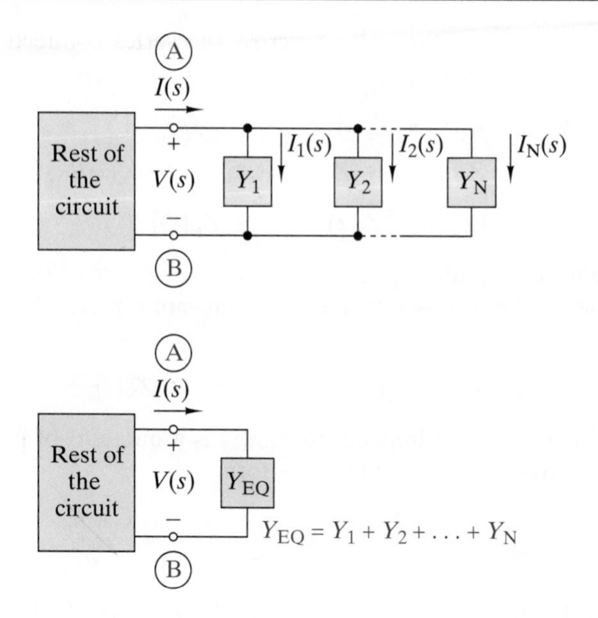

FIGURE 10–10 *Parallel equivalence in the s domain.*

In general, the equivalent admittance $Y_{EQ}(s)$ is a quotient of polynomials in the complex frequency variable s. Since impedance and admittance are reciprocals, it turns out that if $Y_{EQ}(s) = p(s)/q(s)$, then the equivalent impedance at the same pair of terminals has the form $Z_{EQ}(s) = 1/Y_{EQ}(s) = q(s)/p(s)$. That is, at a given pair of terminals the poles of $Z_{EQ}(s)$ are zeros of $Y_{EQ}(s)$, and vice versa.

Combining Eqs. (10–16) and (10–17), we can write the element currents in the form

$$I_1(s) = \frac{Y_1(s)}{Y_{EQ}(s)} I(s) \quad I_2(s) = \frac{Y_2(s)}{Y_{EQ}(s)} I(s) \cdots I_N(s) = \frac{Y_N(s)}{Y_{EQ}(s)} I(s) \tag{10–18}$$

These equations are the s-domain **current division principle**:

> ***Every element current in a parallel connection is equal to its admittance divided by the equivalent admittance of the connection times the current into the parallel circuit.***

This statement is the dual of the results for a series circuit and parallels the current division rule for resistance circuits.

We begin to see that s-domain circuit analysis involves basic concepts that parallel the analysis of resistance circuits in the t domain. Repeated application of series/parallel equivalence and voltage/current division leads to an analysis approach called circuit reduction, discussed in Chapter 2. The major difference here is that we use impedance and admittances rather than resistance and conductance, and the analysis yields voltage and current transforms rather than waveforms.

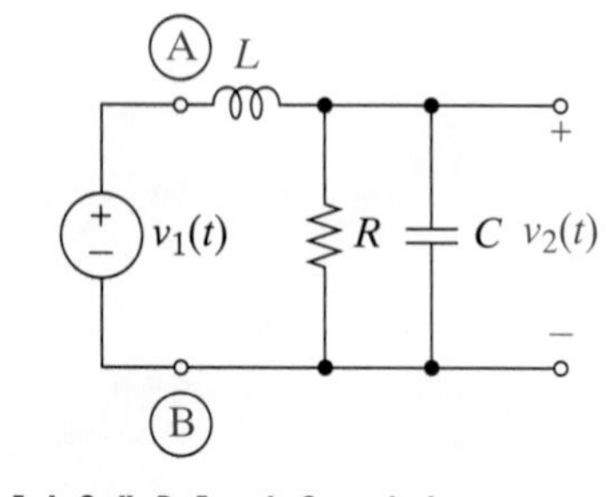

FIGURE 10–11

EXAMPLE 10–3

The inductor current and capacitor voltage in Figure 10–11 are zero at $t = 0$.

(a) Transform the circuit into the s domain and find the equivalent impedance between terminals A and B.
(b) Use voltage division to solve for the output voltage transform $V_2(s)$.

SOLUTION:

(a) Figure 10–12(a) shows the circuit in Figure 10–11 transformed into the *s* domain. As a first step we use parallel equivalence to find the equivalent impedance of the parallel resistor and capacitor.

$$Z_{EQ1}(s) = \frac{1}{Y_{EQ1}(s)} = \frac{1}{\frac{1}{R} + Cs} = \frac{R}{RCs + 1}$$

Figure 10–12(b) shows that the equivalent impedance $Z_{EQ1}(s)$ is connected in series with the inductor. This series combination can be replaced by an equivalent impedance

$$Z_{EQ}(s) = Ls + Z_{EQ1}(s) = Ls + \frac{R}{RCs + 1}$$

$$= \frac{RLCs^2 + Ls + R}{RCs + 1} \quad \Omega$$

as shown in Figure 10–12(c). The rational function $Z_{EQ}(s)$ is the impedance seen between terminals A and B in Figure 10–12(a).

(b) Using voltage division in Figure 10–12(b), we find $V_2(s)$ as

$$V_2(s) = \left[\frac{Z_{EQ1}(s)}{Z_{EQ}(s)}\right]V_1(s) = \left[\frac{R}{RLCs^2 + Ls + R}\right]V_1(s)$$

Note that $Z_{EQ}(s)$ and $V_2(s)$ are rational functions of the complex frequency variable *s*. ■

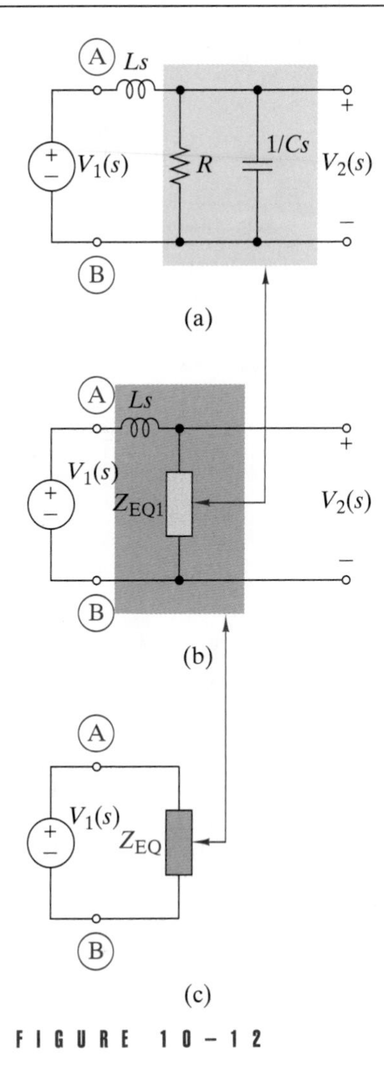

FIGURE 10–12

DESIGN EXAMPLE 10–4

In a circuit analysis problem we are required to find the poles and zeros of a circuit. In circuit design we are required to adjust circuit parameters to place the poles and zeros at specified *s*-plane locations. This example is a simple pole-placement design problem.

(a) Transform the circuit in Figure 10–13(a) into the *s* domain and find the equivalent impedance between terminals A and B.

(b) Select the values of *R* and *C* such that $Z_{EQ}(s)$ has a zero at $s = -5000$ rad/s.

SOLUTION:

(a) Figure 10–13(b) shows the circuit transformed to the *s* domain. The equivalent impedance $Z_{EQ1}(s)$ is

$$Z_{EQ1}(s) = \frac{1}{Y_R + Y_C} = \frac{1}{\frac{1}{R} + Cs}$$

$$= \frac{R}{RCs + 1}$$

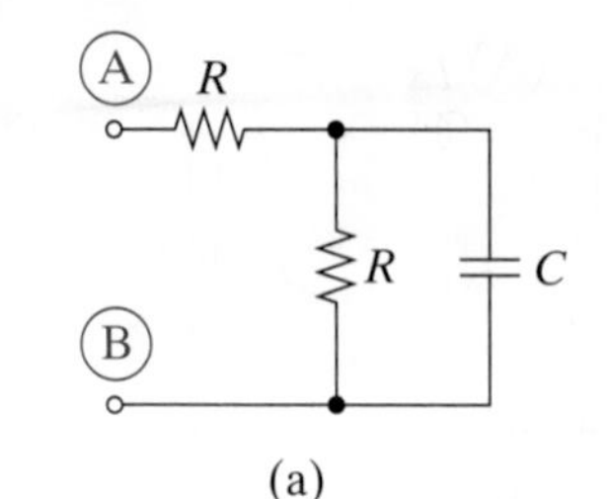

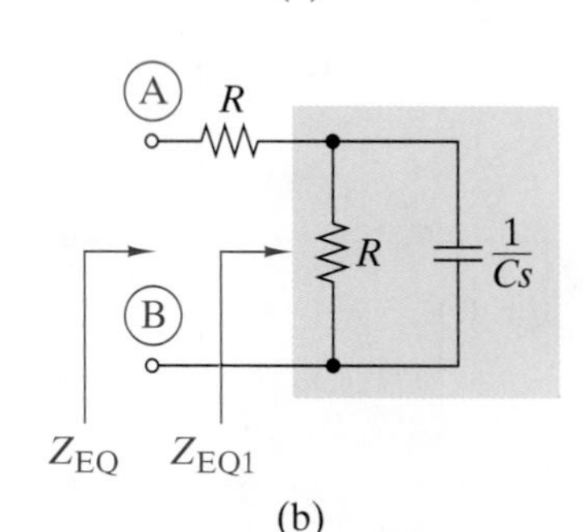

FIGURE 10–13

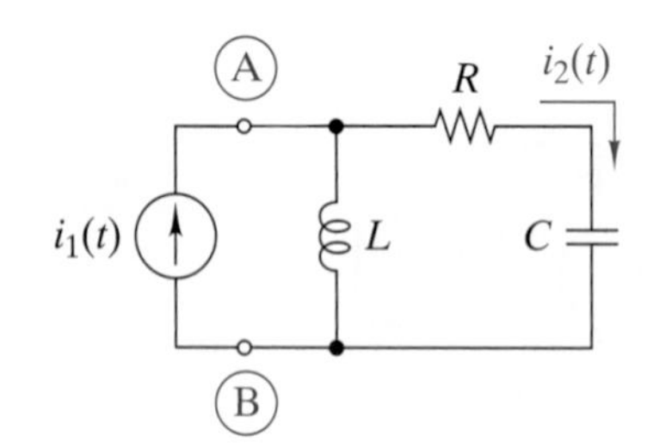

FIGURE 10–14

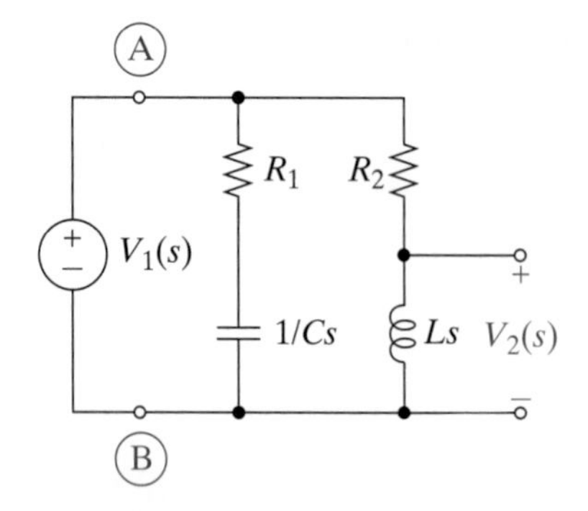

FIGURE 10–15

Hence the equivalent impedance between terminals A and B is

$$Z_{EQ}(s) = R + Z_{EQ1}(s) = R + \frac{R}{RCs + 1}$$

$$= R\frac{RCs + 2}{RCs + 1}\ \Omega$$

(b) For $Z_{EQ}(s)$ to have a zero at $s = -5000$ requires $2/RC = 5000$ or $RC = 4 \times 10^{-4}$. Selecting a standard value for the resistor $R = 10$ kΩ in turn requires $C = 40$ nF. ■

Exercise 10–1

The inductor current and capacitor voltage in Figure 10–14 are zero at $t = 0$.

(a) Transform the circuit into the s domain and find the equivalent admittance between terminals A and B.

(b) Solve for the output current transform $I_2(s)$ in terms of the input current $I_1(s)$.

Answers:

(a) $Y_{EQ}(s) = \dfrac{LCs^2 + RCs + 1}{Ls(RCs + 1)}$

(b) $I_2(s) = \left[\dfrac{LCs^2}{LCs^2 + RCs + 1}\right] I_1(s)$

Exercise 10–2

The inductor current and capacitor voltage in Figure 10–15 are zero at $t = 0$.

(a) Transform the circuit into the s domain and find the equivalent impedance between terminals A and B.

(b) Solve for the output voltage transform $V_2(s)$ in terms of the input voltage $V_1(s)$.

Answers:

(a) $Z_{EQ}(s) = \dfrac{(R_1Cs + 1)(Ls + R_2)}{LCs^2 + (R_1 + R_2)Cs + 1}$

(b) $V_2(s) = \left[\dfrac{Ls}{Ls + R_2}\right] V_1(s)$

10–3 CIRCUIT THEOREMS IN THE s DOMAIN

In this section we study the s-domain versions of proportionality, superposition, and Thévenin/Norton equivalent circuits. These theorems define fundamental properties that provide conceptual tools for the analysis and design of linear circuits. With some modifications, all of the theorems studied in Chapter 3 apply to linear dynamic circuits in the s domain.

PROPORTIONALITY

For linear resistance circuits the **proportionality theorem** states that any output y is proportional to the input x:

$$y = Kx \tag{10–19}$$

The same concept applies to linear dynamic circuits in the s domain except that the proportionality factor is a rational function of s rather than a constant. For instance, in Example 10–3 we found the output voltage $V_2(s)$ to be

$$V_2(s) = \left[\frac{R}{RLCs^2 + Ls + R}\right] V_1(s) \tag{10–20}$$

where $V_1(s)$ is the transform of the input voltage. The quantity inside the brackets is a rational function that serves as the proportionality factor between the input and output transforms.

In the s-domain, rational functions that relate inputs and outputs are called **network functions**. We begin the formal study network functions in Chapter 11. In this chapter we will simply illustrate network functions by an example.

EXAMPLE 10–5

There is no initial energy stored in the circuit in Figure 10–16. Find the network functions relating $I_R(s)$ to $V_1(s)$ and $I_C(s)$ to $V_1(s)$.

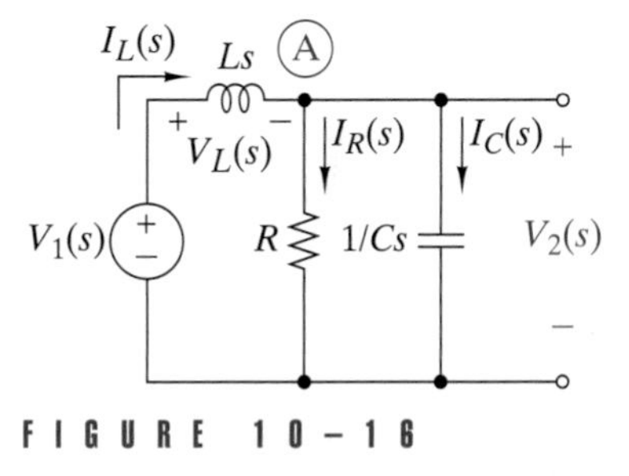

FIGURE 10–16

SOLUTION:

The equivalent impedance seen by the voltage source is

$$Z_{EQ} = Ls + \frac{1}{\dfrac{1}{R} + Cs} = \frac{RLCs^2 + Ls + R}{RCs + 1}$$

Hence we can relate the $I_L(s)$ and $V_1(s)$ as

$$I_L(s) = \frac{V_1(s)}{Z_{EQ}(s)} = \left[\frac{RCs + 1}{RLCs^2 + Ls + R}\right] V_1(s)$$

Using s-domain current division, we can relate $I_R(s)$ and $I_C(s)$ to $I_L(s)$ as

$$I_R(s) = \frac{\dfrac{1}{R}}{\dfrac{1}{R} + Cs} I_L(s) = \left[\frac{1}{RCs + 1}\right] I_L(s)$$

$$I_C(s) = \frac{Cs}{\dfrac{1}{R} + Cs} I_L(s) = \left[\frac{RCs}{RCs + 1}\right] I_L(s)$$

Finally, using these relationships plus the relationship between $I_L(s)$ and $V_1(s)$ derived previously, we obtain the required network functions.

$$I_R(s) = \left[\frac{1}{RLCs^2 + Ls + R}\right] V_1(s)$$

$$I_C(s) = \left[\frac{RCs}{RLCs^2 + Ls + R}\right] V_1(s)$$

■

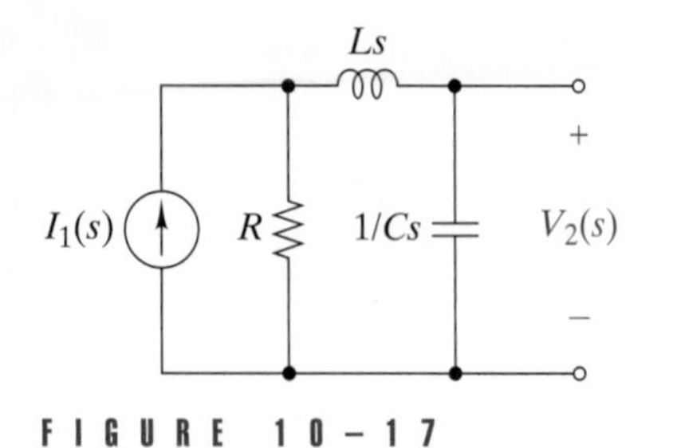

FIGURE 10–17

Exercise 10–3

In Figure 10–17 find the network function relating the output $V_2(s)$ to the input $I_1(s)$.

Answer:
$$V_2(s) = \left[\frac{R}{LCs^2 + RCs + 1}\right] I_1(s)$$

SUPERPOSITION

For linear resistance circuits, the **superposition theorem** states that any output y of a linear circuit can be written as

$$y = K_1x_1 + K_2x_2 + K_3x_3 + \cdots \quad (10\text{–}21)$$

where $x_1, x_2, x_3, \ldots$ are circuit inputs and $K_1, K_2, K_3, \ldots$ are weighting factors that depend on the circuit. The same concept applies to linear dynamic circuits in the s domain except that the weighting factors are rational functions of s rather than constants.

Superposition is usually thought of as a way to find the circuit response by adding the individual responses caused by each input acting alone. However, the principle applies to groups of sources as well. In particular, in the s domain there are two types of independent sources: (1) voltage and current sources representing the external driving forces for $t \geq 0$ and (2) initial condition voltage and current sources representing the energy stored at $t = 0$. As a result, the superposition principle states that the s-domain response can be found as the sum of two components: (1) the **zero-input response** caused by the initial condition sources with the external inputs turned off or (2) the **zero-state response** caused by the external inputs with the initial condition sources turned off. Turning a source off means replacing voltage sources by short circuits $[V_S(s) = 0]$ and current sources by open circuits $[I_S(s) = 0]$.

The zero-input response is the response of a circuit to its initial conditions when the input excitations are set to zero. The zero-state response is the response of a circuit to its input excitations when all of the initial conditions are set to zero. The term *zero input* is self-explanatory. The term *zero state* is used because there is no energy stored in the circuit at $t = 0$.

The result is that voltage and current transform in a linear circuit can be found as the sum of two components of the form

$$V(s) = V_{zs}(s) + V_{zi}(s) \qquad I(s) = I_{zs}(s) + I_{zi}(s) \quad (10\text{–}22)$$

where the subscript zs stands for zero state and zi for zero input. An important corollary is that the time-domain response can also be partitioned into zero-state and zero-input components because the inverse Laplace transformation is a linear operation.

We analyze the circuit treated in Example 10–2 to illustrate the superposition of zero-state and zero-input responses. The transformed circuit in Figure 10–18 has two independent voltage sources: (1) an input voltage source and (2) a voltage source representing the initial inductor current. The resistor and inductor are in series, so these two elements can be replaced by an impedance $Z_{EQ}(s) = Ls + R$.

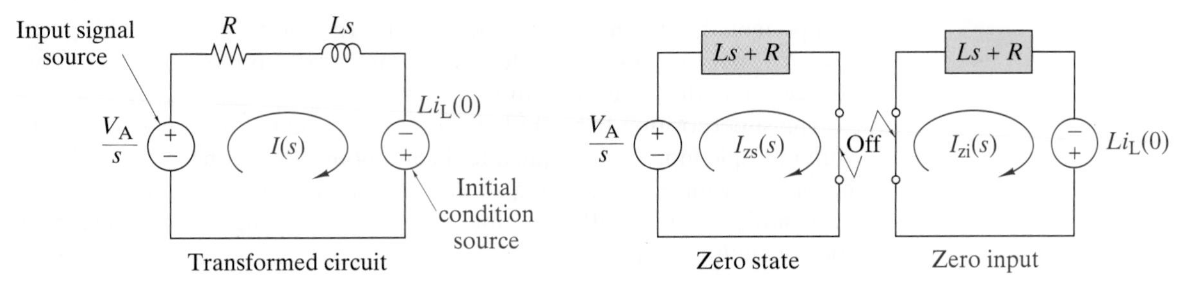

FIGURE 10–18 *Using superposition to find the zero-state and zero-input responses.*

First we turn off the initial condition source and replace its voltage source by a short circuit. Using the resulting zero-state circuit shown in Figure 10–18, we obtain the zero-state response:

$$I_{zs}(s) = \frac{V_A/s}{Z_{EQ}(s)} = \frac{V_A/L}{s(s + R/L)} \tag{10–23}$$

The pole at $s = 0$ comes from the input source and the pole at $s = -R/L$ comes from the circuit. Next, we turn off the input source and use the zero-input circuit shown in Figure 10–18 to obtain the zero-input response:

$$I_{zi}(s) = \frac{Li_L(0)}{Z_{EQ}(s)} = \frac{i_L(0)}{s + R/L} \tag{10–24}$$

The pole at $s = -R/L$ comes from the circuit. The zero-input response does not have a pole at $s = 0$ because the step function input is turned off.

Superposition states that the total response is the sum of the zero-state component in Eqs. (10–23) and the zero-input component in Eq. (10–24).

$$I(s) = I_{zs}(s) + I_{zi}(s) = \frac{V_A/L}{s(s + R/L)} + \frac{i_L(0)}{s + R/L} \tag{10–25}$$

The transform $I(s)$ in this equation is the same as found in Example 10–1. To derive the time-domain response, we expand $I(s)$ by partial fractions:

$$I(s) = \underbrace{\frac{V_A/R}{s} - \frac{V_A/R}{s + R/L}}_{\text{zero state}} + \underbrace{\frac{i_L(0)}{s + R/L}}_{\text{zero input}} \tag{10–26}$$

Performing the inverse transformation on each term yields

$$i(t) = \Bigg[\underbrace{\overbrace{\frac{V_A}{R}}^{\text{forced}} - \frac{V_A}{R}e^{-Rt/L}}_{\text{zero state}} \overbrace{+ \underbrace{i_L(0)\,e^{-Rt/L}}_{\text{zero input}}}^{\text{natural}}\Bigg]u(t) \quad \text{A} \tag{10–27}$$

Using superposition to partition the waveform into zero-state and zero-input components produces the same result as Example 10–1. The zero-state component contains the forced response. The zero-state and zero-input components both contain an exponential term due to the natural pole at $s = -R/L$ because both the external driving force and the initial condition source excite the circuit's natural response.

The superposition theorem helps us understand the response of circuits with multiple inputs, including initial conditions. It is a conceptual tool that helps us organize our thinking about s-domain circuits in general. It is not necessarily the most efficient analysis tool for finding the response of a specific multiple-input circuit.

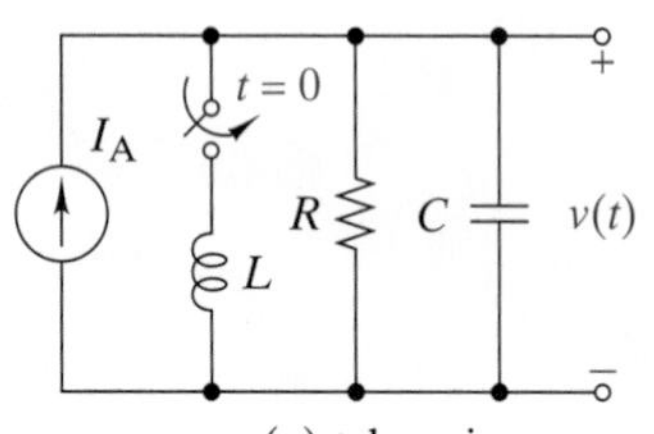

(a) t domain

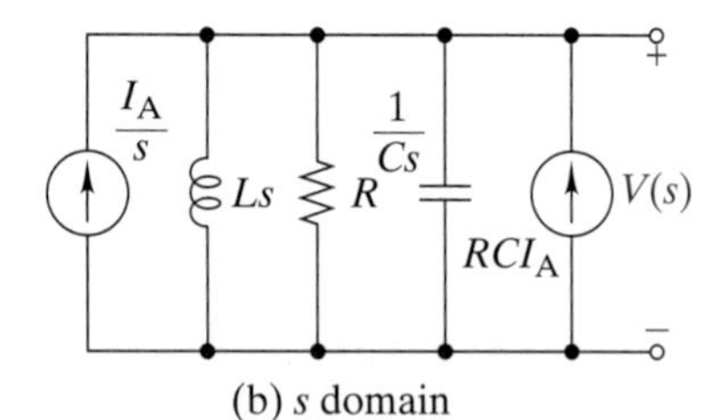

(b) s domain

FIGURE 10–19

EXAMPLE 10–6

The switch in Figure 10–19(a) has been open for a long time and is closed at $t = 0$.

(a) Transform the circuit into the s domain.
(b) Find the zero-state and zero-input components of $V(s)$.
(c) Find $v(t)$ for $I_A = 1$ mA, $L = 2$ H, $R = 1.5$ kΩ, and $C = 1/6$ μF.

SOLUTION:

(a) To transform the circuit into the s domain, we must find the initial inductor current and capacitor voltage. For $t < 0$ the circuit is in a dc steady-state condition with the switch open. The inductor acts like a short circuit, and the capacitor acts like an open circuit. By inspection, the initial conditions at $t = 0-$ are $i_L(0) = 0$ and $v_C(0) = I_A R$. Figure 10–19(b) shows the s-domain circuit for these initial conditions. The current source version for the capacitor's initial condition is used here because the circuit elements are connected in parallel. The switch and constant current source combine to produce a step function $I_A u(t)$ whose transform is I_A/s.

(b) The resistor, capacitor, and inductor can be replaced by an equivalent impedance

$$Z_{EQ} = \frac{1}{Y_{EQ}} = \frac{1}{\frac{1}{Ls} + \frac{1}{R} + Cs}$$

$$= \frac{RLs}{RLCs^2 + Ls + R}$$

The zero-state response is found with the capacitor initial condition source replaced by an open circuit and the step function input source on:

$$V_{zs}(s) = Z_{EQ}(s)\frac{I_A}{s} = \left[\frac{RLs}{RLCs^2 + Ls + R}\right]\frac{I_A}{s} = \frac{I_A/C}{s^2 + \frac{s}{RC} + \frac{1}{LC}}$$

The pole in the input at $s = 0$ is canceled by the zero at the origin in $Z_{EQ}(s)$. As a result, the zero-state response does not have a forced pole at $s = 0$. The zero-input response is found by replacing the input source by an open circuit and turning the capacitor initial condition source on:

$$V_{zi}(s) = [Z_{EQ}(s)]\,[CRI_A] = \frac{RI_A s}{s^2 + \frac{s}{RC} + \frac{1}{LC}}$$

(c) Inserting the given numerical values of the circuit parameters and expanding the zero-state and zero-input response transforms by partial fractions yields

$$V_{zs}(s) = \frac{6000}{(s+1000)(s+3000)} = \frac{3}{s+1000} + \frac{-3}{s+3000} \quad \text{V-s}$$

$$V_{zi}(s) = \frac{1.5\,s}{(s+1000)(s+3000)} = \frac{-0.75}{s+1000} + \frac{2.25}{s+3000} \quad \text{V-s}$$

The inverse transforms of these expansions are

$$v_{zs}(t) = [3e^{-1000t} - 3e^{-3000t}]\,u(t) \quad \text{V}$$

$$v_{zi}(t) = [-0.75e^{-1000t} + 2.25e^{-3000t}]\,u(t) \quad \text{V}$$

Note that the circuit responses contain only transient terms that decay to zero. There is no forced response because in the dc steady state the inductor acts like a short circuit, forcing $v(t)$ to zero for $t \to \infty$. From an *s*-domain viewpoint there is no forced response because the forced pole at $s = 0$ is canceled by a zero in the network function. ■

EXAMPLE 10–7

Use superposition to find the zero-state component of $I(s)$ in the *s*-domain circuit shown in Figure 10–20(a).

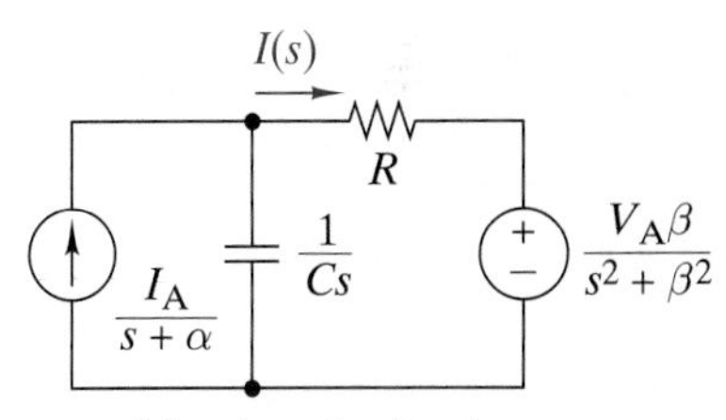

(a) *s*-domain circuit

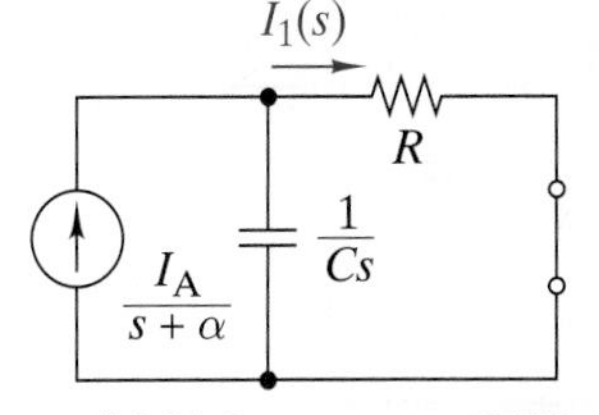

(b) Voltage source OFF

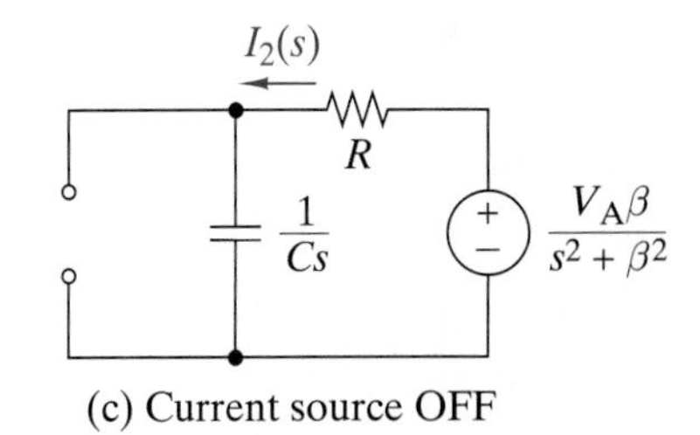

(c) Current source OFF

FIGURE 10–20

SOLUTION:

Turning the voltage source off produces the circuit in Figure 10–20(b). In this circuit the resistor and capacitor are connected in parallel, so current division yields $I_1(s)$ in the form

$$I_1(s) = \frac{Y_R}{Y_C + Y_R}\frac{I_A}{(s+\alpha)} = \frac{I_A}{(RCs+1)(s+\alpha)}$$

Turning the voltage source on and the current source off produces the circuit in Figure 10–20(c). In this case the resistor and capacitor are connected in series, and series equivalence gives the current $I_2(s)$ as

$$I_2(s) = \frac{1}{Z_R + Z_C}\frac{V_A\beta}{s^2+\beta^2} = \frac{CsV_A\beta}{(RCs+1)(s^2+\beta^2)}$$

Using superposition, the total zero-state response is

$$I_{zs}(s) = I_1(s) - I_2(s)$$

$$= \frac{I_A}{(RCs+1)(s+\alpha)} - \frac{CsV_A\beta}{(RCs+1)(s^2+\beta^2)}$$

There is a minus in this equation because $I_1(s)$ and $I_2(s)$ were assigned opposite reference directions in Figures 10–20(b) and 10–20(c). The total zero-state response has four poles. The natural pole at $s = -1/RC$ came from the circuit. The forced pole at $s = -\alpha$ came from the current source, and the two forced poles at $s = \pm j\beta$ came from the voltage source.

In this example the time-domain response would have a transient component $Ke^{-t/RC}$ due to the natural pole, a forced component $Ke^{-\alpha t}$ due to the current source, and a forced component of the form $K_A \cos \beta t + K_B \sin \beta t$ due to the voltage source. We can infer these general conclusions regarding the time-domain response by simply examining the poles of the s-domain response. ■

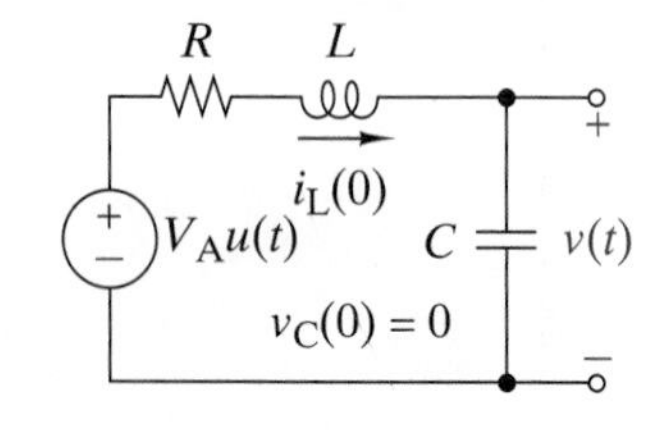

FIGURE 10–21

Exercise 10–4

The initial conditions for the circuit in Figure 10–21 are $v_C(0-) = 0$ and $i_L(0-) = I_0$. Transform the circuit into the s domain and find the zero-state and zero-input components of $V(s)$.

Answers:

$$V_{zs}(s) = \left[\frac{1}{LCs^2 + RCs + 1}\right]\frac{V_A}{s}$$

$$V_{zi}(s) = \frac{LI_0}{LCs^2 + RCs + 1}$$

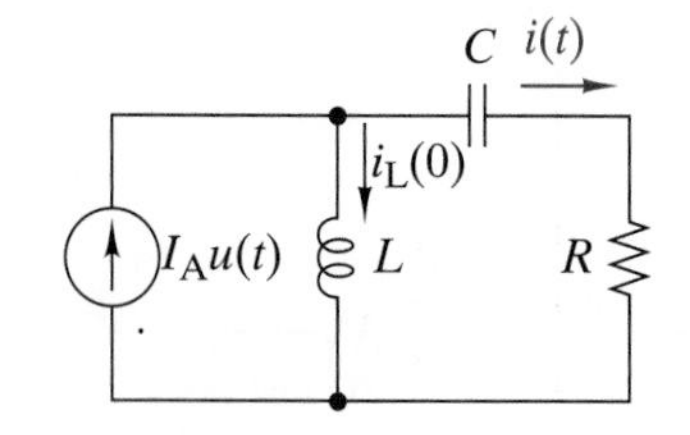

FIGURE 10–22

Exercise 10–5

The initial conditions for the circuit in Figure 10–22 are $v_C(0-) = 0$ and $i_L(0-) = I_0$. Transform the circuit into the s domain and find the zero-state and zero-input components of $I(s)$.

Answers:

$$I_{zs}(s) = \left[\frac{LCs^2}{LCs^2 + RCs + 1}\right]\frac{I_A}{s}$$

$$I_{zi}(s) = \left[\frac{LCs^2}{LCs^2 + RCs + 1}\right]\frac{I_0}{s}$$

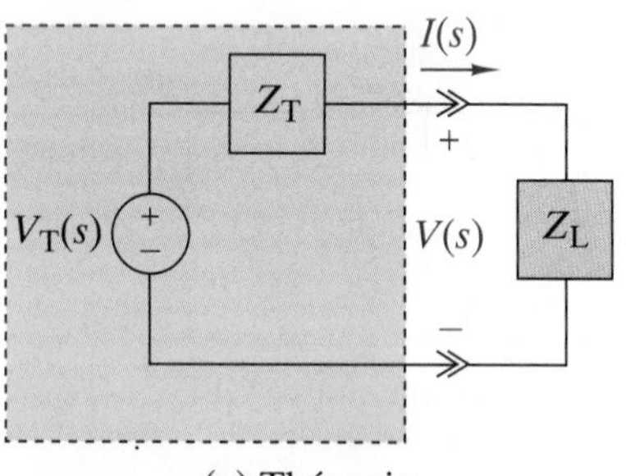

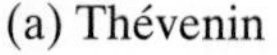

(a) Thévenin

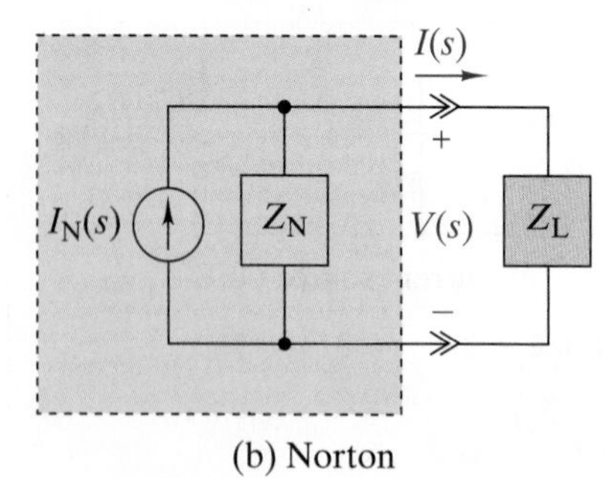

(b) Norton

FIGURE 10–23 *Thévenin and Norton equivalent circuits in the s domain.*

THÉVENIN AND NORTON EQUIVALENT CIRCUITS

In the s domain a two-terminal circuit containing linear elements and sources can be replaced by the Thévenin or Norton equivalent circuits in Figure 10–23. The general concept and restrictions are the same in the s domain as for resistive circuits. The important differences here are that the source terms $V_T(s)$ and $I_N(s)$ are transforms while Z_T and Z_N are s-domain impedances.

To find the Thévenin or Norton equivalent circuit, we use the same process as for resistance circuits, except that now we must manipulate rational functions of s. Since the Thévenin and Norton circuits are equivalent to each other, their circuit parameters are related to the s-domain open-circuit voltage $V_{OC}(s)$ and short-circuit current $I_{SC}(s)$ as

$$\begin{aligned} V_{OC}(s) &= V_T(s) = I_N(s)Z_N \\ I_{SC}(s) &= \frac{V_T(s)}{Z_T} = I_N(s) \\ Z_T &= Z_N = \frac{V_{OC}(s)}{I_{SC}(s)} \end{aligned} \tag{10–28}$$

Algebraically the results in Eq. (10–28) are identical to the corresponding equations for resistance circuits, except that these equations involve transforms and impedances rather than waveforms and resistances. Collectively these equations show that finding a Thévenin or Norton equivalent involves finding any two of the following three quantities: (1) the open-circuit voltage $V_{OC}(s)$, (2) the short-circuit current $I_{SC}(s)$, and (in the absence of dependent sources) (3) the lookback impedance with all independent sources turned off.

The relationships in Eq. (10–28) also define *source transformations* that allow us to convert a voltage source in series with an impedance into a current source in parallel with the same impedance, or vice versa. Performing *s*-domain source transformations may lead to circuit simplifications and can be useful when formulating node-voltage or mesh-current equations, as discussed in the next section.

Thévenin and Norton equivalent circuits should be regarded as important conceptual tools offering insight into how circuits operate in the *s* domain. They are not, in general, important tools for reducing the computational effort involved in *s*-domain circuit analysis.

EXAMPLE 10–8

The circuit in Figure 10–24(a) is in the zero state. Use a source transformation and voltage division to find the *s*-domain relationship between the input $I_1(s)$ and the output $V_2(s)$.

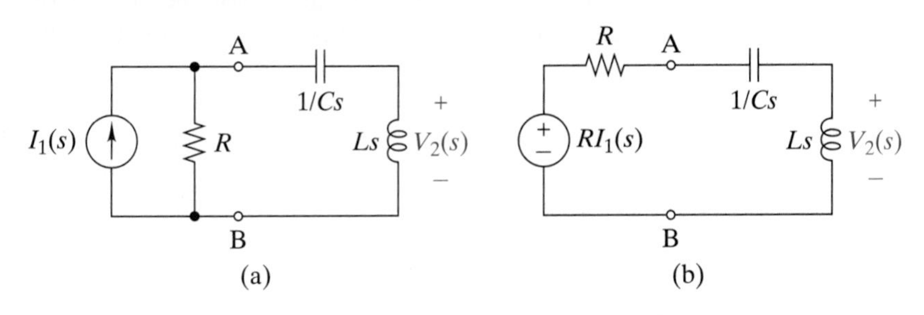

FIGURE 10–24

SOLUTION:

In this example we use a source transformation on the subcircuit to the left of points A and B in Figure 10–24(a). This Norton subcircuit consists of an independent current source $I_N = I_1(s)$ in parallel with an impedance $Z_N = R$. The equivalent Thévenin circuit consists of a voltage source $V_T = I_N Z_N = RI_1(s)$ in series with an impedance $Z_T = Z_N = R$. Figure 10–24(b) shows the circuit after the source transformation. Applying voltage division in the modified circuit yields the required input-output relationship.

$$V_2(s) = \left[\frac{Ls}{R + \dfrac{1}{Cs} + Ls}\right] RI_1(s) = \left[\frac{RLCs^2}{LCs^2 + RCs + 1}\right] I_1(s)$$

EXAMPLE 10–9

The circuit in Figure 10–25(a) is in the zero state. Use a Thévenin equivalent to find the *s*-domain relationship between the input $V_1(s)$ and the output $V_2(s)$.

FIGURE 10–25

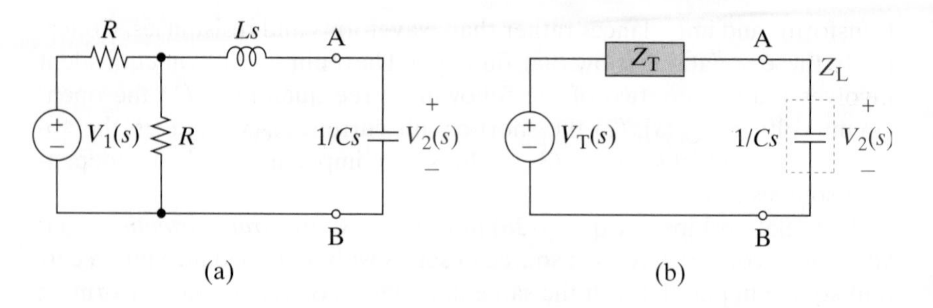

SOLUTION:

In this example we treat the capacitor as a load and find the Thévenin equivalent circuit to the left of points A and B as shown in Figure 10–25(a). To obtain the required Thévenin circuit, we find the open-circuit voltage and lookback impedance. Figure 10–26(a) shows the open-circuit situation. There is no voltage across the inductor, so the open-circuit voltage $V_T(s)$ is the same as the voltage across the second resistor. Using voltage division we have

$$V_T(s) = \frac{R}{R + R} V_1(s) = \frac{V_1(s)}{2}$$

To find the lookback impedance, we turn the input voltage source off [replace $V_1(s)$ by a short circuit] to obtain the situation in Figure 10–26(b). By inspection

$$Z_T = Ls + R \| R = Ls + \frac{R}{2}$$

FIGURE 10–26

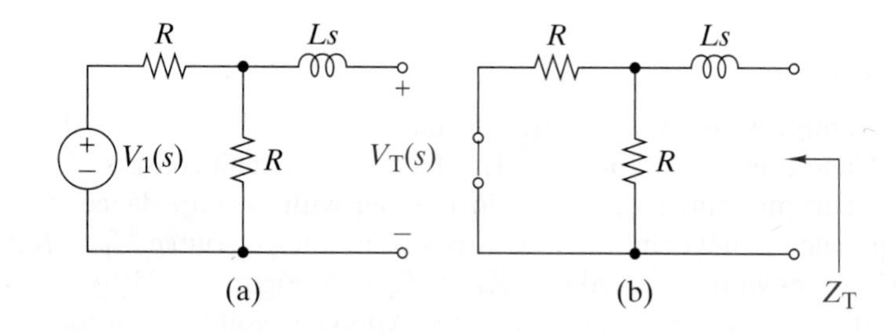

Given the $V_T(s)$ and Z_T, we return to Figure 10–25(b) and use voltage division to obtain the desired input-output relationship.

$$V_2(s) = \left[\frac{Z_L}{Z_T + Z_L}\right] V_T(s) = \left[\frac{\dfrac{1}{Cs}}{Ls + \dfrac{R}{2} + \dfrac{1}{Cs}}\right] \frac{V_1(s)}{2}$$

$$= \left[\frac{1}{2LCs^2 + RCs + 2}\right] V_1(s)$$

Exercise 10–6

Find the Norton equivalent of the s-domain circuits in Figure 10–27.

Answers:

(a) $I_N(s) = \dfrac{I_A}{(RCs + 1)(s + \alpha)} \qquad Z_N(s) = \dfrac{RCs + 1}{Cs}$

(b) $I_N(s) = \dfrac{RI_A}{(Ls + R)(s + \alpha)} \qquad Z_N(s) = \dfrac{Ls + R}{LCs^2 + RCs + 1}$

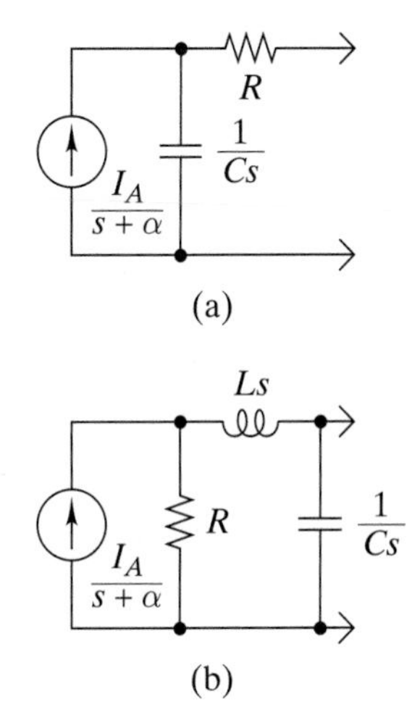

FIGURE 10–27

10–4 NODE-VOLTAGE ANALYSIS IN THE s DOMAIN

The previous sections deal with basic analysis methods using equivalence, reduction, and circuit theorems. These methods are valuable because we work directly with the element impedances and thereby gain insight into s-domain circuit behavior. We also need general methods to deal with more complicated circuits that these basic methods cannot easily handle.

FORMULATING NODE-VOLTAGE EQUATIONS

Formulating node-voltage equations involves selecting a reference node and assigning a node-to-datum voltage to each of the remaining nonreference nodes. Because of KVL, the voltage across any two-terminal element is equal to the difference of two node voltages. This fundamental property of node voltages, together with element impedances, allows us to write KCL constraints at each of the nonreference nodes.

For example, consider the s-domain circuit in Figure 10–28. The sum of currents leaving node A can be written as

$$I_{S2}(s) - I_{S1}(s) + \frac{V_A(s)}{Z_1(s)} + \frac{V_A(s) - V_B(s)}{Z_2(s)} + \frac{V_A(s) - V_C(s)}{Z_3(s)} = 0$$

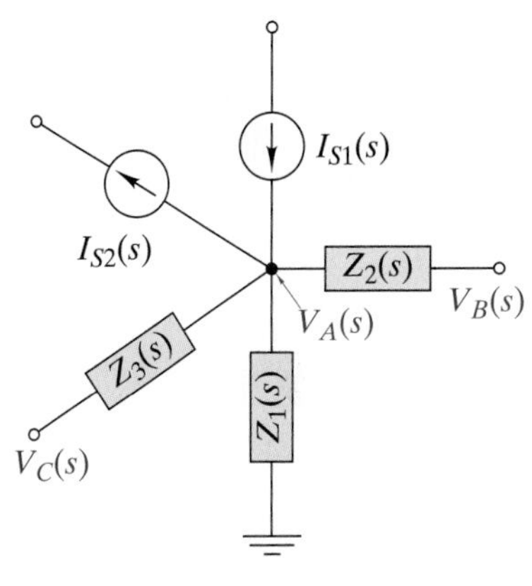

FIGURE 10–28 *An example node.*

Rewriting this equation with unknown node voltages grouped on the left and inputs on the right yields

$$\left[\frac{1}{Z_1(s)} + \frac{1}{Z_2(s)} + \frac{1}{Z_3(s)}\right] V_A(s) - \frac{1}{Z_2(s)} V_B(s) - \frac{1}{Z_3(s)} V_C(s) = I_{S1}(s) - I_{S2}(s)$$

Expressing this result in terms of admittances produces the following equation:

$$[Y_1(s) + Y_2(s) + Y_3(s)]\, V_A(s) - [Y_2(s)]\, V_B(s) - [Y_3(s)] V_C(s) = I_{S1}(s) - I_{S2}(s)$$

This equation has a familiar pattern. The unknowns are the node-voltage transforms $V_A(s)$, $V_B(s)$, and $V_C(s)$. The coefficient $[Y_1(s) + Y_2(s) + Y_3(s)]$ of $V_A(s)$ is the sum of the admittances of the elements connected to node A. The coefficient $[Y_2(s)]$ of $V_B(s)$ is the admittance of the elements connected between nodes A and B, while $[Y_3(s)]$ is the admittance of the elements connected between nodes A and C. Finally, $I_{S1}(s) - I_{S2}(s)$ is the sum of the source currents directed into node A. These observations suggest that we can write node-voltage equations for s-domain circuits by inspection, just as we did with resistive circuits.

The formulation method just outlined assumes that there are no voltage sources in the circuit. When transforming the circuit, we can always select the current source models to represent the initial conditions. However, the circuit may contain dependent or independent voltage sources. If so, they can be treated using the following methods:

Method 1: If there is an impedance in series with the voltage source, use a source transformation to convert it into an equivalent current source.

Method 2: Select the reference node so that one terminal of one or more of the voltage sources is connected to ground. The source voltage then determines the node voltage at the other source terminal, thereby eliminating an unknown.

Method 3: Create a supernode surrounding any voltage source that cannot be handled by method 1 or 2.

Some circuits may require more than one of these methods.

Formulating a set of equilibrium equations in the s domain is a straightforward process involving concepts developed in Chapters 3 and 4 for resistance circuits. The following example illustrates the formulation process.

EXAMPLE 10–10

Formulate s-domain node-voltage equations for the circuit in Figure 10–29(a).

FIGURE 10–29

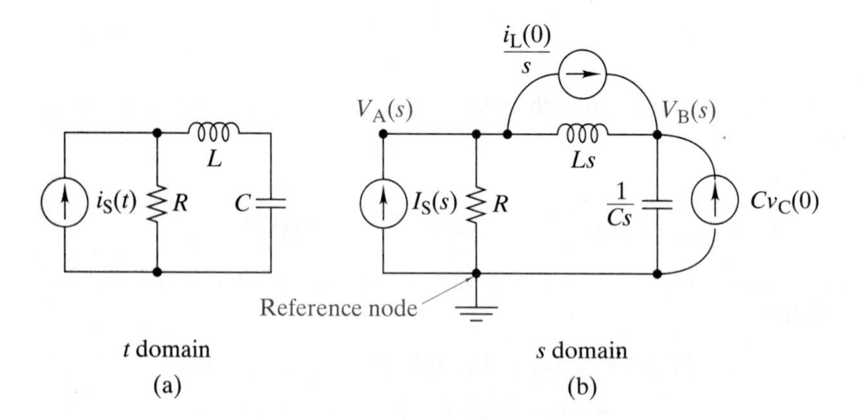

SOLUTION:

Figure 10–29(b) shows the circuit in the s domain. In transforming the circuit, we use current sources to represent the inductor and capacitor initial conditions. This choice facilitates writing node equations since the resulting *s*-domain circuit contains only current sources and fewer nodes. The sum of currents leaving nodes A and B can be written as

$$\text{Node A: } \frac{V_A(s)}{R} + \frac{V_A(s) - V_B(s)}{Ls} - I_S(s) + \frac{i_L(0)}{s} = 0$$

$$\text{Node B: } \frac{V_B(s)}{1/Cs} + \frac{V_B(s) - V_A(s)}{Ls} - \frac{i_L(0)}{s} - Cv_C(0) = 0$$

Rearranging these equations in the standard format with the unknowns on the left and the inputs on the right yields

$$\text{Node A: } \left(G + \frac{1}{Ls}\right) V_A(s) - \left(\frac{1}{Ls}\right) V_B(s) = I_S(s) - \frac{i_L(0)}{s}$$

$$\text{Node B: } -\left(\frac{1}{Ls}\right) V_A(s) + \left(\frac{1}{Ls} + Cs\right) V_B(s) = Cv_C(0) + \frac{i_L(0)}{s}$$

where $G = 1/R$ is the conductance of the resistor. Note that $(G + 1/Ls)$ is the sum of the admittance connected to node A, $(1/Ls + Cs)$ is the sum of the admittances connected to node B, and $1/Ls$ is the admittance connected between nodes A and B. The circuit is driven by an independent current source $I_S(s)$ and two initial condition sources $Cv_C(0)$ and $i_L(0)/s$. The terms on the right side of these equations are the sum of source currents directed into each node. With practice, we learn to write these equations by inspection. ■

SOLVING S-DOMAIN CIRCUIT EQUATIONS

Example 10–10 shows that node-voltage equations are linear algebraic equations in the unknown node voltages. In theory, solving these node equations is accomplished using techniques such as Cramer's rule or perhaps Gaussian reduction. In practice, we quickly lose interest in Gaussian reduction since coefficients in the equations are polynomials, making the algebra rather complicated. For hand analysis, Cramer's rule is the better approach, especially when the element parameters are in symbolic form. For computer-aided analysis we use the symbolic analysis capability of programs like Mathcad, MATLAB, and others to solve these linear equations.

We will illustrate the Cramer's rule solution process using an example from earlier in this section. In Example 10–10 we formulated the following node-voltage equations for the circuit in Figure 10–29:

$$\left(G + \frac{1}{Ls}\right) V_A(s) - \frac{1}{Ls} V_B(s) = I_S(s) - \frac{i_L(0)}{s}$$

$$-\frac{1}{Ls} V_A(s) + \left(Cs + \frac{1}{Ls}\right) V_B(s) = \frac{i_L(0)}{s} + Cv_C(0)$$

When using Cramer's rule, it is convenient first to find the determinant of these equations:

$$\begin{aligned}\Delta(s) &= \begin{vmatrix} G + 1/Ls & -1/Ls \\ -1/Ls & Cs + 1/Ls \end{vmatrix} \\ &= (G + 1/Ls)(Cs + 1/Ls) - (1/Ls)^2 \\ &= \frac{GLCs^2 + Cs + G}{Ls}\end{aligned}$$

We call $\Delta(s)$ the **circuit determinant** because it only depends on the element parameters L, C, and $G = 1/R$. The determinant $\Delta(s)$ characterizes the circuit and does not depend on the input driving forces or initial conditions.

The node voltage $V_A(s)$ is found using Cramer's rule.

$$V_A(s) = \frac{\Delta_A(s)}{\Delta(s)} = \frac{\begin{vmatrix} I_S(s) - \dfrac{i_L(0)}{s} & -\dfrac{1}{Ls} \\ \dfrac{i_L(0)}{s} + Cv_C(0) & Cs + \dfrac{1}{Ls} \end{vmatrix}}{\Delta(s)}$$
$$= \underbrace{\frac{(LCs^2 + 1)\, I_S(s)}{GLCs^2 + Cs + G}}_{\text{zero state}} + \underbrace{\frac{-LCsi_L(0) + Cv_C(0)}{GLCs^2 + Cs + G}}_{\text{zero input}} \tag{10–29}$$

Solving for the other node voltage $V_B(s)$ yields

$$V_B(s) = \frac{\Delta_B(s)}{\Delta(s)} = \frac{\begin{vmatrix} G + \dfrac{1}{Ls} & I_S(s) - \dfrac{i_L(0)}{s} \\ -\dfrac{1}{Ls} & \dfrac{i_L(0)}{s} + Cv_C(0) \end{vmatrix}}{\Delta(s)}$$
$$= \underbrace{\frac{I_S(s)}{GLCs^2 + Cs + G}}_{\text{zero state}} + \underbrace{\frac{GLi_L(0) + (GLs + 1)Cv_C(0)}{GLCs^2 + Cs + G}}_{\text{zero input}} \tag{10–30}$$

Cramer's rule gives both the zero-input and zero-state components of the response transforms $V_A(s)$ and $V_B(s)$.

Cramer's rule yields the node voltages as a ratio of determinants of the form

$$V_X(s) = \frac{\Delta_X(s)}{\Delta(s)} \tag{10–31}$$

The response transform is a rational function of s whose poles are either zeros of the circuit determinant or poles of the determinant $\Delta_X(s)$. That is, $V_X(s)$ in Eq. (10–31) has poles when $\Delta(s) = 0$ or $\Delta_X(s) \rightarrow \infty$. The partial-fraction expansion of $V_X(s)$ will contain terms for each of these poles. We call the zeros of $\Delta(s)$ the **natural poles** because they depend only on the circuit and give rise to the natural response terms in the partial-fraction expansion. We call the poles of $\Delta_X(s)$ the **forced poles** because they depend on the form of the input signal and give rise to the forced response terms in the partial-fraction expansion.

EXAMPLE 10–11

The response transforms $V_A(s)$ and $V_B(s)$ in Eqs. (10–29) and (10–30) include the zero-state and zero-input components. The purpose of this example is to calculate the zero-state components of the waveforms $v_A(t)$ and $v_B(t)$ when $R = 1$ kΩ, $L = 0.5$ H, $C = 0.2$ μF, and $i_S(t) = 10u(t)$ mA.

SOLUTION:
Inserting the specified numerical values into Eqs. (10–29) and (10–30) yields

$$V_A(s) = \left(\frac{10^{-7}s^2 + 1}{10^{-10}s^2 + 0.2 \times 10^{-6}s + 10^{-3}}\right)\frac{10^{-2}}{s}$$

$$= 10\frac{s^2 + 10^7}{s[(s + 1000)^2 + 3000^2]} \quad \text{V-s}$$

$$V_B(s) = \left(\frac{1}{10^{-10}s^2 + 0.2 \times 10^{-6}s + 10^{-3}}\right)\frac{10^{-2}}{s}$$

$$= 10\frac{10^7}{s[(s + 1000)^2 + 3000^2]} \quad \text{V-s}$$

Both response transforms have three poles: a forced pole at $s = 0$ and two natural poles at $s = -1000 \pm j3000$. The forced pole comes from the step function input, and the two natural poles are zeros of the circuit determinant. Expanding these rational functions as

$$V_A(s) = \frac{10}{s} - \frac{20}{3}\left(\frac{3000}{(s + 1000)^2 + 3000^2}\right)$$

$$V_B(s) = \frac{10}{s} - \frac{10}{3}\left(\frac{3000}{(s + 1000)^2 + 3000^2}\right) - 10\left(\frac{s + 1000}{(s + 1000)^2 + 3000^2}\right)$$

and taking the inverse transforms yields the required zero-state response waveforms:

$$v_A(t) = 10u(t) - 20e^{-1000t}\left[\frac{1}{3}\sin(3000\,t)\right]u(t) \quad \text{V}$$

$$v_B(t) = 10u(t) - 10e^{-1000t}\left[\frac{1}{3}\sin(3000\,t) + \cos(3000\,t)\right]u(t) \quad \text{V}$$

The step function in both responses is the forced response caused by the forced pole at $s = 0$. The damped sinusoids are natural responses determined by the natural poles.

Figure 10–30 shows part of an Excel spreadsheet that produces plots of $v_A(t)$ and $v_B(t)$. Spreadsheets are useful for generating graphs, especially when we wish to compare waveforms. The two plots show that the two response waveforms are different even though they have the same poles. In other words, the basic form of a response is determined by the forced and natural poles, but the relative amplitudes are influenced by the zeros as well. ■

APPLICATION EXAMPLE 10–12

In *s*-domain circuit analysis and design, the location of complex poles is often specified in terms of the undamped natural frequency (ω_0) and damping ratio (ζ) parameters introduced in our study of second-order cicuits. Using these parameters, the standard form of a second-order factor is $s^2 + 2\,\zeta\,\omega_0 s + \omega_0^2$, which locates the poles at

$$s_{1,2} = \omega_0(-\zeta \pm \sqrt{\zeta^2 - 1})$$

FIGURE 10–30

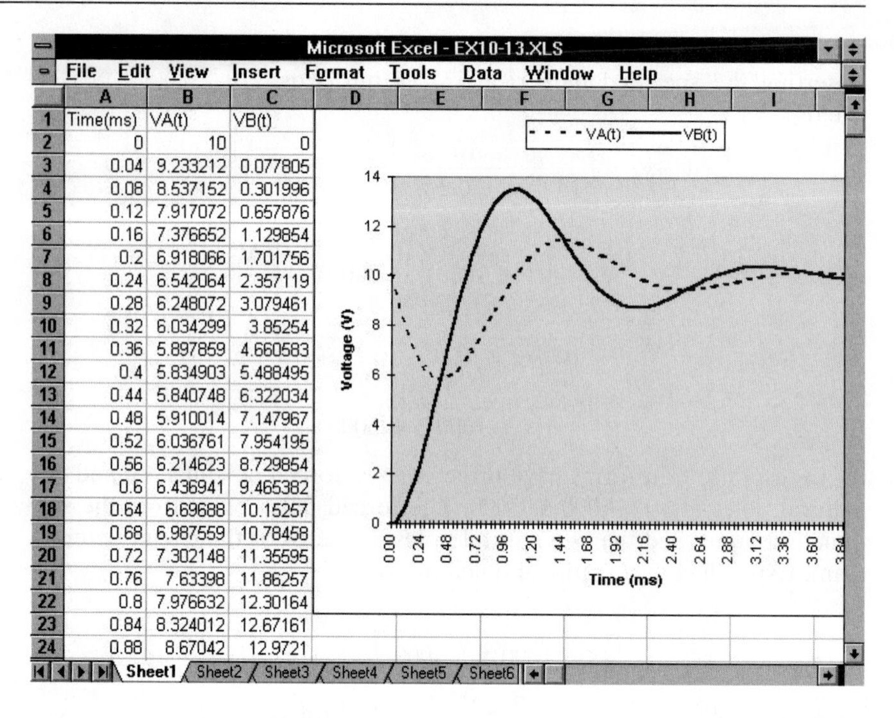

	A	B	C
1	Time(ms)	VA(t)	VB(t)
2	0	10	0
3	0.04	9.233212	0.077805
4	0.08	8.537152	0.301996
5	0.12	7.917072	0.657876
6	0.16	7.376652	1.129854
7	0.2	6.918066	1.701756
8	0.24	6.542064	2.357119
9	0.28	6.248072	3.079461
10	0.32	6.034299	3.85254
11	0.36	5.897859	4.660583
12	0.4	5.834903	5.488495
13	0.44	5.840748	6.322034
14	0.48	5.910014	7.147967
15	0.52	6.036761	7.954195
16	0.56	6.214623	8.729854
17	0.6	6.436941	9.465382
18	0.64	6.69688	10.15257
19	0.68	6.987559	10.78458
20	0.72	7.302148	11.35595
21	0.76	7.63398	11.86257
22	0.8	7.976632	12.30164
23	0.84	8.324012	12.67161
24	0.88	8.67042	12.9721

The quantity under the radical depends only on the damping ratio ζ. When $\zeta > 1$ the quantity is positive and the two poles are real and distinct, and the second-order factor becomes the product of two first-order terms. If $\zeta = 1$ the quantity under the radical vanishes and there is a double pole as $s = -\omega_0$. If $\zeta < 1$ the quantity under the radical is negative and the two roots are complex conjugates.

The location of complex poles is also defined in terms of the two natural frequency parameters α and β. Using these parameters, the poles are at $s_{1,2} = -\alpha \pm j\beta$, and the standard form of a second-order factor is $(s + \alpha)^2 + \beta^2$.

In s-domain circuit design, we often need to convert from one set of parameters to the other. First, equating their standard forms

$$s^2 + 2\alpha s + \alpha^2 + \beta^2 = s^2 + 2\zeta\omega_0 s + \omega_0^2$$

and then equating the coefficients of like powers of s yields

$$\omega_0 = \sqrt{\alpha^2 + \beta^2} \quad \text{and} \quad \zeta = \frac{\alpha}{\sqrt{\alpha^2 + \beta^2}}$$

and conversely

$$\alpha = \zeta\omega_0 \quad \text{and} \quad \beta = \omega_0\sqrt{1 - \zeta^2}$$

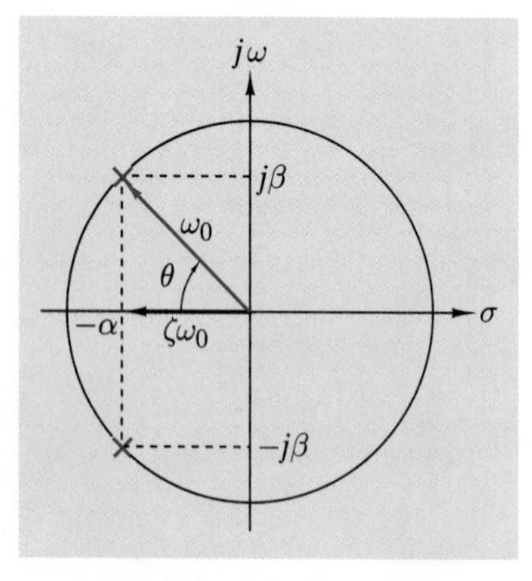

FIGURE 10–31 *s-plane geometry relating α and β to ζ and ω_0.*

Figure 10–31 shows how these parameters define the locations of complex poles in the s plane. The natural frequency parameters α and β define the rectangular coordinates of the poles. In a sense, the parameters ω_0 and ζ define the corresponding polar coordinates. The parameter ω_0 is the radial distance from the origin to the poles. The angle θ is determined by the damping ratio ζ alone, since $\theta = \cos^{-1}\zeta$.

DESIGN EXAMPLE 10–13

The *s*-domain circuit in Figure 10–32 is to be designed to produce a pair of complex poles defined by $\zeta = 0.5$ and $\omega_0 = 1000$ rad/s. To simplify production the design will use equal element values $R_1 = R_2 = R$ and $C_1 = C_2 = C$. Select the values of R, C, and the gain μ so that the circuit has the desired natural poles.

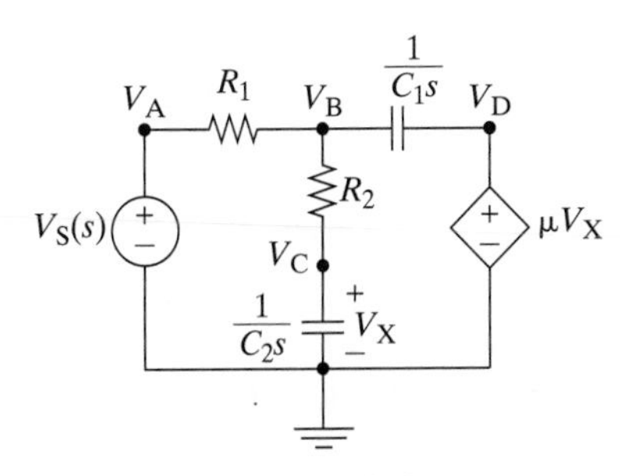

FIGURE 10–32

SOLUTION:

To locate the natural poles, we find the circuit determinant using node-voltage equations. The circuit has four nodes but only two of these involve independent variables. For the indicated reference node, the voltages at nodes A and D are $V_A(s) = V_S(s)$ and $V_D(s) = \mu V_X(s) = \mu V_C(s)$. That is, the two grounded voltage sources specify the voltages at nodes A and D. Consequently, we only need equations at nodes B and C. The sum of currents leaving these nodes are

$$\text{Node B: } \frac{V_B(s) - V_S(s)}{R_1} + \frac{V_B(s) - V_C(s)}{R_2} + \frac{V_B(s) - \mu V_C(s)}{1/C_1 s} = 0$$

$$\text{Node C: } \frac{V_C(s) - V_B(s)}{R_2} + \frac{V_C(s)}{1/C_2 s} = 0$$

Arranging these equations with unknown node voltages on the left and the source terms on the right yields

$$(C_1 s + G_1 + G_2)V_B(s) - (\mu C_1 s + G_2)V_C(s) = G_1 V_S(s)$$
$$-(G_2)V_B(s) + (C_2 s + G_2)V_C(s) = 0$$

where $G_1 = 1/R_1$ and $G_2 = 1/R_2$. The natural poles are zeros of the circuit determinant:

$$\Delta(s) = \begin{vmatrix} (C_1 s + G_1 + G_2) & -(\mu C_2 s + G_2) \\ -(G_2) & (C_2 s + G_2) \end{vmatrix}$$
$$= C_1 C_2 s^2 + (G_2 C_1 + G_1 C_2 + G_2 C_2 - \mu G_2 C_2)s + G_1 G_2$$

For equal resistances $R_1 = R_2 = R$ and equal capacitances $C_1 = C_2 = C$, the circuit determinant reduces to

$$\frac{\Delta(s)}{C^2} = s^2 + \left(\frac{3 - \mu}{RC}\right)s + \left(\frac{1}{RC}\right)^2$$

Comparing this second-order factor to the standard form $s^2 + 2\zeta\omega_0 s + \omega_0^2$ yields the following design constraints:

$$\omega_0 = \frac{1}{RC} = 1000 \text{ and } \zeta = \frac{3 - \mu}{2} = 0.5$$

These constraints lead to the conditions $RC = 10^{-3}$ and $\mu = 3$. Selecting $R = 10$ kΩ makes $C = 100$ nF. For the specified conditions the natural poles are located at $s = \alpha \pm j\beta$, where

$$\alpha = -\zeta\omega_0 = -500 \text{ rad/s} \quad \text{and} \quad \beta = \omega_0\sqrt{1 - \zeta^2} = 866 \text{ rad/s}$$

EXAMPLE 10–14

(a) For the *s*-domain circuit in Figure 10–33, solve for the zero-state output $V_O(s)$ in terms of a general input $V_S(s)$.
(b) Solve for the zero-state output $v_O(t)$ when the input is a unit step function $v_S(t) = u(t)$.

FIGURE 10–33

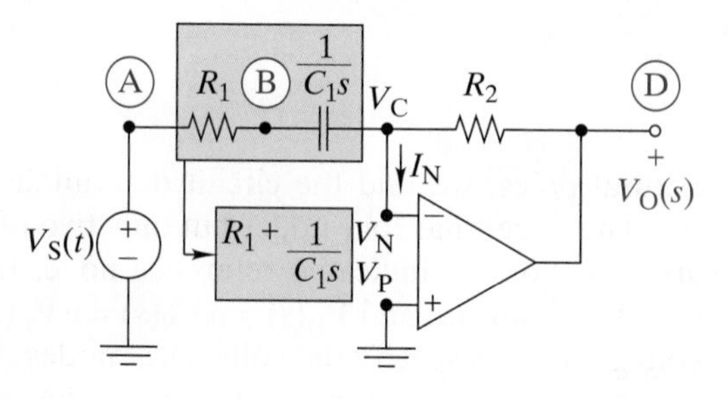

SOLUTION:

(a) We use the node-voltage method to find the OP AMP output. A node equation is not required at node A since the selected reference node makes $V_A(s) = V_S(s)$. Likewise, one is not needed at node D since it is the OP AMP output. Finally, we can avoid writing a node equation at node B by observing that the element impedances R_1 and $1/Cs$ are connected in series. We can treat this series combination as a single element with an equivalent impedance $R_1 + 1/Cs$. Using all of these observations, the sum of the currents leaving Node C is

$$\text{Node C:}\quad \frac{V_C(s) - V_S(s)}{R_1 + 1/C\,s} + \frac{V_C(s) - V_O(s)}{R_2} + I_N(s) = 0$$

In the *s*-domain the ideal OP AMP model in Eq. (10–4) requires $I_N(s) = 0$ and $V_P(s) = V_N(s)$. But $V_P(s) = 0$ since the noninverting input is grounded; hence, $V_N(s) = V_C(s) = 0$. Inserting these conditions in the node C equation and solving for the output voltage yields

$$\begin{aligned} V_O(s) &= \left[-\frac{R_2}{R_1 + 1/C\,s} \right] V_S(s) \\ &= \left[-\frac{R_2}{R_1}\left(\frac{s}{s + 1/R_1C} \right) \right] V_S(s) \quad \text{V-s} \end{aligned}$$

This equation relates the zero-state output to a general input $V_S(s)$. The output transform is proportional to the input transform since the circuit is linear. The proportionality factor within the brackets is called a *network function*. In this case the network function has a natural pole at $s = -1/R_1C$ and a zero at $s = 0$.

(b) A step function input $V_S(s) = 1/s$ produces a forced pole at $s = 0$. However, the zero in the network function cancels the forced pole so that

$$\begin{aligned} v_O(t) &= \mathscr{L}^{-1}\left\{ -\frac{R_2}{R_1}\left(\frac{s}{s + 1/R_1C} \right) \frac{1}{s} \right\} = \mathscr{L}^{-1}\left\{ -\frac{R_2}{R_1}\left(\frac{1}{s + 1/R_1C} \right) \right\} \\ &= \left(-\frac{R_2}{R_1} e^{-t/R_1C} \right) u(t) \quad \text{V} \end{aligned}$$

For a step function input the zero-state output has no forced pole, only a natural pole at $s = -1/R_1C$. The general principle is that the forced response can be zero even when the input is not zero. In the s domain this occurs when the network function relating output to input has zeros at the same location as forced poles.

Exercise 10–7

Formulate node-voltage equations for the circuit in Figure 10–34 and find the circuit determinant. Assume that the initial conditions are zero.

Answer:

The node equations are:

$$\text{Node B:}\quad \left(G_1 + G_2 + \frac{1}{Ls}\right)V_B(s) - G_2V_C(s) = G_1V_S(s)$$

$$\text{Node C:}\quad -G_2V_B(s) + (G_2 + Cs)V_C(s) = CsV_S(s)$$

The circuit determinant is

$$\Delta(s) = \frac{(G_1 + G_2)LCs^2 + (G_1G_2L + C)s + G_2}{Ls}$$

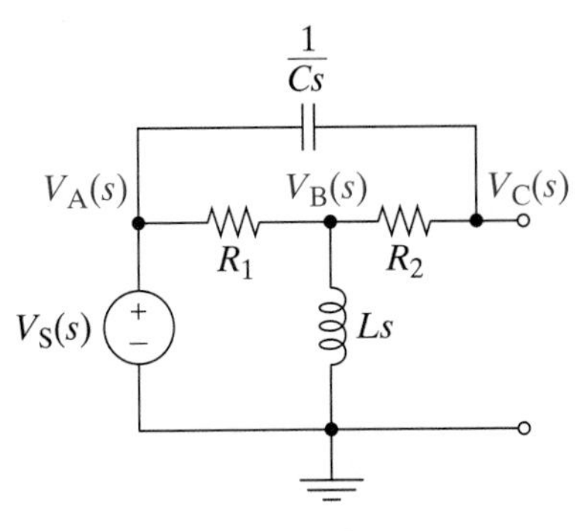

FIGURE 10–34

Exercise 10–8

Formulate node-voltage equations for the circuit in Figure 10–34 when a resistor R_3 is connected between node C and ground. Assume that the initial conditions are zero.

Answer:

$$\text{Node B:}\quad \left(G_1 + G_2 + \frac{1}{Ls}\right)V_B(s) - G_2V_C(s) = G_1V_S(s)$$

$$\text{Node C:}\quad -G_2V_B(s) + (G_2 + G_3 + Cs)V_C(s) = CsV_S(s)$$

10–5 MESH-CURRENT ANALYSIS IN THE *s* DOMAIN

We can use the mesh-current method only when the circuit can be drawn on a flat surface without crossovers. Such planar circuits have special loops called meshes that are defined as closed paths that do not enclose any elements. The mesh-current variables are the loop currents assigned to each mesh in a planar circuit. Because of KCL the current through any two-terminal element can be expressed as the difference of two adjacent mesh currents. This fundamental property of mesh currents, together with the element impedances, allows us to write KVL constraints around each of the meshes.

For example, in Figure 10–35 the sum of voltages around mesh A can be written as

$$Z_1(s)I_A(s) + Z_3[I_A(s) - I_C(s)] - V_{S1}(s) + Z_2[I_A(s) - I_B(s)] + V_{S2}(s) = 0$$

FIGURE 10–35 *An example mesh.*

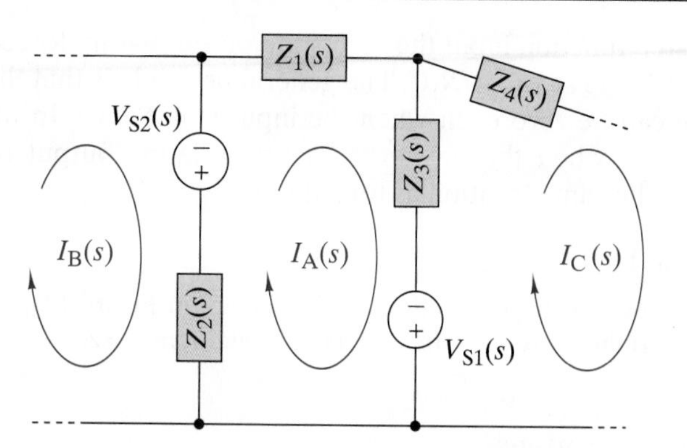

Rewriting this equation with unknown mesh currents grouped on the left and inputs on the right yields

$$(Z_1 + Z_2 + Z_3)I_A(s) - Z_2 I_B(s) - Z_3 I_C(s) = V_{S1}(s) - V_{S2}(s)$$

This equation displays the following pattern. The unknowns are the mesh-current transforms $I_A(s)$, $I_B(s)$, and $I_C(s)$. The coefficient $[Z_1(s) + Z_2(s) + Z_3(s)]$ of $I_A(s)$ is the sum of the impedances of the elements in mesh A. The coefficients $[Z_2(s)]$ of $I_B(s)$ and $[Z_3(s)]$ of $I_C(s)$ are the impedances common to mesh A and the other meshes. Finally, $V_{S1}(s) - V_{S2}(s)$ is the sum of the source voltages around mesh A. These observations suggest that we can write node-voltage equations for *s*-domain circuits by inspection, just as we did with resistive circuits.

The formulation approach just outlined assumes that there are no current sources in the circuit. When writing mesh-current equations, we select the voltage source model to represent the initial conditions. If the circuit contains dependent or independent current sources, they can be treated using the following methods:

> **Method 1:** If there is an admittance in parallel with the current source, use a source transformation to convert it into an equivalent voltage source.
>
> **Method 2:** Draw the circuit diagram so that only one mesh current circulates through the current source. This mesh current is then determined by the source current.
>
> **Method 3:** Create a supermesh for any current source that cannot be handled by methods 1 or 2.

Some circuits may require more than one of these methods.

The following examples illustrate the mesh-current method of *s*-domain circuit analysis.

EXAMPLE 10–15

(a) Formulate mesh-current equations for the circuit in Figure 10–36(a).
(b) Solve for the zero-input component of $I_A(s)$ and $I_B(s)$.
(c) Find the zero-input responses $i_A(t)$ and $i_B(t)$ for $R_1 = 200\ \Omega$, $R_2 = 300\ \Omega$, $L_1 = 50$ mH, and $L_2 = 100$ mH.

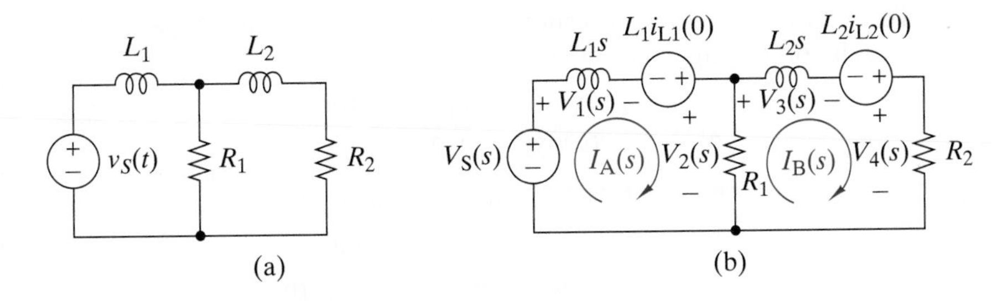

FIGURE 10–36

SOLUTION:

(a) Figure 10–36(b) shows the circuit transformed into the s domain. In transforming the circuit we used the voltage source model for the initial conditions. The net result is that the transformed circuit contains only voltage sources. The sum of voltages around meshes A and B can be written as

$$\text{Mesh A:}\quad -V_S(s) + L_1sI_A(s) - L_1i_{L_1}(0) + R_1[I_A(s) - I_B(s)] = 0$$

$$\text{Mesh B:}\quad R_1[I_B(s) - I_A(s)] + L_2sI_B(s) - L_2i_{L_2}(0) + R_2I_B(s) = 0$$

Rearranging these equations in standard form yields

$$\text{Mesh A:}\quad (L_1s + R_1)I_A(s) - R_1I_B(s) = V_S(s) + L_1i_{L_1}(0)$$

$$\text{Mesh B:}\quad -R_1I_A(s) + (L_2s + R_1 + R_2)I_B(s) = L_2i_{L_2}(0)$$

These s-domain circuit equations are two linear algebraic equations in the two unknown mesh currents $I_A(s)$ and $I_B(s)$.

(b) To solve for the mesh equations, we first find the circuit determinant:

$$\Delta(s) = \begin{vmatrix} L_1s + R_1 & -R_1 \\ -R_1 & L_2s + R_1 + R_2 \end{vmatrix}$$
$$= L_1L_2s^2 + (R_1L_2 + R_1L_1 + R_2L_1)s + R_1R_2$$

To find the zero-input component of $I_A(s)$, we let $V_S(s) = 0$ and use Cramer's rule:

$$I_A(s) = \frac{\begin{vmatrix} L_1i_{L_1}(0) & -R_1 \\ L_2i_{L_2}(0) & L_2s + R_1 + R_2 \end{vmatrix}}{\Delta(s)}$$
$$= \frac{(L_2s + R_1 + R_2)L_1i_{L_1}(0) + R_1L_2i_{L_2}(0)}{L_1L_2s^2 + (R_1L_2 + R_1L_1 + R_2L_1)s + R_1R_2}$$

Similarly, the zero-input component in $I_B(s)$ is

$$I_B(s) = \frac{\begin{vmatrix} L_1s + R_1 & L_1i_{L_1}(0) \\ -R_1 & L_2i_{L_2}(0) \end{vmatrix}}{\Delta(s)}$$
$$= \frac{(L_1s + R_1)L_2i_{L_2}(0) + R_1L_1i_{L_1}(0)}{L_1L_2s^2 + (R_1L_2 + R_1L_1 + R_2L_1)s + R_1R_2}$$

(c) To find the time-domain response, we insert the numerical parameters into the preceding expressions to obtain

$$I_A(s) = \frac{(0.005s + 15)i_{L_1}(0) + 10i_{L_2}(0)}{0.005s^2 + 250s + 20{,}000}$$

$$= \frac{(s + 3000)i_{L_1}(0) + 2000i_{L_2}(0)}{(s + 1000)\,(s + 4000)}$$

$$I_B(s) = \frac{(0.005s + 100)i_{L_2}(0) + 5i_{L_1}(0)}{0.005s^2 + 250s + 20{,}000}$$

$$= \frac{(s + 2000)i_{L_2}(0) + 1000i_{L_1}(0)}{(s + 1000)\,(s + 4000)}$$

The circuit has natural poles at $s = -1000$ and -4000 rad/s. Expanding by partial fractions yields

$$I_A(s) = \frac{2}{3} \times \frac{i_{L_1}(0) + i_{L_2}(0)}{s + 1000} + \frac{1}{3} \times \frac{i_{L_1}(0) - 2i_{L_2}(0)}{s + 4000} \quad \text{A-s}$$

$$I_B(s) = \frac{1}{3} \times \frac{i_{L_1}(0) + i_{L_2}(0)}{s + 1000} - \frac{1}{3} \times \frac{i_{L_1}(0) - 2i_{L_2}(0)}{s + 4000} \quad \text{A-s}$$

The inverse transforms of these expansions are the required zero-state response waveforms:

$$i_A(t) = \left[\frac{2}{3}[i_{L_1}(0) + i_{L_2}(0)]e^{-1000t} + \frac{1}{3}[i_{L_1}(0) - 2i_{L_2}(0)]e^{-4000t}\right]u(t) \quad \text{A}$$

$$i_B(t) = \left[\frac{1}{3}[i_{L_1}(0) + i_{L_2}(0)]e^{-1000t} - \frac{1}{3}[i_{L_1}(0) - 2i_{L_2}(0)]e^{-4000t}\right]u(t) \quad \text{A}$$

Notice that if the initial conditions are $i_{L_1}(0) = -i_{L_2}(0)$, then both $I_A(s)$ and $I_B(s)$ have a zero at $s = -1000$. This zero effectively cancels the natural pole at $s = -1000$. As a result, this pole has zero residue in both partial fraction expansions, and the corresponding terms disappear from the time-domain responses. Likewise, if the initial conditions are $i_{L_1}(0) = 2i_{L_2}(0)$, then both $I_A(s)$ and $I_B(s)$ have a zero at $s = -4000$, and the natural pole at $s = -4000$ disappears in the s-domain responses. The general principle is that all of the circuit's natural poles may not be present in a given response. When this happens the response transform has a zero at the same location as a natural pole, and we say that the natural pole is not observable in the specified response. ■

EXAMPLE 10–16

(a) Formulate mesh-current equations for the circuit in Figure 10–37(a).
(b) Solve for the zero-input component of $i_A(t)$ for $i_L(0) = 0$, $v_C(0) = 10$ V, $L = 250$ mH, $C = 1$ μF, and $R = 1$ kΩ.

SOLUTION:

(a) Figure 10–37(a) is the s-domain circuit used in Example 10–10 to develop node equations. In this circuit each current source is connected in parallel with an impedance. Source transformations convert these

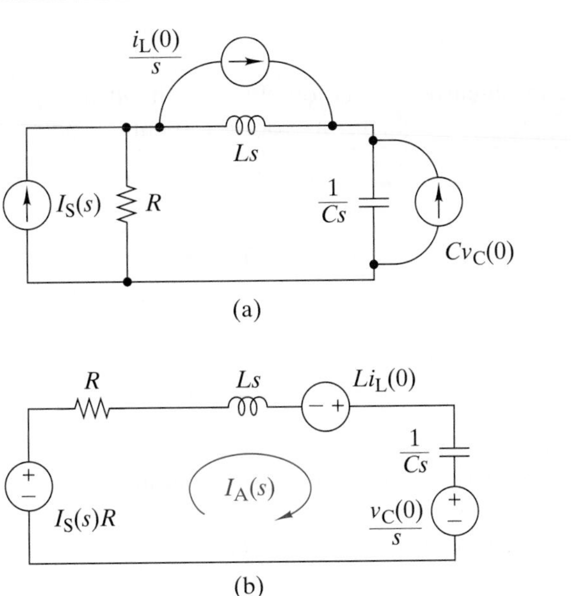

FIGURE 10–37

current sources into the equivalent voltage sources shown in Figure 10–37(b). The circuit in Figure 10–37(b) is a series *RLC* circuit of the type treated in Chapter 7. By inspection, the KVL equation for the single mesh in this circuit is

$$\left[R + Ls + \frac{1}{Cs}\right] I_A(s) = RI_S(s) + Li_L(0) - \frac{v_C(0)}{s}$$

The circuit determinant is the factor $R + Ls + 1/Cs = (LCs^2 + RCs + 1)/Cs$. The zeros of the circuit determinant are roots of the quadratic equation $LCs^2 + RCs + 1 = 0$, which we recognize as the characteristic equation of a series *RLC* circuit. In our study of *RLC* circuits we called these roots natural frequencies. Thus, the natural poles of the circuit are its natural frequencies.

(b) Solving the mesh equation for the zero-input component yields

$$I_A(s) = \frac{LCsi_L(0) - Cv_C(0)}{LCs^2 + RCs + 1} \quad \text{A-s}$$

Inserting the given numerical values produces

$$I_A(s) = -\frac{10^{-5}}{0.25 \times 10^{-6}s^2 + 10^{-3}s + 1} = -\frac{40}{s^2 + 4 \times 10^3 s + 4 \times 10^6}$$

$$= -\frac{40}{(s + 2000)^2} \quad \text{A-s}$$

The zero-state response has two natural poles, both located at $s = -2000$. The inverse transform of $I_A(s)$ is a damped ramp waveform:

$$i_A(t) = -[40te^{-2000t}]\, u(t) \quad \text{A}$$

The damped ramp response indicates a critically damped second-order circuit. The minus sign means the direction of the actual current is opposite to the reference mark assigned to $I_A(s)$ in Figure 10–37. This sign makes sense physically since the capacitor initial condition source in Figure 10–37(b) tends to drive current in a direction opposite to the assigned reference mark. ■

EXAMPLE 10–17

Formulate mesh-current equations for the circuit in Figure 10–38 and solve for $I_B(s)$ in symbolic form. Locate the natural poles of the circuit for $R = 1\ \text{k}\Omega$, $C = 4\ \mu\text{F}$, and $L = 1$ H.

FIGURE 10–38

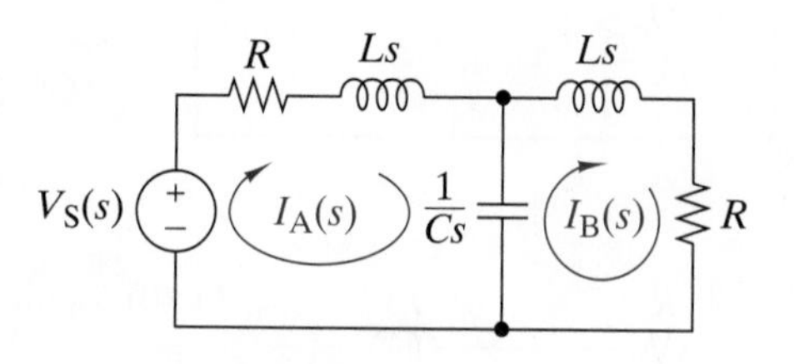

SOLUTION:

By inspection the two mesh-current equations are

$$\text{Mesh A: } \left(R + Ls + \frac{1}{Cs}\right) I_A(s) - \frac{1}{Cs} I_B(s) = V_S(s)$$

$$\text{Mesh B: } -\frac{1}{Cs} I_A(s) + \left(R + Ls + \frac{1}{Cs}\right) I_B(s) = 0$$

The circuit determinant is

$$\Delta(s) = \left(R + Ls + \frac{1}{Cs}\right)^2 - \left(\frac{1}{Cs}\right)^2$$

$$= R^2 + 2RLs + 2\frac{R}{Cs} + L^2s^2 + 2\frac{L}{C}$$

$$= \frac{(Ls + R)(LCs^2 + RCs + 2)}{Cs}$$

and the required mesh current is

$$I_B(s) = \frac{\Delta_B(s)}{\Delta(s)} = \frac{\begin{vmatrix} R + Ls + \frac{1}{Cs} & V_S(s) \\ -\frac{1}{Cs} & 0 \end{vmatrix}}{\Delta(s)}$$

$$= \frac{V_S(s)}{(Ls + R)(LCs^2 + RCs + 2)}$$

The natural poles are roots of the the denominator of $I_B(s)$, namely

$$\left(s + \frac{R}{L}\right)\left(s^2 + \frac{R}{L}s + \frac{2}{LC}\right) = 0$$

For $R = 1\ \text{k}\Omega$, $C = 4\ \mu\text{F}$, and $L = 1$ H, this expression factors as

$$(s + 1000)(s^2 + 1000s + 5 \times 10^5) = (s + 1000)[(s + 500)^2 + 500^2] = 0$$

so the natural poles are located at $s = -1000$ rad/s and $s = -500 \pm j500$ rad/s. ■

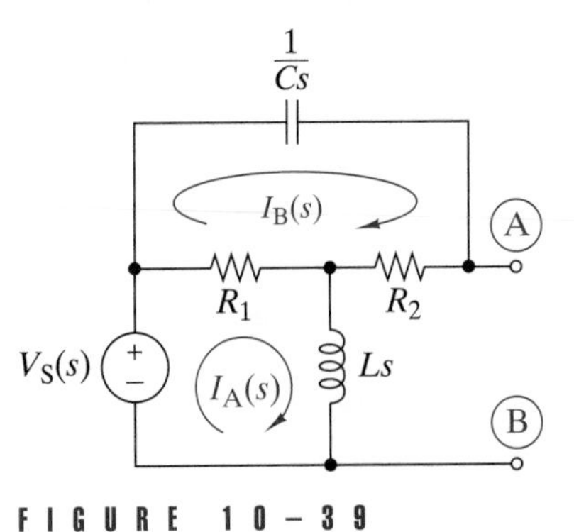

FIGURE 10–39

Exercise 10–9

(a) Formulate mesh-current equations for the circuit in Figure 10–39. Assume that the initial conditions are zero.
(b) Find the circuit determinant.
(c) Solve for the zero-state component of $I_B(s)$.

Answers:

(a)
$$\begin{cases} (R_1 + Ls)I_A(s) - R_1 I_B(s) = V_S(s) \\ -R_1 I_A(s) + (R_1 + R_2 + 1/Cs)I_B(s) = 0 \end{cases}$$

(b)
$$\Delta(s) = \frac{(R_1 + R_2)LCs^2 + (R_1R_2C + L)s + R_1}{Cs}$$

(c)
$$I_B(s) = \frac{R_1CsV_S(s)}{(R_1 + R_2)LCs^2 + (R_1R_2C + L)s + R_1}$$

Exercise 10–10

Formulate mesh-current equations for the circuit in Figure 10–39 when a resistor R_3 is connected between nodes A and B. Assume that the initial conditions are zero.

Answer:

$$(R_1 + Ls)I_A(s) - R_1 I_B(s) - LsI_C(s) = V_S(s)$$

$$-R_1 I_A(s) + \left(R_1 + R_2 + \frac{1}{Cs}\right)I_B(s) - R_2 I_C(s) = 0$$

$$-LsI_A(s) - R_2 I_B(s) + (R_2 + R_3 + Ls)I_C(s) = 0$$

10–6 SUMMARY OF S-DOMAIN CIRCUIT ANALYSIS

At this point we review our progress and put *s*-domain circuit analysis into perspective. We have shown that linear circuits can be transformed from the time domain into the *s* domain. In this domain KCL and KVL apply to transforms and the passive element *i–v* characteristics become impedances with series or parallel initial condition sources. In relatively simple circuits we can use basic analysis methods, such as reduction, superposition, and voltage/current division. For more complicated circuits we use systematic procedures, such as the node-voltage or mesh-current methods, to solve for the circuit response.

In theory, we can perform *s*-domain analysis on circuits of any complexity. In practice, the algebraic burden of hand computations gets out of hand for circuits with more than three or four nodes or meshes. Of what practical use is an analysis method that becomes impractical at such a modest level of circuit complexity? Why not just appeal to computer-aided analysis tools in the first place?

Unquestionably, large-scale circuits are best handled by computer-aided analysis. Computer-aided analysis is probably the right approach even for small-scale circuits when numerical values for all circuit parameters are known and the desired end product is a plot or numerical listing of the response waveform. Simply put, s-domain circuit analysis is not a particularly efficient algorithm for generating numerical response data.

The purpose of s-domain circuit analysis is to gain insight into circuit behavior, not to grind out particular response waveforms. In this regard s-domain circuit analysis complements programs like PSpice. It offers a way of characterizing circuits in very general terms. It provides guidelines that allow us to use computer-aided analysis tools intelligently. Some of the useful general principles derived in this chapter are the following.

The response transform $Y(s)$[1] is a rational function whose partial-fraction expansion leads directly to a response waveform of the form

$$y(t) = \sum_{j=1}^{\text{number of poles}} k_j e^{p_j t}$$

where k_j is the residue of the pole in $Y(s)$ located at $s = -p_j$. The location of the poles tells us a great deal about the form of the response. The pair of conjugate complex poles in Example 10–11 produced a damped sine waveform, the two distinct real poles in Example 10–15 produced exponential waveforms, and the double pole in Example 10–16 led to a damped ramp waveform. The general principle illustrated is as follows:

The poles of* Y(s) *are either real or complex conjugates. Simple real poles lead to exponentials, double real poles lead to a damped ramp, and complex conjugate poles lead to damped sinusoids.

The poles in $Y(s)$ are introduced either by the circuit itself (natural poles) or by the input driving force (forced poles).

The natural poles are zeros of the circuit determinant and lead to the natural response. The forced poles are poles of the input* X(s) *and lead to the forced response.

Stability is a key concept in circuit analysis and design. For our present purposes we say that a linear circuit is **stable** if its natural response decays to zero as $t \to \infty$. Figure 10–40 shows the waveforms of the natural modes corresponding to different pole locations in the s plane. Poles in the left half plane give rise to waveforms that decay to zero as time increases, while those in the right half plane increase without bound. As a result, we can say that

A circuit is stable if all of its natural poles are located in the left half of the* s *plane.

Stability requires *all* of the natural poles to be in the left half plane (LHP). The circuit is *unstable* if even one natural pole falls in the right half plane (RHP).

In Figure 10–40 the $j\omega$-axis is the boundary between the LHP (stable circuits) and RHP (unstable circuits). Poles exactly on this boundary require further discussion. As Figure 10–40 shows, *simple* j-axis poles at $s = 0$

---

1 In this context $Y(s)$ is not an admittance but the Laplace transform of the circuit output $y(t)$.

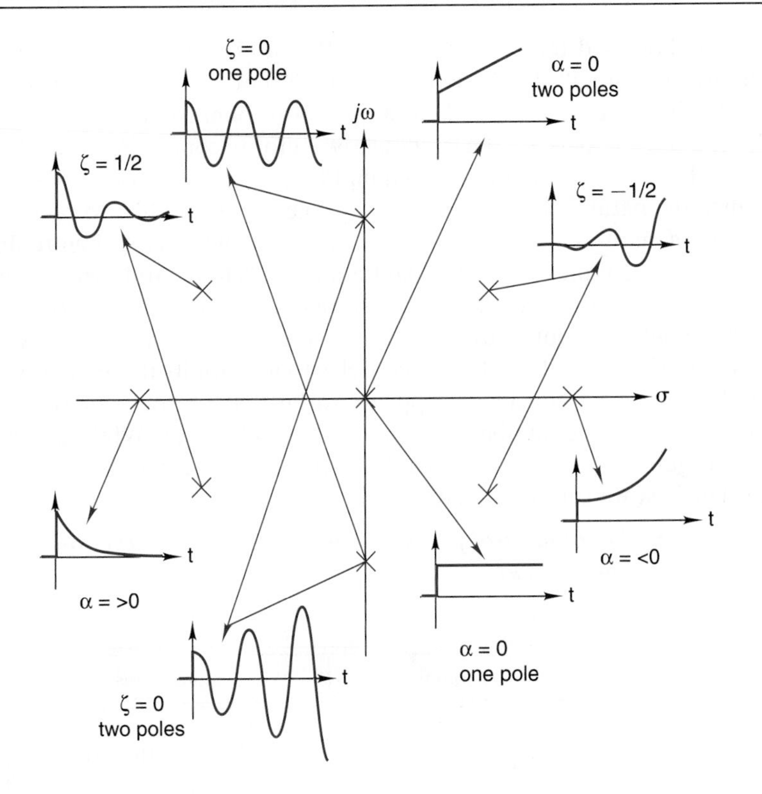

FIGURE 10–40 *Form of the natural response corresponding to different pole locations.*

and $s = \pm j\beta$ lead to natural modes like $u(t)$ and $\cos(\beta t)$ that neither decay to zero nor increase without bound. The figure also shows that *double* poles on the j-axis lead to natural modes like $tu(t)$ and $t\cos(\beta t)$ that increase without bound. Circuits with *simple* poles on the j-axis are sometimes said to be **marginally stable**,[2] while those with *multiple* poles on the j-axis are clearly *unstable*.

Circuit stability is determined by natural poles, not forced poles. For example, suppose an input $x(t) = e^{10t}$ produces an output transform

$$Y(s) = \frac{12s}{\underbrace{(s+2)}_{\text{LHP}}\underbrace{(s-10)}_{\text{RHP}}}$$

This transform has a left half plane (LHP) pole and a right half plane (RHP) pole. The corresponding waveform

$$y(t) = \underbrace{2e^{-2t}}_{\substack{\text{natural}\\\text{bounded}}} + \underbrace{10e^{10t}}_{\substack{\text{forced}\\\text{unbounded}}} \quad t > 0$$

2 They could just as logically be called marginally unstable. The stability status of simple j-axis poles depends on the application. For example, electronic circuits with simple poles firmly rooted on the j-axis are called stable oscillators. On the other hand, j-axis poles in audio amplifiers cause "ringing," a dirty word among audiophiles.

has an unbounded term due to the RHP pole. Even with an unbounded response the circuit is still said to be stable because the natural pole at $s = -2$ is in the LHP and leads to a natural response that decays to zero. The unbounded part of the response waveform comes from the forced RHP pole caused by the unbounded input.

Since the natural response play a key role, we would like to predict the number of natural poles by simply examining the circuit. Figure 10–41 summarizes examples from this chapter and leads to the following observations. Circuits with only one energy storage element (inductor or capacitor) have only one pole, circuits with two independent elements have two poles, and Example 10–17 has three poles to go with its three energy storage elements. The conclusion appears to be that the number of natural poles is equal to the number of energy storage elements. While this rule is a useful guideline, there are exceptions (capacitors in parallel, for example). The best we can say is that

> ***The number of natural poles does not exceed the number of energy storage elements.***

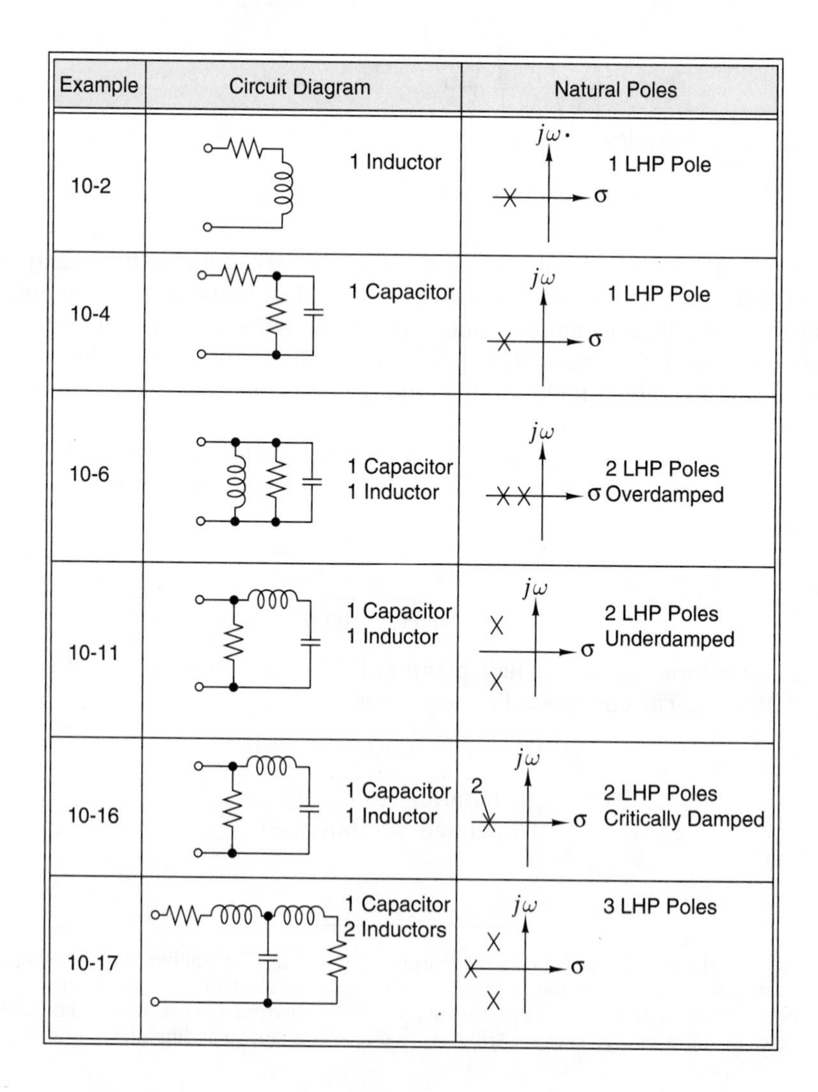

FIGURE 10–41 *Summary of Chapter 10 examples.*

Another implication in Figure 10–41 comes from two additional observations. First, all of the natural poles are in the LHP; hence, all of the circuits are stable. Second, all of the circuits contain only passive resistors, capacitors, and/or inductors; that is, there are no active elements. These observations imply that

Circuits consisting of passive resistors, capacitors, and inductors are inherently stable.

This conclusion makes sense physically since the passive elements can only store or dissipate energy. They cannot produce the energy needed to sustain an unbounded response.

What about circuits with active elements like dependent sources or OP AMPs? Such circuits can be unstable, as we can see by reviewing Example 10–13. In that example we analyzed the active *RC* circuit in Figure 10–32 and found the circuit determinant to be

$$\frac{\Delta(s)}{C^2} = s^2 + \left(\frac{3-\mu}{RC}\right)s + \left(\frac{1}{RC}\right)^2$$

and the two natural poles are defined by

$$\omega_0 = \frac{1}{RC} \quad \text{and} \quad \zeta = \frac{3-\mu}{2}$$

where μ is the gain of the dependent source, the active element in the circuit. For $\mu = 0$ (active element turned off), the damping ratio is $\zeta = 1.5 > 1$ and the circuit is overdamped. For $\mu = 1$, the damping ratio is $\zeta = 1$ and the circuit is critically damped. For $3 > \mu > 1$, the damping ratio is $1 > \zeta > 0$ and the circuit is underdamped. For $\mu = 3$, the damping ratio is $\zeta = 0$ and the circuit is undamped. Finally, for $\mu > 3$, the damping ratio is $\zeta < 0$ and the circuit is said to have negative damping, which is an unstable condition.

Figure 10–42 shows the locus of the natural poles as the gain increases. For $\mu > 3$, the poles move into the RHP and the circuit becomes unstable.

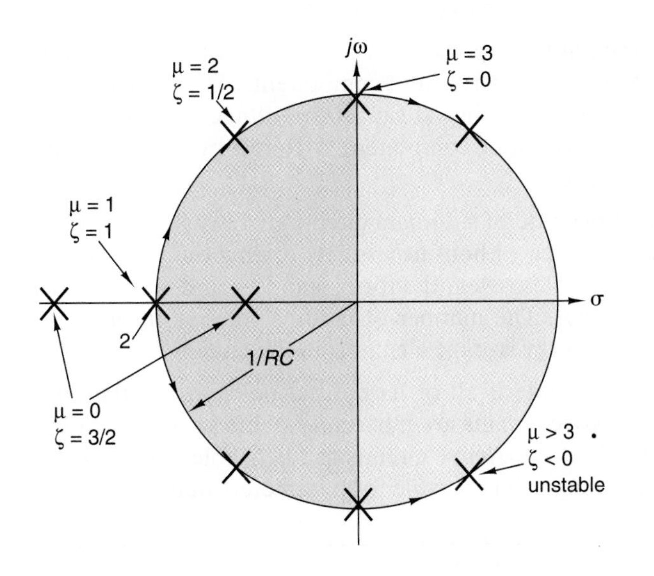

FIGURE 10–42 *Complex pole locus on a circle of radius ω_0.*

This makes sense physically. When the gain is high enough, the active element can produce the energy needed to sustain an unbounded output. Since instability is almost always undesirable, we usually state the conclusion the other way around. That is, this active *RC* circuit is stable provided the gain $\mu < 3$. It is common for active circuits to be stable when circuit parameters are in one range and unstable when they are outside this range. For double-pole circuits the stable range can be found by relating the damping ratio to circuit parameters. For single-pole circuits the stable range ensures that the pole lies on the negative real axis.

SUMMARY

- Kirchhoff's laws apply to voltage and current waveforms in the time domain and to the corresponding transforms in the *s* domain.
- The *s*-domain models for the passive elements include initial condition sources and the element impedance or admittance. Impedance is the proportionality factor in the expression $V(s) = Z(s)I(s)$ relating the voltage and current transforms. Admittance is the reciprocal of impedance.
- The impedances of the three passive elements are $Z_R(s) = R$, $Z_L(s) = Ls$, and $Z_C(s) = 1/Cs$.
- The *s*-domain circuit analysis techniques closely parallel the analysis methods developed for resistance circuits. Basic analysis techniques, such as circuit reduction, Thévenin's and Norton's theorems, the unit output method, or superposition, can be used in simple circuits. More complicated networks require a general approach, such as the node-voltage or mesh-current methods.
- Response transforms are rational functions whose poles are zeros of the circuit determinant or poles of the transform of the input driving forces. Poles introduced by the circuit determinant are called natural poles and lead to the natural response. Poles introduced by the input are called forced poles and lead to the forced response.
- In linear circuits, response transforms and waveforms can be separated into zero-state and zero-input components. The zero-state component is found by setting the initial capacitor voltages and inductor currents to zero. The zero-input component is found by setting all input driving forces to zero.
- The main purpose of *s*-domain circuit analysis is to gain insight into circuit performance without necessarily finding the time-domain response. The natural poles reveal the form, stability, and observability of the circuit's response. The number of natural poles is never greater than the number of energy storage elements in the circuit.
- A circuit is stable if all of its natural poles are in the left half of the *s* plane. Passive circuits are inherently stable so their natural poles are in the left half plane. Active circuits can be stable when circuit parameters are in one range and unstable for parameters outside this range.

PROBLEMS

OBJECTIVE 10–1 EQUIVALENT IMPEDANCE (SECTS. 10–1, 10–2)

Given a linear circuit, use series and parallel equivalences to find the poles and zeros of the equivalent impedance at specified terminal pairs. Select element values to obtain specified pole locations.
See Examples 10–2, 10–4, 10–5, 10–6 and Exercise 10–1, 10–2

10–1 Find $Z_{EQ}(s)$ in Figure P10–1. Express $Z_{EQ}(s)$ as a rational function and locate its poles and zeros.

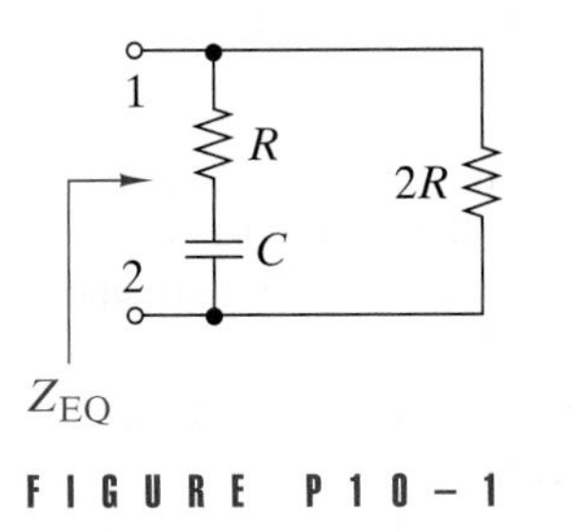

FIGURE P10–1

10–2 Find $Z_{EQ}(s)$ in Figure P10–2. Express $Z_{EQ}(s)$ as a rational function and locate its poles and zeros.

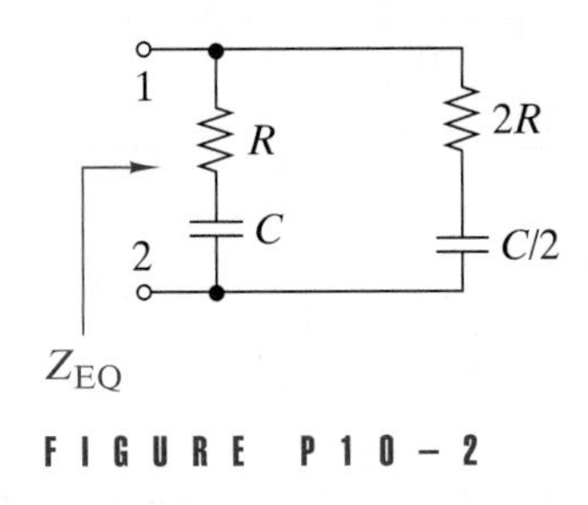

FIGURE P10–2

10–3 Find $Z_{EQ}(s)$ in Figure P10–3. Express $Z_{EQ}(s)$ as a rational function and locate its poles and zeros for $R = 4$ kΩ, $L = 0.4$ H, and $C = 100$ nF.

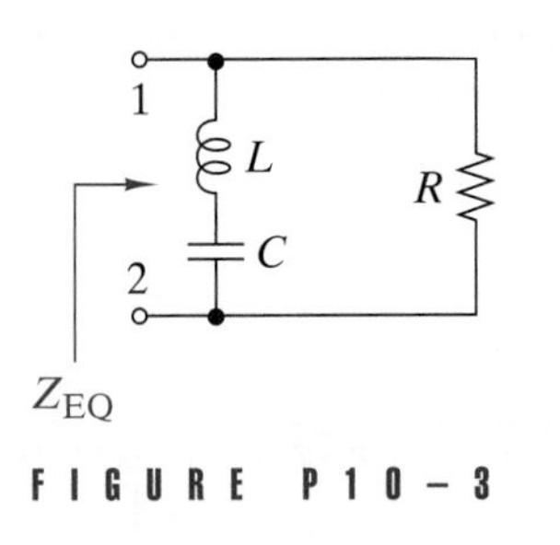

FIGURE P10–3

10–4 Find $Z_{EQ}(s)$ in Figure P10–4. Express $Z_{EQ}(s)$ as a rational function and locate its poles and zeros.

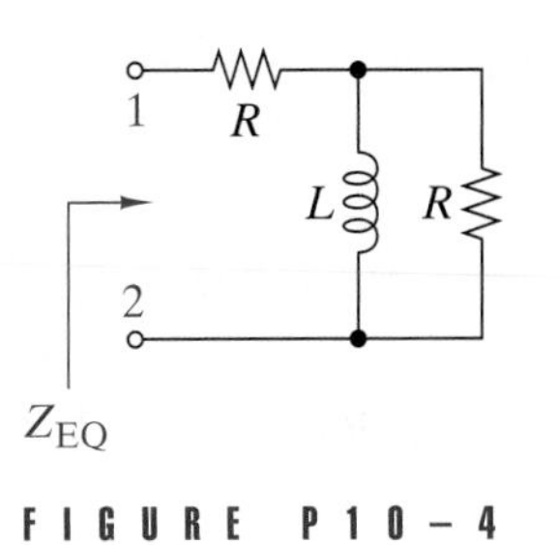

FIGURE P10–4

10–5 Find $Z_{EQ}(s)$ in Figure P10–5. Express $Z_{EQ}(s)$ as a rational function and locate its poles and zeros.

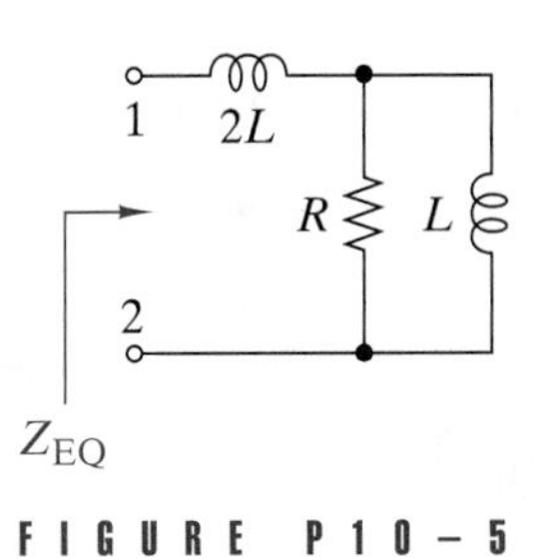

FIGURE P10–5

10–6 Find the poles and zeros of Z_{EQ} in Figure P10–6 for $R = 1$ kΩ, $L = 1$ H, and $C = 500$ nF.

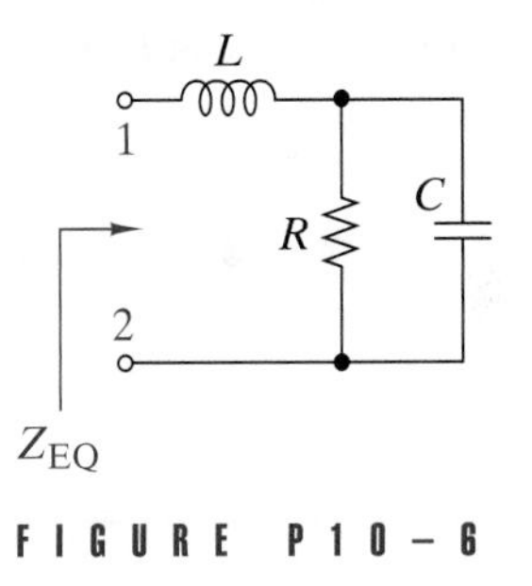

FIGURE P10–6

10–7 Find $Z_{EQ}(s)$ in Figure P10–7. Express $Z_{EQ}(s)$ as a rational function and locate its poles and zeros.

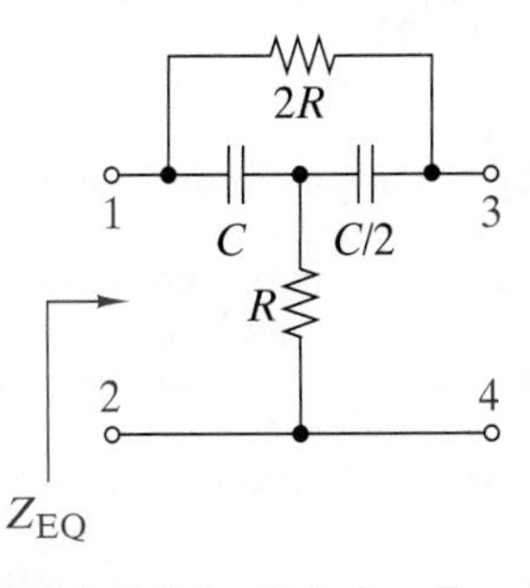

FIGURE P10–7

10–8 Find $Z_{EQ}(s)$ in Figure P10–8. Express $Z_{EQ}(s)$ as a rational function and locate its poles and zeros.

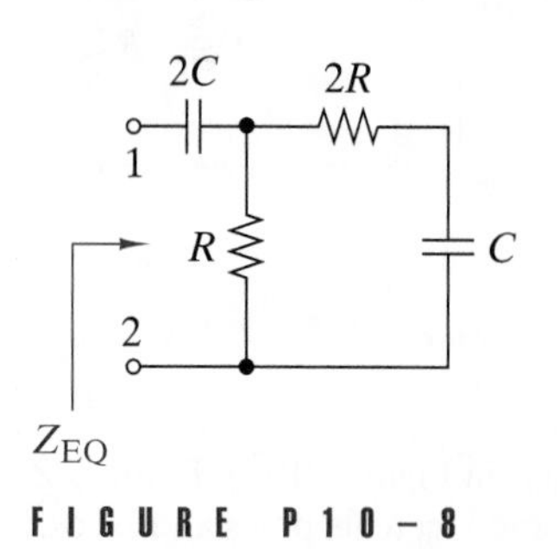

FIGURE P10–8

10–9 Find $Z_{EQ}(s)$ in Figure P10–9. Express $Z_{EQ}(s)$ as a rational function and locate its poles and zeros.

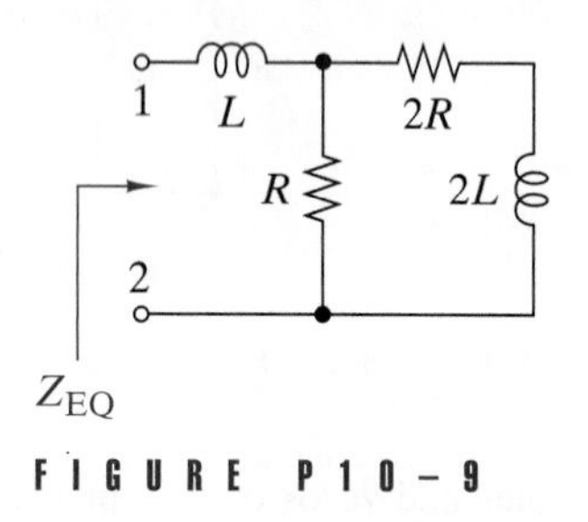

FIGURE P10–9

10–10 Find $Z_{EQ}(s)$ in Figure P10–10. Select values of R and L so that $Z_{EQ}(s)$ has a pole at $s = -1000$ rad/s. Locate the zeros of $Z_{EQ}(s)$ for your choice of R and L.

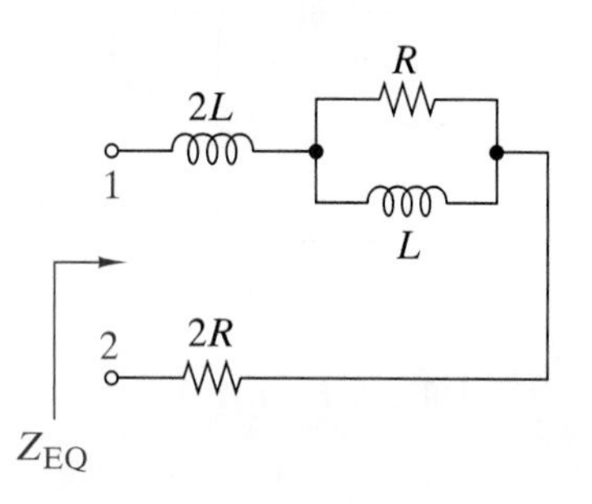

FIGURE P10–10

OBJECTIVE 10–2 BASIC CIRCUIT ANALYSIS TECHNIQUES (SECTS. 10–2, 10–3)

Given a linear circuit:
(a) Determine the initial conditions (if not given) and transform the circuit into the s domain.
(b) Solve for zero-state and/or zero-input response transforms and waveforms using basic analysis methods such as circuit reduction, unit output, Thévenin/Norton equivalent circuits, or superposition.
(c) Locate the forced and natural poles or select circuit parameters to place the natural poles at specified locations.
See Examples 10–2, 10–3, 10–4, 10–5, 10–6, 10–7, 10–8, 10–9

10–11 The switch in Figure P10–11 has been in position A for a long time and is moved to position B at $t = 0$. Transform the circuit into the s domain and solve for $I_L(s)$ and $i_L(t)$ in symbolic form.

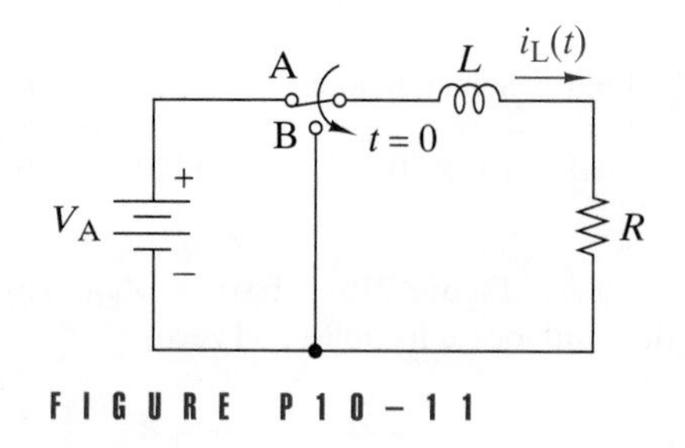

FIGURE P10–11

10–12 The switch in Figure P10–11 has been in position B for a long time and is moved to position A at $t = 0$. Transform the circuit into the s domain and solve for $I_L(s)$ and $i_L(t)$ in symbolic form.

10–13 The switch in Figure P10–13 has been in position A for a long time and is moved to position B at $t = 0$. Transform the circuit into the s domain and solve for $I_C(s)$ and $i_C(t)$ in symbolic form.

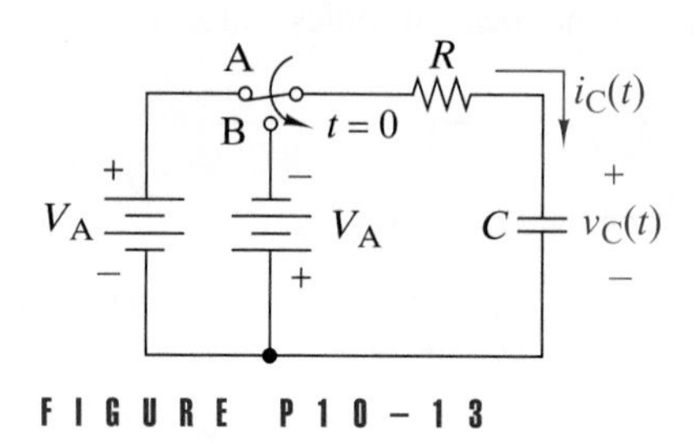

FIGURE P10–13

10–14 The switch in Figure P10–13 has been in position B for a long time and is moved to position A at $t = 0$. Transform the circuit into the s domain and solve for $V_C(s)$ and $v_C(t)$ in symbolic form.

10–15 There is no initial energy stored in the circuit in Figure P10–15. Transform the circuit into the s domain and find $I_L(s)$ and $i_L(t)$ when $v_1(t) = V_A\,u(t)$.

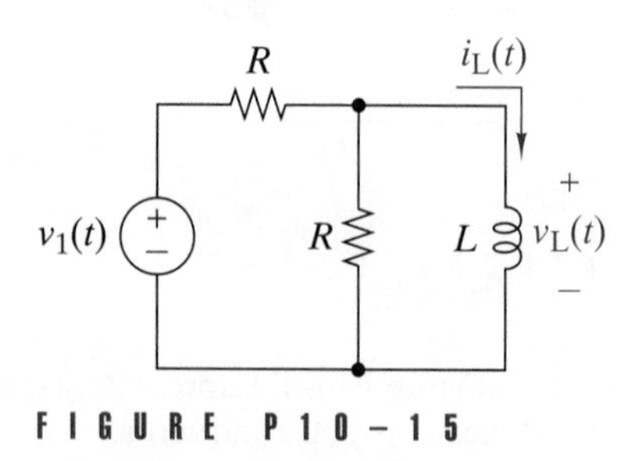

FIGURE P10–15

10–16 There is no initial energy stored in the circuit in Figure P10–15. Transform the circuit into the s domain and find $V_L(s)$ and $v_L(t)$ when $v_1(t) = V_A u(t)$.

10–17 The switch in Figure P10–17 has been in position A for a long time and is moved to position B at $t = 0$.

(a) Transform the circuit into the s domain and solve for $I_L(s)$ in symbolic form.

(b) Find $i_L(t)$ for $R_1 = R_2 = 500\ \Omega$, $L = 250$ mH, $C = 4\ \mu$F, and $V_A = 15$ V.

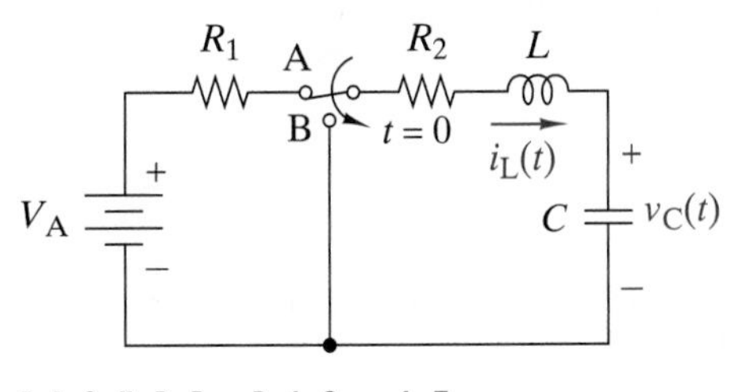

FIGURE P10–17

10–18 The switch in Figure P10–17 has been in position B for a long time and is moved to position A at $t = 0$.

(a) Transform the circuit into the s domain and solve for $V_C(s)$ in symbolic form.

(b) Find $v_C(t)$ for $R_1 = R_2 = 500\ \Omega$, $L = 500$ mH, $C = 1\ \mu$F, and $V_A = 15$ V.

10–19 The circuit in Figure P10–19 is in the zero state. Find the s-domain relationship between the input $I_1(s)$ and the output $I_R(s)$.

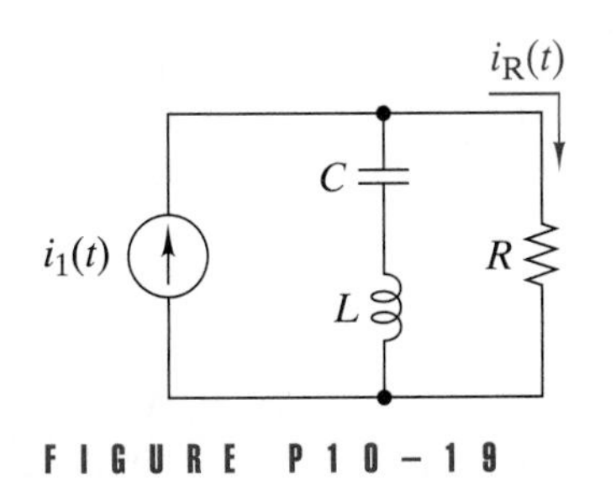

FIGURE P10–19

10–20 The circuit in Figure P10–20 is in the zero state. Find the s-domain relationship between the input $I_1(s)$ and the output $V_O(s)$.

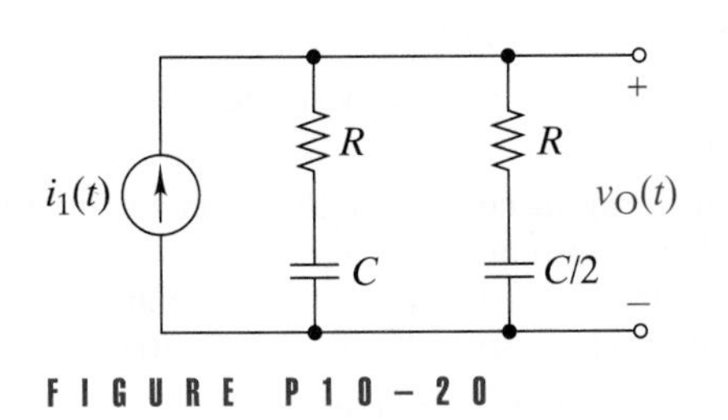

FIGURE P10–20

10–21 The initial conditions in Figure P10–21 are $v_C(0) = V_0$ and $i_L(0) = 0$. Transform the circuit into the s domain and use superposition and voltage division to find the zero-state and zero-input components of $V_C(s)$.

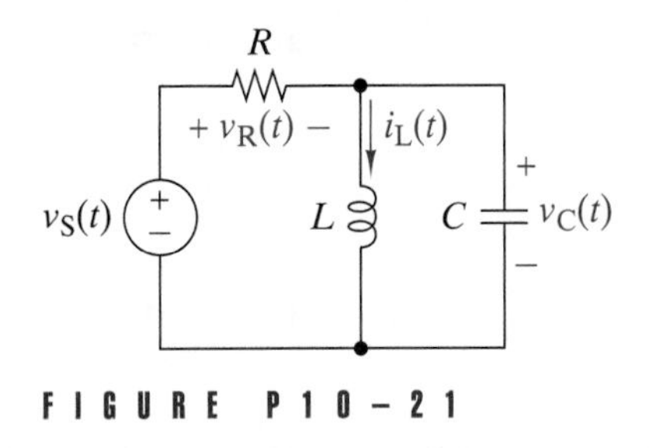

FIGURE P10–21

10–22 The initial conditions in Figure P10–21 are $v_C(0) = 0$ and $i_L(0) = I_0$. Transform the circuit into the s domain and use superposition and voltage division to find the zero-state and zero-input components of $V_R(s)$.

10–23 There is no energy stored in the capacitor in Figure P10–23 at $t = 0$. Transform the circuit into the s domain and use current division to find $v_O(t)$ when the input is $i_S(t) = 2.5\, e^{-100t}u(t)$ mA. Identify the forced and natural poles in $V_O(s)$.

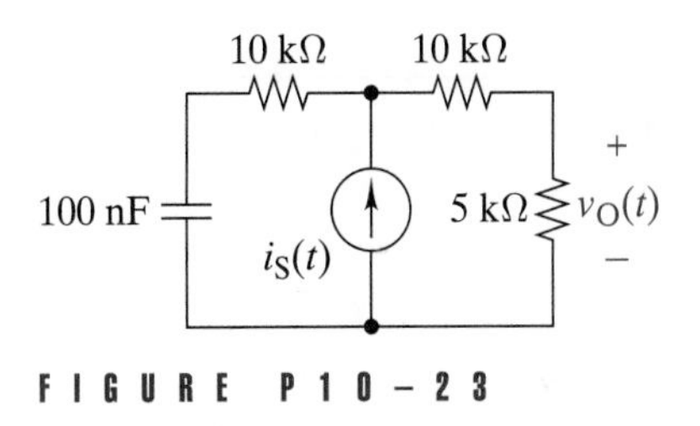

FIGURE P10–23

10–24 Repeat Problem 10–23 when $i_S(t) = 2.5\, e^{-1000t}u(t)$ mA.

10–25 The circuit in Figure P10–25 is in the zero state. Use a Thévenin equivalent to find the s-domain relationship between the input $I_S(s)$ and the interface current $I(s)$.

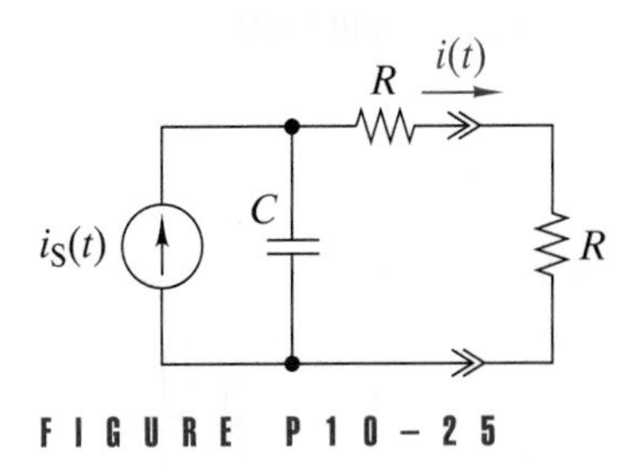

FIGURE P10–25

10–26 Repeat Problem 10–25 using a Norton equivalent circuit.

10–27 The circuit in Figure P10–27 is in the zero state. Find the Thévenin equivalent to the left of the interface.

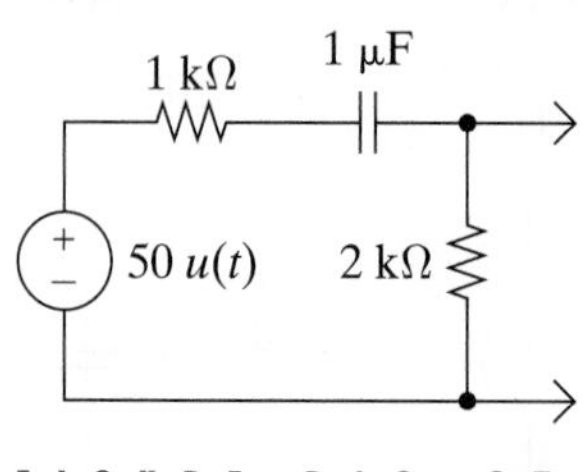

FIGURE P10–27

10–28 There is no initial energy stored in the circuit in Figure P10–28. Transform the circuit into the s domain and use superposition to find $V(s)$. Identify the forced and natural poles in $V(s)$.

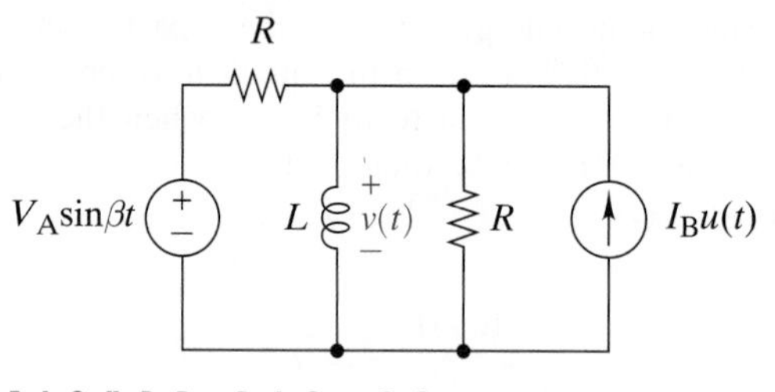

FIGURE P10–28

10–29 The equivalent impedance between a pair of terminals is

$$Z(s) = 1000\left[\frac{s + 2000}{s + 1000}\right]$$

A voltage $v(t) = 10u(t)$ is applied across the terminals. Find the resulting current response $i(t)$.

10–30 There is no initial energy stored in the circuit in Figure P10–30. Use circuit reduction to find the output voltage $V_2(s)$ in terms of the input voltage $V_1(s)$.

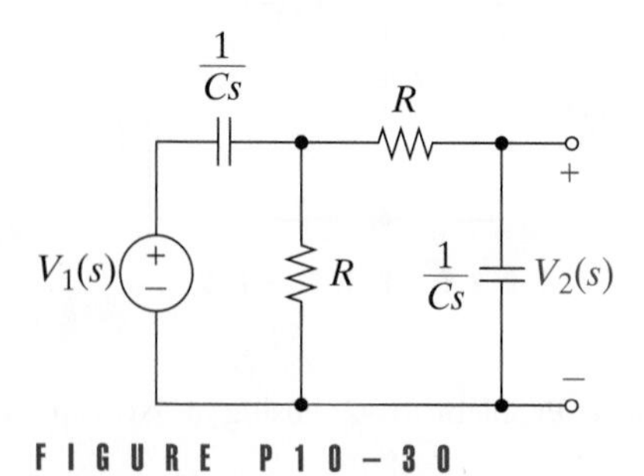

FIGURE P10–30

OBJECTIVE 10–3 GENERAL CIRCUIT ANALYSIS (SECTS. 10–4, 10–5, 10–6)

Given a linear circuit:
(a) Determine the initial conditions (if not given) and transform the circuit into the s domain.
(b) Solve for zero-state and/or zero-input response transforms and waveforms using node-voltage or mesh-current methods.
(c) Identify the forced and natural poles or select circuit parameters to place the natural poles at specified locations.
See Examples 10–10, 10–11, 10–13, 10–14, 10–15, 10–16, 10–17

10–31 There is no initial energy stored in the circuit in Figure P10–31.

(a) Transform the circuit into the s domain and formulate mesh-current equations.
(b) Solve these equations for $I_2(s)$ in symbolic form.
(c) Find $i_2(t)$ for $v_1(t) = 100u(t)$ V, $R_1 = 1$ kΩ, $R_2 = 2$ kΩ, $L = 4$H, and $C = 500$ nF.

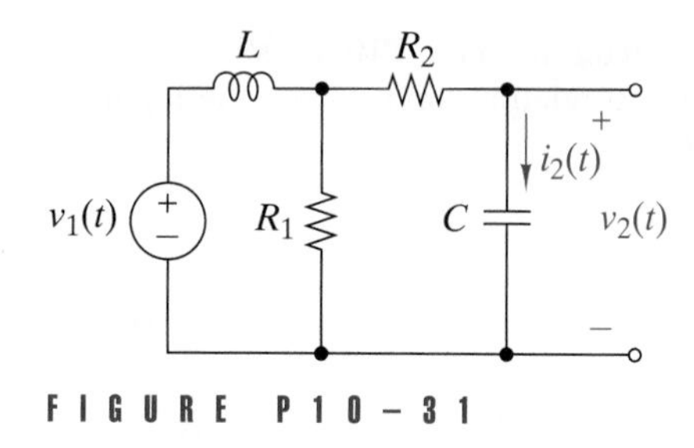

FIGURE P10–31

10–32 There is no initial energy stored in the circuit in Figure P10–31.

(a) Transform the circuit into the s domain and formulate node-voltage equations.
(b) Solve these equations for $V_2(s)$ in symbolic form.
(c) Find $v_2(t)$ for $v_1(t) = 10u(t)$ V, $R_1 = R_2 = 500$ Ω, $L = 0.5$ H, and $C = 2$ μF.

10–33 There is no initial energy stored in the circuit in Figure P10–33.

(a) Transform the circuit into the s domain and formulate node-voltage equations.
(b) Solve these equations for $V_2(s)$ in symbolic form.
(c) Find $v_2(t)$ for $v_1(t) = 30u(t)$ V, $R_1 = 10$ kΩ, $R_2 = 20$ kΩ, $C_1 = 1$ μF, and $C_2 = 0.5$ μF.

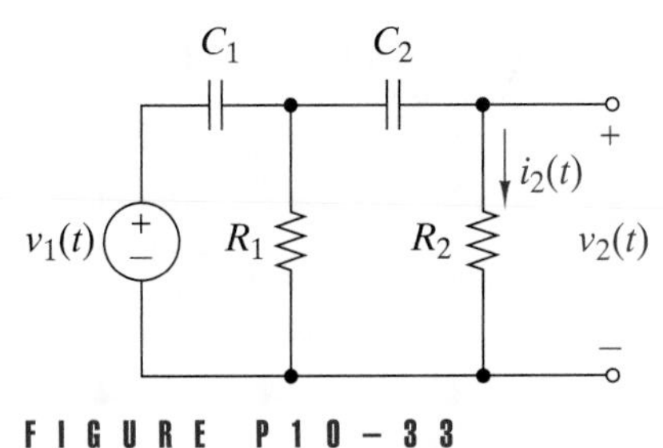

FIGURE P10–33

10–34 There is no initial energy stored in the circuit in Figure P10–33.

(a) Transform the circuit into the s domain and formulate mesh-current equations.

(b) Solve these equations for $I_2(s)$ in symbolic form.

(c) Find $i_2(t)$ for $v_1(t) = 4000r(t)$ V, $R_1 = 5$ kΩ, $R_2 = 20$ kΩ, $C_1 = 5$ μF, and $C_2 = C_1/3$.

10–35 There is no initial energy stored in the bridged-T circuit in Figure P10–35.

(a) Transform the circuit into the s domain and formulate mesh-current equations.

(b) Use the mesh-current equations to find the s-domain relationship between the input $V_1(s)$ and the output $V_2(s)$.

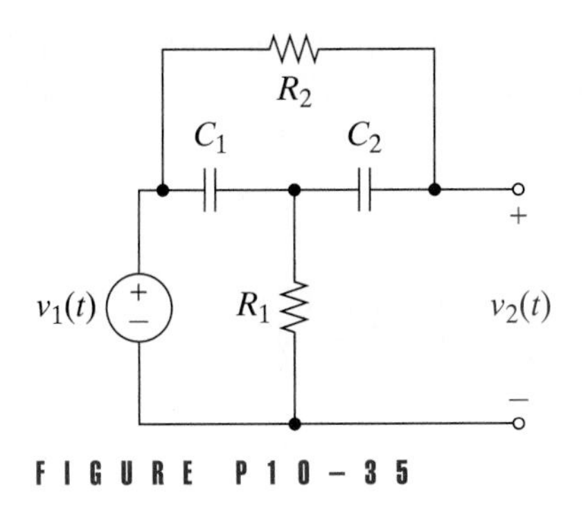

FIGURE P10–35

10–36 There is no initial energy stored in the bridged-T circuit in Figure P10–35.

(a) Transform the circuit into the s domain and formulate node-voltage equations.

(b) Use the node-voltage equations to find the s-domain relationship between the input $V_1(s)$ and the output $V_2(s)$.

10–37 There is no initial energy stored in the circuit shown in Figure P10–37. Find the zero-state mesh currents $i_A(t)$ and $i_B(t)$ when $v_1(t) = 100u(t)$ V.

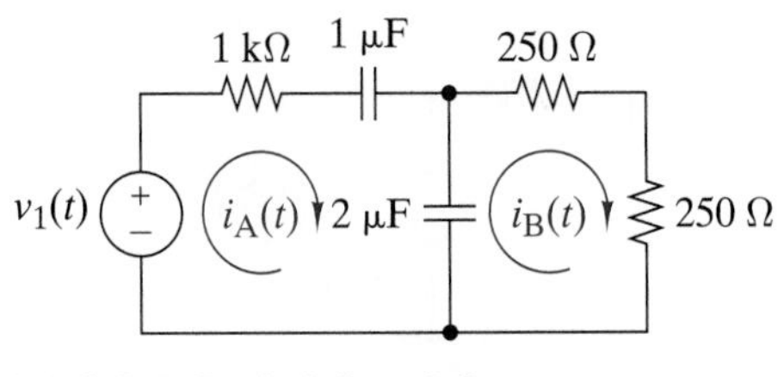

FIGURE P10–37

10–38 There is no external input in the circuit in Figure P10–38. Find the zero-input node voltages $v_A(t)$ and $v_B(t)$ when $v_C(0) = 5$ V and $i_L(0) = 0$.

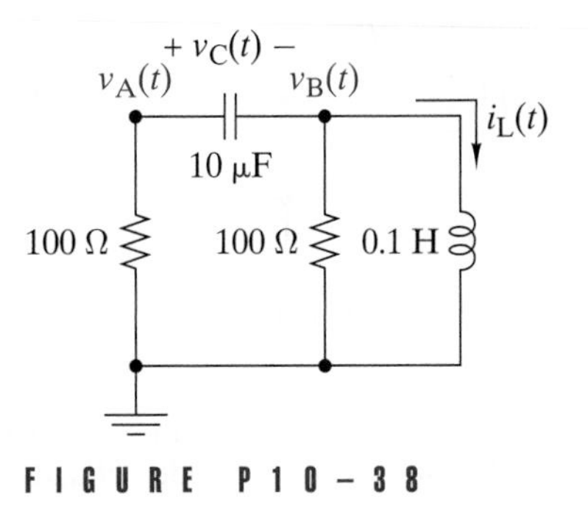

FIGURE P10–38

10–39 **D** The circuit in Figure P10–39 is in the zero state. Use node-voltage equations to find the circuit determinant. Select values of R, C, and μ so that the circuit has $\omega_0 = 5$ krad/s and $\zeta = 1/\sqrt{2}$.

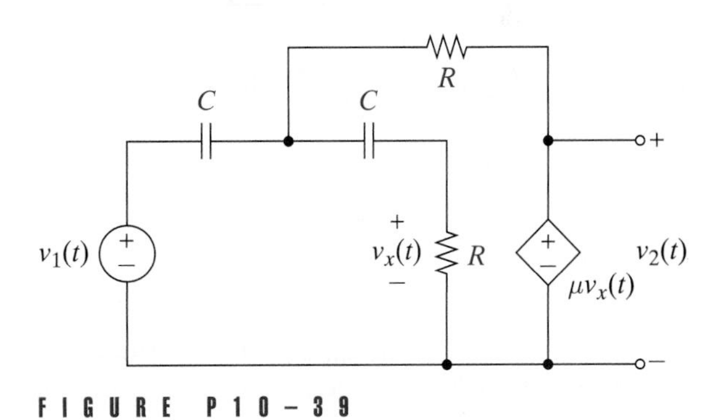

FIGURE P10–39

10–40 **D** The circuit in Figure P10–40 is in the zero state. Use mesh-current equations to find the circuit determinant. Select values of R, L, and C so that the circuit has $\omega_0 = 5$ krad/s and $\zeta = 1/\sqrt{2}$.

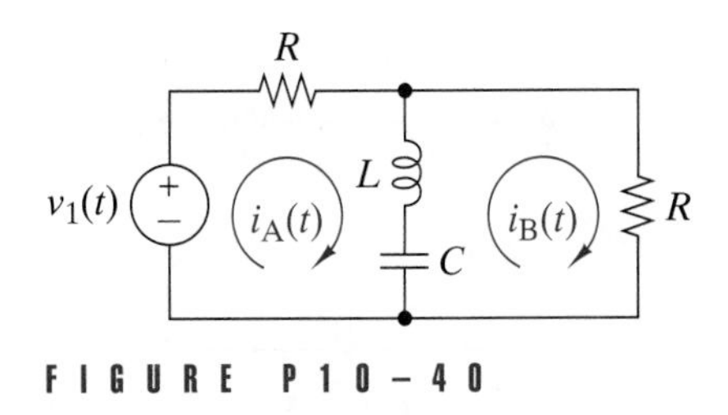

FIGURE P10–40

10–41 **D** The circuit in Figure P10–41 is in the zero state. Use node-voltage equations to find the circuit determinant. Select values of R, L, and C so that the circuit has $\omega_0 = 5$ krad/s and $\zeta = 1/\sqrt{2}$.

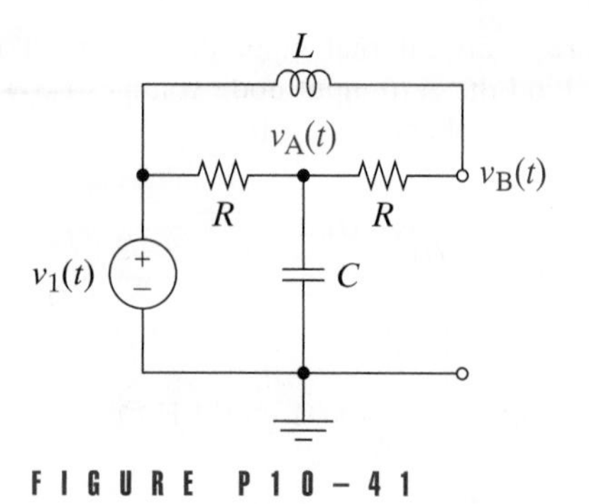

FIGURE P10–41

10–42 Three node voltages are shown in Figure P10–42.

(a) Explain why only one of the node voltages is independent.

(b) Write a node-voltage equation in the independent node voltage.

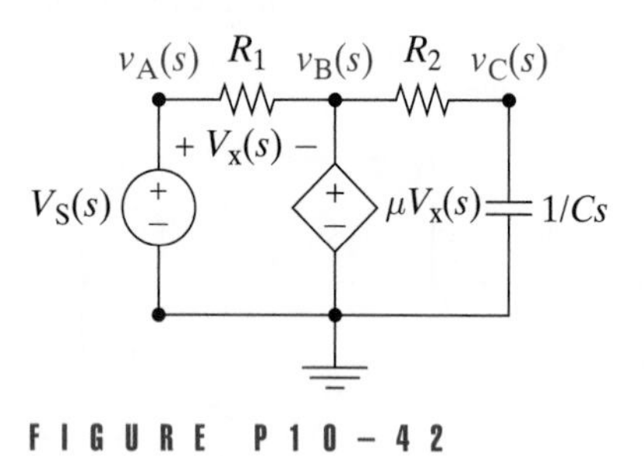

FIGURE P10–42

10–43 Three mesh currents are shown in Figure P10–43.

(a) Explain why only two of these mesh currents are independent.

(b) Write *s*-domain mesh-current equations in two independent mesh currents.

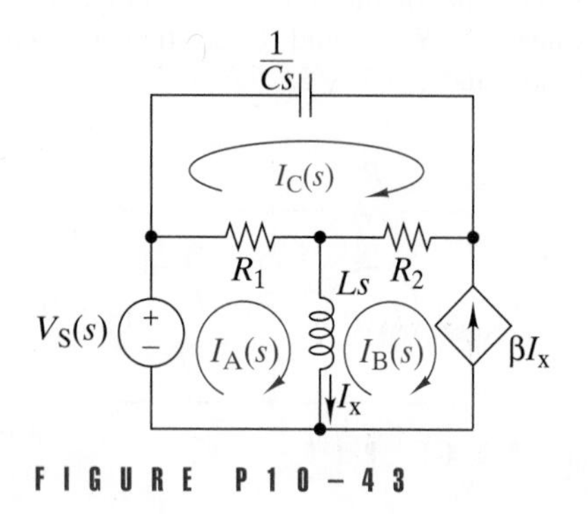

FIGURE P10–43

10–44 The switch in Figure P10–44 has been in position A for a long time and is moved to position B at $t = 0$. Solve for $V_C(s)$ and $v_C(t)$.

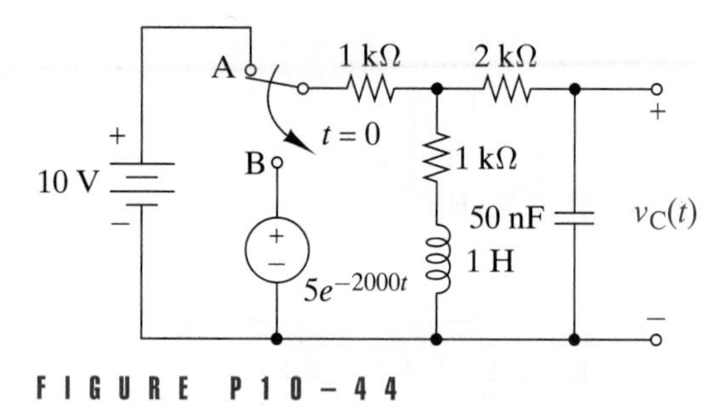

FIGURE P10–44

10–45 There is no energy stored in the circuit in Figure P10–45 at $t = 0$. Transform the circuit into the *s* domain and solve for $V_O(s)$ and $v_O(t)$.

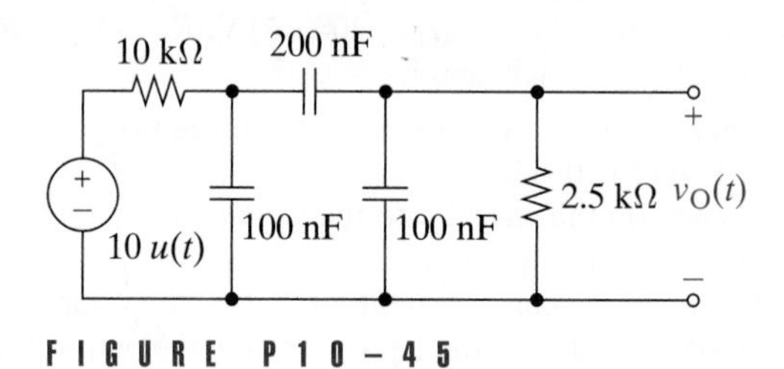

FIGURE P10–45

10–46 The switch in Figure P10–46 has been open for a long time and is closed at $t = 0$. Transform the circuit into the *s* domain and solve for $I_O(s)$ and $i_O(t)$.

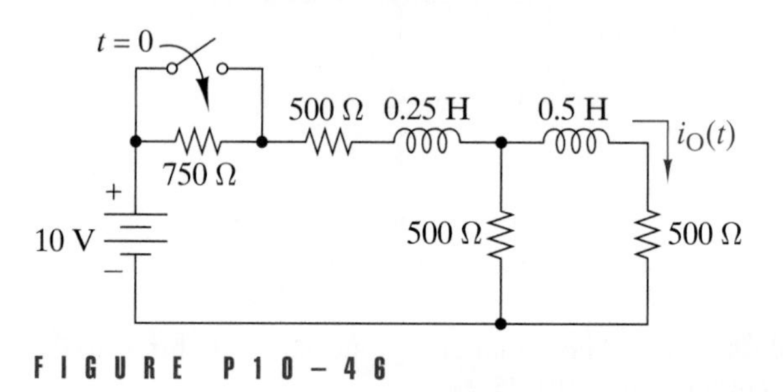

FIGURE P10–46

10–47 There is no initial energy stored in the circuit in Figure P10–47.

(a) Transform the circuit into the *s* domain and solve for $V_O(s)$ in symbolic form.

(b) Find $v_O(t)$ when $v_S(t) = 10^4 t u(t)$ V, $R_1 = R_2 = 2$ kΩ, $C = 0.5$ μF, and $L = 2$ H.

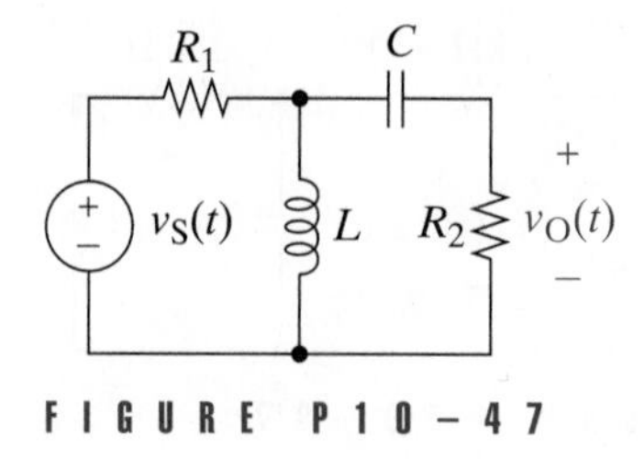

FIGURE P10–47

10–48 Repeat Problem 10–47 with $L = 0.4$ H.

10–49 Show that the circuit in Figure P10–49 has natural poles at $s = -4/RC$ and $s = -2/RC \pm j2/RC$ when $L = R^2C/4$.

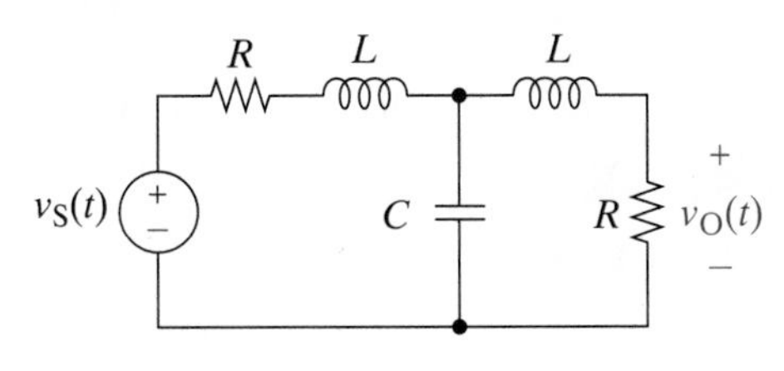

FIGURE P10–49

10–50 Find the range of the gain μ for which the circuit in Figure P10–50 is stable.

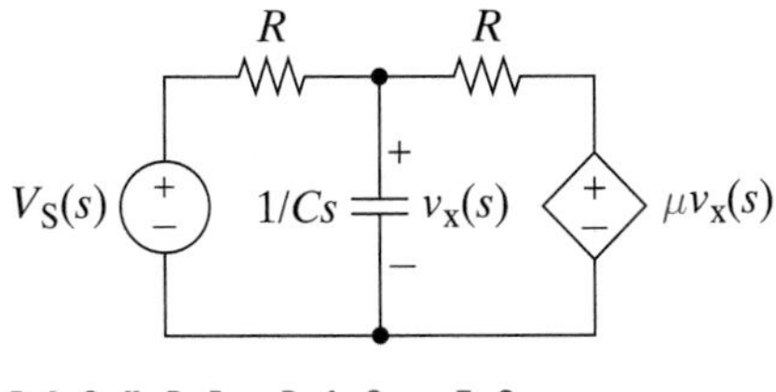

FIGURE P10–50

INTEGRATING PROBLEMS

10–51 **A** Thévenin's Theorem from Time-Domain Data

A black box containing a linear circuit has an ON/OFF switch and a pair of external terminals. When the switch is turned on, the open-circuit voltage between the external terminals is observed to be

$$v_{OC}(t) = (18\, e^{-10t} - 12\, e^{-50t})\, u(t)$$

When a 50-Ω load resistance is connected across the terminals and the switch again turned on, the voltage delivered to the load is observed to be

$$v_L(t) = (5\, e^{-10t} - 5\, e^{-50t})\, u(t)$$

What is the Thévenin's impedance looking into the box?

10–52 **D** Design a Load Impedance

In order to match the Thévenin impedance of a source, the load impedance in Figure P10–52 must be

$$Z_L(s) = \frac{s+5}{s+10}$$

(a) What impedance $Z_2(s)$ is required?

(b) How would you realize $Z_2(s)$ using only resistors, inductors, and/or capacitors?

Hint: Expand $Y_2(s) = 1/Z_2(s)$ by partial fractions.

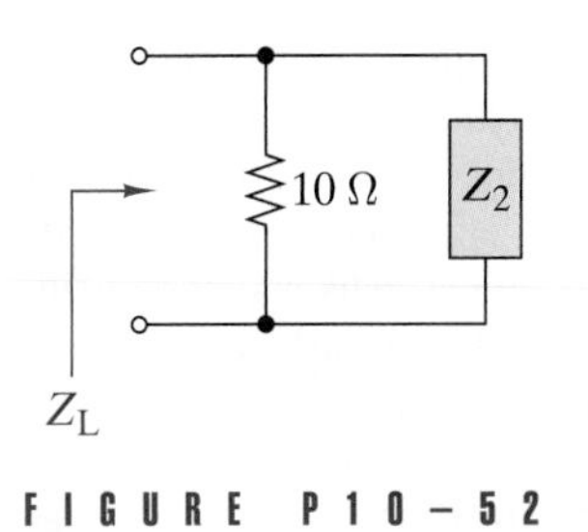

FIGURE P10–52

10–53 **A D** *RC* Circuit Analysis and Design

The *RC* circuits in Figure P10–53 represent the situation at the input to an oscilloscope. The parallel combination of R_1 and C_1 represents the probe used to connect the oscilloscope to a test point. The parallel combination of R_2 and C_2 represents the input impedance of the oscilloscope.

(a) **A** Assuming zero initial conditions, transform the circuit into the *s* domain and find the relationship between the test point voltage $V_S(s)$ and the voltage $V_O(s)$ at the oscilloscope's input.

(b) **D** For $R_2 = 10$ MΩ and $C_2 = 2$ pF, determine the values of R_1 and C_1 that make the input voltage a scaled duplicate of the test point voltage.

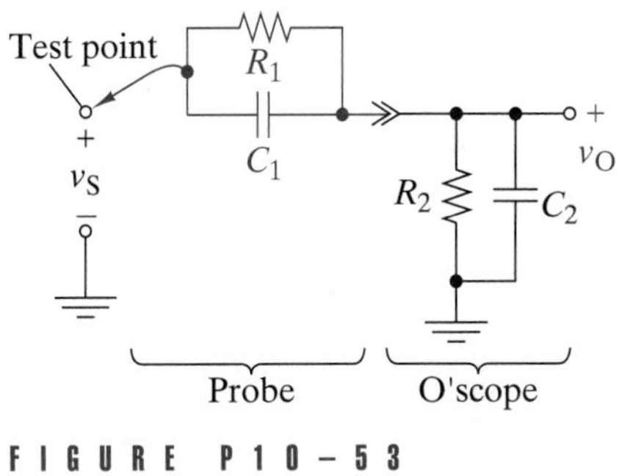

FIGURE P10–53

10–54 **A** *s*-domain OP AMP Circuit Analysis

The OP AMP circuit in Figure P10–54 is in the zero state. Transform the circuit into the *s* domain and use the OP AMP circuit analysis techniques developed in Section 4–5 to find the relationship between the input $V_1(s)$ and the output $V_2(s)$

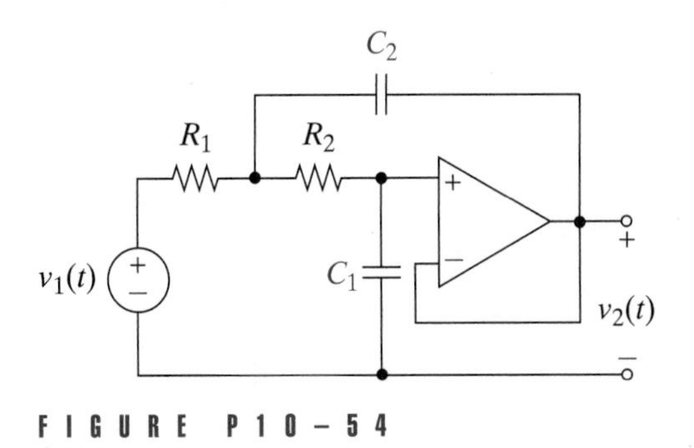

FIGURE P10–54

10–55 E Pulse Conversion Circuit
The purpose of the test setup in Figure P10–55 is to deliver damped sine pulses to the test load. The excitation comes from a 1-Hz square wave generator. The pulse conversion circuit must deliver damped sine waveforms with $\zeta < 0.5$ and $\omega_0 > 10$ krad/s to 50-Ω and 600-Ω loads. The recommended values for the pulse conversion circuit are $L = 10$ mH and $C = 100$ nF. Verify that the test setup meets the specifications.

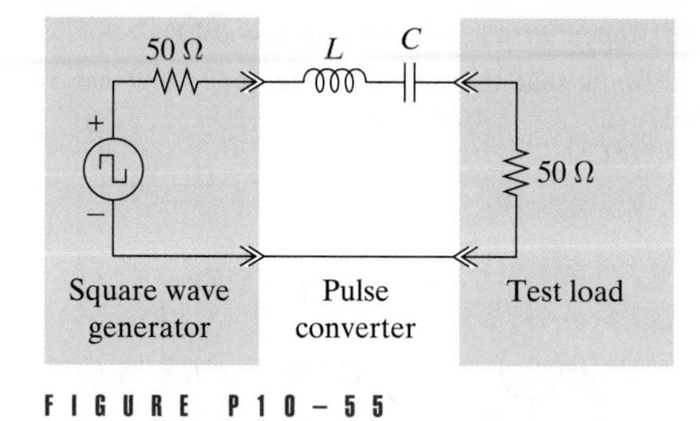

FIGURE P10–55

CHAPTER 11

NETWORK FUNCTIONS

The driving-point impedance of a network is the ratio of an impressed electromotive force at a point in a branch of the network to the resulting current at the same point.

Ronald M. Foster, 1924,
American Engineer

Some History Behind This Chapter

The network function concept emerged in the 1920s during the development of systematic methods of designing electric filters for long-distance telephone systems. The filter design effort eventually evolved into a theory known as *network synthesis.* The purpose of network synthesis is to obtain circuits that produce a desired network function. Ronald Foster along with Sidney Darlington, Hendrik Bode, Wilhelm Cauer, and Otto Brune are generally considered the founders of modern network synthesis.

Why This Chapter Is Important Today

This special chapter introduces one of the most important concepts of electrical engineering—the network function. In this chapter you will learn what network functions are and why they are important descriptors of electric circuits. Most importantly, you will learn how to design circuits that can realize a desired network function. But we don't stop there. Since design can lead to many different answers, we introduce you to the criteria used to evaluate alternative solutions.

Chapter Sections

11–1 Definition of a Network Function

11–2 Network Functions of One- and Two-Port Circuits

11–3 Network Functions and Impulse Response

11–4 Network Functions and Step Response

11–5 Network Functions and the Sinusoidal Steady-State Response

11–6 Impulse Response and Convolution

11–7 Network Function Design

Chapter Learning Objectives

11-1 Network Functions (Sects. 11-1, 11-2)

Given a linear circuit:

(a) Find specified network functions and locate their poles and zeros.
(b) Select the element values to produce specified poles and zeros.

11-2 Network Functions, Impulse Response, and Step Response (Sects. 11-3, 11-4)

(a) Given a first- or second-order linear circuit, find its impulse or step response.
(b) Given the impulse or step response of a linear circuit, find the network function.
(c) Given the impulse or step response of a linear circuit, find the response due to other inputs.

11-3 Network Functions and the Sinusoidal Steady-State Response (Sect. 11-5)

(a) Given a first- or second-order linear circuit with a specified input sinusoid, find the sinusoidal steady-state response.
(b) Given the network function, impulse response, or step response, find the sinusoidal steady-state response for a specified input sinusoid.

11-4 Network Functions and Convolution (Sect. 11-6)

(a) Given the impulse response of a linear circuit, use the convolution integral to find the response to a specified input.
(b) Use the convolution integral to derive properties of linear circuits.

11-5 Network Function Design (Sect. 11-7)

(a) Design circuits that realizes a given network function and meets other stated constraints.
(b) Evaluate alternative designs using stated criteria.

11–1 DEFINITION OF A NETWORK FUNCTION

The proportionality property of linear circuits states that the output is proportional to the input. In Chapter 10 we noted that in the s domain the proportionality factor is a rational function of s called a network function. More formally, a network function is defined as the ratio of a zero-state response transform (output) to the excitation (input) transform.

$$\text{Network Function} = \frac{\text{Zero-state response transform}}{\text{Input signal transform}} \tag{11–1}$$

Note carefully that this definition specifies zero initial conditions and implies only one input.

To study the role of network functions in determining circuit responses, we write the s-domain input-output relationship as

$$Y(s) = T(s)X(s) \tag{11–2}$$

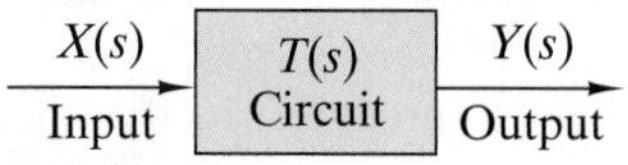

$X(s)$ Input | $T(s)$ Circuit | $Y(s)$ Output

FIGURE 11–1 *Block diagram for an s-domain input-output relationship.*

where $T(s)$ is a network function, $X(s)$ is the input signal transform, and $Y(s)$ is a zero-state response or output.[1] Figure 11–1 shows a block diagram representation of the s-domain input-output relationship in Eq. (11–2).

In an analysis problem, the circuit and input [$X(s)$ or $x(t)$] are specified. We determine $T(s)$ from the circuit, use Eq. (11–2) to find the response transform $Y(s)$, and use the inverse transformation to obtain the response waveform $y(t)$. In a design problem the circuit is unknown. The input and output are specified, or their ratio $T(s) = Y(s)/X(s)$ is given. The objective is to devise a circuit that realizes the specified input-output relationship. A linear circuit analysis problem has a unique solution, but a design problem may have one, many, or even no solutions.

Equation (11–2) points out that the poles of the response $Y(s)$ come from either the network function $T(s)$ or the input signal $X(s)$. When there are no repeated poles, the partial-fraction expansion of the right side of Eq. (11–2) takes the form

$$Y(s) = \underbrace{\sum_{j=1}^{N} \frac{k_j}{s - p_j}}_{\substack{\text{natural}\\ \text{poles}}} + \underbrace{\sum_{\ell=1}^{M} \frac{k_\ell}{s - p_\ell}}_{\substack{\text{forced}\\ \text{poles}}} \tag{11–3}$$

where p_j ($j = 1, 2, \ldots N$) are the poles of $T(s)$ and $s = p_\ell$ ($\ell = 1, 2, \ldots M$) are the poles of $X(s)$. The inverse transform of this expansion is

$$y(t) = \underbrace{\sum_{j=1}^{N} k_j e^{p_j t}}_{\substack{\text{natural}\\ \text{response}}} + \underbrace{\sum_{\ell=1}^{M} k_\ell\, e^{p_\ell t}}_{\substack{\text{forced}\\ \text{response}}} \tag{11–4}$$

1 In this context $Y(s)$ is not an admittance, but the transform of the output waveform $y(t)$.

The poles of $T(s)$ lead to the natural response. In a stable circuit, the natural poles are all in the left half of the s plane, and all of the exponential terms in the natural response eventually decay to zero. The poles of $X(s)$ lead to the forced response. In a stable circuit, those elements in the forced response that do not decay to zero are called the **steady-state response**.

It is important to remember that the complex frequencies in the natural response are determined by the circuit and do not depend on input. Conversely, the complex frequencies in the forced response are determined by the input and do not depend on the circuit. However, the amplitude of its part of the response depends on the residues in the partial-fraction expansion in Eq. (11–3). These residues are influenced by all of the poles and zeros, whether forced or natural. Thus, the amplitudes of the forced and natural responses depend on an interaction between the poles and zeros of $T(s)$ and $X(s)$.

The following example illustrates this discussion.

EXAMPLE 11–1

The transfer function of a circuit is

$$T(s) = \frac{V_2(s)}{V_1(s)} = \frac{2000\,(s + 2000)}{(s + 1000)\,(s + 4000)}$$

Find the zero-state response $v_2(t)$ when the input waveform is $v_1(t) = [20 + 15e^{-5000t}]u(t)$.

SOLUTION:

The transform of the input waveform is

$$V_1(s) = \frac{20}{s} + \frac{15}{s + 5000} = \frac{35\,s + 10^5}{s(s + 5000)}$$

Using the s-domain input-output relationship in Eq. (11–2), the transform of the response is

$$V_2(s) = \frac{10^4(s + 2000)\,(7s + 20000)}{(s + 1000)\,(s + 4000)s(s + 5000)}$$

Expanding by partial fractions,

$$V_2(s) = \underbrace{\frac{k_1}{s + 1000} + \frac{k_2}{s + 4000}}_{\text{natural poles}} + \underbrace{\frac{k_3}{s} + \frac{k_4}{s + 5000}}_{\text{forced poles}}$$

The two natural poles came from the circuit via the network function $T(s)$. The forced poles came from the step function and exponential inputs. Using the cover-up method to evaluate the residues yields

$$k_1 = \left.\frac{10^4(s + 2000)\,(7s + 20000)}{(s + 4000)s(s + 5000)}\right|_{s=-1000} = -\frac{65}{6}$$

$$k_2 = \left.\frac{10^4(s + 2000)\,(7s + 20000)}{(s + 1000)s(s + 5000)}\right|_{s=-4000} = \frac{40}{3}$$

$$k_3 = \left.\frac{10^4(s+2000)(7s+20000)}{(s+1000)(s+4000)(s+5000)}\right|_{s=0} = 20$$

$$k_4 = \left.\frac{10^4(s+2000)(7s+20000)}{(s+1000)(s+4000)s}\right|_{s=-5000} = -\frac{45}{2}$$

Collectively the residues depend on all of the poles and zeros. The inverse transform yields the zero-state response as

$$v_2(t) = \Bigg[\underbrace{-\frac{65}{6}e^{-1000t} + \frac{40}{3}e^{-4000t}}_{\text{natural response}} + \underbrace{20 - \frac{45}{2}e^{-5000t}}_{\text{forced response}}\Bigg]u(t)$$

The natural poles as $s = -1000$ and $s = -4000$ are in the left half of the s plane, so the natural response decays to zero. The forced pole as $s = -5000$ leads to an exponential term that also decays to zero, leaving a steady-state response of $20u(t)$. ■

TEST SIGNALS

While the transfer function is a useful concept, it is clear that we cannot find the circuit response until we are given an input signal. Here, we encounter a central paradox of circuit analysis. In practice, the input signal is a carrier of information and is therefore unpredictable. We could spend a lifetime studying a circuit for various inputs and still not treat all possible signals that might be encountered in practice. What we must do is calculate the responses due to certain standard test signals. Although these test signals may never occur as real input signals, their responses tell us enough to understand the signal-processing capabilities of a circuit.

The two premier test signals used are the pulse and the sinusoid. The study of pulse response divides into two extreme cases, short and long. When the pulse is very short compared to the circuit response time, the sudden injection of energy causes a circuit response long after the input returns to zero. The short pulse is modeled by an impulse, and the resulting *impulse response* is treated in Sect. 11–3. At the other extreme, the long pulse has a duration that greatly exceeds the circuit response time. In this case, the circuit has ample time to be driven from the zero state to a new steady-state condition. The step function is used to model the long pulse input, and the resulting *step response* is studied in Sect. 11–4.

The impuse response is of great importance because it contains all of the information needed to calculate the response due to any other input. The step response is important because it describes how a circuit response transitions from one state to another. The signal transition requirements for circuits and systems are often stated in terms of the step response using partial waveform descriptors such as rise time, fall time, propagation delay, and overshoot.

The unique properties of the sinusoid make it a useful input for characterizing the signal-processing capabilities of linear circuits and systems. When a stable linear circuit is driven by a sinusoidal input, the steady-state output is a sinusoid with the same frequency, but with a different phase angle and amplitude. The frequency-dependent relationship between the sinusoidal input and the steady-state output is called *frequency response,* a signal-processing description that is often used to specify the performance of circuits and systems. The relationship between network functions and the sinusoidal steady-state response is studied in Sect. 11–5.

11–2 NETWORK FUNCTIONS OF ONE- AND TWO-PORT CIRCUITS

The two major types of network functions are driving-point impedance and transfer functions. A **driving-point impedance** relates the voltage and current at a pair of terminals called a port. The driving-point impedance $Z(s)$ of the one-port circuit in Figure 11–2 is defined as

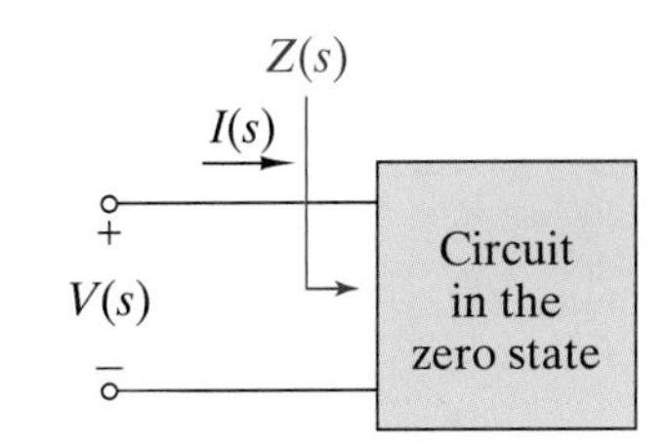

FIGURE 11–2 *A one-port circuit.*

$$Z(s) = \frac{V(s)}{I(s)} \tag{11–5}$$

When the one port is driven by a current source, the response is $V(s) = Z(s)I(s)$ and the natural frequencies in the response are the poles of impedance $Z(s)$. On the other hand, when the one port is driven by a voltage source, the response is $I(s) = [Z(s)]^{-1}V(s)$ and the natural frequencies in the response are the poles of $1/Z(s)$; that is, the zeros of $Z(s)$. In other words, the driving-point impedance is a network function whether upside down or right side up.

The term *driving point* means that the circuit is driven at one port and the response is observed at the same port. The element impedances defined in Sect. 10–1 are elementary examples of driving-point impedances. The equivalent impedances found by combining elements in series and parallel are also driving-point impedances. Driving-point functions are the s-domain generalization of the concept of the input resistance. The terms *driving-point impedance*, *input impedance*, and *equivalent impedance* are synonymous.

The driving-point impedance seen at a pair of terminals determines the loading effects that result when those terminals are connected to another circuit. When two circuits are connected together, these loading effects can profoundly alter the responses observed when the same two circuits operated in isolation. In an analysis situation it is important to be able to predict the response changes that occur when one circuit loads another. In design situations it is important to know when the circuits can be designed separately and then interconnected without encountering loading effects that alter their designed performance. The conditions under which loading can or cannot be ignored will be studied in this and subsequent chapters.

Transfer functions are usually of greater interest in signal-processing applications than driving-point impedances because they describe how a signal is modified by passing through a circuit. A **transfer function** relates an input and response (or output) at different ports in the circuit. Figure 11–3 shows the possible input-output configurations for a two-port circuit. Since

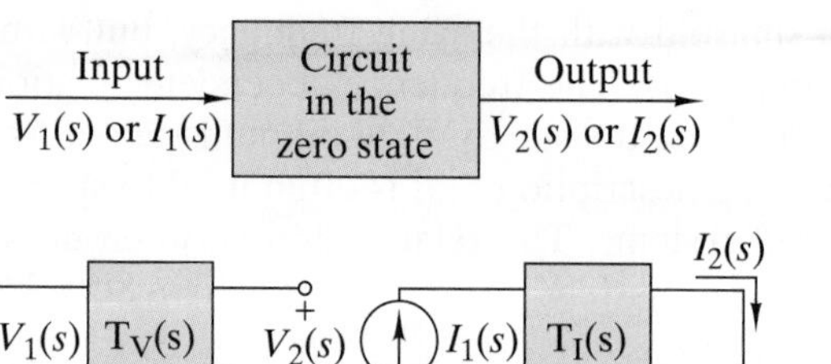

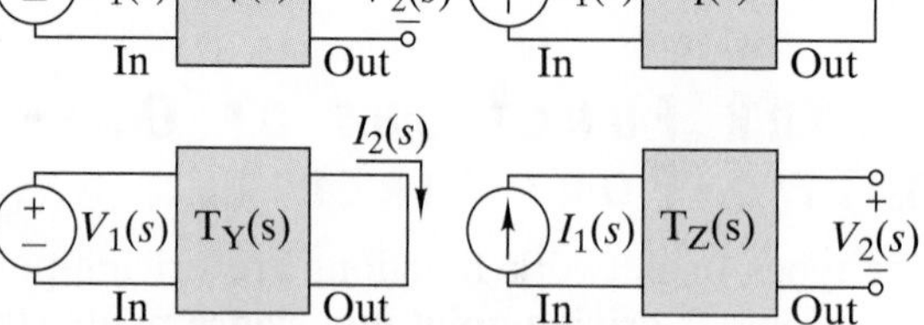

FIGURE 11–3 *Two-port circuits and transfer functions.*

the input and output signals can be either a current or a voltage, we can define four kinds of transfer function:

$$
\begin{aligned}
T_V(s) &= \text{Voltage Transfer Function} = \frac{V_2(s)}{V_1(s)} \\
T_I(s) &= \text{Current Transfer Function} = \frac{I_2(s)}{I_1(s)} \\
T_Y(s) &= \text{Transfer Admittance} = \frac{I_2(s)}{V_1(s)} \\
T_Z(s) &= \text{Transfer Impedance} = \frac{V_2(s)}{I_1(s)}
\end{aligned}
\tag{11–6}
$$

The functions $T_V(s)$ and $T_I(s)$ are dimensionless since the input and output signals have the same units. The function $T_Z(s)$ has units of ohms and $T_Y(s)$ has units of siemens.

The functions in Eq. (11–6) are sometimes called forward transfer functions because they relate inputs applied at port 1 to outputs occurring at port 2. There are, of course, reverse transfer functions that relate inputs at port 2 to outputs at port 1. It is important to realize that a transfer function is only valid for a specified input port and output port. For example, the voltage transfer function $T_V(s) = V_2(s)/ V_1(s)$ relates the input voltage applied at port 1 in Figure 11–3 to the voltage response observed at the output port. The reverse voltage transfer function for signal transmission from output to input is *not* $1/T_V(s)$. Unlike driving-point impedance, transfer functions are not network functions when they are turned upside down.

DETERMINING NETWORK FUNCTIONS

The rest of this section illustrates analysis techniques for deriving network functions. The application of network functions in circuit analysis and design begins in the next section and continues throughout the rest of this book. But first, we illustrate ways to find the network functions of a given circuit.

The divider circuits in Figure 11–4 occur so frequently that it is worth taking time to develop their transfer functions in general terms. Using s-domain voltage division in Figure 11–4(a), we can write

$$V_2(s) = \left[\frac{Z_2(s)}{Z_1(s) + Z_2(s)}\right] V_1(s)$$

Therefore, the voltage transfer function of a voltage divider circuit is

$$T_V(s) = \frac{V_2(s)}{V_1(s)} = \frac{Z_2(s)}{Z_1(s) + Z_2(s)} \quad (11\text{–}7)$$

Similarly, using s-domain current division in Figure 11–4(b) yields the transfer function of a current divider circuit as

$$T_I(s) = \frac{I_2(s)}{I_1(s)} = \frac{Y_2(s)}{Y_1(s) + Y_2(s)} \quad (11\text{–}8)$$

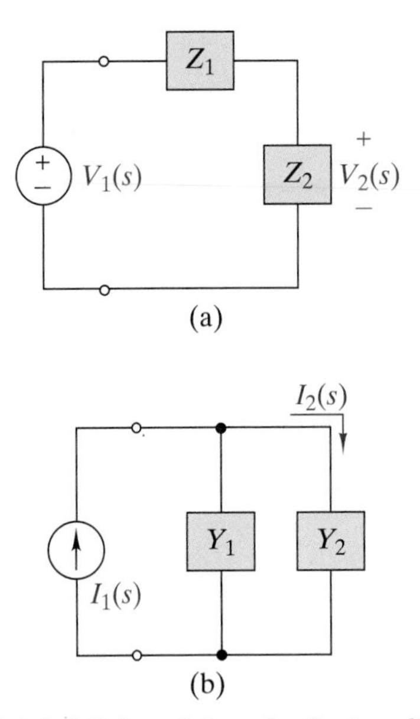

FIGURE 11–4 *Basic divider circuits. (a) Voltage divider. (b) Current divider.*

By series equivalence, the driving-point impedance at the input of the voltage divider is $Z_{EQ}(s) = Z_1(s) + Z_2(s)$. By parallel equivalence the driving-point impedance at the input of the current divider is $Z_{EQ}(s) = 1/(Y_1(s) + Y_2(s))$.

Two other useful circuits are the inverting and noninverting OP AMP configurations shown in Figure 11–5. To determine the voltage transfer function of the inverting circuit in Figure 11–5(a), we write the sum of currents leaving node B:

$$\frac{V_B(s) - V_A(s)}{Z_1(s)} + \frac{V_B(s) - V_C(s)}{Z_2(s)} + I_N(s) = 0$$

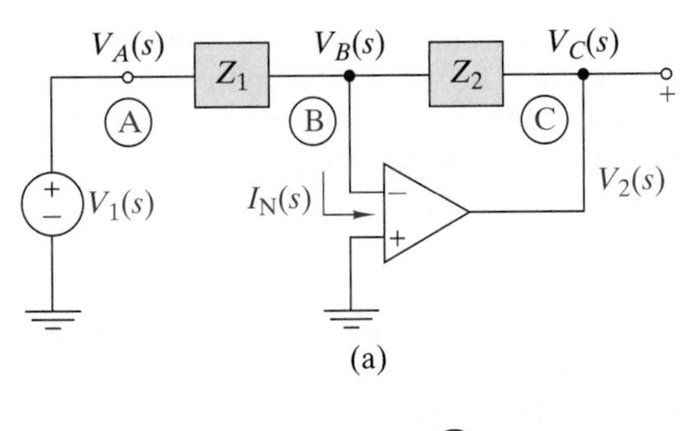

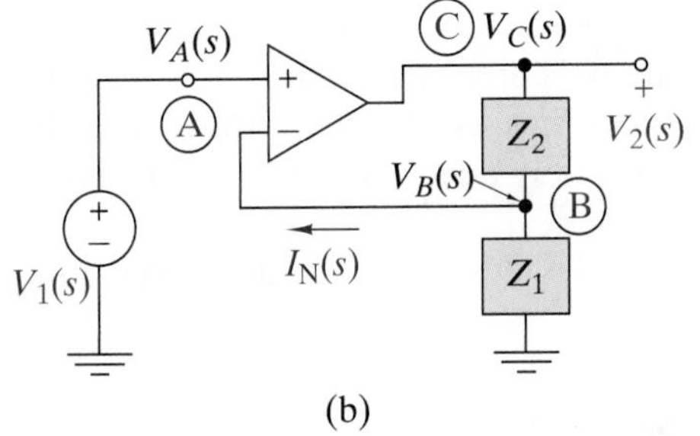

FIGURE 11–5 *Basic OP AMP circuits. (a) Inverting amplifier. (b) Noninverting amplifier.*

But the ideal OP AMP constraints require that $I_N(s) = 0$ and $V_B(s) = 0$ since the noninverting input is grounded. By definition, the output voltage $V_2(s)$ equals node voltage $V_C(s)$ and the voltage source forces $V_A(s)$ to equal the input voltage $V_1(s)$. Inserting all of these considerations into the node equations and solving for the voltage transfer function yields

$$T_V(s) = \frac{V_2(s)}{V_1(s)} = -\frac{Z_2(s)}{Z_1(s)} \tag{11–9}$$

From the study of OP AMP circuits in Chapter 4, you should recognize Eq. (11–9) as the s-domain generalization of the inverting OP AMP circuit gain equation, $K = -R_2/R_1$.

The driving-point impedance at the input to the inverting circuit is

$$Z_{IN}(s) = \frac{V_S(s)}{[V_A(s) - V_B(s)]/Z_1(s)}$$

But $V_A(s) = V_S(s)$ and $V_B(s) = 0$; hence the input impedance is $Z_{IN}(s) = Z_1(s)$.

For the noninverting circuit in Figure 11–5(b), the sum of currents leaving node B is

$$\frac{V_B(s) - V_C(s)}{Z_2(s)} + \frac{V_B(s)}{Z_1(s)} + I_N(s) = 0$$

In the noninverting configuration the ideal OP AMP constraints require that $I_N(s) = 0$ and $V_B(s) = V_1(s)$. By definition, the output voltage $V_2(s)$ equals node voltage $V_C(s)$. Combining all of these considerations and solving for the voltage transfer function yields

$$T_V(s) = \frac{V_2(s)}{V_1(s)} = \frac{Z_1(s) + Z_2(s)}{Z_1(s)} \tag{11–10}$$

Equation (11–10) is the s-domain version of the noninverting amplifier gain equation, $K = (R_1 + R_2)/R_1$. The transfer function of the noninverting configuration is the reciprocal of the transfer function of the voltage divider in the feedback path. The ideal OP AMP draws no current at its input terminals, so theoretically the input impedance of the noninverting circuit is infinite.

The transfer functions of divider circuits and the basic OP AMP configurations are useful analysis and design tools in many practical situations. However, a general method is needed to handle circuits of greater complexity. One general approach is to formulate either node-voltage or mesh-current equations with all initial conditions set to zero. These equations are then solved for network functions using Cramer's rule for hand calculations or symbolic math analysis programs such as Mathcad. The algebra involved can be a bit tedious at times, even with Mathcad. But the tedium is reduced somewhat because we only need the zero-state response for a single input source.

The following examples illustrate methods of calculating network functions.

EXAMPLE 11–2

(a) Find the transfer functions of the circuits in Figure 11–6.
(b) Find the driving-point impedances seen by the input sources in these circuits.

SOLUTION:

(a) These are all divider circuits, so the required transfer functions can be obtained using Eq. (11–7) or (11–8).

For circuit C1: $Z_1 = R$, $Z_2 = 1/Cs$, and $T_V(s) = 1/(RCs + 1)$

For circuit C2: $Z_1 = Ls$, $Z_2 = R$, and $T_V(s) = 1/(GLs + 1)$

For circuit C3: $Y_1 = Cs$, $Y_2 = G$, and $T_I(s) = 1/(RCs + 1)$

These transfer functions are all of the form $1/(\tau s + 1)$, where τ is the circuit time constant.

(b) The driving-point impedances are found by series or parallel equivalence.

For circuit C1: $Z(s) = Z_1 + Z_2 = (RCs + 1)/Cs$

For circuit C2: $Z(s) = Z_1 + Z_2 = Ls + R$

For circuit C3: $Z(s) = 1/(Y_1 + Y_2) = 1/(Cs + G) = R/(RCs + 1)$

The three circuits have different driving-point impedances even though they have the same transfer functions.

The general principle illustrated here is that several different circuits can have the same transfer function. Put differently, a desired transfer function can be realized by several different circuits. This fact is important in design because circuits that produce the same transfer function offer alternatives that may differ in other features. In this example, they all have different input impedances. ■

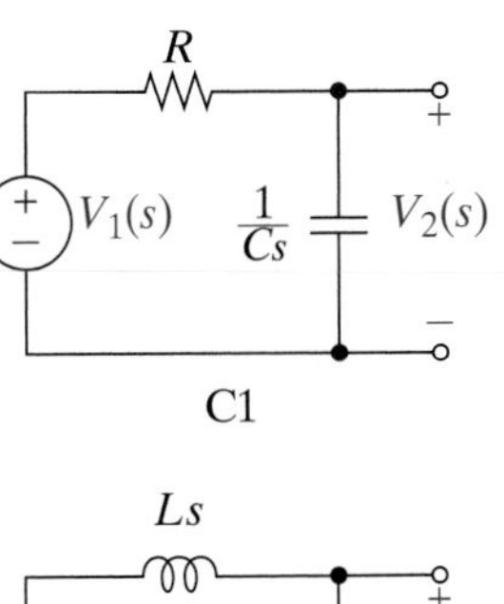

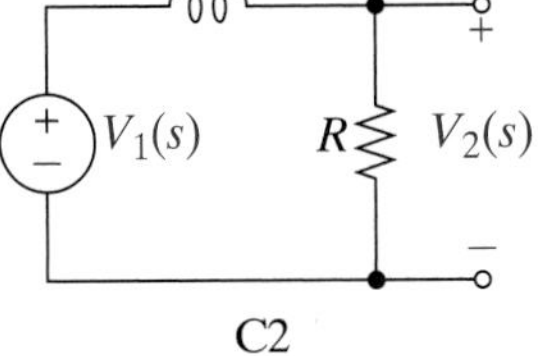

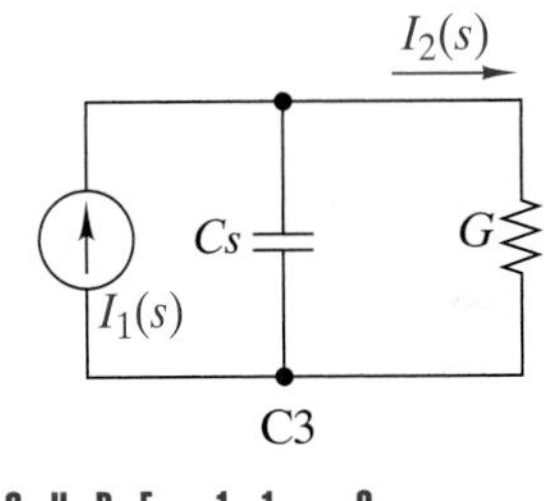

FIGURE 11–6

EXAMPLE 11–3

(a) Find the input impedance seen by the voltage source in Figure 11–7.
(b) Find the voltage transfer function $T_V(s) = V_2(s)/V_1(s)$ of the circuit.
(c) Locate the poles and zeros of $T_V(s)$ for $R_1 = 10$ kΩ, $R_2 = 20$ kΩ, $C_1 = 100$ nF, and $C_2 = 50$ nF.

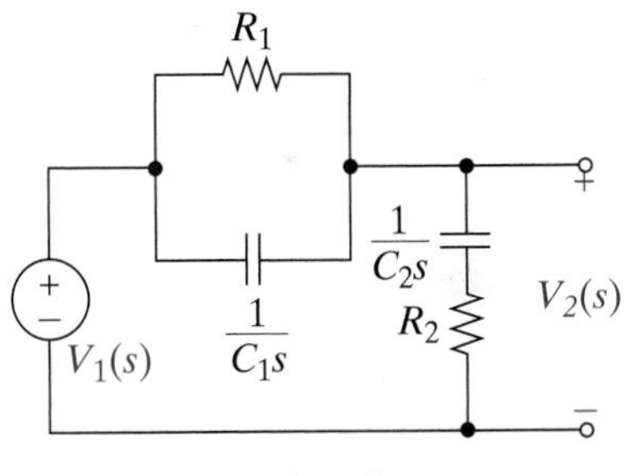

FIGURE 11–7

SOLUTION:

(a) The circuit is a voltage divider. We first calculate the equivalent impedances of the two legs of the divider. The two elements in parallel combine to produce the series leg impedance $Z_1(s)$ as

$$Z_1(s) = \frac{1}{C_1 s + 1/R_1} = \frac{R_1}{R_1 C_1 s + 1}$$

The two elements in series combine to produce shunt leg impedance $Z_2(s)$:

$$Z_2(s) = R_2 + 1/C_2 s = \frac{R_2 C_2 s + 1}{C_2 s}$$

Using series equivalence, the driving-point impedance seen at the input is

$$Z_{EQ}(s) = Z_1(s) + Z_2(s)$$

$$= \frac{R_1 C_1 R_2 C_2 s^2 + (R_1 C_1 + R_2 C_2 + R_1 C_2)s + 1}{C_2 s(R_1 C_1 s + 1)}$$

(b) Using voltage division, the voltage transfer function is

$$T_V(s) = \frac{Z_2(s)}{Z_{EQ}(s)} = \frac{(R_1C_1s + 1)(R_2C_2s + 1)}{R_1C_1R_2C_2s^2 + (R_1C_1 + R_2C_2 + R_1C_2)s + 1}$$

(c) Inserting the specified numerical values into $T_V(s)$ yields

$$T_V(s) = \frac{(10^{-3}s + 1)(10^{-3}s + 1)}{10^{-6}s^2 + 2.5 \times 10^{-3}s + 1} = \frac{(s + 1000)^2}{(s + 500)(s + 2000)}$$

which indicates a double zero at $s = -1000$ rad/s, and simple poles at $s = -500$ rad/s and $s = -2000$ rad/s. ■

EXAMPLE 11–4

Find the driving-point impedance seen by the voltage source in Figure 11–8. Find the voltage transfer function $T_V(s) = V_2(s)/V_1(s)$ of the circuit. The poles of $T_V(s)$ are located at $p_1 = -1000$ rad/s and $p_2 = -5000$ rad/s. If $R_1 = R_2 = 20$ kΩ, what values of C_1 and C_2 are required?

FIGURE 11–8

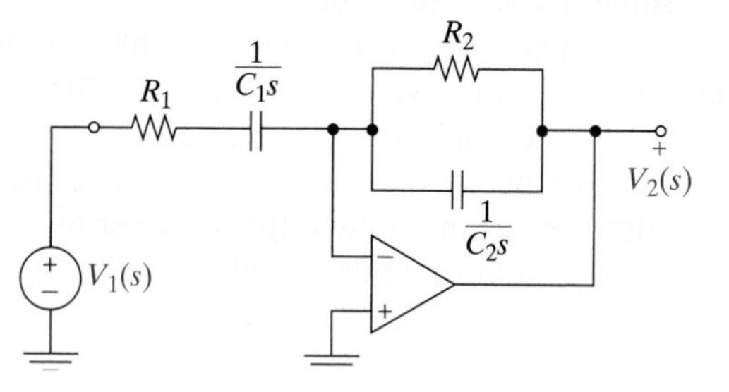

SOLUTION:

The circuit is an inverting OP AMP configuration of the form in Figure 11–5(a). The input impedance of this circuit is

$$Z_1(s) = R_1 + \frac{1}{C_1s} = \frac{R_1C_1s + 1}{C_1s}$$

The impedance Z_2 in the feedback path is

$$Z_2(s) = \frac{1}{C_2s + 1/R_2} = \frac{R_2}{R_2C_2s + 1}$$

and the voltage transfer function is

$$T_V(s) = -\frac{Z_2(s)}{Z_1(s)} = -\frac{R_2C_1s}{(R_1C_1s + 1)(R_2C_2s + 1)}$$

The poles of $T_V(s)$ are located at $p_1 = -1/R_1C_1 = -1000$ and $p_2 = -1/R_2C_2 = -5000$. If $R_1 = R_2 = 20$ kΩ, then $C_1 = 1/1000R_1 = 50$ nF and $C_2 = 1/5000R_2 = 10$ nF. ■

EXAMPLE 11–5

For the circuit in Figure 11–9 find the input impedance $Z(s) = V_1(s)/I_1(s)$, the transfer impedance $T_Z(s) = V_2(s)/I_1(s)$, and the voltage transfer function $T_V(s) = V_2(s)/V_1(s)$.

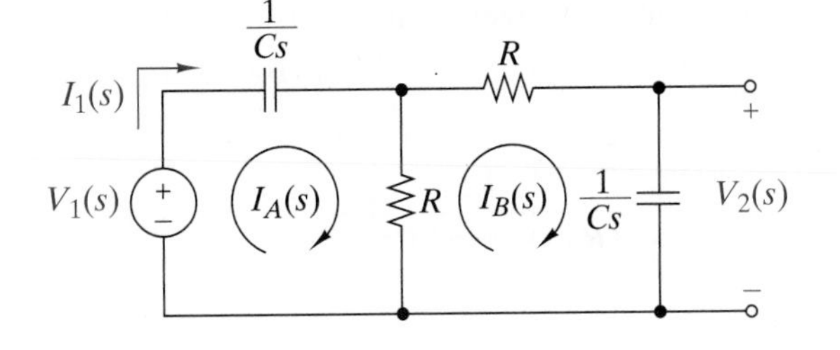

FIGURE 11–9

SOLUTION:

The circuit is not a simple voltage divider, so we use mesh-current equations to illustrate the general approach to finding network functions. By inspection, the mesh-current equations for this ladder circuit are

$$\left(R + \frac{1}{Cs}\right) I_A(s) - RI_B(s) = V_1(s)$$

$$-RI_A(s) + \left(2R + \frac{1}{Cs}\right) I_B(s) = 0$$

In terms of the mesh current, the input impedance is $Z(s) = V_1(s)/I_A(s)$. Using Cramer's rule to solve for $I_A(s)$ yields

$$I_A(s) = \frac{\Delta_A}{\Delta} = \frac{\begin{vmatrix} V_1(s) & -R \\ 0 & 2R + \dfrac{1}{Cs} \end{vmatrix}}{\begin{vmatrix} R + \dfrac{1}{Cs} & -R \\ -R & 2R + \dfrac{1}{Cs} \end{vmatrix}} = \frac{Cs(2RCs + 1)}{(RCs)^2 + 3RCs + 1} V_1(s)$$

The input impedance of the circuit is

$$Z(s) = \frac{V_1(s)}{I_A(s)} = \frac{(RCs)^2 + 3RCs + 1}{Cs(2RCs + 1)}$$

In terms of mesh current, the transfer impedance is $T_Z(s) = V_2(s)/I_A(s)$. The mesh-current equations do not yield the output voltage directly. But since $V_2(s) = I_B(s)Z_C(s)$, we can solve the second mesh equation for $I_B(s)$ in terms of as $I_A(s)$ as

$$I_B(s) = \frac{RCs}{2RCs + 1} I_A(s)$$

and obtain the specified transfer impedance as

$$T_Z(s) = \frac{I_B(s)[1/Cs]}{I_A(s)} = \frac{R}{2RCs + 1}$$

To obtain the specified voltage transfer function, we could use Cramer's rule to solve for $I_B(s)$ in terms of $V_1(s)$ and then use the fact that $V_2(s) = I_B(s)Z_C(s)$. But a moment's reflection reveals that

$$T_V(s) = \frac{V_2(s)}{V_1(s)} = \left[\frac{V_2(s)}{I_1(s)}\right]\left[\frac{I_1(s)}{V_1(s)}\right] = T_Z(s) \times \frac{1}{Z(s)}$$

Hence the specified voltage transfer function is

$$T_V(s) = \frac{R}{2RCs + 1} \times \frac{Cs(2RCs + 1)}{(RCs)^2 + 3RCs + 1}$$
$$= \frac{RCs}{(RCs)^2 + 3RCs + 1}$$

■

EXAMPLE 11–6

Find the voltage transfer function $T_V(s) = V_2(s)/V_1(s)$ of the circuit in Figure 11–10.

FIGURE 11–10

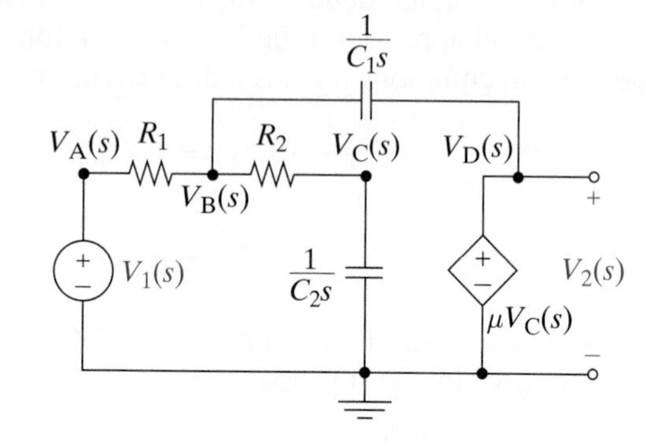

SOLUTION:
The voltage-controlled voltage source makes this an active *RC* circuit. We use node-voltage equations in this problem because the required output is a voltage. The circuit contains two voltage sources connected at a common node. Selecting this common node as the reference eliminates two unknowns since $V_A(s) = V_1(s)$ and $V_D(s) = \mu V_C(s) = V_2(s)$. The sums of currents leaving nodes B and C are

$$\text{Node B:} \quad \frac{V_B(s) - V_1(s)}{R_1} + \frac{V_B(s) - V_C(s)}{R_2} + \frac{V_B(s) - \mu V_C(s)}{1/C_1 s} = 0$$

$$\text{Node C:} \quad \frac{V_C(s) - V_B(s)}{R_2} + \frac{V_C(s)}{1/C_2 s} = 0$$

Multiplying both equations by R_1R_2 and rearranging terms produces

$$\text{Node B:} \quad (R_1 + R_2 + R_1R_2C_1s)V_B(s) - (R_1 + \mu R_1R_2C_1s)V_C(s) = R_2V_1(s)$$

$$\text{Node C:} \quad -V_B(s) + (1 + R_2C_2s)V_C(s) = 0$$

Using the node C equation to eliminate $V_B(s)$ from the node B equation leaves

$$(R_1 + R_2 + R_1R_2C_1s)(1 + R_2C_2s)V_C(s) - (R_1 + \mu R_1R_2C_1s)V_C(s) = R_2V_1(s)$$

Since the output $V_2(s) = \mu V_C(s)$, the required transfer function is

$$T_V(s) = \frac{V_2(s)}{V_1(s)} = \frac{\mu}{R_1R_2C_1C_2s^2 + (R_1C_1 + R_1C_2 + R_2C_2 - \mu R_1C_1)s + 1}$$

This circuit is a member of the Sallen-Key family often used in filter design with $R_1 = R_2 = R$ and $C_1 = C_2 = C$, in which case the transfer function reduces to

$$T_V(s) = \frac{\mu}{(RCs)^2 + (3 - \mu)RCs + 1}$$

We will encounter this result again in later chapters. ■

Exercise 11–1

Find the driving-point impedance seen by the voltage source in Figure 11–11.

Answer:

$$Z(s) = \frac{RLCs^2 + Ls + R}{LCs^2 + 1}$$

Exercise 11–2

Find the voltage transfer function $T_V(s) = V_2(s)/V_1(s)$ in Figure 11–11.

Answer:

$$T_V(s) = \frac{LCs^2 + 1}{LCs^2 + GLs + 1}$$

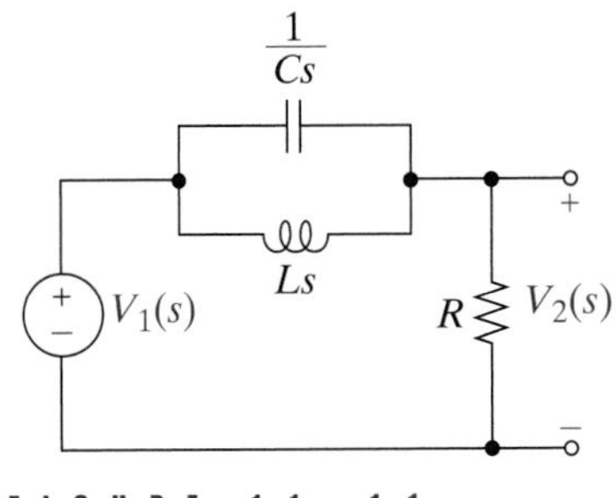
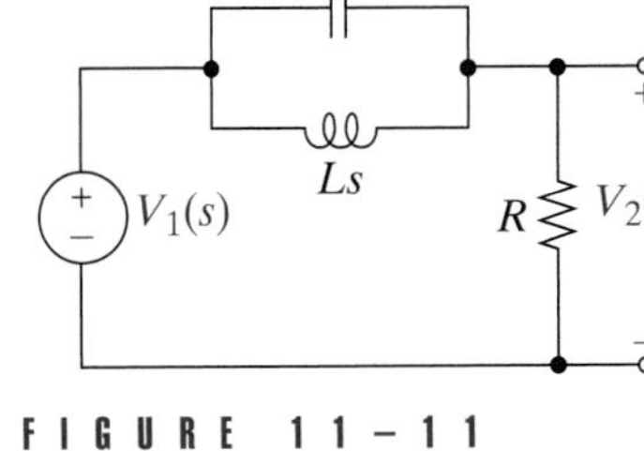

FIGURE 11–11

THE CASCADE CONNECTION AND THE CHAIN RULE

Signal-processing circuits often involve a **cascade connection** in which the output voltage of one circuit serves as the input to the next stage. In some cases, the overall voltage transfer function of the cascade can be related to the transfer functions of the individual stages by a **chain rule**

$$T_V(s) = T_{V1}(s)T_{V2}(s)\cdots T_{Vk}(s) \tag{11–11}$$

where T_{V1}, T_{V2}, . . . and T_{Vk} are the voltage transfer functions of the individual stages when operated separately. It is important to understand when the chain rule applies since it greatly simplifies the analysis and design of cascade circuits.

To illustrate the chain rule concept, consider the two-stage RC circuit in Figure 11–12. When disconnected and operated in isolation, the transfer function of each stage can be found using voltage division as

$$T_{V1}(s) = \frac{R}{R + 1/Cs} = \frac{RCs}{RCs + 1}$$

$$T_{V2}(s) = \frac{1/Cs}{R + 1/Cs} = \frac{1}{RCs + 1}$$

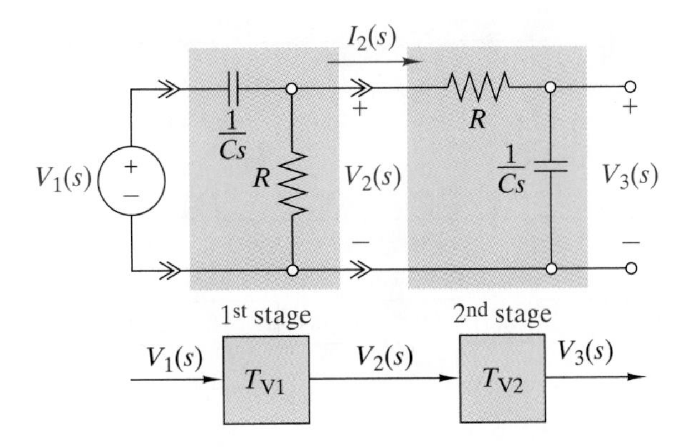

FIGURE 11–12 *Two-port circuits connected in cascade.*

When connected in cascade, the output of the first stage serves as the input to the second stage. If the chain rule applies, we would obtain the overall transfer function as

$$T_V(s) = \frac{V_3(s)}{V_1(s)} = \left(\frac{V_2(s)}{V_1(s)}\right)\left(\frac{V_3(s)}{V_2(s)}\right) = (T_{V1}(s))\,(T_{V2}(s)) \tag{11–12}$$

$$= \underbrace{\left(\frac{RCs}{RCs+1}\right)}_{\substack{\text{first}\\ \text{stage}}}\underbrace{\left(\frac{1}{RCs+1}\right)}_{\substack{\text{second}\\ \text{stage}}} = \underbrace{\frac{RCs}{(RCs)^2 + 2RCs + 1}}_{\text{overall}}$$

However, in Example 11–5, the overall transfer function of this circuit was found to be

$$T_V(s) = \frac{RCs}{(RCs)^2 + 3RCs + 1} \tag{11–13}$$

which disagrees with the chain rule result in Eq. (11–12).

The reason for the discrepancy is that when they are connected in cascade, the second circuit "loads" the first circuit. That is, the voltage-divider rule requires that the interface current $I_2(s)$ in Figure 11–12 be zero. The no-load condition $I_2(s) = 0$ applies when the stages operate separately, but when connected in cascade, the interface current is not zero. The chain rule does not apply here because loading caused by the second stage changes the transfer function of the first stage.

However, Figure 11–13 shows how the loading problem goes away when an OP AMP voltage follower is inserted between the *RC* circuit stages. The follower does not draw any current from the first *RC* circuit [$I_2(s) = 0$] and applies $V_2(s)$ directly across the input of the second *RC* circuit. With this modification the chain rule in Eq. (11–11) applies because the voltage follower isolates the two circuits, thereby solving the loading problem.

In the *s* domain, *loading* causes the transfer function of a circuit to change when it drives the input of another circuit. In a cascade connection, loading does not occur at an interface if (1) the output (Thévenin) impedance of the driving stage is zero or (2) the input impedance of the driven stage is infinite. The voltage follower in Figure 11–13 is an example of a

FIGURE 11–13 *Cascade connection with voltage follower isolation.*

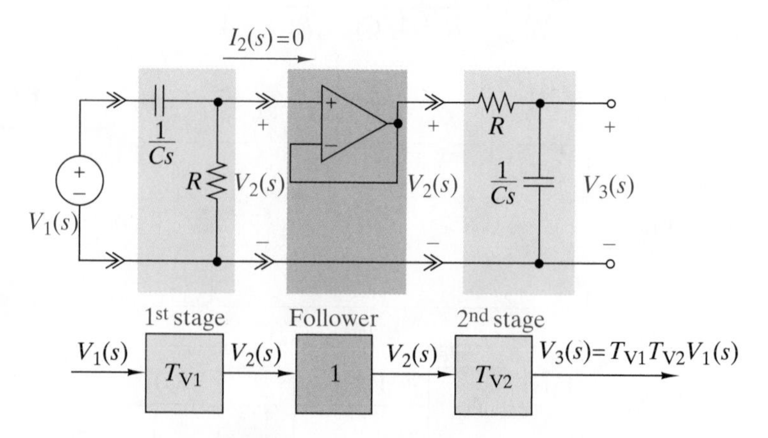

stage that meets both criteria (1) and (2). In general, an inverting OP AMP stage meets criteria (1) but not criteria (2), while a voltage-divider stage meets neither criteria.

When analyzing or designing a cascade connection, it is important to recognize situations in which the chain rule applies. The next example illustrates this point.

EVALUATION EXAMPLE 11–7

Figure 11–14 shows two cascade connections involving the same two stages but with their positions reversed. Do either of these connections involve loading? If not, use the chain rule to find the overall transfer function.

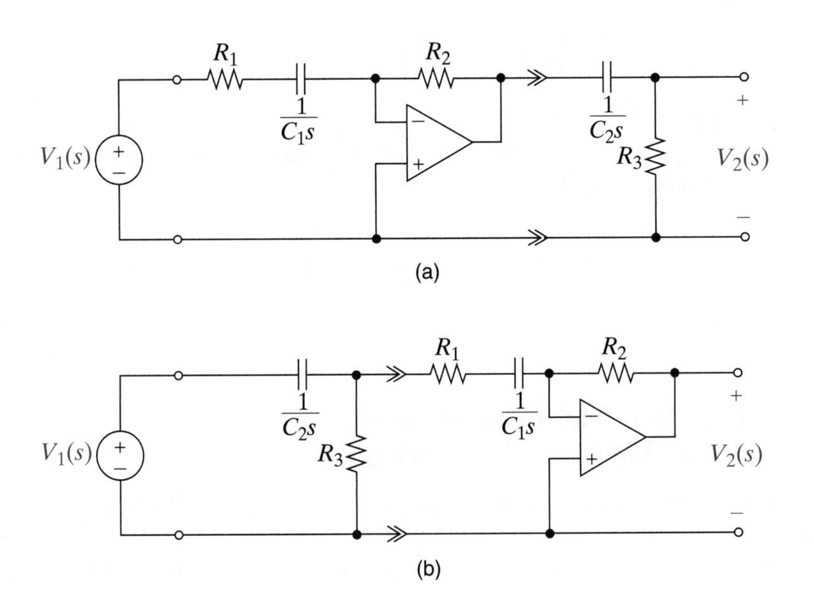

FIGURE 11–14

SOLUTION:

Both circuits involve a cascade connection of a voltage-divider stage and an inverting amplifier stage. The version in Figure 11–14(a) does not involve loading because the output impedance of the first stage is zero. Hence, connecting the second-stage voltage divider does not load the first stage and the chain rule applies. The transfer function of the inverting amplifier stage is

$$T_{V1}(s) = -\frac{Z_2(s)}{Z_1(s)} = -\frac{R_2}{R_1 + 1/C_1 s} = -\frac{R_2 C_1 s}{R_1 C_1 s + 1}$$

The second stage is a voltage divider whose transfer function is

$$T_{V2}(s) = \frac{Z_2(s)}{Z_2(s) + Z_1(s)} = \frac{R_3}{R_3 + 1/C_2 s} = \frac{R_3 C_2 s}{R_3 C_2 s + 1}$$

and the chain rule yields the overall transfer function as

$$T_V(s) = T_{V1}(s) \times T_{V2}(s) = \frac{-R_2C_1s}{R_1C_1s + 1} \times \frac{R_3C_2s}{R_3C_2s + 1}$$

$$= \frac{-R_2C_1R_3C_2s^2}{R_1C_1R_3C_2s^2 + (R_1C_1 + R_3C_2)s + 1}$$

The cascade connection in Figure 11–14(b) interchanges the positions of the two stages. The loading occurs in this case because the first stage is a voltage divider with a nonzero output (Thévenin} impedance of

$$Z_T = \frac{1}{\frac{1}{R_3} + C_2s} = \frac{R_3}{R_3C_2s + 1}$$

and the inverting amplifier in the second stage has a finite input impedance of

$$Z_{IN} = R_1 + \frac{1}{C_1s} = \frac{R_1C_1s + 1}{C_1s}$$

The chain rule does not apply to this connection since the second stage loads the first stage, as can be seen from the transfer function of the connection in Figure 11–14(b):

$$T_V(s) = \frac{R_2C_1R_3C_2s^2}{R_1C_1R_3C_2s^2 + (R_1C_1 + R_3C_2 + R_3C_1)s + 1}$$

which is not equal to $T_{V1}(s) \times T_{V2}(s)$. ■

11–3 NETWORK FUNCTIONS AND IMPULSE RESPONSE

The **impulse response** is the zero-state response of a circuit when the driving force is a unit impulse applied at $t = 0$. When the input signal is $x(t) = \delta(t)$, then $X(s) = \mathcal{L}\{\delta(t)\} = 1$ and the input-output relationship in Eq. (11–2) reduces to

$$Y(s) = T(s) \times 1 = T(s)$$

The impulse response transform equals the network function, and we could treat $T(s)$ as if it is a signal transform. However, to avoid possible confusion between a network function (description of a circuit) and a transform (description of a signal), we denote the impulse response transform as $H(s)$ and use $h(t)$ to denote the corresponding waveform.[2] That is,

Impulse Response

Transform	Waveform	
$H(s) = T(s) \times 1$	$h(t) = \mathcal{L}^{-1}\{H(s)\}$	(11–14)

2 Not all books make this distinction. Books on signals and circuits often use $H(s)$ to represent both a transfer function and the impulse response transform.

When there are no repeated poles, the partial-fraction expansion of $H(s)$ is

$$H(s) = \underbrace{\frac{k_1}{s - p_1} + \frac{k_2}{s - p_2} + \cdots + \frac{k_N}{s - p_N}}_{\text{natural poles}}$$

where $p_1, p_2, \ldots p_N$ are the natural poles in the denominator of the transfer function $T(s)$. All of the poles of $H(s)$ are natural poles since the impulse excitation does not introduce any forced poles. The inverse transform gives the impulse response waveform as

$$h(t) = \underbrace{[k_1 e^{p_1 t} + k_2 e^{p_2 t} + \cdots + k_N e^{p_N t}]}_{\text{natural response}} u(t)$$

When the circuit is stable, all of the natural poles are in the left half plane and the impulse response waveform $h(t)$ decays to zero as $t \to \infty$. A linear circuit whose impulse response ultimately returns to zero is said to be **asymptotically stable**. Asymptotic stability means that the impulse response has a finite time duration. That is, for every $\varepsilon > 0$ there exists a finite time duration T_D such that $|h(t)| < \varepsilon$ for all $t > T_D$.

It is important to note that the impulse response $h(t)$ contains all the information needed to determine the circuit response to any other input. That is, since $\mathcal{L}\{h(t)\} = T(s)$, we can calculate the output $y(t)$ for any Laplace transformable input $x(t)$ as

$$y(t) = \mathcal{L}^{-1}\{H(s)X(s)\}$$

This expression, known as the *convolution theorem,* states that the impulse response can be used to relate the input and output of a linear circuit. Thus, the impulse response $h(t)$ or $H(s)$ can be considered as a mathematical model of a linear circuit. Obviously it is important to be able to find the impulse response and to know how to use the impulse response to find the output for other inputs. The following examples illustrate both of these issues.

EXAMPLE 11–8

Find the response $v_2(t)$ in Figure 11–15 when the input is $v_1(t) = \delta(t)$. Use the element values $R_1 = 10$ kΩ, $R_2 = 12.5$ kΩ, $C_1 = 1$ μF, and $C_2 = 2$ μF.

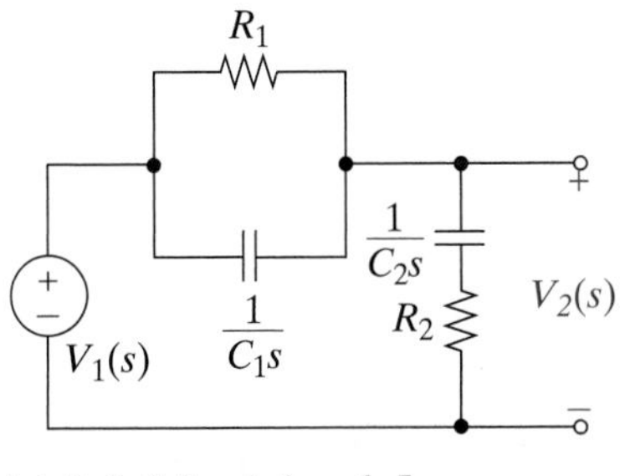

FIGURE 11–15

SOLUTION:

In Example 11–3, the transfer function of this circuit was found to be

$$T_V(s) = \frac{V_2(s)}{V_1(s)} = \frac{(R_1C_1 s + 1)(R_2C_2 s + 1)}{R_1C_1R_2C_2 s^2 + (R_1C_1 + R_2C_2 + R_1C_2)s + 1}$$

For the given element values, the impulse response transform is

$$H(s) = \frac{(s + 100)(s + 40)}{s^2 + 220s + 4000} = \frac{(s + 100)(s + 40)}{(s + 20)(s + 200)}$$

This $H(s)$ is not a proper rational function, so we use one step of a long division plus a partial-fraction expansion to obtain

$$H(s) = 1 + \frac{80/9}{s + 20} - \frac{800/9}{s + 200}$$

and the impulse response is

$$h(t) = \delta(t) + \frac{80}{9}[e^{-20t} - 10e^{-200t}]u(t)$$

In this case, the impulse response contains an impulse because the network function is not a proper rational function. ■

EXAMPLE 11–9

The impulse response of a linear circuit is $h(t) = 200e^{-100t}u(t)$. Find the output when the input is a unit ramp $r(t) = tu(t)$.

SOLUTION:
The circuit impulse response is

$$H(s) = \mathcal{L}\{h(t)\} = \mathcal{L}\{200e^{-100t}u(t)\} = \frac{200}{s + 100}$$

The Laplace transform of the unit ramp input is $1/s^2$; hence, using the convolution theorem the response due to a ramp input is

$$\begin{aligned} y(t) &= \mathcal{L}^{-1}\left\{H(s)\frac{1}{s^2}\right\} = \mathcal{L}^{-1}\left\{\frac{200}{(s + 100)s^2}\right\} \\ &= \mathcal{L}^{-1}\left\{\frac{1/50}{s + 100} + \frac{2}{s^2} - \frac{1/50}{s}\right\} \\ &= \frac{1}{50}(e^{-100t} + 100t - 1)u(t) \end{aligned}$$

This example illustrates that the impulse response $h(t)$ contains all the information needed to calculate the response due to another input. ■

Exercise 11–3

The impulse response of a circuit is $h(t) = 100e^{-20t}u(t)$. Find the output when the input is a step function $x(t) = u(t)$.

Answer: $y(t) = 5\,[1 - e^{-20t}]u(t)$

Exercise 11–4

Find the impulse response of the circuit in Figure 11–16.

Answer: $h(t) = 0.1\delta(t) + [90e^{-100t}]u(t)$

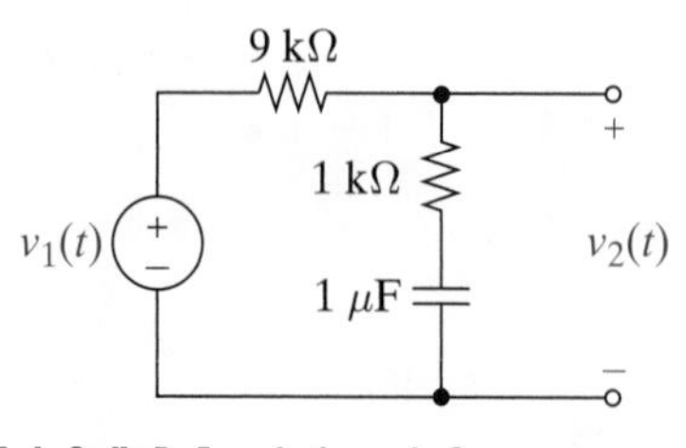

FIGURE 11–16

Exercise 11–5

When the input to a linear circuit is $v_1(t) = \delta(t)$, the output is $v_2(t) = \delta(t) - 10e^{-20t}u(t)$ Find the poles and zeros of the network function $T_V(s) = V_2(s)/V_1(s)$.

Answer: A simple pole at $s = -20$ rad/s and a simple zero at $s = -10$ rad/s.

11–4 NETWORK FUNCTIONS AND STEP RESPONSE

The **step response** is the zero-state response of the circuit output when the driving force is a unit step function applied at $t = 0$. When the input is $x(t) = u(t)$, then $X(s) = \mathcal{L}\{u(t)\} = 1/s$ and the s-domain input-output relationship in Eq. (11–2) yields $Y(s) = T(s)/s$. The step response transform and waveform will be denoted by $G(s)$ and $g(t)$, respectively. That is,

Step Response

Transform	Waveform	
$G(s) = \dfrac{T(s)}{s}$	$g(t) = \mathcal{L}^{-1}\{G(s)\}$	(11–15)

The poles of $G(s)$ are the natural poles contributed by the network function $T(s)$ and a forced pole at $s = 0$ introduced by the step function input. The partial fraction expansion of $G(s)$ takes the form

$$G(s) = \underbrace{\frac{k_0}{s}}_{\text{forced pole}} + \underbrace{\frac{k_1}{s - p_1} + \frac{k_2}{s - p_2} + \cdots + \frac{k_N}{s - p_N}}_{\text{natural poles}}$$

where $p_1, p_2, \ldots p_N$ are the natural poles in $T(s)$. The inverse transformation gives the step response waveform as

$$g(t) = \underbrace{k_0 u(t)}_{\text{forced response}} + \underbrace{[k_1 e^{p_1 t} + k_2 e^{p_2 t} + \cdots + k_N e^{p_N t}]u(t)}_{\text{natural response}}$$

When the circuit is stable, the natural response decays to zero, leaving a forced component called the **dc steady-state response**. The amplitude of the steady-state response is the residue in the partial-fraction expansion of the forced pole at $s = 0$. By the cover-up method, this residue is

$$k_0 = sG(s)\big|_{s=0} = T(0)$$

For a unit step input, the amplitude of the dc steady-state response equals the value of the transfer function at $s = 0$. By linearity, the general principle is that an input $Au(t)$ produces a dc steady-state output whose amplitude is $AT(0)$.

We next show the relationship between the impulse and step responses. First, combining Eqs. (11–14) and (11–15) gives

$$G(s) = \frac{H(s)}{s}$$

The step response transform is the impulse response transform divided by s. The integration property of the Laplace transform tells us that division by s in the s domain corresponds to integration in the time domain. Therefore, in the time domain we can relate the impulse and step response waveforms by integration:

$$g(t) = \int_0^t h(\tau)d\tau \qquad (11\text{–}16)$$

Using the fundamental theorem of calculus, the impulse response waveform is expressed in terms of the step response waveform

$$h(t)(=)\frac{dg(t)}{dt} \tag{11–17}$$

where the symbol (=) means equal almost everywhere, a condition that excludes those points at which $g(t)$ has a discontinuity. In the time domain, the step response waveform is the integral of the impulse response waveform. Conversely, the impulse response waveform is (almost everywhere) the derivative of the step response waveform.

The key idea is that there are relationships between the network function $T(s)$ and the responses $H(s)$, $h(t)$, $G(s)$, and $g(t)$. If any one of these quantities is known, we can obtain any of the other four using relatively simple mathematical operations.

EXAMPLE 11–10

The element values for the circuit in Figure 11–17 are $R_1 = 10$ kΩ, $R_2 = 100$ kΩ, $C_1 = C_2 = 100$ nF, and $v_1(t) = u(t)$ V. Find the response $v_2(t)$.

FIGURE 11–17

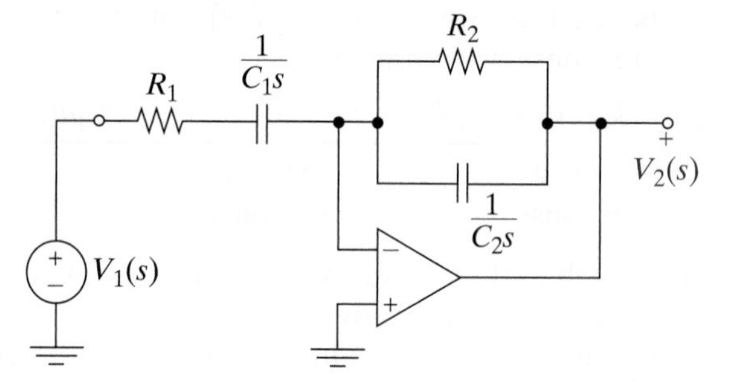

SOLUTION:

For a unit step input $V_1(s) = 1/s$. In Example 11–4 the transfer function of this circuit is shown to be

$$T_V(s) = \frac{V_2(s)}{V_1(s)} = -\frac{R_2C_1s}{(R_1C_1s + 1)(R_2C_2s + 1)}$$

Substituting the numerical element values into this expression yields the step response transform as

$$V_2(s) = \frac{T_V(s)}{s} = -\frac{1000}{(s + 100)(s + 1000)}$$
$$= \frac{-10/9}{s + 100} + \frac{10/9}{s + 1000}$$

The inverse transform yields

$$v_2(t) = \frac{10}{9}(-e^{-100t} + e^{-1000t})u(t)$$

In this case the forced pole at $s = 0$ is canceled by a zero of $T_V(s)$ and the final value of the step response is zero. Using the final value theorem, we have

$$g(\infty) = \underset{t\to\infty}{\text{Limit}}\; g(t) = \underset{s\to 0}{\text{Limit}}\; sG(s) = T(0)$$

the final steady-state value of the step response is equal to the value of the transfer function evaluated at $s = 0$. In the present case $T_V(0) = 0$, so the dc steady-state output of the circuit is zero. Recall from Chapter 6 that in the dc steady state, capacitors can be replaced by open circuits. Replacing C_1 in Figure 11–17 by an open circuit disconnects the input source from the OP AMP, so no dc signals can be transferred through the circuit. A series capacitor that prevents the passage of dc signals is commonly called a *blocking capacitor.* ■

EXAMPLE 11–11

The impulse response of a linear circuit is

$$h(t) = 5000\,(e^{-1000t}\sin 2000\,t)u(t)$$

Find the step response.

SOLUTION:

Using Eq. (11–16) we have

$$\begin{aligned} g(t) &= \int_0^t h(\tau)d\tau = 5000\int_0^t e^{-1000\tau}\sin 2000\,\tau\, d\tau \\ &= (-e^{-1000\tau}\sin 2000\,\tau - 2e^{-1000\tau}\cos 2000\,\tau)\big|_0^t \\ &= (2 - e^{-1000t}\sin 2000\,t - 2e^{-1000t}\cos 2000\,t)u(t) \end{aligned}$$

We could have found $g(t)$ by first finding the Laplace transform of the impulse response $H(s)$ and then calculating the step response as $g(t) = \mathcal{L}^{-1}\{H(s)/s\}$. However, we choose to use Eq. (11–16), which means that the step response calculation took place entirely in the time domain. This is a simple example of a general time-domain method based on the *convolution integral,* a procedure that will be covered in Sect. 11–6. ■

APPLICATION EXAMPLE 11–12

Three time-domain parameters often used to describe the step response are rise time, delay time, and overshoots. **Rise time (T_R)** is the time interval required for the step response to rise from 10% to 90% of its steady-state value $g(\infty)$. **Delay time (T_D)** is the time interval required for the step response to reach 50% of its steady-state value. **Overshoot** is the difference between the peak value of the step response and its steady-state value. Overshoot is usually expressed as a percentage of the steady-state value, namely

$$\text{Overshoot} = \frac{g_{max} - g(\infty)}{g(\infty)} \times 100$$

Figure 11–18 illustrates these descriptors for a typical step response.

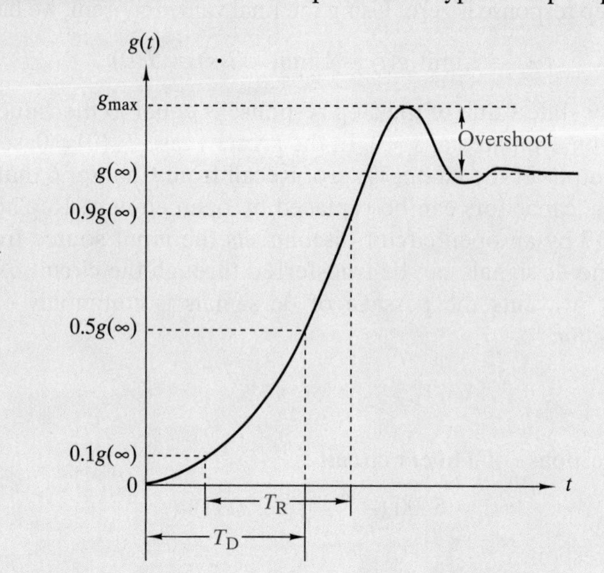

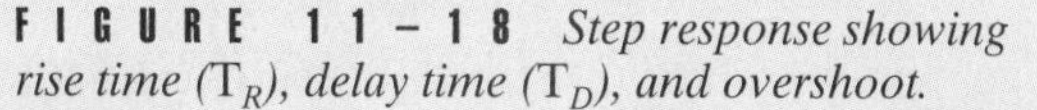
FIGURE 11–18 *Step response showing rise time (T_R), delay time (T_D), and overshoot.*

Step response descriptors are used to specify the performance of both analog and digital systems. Rise time governs how rapidly the system responds to an abrupt change in the input. Delay time controls the time between the application of an abrupt change and the appearance of a significant change in the output. Overshoot indicates the amount of damping present in the system. Lightly damped oscillations produce large overshoots and may cause erroneous state changes in digital systems.

Rise time, delay time, and overshoot can be determined experimentally or calculated using modern computer tools. For example, the MATLAB function `step` calculates and plots the step response of a transfer function of the form $T(s) = n(s)/d(s)$, where $n(s)$ is the numerator polynomial and $d(s)$ is the denominator polynomial. In the MATLAB command window these polynomials are defined by the row matrices `num` and `den` that contain the coefficients of the polynomials $n(s)$ and $d(s)$ in descending powers of s.

For example, suppose we need to evaluate the step response descriptors of a circuit whose transfer function is

$$T(s) = \frac{400(s + 100)}{(s + 100)^2 + 100^2}$$

The two coefficient matrices for this transfer function are entered in the MATLAB command window as

```
num=[400,400*100];
den=[1,2*100,100^2+100^2];
```

The first element in either matrix is the coefficient of the highest power of s, which is followed by the other coefficients in descending powers of s. The MATLAB statement

```
step(num,den)
```

produces the step response plot shown in Figure 11–19.

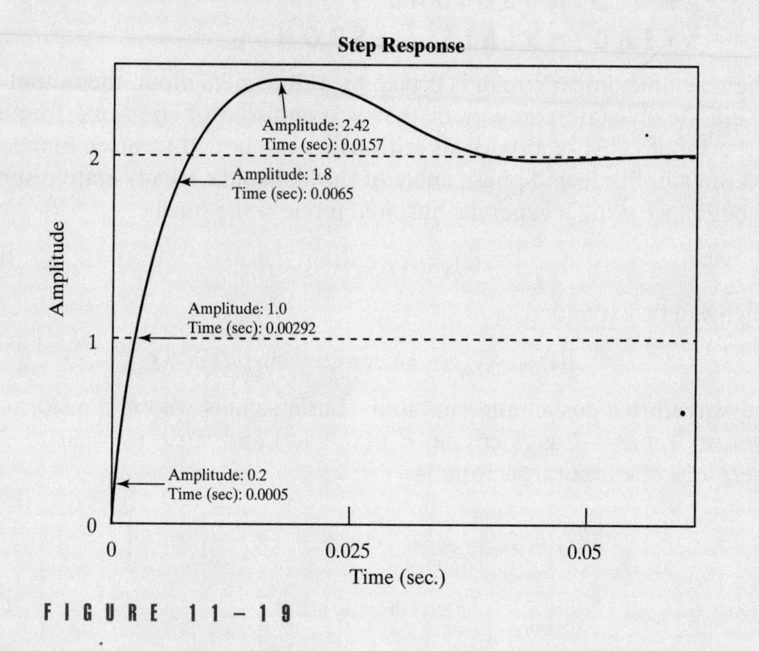

FIGURE 11–19

First note that the steady-state response is $g(\infty) = 2$. Using the cursor in the MATLAB graphics window, we can find the points at which certain amplitudes are reached. The key points shown in the figure are:

1. The 10% rise point (amplitude = 0.2 and time = 0.0005)
2. The 50% rise point (amplitude = 1 and time = 0.00292)
3. The 90% rise point (amplitude = 1.8 and time = 0.0065)
4. The maximum point (amplitude = 2.42 and time = 0.0157)

Hence the step response descriptors for the circuit are:

Rise time: $T_R = 0.0065 - 0.005 = 6$ ms

Delay time: $T_D = 0.00292 = 2.92$ ms

Overshoot = $(2.42 - 2)/2 = 21\%$

Exercise 11–6

The impulse response of a circuit is $h(t) = 0.1\delta(t) + 90e^{-100t}u(t)$. Find the step response.

Answer: $g(t) = [1 - 0.9e^{-1000t}]u(t)$

Exercise 11–7

The step response of a linear circuit is $g(t) = 5[e^{-1000t}\sin(2000t)]u(t)$. Find the circuit transfer function $T(s)$.

Answer: $$T(s) = \frac{10^4 s}{s^2 + 2000s + 5 \times 10^6}$$

11–5 NETWORK FUNCTIONS AND SINUSOIDAL STEADY-STATE RESPONSE

When a stable, linear circuit is driven by a sinusoidal input, the output contains a steady-state component that is a sinusoid of the same frequency as the input. This section deals with using the circuit transfer function to find the amplitude and phase angle of the sinusoidal steady-state response. To begin, we write a general sinusoidal input in the form

$$x(t) = X_A \cos(\omega t + \phi) \tag{11–18}$$

which can be expanded as

$$x(t) = X_A(\cos \omega t \cos \phi - \sin \omega t \sin \phi)$$

The waveforms $\cos \omega t$ and $\sin \omega t$ are basic signals whose transforms are given in Table 9–2 as $\mathscr{L}\{\cos \omega t\} = s/(s^2 + \omega^2)$ and $\mathscr{L}\{\sin \omega t\} = \omega/(s^2 + \omega^2)$. Therefore, the input transform is

$$\begin{aligned} X(s) &= X_A\left[\frac{s}{s^2 + \omega^2}\cos \phi - \frac{\omega}{s^2 + \omega^2}\sin \phi\right] \\ &= X_A\left[\frac{s \cos \phi - \omega \sin \phi}{s^2 + \omega^2}\right] \end{aligned} \tag{11–19}$$

Equation (11–19) is the Laplace transform of the general sinusoidal waveform in Eq. (11–18).

Using Eq. (11–2), we obtain the response transform for a general sinusoidal input:

$$Y(s) = X_A\left[\frac{s \cos \phi - \omega \sin \phi}{(s - j\omega)(s + j\omega)}\right] T(s) \tag{11–20}$$

The response transform contains forced poles at $s = \pm j\omega$ because the input is a sinusoid. Expanding Eq. (11–20) by partial fractions,

$$Y(s) = \underbrace{\frac{k}{s - j\omega} + \frac{k^*}{s + j\omega}}_{\text{forced poles}} + \underbrace{\frac{k_1}{s - p_1} + \frac{k_2}{s - p_2} + \cdots + \frac{k_N}{s - p_N}}_{\text{natural poles}}$$

where $p_1, p_2, \ldots p_N$ are the natural poles contributed by the transfer function $T(s)$. To obtain the response waveform, we perform the inverse transformation:

$$y(t) = \underbrace{ke^{j\omega t} + k^*e^{-j\omega t}}_{\text{forced response}} + \underbrace{k_1e^{p_1t} + k_2e^{p_2t} + \cdots + k_Ne^{p_Nt}}_{\text{natural response}}$$

When the circuit is stable, the natural response decays to zero, leaving a sinusoidal steady-state response due to the forced poles as $s = \pm j\omega$. The steady-state response is

$$y_{SS}(t) = ke^{j\omega t} + k^*e^{-j\omega t}$$

where the subscript SS identifies a steady-state condition.

To determine the amplitude and phase of the steady-state response, we must find the residue k. Using the cover-up method from Chapter 9, we find k to be

$$k = (s - j\omega)X_A\left[\frac{s\cos\phi - \omega\sin\phi}{(s - j\omega)(s + j\omega)}\right]T(s)\bigg|_{s=j\omega}$$

$$= X_A\left[\frac{j\omega\cos\phi - \omega\sin\phi}{2j\omega}\right]T(j\omega)$$

$$= X_A\left[\frac{\cos\phi + j\sin\phi}{2}\right]T(j\omega) = \frac{1}{2}X_A e^{j\phi}T(j\omega)$$

The complex quantity $T(j\omega)$ can be written in magnitude and angle form as $|T(j\omega)|e^{j\theta}$. Using these results, the residue becomes

$$k = \left[\frac{1}{2}X_A e^{j\phi}\right]\left|T(j\omega)\right|e^{j\theta}$$

$$= \frac{1}{2}X_A\left|T(j\omega)\right|e^{j(\phi+\theta)}$$

The inverse transform yields the steady-state response in the form

$$y_{SS}(t) = 2|k|\cos(\omega t + \phi + \angle k)$$

$$= \underbrace{X_A|T(j\omega)|}_{\text{amplitude}}\cos(\omega t + \underbrace{\phi + \theta}_{\text{phase}}) \quad (11\text{–}21)$$

In the development leading to Eq. (11–21), we treat frequency as a general variable where the symbol ω represents all possible input frequencies. In some cases the input frequency has a specific value, which we denote as ω_A. In this case the input is written as

$$x(t) = X_A\cos(\omega_A t + \phi)$$

To obtain the steady-state output, we evaluate the transfer function at the specific frequency (ω_A) of the input sinusoid, namely

$$T(j\omega_A) = |T(j\omega_A)|\angle T(j\omega_A)$$

and the steady-state output is expressed as

$$y_{SS}(t) = X_A|T(j\omega_A)|\cos[\omega_A t + \phi + \angle T(j\omega_A)]$$

This result emphasizes three things about the steady-state response:

(1) Output Frequency = Input Frequency = ω_A

(2) Output Amplitude = Input Amplitude × $|T(j\omega_A)|$ (11–22)

(3) Output Phase = Input Phase + $\angle T(j\omega_A)$

The next two examples illustrate sinusoidal steady-state response calculations. In the first example, frequency is treated as a general variable and we examine the steady-state response as frequency varies over a wide range. In the second example, we evaluate the steady-state response at two specific frequencies.

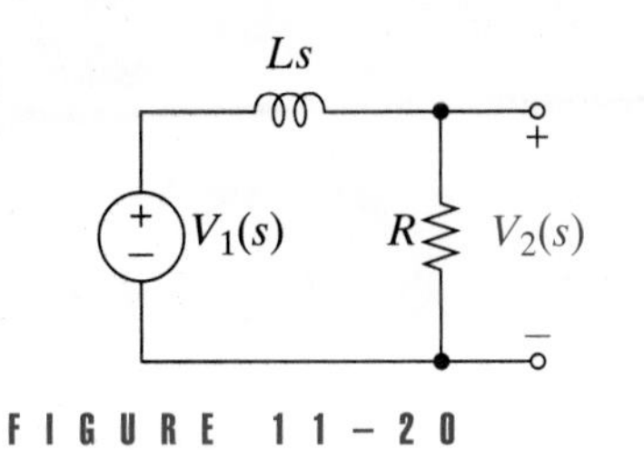

FIGURE 11–20

EXAMPLE 11–13

Find the steady-state output in Figure 11–20 for a general input $v_1(t) = V_A \cos(\omega t + \phi)$.

SOLUTION:
In Example 11–2, the circuit transfer function is shown to be

$$T(s) = \frac{R}{Ls + R}$$

The magnitude and angle of $T(j\omega)$ are

$$|T(j\omega)| = \frac{R}{\sqrt{R^2 + (\omega L)^2}}$$

$$\theta(\omega) = -\tan^{-1}\left(\frac{\omega L}{R}\right)$$

Using Eq. (11–21), the sinusoidal steady-state output is

$$v_{2SS}(t) = \frac{V_A R}{\sqrt{R^2 + (\omega L)^2}} \cos[\omega t + \phi - \tan^{-1}(\omega L/R)]$$

Note that both the amplitude and phase angle of the steady-state response depend on the frequency of the input sinusoid. In particular, at $\omega = 0$ the amplitude of the steady-state output reduces to V_A, which is the same as the amplitude of the input sinusoid. This makes sense because at dc the inductor in Figure 11–20 acts like a short circuit that directly connects the output port to the input port. At very high frequency the amplitude of the steady-state output approaches zero. This also makes sense because at very high frequency the inductor acts like an open circuit that disconnects the output port from the input port. In between these two extremes the output amplitude decreases as the frequency increases. ■

EXAMPLE 11–14

The impulse response of a linear circuit is $h(t) = 5000[2e^{-1000t}\cos 2000t - e^{-1000t}\sin 2000t]u(t)$.

(a) Find the sinusoidal steady-state response when $x(t) = 5 \cos 1000t$.
(b) Repeat (a) when $x(t) = 5 \cos 3000t$.

SOLUTION:
The transfer function corresponding to $h(t)$ is

$$T(s) = H(s) = \mathcal{L}\{h(t)\}$$

$$= 5000\left[\frac{2(s + 1000)}{(s + 1000)^2 + (2000)^2} - \frac{2000}{(s + 1000)^2 + (2000)^2}\right]$$

$$= \frac{10^4 s}{s^2 + 2000s + 5 \times 10^6}$$

(a) At $\omega_A = 1000$ rad/s, the value of $T(j\omega_A)$ is

$$T(j1000) = \frac{10^4(j1000)}{(j1000)^2 + 2000\,(j1000) + 5 \times 10^6}$$

$$= \frac{j10^7}{(5 \times 10^6 - 10^6) + j2 \times 10^6} = \frac{j10}{4 + j2}$$

$$= \frac{10\, e^{j90^\circ}}{\sqrt{20}e^{j26.6^\circ}} = 2.24\, e^{j63.4^\circ}$$

and the steady-state response for $x(t) = 5\cos 1000t$ is

$$y_{SS}(t) = 5 \times 2.24 \cos(1000t + 0^\circ + 63.4^\circ)$$
$$= 11.2\cos(1000t + 63.4^\circ)$$

(b) At $\omega_A = 3000$ rad/s, the value of $T(j\omega_A)$ is

$$T(j3000) = \frac{10^4(j3000)}{(j3000)^2 + 2000\,(j3000) + 5 \times 10^6}$$

$$= \frac{j3 \times 10^7}{5 \times 10^6 - 9 \times 10^6 + j6 \times 10^6}$$

$$= \frac{j30}{-4 + j6} = \frac{30\, e^{j90^\circ}}{\sqrt{52}e^{j123.7^\circ}} = 4.16\, e^{-j33.7^\circ}$$

and the steady-state response for $x(t) = 5\cos 3000t$ is

$$y_{SS}(t) = 5 \times 4.16 \cos(3000t + 0^\circ - 33.7^\circ)$$
$$= 20.8\cos(3000t - 33.7^\circ)$$

Again note that the amplitude and phase angle of the steady-state response depend on the input frequency. ■

Exercise 11–8

The transfer function of a linear circuit is $T(s) = 5(s + 100)/(s + 500)$. Find the steady-state output for
(a) $x(t) = 3\cos 100t$
(b) $x(t) = 2\sin 500t$

Answers:

(a) $y_{SS}(t) = 4.16\cos(100t + 33.7^\circ)$
(b) $y_{SS}(t) = 7.21\cos(500t - 56.3^\circ)$

Exercise 11–9

The impulse response of a linear circuit is $h(t) = \delta(t) - 100[e^{-100t}]u(t)$. Find the steady-state output for
(a) $x(t) = 25\cos 100t$
(b) $x(t) = 50\sin 100t$

Answers:

(a) $y_{SS}(t) = 17.7\cos(100t + 45^\circ)$
(b) $y_{SS}(t) = 35.4\cos(100t - 45^\circ)$

NETWORK FUNCTIONS AND PHASOR CIRCUIT ANALYSIS

In this text we present two methods of finding the sinusoidal steady-state response of a linear circuit. Both methods depend on the fact that in the steady state every voltage and current in a linear circuit is a sinusoid of the same frequency as the input sinusoid. As a result, every method of sinusoidal steady-state analysis boils down to finding the amplitude and phase angle of sinusoidal waveforms, all of which have the same frequency.

The steady-state analysis method developed in this chapter involves first finding the network function $T(s)$ that relates an input and a particular output. We next form the complex quantity $T(j\omega)$ by replacing s by $j\omega$, where ω is the frequency of the input sinusoid. The magnitude $|T(j\omega)|$ and angle $\angle T(j\omega)$ then give us the amplitude and phase angle of the steady-state output via Eq. (11–21). Since $T(j\omega)$ is a function of ω, it can describe the steady-state response for a single frequency or whole range of frequencies.

The steady-state analysis method developed in Chapter 8 involves representing the amplitude and phase angles of a sinusoid by a complex number called a phasor. Although not strictly limited to a single frequency, phasor analysis works best when analyzing circuits driven by a single frequency. This method emphasizes the impedance $Z(s)$ of the individual two-terminal elements in the circuit. The element's phasor impedance $Z(j\omega)$ is found by replacing s by $j\omega$, where ω is the frequency of the input sinusoid. When numerical values of the element impedances are given, the specific value of frequency ω is not required. When the element impedances are not given, the frequency must be given so we can compute the phasor impedances of the three passive elements.

$$\text{Resistor:} \quad Z_R = R$$

$$\text{Inductor:} \quad Z_L = j\omega L$$

$$\text{Capacitor:} \quad Z_C = \frac{1}{j\omega C}$$

The advantage of the phasor method is that it uses familiar analysis tools like voltage or current division, series or parallel equivalence, source transformation, and mesh or node analysis to solve for the complex numbers (phasors) representing the voltages and currents in the circuit. The magnitude and angle of a phasor are the amplitude and phase angle of the corresponding sinusoidal steady-state response. Responses found using the phasor method are *identical* to those found using the network function method developed in this chapter.

Under what circumstances is one method better than the other?

Phasor circuit analysis works best when the circuit is driven at a single frequency and we need to find several voltages and currents, or the average power, such as in the electric power systems studied in Chapter 16. The network function method works best when there is a single output and the circuit is driven at many different frequencies, such as in the frequency-response and filter applications studied in Chapters 12 and 14. The network function method is the only method available when we know the impulse or step response and need to infer the frequency response of a circuit from its time-domain response. Thus, the preferred method depends on how the circuit is driven (single or multiple frequencies), how much we know about the circuit (complete circuit diagram or only the impulse response), and what we need to find out (multiple responses or a single response). Understanding both methods and how they are applied is important to engineers working with electrical circuits.

11–6 IMPULSE RESPONSE AND CONVOLUTION

In signal processing the term **convolution** refers to a process by which the impulse response of a linear system is used to determine the zero-state response due to other inputs. When the impulse response and input are given in the s domain as $H(s)$ and $X(s)$, the zero-state response is obtained from the inverse Laplace transform of their product.

$$y(t) = \mathcal{L}^{-1}\{H(s)X(s)\} \tag{11–23}$$

The purpose of this section is to introduce the notion that the convolution process can also be viewed and carried out entirely in the time domain. Specifically, given a causal impulse response $h(t)$ and a causal input $x(t)$, the zero-state response is obtained from the **convolution integral**[3]

$$y(t) = \int_0^t h(t - \tau)x(\tau)d\tau \tag{11–24}$$

where τ is a dummy variable of integration. The shorthand notation $y(t) = h(t)*x(t)$ is used to represent the t-domain process, where the asterisk indicates a convolution integral, not a multiplication. That is, the expression $h(t)*x(t)$ reads "$h(t)$ convolved with $x(t)$," not "$h(t)$ times $x(t)$."

Figure 11–21 indicates the parallelism between the input-output relationships in the s domain and t domain. In the s domain the impulse response $H(s)$ multiplies the input transform to produce the output transform. In the t domain the impulse response $h(t)$ is convolved with the input waveform to produce the output waveform.

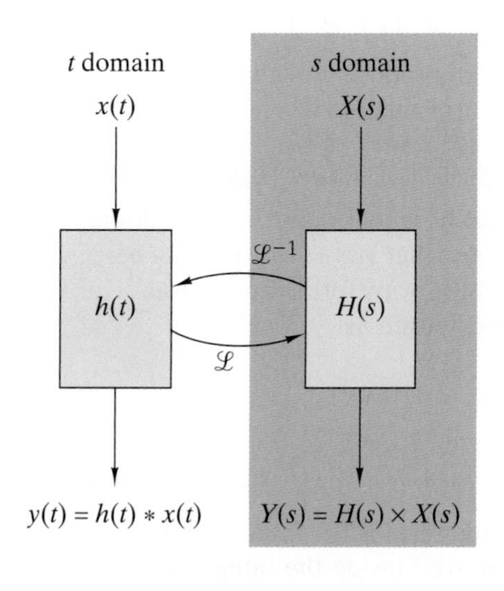

FIGURE 11–21 *Input-output relationships in the* t *domain and the* s *domain.*

3 Recall from Chapter 5 that a waveform is causal if $f(t) = 0$ for all $t < 0$. If the impulse response $h(t)$ is not causal, the upper limit in the convolution integral becomes $+\infty$. If the input $x(t)$ is not casual, the lower limit becomes $-\infty$.

The following example illustrates that the same result is achieved whether convolution is carried out in the s domain or the t domain.

EXAMPLE 11–15

A linear circuit has an impulse response $h(t) = 2e^{-t}u(t)$ and an input $x(t) = e^{-2t}u(t)$. Find the zero-state response using

(a) The s-domain process in Eq. (11–23)
(b) The t-domain convolution integral in Eq. (11–24).

SOLUTION:

(a) Converting $h(t)$ and $x(t)$ to the s domain yields

$$H(s) = \mathcal{L}\{2e^{-t}\} = \frac{2}{s+1} \quad \text{and} \quad X(s) = \mathcal{L}\{e^{-2t}\} = \frac{1}{s+2}$$

Applying the Eq. (11–23) produces $y(t)$ as

$$y(t) = \mathcal{L}^{-1}\{H(s)X(s)\} = \mathcal{L}^{-1}\left\{\frac{1}{(s+1)(s+2)}\right\}$$

$$= \mathcal{L}^{-1}\left\{\frac{2}{s+1} - \frac{2}{s+2}\right\} = 2e^{-t} - 2e^{-2t} \quad \text{for } t > 0$$

(b) Meanwhile, back in the t domain the convolution integral in Eq. (11–24) yields

$$y(t) = \int_0^t h(t-\tau)x(\tau)d\tau = \int_0^t 2e^{-(t-\tau)}e^{-2\tau}d\tau$$

$$= 2e^{-t}\int_0^t e^{\tau}e^{-2\tau}d\tau = 2e^{-t}\int_0^t e^{-\tau}d\tau$$

$$= 2e^{-t}(1 - e^{-t}) = 2\,e^{-t} - 2e^{-2t} \quad \text{for } t > 0$$

The two methods produce the same result. The difference is that the convolution integral evaluation is carried out entirely in the time domain.

EQUIVALENCE OF S-DOMAIN AND T-DOMAIN CONVOLUTION

The approach used here starts out in the s domain with $Y(s) = H(s)X(s)$ and proceeds to show that $y(t) = h(t)*x(t)$. By beginning in the s domain we are assuming that the waveforms are causal. The s-domain input-output relationship can be written as

$$Y(s) = H(s)X(s) = H(s)\underbrace{\left[\int_0^{\infty} x(\tau)e^{-s\tau}d\,\tau\right]}_{\mathcal{L}\{x(t)\}}$$

where the bracketed term is the integral definition of $\mathcal{L}\{x(t)\}$. The impulse response can be moved inside the integration since $H(s)$ does not depend on the dummy variable τ.

$$Y(s) = \int_0^{\infty} [H(s)e^{-s\tau}]x(\tau)d\,\tau \tag{11–25}$$

Using the time translation property from Chapter 9 and the integral definition of the Laplace transformation, the bracketed term can be written as

$$H(s)e^{-s\tau} = \mathcal{L}\{h(t-\tau)u(t-\tau)\}$$
$$= \int_0^\infty h(t-\tau)u(t-\tau)e^{-st}dt$$

When the last line in this result replaces the bracketed term in Eq. (11–25), we obtain

$$Y(s) = \int_0^\infty \Bigg[\underbrace{\int_0^\infty h(t-\tau)u(t-\tau)e^{-st}\,dt}_{H(s)e^{-st}}\Bigg]x(\tau)d\tau$$

Interchanging the order of integration produces

$$Y(s) = \int_0^\infty \left[\int_0^t h(t-\tau)\,x(\tau)\,d\tau\right]e^{-st}\,dt$$

The inner integration is now carried out with respect to the dummy variable τ. The upper limit on this integration need only extend to $\tau = t$ (rather than infinity) since $u(t-\tau) = 0$ for $\tau > t$. By definition, the outer integration (now with respect to t) yields the Laplace transform of the quantity inside the bracket. In other words, this equation is equivalent to the statement

$$Y(s) = \mathcal{L}\Bigg[\underbrace{\int_0^t h(t-\tau)\,x(\tau)\,d\tau}_{h(t)\,*\,x(t)}\Bigg]$$

So finally we have

$$y(t) = \mathcal{L}^{-1}\{Y(s)\} = \mathcal{L}^{-1}\{H(s)Y(s)\} = h(t) * x(t)$$

which establishes the equivalence of Eqs. (11–23) and (11–24). It is important to remember that this equivalence only applies to causal waveforms.

APPLICATIONS OF THE CONVOLUTION INTEGRAL

Given that s-domain and t-domain convolutions are equivalent, why study both methods? The best answer the authors can give at this point will have an "eat your spinach" ring to it. Suffice it to say that in subsequent courses you will encounter signals for which Laplace transforms do not exist; hence, only the t-domain method convolution is possible. Examples of such application are the noncausal waveforms used in communication systems and the discrete-time signals used in digital signal processing. We cannot treat such applications here, but rather only introduce the student to the concept of viewing convolution as a t-domain process.

EXAMPLE 11–16

A linear circuit has an impulse response $h(t) = 2e^{-t}u(t)$. Use the convolution integral to find the zero-state response for $x(t) = tu(t)$.

SOLUTION:

Direct application of Eq. (11–24) produces the following results

$$y(t) = \int_0^t 2e^{-(t-\tau)}\tau \, d\tau = 2e^{-t}\int_0^t \tau \, e^{\tau} \, d\tau$$

$$= 2\, e^{-t}[e^{\tau}(\tau - 1)]_0^t$$

$$= 2(t - 1 + e^{-t}) \quad \text{for} \quad t \geq 0$$

Although evaluation of the integral is straightforward in this case, a geometric interpretation of the process helps us proceed in more complicated situations. ■

Figure 11–22 shows a geometric interpretation of the convolution in Example 11–16. In Figures 11–22(a) and 11–22(b) the input and impulse response waveforms are plotted against the dummy variable τ. Forming $h(-\tau)$ *reflects* the impulse response across the $\tau = 0$ axis, as shown in Figure 11–22(c). Forming $h(t - \tau)$ *shifts* the reflected impulse response to the right by t seconds, as shown in Figure 11–22(d). *Multiplying* the reflected/shifted impulse response by the input produces the product $h(t - \tau) \times x(\tau)$, shown in Figure 11–22(e). The *integrating* from $\tau = 0$ to $\tau = t$ yields the area under this product, which is the value of the zero-state output at time t. At a later instant of time the reflected impulse response $h(t - \tau)$ shifts farther to the right, creating a new product $h(t - \tau) \times x(\tau)$ with a new area and a new value of $y(t)$.

Thus, the geometric interpretation of t-domain convolution involves four operations: reflecting, shifting, multiplying, and integrating. We visualize convolution as a process that reflects the impulse response across the origin and then progressively shifts it to the right as t increases. At any time t the output is the area under the product of the reflected/shifted impulse response and the input. Under this interpretation we can think of the impulse response as a weighting function. That is, when integrating the product $h(t - \tau) \times x(\tau)$, the impulse response tells us how much weight to assign to previous values of the input.

EXAMPLE 11–17

A certain circuit has an impulse response $h(t) = [2e^{-t}]u(t)$. Use the convolution integral to find the zero-state response for $x(t) = 5[u(t) - u(t-2)]$.

SOLUTION:

Evaluation of the convolution integral can be divided into the three situations shown in Figure 11–23. The situation for $t < 0$ is shown in Figure 11–23(a). For this case the reflected impulse response $h(t - \tau)$ and the input $x(\tau)$ do not overlap so the area under their product is zero. Hence $y(t) = 0$ for $t < 0$. This simply says that the zero-state response is causal when the impulse response and input are causal.

For $0 < t < 2$, the reflected impulse response and input overlap as shown in Figure 11–23(b). In this situation the area under the product $h(t - \tau) \times x(\tau)$ is found by integrating $\tau = 0$ to $\tau = t$.

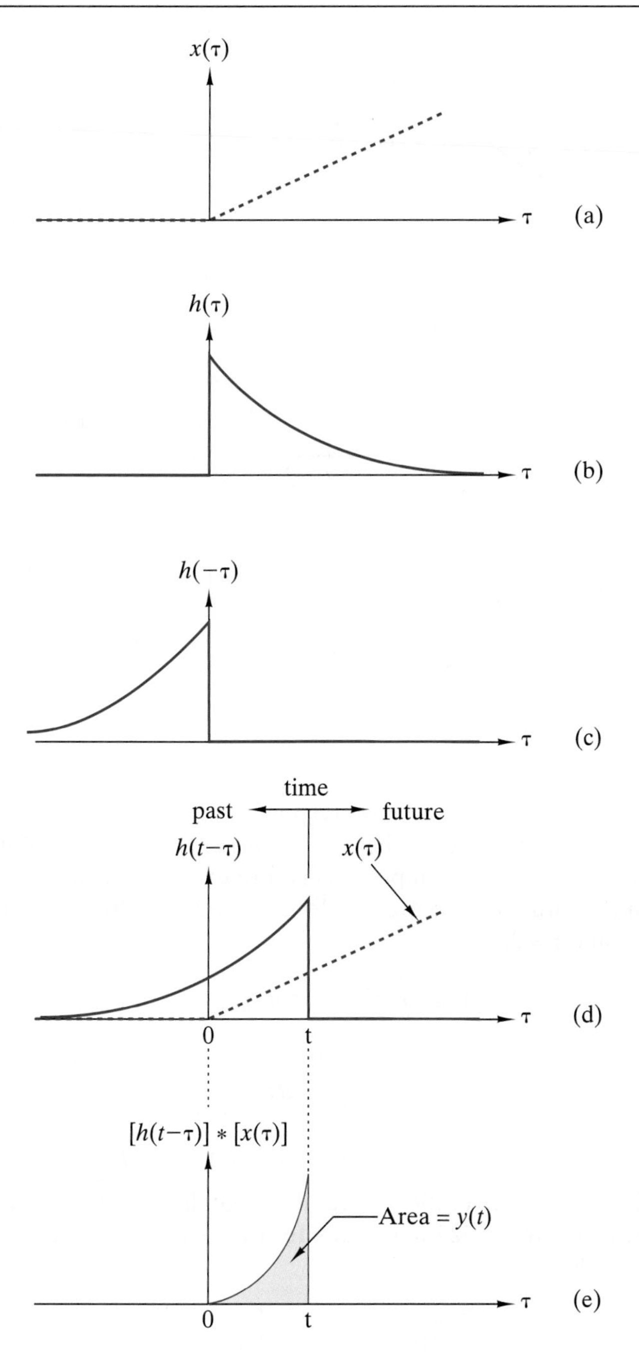

FIGURE 11–22 *Graphical interpretation of convolution.*

$$y(t) = \int_0^t [2e^{-(t-\tau)}][5]d\tau$$

$$= 10\,e^{-t}\int_0^t e^{\tau}d\tau = 10\,e^{-t}(e^{t} - 1)$$

$$= 10(1 - e^{-t}) \quad \text{for} \quad 0 < t < 2$$

FIGURE 11–23

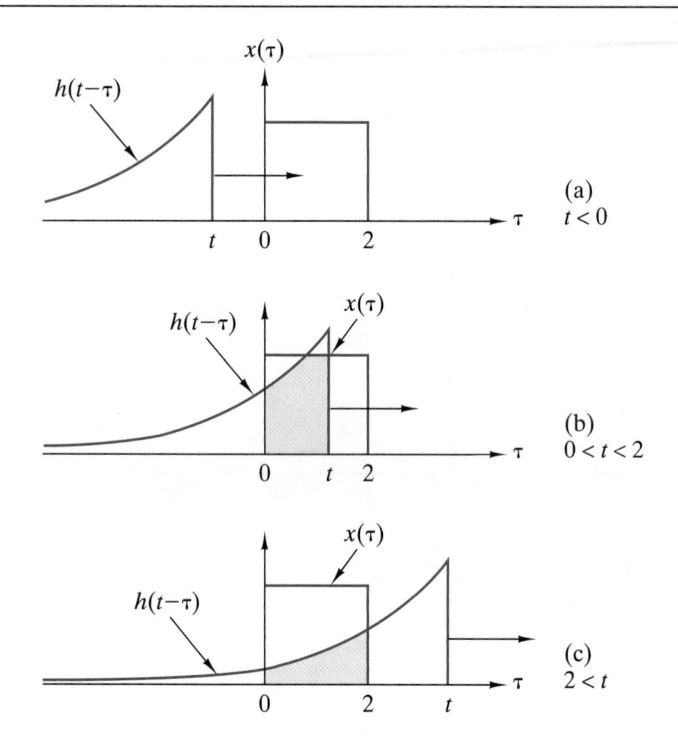

For $t > 2$, the reflected impulse response and input overlap as shown in Figure 11–23(c). In this case the product $h(t - \tau) \times x(\tau)$ is confined to the interval $0 < \tau < 2$ since the input $x(\tau)$ is zero everywhere outside of this interval. In this situation the area under the product is found by integrating from $\tau = 0$ and $\tau = 2$.

$$
\begin{aligned}
y(t) &= \int_0^2 [2e^{-(t-\tau)}][5]d\tau \\
&= 10\, e^{-t} \int_0^2 e^{\tau}\, d\tau \\
&= 10\, e^{-t}(e^2 - 1) \quad \text{for} \quad 2 \leq t
\end{aligned}
$$

Evaluation of the convolution integral was guided by the geometric interpretation in Figure 11–23 and leads to a zero-state response defined on three intervals.

$$
y(t) = \begin{vmatrix} 0 & \text{for} \quad t < 0 \\ 10(1 - e^{-t}) & \text{for} \quad 0 \leq t < 2 \\ 10e^{-t}(e^2 - 1) & \text{for} \quad 2 \leq t \end{vmatrix}
$$

■

EXAMPLE 11–18

Use time-domain and s-domain convolution to find the zero-state response when

$$h(t) = x(t) = [2e^{-t}]u(t)$$

SOLUTION:
Using the convolution integral in Eq. (11–24) produces

$$y(t) = \int_0^t 2e^{-(t-\tau)}2e^{-\tau}d\tau = 4e^{-t}\int_0^t e^{\tau}e^{-\tau}d\tau$$

$$= 4e^{-t}\int_0^t d\tau = 4te^{-t} \quad \text{for} \quad t > 0$$

Using s-domain convolution, we have $H(s) = X(s) = 2/(s+1)$; hence,

$$Y(s) = H(s) \times X(s) = \frac{4}{(s+1)^2}$$

which yields $y(t) = \mathscr{L}^{-1}\{Y(s)\} = [4te^{-t}]u(t)$. ■

Exercise 11–10

Use the convolution integral to find the zero-state response for $h(t) = 2u(t)$ and $x(t) = 5[u(t) - u(t-2)]$.

Answer:

$$y(t) = \begin{vmatrix} 0 & \text{for} \quad t < 0 \\ 10\,t & \text{for} \quad 0 \le t < 2 \\ 20 & \text{for} \quad 2 \le t \end{vmatrix}$$

11–7 NETWORK FUNCTION DESIGN

Finding and using a network function of a given circuit is an s-domain **analysis** problem. An s-domain **synthesis** problem involves finding a circuit that realizes a given network function. For linear circuits, an analysis problem always has a unique solution. In contrast, a synthesis problem may have many solutions because different circuits can have the same network function. A transfer function design problem involves synthesizing several circuits that realize a given function and evaluating the alternative designs, using criteria such as input or output impedance, cost, and power consumption.

The design process discussed here begins with a given transfer function $T_V(s)$. We partition this transfer function into a product of simpler function.

$$T_V(s) = T_{V1}(s)\,T_{V2}(s)\cdots T_{Vk}(s)$$

We then realize each of these simpler functions using basic circuit modules such as voltage dividers, inverting amplifiers, and noninverting amplifiers. The overall transfer function is then achieved by connecting the individual stages in cascade, as indicated in Figure 11–24.

Of course, this approach assumes that the chain rule applies. In other words, we must avoid loading when designing the stages in the cascade realization. This is accomplished by coordinating the input and output impedances of adjacent stages or using OP AMP voltage followers to isolate the individual stages.

Before turning to examples, we discuss the design of simple one-pole modules that serve as the building block stages in a cascade design.

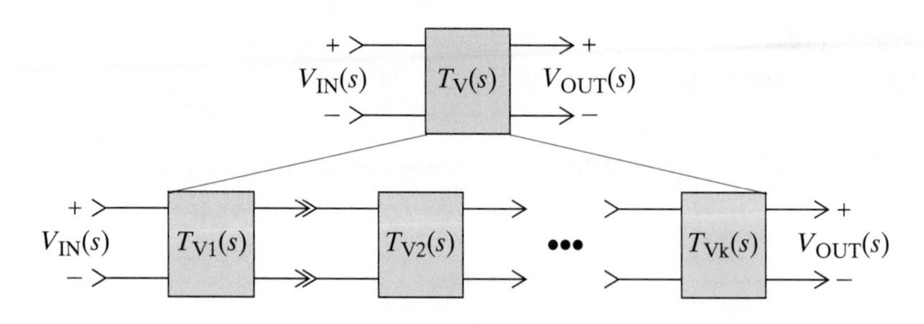

FIGURE 11–24 *Cascade connection transfer functions*

FIRST-ORDER VOLTAGE-DIVIDER CIRCUIT DESIGN

We begin our study of transfer function design by developing a voltage-divider realization of a first-order transfer function of the form $K/(s + \alpha)$. The impedances $Z_1(s)$ and $Z_2(s)$ are related to the given transfer function using the voltage-divider relationship.

$$T_V(s) = \frac{K}{s + \alpha} = \frac{Z_2(s)}{Z_1(s) + Z_2(s)} \tag{11–26}$$

To obtain a circuit realization, we must assign part of the given $T_V(s)$ to $Z_2(s)$ and the remainder to $Z_1(s)$. There are many possible realizations of $Z_1(s)$ and $Z_2(s)$ because there is no unique way to make this assignment. For example, simply equating the numerators and denominators in Eq. (11–26) yields

$$Z_2(s) = K \text{ and } Z_1(s) = s + \alpha - Z_2(s) = s + \alpha - K \tag{11–27}$$

Inspecting this result, we see that $Z_2(s)$ is realizable as a resistance ($R_2 = K\ \Omega$) and $Z_1(s)$ as an inductance ($L_1 = 1$ H) in series with a resistance [$R_1 = (\alpha - K)\ \Omega$]. The resulting circuit diagram is shown in Figure 11–25(a). For $K = \alpha$ the resistance R_1 can be replaced by a short circuit because its resistance is zero. A gain restriction $K \leq \alpha$ is necessary because a negative R_1 is not physically realizable as a single component.

An alternative synthesis approach involves factoring s out of the denominator of the given transfer function. In this case, Eq. (11–26) is rewritten in the form

$$T_V(s) = \frac{K/s}{1 + \alpha/s} = \frac{Z_2(s)}{Z_1(s) + Z_2(s)} \tag{11–28}$$

Equating numerators and denominators yields the branch impedances

$$Z_2(s) = \frac{K}{s} \text{ and } Z_1(s) = 1 + \frac{\alpha}{s} - Z_2(s) = 1 + \frac{\alpha - K}{s} \tag{11–29}$$

In this case we see that $Z_2(s)$ is realizable as a capacitance ($C_2 = 1/K$ F) and $Z_1(s)$ as a resistance ($R_1 = 1\ \Omega$) in series with a capacitance [$C_1 = 1/(\alpha - K)$ F]. The resulting circuit diagram is shown in Figure 11–25(b). For $K = \alpha$, the capacitance C_1 can be replaced by a short circuit because its capacitance is infinite. A gain restriction $K \leq \alpha$ is required to keep C_1 from being negative.

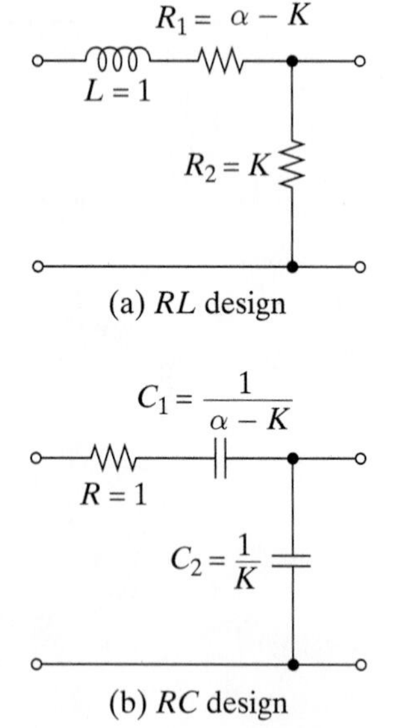

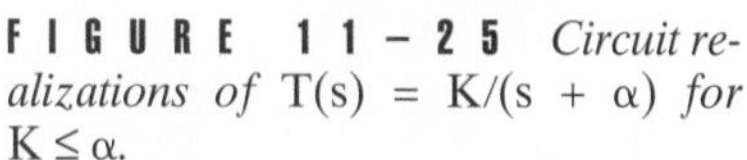
FIGURE 11–25 *Circuit realizations of* T(s) = K/(s + α) *for* K ≤ α.

As a second design example, consider a voltage-divider realization of the transfer function $Ks/(s+\alpha)$. We can find two voltage-divider realizations by writing the specified transfer function in the following two ways:

$$T(s) = \frac{Ks}{s+\alpha} = \frac{Z_2(s)}{Z_1(s)+Z_2(s)} \tag{11–30a}$$

$$T(s) = \frac{K}{1+\alpha/s} = \frac{Z_2(s)}{Z_1(s)+Z_2(s)} \tag{11–30b}$$

Equation (11–30a) uses the transfer function as given, while Eq. (11–30b) factors s out of the numerator and denominator. Equating the numerators and denominators in Eqs. (11–30a) and (11–30b) yields two possible impedance assignments:

$$\text{Using Eq. (11–30a): } Z_2 = Ks \quad \text{and} \quad Z_1 = s + \alpha - Z_2 = (1-K)s + \alpha \tag{11–31a}$$

$$\text{Using Eq. (11–30b): } Z_2 = K \quad \text{and} \quad Z_1 = 1 + \frac{\alpha}{s} - Z_2 = (1-K) + \frac{\alpha}{s} \tag{11–31b}$$

The assignment in Eq. (11–31a) yields $Z_2(s)$ as an inductance ($L_2 = K$ H) and $Z_1(s)$ as an inductance [$L_1 = (1-K)$ H] in series with a resistance ($R_1 = \alpha\ \Omega$). The assignment in Eq. (11–31b) yields $Z_2(s)$ as a resistance ($R_2 = K$) and $Z_1(s)$ as a resistance [$R_1 = (1-K)\ \Omega$] in series with a capacitance ($C_1 = 1/\alpha$ F). The two realizations are shown in Figure 11–26. Both realizations require $K \leq 1$ for the branch impedances to be realizable and both simplify when $K = 1$.

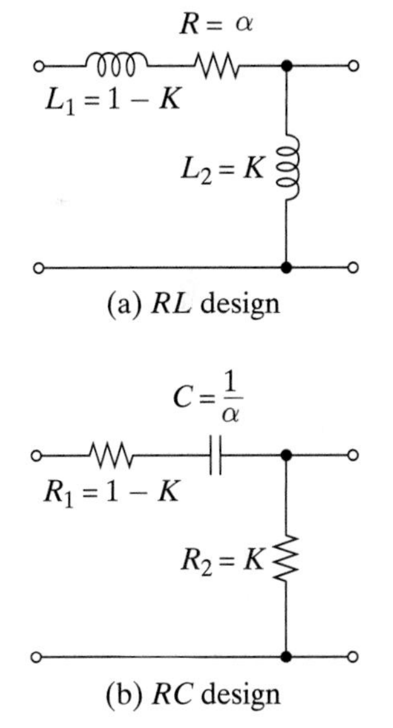

FIGURE 11–26 *Circuit realizations of* $T(s) = Ks/(s+\alpha)$ *for* $K \leq 1$.

VOLTAGE-DIVIDER AND OP AMP CASCADE CIRCUIT DESIGN

The examples in Figures 11–25 and 11–26 illustrate an important feature of voltage-divider realizations. In general, we can write a transfer function as a quotient of polynomials $T(s) = r(s)/q(s)$. A voltage-divider realization requires the impedances $Z_2(s) = r(s)$ and $Z_1(s) = q(s) - r(s)$ to be physically realizable. A voltage-divider circuit usually places limitations on the gain K. This gain limitation can be overcome by using an OP AMP circuit in cascade with divider circuit.

For example, a voltage-divider realization of the transfer function in Eq. (11–26) requires $K \leq \alpha$. When $K > \alpha$, then $T(s)$ is not realizable as a simple voltage divider, since $Z_2(s) = s + \alpha - K$ requires a negative resistance. However, the given transfer function can be written as a two-stage product:

$$T_V(s) = \frac{K}{s+\alpha} = \underbrace{\left[\frac{K}{\alpha}\right]}_{\text{first stage}}\underbrace{\left[\frac{\alpha}{s+\alpha}\right]}_{\text{second stage}}$$

When $K > \alpha$, the first stage has a positive gain greater than unity. This stage can be realized using a noninverting OP AMP circuit with a gain of $(R_1 + R_2)/R_1$. The first-stage design constraint is

$$\frac{K}{\alpha} = \frac{R_1 + R_2}{R_1}$$

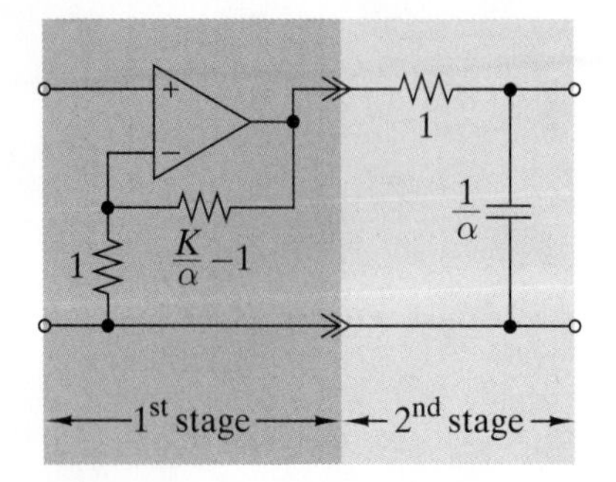

FIGURE 11–27 *Circuit realization of* T(s) = K/(s + α) *for* K > α.

Choosing $R_1 = 1\ \Omega$ requires that $R_2 = (K/\alpha) - 1$. An *RC* voltage-divider realization of the second stage is obtained by factoring an *s* out of the stage transfer function. This leads to the second-stage design constraint

$$\frac{\alpha/s}{1+\alpha/s} = \frac{Z_2(s)}{Z_1(s)+Z_2(s)}$$

Equating numerators and denominators yields $Z_2(s) = \alpha/s$ and $Z_1(s) = 1$. Figure 11–27 shows a cascade connection of a noninverting first stage and the RC divider second stage. The chain rule applies to this circuit, since the first stage has an OP AMP output. The cascade circuit in Figure 11–27 realizes the first-order transfer function $K/(s + \alpha)$ for $K > \alpha$, a gain requirement that cannot be met by the divider circuit alone.

DESIGN EXAMPLE 11–19

Design a circuit to realize the following transfer function using only resistors, capacitors, and OP AMPs:

$$T_V(s) = \frac{3000s}{(s+1000)(s+4000)}$$

SOLUTION:

The given transfer function can be written as a three-stage product.

$$T_V(s) = \underbrace{\left[\frac{K_1}{s+1000}\right]}_{\substack{\text{first}\\ \text{stage}}} \underbrace{[K_2]}_{\substack{\text{second}\\ \text{stage}}} \underbrace{\left[\frac{K_3 s}{s+4000}\right]}_{\substack{\text{third}\\ \text{stage}}}$$

where the stage gains K_1, K_2, and K_3 have yet to be selected. Factoring *s* out of the denominator of the first-stage transfer function leads to an *RC* divider realization:

$$\frac{K_1/s}{1+1000/s} = \frac{Z_2(s)}{Z_1(s)+Z_2(s)}$$

Equating numerators and denominators yields

$$Z_2(s) = K_1/s \quad \text{and} \quad Z_1(s) = 1 + (1000 - K_1)/s$$

The first stage $Z_1(s)$ is simpler when we select $K_1 = 1000$. Factoring *s* out of the denominator of the third-stage transfer function leads to an *RC* divider realization:

$$\frac{K_3}{1+4000/s} = \frac{Z_2(s)}{Z_1(s)+Z_2(s)}$$

Equating numerators and denominators yields

$$Z_2(s) = K_3 \text{ and } Z_1(s) = 1 - K_3 + 4000/s$$

The third stage $Z_1(s)$ is simpler when we select $K_3 = 1$. The stage gains must meet the constraint $K_1 \times K_2 \times K_3 = 3000$ since the overall gain of the given transfer function is 3000. We have selected $K_1 = 1000$ and $K_3 = 1$, which requires $K_2 = 3$. The second stage must have a positive gain greater than 1 and can be realized using a noninverting amplifier with $K_2 = (R_1 + R_2)/R_1 = 3$. Selecting $R_1 = 1\ \Omega$ requires that $R_2 = 2\ \Omega$.

Figure 11–28 shows the three stages connected in cascade. The chain rule applies to this cascade connection because the OP AMP in the second stage isolates the *RC* voltage-divider circuits in the first and third stages. The circuit in Figure 11–28 realizes the given transfer function but is not a realistic design because the values of resistance and capacitance are impractical. For this reason we call this circuit a **prototype** design. We will shortly discuss how to scale a prototype to obtain practical element values. ■

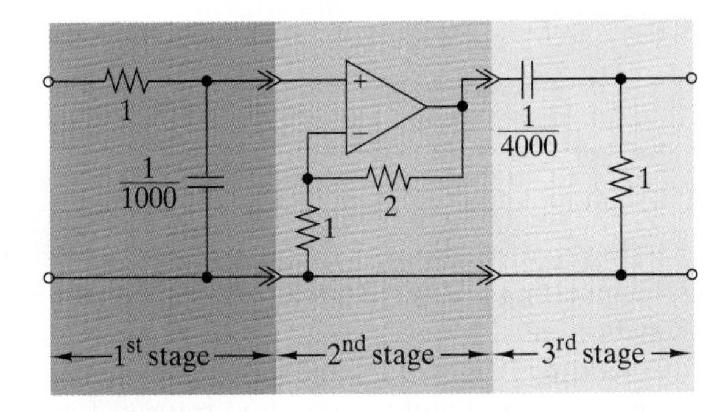

FIGURE 11–28

INVERTING OP AMP CIRCUIT DESIGN

The inverting OP AMP circuit places fewer restrictions on the form of the desired transfer function than does the basic voltage divider. To illustrate this, we will develop two inverting OP AMP designs for a general first-order transfer function of the form

$$T_V(s) = -K\frac{s+\gamma}{s+\alpha}$$

The general transfer function of the inverting OP AMP circuit is $-Z_2(s)/Z_1(s)$, which leads to the general design constraint

$$-K\frac{s+\gamma}{s+\alpha} = -\frac{Z_2(s)}{Z_1(s)} \quad (11\text{–}32)$$

The first design is obtained by equating the numerators and denominators in Eq. (11–32) to obtain the OP AMP circuit impedances as $Z_2(s) = Ks + K\gamma$ and $Z_1(s) = s + \alpha$. Both of these impedances are of the form $Ls + R$ and can be realized by an inductance in series with a resistance, leading to the design realization in Figure 11–29(a).

A second inverting OP AMP realization is obtained by equating $Z_2(s)$ in Eq. (11–32) to the reciprocal of the denominator and equating $Z_1(s)$ to the reciprocal of the numerator. This assignment yields the impedances $Z_1(s) = 1/(Ks + K\gamma)$ and $Z_2(s) = 1/(s + \alpha)$. Both of these impedances are of the form $1/(Cs + G)$, where Cs is the admittance of a capacitor and G is the admittance of a resistor. Both impedances can be realized by a capacitance in parallel with a resistance. These impedance identifications produce the *RC* circuit in Figure 11–29(b).

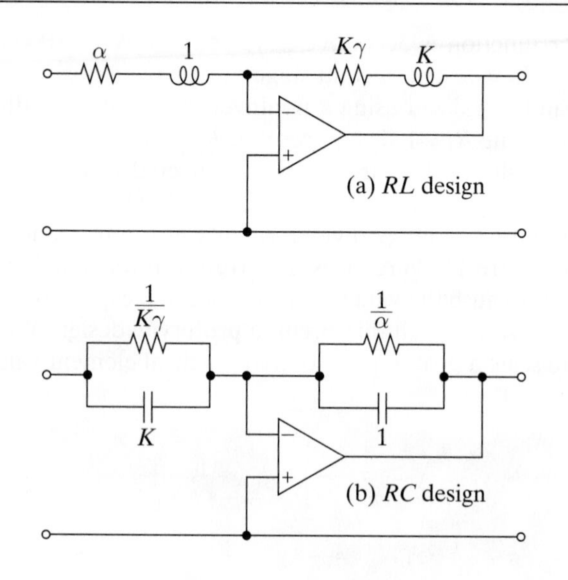

FIGURE 11–29 *Inverting OP AMP circuit realizations of* $T(s) = -K(s+\gamma)/(s+\alpha)$.

Because it has fewer restrictions, it is often easier to realize transfer functions using the inverting OP AMP circuit. To use inverting circuits, the given transfer function must require an inversion or be realized using an even number of inverting stages. In some cases, the sign in front of the transfer function is immaterial and the required transfer function is specified as $\pm T_V(s)$. *Caution:* The input impedance of an inverting OP AMP circuit may load the source circuit.

DESIGN EXAMPLE 11–20

Design a circuit to realize the transfer function given in Example 11–19 using inverting OP AMP circuits.

SOLUTION:

The given transfer function can be expressed as the product of two inverting transfer functions:

$$T_V(s) = \frac{3000s}{(s+1000)\,(s+4000)} = \underbrace{\left[-\frac{K_1}{s+1000}\right]}_{\text{first stage}}\underbrace{\left[-\frac{K_2 s}{s+4000}\right]}_{\text{second stage}}$$

where the stage gains K_1 and K_2 have yet to be selected. The first stage can be realized in an inverting OP AMP circuit since

$$-\frac{K_1}{s+1000} = -\frac{K_1/1000}{1+s/1000} = -\frac{Z_2(s)}{Z_1(s)}$$

Equating the $Z_2(s)$ to the reciprocal of the denominator and $Z_1(s)$ to the reciprocal of the numerator yields

$$Z_2 = \frac{1}{1+s/1000} \quad \text{and} \quad Z_1 = 1000/K_1$$

The impedance $Z_2(s)$ is realizable as a capacitance ($C_2 = 1/1000$ F) in parallel with a resistance ($R_2 = 1\ \Omega$) and $Z_1(s)$ as a resistance ($R_1 = 1000/K_1\ \Omega$). We select $K_1 = 1000$ so that the two resistances in the first stage are equal. Since the overall gain requires $K_1 \times K_2 = 3000$, this means that $K_2 = 3$. The second-stage transfer function can also be produced using an inverting OP AMP circuit:

$$-\frac{3s}{s+4000} = -\frac{3}{1+4000/s} = -\frac{Z_2(s)}{Z_1(s)}$$

Equating numerators and denominators yields $Z_2(s) = R_2 = 3$ and $Z_1(s) = R_1 + 1/C_1 s = 1 + 4000/s$.

Figure 11–30 shows the cascade connection of the *RC* OP AMP circuits that realize each stage. The overall transfer function is noninverting because the cascade uses an even number of inverting stages. The chain rule applies here since the first stage has an OP AMP output. The circuit in Figure 11–30 is a prototype design because the values of resistance and capacitance are impractical. ■

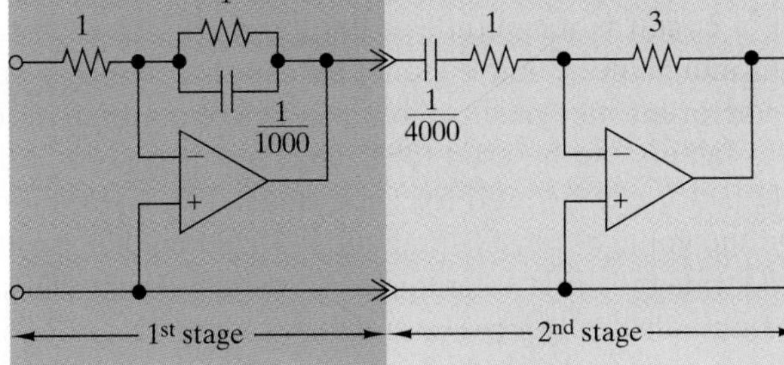

FIGURE 11–30

MAGNITUDE SCALING

The circuits obtained in Examples 11–19 and 11–20 are called prototype designs because the element values are outside of practical ranges. The allowable ranges depend on the fabrication technology used to construct the circuits. For example, monolithic integrated circuit technology limits capacitances to a few hundred picofarads. An OP AMP circuit should have a feedback resistance greater than around 10 kΩ to keep the output current demand within the capabilities of general-purpose OP AMP devices. Other technologies and applications place different constraints on element values.

There are no hard and fast rules here, but, roughly speaking, a circuit is probably realizable by some means if its passive element values fall in the following ranges:

Capacitors: 1 pF to 100 μF

Inductors: 10 μH to 100 mH

Resistors: 10 Ω to 100 MΩ.

The important idea here is that circuit designs like Figure 11–30 are impractical because 1-Ω resistors and 1-mF capacitors are unrealistic values.

It is often possible to scale the magnitude of circuit impedances so that the element values fall into practical ranges. The key is to scale the element values in a way that does not change the transfer function of the circuit. Multiplying the numerator and denominator of the transfer function of a voltage-divider circuit by a scale factor k_m yields

$$T_V(s) = \frac{k_m}{k_m}\frac{Z_2(s)}{Z_1(s) + Z_2(s)} = \frac{k_m Z_2(s)}{k_m Z_1(s) + k_m Z_2(s)} \tag{11–33}$$

Clearly, this modification does not change the transfer function but scales each impedance by a factor of k_m and changes the element values in the following way:

$$R_{after} = k_m R_{before} \quad L_{after} = k_m L_{before} \quad C_{after} = \frac{C_{before}}{k_m} \tag{11–34}$$

Equation (11–34) was derived using the transfer function of a voltage-divider circuit. It is easy to show that we would reach the same conclusion if we had used the transfer functions of inverting or noninverting OP AMP circuits.

In general, a circuit is magnitude scaled by multiplying all resistances, multiplying all inductances, and dividing all capacitances by a scale factor k_m. The scale factor must be positive, but can be greater than or less than 1. Different scale factors can be used for each stage of a cascade design, but only one scale factor can be used for each stage. These scaling operations do not change the voltage transfer function realized by the circuit.

Our design strategy is first to create a prototype circuit whose element values may be unrealistically large or small. Applying magnitude scaling to the prototype produces a design with practical element values. Sometimes there may be no scale factor that brings the prototype element values into a practical range. When this happens, we must seek alternative realizations because the scaling process is telling us that the prototype is not a viable candidate.

EXAMPLE 11–21

Magnitude scale the circuit in Figure 11–30 so all resistances are at least 10 kΩ and all capacitances are less than 1 μF.

SOLUTION:

The resistance constraint requires $k_m R \geq 10^4$ Ω. The smallest resistance in the prototype circuit is 1 Ω; therefore, the resistance constraint requires $k_m \geq 10^4$. The capacitance constraint requires $C/k_m \leq 10^{-6}$ F. The largest capacitance in the prototype is 10^{-3} F; therefore, the capacitance constraint requires $k_m \geq 10^3$. The resistance condition on k_m dominates the two constraints. Selecting $k_m = 10^4$ produces the scaled design in Figure 11–31. This circuit realizes the same transfer function as the prototype in Figure 11–30 but uses practical element values. ■

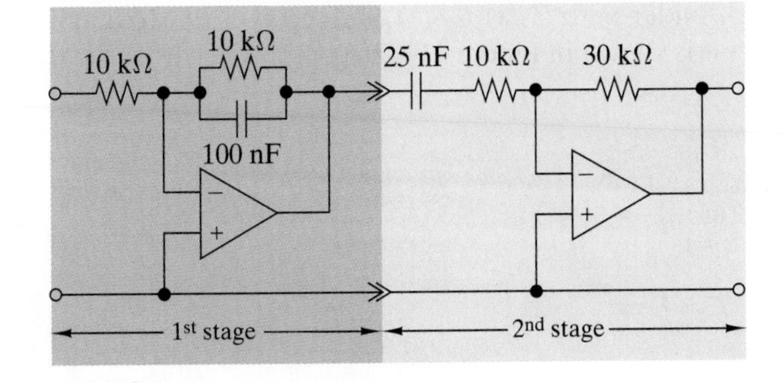

FIGURE 11–31

Exercise 11–11

Select a magnitude scale factor for each stage in Figure 11–28 so that both capacitances are 10 nF and all resistances are greater than 10 kΩ.

Answer: $k_m = 10^5$ for the first stage; $k_m = 10^4$ for the second stage; $k_m = 0.25 \times 10^5$ for the third stage

SECOND-ORDER CIRCUIT DESIGN

An *RLC* voltage divider can also be used to realize second-order transfer functions. For example, the transfer function

$$T_V(s) = \frac{K}{s^2 + 2\zeta\omega_0 s + \omega_0^2}$$

can be realized by factoring *s* out of the denominator and equating the result to the voltage-divider input-output relationship:

$$T_V(s) = \frac{K/s}{s + 2\zeta\omega_0 + \omega_0^2/s} = \frac{Z_2(s)}{Z_1(s) + Z_2(s)}$$

Equating numerators and denominators yields

$$Z_2(s) = \frac{K}{s} \text{ and } Z_1(s) = s + 2\zeta\omega_0 + \frac{\omega_0^2 - K}{s}$$

The impedance $Z_2(s)$ is realizable as a capacitance ($C_2 = 1/K$ F) and $Z_1(s)$ as a series connection of an inductance ($L_1 = 1$ H), resistance ($R_1 = 2\zeta\omega_0$ Ω), and capacitance [$C_1 = 1/(\omega_0^2 - K)$ F]. The resulting voltage-divider circuit is shown in Figure 11–32(a). The impedances in this circuit are physically realizable when $K \leq \omega_0^2$. Note that the resistance controls the damping ratio ζ because it is the element that dissipates energy in the circuit.

When $K > \omega_0^2$, we can partition the transfer function into a two-stage cascade of the form

$$T_V(s) = \underbrace{\left[\frac{K}{\omega_0^2}\right]}_{\text{first stage}} \underbrace{\left[\frac{\omega_0^2/s}{s + 2\zeta\omega_0 + \omega_0^2/s}\right]}_{\text{second stage}}$$

The first stage requires a positive gain greater than unity and can be realized using a noninverting OP AMP circuit. The second stage can be realized as a voltage divider with $Z_2(s) = \omega_0^2/s$ and $Z_1(s) = s + 2\zeta\omega_0$. The resulting cascade circuit is shown in Figure 11–32(b).

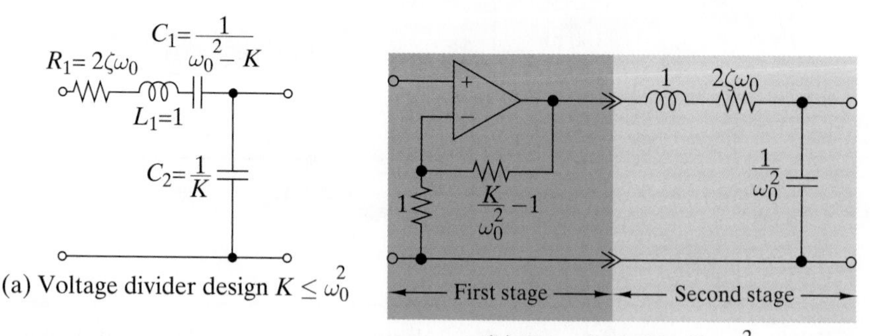

FIGURE 11–32 *Second-order circuit realizations.*

DESIGN EXAMPLE 11–22

Find a second-order realization of the transfer function given in Example 11–19.

SOLUTION:

The given transfer function can be written as

$$T_V(s) = \frac{3000s}{(s + 1000)(s + 4000)} = \frac{3000s}{s^2 + 5000s + 4 \times 10^6}$$

Factoring s out of the denominator and equating the result to the transfer function of a voltage divider gives

$$\frac{3000}{s + 5000 + 4 \times 10^6/s} = \frac{Z_2(s)}{Z_1(s) + Z_2(s)}$$

Equating the numerators and denominators yields

$$Z_2(s) = 3000 \quad \text{and} \quad Z_1(s) = s + 2000 + 4 \times 10^6/s$$

Both of these impedances are realizable, so a single-stage voltage-divider design is possible. The prototype impedance $Z_1(s)$ requires a 1-H inductor, which is a bit large. A more practical value is obtained using a scale factor of $k_m = 0.1$. The resulting scaled voltage divider circuit is shown in Figure 11–33. ■

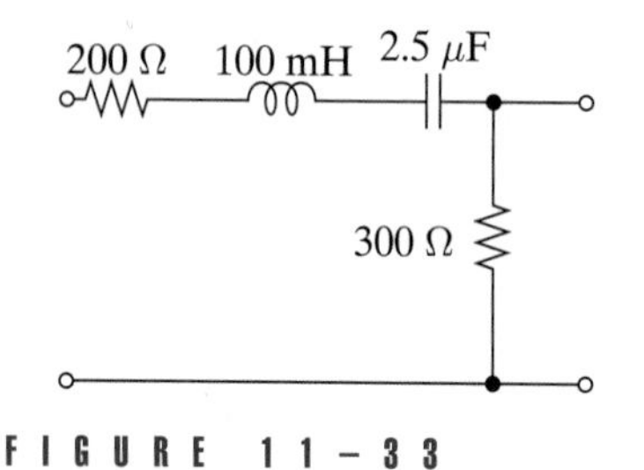

FIGURE 11–33

DESIGN EVALUATION SUMMARY

Examples 11–19, 11–20, and 11–22 show three different ways to realize the transfer function

$$T_V(s) = \frac{3000\,s}{(s + 1000)(s + 4000)}$$

This illustrates that a design requirement can have many solutions. Selecting the best design from among the alternatives involves additional criteria such as element count, power requirements, and output loading effects.

The element counts for each design are shown in the table below. On a pure element count basis, the *RLC* divider in Figure 11–33 is the best design. However, inductors have some serious drawbacks. They are heavy and lossy in low-frequency applications and are not easily fabricated in integrated circuit form. Fortunately inductors are not essential to transfer function design, as shown by the two *RC* OP AMP designs.

			Number of			
Example	Figure	Description	R	L	C	OP AMP
11–19	11–28	RC voltage-divider cascade	4	0	2	1
11–20	11–30	RC inverting cascade	4	0	2	2
11–22	11–33	RLC voltage divider	2	1	1	0

Power Requirements: The two *RC* OP AMP designs require external dc power supplies. The voltage divider cascade in Figure 11–28 requires less power since it uses only one OP AMP, compared with the two-OP-AMP inverting cascade. Thus, power requirements would favor the one-OP-AMP circuit over the two-OP-AMP circuit.

Output Loading: The output impedance of the design is important if the circuit must drive a finite load of, say, 1 kΩ. The resulting loading effects could defeat the basic purpose of the circuit by changing its transfer function. Output loading considerations favor the inverting cascade in Figure 11–30 because it has an OP AMP output that has zero output impedance.

A design problem involves more than simply finding a prototype that realizes a given transfer function. In general, the first step in a design problem involves determining an acceptable transfer function, one that meets performance requirements such as the characteristics of the step or frequency response. In other words, we must first design the transfer function and then design several circuits that realize the transfer function. To deal with transfer function design we must understand how performance characteristics are related to transfer functions. The next two chapters provide some background on this issue.

D E DESIGN AND EVALUATION EXAMPLE 11–23

Given the step response $g(t) = \pm[1 + 4e^{-500t}]u(t)$,

(a) Find the transfer function $T(s)$.
(b) Design two *RC* OP AMP circuits that realize the $T(s)$ found in part (a).
(c) Evaluate the two designs on the basis of element count, input impedance, output impedance.

SOLUTION:

(a) The transform of the step response is

$$G(s) = \pm\mathscr{L}\{[1 + 4e^{-500t}]u(t)\} = \pm\left[\frac{1}{s} + \frac{4}{s+500}\right] = \pm\frac{5s+500}{s(s+500)}$$

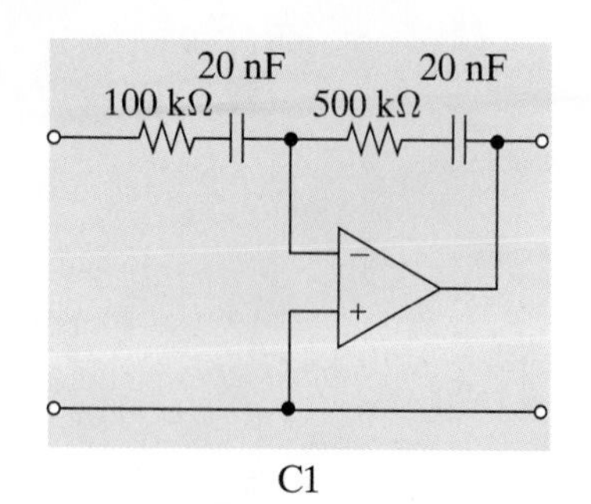

C1

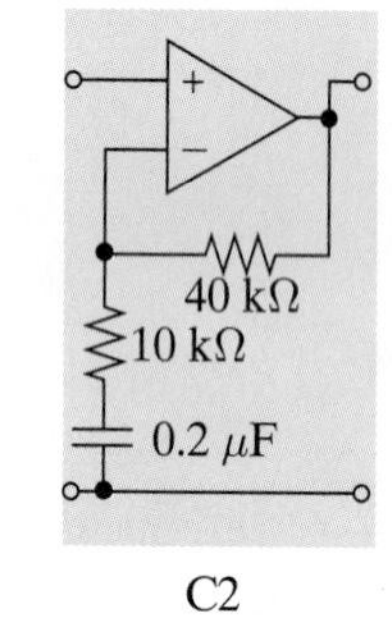

C2

FIGURE 11–34

and the required transfer function is

$$T(s) = H(s) = sG(s) = \pm\frac{5s + 500}{s + 500}$$

(b) The first design uses an inverting OP AMP configuration. Using the minus sign on the transfer function $T(s)$ and factoring an s out of the numerator and denominator yield

$$-T(s) = -\frac{5 + 500/s}{1 + 500/s} = -\frac{Z_2(s)}{Z_1(s)}$$

Equating numerators and denominators yields $Z_2(s) = 5 + 500/s$ and $Z_1(s) = 1 + 500/s$. The impedance $Z_2(s)$ is realizable as a resistance ($R_2 = 5\ \Omega$) in series with a capacitance ($C_2 = 1/500$ F) and $Z_1(s)$ as a resistance ($R_1 = 1\ \Omega$) in series with a capacitance ($C_1 = 1/500$ F). Using a magnitude scale factor $k_m = 10^5$ produces circuit C1 in Figure 11–34.

The second design uses a noninverting OP AMP configuration. Using the plus sign on the transfer function $T(s)$ and factoring an s out of the numerator and denominator yield

$$T(s) = \frac{5 + 500/s}{1 + 500/s} = \frac{Z_1(s) + Z_2(s)}{Z_1(s)}$$

Equating numerators and denominators yields

$$Z_1(s) = 1 + \frac{500}{s} \quad \text{and} \quad Z_2(s) = 5 + \frac{500}{s} - Z_1(s) = 4$$

The impedance $Z_1(s)$ is realizable as a resistance ($R_1 = 1\ \Omega$) in series with a capacitance ($C_1 = 1/500$ F) and $Z_2(s)$ as a resistance ($R_2 = 4\ \Omega$). Using a scale factor of $k_m = 10^4$ produces circuit C2 in Figure 11–34.

(c) Circuit C1 uses one more capacitor than circuit C2. The OP AMP output on both circuits means that they each have almost zero output impedance. The input impedance to circuit C2 is very large, because its input is the noninverting input of the OP AMP. The input impedance of circuit C1 is $Z_1(s) = k_m(1 + 500/s)$; hence, the scale factor must be selected to avoid loading the source circuit. The final design for circuit C1 in Figure 11–34 uses $k_m = 10^5$, which means that $|Z_1| > 100\ \text{k}\Omega$, which should be high enough to avoid loading the source circuit. ■

D DESIGN EXAMPLE 11–24

Verify that circuit C2 in Figure 11–34 meets its design requirements.

SOLUTION:

One of the important uses of computer-aided analysis is to verify that a proposed design meets the performance specifications. The circuit C2 in Figure 11–34 is designed to produce a specified step response

$$g(t) = [1 + 4e^{-500t}]u(t) \quad \text{V}$$

This response jumps from zero to 5 V at $t = 0$ and then decays exponentially to 1 V at large t. The time constant of the exponential is $1/500 = 2$ ms, which means that the final value is effectively reached after about five time constants, or 10 ms.

Figure 11–35 shows the circuit diagram drawn in the Electronics Work-Bench circuit window. From the **Analysis** menu we bring up the **Transient Analysis** dialog box where we set the initial conditions to zero (to get a zero-state response), set the start time at zero, the end time at 10 ms (five time constants), and set node 4 (to OP AMP output) as the node for analysis. The step response calculated by Electronics WorkBench is shown at the bottom of Figure 11–35. The response jumps to 5 V at $t = 0$ and then exponentially decays to 1 V after about 10 ms. The two cursors are set at $x_1 =$ 2.0079 ms (about one time constant) where simulation predicts $y_1 = 2.4657$ and $x_2 = 4.0079$ ms (about two time constants) where simulation predicts y_2 = 1.5392. The theoretical values of $g(t)$ at these points are

$$g(t_1) = 1 + 4e^{-1} = 2.4715$$

$$g(t_2) = 1 + 4e^{-2} = 1.5413$$

Thus, theory and simulation agree to three significant figures, confirming that circuit C2 meets the design requirement. ■

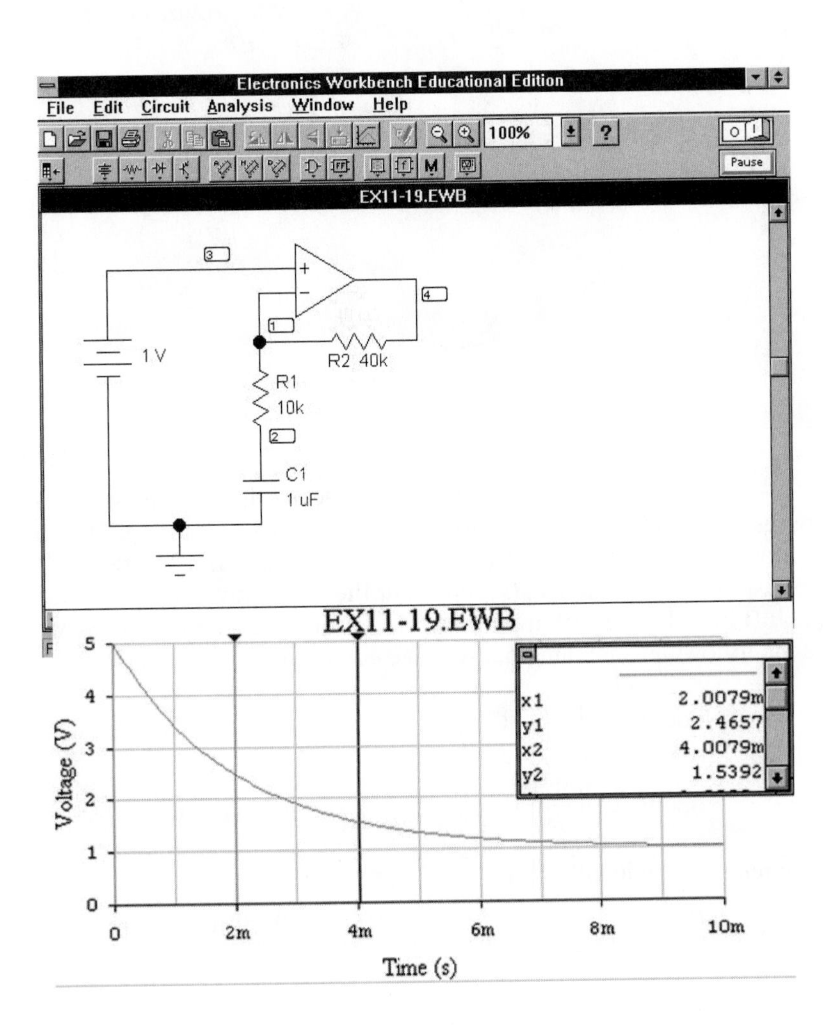

FIGURE 11–35

APPLICATION EXAMPLE 11–25

The operation of a digital system is coordinated and controlled by a periodic waveform called a clock. The *clock waveform* provides a standard timing reference to maintain synchronization between signal processing results that are generated asynchronously. Because of differences in digital circuit delays, there must be agreed-upon instants of time at which circuit outputs can be treated as valid inputs to other circuits.

Figure 11–36 shows a section of the clock distribution network in an integrated circuit. In this network the clock waveform is generated at one point and distributed to other on-chip locations by interconnections that can be modeled as lumped resistors and capacitors. Clock distribution problems arise when the *RC* circuit delays at different locations are not the same. This delay dispersion is called *clock skew*, defined as the time difference between a clock edge at one location and the corresponding edge at another location.

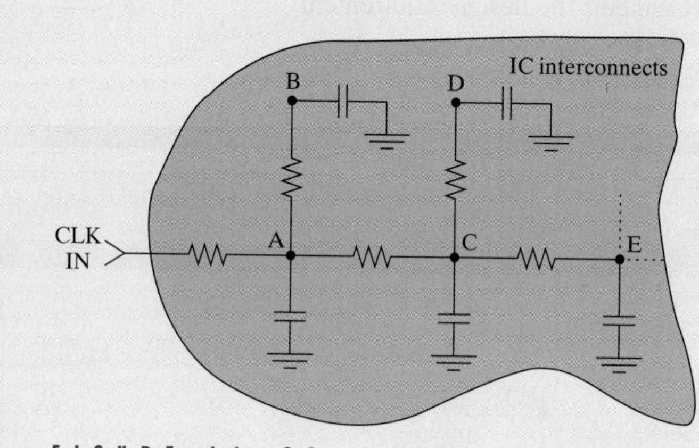

FIGURE 11–36 Clock distribution network.

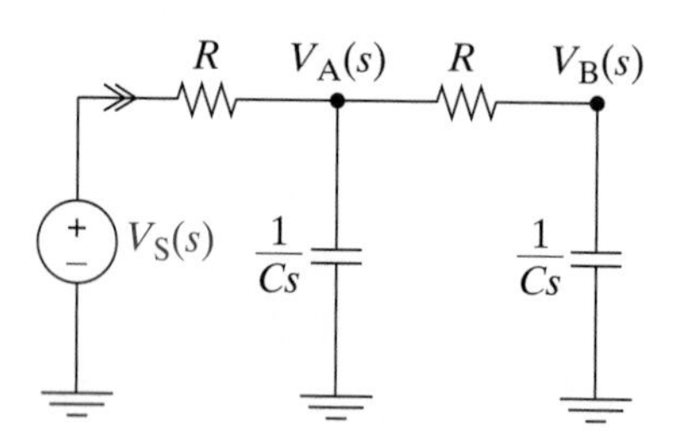

FIGURE 11–37 Two-stage *RC* circuit model.

To qualitatively calculate a clock skew, we will find the step responses in the *RC* circuit in Figure 11–37. The input $V_S(s)$ is a unit step function which simulates the leading edge of a clock pulse. The resulting step responses $V_A(s)$ and $V_B(s)$ represent the clock waveforms at points A and B in a clock distribution network. To find the step responses, we use the following *s*-domain node-voltage equations.

$$\text{Node A:} \quad \left(\frac{2}{R} + Cs\right)V_A(s) - \left(\frac{1}{R}\right)V_B(s) = \frac{V_S(s)}{R}$$

$$\text{Node B:} \quad -\left(\frac{1}{R}\right)V_A(s) + \left(\frac{1}{R} + Cs\right)V_B(s) = 0$$

The circuit determinant is

$$\Delta(s) = \frac{(RCs)^2 + 3(RCs) + 1}{R^2} = \frac{(RCs + 0.382)(RCs + 2.618)}{R^2}$$

which indicates that the circuit has simple poles at $s = -0.382/RC$ and $s = -2.618/RC$. Using the circuit determinant and a unit step input, we can easily solve the node equations for $V_A(s)$ and $V_B(s)$ as

$$V_A(s) = \frac{RCs + 1}{s(RCs + 0.382)(RCs + 2.618)}$$
$$= \frac{1}{s} - \frac{0.7235}{s + 0.382/RC} - \frac{0.2764}{s + 2.618/RC}$$

$$V_B(s) = \frac{1}{s(RCs + 0.382)(RCs + 2.618)}$$
$$= \frac{1}{s} - \frac{1.171}{s + 0.382/RC} + \frac{0.1710}{s + 2.618/RC}$$

From which we obtain the time-domain step responses as

$$v_A(t) = 1 - 0.7235\,e^{-0.382t/RC} - 0.2764\,e^{-2.618t/RC}$$
$$v_B(t) = 1 - 1.171\,e^{-0.382t/RC} + 0.1710\,e^{-2.618t/RC} \quad \text{for} \quad t > 0$$

These two responses are plotted in Figure 11–38. For a unit step input, both responses have a final value of unity. Using the definition of step response *delay time* given in Example 11–12 (time required to reach 50% of the final value), we see that

$$T_{D_A} = 1.06/RC \quad \text{and} \quad T_{D_B} = 2.23/RC$$

The delay time skew is

$$\text{Delay Skew} = T_{D_B} - T_{D_A} = 1.17/RC$$

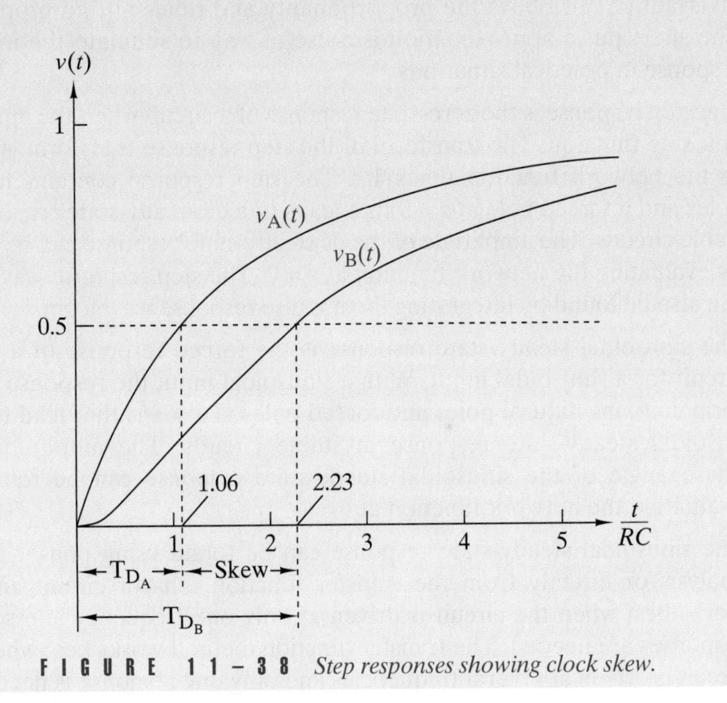

FIGURE 11–38 *Step responses showing clock skew.*

The clock distribution problem is not that the RC elements representing the interconnects produce time delay, but that delays are not all the same. Ideally, digital devices at different locations should operate on their respective digital inputs at exactly the same instant of time. Erroneous results may occur when the clock pulse defining that instant does not arrive at all locations at the same time. Minimizing clock skew is one of the major constraints on the design of the clock distribution network in large-scale integrated circuits.

SUMMARY

- A network function is defined as the ratio of the zero-state response transform to the input transform. Network functions are either driving-point functions or transfer functions. Network functions are rational functions of s with real coefficients whose complex poles and zeros occur in conjugate pairs.
- Network functions for simple circuits like voltage and current dividers and inverting and noninverting OP AMPs are easy to derive and often useful. Node-voltage or mesh-current methods are used to find the network functions for more complicated circuits. The transfer function of a cascade connection obeys the chain rule when each stage does not load the preceding stage in the cascade.
- The impulse response is the zero-state response of a circuit for a unit impulse input. The transform of the impulse response is equal to the network function. The impulse response contains only natural poles and decays to zero in stable circuits. The impulse response of a linear, time-invariant circuit obeys the proportionality and time-shifting properties. The short pulse approximation is a useful way to simulate the impulse response in practical situations.
- The step response is the zero-state response of a circuit when the input is a unit step function. The transform of the step response transform is equal to the network function times $1/s$. The step response contains natural poles and a forced pole at $s = 0$ that leads to a dc steady-state response in stable circuits. The amplitude of the dc steady-state response can be found by evaluating the network function at $s = 0$. The step response waveform can also be found by integrating the impulse response waveform.
- The sinusoidal steady-state response is the forced response of a stable circuit for a sinusoidal input. With a sinusoidal input the response transform contains natural poles and forced poles at $s = \pm j\omega$ that lead to a sinusoidal steady-state response in stable circuits. The amplitude and phase angle of the sinusoidal steady-state response can be found by evaluating the network function at $s = j\omega$.
- The sinusoidal steady-state response can be found using phasor circuit analysis or directly from the transfer function. Phasor circuit analysis works best when the circuit is driven at only one frequency and several responses are needed. The transfer function method works best when the circuit is driven at several frequencies and only one response is needed.

- The convolution integral is a t-domain method relating the impulse response $h(t)$ and input waveform $x(t)$ to the zero-state response $y(t)$. Symbolically the convolution integral is represented by $y(t) = h(t)*x(t)$. Time-domain convolution and s-domain convolution are equivalent; that is, $y(t) = h(t)*x(t) = \mathcal{L}^{-1}\{H(s)X(s)\}$. The geometric interpretation of t-domain convolution involves four operations: reflecting, shifting, multiplying, and integrating.
- First- and second-order transfer functions can be designed using voltage dividers and inverting or noninverting OP AMP circuits. Higher-order transfer functions can be realized using a cascade connection of first- and second-order circuits. Prototype designs usually require magnitude scaling to obtain practical element values.

PROBLEMS

OBJECTIVE 11–1 NETWORK FUNCTIONS (SECTS. 11–1, 11–2)

(a) Given a linear circuit, find specified network functions and locate their poles and zeros.
(b) Given a linear circuit, select the element values to produce specified poles and zeros.
See Examples 11–2, 11–3, 11–4, 11–5, 11–6, 11–7 and Exercises 11–1, 11–2

11–1 Find the driving-point impedance seen by the voltage source in Figure P11–1 and the voltage transfer function $T_V(s) = V_2(s)/V_1(s)$.

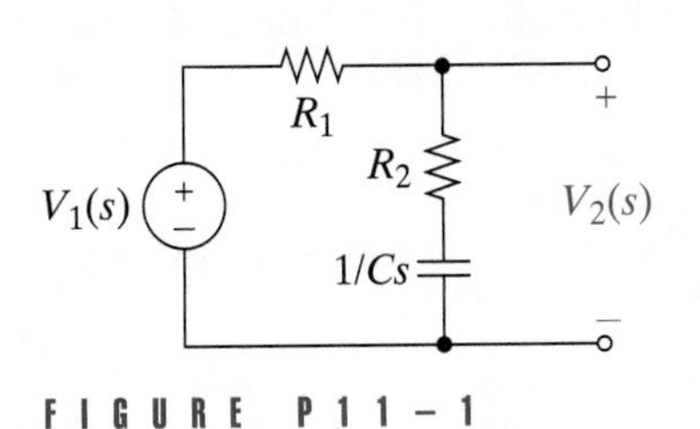

FIGURE P11–1

11–2 Find the driving-point impedance seen by the voltage source in Figure P11–2 and the voltage transfer function $T_V(s) = V_2(s)/V_1(s)$.

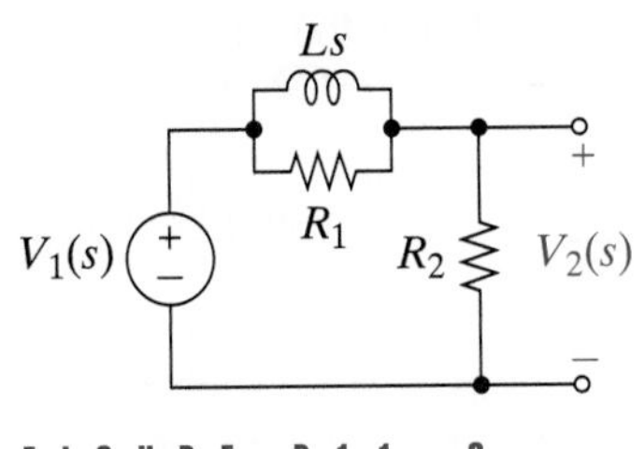

FIGURE P11–2

11–3 Find the driving-point impedance seen by the voltage source in Figure P11–3 and the voltage transfer function $T_V(s) = V_2(s)/V_1(s)$.

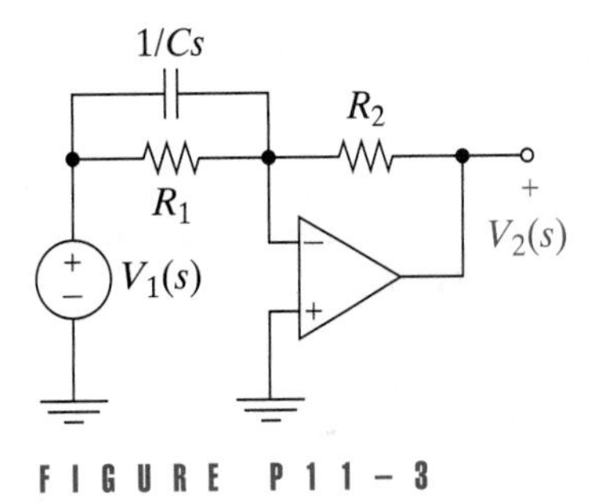

FIGURE P11–3

11–4 Find the driving-point impedance seen by the voltage source in Figure P11–4 and the voltage transfer function $T_V(s) = V_2(s)/V_1(s)$.

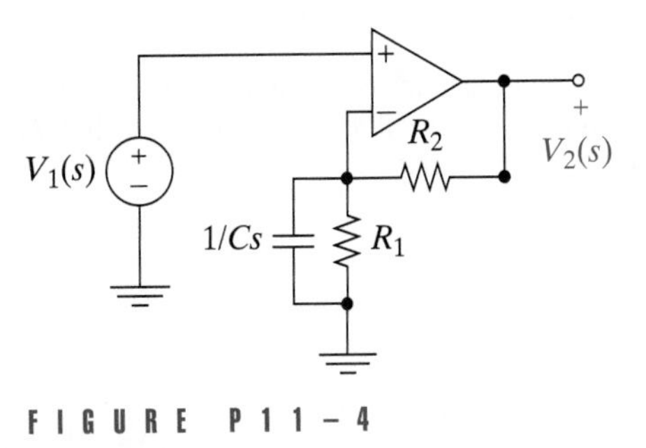

FIGURE P11–4

11–5 Find the voltage transfer function $T_V(s) = V_2(s)/V_1(s)$ in Figure P11–5.

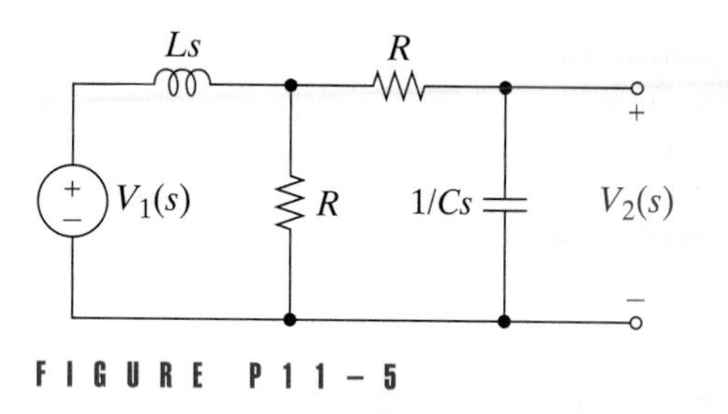

FIGURE P11–5

11–6 Find the driving-point impedance seen by the voltage source in Figure P11–6 and the transfer impedance $T_Z(s) = V_2(s)/I_1(s)$.

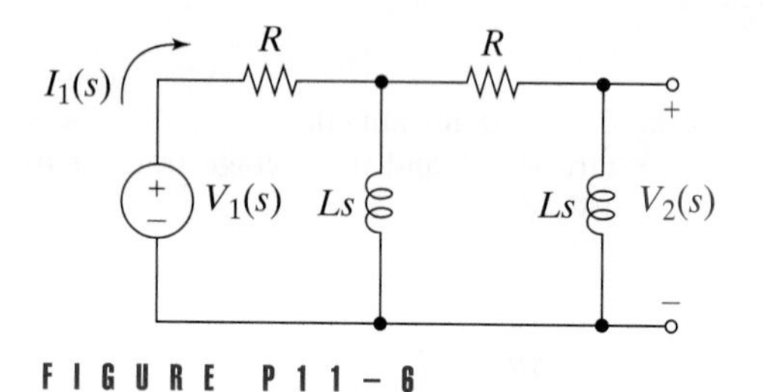

FIGURE P11–6

11–7 Find the voltage transfer function $T_V(s) = V_2(s)/V_1(s)$ in Figure P11–7. Select values of R and C so that $T_V(s)$ has a pole at $s = -500$ rad/s.

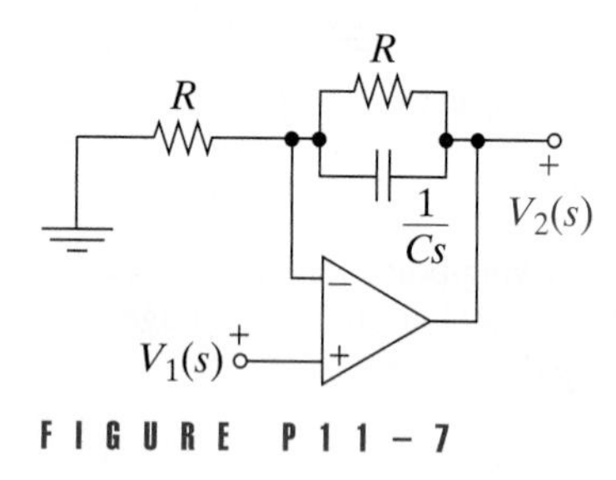

FIGURE P11–7

11–8 Find the current transfer function $T_I(s) = I_2(s)/I_1(s)$ in Figure P11–8. Select values of R and L so that $T_I(s)$ has a pole at $s = -500$ rad/s.

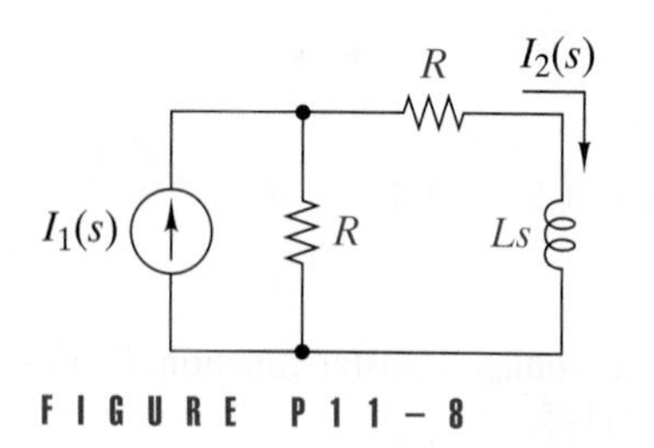

FIGURE P11–8

11–9 Find the voltage transfer function $T_V(s) = V_2(s)/V_1(s)$ of the cascade connection in Figure P11–9. Locate the poles and zeros of the transfer function.

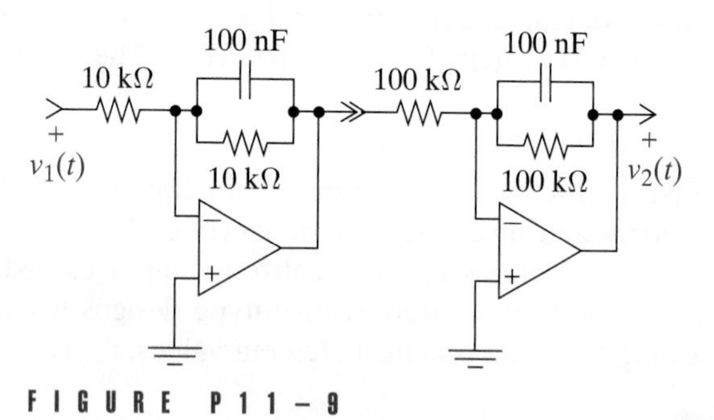

FIGURE P11–9

11–10 Find the voltage transfer function $T_V(s) = V_2(s)/V_1(s)$ of the cascade connection in Figure P11–10. Locate the poles and zeros of the transfer function.

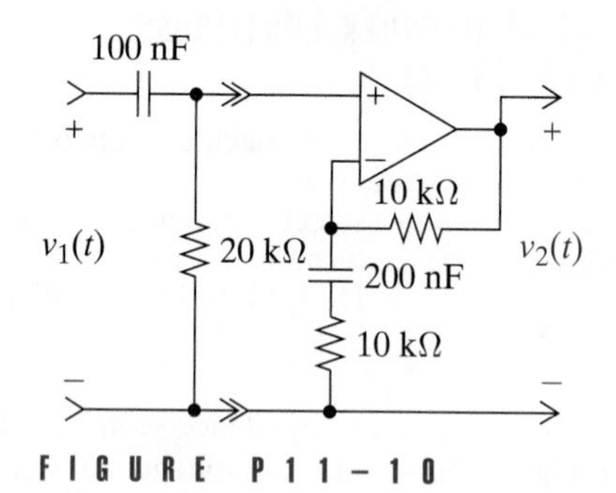

FIGURE P11–10

OBJECTIVE 11–2 NETWORK FUNCTIONS AND IMPULSE AND STEP RESPONSES (SECTS. 11–3, 11–4)

(a) Given a first- or second-order linear circuit, find its impulse or step response.
(b) Given the impulse or step response of a linear circuit, find the network functions.
(c) Given the impulse or step response of a linear circuit, find the response due to other inputs.

See Examples 11–8, 11–9, 11–10, 11–11 and Exercises 11–3, 11–4, 11–5, 11–6, 11–7

11–11 Find $v_2(t)$ in Figure P11–11 when $v_1(t) = \delta(t)$.

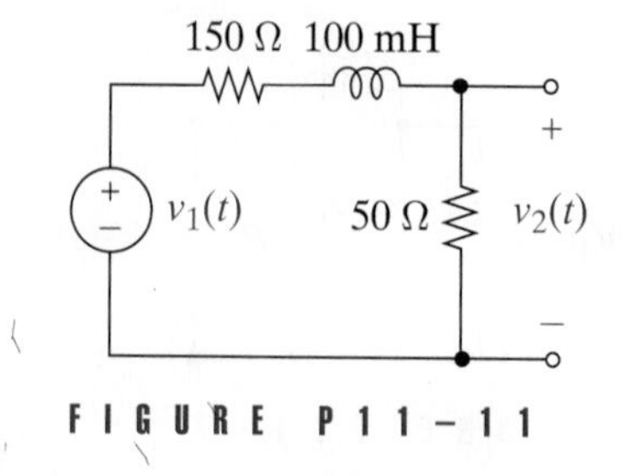

FIGURE P11–11

11–12 Find $v_2(t)$ in Figure P11–12 when $v_1(t) = \delta(t)$.

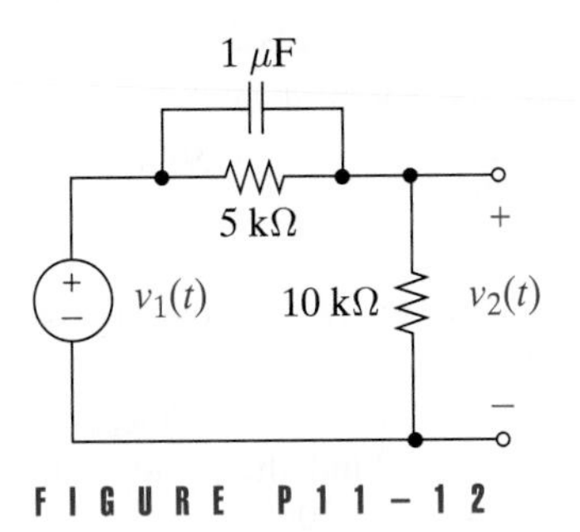

FIGURE P11–12

11–13 Find $v_2(t)$ in Figure P11–13 when $v_1(t) = u(t)$.

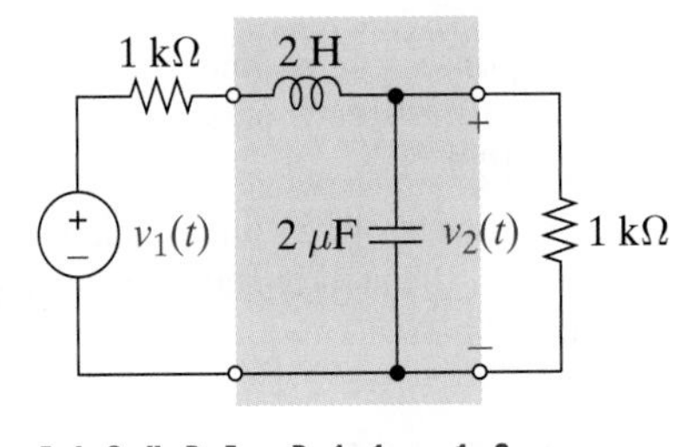

FIGURE P11–13

11–14 Find $v_2(t)$ in Figure P11–14 when $v_1(t) = u(t)$.

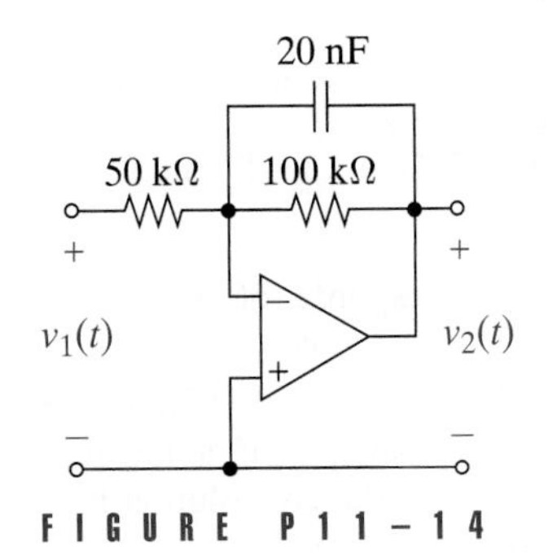

FIGURE P11–14

11–15 Find $v_2(t)$ in Figure P11–15 when $v_1(t) = \delta(t)$.

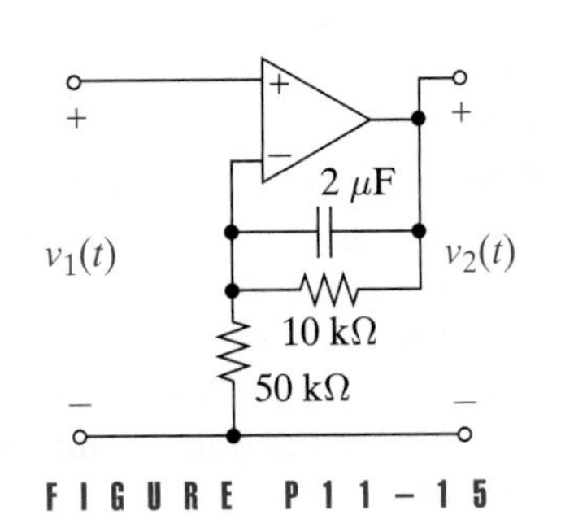

FIGURE P11–15

11–16 The impulse response of a linear circuit is $h(t) = [40e^{-20t} - 30e^{-10t}]u(t)$. Find the circuit step response $g(t)$.

11–17 The impulse response of a linear circuit is $h(t) = \delta(t) - [1000e^{-500t}]u(t)$. Find the circuit step response $g(t)$.

11–18 The step response of a linear circuit is $g(t) = [1 - e^{-500t}]u(t)$. Find the circuit impulse response $h(t)$.

11–19 The step response of a linear circuit is $g(t) = 15[e^{-10t} - e^{-30t}]u(t)$. Find the circuit impulse response $h(t)$.

11–20 Find the circuit transfer function corresponding to each of the following step responses:

(a) $g(t) = [500te^{-250t}]u(t)$

(b) $g(t) = [2 - e^{-250t}]u(t)$

11–21 The impulse response of a linear circuit is $h(t) = 1000[e^{-1000t}]u(t)$. Find the output $y(t)$ when the input is $x(t) = 5tu(t)$ V.

11–22 The step response of a linear circuit is $g(t) = 0.5[1 - e^{-100t}]u(t)$. Find the output $y(t)$ when the input is $x(t) = [10e^{-200t}]u(t)$.

11–23 The step response of a linear circuit is $g(t) = 10[e^{-100t}\sin 200t]u(t)$. Find the circuit impulse response $h(t)$.

11–24 The impulse response of a linear circuit is $H(s) = (s + 2000)/(s + 1000)$. Find the output $y(t)$ when the input is $x(t) = [5e^{-1000t}]u(t)$.

11–25 The impulse response of a linear circuit is $h(t) = 10u(t) + \delta(t)$. Find the output $y(t)$ when the input is $x(t) = [e^{-10t}]u(t)$.

OBJECTIVE 11–3 NETWORK FUNCTIONS AND SINUSOIDAL STEADY-STATE RESPONSE (SECT. 11–5)

(a) Given a first- or second-order linear circuit with a specified input sinusoid, find the sinusoidal steady-state response.

(b) Given a network function, impulse response or step response, find the sinusoidal steady-state response for a specified input sinusoid.

See Examples 11–13, 11–14 and Exercises 11–8, 11–9

11–26 The circuit in Figure P11–26 is in the steady state with $v_1(t) = 5\cos 300t$ V. Find $v_{2SS}(t)$. Repeat with $v_1(t) = 10\cos 600t$ V.

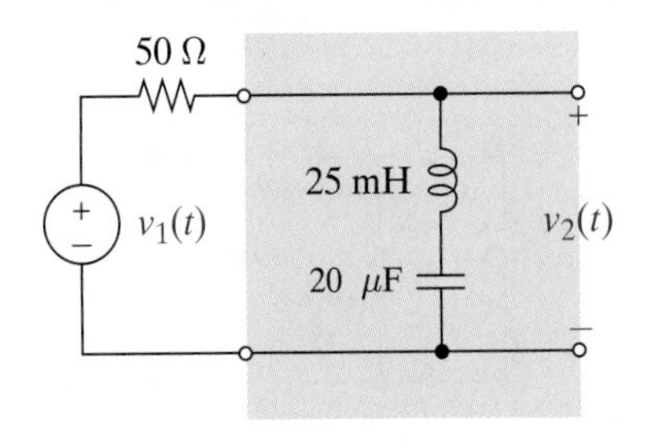

FIGURE P11–26

11–27 The circuit in Figure P11–27 is in the steady state with $v_1(t) = 10 \cos 2000t$ V. Find $v_{2SS}(t)$. Repeat with $v_1(t) = 3 \cos 4000t$ V.

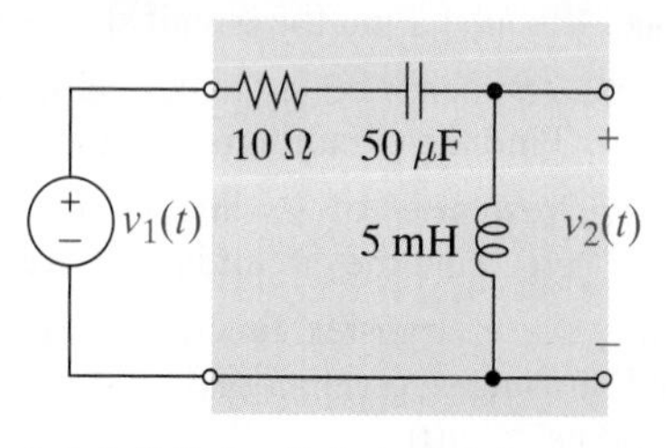

FIGURE P11–27

11–28 The circuit in Figure P11–28 is in the steady state with $i_1(t) = 10 \cos 500t$ mA, $R_1 = 100\ \Omega$, $R_2 = 400\ \Omega$, and $L = 100$ mH. Find $i_{2SS}(t)$. Repeat with $i_1(t) = 10 \cos 5000t$ mA.

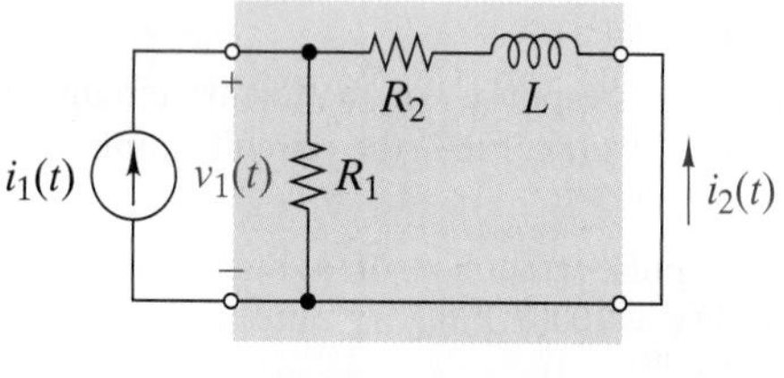

FIGURE P11–28

11–29 The circuit in Figure P11–29 is in the steady state with $i_1(t) = 5 \cos 1000t$ mA, $R = 1$ kΩ, $L = 2$ H, and $C = 500$ nF. Find $i_{2SS}(t)$. Repeat with $i_1(t) = 10 \cos 2000t$ mA.

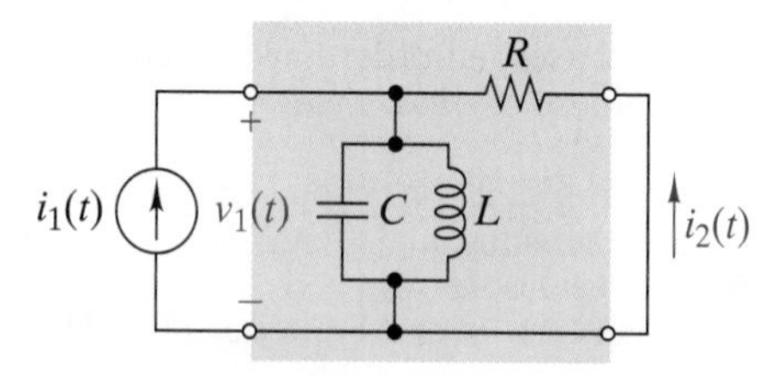

FIGURE P11–29

11–30 The circuit in Figure P11–30 is in the steady state with $i_1(t) = 10 \cos 5000t$ mA. Find the steady-state voltage $v_{2SS}(t)$. Repeat with $i_1(t) = 5 \cos 2500t$ mA.

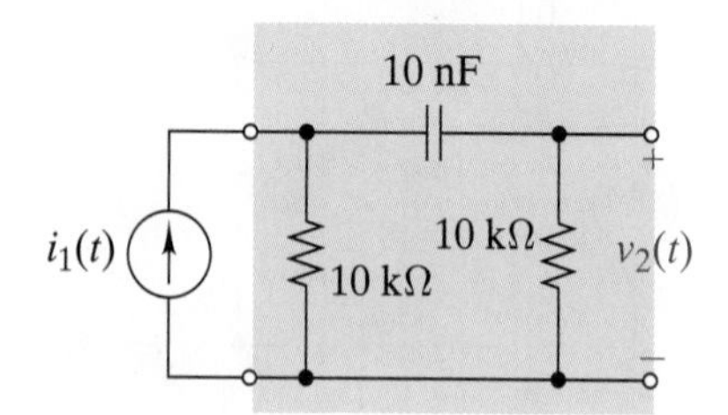

FIGURE P11–30

11–31 The transfer function of a linear circuit is $T(s) = (s + 200)/(s + 50)$. Find the steady-state output $y_{SS}(t)$ for an input $x(t) = 5 \cos 200t$.

11–32 The step response of a linear circuit is $g(t) = [5e^{-500t}]u(t)$. Find the steady-state output $y_{SS}(t)$ for an input $x(t) = 2 \cos 2000t$.

11–33 The step response of a linear circuit is $g(t) = [2e^{100t}]u(t)$. Find the steady-state output $y_{SS}(t)$ for an input $x(t) = 5 \cos 500t$.

11–34 The impulse response of a linear circuit is $h(t) = [800e^{-1000t}]u(t) - \delta(t)$. Find the steady-state output $y_{SS}(t)$ for an input $x(t) = 3 \sin 400t$.

11–35 The impulse response of a linear circuit is $h(t) = 400[e^{-100t} - e^{-400t}]u(t)$. Find the steady-state output $y_{SS}(t)$ for an input $x(t) = 5 \cos 200t$.

11–36 The step response of a linear circuit is $g(t) = [e^{-60t} \sin 80t]u(t)$. Find the steady-state output $y_{SS}(t)$ for an input $x(t) = 20 \cos 100t$.

11–37 The step response of a linear circuit is $g(t) = [1 - 20te^{-10t}]u(t)$. Find the steady-state output $y_{SS}(t)$ for an input $x(t) = 25 \cos 10t$.

OBJECTIVE 11–4 NETWORK FUNCTIONS AND CONVOLUTION (SECT. 11–6)

(a) Given the impulse response of a linear circuit, use the convolution integral to find the response to a specified input.
(b) Use the convolution integral to derive properties of linear circuits.
See Examples 11–15, 11–16, 11–17 and Exercise 11–10

11–38 The impulse response of a linear circuit is $h(t) = [u(t) - u(t - 1)]$. Use the convolution integral to find the response $y(t)$ due to an input $x(t) = u(t - 1)$.

11–39 The impulse response of a linear circuit is $h(t) = [u(t) - u(t - 1)]$. Use the convolution integral to find the response $y(t)$ due to an input $x(t) = u(t) - u(t - 2)$.

11–40 The impulse response of a linear circuit is $h(t) = t[u(t) - u(t - 1)]$. Use the convolution integral to find the response $y(t)$ due to an input $x(t) = u(t - 1)$.

11–41 The impulse response of a linear circuit is $h(t) = e^{-t}u(t)$. Use the convolution integral to find the response $y(t)$ due to an input $x(t) = u(t - 2)$.

11–42 Repeat Problem 11–41 for $x(t) = e^{-t}u(t)$.

11–43 The impulse response of a linear circuit is $h(t) = e^{-t}u(t)$. Use the convolution integral to find the response $y(t)$ due to an input $x(t) = tu(t)$.

11–44 Show that $f(t)*\delta(t) = f(t)$. That is, show that the convolution of a waveform $f(t)$ with an impulse leaves the waveform unchanged.

11–45 Use the convolution integral to show if $h(t) = u(t)$, then the output $y(t)$ for any input $x(t)$ is

$$y(t) = \int_0^t x(\tau)\, d\tau$$

11–46 Use the convolution integral to show that the step response is the integral of the impulse response. That is, show that if $x(t) = u(t)$, then Eq. (11–24) reduces to

$$y(t) = g(t) = \int_0^t h(\tau)\, d\tau$$

11–47 Use the convolution integral to show that the step response is the derivative of the ramp response. That is, show that if $x(t) = tu(t)$, then the derivative of Eq. (11–24) yields

$$\frac{dy(t)}{dt} = \int_0^t h(\tau)\, d\tau = g(t)$$

11–48 The impulse response of a linear circuit is $h(t) = u(t)$. Use the convolution integral to find the response $y(t)$ due to an input $x(t) = t[u(t) - u(t-1)]$.

11–49 The impulse response of a linear circuit is $H(s) = 25/(s+5)$ and $x(t) = tu(t)$. Use s-domain convolution to find the zero-state response $y(t)$.

11–50 The impulse responses of two linear circuits are $h_1(t) = [e^{-2t}]u(t)$ and $h_2(t) = [4e^{-4t}]u(t)$. What is the impulse response of a cascade connection of these two circuits?

OBJECTIVE 11–5 NETWORK FUNCTION DESIGN (SECT. 11–7)

(a) Design a circuit that realizes a given $T(s)$ and meets other stated constraints.
(b) Evaluate alternative designs using stated criteria.

11–51 **D** Design a circuit to realize the transfer function below using only resistors, capacitors, and OP AMPs. Scale the circuit so that all resistors are greater than 10 kΩ and all capacitors are less than 1 μF.

$$T_V(s) = \pm\frac{2 \times 10^4}{(s+250)(s+2500)}$$

11–52 **D** Design a circuit to realize the transfer function below using only resistors, capacitors, and not more than one OP AMP. Scale the circuit so that all capacitors are exactly 10 nF.

$$T_V(s) = \pm\frac{100(s+1000)}{(s+250)(s+2000)}$$

11–53 **D** Design a circuit to realize the transfer function below using only resistors, capacitors, and not more than one OP AMP. Scale the circuit so that the final design uses only 20-kΩ resistors.

$$T_V(s) = \pm\frac{1000\, s}{(s+500)(s+1000)}$$

11–54 **D** Design a circuit to realize the transfer function below using only resistors, capacitors, and inductors (no OP AMPs allowed). Scale the circuit so that all inductors are 50 mH or less.

$$T_V(s) = \frac{s^2}{(s+2000)(s+4000)}$$

11–55 **D** Design a circuit to realize the transfer function listed below using only resistors, capacitors, and not more than one OP AMP. Scale the circuit so that all resistors are greater than 10 kΩ and all capacitors are less than 1 μF.

$$T_V(s) = \pm\frac{(s+100)(s+500)}{(s+200)(s+1000)}$$

11–56 **D** Design a circuit to realize the transfer function below using only resistors, capacitors, and OP AMPs. Scale the circuit so that all capacitors are exactly 100 nF.

$$T_V(s) = -\frac{400(s+100)}{s(s+200)}$$

11–57 **D** Design a circuit to realize the transfer function below using practical element values.

$$T_V(s) = \pm\frac{100s + 10^6}{s^2 + 100s + 10^6}$$

11–58 **E** It is claimed that both circuits in Figure P11–58 realize the transfer function

$$T_V(s) = K\left(\frac{s+2000}{s+1000}\right)$$

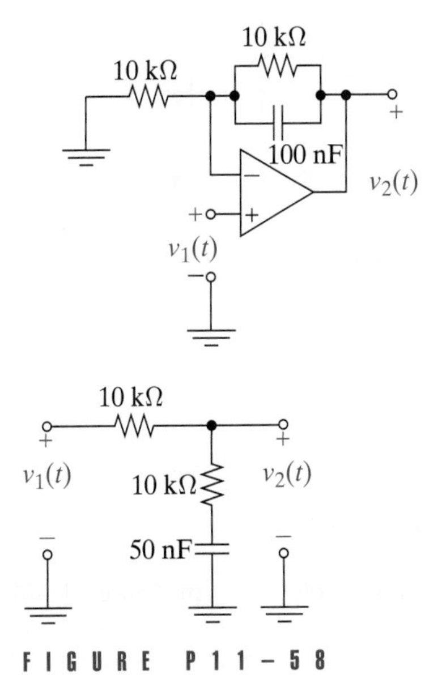

FIGURE P11–58

(a) Verify that both circuits realize the specified $T_V(s)$.
(b) Which circuit would you choose if the output must drive a 1-kΩ load?
(c) Which circuit would you choose if the input comes from a 50-Ω source?
(d) It is further claimed that connecting the two circuits in cascade produces an overall transfer function of $[T_V(s)]^2$ no matter which circuit is the first stage and which is the second stage. Do you agree or disagree? Explain.

11–59 **E** It is claimed that both circuits in Figure P11–59 realize the transfer function

$$T_V(s) = \frac{\pm 1000\, s}{(s + 1000)(s + 4000)}$$

(a) Verify that both circuits realize the specified $T_V(s)$.
(b) Which circuit would you choose if the output must drive a 1-kΩ load?
(c) Which circuit would you choose if the input comes from a 50-Ω source?
(d) It is further claimed that connecting the two circuits in cascade produces an overall transfer function of $[T_V(s)]^2$ no matter which circuit is the first stage and which is the second stage. Do you agree or disagree? Explain.

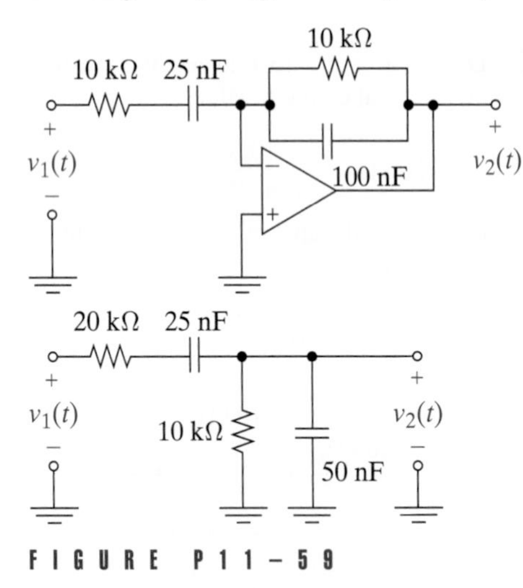

FIGURE P11–59

11–60 **D** Design a circuit that produces the following step response:

$$g(t) = [1 - e^{-50t} - 50\,t\,e^{-50t}]u(t)$$

INTEGRATING PROBLEMS

11–61 **A** **D** First-Order Circuit Impulse and Step Responses
Each row in the table shown in Figure P11–61 refers to a first-order circuit with an impulse response $h(t)$ and a step response $g(t)$. Fill in the missing entries in the table.

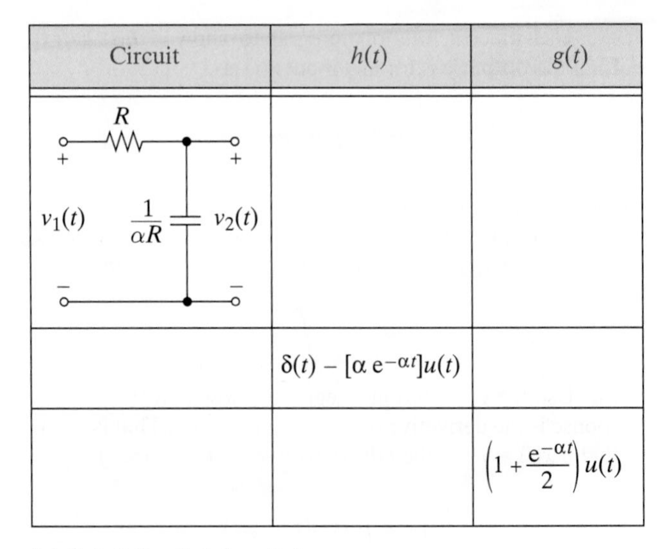

Circuit	$h(t)$	$g(t)$
R; $\frac{1}{\alpha R}$; $v_1(t)$; $v_2(t)$		
	$\delta(t) - [\alpha\, e^{-\alpha t}]u(t)$	
		$\left(1 + \frac{e^{-\alpha t}}{2}\right)u(t)$

FIGURE P11–61

11–62 **A** OP AMP Modules and Loading
Figure P11–62 shows an interconnection of three basic OP AMP modules. Does this interconnection involve loading? Find the overall transfer function of the interconnection and locate its poles and zeros.

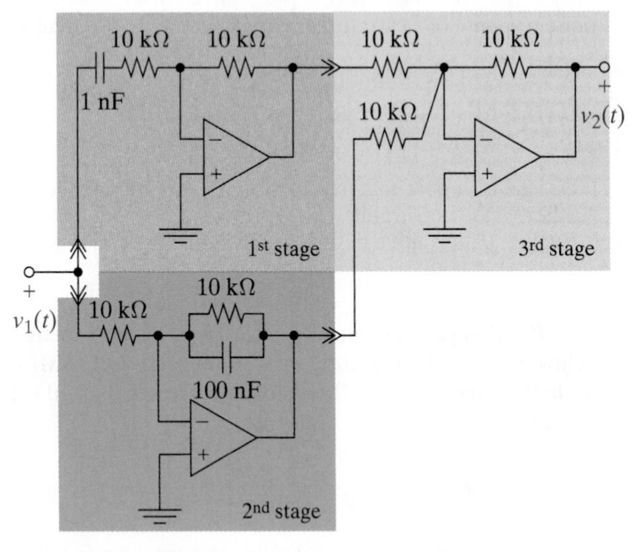

FIGURE P11–62

11–63 **A** OP AMP Modules and Stability
Figure P11–63 shows an interconnection of three basic circuit modules. Does this interconnection involve loading? Find the overall transfer function of the interconnection and locate its poles and zeros. Is the circuit stable?

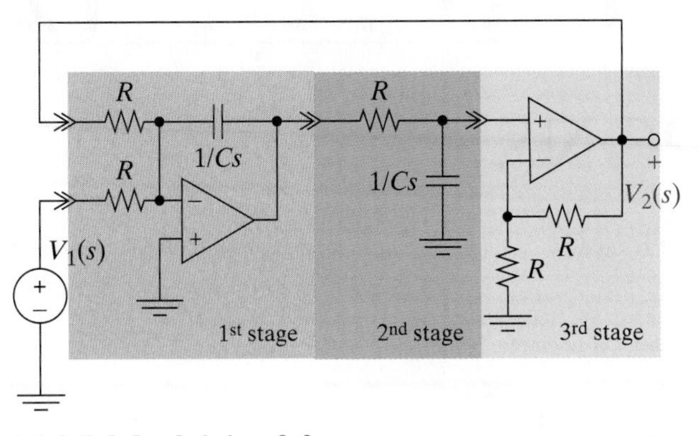

FIGURE P11–63

11–64 A Step Response and Fan-Out

The fan-out of a digital device is defined as the maximum number of inputs to similar devices that can be reliably driven by the device output. Figure P11–64 is a simplified diagram of a device output driving n identical capacitive inputs. To operate reliably, a 5-V step function at the device output must drive the capacitive inputs to 3.7 V in 10 ns or less. Determine the device fan-out for $R = 1\ \text{k}\Omega$ and $C = 3$ pF.

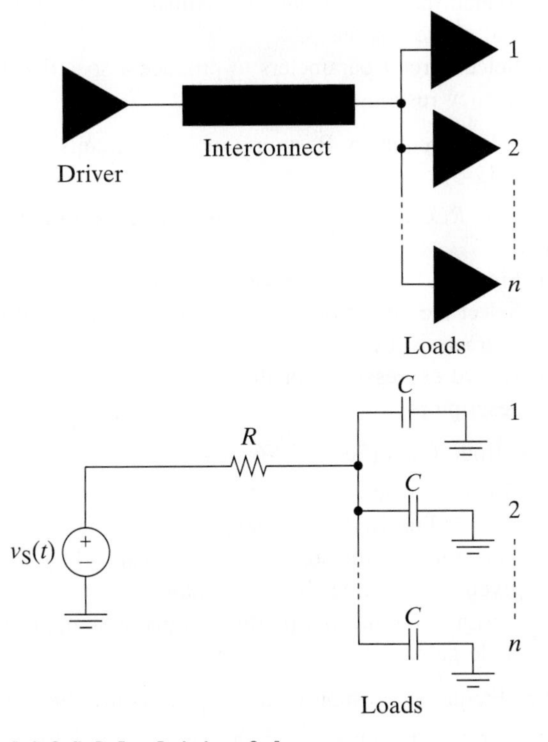

FIGURE P11–64

11–65 A Sinusoidal Steady State Response

The input to the RL circuit in Figure P11–65 is $v_1(t) = V_A \cos \omega t$.

(a) Use the s-domain network function method in this chapter to find the sinusoidal steady-state response $i_{SS}(t)$.

(b) Formulate the circuit differential equation and use the classical methods of Chapter 7 to find the forced component of $i(t)$.

(c) The responses found in parts (a) and (b) should be identical. Which method do you think is easiest to apply? Discuss.

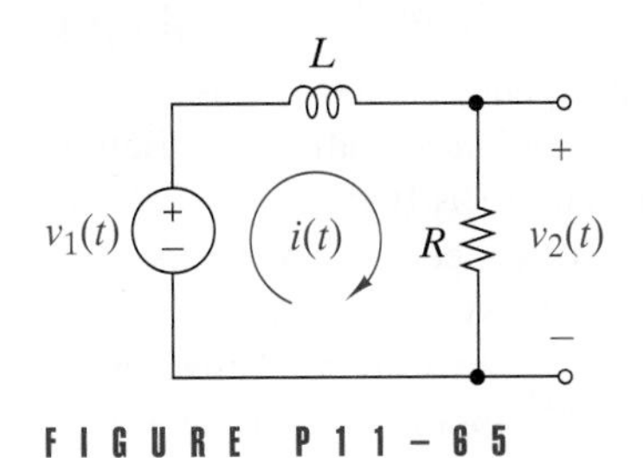

FIGURE P11–65

CHAPTER 12

FREQUENCY RESPONSE

The advantage of the straight-line approximation is, of course, that it reduces the complete characteristics to a sum of elementary characteristics.

Hendrik W. Bode, 1945,
American Engineer

Some History Behind This Chapter

Hendrik Bode spent most of his distinguished career as a member of the technical staff of the Bell Telephone Laboratories. In the 1930s Bode made major contributions to feedback amplifier theory during the development of long-distance telephone systems. His contributions included an approach to frequency response based on logarithmic plots and straight-line approximation. These so called Bode plots are still valuable today and remain the industry-standard way to describe the frequency response of circuits and systems.

Why This Chapter Is Important Today

In Chapter 8 we used phasors to study the steady-state response at a single-frequency. In Chapter 11 we showed that network functions give the steady-state response at any frequency. In this chapter we use network functions and Bode diagrams to describe the steady-state response over a range of frequencies. Bode diagrams are used here because they allow us to quickly visualize how the poles and zeros of a network function affect the frequency response of a circuit.

Chapter Sections

12–1 Frequency-Response Descriptors

12–2 Bode Diagrams

12–3 First-order Low-Pass and High-Pass Responses

12–4 Bandpass and Bandstop Responses

12–5 The Frequency Response of *RLC* Circuits

12–6 Bode Diagrams with Real Poles and Zeros

12–7 Bode Diagrams with Complex Poles and Zeros

12–8 Frequency Response and Step Response

Chapter Learning Objectives

12-1 First-order Circuit Frequency Response (Sects. 12-1, 12-2, 12-3)

Given a first-order circuit or transfer function:

(a) Find and classify the frequency response.
(b) Sketch the straight-line approximations of the gain and phase responses.
(c) Select circuit parameters to produce a specified frequency response.

12-2 Bandpass and Bandstop Responses (Sect. 12-4)

Given a cascade or parallel connection of two first-order circuits:

(a) Find and classify the frequency response.
(b) Sketch the straight-line approximations of the gain and phase responses.
(c) Select circuit parameters to produce a specified frequency response.

12-3 The Frequency Response of *RLC* Circuits (Sect. 12-5)

Given an *RLC* circuit connected as a bandpass or bandstop filter:

(a) Find the frequency-response descriptors.
(b) Select the circuit parameters to achieve specified filter characteristics.
(c) Derive expressions for the frequency-response descriptors.

12-4 Bode Plots (Sects. 12-6, 12-7)

(a) Given a linear circuit or transfer function, construct straight-line Bode gain plots.
(b) Construct the transfer function corresponding to a given straight-line Bode gain plot.
(c) Design a circuit that produces a given straight-line Bode gain plot.

12-5 Frequency Response and Step Response (Sect. 12-8)

Given a step response, find the straight-line Bode gain response, and vice versa.

12–1 FREQUENCY-RESPONSE DESCRIPTORS

In Chapter 11 we learned that the sinusoidal steady-state output can be found by evaluating the transfer function $T(s)$ at $s = j\omega$, where ω is the frequency of the sinusoidal input. The function $T(j\omega)$ determines the amplitude and phase angle of the output through the **gain** function $|T(j\omega)|$ and **phase** function $\theta(\omega) = \angle T(j\omega)$.

$$\text{Output amplitude} = |T(j\omega)| \times (\text{Input amplitude})$$
$$\text{Output phase} = \text{Input phase} + \theta(\omega) \tag{12–1}$$

The gain and phase functions are frequency dependent and together reveal how a circuit responds to input sinusoids of different frequencies. This frequency-dependent relationship between sinusoidal inputs and the resulting steady-state outputs is called the **frequency response** of the circuit.

Frequency-response concepts and techniques find wide applications in communication, control, and instrumentation systems. A key component in these applications is the electric **filter**, a signal processor that modifies or reshapes the frequency content of signals.

The gain and phase functions can be expressed mathematically or presented graphically, as shown in Figure 12–1. Most of the descriptive terminology of frequency response is based on the shape of the gain function. For example, the gain plot in Figure 12–1 is relatively constant at lower frequencies and decreases rapidly at higher frequencies. The range of frequencies with nearly constant gain is called a **passband**. The range of frequencies with significantly reduced gain is called a **stopband**. The frequency associated with the transition from a passband to an adjacent stopband is called the **cutoff frequency**, denoted as ω_C or f_C.

In linear circuits there is a gradual transition from a passband to a stopband, so the location of the cutoff frequency is a matter of definition. The

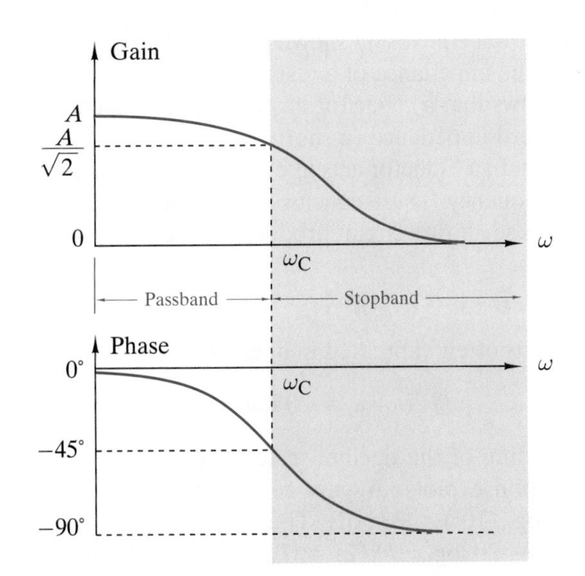

FIGURE 12–1 *Frequency-response plots.*

most widely used definition assigns cutoff to the frequency at which the passband gain decreases by a factor of $1/\sqrt{2}$ from its maximum value. Under this definition ω_C is found from the condition

$$|T(j\omega_C)| = \frac{1}{\sqrt{2}} T_{max} \tag{12–2}$$

where T_{max} is the maximum gain in the passband.

Additional terminology is based on the four basic types of gain responses in Figure 12–2. The figure also shows the sinusoidal input and outputs for each type. The input to all four is a composite signal consisting of three equal-amplitude sinusoids at distinct frequencies ω_1, ω_2, and ω_3. The output signals all contain the same three frequencies but have amplitudes modified by the form of the gain response.

- The *low-pass* gain response has a single passband extending from zero frequency (dc) to ω_C. This type of gain passes the input at ω_1 unchanged and attenuates the inputs at ω_2 and ω_3 since they fall in the stopband above ω_C.
- The *high-pass* response has a single passband extending from ω_C to infinite frequency. This type of gain passes the inputs at ω_2 and ω_3 unchanged and attenuates the input at ω_1 since it falls in the stopband below ω_C.
- The *bandpass* response has a single passband with two adjacent stopbands—one below ω_{C1} and another above ω_{C2}. This gain response passes the input at ω_2 unchanged and attenuates the inputs at ω_1 and ω_3.
- The *bandstop* response has a single stopband with two adjacent passbands—one below ω_{C1} and another above ω_{C2}. This gain attenuates the input at ω_2 and passes the inputs at ω_1 and ω_3 unchanged.

TABLE 12–1 PASSIVE ELEMENT IMPEDANCES AT ZERO AND INFINITE FREQUENCY

Element	Impedance at $\omega = 0$	Impedance at $\omega = \infty$
Resistor $Z_R = R$	R	R
Capacitor $Z_C = 1/j\omega C$	∞ Open circuit	0 Short circuit
Inductor $Z_L = j\omega L$	0 Short circuit	∞ Open circuit

The element impedances $Z(s)$ play a key role in circuit frequency response. Replacing s by $j\omega$ and evaluating $Z(j\omega)$ at $\omega = 0$ (dc) and $\omega = \infty$ leads to the important conclusion shown in Table 12–1. The first row in Table 12–1 shows that the impedance of a resistor does not change with frequency. The next row shows that a capacitor has an infinite impedance (an open circuit) at dc and zero impedance (a short circuit) at infinite frequency. In the last row we see that an inductor acts like a short circuit at dc and an open circuit at infinite frequency. These conclusions are worth remembering because they often help us see how a circuit produces particular gain responses.

12–2 BODE DIAGRAMS

The gain $|T(j\omega)|$ is often expressed in **decibels** (dB), defined as

$$|T(j\omega)|_{dB} = 20 \log_{10}|T(j\omega)| \tag{12–3}$$

Some understanding of the decibel scale is necessary to construct and interpret gain-response plots. A gain of $|T(j\omega)| = 10^n$ expressed in dB is $|T(j\omega)|_{dB} = 20 \log_{10}(10^n) = 20n$ dB. The gain of $T_{max}/\sqrt{2}$ expressed in dB is $20 \log_{10}(T_{max}) - 20 \log_{10}(1/\sqrt{2}) = |T_{max}|_{dB} - 3$ dB. In other words, the

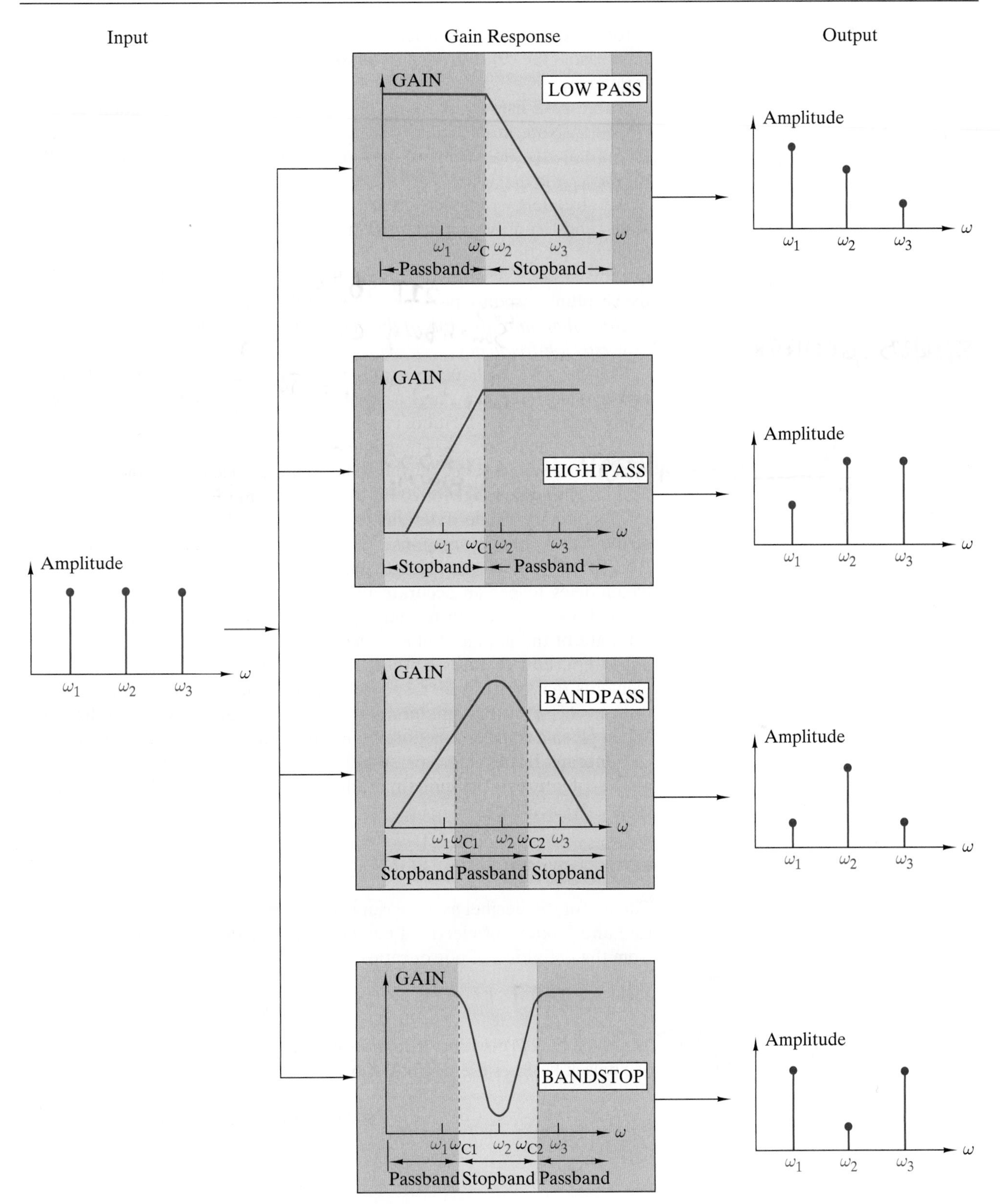

FIGURE 12–2 *Four basic gain responses.*

TABLE 12–2 VALUES OF GAIN AND GAIN IN DB

$\lvert T(j\omega)\rvert$	$\lvert T(j\omega)\rvert_{dB}$
10^{-3}	−60
10^{-2}	−40
10^{-1}	−20
0.5	−6
$1/\sqrt{2}$	−3
1	0
$\sqrt{2}$	3
2	6
10	20
10^2	40
10^3	60

cutoff frequency occurs when the passband gain is reduced by 3 dB. For this reason, the cutoff frequency is sometimes referred to as the 3-dB down frequency.

Table 12–2 lists other values of $|T(j\omega)|$ and the corresponding value of $|T(j\omega)|_{dB}$. Note in particular that a gain of 1 is 0 dB and a gain of 2(0.5) is +6(−6) dB. This means that the gain $2|T(j\omega)|$ expressed in dB is $|T(j\omega)|_{dB} + 6$ dB while $|T(j\omega)|/2$ is $|T(j\omega)|_{dB} - 6$ dB.

The frequency range of interest is often so wide that a linear frequency scale tends to mask important features for the response. For this reason frequency-response plots almost always use a logarithmic scale for the frequency variable. The use of log-frequency scales is a standard practice, and the resulting response plots are called **Bode diagrams**.

Bode diagrams are plots of the gain $|T(j\omega)|_{dB}$ and phase $\theta(\omega)$ versus log-frequency.

The use of log-frequency scales involves some special terminology. An **octave** is any frequency range whose end points have a 2:1 ratio, and a **decade** is any range with a 10:1 ratio. For example, the frequency range from 10 Hz to 20 Hz is one octave, as is the range from 20 MHz to 40 MHz. The standard UHF (ultrahigh frequency) band spans a one-decade range from 0.3 GHz to 3 GHz. The audio range for 20 Hz to 20 kHz spans three decades.

In summary, Bode diagrams (also called Bode plots) are used to describe the frequency response of circuits and systems. To construct a Bode plot, we need to calculate gain and phase at a sufficient number of frequencies to get an accurate picture of frequency response. In many cases, however, we can use straight-line approximations to get a useful estimate of the gain and phase responses. These straight-line plots highlight important features of the gain and can help us identify the circuit elements that control the response. When greater accuracy is needed, we use one of several computer tools to calculate and plot Bode diagrams. When given a transfer function, we use programs like Mathcad or MATLAB to generate Bode diagrams. When given a circuit, we can use the ac analysis features of circuit simulation programs such as PSpice or Electronics WorkBench.

APPLICATION EXAMPLE 12–1

The use of the decibel as a measure of performance pervades the literature and folklore of electrical engineering. The decibel originally came from the definition of power ratios in **bels**.[1]

$$\text{Number of bels} = \log_{10}\frac{P_{OUT}}{P_{IN}}$$

The decibel (dB) is more commonly used in practice. The number of decibels is 10 times the number of bels:

$$\text{Number of dB} = 10 \times (\text{Number of bels}) = 10\log_{10}\frac{P_{OUT}}{P_{IN}}$$

1 The name of the unit honors Alexander Graham Bell (1847–1922), the inventor of the telephone.

When the input and output powers are delivered to equal input and output resistances R, then the power ratio can be expressed in terms of voltages across the resistances.

$$\text{Number of dB} = 10 \log_{10}\frac{v_{\text{OUT}}^2/R}{v_{\text{IN}}^2/R} = 20 \log_{10}\frac{v_{\text{OUT}}}{v_{\text{IN}}}$$

or in terms of currents through the resistances:

$$\text{Number of dB} = 10 \log_{10}\frac{i_{\text{OUT}}^2 \times R}{i_{\text{IN}}^2 \times R} = 20 \log_{10}\frac{i_{\text{OUT}}}{i_{\text{IN}}}$$

The definition of gain in dB in Eq. (12–3) is consistent with these results, since in the sinusoidal steady state the transfer function equals the ratio of output amplitude to input amplitude. The preceding discussion is not a derivation of Eq. (12–3) but simply a summary of its historical origin. In practice Eq. (12–3) is applied when the input and output are not measured across resistances of equal value.

When the chain rule applies to a cascade connection, the overall transfer function is a product

$$T(j\omega) = T_1 \times T_2 \times \cdots T_{\text{N}}$$

where $T_1, T_2, \ldots T_{\text{N}}$ are the transfer functions of the individual stages in the cascade. Expressed in dB, the overall gain is

$$\begin{aligned}|T(j\omega)|_{\text{dB}} &= 20 \log_{10}(|T_1| \times |T_2| \times \cdots |T_{\text{N}}|)\\ &= 20 \log_{10}|T_1| + 20 \log_{10}|T_2| + \cdots + 20 \log_{10}|T_{\text{N}}|\\ &= |T_1|_{\text{dB}} + |T_2|_{\text{dB}} + \cdots + |T_{\text{N}}|_{\text{dB}}\end{aligned}$$

Because of the logarithmic definition, the overall gain (in dB) is the sum of the gains (in dB) of the individual stages in a cascade connection. The effect of altering a stage or adding an additional stage can be calculated by simply adding or subtracting the change in dB. Since summation is simpler than multiplication, the enduring popularity of the dB comes from its logarithmic definition, not its somewhat tenuous relationship to power ratios.

12–3 FIRST-ORDER LOW-PASS AND HIGH-PASS RESPONSES

FIRST-ORDER LOW-PASS RESPONSE

We begin the study of frequency response with the first-order low-pass transfer function:

$$T(s) = \frac{K}{s + \alpha} \tag{12–4}$$

The constants K and α are real. The constant K can be positive or negative, but α must be positive so that the natural pole at $s = -\alpha$ is in the left half of the s plane to ensure that the circuit is stable. Remember, the concepts of sinusoidal steady state and frequency response do not apply to unstable circuits that have poles in the right half of the s plane or on the j-axis.

To describe the frequency response of the low-pass transfer function, we replace s by $j\omega$ in Eq. (12–4)

$$T(j\omega) = \frac{K}{\alpha + j\omega} \tag{12–5}$$

and express the gain and phase functions as

$$|T(j\omega)| = \frac{|K|}{\sqrt{\omega^2 + \alpha^2}}$$

$$\theta(\omega) = \angle K - \tan^{-1}(\omega/\alpha) \tag{12–6}$$

The gain function is a positive number. Since K is real, the angle of K ($\angle K$) is either 0° when $K > 0$ or ±180° when $K < 0$. An example of a negative K occurs in an inverting OP AMP configuration where $T(s) = -Z_2(s)/Z_1(s)$.

Figure 12–3 shows Bode plots of the gain and phase of the first-order low-pass function using a log scale for the normalized frequency ω/α. The gain plot displays a low-pass characteristic with a passband at low

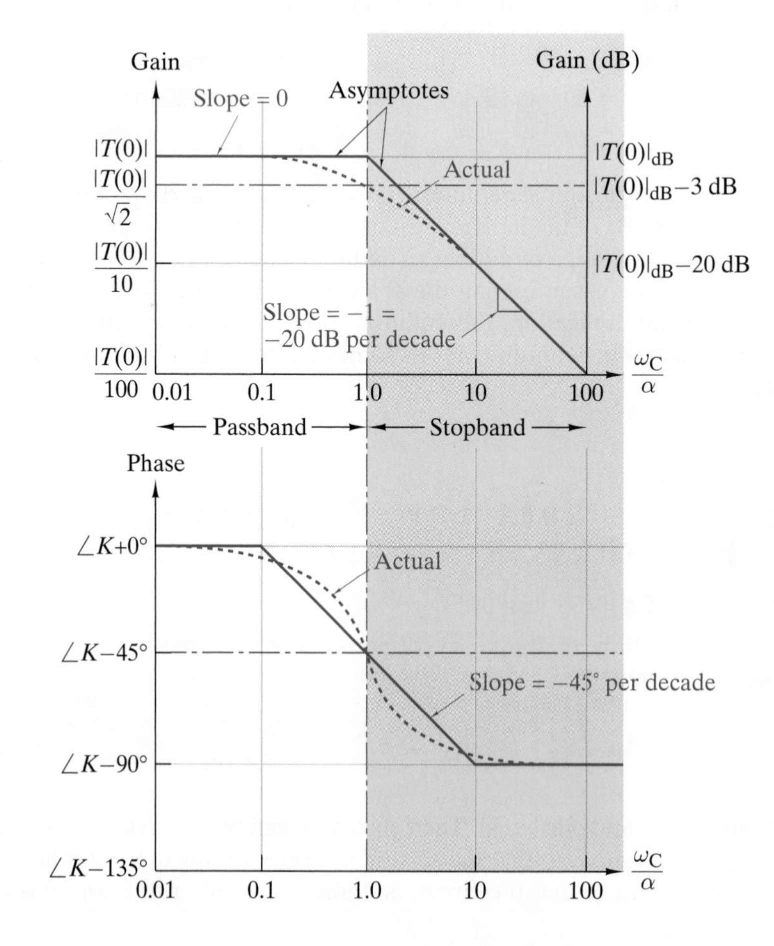

FIGURE 12–3 *First-order low-pass Bode plots.*

frequency and a stopband at high frequency. The maximum passband gain occurs at $\omega = 0$, where $T_{max} = |K|/\alpha$. The gain gradually decreases as frequency increases until at $\omega = \alpha$ we have

$$|T(j\alpha)| = \frac{|K|}{\sqrt{\alpha^2 + \alpha^2}} = \frac{|K|/\alpha}{\sqrt{2}} = \frac{1}{\sqrt{2}} T_{max} \quad (12\text{–}7)$$

Referring to Eq. (12–2), we conclude that $\omega_C = \alpha$ is the cutoff frequency marking the boundary between a low-frequency passband and a high-frequency stopband.

The low- and high-frequency gain asymptotes shown in Figure 12–3 are especially important. The low-frequency asymptote is the horizontal line and the high-frequency asymptote is the sloped line. At low frequencies ($\omega << \alpha$) the gain approaches $|T(j\omega)| \to |K|/\alpha$. At high frequencies ($\omega >> \alpha$) the gain approaches $|T(j\omega)| \to |K|/\omega$. The intersection of the two asymptotes occurs when $|K|/\alpha = |K|/\omega$. The intersection forms a "corner" at $\omega = \alpha$, so the cutoff frequency is also called the **corner frequency**.

The high-frequency gain asymptote decreases by a factor of 10 (–20 dB) whenever the frequency increases by a factor of 10 (one decade). As a result, the high-frequency asymptote has a slope of –1 or –20 dB per decade and the low-frequency asymptote has a slope of 0 or 0 dB/decade. These two asymptotes provide a straight-line approximation to the gain response that differs from the true response by a maximum of 3 dB at the corner frequency.

The semilog plot of the phase shift of the first-order low-pass transfer function is shown in Figure 12–3. At $\omega = \alpha$ the phase angle in Eq. (12–6) is

$$\begin{aligned}\theta(\omega_C) &= \angle K - \tan^{-1}\left(\frac{\alpha}{\alpha}\right)\\ &= \angle K - 45^\circ\end{aligned}$$

At low frequency ($\omega << \alpha$) the phase angle approaches $\angle K$ and at high frequencies ($\omega >> \alpha$) the phase approaches $\angle K - 90^\circ$. Almost all of the -90° phase change occurs in the two-decade range from $\omega/\alpha = 0.1$ to $\omega/\alpha = 10$. The straight-line segments in Figure 12–3 provide an approximation of the phase response. The phase approximation below $\omega/\alpha = 0.1$ is $\theta = \angle K$ and above $\omega/\alpha = 10$ is $\theta = \angle K - 90^\circ$. Between these values the phase approximation is a straight line that begins at $\theta = \angle K$, passes through $\theta = \angle K - 45^\circ$ at the cutoff frequency, and reaches $\theta = \angle K - 90^\circ$ at $\omega/\alpha = 10$. The slope of this line segment is -45°/decade since the total phase change is -90° over a two-decade range.

To construct the straight-line approximations for a first-order low-pass transfer function, we need two parameters, the value of $T(0)$ and α. The parameter α defines the cutoff frequency and the value of $T(0)$ defines the passband gain $|T(0)|$ and the low-frequency phase $\angle T(0)$. The required quantities $T(0)$ and α can be determined directly from the transfer function $T(s)$ and can often be estimated by inspecting the circuit itself.

Using logarithmic scales in Bode plots allows us to make straight-line approximations to both the gain and phase responses. These approximations provide a useful way of visualizing a circuit's frequency response.

Often such graphical estimates are adequate for developing analysis and design approaches. For example, the frequency response of the first-order low-pass function can be characterized by calculating the gain and phase over a two-decade band from one decade below to one decade above the cutoff frequency.

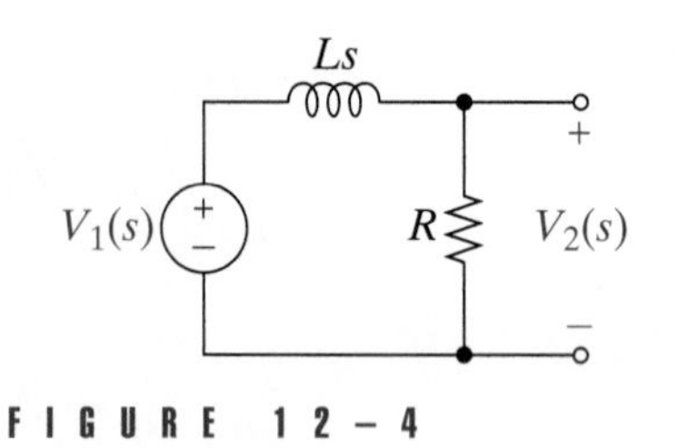

FIGURE 12–4

EXAMPLE 12–2

Consider the circuit in Figure 12–4. Find the transfer function $T(s) = V_2(s)/V_1(s)$ and construct the straight-line approximations to the gain and phase responses.

SOLUTION:

Applying voltage division, the voltage transfer function for the circuit is

$$T(s) = \frac{R}{Ls + R} = \frac{R/L}{s + R/L}$$

Comparing this with Eq. (12–4), we see that the circuit has a low-pass gain response with $\alpha = R/L$ and $T(0) = 1$. Therefore, $|T(0)|_{dB} = 0$ dB, $\omega_C = R/L$, and $\angle K = 0°$. Given these quantities, we construct the straight-line approximations shown in Figure 12–5. Note that the frequency scale in Figure 12–5 is normalized by multiplying ω by $L/R = 1/\alpha$.

Circuit Interpretation: The low-pass response in Figure 12–5 can be explained in terms of circuit behavior. At zero frequency the inductor acts like a short circuit that directly connects the input port to the output port to produce a passband gain of 1 (or 0 dB). At infinite frequency the inductor acts like an open circuit, effectively disconnecting the input and output ports and leading to a gain of zero. Between these two extremes the impedance of the inductor gradually increases, causing the circuit gain to decrease. In particular, at the cutoff frequency we have $\omega L = R$, the impedance of the inductor is $j\omega L = jR$, and the transfer function reduces to

$$T(j\omega_C) = \frac{R}{R + jR} = \frac{1}{\sqrt{2}}\angle -45°$$

In other words, at the cutoff frequency the gain is –3 dB and the phase shift –45°. Obviously, the changing impedance of the inductor gives the circuit its low-pass gain features. ■

DESIGN EXAMPLE 12–3

(a) Show that the transfer function $T(s) = V_2(s)/V_1(s)$ in Figure 12–6 has a low-pass gain characteristic.
(b) Select element values so the passband gain is 4 and the cutoff frequency is 100 rad/s.

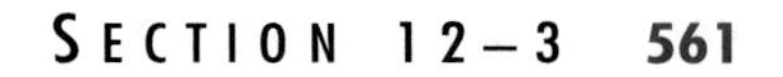

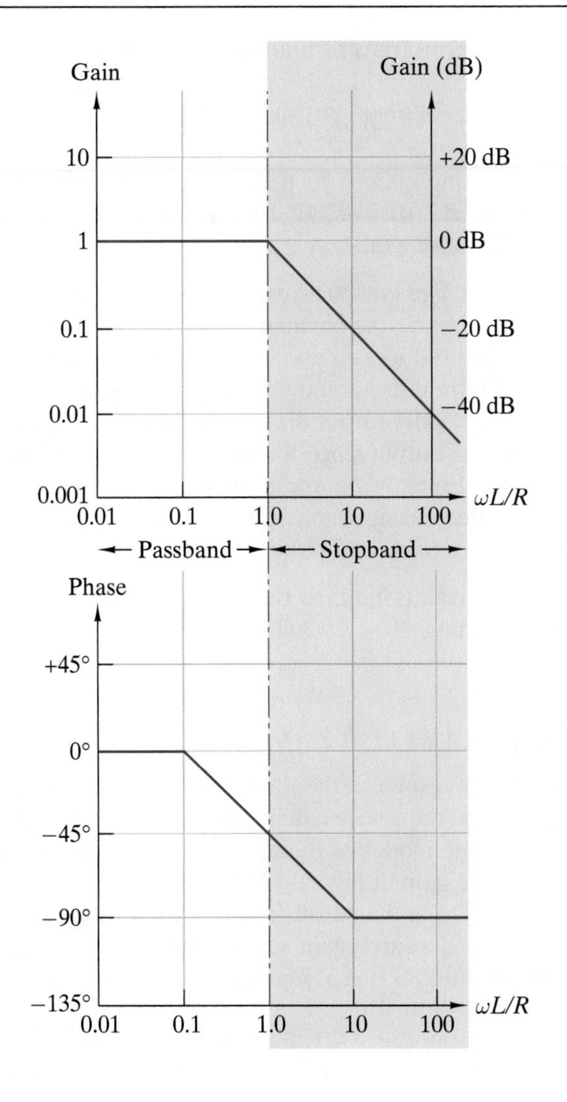

FIGURE 12–5

SOLUTION:

(a) The circuit is an inverting amplifier configuration with

$$Z_1(s) = R_1 \quad \text{and} \quad Z_2(s) = \frac{1}{C_2 s + \dfrac{1}{R_2}} = \frac{R_2}{R_2 C_2 s + 1}$$

The circuit transfer function is found as

$$T(s) = -\frac{Z_2(s)}{Z_1(s)} = -\frac{R_2}{R_1} \times \frac{1}{R_2 C_2 s + 1}$$

Rearranging the standard low-pass form in Eq. (12–4) as

$$T(s) = \frac{K/\alpha}{s/\alpha + 1}$$

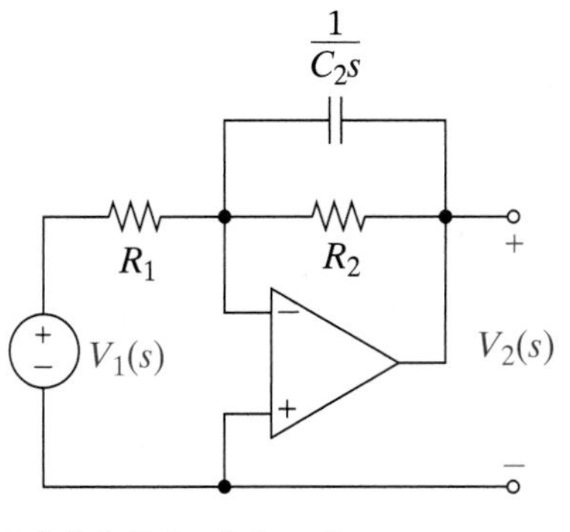

FIGURE 12–6

shows that the circuit transfer function has a low-pass form with

$$\omega_C = \alpha = \frac{1}{R_2C_2} \quad \text{and} \quad T(0) = -\frac{R_2}{R_1}$$

This is an inverting circuit, so the –90° phase swing of the low-pass form runs from $\angle T(0) = -180°$ to $\angle T(\infty) = -270°$, passing through $\angle T(j\omega_C) = -225°$ along the way.

Circuit Interpretation: The low-pass response is easily deduced from known circuit performance. At dc the capacitor acts like an open circuit and the circuit in Figure 12–6 reduces to a resistance inverting amplifier with $K = T(0) = -R_2/R_1$. At infinite frequency the capacitor acts like a short circuit that connects the OP AMP output directly to the inverting input. This connection results in zero output since the node voltage at the inverting input is necessarily zero. In between these two extremes the gain gradually decreases as the decreasing capacitor impedance gradually pulls the OP AMP output down to zero at infinite frequency.

(b) The design constraints require that $\omega_C = 1/R_2C_2 = 100$ and $|T(0)| = R_2/R_1 = 4$. Selecting $R_1 = 10$ kΩ implies that $R_2 = 40$ kΩ and $C_2 = 250$ nF. ■

APPLICATION EXAMPLE 12–4

In terms of frequency response, the ideal OP AMP model introduced in Chapter 4 assumes that the device has an infinite gain and an infinite bandwidth. A more realistic model of the device is shown in Figure 12–7(a). The controlled source gain in Figure 12–7(a) is a low-pass transfer function with a dc gain of A and a cutoff frequency ω_C. The straight-line asymptotes of controlled source gain are shown in Figure 12–7(b). The **gain-bandwidth product** ($G = A\omega_C$) is the basic performance parameter of this model.

With no feedback the OP AMP transfer function is the same as the controlled-source transfer function. The gain-bandwidth product of the open-loop transfer function is

$$G = A\omega_C \quad \text{(open loop)}$$

The closed-loop transfer function of the circuit in Figure 12–7(a) is found by writing the following device and connection equations:

$$\text{Device equation: } V_O = \frac{A}{s/\omega_C + 1}(V_P - V_N)$$

$$\text{Input connection: } V_P = V_S$$

$$\text{Feedback connection: } V_N = V_O$$

Substituting the connection equations into the OP AMP device equation yields

$$V_O(s) = \frac{A}{s/\omega_C + 1}(V_S(s) - V_O(s))$$

FIGURE 12–7

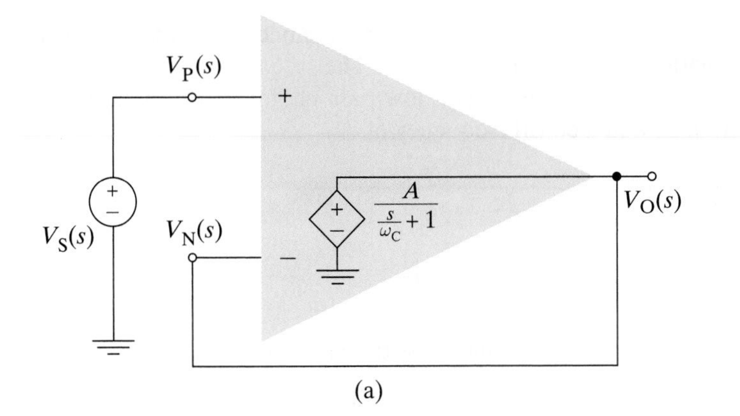

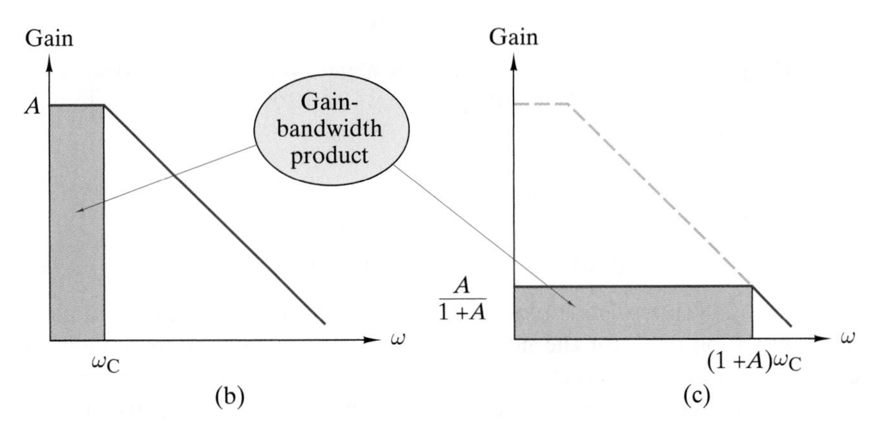

Solving for the closed-loop transfer function produces

$$T(s) = \frac{V_O(s)}{V_S(s)} = \frac{A}{A+1}\left[\frac{1}{\dfrac{s}{(A+1)\omega_C}+1}\right]$$

The straight-line asymptotes of the closed-loop transfer function are shown in Figure 12–7(c).

The closed-loop circuit has a low-pass transfer function with a dc gain of $A/(A+1)$ and a cutoff frequency of $(A+1)\omega_C$. The gain-bandwidth product of the closed-loop circuit is

$$G = \left[\frac{A}{A+1}\right][(A+1)\omega_C] = A\omega_C \quad \text{(closed loop)}$$

which is the same as the open-loop case. In other words, the gain-bandwidth product is invariant and is not changed by feedback. It can be shown that this result is a general one that applies to all linear OP AMP circuits, regardless of the circuit configuration.

Gain-bandwidth product is a fundamental parameter that limits the frequency response of OP AMP circuits. For example, an OP AMP with a gain-bandwidth product of $G = 10^6$ Hz is connected as a noninverting amplifier with a closed-loop gain of 20. The frequency response of the resulting closed-loop circuit has a low-pass characteristic with a passband gain of 20 and a cutoff frequency of

$$f_C = \frac{10^6}{20} = 50 \text{ kHz}$$

FIRST-ORDER HIGH-PASS RESPONSE

We next treat the first-order high-pass transfer function

$$T(s) = \frac{Ks}{s + \alpha} \tag{12–8}$$

The high-pass function differs from the low-pass case by the introduction of a zero at $s = 0$. Replacing s by $j\omega$ in $T(s)$ and solving for the gain and phase functions yields

$$|T(j\omega)| = \frac{|K|\omega}{\sqrt{\omega^2 + \alpha^2}}$$
$$\theta(\omega) = \angle K + 90° - \tan^{-1}(\omega/\alpha) \tag{12–9}$$

Figure 12–8 shows Bode plots for the first-order high-pass function plotted using a log scale for the normalized frequency ω/α. The gain diagram displays a high-pass characteristic with a passband at high frequency and a stopband at low frequency. The maximum passband gain occurs at high frequency($\omega >> \alpha$) where the gain $|T(j\omega)| \rightarrow T_{max} = |K|$. In the passband the gain gradually decreases as frequency decreases until at $\omega = \alpha$ we have

$$|T(j\alpha)| = \frac{|K|\alpha}{\sqrt{\alpha^2 + \alpha^2}} = \frac{|K|}{\sqrt{2}} = \frac{1}{\sqrt{2}} T_{max} \tag{12–10}$$

Again invoking Eq. (12–2), we find that $\omega_C = \alpha$ is the cutoff frequency marking the boundary between a high-frequency passband and a low-frequency stopband.

The low- and high-frequency gain asymptotes approximate the gain response in Figure 12–8. The high-frequency asymptote ($\omega >> \alpha$) is the horizontal line whose ordinate is $|K|$ (slope = 0 or 0 dB/decade). The low-frequency asymptote ($\omega << \alpha$) is a line of the form $|K|\omega/\alpha$ (slope = +1 or +20 dB/decade). The intersection of these two asymptotes occurs when $|K| = |K|\omega/\alpha$, which defines a corner frequency at $\omega = \alpha$.

The semilog plot of the phase shift of the first-order high-pass function is shown in Figure 12–8. The phase shift approaches $\angle K$ at high frequency, passes through $\angle K + 45°$ at the cutoff frequency, and approaches $\angle K + 90°$ at low frequency. Most of the 90° phase change occurs over the two-decade range centered on the cutoff frequency. The phase shift can be approximated by the straight-line segments shown in Figure 12–8. As in the low-pass case, $\angle K$ is 0° when K is positive and ±180° when K is negative.

Like the low-pass function, the first-order high-pass Bode plots can be approximated by straight-line segments. To construct these lines we need two parameters, $T(\infty)$ and α. The parameter α defines the cutoff frequency,

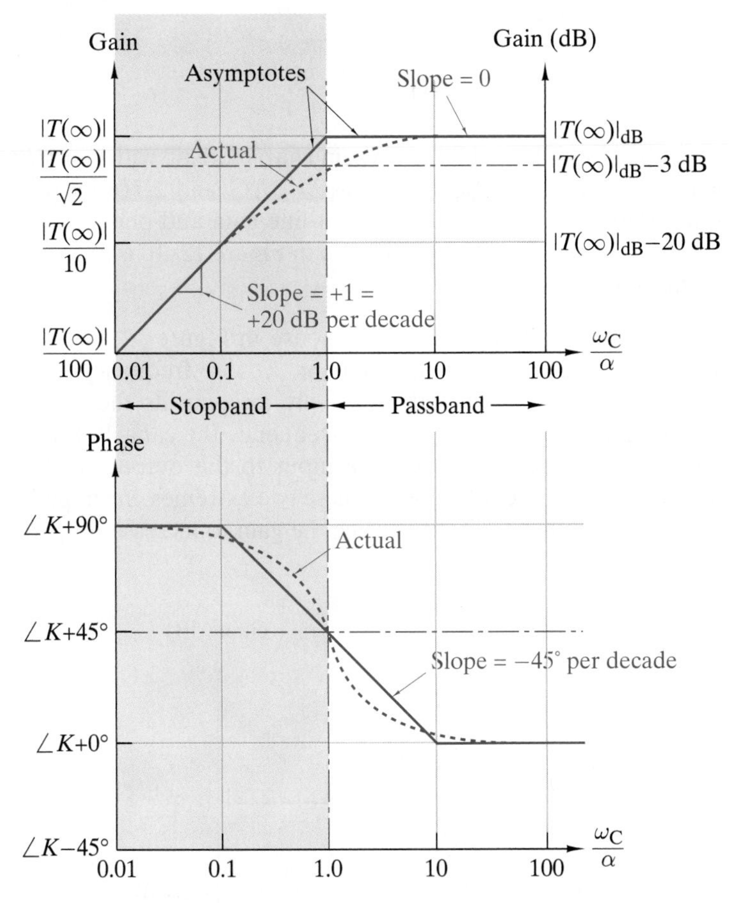

FIGURE 12–8 *First-order high-pass Bode plots.*

and the quantity $T(\infty)$ gives the passband gain $|T(\infty)|$ and the high-frequency phase angle $\angle T(\infty)$. The quantities $T(\infty)$ and α can be determined directly from the transfer function or estimated directly from the circuit in some cases. The straight line shows that the first-order high-pass response can be characterized by calculating the gain and phase over a two-decade band from one decade below to one decade above the cutoff frequency.

EXAMPLE 12–5

Show that the transfer function $T(s) = V_2(s)/V_1(s)$ in Figure 12–9 has a high-pass gain characteristic. Construct the straight-line approximations to the gain and phase responses of the circuit.

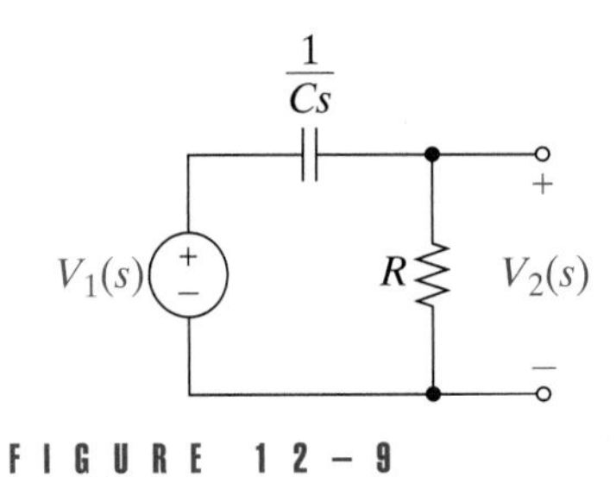

FIGURE 12–9

SOLUTION:

Applying voltage division, the voltage transfer function for the circuit is

$$T(s) = \frac{R}{R + 1/Cs} = \frac{RCs}{RCs + 1}$$

Rearranging Eq. (12–8) as

$$T(s) = \frac{K(s/\alpha)}{s/\alpha + 1}$$

shows that the circuit has a high-pass gain characteristic with $\alpha = 1/RC$ and $T(\infty) = 1$. Therefore, $|T(\infty)|_{\text{dB}} = 0$ dB, $\omega_C = 1/RC$, and $\angle T(\infty) = 0°$. Given these quantities, we construct the straight-line gain and phase approximations in Figure 12–10. The frequency scale in Figure 12–10 is normalized by multiplying ω by $RC = 1/\alpha$.

Circuit Interpretation: The high-pass response in Figure 12–10 can be understood in terms of known circuit behavior. At zero frequency the capacitor acts like an open circuit that effectively disconnects the input signal source, leading to zero gain. At infinite frequency the capacitor acts like a short circuit that directly connects the input to the output, leading to a passband gain of 1 (or 0 dB). Between these two extremes, the impedance of the capacitor gradually decreases, causing the gain to increase. In particular,

FIGURE 12–10

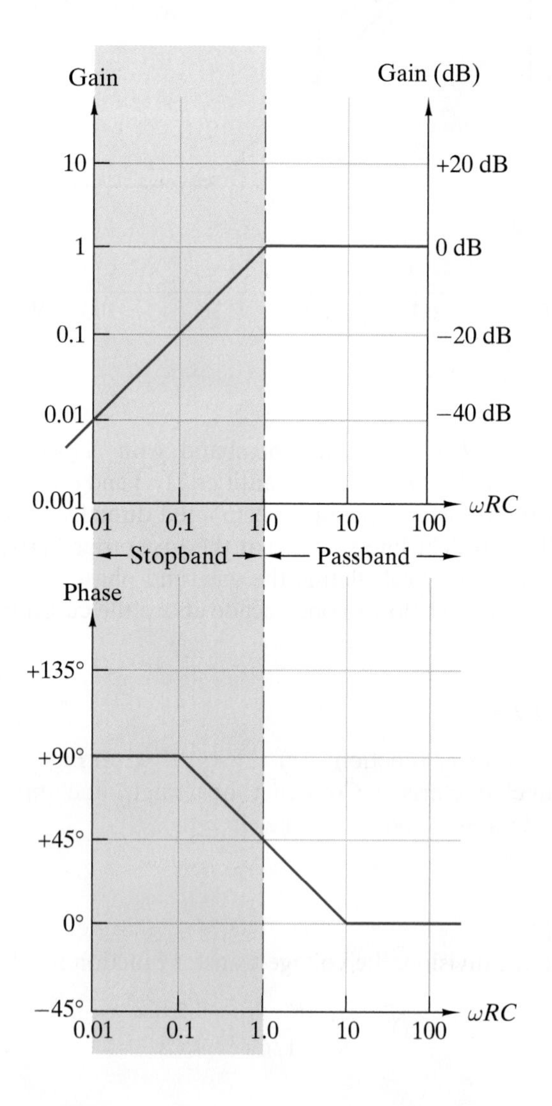

at the cutoff frequency we have $1/\omega C = R$, the impedance of the capacitor is $1/j\omega C = -jR$, and the transfer function is

$$T(j\omega_C) = \frac{R}{R - jR} = \frac{1}{\sqrt{2}}\angle{+45^\circ}$$

In other words, at the cutoff frequency the gain is –3 dB and the phase shift +45°. Obviously, the decreasing impedance of the capacitor gives the circuit its high-pass gain characteristics. ■

DESIGN EXAMPLE 12–6

(a) Show that the transfer function $T(s) = V_2(s)/V_1(s)$ of the circuit in Figure 12–11 has a high-pass gain characteristic.

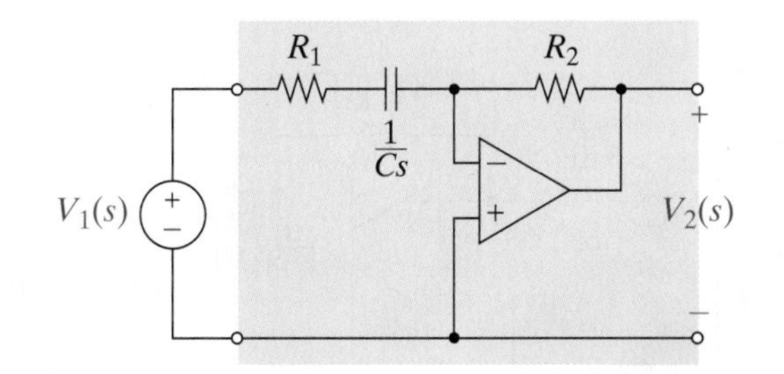

FIGURE 12 – 11

(b) Select the element values to produce a passband gain of 4 and a cutoff frequency of 40 krad/s.

SOLUTION:

(a) The branch impedances of the inverting OP AMP configuration in Figure 12–11 are

$$Z_1(s) = R_1 + \frac{1}{Cs} = \frac{R_1Cs + 1}{Cs} \quad \text{and} \quad Z_2(s) = R_2$$

and the voltage transfer function is

$$T(s) = -\frac{Z_2(s)}{Z_1(s)} = -\frac{R_2Cs}{R_1Cs + 1} = \frac{(-R_2/R_1)s}{s + 1/R_1C}$$

This results in a high-pass transfer function of the form $Ks/(s + \alpha)$ with $K = -R_2/R_1$ and $\alpha = \omega_C = 1/R_1C$.

Circuit Interpretation: The high-pass response of this circuit is easily understood in terms of element impedances. At dc the capacitor in Figure 12–11 acts like an open circuit that effectively disconnects the input source, resulting in zero gain. At infinite frequency the capacitor acts like a short circuit that reduces the circuit to an inverting amplifier with $K = T(\infty) = -R_2/R_1$. As the frequency varies from zero to infinity, the gain gradually increases as the capacitor impedance decreases.

(b) The design requirements specify that $1/R_1C = 4 \times 10^4$ and $R_2/R_1 = 4$. Selecting $R_1 = 10$ kΩ requires $R_2 = 40$ kΩ and $C = 2.5$ nF. ■

EVALUATION EXAMPLE 12–7

Your company issued a request for proposals listing the following design requirements and evaluation criteria.

Design requirements call for a high-pass filter with a passband gain of unity and a cutoff frequency of 150 Hz ± 10%. The filter input is driven by a sensor with a 50-Ω source resistance.

Evaluation criteria are filter performance, parts count, power consumption, and cost. The three vendors have responded with the designs shown in Figure 12–12. As a junior engineer, you have been asked to evaluate the designs and identify the best design. Which vendor would you recommend and why?

FIGURE 12–12

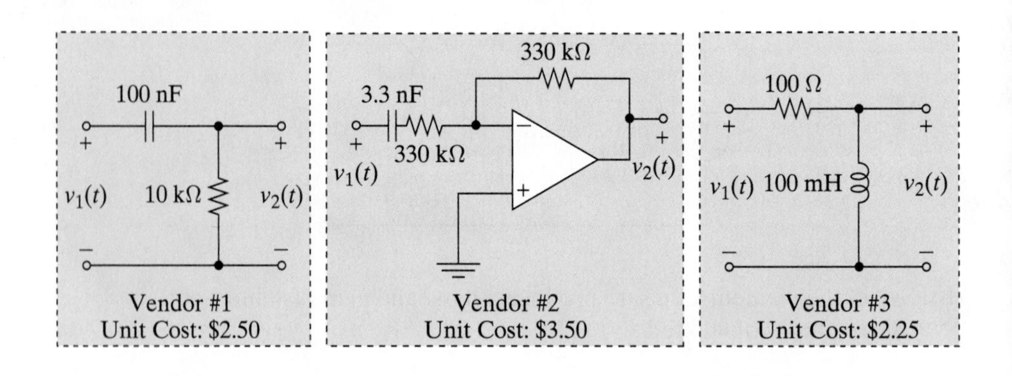

SOLUTION:

The design requirements describe a high-pass transfer function of the form

$$T(s) = \frac{Ks}{s + \alpha} = \frac{\pm 1s}{s + 2\pi 150} = \frac{\pm s}{s + 942.5}$$

The first task is to analyze the performance of the three proposed filters.

Vendor #1: Using voltage division to find $T(s)$ yields

$$T(s) = \frac{10^4}{10^4 + \dfrac{1}{10^{-7}s}} = \frac{10^{-3}s}{10^{-3}s + 1} = \frac{s}{s + 1000}$$

This is a high-pass response with $|T(\infty)| = 1$ and $f_C = 1000/(2\pi) = 159.2$ Hz.

Vendor #2: The circuit is an inverting amplifier whose transfer function is

$$T(s) = \frac{-330 \times 10^3}{330 \times 10^3 + \dfrac{1}{3.3 \times 10^{-9}s}} = \frac{-1.089 \times 10^{-3}s}{1.089 \times 10^{-3}s + 1} = \frac{-s}{s + 918.3}$$

This is a high-pass response with $|T(\infty)| = 1$ and $f_C = 918.3/(2\pi) = 146.2$ Hz.

Vendor #3: Using voltage division to find $T(s)$ yields

$$T(s) = \frac{0.1s}{0.1s + 100} = \frac{s}{s + 1000}$$

This is a high-pass response with $|T(\infty)| = 1$ and $f_C = 1000/(2\pi) =$ 159.2 Hz.

TABLE 12–3 ANALYSIS SUMMARY

	PASSBAND GAIN	CUTOFF FREQ. (HZ)	PARTS COUNT	POWER CONSUMPTION	COST	SOURCE LOADING
Vendor #1	1	159.2	2	Low	\$2.50	No
Vendor #2	−1	146.2	4	Medium	\$3.50	No
Vendor #3	1	159.2	2	Low	\$2.25	Yes

Evaluation Discussion: The analysis summary in Table 12–3 shows that all three filters meet the basic passband gain (±1) and cutoff frequency (150 Hz $\pm$ 10%) requirements. The vendor #2 design can be eliminated because it has the largest part count, requires a dc power supply for its OP AMP, and has the highest cost. The vendor #3 filter has a serious loading problem that offsets its lower cost. This filter has a 100-Ω input resistor that loads the specified 50-Ω input source. The actual cutoff frequency of the source/filter combination is $(R_{\text{source}} + R_{\text{filter}})/L = (50 + 100)/\ 0.1 =$ 1500 rad/s or 238.7 Hz, not the 159.2 Hz found by analyzing the filter in isolation. The input impedance of the vendor #1 design is greater than 10 kΩ, so it does not load the 50-Ω source. All factors considered, the best design is the filter proposed by vendor #1. ■

Exercise 12–1

For each circuit in Figure 12–13, identify whether the gain response has low-pass or high-pass characteristics and find the passband gain and cutoff frequency.

Answers:

(a) High pass, $|T(\infty)| = 1/3$, $\omega_C = 66.7$ rad/s
(b) Low pass, $|T(0)| = 2/3$, $\omega_C = 300$ rad/s
(c) Low pass, $|T(0)| = 1$, $\omega_C = 333$ krad/s
(d) High pass, $|T(\infty)| = 1/3$, $\omega_C = 333$ krad/s

Exercise 12–2

State whether the following transfer functions have low-pass or high-pass gain characteristics and find the passband gain and cutoff frequency.

(a) $T_1(s) = \dfrac{1}{10s^{-1} + 10^{-3}}$

(b) $T_2(s) = \dfrac{10^2}{25s + 10^3}$

(c) $T_3(s) = \dfrac{20/s}{50 + 20/s}$

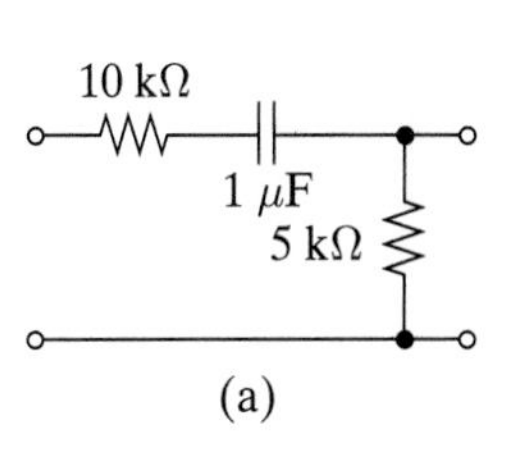

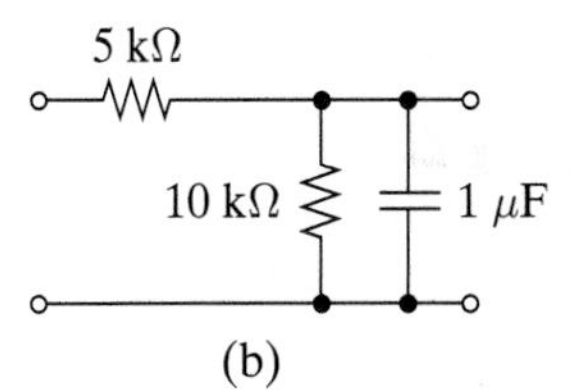

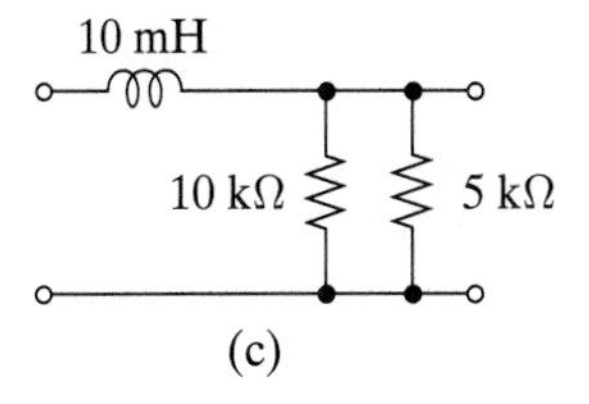

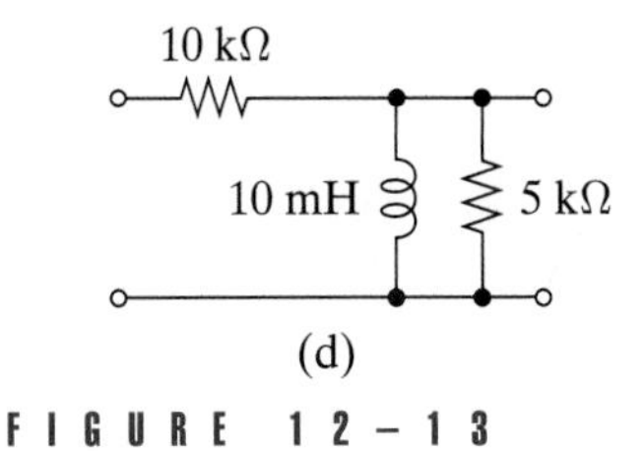

FIGURE 12–13

Answers:

(a) High pass, passband gain = 1000, ω_C = 10 krad/s
(b) Low pass, passband gain = 0.1, ω_C = 40 rad/s
(c) Low pass, passband gain = 1, ω_C = 0.4 rad/s

12–4 BANDPASS AND BANDSTOP RESPONSES

Bandpass and bandstop responses can be obtained using first-order high-pass and low-pass circuits as building blocks. Figure 12–14 shows a cascade connection of first-order high-pass and low-pass circuits. When the second stage does not load the first, the overall transfer function can be found by the chain rule as

$$T(s) = T_1(s) \times T_2(s) = \underbrace{\left(\frac{K_1 s}{s + \alpha_1}\right)}_{\text{high pass}} \underbrace{\left(\frac{K_2}{s + \alpha_2}\right)}_{\text{low pass}} \tag{12–11}$$

Replacing s by $j\omega$ in Eq. (12–11) and solving for the gain response yields

$$|T(j\omega)| = \underbrace{\left(\frac{|K_1|\omega}{\sqrt{\omega^2 + \alpha_1^2}}\right)}_{\text{high pass}} \underbrace{\left(\frac{|K_2|}{\sqrt{\omega^2 + \alpha_2^2}}\right)}_{\text{low pass}} \tag{12–12}$$

Note that the overall gain is zero at $\omega = 0$ and again at infinite frequency. This pattern suggests a bandpass response.

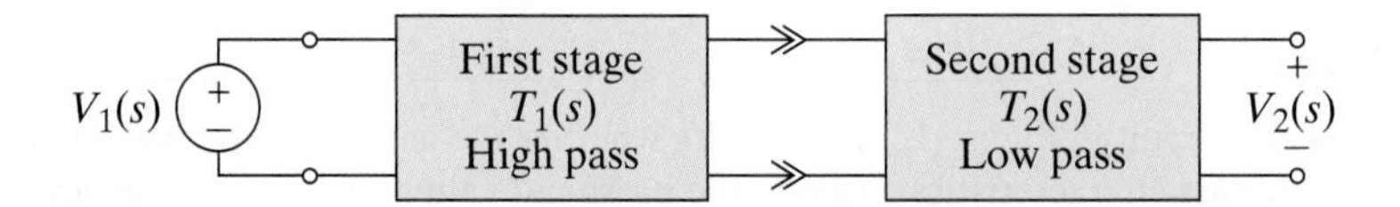

FIGURE 12–14 *Cascade connection of high-pass and low-pass circuits.*

The overall gain in Eq. (12–12) is bandpass when the high-pass cutoff frequency is much lower than the low-pass cutoff frequency ($\alpha_1 << \alpha_2$). To see why, we develop the gain asymptotes for Eq. (12–12) in three frequency ranges.

- **Low frequency** ($\omega << \alpha_1 << \alpha_2$): This range falls in the stopband of the high-pass gain and the passband of the low-pass gain. As a result, the overall gain approaches

$$|T(j\omega)| \rightarrow \underbrace{\left(\frac{|K_1|\omega}{\alpha_1}\right)}_{\text{high pass}} \underbrace{\left(\frac{|K_2|}{\alpha_2}\right)}_{\text{low pass}} = \frac{|K_1||K_2|\omega}{\alpha_1\alpha_2}$$

- **High frequency** ($\alpha_1 << \alpha_2 << \omega$): This range falls in the passband of the high-pass gain and the stopband of the low-pass gain. In this range the overall gain approaches

$$|T(j\omega)| \to \underbrace{\left(\frac{|K_1|\omega}{\omega}\right)}_{\text{high pass}}\underbrace{\left(\frac{|K_2|}{\omega}\right)}_{\text{low pass}} = \frac{|K_1||K_2|}{\omega}$$

- **Mid frequency** ($\alpha_1 << \omega << \alpha_2$): This range falls in the passband of both first-order gains, so the overall gain approaches

$$|T(j\omega)| \to \underbrace{\left(\frac{|K_1|\omega}{\omega}\right)}_{\text{high pass}}\underbrace{\left(\frac{|K_2|}{\alpha_2}\right)}_{\text{low pass}} = \frac{|K_1||K_2|}{\alpha_2}$$

Figure 12–15 shows a plot of the low-, mid-, and high-frequency gain asymptotes. The low-frequency and mid-frequency asymptotes intersect when $|K_1K_2|\omega/\alpha_1\alpha_2 = |K_1K_2|/\alpha_2$. This occurs at $\omega = \alpha_1$, which is the cutoff frequency of the high-pass stage. The high-frequency and mid-frequency asymptotes intersect when $|K_1K_2|/\omega = |K_1K_2|/\alpha_2$. This occurs at $\omega = \alpha_2$, which is the cutoff frequency of the low-pass stage. The straight-line gain plot based on these asymptotes indicates a passband between cutoff frequencies $\omega_{C1} = \alpha_1$ and $\omega_{C2} = \alpha_2$. The mid-frequency gain applies in this passband. Finally, there are two stopbands; one below ω_{C1} and the other above ω_{C2}.

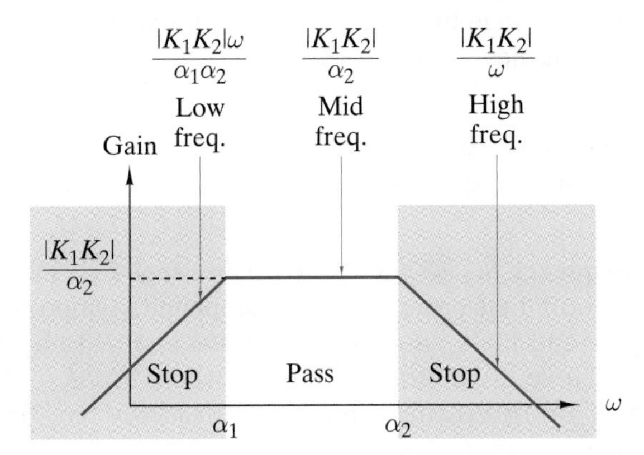

FIGURE 12–15 *Bandpass gain characteristic.*

The input signal to the bandpass cascade must pass through both a high-pass and a low-pass stage to reach the output. In the parallel connection in Figure 12–16, an input can reach the output via either a high-pass or a low-pass path. As a result, the overall transfer function is the sum of the high-pass and low-pass transfer functions.

$$T(s) = T_1(s) + T_2(s) = \underbrace{\left(\frac{K_1 s}{s+\alpha_1}\right)}_{\text{high pass}} + \underbrace{\left(\frac{K_2}{s+\alpha_2}\right)}_{\text{low pass}}$$

Since the overall gain is a sum, we must consider each of these paths separately. Replacing s by $j\omega$ and solving for the individual path gains gives

$$|T_1(j\omega)| = \underbrace{\left(\frac{|K_1|\omega}{\sqrt{\omega^2+\alpha_1^2}}\right)}_{\text{high pass}} \quad \text{and} \quad |T_2(j\omega)| = \underbrace{\left(\frac{|K_2|}{\sqrt{\omega^2+\alpha_2^2}}\right)}_{\text{low pass}} \tag{12–13}$$

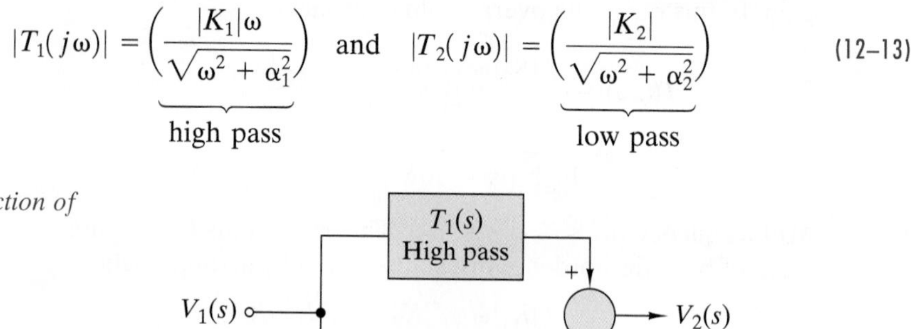

FIGURE 12–16 *Parallel connection of high-pass and low-pass circuits.*

The overall gain has a bandstop characteristic when the high-pass cutoff frequency is much higher than the low-pass cutoff frequency ($\alpha_2 << \alpha_1$). To see why, we develop the gain asymptotes of Eq. (12–13) in three frequency ranges.

- **Low frequency** ($\omega << \alpha_2 << \alpha_1$): This range falls in the stopband of the high-pass gain and the passband of the low-pass gain. In this range the low-pass path dominates, so the overall gain approaches $|T(j\omega)| \to |K_2|/\alpha_2$.
- **High frequency** ($\alpha_2 << \alpha_1 << \omega$): This range falls in the passband of the high-pass gain and the stopband of the low-pass gain. In this range the high-pass path dominates, so the overall gain ap`proaches $|T(j\omega)| \to |K_1|$.

Thus, there is a low-frequency passband and a high-frequency passband. In a bandstop response these passbands normally have the same gain, so that $|K_1| = |K_2|/\alpha_2$. A mid-frequency stopband lies in between these two passbands.

- **Mid frequency** ($\alpha_2 << \omega << \alpha_1$): This range falls in the stopband of both first-order gains. The stopband asymptotes of the low-pass and high-pass gains are $|K_2|/\omega$ and $|K_1|\omega/\alpha_1$, respectively. These asymptotes intersect when $|K_2|/\omega = |K_1|\omega/\alpha_1$. Since $|K_1| = |K_2|/\alpha_2$ the intersection frequency turns out to be $\omega = \sqrt{\alpha_1\alpha_2}$.

Figure 12–17 is a plot of the low-, mid-, and high-frequency gain asymptotes. The low-pass gain dominates at frequencies below the intersection frequency $\sqrt{\alpha_1\alpha_2}$ and the high-pass gain dominates above the intersection. The straight-line gain plot indicates a stopband between cutoff frequencies at $\omega_{C1} = \alpha_2$ and $\omega_{C2} = \alpha_1$ and two passbands: one below ω_{C1} and the other above ω_{C2}.

This analysis shows that bandpass and bandstop responses can be obtained using first-order circuit building blocks. More importantly, the straight-line gain approximations to the first-order gains help us *understand* how the building blocks interact to produce other types of responses.

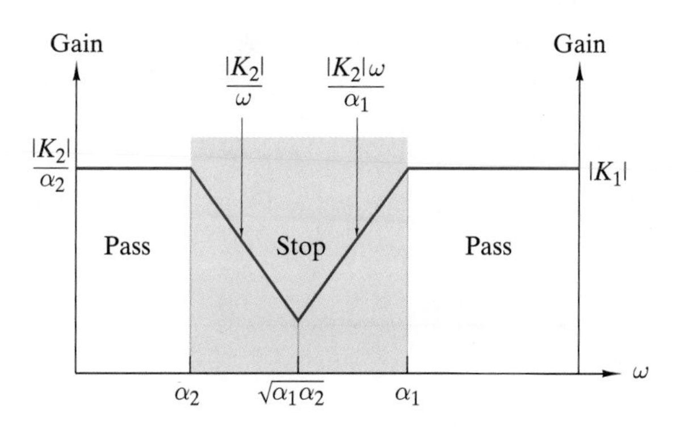

FIGURE 12–17 *Bandstop gain characteristic.*

This understanding explains the enduring utility of Bode plots. The Bode plots in Figures. 12–15 and 12–17 are a reasonably good approximation of the actual gain as long as the two first-order cutoff frequencies are widely separated. Although *widely separated* has no precise definition, the straight-line gain works fairly well when $10\alpha_1 << \alpha_2$ for a bandpass response and $\alpha_1 > 10\alpha_2$ for a bandstop response.

DESIGN EXAMPLE 12–8

Design a bandpass circuit with a passband gain of 10 and cutoff frequencies at 20 Hz and 20 kHz.

SOLUTION:

Our design uses a cascade connection of first-order low- and high-pass building blocks. The required transfer function has the form in Eq. (12–11) with the following constraints

$$\omega_{C1} = \alpha_1 = 2\pi(20) = 40\pi \text{ rad/s}$$

$$\omega_{C2} = \alpha_2 = 2\pi(20 \times 10^3) = 4\pi \times 10^4 \text{ rad/s}$$

$$\frac{|K_1 K_2|}{\alpha_2} = 10$$

Inserting these numerical values in Eq. (12–11) allows us to write the required transfer function as

$$T(s) = \underbrace{\left(\frac{s}{s + 40\pi}\right)}_{\text{high pass}} \underbrace{(10)}_{\text{gain}} \underbrace{\left(\frac{4\pi \times 10^4}{s + 4\pi \times 10^4}\right)}_{\text{low pass}}$$

This transfer function can be realized using the three-stage cascade circuit in Figure 12–18. The first stage is the *RC* high-pass circuit from Example 12–5 and the third stage is the *RL* low-pass circuit from Example 12–2. The noninverting OP AMP circuit in the second stage serves two purposes: (1) It isolates the first and third stages so the chain rule applies, and (2) it supplies the mid-band gain. Using the chain rule, the circuit transfer function is

$$T(s) = \underbrace{\left(\frac{s}{s + 1/R_C C}\right)}_{\text{high pass}} \underbrace{\left(\frac{R_1 + R_2}{R_1}\right)}_{\text{gain}} \underbrace{\left(\frac{R_L/L}{s + R_L/L}\right)}_{\text{low pass}}$$

FIGURE 12–18

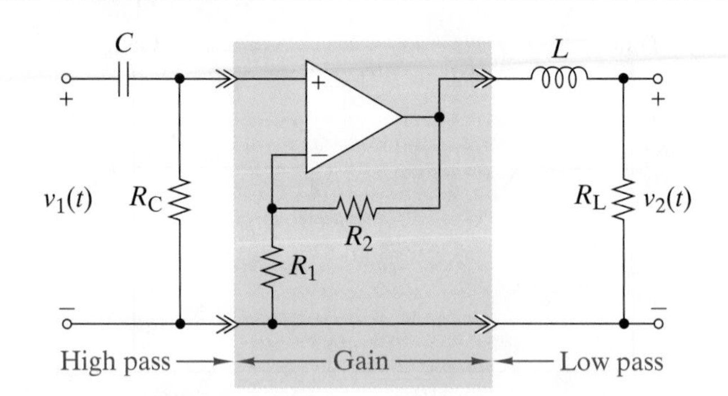

Comparing the circuit transfer function to the required transfer function leads to the following constraints and stage designs. As always, many other solutions are possible.

STAGE	CONSTRAINT	STAGE DESIGN
High pass	$R_C C = 1/40\pi$	Let $R_C = 100\ \text{k}\Omega$; then $C = 79.6$ nF
Gain	$1 + R_2/R_1 = 10$	Let $R_1 = 10\ \text{k}\Omega$; then $R_2 = 90\ \text{k}\Omega$
Low pass	$L/R_L = 1/40000\pi$	Let $R_L = 200\ \text{k}\Omega$; then $L = 0.628$ H

■

Exercise 12–3

Select the element values in Figure 12–18 so that the passband gain is 6 dB and the cutoff frequencies are 1 krad/s and 50 krad/s.

Answers: $R_C = 5\ \text{k}\Omega$; $C = 200$ nF; $R_1 = R_2$; $R_L = 20\ \text{k}\Omega$; $L = 400$ mH

Exercise 12–4

Two first-order circuits in a cascade connection have the following transfer functions.

$$T_1(s) = \frac{20}{\frac{s}{2000} + 1} \quad \text{and} \quad T_2(s) = \frac{s}{s + 40}$$

What are the cutoff frequencies and the passband gain? Assume the chain rule applies.

Answers: $\omega_{C1} = 40$ rad/s; $\omega_{C1} = 2000$ rad/s; passband gain $= 20$

EXAMPLE 12–9

Use computer-aided analysis to show that the circuit in Figure 12–19 implements a bandstop gain response.

SOLUTION:

This circuit implements the block diagram in Figure 12–16. The two first-order circuits have the same input. The upper circuit is the high-pass OP

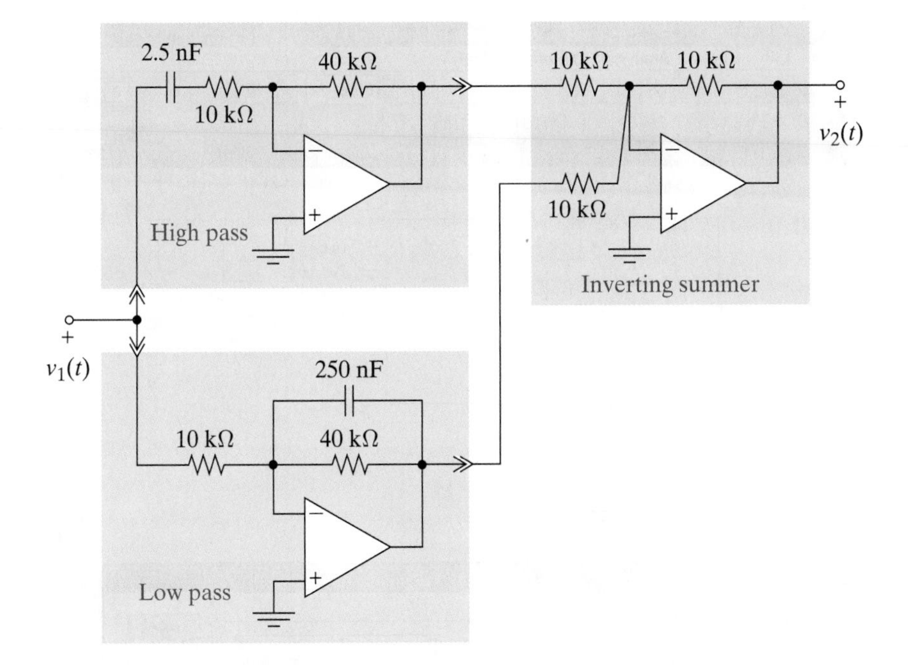

FIGURE 12–19

AMP circuit designed in Example 12–6 to have a passband gain of 4 and a cutoff frequency $\alpha_1 = 40$ krad/s. The lower path is the low-pass OP AMP circuit designed in Example 12–3 to have a passband gain of 4 and a cutoff frequency $\alpha_2 = 100$ rad/s. The inverting summer at the far right implements the summing point function in Figure 12–16. Since the cutoff frequency of the low-pass circuit is much lower than the cutoff frequency of the high-pass circuit, we expect to see two passbands on either side of a stopband centered at

$$\omega = \sqrt{\alpha_1 \alpha_2} = \sqrt{100 \times 4 \times 10^4} = 2000 \text{ rad/s}$$

Figure 12–20 shows the circuit as drawn in the Electronics WorkBench circuit window. The input voltage source $\mathbf{V}_1$ is set to produce 1-V ac, so the circuit gain is equal to the ac voltage at node 8 (the summer output). Selecting **AC Frequency** from the **Analysis** menu brings up the **AC Frequency Analysis** dialog box shown in the figure. Here we must specify the analysis frequency range. This range should extend from at least a decade below the lowest cutoff frequency ($\alpha_2/2\pi$ = 16 Hz) to at least a decade above the highest cutoff frequency ($\alpha_1/2\pi$ = 6.4 kHz). To span this range we select a **Start frequency (FSTART)** of 1 Hz and an **End frequency (FSTOP)** of 100 kHz. To span this range we choose a **Decade** sweep and select the summer output (node 8) as the **Node for analysis.** Clicking the simulate button launches an ac frequency analysis that produces the output voltage plot shown at the bottom of Figure 12–20.

As expected, the frequency response of the overall circuit has two passbands surrounding a stopband. In this plot, one cursor is set in the lower passband and the other at the stopband minimum. The passband output voltage is y_1 = 3.9921 V (gain = 12 dB) at x_1 = 1.0000 = 1 Hz, which confirms that the two passbands have gain = 4 as straight-line analysis predicts.

FIGURE 12–20

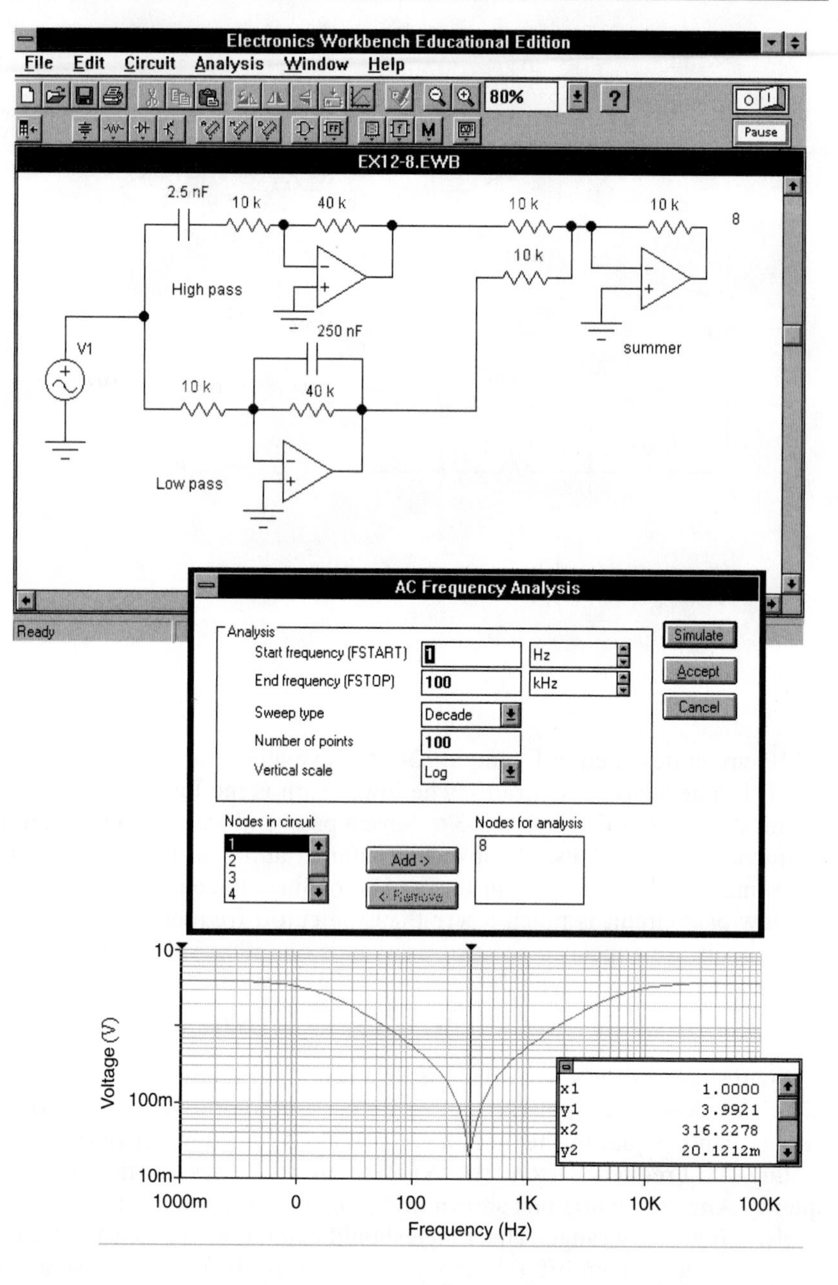

The minimum output is y_2 = 20.1212 mV (gain = −33.9 dB) at x_2 = 316.2278 Hz. This notch frequency corresponds to a radian frequency of $\omega = 2\pi x_2$ = 1987 rad/s, which is very close to the 2000 rad/s predicted by straight-line gain analysis. ■

12–5 THE FREQUENCY RESPONSE OF *RLC* CIRCUITS

A second-order circuit can take many forms as long as it has two energy storing elements. However, the properties of second-order circuits are traditionally introduced using simple series and parallel *RLC* circuits. In Chapter 7 these circuits were used to introduce the role of ζ and ω_0 in the transient response of second-order circuits. In this section they are used to introduce the role of impedance in second-order circuit frequency response. These canonic circuits are useful introductory vehicles because they give us physical insight into the relationship between circuit parameters and circuit response.

SERIES *RLC* BANDPASS CIRCUIT

The transfer function of the *RLC* voltage divider in Figure 12–21 is

$$T(s) = \frac{V_2(s)}{V_1(s)} = \frac{R}{R + Ls + 1/Cs} = \frac{R}{R + Z_{LC}(s)} \quad (12\text{–}14)$$

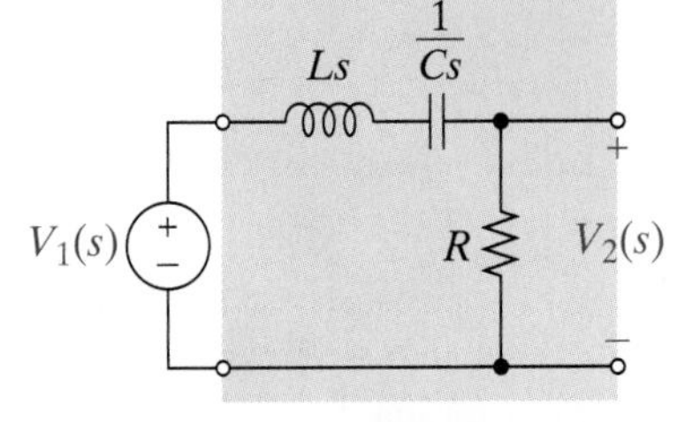

FIGURE 12–21 Series RLC bandpass circuit.

where $Z_{LC}(s) = Ls + 1/Cs$ is the impedance of the series leg of the voltage divider. To describe the frequency response of the circuit, we replace s by $j\omega$ to obtain

$$T(j\omega) = \frac{R}{R + Z_{LC}(j\omega)}$$

The variation of the impedance $Z_{LC}(j\omega)$ is the key to understanding *RLC* circuit frequency response.

Figure 12–22 shows how the impedance $Z_{LC}(j\omega) = j(\omega L - 1/\omega C)$ produces a bandpass gain response. At $\omega = 0$ (dc) the capacitor acts like an open

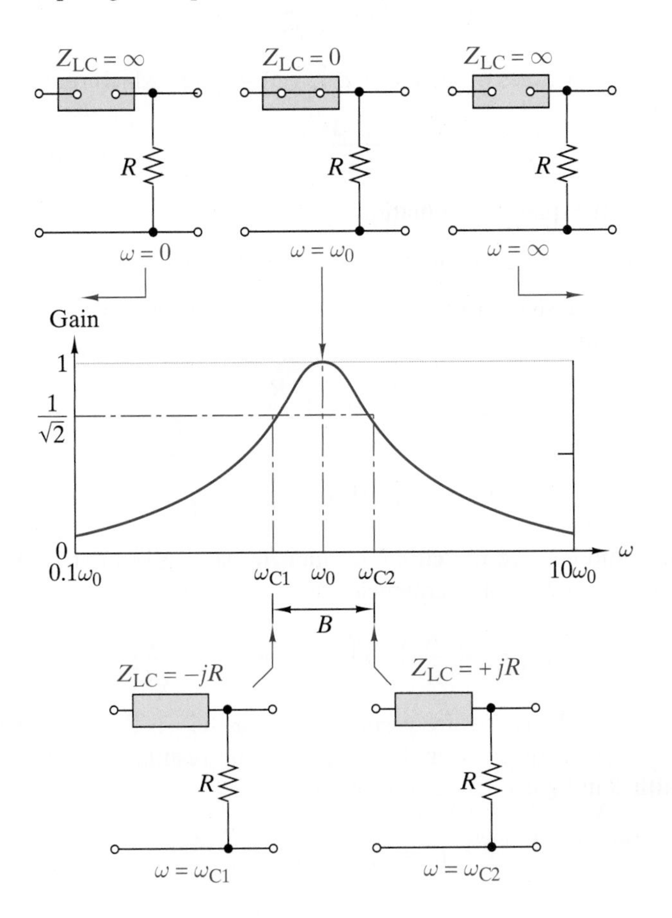

FIGURE 12–22 Effect of Z_{LC} on the series RLC bandpass response.

circuit and $Z_{LC}(0) = \infty$. At $\omega = \infty$ the inductor acts like an open circuit and $Z_{LC}(\infty) = \infty$. In either case, the open circuit effectively disconnects the input source, making $V_2 = 0$ and producing zero-gain stopbands at low and high frequencies.

At $\omega = \omega_0 = 1/\sqrt{LC}$, the impedance $Z_{LC}(j\omega_0) = 0$, since

$$Z_{LC}(j\omega_0) = j\left(\frac{L}{\sqrt{LC}} - \frac{\sqrt{LC}}{C}\right) = j\left(\sqrt{\frac{L}{C}} - \sqrt{\frac{L}{C}}\right) = 0$$

That is, the inductor and capacitor combine to produce a short circuit that makes a direct connection between the input and the output ports. This connection makes $V_2 = V_1$, producing a voltage gain of $|T(j\omega_0)| = T_{max} = 1$. Clearly T_{max} is the maximum voltage gain available from this voltage-divider circuit. The cutoff frequencies occur when $Z_{LC}(j\omega) = \pm jR$, since

$$|T(j\omega)| = \left|\frac{R}{R \pm jR}\right| = \frac{1}{\sqrt{2}} = \frac{1}{\sqrt{2}}T_{max}$$

so that the voltage gain is reduced by a factor of $1/\sqrt{2}$ from its maximum value at $\omega = \omega_0$.

Several parameters are used to describe the bandpass gain response. First, the maximum gain occurs at the **center frequency** ω_0 located at

$$\omega_0 = \frac{1}{\sqrt{LC}} \tag{12–15}$$

The cutoff frequencies occur when $Z_{LC}(j\omega) = \pm jR$, which requires that

$$\left(\omega L - \frac{1}{\omega C}\right) = \pm R$$

and leads to the quadratic equation

$$LC\omega^2 - (\pm R)C\omega - 1 = 0$$

Because of the $\pm$ sign, this quadratic has four roots, only two of which have physical meaning, namely

$$\begin{aligned}\omega_{C1} &= -\frac{R}{2L} + \sqrt{\left(\frac{R}{2L}\right)^2 + \frac{1}{LC}}\\ \omega_{C2} &= +\frac{R}{2L} + \sqrt{\left(\frac{R}{2L}\right)^2 + \frac{1}{LC}}\end{aligned} \tag{12–16}$$

where ω_{C1} and ω_{C2} are the **cutoff frequencies** shown in Figure 12–22. The product of the two cutoff frequencies in Eq. (12–16) is

$$\omega_{C1}\omega_{C2} = -\left(\frac{R}{2L}\right)^2 + \left(\frac{R}{2L}\right)^2 + \frac{1}{LC} = \frac{1}{LC} = \omega_0^2$$

In other words, the center frequency $\omega_0 = \sqrt{\omega_{C1}\omega_{C2}}$ is the geometric mean of the two cutoff frequencies. Finally, using the results in Eq. (12–16), the **bandwidth** B in Figure 12–22 is found to be

$$B = \omega_{C2} - \omega_{C1} = \frac{R}{L} \tag{12–17}$$

In summary, the descriptive parameters ω_0, ω_{C1}, ω_{C2}, and B are related to the series RLC circuit parameters by Eqs. (12–15), (12–16), and (12–17).

For historical reasons, it is traditional to add a fifth descriptive parameter called the **quality factor** Q, defined as the ratio of center frequency over bandwidth, namely

$$Q = \frac{\omega_0}{B} = \frac{\sqrt{L/C}}{R} \qquad (12\text{–}18)$$

Figure 12–23 shows the effect of Q on the bandpass response characteristics. When $Q \gg 1$, the response is said to be *narrow band*, since $B \ll \omega_0$. Conversely, when $Q \ll 1$, the response is said to be *wide band*, since $B \gg \omega_0$. Thus, $Q = 1$ is the dividing point between narrow-band or high-Q responses and wide-band or low-Q responses. But regardless of the value of Q, the maximum gain is always $T_{max} = 1$ (or 0 dB) at $\omega = \omega_0$.

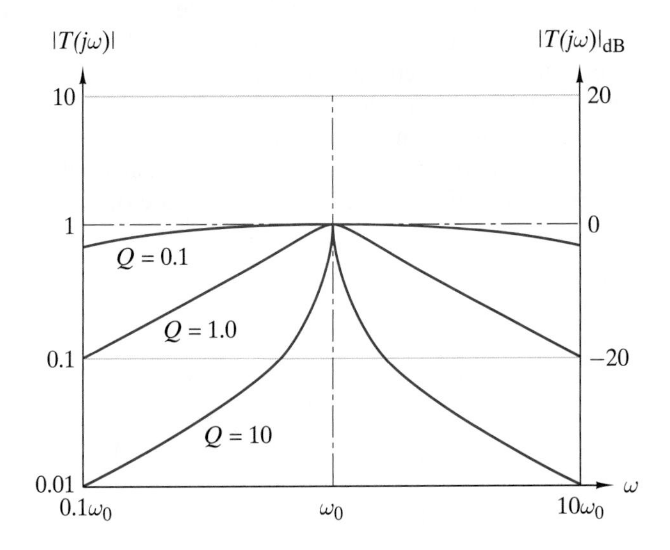

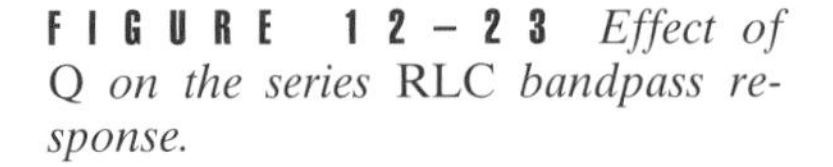
FIGURE 12–23 *Effect of Q on the series* RLC *bandpass response.*

A high-Q, bandpass circuit is sometimes called a *tuned filter.* This terminology applies when the center frequency is carefully adjusted (tuned) to select a narrow band of signal frequencies while rejecting a much broader range of signal frequencies outside of the passband. At the center frequency $Z_{LC}(j\omega_0) = 0$, so the filter input impedance is $R + j0$, a condition known as *resonance.* For this reason the center frequency is also called the *resonant frequency.*

EXAMPLE 12–10

A series RLC circuit has a center frequency of $\omega_0 = 20$ krad/s, a quality factor of $Q = 5$, and a resistance $R = 50\ \Omega$. Find the values of L, C, B, ω_{C1}, and ω_{C2}.

SOLUTION:

Using the definition of Q, the bandwidth is $B = \omega_0/Q = 4$ krad/s. For a series RLC circuit, $B = R/L$ and the inductance is $L = R/B = 50/4000 = 12.5$ mH. Using this inductance and the center frequency, the capacitance is found as $C = 1/\omega_0^2 L = 0.2\ \mu F$. Inserting these results into Eq. (12–16) yields the lower cutoff frequency as

$$\omega_{C1} = -2000 + \sqrt{2000^2 + 20{,}000^2} = 18.1 \text{ krad/s}$$

Using the definition of bandwidth, we get $\omega_{C2} = \omega_{C1} + B = 22.1$ krad/s.

■

D DESIGN EXAMPLE 12–11

Design a bandpass circuit using a series *RLC* circuit that meets the filter requirements in Example 12–8. Compare this *RLC* design with the circuit developed in Example 12–8.

SOLUTION:

The bandpass filter requirements given in Example 12–8 call for passband gain of 10 and cutoff frequencies at 20 Hz and 20 kHz. The series *RLC* circuit can produce the required cutoff frequencies. However, we also need a gain stage since the *RLC* circuit has a maximum gain of 1. The required bandpass response and gain can be obtained using the cascade connection in Figure 12–24. The chain rule applies to this circuit since the gain stage has an infinite input impedance that does not load the output of the *RLC* stage.

FIGURE 12–24

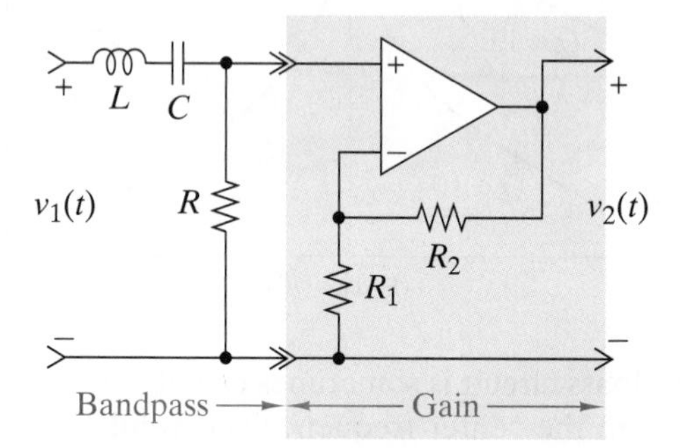

Given that $f_{C1} = 20$ Hz and $f_{C2} = 20$ kHz, the descriptive parameters of the bandpass stage are

$$\omega_0 = 2\pi\sqrt{f_{C1} f_{C2}} = 3.974 \text{ krad/s}$$

$$B = 2\pi(f_{C2} - f_{C1}) = 125.5 \text{ krad/s}$$

$$Q = \frac{\omega_0}{B} = 0.0317$$

Since $Q \ll 1$, the design requirements here and in Example 12–8 describe a wide-band filter. Selecting $R = 10\ k\Omega$, the inductance in the series *RLC* circuit is $L = R/B = 79.66$ mH. Using this inductance and the center frequency gives the capacitance as $C = 1/\omega_0^2 L = 0.795\ \mu F$. The gain stage is a noninverting amplifier for which a gain of 10 requires $1 + R_2/R_1 = 10$. Selecting $R_1 = 10\ k\Omega$ makes $R_2 = 90\ k\Omega$.

Evaluation Discussion: The design developed here uses the series *RLC* cascade in Figure 12–24. The design developed in Example 12–8 uses the cascade of first-order circuits shown in Figure 12–18. The element count for the two designs is summarized below.

FIGURE	DESCRIPTION	*R*	*L*	*C*	OP AMP
12–18	First-order cascade	4	1	1	1
12–24	Series *RLC* cascade	3	1	1	1

The minor difference in element count is not significant. What is significant is the difference in possible output stage loading. The circuit in Figure 12–24 is not subject to loading since the output is an OP AMP whose output impedance is zero. The first-order cascade in Figure 12–18 is susceptible to loading since its output stage is a passive *RL* circuit with a finite output impedance. ■

THE SERIES *RLC* BANDSTOP CIRCUIT

The bandpass circuit in Figure 12–21 and the bandstop circuit in Figure 12–25 are both series *RLC* circuits. The difference is that the bandpass circuit takes its output across the resistor while the bandstop circuit takes its output across the inductor and capacitor in series. The voltage transfer function of the bandstop voltage-divider circuit in Figure 12–25 is

$$T(s) = \frac{V_2(s)}{V_1(s)} = \frac{Ls + 1/Cs}{R + Ls + 1/Cs} = \frac{Z_{LC}(s)}{R + Z_{LC}(s)} \quad (12\text{–}19)$$

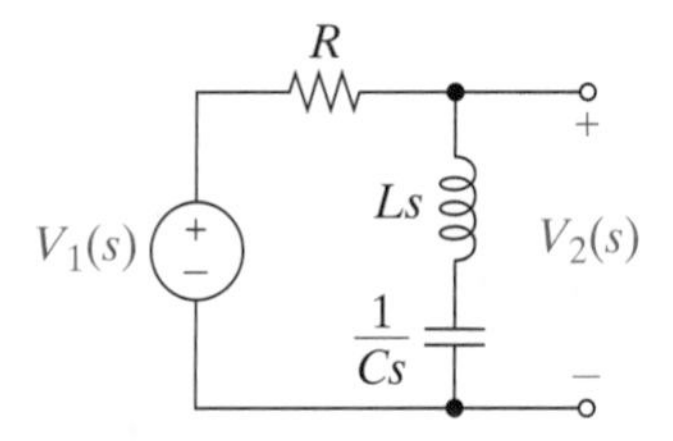

FIGURE 12–25 Series RLC bandstop circuit.

where $Z_{LC}(s) = Ls + 1/Cs$ is the impedance of the shunt leg of the voltage divider. This impedance is the key to understanding the shape of *RLC* circuit frequency response.

Figure 12–26 shows how the impedance $Z_{LC}(j\omega) = j(\omega L - 1/\omega C)$ produces a bandstop gain response. At $\omega = 0$, the capacitor acts like an open circuit so that the output voltage is $V_2 = V_{OC} = V_1$. At $\omega = \infty$, the inductor acts like an open circuit and again we have $V_2 = V_{OC} = V_1$. In either case, the condition $V_2 = V_1$ produces maximum gains of $T_{max} = 1$ in the low- and high-frequency passbands.

At $\omega = \omega_0 = 1/\sqrt{LC}$, the impedance $Z_{LC}(j\omega) = 0$. The resulting short circuit makes $V_2 = 0$, reducing the gain to zero and producing the null or notch in the gain response.

The cutoff frequencies occur when $Z_{LC}(j\omega) = \pm jR$, since

$$|T(j\omega)| = \left|\frac{\pm jR}{R \pm jR}\right| = \frac{1}{\sqrt{2}} = \frac{1}{\sqrt{2}}T_{max}$$

so that the voltage gain is reduced by a factor of $1/\sqrt{2}$ from its maximum value in the two passbands. In sum, the values $Z_{LC}(j\omega)$ that produce the bandpass response in Figure 12–22 produce the bandstop response shown in Figure 12–26.

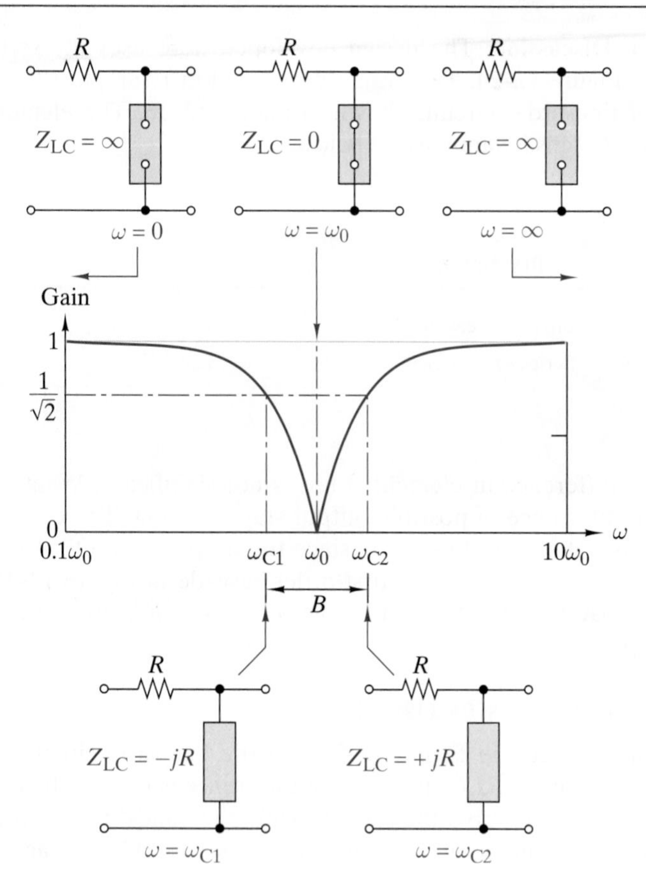

FIGURE 12–26 *Effect of Z_{LC} on the series RLC bandstop response.*

Since Z_{LC} controls bandstop gain response, the equations relating the descriptive parameters to circuit parameters are the same as those governing the bandpass case. Specifically, the notch occurs when $Z_{LC} = 0$ at a frequency of

$$\omega_0 = \frac{1}{\sqrt{LC}}$$

The passband cutoff frequencies occur when $Z_{LC} = \pm jR$ at frequencies of

$$\omega_{C1} = -\frac{R}{2L} + \sqrt{\left(\frac{R}{2L}\right)^2 + \frac{1}{LC}}$$

$$\omega_{C2} = +\frac{R}{2L} + \sqrt{\left(\frac{R}{2L}\right)^2 + \frac{1}{LC}}$$

Finally, the width of the stopband is

$$B = \omega_{C2} - \omega_{C1} = \frac{R}{L}$$

These equations are the same as Eqs. (12–15), (12–16), and (12–17) for the bandpass circuit, except that here they describe the location and width of the stopband rather than the passband.

The *RLC* bandstop circuit is best suited to narrow-band applications aimed at eliminating a single frequency. Narrow-band notch circuits are often used to eliminate power line noise at 60 Hz (50 Hz in many other countries), especially in biomedical instrumentation, where signal levels are usually very low.

D DESIGN EXAMPLE 12–12

A voltage source with a Thévenin resistance of 50 Ω has a spurious (undesirable) conducted emission at 25 krad/s. Connecting an inductor and capacitor in series across the 50-Ω source produces a bandstop filter that can eliminate the troublesome signal. To avoid reducing nearby useful signals, the stopband bandwidth must be less than 1 krad/s. Select values of L and C.

SOLUTION:

The stopband bandwidth limitation requires $B = R/L < 1$ krad/s. Since $R = 50\ \Omega$, this constraint means that $L > R/1000 = 50$ mH. To eliminate the spurious emission, the bandstop notch must be located at

$$\omega_0 = \frac{1}{\sqrt{LC}} = 25 \times 10^3 \text{ rad/s}$$

Selecting $L = 100$ mH makes the notch bandwidth $B = R/L = 0.5$ krad/s < 1 krad/s, which meets the bandwidth limitation. Given this inductance, the required notch frequency calls for a capacitance of $C = 1/\omega_0^2 L =$ 16 nF. ■

PARALLEL *RLC* BANDPASS CIRCUIT

Using current division, the current transfer function of the parallel *RLC* circuit in Figure 12–27 is written as

$$T(s) = \frac{I_2(s)}{I_1(s)} = \frac{1/R}{1/R + Cs + 1/Ls} = \frac{1/R}{1/R + Y_{\text{LC}}(s)} \qquad (12\text{–}20)$$

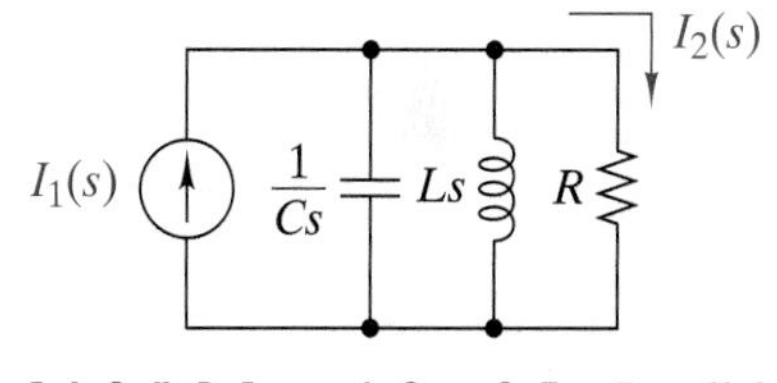

FIGURE 12–27 *Parallel RLC bandpass circuit.*

where $Y_{\text{LC}}(s) = Cs + 1/Ls$ is the admittance of the inductor and capacitor in parallel. The bandpass features of the parallel *RLC* circuit are controlled by this admittance.

The variation of the admittance $Y_{\text{LC}}(j\omega) = j(\omega C - 1/\omega L)$ produces a bandpass response for the following reasons. At $\omega = 0$ and $\omega = \infty$, the admittance $Y_{\text{LC}}(j\omega)$ is infinite, which is equivalent to a zero-impedance short circuit. This short circuit shunts all of the input current around the resistor in Figure 12–27, making $I_2 = 0$ and producing zero-gain stopbands at low and high frequencies.

At $\omega = \omega_0 = 1/\sqrt{LC}$, the admittance $Y_{\text{LC}}(j\omega) = 0$, since

$$Y_{\text{LC}}(j\omega_0) = j\left(\frac{C}{\sqrt{LC}} - \frac{\sqrt{LC}}{L}\right) = j\left(\sqrt{\frac{C}{L}} - \sqrt{\frac{C}{L}}\right) = 0$$

A zero admittance is the same as an infinite-impedance open circuit. Because of this open circuit, all of the input current passes through the resistor, making $I_2 = I_1$. This creates a current gain of $|T(j\omega_0)| = T_{\max} = 1$.

Clearly T_{max} is the maximum current gain available from this current-divider circuit.

The cutoff frequencies occur when $Y_{LC}(j\omega) = \pm j/R$, since

$$|T(j\omega)| = \left|\frac{1/R}{1/R \pm j/R}\right| = \frac{1}{\sqrt{2}} = \frac{1}{\sqrt{2}}T_{max}$$

and the current gain is reduced by a factor of $1/\sqrt{2}$ from its maximum value at $\omega = \omega_0$. Thus, the current gain of the parallel circuit in Figure 12–27 displays a passband response centered at ω_0 with stopbands at both high and low frequencies.

The admittance $Y_{LC}(j\omega)$ plays the same role in the parallel *RLC* circuit that the impedance $Z_{LC}(j\omega)$ plays in the series *RLC* circuit. In fact, comparing the transfer functions in Eqs. (12–14) and (12–20) reveals that the two circuits are duals. That is, one transfer function can be converted into the other using the following duality interchanges.

Series *RLC*		Parallel *RLC*
R	$\leftrightarrow$	$1/R$
L	$\leftrightarrow$	C
C	$\leftrightarrow$	L

Duality also means that these interchanges convert Eqs. (12–15), (12–16), (12–17), and (12–18) for the series *RLC* circuit into the corresponding relationships for the parallel *RLC* circuit.

Starting with Eq. (12–15), the interchange leaves the center frequency unchanged at

$$\omega_0 = \frac{1}{\sqrt{LC}} \tag{12–21}$$

The two cutoff frequencies in Eq. (12–16) become

$$\begin{aligned} \omega_{C1} &= -\frac{1}{2RC} + \sqrt{\left(\frac{1}{2RC}\right)^2 + \frac{1}{LC}} \\ \omega_{C2} &= +\frac{1}{2RC} + \sqrt{\left(\frac{1}{2RC}\right)^2 + \frac{1}{LC}} \end{aligned} \tag{12–22}$$

It is easy to show that the center frequency $\omega_0 = \sqrt{\omega_{C1}\omega_{C2}}$ is the geometric mean of these two cutoff frequencies. Given these cutoff frequencies, the bandwidth and quality factor of the parallel *RLC* circuit are

$$\begin{aligned} B &= \omega_{C2} - \omega_{C1} = \frac{1}{RC} \\ Q &= \frac{\omega_0}{B} = R\sqrt{C/L} \end{aligned} \tag{12–23}$$

Equations (12–21), (12–22), and (12–23) relate the descriptive parameters of a bandpass response to the circuit parameters of the parallel *RLC* circuit.

Finally, the gain plots in Figure 12–23 also apply to the current gain of the parallel *RLC* circuit. As we noted with the series circuit, the response is narrow band when $Q \gg 1$ and is wide band when $Q \ll 1$. Thus, high-Q

and narrow-band are synonymous terms in both the series and parallel *RLC* circuits. But regardless of the value of Q, the maximum current gain is always $T_{max} = 1$ (or 0 dB) at $\omega = \omega_0$.

A high-Q parallel *RLC* circuit is sometimes called a *tank circuit.* This terminology apparently comes from the vacuum tube era, when the tunable *LC* components in frequency-selective amplifiers were packaged in a metal "tanklike" container. The center frequency is also the resonant frequency of the parallel *RLC* circuit, since

$$Z_{IN}(j\omega_0) = \frac{1}{1/R + Y_{LC}(j\omega_0)} = R + j0$$

So, again, the passband center frequency is also called the *resonant frequency.*

EXAMPLE 12–13

A parallel *RLC* circuit has a bandwidth of $B = 12$ krad/s, a quality factor of $Q = 3$, and a 1-mH inductance. Find the values of R, C, ω_0, ω_{C1}, and ω_{C2}.

SOLUTION:

Using the definition of Q, we find the center frequency as $\omega_0 = BQ =$ 36 krad/s. Using the given inductance and the center frequency, we get the capacitance as $C = 1/\omega_0^2 L = 0.772\ \mu\text{F}$. For a parallel *RLC* circuit, $B = 1/RC$ and the resistance is $R = 1/BC = 108\ \Omega$. Inserting these results into Eq. (12–22) yields the lower cutoff frequency as

$$\omega_{C1} = -6000 + \sqrt{6000^2 + 36{,}000^2} = 30.5 \text{ krad/s}$$

Then, using the definition of bandwidth, we get $\omega_{C2} = \omega_{C1} + B =$ 42.5 krad/s. ■

DESIGN EXAMPLE 12–14

Design a parallel *RLC* circuit with cutoff frequencies at 12 kHz and 16 kHz.

SOLUTION:

The descriptive parameters of the required circuit are

$$\omega_0 = 2\pi\sqrt{f_{C1} f_{C2}} = 87.1 \text{ krad/s}$$

$$B = 2\pi(f_{C1} - f_{C2}) = 25.1 \text{ krad/s}$$

Selecting $R = 1\ \text{k}\Omega$ gives $C = 1/RB = 39.8$ nF and $L = 1/\omega_0^2 C = 3.32$ mH. ■

Exercise 12–5

A series *RLC* circuit has cutoff frequencies at 100 rad/s and 10 krad/s. Find the values of B, ω_0, and Q. Does the circuit have a wide-band or narrow-band response?

Answers: $B = 9.9$ krad/s; $\omega_0 = 1$ krad/s; $Q = 0.101$; wide band

Exercise 12–6

A series *RLC* bandstop circuit has a notch bandwidth of 2 krad/s and a notch frequency at 10 krad/s. Find the two cutoff frequencies.

Answers: $\omega_{C1} = 9.05$ krad/s; $\omega_{C2} = 11.05$ krad/s

Exercise 12–7

A parallel *RLC* circuit has a center frequency of 200 krad/s, a bandwidth of 10 krad/s, and a resistance of $R = 10\ \text{k}\Omega$. Find the center frequency and bandwidth when the resistance is increased to 40 kΩ.

Answers: $\omega_0 = 200$ krad/s; $B = 2.5$ krad/s

12–6 BODE DIAGRAMS WITH REAL POLES AND ZEROS

The purpose of this section is to present a method of quickly drawing straight-line approximations to a Bode plot of transfer functions with real poles and zeros. In an analysis situation the straight-line approximations tell us how the poles and zeros of $T(s)$ affect frequency response. In circuit design these straight-line plots serve as a shorthand notation for outlining design approaches or developing design requirements. The straight-line plots help us use computer-aided analysis effectively because they tell us what frequency ranges are important and what features of the response need to be investigated in greater detail.

The straight-line versions of Bode plots are particularly useful when the poles and zeros are located on the real axis in the s plane. As a starting place, consider the transfer function

$$T(s) = \frac{Ks(s + \alpha_1)}{(s + \alpha_2)(s + \alpha_3)} \tag{12–24}$$

where K, α_1, α_2, and α_3 are real. This function has zeros at $s = 0$ and $s = -\alpha_1$, and poles at $s = -\alpha_2$ and $s = -\alpha_3$. All of these critical frequencies lie on the real axis in the s plane. When making Bode plots, we put $T(j\omega)$ in a standard format obtained by factoring out α_1, α_2, and α_3:

$$T(j\omega) = \left(\frac{K\alpha_1}{\alpha_2\alpha_3}\right)\frac{j\omega(1 + j\omega/\alpha_1)}{(1 + j\omega/\alpha_2)(1 + j\omega/\alpha_3)} \tag{12–25}$$

Using the following notation

$$\text{Magnitude} = M = |1 + j\omega/\alpha| = \sqrt{1 + (\omega/\alpha)^2}$$

$$\text{Angle} = \theta = \angle(1 + j\omega/\alpha) = \tan^{-1}(\omega/\alpha) \tag{12–26}$$

$$\text{Scale Factor} = K_0 = \frac{K\alpha_1}{\alpha_2\alpha_3}$$

we can write the transfer function in Eq. (12–25) in the form

$$T(j\omega) = K_0 \frac{(\omega\, e^{j90^\circ})\,(M_1\, e^{j\theta_1})}{(M_2\, e^{j\theta_2})\,(M_3\, e^{j\theta_3})} = \frac{|K_0|\omega M_1}{M_2 M_3} e^{j(\angle K_0 + 90^\circ + \theta_1 - \theta_2 - \theta_3)} \tag{12–27}$$

The gain (in dB) and phase responses are

$$|T(j\omega)|_{\text{dB}} = \underbrace{20\log_{10}|K_0|}_{\text{scale factor}} + \underbrace{20\log_{10}\omega}_{\text{zero}} + \underbrace{20\log_{10}M_1}_{\text{zero}} - \underbrace{20\log_{10}M_2}_{\text{pole}} - \underbrace{20\log_{10}M_3}_{\text{pole}}$$

$$\theta(\omega) = \overbrace{\angle K_0} + \overbrace{90^\circ} + \overbrace{\theta_1} - \overbrace{\theta_2} - \overbrace{\theta_3} \tag{12–28}$$

The terms in Eq. (12–28) caused by zeros have positive signs and increase the gain and phase angle, while the pole terms have negative signs and decrease the gain and phase.

The summations in Eq. (12–28) illustrate a general principle. In a Bode plot, the gain and phase responses are determined by the following types of factors:

1. The scale factor K_0
2. A factor of the form $j\omega$ due to a zero or a pole at the origin
3. Factors of the form $(1 + j\omega/\alpha)$ caused by a zero or pole at $s = -\alpha$

We can construct Bode plots by considering the contributions of these three factors.

The Scale Factor. The gain and phase contributions of the scale factor are constants that are independent of frequency. The gain contribution $20 \log_{10} |K_0|$ is positive when $|K_0| > 1$ and negative when $|K_0| < 1$. The phase contribution $\angle K_0$ is 0° when $K_0 > 0$ and ±180° when $K_0 < 0$.

The Factor jω. A simple zero or pole at the origin contributes $\pm 20 \log_{10}\omega$ to the gain and ±90° to the phase, where the plus sign applies to a zero and the minus to a pole. When $T(s)$ has a factor s^n in the numerator (denominator), it has a zero (pole) of order n at the origin. Multiple zeros or poles at $s = 0$ contribute $\pm 20n \log_{10} \omega$ to the gain and $\pm n90^\circ$ to the phase. Figure 12–28

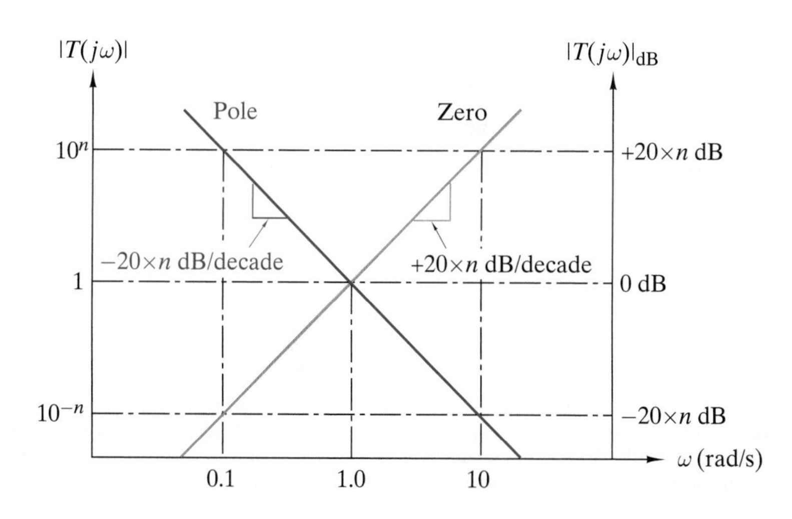

FIGURE 12–28 *Gain responses of poles and zeros at s = 0.*

shows that the gain factors contributed by zeros and poles at the origin are straight lines that pass through a gain of 1 (0 dB) at $\omega = 1$.[2]

The Factor $1 + j\omega/a$. The gain contributions of first-order zeros and poles are shown in Figure 12–29. Like the first-order transfer functions studied earlier in this chapter, these factors produce straight-line gain asymptotes at low and high frequency. In a Bode plot the low-frequency ($\omega << \alpha$) asymptotes are horizontal lines at gain = 1 (0 dB). The high-frequency

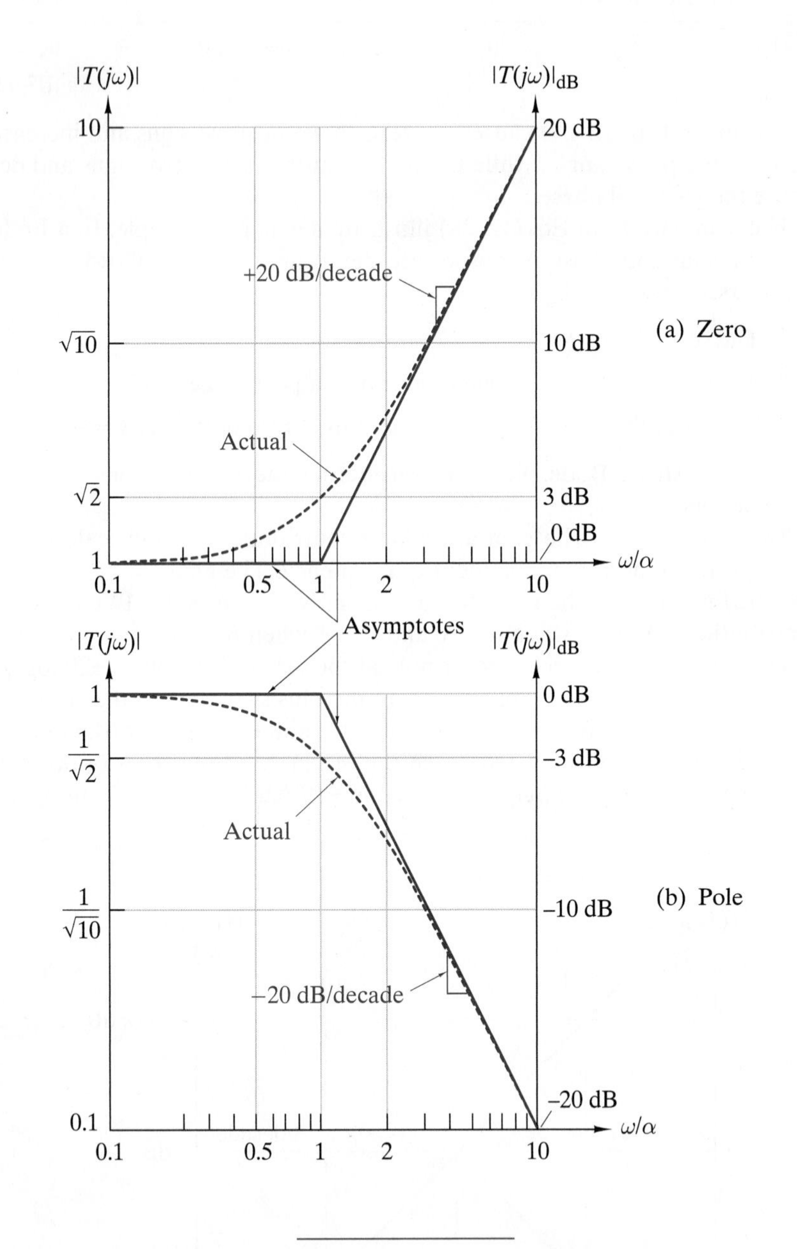

FIGURE 12–29 *Gain responses of poles and zeros at* $s = -\alpha$.

2 Strictly speaking, a circuit with a natural pole at the origin is unstable and does not have a sinusoidal steady-state response. Nevertheless, it is traditional to treat poles at the origin in Bode diagrams because there are practical applications in which such poles are important considerations.

($\omega >> \alpha$) asymptotes are straight lines of the form $\pm\omega/\alpha$ ($\pm 20 \log(\omega/\alpha)$ dB), where the plus sign applies to a zero and the minus to a pole. The high-frequency gain asymptote is proportional to the frequency ω (slope = +1 or +20 dB/decade) for a zero and proportional to $1/\omega$ (slope = –1 or –20 dB/decade) for a pole. In either case, the low- and high-frequency asymptotes intersect at the corner frequency $\omega = \alpha$.

To construct straight-line (SL) gain approximations, we develop a piecewise linear function $|T(j\omega)|_{SL}$ defined by the asymptotes of each of the factors in $T(j\omega)$. The function $|T(j\omega)|_{SL}$ has a corner frequency at each of the critical frequencies of the transfer function. At frequencies below a corner ($\omega < \alpha$), a first-order factor is represented by a gain of 1. Above the corner frequency ($\omega > \alpha$), the factor is represented by its high-frequency asymptote ω/α. To generate $|T(j\omega)|_{SL}$, we start out below the lowest critical frequency with a low-frequency baseline that accounts for the scale factor K_0 and any poles or zeros at the origin. We increase frequency and change the form of $|T(j\omega)|_{SL}$ whenever we pass a corner frequency. We proceed upward in frequency until we have gone beyond the highest critical frequency, at which point we have a complete expression for $|T(j\omega)|_{SL}$.

The following example illustrates the process.

EXAMPLE 12–15

(a) Construct the Bode plot of the straight-line approximation of the gain of the transfer function

$$T(s) = \frac{12{,}500\,(s + 10)}{(s + 50)\,(s + 500)}$$

(b) Find the point at which the high-frequency gain falls below the dc gain.

SOLUTION:

(a) Written in the standard form for a Bode plot, the transfer function is

$$T(j\omega) = \frac{5(1 + j\omega/10)}{(1 + j\omega/50)\,(1 + j\omega/500)}$$

The scale factor is $K_0 = 5$ and the corner frequencies are at $\omega_C = 10$ (zero), 50 (pole), and 500 (pole) rad/s. At low frequency ($\omega < 10$ rad/s), all of the first-order factors are represented by their low-frequency asymptotes. As a result $|T(j\omega)| \approx 5(1)/[(1)(1)] = 5$, so the low-frequency baseline is $|T(j\omega)|_{SL} = 5$ for $\omega \leq 10$. At $\omega = 10$ rad/s we encounter the first critical frequency. Beginning at this point the factor $(1 + j\omega/10)$ is represented by its high-frequency asymptote $\omega/10$ and the straight-line gain becomes $|T(j\omega)|_{SL} = 5(\omega/10) = \omega/2$. This expression applies until we pass the critical frequency at $\omega_C = 50$ due to the pole at $s = -50$. After this point the gain contribution due to the pole factor $1/(1 + j\omega/50)$ is represented by its high-frequency asymptote $\omega/50$ and $|T(j\omega)|_{SL} = 5(\omega/10)/(\omega/50) = 25$. This version applies until we pass the final critical frequency at $\omega_C = 500$, beyond which the gain contribution of the last pole is approximated by $\omega/500$ and the high-frequency gain rolls off as $|T(j\omega)|_{SL} = 5(\omega/10)/[(\omega/50)(\omega/500)] = 12500/\omega$. In summary, the straight-line approximation to the gain is

$$|T(j\omega)|_{SL} = \begin{cases} 5 & \text{if} \quad 0 < \omega \leq 10 \\ \omega/2 & \text{if} \quad 10 < \omega \leq 50 \\ 25 & \text{if} \quad 50 < \omega \leq 500 \\ 12{,}500/\omega & \text{if} \quad 500 < \omega \end{cases}$$

Given this function, we can easily plot the straight-line gain response in Figure 12–30. At low frequency ($\omega < 10$) the gain is flat at a value of 5 (14 dB). At $\omega = 10$ the zero causes the gain to increase as ω (slope = +1 or +20 dB/decade). This increasing gain continues until $\omega = 50$, where the first pole cancels the effect of the zero and the gain is flat at a value of 25 (28 dB). The gain remains flat until the final pole causes a corner at $\omega = 500$. Thereafter the gain falls off as $1/\omega$ (slope = –1 or –20 dB/decade).

(b) The dc gain is 5. A quick look at the sketch in Figure 12–30 shows that the high-frequency gain falls below the dc gain in the region above $\omega = 500$, where the straight-line gain is $12{,}500/\omega$. Hence we estimate the required frequency to be $\omega = 12{,}500/5 = 2500$ rad/s. ■

FIGURE 12–30

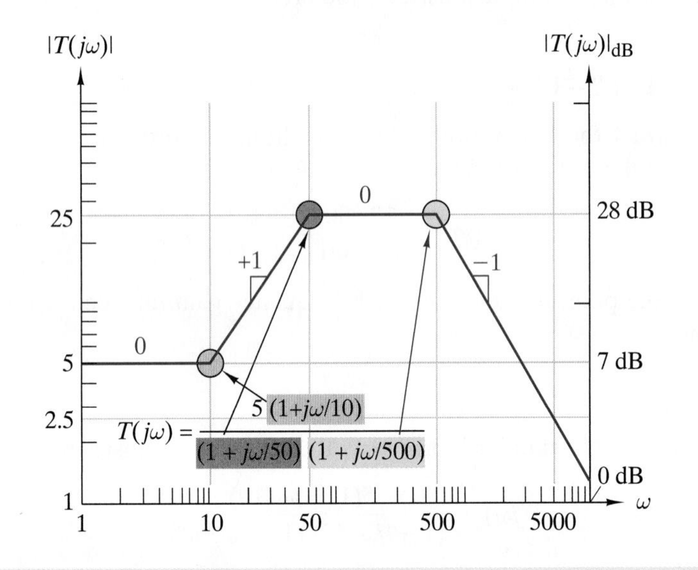

EXAMPLE 12–16

(a) Construct the Bode plot of the straight-line approximation of the gain of the transfer function:

$$T(s) = \frac{10s^2}{(s + 40)(s + 200)}$$

(b) Find the point at which the low-frequency gain falls –40 dB below the passband gain.

SOLUTION:

(a) Writing $T(j\omega)$ in standard form produces

$$T(j\omega) = \frac{1}{800}\left[\frac{(j\omega)^2}{(1 + j\omega/40)(1 + j\omega/200)}\right]$$

In this form $T(j\omega)$ has a scale factor of $K_0 = 1/800$, a double zero ($n = 2$) at the origin, and finite critical frequencies at $\omega = 40$ and 200, both due to poles. At low frequency ($\omega < 40$) the two first-order factors are represented by their low-frequency asymptote. As a result, the low-frequency gain baseline is $|T(j\omega)|_{SL} = \omega^2/800$. This trend continues until we pass the critical frequency at $\omega = 40$. After this point the straight-line gain is

$$|T(j\omega)|_{SL} = (\omega^2/800)/(\omega/40) = \omega/20$$

The gain continues to increase at a reduced rate until we pass the final critical frequency at $\omega = 200$. Beyond this point the straight-line gain is constant at the high frequency of

$$|T(j\omega)|_{SL} = (\omega^2/800)/[(\omega/40)(\omega/200)] = 10$$

In summary, the straight-line approximation to the gain response is

$$|T(j\omega)|_{SL} = \begin{vmatrix} \omega^2/800 & \text{if} & 0 < \omega \leqslant 40 \\ \omega/20 & \text{if} & 40 < \omega \leqslant 200 \\ 10 & \text{if} & 200 < \omega \end{vmatrix}$$

We can easily plot the straight-line gain response shown in Figure 12–31. At low frequency ($\omega < 20$) the gain is increasing as ω^2 (slope = +2 or +40 dB/decade). This upward trend lasts until the critical frequency (a pole) at $\omega = 40$, at which point the gain is $(40)^2/800 = 2$. Thereafter the gain increases as ω (slope = +1 or +20 dB/decade) until the final critical frequency at $\omega = 200$ reduces the slope to zero at a flat gain of 10 (20 dB).

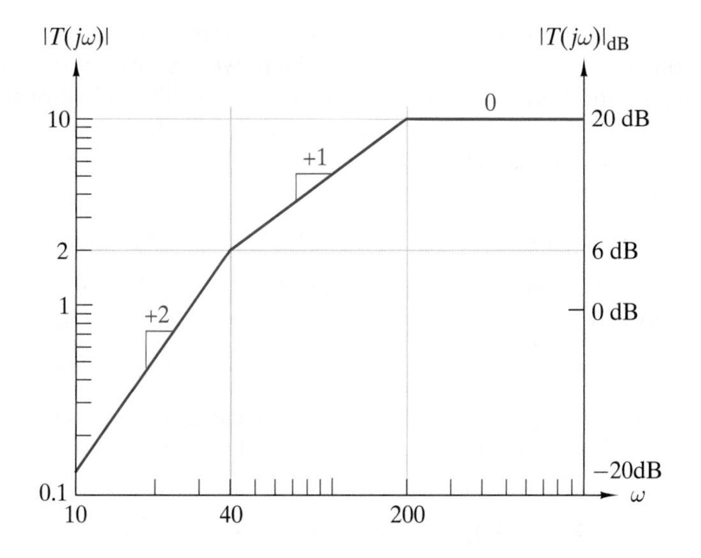

FIGURE 12–31

(b) A quick look at the plot in Figure 12–31 shows that the gain response has a high-pass characteristic with a passband above $\omega = 200$, where the passband gain is 20 dB. The low-frequency gain will be –40 dB below this level in the frequency range below $\omega = 40$, where the straight-line gain is $\omega^2/800$. The actual gain must be 20 dB – 40 dB = –20 dB or 0.1. This gain occurs when $\omega^2/800 = 0.1$ at $\omega = 8.9$ rad/s. ■

D DESIGN EXAMPLE 12–17

(a) Find the transfer function $T(s)$ corresponding to the straight-line gain response in Fig. 12–32.

(b) Design a cascade circuit that realizes the $T(s)$ found in (a).

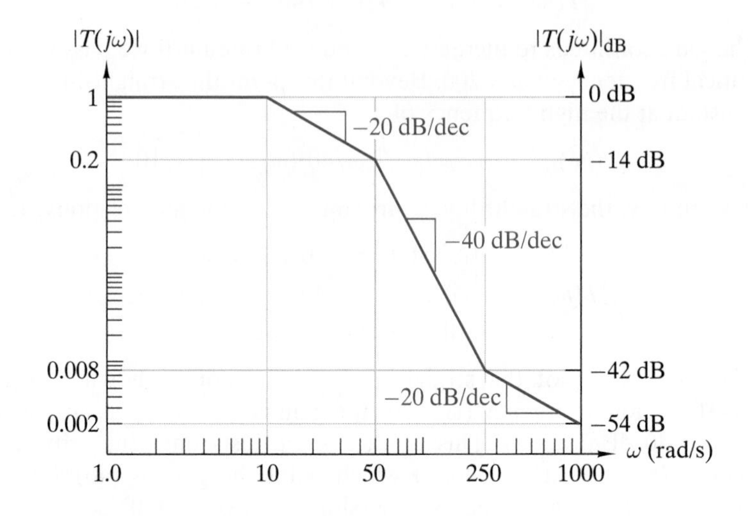

FIGURE 12–32

SOLUTION:

(a) In previous examples we were given a transfer function and required to construct a straight-line gain plot. Here we are given a straight-line gain plot and asked to find the transfer function. The gain plot shows a finite gain at dc and corner frequencies at $\omega = 10$, 50, and 250 rad/s. The slope of the straight-line gain response between the first corner at $\omega = 10$ and 50 rad/s is

$$m = (-14 - 0)/(\log 50 - \log 10) = -20 \text{ dB/decade}$$

The -20 dB/decade slope means that $T(s)$ has a pole at $s = -10$. The gain slope between the second corner at $\omega = 50$ and 250 rad/s is

$$m = [-42 - (-14)]/(\log 250 - \log 50) = -40 \text{ dB/decade}$$

The slope decreases by another -20 dB/decade, so $T(s)$ also has a pole at $s = -50$. The gain slope between the last corner at $\omega = 250$ and 1000 rad/s is

$$m = [-54 - (-42)]/(\log 1000 - \log 250) = -20 \text{ dB/decade}$$

The gain slope increases by $+20$ dB/decade, so $T(s)$ must have zero at $s = -250$. These critical frequencies account for all of the corner frequencies shown in the figure, and $T(s)$ has the form

$$T(s) = \frac{K(s + 250)}{(s + 10)(s + 50)}$$

The dc gain of this transfer function is $T(0) = K/2$. The dc gain in Fig. 12–32 is $T(0) = 1$, so $K = 2$.

(b) The transfer function found in (a) can be expressed as a product of two transfer functions:

$$T(s) = \underbrace{\left(\frac{1}{\dfrac{s}{10}+1}\right)}_{T_1(s)}\underbrace{\left(\frac{\dfrac{s}{250}+1}{\dfrac{s}{50}+1}\right)}_{T_2(s)}$$

Both of these transfer functions can be realized by voltage dividers. Factoring s out of the denominator of $T_1(s)$ and equating the result to a voltage divider yields

$$\underbrace{\frac{1/s}{1/10 + 1/s}}_{T_1(s)} = \underbrace{\frac{Z_2(s)}{Z_1(s) + Z_2(s)}}_{\text{voltage divider}}$$

Equating the numerators and denominators yields a voltage divider with $Z_2 = 1/C_1s$ and $Z_1 = R_1$, where $C_1 = 1$ F and $R_1 = 1/10\ \Omega$.

Factoring s out of the numerator and denominator of $T_2(s)$ and equating the result to a voltage divider yields

$$\underbrace{\frac{1/250 + 1/s}{1/50 + 1/s}}_{T_2(s)} = \underbrace{\frac{Z_2(s)}{Z_1(s) + Z_2(s)}}_{\text{voltage divider}}$$

Equating the numerators and denominators leads to a voltage divider with $Z_2 = R_2 + 1/C_1s$ and $Z_1 = R_3$, where $R_2 = 1/250\ \Omega$, $C_2 = 1$ F, and $R_3 = 4/250\ \Omega$.

Figure 12–33 shows a cascade circuit using the two voltage dividers separated by a voltage follower to ensure that the chain rule applies. The element values in the prototype version of this circuit are $R_1 = 1/10\ \Omega$, $C_1 = C_2 = 1$ F, $R_2 = 1/250\ \Omega$, and $R_3 = 4/250\ \Omega$. Using a magnitude scale factor of $k_m = 10^6$ to get practical values produces a final design with $R_1 = 100$ kΩ, $C_1 = C_2 = 1\ \mu$F, $R_2 = 4$ kΩ, and $R_3 = 16$ kΩ. Many other designs are possible.

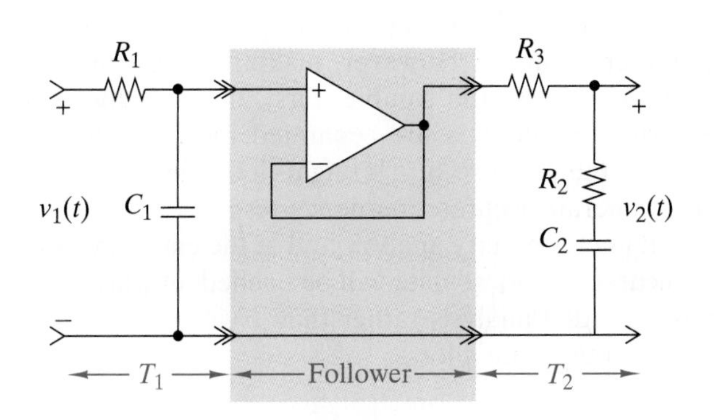

FIGURE 12–33

Exercise 12–8

(a) Derive an expression for the straight-line approximation to the gain response of the following transfer function:

$$T(s) = \frac{500\,(s + 50)}{(s + 20)(s + 500)}$$

(b) Find the straight-line gains at $\omega = 10$, 30, and 100 rad/s.
(c) Find the frequency at which the high-frequency gain asymptote falls below –20 dB.

Answers:

(a) $$|T(j\omega)|_{SL} = \begin{cases} 2.5 & \text{if} \quad 0 < \omega \leq 2 \\ 50/\omega & \text{if} \quad 20 < \omega \leq 50 \\ 1 & \text{if} \quad 50 < \omega \leq 500 \\ 500/\omega & \text{if} \quad 500 < \omega \end{cases}$$

(b) 8 dB, 4.4 dB, 0 dB
(c) 5 krad/s

Exercise 12–9

Given the following transfer function

$$T(s) = \frac{4000s}{(s + 100)(s + 2000)}$$

(a) Find the straight-line approximation to the gain at $\omega = 50$, 100, 500, 2000, and 4000 rad/s.
(b) Estimate the actual gain at the frequencies in (a).

Answers:

(a) 0 dB; 6 dB; 6 dB; 6 dB; 0 dB
(b) –1 dB; 3 dB; 6 dB; 3 dB; –1 dB

In many situations the straight-line Bode plot tells us all we need to know. When greater accuracy is needed, the straight-line gain plot can be refined by adding gain corrections in the neighborhood of the corner frequency. Figure 12–29 shows that the actual gains and the straight-line approximations differ by ±3 dB at the corner frequency. They differ by roughly ±1 dB an octave above or below the corner frequency. When these "corrections" are included, we can sketch the actual response and achieve somewhat greater accuracy. However, making the graphical gain corrections is usually not worth the trouble. First, the gain corrections overlap unless the corner frequencies are separated by more than two octaves. More important, the purpose of a straight-line gain analysis is to provide insight, not to generate accurate frequency-response data. The straight-line plots are useful in preliminary analysis and in the early stages of design. At some point accurate response data will be needed, in which case it is better to use computer-aided analysis rather than trying to "correct the errors" graphically in a straight-line plot.

STRAIGHT-LINE PHASE ANGLE PLOTS

Figure 12–34 shows the phase contributions of first-order zeros and poles. The straight-line approximations are similar to the gain asymptotes except that there are two slope changes. The first occurs a decade below the gain corner frequency, and the second occurs a decade above. The total phase

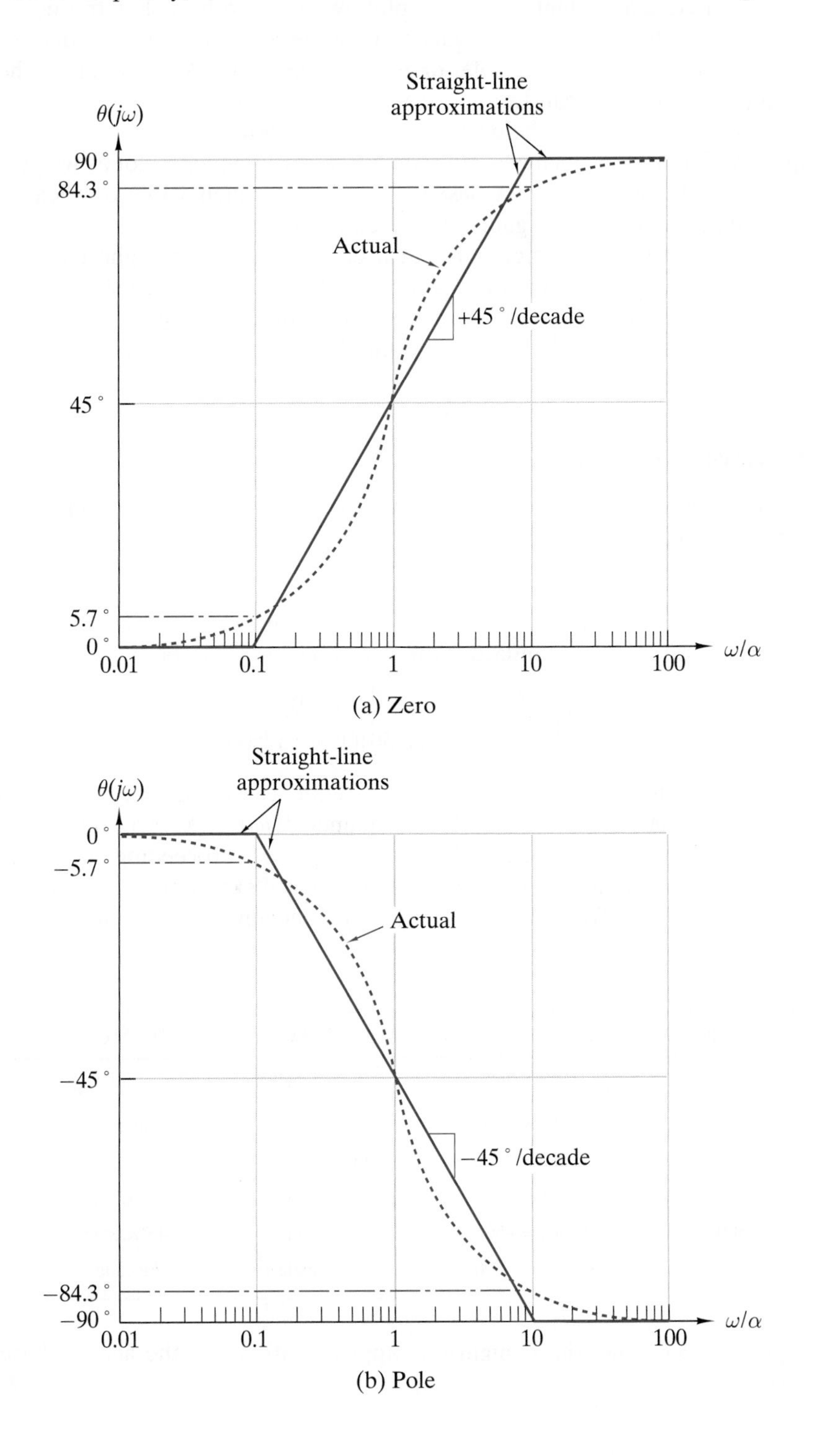

FIGURE 12–34 *Phase responses of poles and zeros at* s = −α.

changes by 90° over this two-decade range, so the straight-line approximations have slopes of ±45° per decade, where the plus sign applies to a zero and the minus to a pole. Poles and zeros at the origin contribute a constant phase angle of $\pm n90°$, where n is the order of the critical frequency and the plus (minus) sign applies to zeros (poles).

To generate a straight-line phase plot, we begin with the low-frequency phase asymptote. This low-frequency baseline accounts for the effect of the scale factor K_0 and any poles or zeros at the origin. We account for the effect of other critical frequencies by introducing a slope change of ±45°/decade one decade below and one decade above each gain corner frequency. These slope changes generate a straight-line phase plot as we proceed from the low-frequency baseline to a high frequency that is at least a decade above the highest gain corner frequency.

It is important to remember that a decade above the highest corner frequency the phase asymptote is a constant value with zero slope. That is, at high frequency the straight-line phase plot is a horizontal line at $\theta(j\omega) = (m - n)90°$, where m is the number of finite zeros and n is the number of poles.

EXAMPLE 12–18

Find the straight-line approximation to the phase response of the transfer function in Example 12–15.

SOLUTION:

In Example 12–15 the standard form of $T(j\omega)$ is shown to be

$$T(j\omega) = \frac{5(1 + j\omega/10)}{(1 + j\omega/50)(1 + j\omega/500)}$$

The scale factor is $K_0 = 5$ and the corner frequencies are $\omega_C = 10$ (zero), 50 (pole), and 500 (pole) rad/s. At low frequency $T(j\omega) \to K_0 = 5$, so the low-frequency phase asymptote is $\theta(\omega) \to \angle K_0 = 0°$. Proceeding from one decade below the lowest corner frequency (1 rad/s) to one decade above the highest corner frequency (5000 rad/s), we encounter the following slope changes:

FREQUENCY	CAUSED BY	SLOPE CHANGE	NET SLOPE
1	zero at $s = -10$	+45°/decade	+45°/decade
5	pole at $s = -50$	–45°/decade	0°/decade
50	pole at $s = -500$	–45°/decade	–45°/decade
100	zero at $s = -10$	–45°/decade	–90°/decade
500	pole at $s = -50$	+45°/decade	–45°/decade
5000	pole at $s = -500$	+45°/decade	0°/decade

Figure 12–35 shows the straight-line approximation and the actual phase response. ■

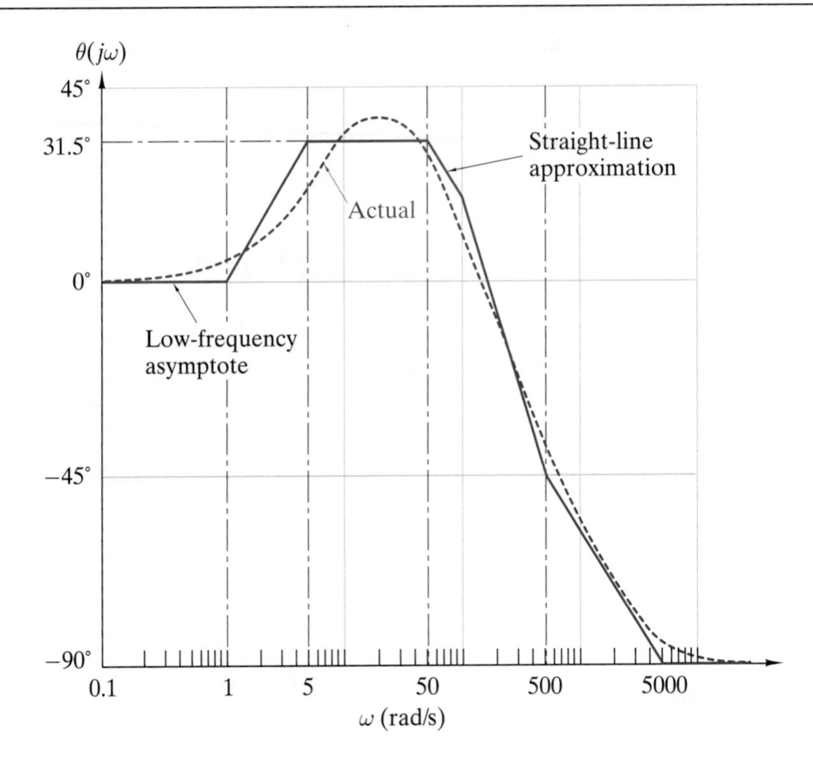

FIGURE 12–35

Exercise 12–10

Construct a Bode plot of the straight-line approximation to the phase response of the transfer function in Exercise 12–8. Use the plot to estimate the phase angles at $\omega = 1, 15, 300$, and 10^4 rad/s.

Answers: 0°, −18°, −45°, −90°

12–7 BODE DIAGRAMS WITH COMPLEX POLES AND ZEROS

The frequency response of transfer functions with complex poles and zeros can be analyzed using straight-line gain plots. However, complex critical frequencies may produce resonant peaks (or valleys) where the actual gain response departs significantly from the straight-line approximation. Straight-line plots can be used to define a starting place for describing the frequency response of these highly resonant circuits.

Complex poles and zeros occur in conjugate pairs that appear as quadratic factors of the form

$$s^2 + 2\zeta\omega_0 s + \omega_0^2 \tag{12–29}$$

where ζ and ω_0 are the damping ratio and undamped natural frequency. In a Bode diagram the appropriate standard form of the quadratic factor is obtained by factoring out ω_0^2 and replacing s by $j\omega$ to obtain

$$1 - (\omega/\omega_0)^2 + j2\zeta(\omega/\omega_0) \tag{12–30}$$

In a Bode diagram this quadratic factor introduces gain and phase terms of the following form:

$$|T(j\omega)|_{dB} = \pm 20 \log_{10} \sqrt{[1 - (\omega/\omega_0)^2]^2 + (2\zeta\omega/\omega_0)^2} \tag{12–31a}$$

$$\theta(\omega) = \pm \tan^{-1} \frac{2\zeta\omega/\omega_0}{1 - (\omega/\omega_0)^2} \tag{12–31b}$$

where the plus sign applies to complex zeros of $T(s)$ and the minus sign to complex poles.

Figure 12–36 shows the gain contribution of complex poles and zeros for several values of the damping ratio ζ. The low-frequency ($\omega << \omega_0$) gain asymptotes for these plots are unity (0 dB). The high-frequency ($\omega >> \omega_0$)

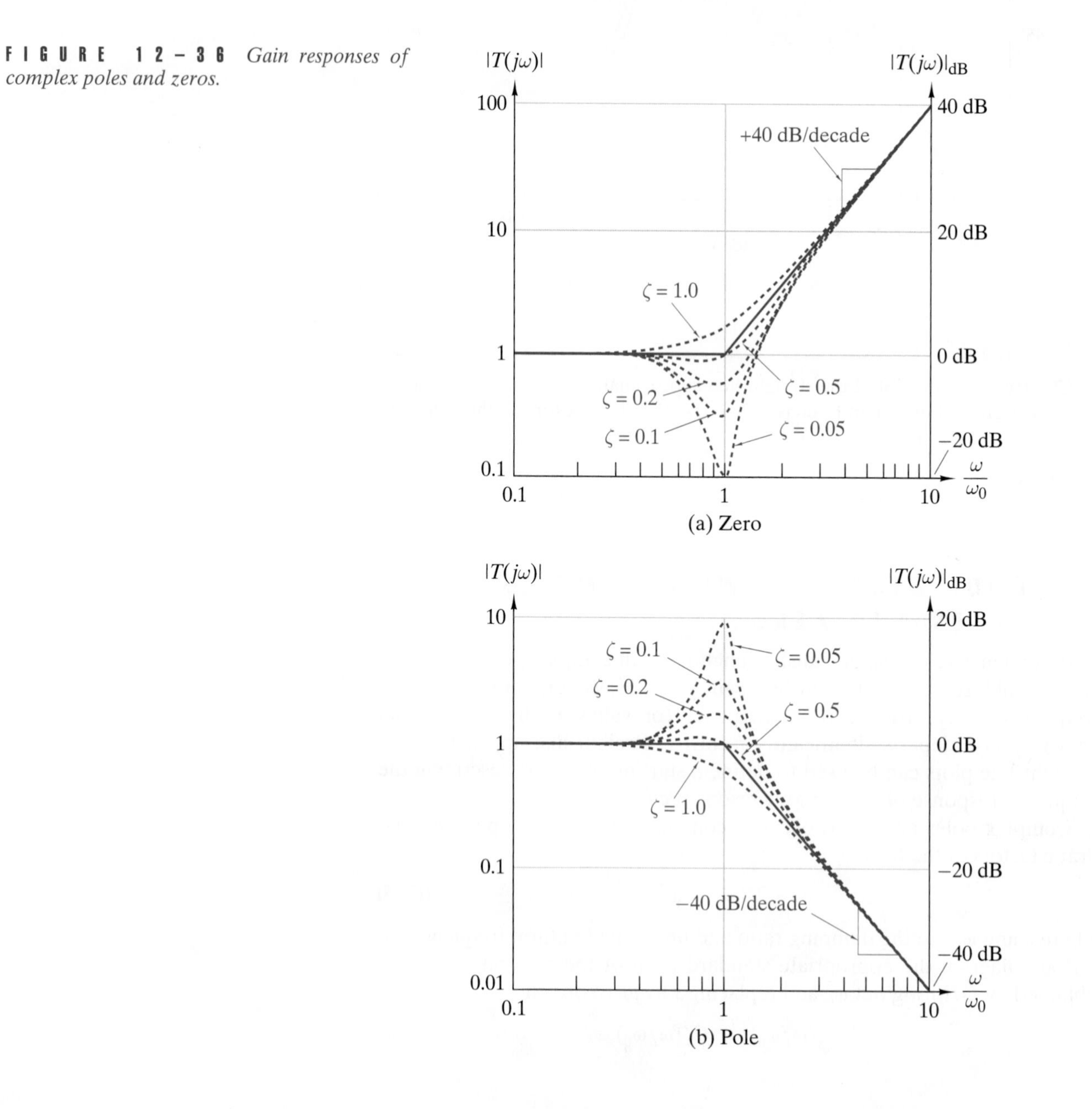

FIGURE 12–36 *Gain responses of complex poles and zeros.*

gain asymptotes are of the form $(\omega/\omega_0)^{\pm 2}$. Expressed in dB, the high-frequency asymptote is $\pm 40\ \log_{10}(\omega/\omega_0)$, which in a Bode diagram is a straight line with a slope of ±2 or ±40 dB/decade, where again the plus sign applies to zeros and the minus to poles.

These asymptotes intersect at a corner frequency of $\omega = \omega_0$. The gains in the neighborhood of the corner frequency are strong functions of the damping ratio. For $\zeta > 1/\sqrt{2}$ the actual gain lies entirely above the asymptotes for complex zeros and entirely below the asymptotes for complex poles. For $\zeta < 1/\sqrt{2}$ the gain is a minimum at $\omega = \omega_0\sqrt{1 - 2\zeta^2}$ for complex zeros and a maximum for complex poles. These valleys (for zeros) and peaks (for poles) are not particularly conspicuous until $\zeta < 0.5$.

To develop a straight-line gain plot for complex critical frequencies we insert a corner frequency at $\omega = \omega_0$. Below this corner frequency we use the low-frequency asymptote to approximate the gain, and above the corner we use the high-frequency asymptote. The actual gain around the corner frequency depends on ζ. But generally speaking, the straight-line gain is within ±3 dB of the actual gain for ζ in the range from about 0.3 to about 0.7. When ζ falls outside this range, we can calculate the actual gain at the corner frequency and perhaps a few points on either side of the corner frequency. These gains may give us a better picture of the gain plot in the vicinity of the corner frequency.

However, we should keep in mind that the purpose of straight-line gain analysis is insight into the major features of a circuit's frequency response. If greater accuracy is required, then computer-aided analysis is the best approach. The straight-line gain gives useful results when the resonant peaks and valleys are not too abrupt. When a circuit has lightly damped critical frequencies, the straight-line approach may not be particularly helpful. The following examples illustrate both of these cases.

EXAMPLE 12–19

(a) Construct the straight-line gain plot for the transfer function

$$T(s) = \frac{5000(s + 100)}{s^2 + 400s + (500)^2}$$

(b) Use the straight-line plot to estimate the maximum gain and the frequency at which it occurs.

SOLUTION:

(a) The transfer function has a real zero at $s = -100$ rad/s and a pair of complex poles with $\zeta = 0.4$ and $\omega_0 = 500$ rad/s. This damping ratio falls in the range (0.3 to 0.7) in which the resonant peak due to the complex poles is not too pronounced. Hence we expect the straight-line gain to give a useful approximation. Written in standard form, $T(j\omega)$ is

$$T(j\omega) = 2\left(\frac{1 + j\omega/100}{1 - (\omega/500)^2 + j400\omega/500}\right)$$

The scale factor is $K_0 = 2$, and there are corner frequencies at $\omega = 100$ rad/s due to the zero and $\omega = 500$ rad/s due to the pair of complex poles. At low frequency $T(j\omega) \rightarrow 2$, so the low-frequency $(\omega < 100)$

baseline is $|T(j\omega)|_{SL} = 2$. This gain applies until we pass the first critical frequency at $\omega = 100$. Beginning at that point, the zero is represented by its high-frequency asymptote $(\omega/100)$ and the straight-line gain becomes

$$|T(j\omega)|_{SL} = 2(\omega/100) = \omega/50$$

This linearly increasing gain applies until we pass the critical frequency at $\omega_0 = 500$. Thereafter the complex poles are represented by their high-frequency asymptote $(500^2/\omega^2)$. After this point the gain rolls off as

$$|T(j\omega)|_{SL} = 2(\omega/100)(500^2/\omega^2) = 5000/\omega$$

In summary, the straight-line gain function is

$$|T(j\omega)|_{SL} = \begin{vmatrix} 2 & \text{if} \quad 0 < \omega \leq 100 \\ \omega/50 & \text{if} \quad 100 < \omega \leq 500 \\ 5000/\omega & \text{if} \quad 500 < \omega \end{vmatrix}$$

Figure 12–37 shows a plot of the straight-line gain. We expect to see a gain peak around $\omega = 500$ rad/s due to the complex poles. The plot in Figure 12–37 shows that the zero at $s = -100$ rad/s causes the gain to bend upward prior to the corner frequency at $\omega = 500$ rad/s. This upward bend enhances the height of the resonant peak caused by the complex poles.

FIGURE 12–37

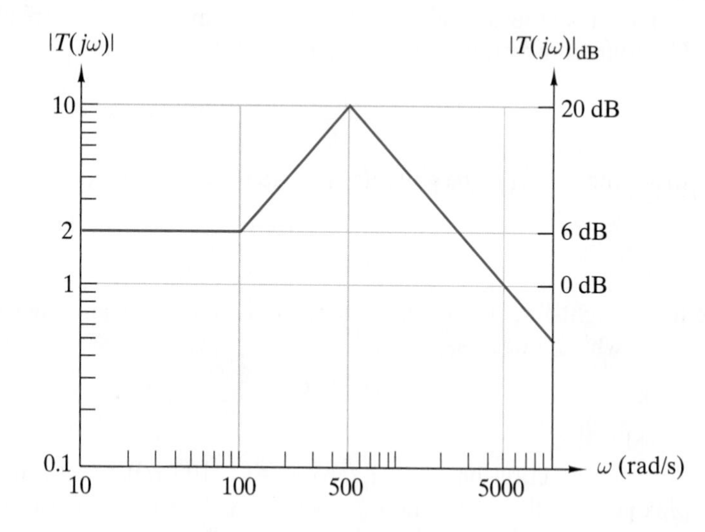

(b) The straight-line gain predicts a maximum gain of 10 (20 dB) at $\omega =$ 500 rad/s. The actual gain at that point is

$$|T(j500)| = 2\left|\frac{1 + j5}{j0.8}\right| = 12.7 \text{ (22 dB)}$$

In this case the straight-line plot gives us a useful approximation of the gain in the vicinity of the resonant peak. ■

EXAMPLE 12–20

Plot the straight-line gain and actual gain of the transfer function

$$T(s) = \frac{20\,s}{s^2 + 2\,s + 2500}$$

SOLUTION:

The denominator is a quadratic factor with $\omega_0 = \sqrt{2500} = 50$ and $2\zeta\omega_0 = 2$; hence $\zeta = 0.02$. Replacing s by $j\omega$ leads to a gain function

$$|T(j\omega)| = \frac{20\,\omega}{|2500 - \omega^2 + j2\omega|}$$

At low frequency ($\omega << \omega_0 = 50$), the gain asymptote is $|T(j\omega)| \to 20\omega/2500$. At high frequency ($\omega >> \omega_0 = 50$), the gain asymptote is $|T(j\omega)| \to 20/\omega$. These asymptotes intersect when $20\omega/2500 = 20/\omega$, which occurs at a corner frequency of $\omega = \omega_0 = 50$. Using these asymptotes, the straight-line gain for this transfer function is written as

$$|T(j\omega)|_{SL} = \begin{cases} 20\omega/2500 & \text{if} \quad 0 < \omega \le 50 \\ 20/\omega & \text{if} \quad 50 < \omega \end{cases}$$

Figure 12–38 shows a Mathcad worksheet that generates plots of the straight-line gain (solid lines) and the actual gain (dashed curves). The actual gain has a narrow-band bandpass response. The straight-line gain plot also indicates a bandpass gain but does not tell us anything about the bandwidth. The straight-line gain at the corner frequency is $|T(j50)|_{SL} = 0.2$. The actual gain at the corner frequency is $|T(j50)| = 10$, a difference of

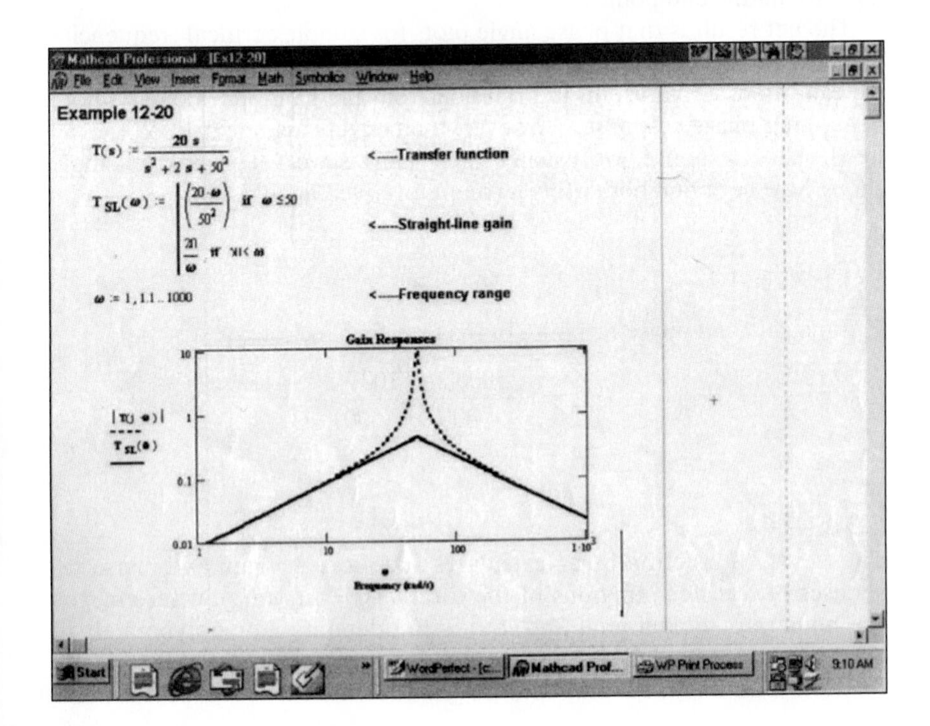

FIGURE 12–38

about 34 dB. The straight-line gain provides a rough starting place, but beyond that it is not particularly helpful in this case since the poles have a low ζ or, equivalently, a high Q.

Exercise 12–11

(a) Construct the straight-line gain function for the transfer function

$$T(s) = \frac{10^5}{[s^2 + 100s + 10^4](s + 10)}$$

(b) Find the straight-line gain at $\omega = 1, 10, 100$, and 1000 rad/s.

Answers:

(a) $|T(j\omega)|_{SL} = \begin{vmatrix} 1 & \text{if} & 0 \le \omega \le 10 \\ 10/\omega & \text{if} & 10 \le \omega \le 100 \\ 10^5/\omega^3 & \text{if} & 100 \le \omega \end{vmatrix}$

(b) 0 dB; 0 dB; –20 dB; –80 dB

PHASE PLOTS FOR COMPLEX CRITICAL FREQUENCIES

Figure 12–39 shows the phase contribution from complex poles or zeros for several values of ζ. The low-frequency phase asymptotes are 0° and the high-frequency limits are ±180°. The phase is always ±90° at $\omega = \omega_0$ regardless of the value of the damping ratio. The total phase change is ±180°, and most of this change occurs in a two-decade range from $\omega_0/10$ to $10\omega_0$. As a result, the straight lines in Figure 12–39 offer crude approximations to the phase shift. The shape of the phase curves change radically with the damping ratio, so these straight-line approximations are of little use except at ω_0 and around the end points.

The net result is that phase angle plots for complex critical frequencies are best generated by computer-aided analysis. In practical applications we can often derive useful information from the gain plot alone without generating phase response. However, the converse is not true. When we need phase response, we usually need gain as well. The next example shows how to obtain both using a computer-aided analysis tool.

EXAMPLE 12–21

Plot the gain and phase response of the transfer function

$$T(s) = \frac{5000(s + 100)}{s^2 + 400s + (500)^2}$$

SOLUTION:

The MATLAB function `bode` calculates and plots the gain and phase responses of transfer functions of the form $T(s) = n(s)/d(s)$, where $n(s)$ is the numerator polynomial and $d(s)$ is the denominator polynomial. In the MATLAB command window these polynomials are defined by the row matrices `num` and `den` that contain the coefficients of the polynomials

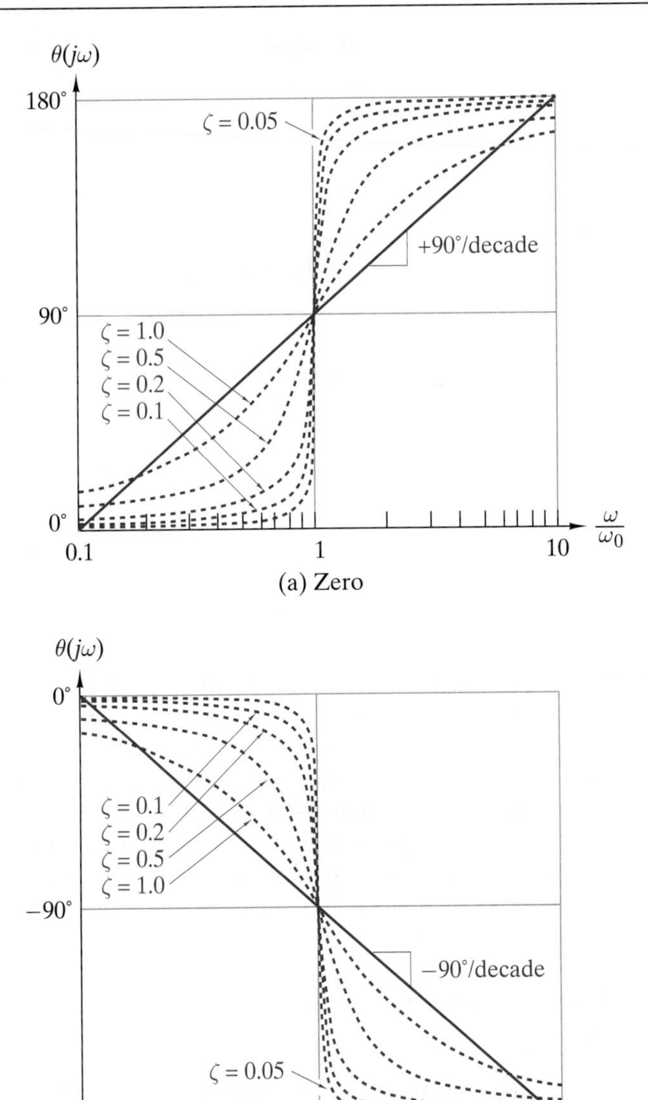

FIGURE 12–39 *Phase responses of complex poles and zeros.*

$n(s)$ and $d(s)$ in descending powers of s. In the present case these matrices are

```
num=[5000,5000*100];
den=[1,400,500^2];
```

The first element in both matrices is the coefficient of the highest power of s followed by the other coefficients in descending powers of s. The order of the polynomial is determined by the total number of entries in the matrix. This means that every polynomial coefficient must be entered in the matrix even if it is zero. The MATLAB statement

```
bode (num,den)
```

produces the Bode plots shown in Figure 12–40.

FIGURE 12–40

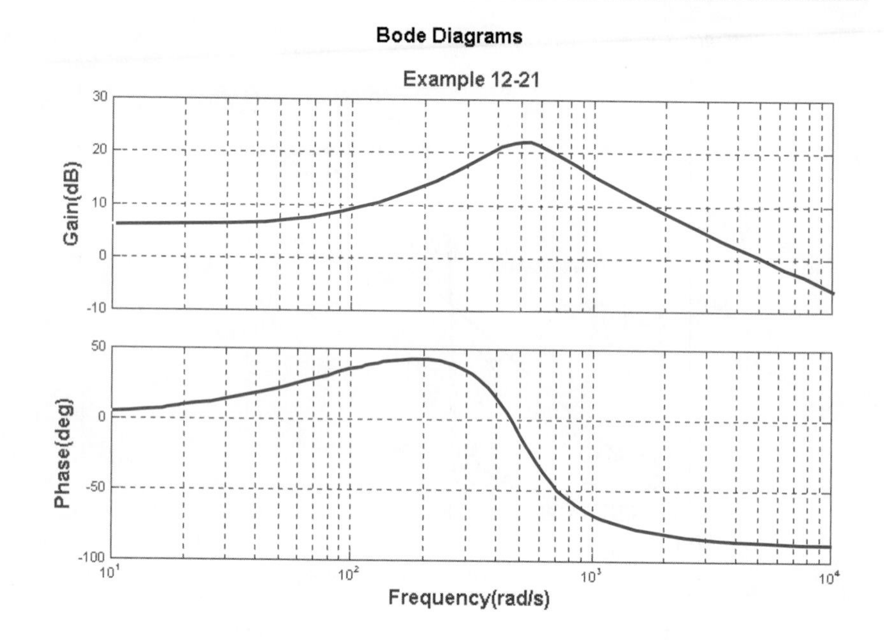

At low frequencies the phase angle gradually increases under the influence of the zero at $s = -100$ rad/s. As the input frequency approaches the natural frequency of the complex poles, the phase begins to decrease. As the frequency passes natural frequency at $\omega = 500$ rad/s, the phase rapidly decreases and eventually approaches –90° at high frequency. The net effect of the zero and the two poles is that the phase initially increases and then decreases as the poles come into play. The result is that the phase has a maximum of about 42° at around 200 rad/s. ■

12–8 FREQUENCY RESPONSE AND STEP RESPONSE

The step response $g(t)$ and frequency response $|T(j\omega)|$ are alternative ways to describe the performance of a filter. The transient and frequency responses are not independent. In fact, one can be derived from the other. Filter performance is usually specified in terms of frequency response. Hence, it is important to understand the time-domain consequences of frequency-domain specifications, and vice versa.

As a starting place, consider the first-order low-pass filter defined by

$$T(s) = \frac{K}{s + \alpha}$$

Section 12–3 shows that this filter has a passband (dc) gain of K/α and a cutoff frequency of $\omega_C = \alpha$. Recall from Chapter 11 that the step response of a circuit is $G(s) = T(s)/s$. Hence, the step response of this low-pass filter is

$$G(s) = \frac{K}{s(s + \alpha)} = \frac{K/\alpha}{s} - \frac{K/\alpha}{s + \alpha}$$

The time-domain step response $g(t)$ is

$$g(t) = \mathcal{L}^{-1}\{G(s)\} = \frac{K}{\alpha}(1 - e^{-\alpha t}) \text{ for } t \geq 0$$

The time-domain response has a final value ($t \rightarrow \infty$) of K/α and a time constant of $T_C = 1/\alpha$.

Specifying the gain and cutoff frequency of the frequency response also determines the amplitude and duration of the transient response, and vice versa. For instance, if the duration of the transient response is specified to be less than 20 μs, then $5T_C = 5/\alpha = 5/\omega_C < 20 \times 10^{-6}$ and hence $\omega_C > 250$ krad/s or 39.8 kHz. This cutoff frequency might not satisfy the frequency-response requirements. In any case, we need to look at the response in both domains.

Today's users of communication technology generally recognize that wide band and high speed are synonymous. The first-order low-pass filter illustrates the connection, since $\omega_C T_C = 1$. If we want a short response time (small T_C), then a wide bandwidth (large ω_C) is required. On the other hand, a fixed bandwidth (given ω_C) limits the speed of response, since $T_C = 1/\omega_C$. This illustrates a signal-processing principle called *reciprocal spreading*. Shrinking the response in one domain causes the response in the other domain to spread out.

The important point is that evaluating filter performance involves both the frequency domain and the time domain. We will encounter this idea again in Chapter 14, where we design much more complex filters. In any case is it clear that it is important to be able to go from one domain to the other. Changing domains involves concepts learned in several previous chapters, as illustrated in the following examples.

EXAMPLE 12–22

For $t \geq 0$, the step response of a filter is

$$g(t) = -5e^{-40t} + 25e^{-200t}$$

Is this a low-pass, high-pass, bandpass, or bandstop filter? What is the passband gain and the approximate cutoff frequency?

SOLUTION:

Recall from Chapter 11 that $T(s) = sG(s)$ where $G(s) = \mathcal{L}\{g(t)\}$. The filter transfer function is found as follows:

$$T(s) = s\,\mathcal{L}\{g(t)\} = -\frac{5s}{s+40} + \frac{25s}{s+200}$$

$$= \frac{20\,s^2}{(s+40)(s+200)}$$

This is a high-pass filter with $T(0) = 0$ and $T(\infty) = 20$. The high-frequency passband gain is 20 or, equivalently, 26 dB. The filter has a corner frequency (pole) at $\omega = 40$ rad/s and another at $\omega = 200$ rad/s. The low-frequency gain asymptote is $20\omega^2/(40 \times 200) = \omega^2/400$. This asymptote has a slope of −40 dB/decade and a value of 4 (14 dB) at the first corner. Above the first corner, the gain asymptote is $20\omega/(200) = \omega/10$, which has a slope of −20 dB/decade and a value of 20 (26 dB) at the second corner. Given these results, we can easily sketch the straight-line gain shown in Figure 12–41. Clearly the cutoff frequency is in the vicinity of the second corner frequency, hence $\omega_C \approx 200$ rad/s.

FIGURE 12–41

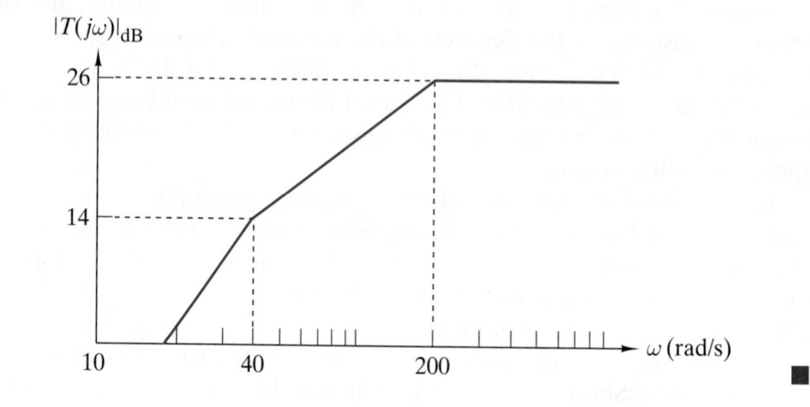

EXAMPLE 12–23

The straight-line gain response of a filter is shown in Figure 12–42. What are the initial and final values of the step response? What is the approximate duration of the transient response?

FIGURE 12–42

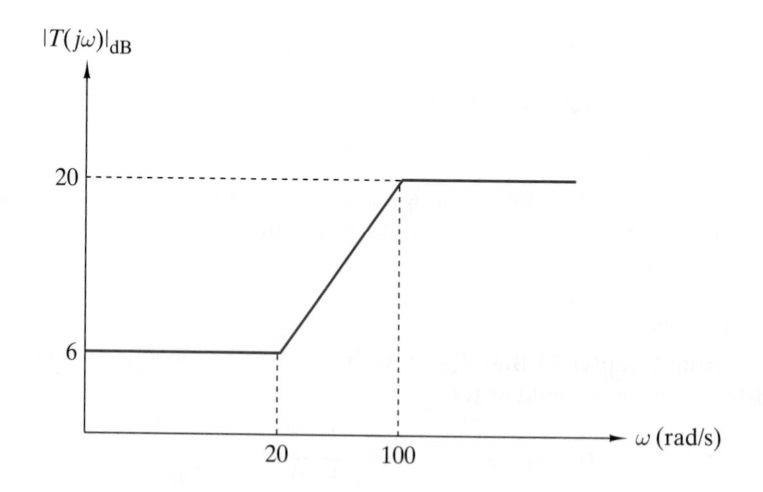

SOLUTION:

In this case we must derive the transfer function from the gain plot. The straight-line gain has corner frequencies at $\omega = 20$ rad/s and another at $\omega = 100$ rad/s. The first corner increases the gain slope by $(20 - 6)/(\log 100 - \log 20) = +20$ dB/decade, so there is a simple zero at $s = -20$. The second corner decreases the slope by −20 dB/decade, so there is a simple pole at $s = -100$. The two critical frequencies account for all of the corner frequencies shown in the figure, so the necessary transfer function has the form

$$T(s) = K\frac{s + 20}{s + 100}$$

The infinite frequency gain of this transfer function is $T(\infty) = K$. The high-frequency gain in Figure 12–42 is +20 dB; hence, $K = 10$. In general the step response is $G(s) = T(s)/s$; hence, for this filter we have

$$G(s) = \frac{10}{s}\frac{s + 20}{s + 100} = \frac{2}{s} + \frac{8}{s + 100}$$

The time-domain step response $g(t)$ is

$$g(t) = \mathscr{L}^{-1}\{G(s)\} = 2 + 8e^{-100t} \quad t > 0$$

The initial value is $g(0) = 10$, the final value is $g(\infty) = 2$, and the time constant of the transient term is $T_C = 1/100$ s. The duration of the transient response is about $5T_C = 50$ ms. ■

EVALUATION EXAMPLE 12–24

Modern piezoelectric pressure transducers are fabricated as an integrated circuit that includes both the sensing crystal and the signal-conditioning electronics. The step response of such a transducer takes the form

$$v_{TR}(t) = K_V e^{-t/T_C} p(t)$$

where $v_{TR}(t)$ is the transducer output voltage caused by an abrupt change in pressure $p(t) = P_A u(t)$. The transducer transient response is defined by the sensitivity K_V (usually in mV/psi), and the discharge time constant T_C (usually in seconds). A transducer is to be installed in an environment with sinusoidal pressure variations in the range from 2 mHz to 2 kHz. A vendor offers transducers with the following characteristic.

MODEL	261	265	266	268	269
K_V (mV/psi)	50	10	5	1	0.5
T_C (s)	10	50	100	500	1000

Select the model that meets the frequency-response requirements and produces the largest output signal for any given pressure change.

SOLUTION:

The specified response $v_{TR}(t)$ is for an input $P_A u(t)$. The unit step response $g(t)$ is derived from this response by dividing out the amplitude of the input step P_A.

$$g(t) = \frac{1}{P_A} v_{TR}(t) = K_V e^{-t/T_C} u(t)$$

In the s domain the step response is

$$G(s) = \frac{K_V}{s + 1/T_C}$$

and the pressure-to-voltage transfer function is found to be

$$T(s) = sG(s) = \frac{K_V s}{s + 1/T_C}$$

This transfer function has a first-order high-pass characteristic with a passband gain of K_V and a cutoff frequency of $\omega_C = 1/T_C$. The specified frequency range from 2 mHz to 2 kHz must fall in the transducer passband, which means that the cutoff frequency must be less than 2 mHz. In equation form this requires

$$\omega_C = \frac{1}{T_C} < 2\pi \times 0.002 \quad \text{or} \quad T_C > \frac{250}{\pi} = 79.6 \text{ s}$$

Models 266, 268, and 269 meet the frequency-response requirement, since they have discharge time constants T_C greater than 79.6 s. Model 266 is the best choice because it has the most sensitivity (passband gain) at $K_V =$ 5 mV/psi. ■

SUMMARY

- The frequency response of a circuit is defined by the variation of the gain $|T(j\omega)|$ and phase $\angle T(j\omega)$ with frequency. The gain function is usually expressed in dB in frequency-response plots. Logarithmic frequency scales are used on frequency-response plots of the gain and phase functions.
- A passband is a range of frequencies over which the steady-state output is essentially constant with very little attenuation. A stopband is a range of frequencies over which the steady-state output is significantly attenuated. The cutoff frequency is the boundary between a passband and the adjacent stopband.
- Circuit gain responses are classified as low pass, high pass, bandpass, and bandstop depending on the number and location of the stop and passbands. The performance of devices and circuits is often specified in terms of frequency-response descriptors such as bandwidth, passband gain, and cutoff frequency.
- The low- and high-frequency gain asymptotes of a first-order circuit intersect at a corner frequency determined by the location of its pole. The total phase change from low to high frequency is ±90°. First-order circuits can be connected to produce bandpass and bandstop responses.
- Series and parallel *RLC* circuits provide bandpass and bandstop gain characteristics that are easily related to the circuit parameters.
- Bode plots are graphs of the gain (in dB) and phase angle (in degrees) versus log-frequency scales. Straight-line approximations to the gain and phase can be constructed using the corner frequencies defined by

the poles and zeros of $T(s)$. The purpose of the straight-line approximations is to develop a conceptual understanding of frequency response. The straight-line plots do not necessarily provide accurate data at all frequencies, especially for circuits with complex poles.

- Computer-aided circuit analysis programs can accurately generate and plot frequency-response data. The user must have a rough idea of the gain and frequency ranges of interest to use these tools intelligently.
- It is often necessary to relate the time-domain characteristics of a circuit to its frequency response, and vice versa. The transfer function provides a link between the frequency-domain and time-domain responses.

PROBLEMS

OBJECTIVE 12–1 FIRST-ORDER CIRCUIT FREQUENCY RESPONSE (SECTS. 12–1, 12–2, 12–3)

Given a first-order circuit or transfer function:
(a) Determine frequency-response descriptors and classify the response.
(b) Sketch the straight-line approximations of the gain and phases responses.
(c) Select circuit parameters to produce a specified frequency response.
See Examples 12–2, 12–3, 12–4, 12–5, 12–6, 12–7 and Exercise 12–1, 12–2

12–1 Find the transfer function $T_V(s) = V_2(s)/V_1(s)$ of the circuit in Figure P12–1.

(a) Find the dc gain, infinite frequency gain, and cutoff frequency. Identify the type of gain response.
(b) Sketch the straight-line approximations of the gain and phase responses.
(c) Calculate the gain at $\omega = 0.5\omega_C$, ω_C, and $2\omega_C$.

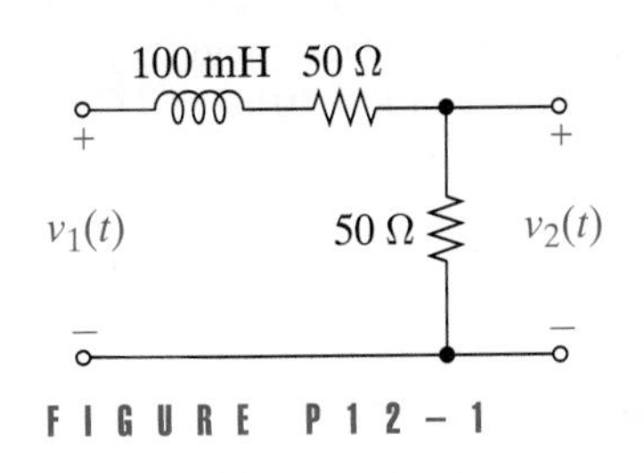

FIGURE P12–1

12–2 Find the transfer function $T_V(s) = V_2(s)/V_1(s)$ of the circuit in Figure P12–2.

(a) Find the dc gain, infinite frequency gain, and cutoff frequency. Identify the type of gain response.
(b) Sketch the straight-line approximations of the gain response.
(c) Calculate the gain at $\omega = 0.5\omega_C$, ω_C, and $2\omega_C$.

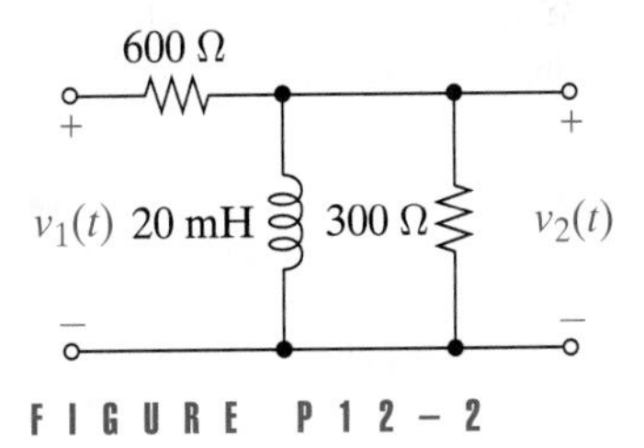

FIGURE P12–2

12–3 Find the transfer function $T_V(s) = V_2(s)/V_1(s)$ of the circuit in Figure P12–3.

(a) Find the dc gain, infinite frequency gain, and cutoff frequency. Identify the type of gain response.
(b) Sketch the straight-line approximation of the gain response.
(c) What element values would you change to increase the passband gain to 10?

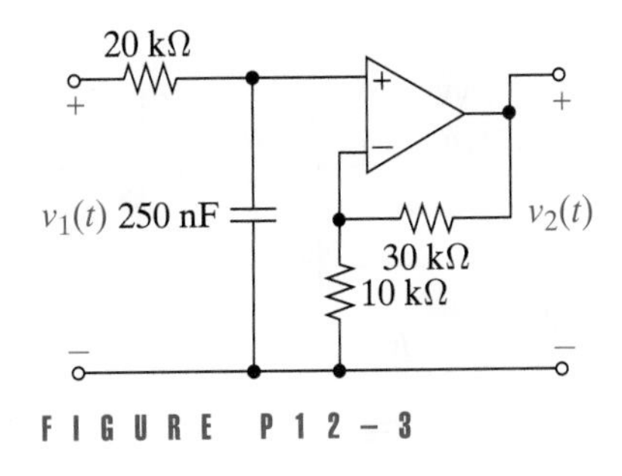

FIGURE P12–3

12–4 Find the transfer function $T_V(s) = V_2(s)/V_1(s)$ of the circuit in Figure P12–4.

(a) Find the dc gain, infinite frequency gain, and cutoff frequency. Identify the type of gain response.

(b) Draw the straight-line approximations of the gain and phase of $T_V(j\omega)$.
(c) What element values would you change to double the passband gain without changing the cutoff frequency?

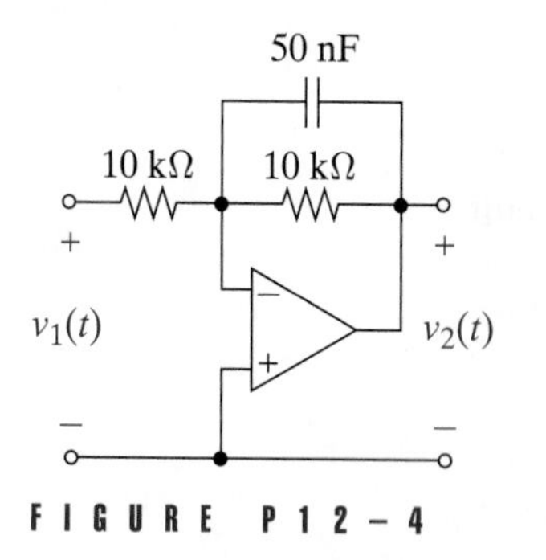

FIGURE P12–4

12–5 Find the transfer function $T_V(s) = V_2(s)/V_1(s)$ of the circuit in Figure P12–5.
(a) Find the dc gain, infinite frequency gain, and cutoff frequency. Identify the type of gain response.
(b) What element value would you change to increase the cutoff frequency by one decade?

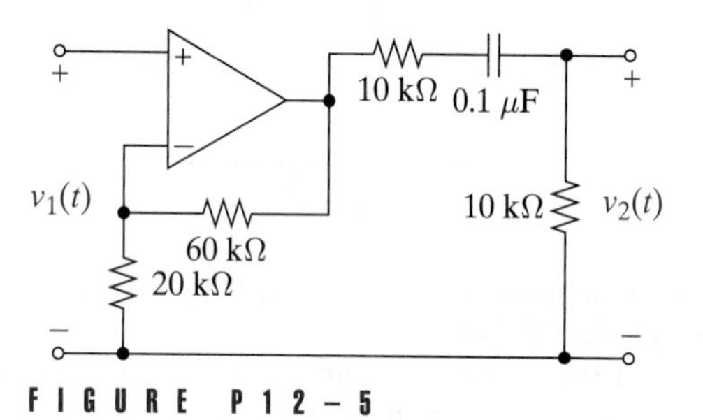

FIGURE P12–5

12–6 Find the transfer function $T_V(s) = V_2(s)/V_1(s)$ of the circuit in Figure P12–6. What type of gain response does the circuit have? What is the passband gain? Select values of R and C so that the cutoff frequency is 5 kHz.

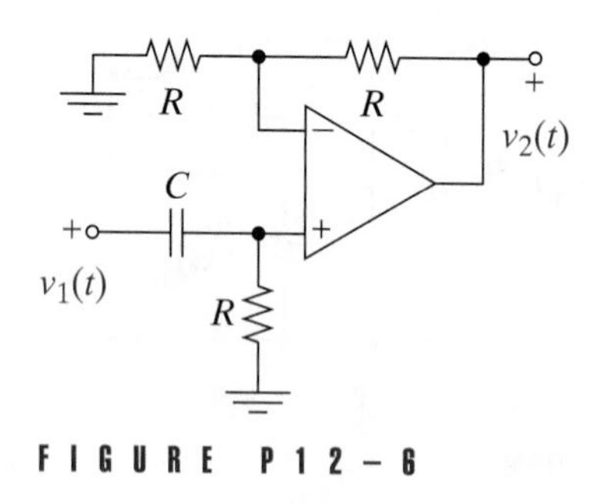

FIGURE P12–6

12–7 A first-order high-pass circuit has a passband gain of 20 dB and a cutoff frequency of 350 rad/s. Find the gain (in dB) at ω = 200, 400, and 800 rad/s.

12–8 A first-order low-pass circuit has a passband gain of 5 dB and a cutoff frequency of 2 krad/s. Find the gain at $\omega = 0.5\omega_C$, ω_C, and $2\omega_C$.

12–9 The transfer function of a first-order circuit is

$$T(s) = \frac{2}{10^{-2} + 20/s}$$

(a) Identify the type of gain response. Find the cutoff frequency and the passband gain.
(b) Sketch the straight-line gain response. Use the straight-line gain to estimate the gain at $\omega = 0.5\omega_C$, ω_C, and $2\omega_C$.

12–10 Repeat Problem 12–9 for

$$T(s) = \frac{10/s}{100/s + 1/20}$$

12–11 **D** Design a passive first-order low-pass filter with a dc gain of −6 dB and a cutoff frequency of 500 Hz.

12–12 **D** Design a first-order high-pass filter with a high-frequency gain of 5 and a cutoff frequency of 1.5 kHz.

12–13 **D** Design a first-order low-pass filter with a dc gain of 2 and a gain of −3 dB at 400 rad/sec.

OBJECTIVE 12–2 BANDPASS AND BANDSTOP RESPONSES (SECT. 12–4)

Given a cascade or parallel connection of two first-order circuits or their transfer functions:
(a) Relate the cutoff frequencies and passband gain to circuit elements.
(b) Select circuit parameters to produce specified cutoff frequencies and passband gain.
(c) Sketch the straight-line approximations of the gain responses.
See Examples 12–8, 12–9 and Exercises 12–3, 12–4

12–14 The circuit in Figure P12–14 produces a bandpass response for a suitable choice of element values. Identify the elements that control the two cutoff frequencies. Select the element values so that the passband gain is 10 and the cutoff frequencies are 100 rad/s and 2500 rad/s. Use practical element values with $R \geq 10\ \text{k}\Omega$ and $C \leq 1\ \mu\text{F}$.

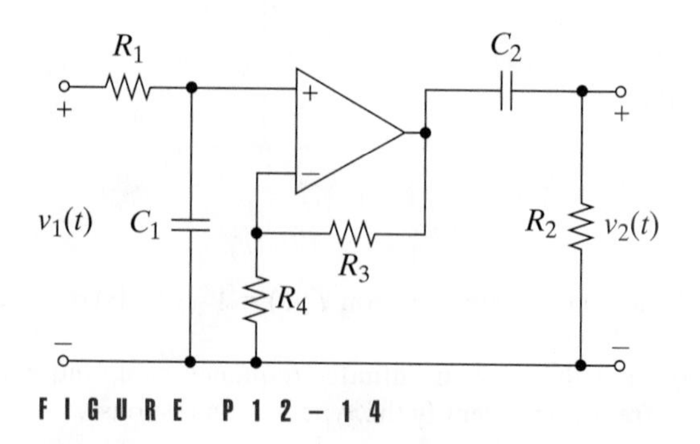

FIGURE P12–14

12–15 Repeat Problem 12–14 for the circuit in Figure P12–15.

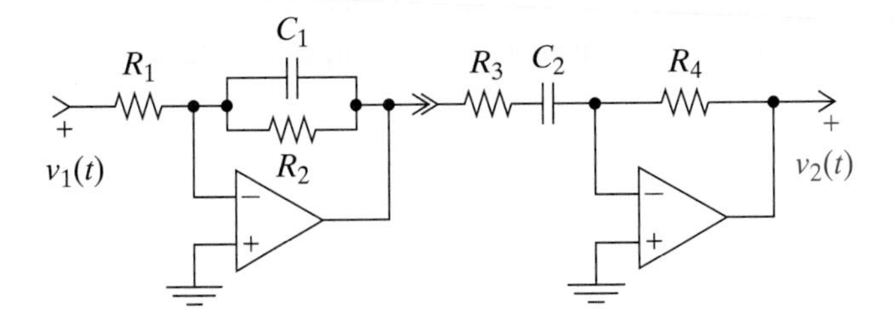

FIGURE P12–15

12–16 The circuit in Figure P12–16 produces a bandstop response for a suitable choice of element values. Identify the elements that control the two cutoff frequencies. Select the element values so that the cutoff frequencies are 10 rad/s and 500 rad/s. What is the passband gain for your choices? Use practical element values with $R \geq 10\text{ k}\Omega$, $L \leq 100\text{ mH}$, and $C \leq 1\text{ }\mu\text{F}$.

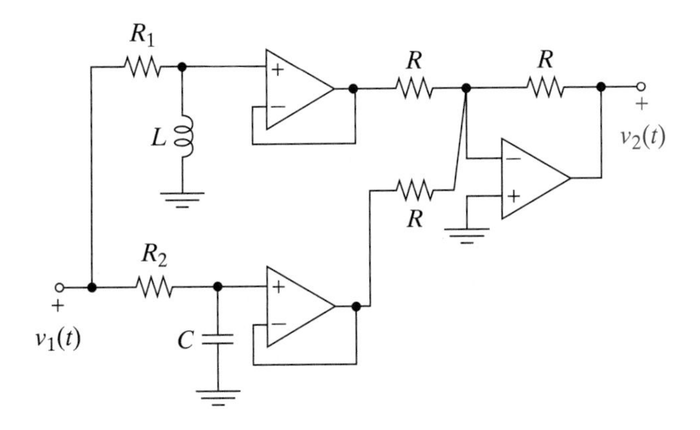

FIGURE P12–16

Each of the following problems involve a cascade connection of two first-order circuits. Sketch the straight-line Bode plot of the overall gain of each connection.

Problem	First Circuit	Second Circuit
12–17	$T_1(s) = \dfrac{s}{s + 100}$	$T_2(s) = \dfrac{\sqrt{10}}{s/2000 + 1}$
12–18	$T_1(s) = \dfrac{\sqrt{10}s}{s + 1500}$	$T_2(s) = \dfrac{\sqrt{10}}{s/30000 + 1}$
12–19	$T_1(s) = \dfrac{500}{s + 500}$	$T_2(s) = \dfrac{10}{s/2000 + 1}$
12–20	$T_1(s) = \dfrac{4s}{s + 200}$	$T_2(s) = \dfrac{4s}{s + 500}$

OBJECTIVE 12–3 THE FREQUENCY RESPONSE OF *RLC* CIRCUITS (SECT. 12–5)

Given an *RLC* circuit connected as a bandpass or bandstop filter:

(a) Find the circuit parameters or frequency-response descriptors.

(b) Select the circuit parameters to achieve specified filter characteristics.

(c) Derive expressions for the frequency-response descriptors.

See Examples 12–10, 12–11, 12–12, 12–13, 12–14 and Exercises 12–5, 12–6, 12–7

12–21 A series *RLC* bandpass circuit with $R = 24$ is designed to have a bandwidth of 4 Mrad/s and a center frequency of 40 Mrad/s. Find L, C, Q, and the two cutoff frequencies.

12–22 A parallel *RLC* bandpass circuit with $C = 20\text{ nF}$ and $Q = 10$ has a center frequency of 1 Mrad/s. Find R, L, and the two cutoff frequencies.

12–23 A 20-mH inductor with an internal series resistance of 20 Ω is connected in series with a capacitor and a voltage source with a Thévenin resistance of 60 Ω.

(a) What value of C is needed to produce $\omega_0 = 5$ krad/s?

(b) Find the bandwidth and quality factor of the circuit.

12–24 In a series *RLC* circuit, which element would you adjust (and by how much) to do the following?

(a) Double the bandwidth without changing the center frequency

(b) Double the center frequency without changing the bandwidth.

Repeat for a parallel *RLC* circuit.

12–25 A parallel *RLC* circuit with $R = 2\text{ k}\Omega$ has a center frequency of 1 MHz and a bandwidth of 10 kHz. Find the values of L and C.

12–26 **D** A series *RLC* bandpass filter is required to have resonance at $f_0 = 30$ kHz. The series connected L and C are driven by a sinusoidal source with a Thévenin resistance of 60 Ω. The following standard capacitors are available in the stock room: 1 μF, 680 nF, 470 nF, 330 nF, 200 nF, and 120 nF. The inductor will be custom-designed to match the capacitor used. Select the available capacitor that minimizes the filter bandwidth.

12–27 **D** A series *RLC* bandstop circuit is to be used as a notch filter to eliminate a bothersome 60-Hz hum in an audio channel. The signal source has a Thévenin resistance of 600 Ω. Select values of L and C so the notch frequency is 60 Hz and the upper cutoff frequency is below 100 Hz.

12–28 Find the transfer function $T_V(s) = V_2(s)/V_1(s)$ for the bandpass circuit in Figure P12–28. Find the parameters B, Q, ω_{C1}, and ω_{C2} for $R = 50\ \Omega$, $L = 50\ \mu H$, and $C = 20$ nF.

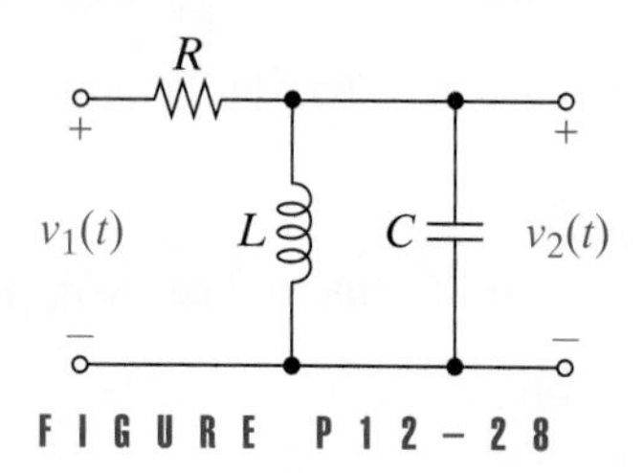

FIGURE P12–28

12–29 Show that the transfer function $T_V(s) = V_2(s)/V_1(s)$ of the circuit in Figure P12–29 has a bandstop filter characteristic. Derive expressions relating the passband gains, the notch frequency, and the cutoff frequencies to R, L, and C.

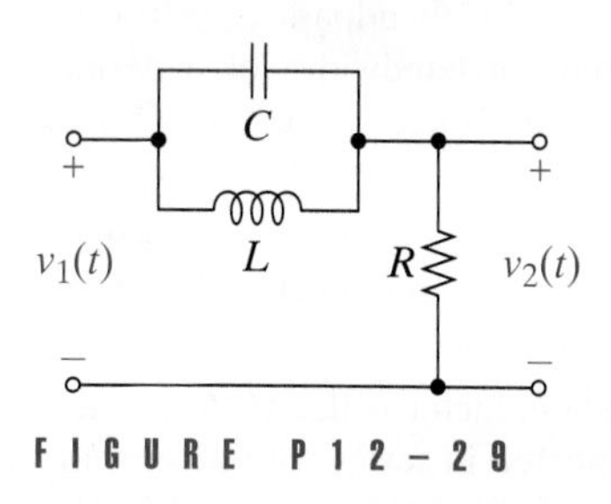

FIGURE P12–29

12–30 Figure P12–30 shows an RLC filter with an input current and an output voltage. The purpose of this problem is to determine the filter type using informal circuit analysis. Use the element impedances and basic analysis tools to find the magnitude of the output voltage $|V_2(j\omega)|$ at $\omega = 0$, $\omega = \infty$, and $\omega = \omega_0 = 1/\sqrt{LC}$. What is the filter type?

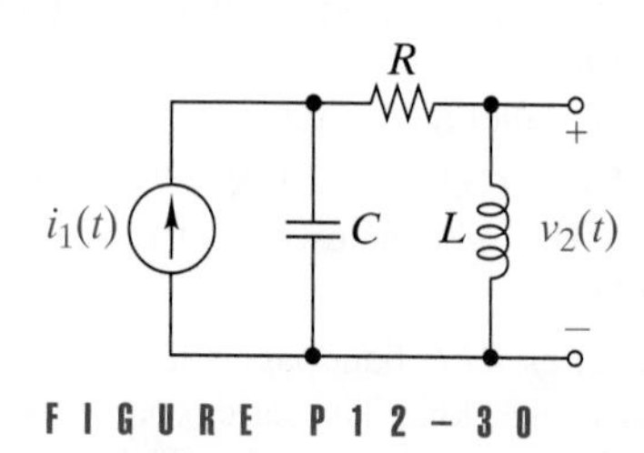

FIGURE P12–30

OBJECTIVE 12–4 BODE PLOTS (SECTS. 12–6, 12–7)

(a) Construct Bode plots of the straight-line approximations of gain and phase responses of a given circuit or transfer function.
(b) Construct the transfer function corresponding to a given straight-line gain plot.
(c) Design a circuit that produces a given straight-line gain plot.

See Examples 12–15, 12–16, 12–17, 12–18, 12–19, 12–20, 12–21 and Exercises 12–8, 12–9, 12–10, 12–11

12–31 Construct the straight-line Bode plot of the gain of the transfer function $T_V(s) = V_2(s)/V_1(s)$ for the circuit in Figure P12–31. Use the straight-line gain plot to estimate the amplitude of the steady-state output for $v_1(t) = 10 \sin 30t$ V. Calculate the actual output amplitude for this input and compare the two results.

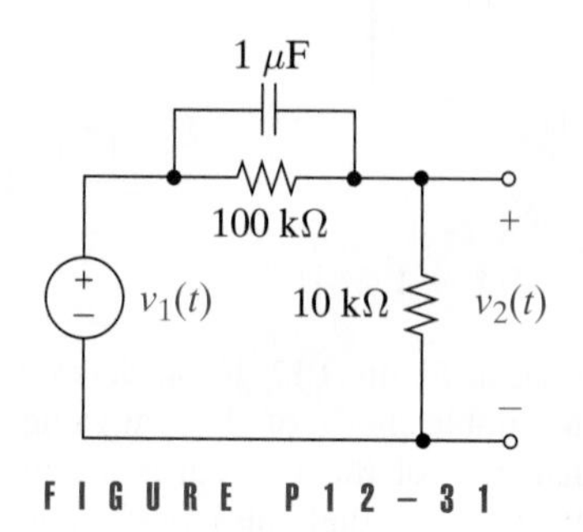

FIGURE P12–31

12–32 Repeat Problem 12–31 using the circuit in Figure P12–32.

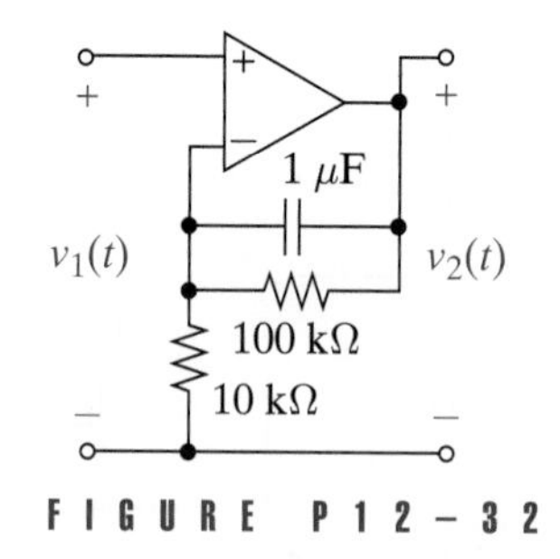

FIGURE P12–32

12–33 Construct the straight-line Bode plot of the gain of the following transfer function. Is this a low-pass, high-pass, bandpass, or bandstop function? Estimate the cutoff frequency and passband gain.

$$T(s) = \frac{(s + 20)}{(10s + 4)(s + 5)}$$

12–34 Construct the straight-line Bode plot of the gain of the following transfer function. Is this a low-pass, high-pass, bandpass, or bandstop function? Estimate the cutoff frequency and passband gain.

$$T(s) = \frac{s(s + 200)}{(s + 1)(s + 5)}$$

12–35 Construct the straight-line Bode plot of the gain of the following transfer function. Is this a low-pass, high-pass, bandpass, or bandstop function? Estimate the cutoff frequency and passband gain.

$$T(s) = \frac{25s(s + 20)}{(s + 5)(s + 100)}$$

12–36 Construct the straight-line Bode plot of the gain of the following transfer function. Is this a low-pass, high-pass, bandpass, or bandstop function? Estimate the cutoff frequency and passband gain.

$$T(s) = \frac{10(s+5)(s+20)}{(s+1)(s+100)}$$

12–37 Construct the straight-line Bode plot of the gain of the following transfer function. Is this a low-pass, high-pass, bandpass, or bandstop function? Estimate the cutoff frequency and passband gain.

$$T(s) = \frac{8s^2}{(0.4s+1)^2}$$

12–38 Construct the straight-line Bode plot of the gain of the following transfer function. Is this a low-pass, high-pass, bandpass, or bandstop function? Use the straight-line plot to estimate the maximum gain and the frequency at which it occurs.

$$T(s) = \frac{4s}{0.04s^2 + 0.2s + 1}$$

12–39 Find the transfer function corresponding to the straight-line gain plot in Figure P12–39. Compare the straight-line gain and the actual gain at ω = 10 and 400 rad/s.

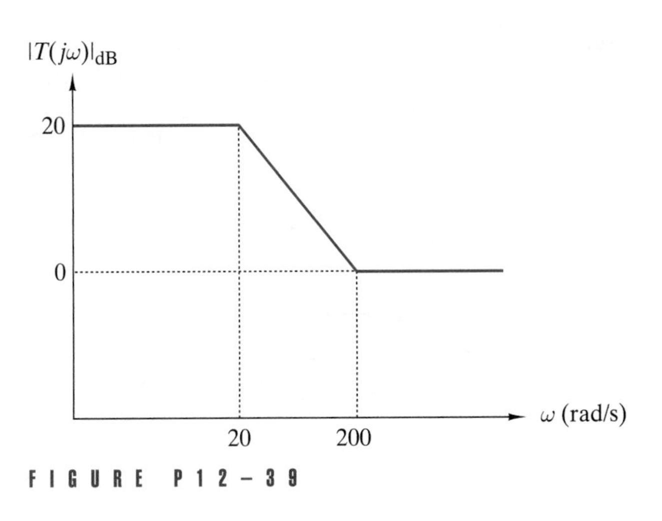

FIGURE P12–39

12–40 Find the transfer function corresponding to the straight-line gain plot in Figure P12–40. Compare the straight-line gain and the actual gain at ω = 10 and 500 rad/s.

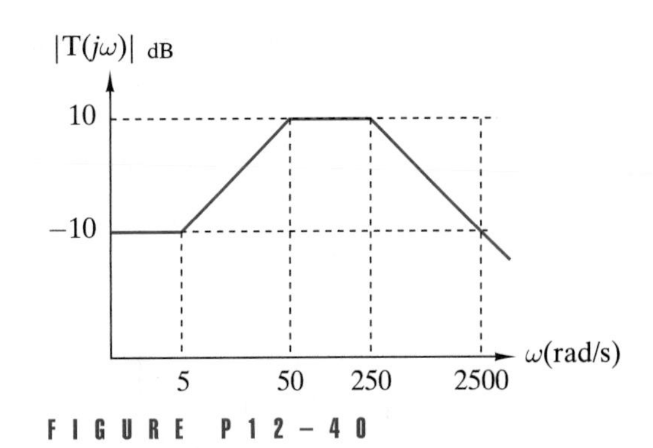

FIGURE P12–40

12–41 Find the transfer function corresponding to the straight-line gain plot in Figure P12–41. Compare the straight-line gain and the actual gain at ω = 50 rad/s and 100 krad/s.

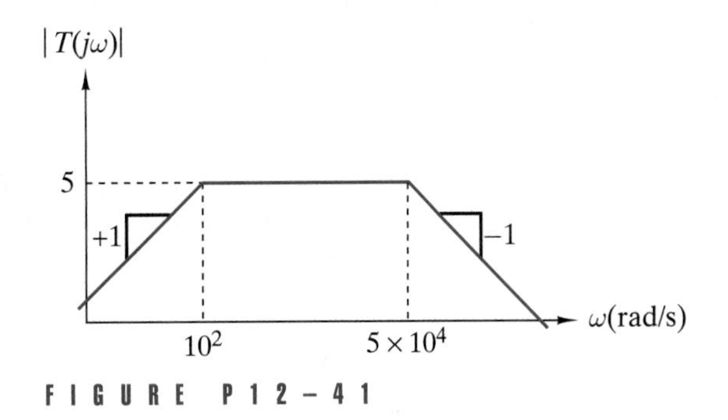

FIGURE P12–41

12–42 D Design a circuit that produces the straight-line gain plot in Figure P12–42.

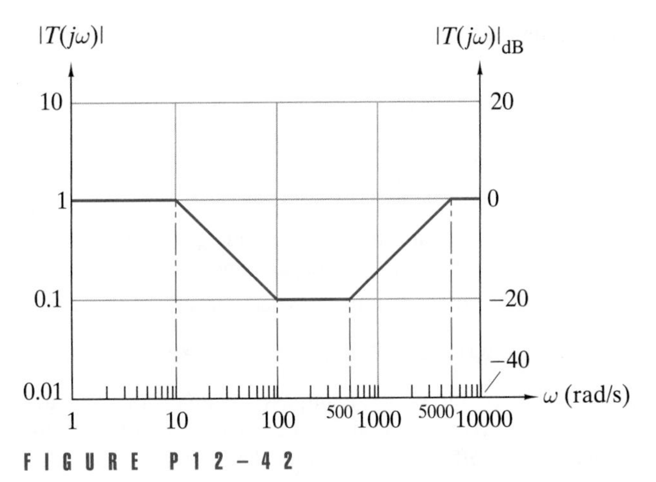

FIGURE P12–42

12–43 **D** Design a circuit that produces the straight-line gain plot in Figure P12–43.

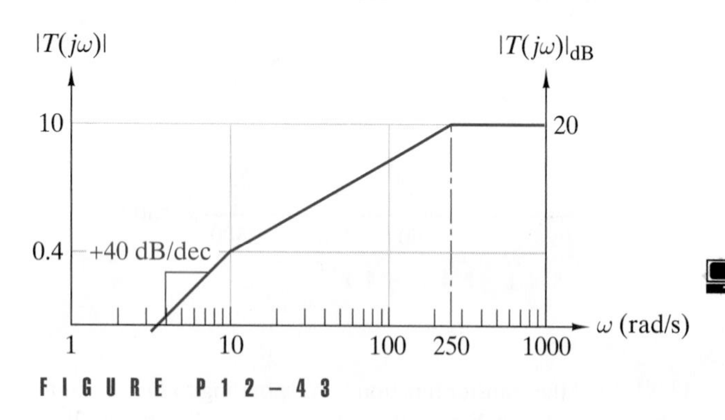

FIGURE P12–43

12–44 Construct the straight-line Bode plot of the phase of the following transfer function. Use the straight-line phase to estimate the actual phase at ω = 1, 10, and 100 rad/s.

$$T(s) = \frac{10(s + 20)}{s + 1}$$

12–45 Construct the straight-line Bode plot of the phase of the following transfer function. Use the straight-line phase to estimate the actual phase at ω = 1, 10, and 100 rad/s.

$$T(s) = \frac{10(s + 5)}{(s + 1)(s + 100)}$$

OBJECTIVE 12–5 FREQUENCY RESPONSE AND STEP RESPONSE (SECT. 12–8)

(a) Find the gain response corresponding to a given step response.
(b) Find the step response corresponding to a given gain response.
See Examples 12–22, 12–23, 12–24

12–46 The step response of a linear circuit is

$$g(t) = 12 - 10e^{-10t} - 2e^{-100t} \quad t > 0$$

Is the circuit a low-pass, high-pass, bandpass, or bandstop filter? Construct the straight-line Bode gain plot and estimate the cutoff frequency and passband gain.

12–47 The step response of a linear circuit is

$$g(t) = 3 - 2e^{-20t} + 2e^{-500t} \quad t > 0$$

Is the circuit a low-pass, high-pass, bandpass, or bandstop filter? Construct the straight-line Bode gain plot and estimate the cutoff frequency and passband gain.

12–48 The step response of a linear circuit is

$$g(t) = 4\left[e^{-50t} - e^{-2000t}\right] \quad t > 0$$

Is the circuit a low-pass, high-pass, bandpass, or bandstop filter? Construct the straight-line Bode gain plot and estimate the cutoff frequency and passband gain.

12–49 The straight-line gain response of a linear circuit is shown in Figure P12–39. What are the initial and final values of the circuit step response? What is the approximate duration of the transient response?

12–50 The straight-line gain response of a linear circuit is shown in Figure P12–41. What are the initial and final values of the circuit step response? What is the approximate duration of the transient response?

INTEGRATING PROBLEMS

12–51 **A** Step Response of an *RLC* Bandpass Circuit
The step response of a series *RLC* bandpass circuit is

$$g(t) = \frac{4}{5}e^{-200t}\sin(500t) \quad t > 0$$

Find the passband center frequency and the two cutoff frequencies.

12–52 **A** A Tunable Tank Circuit
The *RLC* circuit in Figure P12–52 (often called a tank circuit) has $R = 10\text{ k}\Omega$, $C = 125$ pF, and an adjustable (tunable) L ranging from 64 to 640 μH.
(a) Show that the circuit is a bandpass filter.
(b) Find the frequency range (in Hz) over which the center frequency can be tuned.
(c) Find the bandwidth (in Hz) at the end points of this range.

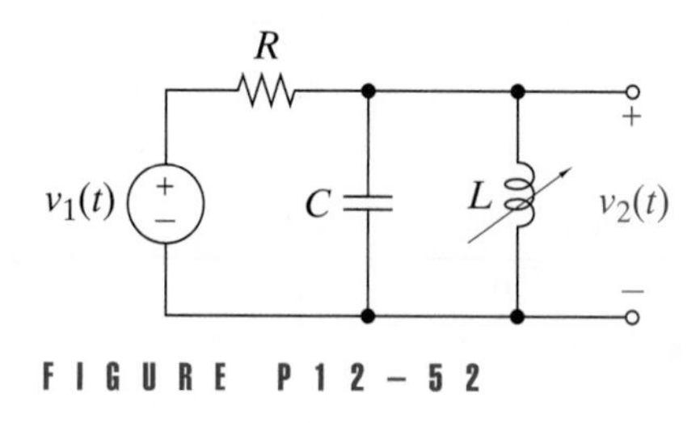

FIGURE P12–52

12–53 **D** Filter Design Specification
Construct a transfer function whose gain response lies entirely within the unshaded region in Figure P12–53.

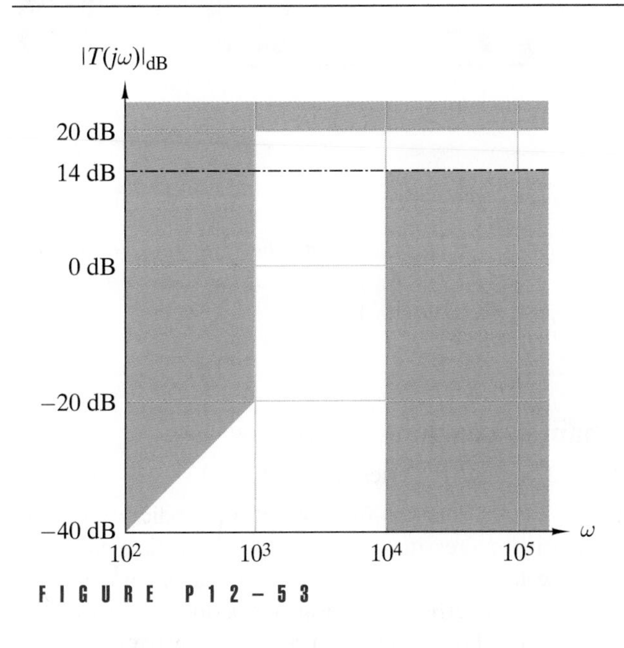

FIGURE P12–53

12–54 Chip *RC* Networks

Integrated circuit (chip) *RC* networks are used at parallel data ports to suppress radio frequency (RF) noise. In a certain application RF noise at 3.2 MHz is interfering with a four-bit parallel data signal operating at 1.1 MHz. A chip *RC* network is to be used to reduce the RF noise on the parallel bus by at least 7 dB without reducing the data signals by more than 2 dB. A vendor offers a family of chip *RC* networks connected as shown in Figure P12–54. The available circuit parameters are shown in Table P12–54. Select the part number that best meets the noise suppression requirements.

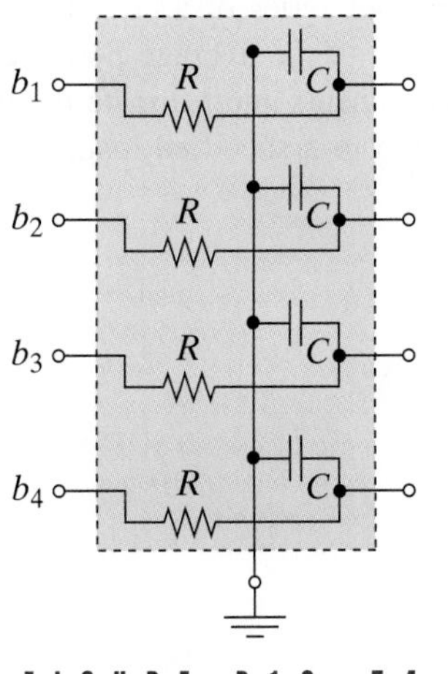

FIGURE P12–54

TABLE P12–54

$R(\Omega)$	C (pF)	PART NO.	$R(\Omega)$	C (pF)	PART NO.
47	220	ZAEN14	220	220	ZAEN34
47	470	ZAEN15	220	470	ZAEN35
47	1000	ZAEN16	220	1000	ZAEN36
100	220	ZAEN24	470	220	ZAEN44
100	470	ZAEN25	470	470	ZAEN45
100	1000	ZAEN26	470	1000	ZAEN46

12–55 Design Evaluation

Your company issued a request for proposals listing the following design requirements and evaluation criteria.

Design requirements: Design a low-pass filter with a passband gain of 9 ± 10% and a cutoff frequency of 90 ± 10% krad/s. The filter input is driven by a sensor with a 1-kΩ source resistance and an open-circuit voltage range of ±1.6 V.

Evaluation criteria: Evaluate filter performance, parts count, use of standard parts, and cost. The two vendors have responded with the designs shown in Figure P12–55. As a junior engineer, you are asked to evaluate the designs and recommend a vendor. Which vendor would you recommend and why?

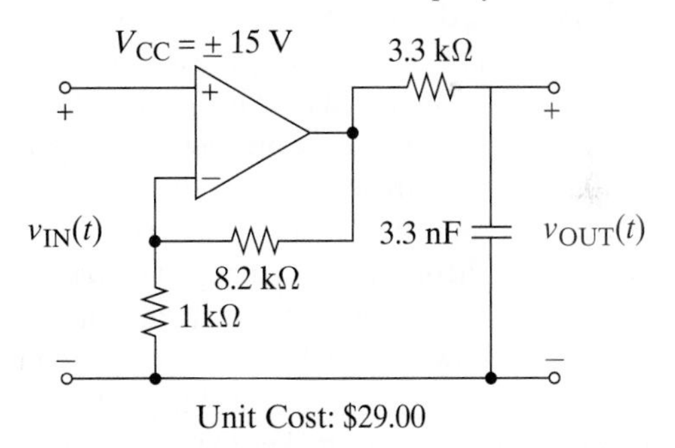

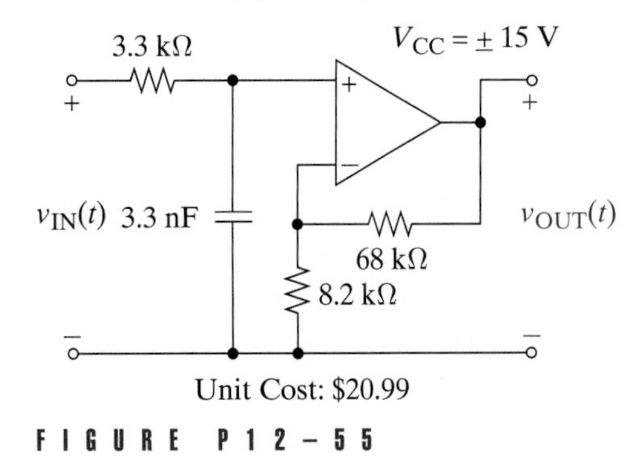

FIGURE P12–55

CHAPTER 13

FOURIER SERIES

The series formed of sines or cosines of multiple arcs are therefore adapted to represent between definite limits all possible functions, and the ordinates of lines or surfaces whose form is discontinuous.

Jean Baptiste Joseph Fourier, 1822,
French Physicist

Some History Behind This Chapter

The analysis techniques in this chapter have their roots in the works of the French physicist J. J. B. Fourier (1768–1830). In his studies Fourier found that discontinuous functions could be represented by an infinite series of harmonic sinusoids. Fourier was guided by his physical intuition and presented empirical evidence to support his claim but left the question of series convergence unanswered. This question was later settled by the German mathematician P. G. L. Dirichlet (1805–1859). Fourier's methods were eventually accepted and are now widely used in many branches of science and engineering.

Why This Chapter Is Important Today

The word *spectrum* denotes both a natural resource and a mathematical concept. As a natural resource, it is critical to modern communications and its use is regulated by national and international agencies. As a mathematical concept, it describes the range of frequencies needed to reproduce a signal. Perfect reproduction is not always possible or even necessary. Fourier theory suggests that perfect reproduction requires an infinite spectrum, a practical impossibility. The theory also shows that a finite range of frequencies carry most of the energy in periodic signals. Intentionally limiting a signal to this range makes the spectrum available to other users and allows today's profusion of communication systems.

Chapter Sections

13–1 Overview of Fourier Analysis

13–2 Fourier Coefficients

13–3 Waveform Symmetries

13–4 Circuit Analysis Using the Fourier Series

13–5 RMS Value and Average Power

Chapter Learning Objectives

13-1 The Fourier Series (Sects. 13-1, 13-2, 13-3)

(a) Given an equation or graph of a periodic waveform, derive expressions for the Fourier coefficients.
(b) Use the results in Figure 13–3 to calculate the Fourier coefficients of a given periodic waveform.
(c) Given a Fourier series of a periodic waveform, determine properties of the waveform, and plot its amplitude and phase spectra.

13-2 Fourier Series and Circuit Analysis (Sect. 13-4)

(a) Given a linear circuit with a periodic input waveform, find the Fourier series of a steady-state response.
(b) Given a network function with a periodic input, find the amplitude and phase spectra of the steady-state output.

13-3 rms Value and Average Power (Sect. 13-5)

(a) Given a periodic waveform, find the rms value of the waveform and the average power delivered to a specified load.
(b) Given the Fourier series of a periodic waveform, find the fraction of the average power carried by specified components and estimate the average power delivered to a specified load.

13–1 OVERVIEW OF FOURIER ANALYSIS

In this chapter we develop a method of finding the steady-state response of circuits to periodic inputs such as the waveforms in Figure 13–1. These periodic waveforms can be written as a Fourier series consisting of an infinite sum of harmonically related sinusoids. More specifically, if $f(t)$ is periodic with period T_0 and is reasonably well behaved, then $f(t)$ can be expressed as a **Fourier series** of the form

$$\begin{aligned} f(t) = a_0 &+ a_1\cos(2\pi f_0 t) + a_2\cos(2\pi 2 f_0 t) + \ldots + a_n\cos(2\pi n f_0 t) + \ldots \\ &+ b_1\sin(2\pi f_0 t) + b_2\sin(2\pi 2 f_0 t) + \ldots + b_n\sin(2\pi n f_0 t) + \ldots \end{aligned} \tag{13–1}$$

or, more compactly,

$$f(t) = \underbrace{a_0}_{\text{dc}} + \underbrace{\sum_{n=1}^{\infty}[a_n\cos(2\pi n f_0 t) + b_n\sin(2\pi n f_0 t)]}_{\text{ac}} \tag{13–2}$$

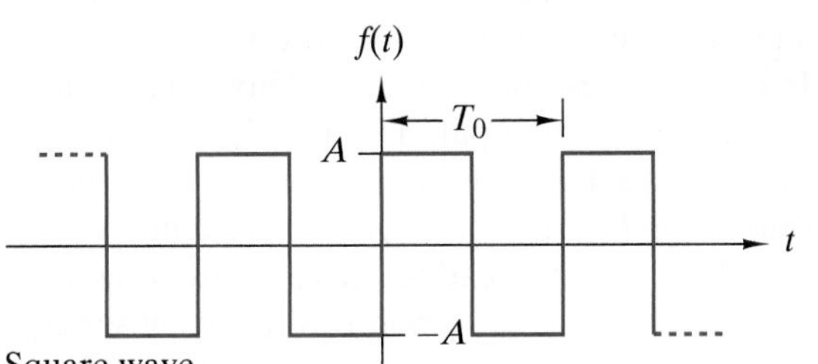

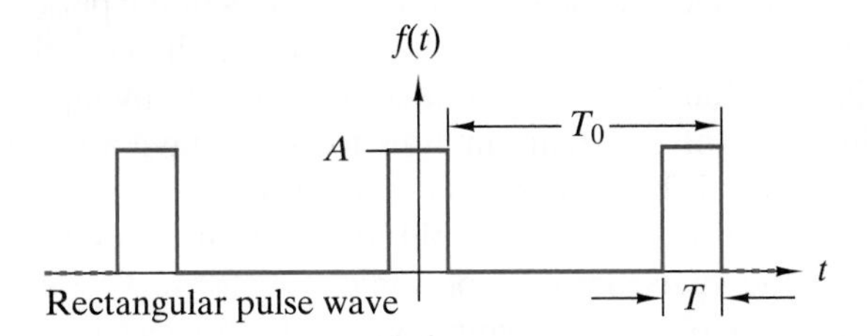

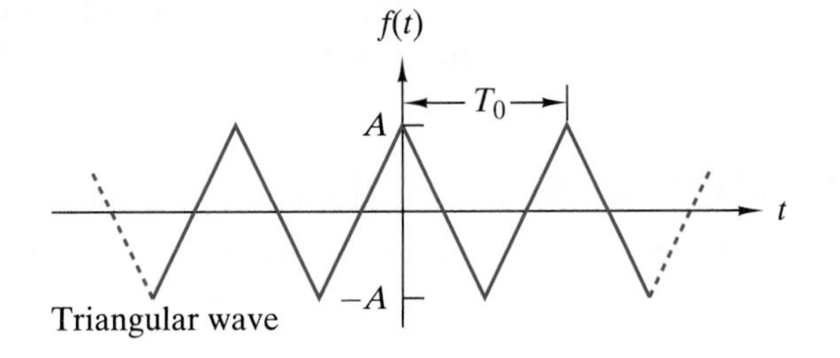

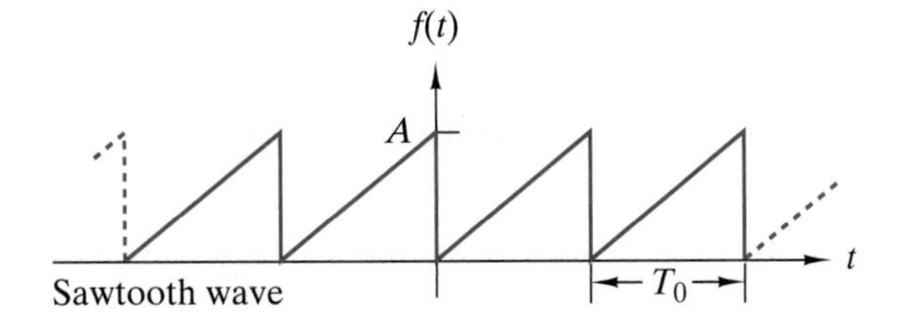

FIGURE 13–1 *Examples of periodic waveforms.*

The coefficient a_0 is the dc component or average value of $f(t)$. The constants a_n and b_n ($n = 1, 2, 3, \ldots$) are the **Fourier coefficients** of the sinusoids in the ac component. The lowest frequency in the ac component occurs for $n = 1$ and is called the **fundamental frequency**, defined as $f_0 = 1/T_0$. The other frequencies are integer multiples of the fundamental called the second harmonic ($2f_0$), third harmonic ($3f_0$) and, in general, the nth harmonic (nf_0).

Since Eq. (13–2) is an infinite series, there is always a question of convergence. We have said that the series converges as long as $f(t)$ is reasonably well behaved. Basically, this means that $f(t)$ is single valued, the integral of $|f(t)|$ over a period is finite, and $f(t)$ has a finite number of discontinuities in any one period. These requirements, called the **Dirichlet conditions**, are sufficient to assure convergence. Every periodic waveform that meets the Dirichlet conditions has a convergent Fourier series. However, there are waveforms that do not meet the Dirichlet conditions that also have convergent Fourier series. That is, while the Dirichlet conditions are sufficient, they are not necessary and sufficient. This limitation does not present a serious problem because the Dirichlet conditions are satisfied by the waveforms generated in physical systems. All of the periodic waveforms in Figure 13–1 meet the Dirichlet requirements.

It is important to have an overview of how a Fourier series is used in circuit analysis. In our previous study we learned how to find the steady-state responses due to a dc or an ac input. The Fourier series resolves a periodic input into a dc component and an infinite sum of ac components. We treat each of the terms in the Fourier series as a separate input source and use our circuit analysis tools to find the steady-state responses due to each term acting alone. The complete steady-state response is found by superposition, that is, adding up the responses due to each term acting alone. The net result is that the output is a modified version of the Fourier series of the periodic input.

At first glance it may seem complicated to find the complete response by finding the individual responses due to infinitely many inputs. However, each step simply involves circuit analysis tools we have already mastered. The end result tells us how the circuit transforms the input series into the output series. The distribution of amplitudes and phase angles in a Fourier series is called the **spectrum** of a periodic waveform. Frequency-domain signal processing involves modifying a given input spectrum to produce a desired output spectrum. Thus, finding the Fourier series of periodic waveforms is not an end in itself, but the first step in the study of frequency-domain signal processing.

13–2 FOURIER COEFFICIENTS

The Fourier coefficients for any periodic waveform $f(t)$ satisfying the Dirichlet conditions can be obtained from the equations

$$
\begin{aligned}
a_0 &= \frac{1}{T_0}\int_{-T_0/2}^{+T_0/2} f(t)dt \\
a_n &= \frac{2}{T_0}\int_{-T_0/2}^{+T_0/2} f(t)\cos(2\pi nt/T_0)dt \\
b_n &= \frac{2}{T_0}\int_{-T_0/2}^{+T_0/2} f(t)\sin(2\pi nt/T_0)dt
\end{aligned}
\tag{13–3}
$$

The integration limits in these equations extend from $-T_0/2$ to $+T_0/2$. However, the limits can span any convenient interval as long as it is exactly one period. For example, the limits could be from 0 to T_0 or $-T_0/4$ to $3T_0/4$. We will show where Eq. (13–3) comes from in a moment, but first we use these equations to obtain the Fourier coefficients of the sawtooth wave.

EXAMPLE 13–1

Find the Fourier coefficients for the sawtooth wave in Figure 13–1.

SOLUTION:

An expression for a sawtooth wave on the interval $0 \leq t \leq T_0$ is

$$f(t) = \frac{At}{T_0} \qquad 0 \leqslant t < T_0$$

For this definition of $f(t)$ we use 0 and T_0 as the limits in Eq. (13–3). The first expression in Eq. (13–3) yields a_0 as

$$a_0 = \frac{1}{T_0}\int_0^{T_0} \frac{At}{T_0}dt = \left.\frac{At^2}{2T_0^2}\right|_0^{T_0} = \frac{A}{2}$$

This result states that the average or dc value is $A/2$, which is easy to see because the area under one cycle of the sawtooth wave is $AT_0/2$. The second expression in Eq. (13–3) yields a_n as

$$\begin{aligned}
a_n &= \frac{2}{T_0}\int_0^{T_0} \frac{At}{T_0}\cos(2\pi nt/T_0)dt \\
&= \frac{2A}{T_0^2}\left[\frac{\cos(2\pi nt/T_0)}{(2\pi n/T_0)^2} + \frac{t \times \sin(2\pi nt/T_0)}{(2\pi n/T_0)}\right]_0^{T_0} \\
&= \frac{2A}{T_0^2}\left[\frac{\cos(2\pi n) - \cos(0)}{(2\pi n/T_0)^2}\right] = 0 \qquad \text{for all } n
\end{aligned}$$

Since $a_n = 0$ for all n, there are no cosine terms in the series. The b_n coefficients are found using the third expression in Eq. (13–3):

$$\begin{aligned}
b_n &= \frac{2}{T_0}\int_0^{T_0} \frac{At}{T_0}\sin(2\pi nt/T_0)dt \\
&= \frac{2A}{T_0^2}\left[\frac{\sin(2\pi nt/T_0)}{(2\pi n/T_0)^2} - \frac{t \times \cos(2\pi nt/T_0)}{(2\pi n/T_0)}\right]_0^{T_0} \\
&= \frac{2A}{T_0^2}\left[\frac{T_0\cos(2\pi n)}{(2\pi n/T_0)}\right] = -\frac{A}{n\pi} \qquad \text{for all } n
\end{aligned}$$

Given the coefficients a_n and b_n found above, the Fourier series for the sawtooth wave is

$$f(t) = \frac{A}{2} + \sum_{n=1}^{\infty}\left[-\frac{A}{n\pi}\right]\sin(2\pi n f_0 t)$$

■

EXAMPLE 13–2

In this example we use a computer tool to show that a truncated Fourier series approximates a periodic waveform. The waveform is a sawtooth with $A = 10$ and $T_0 = 2$ ms. Calculate the Fourier coefficients of the first 20 harmonics and plot the truncated series representation of the waveform using the first 5 harmonics and the first 10 harmonics.

SOLUTION:

From Example 13–1 the Fourier coefficients for the sawtooth wave are

$$a_0 = A/2 \quad a_n = 0 \quad \text{and} \quad b_n = -\frac{A}{n\pi} \quad \text{for all } n$$

Figure 13–2 shows a Mathcad worksheet that generates truncated Fourier series. The first line defines the waveform amplitude and period. The Fourier coefficients of the first 20 harmonics are calculated in the second line using the index variable $n = 1, 2, \ldots 20$. In Mathcad syntax $n = I, J, \ldots K$ means FOR $n = I$ TO K STEP $J - I$ DO. The next line defines a truncated Fourier series $f(k,t)$ consisting of the dc component a_0 plus the sum of the first k harmonics. Specifically, the function $f(5, t)$ is the sum of the dc component plus the first 5 harmonics and $f(10, t)$ is the sum of the dc component plus the first 10 harmonics. The plots of these two functions at the bottom of the worksheet show how function $f(k, t)$ approaches the sawtooth

FIGURE 13–2

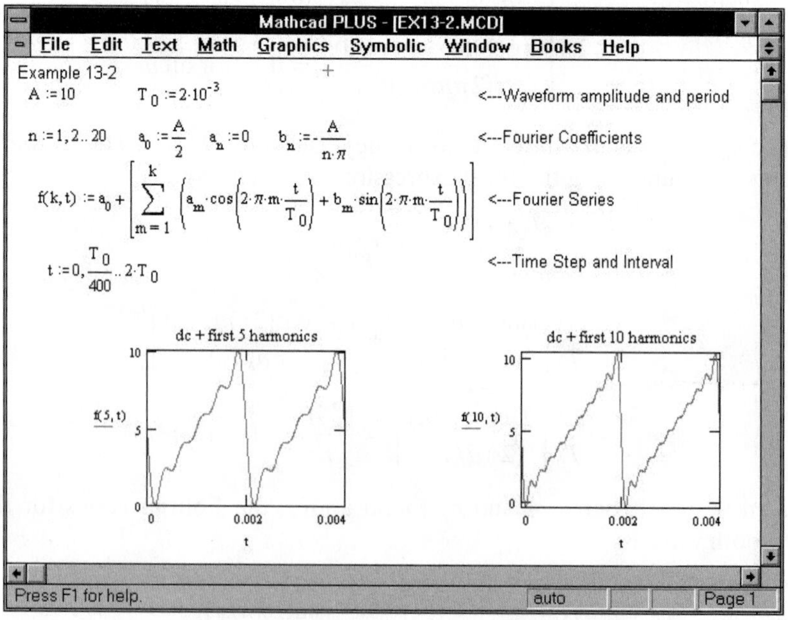

Mathcad PLUS - [EX13-2.MCD]

File Edit Text Math Graphics Symbolic Window Books Help

Example 13-2

$A := 10 \quad T_0 := 2\cdot 10^{-3}$ <---Waveform amplitude and period

$n := 1, 2 .. 20 \quad a_0 := \frac{A}{2} \quad a_n := 0 \quad b_n := -\frac{A}{n\cdot\pi}$ <---Fourier Coefficients

$f(k,t) := a_0 + \left[\sum_{m=1}^{k}\left(a_m\cdot\cos\left(2\cdot\pi\cdot m\cdot\frac{t}{T_0}\right) + b_m\cdot\sin\left(2\cdot\pi\cdot m\cdot\frac{t}{T_0}\right)\right)\right]$ <---Fourier Series

$t := 0, \frac{T_0}{400} .. 2\cdot T_0$ <---Time Step and Interval

dc + first 5 harmonics

f(5, t)

10 5 0

0 0.002 0.004

t

dc + first 10 harmonics

f(10, t)

10 5 0

0 0.002 0.004

t

Press F1 for help. auto Page 1

wave as more harmonics are added. Infinitely many harmonics are needed to represent a sawtooth wave exactly. However, the plots suggest that a relatively small number, say 5 or 10, provides a reasonable approximation of the major features of the sawtooth wave.

The Mathcad worksheet in Figure 13–2 is a template for generating truncated Fourier series. Changing the amplitude and period in the first line and the Fourier coefficients in the second line generates plots of the truncated Fourier series for other periodic waveforms. The index for n in the second line may need to be changed as well. For example, if only odd harmonics are present in the series, then the index should be $n = 1, 3, \ldots 21$.

■

DERIVING EQUATIONS FOR a_n AND b_n

The sawtooth wave example shows how to calculate the Fourier coefficients using Eq. (13–3). We now turn to the derivation of these equations. An equation for a Fourier coefficient is derived by multiplying both sides of Eq. (13–2) by the sinusoid associated with the coefficient and then integrating the result over one period. This multiply and integrate process isolates one coefficient because it turns out that all of the integrations produce zero except one.

The following derivation makes use of the fact that the area under a sine or cosine wave over an integer number of cycles is zero. That is,

$$\int_{-T_0/2}^{+T_0/2} \sin(2\pi k f_0 t)dt = 0 \qquad \text{for all } k$$

$$\begin{aligned} \int_{-T_0/2}^{+T_0/2} \cos(2\pi k f_0 t)dt &= 0 \qquad \text{for } k \neq 0 \\ &= T_0 \qquad \text{for } k = 0 \end{aligned} \tag{13–4}$$

where k is an integer. These equations state that integrating a sinusoid over $k \neq 0$ cycles produces zero, since the areas under successive half-cycles cancel. The single exception occurs when $k = 0$, in which case the cosine function reduces to one and the net area for one period is T_0.

We derive the equation for the amplitude of the dc component a_0 by integrating both sides of Eq. (13–2):

$$\int_{-T_0/2}^{+T_0/2} f(t)dt = \tag{13–5}$$

$$\begin{aligned} & a_0 \int_{-T_0/2}^{+T_0/2} dt + \sum_{n=1}^{\infty} \left[a_n \int_{-T_0/2}^{+T_0/2} \cos(2\pi n f_0 t)dt + b_n \int_{-T_0/2}^{+T_0/2} \sin(2\pi n f_0 t)dt \right] \\ & = a_0 T_0 + \qquad 0 \qquad + \qquad 0 \end{aligned}$$

The integrals of the ac components vanish because of the properties in Eq. (13–4), and the right side of this expression reduces to $a_0 T_0$. Solving for a_0 yields the first expression in Eq. (13–3).

To derive the expression for a_n we multiply Eq. (13–2) by $\cos(2\pi m f_0 t)$ and integrate over the interval from $-T_0/2$ to $+T_0/2$:

$$\int_{-T_0/2}^{+T_0/2} f(t)\cos(2\pi m f_0 t)dt = a_0\int_{-T_0/2}^{+T_0/2}\cos(2\pi m f_0 t)dt$$
$$+\sum_{n=1}^{\infty}\left[a_n\int_{-T_0/2}^{+T_0/2}\cos(2\pi m f_0 t)\cos(2\pi n f_0 t)dt + b_n\int_{-T_0/2}^{+T_0/2}\cos(2\pi m f_0 t)\sin(2\pi n f_0 t)dt\right] \tag{13–6}$$

All of the integrals on the right side of this equation are zero except one. To show this we use identities

$$\cos(x)\cos(y) = \frac{1}{2}\cos(x-y) + \frac{1}{2}\cos(x+y)$$
$$\cos(x)\sin(y) = \frac{1}{2}\sin(x-y) + \frac{1}{2}\sin(x+y)$$

to change Eq. (13–6) into the following form:

$$\int_{-T_0/2}^{+T_0/2} f(t)\cos(2\pi m f_0 t)dt = a_0\int_{-T_0/2}^{+T_0/2}\cos(2\pi m f_0 t)dt$$
$$+\sum_{n=1}^{\infty}\left\{\frac{a_n}{2}\left[\int_{-T_0/2}^{+T_0/2}\cos[2\pi(m-n)f_0 t]dt + \int_{-T_0/2}^{+T_0/2}\cos[2\pi(m+n)f_0 t]dt\right]\right\}$$
$$+\sum_{n=1}^{\infty}\left\{\frac{b_n}{2}\left[\int_{-T_0/2}^{+T_0/2}\sin[2\pi(m-n)f_0 t]dt + \int_{-T_0/2}^{+T_0/2}\sin[2\pi(m+n)f_0 t]dt\right]\right\} \tag{13–7}$$

All of the integrals are now in the form of expressions in Eq. (13–4). Consequently, we see all of the integrals on the right side of Eq. (13–7) vanish, except for one cosine integral when $m = n$. This one survivor corresponds to the $k = 0$ case in Eq. (13–4), and the right side of Eq. (13–7) reduces to

$$\int_{-T_0/2}^{T_0/2} f(t)\cos(2\pi n f_0 t)dt = \frac{a_n}{2}\int_{-T_0/2}^{T_0/2}\cos[2\pi(n-n)f_0 t]dt$$
$$= \frac{a_n}{2}T_0$$

Solving Eq. (13–7) for a_n yields the second expression in Eq. (13–3).

To obtain the expression for b_n we multiply Eq. (13–2) by $\sin(2\pi m f_0 t)$ and integrate over the interval $t = -T_0/2$ to $+T_0/2$. The derivation steps then parallel the approach used to find a_n. The end result is that the dc component integral vanishes and the ac component integrals reduce to $b_n T_0/2$, which yields the expression for b_n in Eq. (13–3).

The derivation of Eq. (13–3) focuses on the problem of finding the Fourier coefficients of a given periodic waveform. Some experience and practice are necessary to understand the implications of this procedure. On the other hand, it is not necessary to go through these mechanics for every newly encountered periodic waveform because tables of Fourier series expansions are available. For our purposes the listing in Figure 13–3 will suffice. For each waveform defined graphically, the figure lists the expressions for a_0, a_n, and b_n as well as restrictions on the integer n.

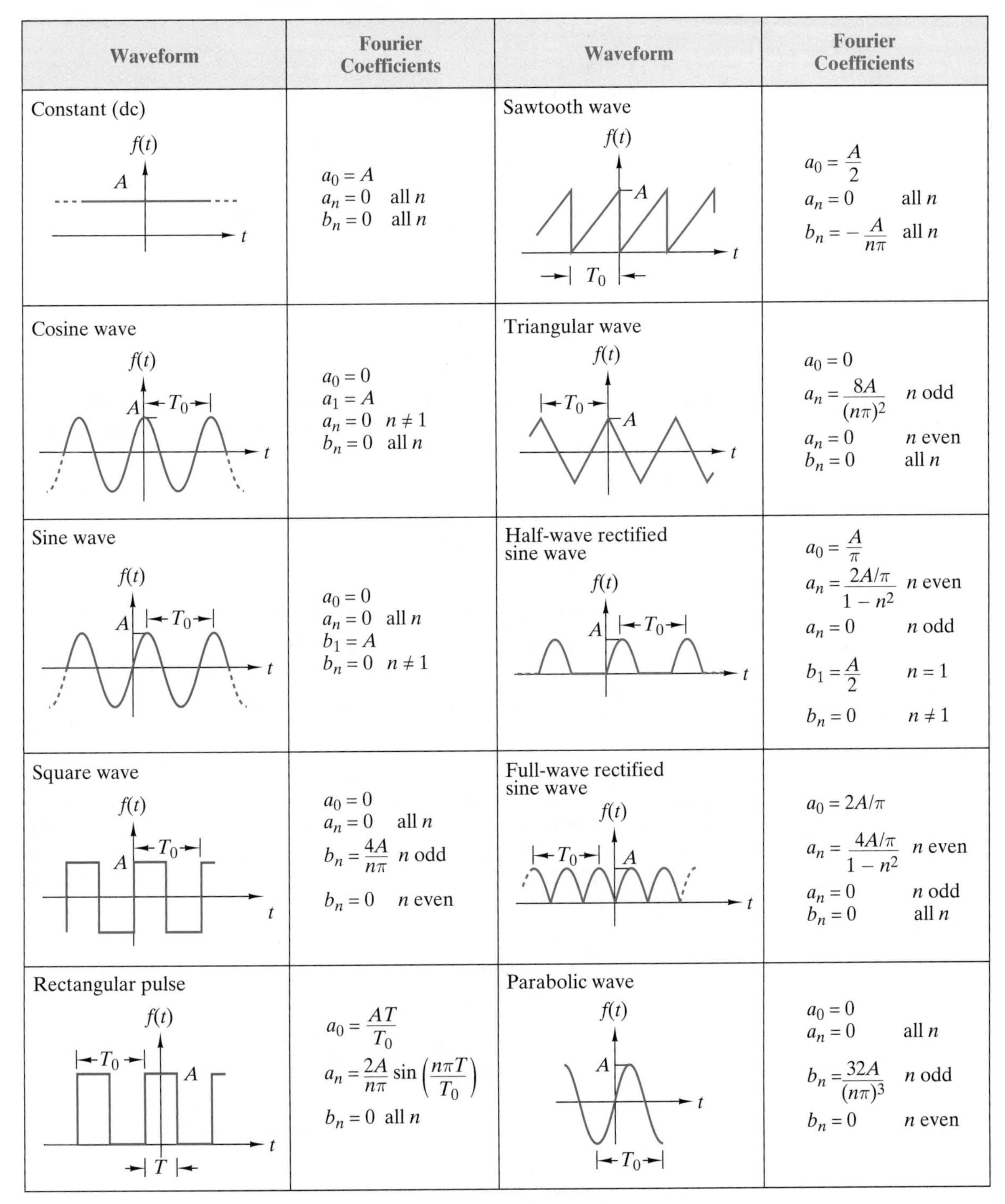

Waveform	Fourier Coefficients	Waveform	Fourier Coefficients
Constant (dc)	$a_0 = A$ $a_n = 0$ all n $b_n = 0$ all n	Sawtooth wave	$a_0 = \frac{A}{2}$ $a_n = 0$ all n $b_n = -\frac{A}{n\pi}$ all n
Cosine wave	$a_0 = 0$ $a_1 = A$ $a_n = 0$ $n \neq 1$ $b_n = 0$ all n	Triangular wave	$a_0 = 0$ $a_n = \frac{8A}{(n\pi)^2}$ n odd $a_n = 0$ n even $b_n = 0$ all n
Sine wave	$a_0 = 0$ $a_n = 0$ all n $b_1 = A$ $b_n = 0$ $n \neq 1$	Half-wave rectified sine wave	$a_0 = \frac{A}{\pi}$ $a_n = \frac{2A/\pi}{1-n^2}$ n even $a_n = 0$ n odd $b_1 = \frac{A}{2}$ $n = 1$ $b_n = 0$ $n \neq 1$
Square wave	$a_0 = 0$ $a_n = 0$ all n $b_n = \frac{4A}{n\pi}$ n odd $b_n = 0$ n even	Full-wave rectified sine wave	$a_0 = 2A/\pi$ $a_n = \frac{4A/\pi}{1-n^2}$ n even $a_n = 0$ n odd $b_n = 0$ all n
Rectangular pulse	$a_0 = \frac{AT}{T_0}$ $a_n = \frac{2A}{n\pi}\sin\left(\frac{n\pi T}{T_0}\right)$ $b_n = 0$ all n	Parabolic wave	$a_0 = 0$ $a_n = 0$ all n $b_n = \frac{32A}{(n\pi)^3}$ n odd $b_n = 0$ n even

FIGURE 13–3 *Fourier coefficients for some periodic waveforms.*

EXAMPLE 13–3

Verify the Fourier coefficients given for the square wave in Figure 13–3 and write the first three nonzero terms in its Fourier series.

SOLUTION:

An expression for a square wave on the interval $0 < t <$ to T_0 is

$$f(t) = \begin{cases} A & 0 < t < T_0/2 \\ -A & T_0/2 < t < T_0 \end{cases}$$

Using the first expression in Eq. (13–3) to find a_0 yields

$$\begin{aligned} a_0 &= \frac{1}{T_0}\int_0^{T_0/2} A\,dt + \frac{1}{T_0}\int_{T_0/2}^{T_0} (-A)dt \\ &= \frac{A}{T_0}\left[\frac{T_0}{2} - 0 - T_0 + \frac{T_0}{2}\right] = 0 \end{aligned}$$

The result $a_0 = 0$ means that the dc value of the square wave is zero, which is easy to see because the area under a positive half-cycle cancels the area under a negative half-cycle. Using the second expression in Eq. (13–3) to find a_n produces

$$\begin{aligned} a_n &= \frac{2}{T_0}\int_0^{T_0/2} A\cos(2\pi nt/T_0)dt + \frac{2}{T_0}\int_{T_0/2}^{T_0} (-A)\cos(2\pi nt/T_0)dt \\ &= \frac{2A}{T_0}\left[\frac{\sin(2\pi nt/T_0)}{2\pi n/T_0}\right]_0^{T_0/2} - \frac{2A}{T_0}\left[\frac{\sin(2\pi nt/T_0)}{2\pi n/T_0}\right]_{T_0/2}^{T_0} \\ &= \frac{A}{n\pi}[\sin(n\pi) - \sin(0) - \sin(2n\pi) + \sin(n\pi)] = 0 \end{aligned}$$

Since $a_n = 0$ for all n, there are no cosine terms in the series. This makes some intuitive sense because a sine wave with the same fundamental frequency as the square wave fits nicely inside the square wave with zeros crossing at the same points, whereas a cosine with the same frequency does not fit at all. The b_n coefficients for the sine terms are found using the third expression in Eq. (13–3):

$$\begin{aligned} b_n &= \frac{2}{T_0}\int_0^{T_0/2} A\sin(2\pi nt/T_0)dt + \frac{2}{T_0}\int_{T_0/2}^{T_0} (-A)\sin(2\pi nt/T_0)dt \\ &= \frac{2A}{T_0}\left[-\frac{\cos(2\pi nt/T_0)}{2\pi n/T_0}\right]_0^{T_0/2} - \frac{2A}{T_0}\left[-\frac{\cos(2\pi nt/T_0)}{2\pi n/T_0}\right]_{T_0/2}^{T_0} \\ &= \frac{A}{n\pi}[-\cos(n\pi) + \cos(0) + \cos(2n\pi) - \cos(n\pi)] \\ &= \frac{2A}{n\pi}[1 - \cos(n\pi)] \end{aligned}$$

The term $[1 - \cos(n\pi)] = 2$ if n is odd and zero if n is even. Hence b_n can be written as

$$b_n = \begin{cases} \dfrac{4A}{n\pi} & n \text{ odd} \\ 0 & n \text{ even} \end{cases}$$

The first three nonzero terms in the Fourier series of the square wave are

$$f(t) = \frac{4A}{\pi}\left[\sin 2\pi f_0 t + \frac{1}{3}\sin 2\pi 3 f_0 t + \frac{1}{5}\sin 2\pi 5 f_0 t + \cdots\right]$$

Note that this series contains only odd harmonic terms. ■

Exercise 13–1

The triangular wave in Figure 13–3 has a peak amplitude of $A = 10$ and $T_0 = 2$ ms. Calculate the Fourier coefficients of the first nine harmonics.

Answers: $a_1 = 8.11$, $a_2 = 0$, $a_3 = 0.901$, $a_4 = 0$, $a_5 = 0.324$, $a_6 = 0$, $a_7 = 0.165$, $a_8 = 0$, $a_9 = 0.100$, $b_n = 0$ for all n.

ALTERNATIVE FORM OF THE FOURIER SERIES

The series in Eq. (13–1) can be written in several alternative yet equivalent forms. From our study of sinusoids in Chapter 5, we recall that the Fourier coefficients determine the amplitude and phase angle of the general sinusoid. Thus, we can write a general Fourier series in the form

$$f(t) = A_0 + A_1\cos(2\pi f_0 t + \phi_1) + A_2\cos(2\pi 2 f_0 t + \phi_2) + \cdots + A_n\cos(2\pi n f_0 t + \phi_n) + \cdots \quad (13\text{–}8)$$

where

$$A_n = \sqrt{a_n^2 + b_n^2} \quad \text{and} \quad \phi_n = \tan^{-1}\frac{-b_n}{a_n} \quad (13\text{–}9)$$

The coefficient A_n is the amplitude of the nth harmonic and ϕ_n is its phase angle.[1]

Note that the amplitude A_n and phase angle ϕ_n contain all of the information needed to construct the Fourier series in the form of Eq. (13–8). Figure 13–4 shows how plots of this information are used to display the spectral content of a periodic waveform $f(t)$. The plot of A_n versus nf_0 (or $n\omega_0$) is called the **amplitude spectrum**, while the plot of ϕ_n versus nf_0 (or $n\omega_0$) is called the **phase spectrum**. Both plots are **line spectra** because spectral content can be represented as a line at discrete frequencies.

In theory, a Fourier series includes infinitely many harmonics, although the harmonics tend to decrease in amplitude at high frequency. For example,

---

1 There is a 180° ambiguity in the value returned by the inverse tangent function in most computational tools. The ambiguity is resolved by the following rule: $b_n < 0$ implies that the angle is in the range 0 to –180°, while $b_n > 0$ implies the 0 to +180° range.

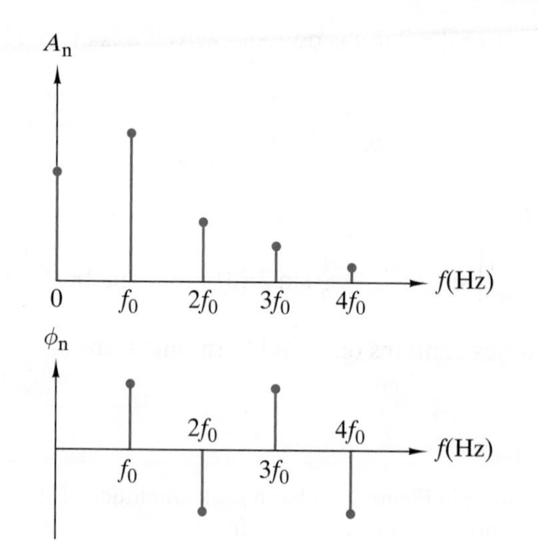

FIGURE 13–4 *Amplitude and phase spectra.*

the summary in Figure 13–3 shows that the amplitudes of the square wave decrease as $1/n$, the triangular wave as $1/n^2$, and the parabolic wave as $1/n^3$. The $1/n^3$ dependence means that the amplitude of the fifth harmonic in a parabolic wave is less than 1% of the amplitude of the fundamental (actually 1/125th of the fundamental). In practical signals the harmonic amplitudes decrease at high frequency so that at some point the higher-order components become negligibly small. This means that we can truncate the series at some finite frequency and still retain the important features of the signal. This is an important consideration in systems with finite bandwidth.

EXAMPLE 13–4

Derive expressions for the amplitude A_n and phase angle ϕ_n of the Fourier series of the sawtooth wave in Figure 13–3. Sketch the amplitude and phase spectra of a sawtooth wave with $A = 5$ and $T_0 = 4$ ms.

SOLUTION:

Figure 13–3 gives the Fourier coefficients of the sawtooth wave as

$$a_0 = \frac{A}{2} \qquad a_n = 0 \qquad b_n = -\frac{A}{n\pi} \qquad \text{for all } n$$

Using Eq. (13–9) yields

$$A_n = \sqrt{a_n^2 + b_n^2} = \begin{cases} \dfrac{A}{2} & n = 0 \\ \dfrac{A}{n\pi} & n > 0 \end{cases}$$

and

$$\phi_n = \tan^{-1}\frac{-b_n}{a_n} = \begin{cases} \text{undefined} & n = 0 \\ 90^\circ & n > 0 \end{cases}$$

For $A = 5$ and $f_0 = 1/T_0 = 250$ Hz the first four nonzero terms in the series are

$$f(t) = 2.5 + 1.59\cos(2\pi 250t + 90^\circ) + 0.796\cos(2\pi 500t + 90^\circ) + 0.531\cos(2\pi 750t + 90^\circ) + \cdots$$

Figure 13–5 shows the amplitude and phase spectra for this signal. Note that the lines in the amplitude spectrum are inversely proportional to frequency. As frequency increases, the amplitudes decrease so that the high-frequency components become negligible. ■

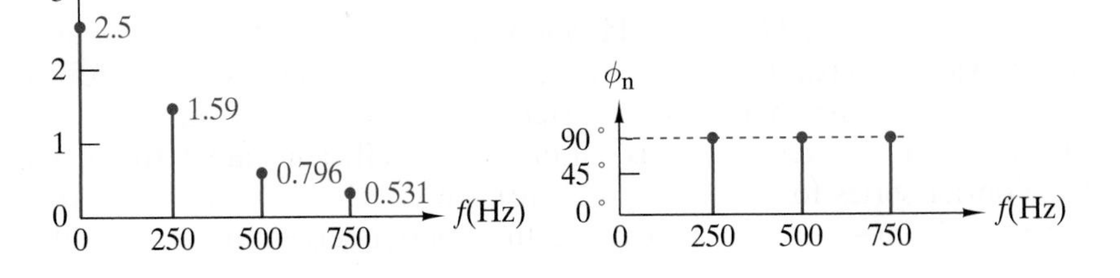

FIGURE 13–5

Exercise 13–2

Derive expressions for the amplitude A_n and phase angle ϕ_n for the triangular wave in Figure 13–3 and write an expression for the first three nonzero terms in the Fourier series with $A = \pi^2/8$ and $T_0 = 2\pi/5000$ s.

Answers:

$$A_n = \frac{8A}{(n\pi)^2} \quad \phi_n = 0^\circ \quad n \text{ odd}$$

$$A_n = 0 \quad \phi_n \text{ undefined} \quad n \text{ even}$$

$$f(t) = \cos(5000t) + \frac{1}{9}\cos(15{,}000t) + \frac{1}{25}\cos(25{,}000t) + \cdots$$

13–3 WAVEFORM SYMMETRIES

Many of the Fourier coefficients are zero when a periodic waveform has certain types of symmetries. It is helpful to recognize these symmetries, since they may simplify the calculation of the Fourier coefficients.

The first expression in Eq. (13–3) shows that the amplitude of the dc component a_0 is the average value of the periodic waveform $f(t)$. If the waveform has equal area above and below the time axis, then the integral over one cycle vanishes, the average value is zero, and $a_0 = 0$. The square wave, triangular wave, and parabolic wave in Figure 13–3 are examples of periodic waveforms with zero average value.

A waveform is said to have **even symmetry** if $f(-t) = f(t)$. The cosine wave, rectangular pulse, and triangular wave in Figure 13–3 are examples of waveforms with even symmetry. The Fourier series of an even waveform is made up entirely of cosine terms: that is, all of the b_n coefficients are zero. To show this we write the Fourier series for $f(t)$ in the form

$$f(t) = a_0 + \sum_{n=1}^{\infty}[a_n \cos(2\pi n f_0 t) + b_n \sin(2\pi n f_0 t)] \tag{13–10}$$

Given the Fourier series for $f(t)$, we use the identities $\cos(-x) = \cos(x)$ and $\sin(-x) = -\sin(x)$ to write the Fourier series for $f(-t)$ as follows:

$$f(-t) = a_0 + \sum_{n=1}^{\infty}[a_n \cos(2\pi n f_0 t) - b_n \sin(2\pi n f_0 t)] \tag{13–11}$$

For even symmetry $f(t) = f(-t)$ and the right sides of Eqs. (13–10) and (13–11) must be equal. Comparing the Fourier coefficients term by term, we find that $f(t) = f(-t)$ requires $b_n = -b_n$. The only way this can happen is for $b_n = 0$ for all n.

A waveform is said to have **odd symmetry** if $-f(-t) = f(t)$. The sine wave, square wave, and parabolic wave in Figure 13–3 are examples of waveforms with this type of symmetry. The Fourier series of odd waveforms are made up entirely of sine terms: that is, all of the a_n coefficients are zero. Given the Fourier series for $f(t)$ in Eq. (13–10), we use the identities $\cos(-x) = \cos(x)$ and $\sin(-x) = -\sin(x)$ to write the Fourier series for $-f(-t)$ in the form

$$-f(-t) = -a_0 + \sum_{n=1}^{\infty}[-a_n \cos(2\pi n f_0 t) + b_n \sin(2\pi n f_0 t)] \tag{13–12}$$

With odd symmetry $f(t) = -f(-t)$ and the right sides of Eq. (13–10) and (13–12) must be equal. Comparing the Fourier coefficients term by term, we find that odd symmetry requires $a_0 = -a_0$ and $a_n = -a_n$. The only way this can happen is for $a_n = 0$ for all n, including $n = 0$.

A waveform is said to have **half-wave symmetry** if $-f(t - T_0/2) = f(t)$. This requirement states that inverting the waveform $[-f(t)]$ and then time shifting by half a cycle $(T_0/2)$ must produce the same waveform. Basically, this means that successive half-cycles have the same waveshape but opposite polarities. In Figure 13–3 the sine wave, cosine wave, square wave, triangular wave, and parabolic wave have half-wave symmetry. The sawtooth wave, half-wave sine, rectangular pulse train, and full-wave sine do not have this symmetry.

With half-wave symmetry the amplitudes of all even harmonics are zero. To show this we use the identities $\cos(x - n\pi) = (-1)^n \cos(x)$ and $\sin(x - n\pi) = (-1)^n \sin(x)$ to write the Fourier series of $-f(t - T_0/2)$ in the form

$$\begin{aligned} -f(t - T_0/2) = -a_0 + \sum_{n=1}^{\infty}[&-(-1)_n a_n \cos(2\pi n f_0 t) \\ &-(-1)^n b_n \sin(2\pi n f_0 t)] \end{aligned} \tag{13–13}$$

For half-wave symmetry the right sides of Eqs. (13–10) and (13–13) must be equal. Comparing the coefficients term by term, we find that equality requires $a_0 = -a_0$, $a_n = -(-1)^n a_n$, and $b_n = -(-1)^n b_n$. The only way this can happen is for $a_0 = 0$ and for $a_n = b_n = 0$ when n is even. In other words, the only nonzero Fourier coefficients occur when n is odd.

A waveform may have more than one symmetry. For example, the triangular wave in Figure 13–3 has even symmetry and half-wave symmetry, while the square wave has both odd and half-wave symmetries. The sawtooth wave in Figure 13–3 is an example where an underlying odd symmetry is masked by a dc component. A symmetry that is not apparent until the dc component is removed is sometimes called a **hidden symmetry**.

Finally, whether a waveform has even or odd symmetry (or neither) depends on where we choose to define $t = 0$. For example, the triangular wave in Figure 13–3 has even symmetry because the $t = 0$ vertical axis is located at a local maximum. If the axis is shifted to a zero crossing, the waveform has odd symmetry and the cosine terms in the series are replaced by sine terms. If the vertical axis is shifted to a point between a zero cross and a maximum, then the resulting waveform is neither even nor odd and its Fourier series contains both sine and cosine terms.

EXAMPLE 13–5

Given that $f(t)$ is a square wave of amplitude A and period T_0, use the Fourier coefficients in Figure 13–3 to find the Fourier coefficients of $g(t) = f(t + T_0/4)$.

SOLUTION:

Figure 13–6 compares the square waves $f(t)$ and $g(t) = f(t + T_0/4)$. The square wave $f(t)$ has odd symmetry (sine terms only) and half-wave symmetry (odd harmonics only). Using the coefficients in Figure 13–3, the Fourier series for $f(t)$ is

$$f(t) = \sum \frac{4A}{n\pi} \sin(2\pi nt/T_0) \quad n \text{ odd}$$

The Fourier series for $g(t) = f(t + T_0/4)$ can be written in the form

$$g(t) = f(t + T_0/4) = \sum \frac{4A}{n\pi} \sin[2\pi n(t + T_0/4)/T_0] \quad n \text{ odd}$$

$$= \sum \frac{4A}{n\pi} \sin(2\pi nt/T_0 + n\pi/2)$$

$$= \sum \frac{4A}{n\pi} \cos(2\pi nt/T_0)\sin(n\pi/2)$$

$$= \sum \frac{4A}{n\pi} \cos(2\pi nt/T_0)(-1)^{\frac{n-1}{2}} \quad n \text{ odd}$$

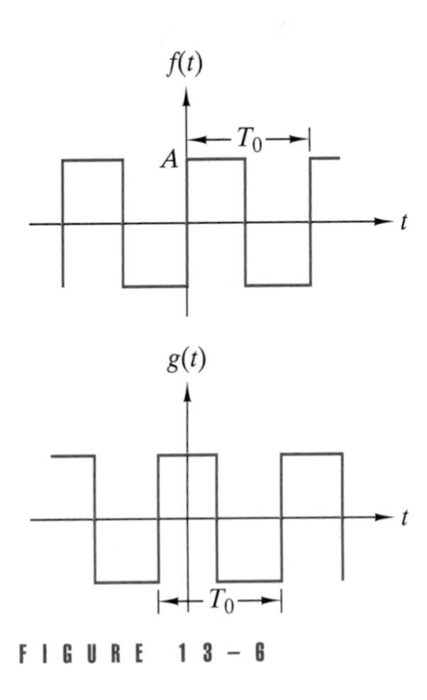

FIGURE 13–6

Figure 13–6 shows that $g(t)$ has even and half-wave symmetry, so its Fourier series has only cosine terms and odd harmonics. The Fourier coefficients for $g(t)$ are

$$a_0 = 0 \quad a_n = \begin{cases} 0 & n \text{ even} \\ \left[\dfrac{4A}{n\pi}\right](-1)^{\frac{n-1}{2}} & n \text{ odd} \end{cases}$$

$$b_n = 0 \quad \text{all } n$$

Shifting the time origin alters the even or odd symmetry properties of a periodic waveform because these symmetries depend on values of $f(t)$ on opposite sides of the vertical axis at $t = 0$. The half-wave symmetry of a waveform is not changed by time shifting because this symmetry only requires successive half-cycles to have the same form but opposite polarities. ■

Exercise 13–3

(a) Identify the symmetries in the waveform $f(t)$ whose Fourier series is

$$f(t) = \frac{2\sqrt{3}A}{\pi}\left[\cos(\omega_0 t) - \frac{1}{5}\cos(5\,\omega_0 t) + \frac{1}{7}\cos(7\,\omega_0 t) - \frac{1}{11}\cos(11\,\omega_0 t) + \frac{1}{13}\cos(13\,\omega_0 t) + \cdots\right]$$

(b) Write the corresponding terms of the function $g(t) = f(t - T_0/4)$.

Answers:

(a) Even symmetry; half-wave symmetry; zero average value.

(b) $$g(t) = \frac{2\sqrt{3}A}{\pi}\left[\sin(\omega_0 t) - \frac{1}{5}\sin(5\,\omega_0 t) - \frac{1}{7}\sin(7\,\omega_0 t) + \frac{1}{11}\sin(11\,\omega_0 t) + \frac{1}{13}\sin(13\,\omega_0 t) + \ldots\right]$$

13–4 CIRCUIT ANALYSIS USING THE FOURIER SERIES

Up to this point we have concentrated on finding the Fourier series description of periodic waveforms. We are now in a position to address circuit analysis problems of the type illustrated in Figure 13–7. This first-order *RL* circuit is driven by a periodic sawtooth voltage, and the objective is to find the steady-state current $i(t)$.

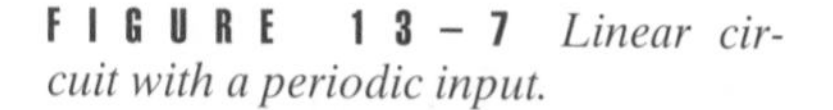

FIGURE 13–7 *Linear circuit with a periodic input.*

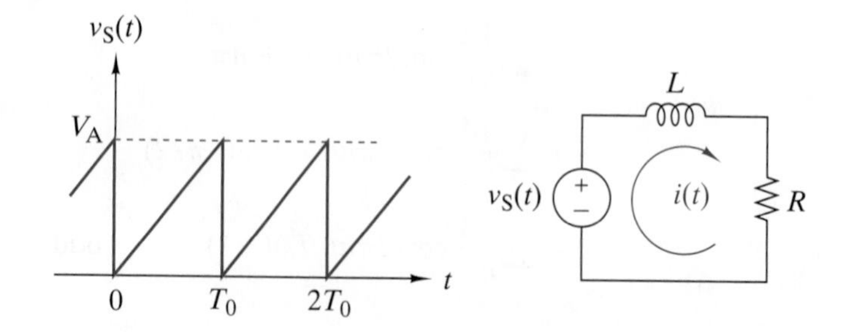

We begin by using the results in Example 13–4 to express the input voltage as a Fourier series in the form

$$v_S(t) = \underbrace{\frac{V_A}{2}}_{\text{dc}} + \underbrace{\sum_{n=1}^{\infty} \frac{V_A}{n\pi}\cos(n\omega_0 t + 90°)}_{\text{ac}} \tag{13–14}$$

This result expresses the input driving force as the sum of a dc component plus ac components at harmonic frequencies $n\omega_0 = 2\pi n f_0$, $n = 1, 2, 3, \ldots$. Since the circuit is linear, we find the steady-state response caused by each component acting alone and then obtain the total response by superposition.

In the dc steady state the inductor acts like a short circuit, so the steady-state current due to the dc input $V_A/2$ is simply $I_0 = V_A/(2R)$. The amplitude and phase angle of the nth harmonic of the sawtooth input are

$$V_n = \frac{V_A}{n\pi} \quad \text{and} \quad \phi_n = 90°$$

In Chapter 11 we found that the sinusoidal steady-state response at frequency ω_A can be found directly from a network function $T(s)$ as

$$\text{Output amplitude} = \text{Input amplitude} \times |T(j\omega_A)|$$

$$\text{Output phase} = \text{Input phase} + \angle T(j\omega_A)$$

In the present case the input is the nth harmonic at $\omega_A = n\omega_0$, the output is the current $I(s) = V(s)/Z(s)$, and hence the network function of interest is $T(s) = 1/(Ls + R)$. Applying the Chapter 11 method here yields the amplitude and phase of the nth harmonic current as

$$\text{Amplitude} = V_n \times \left|\frac{1}{jn\omega_0 L + R}\right| = \frac{V_A}{n\pi R}\frac{1}{\sqrt{1 + jn\omega_0 L/R}}$$

$$\text{Phase} = \phi_n + \text{angle}\left(\frac{1}{jn\omega_0 L + R}\right) = 90° - \tan^{-1}(n\omega_0 L/R)$$

Defining $\theta_n = \tan^{-1}(n\omega_0 L/R)$, we get the waveform of the steady-state response to the nth harmonic input.

$$i_n(t) = \underbrace{\frac{V_A}{n\pi R}\frac{1}{\sqrt{1 + (n\omega_0 L/R)^2}}}_{\text{amplitude}}\cos(\underbrace{n\omega_0 t + 90° - \theta_n}_{\text{phase}})$$

We have now found the steady-state response of the circuit due to the dc component acting alone and the nth harmonic ac component acting alone. Since the circuit is linear, superposition applies and we find that the steady-state response caused the sawtooth input triangular by summing the contributions of each of these sources:

$$\begin{aligned} i(t) &= I_0 + \sum_{n=1}^{\infty} i_n(t) \\ &= \frac{V_A}{2R} + \frac{V_A}{R}\sum_{n=1}^{\infty}\frac{1}{n\pi\sqrt{1 + (n\omega_0 L/R)^2}}\cos(n\omega_0 t + 90° - \theta_n) \end{aligned} \tag{13–15}$$

The Fourier series in Eq. (13–15) represents the steady-state current due to a sawtooth driving force whose Fourier series is in Eq. (13–14).

EXAMPLE 13–6

Find the first four nonzero terms of the Fourier series in Eq. (13–15) for $V_A = 25$ V, $R = 50\ \Omega$, $L = 40\ \mu$H, $\omega_0 = 1$ Mrad/s.

SOLUTION:

Equation (13–15) gives a Fourier series of the form

$$i(t) = I_0 + \sum_{n=1}^{\infty} I_n \cos(n\,\omega_0 t + \psi_n)$$

where

$$I_0 = \frac{V_A}{2R} \qquad I_n = \frac{V_A}{R}\frac{1}{n\pi\sqrt{1 + (n\omega_0 L/R)^2}} \qquad \psi_n = \frac{\pi}{2} - \tan^{-1}(n\omega_0 L/R)$$

Inserting the numerical values leads to

$$I_0 = \frac{1}{4} \qquad I_n = \frac{1}{2n\,\pi\sqrt{1 + (n \times 0.8)^2}} \qquad \psi_n = \frac{\pi}{2} - \tan^{-1}(n \times 0.8)$$

The first four nonzero terms in the Fourier series include the dc component plus the first three harmonics. For $n = 0, 1, 2, 3$, we have

$$\begin{aligned}
I_0 &= 0.25 \\
I_1 &= 0.124 \quad \psi_1 = 0.896 \text{ rad} = 51.3^\circ \\
I_2 &= 0.0422 \quad \psi_2 = 0.559 \text{ rad} = 32.0^\circ \\
I_3 &= 0.0204 \quad \psi_3 = 0.395 \text{ rad} = 22.6^\circ
\end{aligned}$$

hence, the desired expression is

$$\begin{aligned}
i(t) &= 0.25 + 0.124\cos(10^6 t + 51.3^\circ) \\
&\quad + 0.0422\cos(2 \times 10^6 t + 32^\circ) + 0.0204\cos(3 \times 10^6 t + 22.6^\circ) \ldots
\end{aligned}$$

■

Given the Fourier series of a periodic input, it is a straightforward procedure to obtain the Fourier series of the steady-state response. But interpreting the analysis result requires some thought since the response is presented as an infinite series. In practice the series converges rather rapidly, so we can calculate specific values or generate plots using computer tools to sum a truncated version of the infinite series. Before doing so, let us look closely at Eq. (13–15) to see what we can infer about the response.

First note that if $L = 0$, Eq. (13–15) reduces to

$$i(t) = \frac{V_A/R}{2} + \sum_{n=1}^{\infty} \frac{V_A/R}{n\pi}\cos(n\omega_0 t + 90^\circ)$$

which is the Fourier series of a sawtooth wave of amplitude V_A/R. This makes sense because without the inductor the circuit in Figure 13–7 is a simple resistive circuit in which $i(t) = v_S(t)/R$, so the input and response must have the same waveform.

When $L \neq 0$ the response is not a sawtooth, but we can infer some features of its waveform if we examine the amplitude spectrum. At high frequency $[(n\omega_0 L/R) >> 1]$ Eq. (13–15) points out that the amplitudes of the ac components are approximately

$$I_n \approx \frac{V_A}{R} \frac{1}{n^2 \pi \omega_0 L/R} \tag{13–16}$$

In the steady-state response the amplitudes of the high-frequency ac components decrease as $1/n^2$, whereas the ac components in the input sawtooth decrease as $1/n$. In other words, the relative amplitudes of the high-frequency components are much smaller in the response than in the input. This makes sense because the inductor's impedance increases with frequency and thereby reduces the amplitudes of the high-frequency ac currents. We would expect the circuit to filter out the high-frequency components in the input and produce a response without the sharp corners and discontinuities in the input sawtooth.

The next example examines this thought for a specific set of parameters.

EXAMPLE 13–7

The parameters of the steady-state waveform in Eq. (13–15) are V_A = 25 V, T_0 = 5 μs, L = 40 μH, and R = 50 Ω. Calculate and plot a truncated Fourier series representation of the steady-state current using the first 5 harmonics and the first 10 harmonics.

SOLUTION:

Figure 13–8 shows a Mathcad worksheet that calculates truncated Fourier series of the steady-state current in Eq. (13–15). The first two lines define the waveform and circuit parameters. The Fourier parameters for the first 20 harmonics are calculated in the third line using the index variable n = 1, 2, . . . 20. The next line defines a truncated Fourier series $i(k, t)$ consisting of the dc component I_0 plus the sum of the first k harmonics. The responses $I(5, t)$ and $I(10, t)$ are plotted at the bottom of the worksheet. There is not much change between the two plots, suggesting that a truncated series converges rather rapidly and gives a reasonable approximation to the steady-state current.

The plots in Figure 13–8 show that the steady-state current is indeed smoother than the input driving force. The reason is that the inductor suppresses the high-frequency components that are important contributors to the discontinuities in the sawtooth waveform. As a result, the response does not have the abrupt changes and sharp corners present in the sawtooth input. ■

FIGURE 13–8

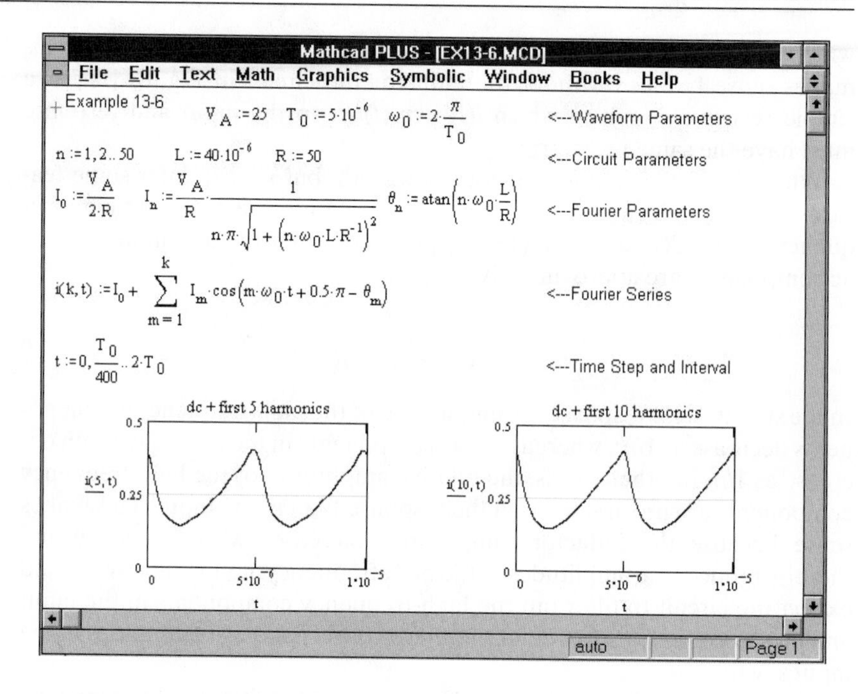

Exercise 13–4

Derive an expression for the steady-state current of the *RL* circuit in Figure 13–7 when the input voltage is a triangular wave with amplitude V_A and frequency ω_0.

Answer:

$$i(t) = \sum_{n=1}^{\infty} \frac{8V_A/R}{(n\pi)^2\sqrt{1 + (n\omega_0 L/R)^2}} \cos(n\omega_0 t - \theta_n) \qquad n \text{ odd}$$

where

$$\theta_n = \tan^{-1}(n\omega_0 L/R)$$

EXAMPLE 13–8

A sawtooth wave with $V_A = 12$ V and $\omega_0 = 40$ rad/s drives a circuit with a transfer function $T(s) = 100/(s + 50)$. Find the amplitude of the first four nonzero terms in the Fourier series of the steady-state output.

SOLUTION:

The Fourier coefficients of the input are

$$a_0 = \frac{V_A}{2} \qquad a_n = 0 \qquad b_n = -\frac{V_A}{n\pi} \qquad \text{for all } n$$

The amplitudes of the Fourier series for the steady-state output are found from

$$A_0 = |a_0 T(0)|$$

$$A_n = |b_n T(jn\omega_0)| = \left| \frac{12}{n\pi} \frac{100}{\sqrt{50^2 + (n\,40)^2}} \right|$$

Which yields $A_0 = 12$, $A_1 = 5.97$, $A_2 = 2.02$, and $A_3 = 0.979$. ■

EXAMPLE 13–9

Figure 13–9 shows a block diagram of a dc power supply. The ac input is a sinusoid that is converted to a full-wave sine by the rectifier. The filter passes the dc component in the rectified sine and suppresses the ac components. The result is an output consisting of a small residual ac ripple riding on top of a much larger dc signal.

Calculate and plot the first 10 harmonics in amplitude spectra of the filter input and output for $V_A = 23.6$ V, $T_0 = 1/60$ s, and a low-pass filter transfer function of

$$T(s) = \frac{(200)^2}{s^2 + 70s + (200)^2}$$

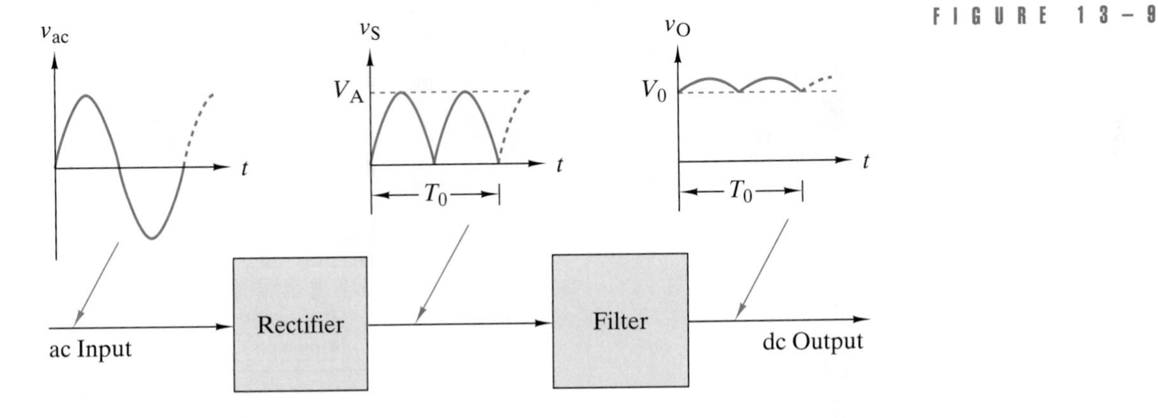

FIGURE 13–9

SOLUTION:

The amplitude spectrum of the filter input is obtained using the Fourier coefficients for the full-wave rectified sine in Figure 13–3:

$$V_0 = 2V_A/\pi = 15.02 \text{ V}$$

$$V_n = \begin{cases} 0 & n \text{ odd} \\ \left|\dfrac{4V_A/\pi}{1-n^2}\right| = \dfrac{30.04}{n^2-1} & n \text{ even} \end{cases} \tag{13–17}$$

The magnitude of the transfer function at each of these discrete frequencies is

$$|T(jn\omega_0)| = \frac{(200)^2}{\sqrt{[(200)^2 - (n\omega_0)^2]^2 + (70\,n\omega_0)^2}} \tag{13–18}$$

To obtain the specified output spectrum, we must generate the product of the input amplitude times the transfer function magnitude for $n = 0, 1, 2, 3, \ldots 10$.

Spreadsheets are ideally suited to making repetitive calculations of this type. Figure 13–10 shows an Excel spreadsheet that implements the required calculations. Column A gives the index n and column B gives the corresponding frequencies in Hz. The input amplitudes in column C are calculated using Eq. (13–17), while the entries in column D are based on Eq. (13–18). Finally, the entries in column E are the product of those in columns C and D. Since the lowpass filter has unity gain at zero frequency, the dc components in the input and output are equal. The first nonzero ac component is the second harmonic, which has an amplitude of 10 V in the input but less than 1 V in the output. By the time we get to the next nonzero harmonic at 240 Hz, the ac amplitudes in the output are entirely negligible. ■

FIGURE 13–10

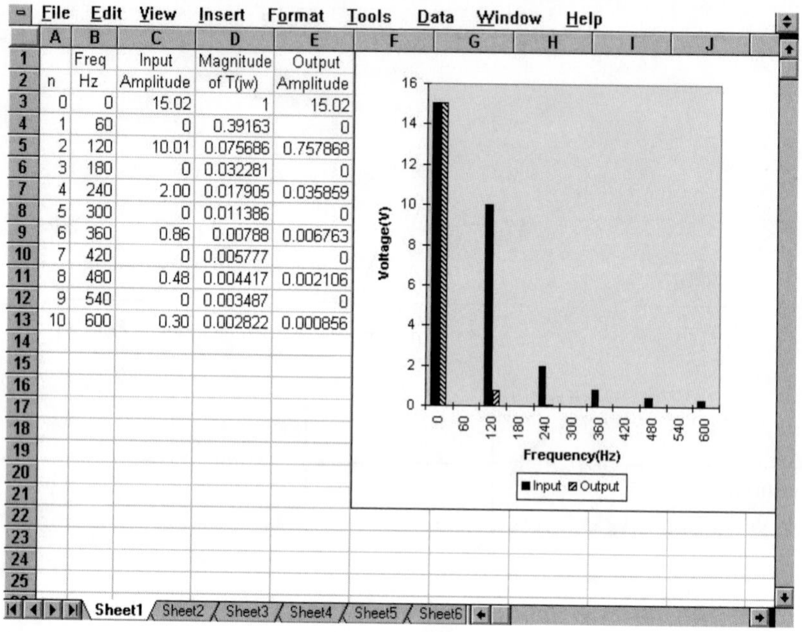

n	Freq Hz	Input Amplitude	Magnitude of T(jw)	Output Amplitude
0	0	15.02	1	15.02
1	60	0	0.39163	0
2	120	10.01	0.075686	0.757868
3	180	0	0.032281	0
4	240	2.00	0.017905	0.035859
5	300	0	0.011386	0
6	360	0.86	0.00788	0.006763
7	420	0	0.005777	0
8	480	0.48	0.004417	0.002106
9	540	0	0.003487	0
10	600	0.30	0.002822	0.000856

Exercise 13–5

Derive an expression for the first three nonzero terms in the Fourier series of the steady-state output voltage in Example 13–9.

Answer: $v_O(t) = 15.02 + 0.758 \cos(2\pi\, 120t + 5.7°) + 0.036 \cos(2\pi\, 240t + 2.7°)$ V

13–5 RMS VALUE AND AVERAGE POWER

In Chapter 5 we introduced the rms value of a periodic waveform as a descriptor of the average power carried by a signal. In this section we relate the rms value of the waveform to the amplitudes of the dc and ac components in its Fourier series. The rms value of a periodic waveform is defined as

$$F_{\text{rms}} = \sqrt{\frac{1}{T_0}\int_0^{T_0} [f(t)]^2\, dt} \tag{13–19}$$

The waveform $f(t)$ can be expressed as a Fourier series in the amplitude and phase form as

$$f(t) = A_0 + \sum_{n=1}^{\infty} A_n \cos(n\omega_0 t + \phi_n)$$

Substituting this expression into Eq. (13–19), we can write F^2_{rms} as

$$F^2_{rms} = \frac{1}{T_0}\int_0^{T_0}\left[A_0 + \sum_{n=1}^{\infty} A_n \cos(n\omega_0 t + \phi_n)\right]^2 dt \tag{13–20}$$

Squaring and expanding the integrand on the right side of this equation produces three types of terms. The first is the square of the dc component:

$$\frac{1}{T_0}\int_0^{T_0}[A_0]^2\,dt = A_0^2 \tag{13–21}$$

The second is the cross product of the dc and ac components, which takes the form

$$\frac{1}{T_0}\sum_{n=1}^{\infty} 2A_0\int_0^{T_0} A_n \cos(n\omega_0 t + \phi_n)dt = 0 \tag{13–22}$$

These terms all vanish because they involve integrals of sinusoids over an integer number of cycles. The third and final type of term is the square of the ac components, which can be written as

$$\frac{1}{T_0}\sum_{n=1}^{\infty}\sum_{m=1}^{\infty}\int_0^{T_0} A_n \cos(n\omega_0 t + \phi_n)A_m \cos(m\omega_0 t + \phi_m)dt = \frac{1}{2}\sum_{n=1}^{\infty} A_n^2 \tag{13–23}$$

This rather formidable expression boils down to a simple sum of squares because all of the integrals vanish except when $m = n$.

Combining Eqs. (13–19) through (13–23), we obtain the rms value as

$$F_{rms} = \sqrt{A_0^2 + \sum_{n=1}^{\infty}\frac{A_n^2}{2}}$$

$$= \sqrt{A_0^2 + \sum_{n=1}^{\infty}\left(\frac{A_n}{\sqrt{2}}\right)^2} \tag{13–24}$$

Since the rms value of a sinusoid of amplitude A is $A/\sqrt{2}$, we conclude that

The rms value of a periodic waveform is equal to the square root of the sum of the square of the dc value and the square of the rms value of each of the ac components.

In Chapter 5 we found that the average power delivered to a resistor is related to its rms voltage or current as

$$P = \frac{V^2_{rms}}{R} = I^2_{rms}R$$

Combining these expressions with the result in Eq. (13–24), we can write the average power delivered by a periodic waveform to the average power delivered by each of its Fourier components:

$$P = \frac{V_0^2}{R} + \sum_{n=1}^{\infty} \frac{V_n^2}{2R} = I_0^2 R + \sum_{n=1}^{\infty} \frac{I_n^2}{2} R$$
$$= P_0 + \sum_{n=1}^{\infty} P_n \tag{13–25}$$

where P_0 is the average power delivered by the dc component and P_n is the average power delivered by the nth ac component. This additive feature is important because it means we can find the total average power by adding the average power carried by the dc plus that carried by each of the ac components.

Caution: In general, we cannot find the total power by adding the power delivered by each component acting alone because the superposition principle does not apply to power. However, the average power carried by harmonic sinusoids is additive because they belong to a special class called orthogonal signals.

EXAMPLE 13–10

Derive an expression for the average power delivered to a resistor by a sawtooth voltage of amplitude V_A and period T_0. Then calculate the fraction of the average power carried by the dc component plus the first three ac components.

SOLUTION:

An equation for the sawtooth voltage is $v(t) = V_A t/T_0$ for the range $0 < t < T_0$. The square of the rms value of a sawtooth is

$$V_{rms}^2 = \frac{1}{T_0}\int_0^{T_0}\left(\frac{V_A t}{T_0}\right)^2 dt = V_A^2\left[\frac{t^3}{3T_0^3}\right]_0^{T_0} = \frac{V_A^2}{3}$$

The average power delivered to a resistor is

$$P = \frac{V_{rms}^2}{R} = \frac{V_A^2}{3R} = 0.333\frac{V_A^2}{R}$$

This result is obtained directly from the sawtooth waveform without having to sum an infinite series. The same answer could be obtained by summing an infinite series. The question this example asks is, How much of the average power is carried by the first four components in the Fourier series of the sawtooth wave? The amplitude spectrum of the sawtooth wave (see Example 13–4) is

$$V_n = \begin{cases} \dfrac{V_A}{2} & n = 0 \\ \dfrac{V_A}{n\pi} & n > 0 \end{cases}$$

From Eq. (13–25), the average power in terms of amplitude spectrum is

$$P = \frac{(V_A/2)^2}{R} + \sum_{n=1}^{\infty}\frac{(V_A/n\pi)^2}{2R}$$

which can be arranged in the form

$$P = \frac{V_A^2}{R}\Bigg\{\underbrace{\overbrace{\frac{1}{(2)^2} + \frac{1}{2(\pi)^2} + \frac{1}{2(2\pi)^2} + \frac{1}{2(3\pi)^2}}^{}}_{0.319(96\%)} + \frac{1}{2(4\pi)^2} + \cdots\Bigg\}$$

$$\underbrace{\qquad\qquad\qquad\qquad\qquad\qquad}_{0.333}$$

The infinite series within the braces must sum to 0.333 to match the average power we calculated directly from the waveform itself. The dc component plus the first three ac terms contribute 0.319 to the infinite sum. In other words, these four components alone deliver 96% of the average power carried by the sawtooth wave. ■

Exercise 13–6

The full-wave sine shown in Figure 13–3 has an rms value of $A/\sqrt{2}$. What fraction of the average power that the waveform delivers to a resistor is carried by the first two nonzero terms in its Fourier series?

Answer: Fraction = $88/9\pi^2$ or 99.07%

EXAMPLE 13–11

Figure 13–7 shows a series *RL* circuit driven by a sawtooth voltage source. Estimate the average power delivered to the resistor for V_A = 25 V, $R = 50\ \Omega$, $L = 40\ \mu H$, $T_0 = 5\ \mu s$, and $\omega_0 = 2\pi/T_0 = 1.26$ Mrad/s.

SOLUTION:

The Fourier series of the current in a series *RL* circuit is given in Eq. (13–15) as

$$i(t) = I_0 + \sum_{n=1}^{\infty} I_n \cos(n\omega_0 t + \psi_n)$$

where

$$I_0 = \frac{V_A}{2R} \qquad I_n = \frac{V_A}{R}\frac{1}{n\pi\sqrt{1 + (n\omega_0 L/R)^2}}$$

We cannot directly calculate the rms value of this current without a closed form expression for its waveform. However, we can get an estimate of the average power from the Fourier series. In terms of its Fourier series, the average power delivered by the current is

$$P = I_0^2 R + \sum_{n=1}^{\infty} \frac{I_n^2}{2} R$$

When $n\omega_0 L/R >> 1$, the amplitude I_n decreases as $1/n^2$ and its contribution to the average power falls off as $1/n^4$. In other words, the infinite series for the average power converges very rapidly. To show how rapidly the series converges, we calculate the first few terms using the specified numerical values

$$P = I_0^2 R + \frac{I_1^2}{2} R + \frac{I_2^2}{2} R + \frac{I_3^2}{2} R + \cdots$$

$$= 3.125 + 0.315 + 0.0314 + 0.00697 + \cdots$$

$$\longleftarrow 3.44 \longrightarrow$$
$$\longleftarrow 3.471 \longrightarrow$$
$$\longleftarrow 3.478 \longrightarrow$$

These results indicate that $P = 3.48$ W is a reasonable estimate of the average power, an estimate obtained using only four terms in the infinite series. The important point is that the high-frequency ac components are not important contributors to the average power carried by a signal. ■

APPLICATION EXAMPLE 13–12

The Fourier series serves as an introduction to the concept that a signal can be described by a spectrum that gives the distribution of amplitudes (and sometimes phases) of the sinusoidal components in a waveform. Radio, television, satellite communication, and radar systems must confine their signal spectra to an allocated portion of the available electromagnetic spectrum. Spectral allocations are regulated by various governmental agencies since there are many potential users and a limited spectral resource. As a result, users must design their systems to operate within specified spectral limitations.

These limitations are often specified using the concept of **signal bandwidth,** defined as the frequency interval outside of which the amplitude spectrum is less than some specified value. Accordingly, signal bandwidth (B) can be quantified by the expression

$$B = f_U - f_L$$

where f_U and f_L are the upper and lower limits on the bandwidth interval. Bandwidth is a partial signal descriptor that places an upper bound on the spectral content outside of the interval $f_L < f < f_U$. *Caution:* The spectral content inside the interval can be less than this upper bound at some frequencies but must be less at every frequency outside the interval.

For a periodic waveform the lower frequency bound is $f_L = f_0$ when there is no dc component ($a_0 = 0$) and is $f_L = 0$ when a_0 is not zero. The upper bound is another matter. In general, the Fourier series is an infinite series, so that in theory its spectral content extends to infinite frequency. However, we have seen the amplitudes of the higher harmonic become negligibly small and make little contribution to the waveform's energy content. In other words, an acceptable approximation to a periodic waveform is obtained when its Fourier series is terminated at some designated $f_U = nf_0$.

One simple way to select nf_0 is to compare the amplitudes of the high-frequency harmonics with the amplitude of the largest ac component, usually the fundamental. Harmonics whose amplitudes are less than a specified fraction (say 5%) of the fundamental are ignored. Applying a 5% criteria to waveforms whose Fourier coefficients decrease as $1/n$ (square wave, rectangular pulse, sawtooth wave; see Figure 13–3) yields $f_U = 20f_0$. Waveforms whose harmonics fall off as $1/n^2$ (triangular wave, rectified sine waves) meet a 5% bound at about $f_U = 5f_0$. Regardless of criteria used, the basic idea is that signal bandwidth is an important design consideration that in turn influences the bandwidth required in system components.

For example, accurate representation of arterial blood pressure waveforms requires the first eight harmonics of the pressure signal. To get a useful record of the arterial pressure waveform, the bandwidth (in Hz) of a measurement system must be at least eight times the maximum heart rate (in beats/second). Thus, designing a recording system for a maximum heart rate of 180 beats/minute requires a minimum bandwidth of 24 Hz.

APPLICATION EXAMPLE 13–13

Digital signal processing uses samples of an analog waveform, as contrasted with analog processing, which operates on the entire waveform. Sampling refers to the process of selecting discrete values of a time-varying analog waveform for further processing. By far the most common method of sampling is to record the waveform amplitudes at equally spaced time intervals. The set of samples $v_k\{k = 1, 2, 3, \cdots\}$ of an analog waveform $v(t)$ is defined as

$$v_k = v(kT_S)$$

where k is an integer and T_S is the time interval between successive samples. Figure 13–11 shows an example of sampling an analog waveform.

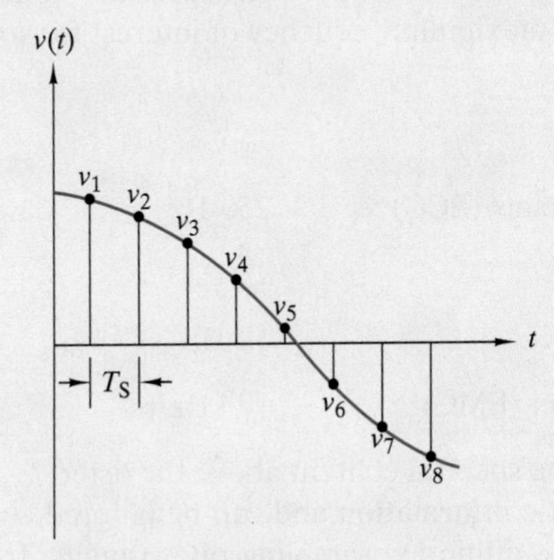

FIGURE 13–11 *Sampled signal.*

The digitized samples can be stored in a computer or some other medium such as a compact disc. These samples become the only means of describing the original analog signal. How good of a description can they be? From our experience we know that a reasonable facsimile can be obtained if the samples are closely spaced. Intuitively it might seem that exact reconstruction of the analog waveform would require the time between samples to approach zero. Surprisingly, exact reproduction is possible from samples taken at finite intervals.

> ***An analog waveform $v(t)$ whose spectral content falls below f_{max} can be reproduced exactly from samples $v_k = v(kT_S)$, if the sampling rate $f_S = 1/T_S$ is greater than $2f_{max}$.***

This statement, known as the **sampling theorem**, is one of the key principles of signal processing. The theorem states that exact recovery requires a minimum sampling rate of $2f_{max}$.[2] Analog waveforms are usually sampled at a rate higher than the minimum. For example, the industry standard sampling rate for recording digital audio signals is $f_s =$ 44.1 kHz, which is slightly more than twice the generally accepted upper limit on human hearing at $f_{max} = 20$ kHz.

The important point is that the minimum sampling rate is determined by the spectral content of the original analog waveform. Waveforms whose sinusoidal components all fall in the band $0 \leq f \leq f_{max}$ are said to be **strictly band-limited**. For example, the waveform

$$v(t) = 8 + 5\cos(2\pi 200t) + 3\cos(2\pi 400t) + 1.5\cos(2\pi 1200t)$$

is strictly band-limited with $f_{max} = 1.2$ kHz and can be reproduced exactly from samples taken at any $f_S > 2.4$ kHz.

Most analog signals are not strictly band-limited but usually have an upper frequency limit beyond which the spectral content is of no interest. For example, a periodic waveform is not strictly band-limited because its Fourier series is an infinite sum of harmonics at nf_0. However, the amplitudes of the higher harmonic become negligibly small as $n \to \infty$. A band-limited approximation to a periodic waveform is obtained by defining $f_{max} = nf_0$, where nf_0 is the harmonic beyond which the spectral amplitudes are less than some specified value.

Examples of the maximum frequency of interest for some biomedical signals are:

Type	f_{max}
Electrocardiograms (ECG)	250 Hz
Blood flow	25 Hz
Respiratory rate	10 Hz
Electromyogram (EMG)	10 kHz

For these signals, the spectral content above the listed f_{max} does not contain useful diagnostic information and can be ignored. As a result, biomedical signals have minimum sampling rates ranging from $f_S = 20$ Hz to $f_S = 20$ kHz.

2 This minimum sampling rate is called the *Nyquist rate*. The name honors Harry Nyquist, who along with Claude Shannon and others made several key breakthroughs in signal processing in the era from 1920 to 1950 while working at the Bell Telephone Laboratories.

But we cannot simply ignore the spectral content above f_{max}. It turns out that sampling signals that are not strictly band-limited causes *aliasing,* a process by which seemingly negligible out-of-band spectral content reappears as in-band distortion. The answer to the aliasing problem is to filter the analog signal prior to sampling. These *anti-aliasing filters* must pass spectral content up to the *highest frequency of interest,* f_{max}, and suppress the spectral content above the *lowest aliasing frequency* at $f_S - f_{max}$, where f_S is the sampling frequency.

Figure 13–12 shows the gain response of an anti-aliasing filter used in telecommunication. This low-pass filter has 0 dB gain at $f_{max} = 3.3$ kHz and less than -60 dB gain at $2\,f_{max}$. This type of gain response cannot be achieved by the first- and second-order filters studied in Chapter 12. This type of performance calls for higher-order filters of the type discussed in Chapter 14.

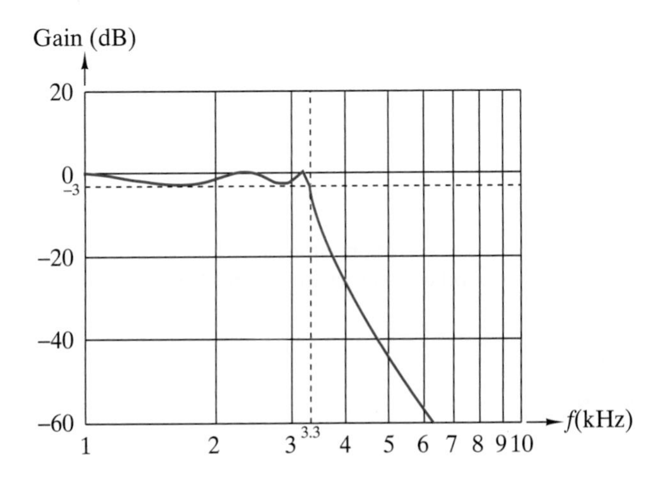

FIGURE 13–12 *Anti-aliasing filter.*

SUMMARY

- The Fourier series resolves a periodic waveform into a dc component plus an ac component containing an infinite sum of harmonic sinusoids. The dc component is equal to the average value of the waveform. The amplitudes of the sine and cosine terms in the ac component are called Fourier coefficients.
- The fundamental frequency of the ac component is determined by the period T_0 of the waveform ($f_0 = 1/T_0$). The harmonic frequencies in the ac component are integer multiples of the fundamental frequency.
- Waveform symmetries cause the amplitudes of some terms in a Fourier series to be zero. Even symmetry causes all of the sine terms in the ac component to be zero. Odd symmetry causes all of the cosine terms to be zero. Half-wave symmetry causes all of the even harmonics to be zero.
- An alternative form of the Fourier series represents each harmonic in the ac component by its amplitude and phase angle. A plot of amplitudes

versus frequency is called the amplitude spectrum. A plot of phase angles versus frequency is called the phase spectrum. A periodic waveform has spectral components at the discrete frequencies present in its Fourier series.

- The steady-state response of a linear circuit for a periodic driving force can be found by first finding the steady-state response due to each term in the Fourier series of the input. The Fourier series of the steady-state response is then found by adding (superposing) responses due to each term acting alone. The individual responses can be found using either phasor or s-domain analysis.
- The rms value of a periodic waveform is equal to the square root of the sum of the square of the dc value and the square of the rms value of each of the ac components. The average power delivered by a periodic waveform is equal to the average power delivered by the dc component plus the sum of average power delivered by each of the ac components.

PROBLEMS

OBJECTIVE 13–1 THE FOURIER SERIES (SECTS. 13–1, 13–2, 13–3)

(a) Given an equation or graph of a periodic waveform, derive expressions for the Fourier coefficients.
(b) Use the results in Figure 13–3 to calculate the Fourier coefficients of a given periodic waveform.
(c) Given a Fourier series of a periodic waveform, determine properties of the waveform, plot its amplitude and phase spectra, and identify the waveform.
See Examples 13–1, 13–3, 13–4, 13–5 and Exercises 13–1, 13–2, 13–3

13–1 The equation for the first cycle ($0 \leq t \leq T_0$) of a periodic waveform is

$$v(t) = V_A(-0.5 + t/T_0)$$

(a) Sketch the first two cycles of the waveform.
(b) Derive expressions for the Fourier coefficients a_n and b_n.

13–2 The equation for the first cycle ($0 \leq t \leq T_0$) of a periodic pulse train is

$$v(t) = V_A[u(t) + 2u(t - T_0/2)]$$

(a) Sketch the first two cycles of the waveform.
(b) Derive expressions for the Fourier coefficients a_n and b_n.

13–3 Derive expressions for the Fourier coefficients of the periodic waveform in Figure P13–3.

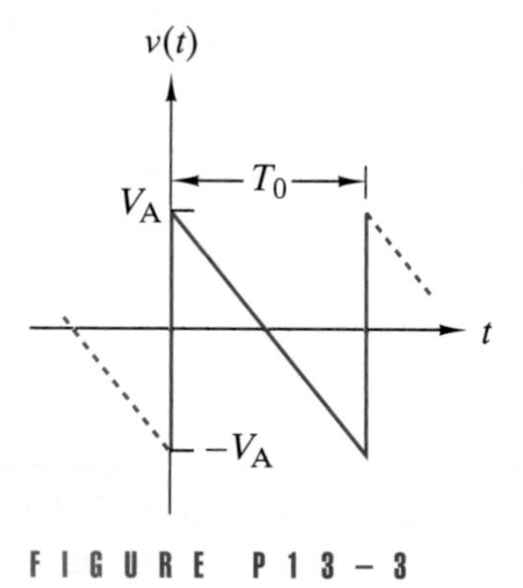

FIGURE P13–3

13–4 Derive expressions for the Fourier coefficients of the periodic waveform in Figure P13–4.

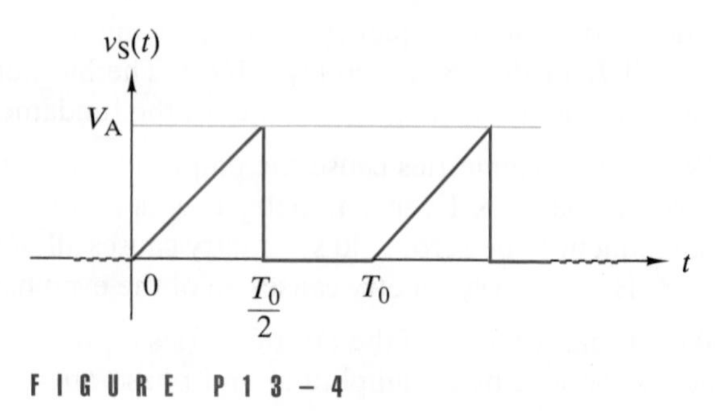

FIGURE P13–4

13–5 Use the results in Figure 13–3 to calculate Fourier coefficients of the square wave in Figure P13–5. Write an expression for the first four nonzero terms in the Fourier series.

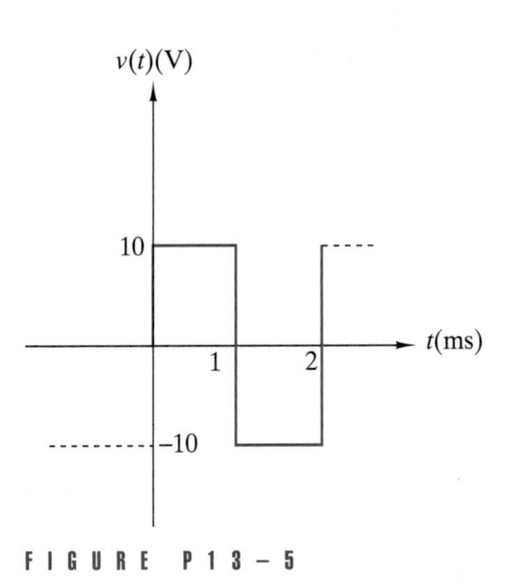

FIGURE P13–5

13–6 Use the results in Figure 13–3 to calculate the Fourier coefficients of the shifted triangular wave in Figure P13–6. Write an expression for the first four nonzero terms in the Fourier series.

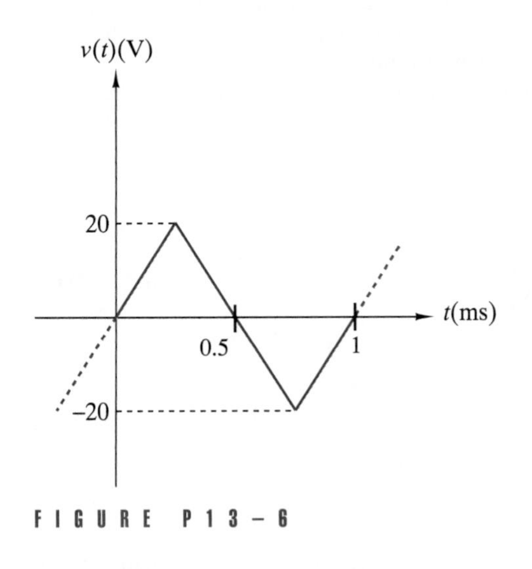

FIGURE P13–6

13–7 Use the results in Figure 13–3 to calculate the Fourier coefficients of the full-wave rectified sine wave in Figure P13–7. Write an expression for the first four nonzero terms in the Fourier series.

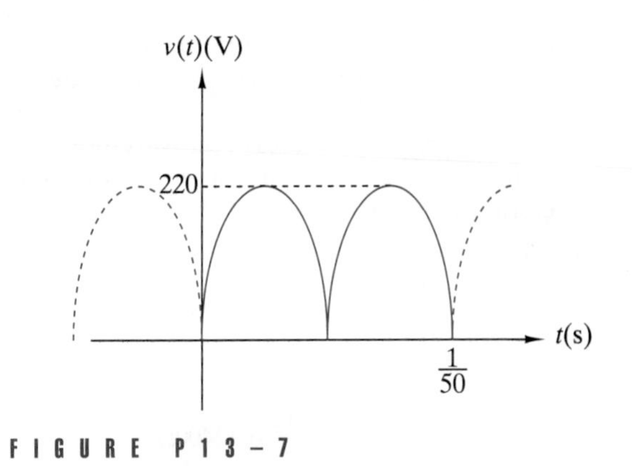

FIGURE P13–7

13–8 A half-wave rectified sine wave has an amplitude of 60 V and a fundamental frequency of 50 Hz. Use the results in Figure 13–3 to write an expression for the first four nonzero terms in the Fourier series and plot the amplitude spectrum of the signal.

13–9 The waveform $f(t)$ is a 1-kHz triangular wave with a *peak-to-peak* amplitude of 10 V. Use the results in Figure 13–3 to write an expression for the first four nonzero terms in the Fourier series of $g(t) = 10 + f(t)$ and plot its amplitude spectrum.

13–10 A sawtooth wave has a *peak-to-peak* amplitude of 15 V and a fundamental frequency of 2 kHz. Use the results in Figure 13–3 to write an expression for the first four nonzero terms in the Fourier series and plot the amplitude spectrum of the signal.

13–11 The equation for the first cycle ($0 \leq t \leq T_0$) of a periodic pulse train is

$$v(t) = V_A[u(3t - T_0) - u(3t - 2T_0)]$$

(a) Sketch the first two cycles of the waveform and identify a related signal in Figure 13–3.
(b) Use the Fourier series of the related signal to find the Fourier coefficients of $v(t)$.

13–12 The equation for a periodic waveform is

$$v(t) = V_A[\sin(2\pi t/T_0) + |\sin(2\pi t/T_0)|]$$

(a) Sketch the first two cycles of the waveform and identify a related signal in Figure 13–3.
(b) Use the Fourier series of the related signal to find the Fourier coefficients of $v(t)$.

13–13 The first four terms in the Fourier series of a periodic waveform are

$$v(t) = 15\left[\sin(20\pi t) - \frac{1}{9}\sin(60\pi t) + \frac{1}{25}\sin(100\pi t) - \frac{1}{49}\sin(140\pi t) + \cdots\right]$$

(a) Find the period and fundamental frequency in rad/s and Hz. Identify the harmonics present in the first four terms.
(b) Identify the symmetry features of the waveform.
(c) Write the first four terms in Fourier series for the waveform $v(t - T_0/4)$.

13–14 The first five terms in the Fourier series of a periodic waveform are

$$v(t) = 20 + 40\left[\frac{\pi}{4}\sin(400t) - \frac{1}{3}\cos(800t) - \frac{1}{15}\cos(1600t) - \frac{1}{35}\cos(2400t) + \cdots\right]$$

(a) Find the period and fundamental frequency in rad/s and Hz. Identify the harmonics present in the first five terms.
(b) Identify the symmetry features of the waveform.
(c) Write the first five terms of the Fourier series for the waveform $v(-t)$.

13–15 The equation for a full-wave rectified cosine is $v(t) = V_A\,|\cos(\pi t/T_0)|$. Sketch two cycles of the waveform and calculate the dc component of the waveform.

OBJECTIVE 13–2 FOURIER SERIES AND CIRCUIT ANALYSIS (SECT. 13–4)

(a) Given a linear circuit with a periodic input waveform, find the Fourier series representation of a steady-state response.
(b) Given a network function with a periodic input, find the amplitude and/or phase spectrum of the steady-state output.
See Examples 13–6, 13–7, 13–8, 13–9 and Exercises 13–4, 13–5

13–16 The periodic pulse train in Figure P13–16 is applied to the RL circuit shown in the figure.

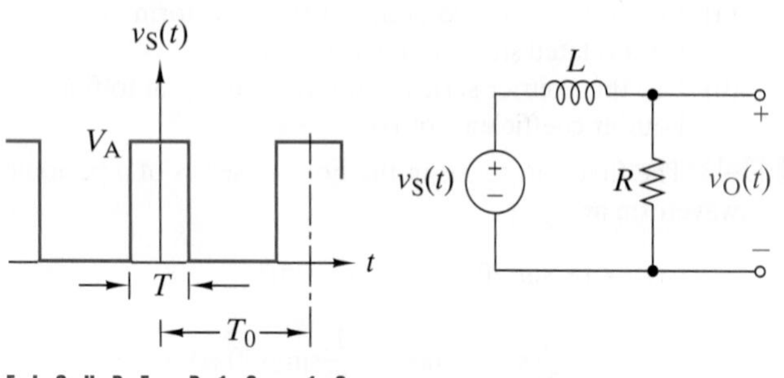

FIGURE P13–16

(a) Use the results in Figure 13–3 to find the Fourier coefficients of the input for $V_A = 5$ V, $T_0 = 2\pi$ ms, and $T = T_0/5$.
(b) Find the first four nonzero terms in the Fourier series of $v_O(t)$ for $R = 50\ \Omega$ and $L = 100$ mH.

13–17 The periodic triangular wave in Figure P13–17 is applied to the RC circuit shown in the figure. The Fourier coefficients of the input are:

$$a_0 = 0 \quad a_n = 0 \quad b_n = \frac{8\,V_A}{(n\pi)^2}\sin\left(n\frac{\pi}{2}\right)$$

If $V_A = 20$ V and $T_0 = 10\pi$ ms, find the first four nonzero terms in the Fourier series of $v_O(t)$ for $R = 50$ kΩ and $C = 50$ nF.

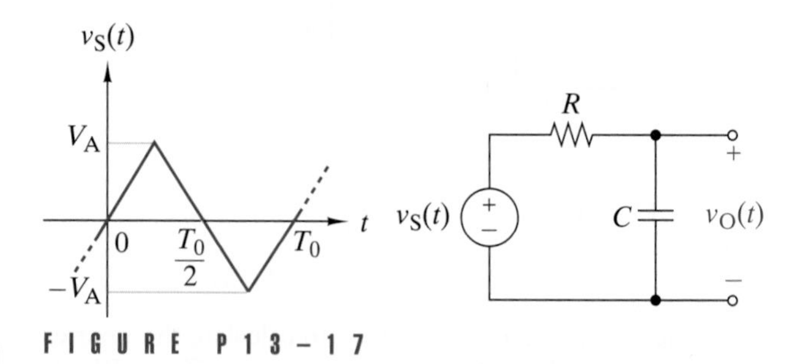

FIGURE P13–17

13–18 The periodic sawtooth wave in Figure P13–18 drives the OP AMP circuit shown in the figure.
(a) Use the results in Figure 13–3 to find the Fourier coefficients of the input for $V_A = 5$ V and $T_0 = 4\pi$ ms.
(b) Find the first four nonzero terms in the Fourier series of $v_O(t)$ for $R_1 = 10$ kΩ, $R_2 = 50$ kΩ, and $C = 100$ nF.

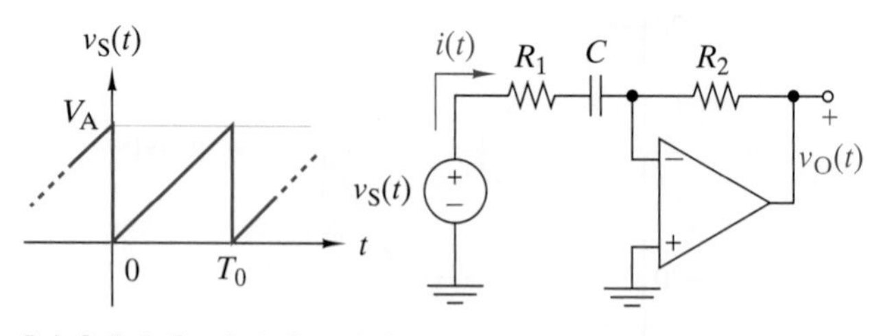

FIGURE P13–18

13–19 The periodic sawtooth wave in Figure P13–18 drives the OP AMP circuit shown in the figure.
(a) Use the results in Figure 13–3 to find the Fourier coefficients of the input for $V_A = 4$ V and $T_0 = 8\pi$ ms.
(b) Find the first four nonzero terms in the Fourier series of $i(t)$ for $R_1 = 100$ kΩ, $R_2 = 50$ kΩ, and $C = 40$ nF.

13–20 The periodic triangular wave in Figure P13–20 is applied to the *RLC* circuit shown in the figure.

(a) Use the results in Figure 13–3 to find the Fourier coefficients of the input for $V_A = 5$ V and $T_0 = 800\pi$ μs.

(b) Find the amplitude of the first five nonzero terms in the Fourier series for $i(t)$ when $R = 1\ \Omega$, $L = 8$ mH, and $C = 800$ nF. What term in the Fourier series tends to dominate the response? Explain.

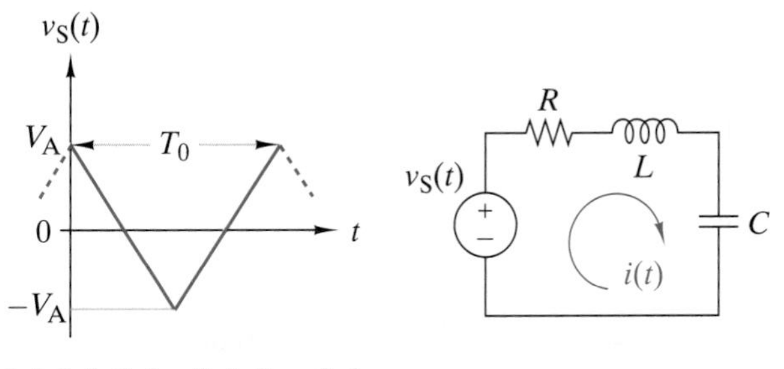

FIGURE P13–20

13–21 A sawtooth wave with $V_A = 15$ V and $T_0 = 50\pi$ ms drives a circuit with a transfer function $T(s) = s/(s + 100)$. Find the amplitude of the first four nonzero terms in the Fourier series of the steady-state output. Construct plots of the amplitude spectra for the input and output waveforms and comment on any differences.

13–22 Repeat Problem 13–21 for $T(s) = 100/(s + 100)$.

13–23 The voltage across a 1-nF capacitor is a triangular wave with $V_A = 15$ V and $f_0 = 1$ kHz. Construct plots of the amplitude spectra of the capacitor voltage and current. Discuss any differences in spectral content.

13–24 An ideal time delay is a signal processor whose output is $v_O(t) = v_{IN}(t - T_D)$. Write an expression for $v_O(t)$ for $T_D = 1$ ms and

$$v_{IN}(t) = 10 + 10\cos(2\pi 500t) + 3\cos(2\pi 1000t) + 2\cos(2\pi 4000t)$$

Discuss the spectral changes caused by the time delay.

13–25 A sawtooth wave with $V_A = 5\pi$ V and $T_0 = 10\pi$ ms drives a circuit whose transfer function is

$$T(s) = \frac{100\,s}{(s+50)^2 + 400^2}$$

Find the amplitude of the first four nonzero terms in the Fourier series of the steady-state output. What term in the Fourier series tends to dominate the response? Explain.

OBJECTIVE 13–3 RMS VALUE AND AVERAGE POWER (SECT. 13–5)

(a) Given a periodic waveform, find the rms value of the waveform and the average power delivered to a specified load.

(b) Given the Fourier series of a periodic waveform, find the fraction of the average power carried by specified components and estimate the average power delivered to a specified load.

See Examples 13–10, 13–11 and Exercise 13–6

13–26 The current through a 500-Ω resistor is

$$i(t) = 30 + 15\cos(120\pi t - 30^\circ) - 10\cos(360\pi t + 45^\circ)\text{ mA}$$

Find the average power delivered to the resistor and the rms value of the current.

13–27 The voltage across a 50-Ω resistor is

$$v(t) = 10 + 15\cos(300\pi t) - 20\sin(900\pi t) + 5\sin(1200\pi t)\text{ V}$$

Find the average power delivered to the resistor and the rms value of the voltage.

13–28 Find the rms value of a full-wave rectified sine wave and the fraction of the total average power carried by the dc component plus the first two nonzero ac components in the Fourier series.

13–29 Find the rms value of the periodic waveform in Figure P13–29 and the average power the waveform delivers to a resistor. Find the dc component of the waveform and the average power carried by the dc component. What fraction of the total average power is carried by the dc component? What fraction is carried by the ac components?

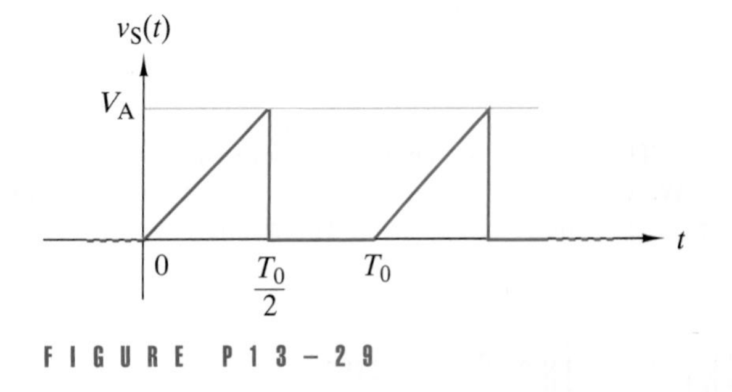

FIGURE P13–29

13–30 Repeat Problem 13–29 for the periodic waveform in Figure P13–30.

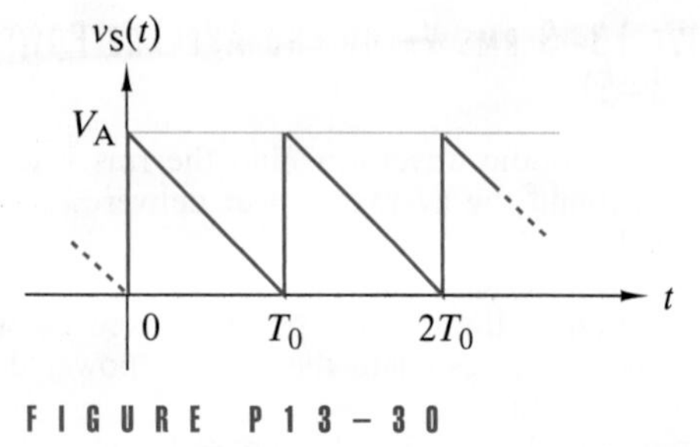

FIGURE P13–30

13–31 A first-order low-pass filter has a cutoff frequency of 250 rad/s and a passband gain of 20 dB. The input to the filter is $x(t) = 5\sin 500t + 2\cos 2000t$ V. Find the rms value of the steady-state output.

13–32 Repeat Problem 13–31 for a first-order high-pass filter with the same cutoff frequency and passband gain.

13–33 Estimate the rms value of the periodic voltage

$$v(t) = V_A\left[\frac{\pi^2}{12} - \cos\omega_0 t + \frac{1}{4}\cos 2\omega_0 t - \frac{1}{9}\cos 3\omega_0 t + \cdots\right]$$

13–34 The input to the circuit in Figure P13–34 is the voltage

$$v_S(t) = 35\cos(2\pi 3000t) + 15\sin(2\pi 6000t)\ \text{V}$$

Calculate the average power delivered to the 50-Ω resistor.

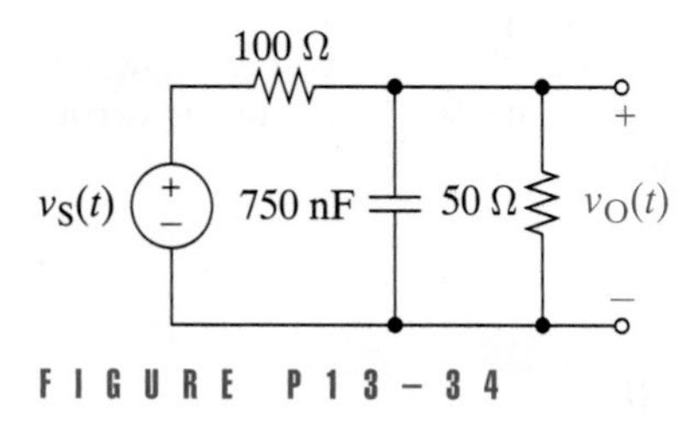

FIGURE P13–34

13–35 The input to the circuit in Figure P13–34 is a square wave with $V_A = 5$ V and $T_0 = 50\pi$ μs. Estimate the average power delivered to the 50-Ω resistor.

INTEGRATING PROBLEMS

13–36 **A** Fourier Series from a Bode Plot

The transfer function of a linear circuit has the straight-line gain and phase Bode plots in Figure P13–36. The first four terms in the Fourier series of a periodic input $v_1(t)$ to the circuit are

$$v_1(t) = 14\cos(10t) + 4.667\cos(30t) + 2.8\cos(50t) + 2\cos(70t)\ \text{V}$$

Estimate the amplitudes and phase angles of the first four terms in the Fourier series of the steady-state output $v_2(t)$.

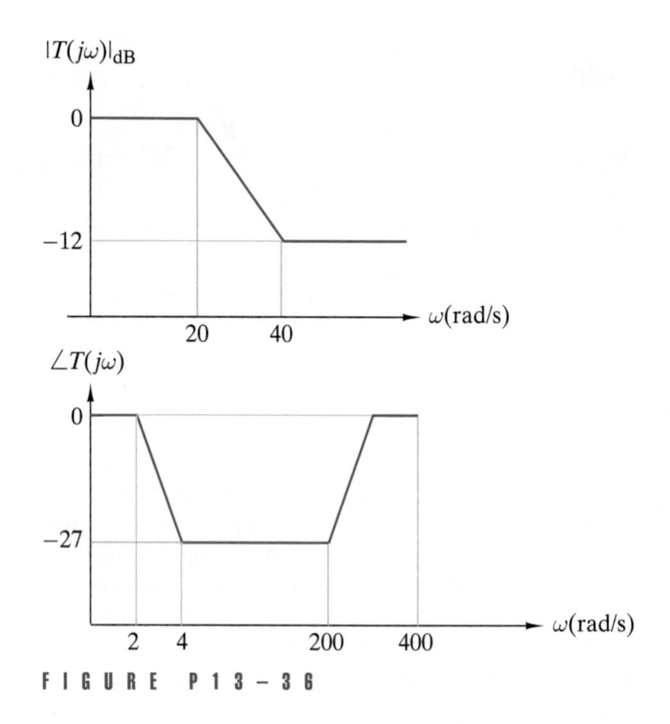

FIGURE P13–36

13–37 **A** Spectrum of a Periodic Impulse Train

A periodic impulse train can be written as

$$x(t) = T_0 \sum_{n=-\infty}^{\infty} \delta(t - nT_0)$$

Find the Fourier coefficients of $x(t)$. Plot the amplitude spectrum and comment on the frequencies contained in the impulse train.

13–38 **D** Power Supply Filter Design

The input to a power supply filter is a full-wave rectified sine wave with $f_0 = 60$ Hz. The filter is a first-order low pass with unity dc gain. Select the cutoff frequency of the filter so that all of the ac components in the filter output are all less than 1% of the dc component.

13–39 **D** Periodic Signal Design

The first half-cycle $(0 \leq t \leq T_0/2)$ of a periodic waveform is defined as $v(t) = 2V_A t/T_0$. Define a half-cycle on $-T_0/2 \leq t \leq 0$ such that the Fourier series of $v(t)$ contains only sine terms. Repeat for a series with only cosine terms. Repeat again for a series with only odd harmonics.

13–40 E Spectrum Analyzer Calibration
A certain spectrum analyzer measures the average power delivered to a calibrated resistor by the individual harmonics of periodic waveforms. The calibration of the analyzer has been checked by applying a 1-MHz square wave and the following results reported:

f(MHz)	1	3	5	7	9	11
P(dBm)	12.1	2.56	−1.88	−4.80	−6.98	−8.73

The reported power in dBm is $P = 10\log(P_n)$, where P_n is the average power delivered by the nth harmonic in mW. Is the spectrum analyzer correctly calibrated?

CHAPTER 14

ACTIVE FILTER DESIGN

In its usual form the electric wave-filter transmits currents of all frequencies lying within one or more specified ranges, and excludes currents of all other frequencies.

George A. Campbell, 1922,
American Engineer

Some History Behind This Chapter

The electric filter was independently invented during World War I by George Campbell in the United States and by K. W. Wagner in Germany. Electric filters and vacuum tube amplifiers were key technologies that triggered the growth of telephone and radio communication systems in the 1920s and 1930s. The emergence of semiconductor electronics in the 1960s, especially the integrated circuit OP AMP, allowed the functions of filtering and amplification to be combined into what are now called active filters.

Why This Chapter Is Important Today

This is a fun chapter dedicated to the design of practical filters. After studying this chapter, you will be able to design a wide range of analog filters that find applications in instrumentation systems, audio systems, communication systems, and even digital systems. You will also learn how to evaluate different solutions to a filter design problem.

Chapter Sections

14–1 Active Filters

14–2 Second-Order Low-Pass and High-Pass Filters

14–3 Second-Order Bandpass and Bandstop Filters

14–4 Low-Pass Filter Design

14–5 Low-Pass Filter Evaluation

14–6 High-Pass Filter Design

14–7 Bandpass and Bandstop Filter Design

Chapter Objectives

14-1 Second-Order Filter Analysis (Sects. 14-2, 14-3)

(a) Given a second-order filter circuit, find a specified transfer function.
(b) Given the transfer function of a second-order circuit, develop a method of selecting the element values to achieve specified filter characteristics.

14-2 Second-Order Filter Design (Sects. 14-2, 14-3)

(a) Construct a second-order transfer function with specified filter characteristics.
(b) Design a second-order circuit with specified filter characteristics.

14-3 Low-Pass Filter Design (Sect. 14-4)

Given a low-pass filter specification:
(a) Construct a transfer function that meets the specification.
(b) Design a cascade of first- and second-order circuits that implements a given transfer function.

14-4 High-Pass, Bandpass and Bandstop Filter Design (Sects. 14-6, 14-7)

Given a high-pass, bandpass, or bandstop filter specification:
(a) Construct a transfer function that meets the specification.
(b) Design a cascade or parallel connection of first- and second-order circuits that implements a given transfer function.

14–1 ACTIVE FILTERS

An electric **filter** is a signal processor that modifies or reshapes the frequency content of input signals. Filters are found in a multitude of applications. In communication systems filters are used for noise suppression and to detect or isolate a desired signal. They are used in digital systems to bandlimit analog inputs prior to sampling and to convert D/A outputs to continuous analog signals. The crossover networks in audio playback systems are filters that split the input spectrum into two or more bands for distribution to different speakers. Biomedical systems use filters to interface physiological sensors with data-logging and diagnostic equipment.

Filters are classified as passive or active depending on the components used in their physical realization. **Passive filter** circuits contain only resistors, capacitors, and inductors. The *RLC* bandpass and bandstop circuits studied in Chapter 12 are passive filters. These circuits can be highly selective when losses are low, and the response is highly resonant. However, they can not supply passband gains greater than one and suffer from loading effects that can nullify the chain rule in a cascade design.

In this chapter we define an **active filter** as a circuit that contains only resistors, capacitors, and OP AMPs. These filters offer several advantages:

- They provide frequency selectivity comparable to passive *RLC* circuits plus passband gains greater than one.
- They have OP AMP outputs, which means that the chain rule applies in a cascade design.
- They do not require inductors, which can be large, lossy, and expensive in low-frequency application.

Active filter design involves devising circuits that realize a given transfer function $T(s)$. Our design strategy is based on the familiar cascade connection in Figure 14–1. Under the chain rule, the overall transfer function is

$$T(s) = T_1(s) \times T_2(s) \times T_3(s) \ldots T_n(s)$$

The stages in the cascade are either first-order or second-order active filters. The real poles in $T(s)$ are produced using the first-order building block developed in Chapter 12. The complex poles are produced by second-order building blocks developed in the next two sections. The complex pole locations are specified by the customary second-order parameters—our old friends the damping ratio ζ and undamped natural frequency ω_0. Thus, we design an active filter by controlling the poles introduced by each stage in a cascade connection.

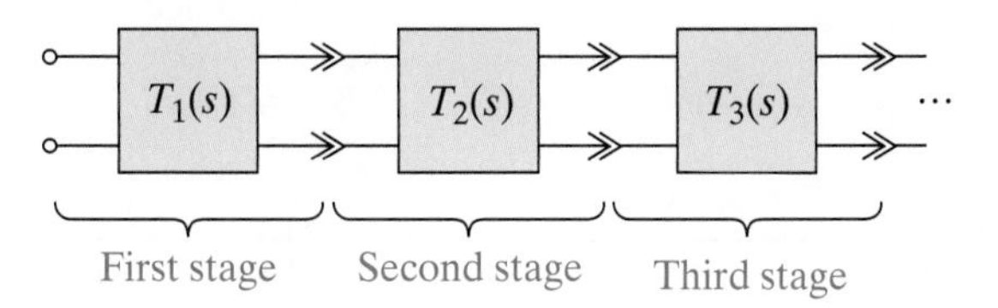

FIGURE 14–1 *A cascade connection.*

14–2 SECOND-ORDER LOW-PASS AND HIGH-PASS FILTERS

The second-order building blocks developed in this section are the counterparts of the first-order low-pass and high-pass filters in Chapter 12. These second-order circuits all have the following features: (1) an OP AMP output (to avoid loading), (2) two capacitors (to get two poles), and (3) at least one feedback path (to make the poles complex). We begin our development with the low-pass case.

SECOND-ORDER LOW-PASS FILTERS

The transfer function of a second-order low-pass filter has the form

$$T(s) = \frac{K}{(s/\omega_0)^2 + 2\zeta(s/\omega_0) + 1} \tag{14–1}$$

The gain response is found by substituting $j\omega$ for s to obtain

$$|T(j\omega)| = \frac{|K|}{\sqrt{[1 - (\omega/\omega_0)^2]^2 + 4\zeta^2(\omega/\omega_0)^2}} \tag{14–2}$$

At low frequency ($\omega << \omega_0$) the gain approaches $|T(j\omega)| \rightarrow |K| = |T(0)|$. At high frequency ($\omega >> \omega_0$) the gain approaches $|T(j\omega)| \rightarrow |K|(\omega_0/\omega)^2$. Figure 14–2 presents Bode plots of these gain asymptotes (solid lines) and the actual gain (dashed curves) for several values of ζ. The key points to remember are the following:

- The low- and high-frequency gain asymptotes intersect at $\omega = \omega_0$.
- For $\omega < \omega_0$ the asymptotic gain equals the dc gain $|T(0)| = |K|$.
- For $\omega > \omega_0$ the slope of the asymptotic gain is -40 dB/decade.
- The actual gain at $\omega = \omega_0$ is $|T(j\omega_0)| = |K|/2\zeta$.

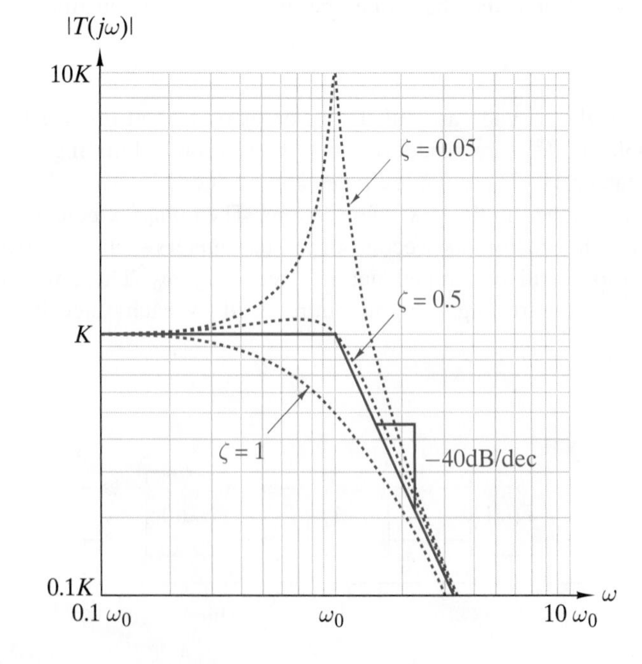

FIGURE 14–2 *Second-order low-pass gain responses.*

From a design perspective, we say that ω_0 locates the corner frequency and ζ controls the gain in the vicinity of the corner.

The circuit in Figure 14–3(a) is analyzed in Chapter 11 (Example 11–6), where its voltage transfer function is shown to be

$$T(s) = \frac{V_2(s)}{V_1(s)} = \frac{\mu}{R_1R_2C_1C_2s^2 + (R_1C_1 + R_1C_2 + R_2C_2 - \mu R_1C_1)s + 1} \tag{14–3}$$

This is a second-order low-pass function with a dc gain of $|T(0)| = \mu$.

An OP AMP RC filter is obtained by replacing the dependent source in Figure 14–3(a) by the OP AMP circuit in Figure 14–3(b). The gain of this noninverting amplifier gain is $1 + R_A/R_B$. Thus, modification simulates the dependent source when $1 + R_A/R_B = \mu$ or, equivalently, $R_A = (\mu - 1)R_B$. The modified circuit is our first second-order building block. Note that it has an OP AMP output, two capacitors, and a feedback path.

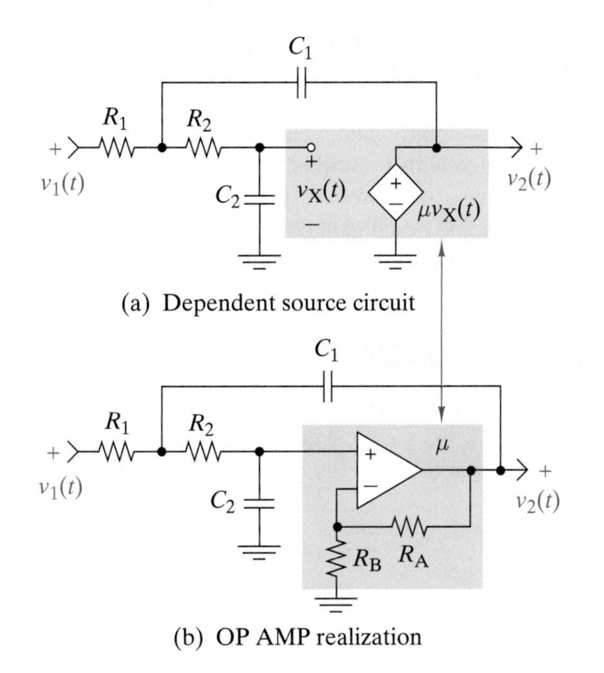

FIGURE 14–3 Second-order low-pass circuits.

Our objective is to find methods of determining the gain μ and the four element values that produce specified values of ω_0 and ζ. Comparing the denominators in Eqs. (14–1) and (14–3) gives

$$\sqrt{R_1R_2C_1C_2} = 1/\omega_0 \quad \text{and} \quad R_1C_2 + R_2C_2 + (1 - \mu)R_1C_1 = 2\zeta/\omega_0$$

Using the first equation to eliminate ω_0 from the second equation leads to

$$\sqrt{R_1R_2C_1C_2} = \frac{1}{\omega_0} \quad \text{and} \quad \sqrt{\frac{R_1C_2}{R_2C_1}} + \sqrt{\frac{R_2C_2}{R_1C_1}} + (1 - \mu)\sqrt{\frac{R_1C_1}{R_2C_2}} = 2\zeta \tag{14–4}$$

Two methods of selecting element values to meet these design goals are discussed below.

The *equal element* method requires that $R_1 = R_2 = R$ and $C_1 = C_2 = C$. Inserting these conditions into Eq. (14–4) leads to

$$RC = \frac{1}{\omega_0} \quad \text{and} \quad \mu = 3 - 2\zeta \tag{14–5}$$

Using this method, we select values of R (or C) and R_B, then solve for C (or R) and $R_A = (\mu - 1)R_B$. The dc gain achieved by this method is $|T(0)| = \mu = 3 - 2\zeta$.

The *unity gain* method requires that $R_1 = R_2 = R$ and $\mu = 1$. Inserting these conditions in Eq. (14–4) leads to

$$R\sqrt{C_1C_2} = \frac{1}{\omega_0} \quad \text{and} \quad \frac{C_2}{C_1} = \zeta^2 \tag{14–6}$$

Using this method, we select a value of C_1 and calculate $C_2 = \zeta^2C_1$ and $R = (\omega_0\sqrt{C_1C_2})^{-1}$. To get a gain $\mu = 1$, we make the noninverting OP AMP circuit a voltage follower. That is, we replace R_A by a short circuit and R_B by an open circuit. This eliminates the need for R_A and R_B but requires two different capacitors. Obviously the dc gain achieved by this design method is $|T(0)| = \mu = 1$.

The equal element and unity gain methods provide alternative ways to design an active low-pass filter with prescribed values of ω_0 and ζ. However, the dc gains produced by these methods are predetermined and are not adjustable design parameters. An additional gain correction stage may be needed when ω_0, ζ, and the dc gain are all three prescribed.

D DESIGN EXAMPLE 14–1

Construct a second-order low-pass transfer function with a corner frequency at $\omega_0 = 1$ krad/s and a corner frequency gain equal to the dc gain. Design a second-order low-pass circuit to realize the transfer function.

SOLUTION:

The required transfer function has the form

$$T(s) = \frac{K}{(s/1000)^2 + 2\zeta(s/1000) + 1}$$

The dc gain is $T(0) = K$ and the corner frequency gain is $|T(j1000)| = K/2\zeta$. These two gains are equal when $\zeta = 0.5$.

Equal Element Design: Inserting $\omega_0 = 1000$ rad/s and $\zeta = 0.5$ into Eq. (14–5) yields $RC = 10^{-3}$ s and $\mu = 2$. Selecting $R = R_B = 10\text{ k}\Omega$ requires $C = 100$ nF and $R_A = (\mu - 1)R_B = 10\text{ k}\Omega$. The resulting circuit in Figure 14–4(a) has a dc gain of $|T(0)| = \mu = 2$.

Unity Gain Design: Inserting $\omega_0 = 1000$ rad/s and $\zeta = 0.5$ into Eq. (14–6) yields $R\sqrt{C_1C_2} = 10^{-3}$ s and $C_2 = \zeta^2C_1 = 0.25C_1$. Selecting $C_1 = 100$ nF

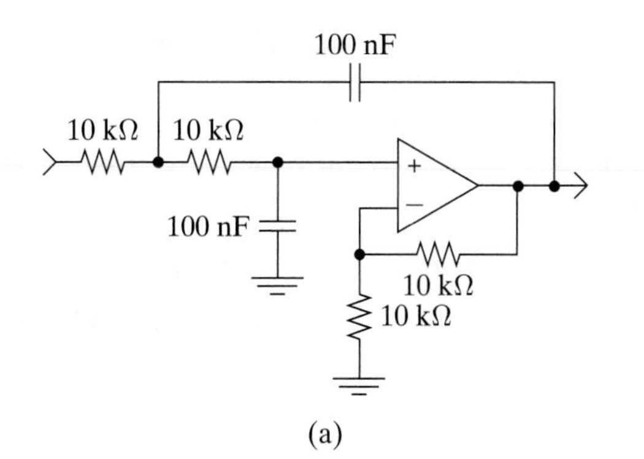

(a)

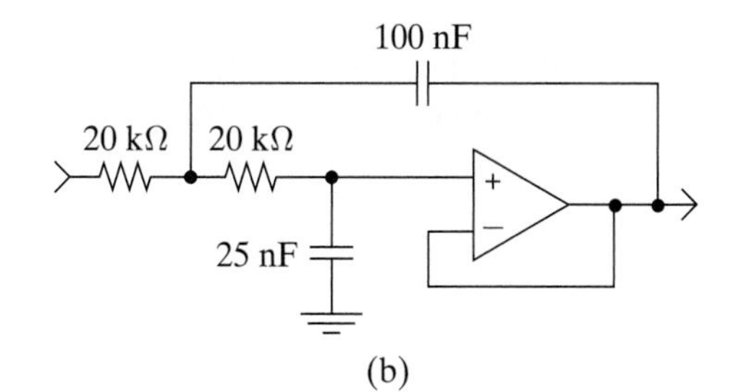

(b)

FIGURE 14–4

dictates that $C_2 = 25$ nF and $R = 20$ kΩ. The resulting circuit in Figure 14–4(b) uses fewer resistors but has less dc gain since $|T(0)| = 1$ and requires two different capacitors. ■

Exercise 14–1

Rework the unity gain design in Example 14–1 with $R = 40$ kΩ.

Answer: $C_1 = 50$ nF; $C_2 = 12.5$ nF

Exercise 14–2

Construct a second-order low-pass transfer function with a corner frequency of 50 rad/s, a dc gain of 2, and a gain of 4 at the corner frequency.

Answer:

$$T(s) = \frac{5000}{s^2 + 25s + 2500}$$

SECOND-ORDER HIGH-PASS FILTERS

The transfer function of a second-order high-pass filter has the form

$$T(s) = \frac{K(s/\omega_0)^2}{(s/\omega_0)^2 + 2\zeta(s/\omega_0) + 1} \tag{14–7}$$

This transfer function has two poles and a double zero at $s = 0$. Substituting $j\omega$ for s yields the gain response as

$$|T(j\omega)| = \frac{|K|(\omega/\omega_0)^2}{\sqrt{[1 - (\omega/\omega_0)^2]^2 + 4\zeta^2(\omega/\omega_0)^2}} \tag{14–8}$$

At high frequency ($\omega >> \omega_0$) the gain approaches $|T(j\omega)| \rightarrow |K| = |T(\infty)|$. At low frequency ($\omega << \omega_0$) the gain approaches $|T(j\omega)| \rightarrow |K|(\omega/\omega_0)^2$. Figure 14–5 presents Bode plots of these gain asymptotes (solid lines) and the actual gain (dashed curves) for several values of ζ. The key points to remember are the following:

- The high- and low-frequency gain asymptotes intersect at $\omega = \omega_0$.
- For $\omega > \omega_0$ the asymptotic gain equals the infinite frequency gain $|T(\infty)| = |K|$.
- For $\omega < \omega_0$ the slope of the asymptotic gain is +40 dB/decade.
- The actual gain at $\omega = \omega_0$ is $|T(j\omega_0)| = |K|/2\zeta$.

As in the low-pass case, the design parameter ω_0 locates the corner frequency and ζ controls the actual gain around the corner frequency.

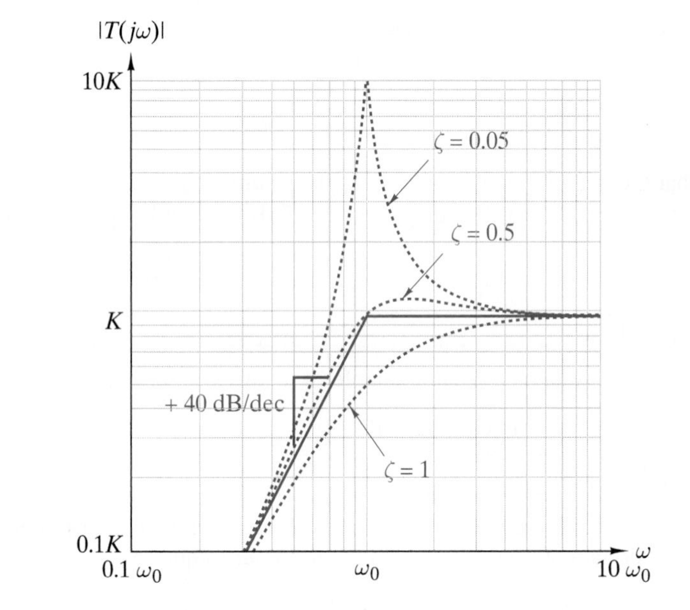

FIGURE 14–5 *Second-order high-pass gain responses.*

For future reference, note that replacing s/ω_0 by ω_0/s converts the low-pass $T(s)$ in Eq. (14–1) into the high-pass $T(s)$ in Eq. (14–7). We will make good use of this *low-pass to high-pass transformation* in a later section.

The circuit in Figure 14–6(a) has a transfer function of the form

$$T(s) = \frac{V_2(s)}{V_1(s)} = \frac{\mu R_1 R_2 C_1 C_2 s^2}{R_1 R_2 C_1 C_2 s^2 + (R_2 C_2 + R_1 C_1 + R_1 C_2 - \mu R_2 C_2)s + 1} \tag{14–9}$$

This is a second-order high-pass function with an infinite-frequency gain of $|T(\infty)| = \mu$. The high-pass circuit in Figure 14–6(a) is obtained from the low-pass circuit in Figure 14–3(a) by interchanging the locations of the resistors

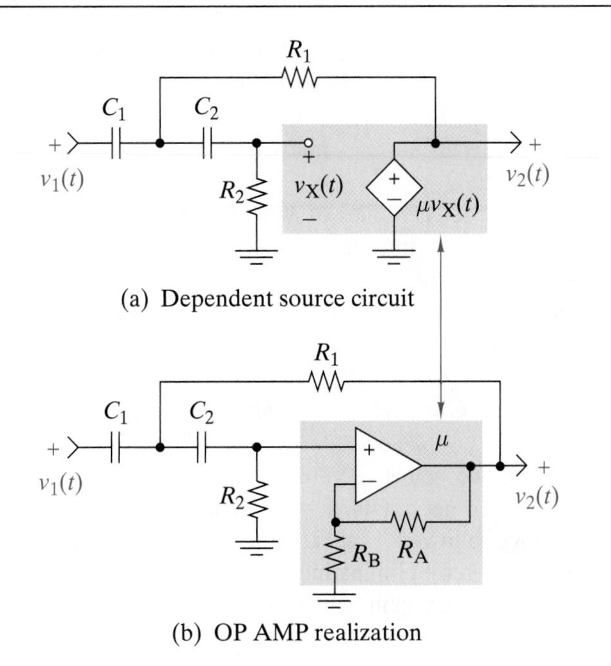

FIGURE 14–6 *Second-order high-pass circuits.*

and capacitors.[1] In Problem 14–1 the student will discover how this interchange converts the low-pass $T(s)$ in Eq. (14–3) into the high-pass $T(s)$ in Eq. (14–9).

As in the low-pass case, we replace the dependent source in Figure 14–6(a) by the noninverting OP AMP circuit in Figure 14–6(b). The OP AMP circuit simulates the dependent source when $1 + R_A/R_B = \mu$ or, equivalently, $R_A = (\mu - 1)R_B$. The modified high-pass circuit is a second-order building block with an OP AMP output, two capacitors, and a feedback path.

Again our goal is to find methods of determining the gain μ and the four element values that produce specified values of ω_0 and ζ. Comparing the denominators in Eqs. (14–7) and (14–9) gives

$$\sqrt{R_1R_2C_1C_2} = 1/\omega_0 \quad \text{and} \quad R_1C_1 + R_1C_2 + (1 - \mu)R_2C_2 = 2\zeta/\omega_0$$

Using the first equation to eliminate ω_0 from the second equation leads to

$$\sqrt{R_1R_2C_1C_2} = \frac{1}{\omega_0} \quad \text{and} \quad \sqrt{\frac{R_1C_1}{R_2C_2}} + \sqrt{\frac{R_1C_2}{R_2C_1}} + (1 - \mu)\sqrt{\frac{R_2C_2}{R_1C_1}} = 2\zeta \tag{14–10}$$

To select the element values in Eq. (14–10), we use methods similar to those used to design our low-pass building block.

---

1 Both circuits belong to a family of circuits originally proposed by R. P. Sallen and E. L. Key in "A Practical Method of Designing *RC* Active Filters," *IRE Transactions on Circuit Theory,* Vol. CT-2, pp. 74–85, 1955. In 1955 the controlled sources were obtained using vacuum tubes.

The *equal element method* requires that $R_1 = R_2 = R$ and $C_1 = C_2 = C$. Inserting these conditions in Eq. (14–10) leads to

$$RC = \frac{1}{\omega_0} \quad \text{and} \quad \mu = 3 - 2\zeta \tag{14–11}$$

Under this method we select values of R (or C) and R_B, and solve for C (or R) and $R_A = (\mu - 1)R_B$. The infinite-frequency gain achieved by this method is $|T(\infty)| = \mu = 3 - 2\zeta$.

The *unity gain method* requires that $C_1 = C_2 = C$ and $\mu = 1$. Inserting these conditions in Eq. (14–10) gives

$$C\sqrt{R_1 R_2} = \frac{1}{\omega_0} \quad \text{and} \quad \frac{R_1}{R_2} = \zeta^2 \tag{14–12}$$

Using this method, we select a value of R_2 and calculate $R_1 = \zeta^2 R_2$ and $C = (\omega_0\sqrt{R_1 R_2})^{-1}$. To get a gain $\mu = 1$, we make the noninverting OP AMP circuit into a voltage follower. That is, we replace R_A by a short circuit and R_B by an open circuit, thereby eliminating the need for these two resistors. Obviously the infinite-frequency gain achieved by this method is $|T(\infty)| = \mu = 1$.

The equal element and unity gain methods are alternative ways to design an active high-pass filter with prescribed values of ω_0 and ζ. As we found in the low-pass case, the passband gains produced by these methods are predetermined and are not adjustable design parameters. An additional gain correction stage may be needed when ω_0, ζ, and the infinite-frequency gain are all three prescribed.

DESIGN EXAMPLE 14–2

Construct a second-order high-pass transfer function with a corner frequency at $\omega_0 = 20$ krad/s, an infinite frequency gain of 0 dB, and a corner frequency gain of -3 dB. Design a second-order circuit that realizes this transfer function.

SOLUTION:

The required transfer function has the form

$$T(s) = \frac{K(s/20000)^2}{(s/20000)^2 + 2\zeta(s/20000) + 1}$$

The infinite frequency gain is $T(\infty) = K$ and the corner frequency gain is $|T(j\omega_0)| = K/2\zeta$. A gain of 0 dB at infinite frequency requires $K = 1$. A gain of -3 dB at the corner frequency requires $|T(j\omega_0)| = 1/\sqrt{2}$, which in turn requires $\zeta = 1/\sqrt{2}$.

Equal Element Design: Substituting $\omega_0 = 20$ krad/s and $\zeta = 1/\sqrt{2}$ into Eq. (14–11) yields $RC = 5 \times 10^{-5}$ s and $\mu = 3 - \sqrt{2} = 1.586$. Selecting $C =$ 5 nF and $R_B = 50$ kΩ requires $R = 10$ kΩ and $R_A = (\mu - 1)R_B = 29.3$ kΩ. The high-frequency gain of this design is $|T(\infty)| = \mu = 1.586$, which is more

FIGURE 14–7

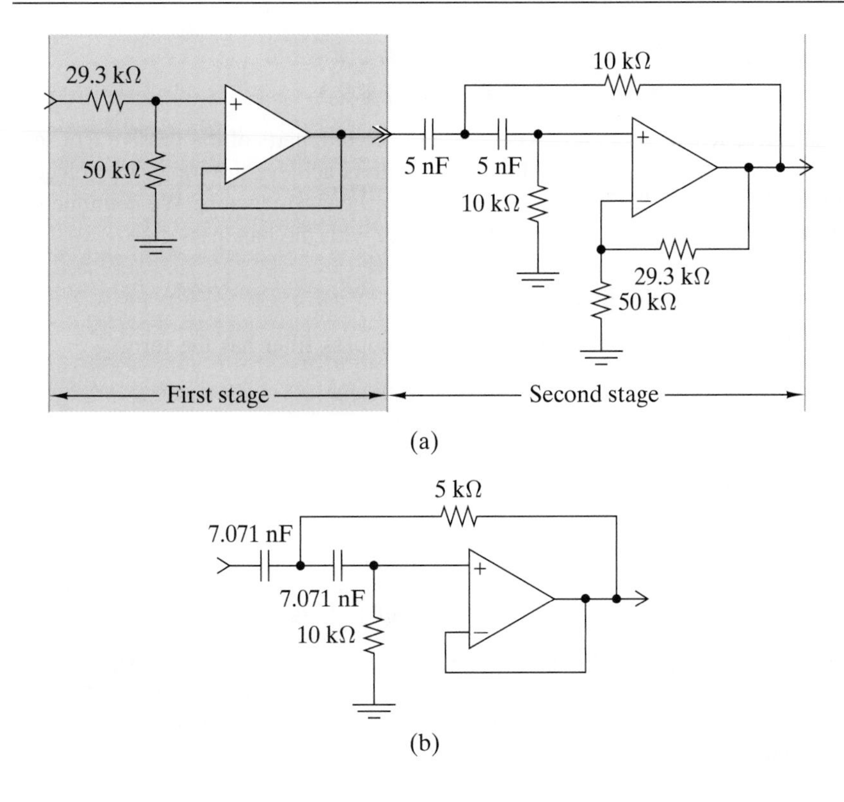

than the prescribed value of one (0 dB). We add a gain correction stage with a gain of $1/1.586 = 0.6305$ to bring the overall gain down to 0 dB. Figure 14–7(a) shows the resulting two-stage design that meets all three parts of the specification.

Unity Gain Design: Substituting $\omega_0 = 20$ krad/s and $\zeta = 1/\sqrt{2}$ into Eq. (14–12) yields $C\sqrt{R_1R_2} = 5 \times 10^{-5}$ s and $R_1 = 0.5R_2$. Selecting $R_2 = 10$ kΩ dictates that $R_1 = 5$ kΩ and $C = 7.071$ nF. The high-frequency gain of this design is $|T(\infty)| = 1$, which matches the prescribed 0 dB. The single-stage design in Figure 14–7(b) meets all three parts of the specification without a gain correction stage. ■

Exercise 14–3

Rework the unity gain design in Example 14–2 with $C = 20$ nF.

Answers: $R_1 = 1.77$ kΩ; $R_2 = 3.54$ kΩ

Exercise 14–4

Construct a second-order high-pass transfer function with a corner frequency of 20 rad/s, an infinite-frequency gain of 4, and a gain of 2 at the corner frequency.

Answer:

$$T(s) = \frac{4s^2}{s^2 + 40s + 400}$$

14–3 SECOND-ORDER BANDPASS AND BANDSTOP FILTERS

The active *RC* filters in this section are the counterparts of the passive *RLC* circuits studied in Chapter 12. These active filters achieve the frequency selectivity of the passive *RLC* circuits without the need for inductors. We continue our development of active filter building blocks with the bandpass case.

SECOND-ORDER BANDPASS FILTERS

The transfer function of a second-order bandpass filter has the form

$$T(s) = \frac{K(s/\omega_0)}{(s/\omega_0)^2 + 2\zeta(s/\omega_0) + 1} \tag{14–13}$$

This transfer function has two poles and a zero at $s = 0$. The gain response is found in the usual way as

$$|T(j\omega)| = \frac{|K||(\omega/\omega_0)|}{\sqrt{[1 - (\omega/\omega_0)^2]^2 + 4\zeta^2(\omega/\omega_0)^2}} \tag{14–14}$$

At low frequency ($\omega << \omega_0$) the gain approaches $|T(j\omega)| \to |K|(\omega/\omega_0)$. At high frequency ($\omega >> \omega_0$) the gain approaches $|T(j\omega)| \to |K|(\omega_0/\omega)$. Figure 14–8 presents Bode plots of these gain asymptotes (solid lines) and the actual gain (dashed curves) for several values of ζ. The key points to remember are the following:

- The low- and high-frequency gain asymptotes intersect at $\omega = \omega_0$.
- For $\omega < \omega_0$ the slope of the asymptotic gain is +20 dB/decade.
- For $\omega > \omega_0$ the slope of the asymptotic gain is −20 dB/decade.
- The actual gain at $\omega = \omega_0$ is $|T(j\omega_0)| = |K|/2\zeta$.

Note again that the design parameter ω_0 locates the center frequency and ζ controls the actual gain near the center frequency.

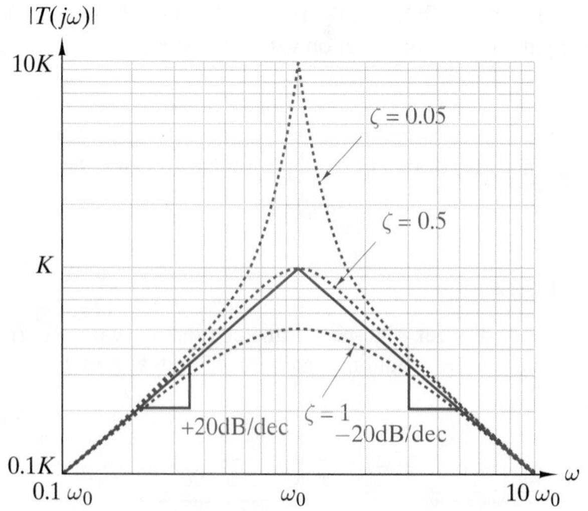

FIGURE 14–8 Second-order bandpass gain responses.

The circuit in Figure 14–9 has a transfer function of the form

$$T(s) = \frac{V_2(s)}{V_1(s)} = \frac{-R_2C_2s}{R_1R_2C_1C_2s^2 + (R_1C_1 + R_1C_2)s + 1} \qquad (14\text{–}15)$$

This circuit produces a second-order bandpass function. It has an OP AMP output, two capacitors, and two feedback paths—one via resistor R_2 and the other through the capacitor C_2. The dual feedback identifies this circuit as a member of the multiple-feedback family of active filters.[2]

Derivation of Eq. (14–15) using node analysis is easy, as the student can readily verify by working Problem 14–2.

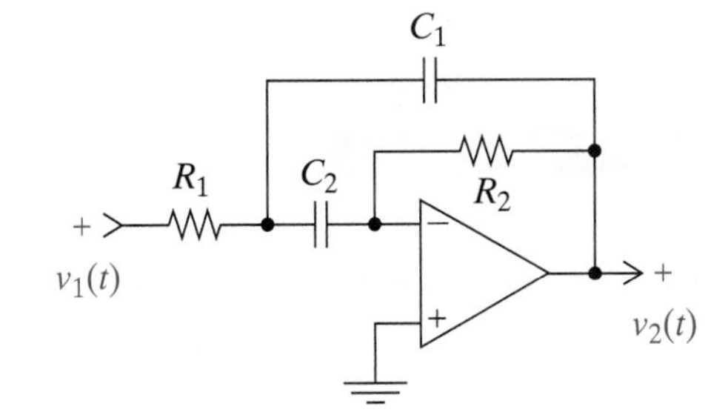

FIGURE 14–9 *Second-order bandpass circuit.*

Our design goal is to select element values to achieve specified values of ω_0 and ζ. Comparing the denominators in Eqs. (14–13) and (14–15) gives

$$\sqrt{R_1R_2C_1C_2} = 1/\omega_0 \quad \text{and} \quad R_1C_1 + R_1C_2 = 2\zeta/\omega_0$$

Using the first equation to eliminate ω_0 from the second equation produces

$$\sqrt{R_1R_2C_1C_2} = \frac{1}{\omega_0} \quad \text{and} \quad \sqrt{\frac{R_1C_1}{R_2C_2}} + \sqrt{\frac{R_1C_2}{R_2C_1}} = 2\zeta \qquad (14\text{–}16)$$

In this case there are two design constraints and four unknown element values.

An *equal-capacitor method* ($C_1 = C_2 = C$) can be used to reduce the number of unknowns. Inserting $C_1 = C_2 = C$ into Eq. (14–16) yields

$$\sqrt{R_1R_2}C = \frac{1}{\omega_0} \quad \text{and} \quad \frac{R_1}{R_2} = \zeta^2 \qquad (14\text{–}17)$$

Under this method we select a value of R_2 and solve for $R_1 = \zeta^2R_2$ and $C = (\omega_0\sqrt{R_1R_2})^{-1}$. Since this method uses $C_1 = C_2$, the center frequency gain found from Eq. (14–15) is $|T(j\omega_0)| = R_2/2R_1 = 1/2\zeta^2$. Note that the center frequency gain is greater than 1 when $\zeta < 1/\sqrt{2}$. For example, when $\zeta = 0.1$, the

2 For an extensive discussion of the multiple-feedback family, see Wai-Kai Chen, Ed., *The Circuits and Filters Handbook,* CRC Press, 1995, Chapter 76, pp. 2372 ff. The bandpass circuit in Figure 14–9 is sometimes called the Delyannis-Friend circuit. See M. E. Van Valkenburg, *Analog Filter Design*, Holt, Rinehart & Winston, Inc, 1982, p. 203.

gain is 50. This contrasts with the passive *RLC* bandpass circuit, whose center frequency gain is always 1.

The key descriptive parameters of a second-order bandpass filter are its **center frequency** ω_0 and **bandwidth** $B = 2\zeta\omega_0$. It is customary to add a third parameter called the **quality factor**, defined as $Q = \omega_0/B$. From this definition it is clear that Q and ζ are both dimensionless parameters related as $Q = 1/2\zeta$. Either parameter can be used to characterize filter bandwidth. When $Q > 1$ ($\zeta < 0.5$), the filter is said to be narrow-band because the bandwidth is less than the center frequency. When $Q < 1$ ($\zeta > 0.5$), the filter is said to be wide-band. The active bandpass building block developed here is best suited to narrow-band applications. The design of wide-band active bandpass filters is discussed in Sect. 14–7.

DESIGN EXAMPLE 14–3

Use the active *RC* circuit in Figure 14–9 to design a bandpass filter with a center frequency at 10 kHz and a bandwidth of 4 kHz. Find the center frequency gain for the design.

SOLUTION:

The specified filter parameters define the center frequency and bandwidth as

$$\omega_0 = 2\pi \times 10^4 = 62.8 \text{ krad/s} \quad \text{and} \quad B = 2\pi \times 4000 = 25.1 \text{ krad/s}$$

The required damping ratio is $\zeta = B/2\omega_0 = 0.2$. Inserting the values of ω_0 and ζ into Eq. (14–17) yields $R_1 = 0.04R_2$ and $C = 1.592 \times 10^{-5}/\sqrt{R_1R_2}$. Selecting $R_2 = 100\ \text{k}\Omega$ requires $R_1 = 4\ \text{k}\Omega$ and $C = 0.796$ nF. The center frequency gain for this design is $|T(j\omega_0)| = 1/2\zeta^2 = 12.5$. ■

Exercise 14–5

Rework the design in Example 14–3, starting with $C = 2$ nF.

Answers: $R_1 = 1.59\ \text{k}\Omega$; $R_2 = 39.8\ \text{k}\Omega$

Exercise 14–6

Construct a second-order bandpass transfer function with a corner frequency of 50 rad/s, a bandwidth of 10 rad/sec. and a center frequency gain of 4.

Answer:

$$T(s) = \frac{40s}{s^2 + 10s + 2500}$$

SECOND-ORDER BANDSTOP FILTERS

The transfer function of a second-order bandstop filter has the form

$$T(s) = \frac{K[(s/\omega_0)^2 + 1]}{(s/\omega_0)^2 + 2\zeta(s/\omega_0) + 1} \tag{14–18}$$

This transfer function has two zeros at $s = \pm j\omega_0$ along with the two poles defined by ζ and ω_0. The resulting gain response is

$$|T(j\omega)| = \frac{|K||1 - (\omega/\omega_0)^2|}{\sqrt{[1 - (\omega/\omega_0)^2]^2 + 4\zeta^2(\omega/\omega_0)^2}} \qquad (14\text{–}19)$$

At low frequency ($\omega \ll \omega_0$) and high frequency ($\omega \gg \omega_0$) the gain approaches $|T(0)| = |T(\infty)| \rightarrow |K|$. At $\omega = \omega_0$ the gain is $|T(j\omega_0)| = 0$. Figure 14–10 shows the gain predicted by Eq. (14–19) for some representative values of ζ. Key points to remember are the following:

- The low- and high-frequency gains are $|T(0)| = |T(\infty)| = |K|$.
- At $\omega = \omega_0$ there is a zero-gain notch in the gain response.

The notch is caused by the j-axis zeros in $T(s)$. From a design perspective, we say that the zeros at $s = \pm j\omega_0$ locate the notch and ζ controls the width of the notch.

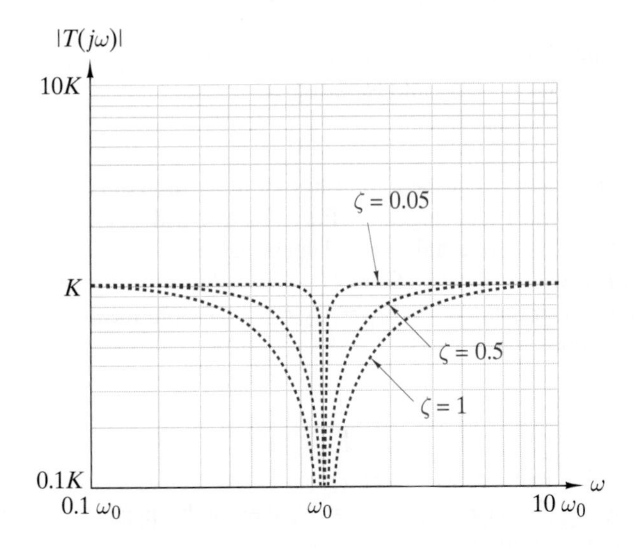

FIGURE 14–10 *Second-order bandstop gain responses.*

The circuit in Figure 14–11 is a modification of the bandpass circuit in Figure 14–9. The modified circuit retains the two feedback paths and adds a second input path via the voltage divider R_A and R_B. The two input and feedback paths combine to produce an overall circuit transfer function of

$$T(s) = \frac{V_2(s)}{V_1(s)} = \frac{R_B}{R_A + R_B}\left[\frac{R_1R_2C_1C_2s^2 + (R_1C_1 + R_1C_2 - R_2C_2R_A/R_B)s + 1}{R_1R_2C_1C_2s^2 + (R_1C_1 + R_1C_2)s + 1}\right] \qquad (14\text{–}20)$$

Derivation of this transfer function involves straightforward (if laborious) node analysis and is left for the enjoyment of the student in Problem 14–3.

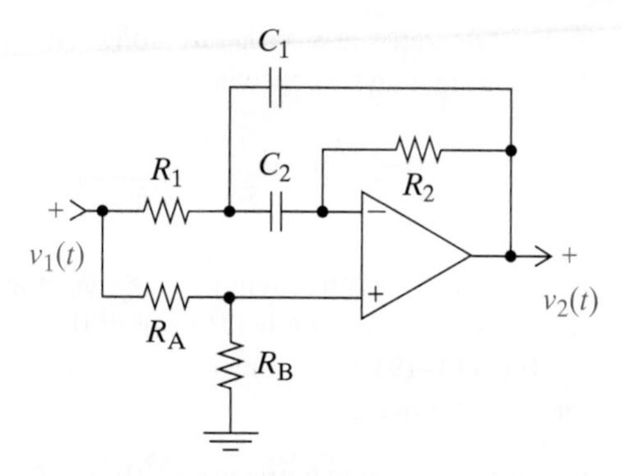

FIGURE 14–11 *Second-order bandstop circuit.*

The denominator in Eq. (14–20) is the same as the denominator of the bandpass filter in Eq. (14–15). Hence, the design constraints in Eq. (14–16) apply here as well, namely

$$\sqrt{R_1R_2C_1C_2} = \frac{1}{\omega_0} \quad \text{and} \quad \sqrt{\frac{R_1C_1}{R_2C_2}} + \sqrt{\frac{R_1C_2}{R_2C_1}} = 2\zeta \tag{14–21}$$

These constraints control the locations of the bandstop filter poles. Comparing the numerators in Eqs. (14–18) and (14–20) gives

$$R_1(C_1 + C_2) - R_2C_2R_A/R_B = 0 \tag{14–22}$$

This constraint ensures that the circuit has the requisite j-axis zeros to produce the bandstop notch. Taken together, Eqs. (14–21) and (14–22) give us three design constraints and five unknown element values.

The *equal-capacitor method* ($C_1 = C_2 = C$) reduces the number of unknowns in the design constraints. When $C_1 = C_2 = C$, the pole location constraints in Eq. (14–21) reduce to

$$\sqrt{R_1R_2}C = \frac{1}{\omega_0} \quad \text{and} \quad \frac{R_1}{R_2} = \zeta^2 \tag{14–23}$$

Using this design method, we select a value of R_2 and solve for $R_1 = \zeta^2R_2$ and $C = (\omega_0\sqrt{R_1R_2})^{-1}$. When $C_1 = C_2 = C$, the bandstop notch requirement in Eq. (14–22) reduces to

$$\frac{R_A}{R_B} = \frac{2R_1}{R_2} \tag{14–24}$$

An obvious way to meet this requirement is to let $R_A = 2R_1$ and $R_B = R_2$.

Second-order bandstop filters are customarily described in terms of the **notch frequency** ω_0 and the **notch bandwidth** $B = 2\zeta\omega_0$. The bandstop circuit in Figure 14–11 is best suited for narrow-band applications aimed at eliminating a narrow range of frequencies or even a single frequency. The design of wide-band bandstop filters is discussed in Sect. 14–7.

Narrow-band notch circuits are often used to eliminate power line noise at 60 Hz (50 Hz in many other countries). The next example illustrates such an application.

DESIGN EXAMPLE 14–4

Use the active *RC* circuit in Figure 14–11 to design a bandstop filter with a notch frequency at 60 Hz and a notch bandwidth of 12 Hz.

SOLUTION:

The specified filter parameters define the notch frequency and bandwidth as

$$\omega_0 = 2\pi \times 60 = 377 \text{ rad/s} \quad \text{and} \quad B = 2\pi \times 12 = 75.4 \text{ rad/s}$$

The required damping ratio is $\zeta = B/2\omega_0 = 0.1$. Using equal capacitors ($C_1 = C_2 = C$) and selecting $R_2 = 100\text{ k}\Omega$, we use Eqs. (14–23) and (14–24) to calculate the remaining values as follows:

$$R_1 = \zeta^2 R_2 = 1\text{ k}\Omega$$

$$C = 1/\sqrt{R_1 R_2}\omega_0 = 265\text{ nF}$$

$$R_A = 2R_1 = 2\text{ k}\Omega$$

$$R_B = R_2 = 100\text{ k}\Omega$$

■

Exercise 14–7

Rework the circuit design in Example 14–4 starting with $C_1 = C_2 = C = 200\text{ nF}$.

Answers: $R_1 = 1.33\text{ k}\Omega$; $R_2 = 133\text{ k}\Omega$; $R_A = 2.66\text{ k}\Omega$; $R_B = 133\text{ k}\Omega$

Exercise 14–8

Construct a second-order bandstop transfer function with a notch frequency of 50 rad/s, a notch bandwidth of 10 rad/sec, and passband gains of 5.

Answer:

$$T(s) = \frac{5(s^2 + 50^2)}{s^2 + 10s + 50^2}$$

14–4 LOW-PASS FILTER DESIGN

In this section we learn how to design multi-pole filters with prescribed low-pass characteristics. The Bode plot in Figure 14–12 shows how low-pass filter characteristics are specified. To satisfy the specification, the filter must be designed so that its gain response lies within the unshaded region. There are many different gain responses that meet this requirement, as illustrated by the two shown in the figure. All such gain responses are constrained by the conditions imposed in three adjacent frequency bands.

In the **passband** ($0 \leq \omega \leq \omega_C$) the gain response must be in the range

$$\frac{T_{MAX}}{\sqrt{2}} \leq |T(j\omega)| \leq T_{MAX}$$

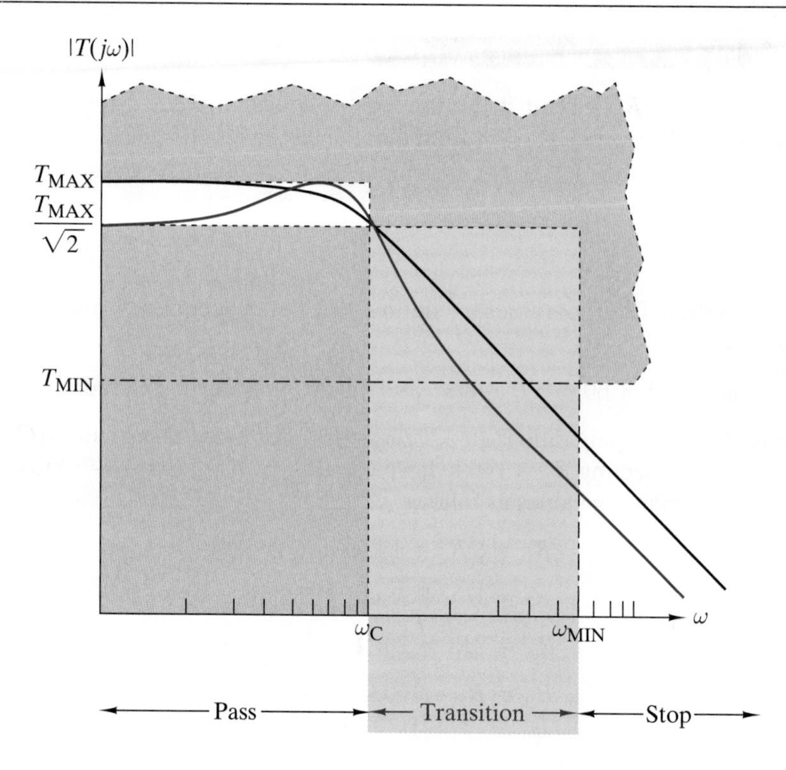

FIGURE 14–12 *Low-pass filter specification and responses.*

In the **stopband** ($\omega_{MIN} \leq \omega$) the gain response must be in the range $|T(j\omega)| \leq T_{MIN}$. In the **transition band** ($\omega_C \leq \omega \leq \omega_{MIN}$) between the passband and stopband the gain response must change dramatically. The required change is defined by two ratios: T_{MAX}/T_{MIN} defines how much the gain must change and ω_{MIN}/ω_C specifies how rapidly the change must occur.

In summary, a low-pass filter design problem is defined by specifying the four parameters: T_{MAX}, ω_C, T_{MIN}, and ω_{MIN}. Transfer functions (there may be several) that meet these requirements are of the form

$$T(s) = \frac{K}{q_n(s)}$$

where $q_n(s)$ is an nth-order polynomial whose roots define the poles of $T(s)$. From our previous design experience, we can see that filter design involves two related tasks: (1) selecting $q_n(s)$ and K so that the $|T(j\omega)|$ meets the filter specification and (2) devising a circuit that realizes the transfer function $T(s)$. We will discuss three methods of dealing with the first task. The second task is accomplished using our active filter building blocks to design a cascade circuit.

FIRST-ORDER CASCADE FILTERS

A simple way to produce a multi-pole low-pass filter is to connect n identical first-order low-pass filters in cascade. The transfer function of such a cascade connection is written as

$$T(s) = \underbrace{\left[\frac{K}{s/\alpha + 1}\right] \times \left[\frac{K}{s/\alpha + 1}\right] \times \ldots \left[\frac{K}{s/\alpha + 1}\right]}_{n \text{ stages}} = \frac{K^n}{(s/\alpha + 1)^n} \quad (14\text{–}25)$$

To design such a filter, we must select K, α, and n such that the gain response of this transfer function meets a filter specification defined by T_{MAX}, T_{MIN}, ω_C, and ω_{MIN}.

The gain response of the transfer function in Eq. (14–25) is

$$|T(j\omega)| = \frac{|K|^n}{\left[\sqrt{1 + (\omega/\alpha)^2}\right]^n} \tag{14–26}$$

The maximum gain in Eq. (14–26) occurs at $\omega = 0$, where the gain is $|T(0)| = |K|^n$. To meet the passband requirement of the specification, we set $|K|^n = T_{MAX}$ and evaluate the gain in Eq. (14–26) at the prescribed cutoff frequency ω_C.

$$|T(j\omega_C)| = \frac{T_{MAX}}{\left[\sqrt{1 + (\omega_C/\alpha)^2}\right]^n} = \frac{T_{MAX}}{\sqrt{2}}$$

Equating the denominators in this equation and then solving for α yields

$$\alpha = \frac{\omega_C}{\sqrt{2^{1/n} - 1}} \tag{14–27}$$

This equation relates the cutoff frequency of each first-order stage (α) to the cutoff frequency of a cascade connection (ω_C) of n first-order stages. Each of these stages has a gain of $K = (T_{MAX})^{1/n}$ and a cutoff frequency of α. To complete a design, we need to know the number of stages required.

Figure 14–13 shows normalized gain responses defined by Eq. (14–26) for $n = 1$ to $n = 8$, where $T_{MAX} = |K|^n$ and α is given by Eq. (14–27). All of these responses meet the passband requirements, and, more importantly, the

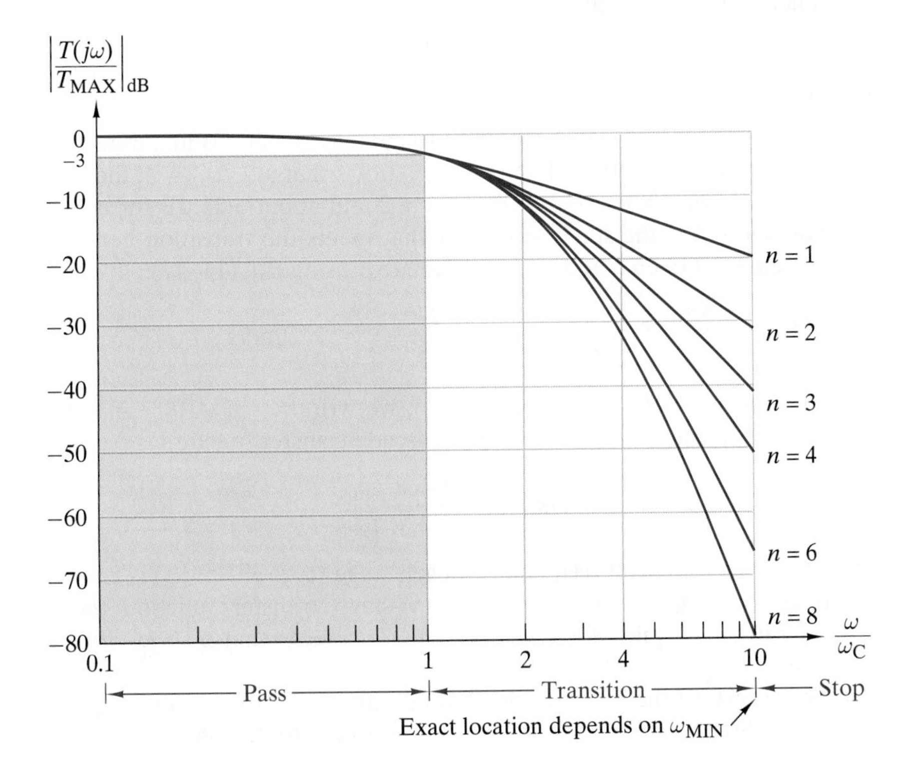

FIGURE 14–13 *First-order low-pass cascade filter responses.*

transition band gain decreases as we increase n. Obviously, we can improve the transition band performance of $T(s)$ by increasing n, but this increase adds more stages to the cascade circuit that realizes $T(s)$. There is a trade-off between transition band performance and circuit complexity. What we need to know is the smallest value of n that meets the filter specification.

We can estimate the smallest value of n from the gain plots in Figure 14–13. For example, suppose the transition band gain must decrease by 40 dB ($T_{MAX}/T_{MIN} = 100$) in the decade above cutoff ($\omega_{MIN}/\omega_C = 10$). In Figure 14–13 we see that at $\omega/\omega_C = 10$, the normalized gain $|T(j\omega)|/T_{MAX}$ is -32 dB for $n = 2$ and -42 dB for $n = 3$. The $n = 3$ curve identifies the smallest value of n that reduces the gain by at least 40 dB in a one-decade transition band.

In summary, given values of T_{MAX}, T_{MIN}, ω_C, and ω_{MIN}, we construct a first-order cascade transfer function as follows. We use Figure 14–13 (or trial and error) to find the smallest integer n that meets the transition band requirements defined by T_{MAX}/T_{MIN} and ω_{MIN}/ω_C. Given n and ω_C, we calculate α using Eq. (14–27), $K = (T_{MAX})^{1/n}$ and use Eq. (14–25) to get the required transfer function. The transfer function is partitioned into a product of n identical first-order functions, each of which is realized using a first-order low-pass circuit. The next example illustrates this design procedure.

DESIGN EXAMPLE 14–5

(a) Construct a first-order cascade transfer function that meets the following requirements: $T_{MAX} = 10$ dB, $\omega_C = 200$ rad/s, $T_{MIN} = -10$ dB, and $\omega_{MIN} = 800$ rad/s.

(b) Design a cascade of active RC circuits that realizes the transfer function developed in (a).

SOLUTION:

(a) The specification requires the gain to decrease by 20 dB in a transition band with $\omega_{MIN}/\omega_C = 4$. Figure 14–13 shows that at $\omega/\omega_C = 4$, the normalized gain is about -17 dB for $n = 2$ and about -22 dB for $n = 3$. Thus, $n = 3$ is the smallest integer that meets the transition band requirement. Given n and ω_C, we calculate α using Eq. (14–27):

$$\alpha = \frac{\omega_C}{\sqrt{2^{1/n} - 1}} = \frac{200}{\sqrt{2^{1/3} - 1}} = 392 \text{ rad/s}$$

Since $T_{MAX} = 10$ dB (factor of $\sqrt{10}$), we write $K = (\sqrt{10})^{1/3} = 1.468$. So, finally, the required first-order cascade transfer function is

$$T(s) = \left(\frac{1.468}{s/392 + 1}\right)^3$$

Note that the cutoff frequency of each stage ($\alpha = 392$ rad/s) is greater than the cutoff frequency of the n-stage transfer function ($\omega_C = 200$ rad/s). A quick look at Eq. (14–27) reveals that $\alpha > \omega_C$ for all $n > 1$.

(b) The transfer function developed in (a) can be partitioned into a product of three identical first-order functions of the form $1.468/(s/392 + 1)$.

Each of these can be realized using the first-order low-pass circuit in Figure 14–14. The transfer function of this circuit is

$$T(s) = \frac{1 + R_A/R_B}{RCs + 1}$$

The design constraints for each stage are $1 + R_A/R_B = 1.468$ and $RC = 1/392$. Selecting $R_B = R = 100\ \text{k}\Omega$ leads to $R_A = 46.8\ \text{k}\Omega$ and $C = 25.5\ \text{nF}$. A cascade connection of three such circuits produces a low-pass filter that meets the design requirements.

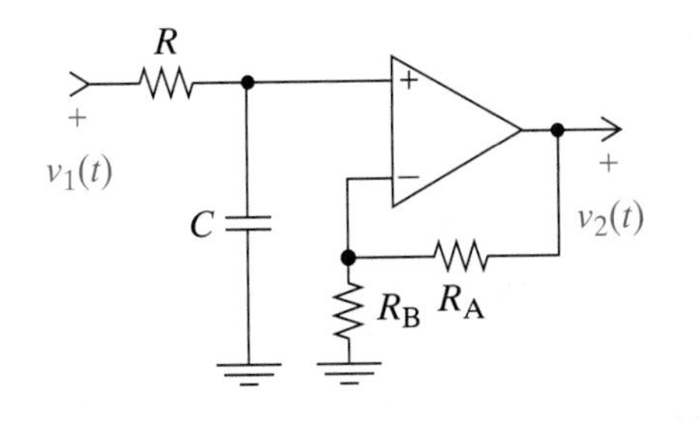

FIGURE 14–14 *First-order low-pass circuit.*

■

Exercise 14–9

Construct a first-order cascade transfer function that meets the following requirements: $T_{MAX} = 0$ dB, $T_{MIN} = -30$ dB, $\omega_C = 200$ rad/s, and $\omega_{MIN} = 1$ krad/s.

Answer:

$$T(s) = \left(\frac{460}{s + 460}\right)^4$$

BUTTERWORTH LOW-PASS FILTERS

All Butterworth low-pass filters have transfer functions of the form $T(s) = K/q_n(s)$, whose gain response is

$$|T(j\omega)| = \frac{|K|}{\sqrt{1 + (\omega/\omega_C)^{2n}}} \tag{14–28}$$

where ω_C is the cutoff frequency and n is the order of the denominator polynomial, that is, the number of poles. By inspection, the maximum gain occurs at dc, where $|T(0)| = |K| = T_{MAX}$. At the cutoff frequency the gain is $|T(j\omega_C)| = |K|/\sqrt{2} = T_{MAX}/\sqrt{2}$ for all values of n. When we make $K = T_{MAX}$, the gain response in Eq. (14–28) meets the passband requirements of a filter specification for all values of n. At high frequency ($\omega \gg \omega_C$) the gain approaches an asymptote of $|K|(\omega_C/\omega)^n$, which has a slope of $-20n$ dB/dec. Thus, we can decrease the gain in the transition band by increasing the number of poles.

Figure 14–15 compares Butterworth and first-order cascade gain responses for $n = 4$. Both responses have high-frequency asymptotes whose slopes are -80 dB/dec. However, the Butterworth gain approaches its asymptote at a lower frequency, so it has less gain in the transition band. The reduced gain means that the Butterworth response has better transition band performance than the first-order cascade.

Figure 14–16 shows normalized plots of the Butterworth gain responses defined by Eq. (14–28), where $T_{MAX} = |K|$. Clearly the transition band gain decreases

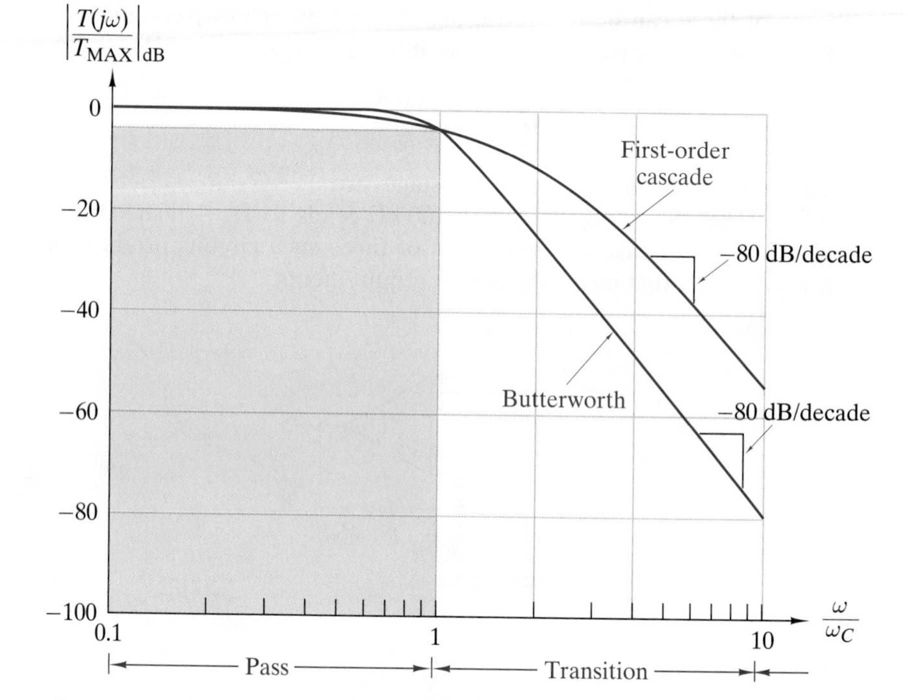

FIGURE 14–15 *First-order cascade and Butterworth low-pass filter responses for* n = 4.

rapidly as we increase n, the filter order. Increasing n means that $T(s) = K/q_n(s)$ has more poles, which in turn means more stages in a cascade circuit realizing $T(s)$. As we discovered with the first-order cascade filter, there is a trade-off between reducing the transition band gain and circuit complexity. What we need to know is the smallest value of n that meets a given filter specification.

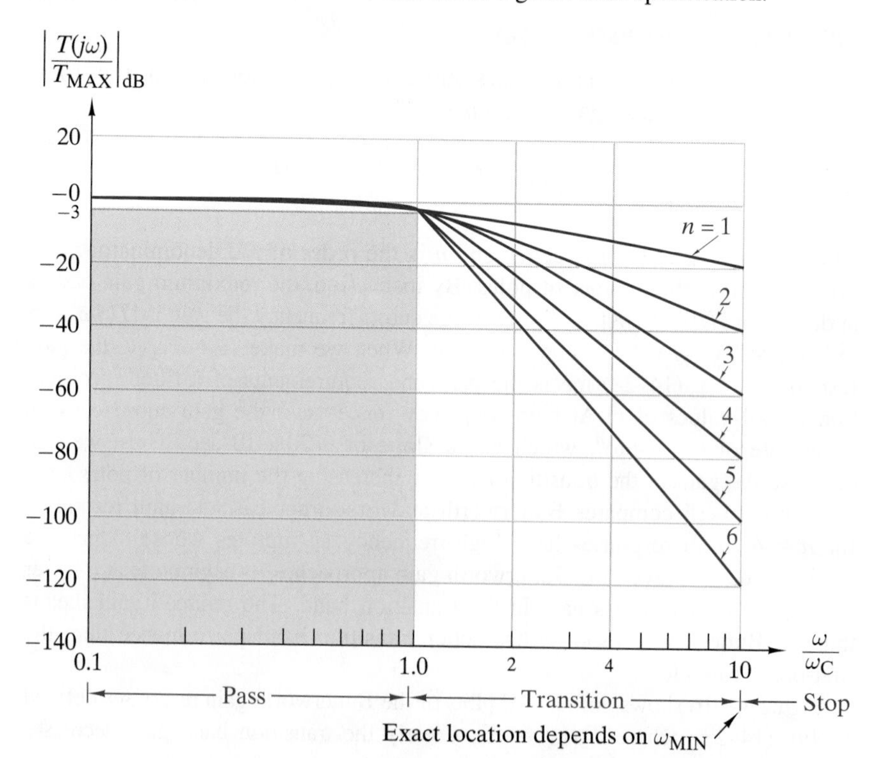

FIGURE 14–16 *Butterworth low-pass filter responses.*

Evaluating the gain in Eq. (14–28) at $\omega = \omega_{MIN}$ with $T_{MAX} = |K|$ produces the inequality

$$|T(j\omega_{MIN})| = \frac{T_{MAX}}{\sqrt{1 + (\omega_{MIN}/\omega_C)^{2n}}} \leq T_{MIN}$$

Solving the inequality for n yields the constraint

$$n \geq \frac{1}{2} \frac{\ln\left[(T_{MAX}/T_{MIN})^2 - 1\right]}{\ln\left[\omega_{MIN}/\omega_C\right]} \tag{14–29}$$

The right side of this equation is a lower bound on the filter order n. Note that the lower bound is determined by the transition band ratios T_{MAX}/T_{MIN} and ω_{MIN}/ω_C. In other words, the smallest n depends on how much and how rapidly the gain must change in the transition band.

For example, if the transition band gain must decrease by 30 dB ($T_{MAX}/T_{MIN} = 10^{3/2}$) in the two octaves above cutoff ($\omega_{MIN}/\omega_C = 4$), then Eq. (14–29) yields

$$n \geq \frac{1}{2} \frac{\ln\left[(10^{3/2})^2 - 1\right]}{\ln[4]} = 2.49$$

Since filter order must be an integer, the smallest value is $n = 3$. The gain plots in Figure 14–16 confirm this result, since a normalized gain below -30 dB at $\omega/\omega_C = 4$ cannot be achieved by any order less than $n = 3$. For Butterworth low-pass filters, the smallest value of n can be determined analytically using Eq. (14–29) or graphically using Figure 14–16.

Once the required value of n is known, we need an nth-order polynomial $q_n(s)$ to construct a transfer function with Butterworth gain characteristics. A method of obtaining such polynomials is given in Appendix D, and using that method we generated the normalized ($\omega_C = 1$) polynomials $q_n(s)$ in Table 14–1. All of these polynomials have the property that $q_n(0) = 1$.

TABLE 14–1 **NORMALIZED POLYNOMIALS THAT PRODUCE BUTTERWORTH RESPONSES**

ORDER	NORMALIZED DENOMINATOR POLYNOMIALS
1	$(s + 1)$
2	$(s^2 + 1.414s + 1)$
3	$(s + 1)(s^2 + s + 1)$
4	$(s^2 + 0.7654s + 1)(s^2 + 1.848s + 1)$
5	$(s + 1)(s^2 + 0.6180s + 1)(s^2 + 1.618s + 1)$
6	$(s^2 + 0.5176s + 1)(s^2 + 1.414s + 1)(s^2 + 1.932s + 1)$

Using these polynomials, we obtain an nth-order Butterworth transfer function with a dc gain of K and cutoff frequency of ω_C as

$$T(s) = \frac{K}{q_n(s/\omega_C)} \tag{14–30}$$

where $q_n(s/\omega_C)$ is the nth-order polynomial in Table 14–1 with s replaced by s/ω_C. Since $q_n(0) = 1$, the dc gain of this transfer function is $|T(0)| = |K|$ and the poles of $T(s)$ are roots of the equation $q_n(s/\omega_C) = 0$.

For example, for $n = 3$ the polynomial in Table 14–1 is $q_3(s) = (s + 1)(s^2 + s + 1)$. A third-order Butterworth low-pass transfer function with a dc gain of 20 dB ($K = 10$) and a cutoff frequency of $\omega_C = 1$ krad/s is written as

$$T(s) = \frac{10}{q_3(s/1000)} = \frac{10}{(s/1000 + 1)[(s/1000)^2 + s/1000 + 1]}$$

This transfer function has a real pole at $s = -1000$ rad/s and a pair of complex poles with $\omega_0 = 1000$ rad/s and $\zeta = 0.5$. The three poles are all located a distance of $\omega_C = 1000$ from the origin in the s plane. This illustrates a general principle of Butterworth poles—*Butterworth poles are all located on a circle of radius ω_C in the left-half of the s plane.*

Once we have a transfer function $T(s)$ whose gain meets the filter specification, we partition it into a product of first- and second-order functions. Each of these functions is then realized using one of our active filter building blocks. A cascade connection of these building blocks then produces the required transfer function. The next example illustrates the procedure.

DESIGN EXAMPLE 14–6

(a) Construct a Butterworth low-pass transfer function that meets the following requirements:

$T_{MAX} = 20$ dB, $\omega_C = 1$ krad/s, $T_{MIN} = -20$ dB, and $\omega_{MIN} = 4$ krad/s.

(b) Design a cascade of active RC circuits that realizes the transfer function found in (a).

SOLUTION:

(a) The specification requires the gain to decrease by 40 dB in the transition band between $\omega_C = 1$ krad/s and $\omega_{MIN} = 4$ krad/s. The transition band ratios for this specification are $T_{MAX}/T_{MIN} = 100$ and $\omega_{MIN}/\omega_C = 4$. Inserting these ratios in Eq. (14–29) yields

$$n \geq \frac{1}{2}\frac{\ln[100^2 - 1]}{\ln[4]} = 3.32$$

Thus, $n = 4$ is the lowest-order Butterworth filter that meets the filter specification. Using $K = T_{MAX} = 10$ and $q_4(s)$ from Table 14–1, the required fourth-order Butterworth low-pass transfer function is

$$T(s) = \frac{K}{q_4(s/1000)}$$

$$= \frac{10}{\left[\left(\frac{s}{1000}\right)^2 + 0.7654\left(\frac{s}{1000}\right) + 1\right]\left[\left(\frac{s}{1000}\right)^2 + 1.848\left(\frac{s}{1000}\right) + 1\right]}$$

(b) The fourth-order Butterworth transfer function in (a) can be partitioned as

$$T(s) = T_1(s)T_2(s)$$

$$= \left[\frac{K_1}{\left(\frac{s}{1000}\right)^2 + 0.7654\left(\frac{s}{1000}\right) + 1}\right]\left[\frac{K_2}{\left(\frac{s}{1000}\right)^2 + 1.848\left(\frac{s}{1000}\right) + 1}\right]$$

where K_1 and K_2 are to be determined. This partitioning calls for a cascade of two second-order low-pass building blocks.

Figure 14–17 shows a design sequence for circuits that realize these transfer functions. The first two rows in the figure show the required transfer functions and the stage parameters ω_0 and ζ. The stage protoype in the third row is the low-pass building block circuit in Figure 14–3. Using the equal element design method for this circuit [see Eq. (14–5)] leads to the design constraints in the fourth row. The element values selections in the fifth row produce the two-stage design in the last row of Figure 14–17.

Item	First Stage	Second Stage
Prototype transfer function	$\dfrac{K_1}{(s/1000)^2 + 0.7654(s/1000) + 1}$	$\dfrac{K_2}{(s/1000)^2 + 1.848(s/1000) + 1}$
Stage parameters	$\omega_0 = 1000 \quad \zeta = 0.7654/2 = 0.3827$	$\omega_0 = 1000 \quad \zeta = 1.848/2 = 0.924$
Stage prototype	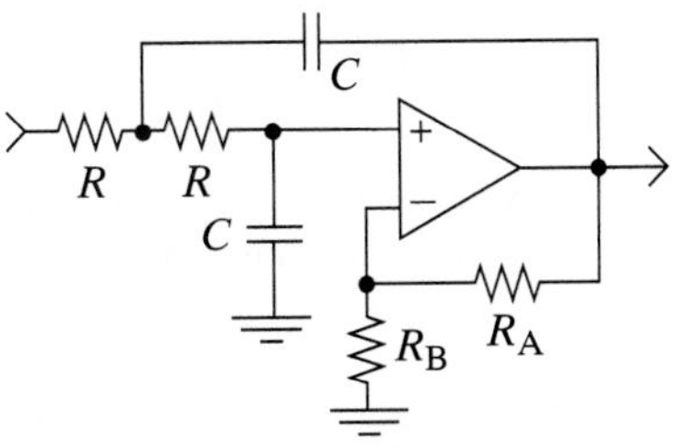 R, R, C, C, R_B, R_A	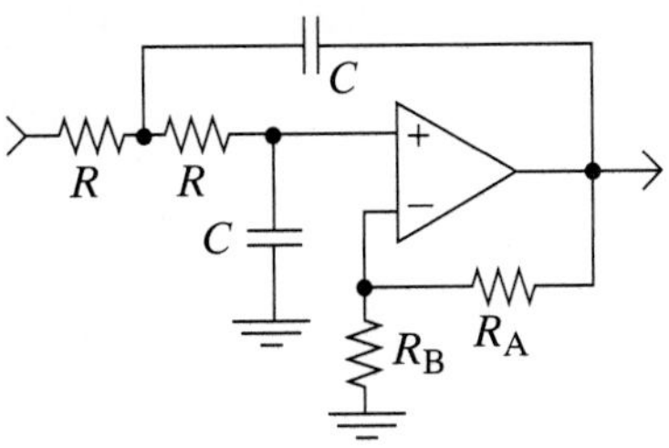 R, R, C, C, R_B, R_A
Design constraints	$RC = \dfrac{1}{\omega_0} = 0.001$ $K_1 = 3 - 2\zeta = 2.2346$ $R_A = (K_1 - 1)R_B$	$RC = \dfrac{1}{\omega_0} = 0.001$ $K_2 = 3 - 2\zeta = 1.152$ $R_A = (K_2 - 1)R_B$
Element values	Let $R = 100$ kΩ, then $C = 10$ nF Let $R_B = 100$ kΩ, then $R_A = 123$ kΩ	Let $R = 100$ kΩ, then $C = 10$ nF Let $R_B = 100$ kΩ, then $R_A = 15.2$ kΩ
Final designs	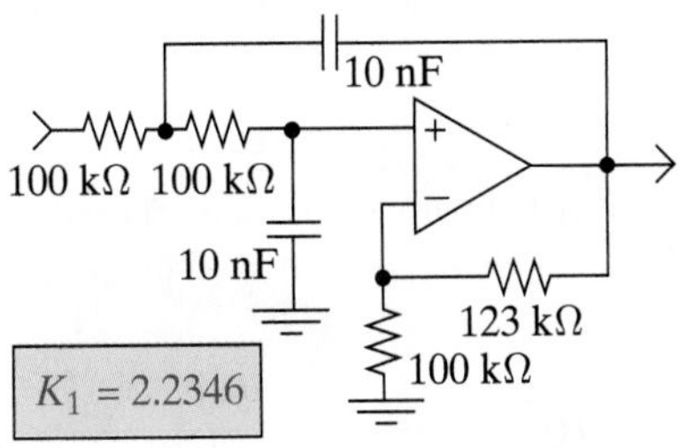 10 nF, 100 kΩ, 100 kΩ, 10 nF, 123 kΩ, 100 kΩ $K_1 = 2.2346$	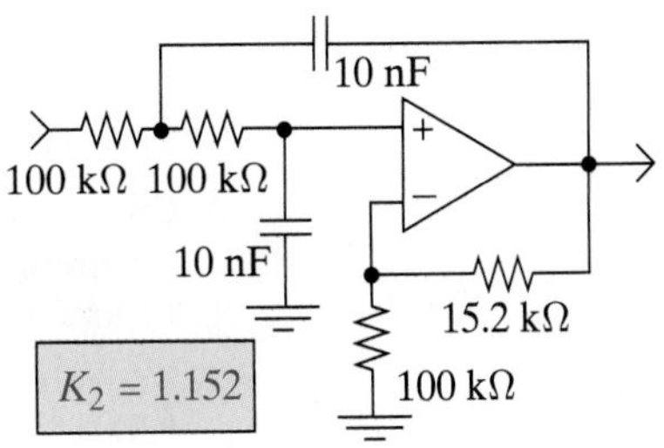 10 nF, 100 kΩ, 100 kΩ, 10 nF, 15.2 kΩ, 100 kΩ $K_2 = 1.152$

FIGURE 14–17 *Design sequence for Example 14–6.*

As discussed in Sect. 14–2, the equal element design method does not allow us to control K_1 and K_2. The values achieved in our two-stage design are $K_1 = 2.235$ and $K_2 = 1.152$. Together the two stages produce an overall dc gain of $K = K_1K_2 = 2.235 \times 1.152 = 2.575$, which is less than the $K = 10$ required by the specification. We deal with this shortfall by adding a gain correction stage with a gain of $10/2.575 = 3.883$. Figure 14–18 shows the final three-stage design with the added gain correction stage. ■

FIGURE 14–18

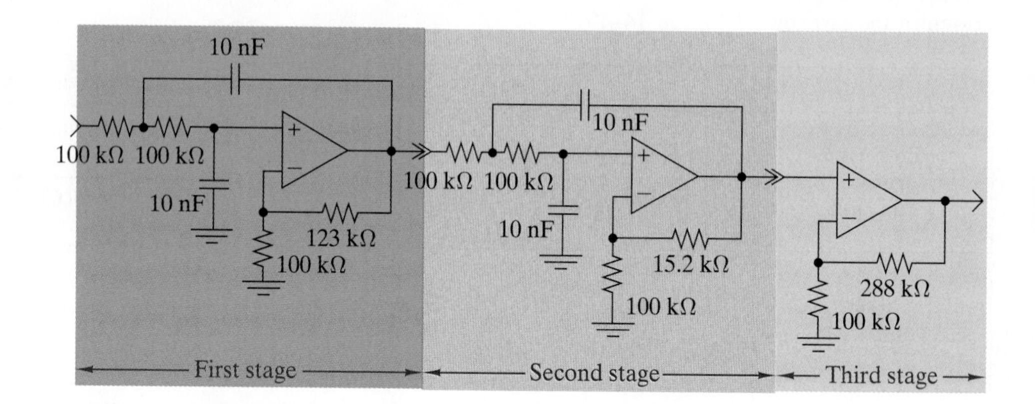

Exercise 14–10

Rework the design in Example 14–6 using the unity gain method discussed in Sect. 14–2 [see Eq. (14–6)] to design the two second-order low-pass circuits.

Answer: Figure 14–19 shows the unity gain method redesign of the second-order stages in Figure 14–18.

FIGURE 14–19

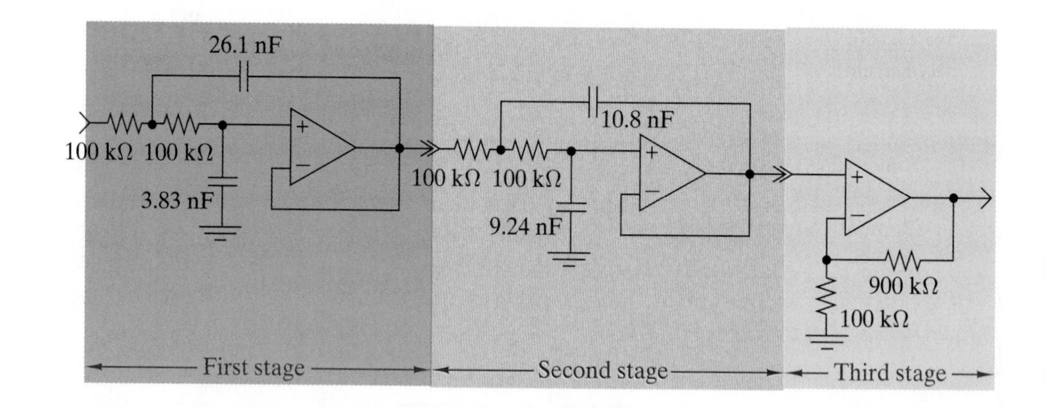

Exercise 14–11

Construct a Butterworth low-pass transfer function that meets the following requirements: $T_{MAX} = 0$ dB, $T_{MIN} = -40$ dB, $\omega_C = 250$ rad/s, and $\omega_{MIN} =$ 1.5 krad/s.

Answer:

$$T(s) = \frac{250^3}{(s + 250)(s + 250s + 250^2)}$$

CHEBYCHEV LOW-PASS FILTERS

All Chebychev low-pass filters have a gain response of the form

$$|T_n(j\omega)| = \frac{|K|}{\sqrt{1 + C_n^2(\omega/\omega_C)}} \tag{14–31}$$

where $C_n(x)$ is an nth-order Chebychev polynomial defined by

$$C_n(x) = \cos[n \times \cos^{-1}(x)] \quad x \leq 1 \tag{14–32a}$$

and

$$C_n(x) = \cosh[n \times \cosh^{-1}(x)] \quad x > 1 \tag{14–32b}$$

In the passband ($x = \omega/\omega_C \leq 1$) $C_n(x)$ is the cosine function in Eq. (14–32a) and the term $1 + C_n^2(\omega/\omega_C)$ in the denominator of Eq. (14–31) varies between 1 (when $C_n = 0$) and 2 (when $C_n = \pm 1$). This means that in the passband the gain $|T(j\omega)|$ in Eq. (14–31) varies between $|K|$ and $|K|/\sqrt{2}$. Thus, the Chebychev gain variation remains within standard passband bounds for all values of n. However, the passband gain variation is oscillatory rather than smooth and steady like the Butterworth response.

Figure 14–20 shows plots of Eq. (14–31) for $n = 4$ and $n = 5$. The oscillatory variation of the term $1 + C_n^2(\omega/\omega_C)$ produces a sequence of resonant peaks and valleys given the descriptive name *ripple*. The gain at the top of every peak $|K|$ and the gain at the bottom of every valley is $|K|/\sqrt{2}$. The Chebychev passband gain is called an *equal-ripple response* because the upper bound is always $|K|$ and the lower bound $|K|/\sqrt{2}$. Because of the equal-ripple response, the Chebychev gain has $T_{MAX} = |K|$ and $|T(j\omega_C)| = |K|/\sqrt{2}$ for all values of n.

However, the Chebychev dc gain is $|K|$ when n is odd and $|K|/\sqrt{2}$ when n is even. In other words, T_{MAX} does not occur at dc when n is even. We must account for this difference in dc gain when constructing low-pass transfer functions with Chebychev poles.

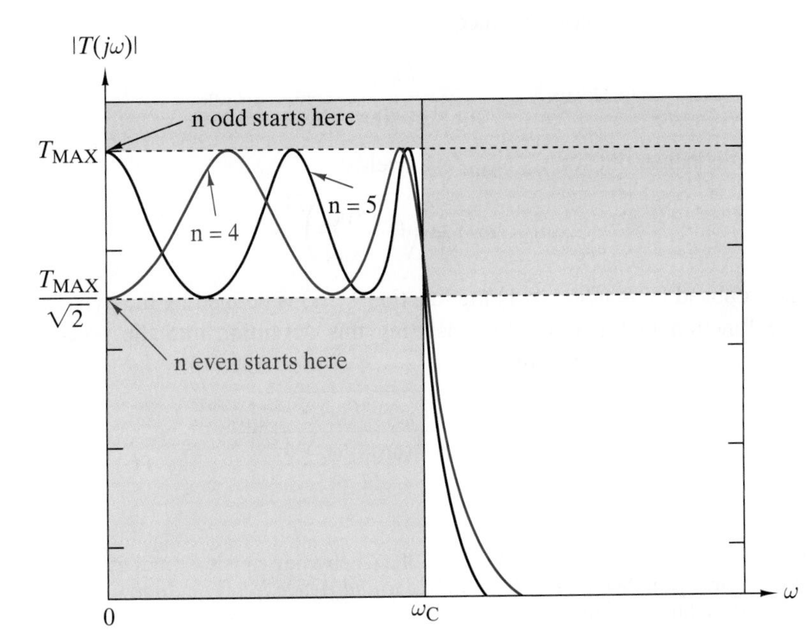

FIGURE 14–20 *Chebychev passband gain responses.*

Figure 14–21 compares the Butterworth and Chebychev gain responses for $n = 4$. Both responses have high-frequency asymptotes with slopes of -80 dB/dec. The Butterworth response is relatively flat in the passband and sedately approaches its high-frequency asymptote in the transition band. In contrast, the last resonant peak in the Chebychev passband produces an initial slope in the transition band that is steeper than -80 dB/dec.[3] As a result, a Chebychev response has less gain in the transition band than a Butterworth response of the same order.

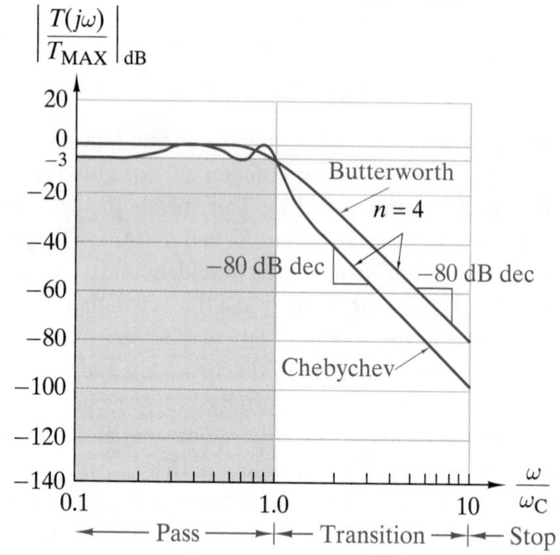

FIGURE 14–21 *Butterworth and Chebychev low-pass filter responses for* n = 4.

Figure 14–22 shows normalized plots of the Chebychev gain responses with $T_{MAX} = |K|$. As with Butterworth responses, the Chebychev transition band gain decreases rapidly as we increase *n*, the filter order. Again to minimize circuit complexity, we need to know the smallest value of *n* that meets a given filter specification. Evaluating the Chebychev gain in Eq. (14–31) at $\omega = \omega_{MIN}$ with $T_{MAX} = |K|$ produces the inequality

$$|T(j\omega_{MIN})| = \frac{T_{MAX}}{\sqrt{1 + C_n^2(\omega_{MIN}/\omega_C)}} \leq T_{MIN}$$

Solving this constraint for $C_n(\omega_{MIN}/\omega_C)$ yields

$$C_n(\omega_{MIN}/\omega_C) \geq \sqrt{\left(\frac{T_{MAX}}{T_{MIN}}\right)^2 - 1}$$

In the stopband ($x = \omega/\omega_C \geq 1$) the function $C_n(x)$ is defined by the hyperbolic cosine function in Eq. (14–32b). Inserting this definition into the preceding inequality and then solving for *n* yields

$$n \geq \frac{\cosh^{-1}\left(\sqrt{(T_{MAX}/T_{MIN})^2 - 1}\right)}{\cosh^{-1}(\omega_{MIN}/\omega_C)} \tag{14–33}$$

---

3 It can be shown that at $\omega = \omega_C$ the slope of a Chebychev gain is *n* times as steep as a Butterworth gain. See Aram Budak, *Passive and Active Network Analysis and Synthesis,* Houghton Mifflin, 1974, p. 516.

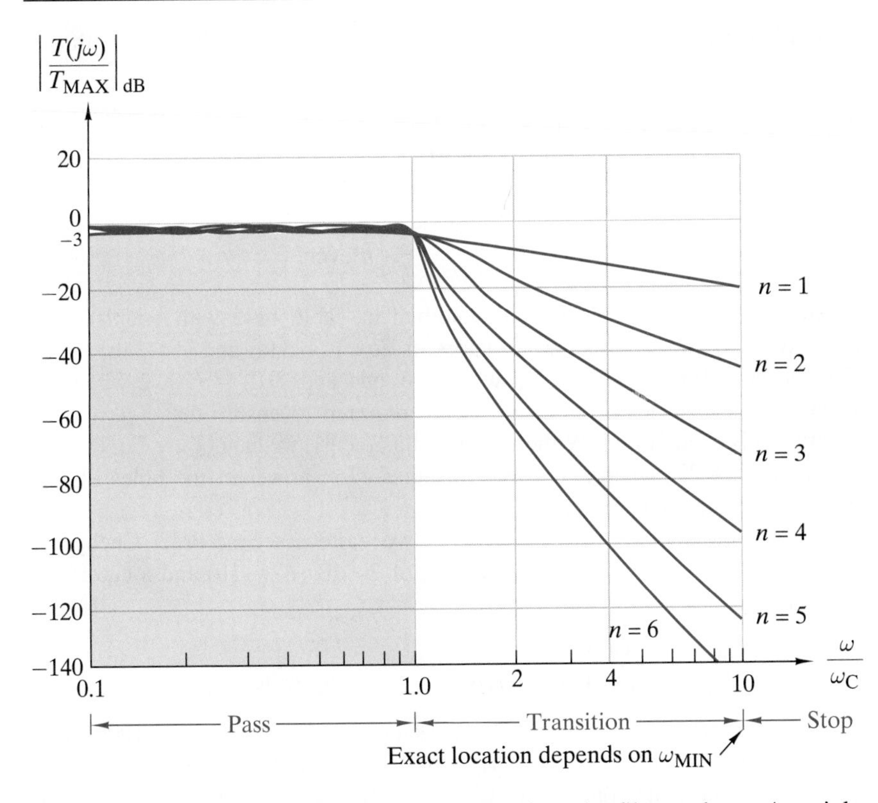

FIGURE 14–22 *Chebychev low-pass filter responses.*

The right side of this equation is a lower bound on the filter order n. As might be expected, the lower bound is determined by the transition band ratios T_{MAX}/T_{MIN} and ω_{MIN}/ω_C. Based on our experience with first-order cascade, Butterworth, and now Chebychev filters, it seems reasonable to infer the following general principle of filter design.

Filter order is determined by how much and how rapidly the gain must decrease in the transition band.

The smallest order for a Chebychev filter can be determined analytically using Eq. (14–33) or graphically using the gain plots in Figure 14–22. Once the required order is known, we need an nth-order polynomial $q_n(s)$ to construct a transfer function with Chebychev gain response. A method of obtaining such a polynomial is given in Appendix D. Using the method given there, we can generate the normalized ($\omega_C = 1$) polynomials $q_n(s)$ in Table 14–2. Note that these polynomials all have the property that $q_n(0) = 1$.

TABLE 14–2 **NORMALIZED POLYNOMIALS THAT PRODUCE CHEBYCHEV RESPONSES**

ORDER	NORMALIZED DENOMINATOR POLYNOMIALS
1	$(s + 1)$
2	$[(s/0.8409)^2 + 0.7654(s/0.8409) + 1]$
3	$[(s/0.2980) + 1][(s/0.9159)^2 + 0.3254(s/0.9159) + 1]$
4	$[(s/0.9502)^2 + 0.1789(s/0.9502) + 1)][(s/0.4425)^2 + 0.9276(s/0.4425) + 1]$
5	$[(s/0.1772) + 1][(s/0.9674)^2 + 0.1132(s/0.9674) + 1][(s/0.6139)^2 + 0.4670(s/0.6139) + 1]$
6	$[(s/0.9771)^2 + 0.0781(s/0.9771) + 1][(s/0.7223)^2 + 0.2886(s/0.7223) + 1][(s/0.2978)^2 + 0.9562(s/0.2978) + 1]$

Using these polynomials, an nth-order Chebychev transfer function with a dc gain of K and cutoff frequency of ω_C is written as

$$T(s) = \frac{K}{q_n(s/\omega_C)} \quad n \text{ odd} \tag{14–34a}$$

or

$$T(s) = \frac{K/\sqrt{2}}{q_n(s/\omega_C)} \quad n \text{ even} \tag{14–34b}$$

where $q_n(s/\omega_C)$ is the nth-order polynomial in Table 14–2 with s replaced by s/ω_C. Since $q_n(0) = 1$, the scale factors in Eqs. (14–34a) and (14–34b) ensure that the dc gains are $T(0) = |K|$ when n is odd and $T(0) = K/\sqrt{2}$ when n is even. These scale factor adjustments are required to ensure that $T_{MAX} = |K|$ whether n is even or odd. Without these adjustments, we cannot use Eq. (14–33) or Figure (14–22) when n is even. In either case, however, the poles of $T(s)$ are roots of the equation $q_n(s/\omega_C) = 0$.

As an example of using Table 14–2, we write a second-order Chebychev low-pass transfer function with a dc gain of 20 dB ($K = 10$) and a cutoff frequency of $\omega_C = 1$ krad/s as

$$T(s) = \frac{10/\sqrt{2}}{q_2(s/1000)} = \frac{7.07}{(s/840.9)^2 + 0.7654(s/840.9) + 1}$$

This transfer function has $T_{MAX} = 10$, a cutoff frequency at $\omega_C = 1000$ rad/s, and a pair of complex poles with $\omega_0 = 840.9$ rad/s and $\zeta = 0.3827$. This illustrates two important properties of Chebychev poles: (1) Their natural frequencies are less than the filter cutoff frequency ($\omega_0 < \omega_C$), and (2) their damping ratios are less than those of Butterworth poles of the same order. As a result, the Chebychev poles produce resonant peaks in the gain response at frequencies below ω_C. Put differently, Chebychev poles are specifically located to produce an equal-ripple response in the passband.

Once we have a $T(s)$ that meets the filter specification, we realize it using a cascade of first- and second-order building blocks. The next example illustrates the design of a Chebychev low-pass filter.

D DESIGN EXAMPLE 14–7

(a) Construct a Chebychev low-pass transfer function that meets the requirements: $T_{MAX} = 20$ dB, $\omega_C = 10$ rad/s, $T_{MIN} = -30$ dB, and $\omega_{MIN} = 50$ rad/s.

(b) Design a cascade of active RC circuits that produces the transfer function found in (a).

SOLUTION:

(a) This specification requires the gain to decrease by 50 dB in the transition band between $\omega_C = 10$ rad/s and $\omega_{MIN} = 50$ rad/s. The corresponding transition band ratios are $T_{MAX}/T_{MIN} = 10^{5/2}$ and $\omega_{MIN}/\omega_C = 5$. Using these ratios in Eq. (14–33) yields

$$n \geq \frac{\cosh^{-1}(\sqrt{(10^{5/2})^2 - 1})}{\cosh^{-1}(5)} = 2.81$$

The smallest integer meeting the specification is $n = 3$. Using the polynomial $q_3(s)$ from Table 14–2, we construct the required Chebychev low-pass transfer function using Eq. (14–34a).

$$T(s) = \frac{K}{q_3(s/10)}$$

$$= \frac{10}{\left[\left(\frac{s}{2.98}\right) + 1\right]\left[\left(\frac{s}{9.159}\right)^2 + 0.3254\left(\frac{s}{9.159}\right) + 1\right]}$$

No scale factor adjustment is needed here because $n = 3$ is odd.

(b) The Chebychev transfer function in (a) can be partitioned as

$$T(s) = T_1(s)T_2(s)$$

$$= \left[\frac{K_1}{\frac{s}{2.98} + 1}\right]\left[\frac{K_2}{\left(\frac{s}{9.159}\right)^2 + 0.3254\left(\frac{s}{9.159}\right) + 1}\right]$$

where K_1 and K_2 are to be determined. This partition calls for a cascade of a first-order and a second-order low-pass filter.

The first row in the design sequence in Figure 14–23 shows the required transfer functions. The second, third, and fourth rows show the stage parameters, stage prototypes, and design constraints. The second-order prototype is the low-pass circuit in Figure 14–3. The design constraints for this circuit use the equal element design method [see Eq. (14–5)]. The equal element method produces a dc gain of $K_2 = 3 - 2\zeta = 2.675$. The first-order prototype is the low-pass circuit in Figure 14–14. Its pole location is determined by the RC product and its gain K_1 can be adjusted without changing the pole location, so we are free to set $K_1 = 10/K_2 = 3.738$, which makes $K = K_1K_2 = 10$ as required. The assigned element values in the fifth row produce the final design in the last row. A cascade connection of these two stages meets all design requirements without an added gain correction stage. ■

Exercise 14–12

Rework the design in Example 14–7 using the unity gain method in Sect. 14–2 to design the required second-order low-pass circuit.

Answer: Figure 14–24 shows the unity gain method redesigns of first- and second-order stages in Figure 14–23.

Exercise 14–13

Construct a Chebychev low-pass transfer function that meets the following requirements: $T_{MAX} = 0$ dB, $T_{MIN} = -30$ dB, $\omega_C = 250$ rad/s, and $\omega_{MIN} = 1.5$ krad/s.

Answer:

$$T(s) = \frac{210^2/\sqrt{2}}{s^2 + 161s + 210^2}$$

Item	First Stage	Second Stage
Prototype transfer function	$\dfrac{K_1}{(s/2.98)+1}$	$\dfrac{K_2}{(s/9.159)^2+0.3254(s/9.159)+1}$
Stage parameters	$\omega_0 = 2.980$	$\omega_0 = 9.159 \quad \zeta = 0.3254/2 = 0.1627$
Stage prototype	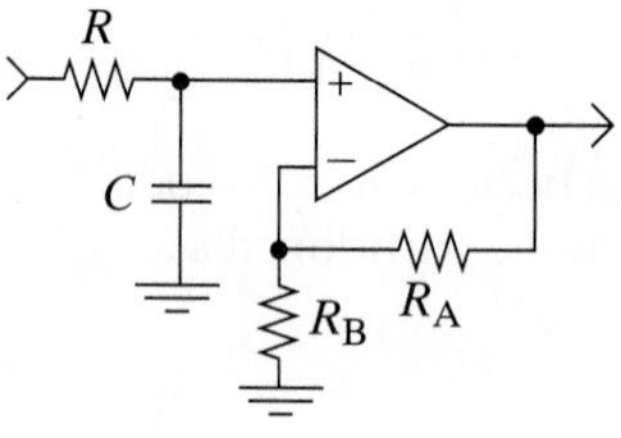 R, C, R_B, R_A	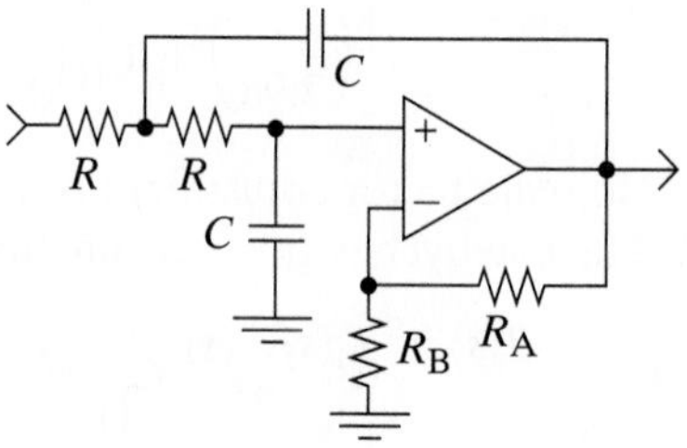 C, R, R, C, R_B, R_A
Design constraints	$RC = \dfrac{1}{\omega_0} = 0.3357$ $K_1 = 10/K_2 = 3.73$ $R_A = (K_1 - 1)R_B$	$RC = \dfrac{1}{\omega_0} = 0.1092$ $K_2 = 3 - 2\zeta = 2.675$ $R_A = (K_2 - 1)R_B$
Element values	Let $R = 100$ kΩ, then $C = 3.36\ \mu$F Let $R_B = 100$ kΩ, then $R_A = 274$ kΩ	Let $R = 100$ kΩ, then $C = 1.092\ \mu$F Let $R_B = 100$ kΩ, then $R_A = 167$ kΩ
Final designs	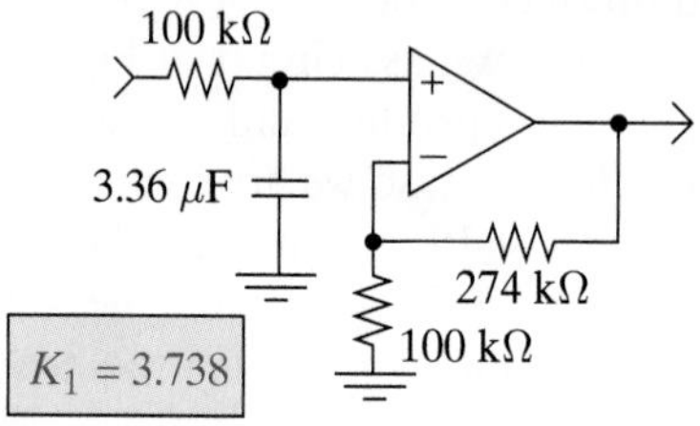 100 kΩ, 3.36 μF, 274 kΩ, 100 kΩ $K_1 = 3.738$	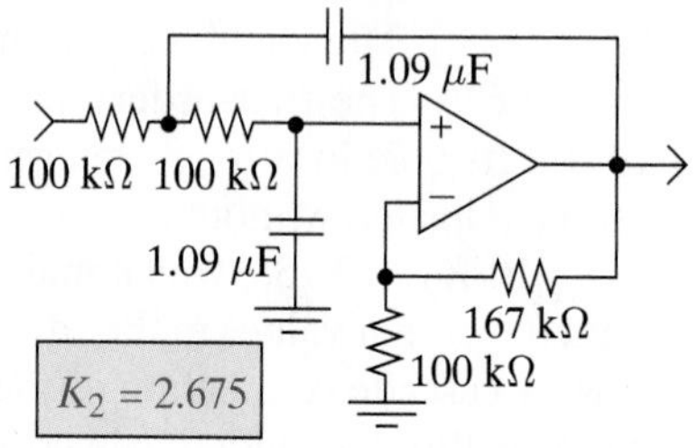 1.09 μF, 100 kΩ, 100 kΩ, 1.09 μF, 167 kΩ, 100 kΩ $K_2 = 2.675$

FIGURE 14–23 *Design sequence for Example 14–7.*

FIGURE 14–24

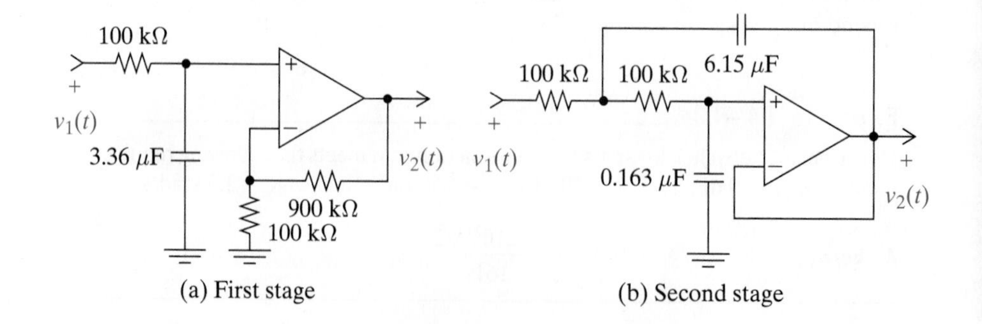

(a) First stage

(b) Second stage

14–5 LOW-PASS FILTER EVALUATION

We have described low-pass filter design methods for three responses, the first-order cascade, the Butterworth, and the Chebychev. At this point we want to compare the methods and discuss how we might choose between them. In filter applications the gain response is obviously important. Figure 14–25 shows the three straight-line asymptotes (solid lines) and the actual gain responses (dashed curves) for $n = 4$. All three responses meet the same passband requirements, have the same cutoff frequency, and have high-frequency asymptotes with slopes of $-20n = -80$ dB/decade. However, the three responses have different corner frequencies. At one extreme the corner frequency of the first-order cascade response is above the cutoff frequency, so its actual gain response approaches its asymptote very gradually. At the other extreme the Chebychev corner frequency lies below the cutoff frequency and the actual response has a resonant peak that fills the gap between the corner frequency and the cutoff frequency. This resonance causes the Chebychev response to decrease rapidly in the neighborhood of the cutoff frequency. The Butterworth response has its corner frequency at the cutoff frequency, so its gain response falls between these two extremes.

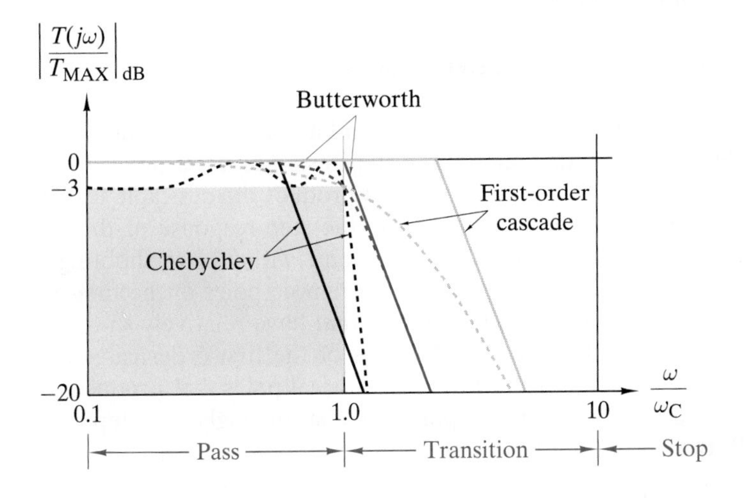

FIGURE 14–25 *First-order cascade, Butterworth, and Chebychev gain responses for* n = 4.

The differences in gain response can be understood by examining the pole-zero diagram in Figure 14–26. The Butterworth poles are evenly distributed on a circle of radius ω_C. The Chebychev poles lie on an ellipse whose minor axis is much smaller than ω_C. As a result, the Chebychev poles are closer to the *j*-axis, have lower damping ratios, and produce a gain response with pronounced resonant peaks. These resonant peaks lead to the equal-ripple response in the passband and the steep gain slope in the neighborhood of the cutoff frequency. At the other extreme the first-order cascade response has a fourth-order pole (quadruple pole) located on the negative real axis. The distance from the *j*-axis to the first-order cascade poles is much larger than ω_C, which explains the rather leisurely way its gain response transitions from the passband to stopband asymptote. As might be expected, the Butterworth poles fall between these two extremes.

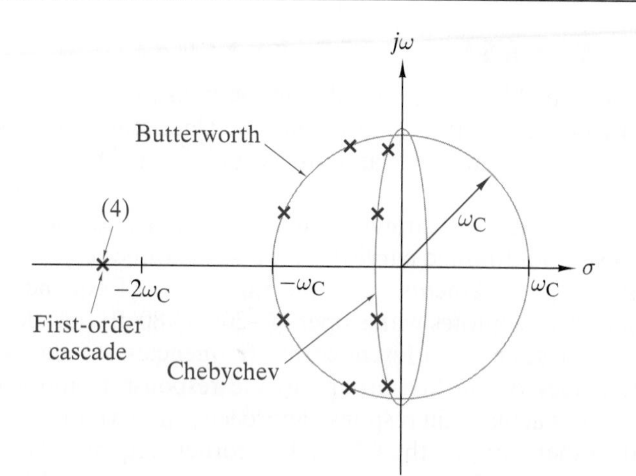

FIGURE 14–26 *First-order cascade, Butterworth, and Chebychev pole locations for* n = 4.

This discussion illustrates the following principle. For any given value of n, the Chebychev response produces more transition band attenuation than the Butterworth response, which, in turn, produces more than the first-order cascade response. If transition band performance is the only consideration, then we should choose the Chebychev response. However, the Chebychev response comes at a price.

Figure 14–27 shows the step response of these three low-pass filters for $n = 4$. The step response of the Chebychev filter has lightly damped oscillations that produce a large overshoot and a long settling time. These undesirable features of the Chebychev step response are a direct result of the low-damping-ratio complex poles that produce the desirable features of its gain response. At the other extreme the step response of the first-order cascade filter rises rapidly to its final value without overshooting. This result should not be surprising since the remote poles of the first-order cascade produce exponential waveforms that have relatively short durations. In other words, the desirable features of the first-order cascade step response are a direct result of the remote real poles that produce the undesirable features of its gain response. Not surprisingly, the step response of the Butterworth filter lies between these two extremes.

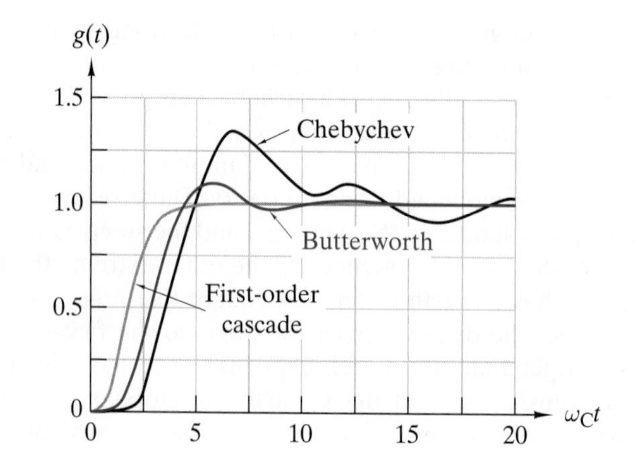

FIGURE 14–27 *First-order cascade, Butterworth, and Chebychev step responses for* n = 4.

Finally, consider the element values in the circuit realizations of these filters. Examination of Table 14–2 reveals that each pair of complex poles in a Chebychev filter has a different ω_0 and a different ζ. These parameters define the constraints on the element values for each stage in the filter. As a result, each stage in a cascade realization of a Chebychev filter has a different set of element values. In contrast, the stages in a first-order cascade filter can be exactly the same. From a manufacturing point of view, it may be better to produce and stock identical circuits rather than uniquely different circuits.

The essential point is that transition band attenuation is important, but it does not tell the whole story. Filter design, and indeed all design, involves trade-offs between conflicting requirements. The choice of a design approach is driven by the weight assigned to conflicting requirements.

14–6 HIGH-PASS FILTER DESIGN

An ideal high-pass filter has a constant gain above a cutoff frequency ω_C and zero gain below ω_C. Filter design specifications define the degree to which real filters are required to approach the ideal. Figure 14–28 is a Bode plot specifying the allowable region for high-pass filter gain responses. The allowable region is defined by four familiar parameters: T_{MAX}, T_{MIN}, ω_C, and ω_{MIN}. We are acquainted with the definitions and use of these parameters from our study of low-pass filter design.

The difference here is that $\omega_{MIN} < \omega_C$, that is, the transition band and stopband are below the passband. In fact, if we look carefully we see that Figure 14–28 is the mirror image of Figure 14–12 for low-pass filters. This suggests that low-pass filter design methods apply to high-pass filter design when we interchange the positions of the passband and the stopband. Mathematically the

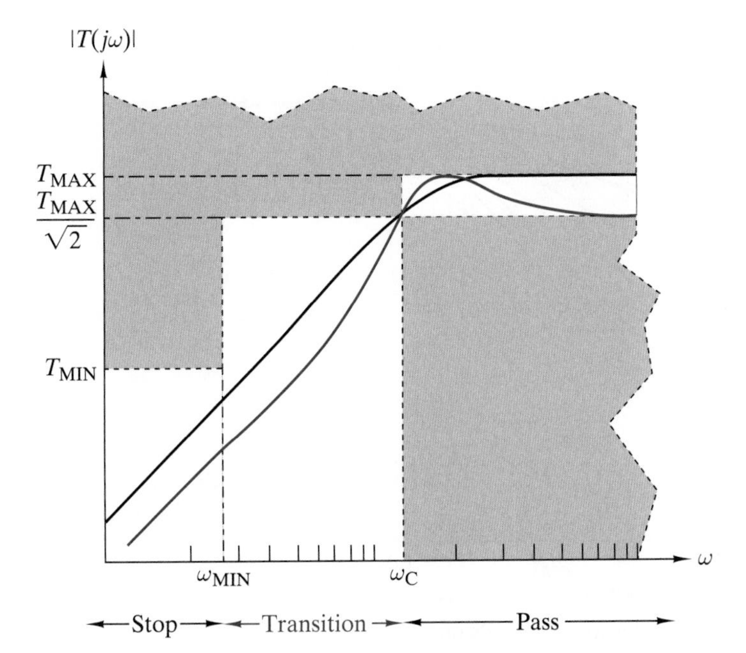

FIGURE 14–28 *High-pass filter specification and responses.*

interchange is called a low-pass to high-pass transformation. The transformation is carried out by inverting the frequency variables in the low-pass filter relationships. We demonstrate this first for Butterworth responses.

BUTTERWORTH HIGH-PASS FILTERS

Butterworth low-pass gain responses are defined in Eq. (14–28) using the normalized frequency variable ω/ω_C. Using an inverted frequency variable ω_C/ω in Eq. (14–28) produces the definition of Butterworth high-pass gain responses:

$$|T(j\omega)| = \frac{|K|}{\sqrt{1 + (\omega_C/\omega)^{2n}}} \quad \text{Butterworth high-pass} \tag{14–35}$$

where n is the filter order. Figure 14–29 shows normalized plots of Eq. (14–35) for $n = 1$ to 6, where $T_{\mathrm{MAX}} = |K|$. These plots confirm that the inverted variable ω_C/ω does in fact interchange the passband and stopband while leaving the cutoff frequency unchanged at $\omega/\omega_C = 1$. The inversion causes low-pass gain responses that occur below ω_C to occur as high-pass responses above ω_C, and vice versa. In particular, Figure 14–29 shows that gains in the high-pass transition band below ω_C decrease rapidly as n increases.

To limit circuit complexity, we need to know the smallest n that meets a given high-pass filter specification. For Butterworth low-pass filters, Eq. (14–29) provides a lower bound on n in terms of the transition band ratios. For Butterworth high-pass filters, the lower bound is found by substituting the inverted

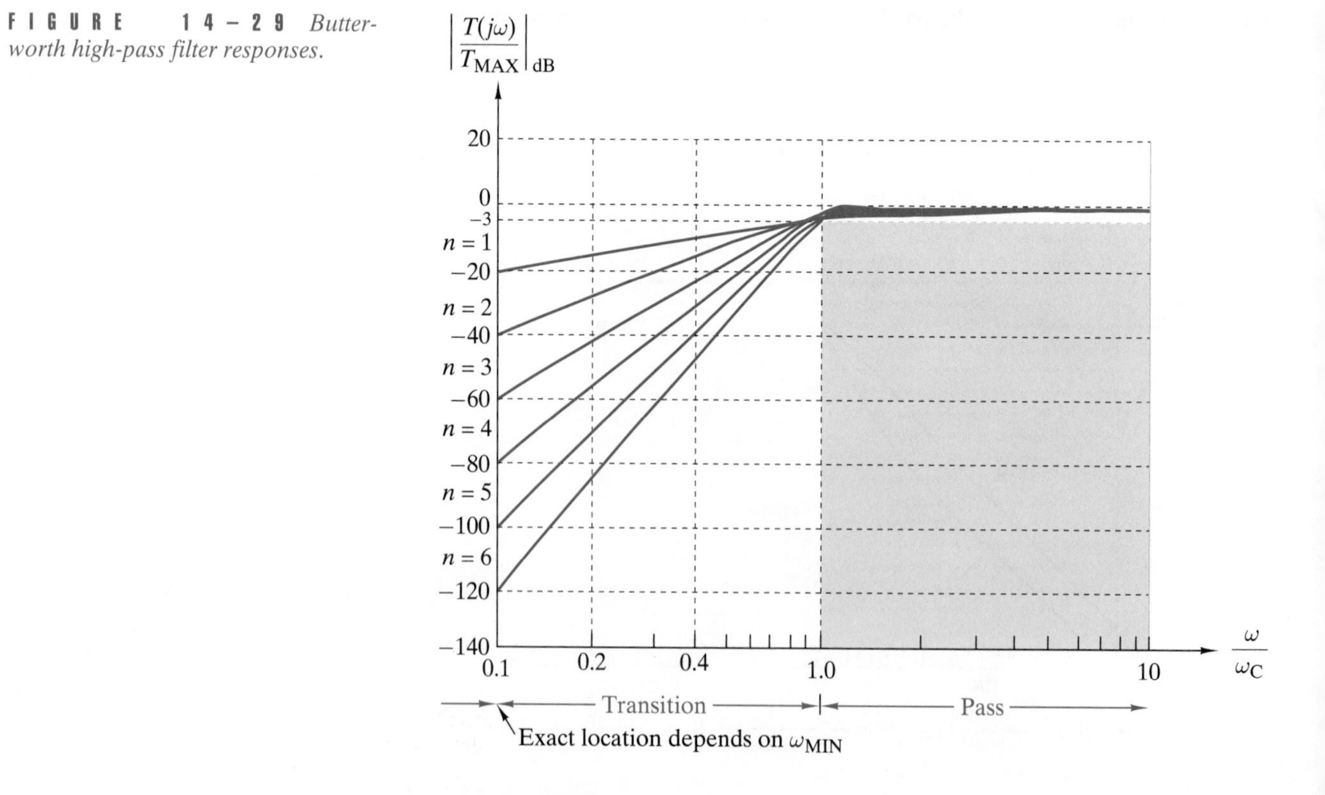

FIGURE 14–29 Butterworth high-pass filter responses.

variable ω_C/ω in the derivation of Eq. (14–29). The modified derivation is straightforward and leads to the following equation:

$$n \geq \frac{1}{2}\frac{\ln[(T_{MAX}/T_{MIN})^2 - 1]}{\ln[\omega_C/\omega_{MIN}]} \quad \text{High-pass order} \qquad (14\text{–}36)$$

The lower bound on the right side of this equation depends on two transition band ratios, namely T_{MAX}/T_{MIN} and ω_C/ω_{MIN}. There is no real surprise here. We have seen several times that filter order depends on how much and how abruptly the transition band gain must decrease to meet a given specification.

To give a concrete example, suppose the transition band gain of a Butterworth high-pass filter must decrease by 40 dB ($T_{MAX}/T_{MIN} = 10^2$) when $\omega_C/\omega_{MIN} = 5$. Equation (14–36) yields a lower bound of

$$n \geq \frac{1}{2}\frac{\ln[(10^2)^2 - 1]}{\ln[5]} = 2.86$$

The smallest integer meeting this bound is $n = 3$. This result is confirmed by the plots in Figure 14–29, which show that a normalized gain below −40 dB at $\omega/\omega_C = 0.2$ cannot be produced by any order less than $n = 3$.

Given the value of n, we need to construct a high-pass transfer function with Butterworth gain response. A Butterworth low-pass transfer function is constructed using a polynomial $q_n(s)$ from Table 14–1 in Eq. (14–30). An nth-order high-pass transfer function is found by inverting the frequency variable in Eq. (14–30) to get

$$T(s) = \frac{K}{q_n(\omega_C/s)} \quad \text{High-pass transfer function} \qquad (14\text{–}37)$$

where $q_n(\omega_C/s)$ is a normalized Butterworth polynomial in Table 14–1 with s replaced by ω_C/s.

To see how the inversion process actually works, let us develop a second-order Butterworth high-pass transfer function with $\omega_C = 1$ krad/s and $T_{MAX} = 20$ dB. For $n = 2$ the normalized polynomial in Table 14–1 is $q_2(s) = s^2 + 1.41s + 1$. Using $\omega_C = 1000$ rad/s and $K = 10$ (20 dB) in Eq. (14–37) produces

$$T(s) = \frac{10}{\left(\frac{1000}{s}\right)^2 + 1.414\left(\frac{1000}{s}\right) + 1} = \frac{10\left(\frac{s}{1000}\right)^2}{\left(\frac{s}{1000}\right)^2 + 1.414\left(\frac{s}{1000}\right) + 1}$$

This transfer function has a double zero at $s = 0$ and a pair of Butterworth poles with $\omega_0 = 1000$ rad/s and $\zeta = 1.414/2 = 0.707$. The double zero makes the dc gain $|T(0)| = 0$ and the infinite-frequency gain $|T(\infty)| = 10$. The gain at the cutoff frequency $\omega_C = 1$ krad/s is

$$|T(j1000)| = 10/1.414 = |T(\infty)|/\sqrt{2}$$

as required. In sum, the transfer function $T(s)$ has the properties of a second-order Butterworth high-pass filter.

Designing an nth-order Butterworth high-pass filter to meet prescribed values of T_{MAX}, T_{MIN}, ω_C, and ω_{MIN} involves the following steps. We first determine the smallest n that meets the specification using Eq. (14–36) or the graphs in Figure 14–29. Taking the polynomial $q_n(s)$ from Table 14–1 and using

$K = T_{MAX}$, we obtain $T(s)$ using Eq. (14–37). The polynomial $q_n(\omega_C/s)$ in the denominator of $T(s)$ supplies n Butterworth poles and n zeros at the origin. We then partition $T(s)$ into a product of first- and second-order high-pass functions and use a cascade of high-pass building blocks to get the overall filter circuit.

The next example illustrates the Butterworth high-pass design process.

DESIGN EXAMPLE 14–8

(a) Construct a Butterworth high-pass transfer function that meets the following requirements: $T_{MAX} = 20$ dB, $\omega_C = 10$ rad/s, $T_{MIN} = -10$ dB, and $\omega_{MIN} = 3$ rad/s.
(b) Design a cascade of active RC circuits that realizes the transfer function found in (a).

SOLUTION:

(a) The specification requires the gain to decrease by 30 dB in the transition band between $\omega_{MIN} = 3$ rad/s and $\omega_C = 10$ krad/s. The transition band ratios for this specification are $T_{MAX}/T_{MIN} = 10^{3/2}$ and $\omega_C/\omega_{MIN} = 3.33$. Inserting these ratios in Eq. (14–36) yields

$$n \geq \frac{1}{2}\frac{\ln[(10^{3/2})^2 - 1]}{\ln(3.33)} = 2.87$$

Hence, $n = 3$ is the lowest-order Butterworth response that meets the transition band requirement. For $n = 3$, Table 14–1 lists $q_3(s) = (s + 1)(s^2 + s + 1)$. Using $\omega_C = 10$ and $K = 10$ (20 dB) in Eq. (14–37) gives the required high-pass transfer function:

$$T(s) = \frac{10}{q_3(10/s)} = \frac{10}{\left[\left(\frac{10}{s}\right) + 1\right]\left[\left(\frac{10}{s}\right)^2 + \left(\frac{10}{s}\right) + 1\right]}$$

$$= \frac{10\left(\frac{s}{10}\right)^3}{\left[\left(\frac{s}{10}\right) + 1\right]\left[\left(\frac{s}{10}\right)^2 + \left(\frac{s}{10}\right) + 1\right]}$$

This third-order high-pass function has three zeros at $s = 0$, a real pole at $s = -10$ rad/s, and a pair of complex poles with $\omega_0 = 10$ rad/s and $\zeta = 0.5$. Note that these high-pass poles all lie on a circle of radius ω_C, as did the poles in Butterworth low-pass filters.

(b) The Butterworth high-pass function developed in (a) can be partitioned as follows:

$$T(s) = T_1(s)T_2(s)$$
$$= \left[\frac{(s/10)^2}{(s/10)^2 + s/10 + 1}\right]\left[\frac{10(s/10)}{(s/10) + 1}\right]$$

In this partition all of the passband gain $T_{MAX} = 10$ (20 dB) has been assigned to the first-order transfer function. This makes the passband gain of the second-order transfer function 1 (0 dB) so that it can be realized using the unity gain design method [see Eq. (14–12)].

Figure 14–30 shows a design sequence based on this partitioning. The stage transfer functions in the first row lead to the stage parameters in the second row. The two stage prototypes are a second-order high-pass circuit with unity gain and a first-order high-pass circuit with an adjustable gain. The design constraints for the second-order stage use the unity gain design method. The design constraints for the first-order stage locate the real pole using the *RC* product and adjust the OP AMP's feedback to get the required

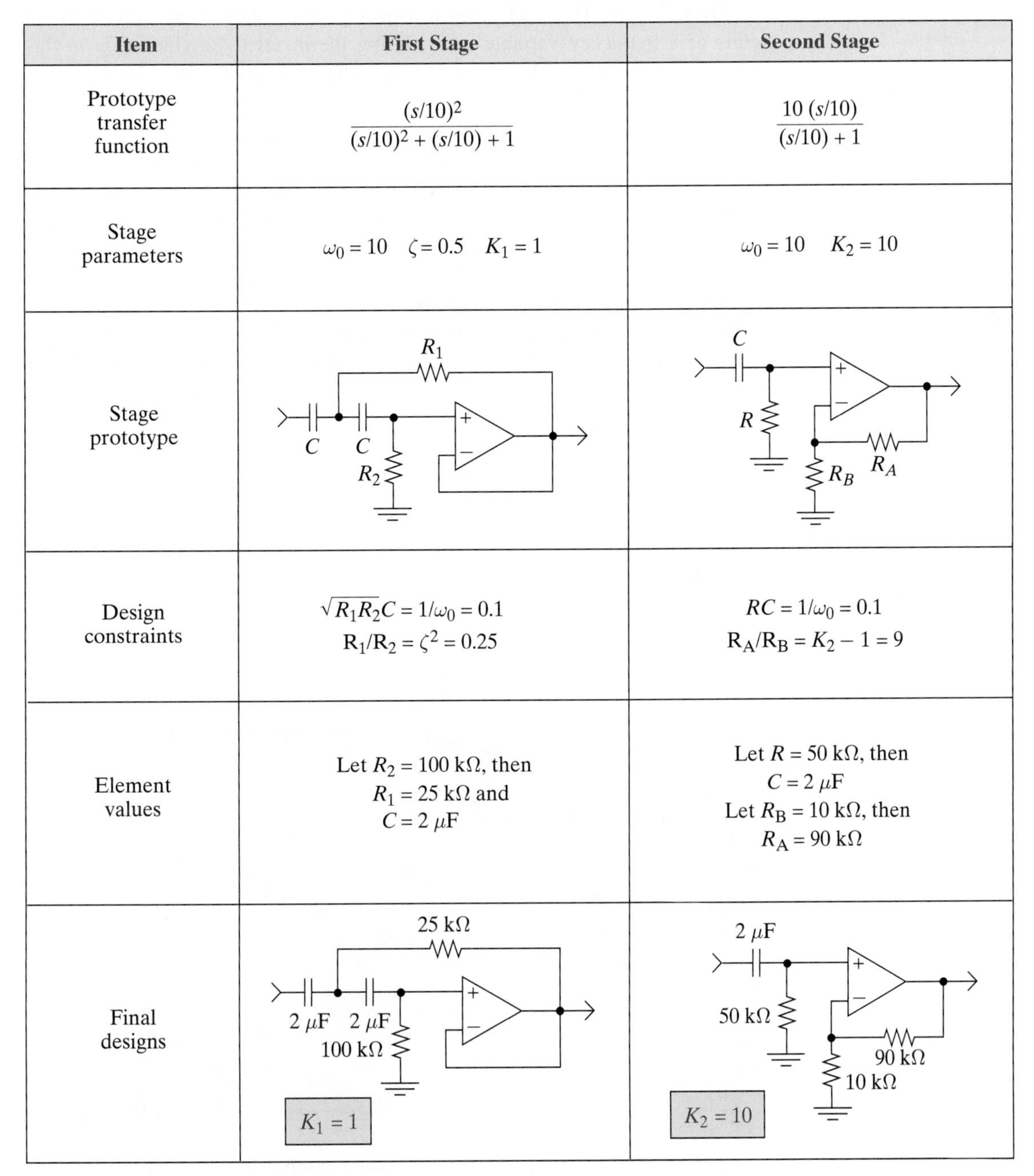

Item	First Stage	Second Stage
Prototype transfer function	$\dfrac{(s/10)^2}{(s/10)^2 + (s/10) + 1}$	$\dfrac{10\,(s/10)}{(s/10) + 1}$
Stage parameters	$\omega_0 = 10 \quad \zeta = 0.5 \quad K_1 = 1$	$\omega_0 = 10 \quad K_2 = 10$
Stage prototype	R_1, C, C, R_2	C, R, R_B, R_A
Design constraints	$\sqrt{R_1 R_2}C = 1/\omega_0 = 0.1$ $R_1/R_2 = \zeta^2 = 0.25$	$RC = 1/\omega_0 = 0.1$ $R_A/R_B = K_2 - 1 = 9$
Element values	Let $R_2 = 100$ kΩ, then $R_1 = 25$ kΩ and $C = 2\ \mu$F	Let $R = 50$ kΩ, then $C = 2\ \mu$F Let $R_B = 10$ kΩ, then $R_A = 90$ kΩ
Final designs	25 kΩ, 2 μF, 2 μF, 100 kΩ $K_1 = 1$	2 μF, 50 kΩ, 90 kΩ, 10 kΩ $K_2 = 10$

FIGURE 14–30 *Design sequence for Example 14–8.*

gain. Using these constraints together with the element value assignments in the fifth row produces the final designs in the last row of Figure 14–30. The required passband gain is provided by the first-order stage, so no additional gain correction stage is needed in this example.

CHEBYCHEV HIGH-PASS FILTERS

Our treatment of Chebychev high-pass filters can be brief because the low-pass to high-pass transformation used to develop Butterworth high-pass filters works equally well here. The Chebychev low-pass gains are defined in Eq. (14–31) in terms of a frequency variable ω/ω_C. Using the inverted variable ω_C/ω in this definition produces the definition of Chebychev high-pass gain responses.

$$|T_n(j\omega)| = \frac{|K|}{\sqrt{1 + C_n^2(\omega_C/\omega)}} \quad \text{Chebyschev high-pass} \qquad (14\text{–}38)$$

where $C_n(x)$ is an nth-order Chebychev polynomial. The normalized plots of the Chebychev high-pass gains in Figure 14–31 show that the passband and stopband have been interchanged. These Chebychev high-pass responses have equal-ripple responses in the passband above ω_C and transition band gains below ω_C that decrease rapidly as n increases.

Equation (14–33) provides a lower bound on n for Chebychev low-pass filters. Using the inverted variable ω_C/ω in the derivation of this equation yields a lower bound on the Chebychev high-pass order as

$$n \geq \frac{\cosh^{-1}\left(\sqrt{(T_{MAX}/T_{MIN})^2 - 1}\right)}{\cosh^{-1}(\omega_C/\omega_{MIN})} \quad \text{High-pass order} \qquad (14\text{–}39)$$

FIGURE 14–31 Chebychev high-pass filter responses.

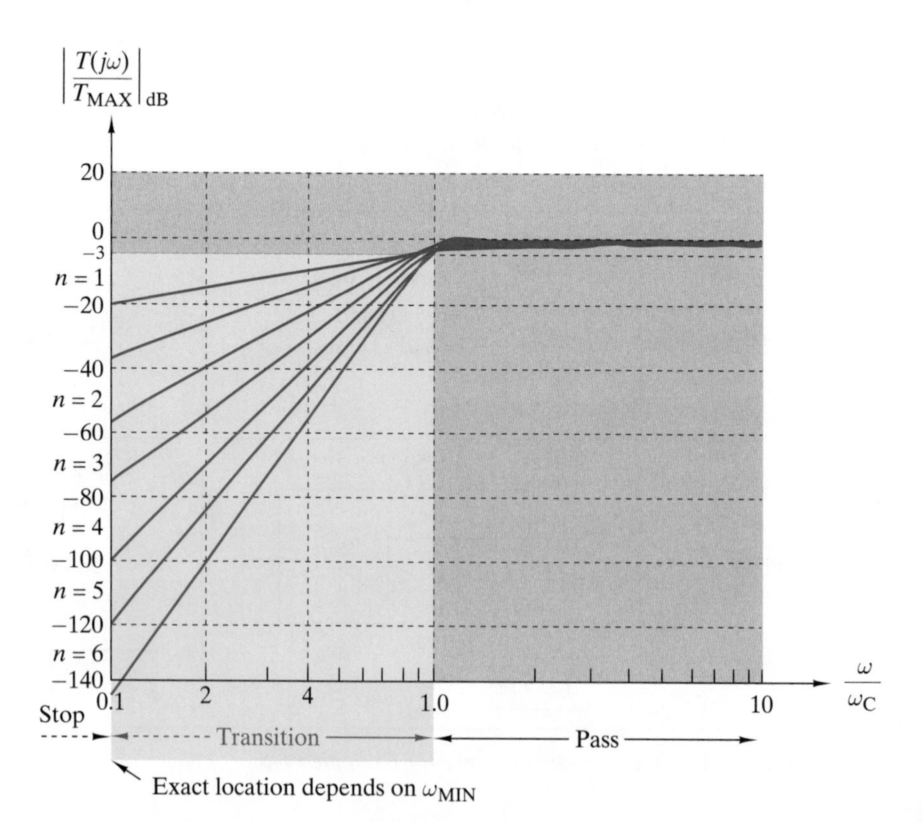

Once again we see that the lower bound on filter order depends on how much (T_{MAX}/T_{MIN}) and how abruptly (ω_C/ω_{MIN}) the transition band gain must decrease to meet the specification.

For any given n, a Chebychev low-pass transfer function is obtained using a polynomial $q_n(s/\omega_C)$ in Eqs. (14–34a) or (14–34b). To obtain a Chebychev high-pass transfer function, we invert the frequency variable and write

$$T(s) = \frac{K}{q_n(\omega_C/s)} \quad n \text{ odd} \tag{14–40a}$$

or

$$T(s) = \frac{K/\sqrt{2}}{q_n(\omega_C/s)} \quad n \text{ even} \tag{14–40b}$$

where $q_n(\omega_C/s)$ is a normalized Chebychev denominator polynomial in Table 14–2 with s replaced by ω_C/s. Thus, the polynomials in Table 14–1 and Table 14–2 serve two purposes. They are used as $q_n(s/\omega_C)$ in the design of low-pass filters and as $q_n(\omega_C/s)$ in the design of high-pass filters. For the Chebychev high-pass case, the order of the polynomial is found analytically using Eq. (14–39) or graphically using the normalized gain plots in Figure 14–31.

The next example illustrates constructing Chebychev high-pass transfer functions to meet a given specification.

DESIGN EXAMPLE 14–9

Construct a Chebychev high-pass transfer function that meets requirements of Example 14–8.

SOLUTION:

The specification requirements in Example 14–8 are $T_{MAX} = 20$ dB, $\omega_C = 10$ rad/s, $T_{MIN} = -10$ dB, and $\omega_{MIN} = 3$ rad/s. The transition band gain must decrease by 30 dB $(T_{MAX}/T_{MIN} = 10^{3/2})$ on the range $\omega_C/\omega_{MIN} = 3.33$. Inserting these ratios in Eq. (14–39) yields

$$n \geq \frac{\cosh^{-1}\left[\sqrt{(10^{3/2})^2 - 1}\right]}{\cosh^{-1}(3.33)} = 2.21$$

Hence, $n = 3$ is the lowest-order Chebychev response that meets the design requirement. The third-order polynomial in Table 14–2 is

$$q_3(s) = (s/0.2980 + 1)[(s/0.9159)^2 + 0.3254(s/0.9159) + 1]$$

Using $\omega_C = 10$ and $K = 10$ $(T_{MAX} = 20$ dB) in Eq. (14–40a) gives the required high-pass transfer function:

$$T_{HP}(s) = \frac{10}{q_3(10/s)}$$

$$= \frac{10}{\left[\left(\frac{10}{0.2980s}\right) + 1\right]\left[\left(\frac{10}{0.9159s}\right)^2 + 0.3254\left(\frac{10}{0.9159s}\right) + 1\right]}$$

$$= \frac{10\,s^3}{(s + 33.55)(s^2 + 3.553s + 119.2)}$$

In Example 14–8 we meet this specification using a third-order Butterworth high pass with poles located on a circle of radius $\omega_C = 10$ rad/s. The third-order Chebychev high-pass function constructed here has three zeros at $s = 0$, a real pole at $s = -33.55$ rad/s, and a pair of complex poles with $\omega_0 = \sqrt{119.2} = 10.92$ rad/s and $\zeta = 0.1627$. These poles all lie outside of a circle of radius $\omega_C = 10$ rad/s. The reason for this is easy to understand. A Chebychev high-pass function has an equal-ripple response in the passband *above* ω_C. Hence, the Chebychev poles that produce the ripple must have natural frequencies greater than ω_C. ■

Exercise 14–14

Construct Butterworth and Chebychev high-pass transfer functions that meet the following requirements: $T_{MAX} = 10$ dB, $\omega_C = 50$ rad/s, $T_{MIN} = -40$ dB, and $\omega_{MIN} = 10$ rad/s.

Answer:

$$T_{BU}(s) = \frac{\sqrt{10}s^4}{(s^2 + 38.3s + 50^2)(s^2 + 92.4s + 50^2)}$$

$$T_{CH}(s) = \frac{\sqrt{10}s^3}{(s + 168)(s^2 + 17.8s + 54.6^2)}$$

14–7 BANDPASS AND BANDSTOP FILTER DESIGN

In Chapter 12 we found that the cascade connection in Figure 14–32 can produce a bandpass filter. When the cutoff frequency of the low-pass filter (ω_{CLP}) is higher than the cutoff frequency of the high-pass filter (ω_{CHP}), the interval between the two frequencies is a passband separating two stopbands. The low-pass filter provides the high-frequency stopband and the high-pass filter the low-frequency stopband. Frequencies between the two cutoffs fall in the passband of both filters and are transmitted through the cascade connection, producing the passband of the resulting bandpass filter.

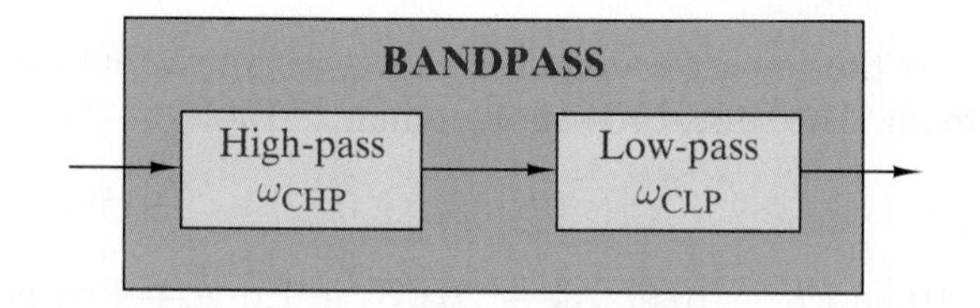

FIGURE 14–32 *Cascade connection of high-pass and low-pass filters.*

When $\omega_{CLP} >> \omega_{CHP}$, two bandpass cutoff frequencies are approximately $\omega_{C1} \approx \omega_{CHP}$ and $\omega_{C2} \approx \omega_{CLP}$. Under these conditions the center frequency and bandwidth of the bandpass filter are

$$\omega_0 = \sqrt{\omega_{CHP}\omega_{CLP}} \quad \text{and} \quad B = \omega_{CLP} - \omega_{CHP}$$

and the ratio of the center frequency over the bandwidth is approximately

$$\frac{\omega_0}{B} = Q \approx \sqrt{\frac{\omega_{CHP}}{\omega_{CLP}}} << 1$$

Since the quality factor is less than 1, this method of bandpass filter design produces a wide-band filter, as contrasted with the narrow-band response ($Q > 1$) produced by the active RC bandpass circuit studied earlier in Sect. 14–3.

DESIGN EXAMPLE 14–10

Use second-order Butterworth low-pass and high-pass functions to obtain a fourth-order bandpass function with a passband gain of 0 dB and cutoff frequencies at $\omega_{C1} = 10$ rad/s and $\omega_{C2} = 50$ rad/s.

SOLUTION:

The upper cutoff frequency at 50 rad/s is produced by a low-pass function. Using second-order Butterworth poles, the required low-pass function is

$$T_{LP}(s) = \frac{1}{(s/50)^2 + \sqrt{2}(s/50) + 1}$$

The lower cutoff frequency at 10 rad/s is to be produced by a second-order high-pass function. Using second-order Butterworth poles, the required high-pass function is

$$T_{HP}(s) = \frac{(s/10)^2}{(s/10)^2 + \sqrt{2}(s/10) + 1}$$

When circuits realizing these two transfer functions are connected in cascade, the overall transfer function is

$$\begin{aligned} T_{HP}(s) \times T_{LP}(s) &= \left[\frac{(s/10)^2}{(s/10)^2 + \sqrt{2}(s/10) + 1}\right] \times \left[\frac{1}{(s/50)^2 + \sqrt{2}(s/50) + 1}\right] \\ &= \frac{2500\, s^2}{s^4 + 60\sqrt{2}s^3 + 3600s^2 + 30{,}000\sqrt{2}s + 250{,}000} \end{aligned}$$

With the transfer function expressed as a quotient of polynomials, we use MATLAB to generate Bode plots of the frequency response. In the MATLAB command window, we define the numerator and denominator polynomials using the row matrices `num` and `den`.

```
num=[2500 0 0];
den=[1 60*sqrt(2) 3600 30000*sqrt(2) 250000];
```

These matrices list the polynomial coefficients in descending order. The MATLAB command

```
bode(num,den)
```

produces the Bode plots shown in Figure 14–33. The gain response displays a bandpass characteristic with a lower cutoff frequency at 10 rad/s and an upper cutoff frequency at 50 rad/s. These two cutoff frequencies come from the high-pass and low-pass functions, respectively. The phase shift swings from +180° at low frequency to –180° at high frequency, passing through zero at the passband center frequency of $\sqrt{10 \times 50} = 22.36$ rad/s. ■

FIGURE 14–33

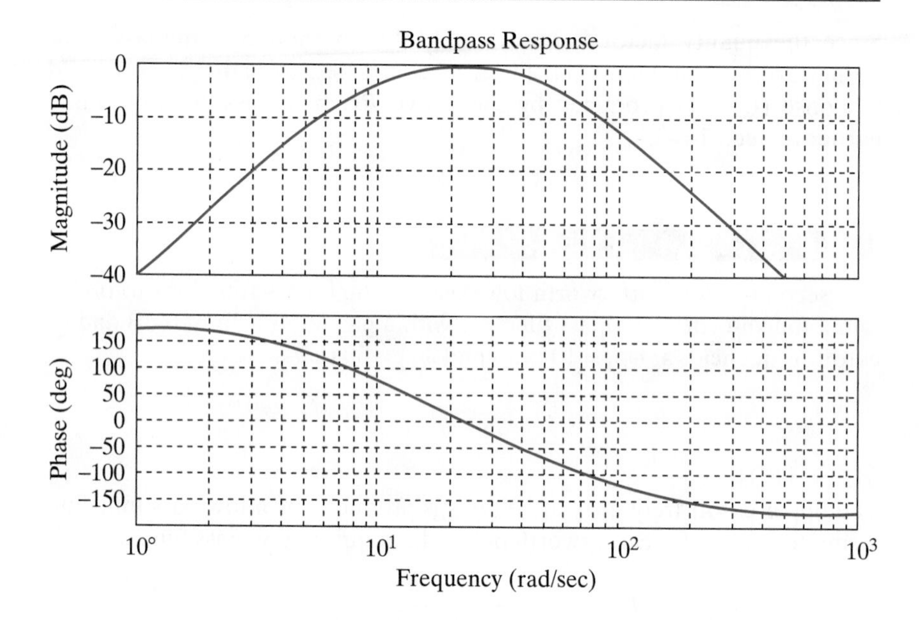

Figure 14–34 shows the dual situation in which a high-pass and a low-pass filter are connected in parallel to produce a bandstop filter. When $\omega_{CLP} << \omega_{CHP}$, the region between the two cutoff frequencies is a stopband separating two passbands. The low-pass filter provides the low-frequency passband via the lower path and the high-pass filter provides the high-frequency passband via the upper path. Frequencies between the two cutoffs fall in the stopband of both filters and are not transmitted through either path in the parallel connection. As a result, the two filters produce the stopband of the resulting bandstop filter. When $\omega_{CHP} >> \omega_{CLP}$, two cutoff frequencies are approximately $\omega_{C1} \approx \omega_{CLP}$ and $\omega_{C2} \approx \omega_{CHP}$.

FIGURE 14–34 *Parallel connection of high-pass and low-pass filters.*

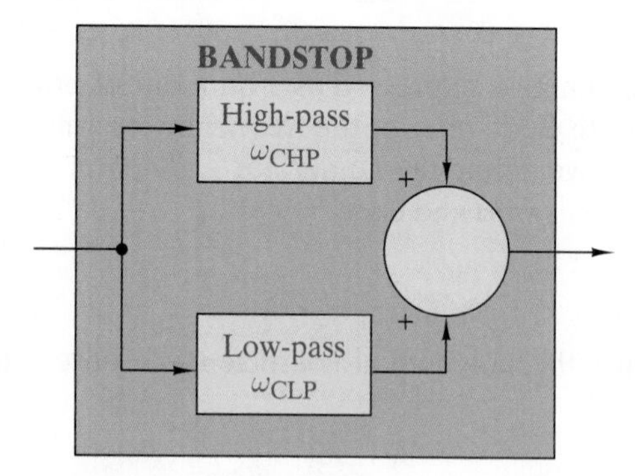

DESIGN EXAMPLE 14–11

Use second-order Butterworth low-pass and high-pass functions to obtain a fourth-order bandstop function with passband gains of 0 dB and cutoff frequencies at $\omega_{C1} = 10$ rad/s and $\omega_{C2} = 50$ rad/s.

SOLUTION:

In Example 14–10 we obtained a Butterworth bandpass response using a cascade connection of a second-order high-pass function and a low-pass function with $\omega_{CLP} = 50 >> \omega_{CHP} = 10$. To obtain a bandstop response, we interchange the cutoff frequencies and write the two functions as

$$T_{LP}(s) = \frac{1}{(s/10)^2 + \sqrt{2}(s/10) + 1}$$

$$T_{HP}(s) = \frac{(s/50)^2}{(s/50)^2 + \sqrt{2}(s/50) + 1}$$

We now have $\omega_{CLP} = 10 << \omega_{CHP} = 50$, which leads to a bandstop response. For the parallel connection the overall transfer function is the sum of the two transfer functions.

$$T_{LP}(s) + T_{HP}(s) = \frac{1}{(s/10)^2 + \sqrt{2}(s/10) + 1} + \frac{(s/50)^2}{(s/50)^2 + \sqrt{2}(s/50) + 1}$$

$$= \frac{s^4 + 10\sqrt{2}s^3 + 200s^2 + 5000\sqrt{2}s + 250{,}000}{s^4 + 60\sqrt{2}s^3 + 3600s^2 + 30{,}000\sqrt{2}s + 250{,}000}$$

With the transfer function expressed as a quotient of polynomials, we use MATLAB to generate Bode plots of the frequency response. In the MATLAB command window, we define the numerator and denominator polynomials using the row matrices `num` and `den`.

```
num=[1 10*sqrt(2) 200 5000*sqrt(2) 250000];
den=[1 60*sqrt(2) 3600 30000*sqrt(2) 250000];
```

These matrices list the polynomial coefficients in descending order. The MATLAB command

```
bode(num,den)
```

produces the Bode plots shown in Figure 14–35. The gain displays a bandstop response with a lower cutoff frequency at 10 rad/s and an upper cutoff

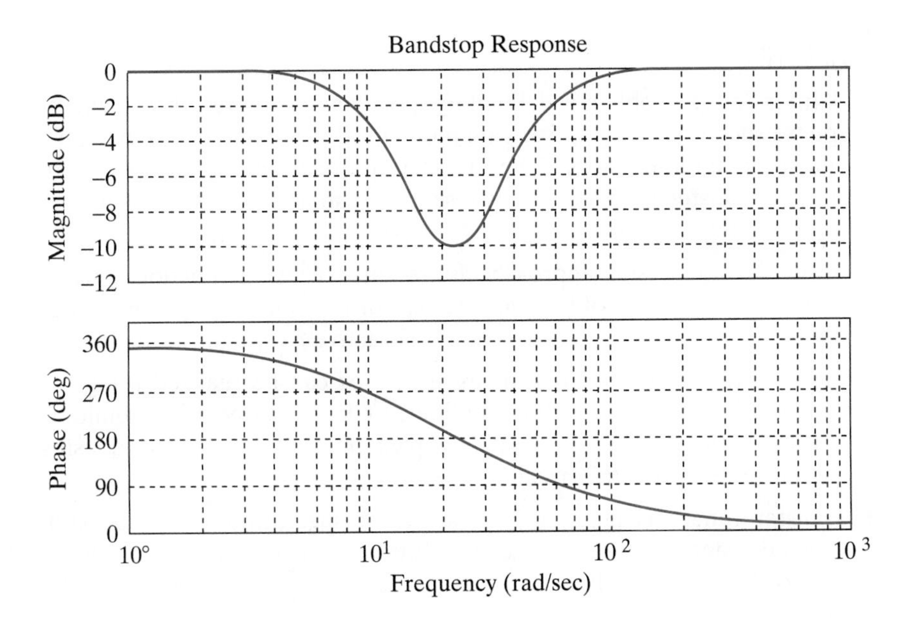

FIGURE 14–35

frequency at 50 rad/s. These two cutoff frequencies come from the low-pass and high-pass functions, respectively. The phase shift changes from +0° at low frequency to –360° at high frequency, passing through –180° at the stopband center frequency of $\sqrt{10 \times 50} = 22.36$ rad/s. ■

When the two cutoff frequencies are widely separated, we can realize wide-band bandpass and bandstop filters using a cascade or parallel connection of low-pass and high-pass filters. The design problem reduces to designing separate low-pass and high-pass filters and then connecting them in cascade or parallel to obtain the required overall response.

Exercise 14–15

Construct Butterworth low-pass and high-pass transfer functions whose cascade connection produces a bandpass function with cutoff frequencies at 20 rad/s and 500 rad/s, a passband gain of 0 dB, and a stopband gain less than –20 dB at 5 rad/s and 2000 rad/s.

Answer:

$$T(s) = \left[\frac{500^2}{s^2 + 707s + 500^2}\right]\left[\frac{s^2}{s^2 + 28.3s + 400}\right]$$

Exercise 14–16

Develop Butterworth low-pass and high-pass transfer functions whose parallel connection produces a bandstop filter with cutoff frequencies at 2 rad/s and 800 rad/s, passband gains of 20 dB, and stopband gains less than –30 dB at 20 rad/s and 80 rad/s.

Answer:

$$T_{LP}(s) = \frac{80}{(s^2 + 2s + 4)(s + 2)}$$

$$T_{HP}(s) = \frac{10s^3}{(s^2 + 800s + 800^2)(s + 800)}$$

SUMMARY

- A filter design problem is defined by specifying attributes of the gain response such as a straight-line gain plot, cutoff frequency, passband gain, and stopband attenuation. The first step in the design process is to construct a transfer function $T(s)$ whose gain response meets the specification requirements.
- In the cascade design approach, the required transfer function is partitioned into a product of first- and second-order transfer functions, which can be independently realized using basic active *RC* building blocks.
- Transfer functions with real poles and zeros can be realized using the voltage divider, noninverting amplifier, or inverting amplifier building blocks. Transfer functions with complex poles can be realized using second-order active *RC* circuits.
- Transfer functions meeting low-pass filter specifications can be constructed using first-order cascade, Butterworth, or Chebychev poles. First-order cascade filters are easy to design but have poor stopband

performance. Butterworth responses produce maximally flat passband responses and more stopband attenuation than a first-order cascade with the same number of poles. The Chebychev responses produce equal-ripple passband responses and more stopband attenuation than the Butterworth response with the same number of poles.

- A high-pass transfer function can be constructed from a low-pass prototype by replacing s with ω_C^2/s. Bandpass (bandstop) filters can be constructed using a cascade (parallel) connection of a low-pass and a high-pass filter.

PROBLEMS

OBJECTIVE 14–1 SECOND-ORDER FILTER ANALYSIS (SECTS. 14–1, 14–2)

(a) Given a second-order filter circuit, find a specified transfer function.
(b) Given the transfer function of a second-order circuit, develop a method of selecting the element values to achieve specified values of ζ and ω_0.
See Examples 14–1, 14–2, 14–3, 14–4

14–1 Interchanging the positions of the resistors and capacitors converts the low-pass filter in Figure 14–3(a) into the high-pass filter in Figure 14–6(a). This *CR-RC* interchange involves replacing R_k by $1/C_k s$ and $C_k s$ by $1/R_k$. Show that this interchange converts the low-pass transfer function in Eq. (14–3) into the high-pass function in Eq. (14–9).

14–2 Show that the circuit in Figure 14–9 has the bandpass transfer function in Eq. (14–15).

14–3 Show that the circuit in Figure 14–11 has the bandstop transfer function in Eq. (14–20).

14–4 Show that the active filter in Figure P14–4 has a transfer function of the form

$$T(s) = \frac{V_2(s)}{V_1(s)} = \frac{1}{R_1R_2C_1C_2s^2 + R_2C_2s + 1}$$

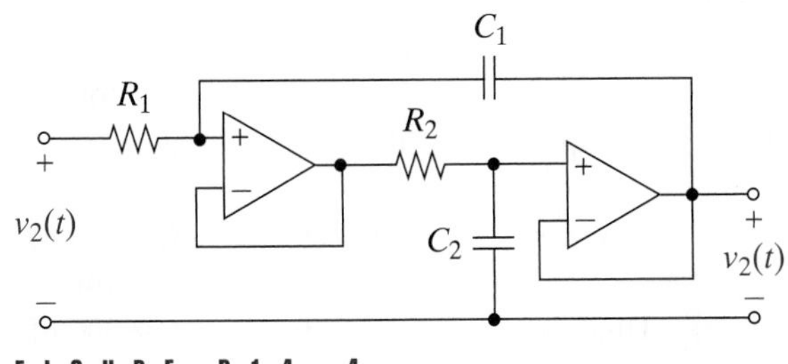

FIGURE P14–4

Using $C_1 = C_2 = C$, develop a method of selecting the element values to achieve given values of ζ and ω_0.

14–5 The circuit in Figure 14–3(a) has a low-pass transfer function given in Eq. (14–3) and repeated below:

$$T(s) = \frac{V_2(s)}{V_1(s)} = \frac{\mu}{R_1R_2C_1C_2s^2 + (R_1C_1 + R_1C_2 + R_2C_2 - \mu R_1C_1)s + 1}$$

In Sect. 14–2 we developed *equal element* and *unity gain* design methods for this circuit. This problem explores an *equal time constant* design method. Using $R_1C_1 = R_2C_2$ and $\mu = 2$, develop a method of selecting the element values to achieve given values of ζ and ω_0.

14–6 The active filter in Figure P14–6 has a transfer function of the form

$$T(s) = \frac{V_2(s)}{V_1(s)} = \frac{-R_2/R_1}{R_2R_3C_1C_2s^2 + [R_3C_2 + R_2C_2(1 + R_3/R_1)]s + 1}$$

Using $R_1 = R_2 = R_3 = R$, develop a method of selecting element values to achieve given values of ζ and ω_0.

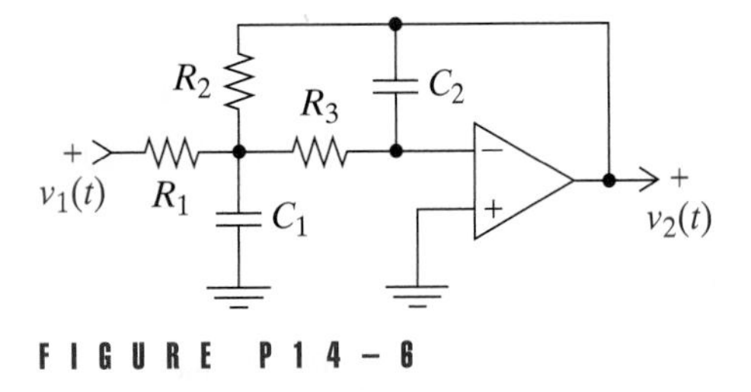

FIGURE P14–6

14–7 Show that the active filter in Figure P14–7 has a transfer function of the form

$$T(s) = \frac{V_2(s)}{V_1(s)} = \frac{R_1R_2C_1C_2s^2}{R_1R_2C_1C_2s^2 + R_1C_1s + 1}$$

Using $C_1 = C_2 = C$, develop a method of selecting the element values to achieve given values of ζ and ω_0.

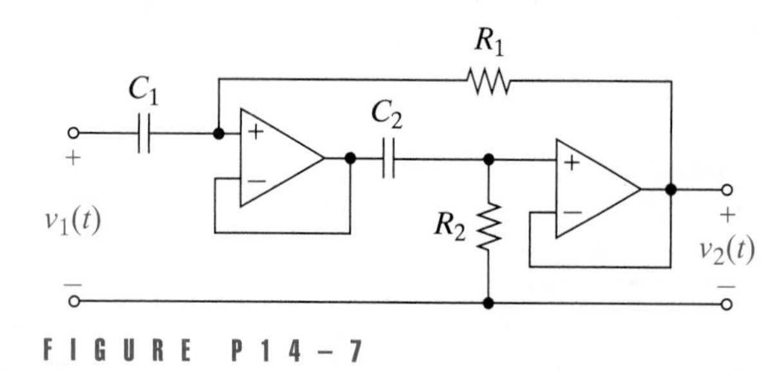

FIGURE P14–7

14–8 The circuit in Figure 14–6(a) has a high-pass transfer function given in Eq. (14–9) and repeated below:

$$T(s) = \frac{V_2(s)}{V_1(s)}$$
$$= \frac{\mu R_1R_2C_1C_2s^2}{R_1R_2C_1C_2s^2 + (R_2C_2 + R_1C_1 + R_1C_2 - \mu R_2C_2)s + 1}$$

In Sect. 14–2 we developed *equal element* and *unity gain* design methods for this circuit. This problem explores an *equal time constant* design method. Using $R_1C_1 = R_2C_2$ and $\mu = 2$, develop a method of selecting the element values to achieve given values of ζ and ω_0.

14–9 Show that the active filter in Figure P14–9 has a transfer function of the form

$$T(s) = \frac{V_2(s)}{V_1(s)} = \frac{-R_1C_1s}{R_1R_2C_1C_2s^2 + (R_1C_2 + R_2C_2)s + 1}$$

Using $R_1 = R_2 = R$, develop a method of selecting the element values to achieve given values of ζ and ω_0.

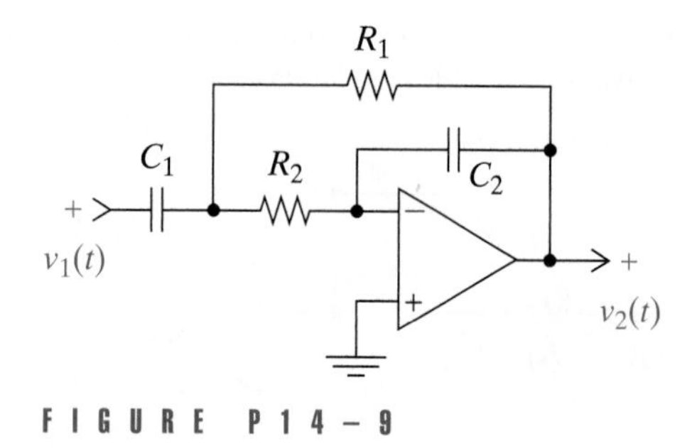

FIGURE P14–9

14–10 The active filter in Figure P14–10 has a transfer function of the form

$$T(s) = \frac{V_2(s)}{V_1(s)}$$
$$= \frac{-R_3C_2s}{R_1R_3C_1C_2s^2 + (R_1C_1 + R_1C_2)s + 1 + R_1/R_2}$$

Using $C_1 = C_2 = C$ and $R_1 = R_2 = R$, develop a method of selecting the element values to achieve given values of ζ and ω_0.

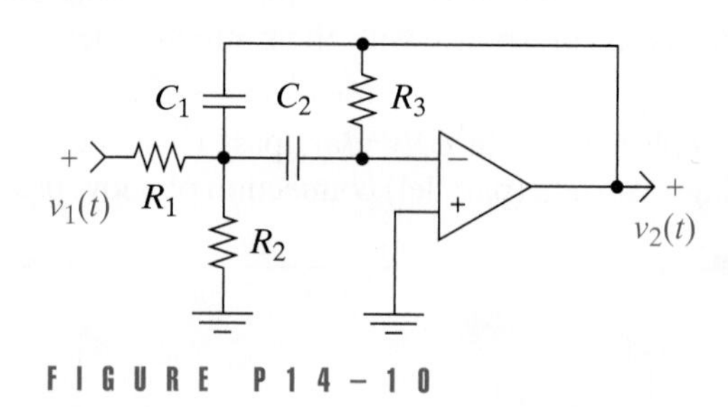

FIGURE P14–10

OBJECTIVE 14–2 SECOND-ORDER FILTER DESIGN (SECTS. 14–2, 14–3)

(a) Construct a second-order transfer function with prescribed filter characteristics.
(b) Design a second-order circuit with prescribed filter characteristics.
See Examples 14–1, 14–2, 14–3, 14–4 and Exercises 14–1, 14–2, 14–3, 14–4, 14–5, 14–6, 14–7, 14–8

Construct second-order transfer functions that meet the following requirements:

Problem	Type	ω_0 (rad/s)	ζ	$\lvert T(j\omega_0)\rvert$	Constraints
14–11	Low pass	500	?	5	dc gain of 10
14–12	High pass	500	0.5	?	Infinite frequency gain of 10
14–13	Bandpass	1000	?	5	B = 200 rad/s
14–14	Bandpass	2000	0.25	10	dc gain of zero
14–15	Bandstop	1000	0.1	0	dc gain of 10

Design second-order active filters that meet the following requirements:

Problem	Type	ω_0 (rad/s)	ζ	Constraints
14–16 D	Low pass	2000	0.5	Use 10-kΩ resistors
14–17 D	Low pass	2500	0.8	dc gain of 20 dB
14–18 D	Low pass	2000	0.25	Use 200-nF capacitors
14–19 D	High pass	2500	0.75	Use 200-nF capacitors

14–20 D	High pass	2000	?	High-frequency gain of 2
14–21 D	High pass	2500	0.25	High-frequency gain of 20 dB
14–22 D	Bandpass	5000	?	Center-frequency gain of 20 dB
14–23 D	Bandpass	5000	?	Bandwidth of 500 rad/s
14–24 D	Bandstop	5000	0.2	Resistors between 1kΩ & 100kΩ
14–25 D	Bandstop	5000	0.1	Use 200-nF capacitors

OBJECTIVE 14–3 LOW-PASS FILTER DESIGN (SECT. 14–4)

Given a low-pass filter specification:
(a) Construct a transfer function $T(s)$ that meets the specification.
(b) Design a cascade of first- and second-order circuits that meets the specification.
See Examples 14–5, 14–6, 14–7 and Exercises 14–9, 14–10, 14–11, 14–12, 14–13

Construct the lowest-order transfer functions that meet the following low-pass filter specifications. Calculate the gain (in dB) of the transfer function at $\omega = \omega_C$ and ω_{MIN}.

Problem	Pole Type	ω_C (rad/s)	T_{MAX}	ω_{MIN} (rad/s)	T_{MIN}
14–26	First-order cascade	200	0 dB	1000	−30 dB
14–27	Butterworth	200	0 dB	1000	−30 dB
14–28	Chebychev	200	0 dB	1000	−30 dB
14–29	Butterworth	100	10 dB	200	−10 dB
14–30	Chebychev	1000	10 dB	3500	−20 dB

14–31 D Design an active low-pass filter to meet the specification in Problem 14–26.

14–32 D Design an active low-pass filter to meet the specification in Problem 14–27.

14–33 D Design an active low-pass filter to meet the specification in Problem 14–28.

14–34 D Design an active low-pass filter to meet the specification in Problem 14–29.

14–35 D Design an active low-pass filter to meet the specification in Problem 14–30.

14–36 D Design a third-order Butterworth low-pass filter with $\omega_C = 250$ rad/s and a passband gain of 5 dB.

14–37 D Design a third-order Chebychev low-pass filter with $\omega_C = 5$ krad/s and a passband gain of 10 dB.

14–38 D A low-pass filter is needed to suppress the harmonics in a periodic waveform with $f_0 = 1$ kHz. The filter must have unity passband gain, less than −25 dB gain at the third harmonic, and less than −40 dB gain at the fifth harmonic. Design a Butterworth filter that meets these requirements.

14–39 D Design a low-pass filter with 6-dB passband gain, a cutoff frequency of 3.2 kHz, and stopband gains less than −20 dB at 6.4 kHz and −40 dB at 12.8 kHz. Calculate the gains realized by your design at 3.2 kHz, 6.4 kHz, and 12.8 kHz.

14–40 D A pesky signal at 200 kHz is interfering with a desired signal at 25 kHz. A careful analysis suggests that reducing the interfering signal by 30 dB will eliminate the problem, provided the desired signal is not reduced by more than 3 dB. Design an active *RC* filter that meets these requirements.

OBJECTIVE 14–4 HIGH-PASS, BANDPASS, AND BANDSTOP FILTER DESIGN (SECTS. 14–5, 14–6)

Given a high-pass, wide-band bandpass, or bandstop filter specification:
(a) Construct a transfer function $T(s)$ that meets the specification.
(b) Design a cascade or parallel connection of first- and second-order circuits that meets the specification.
See Examples 14–8, 14–9, 14–10, 14–11 and Exercises 14–14, 14–15, 14–16

Construct the lowest-order high-pass transfer functions that meet the following filter specifications. Calculate the gain (in dB) of the transfer function at $\omega = \omega_C$ and ω_{MIN}.

Problem	Pole Type	ω_C (rad/s)	T_{MAX}	ω_{MIN} (rad/s)	T_{MIN}
14–41	Butterworth	2000	20 dB	200	−20 dB
14–42	Butterworth	2000	0 dB	500	−30 dB
14–43	Chebychev	2000	0 dB	500	−25 dB
14–44	Butterworth	1000	10 dB	500	−10 dB

14–45 D Design an active high-pass filter to meet the specification in Problem 14–41.

14–46 D Design an active high-pass filter to meet the specification in Problem 14–42.

14–47 D Design an active high-pass filter to meet the specification in Problem 14–43.

14–48 D Design an active high-pass filter to meet the specification in Problem 14–44.

14–49 **D** Design a fourth-order Butterworth bandpass filter with unity passband gain and cutoff frequencies at 400 and 2500 rad/s.

14–50 **D** A bandstop filter specification requires cutoff frequencies at 3 krad/s and 40 krad/s, passband gains of 0 dB, and a stopband gain less than −15 dB at 10 krad/s.

(a) Construct a transfer function $T(s)$ using a second-order Butterworth low-pass function and a second-order Butterworth high-pass function.

(b) Calculate the actual gain of $T(s)$ at $\omega = 3$ krad/s, 10 krad/s, and 40 krad/s.

INTEGRATING PROBLEMS

14–51 **E** Design Evaluation

A need exists for a third-order Butterworth low-pass filter with a cutoff frequency of 2 krad and a dc gain of 0 dB. The design department has proposed the circuit in Figure P14–51. As a junior engineer in the manufacturing department, you have been asked to verify the design and suggest modifications that would simplify production.

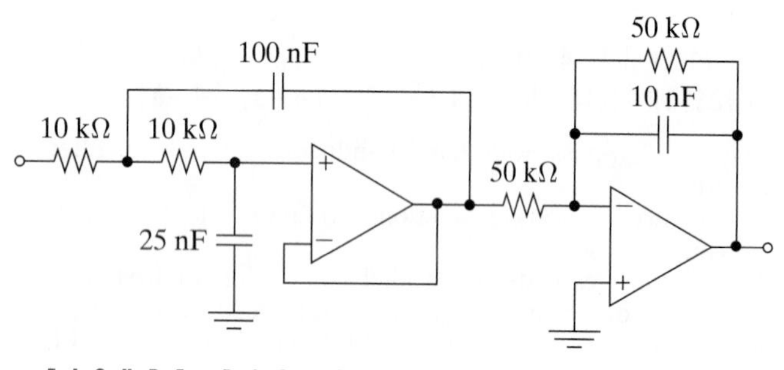

FIGURE P14–51

14–52 **D** Modifying an Existing Circuit

One of your company's products includes the passive *RLC* filter and OP AMP buffer circuit in Figure P14–52. The supplier of the inductor is no longer in business and a suitable replacement is not available, even on eBay. You have been asked to design a suitable inductorless replacement. To minimize production changes, your design must use the existing OP AMP as is and either the 1-kΩ resistor or the 100-nF capacitor or both, if possible.

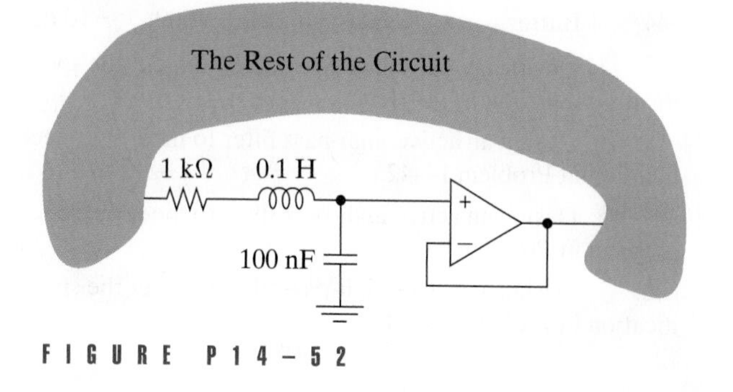

FIGURE P14–52

14–53 **A** What's a High-Pass Filter?

Ten years after earning a BSEE, you return for a master's degree and sign on as the laboratory instructor for the basic circuit analysis course. One experiment asks the students to build the active filter in Figure P14–53 and measure its gain response over the range from 150 Hz to 15 kHz. The lab instructions say the circuit is a high-pass filter with $\omega_0 = 10$ krad/s and an infinite frequency gain of 0 dB. Everything goes well until a student, intrigued by the concept of infinite frequency, inputs 1 MHz and measures a gain of only 0.7. The student then inputs 2 MHz and measures a gain of 0.45. The high-pass filter appears to be a bandpass filter! Motivated by an insatiable thirst for understanding, the student asks you for an explanation. You first check the student's circuit and find it to be correct. You next replace the OP AMP and get almost the same results. Desperate for an explanation (your credibility is on the line here), you read the course textbook (it's Thomas and Rosa) and find the answer in Chapter 12 (Example 12–4). What do you tell the student?

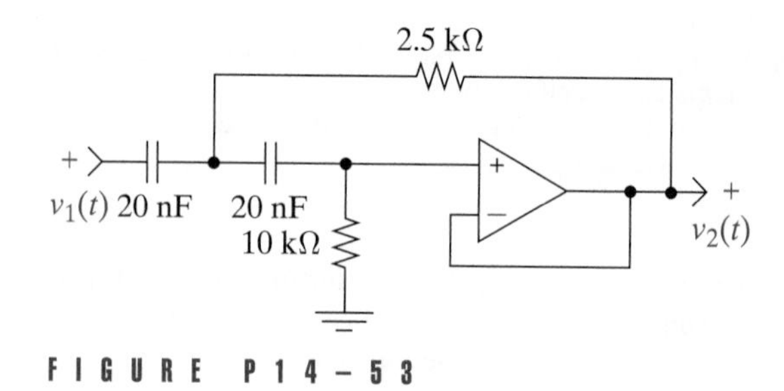

FIGURE P14–53

14–54 **A** Bandpass to Bandstop Transformation

The three-terminal circuit in Figure 14–54(a) has a bandpass transfer function of the form

$$T(s) = \frac{V_O(s)}{V_S(s)} = \frac{2\zeta(s/\omega_0)}{(s/\omega_0)^2 + 2\zeta(s/\omega_0) + 1}$$

Show that the circuit in Figure 14–54(b) has a bandstop transfer function of the form

$$T(s) = \frac{V_O(s)}{V_S(s)} = \frac{(s/\omega_0)^2 + 1}{(s/\omega_0)^2 + 2\zeta(s/\omega_0) + 1}$$

That is, show that interchanging the input and ground terminals changes a unity gain bandpass circuit into a unity gain bandstop circuit.

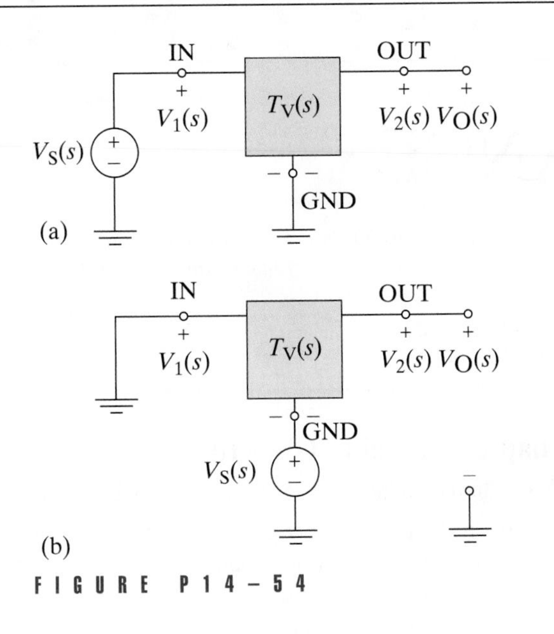

FIGURE P14–54

14–55 A Third-Order Butterworth Circuit

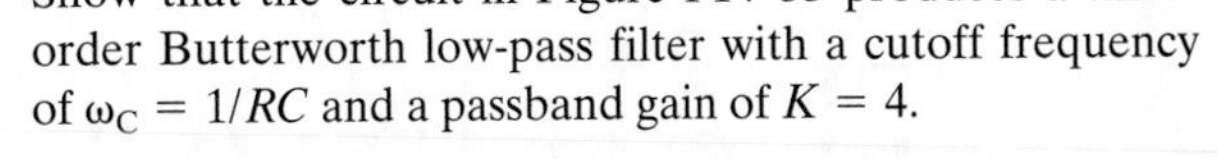

Show that the circuit in Figure P14–55 produces a third-order Butterworth low-pass filter with a cutoff frequency of $\omega_C = 1/RC$ and a passband gain of $K = 4$.

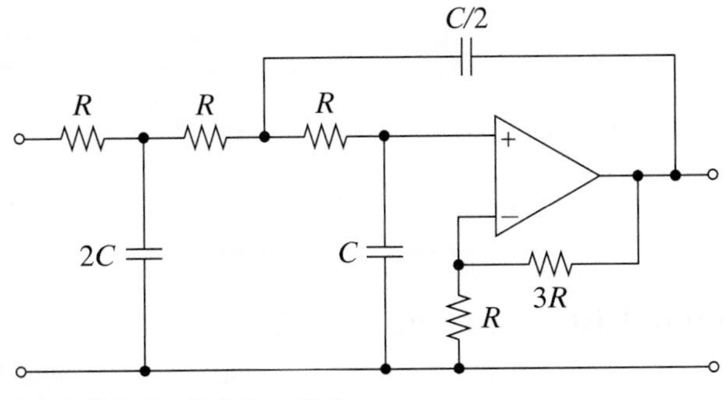

FIGURE P14–55

CHAPTER 15

MUTUAL INDUCTANCE

From the foregoing facts, it appears that a current of electricity is produced, for an instant, in a helix of copper wire surrounding a piece of soft iron whenever magnetism is induced in the iron; also that an instantaneous current in one or the other direction accompanies every change in the magnetic intensity of the iron.

Joseph Henry, 1831,
American Physicist

Some History Behind This Chapter

The discovery of electromagnetic induction led Michael Faraday to wind two separate insulated wires around an iron ring. One winding was connected to a battery via a switch and the other to a galvanometer. He observed that the current in the first winding did indeed induce a current in the second, but only when the switch was opened or closed. From this he correctly deduced that it was the change in current that produced the inductive effect. Faraday's coils and iron ring were actually a crude transformer. Practical ac power transformers were perfected by the British engineers Lucien Gauland and Josiah Gibbs in the early 1880s.

Why This Chapter Is Important Today

Coupled coils are found in applications such as power systems, communications, and high-quality audio. In this chapter you will learn how to model the magnetic coupling between coils using the concept of mutual inductance. This concept leads to a new circuit element called a transformer, which provides impedance matching, electrical isolation, and changes in the voltage level in power systems. The large power transformers used in such systems are very efficient, with internal losses often less than 1%.

Chapter Sections

15–1 Coupled Inductors

15–2 The Dot Conventions

15–3 Energy Analysis

15–4 The Ideal Transformer

15–5 Transformers in the Sinusoidal Steady State

15–6 Transformer Equivalent Circuits

Chapter Learning Objectives

15-1 Mutual Inductance (Sects. 15-1, 15-2, 15-3)

(a) Given the current through or voltage across two coupled inductors, find other currents or voltages.

(b) Given device and circuit parameters, find characteristics of coupled inductors.

15-2 The Ideal Transformer (Sect. 15-4)

Given a circuit containing ideal transformers:

(a) Find specified voltages, currents, powers, and equivalent circuits.

(b) Select the turns ratio to meet prescribed conditions.

15-3 Transformers in the Sinusoidal Steady State (Sects. 15-5, 15-6)

Given a linear circuit with a transformer operating in the sinusoidal steady-state, find phasor voltages and currents, average powers, and equivalent impedances.

15–1 COUPLED INDUCTORS

The i–v characteristic of the inductor results from the magnetic field produced by current through a coil of wire. A constant current produces a constant magnetic field that forms closed loops of magnetic flux lines in the vicinity of the inductor. A changing current causes these closed loops to expand or contract, thereby cutting the turns in the winding that makes up the inductor. Faraday's law states that voltage across the inductor is equal to the time rate of change of the total flux linkage. In earlier chapters we expressed this relationship between a time-varying current and an induced voltage in terms of a circuit parameter called inductance L.

Now suppose that a second inductor is brought close to the first so that the flux from the first inductor links with the turns of the second inductor. If the current in the first inductor is changing, then this flux linkage will generate a voltage in the second inductor. The magnetic coupling between the changing current in one inductor and the voltage generated in a second inductor produces **mutual inductance**.

i–v CHARACTERISTICS

The i–v characteristics of coupled inductors unavoidably involve describing the effects observed in one inductor due to causes occurring in the other. We will use a double-subscript notation because it clearly identifies the various cause-and-effect relationships. The first subscript indicates the inductor in which the effect takes place, and the second identifies the inductor in which the cause occurs. For example, $v_{12}(t)$ is the voltage across inductor 1 due to causes occurring in inductor 2, whereas $v_{11}(t)$ is the voltage across inductor 1 due to causes occurring in inductor 1 itself.

We begin by assuming that the inductors are far apart, as shown in Figure 15–1(a). Under these circumstances there is no magnetic coupling between the two. A current $i_1(t)$ passes through the N_1 turns of the first inductor and $i_2(t)$ through N_2 turns in the second. Each inductor produces a flux

$$\begin{array}{cc} \text{Inductor 1} & \text{Inductor 2} \\ \phi_1(t) = k_1 N_1 i_1(t) & \phi_2(t) = k_2 N_2 i_2(t) \end{array} \tag{15–1}$$

where k_1 and k_2 are proportionality constants. The flux linkage in each inductor is proportional to the number of turns:

$$\begin{array}{cc} \text{Inductor 1} & \text{Inductor 2} \\ \lambda_{11}(t) = N_1 \phi_1(t) & \lambda_{22}(t) = N_2 \phi_2(t) \end{array} \tag{15–2}$$

By Faraday's law the voltage across an inductor is equal to the time rate of change of the flux linkage. Using Eqs. (15–1) and (15–2) together with relationship between voltage and time rate of change of flux linkage gives

$$\begin{aligned} \text{Inductor 1:} \quad v_{11}(t) &= \frac{d\lambda_{11}(t)}{dt} = N_1 \frac{d\phi_1(t)}{dt} = [k_1 N_1^2] \frac{di_1(t)}{dt} \\ \text{Inductor 2:} \quad v_{22}(t) &= \frac{d\lambda_{22}(t)}{dt} = N_2 \frac{d\phi_2(t)}{dt} = [k_2 N_2^2] \frac{di_2(t)}{dt} \end{aligned} \tag{15–3}$$

Equations (15–3) provide the i–v relationships for the inductors when there is no mutual coupling. These results are the same as previously found in Chapter 6.

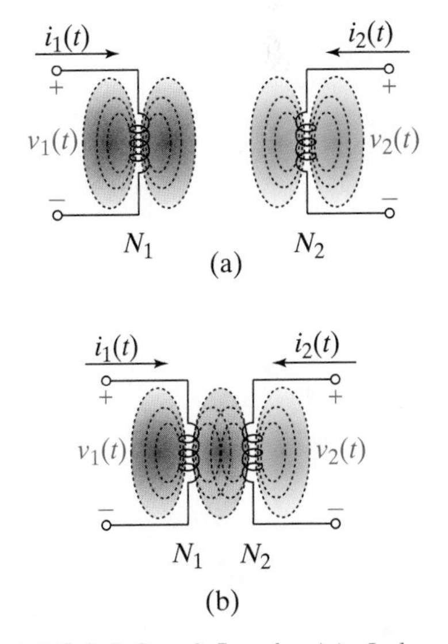

FIGURE 15–1 (a) Inductors separated, only self-inductance present. (b) Inductors coupled, both self- and mutual inductance present.

Now suppose the inductors are brought close together so that part of the flux produced by each inductor intercepts the other, as indicated in Figure 15–1(b). That is, part (but not necessarily all) of the fluxes $\phi_1(t)$ and $\phi_2(t)$ in Eq. (15–1) intercept the opposite inductor. We describe the cross coupling using the double-subscript notation:

$$\begin{array}{cc} \text{Inductor 1} & \text{Inductor 2} \\ \phi_{12}(t) = k_{12}N_2 i_2(t) & \phi_{21}(t) = k_{21}N_1 i_1(t) \end{array} \tag{15–4}$$

The quantity $\phi_{12}(t)$ is the flux intercepting inductor 1 due to the current in inductor 2, and $\phi_{21}(t)$ is the flux intercepting inductor 2 due to the current in inductor 1. The total flux linkage in each inductor is proportional to the number of turns:

$$\begin{array}{cc} \text{Inductor 1} & \text{Inductor 2} \\ \lambda_{12}(t) = N_1\phi_{12}(t) & \lambda_{21}(t) = N_2\phi_{21}(t) \end{array} \tag{15–5}$$

By Faraday's law, the voltage across a winding is equal to the time rate of change of the flux linkage. Using Eqs. (15–4) and (15–5) together with derivative relationship, the time rate of change of flux linkages and voltages gives

$$\begin{array}{ll} \text{Inductor 1:} & v_{12}(t) = \dfrac{d\lambda_{12}(t)}{dt} = N_1\dfrac{d\phi_{12}(t)}{dt} = [k_{12}N_1N_2]\dfrac{di_2(t)}{dt} \\ \text{Inductor 2:} & v_{21}(t) = \dfrac{d\lambda_{21}(t)}{dt} = N_2\dfrac{d\phi_{21}(t)}{dt} = [k_{21}N_1N_2]\dfrac{di_1(t)}{dt} \end{array} \tag{15–6}$$

The expressions in Eq. (15–6) are the i–v relationships describing the cross coupling between inductors when there is mutual coupling.

When the magnetic medium supporting the fluxes is linear, the superposition principle applies, and the total voltage across the inductors is the sum of the results in Eqs. (15–3) and (15–6):

$$\begin{array}{ll} \text{Inductor 1:} & v_1(t) = v_{11}(t) + v_{12}(t) \\ & \quad = [k_1N_1^2]\dfrac{di_1(t)}{dt} + [k_{12}N_1N_2]\dfrac{di_2(t)}{dt} \\ \text{Inductor 2:} & v_2(t) = v_{21}(t) + v_{22}(t) \\ & \quad = [k_{21}N_1N_2]\dfrac{di_1(t)}{dt} + [k_2N_2^2]\dfrac{di_2(t)}{dt} \end{array} \tag{15–7}$$

We can identify four inductance parameters in these equations:

$$L_1 = k_1N_1^2 \qquad L_2 = k_2N_2^2 \tag{15–8}$$

and

$$M_{12} = k_{12}N_1N_2 \qquad M_{21} = k_{21}N_2N_1 \tag{15–9}$$

The two inductance parameters in Eq. (15–8) are the **self-inductance** of the inductors. The two parameters in Eq. (15–9) are the **mutual inductances** between the two inductors. In a linear magnetic medium, $k_{12} = k_{21} = k_M$. As a result, we can define a single mutual inductance parameter M as

$$M = M_{12} = M_{21} = k_MN_1N_2 \tag{15–10}$$

Using the definitions in Eqs. (15–8) and (15–10), the *i–v* characteristics of two coupled inductors are

$$\begin{aligned} \text{Inductor 1:} \quad v_1(t) &= L_1 \frac{di_1(t)}{dt} + M \frac{di_2(t)}{dt} \\ \text{Inductor 2:} \quad v_2(t) &= M \frac{di_1(t)}{dt} + L_2 \frac{di_2(t)}{dt} \end{aligned} \tag{15–11}$$

Coupled inductors involve three inductance parameters, the two self-inductances L_1 and L_2 and the mutual inductance M.

The preceding development assumes that the cross coupling is additive. Additive coupling means that a positive rate of change of current in inductor 2 induces a positive voltage in inductor 1, and vice versa. The additive assumption produces the positive sign on the mutual inductance terms in Eq. (15–11). Unhappily, it is possible for a positive rate of change of current in one inductor to induce a negative voltage in the other. To account for additive and subtractive coupling, the general form of the coupled inductor *i–v* characteristics includes a ± sign on the mutual inductance terms:

$$\begin{aligned} \text{Inductor 1:} \quad v_1(t) &= L_1 \frac{di_1(t)}{dt} \pm M \frac{di_2(t)}{dt} \\ \text{Inductor 2:} \quad v_2(t) &= \pm M \frac{di_1(t)}{dt} + L_2 \frac{di_2(t)}{dt} \end{aligned} \tag{15–12}$$

When applying these element equations, it is necessary to know when to use a plus sign and when to use a minus sign.

15–2 THE DOT CONVENTION

The parameter M is positive, so the question is, What sign should be placed in front of this positive parameter in the *i–v* relationships in Eq. (15–12)? The correct sign depends on two things: (1) the spatial orientation of the two windings and (2) the reference marks given to the currents and voltages.

Figure 15–2 shows the additive and subtractive spatial orientation of two coupled inductors. In either case, the direction of the flux produced by a current is found using the right-hand rule treated in physics courses. In the additive case, currents i_1 and i_2 both produce clockwise fluxes ϕ_1 and ϕ_2. In the subtractive case, the currents produce opposing fluxes because ϕ_1 is clockwise and ϕ_2 is counterclockwise. The sign for the mutual inductance term is positive for the additive orientation and negative for the subtractive case.

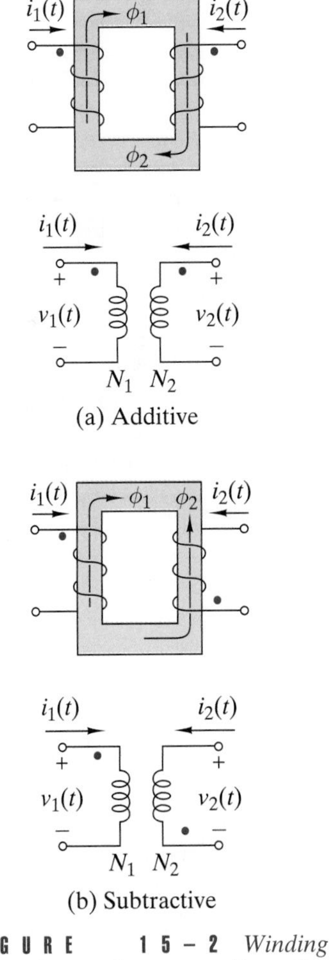

FIGURE 15–2 *Winding orientations and corresponding reference dots.*

In general, it is awkward to show the spatial features of the windings in circuit diagrams. The dots shown near one terminal of each winding in Figure 15–2 are special reference marks indicating the relative orientation of the windings. The reference directions for the currents and voltages are arbitrary. They can be changed as long as we follow the passive sign convention. However, the dots indicate physical attributes of the windings that make up the coupled inductors. They are not arbitrary. They cannot be changed.

The correct sign for the mutual inductance term hinges on how the reference marks for currents and voltages are assigned relative to the dots.

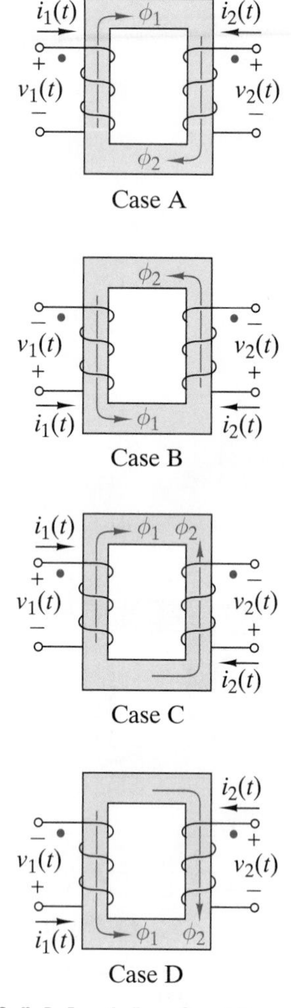

FIGURE 15-3 *All possible current and voltage reference marks for a fixed winding orientation.*

For a given winding orientation, Figure 15–3 shows all four possible current and voltage reference assignments under the passive sign convention. In cases A and B the fluxes are additive, so the mutual inductance term is positive. In cases C and D the fluxes are subtractive and the mutual inductance term is negative. From these results we derive the following rule:

> ***Mutual inductance is additive when both current reference directions point toward or both point away from dotted terminals; otherwise, it is subtractive.***

Because we always use the passive sign convention, the rule can be stated in terms of voltages as follows:

> ***Mutual inductance is additive when the voltage reference marks are both positive or both negative at the dotted terminals; otherwise, it is subtractive.***

Because the current reference directions or voltage polarity marks can be changed, a corollary of this rule is that we can always assign reference directions so that the positive sign applies to the mutual inductance. This corollary is important because a positive sign is built into the mutual inductance models in circuit analysis programs like SPICE.

It may seem that all of this discussion about signs and dots is much ado about nothing. Not so. First, selecting the wrong sign can have nontrivial consequences because the polarity of the output signal is reversed. If the signal is a command to your car's autopilot, then you really need to know whether stepping on the brake pedal will slow the car down or speed it up. Another problem is that the two coupled inductors may appear in different parts of a circuit diagram and may be assigned voltage and current reference marks for other reasons. In such circumstances it is important to understand the underlying principle to select the correct sign for the mutual inductance term.

The following examples and exercises illustrate selecting the correct sign and applying the i–v characteristics in Eq. (15–12).

EXAMPLE 15–1

The source voltage in Figure 15–4 is $v_S(t) = 10 \cos 100t$ V. Find the output voltage $v_2(t)$.

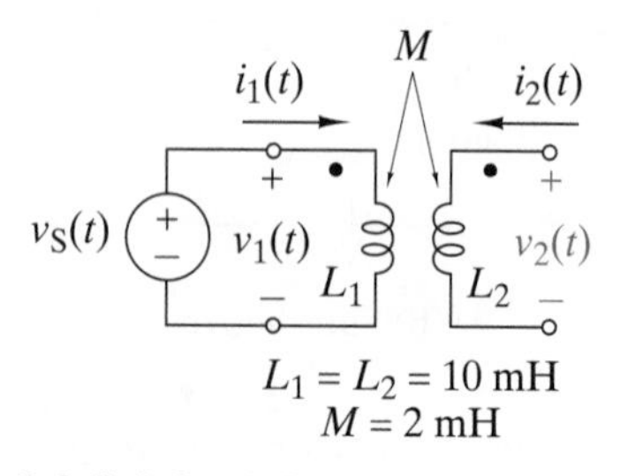

FIGURE 15-4

SOLUTION:

The sign of the mutual inductance term is positive because both current reference directions are toward the dots. Since the load connected across inductor 2 is an open circuit, $i_2(t) = 0$ and the i–v equations of the coupled inductors reduce to

$$\text{Inductor 1:} \quad 10 \cos 100t = 0.01 \frac{di_1(t)}{dt} + 0$$

$$\text{Inductor 2:} \quad v_2(t) = 0.002 \frac{di_1(t)}{dt} + 0$$

Solving the first equation for di_1/dt yields

$$\frac{di_1(t)}{dt} = 1000 \cos 100t$$

Substituting this equation for $i_1(t)$ into the second equation yields

$$v_2(t) = 2 \cos 100t \text{ V (correct)}$$

If we had incorrectly chosen a minus sign, the mutual inductance would have produced a voltage of

$$v_2(t) = -2 \cos 100t \text{ V (incorrect)}$$

which differs from the correct answer by a signal inversion. ■

EXAMPLE 15–2

Solve for $v_x(t)$ in terms of $i_1(t)$ for the coupled inductors in Figure 15–5.

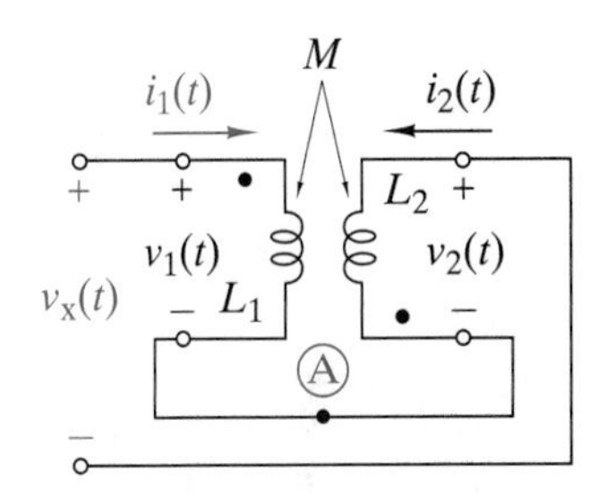

FIGURE 15–5

SOLUTION:

In this case the signs of mutual inductance terms are negative because the reference direction for $i_1(t)$ points toward the dot for inductor 1 and the reference direction for $i_2(t)$ points away from the dot for inductor 2. The coupled inductor i–v equations are

$$\text{Inductor 1:} \quad v_1(t) = L_1 \frac{di_1(t)}{dt} - M \frac{di_2(t)}{dt}$$

$$\text{Inductor 2:} \quad v_2(t) = -M \frac{di_1(t)}{dt} + L_2 \frac{di_2(t)}{dt}$$

A KCL constraint at node A requires that $i_1(t) = -i_2(t)$. Therefore, these i–v equations can be written in the form

$$\text{Inductor 1:} \quad v_1(t) = L_1 \frac{di_1(t)}{dt} - M \frac{d[-i_1(t)]}{dt}$$

$$\text{Inductor 2:} \quad v_2(t) = -M \frac{di_1(t)}{dt} + L_2 \frac{d[-i_1(t)]}{dt}$$

A KVL constraint around the loop requires that $v_x(t) = v_1(t) - v_2(t)$. Subtracting the second equation from the first yields

$$v_x(t) = L_1 \frac{di_1(t)}{dt} + M \frac{di_1(t)}{dt} + M \frac{di_1(t)}{dt} + L_2 \frac{di_1(t)}{dt}$$

$$= (L_1 + L_2 + 2M) \frac{di_1(t)}{dt}$$

In this case the two mutual inductance terms add to produce an equivalent inductance of

$$L_{EQ} = L_1 + L_2 + 2M$$

If the plus sign in the coupled inductor i–v relationships was appropriate, the mutual inductance terms would subtract to produce an equivalent inductance of $L_{EQ} = L_1 + L_2 - 2M$. Clearly it is important to have the right sign in the i–v relationships. ■

Exercise 15–1

Find $v_1(t)$ and $v_2(t)$ for the circuit in Figure 15–6.

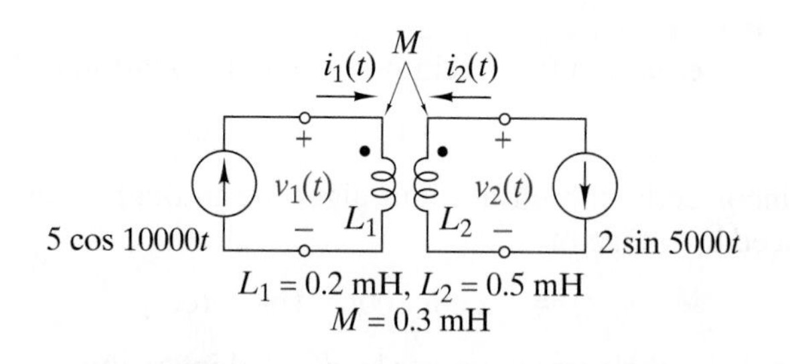

Answers:

$v_1(t) = -10 \sin 10^4 t - 3 \cos 5 \times 10^3 t$ V
$v_2(t) = -15 \sin 10^4 t - 5 \cos 5 \times 10^3 t$ V

15–3 ENERGY ANALYSIS

Calculating the total energy stored in a pair of coupled inductors reveals a fundamental limitation on allowable values of the self- and mutual inductances. To uncover this limitation, we first calculate the total power absorbed. Multiplying the first equation in Eq. (15–12) by $i_1(t)$ and the second equation by $i_2(t)$ produces

$$\begin{aligned} p_1(t) &= v_1(t)i_1(t) = L_1 i_1(t)\frac{di_1(t)}{dt} \pm Mi_1(t)\frac{di_2(t)}{dt} \\ p_2(t) &= v_2(t)i_2(t) = \pm Mi_2(t)\frac{di_1(t)}{dt} + L_2 i_2(t)\frac{di_2(t)}{dt} \end{aligned} \quad (15\text{–}13)$$

The quantities $p_1(t)$ and $p_2(t)$ are the powers absorbed with inductors 1 and 2. The total power is the sum of the individual inductor powers:

$$\begin{aligned} p(t) &= p_1(t) + p_2(t) \\ &= L_1\left[i_1(t)\frac{di_1(t)}{dt}\right] \pm M\left[i_1(t)\frac{di_2(t)}{dt} + i_2(t)\frac{di_1(t)}{dt}\right] + L_2\left[i_2(t)\frac{di_2(t)}{dt}\right] \end{aligned} \quad (15\text{–}14)$$

Each of the bracketed terms in Eq. (15–14) is a perfect derivative. Specifically,

$$\begin{aligned} i_1(t)\frac{di_1(t)}{dt} &= \frac{1}{2}\frac{di_1^2(t)}{dt} \\ i_2(t)\frac{di_2(t)}{dt} &= \frac{1}{2}\frac{di_2^2(t)}{dt} \\ i_1(t)\frac{di_2(t)}{dt} + i_2(t)\frac{di_1(t)}{dt} &= \frac{di_1(t)i_2(t)}{dt} \end{aligned} \quad (15\text{–}15)$$

Therefore, the total power in Eq. (15–14) is

$$p(t) = \frac{d}{dt}\left[\frac{1}{2}L_1 i_1^2(t) \pm M\, i_1(t)i_2(t) + \frac{1}{2}L_2\, i_2^2(t)\right] \quad (15\text{–}16)$$

Because power is the time rate of change of energy, the quantity inside the brackets in Eq. (15–16) is the total energy stored in the two inductors. That is,

$$w(t) = \frac{1}{2}L_1 i_1^2(t) \pm M\, i_1(t) i_2(t) + \frac{1}{2}L_2 i_2^2(t) \qquad (15\text{–}17)$$

In Eq. (15–17) the self-inductance terms are always positive. However, the mutual inductance term can be either positive or negative. At first glance it appears that the total energy could be negative. But the total energy must be positive; otherwise, the coupled inductors could deliver net energy to the rest of the circuit.

The condition $w(t) \geq 0$ places a constraint on the values of the self- and mutual inductances. First, if $i_2(t) = 0$, then $w(t) \geq 0$ in Eq. (15–17) requires $L_1 > 0$. Next, if $i_1(t) = 0$, then $w(t) \geq 0$ in Eq. (15–17) requires $L_2 > 0$. Finally, if $i_1(t) \neq 0$ and $i_2(t) \neq 0$, then we divide Eq. (15–17) by $[i_2(t)]^2$ and defined a variable $x = i_1/i_2$. With these changes, the energy constraint $w(t) > 0$ becomes

$$\frac{w(t)}{i_2^2(t)} = f(x) = \frac{1}{2}L_1 x^2 \pm Mx + \frac{1}{2}L_2 \geq 0 \qquad (15\text{–}18)$$

The minimum value of $f(x)$ occurs when

$$\frac{df(x)}{dx} = L_1 x \pm M = 0 \quad \text{hence} \quad x_{\text{min}} = \mp \frac{M}{L_1} \qquad (15\text{–}19)$$

The value x_{min} yields the minimum of $f(x)$ because the second derivative of $f(x)$ is positive. Substituting x_{min} back into Eq. (15–18) yields the condition

$$f(x_{\text{min}}) = \frac{1}{2}L_1\frac{M^2}{L_1^2} - \frac{M^2}{L_1} + \frac{1}{2}L_2 = \frac{1}{2}\left[-\frac{M^2}{L_1} + L_2\right] \geq 0 \qquad (15\text{–}20)$$

The constraint in Eq. (15–20) means that the stored energy in a pair of coupled inductors is positive if

$$L_1 L_2 \geq M^2 \qquad (15\text{–}21)$$

Energy considerations dictate that in any pair of coupled inductors, the product of the self-inductances must exceed the square of the mutual inductance.

The constraint in Eq. (15–21) is usually written in terms of a new parameter called the **coupling coefficient** k:

$$k = \frac{M}{\sqrt{L_1 L_2}} \leq 1 \qquad (15\text{–}22)$$

The parameter k ranges from 0 to 1. If $M = 0$, then $k = 0$ and the coupling between the inductors is zero. The condition $k = 1$ requires **perfect coupling** in which all of the flux produced by one inductor links the other. Perfect coupling is physically impossible, although careful design can produce coupling coefficients of 0.99 and higher.

The next section discusses a transformer model that assumes perfect coupling ($k = 1$). Computer-aided circuit analysis programs like SPICE specify the parameters of a pair of coupled inductors in terms of the self-inductances L_1 and L_2 and the coupling coefficient k.

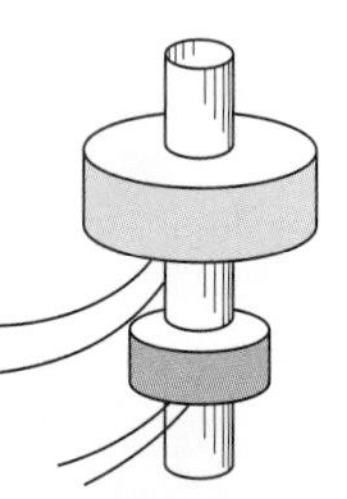

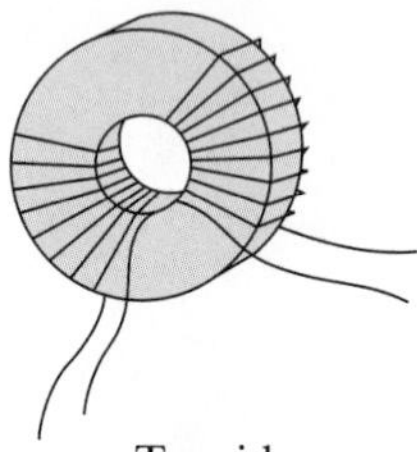

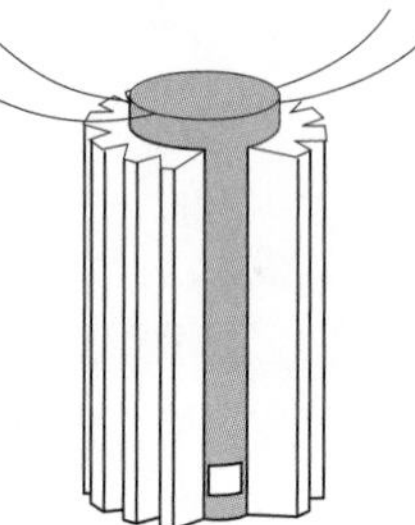

FIGURE 15–7 *Examples of transformer devices.*

Exercise 15–2

What is the coupling coefficient of the coupled inductors in Figure 15–6.

Answer: $k = 0.949$

15–4 THE IDEAL TRANSFORMER

A **transformer** is an electrical device that utilizes magnetic coupling between two inductors. Transformers find application in virtually every type of electrical system, but especially in power supplies and commercial power grids. Some example devices from these applications are shown in Figure 15–7.

In Figure 15–8 the transformer is shown as an interface device between a source and a load. The winding connected to the source is called the **primary winding**, and the one connected to the load is called the **secondary winding**. In most applications the transformer is a coupling device that transfers signals (especially power) from the source to the load. The basic purpose of the device is to change voltage and current levels so that the conditions at the source and load are compatible.

Transformer design involves two primary goals: (1) to maximize the magnetic coupling between the two windings and (2) to minimize the power loss in the windings. The first goal produces nearly perfect coupling ($k \approx 1$) so that almost all of the flux in one winding links the other. The second goal produces nearly zero power loss so that almost all of the power delivered to the primary winding transfers to the load. The **ideal transformer** is a circuit element in which coupled inductors are assumed to have perfect coupling and zero power loss. Using these two idealizations, we can derive the i–v characteristics of an ideal transformer.

PERFECT COUPLING

Perfect coupling means that all of the flux in the first winding links the second, and vice versa. Equation (15–1) defines the total flux in each winding as

$$\begin{array}{cc} \text{Winding 1} & \text{Winding 2} \\ \phi_1(t) = k_1 N_1 i_1(t) & \phi_2(t) = k_2 N_2 i_2(t) \end{array} \tag{15–23}$$

where k_1 and k_2 are proportionality constants. Equation (15–4) defines the cross coupling using the double subscript notation:

$$\begin{array}{cc} \text{Winding 1} & \text{Winding 2} \\ \phi_{12}(t) = k_{12} N_2 i_2(t) & \phi_{21}(t) = k_{21} N_1 i_1(t) \end{array} \tag{15–24}$$

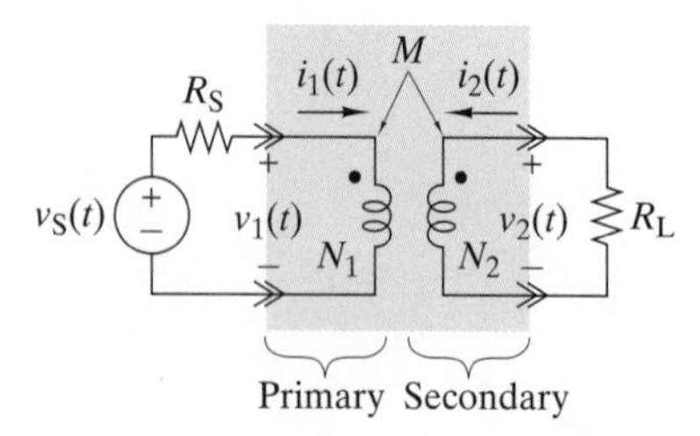

FIGURE 15–8 *Transformer connected at a source-load interface.*

In this equation $\phi_{12}(t)$ is the flux intercepting winding 1 due to the current in winding 2, and $\phi_{21}(t)$ is the flux intercepting winding 2 due to the current in winding 1. Perfect coupling means that

$$\phi_{21}(t) = \phi_1(t) \quad \text{and} \quad \phi_{12}(t) = \phi_2(t) \tag{15–25}$$

Comparing Eqs. (15–23) and (15–24) shows that perfect coupling requires $k_1 = k_{21}$ and $k_2 = k_{12}$. But in a linear magnetic medium $k_{12} = k_{21} = k_M$, so perfect coupling implies

$$k_1 = k_2 = k_{12} = k_{21} = k_M \tag{15–26}$$

Substituting the perfect coupling conditions in Eq. (15–26) into the i–v characteristics in Eq. (15–7) gives

$$\begin{aligned} v_1(t) &= [k_M N_1^2]\frac{di_1(t)}{dt} \pm [k_M N_1 N_2]\frac{di_2(t)}{dt} \\ v_2(t) &= \pm [k_M N_1 N_2]\frac{di_1(t)}{dt} + [k_M N_2^2]\frac{di_2(t)}{dt} \end{aligned} \tag{15–27}$$

Factoring N_1 out of the first equation and $\pm N_2$ out of the second produces

$$\begin{aligned} v_1(t) &= N_1\left([k_M N_1]\frac{di_1(t)}{dt} \pm [k_M N_2]\frac{di_2(t)}{dt}\right) \\ v_2(t) &= \pm N_2\left([k_M N_1]\frac{di_1(t)}{dt} \pm [k_M N_2]\frac{di_2(t)}{dt}\right) \end{aligned} \tag{15–28}$$

Dividing the second equation by the first shows that perfect coupling implies

$$\frac{v_2(t)}{v_1(t)} = \pm\frac{N_2}{N_1} = \pm n \tag{15–29}$$

where the parameter n is called the **turns ratio**.

With perfect coupling the secondary voltage is proportional to the primary voltage, so they have the same waveshape. For example, when the primary voltage is $v_1(t) = V_A \sin \omega t$, the secondary voltage is $v_2(t) = \pm nV_A \sin \omega t$. When the turns ratio $n > 1$, the secondary voltage amplitude is larger than the primary and the device is called a **step-up transformer**. Conversely, when $n < 1$, the secondary voltage is smaller than the primary and the device is called a **step-down transformer**. The ability to increase or decrease ac voltages is a basic feature of transformers. Commercial power systems use transmission voltages of several hundred kilovolts. For residential applications the transmission voltage is reduced to safer levels (typically 220/110 V_{rms}) using step-down transformers.

The $\pm$ sign in Eq. (15–29) reminds us that mutual inductance can be additive or subtractive. Selecting the correct sign is important because of signal inversion. The sign depends on the reference marks given the primary and secondary currents relative to the dots indicating the relative winding orientations. The rule for the ideal transformer is a corollary of the rule for selecting the sign of the mutual inductance term for coupled inductors.

The sign in Eq. (15–29) is positive when the reference directions for both currents point toward or both point away from a dotted terminal; otherwise it is negative.

Exercise 15–3

The transformer in Figure 15–9 has perfect coupling and a turns ratio of $n = 0.1$. The input voltage is $v_S(t) = 120 \sin 377t$ V.

(a) What is the secondary voltage?
(b) What is the secondary current for a 50-Ω load?
(c) Is this a step-up or step-down transformer?

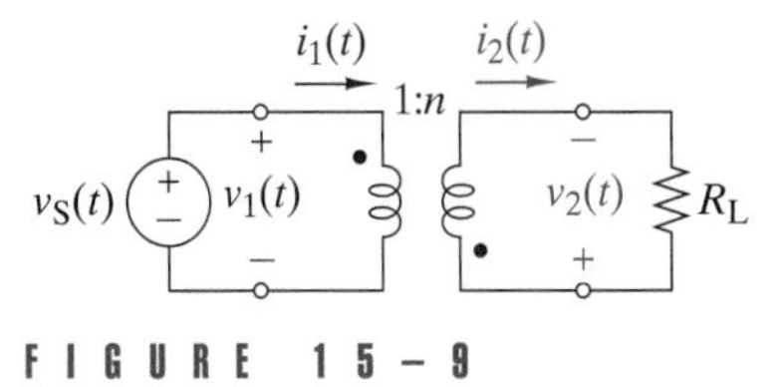

FIGURE 15–9

Answers:

(a) $v_2(t) = +12 \sin 377t$ V
(b) $i_2(t) = -0.24 \sin 377t$ A
(c) Step down

ZERO POWER LOSS

The ideal transformer model also assumes that there is no power loss in the transformer. With the passive sign convention, the quantity $v_1(t)i_1(t)$ is the power in the primary winding and $v_2(t)i_2(t)$ is the power in the secondary winding. Zero power loss requires

$$v_1(t)i_1(t) + v_2(t)i_2(t) = 0 \tag{15–30}$$

This equation states that whatever power enters the transformer on one winding immediately leaves via the other winding. This not only means that there is no energy lost in the ideal transformer, but also that there is no energy stored within the element. The zero-power-loss constraint can be rearranged as

$$\frac{i_2(t)}{i_1(t)} = -\frac{v_1(t)}{v_2(t)} \tag{15–31}$$

But under the perfect coupling assumption, $v_2(t)/v_1(t) = \pm n$. With zero power loss and perfect coupling the primary and secondary currents are related as

$$\frac{i_2(t)}{i_1(t)} = \mp\frac{1}{n} \tag{15–32}$$

The correct sign in this equation depends on the orientation of the current reference directions relative to the dots describing the transformer structure.

With both perfect coupling and zero power loss, the secondary current is inversely proportional to the turns ratio. A step-up transformer ($n > 1$) increases the voltage and decreases the current, which improves transmission line efficiency because the i^2R losses in the conductors are smaller.

i–v CHARACTERISTICS

Equations (15–29) and (15–32) define the i–v characteristics of the ideal transformer circuit element.

$$\begin{aligned} v_2(t) &= \pm\, n\, v_1(t) \\ i_2(t) &= \mp\frac{1}{n}\, i_1(t) \end{aligned} \tag{15–33}$$

where $n = N_2/N_1$ is the turns ratio. The correct sign in these equations depends on the assigned reference directions and transformer dots, as previously discussed.

Because the turns ratio is a key transformer parameter, it is worth relating it to the inductance parameters. Looking back at Eq. (15–27), we see that in a linear magnetic medium with perfect coupling the inductance parameters are $L_1 = k_M N_1^2$, $L_2 = k_M N_2^2$, and $M = k_M N_1 N_2$. As a result, the turns ratio can be expressed in terms of the inductance parameters as

$$n = \frac{N_2}{N_1} = \sqrt{\frac{L_2}{L_1}} = \frac{M}{L_1} = \frac{L_2}{M} \quad (15\text{–}34)$$

Although actual transformer devices are not ideal, they do approach perfect coupling. Hence, these relations are useful approximations with transformers that are tightly coupled ($k \approx 1$).

Using the ideal transformer model requires some caution. The relationships in Eq. (15–33) state that the voltages are independent of the currents. For example, if the secondary winding is connected to an open circuit, then $i_2(t) = 0$. The ideal transformer characteristics require that $i_1(t) = 0$ regardless of the value of $v_1(t)$. In effect this means that the inductance parameters are all infinite, but infinite in such a way that their ratios in Eq. (15–34) are still equal to the turns ratio. Equally important is the fact that the element equations appear to apply to all signals, including constant (dc) signals. In theory, a transformer with perfect coupling and zero power loss would pass dc signals. In practice, some transformers approach these conditions but no real transformer actually achieves this state of perfection. Remember that the element equations of the ideal transformer are an idealization of mutual coupling between inductors and that time-varying signals are necessary; otherwise, the coupling terms Mdi_1/dt and Mdi_2/dt are zero.

EXAMPLE 15–3

The input to the primary of an ideal transformer with a turns ratio of $n = 10$ is $v_1(t) = 120 \sin 377t$ V. The load connected to the secondary winding is a 50-Ω resistor. Assuming additive coupling, find the secondary voltage, the secondary current, the secondary power, the primary current, and the primary power.

SOLUTION:
With additive coupling the secondary voltage is

$$v_2(t) = +nv_1(t) = 10 \times 120 \sin 377\,t = 1200 \sin 377t \text{ V}$$

From Ohm's law the secondary current is

$$i_2(t) = -\frac{v_2(t)}{R_L} = -\frac{1200}{50} \sin 377\,t = -24 \sin 377t \text{ A}$$

The minus sign in this expression comes about because the reference marks for $v_2(t)$ and $i_2(t)$ follow the passive sign convention with respect to the secondary winding and not the load resistance. The power in the secondary winding is

$$p_2(t) = v_2(t)i_2(t) = -28.8 \sin^2 377t \text{ kW}$$

The minus sign here means that the secondary delivers power to the load resistor. With additive coupling the primary current is

$$i_1 = -ni_2(t) = -10 \times (-24 \sin 377\,t) = 240 \sin 377t \text{ A}$$

and the primary power is

$$p_1(t) = v_1(t)i_1(t) = 28.8 \sin^2 377t \text{ kW}$$

The plus sign here means that the primary winding is absorbing power from the source. Note that $p_1(t) + p_2(t) = 0$ as required by the zero-power-loss assumption. ■

EQUIVALENT INPUT RESISTANCE

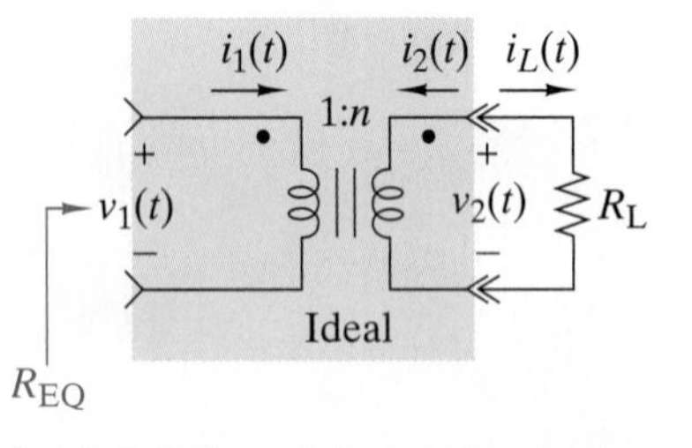

FIGURE 15–10 *Equivalent resistance seen on the primary winding.*

Because a transformer changes the voltage and current levels, it effectively changes the load resistance seen by a source in the primary circuit. To derive the equivalent input resistance, we write the device equations for the ideal transformer shown in Figure 15–10.

$$\text{Resistor:} \quad v_2(t) = R_L i_L(t)$$

$$\text{Transformer:} \quad v_2(t) = n\, v_1(t) \text{ and } i_2(t) = -\frac{1}{n} i_1(t)$$

Dividing the first transformer equation by the second and inserting the load resistance constraint yields

$$\frac{v_2(t)}{i_2(t)} = \frac{i_L(t)R_L}{i_2(t)} = -n^2 \frac{v_1(t)}{i_1(t)}$$

Applying KCL at the output interface tells us that $i_L(t) = -i_2(t)$. Therefore, the equivalent resistance seen on the primary side is

$$R_{EQ} = \frac{v_1(t)}{i_1(t)} = \frac{1}{n^2} R_L \tag{15–35}$$

The equivalent load resistance seen on the primary side depends on the turns ratio and the load resistance. Adjusting the turns ratio can make R_{EQ} equal to the source resistance. Transformer coupling can produce the conditions for maximum power transfer when the source and load resistances are not equal.

The derivation leading to Eq. (15–35) used the ideal transformer with the dot markings and reference directions in Figure 15–10. However, the final result does not depend on the location of the dot marks relative to voltage and current reference directions. In other words, Eq. (15–35) yields the input resistance for any ideal transformer with a turns ratio of n and a load R_L.

EXAMPLE 15–4

A stereo amplifier has an output resistance of 600 Ω. The input resistance of the speaker (the load) is 8 Ω. Select the turns ratio of a transformer to obtain maximum power transfer.

SOLUTION:

The maximum power transfer theorem in Chapter 3 states that the source and load resistance must be matched (equal) to achieve maximum power. Directly connecting the amplifier (600 Ω) to the speaker (8 Ω) produces a mismatch. If a transformer is inserted as shown in Figure 15–11, then the equivalent load resistance seen by the amplifier is

$$R_{EQ} = \frac{1}{n^2} R_L = \frac{1}{n^2} 8$$

To produce a resistance match, we need $R_{EQ} = 600 = 8/n^2$ or a turns ratio of $n = 1/8.66$. ■

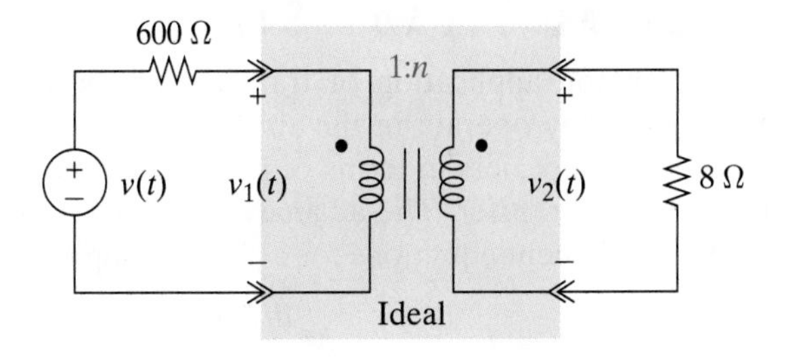

FIGURE 15–11

Exercise 15–4

The load resistor in Figure 15–10 is 50 Ω, and the numbers of turns on the primary and secondary windings are $N_1 = 540$ and $N_2 = 108$. Find the equivalent resistance on the primary side.

Answer: $R_{EQ} = 1250\ \Omega$

APPLICATION EXAMPLE 15–5

In a transformer the primary and secondary windings are magnetically coupled but are usually electrically isolated. Transformer performance in some applications can be improved by electrically connecting the two magnetically coupled windings in a configuration called an **autotransformer**. Figure 15–12 shows a two-winding ideal transformer connected in an autotransformer step-up configuration.

The rating of the ideal transformer when connected in the usual electrically isolated two-winding configuration is $P_{rated} = v_1 i_1 = -v_2 i_2$. Connected in the autotransformer configuration in Figure 15–12, the power delivered to the load is

$$P_{load} = (v_1 + v_2)(-i_2)$$

For an ideal transformer $v_2 = nv_1$ and $i_2 = -i_1/n$, where $n = N_2/N_1$. Hence the power delivered to the load is

$$P_{load} = (v_1 + nv_1)(i_1/n) = \left(1 + \frac{1}{n}\right) v_1 i_1$$

$$= \left(1 + \frac{1}{n}\right) P_{rated}$$

The autotransformer configuration delivers more power to the load than the power rating of the two-winding transformer. Put differently,

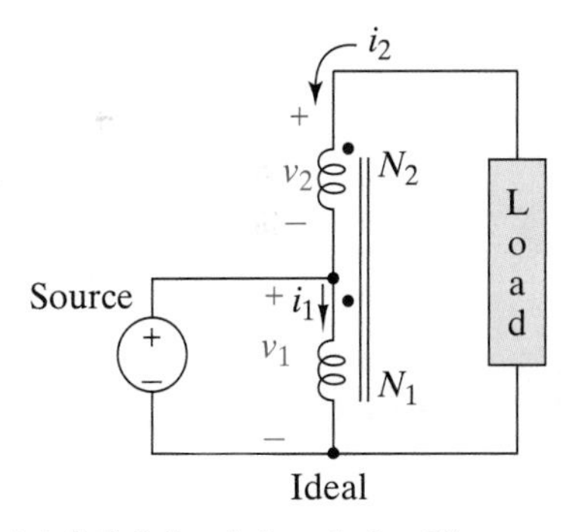

FIGURE 15–12 *The autotransformer connection.*

the autotransformer can supply a specified load power using a transformer with a lower power rating. Autotransformers are normally used when the turns ratio is less than 3:1, so this advantage can be significant. A disadvantage is that the electrical isolation provided by the usual transformer configuration is lost.

15–5 TRANSFORMERS IN THE SINUSOIDAL STEADY STATE

By far the most common application of transformers occurs in electric power systems, where they operate in the sinusoidal steady state. In this context we describe transformers in terms of phasors and impedances and deal with average power transfer. The ac analysis of a transformer begins with the time-domain element equations for a pair of coupled inductors

$$v_1(t) = L_1 \frac{di_1(t)}{dt} \pm M \frac{di_2(t)}{dt}$$

$$v_2(t) = \pm M \frac{di_1(t)}{dt} + L_2 \frac{di_2(t)}{dt}$$

Transforming these equations into the phasor domain involves replacing waveforms by phasors using the derivative property to obtain the phasors for di_1/dt and di_2/dt:

$$\begin{aligned} \mathbf{V}_1 &= j\omega L_1 \mathbf{I}_1 \pm j\omega M \mathbf{I}_2 \\ \mathbf{V}_2 &= \pm j\omega M \mathbf{I}_1 + j\omega L_2 \mathbf{I}_2 \end{aligned} \tag{15–36}$$

where

1. $\mathbf{V}_1$ and $\mathbf{I}_1$ are the phasors representing ac voltage and current of the first winding, and $j\omega L_1$ is the impedance of the self-inductance of first winding.
2. $\mathbf{V}_2$ and $\mathbf{I}_2$ are the phasors representing ac voltage and current of the second winding, and $j\omega L_2$ is the impedance of the self-inductance of second winding.
3. $j\omega M$ is the impedance of the mutual inductance between the two windings.

The self-inductance impedances relate phasor voltage and current at the same pair of terminals, while the mutual inductance impedance relates the phasor voltage at one pair of terminals to the phasor current at the other pair. The phasor-domain equations can also be written in terms of reactances as

$$\mathbf{V}_1 = jX_1\mathbf{I}_1 \pm jX_M\mathbf{I}_2$$

$$\mathbf{V}_2 = \pm jX_M\mathbf{I}_1 + jX_2\mathbf{I}_2$$

where the three reactances are $X_1 = \omega L_1$, $X_2 = \omega L_2$, and $X_M = \omega M$. The degree of coupling between windings is indicated by the coupling coefficient, which can be written in terms of reactances as

$$k = \frac{M}{\sqrt{L_1 L_2}} = \frac{X_M}{\sqrt{X_1 X_2}}$$

In either form, energy considerations dictate that $0 \leq k \leq 1$.

Figure 15–13 shows the phasor-domain version of transformer coupling between a source and a load. Following our previous notation, the winding connected to the source is called the **primary**, and the winding connected to the load is called the **secondary**. Although the transformer is a bilateral device, we normally think of signal and power transfer as passing from the primary to the secondary winding.

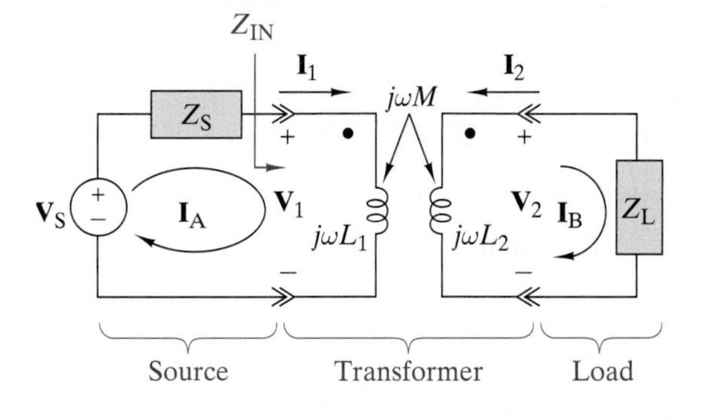

FIGURE 15–13 *Phasor circuit model of the two-winding transformer.*

Our immediate objective is to write circuit equations for the transformer using the mesh currents $\mathbf{I}_A$ and $\mathbf{I}_B$ in Figure 15–13. Applying KVL around the primary circuit (mesh A) and secondary circuit (mesh B), we obtain the following equations:

$$\begin{aligned} &\text{Mesh A:} \quad Z_S\mathbf{I}_A + \mathbf{V}_1 = \mathbf{V}_S \\ &\text{Mesh B:} \quad -\mathbf{V}_2 + Z_L\mathbf{I}_B = 0 \end{aligned} \tag{15–37}$$

The reference directions for the inductor currents in Figure 15–13 are both directed in at dotted terminals, so the mutual inductance coupling is additive and the plus signs in Eq. (15–36) apply. Using KCL we see that the reference directions for the currents lead to the relations $\mathbf{I}_1 = \mathbf{I}_A$ and $\mathbf{I}_2 = -\mathbf{I}_B$. The **I–V** relationships of the coupled inductors in terms of the mesh currents are

$$\begin{aligned} \mathbf{V}_1 &= +j\omega L_1\mathbf{I}_A + j\omega M(-\mathbf{I}_B) \\ \mathbf{V}_2 &= +j\omega M\mathbf{I}_A + j\omega L_2(-\mathbf{I}_B) \end{aligned} \tag{15–38}$$

Substituting the inductor voltages from Eq. (15–38) into the KVL equations in Eq. (15–37) yields

$$\begin{aligned} &\text{Mesh A:} \quad (Z_S + j\omega L_1)\mathbf{I}_A - j\omega M\mathbf{I}_B = \mathbf{V}_S \\ &\text{Mesh B:} \quad -j\omega M\mathbf{I}_A + (Z_L + j\omega L_2)\mathbf{I}_B = 0 \end{aligned} \tag{15–39}$$

This set of mesh equations provides a complete description of the circuit ac response. Once we solve for the mesh currents, we can calculate every phasor voltage and current using Kirchhoff's laws and element equations.

EXAMPLE 15–6

The source circuit in Figure 15–13 has $Z_S = 0 + j20\ \Omega$ and $\mathbf{V}_S = 2500\angle 0°$ V at $\omega = 377$ rad/s. The transformer has $L_1 = 2$ H, $L_2 = 0.2$ H, and $M = 0.6$ H. The load impedance is $Z_L = 25 + j15\ \Omega$. Find $\mathbf{I}_A$, $\mathbf{I}_B$, $\mathbf{V}_1$, $\mathbf{V}_2$, Z_{IN}, and the average power delivered by the source at input interface.

SOLUTION:

The transformer impedances are

$$j\omega L_1 = j377 \times 2 = j754 \ \Omega$$

$$j\omega L_2 = j377 \times 0.2 = j75.4 \ \Omega$$

$$j\omega M = j377 \times 0.6 = j226 \ \Omega$$

Using these impedances in Eq. (15–39) yields the following mesh equations:

$$\text{Mesh A:} \quad (j20 + j754)\mathbf{I}_A - j226\mathbf{I}_B = 2500 \angle 0^\circ$$

$$\text{Mesh B:} \quad -j226\mathbf{I}_A + (25 + j15 + j75.4)\mathbf{I}_B = 0$$

Solving these equations for the mesh currents yields

$$\mathbf{I}_A = 4.36 - j7.49 = 8.67 \; {-}{-}59.8^\circ \text{ A}$$

$$\mathbf{I}_B = 15.0 - j14.6 = 20.9 \; {-}{-}44.2^\circ \text{ A}$$

The winding voltages are found to be

$$\mathbf{V}_1 = \mathbf{V}_S - Z_S\mathbf{I}_A = 2350 - j87.2 = 2350 \angle -2.1^\circ \text{ V}$$

$$\mathbf{V}_2 = Z_L\mathbf{I}_B = 594 - j140 = 610 \angle -13.2^\circ \text{ V}$$

The input impedance seen by the source circuit is

$$Z_{IN} = \frac{\mathbf{V}_1}{\mathbf{I}_A} = R_{IN} + jX_{IN} = 145 + j229 \ \Omega$$

The average power delivered by the source at the input interface is

$$P_{IN} = \frac{|\mathbf{I}_A|^2}{2} R_{IN} = \frac{(8.67)^2}{2} 145 = 5.45 \text{ kW}$$

Exercise 15–5

Repeat Example 15–6 when the dot on the secondary winding in Figure 15–13 is moved to the bottom terminal and all other reference marks stay the same.

Answers:

$\mathbf{I}_A = 4.36 - j7.49$ A; $\mathbf{I}_B = -15.0 + j14.6$ A

$\mathbf{V}_1 = 2350 - j87.2$ V; $\mathbf{V}_2 = -594 + j140$ V; $Z_{IN} = 145 + j229 \ \Omega$; $P_{IN} = 5.45$ kW

The method used in the preceding example illustrates a general approach to the analysis of transformer circuits. The steps in the method are as follows:

STEP 1 Write KVL equations around the primary and secondary circuits using assigned mesh currents, source voltages, and inductor voltages.

STEP 2 Write the **I–V** characteristics of the coupled inductors in terms of the mesh currents using the dot convention to determine whether the coupling is additive or subtractive.

STEP 3 Use the **I–V** relationships from step 2 to eliminate the inductor voltages from the KVL equations obtained in step 1 to obtain mesh-current equations.

The next two examples illustrate this method of formulating mesh equations.

EXAMPLE 15–7

Find $\mathbf{I}_A$, $\mathbf{I}_B$, $\mathbf{V}_1$, $\mathbf{V}_2$, and the impedance seen by the voltage source in Figure 15–14.

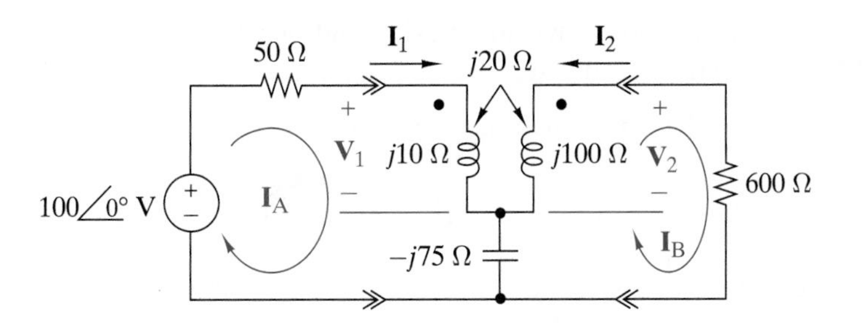

FIGURE 15–14

SOLUTION:

STEP 1 The KVL equations around meshes A and B are

$$\text{Mesh A:}\quad (50 - j75)\mathbf{I}_A - (-j75)\mathbf{I}_B + \mathbf{V}_1 = 100\angle 0°$$
$$\text{Mesh B:}\quad -(-j75)\mathbf{I}_A + (600 - j75)\mathbf{I}_B - \mathbf{V}_2 = 0$$

STEP 2 For the assigned reference directions, the coupling is additive. By KCL we have $\mathbf{I}_1 = \mathbf{I}_A$ and $\mathbf{I}_2 = -\mathbf{I}_B$. Hence, the element equations for the coupled inductors in terms of the mesh currents are

$$\mathbf{V}_1 = j10\mathbf{I}_A + j20(-\mathbf{I}_B)$$
$$\mathbf{V}_2 = j20\mathbf{I}_A + j100(-\mathbf{I}_B)$$

STEP 3 Using these equations to eliminate the inductor voltages from the KVL equations found in step 1 yields the following mesh equations:

$$\text{Mesh A:}\quad (50 - j75 + j10)\mathbf{I}_A + (j75 - j20)\mathbf{I}_B = 100\angle 0°$$
$$\text{Mesh B:}\quad (j75 - j20)\mathbf{I}_A + (600 - j75 + j100)\mathbf{I}_B = 0$$

Solving these equations for the mesh currents produces

$$\mathbf{I}_A = 0.756 + j0.896 = 1.17\angle 49.8°\ \text{A}$$
$$\mathbf{I}_B = 0.0791 - j0.0726 = 0.107\angle -42.5°\ \text{A}$$

Given the mesh currents, we find the inductor voltages from the **I–V** relations:

$$\mathbf{V}_1 = j10\mathbf{I}_A - j20\mathbf{I}_B = -10.4 + j5.98 = 12.0\angle 150°\ \text{V}$$
$$\mathbf{V}_2 = j20\mathbf{I}_A - j100\mathbf{I}_B = -25.2 + j7.21 = 26.2\angle 164°\ \text{V}$$

Finally, the impedance seen by the input voltage source is

$$Z_{IN} = \frac{\mathbf{V}_S}{\mathbf{I}_A} = 55.0 - j65.2\ \Omega$$

■

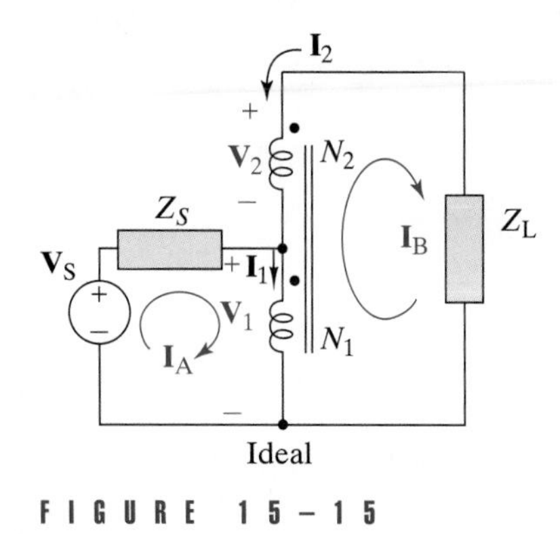

FIGURE 15–15

EXAMPLE 15–8

Figure 15–15 shows an ideal transformer connected as an autotransformer. Find the voltage and average power delivered to the load for $\mathbf{V}_S = 500\angle 0°$ V, $Z_S = j10\ \Omega$, $Z_L = 50 + j0\ \Omega$, $N_1 = 200$, and $N_1 = 280$.

SOLUTION:

The three-step method of writing mesh equations can be used here with a modification to the second step:

STEP 1 The KVL equations around meshes A and B are

$$\text{Mesh A:} \quad Z_S\mathbf{I}_A + \mathbf{V}_1 = \mathbf{V}_S$$
$$\text{Mesh B:} \quad -\mathbf{V}_1 - \mathbf{V}_2 + Z_L\mathbf{I}_B = 0$$

STEP 2 For the assigned reference directions, the coupling is additive so the ideal transformer voltages and currents are related as

$$\mathbf{V}_2 = n\mathbf{V}_1 \quad \text{and} \quad \mathbf{I}_1 = -n\mathbf{I}_2$$

By KCL we have $\mathbf{I}_1 = \mathbf{I}_A - \mathbf{I}_B$ and $\mathbf{I}_2 = -\mathbf{I}_B$. Hence, the ideal transformer constrains the two mesh currents as

$$\mathbf{I}_A = (n+1)\mathbf{I}_B$$

STEP 3 Using these results to eliminate $\mathbf{I}_A$ and $\mathbf{V}_2$ from the KVL equation in step 1 yields

$$\text{Mesh A:} \quad (n+1)Z_S\mathbf{I}_B + \mathbf{V}_1 = \mathbf{V}_S$$
$$\text{Mesh B:} \quad -(n+1)\mathbf{V}_1 + Z_L\mathbf{I}_B = 0$$

Notice that we cannot eliminate both $\mathbf{V}_1$ and $\mathbf{V}_2$ from the KVL equations since voltage and current are independent in an ideal transformer. Substituting the numerical values produces

$$\text{Mesh A:} \quad j24\mathbf{I}_B + \mathbf{V}_1 = 500\angle 0°$$
$$\text{Mesh B:} \quad -2.4\mathbf{V}_1 + 50\mathbf{I}_B = 0$$

Solving these equations for $\mathbf{I}_B$ and $\mathbf{V}_1$ yields

$$\mathbf{I}_B = 10.3 - j11.9 = 15.7\angle -49.1° \text{ A}$$
$$\mathbf{V}_1 = 215 - j247 = 328\angle -49.0° \text{ A}$$

Given the $\mathbf{I}_B$, we find the output quantities as

$$\mathbf{V}_L = Z_L\mathbf{I}_B = 515 - j595 = 787\angle -49.1° \text{ A}$$
$$P_L = \frac{|\mathbf{I}_B|^2}{2}R_L = \frac{15.7^2}{2}50 = 6.16 \text{ kW}$$

■

Exercise 15–6

Find $\mathbf{V}_L$, $\mathbf{V}_1$, $\mathbf{V}_2$, and Z_{IN} in the circuit shown in Figure 15–16.

Answers: $\mathbf{V}_L = 31.6\angle -108°$ V, $\mathbf{V}_1 = 23.7\angle 71.6°$ V, $\mathbf{V}_2 = 23.7\angle 71.6°$ V, $Z_{IN} = 0 + j20\ \Omega$.

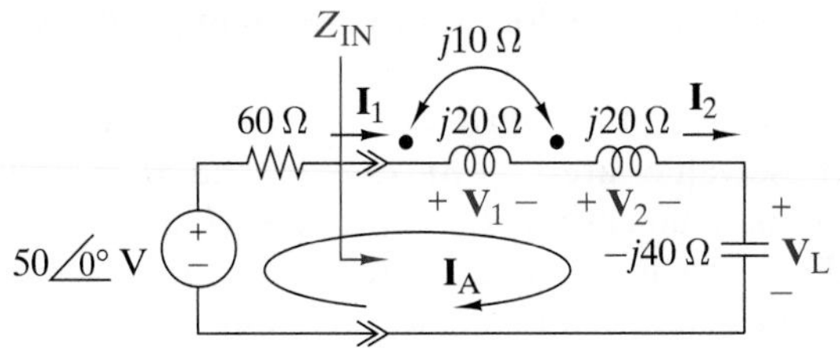

FIGURE 15–16

15–6 TRANSFORMER EQUIVALENT CIRCUITS

The circuit analysis may be simplified when coupled inductors are replaced by an equivalent circuit made up of uncoupled inductors. The inductive T-circuit in Figure 15–17 is one example of such a circuit. By definition, two circuits are equivalent if they have the same **I–V** characteristics at specified terminals. Equation (15–36) gives the **I–V** relations for coupled inductors at the input and output ports. To establish an equivalence, we must find the values of L_A, L_B, and L_C in Figure 15–17 that produce the same **I–V** properties as Eq. (15–36).

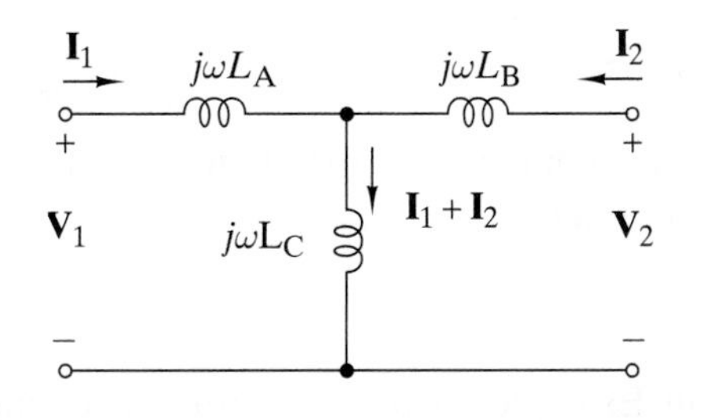

FIGURE 15–17 *T-equivalent circuit of a two-winding transformer.*

Applying KCL in Figure 15–17 shows that the current through L_C is $\mathbf{I}_1 + \mathbf{I}_2$. As a result, applying KVL around the circuit's input and output loops yields

$$-\mathbf{V}_1 + j\omega L_A\mathbf{I}_1 + j\omega L_C(\mathbf{I}_1 + \mathbf{I}_2) = 0$$

$$-j\omega L_C(\mathbf{I}_1 + \mathbf{I}_2) - j\omega L_B\mathbf{I}_2 + \mathbf{V}_2 = 0$$

These equations can be rearranged in the form

$$\mathbf{V}_1 = j\omega(L_A + L_C)\mathbf{I}_1 + j\omega L_C\mathbf{I}_2$$

$$\mathbf{V}_2 = j\omega L_C\mathbf{I}_1 + j\omega(L_B + L_C)\mathbf{I}_2$$

Comparing these equations to the **I–V** relations in Eq. (15–36), we conclude that equivalence requires

$$L_A + L_C = L_1$$

$$L_B + L_C = L_2$$

$$L_C = \pm M$$

Using the third equation to eliminate L_C from the first two equations yields the required values of L_A, L_B, and L_C.

$$L_A = L_1 \mp M$$

$$L_B = L_2 \mp M \qquad (15\text{–}40)$$

$$L_C = \pm M$$

where the upper sign applies for additive coupling and the lower for subtractive.

For additive coupling, the equivalent T-circuit involves three uncoupled inductances whose values are $L_1 - M$, $L_2 - M$, and M. For subtractive coupling, the three inductances are $L_1 + M$, $L_2 + M$, and $-M$. In either case, one of the inductances is negative. A negative inductance is not a passive element, which means we can't actually build an equivalent T-circuit out of ordinary inductors.

The fact that we cannot actually build the equivalent T-circuit in the laboratory does not invalidate the equivalence. Equivalence simply means that a pair of coupled inductors and the T-circuit defined by Eq. (15–40) have the same **I–V** properties at the input and output ports. Coupled inductors interact with the rest of the circuit only via these ports. The equivalent T-circuit will correctly predict external interaction even though one of its inductances is negative.

One final comment is that the T-circuit in Figure 15–17 shows why coupled inductors do not pass steady-state dc signals. Recall that in the dc steady state, inductors act like short circuits. Applying this idea in Figure 15–17 shows that the input and output ports are short-circuited, making it impossible for dc signals to pass from input to output.

EXAMPLE 15–9

Use the equivalent T-circuit in Figure 15–17 to find the input impedance of the transformer in Example 15–6.

SOLUTION:

The transformer in Example 15–6 had additive coupling with $L_1 = 2$ H, $L_2 = 0.2$ H, $M = 0.6$ H, $\omega = 377$ rad/s, and $Z_L = 25 + j15\ \Omega$. Applying Eq. (15–40) using the upper sign yields the impedances of the equivalent T-circuit as

$$j\omega L_A = j\omega(L_1 - M) = j528 \quad \Omega$$

$$j\omega L_B = j\omega(L_2 - M) = -j151 \quad \Omega$$

$$j\omega L_C = j\omega M = j226 \quad \Omega$$

In this case the inductance $L_B = L_2 - M = -0.4$ H is negative. Figure 15–18 shows the equivalent T-circuit together with the load impedance. The input impedance of this equivalent circuit is easily calculated by circuit reduction.

$$Z_{IN} = j528 + j226 \,||(-j151 + Z_L)$$

$$= j528 + \cfrac{1}{\cfrac{1}{j226} + \cfrac{1}{25 - j136}}$$

$$= 145 + j228 \ \Omega$$

This result agrees with the answer in Example 15–6, where the input impedance was found using mesh-current analysis. ■

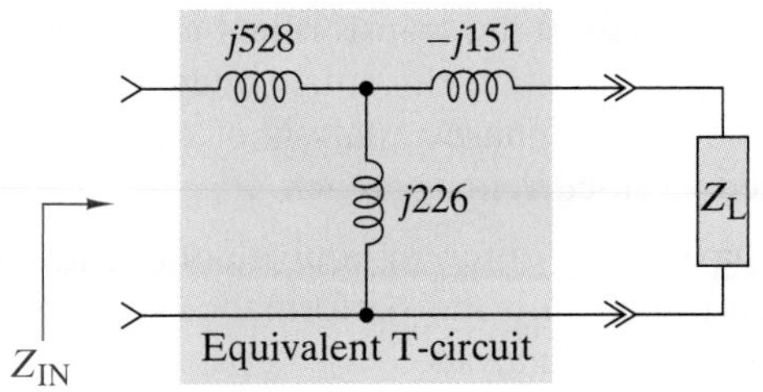

FIGURE 15–18

Exercise 15–7

Use an equivalent T-circuit to find the input impedance of a transformer with subtractive coupling and $L_1 = 20$ mH, $L_2 = 250$ mH, $M = 70$ mH, $\omega = 2000$ rad/s, and $Z_L = 500 + j100\ \Omega$.

Answer: $Z_{IN} = 16.1 + j20.7\ \Omega$

SUMMARY

- Mutual inductance describes magnetic coupling between two inductors. The mutual inductance parameter relates the voltage induced in one inductor to the rate of change of current in the other inductor. The induced voltage can be either positive (additive coupling) or negative (subtractive coupling).
- The dot convention describes the physical orientation of the two magnetically coupled inductors. Mutual inductance coupling is additive when both current reference arrows point toward or away from dotted terminals; otherwise, it is subtractive. The reference directions for the currents can always be selected so that the coupling is additive.
- The degree of coupling is indicated by the coupling coefficient. Energy analysis shows that the coupling coefficient must lie between zero and one. A coupling coefficient of unity is called perfect coupling and means that all of the flux produced by first winding links with a second winding, and vice versa. A coupling coefficient of zero indicates no magnetic linkage between the windings.
- A transformer is an electrical device based on the mutual inductance coupling between two windings. Transformers find applications in almost all electrical systems, especially in power supplies and the electrical power grid. The transformer winding connected to the power source is called the primary winding, and the winding connected to the load is called the secondary winding.
- The ideal transformer is a circuit element in which the primary and secondary windings are assumed to be perfectly coupled and to have no power loss. In an ideal transformer the voltages and currents in the primary and secondary windings are related by the turns ratio, which is the ratio of the number of turns in the secondary winding to that in the primary winding.

- In the sinusoidal steady state, transformers can be analyzed using phasors to determine steady-state currents, voltages, impedances, and average power transfer. The phasor analysis of transformers is carried out using a modified mesh-current approach.
- Transformers have many different equivalent circuits. These equivalent circuits provide insight into the performance of transformers and may simplify certain types of analysis.

PROBLEMS

OBJECTIVE 15–1 MUTUAL INDUCTANCE (SECTS. 15–1, 15–2, 15–3)

(a) Given the current through or voltage across two coupled inductors, find the unspecified currents or voltages.
(b) Find characteristics of coupled inductors when given some device parameters and circuit connections.
See Examples 15–1, 15–2 and Exercises 15–1, 15–2

15–1 In Figure P15–1 $L_1 = 10$ mH, $L_2 = 5$ mH, $M = 7$ mH, and $v_S(t) = 200 \sin 100t$ V.
(a) Write the i–v relationships for the coupled inductors using the reference marks in the figure.
(b) Solve for $v_2(t)$ when the output terminals are open-circuited ($i_2 = 0$).

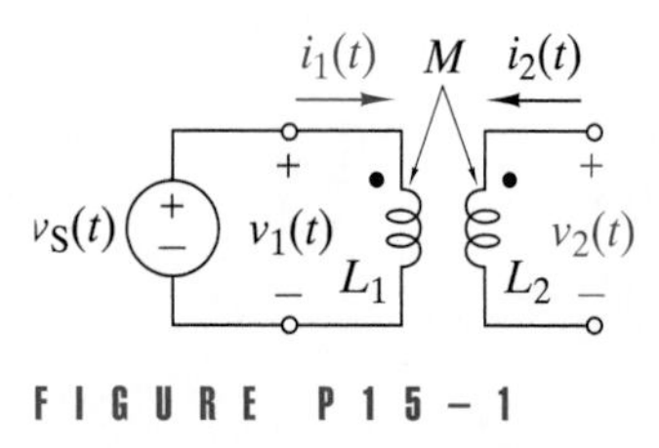

FIGURE P15–1

15–2 In Figure P15–1 $L_1 = 10$ mH, $L_2 = 5$ mH, $M = 7$ mH, and $v_S(t) = 100 \sin 1000t$ V.
(a) Write the i–v relationships for the coupled inductors using the reference marks in the figure.
(b) Solve for $i_1(t)$ and $i_2(t)$ when the output terminals are short-circuited ($v_2 = 0$).

15–3 In Figure P15–1 $L_1 = 10$ mH, $L_2 = 5$ mH, $M = 7$ mH, and the outputs are $v_2(t) = 0$ and $i_2(t) = 35 \sin 1000t$ A.
(a) Write the i–v relationships for the coupled inductors using the reference marks given.
(b) Solve for the source voltage $v_S(t)$.

15–4 In Figure P15–4 $L_1 = L_2 = 3$ mH, $M = 2$ mH, and $i_S(t) = 50 \sin 100t$ A. Solve for $v_1(t)$ and $v_2(t)$ when the output terminals are open-circuited ($i_2 = 0$).

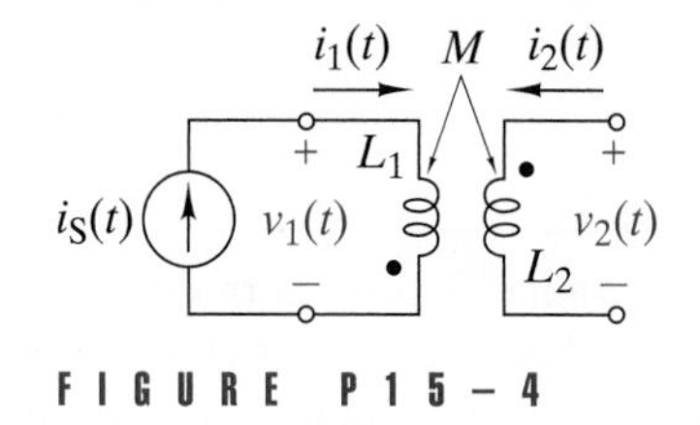

FIGURE P15–4

15–5 In Figure P15–4 $L_1 = L_2 = 3$ mH, $M = 2$ mH, and $i_2(t) = 0.5 \sin 1000t$ A when the output terminals are short-circuited. Solve for $v_1(t)$ and $i_1(t)$.

15–6 A pair of coupled inductors have $L_1 = 3.6$ H and $L_2 = 2.4$ H. With the output open-circuited ($i_2 = 0$), the coil voltages are observed to be $v_1(t) = 4 \sin 1000t$ and $v_2(t) = 3 \sin 1000t$. Find the mutual inductance and the coupling coefficient. Is the coupling additive or subtractive?

15–7 In Figure P15–7 $L_1 = L_2 = 6$ mH, $M = 4$ mH, and $i_1(t) = 5 \cos(1000t)$ A. Find the input voltage $v_x(t)$.

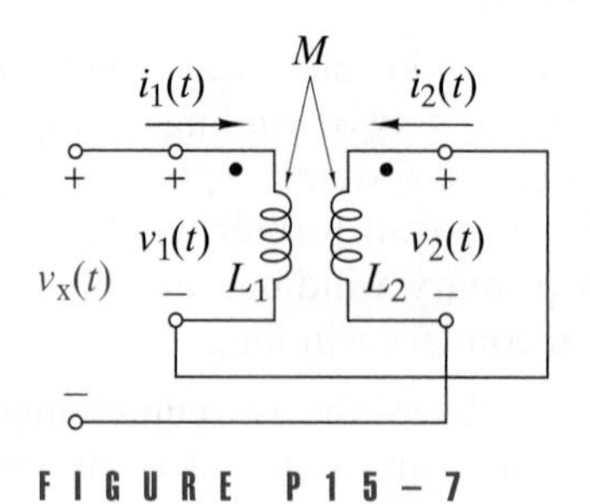

FIGURE P15–7

15–8 In Figure P15–8 $L_1 = 80$ mH, $L_2 = 400$ mH, and $k = 0.5$. Find the equivalent inductance L_{EQ}.

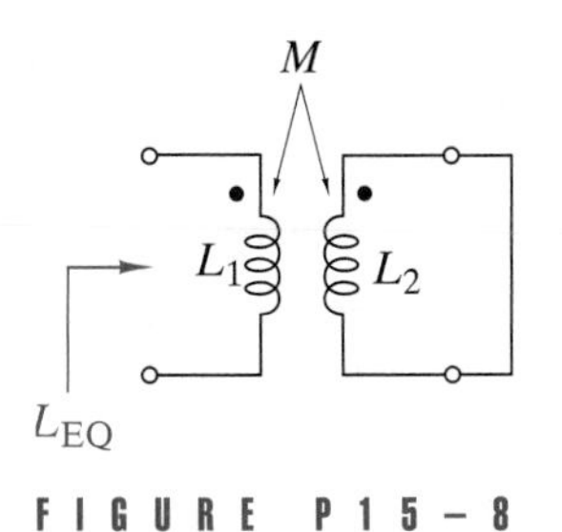

FIGURE P15–8

15–9 A pair of coupled coils have additive coupling. When the output terminals are open-circuited ($i_2 = 0$), the equivalent inductance seen at the input is 50 mH. When the output terminals are short-circuited ($v_2 = 0$), the equivalent inductance seen at the input is 10 mH. Find the coupling coefficient.

15–10 The self-inductances of two coupled inductors are found to be 220 mH and 260 mH. When the two coupled inductors are connected in series, the total inductance is found to be 120 mH. Find the mutual inductance and the coupling coefficient.

OBJECTIVE 15–2 THE IDEAL TRANSFORMER (SECT. 15–4)

Given a circuit containing ideal transformers:
(a) Find specified voltages, currents, powers, and equivalent circuits.
(b) Select the turns ratio to meet prescribed conditions.
See Examples 15–3, 15–4, 15–5 and Exercises 15–3, 15–4

15–11 The primary voltage in an ideal transformer is a 120-V, 60-Hz sinusoid. The secondary voltage is a 24-V, 60-Hz sinusoid. The secondary winding is connected to an 800-Ω resistive load.
(a) Find the transformer turns ratio.
(b) Write expressions for the primary current and voltage.

15–12 The number of turns in primary and secondary of an ideal transformer are $N_1 = 50$ and $N_2 = 400$. The primary winding is connected to a 120-V, 60-Hz source with a source resistance of 50 Ω. The secondary winding is connected to a 1600-Ω load. Find the primary and secondary currents.

15–13 The turns ratio of the second ideal transformer in Figure P15–13 is $n = 5$. Find the equivalent resistance indicated in the figure.

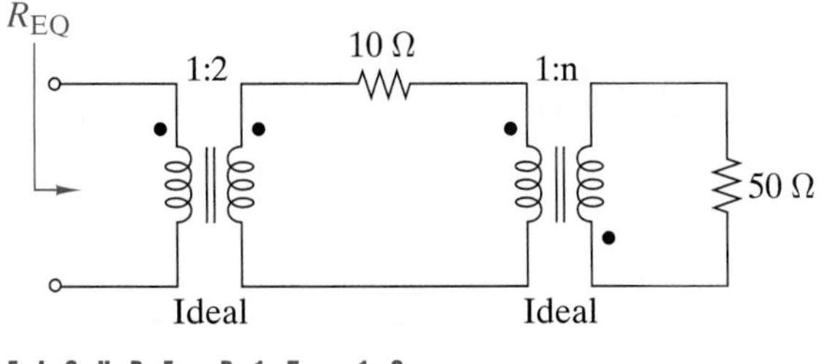

FIGURE P15–13

15–14 The equivalent resistance in Figure P15–13 is $R_{EQ} = 15\ \Omega$. Find the turns ratio of the second ideal transformer.

15–15 Figure P15–15 shows an ideal transformer connected as an autotransformer. Find $i_L(t)$ and $i_S(t)$ when $v_S(t) = 120 \sin 400t$ and $R_L = 60\ \Omega$.

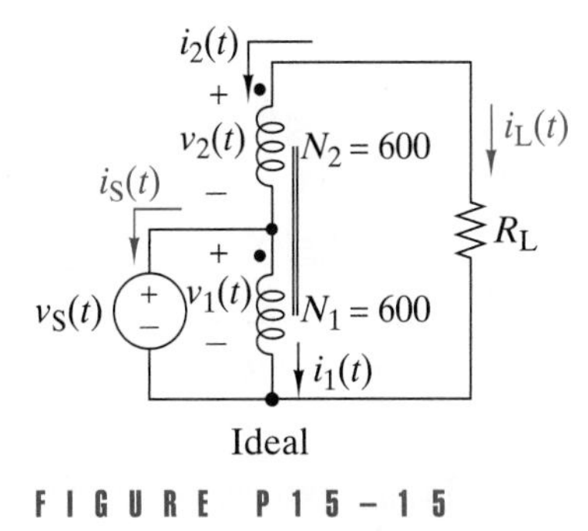

FIGURE P15–15

15–16 An ideal transformer has $n = 1/5$. The primary winding is connected to a voltage source with a source resistance of 2 kΩ. Find the load resistance connected across the secondary winding that will draw maximum power from the source.

15–17 Select the turns ratio of an ideal transformer in the interface circuit shown in Figure P15–17 so that the input resistance seen by the voltage source is 150 Ω.

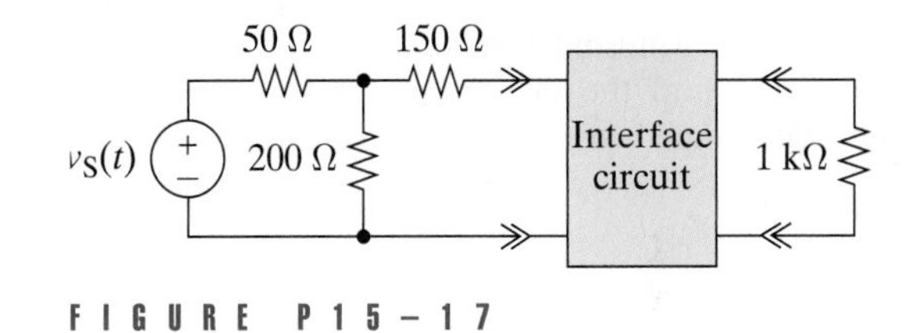

FIGURE P15–17

15–18 A voltage source with Thévenin parameters $v_T = 5 \sin 1000t$ V and $R_T = 25\ \Omega$ drives the primary winding of an ideal transformer with $n = 2$. Find the power delivered to a 300-Ω load connected across the secondary winding.

15–19 A voltage source with $v_S = 15 \sin 377t$ V drives the primary windings of the ideal transformer in Figure P15–19. Find the input resistance seen by the voltage source.

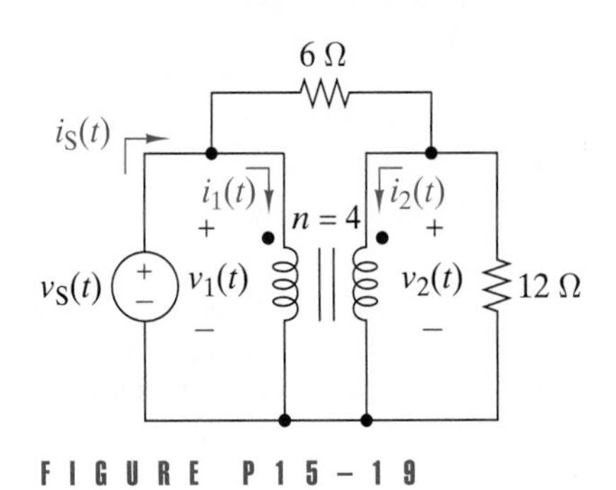

FIGURE P15–19

15–20 Show that the equivalent resistance in Figure P15–20 is

$$R_{EQ} = \left(\frac{N_1 + N_2}{N_2}\right)^2 R_L$$

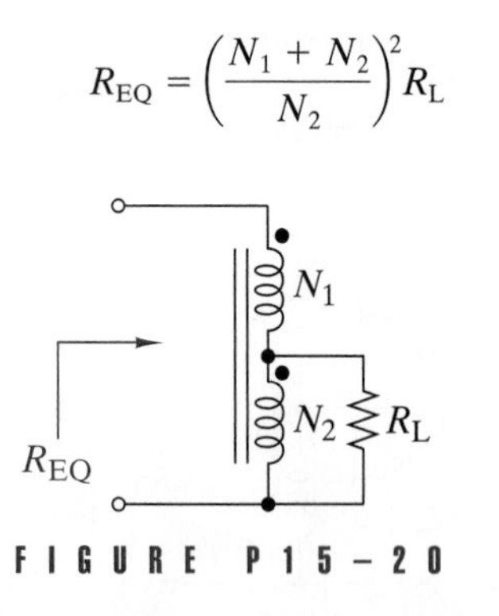

FIGURE P15–20

OBJECTIVE 15–3 TRANSFORMERS IN THE SINUSOIDAL STEADY STATE (SECTS. 15–5, 15–6)

Given a linear transformer in a linear circuit operating in the sinusoidal steady state, find phasor voltages and currents, average powers, and equivalent impedances.

See Examples 15–6, 15–7, 15–8, 15–9 and Exercises 15–5, 15–6, 15–7

15–21 The input voltage to the transformer in Figure P15–21 is a sinusoid $v_1(t) = 60 \cos 2500t$ V. With the circuit operating in the sinusoidal steady state, use mesh-current analysis to find the phasor output voltage $\mathbf{V}_2$ and the equivalent impedance seen by the voltage source.

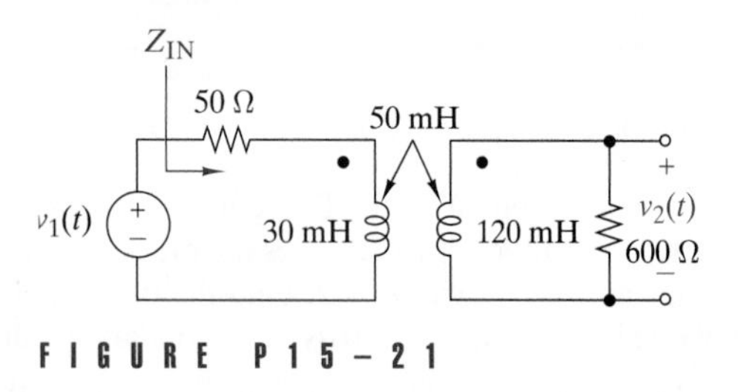

FIGURE P15–21

15–22 Repeat Problem 15–21 with $v_1(t) = 60 \cos 500t$ V.

15–23 A transformer operating in the sinusoidal steady state with $\omega = 377$ rad/s has self-inductances $L_1 = 200$ mH, $L_2 = 400$ mH, and $k = 0.95$. The load connected to the secondary winding is a 75-Ω resistor. Find the transformer input impedance. Assume additive coupling.

15–24 Repeat Problem 15–23 when the load is $Z_L = 75 + j150$ Ω.

15–25 The circuit in Figure P15–25 is operating in the sinusoidal steady state with $v_S(t) = 150 \sin 2000t$ V. The resistors in series with the coupled inductors simulate winding resistances. Find input impedance Z_{IN} and the phasor responses $\mathbf{V}_1$ and $\mathbf{V}_L$.

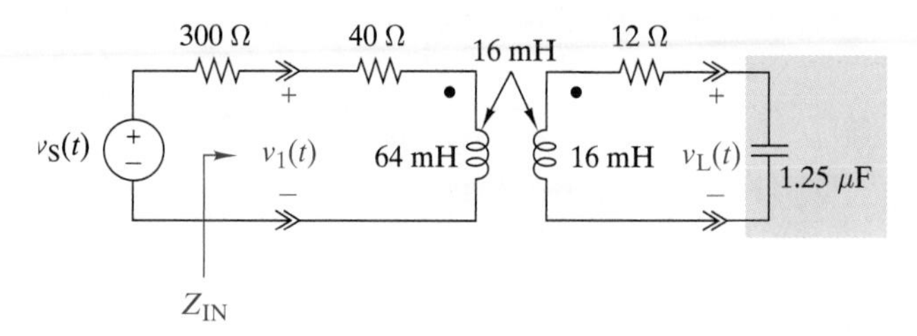

FIGURE P15–25

15–26 Repeat Problem 15–25 when the capacitor load is replaced by $Z_L = 200 + j200$ Ω.

15–27 The circuit in Figure P15–27 is in the sinusoidal steady state with $v_S(t) = 200 \cos 2500t$ V. Find the input current $i(t)$, the impedance seen by the voltage source, and the average power supplied by the voltage source.

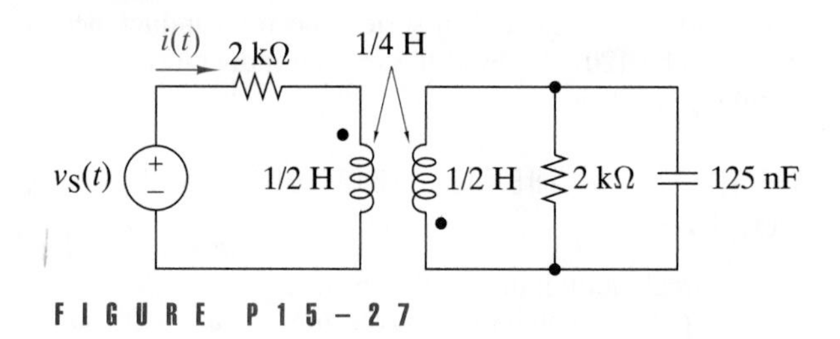

FIGURE P15–27

15–28 Find the equivalent input impedance in Figure P15–28.

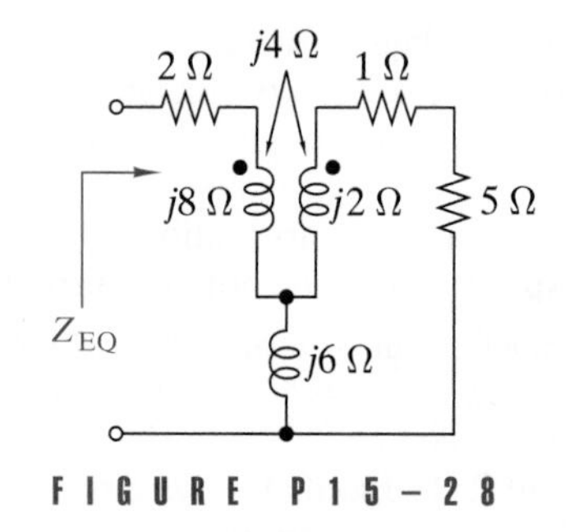

FIGURE P15–28

15–29 Find $\mathbf{V}_1$ and $\mathbf{V}_2$ in Figure P15–29.

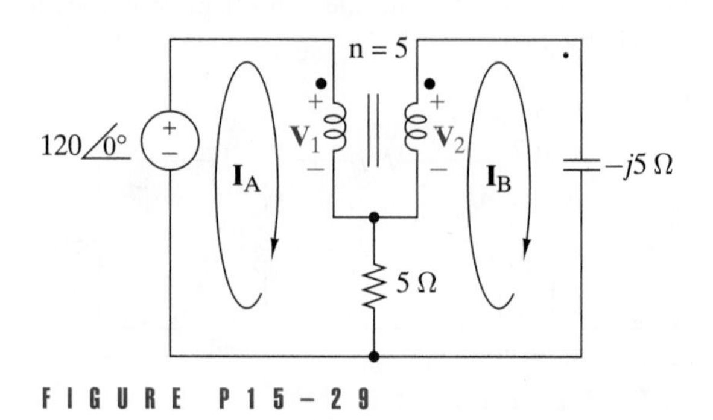

FIGURE P15–29

15–30 A transformer operating in the sinusoidal steady state has inductances $L_1 = 5$ mH, $L_2 = 20$ mH, and $M = 10$ mH. The load connected across the secondary winding is $Z_L = 100 + j200\ \Omega$. The voltage source connected to the primary winding has a frequency of $\omega = 10$ krad/s and a peak amplitude of 200 V. Find the primary and secondary currents. Assume additive coupling.

15–31 The self- and mutual inductances of a transformer can be calculated from measurements of the steady-state ac voltages and currents with the secondary winding open-circuited and short-circuited. Suppose the measurements are $|\mathbf{V}_1| = 120$ V, $|\mathbf{I}_1| = 120$ mA, and $|\mathbf{V}_2| = 240$ V when the secondary is open and $|\mathbf{I}_1| = 10$ A and $|\mathbf{I}_2| = 2.2$ A when the secondary is short-circuited. All measurements were made at $f = 400$ Hz. Find L_1, L_2, and M.

15–32 An ideal transformer has a turns ratio of $n = 10$. The secondary winding is connected to a load $Z_L = 300 + j100\ \Omega$. The primary is connected to a voltage source with a peak amplitude of 300 V and an internal impedance of $Z_S = j2\ \Omega$. Find the average power delivered to the load.

15–33 A certain resistive load requires 25 kW at a voltage of 480 V rms. The available supply is 2400 V rms. A transformer is to be used to reduce the supply voltage to 480 V. A circuit breaker is to be installed on the primary side to protect the equipment. The available circuit breakers have tripping currents of 5 A, 10 A, 15 A, and 20 A. Which breaker would you choose in this application?

15–34 An ideal transformer with $N_1 = 100$ and $N_2 = 250$ is connected as an autotransformer. The primary winding is connected to a 480-V source. A 100-Ω load is connected across the series-connected primary and secondary windings. Find the amplitudes of primary and secondary currents.

15–35 A transformer that can be treated as ideal has 480 turns in the primary winding and 240 turns in the secondary winding. The primary is connected to a 60-Hz source with a peak amplitude of 440 V. The secondary winding delivers 5 kW to a resistive load. Find the primary and secondary currents and the impedance seen by the source.

INTEGRATING PROBLEMS

15–36 A Transformer Step Response
Figure P15–36 shows a transformer circuit with a source resistance of R_1 and a resistive load of R_2. Use s-domain circuit analysis to find the zero-state step response of $v_2(t)$ for $L_1 = 0.1$ H, $L_2 = 0.4$ H, $M = 0.1$ H, $R_1 = 10\ \Omega$, and $R_2 = 40\ \Omega$.

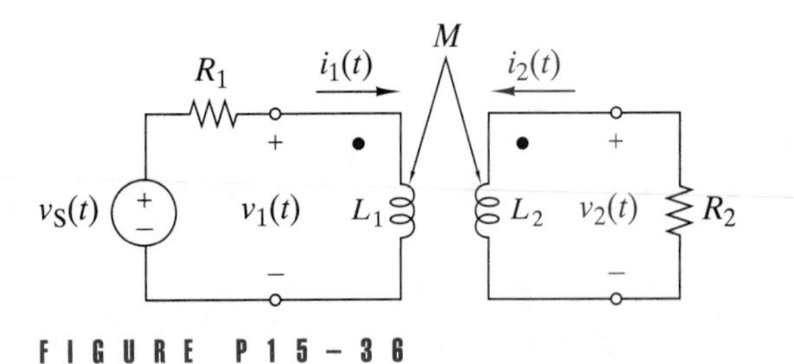

FIGURE P15–36

15–37 A Transformer Frequency Response
Figure P15–36 shows a transformer circuit with a source resistance of R_1 and a resistive load of R_2. Use s-domain circuit analysis to find the transfer function $T_V(s) = V_2(s)/V_S(s)$ for $L_1 = 0.25$ H, $L_2 = 1$ H, $M = 0.3$ H, $R_1 = 50\ \Omega$, and $R_2 = 200\ \Omega$. Is the transformer circuit a low-pass, high-pass, or bandpass filter? Find the cutoff frequencies and passband gain.

15–38 A Three-Winding Ideal Transformer
The i–v characteristics of the three-winding ideal transformer in Figure P15–38 are

$$\frac{v_1(t)}{N_1} = \frac{v_2(t)}{N_2} = \frac{v_3(t)}{N_3}$$

and

$$N_1 i_1(t) + N_2 i_2(t) + N_3 i_3(t) = 0$$

The total instantaneous power delivered to the transformer is

$$p(t) = v_1(t)i_1(t) + v_2(t)i_2(t) + v_3(t)i_3(t)$$

Show that the i–v characteristics imply $p(t) = 0$, that is, zero power loss.

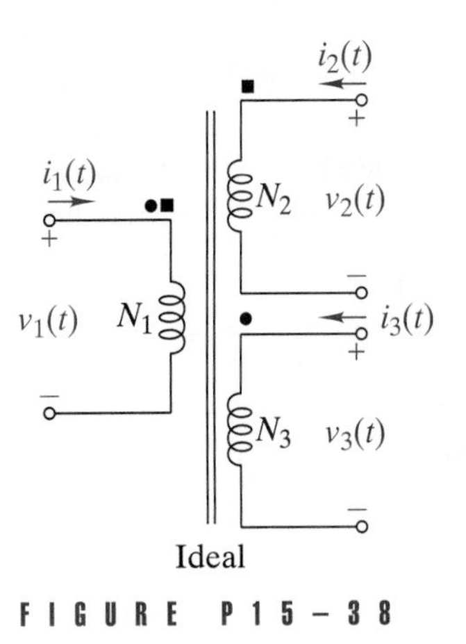

FIGURE P15–38

CHAPTER 16

POWER IN THE SINUSOIDAL STEADY STATE

George Westinghouse was in my opinion, the only man on the globe who could take my alternating current (power) system under the circumstances then existing and win the battle against prejudice and money power.

Nikola Tesla, 1932,
American Engineer

Some History Behind This Chapter

The 1890s saw a competition between the dc power system developed by Thomas Edison and the newly emerging ac system. Initially the main drawback of the ac approach was a lack of practical motors. The three-phase ac induction motor that met this need was invented by the Serbian immigrant Nikola Tesla (1856–1943). In 1887 Tesla founded a company to develop his inventions, eventually producing some 40 patents on three-phase equipment. George Westinghouse recognized the importance of this work and purchased the rights to Tesla's patents. The ac versus dc competition was settled when an ac system was chosen for a large hydroelectric power station at Niagara Falls, New York.

Why This Chapter Is Important Today

Rolling blackouts and collapsing power grids remind us that the reliable flow of electrical power is essential in a modern society. Although you may never work directly in the electrical power field, some understanding of its concepts and limitations is important in many adjacent areas of technology. Among these concepts are notion complex power and three-phase circuits and systems. Have you ever wondered why the electric transmission lines marching across the countryside have three wires each attached to a large insulator? You'll find the answer here.

Chapter Sections

16–1 Average and Reactive Power

16–2 Complex Power

16–3 ac Power Analysis

16–4 Load-Flow Analysis

16–5 Three-Phase Circuits

16–6 Three-Phase ac Power Analysis

Chapter Learning Objectives

16-1 Complex Power (Sects. 16-1, 16-2)

Given a linear circuit operating in the sinusoidal steady-state:

(a) Find the complex power delivered at a specified interface.

(b) Find the phasor voltage or current required to deliver a specified complex power at an interface.

16-2 ac Power Analysis (Sect. 16-3)

Given a linear circuit operating in the sinusoidal steady state:

(a) Find the complex power associated with any element.

(b) Find the load impedance required to draw a given complex power from a source.

16-3 Load-Flow Analysis (Sect. 16-4)

Find unknown voltages or currents required to produce a specified power flow.

16-4 Three-Phase Power (Sects. 16-5, 16-6)

Given a balanced three-phase circuit operating in the sinusoidal steady state:

(a) Find the line (or phase) voltages or currents when given the phase sequence and the phase (or line) voltages or currents.

(b) Find the line current and total complex power when given the line or phase voltage and the load impedance.

(c) Find unknown voltages or currents required to produce a specified power flow.

16–1 AVERAGE AND REACTIVE POWER

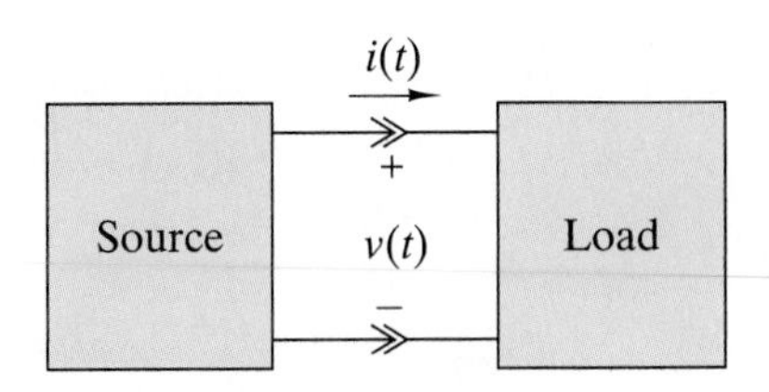

FIGURE 16–1 *A two-terminal interface.*

We begin our study of electric power circuits with the two-terminal interface in Figure 16–1. In power applications we normally think of one circuit as the source and the other as the load. Our objective is to describe the flow of power across the interface when the circuit is operating in the sinusoidal steady state. To this end, we write the interface voltage and current in the time domain as sinusoids of the form

$$v(t) = V_A \cos(\omega t + \theta)$$
$$i(t) = I_A \cos \omega t \qquad (16\text{–}1)$$

In Eq. (16–1) V_A and I_A are real, positive numbers representing the peak amplitudes of the voltage and current, respectively.

In Eq. (16–1) we have selected the $t = 0$ reference at the positive maximum of the current $i(t)$ and assigned a phase angle to $v(t)$ to account for the fact that the voltage maximum may not occur at the same time. In the phasor domain the angle $\theta = \phi_V - \phi_I$ is the angle between the phasors $\mathbf{V} = V_A\angle\phi_V$ and $\mathbf{I} = I_A\angle\phi_I$. In effect, choosing $t = 0$ at the current maximum shifts the phase reference by an amount $-\phi_I$ so that the voltage and current phasors become $\mathbf{V} = V_A\angle\,\theta$ and $\mathbf{I} = I_A\angle 0$.

A method of relating power to phasor voltage and current will be presented in the next section, but at the moment we write the instantaneous power in the time domain.

$$p(t) = v(t) \times i(t)$$
$$= V_A I_A \cos(\omega t + \theta)\cos \omega t \qquad (16\text{–}2)$$

This expression for instantaneous power contains dc and ac components. To separate the components, we first use the identity $\cos(x + y) = \cos x \cos y - \sin x \sin y$ to write $p(t)$ in the form

$$p(t) = V_A I_A[\cos \omega t \cos \theta - \sin \omega t \sin \theta] \cos \omega t$$
$$= [V_A I_A \cos \theta]\cos^2 \omega t - [V_A I_A \sin \theta]\cos \omega t \sin \omega t \qquad (16\text{–}3)$$

Using the identities $\cos^2 x = ½(1 + \cos 2x)$ and $\cos x \sin x = ½ \sin 2x$, we write $p(t)$ in the form

$$p(t) = \underbrace{\left[\frac{V_A I_A}{2}\cos\theta\right]}_{\text{dc component}} + \underbrace{\left[\frac{V_A I_A}{2}\cos\theta\right]\cos 2\omega t - \left[\frac{V_A I_A}{2}\sin\theta\right]\sin 2\omega t}_{\text{ac component}} \qquad (16\text{–}4)$$

Written this way, we see that the instantaneous power is the sum of a dc component and a double-frequency ac component. That is, the instantaneous power is the sum of a constant plus a sinusoid whose frequency is 2ω, which is twice the angular frequency of the voltage and current in Eq. (16–1).

Note that instantaneous power in Eq. (16–4) is periodic. In Chapter 5 we defined the average value of a periodic waveform as

$$P = \frac{1}{T}\int_0^T p(t)dt$$

where $T = 2\pi/2\omega$ is the period of $p(t)$. In Chapter 5 we also showed that the average value of a sinusoid is zero, since the area under the waveform during a positive half-cycle is canceled by the area under a subsequent negative half-cycle. Therefore, the **average value** of $p(t)$, denoted as P, is equal to the constant or dc term in Eq. (16–4):

$$P = \frac{V_A I_A}{2} \cos\theta \tag{16–5}$$

The amplitude of the sin $2\omega t$ term in Eq. (16–4) has a form much like the average power in Eq. (16–5), except it involves sin θ rather than cos θ. This amplitude factor is called the **reactive power** of $p(t)$, where reactive power Q is defined as

$$Q = \frac{V_A I_A}{2} \sin\theta \tag{16–6}$$

Substituting Eqs. (16–5) and (16–6) into Eq. (16–4) yields the instantaneous power in terms of the average power and reactive power:

$$p(t) = \underbrace{P(1 + \cos 2\omega t)}_{\text{unipolar}} - \underbrace{Q \sin 2\omega t}_{\text{bipolar}} \tag{16–7}$$

The first term in Eq. (16–7) is said to be unipolar because the factor 1 + cos $2\omega t$ never changes sign. As a result, the first term is either always positive or always negative depending on the sign of P. The second term is said to be bipolar because the factor sin $2\omega t$ alternates signs every half-cycle.

The energy transferred across the interface during one cycle $T = 2\pi/2\omega$ of $p(t)$ is

$$\begin{aligned} W &= \int_0^T p(t)dt \\ &= P\underbrace{\int_0^T (1 + \cos 2\omega t)dt}_{\text{net energy}} - Q\underbrace{\int_0^T \sin 2\omega t\, dt}_{\text{no net energy}} \\ &= \quad P \times T \quad - \quad 0 \end{aligned} \tag{16–8}$$

Only the unipolar term in Eq. (16–7) provides any net energy transfer, and that energy is proportional to the average power P. With the passive sign convention the energy flows from source to load when $W > 0$. Equation (16–8) shows that the net energy will be positive if the average power $P > 0$. Equation (16–5) points out that the average power P is positive when cos $\theta > 0$, which in turn means $|\theta| < 90°$. We conclude that

The net energy flow in Figure 16–1 is from source to load when the angle between the interface voltage and current is bounded by $-90° < \theta < 90°$; otherwise the net energy flow is from load to source.

The bipolar term in Eq. (16–7) is a power oscillation that transfers no net energy across the interface. In the sinusoidal steady state the load in Figure 16–1 borrows energy from the source circuit during part of a cycle and temporarily stores it in the load inductance or capacitance. In another part of the cycle the borrowed energy is returned to the source unscathed.

The amplitude of the power oscillation is called reactive power because it involves periodic energy storage and retrieval from the reactive elements of the load. The reactive power can be either positive or negative depending on the sign of sin θ. However, the sign of Q tells us nothing about the net energy transfer, which is controlled by the sign of P.

We are obviously interested in average power, since this component carries net energy from source to load. For most power system customers the basic cost of electrical service is proportional to the net energy delivered to the load. Large industrial users may also pay a service charge for their reactive power. This may seem unfair, since reactive power transfers no net energy. However, the electric energy borrowed and returned by the load is generated within a power system that has losses. From a power company's viewpoint, the reactive power is not free because there are losses in the system connecting the generators in the power plant to the source-load interface at which the lossless interchange of energy occurs.

In ac power circuit analysis, it is necessary to keep track of both the average power and reactive power. These two components of power have the same dimensions, but because they represent quite different effects they traditionally are given different units. The average power is expressed in watts (W) while reactive power is expressed in VARs, which is an acronym for "volt-amperes reactive."

Exercise 16–1

Using the reference marks in Figure 16–1, calculate the average and reactive power for the following voltages and currents. State whether the load is absorbing or delivering net energy.

(a) $v(t) = 168 \cos(377t + 45°)$ V, $i(t) = 0.88 \cos 377t$ A
(b) $v(t) = 285 \cos(2500t - 68°)$ V, $i(t) = 0.66 \cos 2500t$ A
(c) $v(t) = 168 \cos(377t + 45°)$ V, $i(t) = 0.88 \cos(377t - 60°)$ A
(d) $v(t) = 285 \cos(2500t - 68°)$ V, $i(t) = 0.66 \sin 2500t$ A

Answers:

(a) $P = +52.3$ W, $Q = +52.3$ VAR, absorbing
(b) $P = +35.2$ W, $Q = -87.2$ VAR, absorbing
(c) $P = -19.1$ W, $Q = +71.4$ VAR, delivering
(d) $P = +87.2$ W, $Q = +35.2$ VAR, absorbing

16–2 COMPLEX POWER

It is important to relate average and reactive power to phasor quantities because steady-state analysis is conveniently carried out using phasors. In our previous work the magnitude of a phasor represented the peak amplitude of a sinusoid. However, in power circuit analysis it is convenient to express phasor magnitudes in rms (root-mean-square) values. In this chapter phasor voltages and currents are expressed as

$$\mathbf{V} = V_{\text{rms}} e^{j\phi_V} \quad \text{and} \quad \mathbf{I} = I_{\text{rms}} e^{j\phi_I} \tag{16–9}$$

Notice that the phasor magnitudes are the rms amplitude of the corresponding sinusoid.

Equations (16–5) and (16–6) express average and reactive power in terms of peak amplitudes V_A and I_A. In Chapter 5 we showed that the peak and rms values of a sinusoid are related by $V_{rms} = V_A/\sqrt{2}$. The expression for average power easily can be converted to rms amplitudes since we can write Eq. (16–5) as

$$P = \frac{V_A I_A}{2}\cos\theta = \frac{V_A}{\sqrt{2}}\frac{I_A}{\sqrt{2}}\cos\theta$$
$$= V_{rms} I_{rms} \cos\theta \qquad (16\text{–}10)$$

where $\theta = \phi_V - \phi_I$ is the angle between the voltage and current phasors. By similar reasoning, Eq. (16–6) becomes

$$Q = V_{rms} I_{rms} \sin\theta \qquad (16\text{–}11)$$

Using rms phasors, the **complex power** (S) at a two-terminal interface is defined as follows:

$$S = \mathbf{VI}^* \qquad (16\text{–}12)$$

That is, the complex power at an interface is the product of the voltage phasor times the conjugate of the current phasor. Substituting Eq. (16–9) into this definition yields

$$S = \mathbf{VI}^* = V_{rms}e^{j\phi_V} I_{rms}e^{-j\phi_I}$$
$$= [V_{rms} I_{rms}]e^{j(\phi_V - \phi_I)} \qquad (16\text{–}13)$$

Using Euler's relationship and the fact that the angle is $\theta = \phi_V - \phi_I$, we can write complex power as

$$S = [V_{rms}I_{rms}]e^{j\theta}$$
$$= [V_{rms}I_{rms}]\cos\theta + j[V_{rms}I_{rms}]\sin\theta \qquad (16\text{–}14)$$
$$= P + jQ$$

The real part of the complex power S is the average power, and the imaginary part is the reactive power. Although S is a complex number, it is not a phasor. However, it is a convenient variable for keeping track of the two components of power when the voltage and currents are expressed as phasors.

The power triangles in Figure 16–2 provide a convenient way to remember complex power relationships and terminology. We confine our study to cases in which net energy is transferred from source to load. In such cases $P > 0$ and the power triangles fall in the first or fourth quadrant, as indicated in Figure 16–2.

The magnitude $|S| = V_{rms}I_{rms}$ is called **apparent power** and is expressed using the unit volt-ampere (VA). The ratio of the average power to the apparent power is called the **power factor** (pf). Using Eq. (16–10), we see that the power factor is

$$\text{pf} = \frac{P}{|S|} = \frac{V_{rms}I_{rms}\cos\theta}{V_{rms}I_{rms}} = \cos\theta \qquad (16\text{–}15)$$

Since pf = cos θ, the angle θ is called the **power factor angle**.

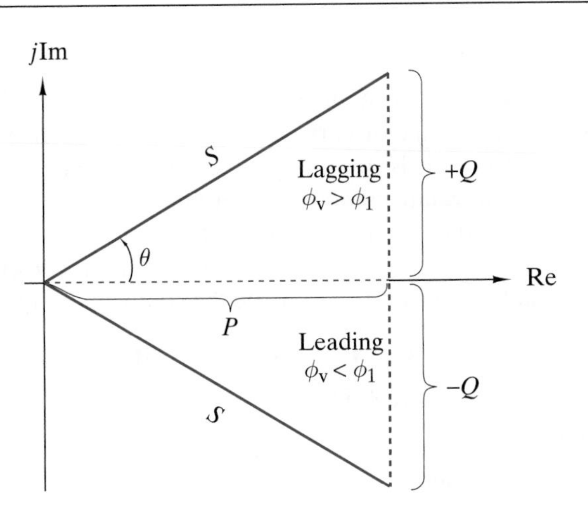

FIGURE 16–2 *Power triangles.*

When the power factor is unity, the phasors **V** and **I** are in phase ($\theta = 0°$) and the reactive power is zero since $\sin\theta = 0$. When the power factor is less than unity, the reactive power is not zero and its sign is indicated by the modifiers *lagging* or *leading*. The term *lagging power factor* means the current phasor lags the voltage phasor so that $\theta = \phi_V - \phi_I > 0$. For a lagging power factor, S falls in the first quadrant in Figure 16–2 and the reactive power is positive since $\sin\theta > 0$. The term *leading power factor* means the current phasor leads the voltage phasor so that $\theta = \phi_V - \phi_I < 0$. In this case S falls in the fourth quadrant in Figure 16–2 and the reactive power is negative since $\sin\theta < 0$. Most industrial and residential loads have lagging power factors.

The apparent power rating of electrical power equipment is an important design consideration. The ratings of generators, transformers, and transmission lines are normally stated in kVA. The ratings of most loads are stated in kW and power factor. The wiring must be large enough to carry the required current and insulated well enough to withstand the rated voltage. However, only the average power is potentially available as useful output, since the reactive power represents a lossless interchange between the source and device. Because reactive power increases the apparent power rating without increasing the net energy output, it is desirable for electrical devices to operate with zero reative power or, equivalently, at unity power factor.

Exercise 16–2

Determine the average power, reactive power, and apparent power for the following voltage and current phasors. State whether the power factor is lagging or leading.

(a) $\mathbf{V} = 208\angle{-90°}$ V (rms), $\mathbf{I} = 1.75\angle{-75°}$ A (rms)
(b) $\mathbf{V} = 277\angle{+90°}$ V (rms), $\mathbf{I} = 11.3\angle 0°$ A (rms)
(c) $\mathbf{V} = 120\angle{-30°}$ V (rms), $\mathbf{I} = 0.30\angle{-90°}$ A (rms)
(d) $\mathbf{V} = 480\angle{+75°}$ V (rms), $\mathbf{I} = 8.75\angle{+105°}$ A (rms)

Answers:

(a) $P = 352$ W; $Q = -94.2$ VAR; $|S| = 364$ VA; leading
(b) $P = 0$ W; $Q = +3.13$ kVAR; $|S| = 3.13$ kVA; lagging
(c) $P = 18$ W; $Q = +31.2$ VAR; $|S| = 36$ VA; lagging
(d) $P = 3.64$ kW; $Q = -2.1$ kVAR; $|S| = 4.20$ kVA; leading

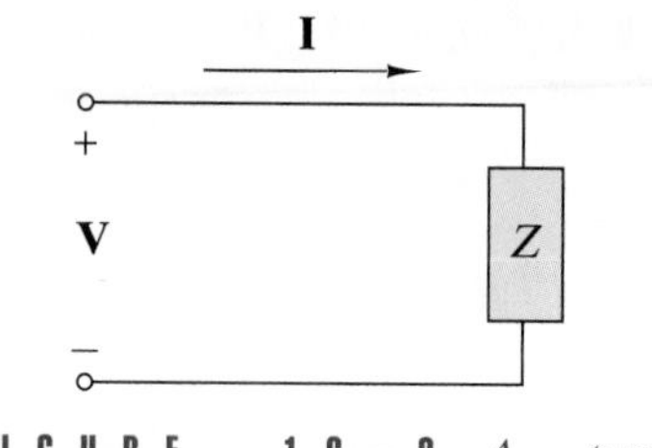

FIGURE 16–3 A two-terminal impedance.

COMPLEX POWER AND LOAD IMPEDANCE

In many cases power circuit loads are described in terms of their power ratings at a specified voltage or current level. To find voltages and current elsewhere in the circuit, it is necessary to know the load impedance. For this reason we need to relate complex power and load impedance.

Figure 16–3 shows the general case for a two-terminal load. For the assigned reference directions the load produces the element constraint $\mathbf{V} = Z\mathbf{I}$. Using this constraint in Eq. (16–12), we write the complex power of the load as

$$\begin{aligned} S &= \mathbf{V} \times \mathbf{I}^* = Z\mathbf{I} \times \mathbf{I}^* = Z|\mathbf{I}|^2 \\ &= (R + jX)\, I_{\text{rms}}^2 \end{aligned}$$

where R and X are the resistance and reactance of the load, respectively. Since $S = P + jQ$, we conclude that

$$R = \frac{P}{I_{\text{rms}}^2} \quad \text{and} \quad X = \frac{Q}{I_{\text{rms}}^2} \tag{16–16}$$

The load resistance and reactance are proportional to the average and reactive power of the load, respectively.

The first condition in Eq. (16–16) demonstrates that resistance cannot be negative, since P cannot be negative for a passive circuit. That is, in the sinusoidal steady state a passive circuit cannot produce average power; otherwise, perpetual motion would be possible and the energy crisis would be a small footnote in the great sweep of human history. The second condition in Eq. (16–16) points out that when the reactive power is positive the load is inductive, since $X_L = \omega L$ is positive. Conversely, when the reactive power is negative the load is capacitive, since $X_C = -1/\omega C$ is negative. The terms *inductive load, lagging power factor,* and *positive reactive power* are synonymous, as are the terms *capacitive load, leading power factor,* and *negative reactive power.*

EXAMPLE 16–1

At 440 V (rms) a two-terminal load draws 3 kVA of apparent power at a lagging power factor of 0.9. Find

(a) I_{rms}
(b) P
(c) Q
(d) the load impedance

Draw the power triangle for the load.

SOLUTION:

(a) $I_{\text{rms}} = |S|/V_{\text{rms}} = 3000/440 = 6.82$ A (rms)
(b) $P = V_{\text{rms}} I_{\text{rms}} \cos\theta = 3000 \times 0.9 = 2.7$ kW
(c) For $\cos\theta = 0.9$ lagging, $\sin\theta = 0.436$ and $Q = V_{\text{rms}} I_{\text{rms}} \sin\theta = 1.31$ kVAR
(d) $Z = (P + jQ)/(I_{\text{rms}})^2 = (2700 + j1310)/46.5 = 58.0 + j28.2\ \Omega$.

Figure 16–4 shows the power triangle for this load. ■

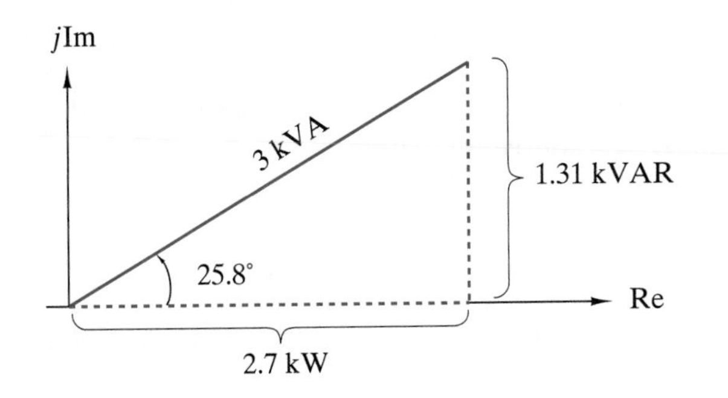

FIGURE 16–4

Exercise 16–3

Find the impedance of a two-terminal load under the following conditions.

(a) $\mathbf{V} = 120\angle 30°$ V (rms) and $\mathbf{I} = 20\angle 75°$ A (rms)
(b) $|S| = 3.3$ kVA, $Q = -1.8$ kVAR, and $I_{\text{rms}} = 7.5$ A
(c) $P = 3$ kW, $Q = 4$ kVAR, and $V_{\text{rms}} = 880$ V
(d) $V_{\text{rms}} = 208$ V, $I_{\text{rms}} = 17.8$ A, and $P = 3$ kW

Answers:

(a) $Z = 4.24 - j4.24\ \Omega$
(b) $Z = 49.2 - j32\ \Omega$
(c) $Z = 92.9 + j124\ \Omega$
(d) $Z = 9.47 \pm j6.85\ \Omega$

16–3 AC POWER ANALYSIS

The nature of ac power analysis can be modeled in terms of the phasor voltage, phasor current, and complex power at the source-load interface in Figure 16–1. Two different types of problems are treated using this model. In the **direct analysis** problem the source and load circuit are given and we are required to find the steady-state responses at one or more interfaces. This type of problem is essentially the same as the phasor circuit analysis problems in Chapter 8, except that we calculate complex powers as well as phasor responses. In the **load-flow** problem we are required to adjust the source so that a prescribed complex power is delivered to the load at a specified interface voltage magnitude. This type of problem arises in electrical power systems where the objective is to supply changing energy demands at a fixed voltage level.

The following examples are direct analysis problems that illustrate the computational tools needed to deal with the load-flow problems discussed in the next section. One of the useful tools is the principle of the **conservation of complex power**, which can be stated as follows:

> ***In a linear circuit operating in the sinusoidal steady state, the sum of the complex powers produced by each independent source is equal to the sum of the complex power absorbed by all other two-terminal elements in the circuit.***

To apply this principle, it is important to distinguish between the complex power "produced by" and "absorbed by" a two-terminal element. To do so, we modify our practice of using the passive sign convention. We continue to use the passive convention (current reference directed in at the + voltage mark) for loads, but use the active convention (current reference directed out at the + voltage mark) for sources. In either case, the complex power is always calculated as $S = \mathbf{VI}^* = P + jQ$. The net result is that the average power produced by a source is positive and the average power absorbed by a load is positive.

EXAMPLE 16–2

(a) Calculate the complex power absorbed by each parallel branch in Figure 16–5.

FIGURE 16–5

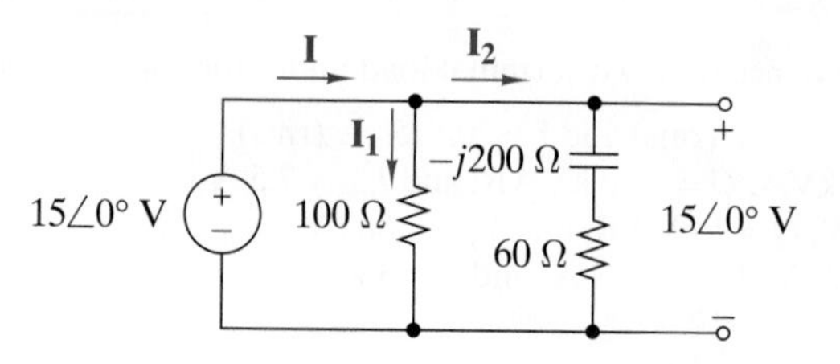

(b) Calculate the complex power produced by the source and the power factor of the load seen by the source.

SOLUTION:

(a) The voltage across each branch is 15∠0° V and the branch impedances are $Z_1 = 100$ and $Z_2 = 60 - j200$. Therefore, the branch currents are

$$\mathbf{I}_1 = \frac{15\angle 0^\circ}{100} = 0.15\angle 0^\circ \text{ A}$$

$$\mathbf{I}_2 = \frac{15\angle 0^\circ}{60 - j200} = 0.0718\angle 73.3^\circ \text{ A}$$

The reference marks for the voltage across and currents through these loads follow the passive sign convention. Hence the complex powers absorbed by the loads are

$$S_1 = (15\angle 0^\circ)\mathbf{I}_1^* = (15\angle 0^\circ)(0.15\angle -0^\circ) = 2.25\angle 0^\circ \text{ VA}$$

$$S_2 = (15\angle 0^\circ)\mathbf{I}_2^* = (15\angle 0^\circ)(0.0718\angle -73.3^\circ) = 1.08\angle -73.3^\circ \text{ VA}$$

(b) Using KCL, the source current **I** is

$$\mathbf{I} = (\mathbf{I}_1 + \mathbf{I}_2) = (0.15\angle 0^\circ + 0.0718\angle 73.3^\circ)$$
$$= 0.171 + j0.0688 = 0.184\angle 21.9^\circ \text{ A}$$

The reference marks for the source current **I** and source voltage 15∠0° conform to the active sign convention, and the complex power produced by the source is

$$S = (15\angle 0^\circ)\mathbf{I}^* = (15\angle 0^\circ)(0.184\angle -21.9^\circ)$$
$$= 2.76\angle -21.9^\circ \text{ VA}$$

The power factor is cos(−21.9°) = 0.928 leading. Alternatively, we can use the conservation of complex power to obtain the complex power produced by the source as the sum of the complex powers delivered to the passive elements:

$$
\begin{aligned}
S_1 + S_2 &= 2.25\angle 0° + 1.08\angle -73.3° \\
&= 2.25 + j0 + 0.310 - j1.03 \\
&= 2.56 - j1.03 = 2.76\angle -21.9° \text{ VA}
\end{aligned}
$$

This result is the same as the answer obtained previously. ■

EXAMPLE 16–3

Find the complex power produced by each source in Figure 16–6 when $\mathbf{V}_{S1} = 440\angle 0°$ V (rms) and $\mathbf{V}_{S2} = 500\angle 0°$ V (rms).

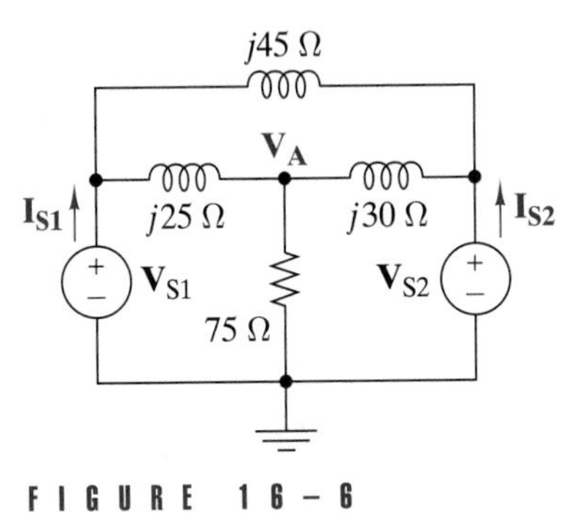

FIGURE 16–6

SOLUTION:

Since both voltage sources are connected to ground, the voltage at node A is the only unknown node voltage in the circuit. By inspection, the node equation at node A is

$$
\left[\frac{1}{75} + \frac{1}{j25} + \frac{1}{j30}\right]\mathbf{V}_A = \frac{\mathbf{V}_{S1}}{j25} + \frac{\mathbf{V}_{S2}}{j30} = -j34.27
$$

which yields $\mathbf{V}_A = 452 - j82.2$ V. The current supplied by source 1 is

$$
\mathbf{I}_{S1} = \frac{\mathbf{V}_{S1} - \mathbf{V}_A}{j25} + \frac{\mathbf{V}_{S1} - \mathbf{V}_{S2}}{j45} = 3.29 + j1.81 \text{ A (rms)}
$$

The complex power supplied by source 1 is

$$
\begin{aligned}
S_{S1} &= \mathbf{V}_{S1}\mathbf{I}_{S1}^* = 440 \times (3.29 - j1.83) \\
&= 1450 - j796 \quad \text{VA}
\end{aligned}
$$

The current supplied by source 2 is

$$
\mathbf{I}_{S2} = \frac{\mathbf{V}_{S2} - \mathbf{V}_A}{j30} + \frac{\mathbf{V}_{S2} - \mathbf{V}_{S1}}{j45} = 2.74 - j2.93 \quad \text{A (rms)}
$$

and the complex power supplied by source 2 is

$$
\begin{aligned}
S_{S2} &= \mathbf{V}_{S2}\mathbf{I}_{S2}^* = 500 \times (2.74 + j2.93) \\
&= 1370 + j1460 \quad \text{VA}
\end{aligned}
$$

■

EXAMPLE 16–4

Given the transformer circuit in Figure 16–7, find the complex power produced by the input voltage source and the power absorbed by the 100-Ω load resistor.

FIGURE 16–7

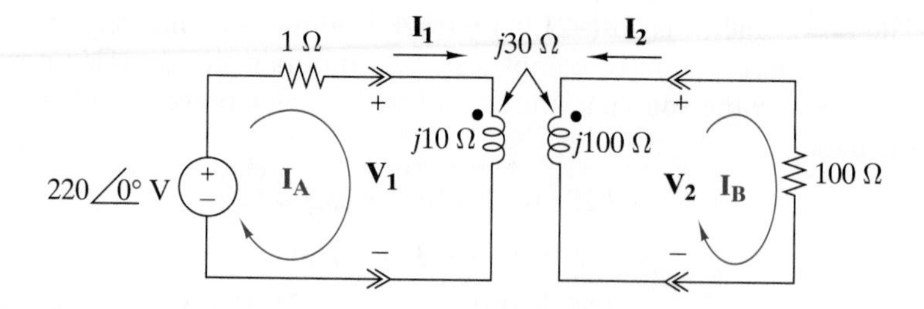

SOLUTION:
Using the voltages and mesh currents in Figure 16–7, the KVL equations around the primary and secondary circuits are

$$1 \times \mathbf{I}_A + \mathbf{V}_1 = 220\angle 0°$$
$$100\,\mathbf{I}_B - \mathbf{V}_2 = 0$$

The coil currents are related to the mesh currents as $\mathbf{I}_1 = \mathbf{I}_A$ and $\mathbf{I}_2 = -\mathbf{I}_B$. Both coil currents are directed inward at the winding dots, so the mutual inductive coupling is additive and the i–v characteristics of the transformer in terms of the mesh currents are

$$\mathbf{V}_1 = j10\mathbf{I}_A + j30(-\mathbf{I}_B)$$
$$\mathbf{V}_2 = j30\mathbf{I}_A + j100(-\mathbf{I}_B)$$

Substituting these i–v characteristics into the KVL equation produces the two mesh-current equations for the transformer:

$$(1 + j10)\mathbf{I}_A - j30\mathbf{I}_B = 220$$
$$-j30\mathbf{I}_A + (100 + j100)\mathbf{I}_B = 0$$

Solving for the two mesh currents produces

$$\mathbf{I}_A = 20 - j20 \text{ A (rms)}$$
$$\mathbf{I}_B = 6 + j0 \text{ A (rms)}$$

Using mesh current $\mathbf{I}_A$, the complex power produced by the source is

$$S_{IN} = (220\angle 0°)\mathbf{I}_A^* = 4400 + j4400 \quad \text{VA}$$

and the power absorbed by the 100-Ω output resistor is

$$S_{OUT} = (100\mathbf{I}_B)\mathbf{I}_B^* = 3600 + j0 \text{ VA}$$

■

Exercise 16–4

Calculate the complex power delivered by each source in Figure 16–8.

FIGURE 16–8

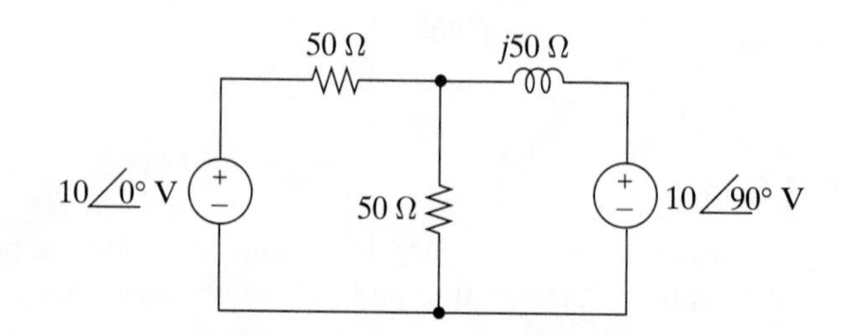

Answers: $S_1 = 0.4 + j0.8$ VA; $S_2 = 1.6 + j1.2$ VA

16–4 LOAD-FLOW ANALYSIS

The analysis of ac electrical power systems is one of the major applications of phasor circuit analysis. Although the loads on power systems change throughout the day, these variations are extremely slow compared with the period of the 50/60-Hz sinusoid involved.[1] Consequently, electrical power system analysis can be carried out using steady-state concepts and phasors. In fact, it was the study of the steady-state performance of ac power equipment that led Charles Steinmetz to advocate phasors in the first place.

In this section we treat ac power analysis using the simple model of an electrical power system in Figure 16–9. This model is a series circuit with an ac source connected to a load via power lines whose wire impedances are Z_W. In a **load-flow problem** the complex power delivered to the load is specified and we are asked to find either the source voltage for a given load voltage or the load voltage for a given source voltage. The analysis approach is similar to the unit output method. That is, we begin with conditions at the load and work backward through the circuit to establish the required source voltage. The load-flow problem is different from the maximum power transfer problem studied in Chapter 8. In a maximum power transfer problem the source is fixed and the load is adjusted to achieve a conjugate match. Conjugate matching does not apply to electrical power systems because the load power is fixed and the source is adjusted to meet customer demands.

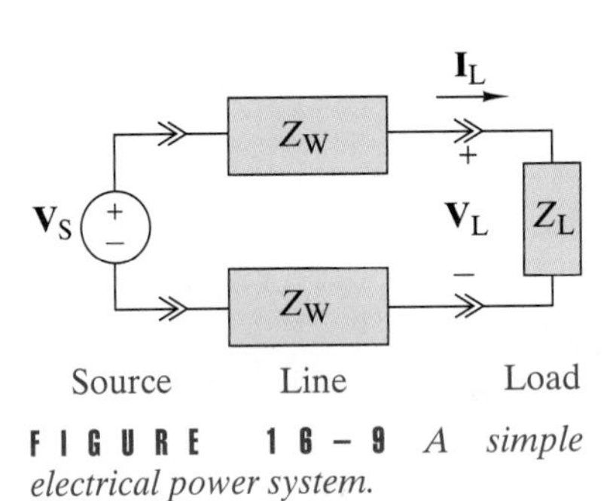

FIGURE 16–9 *A simple electrical power system.*

Large industrial customers are charged for their reactive power, so in some cases it is desirable to reduce the load reactance. Since power system loads are normally inductive, the net reactive power of the load can be reduced by adding a capacitor in parallel with the load. The amount of the negative reactive power taken by the capacitor is selected to cancel some or all of the positive reactance power drawn by the inductive load. Physically, this means that the oscillatory interchange of energy represented by reactive power takes place between the capacitor and inductance in the load, rather than between the load inductance and the lossy power system.

Adding parallel capacitance is called **power factor correction**, since the net power factor of the composite load is increased. If the power factor is increased to unity, then the net reactance is zero and the load is in resonance. Power factor correction reduces the reactive power drain on the power system but does not change the average power delivered to the load.

The following examples illustrate ac power analysis problems, including load flow and power factor correction.

EXAMPLE 16–5

In this problem the two parallel connected impedances in Figure 16–10 are the load in the power system model in Figure 16–9. With $V_L = 480$ V (rms), load Z_1 draws an average power of 10 kW at a lagging power factor of 0.8 and load Z_2 draws 12 kW at a lagging power factor of 0.75. The line impedances shown in Figure 16–9 are $Z_W = 0.35 + j1.5\ \Omega$.

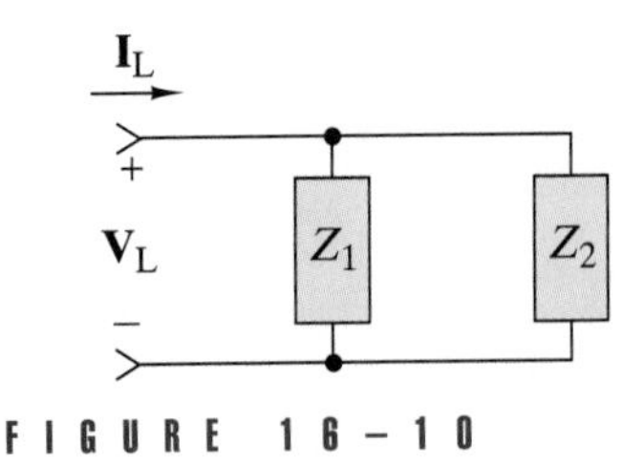

FIGURE 16–10

---

1 In the United States, commercial ac power systems operate at 60 Hz. In most of the rest of the world the standard operating frequency is 50 Hz.

(a) Find the total complex power delivered to the composite load.
(b) Find the apparent power delivered by the source and the source voltage phasor $\mathbf{V}_S$.
(c) Calculate the transmission efficiency of the system.

SOLUTION:

(a) In this example we are given the power factor (pf) and the average power (P) for each load. Because the power factor angle is $\theta = \cos^{-1}(\text{pf})$ and $Q = P\tan\theta$, we can find the complex power for each load as $S = P + P\tan(\cos^{-1}\text{pf})$; hence

$$S_{L1} = 10 + j10\tan(\cos^{-1}0.8) = 10 + j7.5\text{ kVA}$$

$$S_{L2} = 12 + j12\tan(\cos^{-1}0.75) = 12 + j10.6\text{ kVA}$$

The total complex power delivered to the composite load is

$$S_L = S_{L1} + S_{L2} = 22 + j18.1 = 28.5\angle 39.4°\text{ kVA}$$

(b) Using the load voltage as the phase reference, we find the load current to be

$$\mathbf{I}_L^* = \frac{S_L}{\mathbf{V}_L} = \frac{28500\angle 39.4°}{480\angle 0°} = 59.4\angle 39.4°\quad\text{A (rms)}$$

or $\mathbf{I}_L = 59.4\angle{-39.4°}$ A (rms). To find the source power, we need to find the complex power lost in the transmission to the line:

$$S_W = 2|\mathbf{I}_L|^2(R_W + jX_W) = 2 \times (59.4)^2\,(0.35 + j1.5)$$
$$= 2.47 + j10.6\text{ kVA}$$

Using the conservation of complex power, we find that the source must produce

$$S_S = S_L + S_W = 24.5 + j28.7 = 37.7\angle 49.5°\quad\text{kVA}$$

For the series model in Figure 16–9 the source and load currents are equal, so the source power is $S_S = \mathbf{V}_S(\mathbf{I}_L)^*$. Given the source power and load current, we find that the required source voltage is

$$\mathbf{V}_S = \frac{S_S}{\mathbf{I}^*_L} = \frac{37700\angle 49.5°}{59.4\angle 39.4°} = 635\angle 10.1°\quad\text{V (rms)}$$

(c) The transmission efficiency is defined in terms of source and load average power as

$$\eta = \frac{P_L}{P_S} \times 100\% = \frac{22}{24.5} \times 100 = 89.8\%$$ ■

EXAMPLE 16–6

A power system modeled by the circuit in Figure 16–9 delivers a load power of $S_L = 25 + j10$ kVA when the source voltage is $\mathbf{V}_S = 600\angle 0°$. The line impedances are $Z_W = 0.35 + j1.5\ \Omega$. Find the load current, load voltage, and transmission efficiency.

SOLUTION:

In the preceding example the known quantities are the load power and load voltage. In this example the load power and the source voltage are given. As a result, both the load current and the load voltage are unknowns. We can write two constraints on these unknowns. First, the specified complex power delivered to the load requires that

$$S_L = 25000 + j10000 = \mathbf{V}_L\mathbf{I}_L^*$$

Next, writing a KVL equation around the loop in Figure 16–9 yields $\mathbf{V}_S - \mathbf{V}_L - 2Z_W\mathbf{I}_L = 0$. Inserting the known quantities and solving for $\mathbf{V}_L$ produces

$$\mathbf{V}_L = 600 - (0.7 + j3)\mathbf{I}_L$$

Substituting this result into the complex power constraint yields

$$25000 + j1000 - 600\,\mathbf{I}_L^* + (0.7 + j3)\mathbf{I}_L\mathbf{I}_L^* = 0$$

We have a single complex constraint in which the only unknown is the load current phasor $\mathbf{I}_L$. This condition requires both the real part and the imaginary part of the constraint to vanish. Hence, the constraint can be written in terms of two real value functions. Writing the load current phasor as $\mathbf{I}_L = I_1 + jI_2$, hence $\mathbf{I}_L\mathbf{I}_L^* = I_1^2 + I_2^2$, the constraints on the real part and the imaginary part are

$$\text{Real part:} \quad 25000 - 600\,I_1 + 0.7(I_1^2 + I_2^2) = 0$$
$$\text{Imaginary part:} \quad 10000 + 600\,I_2 + 3(I_1^2 + I_2^2) = 0$$

We have now set up the equations so that MATLAB can be used to solve the complex constraints. In the MATLAB command window we write the following statements:

```
syms I1 I2

FR=25000-600*I1+0.7*(I1^2+I^2);

FI=10000+600*I2+3*(I1^2+I2^2);
```

The first statement declares `I1` and `I2` to be symbolic variables. The next two statements define the real part constraint `FR` and the imaginary part constraint `FI` as functions of the form $f(x, y) = 0$. The MATLAB `solve` command is set up to solve equations of this type. Accordingly, we write the command

```
[I1,I2]=solve(FR,FI)
```

and MATLAB responds with the values of `I1` and `I2` that satisfy the conditions `FR=0` and `FI=0`, namely,

```
I1=

[55000/949-700/2847*2669^(1/2)]

[55000/949+700/2847*2669^(1/2)]


I2=

[-246200/2847+1000/949*2669^(1/2)]

[-246200/2847-1000/949*2669^(1/2)]
```

MATLAB returns two possible solutions for the load current. The first solution is

$$\begin{aligned}\mathbf{I}_L &= I_1 + jI_2 \\ &= \left(\frac{55000}{949} - \frac{700}{2847}\sqrt{2669}\right) + j\left(\frac{-246200}{2847} + \frac{1000}{949}\sqrt{2669}\right) \\ &= 45.253 - j32.083 \text{ A}\end{aligned}$$

which produces a load voltage phasor of

$$\mathbf{V}_L = 600 - (0.7 + j3)\mathbf{I}_L = 472.208 - j113.333 \text{ V}$$

The apparent power delivered by the source is

$$\begin{aligned}S_S &= \mathbf{V}_S\mathbf{I}_L{}^* = (600 + j0)(45.253 + j32.038) \\ &= 27.2 + j19.2 \text{ kVA}\end{aligned}$$

and the transmission efficiency for the first solution is

$$\eta = \frac{P_L}{P_S} \times 100\% = \frac{25}{27.2} \times 100 = 91.9\%$$

The second MATLAB solution is

$$\begin{aligned}\mathbf{I}_L &= I_1 + jI_2 \\ &= \left(\frac{55000}{949} + \frac{700}{2847}\sqrt{2669}\right) + j\left(\frac{-246200}{2847} - \frac{1000}{949}\sqrt{2669}\right) \\ &= 70.658 - j140.916 \text{ A}\end{aligned}$$

which produces a load voltage phasor of

$$\mathbf{V}_L = 600 - (0.7 + j3)\mathbf{I}_L = 127.790 - j113.333 \text{ V}$$

The apparent power delivered by the source is

$$\begin{aligned}S_S &= \mathbf{V}_S\mathbf{I}_L{}^* = (600 + j0)(70.658 + j140.915) \\ &= 42.39 + j84.55 \text{ kVA}\end{aligned}$$

and the transmission efficiency for the second solution is

$$\eta = \frac{P_L}{P_S} \times 100\% = \frac{25}{42.395} \times 100 = 59.0\%$$

Several comments are in order. This type of load-flow problem may have several solutions, two in this case. On the basis of transmission efficiency we would choose the first solution over the second. This simple example illustrates some features of the general load-flow problem in large power systems. Such systems have a number of generating stations with different efficiencies and many geographically separated load centers. The load-flow problem involves controlling the output of the different generating stations to meet the total load requirements in a way that maximizes the system efficiency. ■

EXAMPLE 16–7

Using the loads and line impedances defined in Example 16–5:

(a) Find the parallel capacitance needed so that the load power factor is at least 0.95. The power system frequency is 60 Hz.
(b) Find the transmission efficiency with the capacitor connected.

SOLUTION:

(a) To determine the capacitance, it is necessary to relate the reactive power of a capacitor to the load voltage. Since the capacitor is in parallel with the load, the current through it is $\mathbf{I}_C = j\omega C\mathbf{V}_L$. Therefore, the reactive power of the capacitor can be written as

$$Q_C = |\mathbf{I}_C|^2 X_C = |j\omega C\mathbf{V}_L|^2\left(\frac{-1}{\omega C}\right)$$
$$= |\mathbf{V}_L|^2(-\omega C)$$

Note that Q_C is negative. In Example 16–5 the complex power of the two parallel loads is found to be

$$S_L = P_L + jQ_L = 22 + j18.1 \quad \text{kVA}$$

When the capacitor is connected, the complex power delivered to the composite load becomes

$$S_L = P_L + j(Q_L + Q_C)$$

Since the capacitor is a reactive element, it only changes the reactive power, not the average power. To correct the power factor to at least 0.95 requires that

$$\cos\theta = \frac{P_L}{\sqrt{P_L^2 + (Q_L + Q_C)^2}} \geqslant 0.95$$

Solving for $-Q_C$ yields

$$-Q_C \geqslant Q_L - P_L\sqrt{1/(0.95)^2 - 1} = 10.9 \quad \text{kVAR}$$

which means that a lower bound on the capacitance is

$$C = \frac{-Q_C}{V_L^2\omega} \geqslant \frac{10{,}900}{(480)^2(2\pi 60)} = 125\ \mu\text{F}$$

This yields a lower bound on the required capacitance. An upper bound is found by increasing the load-power factor to unity ($\cos\theta = 1$). The required capacitive reactance is

$$-Q_C = Q_L = 18.1 \quad \text{kVAR}$$

which yields an upper bound on the capacitance of

$$C = \frac{-Q_C}{V_L^2\omega} \leqslant \frac{18{,}100}{(480)^2(2\pi 60)} = 208\ \mu\text{F}$$

Thus, the design requirement can be met by any capacitance in the range from 125 μF to 208 μF.

(b) With the minimum capacitance of $C = 125\ \mu F$ connected in parallel with the load, the apparent power delivered to the composite load is $S_L = 22 + j7.2$ kVA. Using the load voltage for the phase reference, we find that the load current is

$$\mathbf{I}_L^* = \frac{S_L}{\mathbf{V}_L} = \frac{22{,}000 + j7200}{480\angle 0^\circ}$$
$$= 48.2\angle 18.1^\circ \quad \text{A (rms)}$$

The apparent power lost in the line is

$$S_W = |\mathbf{I}_L|^2\, 2(R_W + jX_W) = 1.62 + j6.97 \quad \text{kVA}$$

Hence with power factor correction the source produces

$$S_S = S_L + S_W = 23.6 + j14.2 \quad \text{kVA}$$

and the transmission efficiency is

$$\eta = \frac{P_L}{P_S} \times 100\% = \frac{22}{23.6} \times 100 = 93.2\%$$

In Example 16–5 we found the transmission efficiency without power factor correction to be 89.8%. With power factor correction the source delivers the same average power to the load with an increase in efficiency. Reactive power is a burden to a power system even though it represents a lossless interchange of energy at the terminals of the load. ■

Exercise 16–5

Find the source voltage and apparent power required to deliver 2400 V (rms) to a load that draws 25 kVA at a 0.85 lagging power factor from a line with a total line impedance of $2Z_W = 4 + j20\ \Omega$.

Answers: 2.55 kV (rms) and 26.6 kVA at a lagging power factor of 0.82

APPLICATION EXAMPLE 16–8

The electrical power for most residential customers in the United States is supplied by the 60-Hz, 110/220-V (rms) single-phase, three-wire system modeled in Figure 16–11. The term *single phase* means that the phasors representing the two source voltages are in phase. The three lines connecting the sources and loads are labeled A, B, and N (for neutral). The impedances Z_W and Z_N are small compared with the load impedances, so the load voltages differ from the source voltages by only a few percent. The impedances Z_1 and Z_2 connected from lines A or B to neutral represent small appliance and lighting loads which require 110 V (rms) service. The impedances Z_3 connected between lines A and B are heavier loads that require 220 V (rms) service, such as water heaters or clothes dryers.

When the two source voltages are exactly equal and $Z_1 = Z_2$ the system is said to be balanced. Under balanced conditions the current in

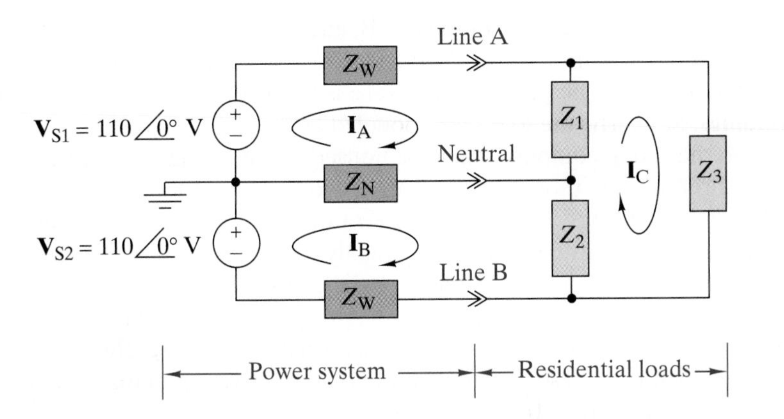

FIGURE 16–11 *Residential power distribution circuit.*

the neutral wire is zero. To show why the neutral current is zero, we write two mesh-current equations:

$$\text{Mesh A: } (Z_W + Z_1 + Z_N)\mathbf{I}_A - Z_N\mathbf{I}_B - Z_1\mathbf{I}_C = \mathbf{V}_{S1}$$

$$\text{Mesh B: } -Z_N\mathbf{I}_A + (Z_W + Z_2 + Z_N)\mathbf{I}_B - Z_2\mathbf{I}_C = \mathbf{V}_{S2}$$

For balanced conditions $\mathbf{V}_{S1} = \mathbf{V}_{S2}$ and $Z_1 = Z_2 = Z_L$. Subtracting the mesh B equation from the mesh A equation yields the condition

$$(Z_W + Z_L + 2Z_N)(\mathbf{I}_A - \mathbf{I}_B) = 0$$

This condition requires $\mathbf{I}_A - \mathbf{I}_B = 0$, since the impedance sum cannot be zero for all loads. Therefore, the net current in the neutral line is zero and theoretically the neutral wire can be disconnected. In practice, the balance is never perfect and the neutral line is included for safety reasons. But even so, the current in the neutral is usually less than the line currents, so losses in the feeder lines are reduced.

16–5 THREE-PHASE CIRCUITS

The three-phase system shown in Figure 16–12 is the predominant method of generating and distributing ac electrical power. The system uses four lines (A, B, C, N) to transmit power from the source to the loads. The

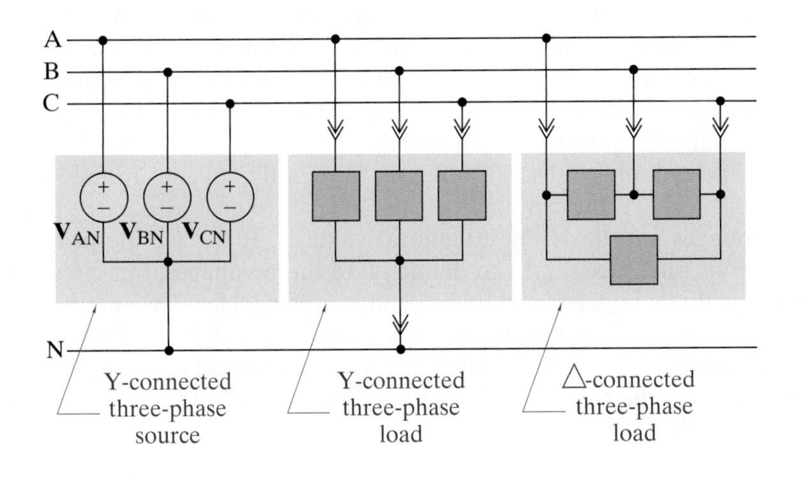

FIGURE 16–12 *A three-phase ac electrical power system.*

symbols stand for the three phases A, B, and C, and a neutral line labeled N. The three-phase generator in Figure 16–12 is modeled as three independent sources, although the physical hardware is a single unit with three separate windings. Similarly, the loads are modeled as three separate impedances, although the actual equipment may be housed within a single container.

A three-phase system involves three voltages and three currents. In a balanced three-phase system the generator produces phasor voltages that are equal in magnitude and symmetrically displaced in phase at 120° intervals. When the three-phase load is balanced (equal impedances), the resulting phasor currents have equal magnitude and are symmetrically displaced in phase at 120° intervals. Thus, a **balanced three-phase** system is one in which the phasor currents and voltages have equal magnitudes and phase differences of 120°.

The terminology *Y-connected* and *Δ-connected* refers to the two ways the source and loads can be electrically connected. Figure 16–13 shows the same electrical arrangement as Figure 16–12 with the elements rearranged to show the Y and Δ nature of the connections (the Δ is upside down in the figure). The circuit diagrams in the two figures are electrically equivalent, but we will use the form in Figure 16–12 because it highlights the purpose of the system. You need only remember that in a Y-connection the three elements are connected from line to neutral, while in the Δ-connection they are connected from line to line. In most systems the source is Y-connected while the loads can be either Y or Δ, although the latter is more common.

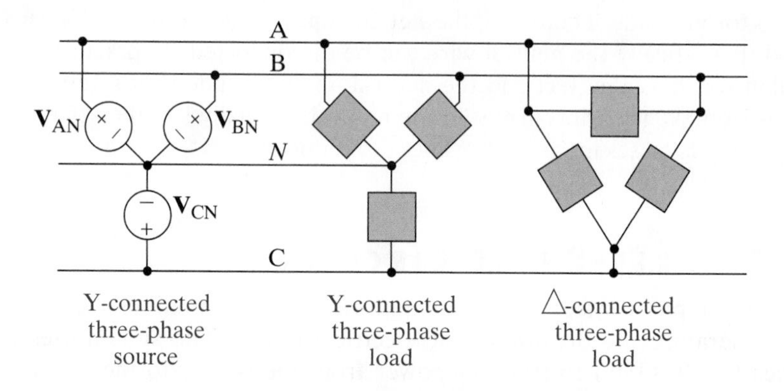

FIGURE 16–13 *A three-phase power system with the loads rearranged.*

Three-phase sources usually are Y-connected because the Δ-connection involves a loop of voltage sources. Large currents may circulate in this loop if the three voltages do not exactly sum to zero. In analysis situations, a Δ-connection of ideal voltage sources is awkward because it is impossible to determine the current in each source.

We use a double subscript notation to identify voltages in the system. The reason is that there are at least six voltages to deal with: three line-to-line voltages and three line-to-neutral voltages. If we use the usual plus and minus reference marks to define all of these voltages, our circuit diagram would be hopelessly cluttered and confusing. Hence we use two subscripts to define the points across which a voltage is defined. For example, $\mathbf{V}_{XY}$ means the voltage between points X and Y, with an implied plus reference mark at the first subscript (X) and an implied minus at the second subscript (Y).

The three line-to-neutral voltages are called the **phase voltages** and are written in double-subscript notation as $\mathbf{V}_{AN}$, $\mathbf{V}_{BN}$, and $\mathbf{V}_{CN}$. Similarly, the three line-to-line voltages, called simply the **line voltages**, are identified as $\mathbf{V}_{AB}$, $\mathbf{V}_{BC}$, and $\mathbf{V}_{CA}$. From the definition of the double-subscript notation it follows that $\mathbf{V}_{XY} = -\mathbf{V}_{YX}$. Using this result and KVL, we derive the relationships between the line voltages and phase voltages:

$$\begin{aligned} \mathbf{V}_{AB} &= \mathbf{V}_{AN} + \mathbf{V}_{NB} = \mathbf{V}_{AN} - \mathbf{V}_{BN} \\ \mathbf{V}_{BC} &= \mathbf{V}_{BN} + \mathbf{V}_{NC} = \mathbf{V}_{BN} - \mathbf{V}_{CN} \\ \mathbf{V}_{CA} &= \mathbf{V}_{CN} + \mathbf{V}_{NA} = \mathbf{V}_{CN} - \mathbf{V}_{AN} \end{aligned} \tag{16–17}$$

A balanced three-phase source produces phase voltages that obey the following two constraints:

$$\begin{aligned} |\mathbf{V}_{AN}| = |\mathbf{V}_{BN}| = |\mathbf{V}_{CN}| &= V_P \\ \mathbf{V}_{AN} + \mathbf{V}_{BN} + \mathbf{V}_{CN} &= 0 + j0 \end{aligned}$$

That is, the phase voltages have equal amplitudes (V_P) and sum to zero. There are two ways to satisfy these constraints:

Positive Phase Sequence	Negative Phase Sequence
$\mathbf{V}_{AN} = V_P\angle 0°$	$\mathbf{V}_{AN} = V_P\angle 0°$
$\mathbf{V}_{BN} = V_P\angle -120°$	$\mathbf{V}_{BN} = V_P\angle -240°$
$\mathbf{V}_{CN} = V_P\angle -240°$	$\mathbf{V}_{CN} = V_P\angle -120°$

(16–18)

Figure 16–14 shows the phasor diagrams for the positive and negative phase sequences. It is apparent that both sequences involve three equal-length phasors that are separated by angles of 120°. As a result, the sum of any two phasors cancels the third. In the positive sequence the phase B voltage lags the phase A voltage by 120°. In the negative sequence phase B lags by 240°. It also is apparent that we can convert one phase sequence into the other by simply interchanging the labels on lines B and C. From a circuit analysis viewpoint, there is no conceptual difference between the two sequences. Consequently, in analysis problems we will use the positive phase sequence unless otherwise stated.

However, the phrase *no conceptual difference* does not mean that phase sequence is unimportant. It turns out that three-phase motors run in one direction when the positive sequence is applied, and in the opposite direction for the negative sequence. This could be a matter of some importance if the motor is driving a conveyor belt at a sewage treatment facility. In practice, it is essential that there be no confusion about which is line A, B, and C and whether the source phase sequence is positive or negative.

A simple relationship between the line and phase voltages is obtained by substituting the positive phase sequence voltages from Eq. (16–18) into the phasor sums in Eq. (16–17). For the first sum

$$\begin{aligned} \mathbf{V}_{AB} &= \mathbf{V}_{AN} - \mathbf{V}_{BN} \\ &= V_P\angle 0° - V_P\angle -120° \\ &= V_P(1 + j0) - V_P(-1/2 - j\sqrt{3}/2) \\ &= V_P(3/2 + j\sqrt{3}/2) \\ &= \sqrt{3}V_P\angle 30° \end{aligned} \tag{16–19}$$

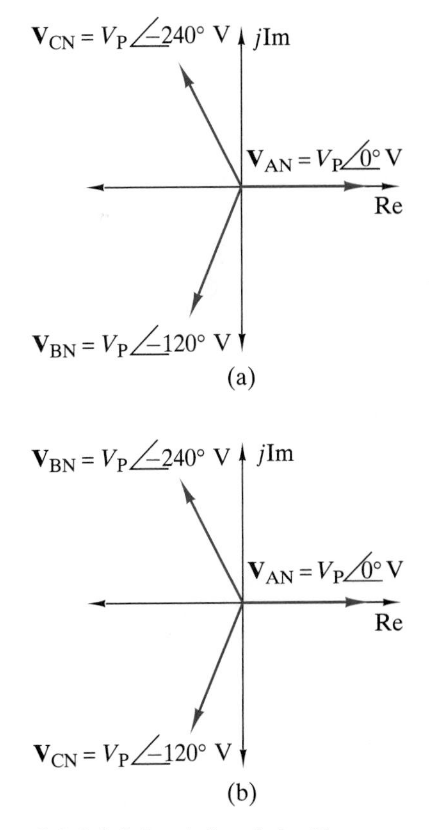

FIGURE 16–14 *Two possible phase sequences: (a) Positive. (b) Negative.*

Using the other sums, we find the other two positive sequence line voltages as

$$\mathbf{V}_{BC} = \sqrt{3}V_P\angle-90^\circ$$
$$\mathbf{V}_{CA} = \sqrt{3}V_P\angle-210^\circ \tag{16–20}$$

Figure 16–15 shows the phasor diagram of these results. The line voltage phasors have the same amplitude and are displaced from each other by 120°. Hence they obey equal-amplitude and zero-sum constraints like the phase voltages.

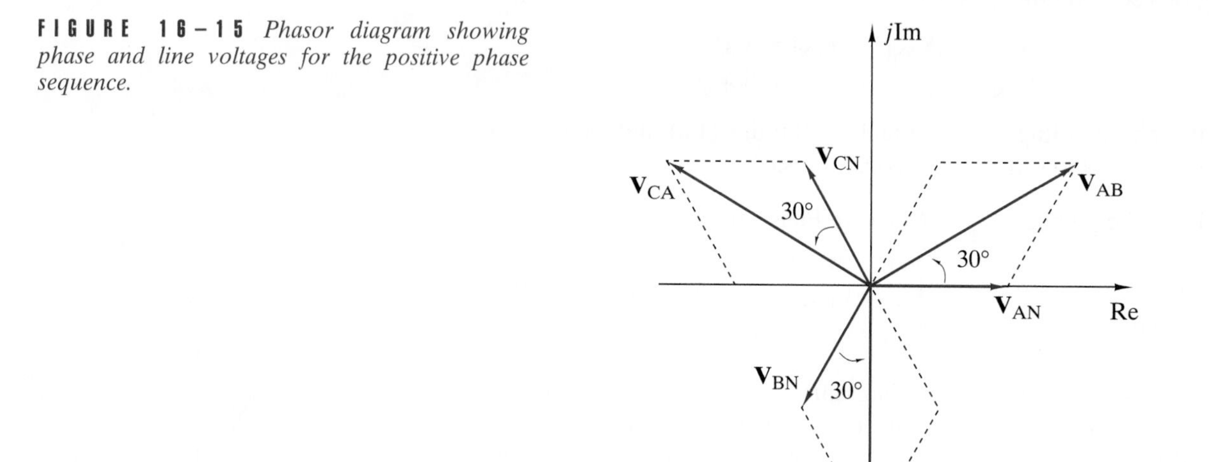

FIGURE 16–15 *Phasor diagram showing phase and line voltages for the positive phase sequence.*

If we denote the amplitude of the line voltages as V_L, then

$$V_L = \sqrt{3}V_P \tag{16–21}$$

In a balanced three-phase system the line voltage amplitude is $\sqrt{3}$ times the phase voltage amplitude. This ratio appears in equipment descriptions such as 277/480 V three phase, where 277 is the phase voltage and 480 the line voltage.

It is necessary to choose one of the phasors as the zero-phase reference when defining three-phase voltages and currents. Usually the reference is the line A phase voltage (i.e., $\mathbf{V}_{AN} = V_P\angle 0^\circ$), as illustrated in Figures 16–14 and 16–15. Unless otherwise stated, $\mathbf{V}_{AN}$ will be used as the phase reference in this chapter.

Exercise 16–6

A balanced Y-connected three-phase source produces positive sequence phase voltages with 2400-V (rms) amplitudes. Write expressions for the phase and line voltage phasors.

Answers:

$\mathbf{V}_{AN} = 2400\angle 0°$ $\mathbf{V}_{BN} = 2400\angle -120°$ $\mathbf{V}_{CN} = 2400\angle -240°$

$\mathbf{V}_{AB} = 4160\angle +30°$ $\mathbf{V}_{BC} = 4160\angle -90°$ $\mathbf{V}_{CA} = 4160\angle -210°$

Exercise 16–7

Given that $\mathbf{V}_{BC} = 480\angle +135°$ in a balanced, positive sequence, three-phase system, write expressions for the three phase-voltage phasors.

Answer:

$\mathbf{V}_{AN} = 277\angle -135°$ $\mathbf{V}_{BN} = 277\angle +105°$ $\mathbf{V}_{CN} = 277\angle -15°$

16–6 THREE-PHASE AC POWER ANALYSIS

This section treats the analysis of balanced three-phase circuits. We first treat the direct analysis problem beginning with the Y-connected source and load shown in Figure 16–16. In a direct analysis problem we are given the source phase voltages $\mathbf{V}_{AN}$, $\mathbf{V}_{BN}$, and $\mathbf{V}_{CN}$ and the load impedances Z. Our objective is to determine the three line currents $\mathbf{I}_A$, $\mathbf{I}_B$, and $\mathbf{I}_C$ and the total complex power delivered to the load.

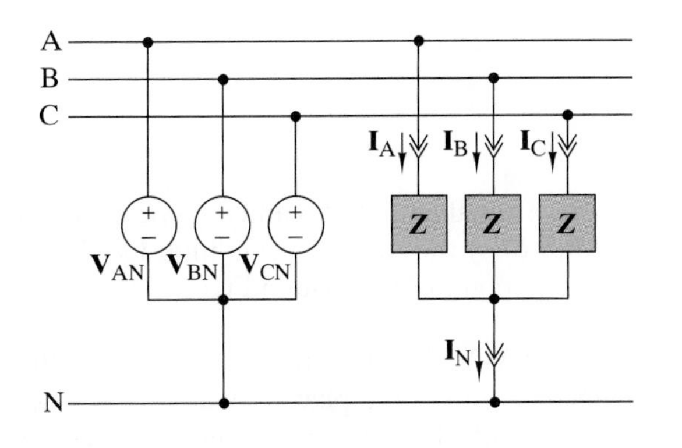

FIGURE 16–16 *A balanced three-phase system with a Y-connected source and load.*

Y-CONNECTED SOURCE AND Y-CONNECTED LOAD

The load in Figure 16–16 is balanced because the phase impedances in the legs of the Y are equal. With the neutral point at the load connected, the voltage across each phase impedance is a phase voltage. Using $\mathbf{V}_{AN}$ as the phase reference, we find that the line currents are

$$\mathbf{I}_A = \frac{\mathbf{V}_{AN}}{Z} = \frac{V_P\angle 0°}{|Z|\angle\theta} = \frac{V_P}{|Z|}\angle -\theta$$

$$\mathbf{I}_B = \frac{\mathbf{V}_{BN}}{Z} = \frac{V_P\angle -120°}{|Z|\angle\theta} = \frac{V_P}{|Z|}\angle -120° - \theta \tag{16–22}$$

$$\mathbf{I}_C = \frac{\mathbf{V}_{CN}}{Z} = \frac{V_P\angle -240°}{|Z|\angle\theta} = \frac{V_P}{|Z|}\angle -240° - \theta$$

Figure 16–17 shows the phasor diagram of the line currents and phase voltages.

The line current phasors in Eq. (16–22) and Figure 16–17 have the same amplitude I_L, where

$$I_L = \frac{V_P}{|Z|} \text{ (Y-connected load)} \tag{16–23}$$

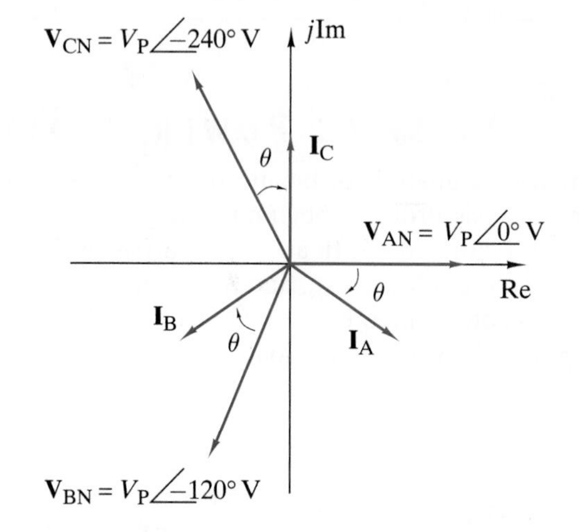

FIGURE 16–17 *Line currents and phase voltages in a balanced three-phase system.*

The line currents have equal amplitudes and are symmetrically disposed at 120° intervals, so they obey the zero-sum condition $\mathbf{I}_A + \mathbf{I}_B + \mathbf{I}_C = 0$. Applying KCL at the neutral point of the load in Figure 16–16, we find that $\mathbf{I}_N = \mathbf{I}_A + \mathbf{I}_B + \mathbf{I}_C = 0$.

Thus, in a balanced Y-Y circuit there is no current in the neutral line. The neutral connection could be replaced by any impedance whatsoever, including infinity, without affecting the power delivered to the load. In other words, the neutral wire can be disconnected without changing the circuit response. Real systems may or may not have a neutral wire, but in solving three-phase problems it is helpful to draw the neutral line because it serves as a reference point for the phase voltages.

The total complex power delivered to the load is

$$\begin{aligned} S_L &= \mathbf{V}_{AN}\mathbf{I}_A^* + \mathbf{V}_{BN}\mathbf{I}_B^* + \mathbf{V}_{CN}\mathbf{I}_C^* \\ &= (V_P\angle 0)(I_L\angle\theta) + (V_P\angle{-120°})(I_L\angle 120° + \theta) \\ &\quad + (V_P\angle{-240°})(I_L\angle 240° + \theta) \\ &= 3V_PI_L\angle\theta \end{aligned} \tag{16–24}$$

Since $V_P = V_L/\sqrt{3}$, the expression for complex power can also be written using the line voltage:

$$S_L = \sqrt{3}V_LI_L\angle\theta \tag{16–25}$$

In either Eq. (16–24) or (16–25) the power factor angle θ is the angle of the per-phase impedance of the Y-connected load.

EXAMPLE 16–9

When the line voltage is 480 V (rms), a balanced Y-connected load draws an apparent power of 40 kVA at a lagging power factor of 0.9.

(a) Find the per-phase impedance of the load.
(b) Find the three line current phasors using $\mathbf{V}_{AN}$ as the phase reference.

SOLUTION:

(a) For the given conditions the phase voltage, line current, and power factor angle are

$$V_P = \frac{V_L}{\sqrt{3}} = \frac{480}{\sqrt{3}} = 277 \quad \text{V (rms)}$$

$$I_L = \frac{|S_L|}{\sqrt{3}V_L} = \frac{4 \times 10^4}{\sqrt{3} \times 480} = 48.1 \text{ A (rms)}$$

$$\theta = \cos^{-1}(0.9) = 25.8^\circ$$

and the per-phase impedance of the Y-connected load is

$$Z_L = \frac{V_P}{I_L}\angle\theta = \frac{277}{48.1}\angle 25.8^\circ$$
$$= 5.18 + j2.51 \quad \Omega$$

(b) With $\mathbf{V}_{AN} = 277\angle 0^\circ$ as the phase reference, we calculate the line current $\mathbf{I}_A$ as

$$\mathbf{I}_A = \frac{\mathbf{V}_{AN}}{Z_L} = I_L\angle -\theta = 48.1\angle -25.8^\circ \text{ A (rms)}$$

Hence for a balanced load the other two line current phasors are

$$\mathbf{I}_B = I_L\angle -120^\circ - \theta = 48.1\angle -145.8^\circ \text{ A (rms)}$$
$$\mathbf{I}_C = I_L\angle -240^\circ - \theta = 48.1\angle -265.8^\circ \text{ A (rms)}$$

■

Exercise 16–8

A Y-connected load with $Z = 10 + j4$ Ω/phase is driven by a balanced, positive sequence three-phase generator with $V_L = 4.16$ kV (rms). Using $\mathbf{V}_{AN}$ as the phase reference:

(a) Find the line currents.
(b) Find the average and reactive power delivered to the load.

Answers:

(a) $\mathbf{I}_A = 223\angle -21.8^\circ$ A $\mathbf{I}_B = 223\angle -141.8^\circ$ A $\mathbf{I}_C = 223\angle -261.8^\circ$ A
(b) $P_L = 1.49$ MW $Q_L = 0.597$ MVAR

Y-CONNECTED SOURCE AND Δ-CONNECTED LOAD

We now turn to the balanced Δ-connected load shown in Figure 16–18. When the objective is to determine the line currents and total complex power, it is convenient to replace the Δ-connected load by an equivalent Y-connected load. Using Figure 16–19, we can easily determine the required transformation. Looking between any two terminals in the Δ-connected load, we see an impedance $Z \| 2Z$. Similarly, looking between any two terminals in the Y-connected load, we see an impedance $2Z_Y$. For these two circuits to be equivalent, we must see the same equivalent impedance between any two terminals. Hence, equivalence requires

$$Z \| 2Z = \frac{2Z^2}{Z + 2Z} = \frac{2}{3}Z = 2Z_Y$$

or

$$Z_Y = \frac{Z}{3} \tag{16–26}$$

FIGURE 16–18 *A balanced three-phase system with a Y-connected source and a Δ-connected load.*

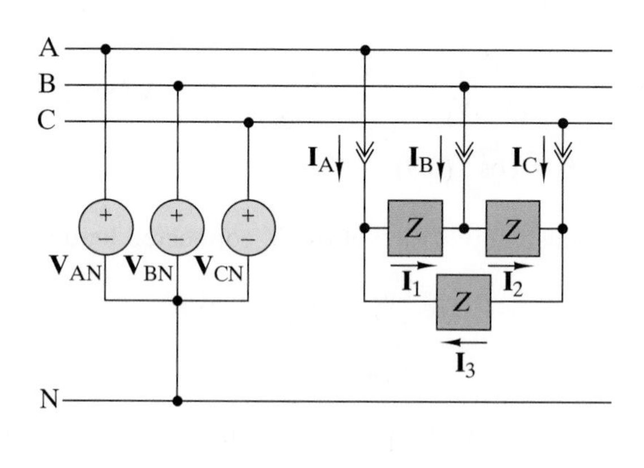

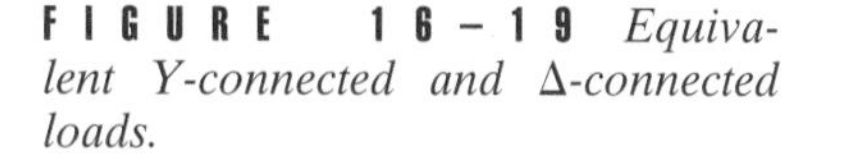
FIGURE 16–19 *Equivalent Y-connected and Δ-connected loads.*

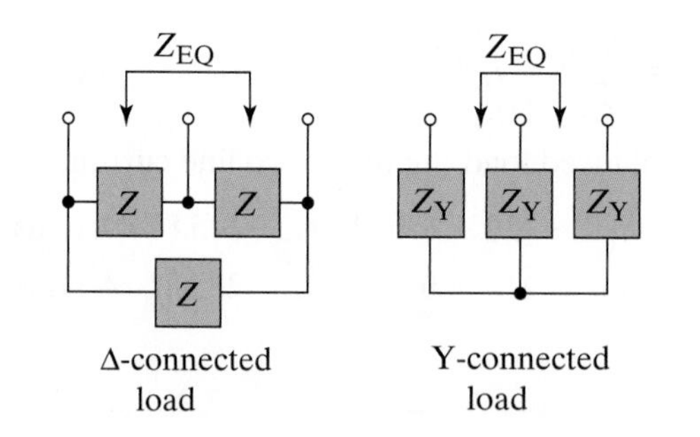

Using the Δ-Y transformation, we reduce the problem to a circuit in which the source and load are Y-connected. The resulting Y-Y configuration can then be analyzed to determine line currents and total power, as discussed previously.

However, when we need to know the current or power delivered to each leg of the Δ load, we must determine the phase currents $\mathbf{I}_1$, $\mathbf{I}_2$, and $\mathbf{I}_3$ shown in Figure 16–18. The phase currents can be expressed in terms of the phase impedances and the line voltage. Assuming the positive phase sequence and using $\mathbf{V}_{AN}$ as the phase reference, these expressions are

$$\mathbf{I}_1 = \frac{\mathbf{V}_{AB}}{Z} = \frac{V_L\angle 30^\circ}{|Z|\angle\theta} = \frac{V_L}{|Z|}\angle 30^\circ - \theta$$

$$\mathbf{I}_2 = \frac{\mathbf{V}_{BC}}{Z} = \frac{V_L\angle -90^\circ}{|Z|\angle\theta} = \frac{V_L}{|Z|}\angle -90^\circ - \theta \quad (16\text{–}27)$$

$$\mathbf{I}_3 = \frac{\mathbf{V}_{CA}}{Z} = \frac{V_L\angle -210^\circ}{|Z|\angle\theta} = \frac{V_L}{|Z|}\angle -210^\circ - \theta$$

The phase currents have the same amplitude I_P, defined as

$$I_P = \frac{V_L}{|Z|} \quad (\Delta\text{-connected load}) \quad (16\text{–}28)$$

Although of no immediate physical importance, note that the phase currents sum to zero because they have equal amplitudes and are symmetrically disposed at 120° intervals.

Using the results in Eq. (16–27), the complex power delivered to each leg of the delta load is

$$S_1 = \mathbf{V}_{AB} \times \mathbf{I}_1^* = (V_L\angle 30^\circ)(I_P\angle\theta - 30^\circ) = V_L I_P \angle\theta$$

$$S_2 = \mathbf{V}_{BC} \times \mathbf{I}_2^* = (V_L\angle -90^\circ)(I_P\angle\theta + 90^\circ) = V_L I_P \angle\theta \quad (16\text{–}29)$$

$$S_3 = \mathbf{V}_{CA} \times \mathbf{I}_3^* = (V_L\angle -210^\circ)(I_P\angle\theta + 210^\circ) = V_L I_P \angle\theta$$

The total complex power delivered to the Δ-connected load is

$$S_L = S_1 + S_2 + S_3 = 3V_L I_P \angle\theta \quad (16\text{–}30)$$

where the power factor angle θ is the angle of the per-phase impedance of the Δ-connected load.

The result in Eq. (16–30) can be put in the same form as Eq. (16–25) by replacing the phase current I_P by the line current I_L. As noted previously, the line currents for a balanced Δ-connected load can be calculated using a Δ-to-Y transformation. In the transformed circuit the line current amplitude is $I_L = V_P/|Z_Y|$, where Z_Y is the impedance in each leg of the equivalent Y-connected load. But in a balanced system $V_P = V_L/\sqrt{3}$, and according to Eq. (16–26) $Z_Y = Z/3$. Therefore, the amplitudes of the line and phase currents in a Δ-connected load are related as follows:

$$I_L = \frac{V_L/\sqrt{3}}{|Z/3|} = \sqrt{3}\frac{V_L}{|Z|} = \sqrt{3}I_P \quad (16\text{–}31)$$

When $I_P = I_L/\sqrt{3}$ is substituted into Eq. (16–30), we obtain

$$S_L = \sqrt{3}V_L I_L \angle\theta \quad (16\text{–}32)$$

Equations (16–25) and (16–32) are identical, which means that the relationship applies to balanced three-phase loads whether Y- or Δ-connected. In either case the power factor angle θ is the angle of the per-phase impedance of the load because the transformation $Z_Y = Z/3$ does not alter the phase angle of the phase impedance.

EXAMPLE 16–10

A Δ-connected load with $Z = 40 + j30$ Ω/phase is driven by a balanced, positive sequence three-phase generator with $V_L = 2400$ V (rms). Using $\mathbf{V}_{AN}$ as the phase reference:

(a) Find the phase currents.
(b) Find the line currents.
(c) Find the average and reactive power delivered to the load.

SOLUTION:

(a) The first phase current is

$$\mathbf{I}_1 = \frac{\mathbf{V}_{AB}}{Z} = \frac{2400\angle 30°}{40 + j30} = 48.0\angle -6.87° \text{ A (rms)}$$

Since the circuit is balanced, the other phase currents all have amplitudes of $I_P = 48.0$ A and are displaced at 120° intervals.

$$\mathbf{I}_2 = 48.0\angle -126.87° \text{ A (rms)}$$
$$\mathbf{I}_3 = 48.0\angle -246.87° \text{ A (rms)}$$

(b) Using $V_P = 2400/\sqrt{3}$ and Δ-Y transformation with $Z_Y = (40 + j30)/3$, we find that the phase A line current is

$$\mathbf{I}_A = \frac{\mathbf{V}_{AN}}{Z_Y} = \frac{2400/\sqrt{3}}{(40 + j30)/3} = 83.1\angle -36.9° \text{ A (rms)}$$

Since the circuit is balanced, the other line currents have amplitudes of $I_L = 83.1$ A and are displaced at 120° intervals.

$$\mathbf{I}_B = 83.1\angle -156.9° \text{ A (rms)}$$
$$\mathbf{I}_C = 83.1\angle -276.9° \text{ A (rms)}$$

(c) The complex power delivered to the load is

$$S_L = \sqrt{3}V_L I_L \angle\theta = \sqrt{3} \times 2400 \times 83.1\angle 36.9°$$
$$= 345\angle 36.9° \text{ kVA}$$

Therefore, $P_L = 276$ kW and $Q_L = 207$ kVAR. ■

Exercise 16–9

Exercise 16–8 involved a balanced Y-connected load with $Z = 10 + j4$ Ω/phase and a line voltage of $V_L = 4.16$ kV (rms). In this exercise these same parameters apply to a Δ-connected load. Using $\mathbf{V}_{AN}$ as the phase reference:

(a) Find phase currents.
(b) Find the line currents.
(c) Find the average and reactive power delivered to the load.

Answers:

(a) $\mathbf{I}_1 = 386\angle +8.20°$ A; $\mathbf{I}_2 = 386\angle -111.8°$ A; $\mathbf{I}_3 = 386\angle -231.8°$ A
(b) $\mathbf{I}_A = 669\angle -21.8°$ A; $\mathbf{I}_B = 669\angle -141.8°$ A; $\mathbf{I}_C = 669\angle -261.8°$ A
(c) $P_L = 4.47$ MW; $Q_L = 1.79$ MVAR

THREE-PHASE LOAD-FLOW ANALYSIS

The analysis and examples given thus far treat direct analysis problems in which the source parameters are given and the power flow is unknown. In a load-flow problem we are required to find the source or load voltages when the power flow is given.

A simple model of the three-phase circuit for the load-flow problem is shown in Figure 16–20. The impedance Z_W represents the wire impedances of the power lines connecting the source and load. It is clear even in this very simple case that including all three phases in a circuit diagram is unwieldy. In more complicated situations, including all three phases in circuit diagrams tends to obscure the working of the system. In a balanced three-phase system we have seen that once we find one of the line currents or voltages, the others are easily derived by shifting the known response at 120° intervals. Thus, in effect, we really do not need all three phases in the circuit diagram to analyze balanced three-phase systems.

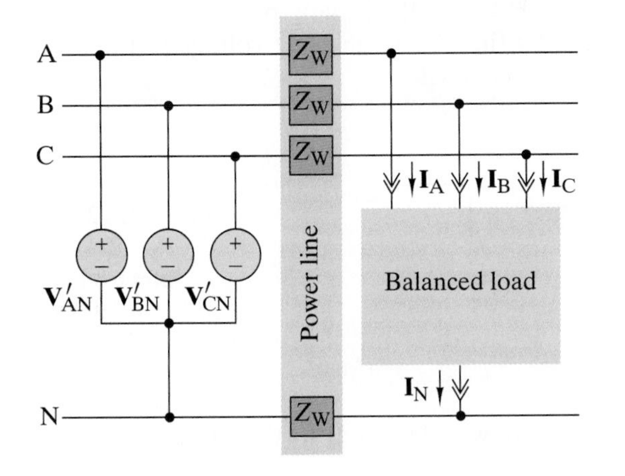

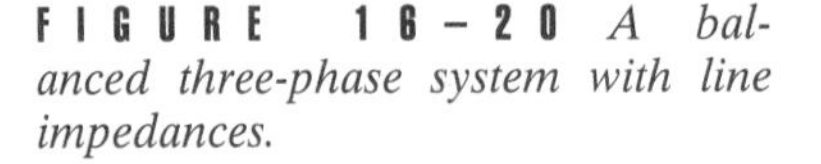
FIGURE 16–20 *A balanced three-phase system with line impedances.*

A simpler representation of a balanced three-phase system is obtained by omitting the neutral and using one line to represent all three phases. Figure 16–21 is a single-line representation of the circuit in Figure 16–20. In power system terminology a **bus** is a group of conductors that serve as a common connection for two or more circuits. In the single-line diagram in Figure 16–21 the buses are represented by short horizontal lines. Bus 1 is a generator bus connecting a three-phase source represented by the circle and a transmission line represented by Z_W. Bus 2 is a load bus connecting the transmission line to a balanced three-phase load represented by an arrow indicating the delivery of complex power. The line voltages and

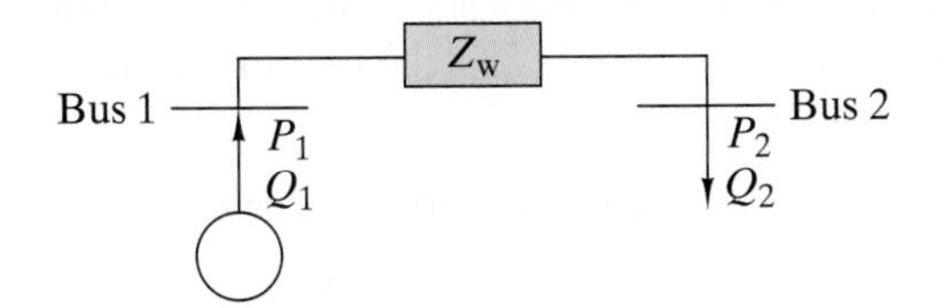

FIGURE 16–21 *Single-line representation of a three-phase system.*

complex power flow are written beside the buses using a subscript that identifies the bus.

Using single-line diagrams makes three-phase load-flow analysis similar to the single-phase two-wire load-flow problems in Sect. 16–5. The only complication here is that we must account for the power in all three phases and we must distinguish between the line and phase voltages. In a three-phase load-flow problem the bus voltages and line currents are represented by their phasor magnitudes without reference to their phase angles. A load-flow problem does not require that we find these phase angles since the required phase information is carried by the specified complex power.

EXAMPLE 16–11

The single-line diagram in Figure 16–22 shows a power system with a generator bus connected to a load bus via power lines with $Z_W = 4 + j16\ \Omega$. The line voltage at the load bus is $V_{L2} = 4800$ V (rms) and the load draws $P_2 = 100$ kW at 0.8 lagging power factor. Find the complex power produced by the source, the line voltage at the source bus, and the transmission efficiency.

FIGURE 16–22

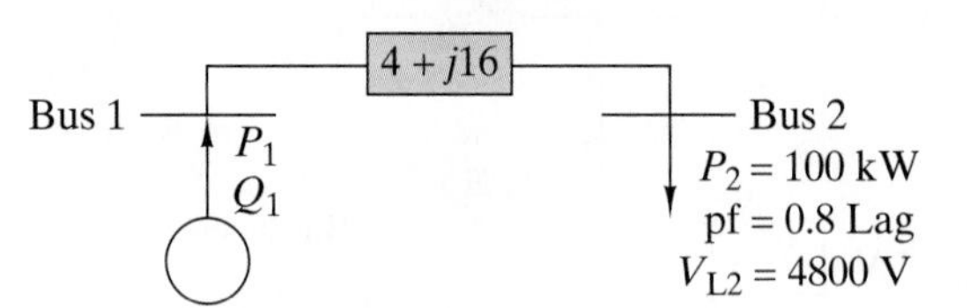

SOLUTION:

The total complex power delivered to the three-phase load connected to bus 2 is

$$S_2 = 100 + j100\tan(\cos^{-1} 0.8) = 100 + j75$$
$$= 125\angle 36.9^\circ \text{ kVA}$$

Using Eq. (16–32) to calculate the line current at the load bus yields

$$I_L = \frac{|S_2|}{\sqrt{3}\,V_{L2}} = \frac{125{,}000}{\sqrt{3} \times 4800} = 15.04 \text{ A (rms)}$$

The total complex power absorbed by the power line is

$$S_W = 3I_L^2(R_W + jX_W) = 3(15.04)^2(4 + j16)$$
$$= 2.71 + j10.8 \text{ kVA}$$

The preceding computation includes the three lines for phases A, B, and C but does not include the neutral wire since the circuit is balanced and there is no current in the neutral line. Using the conservation of complex power yields the source power as

$$S_1 = S_2 + S_W = 102.7 + j85.8$$
$$= 133.8\angle 39.9^\circ \text{ kVA}$$

We now use Eq. (16–32) again to find the line voltage at the source bus as

$$V_{L1} = \frac{|S_1|}{\sqrt{3}I_L} = \frac{133800}{\sqrt{3} \times 15.04}$$
$$= 5.14 \text{ kV (rms)}$$

The transmission efficiency is

$$\eta = \frac{P_2}{P_1} \times 100\% = \frac{100}{102.7} \times 100 = 97.4\%$$ ■

EXAMPLE 16–12

The transmission line in Figure 16–23 has a maximum rated capacity of 250 kVA at line voltage of 4160 V (rms). When operating at rated capacity, the resistive and reactive voltage drops in the line are 2.5% and 6% of the rated voltage, respectively.

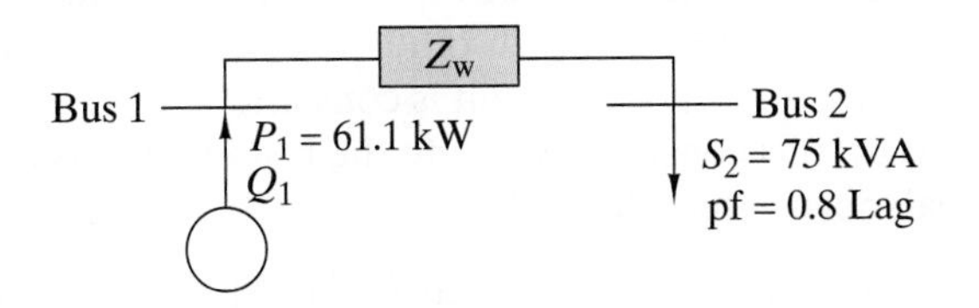

FIGURE 16–23

(a) Find the wire impedance Z_W.
(b) The system is operating with $P_1 = 61.1$ kW and $S_2 = 75$ kVA at pf = 0.8 lagging. Find the line voltage at the load bus.

SOLUTION:

(a) When the power line is operating at maximum rated capacity, the line current is

$$I_L = \frac{|S_L|}{\sqrt{3}\, V_L} = \frac{250{,}000}{\sqrt{3} \times 4160} = 34.7 \text{ A (rms)}$$

The magnitudes of the voltage across line resistance and reactance are

$$I_L R_W = 4160 \times 0.025 = 104 \text{ V}$$
$$I_L X_W = 4160 \times 0.060 = 250 \text{ V}$$

Hence, the wire impedance of the line is

$$Z_W = \frac{104 + j250}{I_L} = 3 + j7.2\ \Omega$$

(b) When the system operates with $P_1 = 61.1$ kW and $S_2 = 75$ kVA at pf = 0.8 lagging, the average power lost in the line is $P_W = P_1 - S_2 \times 0.8 =$ 1.1 kW. The line current is

$$I_L = \sqrt{\frac{P_W}{3\, R_W}} = \sqrt{\frac{1100}{3 \times 3}} = 11.06 \text{ A (rms)}$$

and the line voltage at load bus is

$$V_{L2} = \frac{|S_2|}{\sqrt{3}I_L} = 3.92 \text{ kV (rms)}$$ ■

Exercise 16–10

A balanced three-phase load draws 8 kVA at lagging power factor of 0.9 when the phase voltage is 277 V (rms). The load is fed by a balanced three-phase source via lines with $Z_W = 0.3 + j5\ \Omega$.

(a) Find the line current.
(b) Find the line voltage at the source.

Answers:

(a) $I_L = 9.63$ A
(b) $V_L = 526$ V

INSTANTANEOUS THREE-PHASE POWER

We began this chapter by showing that the instantaneous power in a single-phase circuit consists of a constant dc component (the average power) plus a double-frequency ac component that makes no net contribution to energy transfer. One of the advantages of three-phase operation is that the net ac component is zero. In other words, the total instantaneous power in a balanced three-phase circuit is constant.

To show that $p_T(t)$ is constant, we write the instantaneous power in each phase of the balanced three-phase circuit:

$$p_A(t) = v_{AN}(t) \times i_A(t) = [\sqrt{2}\,V_P \cos(\omega t)] \times [\sqrt{2}\,I_L \cos(\omega t - \theta)]$$

$$p_B(t) = v_{BN}(t) \times i_B(t) = [\sqrt{2}\,V_p \cos(\omega t - 120°)] \times [\sqrt{2}\,I_L \cos(\omega t - 120° - \theta)]$$

$$p_C(t) = v_{CN}(t) \times i_C(t) = [\sqrt{2}\,V_P \cos(\omega t - 240°)] \times [\sqrt{2}\,I_L \cos(\omega t - 240° - \theta)]$$

where $\sqrt{2}\,V_P$ is the peak amplitude of each line-to-neutral voltage and $\sqrt{2}\,I_L$ is the peak amplitude of each line current. Using the trigonometric identity,

$$[\cos x] \times [\cos y] = \frac{1}{2}\cos(x - y) + \frac{1}{2}\cos(x + y)$$

the individual phase powers can be put into the form

$$p_A(t) = V_P I_L \cos\theta + V_P I_L \cos(2\,\omega\,t - \theta)$$

$$p_B(t) = V_P I_L \cos\theta + V_P I_L \cos(2\,\omega\,t - 240° - \theta)$$

$$p_C(t) = V_P I_L \cos\theta + V_P I_L \cos(2\,\omega\,t - 480° - \theta)$$

Each phase power has a constant dc term $V_P I_L \cos\theta$ plus a double-frequency ac term. The double-frequency terms all have the same amplitude $V_P I_L$ and are symmetrically disposed at 120° intervals because –480° is the same as –120°. When viewed as phasors, it is easy to see that the double-frequency sinusoidal terms sum to zero and the total instantaneous power is

$$p_T(t) = p_A(t) + p_B(t) + p_C(t) = 3\,V_P I_L \cos\theta$$

The fact that the total instantaneous power is constant means, among other things, that three-phase motors produce constant mechanical output and the three-phase generators require constant mechanical input. As a result, there is smoother operation with less vibration at the electromechanical interfaces of the system.

APPLICATION EXAMPLE 16–13

The purpose of this example is to use conventional phasor circuit analysis to analyze a three-phase power system. Figure 16–24 shows a single-line diagram of a power system in which a three-phase source with $V_L = 250$ kV supplies power to two power distribution centers through a radial network of transmission lines. The two transmission lines can be modeled by series impedances of

$$Z_{W1} = 6.5 + j14 \quad \Omega \quad \text{and} \quad Z_{W2} = 8.6 + j19 \quad \Omega$$

The two load centers can be modeled as Y-connected loads with per-phase impedance of

$$Z_{Y1} = 1000 \angle 10° \quad \Omega \quad \text{and} \quad Z_{Y2} = 2400 \angle 20° \quad \Omega$$

Find the complex power delivered by the source and the magnitude of the line voltage delivered to each distribution center.

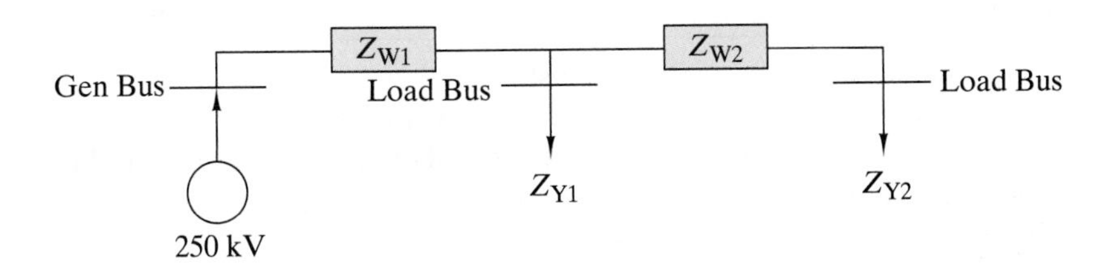

FIGURE 16–24

SOLUTION:

We can analyze a balanced three-phase system on a per-phase basis. If we look at just one phase, say phase A, we get the single-phase equivalent circuit in Figure 16–25. We analyze this circuit to get a solution for phase A. If need be, we use the phase sequence to shift the solution to get phase B and phase C results. The single-phase equivalent circuit includes a neutral connection. We may do this even if the system does not have a neutral wire. In a balanced system, neutral points in Y-connections have the same voltage. Thus, for analysis purposes we can tie these points together with a "virtual" neutral wire without altering the response of the system.

The source voltage in Figure 16–25 is the phase A voltage of the 250-kV source.

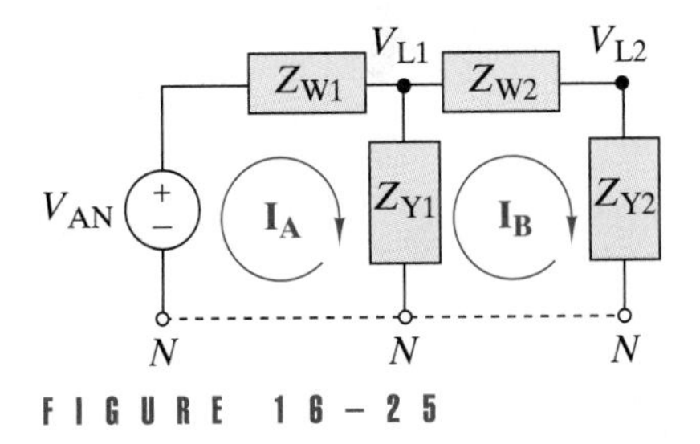

FIGURE 16–25

$$\mathbf{V}_{AN} = \frac{V_L}{\sqrt{3}} = \frac{250}{\sqrt{3}} = 144.3 \angle 0° \quad \text{kV}$$

In matrix format the mesh-current equations for the equivalent circuit are

$$\begin{bmatrix} Z_{W1} + Z_{Y1} & -Z_{Y1} \\ -Z_{Y1} & Z_{W2} + Z_{Y1} + Z_{Y2} \end{bmatrix} \begin{bmatrix} \mathbf{I_A} \\ \mathbf{I_B} \end{bmatrix} = \begin{bmatrix} \mathbf{V_{AN}} \\ 0 \end{bmatrix}$$

Note that in this context $\mathbf{I}_A$ and $\mathbf{I}_B$ are the mesh currents in Figure 16–25, not line currents. Inserting the preceding numerical values and solving for these mesh currents yield

$$\mathbf{I}_A = 194.8 - j48.5 \text{ A} \quad \text{and} \quad \mathbf{I}_B = 54.94 - j21.4 \text{ A}$$

We can now calculate the total complex power delivered by the 25-kV source as

$$S = 3\,S_A = 3\,\mathbf{V}_{AN}\mathbf{I}_A^* = 82.33 + j21.02 \text{ kVA}$$

The $3S_A$ in this equation accounts for the complex power delivered by all three phases. The phase A voltage at each load center is

$$\mathbf{V}_{AN1} = \mathbf{V}_{AN} - \mathbf{I}_A Z_{W1} = 142.4 - j2.41 \text{ kV}$$

$$\mathbf{V}_{AN2} = \mathbf{V}_{AN1} - \mathbf{I}_B Z_{W2} = 141.5 - j3.27 \text{ kV}$$

from which we get the line voltage magnitude at each load center.

$$V_{L1} = \sqrt{3}\,|\mathbf{V}_{AN1}| = 246.6 \text{ kV}$$

$$V_{L2} = \sqrt{3}\,|\mathbf{V}_{AN2}| = 245.1 \text{ kV}$$

■

SUMMARY

- In the sinusoidal steady state the instantaneous power at a two-terminal interface contains average and reactive power components. The average power represents a unidirectional transfer of energy from source to load. The reactive power represents a lossless interchange of energy between the source and load.
- In ac power analysis problems, amplitudes of voltage and current phasors are expressed in rms values. Complex power is defined as the product of the voltage phasor and the conjugate of the current phasor. The real part of the complex power is the average power in watts (W), the imaginary part is the reactive power in volt-amperes reactive (VAR), and the magnitude is the apparent power in volt-amperes (VA).
- In a direct analysis problem the load impedance, line impedances, and source voltage are given, and the load voltage, current, and power are the unknowns. In a load-flow problem the source and load powers are given, and the unknowns are line voltages and currents at different points.
- Three-phase systems transmit power from source to load on four lines labeled A, B, C, and N. The three line-to-neutral voltages $\mathbf{V}_{AN}$, $\mathbf{V}_{BN}$, and $\mathbf{V}_{CN}$ are called the phase voltages. The three line-to-line voltages $\mathbf{V}_{AB}$, $\mathbf{V}_{BC}$, and $\mathbf{V}_{CA}$ are called line voltages.
- In a balanced three-phase system, (1) the neutral wire carries no current, (2) the line voltage amplitude V_L is related to the phase voltage amplitude V_P by $V_L = \sqrt{3}\,V_P$, (3) the three line currents $\mathbf{I}_A$, $\mathbf{I}_B$, and $\mathbf{I}_C$ have the same amplitude I_L, and (4) the total complex power delivered to a Y- or a Δ-connected load is $\sqrt{3}V_L I_L \angle\theta$, where θ is the angle of the phase impedance.

PROBLEMS

OBJECTIVE 16–1 COMPLEX POWER (SECTS. 16–1, 16–2)

Given a linear circuit operating in the sinusoidal steady-state:
(a) Find the complex power delivered at a specified interface.
(b) Find the phasor voltage or current required to deliver a specified complex power at an interface.
(c) Find the load impedance required to draw a specified complex power at a given voltage level.
See Example 16–1 and Exercises 16–1, 16–2

16–1 The following sets of $v(t)$ and $i(t)$ apply to the load circuit in Figure P16–1. Calculate the average power and the reactive power. State whether the load circuit is absorbing or delivering net energy.
(a) $v(t) = 1500\cos(\omega t - 45°)$ V, $i(t) = 2\cos(\omega t + 50°)$ A
(b) $v(t) = 90\cos(\omega t + 60°)$ V, $i(t) = 10.5\cos(\omega t - 20°)$ A

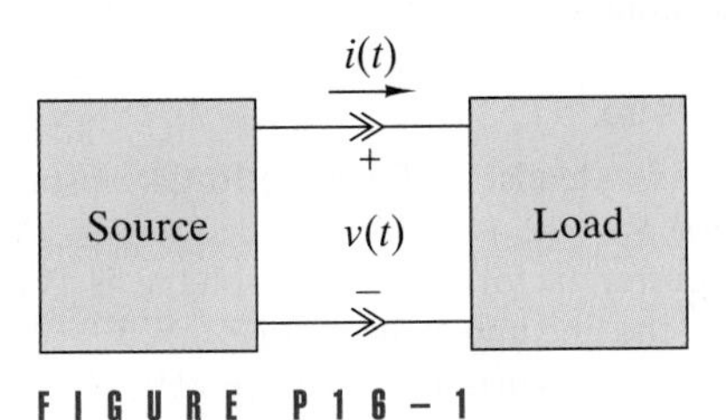

FIGURE P16–1

16–2 The following sets of $v(t)$ and $i(t)$ apply to the load circuit in Figure 16–1. Calculate the average power and the reactive power. State whether the load circuit is absorbing or delivering net energy.
(a) $v(t) = 135\cos(\omega t)$ V, $i(t) = 2\cos(\omega t + 30°)$ A
(b) $v(t) = 370\sin(\omega t)$ V, $i(t) = 10\cos(\omega t + 20°)$ A

16–3 The following sets of **V** and **I** apply to the circuit in Figure P16–3. Calculate the complex power, the average power, the reactive power, and the power factor. State whether the power factor is lagging or leading.
(a) $\mathbf{V} = 240\angle 0°$ V (rms), $\mathbf{I} = 20\angle -75°$ A (rms)
(b) $\mathbf{V} = 120\angle 35°$V (rms), $\mathbf{I} = 12.5\angle 115°$ A (rms)

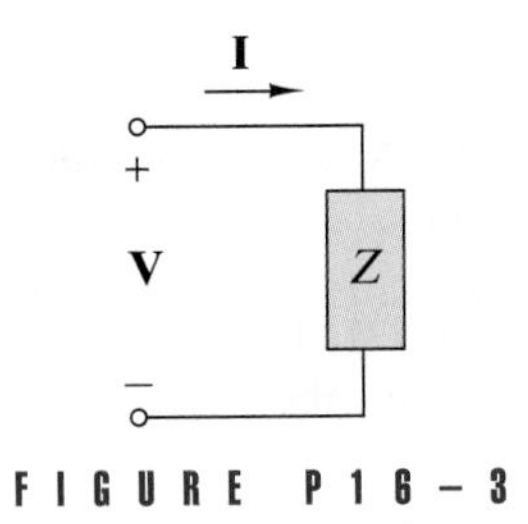

FIGURE P16–3

16–4 The following sets of **V** and **I** apply to the circuit in Figure 16–3. Calculate the complex power, the average power, the reactive power, and the power factor. State whether the power factor is lagging or leading.
(a) $\mathbf{V} = 120\angle 30°$ V (rms), $\mathbf{I} = 3.3\angle -15°$ A (rms)
(b) $\mathbf{V} = 480\angle 45°$ V (rms), $\mathbf{I} = 8.5\angle 90°$ A (rms)

16–5 The conditions in this problem apply to the circuit in Figure 16–3. Calculate the complex power for each condition given below.
(a) $\mathbf{V} = 15\angle 45°$kV (rms), $Z = 500\angle -15°\ \Omega$
(b) $Z = 40 - j30\ \Omega$, $\mathbf{I} = 10\angle 25°$ A (rms)

16–6 Find the power factor under the following conditions. State whether the power factor is lagging or leading.
(a) $S = 1000 + j250$ kVA
(b) $|S| = 15$ kVA, $P = 12$ kW, $Q < 0$

16–7 A load draws an apparent power of 30 kVA at a power factor of 0.8 lagging from a 2400-V (rms) source. Find P, Q, I_{rms}, and the load impedance.

16–8 A load draws 80 kW at a power factor of 0.8 leading from a 880-V (rms) source. Find Q, I_{rms}, and the load impedance.

16–9 A load draws 15 A (rms), 5 kW, and 2.5 kVARS (lagging) from a 60-Hz source. Find the load power factor and impedance.

16–10 Find the impedance of a load that is rated at 440 V (rms), 5 A (rms), and 2.2 kW.

OBJECTIVE 16–2 AC POWER ANALYSIS (SECT. 16–3)

Given a linear circuit operating in the sinusoidal steady state:
(a) Find the complex power associated with any element.
(b) Find the load impedance required to draw a given complex power from a source.
See Examples 16–2, 16–3, 16–4 and Exercise 16–3

16–11 A load consisting of a 100-Ω resistor in series with a 150-mH inductor is connected across a 60-Hz voltage source that delivers 240 V (rms). Find the complex power delivered to the load.

16–12 A load consisting of a 50-Ω resistor in parallel with a 10-μF capacitor is connected across a 400-Hz voltage source that delivers 110 V (rms). Find the complex power delivered to the load.

16–13 A load consisting of a resistor and inductor connected in series absorbs a complex power $S = 1200 + j800$ VA when connected to a 440-V (rms) 60-Hz line. Find the values of R and L.

16–14 The load in Figure P16–14 consists of a 200-Ω resistor in series with an inductor whose reactance is 50 Ω. The source voltage is 2.5 kV and the line impedance is $Z_W = 1 + j10\ \Omega$. Find the load current and the complex power absorbed by the load.

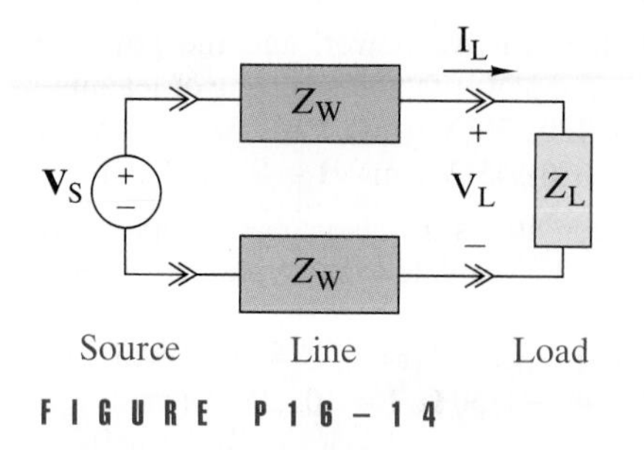

FIGURE P16–14

16–15 The load in Figure P16–14 consists of a 60-Ω resistor in series with a capacitor whose reactance is 30 Ω. The source voltage is 440 V (rms) at 60 Hz, and the line impedance is $Z_W = 2 + j5\ \Omega$. Find the load current and the complex power absorbed by the load.

16–16 A load consists of a 5-H inductor connected in parallel with a 2-kΩ resistor and a 10-μF capacitor in series. The load is connected across a 240-V (rms), 60-Hz voltage source. Find the complex power delivered to the circuit.

16–17 A two-terminal circuit is formed by connecting a 20-Ω resistor in parallel with an impedance of $8 - j6\ \Omega$. The circuit is driven by a 50-V (rms) source. Find the complex power delivered to the circuit and the power factor.

16–18 The three load impedances in Figure P16–18 are: $Z_1 = 25 + j6\ \Omega$, $Z_2 = 16 + j4\ \Omega$, and $Z_3 = 100 + j20\ \Omega$.
(a) Find the current in lines A, B, and N.
(b) Find the complex power produced by each source.

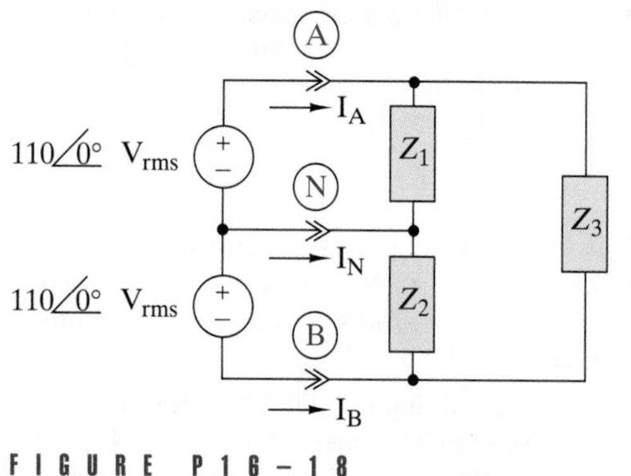

FIGURE P16–18

16–19 The three loads in Figure P16–18 draw complex powers $S_1 = 1250 + j500$ VA, $S_2 = 800 + j0$ VA, and $S_3 = 2000 + j400$ VA.
(a) Find the current in lines A, B, and N.
(b) Find the complex power produced by each source.

16–20 A load rated at 20 kW with a lagging power factor of 0.8 is connected in parallel with a load rated at 16 kW with a lagging power factor of 0.9. Find the power factor of the parallel combination.

OBJECTIVE 16–3 LOAD-FLOW ANALYSIS (SECT. 16–4)

Find unknown voltages or currents required to produce a specified power flow.
See Examples 16–5, 16–6, 16–7 and Exercise 16–5

16–21 The load in Figure P16–14 draws 25 kW at a power factor of 0.75 lagging. The load voltage is 2.4 kV (rms) and the line impedance is $Z_W = 1 + j8\ \Omega$. Find the required source voltage and the complex power produced by the source.

16–22 The load in Figure P16–14 draws an apparent power of 60 kVA at a power factor of 0.85 lagging. The load voltage is 4.4 kV (rms) and the line impedance is $Z_W = 2 + j8\ \Omega$. Find the required source voltage and the source power factor.

16–23 The source in Figure P16–14 delivers 21 kW when the apparent power delivered to the load is $20 + j15$ kVA. The line impedance is $Z_W = 2.1 + j12\ \Omega$. Find the load and source voltages. Assume the phase angle of the line current is zero.

16–24 The two loads in Figure P16–24 absorb complex powers of $S_1 = 12 + j6$ kVA and $|S_2| = 15$ kVA at 0.75 lagging power factor. The load voltage is $|\mathbf{V}_L| = 4.4$ kV (rms) and the line impedances are $Z_W = 3 + j8\ \Omega$.
(a) Find the line current and source voltage.
(b) Find the complex power produced by the source.
(c) Calculate the transmission efficiency (η).

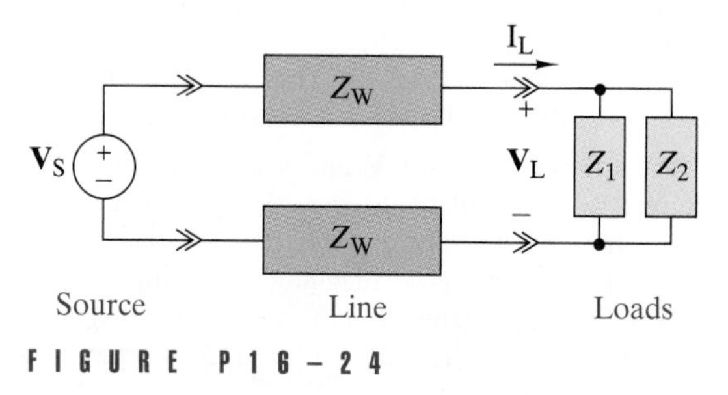

FIGURE P16–24

16–25 The two loads in Figure P16–24 absorb complex powers of $|S_1| = 16$ kVA at 0.8 lagging power factor and $|S_2| = 25$ kVA at unity power factor. The source delivers 50 kVA at 0.8 lagging power factor. The line impedances are $Z_W = 0.1 + j0.5\ \Omega$. Find the load voltage and the source voltage. Assume the phase angle of the line current is zero.

16–26 A load draws 10 A (rms) and 4 kW at a power factor 0.75 (lagging) from a 60-Hz source. Find the capacitance needed in parallel with the load to raise the power factor to unity.

16–27 The load in Figure P16–27 operates at 440 V (rms), 60 Hz, and draws 33 kVA at a power factor of 0.75 lagging.
(a) Find the capacitance needed in parallel with the load to raise the load power factor to 0.95.
(b) Calculate the transmission efficiency with and without the capacitor found in part (a).

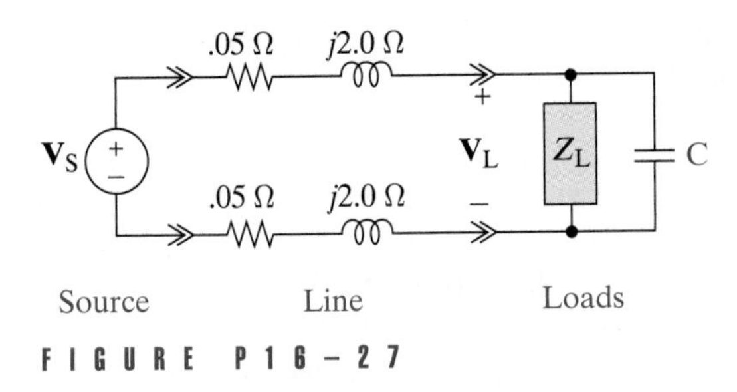

FIGURE P16–27

16–28 An 880-VA load operating at 440-V (rms) and 60 Hz has a power factor of 0.7 lagging. Find the capacitance needed in parallel with the load to raise the power factor to (a) 0.95 and (b) unity.

16–29 The load in Figure P16–29 is supplied by two 60-Hz sources. When $\mathbf{V}_L = 10\angle 0°$ kV (rms), the load draws 1.2 MVA at 0.8 power factor lagging. The first source voltage is $\mathbf{V}_{S1} = 11 + j1.1$kV (rms). Find the second source voltage.

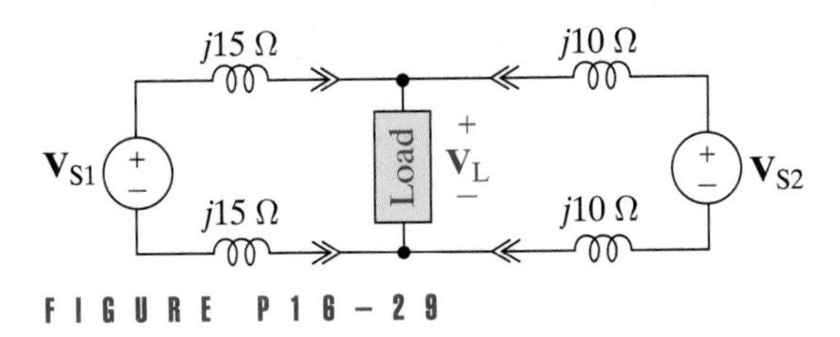

FIGURE P16–29

16–30 The load in Figure P16–14 draws a current of 10 A (rms) and 20 kW at a lagging power factor of 0.8. The line impedance is $Z_W = 2 + j9\ \Omega$. Find the source voltage.

OBJECTIVE 16–4 THREE-PHASE POWER (SECTS. 16–5, 16–6)

Given a balanced three-phase circuit operating in the sinusoidal steady state:
(a) Find the line (phase) voltages or currents when given the phase sequence and the phase (line) voltages or currents.
(b) Find the line current and total complex power (load impedance) when given the line or phase voltage and the load impedance (total complex power).
(c) Find unknown voltages or currents required to produce a specified power flow.
See Examples 16–9, 16–10, 16–11, 16–12, 16–13 and Exercises 16–7, 16–8, 16–9, 16–10

16–31 In a balanced Y-connected three-phase circuit the magnitude of the line voltage is 480 V (rms) and the phase sequence is positive.
(a) Write the line and phase voltage phasors in polar form using $\mathbf{V}_{AN}$ as the reference phasor.
(b) Draw a phasor diagram of the line and phase voltages.

16–32 In a balanced Δ-connected three-phase circuit the magnitude of the line voltage is 2400 V (rms) and the phase sequence is positive.
(a) Write the line and phase voltage phasors in polar form using $\mathbf{V}_{AB}$ as the reference phasor.
(b) Draw a phasor diagram of the line and phase voltages.

16–33 In a balanced Δ-connected three-phase circuit, the magnitude of the phase current is $I_P = 3.5$ A (rms) and the phase sequence is negative.
(a) Write the line and phase current phasors in polar form using $\mathbf{I}_A$ as the reference phasor.
(b) Draw a phasor diagram of the line and phase currents.

16–34 A balanced Y-connected three-phase load with a per-phase impedance of $20 + j15\ \Omega$ operates with a line voltage magnitude of 480 V (rms) using a positive phase sequence. Using $\mathbf{V}_{AN}$ as the reference phasor:
(a) Find the line current phasors in rectangular form.
(b) Calculate the total complex power delivered to the load.
(c) Draw a phasor diagram showing the line currents and phase voltages.

16–35 A balanced Δ-connected three-phase load with a per-phase impedance of $15 + j12\ \Omega$ operates with a line voltage magnitude of 480 V (rms) using a negative phase sequence. Using $\mathbf{V}_{AB}$ as the reference:
(a) Find the line current phasors in rectangular form.
(b) Calculate the total complex power delivered to the load.
(c) Draw a phasor diagram showing the line currents and voltages.

16–36 A balanced Y-connected three-phase load with a per-phase impedance of $50 + j15\ \Omega$ operates with a line voltage magnitude of 480 V (rms) using a positive phase sequence. Using $\mathbf{I}_A$ as the phase reference:
(a) Find the line current phasors in rectangular form.
(b) Calculate the total complex power delivered to the load.

16–37 A balanced Y-connected three-phase load with a per-phase impedance of $8 - j6\ \Omega$ is connected in parallel with a balanced Δ-connected three-phase load with a per-phase impedance of $12\angle 40°\ \Omega$. The line voltage is $V_L = 480$ V (rms) Find the magnitude of the line current and the power factor of the combined loads.

16–38 In a balanced Y-connected three-phase load, the phase A line current is $\mathbf{I}_A = 25\angle -40°$ A (rms) with a line voltage of $\mathbf{V}_{AB} = 480\angle 30°$ V (rms). Find the phase impedance of the load assuming a positive phase sequence.

16–39 A balanced Δ-connected three-phase load absorbs 30 kVA at a power factor of 0.75 lagging when the line voltage magnitude is 480 V (rms).
(a) Find the magnitude of the line current.
(b) Calculate the resistance and reactance of the per-phase impedance.

16–40 A balanced Δ-connected three-phase load absorbs 20 kW when the line current magnitude is 32 A (rms) and the line voltage magnitude is 480 V (rms). Find the resistance and reactance of the per-phase impedance assuming a lagging power factor.

16–41 A balanced Y-connected three-phase load absorbs 30 kVA at a power factor of 0.75 lagging when the line voltage magnitude is 2400 V (rms).
(a) Find the magnitude of the line current.
(b) Calculate the resistance and reactance of the per-phase impedance.

16–42 Two three-phase loads are connected in parallel. The first is a balanced Y-connected circuit absorbing 30 kVA at a power factor of 0.85 lagging. The second is a balanced Δ-connected load with a per-phase impedance of $20 + j15\ \Omega$. The magnitude of the line voltage at the loads is 480 V (rms). Find the magnitude of the total line current and the total complex power delivered to the two loads.

16–43 The line impedance in the single-line diagram of Figure P16–43 is $Z_W = 0.6 + j4\ \Omega$/phase. The balanced load connected to bus 2 is rated at $S_2 = 16 + j8$ kVA when the line voltage is 2400 V (rms). Find the line voltage and the complex power delivered at bus 2 when the generator at bus 1 produces a line voltage is 2400 V (rms).

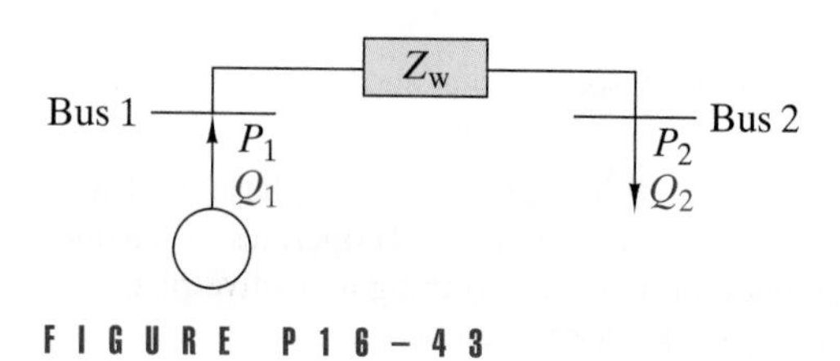

FIGURE P16–43

16–44 The source at bus 1 in Figure P16–43 is a balanced Y-connected generator with an internal impedance of $j2.2\ \Omega$/phase and a Thévenin voltage of 4160 V/phase. The load connected to bus 2 is a balanced Y-connected three-phase load with a phase impedance of $85 + j45\ \Omega$/phase. The line impedance is $Z_W = 2 + j5\ \Omega$/phase.
(a) Find the line current.
(b) Find the complex power delivered to the load and phase voltage at bus 2.
(c) Find the total complex power produced by the source.
(d) Find the transmission efficiency (η).

16–45 The line impedance in Figure P16–43 is $Z_W = 1 + j8\ \Omega$/phase. The line voltage at bus 2 is 7200 V (rms) and the load absorbs 600 kVA at 0.8 power factor lagging. Find the line current, the line voltage at bus 1, and the complex power produced by the source.

16–46 The line impedance in Figure P16–43 is $Z_W = 3 + j6\ \Omega$/phase. The line voltage at bus 2 is 7.2 kV (rms) and the load absorbs 300 kW at 0.8 power factor lagging. Find the magnitude of the line current and the line voltage at bus 1.

16–47 The line impedance in Figure P16–43 is $Z_W = 0.5 + j2\ \Omega$/phase. The load at bus 2 absorbs an apparent power of $S_2 = 5$ kVA at a lagging power factor of 0.7. The line voltage at bus 2 is 440 V (rms). Find the magnitude of the line current and the line voltage at bus 1.

16–48 The three-phase load connected to bus 2 in Figure P16–48 draws an average power of $P_2 = 100$ MW at a power factor of 0.8 lagging. At bus 2 the phase A voltage is $\mathbf{V}_{AN} = 90\angle 0°$ kV (rms). At bus 3 the phase A voltage is $\mathbf{V}_{AN} = 110\angle 10°$ kV (rms). Find the complex power produced by each source.

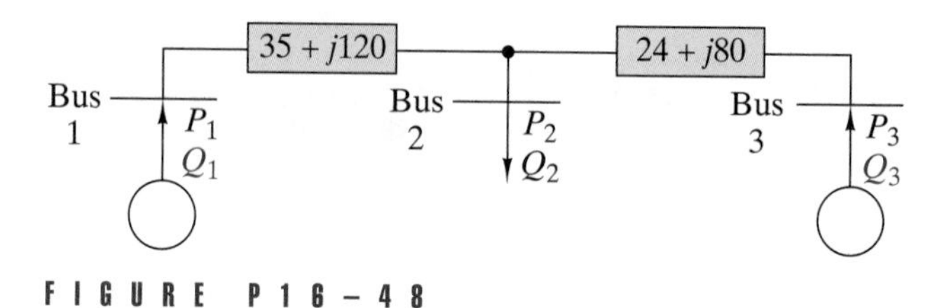

FIGURE P16–48

16–49 Three balanced three-phase loads are connected in parallel and connected to a source by lines with $Z_W = 1 + j10\ \Omega$/phase. The first load is Y-connected with an impedance of $200 + j100\ \Omega$/phase. The second load is Δ-connected with an impedance of $2700 - j1200\ \Omega$/phase. The third load draws a complex power of $110 + j95$ kVA. The phase voltage at the load is 7.2 kV (rms). Find the complex power delivered to the load by the source.

16–50 Three balanced three-phase loads are connected in parallel and have a composite power factor of 0.8 lagging. The first load draws an apparent power of 20 kVA at a power factor of 0.8 lagging. The second load draws an apparent power of 50 kVA and a power factor of 0.88 lagging. The line voltage is 2.4 kV (rms) and the line current magnitude is 25 A (rms). Find the apparent power delivered to the third load.

INTEGRATING PROBLEMS

16–51 **A** Three-Phase Heater Verification

Electric heaters use resistive elements to provide heat in commercial and industrial applications. A certain vendor lists the following ratings on three-phase, Y-connected electric heaters.

Power	Voltage	Current
18 kW	480/227 V	21.7 A/phase
24 kW	480/227 V	28.9 A/phase
32 kW	480/227 V	38.5 A/phase
18 kW	208/120 V	50.0 A/phase
24 kW	208/120 V	67.0 A/phase

Are these rating consistent with resistive three-phase loads?

16–52 **A** **D** Three-Phase Power Factor Corrections

The line voltage in Figure P16–52 is 480 V (rms). The balanced three-phase load operates 8 hours/day and draws 100 kVA at a lagging power factor of 0.8. The electric power supplier charges 7¢/kW-hr when the load power factor is greater than 0.95 and 10¢/kW-hr when the power factor is less than 0.95. The following three-phase capacitor banks are commercially available.

480-V THREE-PHASE CAPACITOR EQUIPMENT

KVAR	PART NUMBER	UNIT PRICE	KVAR	PART NUMBER	UNIT PRICE
10	1N0240A05	$550	40	1N0240A17	$950
20	1N0240A09	$700	50	1N0240A19	$1100
30	1N0240A11	$850	60	1N0240A23	$1500

If one or more capacitor banks are purchased to increase the power factor above 0.95, estimate the time it will take for the accumulated savings in operating costs to equal the capital investment in new equipment.

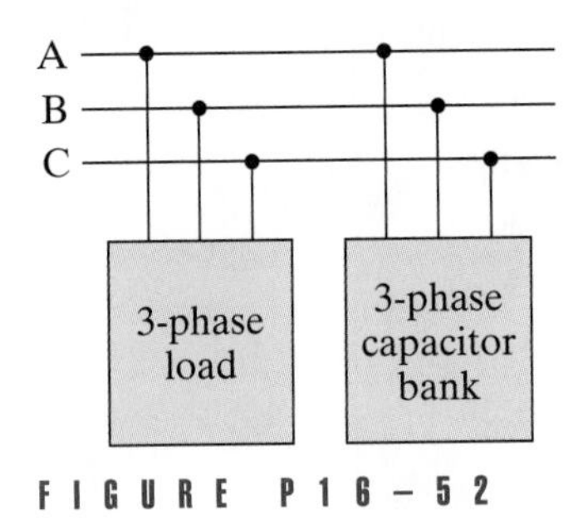

FIGURE P16–52

16–53 **A** Three-Phase Transformer Calculations

The line voltage and line current on the primary side of a three-phase transformer are $2400\angle -30^\circ$ V (rms) and $45\angle -56^\circ$ A (rms). The line voltage and line current on the secondary side are $480\angle 0^\circ$ V (rms) and $225\angle -30^\circ$ A (rms).

(a) What is the complex power and power factor on the primary side?

(b) What is the complex power and power factor on the secondary side?

(c) Is this an ideal transformer?

(d) What is the efficiency of the transformer?

16–54 **A** Phase Converter Efficiency

Three-phase motors are often used in equipment because they are more efficient and reliable than single-phase motors. Such equipment may be installed in locations where only single-phase power is available and the cost of installing three-phase service is prohibitive. The rotary phase converter in Figure P16–54 is one way of providing three-phase power from a single-phase source. Basically the converter is a rotating transformer that shifts the phase of a portion of the single-phase input to produce outputs with the same amplitude but shifted by $\pm 120^\circ$. In an application, a converter supplies three-phase power to a 30-hp motor that is 85% efficiency at full load. The single-phase input at full load output is 220V (rms) and 130 A (rms) at a power factor of 0.95 lagging. Find the efficiency of the converter. *Hint:* 1 hp = 746 W.

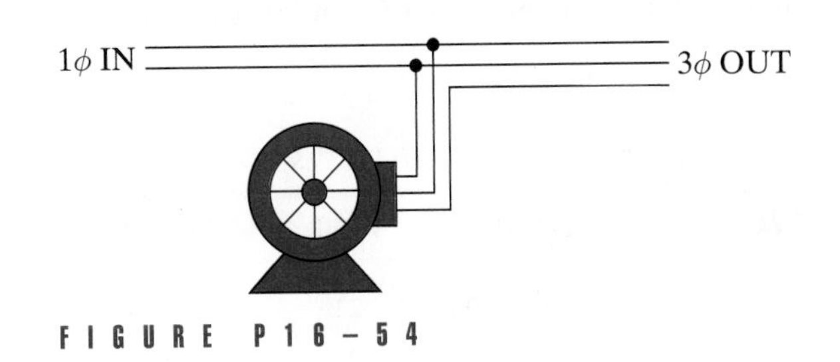

FIGURE P16–54

APPENDIX A

STANDARD VALUES

To reduce inventory costs the electrical industry has agreed on standard values and tolerances for commonly used discrete components such as resistors, capacitors, and zener diodes. The standard values for resistors and capacitors with 5%, 10%, and 20% tolerances are shown in Table A–1. Standard resistances are obtained by multiplying the values in the table by different powers of 10. For example, multiplying the values for ±20% tolerance by 10^4 yields a decade range of standard resistances of 100 kΩ, 150 kΩ, 220 kΩ, 330 kΩ, 470 kΩ, and 680 kΩ. Standard values for tolerances down to ±0.1% are defined, although the resulting proliferation of values tends to defeat the purpose of standardization.

TABLE A–1 STANDARD VALUES FOR RESISTORS AND CAPACITORS

VALUE	TOLERANCES	VALUE	TOLERANCES	VALUE	TOLERANCES
10	±5%, ±10%, ±20%	22	±5%, ±10%, ±20%	47	±5%, ±10%, ±20%
11	±5%	24	±5%	51	±5%
12	±5%, ±10%	27	±5%, ±10%	56	±5%, ±10%
13	±5%	30	±5%	62	±5%
15	±5%, ±10%, ±20%	33	±5%, ±10%, ±20%	68	±5%, ±10%, ±20%
16	±5%	36	±5%	75	±5%
18	±5%, ±10%	39	±5%, ±10%	82	±5%, ±10%
20	±5%	43	±5%	91	±5%

APPENDIX B

SOLUTION OF LINEAR EQUATIONS

The purpose of this appendix is to review the methods of solving systems of linear algebraic equations. Circuit analysis often requires solving linear algebraic equations of the type

$$\begin{aligned} 5x_1 - 2x_2 - 3x_3 &= 4 \\ -5x_1 + 7x_2 - 2x_3 &= -10 \\ -3x_1 - 3x_2 + 8x_3 &= 6 \end{aligned} \qquad \text{(B–1)}$$

where x_1, x_2, and x_3 are unknown voltages or currents. Often some of the unknowns may be missing from one or more of the equations. For example, the equations

$$\begin{aligned} 5x_1 - 2x_2 &= 5 \\ -4x_1 + 7x_2 &= 0 \\ -3x_2 + 8x_3 &= 0 \end{aligned}$$

involve three unknowns with one variable missing in each equation. Such equations can always be put in the standard square form by inserting the missing unknowns with a coefficient of zero.

$$\begin{aligned} 5x_1 - 2x_2 - 0x_3 &= 5 \\ -4x_1 + 7x_2 - 0x_3 &= 0 \\ 0x_1 - 3x_2 + 8x_3 &= 0 \end{aligned} \qquad \text{(B–2)}$$

Equations (B–1) and (B–2) will be used to illustrate the different methods of solving linear equations.

CRAMER'S RULE

Cramer's rule states that the solution of a system of linear equations for any unknown x_k is found as the ratio of two determinants

$$x_k = \frac{\Delta_k}{\Delta} \tag{B–3}$$

where Δ and Δ_k are determinants derived from the given set of equations. A **determinant** is a square array of numbers or symbols called **elements**. The elements are arranged in horizontal rows and vertical columns and are bordered by two vertical straight lines. In general, a determinant contains n^2 elements arranged in n rows and n columns. The value of the determinant is a function of the value and position of its n^2 elements.

The **system determinant** Δ in Eq. (B–3) is made up of the coefficients of the unknowns in the given system of equations. For example, the system determinant for Eq. (B–1) is

$$\Delta = \begin{vmatrix} 5 & -2 & -3 \\ -5 & 7 & -2 \\ -3 & -3 & 8 \end{vmatrix}$$

and for Eq. (B–2) is

$$.\Delta = \begin{vmatrix} 5 & -2 & 0 \\ -4 & 7 & 0 \\ 0 & -3 & 8 \end{vmatrix}$$

These two equations are examples of the general 3×3 determinant

$$\Delta = \begin{vmatrix} a_{11} & a_{12} & a_{13} \\ a_{21} & a_{22} & a_{23} \\ a_{31} & a_{32} & a_{33} \end{vmatrix} \tag{B–4}$$

where a_{ij} is the element in the ith row and jth column.

The determinant Δ_k in Eq. (B–3) is derived from the system determinant by replacing the kth column by the numbers on the right side of the system of equations. For example, Δ_1 for Eq. (B–1) is

$$\Delta_1 = \begin{vmatrix} 4 & -2 & -3 \\ -10 & 7 & -2 \\ 6 & -3 & 8 \end{vmatrix}$$

and Δ_3 for Eq. (B–2) is

$$\Delta_3 = \begin{vmatrix} 5 & -2 & 5 \\ -4 & 7 & 0 \\ 0 & -3 & 0 \end{vmatrix}$$

These examples are 3×3 determinants because the system determinants from which they are derived are 3×3.

In summary, using Cramer's rule to solve linear equations boils down to evaluating the determinants formed using the coefficients of the unknowns and the right side of the system of equations.

EVALUATING DETERMINANTS

The **diagonal rule** gives the value of a 2×2 determinant as the difference in the product of the elements on the main diagonal ($a_{11}a_{22}$) and the product of the elements on the off diagonal ($a_{21}a_{12}$). That is, for a 2×2 determinant

$$\Delta = \begin{vmatrix} a_{11} & a_{12} \\ a_{21} & a_{22} \end{vmatrix} = a_{11}a_{22} - a_{21}a_{12} \tag{B–5}$$

The value of 3×3 and higher-order determinants can be found using the method of minors. Every element a_{ij} has a **minor** M_{ij}, which is formed by deleting the row and column containing a_{ij}. For example, the minor M_{21} of the general 3×3 determinant in Eq. (B–4) is

$$M_{21} = \begin{vmatrix} a_{12} & a_{13} \\ a_{32} & a_{33} \end{vmatrix} = a_{12}a_{33} - a_{32}a_{13}$$

The **cofactor** C_{ij} of the element a_{ij} is its minor M_{ij} multiplied by $(-1)^{i+j}$.

$$C_{ij} = (-1)^{i+j}M_{ij}$$

The signs of the cofactors alternate along any row or column. The appropriate sign for cofactor C_{ij} is found by starting in position a_{11} and counting plus, minus, plus, minus . . . along any combination of rows or columns leading to the position a_{ij}.

To use the **method of minors** we select one (and only one) row or column. The determinant is the sum of the products of the elements in the selected row or column and their cofactors. For example, selecting the first column in Eq. (B–4), we obtain Δ as follows:

$$\begin{aligned} \Delta &= a_{11}C_{11} + a_{21}C_{21} + a_{31}C_{31} \\ &= a_{11}(-1)^2\begin{vmatrix} a_{22} & a_{23} \\ a_{32} & a_{33} \end{vmatrix} + a_{21}(-1)^3\begin{vmatrix} a_{12} & a_{13} \\ a_{32} & a_{33} \end{vmatrix} + a_{31}(-1)^4\begin{vmatrix} a_{12} & a_{13} \\ a_{22} & a_{23} \end{vmatrix} \\ &= a_{11}(a_{22}a_{33} - a_{32}a_{23}) - a_{21}(a_{12}a_{33} - a_{32}a_{13}) + a_{31}(a_{12}a_{23} - a_{22}a_{13}) \end{aligned}$$

An identical expression for Δ is obtained using any other row or column. For determinants greater than 3×3 the minors themselves can be evaluated using this approach. However, a system of equations leading to determinants larger than 3×3 is probably better handled using computer tools.

EXAMPLE B–1

Solve for the three unknowns in Eq. (B–1) using Cramer's rule.

SOLUTION:

Expanding the system determinant about the first column yields

$$\Delta = \begin{vmatrix} 5 & -2 & -3 \\ -5 & 7 & -2 \\ -3 & -3 & 8 \end{vmatrix} = 5\begin{vmatrix} 7 & -2 \\ -3 & 8 \end{vmatrix} - (-5)\begin{vmatrix} -2 & -3 \\ -3 & 8 \end{vmatrix} + (-3)\begin{vmatrix} -2 & -3 \\ 7 & -2 \end{vmatrix}$$

$$\begin{aligned} &= 5[7 \times 8 - (-2)(-3)] - (-5)[(-2) \times 8 - (-3)(-3)] \\ &\quad + (-3)[(-2)(-2) - (7)(-3)] \\ &= 250 - 125 - 75 = 50 \end{aligned}$$

Expanding Δ_1 about the first column yields

$$\Delta_1 = \begin{vmatrix} 4 & -2 & -3 \\ -10 & 7 & -2 \\ 6 & -3 & 8 \end{vmatrix} = 4\begin{vmatrix} 7 & -2 \\ -3 & 8 \end{vmatrix} - (-10)\begin{vmatrix} -2 & -3 \\ -3 & 8 \end{vmatrix} + (6)\begin{vmatrix} -2 & -3 \\ 7 & -2 \end{vmatrix}$$

$$= 200 - 250 + 150 = 100$$

Expanding Δ_2 about the first column yields

$$\Delta_2 = \begin{vmatrix} 5 & 4 & -3 \\ -5 & -10 & -2 \\ -3 & 6 & 8 \end{vmatrix} = 5\begin{vmatrix} -10 & -2 \\ 6 & 8 \end{vmatrix} - (-5)\begin{vmatrix} 4 & -3 \\ 6 & 8 \end{vmatrix} + (-3)\begin{vmatrix} 4 & -3 \\ -10 & -2 \end{vmatrix}$$

$$= -340 + 250 + 114 = 24$$

Expanding Δ_3 about the first column yields

$$\Delta_3 = \begin{vmatrix} 5 & -2 & 4 \\ -5 & 7 & -10 \\ -3 & -3 & 6 \end{vmatrix} = 5\begin{vmatrix} 7 & -10 \\ -3 & 6 \end{vmatrix} - (-5)\begin{vmatrix} -2 & 4 \\ -3 & 6 \end{vmatrix} + (-3)\begin{vmatrix} -2 & 4 \\ 7 & -10 \end{vmatrix}$$

$$= 60 - 0 + 24 = 84$$

Now, applying Cramer's rule, we solve for the three unknowns.

$$x_1 = \frac{\Delta_1}{\Delta} = \frac{100}{50} = 2$$

$$x_2 = \frac{\Delta_2}{\Delta} = \frac{24}{50} = 0.48$$

$$x_3 = \frac{\Delta_3}{\Delta} = \frac{84}{50} = 1.68$$

■

Exercise B–1

Evaluate Δ, Δ_1, Δ_2, and Δ_3 for Eq. (B–2).

Answer: 216; 280; 160; 60

MATRICES AND LINEAR EQUATIONS

Circuit equations can be formulated and solved in matrix format. By definition, a **matrix** is a rectangular array written as

$$\mathbf{A} = \begin{bmatrix} a_{11} & a_{12} & a_{13} & \dots & a_{1n} \\ a_{21} & a_{22} & a_{23} & \dots & a_{2n} \\ \dots & \dots & \dots & \dots & \dots \\ a_{m1} & a_{m2} & a_{m3} & \dots & a_{mn} \end{bmatrix} \tag{B–6}$$

The matrix **A** in Eq. (B–6) contains m rows and n columns and is said to be of order m *by* n (or $m \times n$). The matrix notation in Eq. (B–6) can be abbreviated as follows:

$$\mathbf{A} = [a_{ij}]_{mn} \tag{B–7}$$

where a_{ij} is the element in the ith row and jth column.

SOME DEFINITIONS

Different types of matrices have special names. A **row matrix** has only one row ($m = 1$) and any number of columns. A **column matrix** has only one column ($n = 1$) and any number of rows. A **square matrix** has the same number of rows as columns ($m = n$). A **diagonal matrix** is a square matrix in which all elements not on the main diagonal are zero ($a_{ij} = 0$ for $i \neq j$). An **identity matrix** is a diagonal matrix for which the main diagonal elements are all unity ($a_{ii} = 1$).

For example, given

$$\mathbf{A} = [1 \quad -2 \quad 0 \quad 4] \quad \mathbf{B} = \begin{bmatrix} 3 \\ -2 \\ 6 \\ 0 \end{bmatrix} \quad \mathbf{C} = \begin{bmatrix} 1 & 0 & -7 \\ -3 & 12 & 0 \\ 0 & 0 & -4 \end{bmatrix} \quad \mathbf{U} = \begin{bmatrix} 1 & 0 & 0 & 0 \\ 0 & 1 & 0 & 0 \\ 0 & 0 & 1 & 0 \\ 0 & 0 & 0 & 1 \end{bmatrix}$$

we say that **A** is a 1×4 row matrix, **B** is a 4×1 column matrix, **C** is a 3×3 square matrix, and **U** is a 4×4 identity matrix.

The **determinant** of a square matrix **A** (denoted det **A**) has the same elements as the matrix itself. For example, given

$$\mathbf{A} = \begin{bmatrix} 4 & -6 \\ 1 & -2 \end{bmatrix} \text{ then } \det \mathbf{A} = \begin{vmatrix} 4 & -6 \\ 1 & -2 \end{vmatrix} = -8 + 6 = -2$$

The **transpose** of a matrix **A** (denoted $\mathbf{A}^T$) is formed by interchanging the rows and columns. For example, given

$$\mathbf{A} = \begin{bmatrix} 1 & 2 & 0 & 8 \\ 4 & 7 & -1 & -3 \end{bmatrix} \text{ then } \mathbf{A}^T = \begin{bmatrix} 1 & 4 \\ 2 & 7 \\ 0 & -1 \\ 8 & -3 \end{bmatrix}$$

The **adjoint** of a square matrix **A** (denoted adj **A**) is formed by replacing each element a_{ij} by its cofactor C_{ij} and then transposing.

$$\text{adj } \mathbf{A} = [C_{ij}]^T \tag{B–8}$$

For example, if

$$\mathbf{A} = \begin{bmatrix} -3 & 2 \\ 0 & 5 \end{bmatrix} \text{ then } C_{11} = 5 \quad C_{12} = 0 \quad C_{21} = -2 \quad C_{22} = -3$$

and therefore

$$\text{adj } \mathbf{A} = \begin{bmatrix} 5 & 0 \\ -2 & -3 \end{bmatrix}^T = \begin{bmatrix} 5 & -2 \\ 0 & -3 \end{bmatrix}$$

MATRIX ALGEBRA

The matrices **A** and **B** are equal if and only if they have the same number of rows and columns, and $a_{ij} = b_{ij}$ for all i and j. Matrix addition is only possible when two matrices have the same number of rows and columns.

When two matrices are of the same order, their sum is obtained by adding the corresponding elements: that is,

$$\text{If } \mathbf{C} = \mathbf{A} + \mathbf{B} \text{ then } c_{ij} = a_{ij} + b_{ij} \tag{B–9}$$

For example, given

$$\mathbf{A} = \begin{bmatrix} -1 & 4 \\ -3 & -2 \end{bmatrix} \text{ and } \mathbf{B} = \begin{bmatrix} 3 & 0 \\ 2 & -4 \end{bmatrix} \text{ then } \mathbf{C} = \mathbf{A} + \mathbf{B} = \begin{bmatrix} 2 & 4 \\ -1 & -6 \end{bmatrix}$$

Multiplying a matrix **A** by a scalar constant k is accomplished by multiplying every element by k; that is, $k\mathbf{A} = [ka_{ij}]$. In particular, if $k = -1$ then $-\mathbf{B} = [-b_{ij}]$, and applying the matrix addition rule yields matrix **subtraction**.

$$\text{If } \mathbf{C} = \mathbf{A} - \mathbf{B} \text{ then } c_{ij} = a_{ij} - b_{ij} \tag{B–10}$$

Multiplication of two matrices **AB** is defined only if the number of columns in **A** equals the number of rows in **B**. In general, if **A** is of order $m \times n$ and **B** is of order $n \times r$, then the product $\mathbf{C} = \mathbf{AB}$ is a matrix of order $m \times r$. The element c_{ij} is found by summing the products of the elements in the ith row of **A** and the jth column of **B**.

$$c_{ij} = [a_{i1}\ a_{i2}\ \dots\ a_{in}] \begin{bmatrix} b_{1j} \\ b_{2j} \\ \vdots \\ \vdots \\ \vdots \\ b_{nj} \end{bmatrix} = a_{i1}b_{1j} + a_{i2}b_{2j} + \cdots a_{in}b_{nj}$$

$$= \sum_{k=1}^{n} a_{ik}b_{kj} \tag{B–11}$$

In other words, matrix multiplication is a row by column operation.

Matrix multiplication is not commutative, so usually $\mathbf{AB} \neq \mathbf{BA}$. Two important exceptions are (1) the product of a square matrix **A** and an identity matrix **U** for which $\mathbf{UA} = \mathbf{AU} = \mathbf{A}$, and (2) the product of a square matrix **A** and its **inverse** (denoted $\mathbf{A}^{-1}$) for which $\mathbf{A}^{-1}\mathbf{A} = \mathbf{AA}^{-1} = \mathbf{U}$. A closed-form formula for the inverse of a square matrix is

$$\mathbf{A}^{-1} = \frac{\text{adj } \mathbf{A}}{\det \mathbf{A}} \tag{B–12}$$

That is, the inverse can be found by multiplying the adjoint matrix of **A** by the scalar 1/det **A**. If det **A** = 0, then **A** is said to be **singular** and $\mathbf{A}^{-1}$ does not exist. Equation (B–12) is useful for deriving properties of the inverse of a matrix. It is not, however, a very efficient way to calculate the inverse of a matrix of order greater than 3×3.

Exercise B–2

Given:

$$\mathbf{A} = \begin{bmatrix} -5 & 7 \\ 7 & 11 \end{bmatrix} \text{ and } \mathbf{B} = \begin{bmatrix} 3 & -1 \\ 6 & -2 \end{bmatrix}$$

Calculate AB, BA, A^{-1}, and B^{-1}.

Answers:

$$\mathbf{AB} = \begin{bmatrix} 27 & -9 \\ 87 & -29 \end{bmatrix} \quad \mathbf{BA} = \begin{bmatrix} -22 & 10 \\ -44 & 20 \end{bmatrix}$$

$$\mathbf{A}^{-1} = \frac{1}{104}\begin{bmatrix} -11 & 7 \\ 7 & 5 \end{bmatrix} \quad \mathbf{B}^{-1} \text{ does not exist}$$

MATRIX SOLUTION OF LINEAR EQUATIONS

The three linear equations in Eq. (B–1) are

$$\begin{aligned} 5x_1 - 2x_2 - 3x_3 &= 4 \\ -5x_1 + 7x_2 - 2x_3 &= -10 \\ -3x_1 - 3x_2 + 8x_3 &= 6 \end{aligned}$$

These equations are expressed in matrix form as follows:

$$\begin{bmatrix} 5 & -2 & -3 \\ -5 & 7 & -2 \\ -3 & -3 & 8 \end{bmatrix} \begin{bmatrix} x_1 \\ x_2 \\ x_3 \end{bmatrix} = \begin{bmatrix} 4 \\ -10 \\ 6 \end{bmatrix} \tag{B–13}$$

The left side of Eq. (B–13) is the product of a 3×3 square matrix and a 3×1 column matrix of unknowns. The elements in the square matrix are the coefficients of the unknown in the given equations. The matrix product on the left side in Eq. (B–13) produces a 3×1 matrix, which equals the 3×1 column matrix on the right side. The elements of the 3×1 on the right side are the constants on the right sides of the given equations.

In symbolic form we write the matrix equation in Eq. (B–13) as

$$\mathbf{AX} = \mathbf{B} \tag{B–14}$$

where

$$\mathbf{A} = \begin{bmatrix} 5 & -2 & -3 \\ -5 & 7 & -2 \\ -3 & -3 & 8 \end{bmatrix}, \quad \mathbf{X} = \begin{bmatrix} x_1 \\ x_2 \\ x_3 \end{bmatrix}, \text{ and } \mathbf{B} = \begin{bmatrix} 4 \\ -10 \\ 6 \end{bmatrix}$$

Left multiplying Eq. (B–14) by $\mathbf{A}^{-1}$ yields

$$\mathbf{A}^{-1}\mathbf{AX} = \mathbf{A}^{-1}\mathbf{B}$$

But by definition $\mathbf{A}^{-1}\mathbf{A} = \mathbf{U}$ and $\mathbf{UX} = \mathbf{X}$; therefore

$$\mathbf{X} = \mathbf{A}^{-1}\mathbf{B} \tag{B–15}$$

To solve linear equations by matrix methods, we calculate the product $\mathbf{A}^{-1}\mathbf{B}$.

To implement the matrix approach, we must first find $\mathbf{A}^{-1}$ using Eq. (B–12). The determinant of the coefficient matrix is

$$\det \mathbf{A} = \begin{vmatrix} 5 & -2 & -3 \\ -5 & 7 & -2 \\ -3 & -3 & 8 \end{vmatrix} = 50$$

The cofactors of the first row of the coefficient matrix are

$$C_{11} = -\begin{vmatrix} +7 & -2 \\ -3 & 8 \end{vmatrix} = 50 \quad C_{12} = \begin{vmatrix} -5 & -2 \\ -3 & 8 \end{vmatrix} = 46$$

$$C_{13} = -\begin{vmatrix} -5 & +7 \\ -3 & -3 \end{vmatrix} = 36$$

The cofactors for the second and third rows are

$$C_{21} = 25 \quad C_{22} = 31 \quad C_{23} = 21$$

$$C_{31} = 25 \quad C_{32} = 25 \quad C_{33} = 25$$

Now, using Eq. (B–12), we obtain $\mathbf{A}^{-1}$ as

$$\mathbf{A}^{-1} = \frac{\text{adj } \mathbf{A}}{\det \mathbf{A}} = \frac{1}{50}\begin{bmatrix} 50 & 46 & 36 \\ 25 & 31 & 21 \\ 25 & 25 & 25 \end{bmatrix}^T = \frac{1}{50}\begin{bmatrix} 50 & 25 & 25 \\ 46 & 31 & 25 \\ 36 & 21 & 25 \end{bmatrix}$$

Using Eq. (B–15), we solve for the column matrix of unknowns as

$$\begin{bmatrix} x_1 \\ x_2 \\ x_3 \end{bmatrix} = \mathbf{X} = \mathbf{A}^{-1}\mathbf{B} = \frac{1}{50}\begin{bmatrix} 50 & 25 & 25 \\ 46 & 31 & 25 \\ 36 & 21 & 25 \end{bmatrix}\begin{bmatrix} 4 \\ -10 \\ 6 \end{bmatrix} = \frac{1}{50}\begin{bmatrix} 100 \\ 24 \\ 84 \end{bmatrix}$$

which yields $x_1 = 2$, $x_2 = 24/50$, and $x_3 = 84/50$. These are, of course, the same results previously obtained using Cramer's rule.

USING COMPUTER TOOLS

Computer tools for solving linear equations range from inexpensive hand-held calculators to sophisticated software packages capable of solving hundreds of equations. At the intermediate level are math analysis software packages such as MATLAB and Mathcad. Under what circumstances should you consider using these computer tools in linear circuit analysis?

There are no hard and fast rules here. Somewhere around three or four equations, the burden of hand calculations becomes mildly excruciating. Sets of equations with N up to 20 or 30 are routinely solved using computer tools, except when the equations are ill conditioned (several equations are almost linearly dependent). Well-conditioned systems of equations with $N = 50$ or more can be solved using sophisticated numerical methods. On the other hand, these sophisticated computer tools probably don't buy you very much in linear circuit applications. If you encounter a problem that requires solving, say, 20 or more linear equations, you should redefine the problem so that it can be partitioned into smaller pieces.

MATLAB is a software package for matrix-based computations and data analysis. The original MATLAB was a mainframe resident software written in the 1970s to provide a "matrix laboratory" for linear algebra and matrix theory courses. Since then, personal computer versions have been developed that fit on modest computer platforms. In its present form, MATLAB is an interactive system and programming language whose capabilities extend far beyond the original matrix laboratory application.

To solve the matrix equation in Eq. (B–13) using MATLAB, we first enter the **A** and **B** matrices in the MATLAB command window. In this window the command line prompt is either the character string `<< or EDU<<`, depending on the version in use. This prompt indicates that MATLAB is ready to accept data and commands entered via the keyboard. For example, a matrix can be entered by typing its elements one row at a time. Elements in the same row are separated by spaces. The end of a row is indicated by a semicolon or an ↵enter keystroke. The row-by-row entry is enclosed by a left bracket [at the beginning and a right bracket] at the end of the last row.

For example, the **A** matrix in Eq. (B–13) can be entered by typing

```
A=[5 -2 -3
  -5  7 -2
  -3 -3  8]
```

After the last ↵enter keystroke, MATLAB responds by listing the elements of **A**.

```
A=
     5 -2 -3
    -5  7 -2
    -3 -3  8
```

This echo check allows you to verify that the elements of **A** have been entered correctly. The **A** can also be entered by typing

```
A=[5 -2 -3;-5 7 -2;-3 -3 8];
```

The semicolon following the closing right bracket suppresses the MATLAB echo check if you have supreme confidence that you have entered the matrix elements without error. An individual matrix element can be referenced by enclosing its subscripts in parenthesis. For instance, the command line query

```
A(3,1)
```

asks a_{31} = ? and produces the MATLAB response

```
A(3,1) =
        -3
```

The **B** matrix for Eq. (B–13) can be entered as

```
B = [4; -10; 6];
```

The matrix equations in Eq. (B–13) can be solved for the unknown column matrix **X** by forming the matrix product $\mathbf{A}^{-1}\mathbf{B}$. In the MATLAB command window we enter the statement

```
X=inv(A)*B
```

to which MATLAB responds with

```
X=
       2.0000
       0.4800
       1.6800
```

We can verify this result by checking to see that the matrix product **AX** is equal to the matrix **B**. In the command window we type

```
A*X
```

to which MATLAB responds with

```
ans=
         4
       -10
         6
```

which is the matrix **B** on the right side of Eq. (B–13).

Figure B–1 shows a Mathcad worksheet that performs the same matrix operations. The format and appearance of the mathematical operations in the worksheet are virtually self-defining. Again we have checked the answer by forming the matrix product **AX** in the final step.

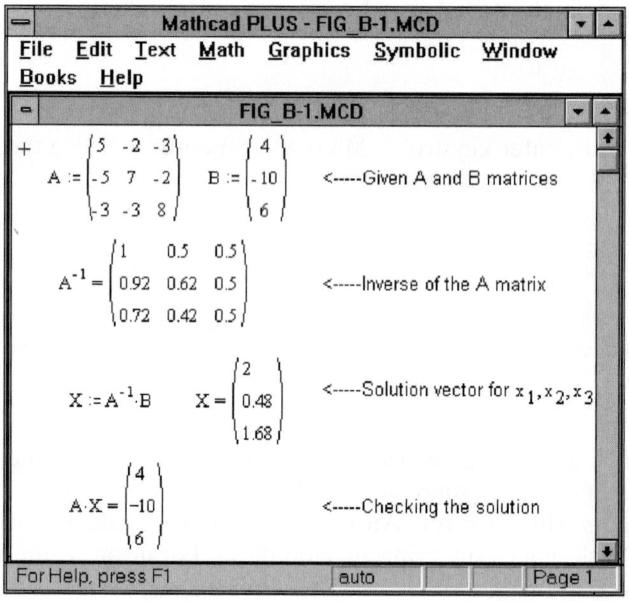

FIGURE B–1 *Mathcad worksheet.*

Mathcad PLUS - FIG_B-1.MCD

File Edit Text Math Graphics Symbolic Window Books Help

FIG_B-1.MCD

$$A := \begin{pmatrix} 5 & -2 & -3 \\ -5 & 7 & -2 \\ -3 & -3 & 8 \end{pmatrix} \quad B := \begin{pmatrix} 4 \\ -10 \\ 6 \end{pmatrix}$$ <-----Given A and B matrices

$$A^{-1} = \begin{pmatrix} 1 & 0.5 & 0.5 \\ 0.92 & 0.62 & 0.5 \\ 0.72 & 0.42 & 0.5 \end{pmatrix}$$ <-----Inverse of the A matrix

$$X := A^{-1} \cdot B \quad X = \begin{pmatrix} 2 \\ 0.48 \\ 1.68 \end{pmatrix}$$ <-----Solution vector for x_1, x_2, x_3

$$A \cdot X = \begin{pmatrix} 4 \\ -10 \\ 6 \end{pmatrix}$$ <-----Checking the solution

For Help, press F1 | auto | Page 1

APPENDIX C

COMPLEX NUMBERS

Using complex numbers to represent signals and circuits is a fundamental tool in electrical engineering. This appendix reviews complex-number representations and arithmetic operations. These procedures, though rudimentary, must be second nature to all who aspire to be electrical engineers. Exercises are provided to confirm your mastery of these basic skills.

COMPLEX-NUMBER REPRESENTATIONS

A complex number z can be written in rectangular form as

$$z = x + jy \tag{C–1}$$

where j represents $\sqrt{-1}$. Mathematicians customarily use i to represent $\sqrt{-1}$, but i represents current in electrical engineering, so we use the symbol j instead.

The quantity z is a two-dimensional number represented as a point in the complex plane, as shown in Figure C–1. The x component is called the **real part** and y (not jy) the **imaginary part** of z. A special notation is sometimes used to indicate these two components:

$$x = \text{Re}\{z\} \quad \text{and} \quad y = \text{Im}\{z\} \tag{C–2}$$

where Re{z} means the real part and Im{z} the imaginary part of z.

Figure C–1 also shows the polar representation of the complex number z. In polar form a complex number is written

$$z = M\angle\theta \tag{C–3}$$

where M is called the **magnitude** and θ the **angle** of z. A special notation is also used to indicate these two components.

$$|z| = M \quad \text{and} \quad \angle z = \theta \tag{C–4}$$

where $|z|$ means the magnitude and $\angle z$ the angle of z.

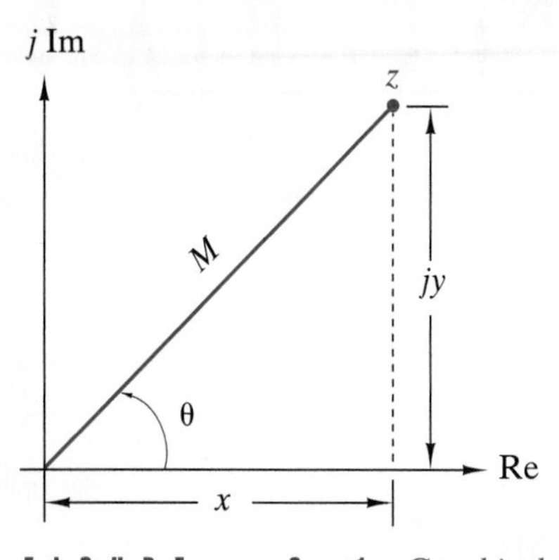

FIGURE C – 1 *Graphical representation of complex numbers.*

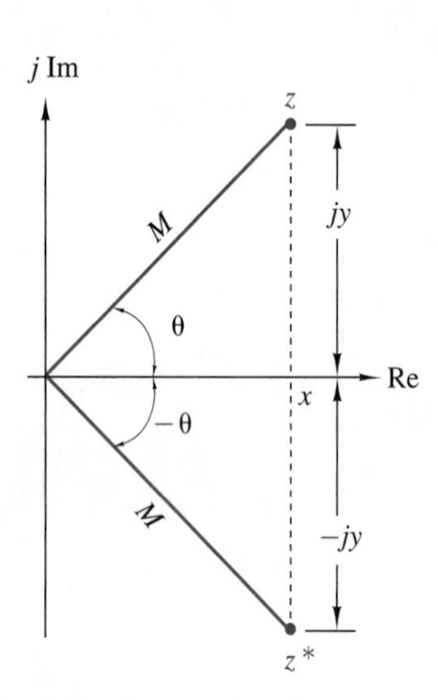

FIGURE C – 2 *Graphical representation of conjugate complex numbers.*

The real and imaginary parts and magnitude and angle of z are all shown geometrically in Figure C–1. The relationships between the rectangular and polar forms are easily derived from the geometry in Figure C–1:

$$\text{Rectangular to polar} \quad M = \sqrt{x^2 + y^2} \quad \theta = \tan^{-1}\frac{y}{x}$$

$$\text{Polar to rectangular} \quad x = M\cos\theta \quad y = M\sin\theta \tag{C–5}$$

The inverse tangent relation for θ involves an ambiguity that can be resolved by identifying the correct quadrant in the z-plane using the signs of the two rectangular components. [See Exercise C–1(b) and (c).]

Another version of the polar form is obtained using Euler's relationship:

$$e^{j\theta} = \cos\theta + j\sin\theta \tag{C–6}$$

We can write the polar form as

$$z = Me^{j\theta} = M\cos\theta + jM\sin\theta \tag{C–7}$$

This polar form is equivalent to Eq. (C–3), since the right side yields the same polar-to-rectangular relationships as Eq. (C–5). Thus, a complex number can be represented in three ways:

$$z = x + jy \quad z = M\angle\theta \quad z = Me^{j\theta} \tag{C–8}$$

The relationships between these forms are given in Eq. (C–5).

The quantity z^* is called the conjugate of the complex number z. The asterisk indicates the **conjugate** of a complex number formed by reversing the sign of the imaginary component. In rectangular form the conjugate of $z = x + jy$ is written as $z^* = x - jy$. In polar form the conjugate is obtained by reversing the sign of the angle of z, $z^* = Me^{-j\theta}$. The geometric interpretation in Figure C–2 shows that conjugation simply reflects a complex number across the real axis in the complex plane.

Exercise C–1

Convert the following complex numbers to polar form:

(a) $1 + j\sqrt{3}$ (b) $-10 + j20$ (c) $-2000 - j8000$ (d) $60 - j80$

Answers: (a) $2e^{j60°}$ (b) $22.4e^{j117°}$ (c) $8246e^{j256°}$ (d) $100e^{j307°}$

Exercise C–2

Convert the following complex numbers to rectangular form:

(a) $12e^{j90°}$ (b) $3e^{j45°}$ (c) $400\angle\pi$ (d) $8e^{-j60°}$ (e) $15e^{j\pi/6}$

Answers: (a) $0 + j12$ (b) $2.12 + j2.12$ (c) $-400 + j0$ (d) $4 - j6.93$ (e) $13 + j7.5$

Exercise C–3

Evaluate the following expressions:

(a) $\text{Re}(12e^{j\pi})$ (b) $\text{Im}(100\angle 60°)$ (c) $\angle(-2 + j6)$ (d) $\text{Im}[(4e^{j\frac{\pi}{4}})^*]$

Answers: (a) -12 (b) 86.6 (c) $108.4°$ (d) -2.83

ARITHMETIC OPERATIONS: ADDITION AND SUBTRACTION

Addition and subtraction are defined in terms of complex numbers in rectangular form. Two complex numbers

$$z_1 = x_1 + jy_1 \quad \text{and} \quad z_2 = x_2 + jy_2 \tag{C–9}$$

are added by separately adding the real parts and imaginary parts. The sum $z_1 + z_2$ is defined as

$$z_1 + z_2 = (x_1 + x_2) + j(y_1 + y_2) \tag{C–10}$$

Subtraction follows the same pattern except that the components are subtracted:

$$z_1 - z_2 = (x_1 - x_2) + j(y_1 - y_2) \tag{C–11}$$

Figure C–3 shows a geometric interpretation of addition and subtraction. In particular, note that $z + z^* = 2x$ and $z - z^* = j2y$.

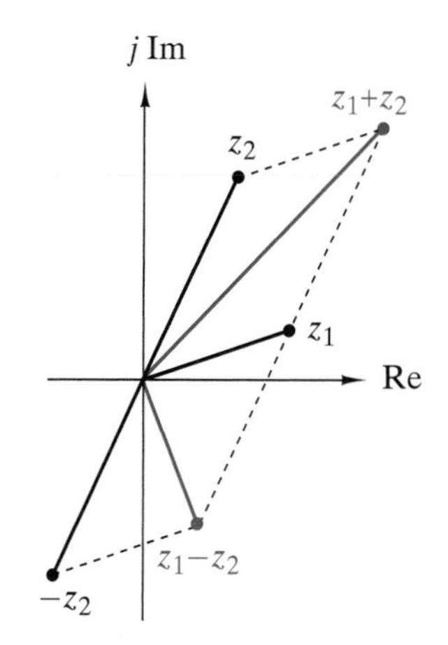

FIGURE C – 3 *Graphical representation of addition and subtraction of complex numbers.*

MULTIPLICATION AND DIVISION

Multiplication and division of complex numbers can be accomplished with the numbers in either rectangular or polar form. For complex numbers in rectangular form, the multiplication operation yields

$$\begin{aligned} z_1 z_2 &= (x_1 + jy_1)(x_2 + jy_2) \\ &= (x_1x_2 + j^2y_1y_2) + j(x_1y_2 + x_2y_1) \\ &= (x_1x_2 - y_1y_2) + j(x_1y_2 + x_2y_1) \end{aligned} \tag{C–12}$$

For numbers in polar form, the product is

$$\begin{aligned} z_1 z_2 &= (M_1e^{j\theta_1})(M_2e^{j\theta_2}) \\ &= (M_1M_2)e^{j(\theta_1+\theta_2)} \end{aligned} \tag{C–13}$$

Multiplication is somewhat easier to carry out with the numbers in polar form, although both methods should be understood. In particular, the product of a complex number z and it conjugate z^* is the square of its magnitude, which is always positive.

$$zz^* = (Me^{j\theta})(Me^{-j\theta}) = M^2 \tag{C–14}$$

For complex numbers in polar form the division operation yields

$$\begin{aligned} \frac{z_1}{z_2} &= \frac{Me^{j\theta_1}}{M_2e^{j\theta_2}} \\ &= \left(\frac{M_1}{M_2}\right)e^{j(\theta_1-\theta_2)} \end{aligned} \tag{C–15}$$

When the numbers are in rectangular form, the numerator and denominator of the quotient are multiplied by the conjugate of the denominator.

$$\frac{z_1}{z_2}\frac{z_2^*}{z_2^*} = \frac{(x_1 + jy_1)(x_2 - jy_2)}{(x_2 + jy_2)(x_2 - jy_2)}$$

Applying the multiplication rule from Eq. (C–12) to the numerator and denominator yields

$$\frac{z_1}{z_2} = \frac{(x_1x_2 + y_1y_2) + j(x_2y_1 - x_1y_2)}{x_2^2 + y_2^2} \tag{C–16}$$

Complex division is easier to carry out with the numbers in polar form, although both methods should be understood.

Exercise C–4

Evaluate the following expressions using $z_1 = 3 + j4$, $z_2 = 5 - j7$, $z_3 = -2 + j3$, and $z_4 = 5\angle -30°$:

(a) z_1z_2 (b) $z_3 + z_4$ (c) z_2z_3/z_4 (d) $z_1^* + z_3z_1$ (e) $z_2 + (z_1z_4)^*$

Answers:

(a) $43 - j$ (b) $2.33 + j0.5$ (c) $-0.995 + j6.12$ (d) $-15 - j3$ (e) $28 - j16.8$

Exercise C–5

Given $z = x + jy = Me^{j\theta}$, evaluate the following statements:

(a) $z + z^*$ (b) $z - z^*$ (c) z/z^* (d) z^2 (e) $(z^*)^2$ (f) zz^*

Answers:

(a) $2x$ (b) $j2y$ (c) $e^{j2\theta}$ (d) $x^2 - y^2 + j2xy$ (e) $x^2 - y^2 - j2xy$ (f) $x^2 + y^2$

Exercise C–6

Given $z_1 = 1$, $z_2 = -1$, $z_3 = j$, and $z_4 = -j$, evaluate (a) z_1/z_3 (b) z_1/z_4 (c) z_3z_4 (d) z_3z_3 (e) z_4z_4 (f) $z_2z_3^*$.

Answers:

(a) $-j$ (b) $+j$ (c) 1 (d) -1 (e) -1 (f) j

Exercise C–7

Evaluate the expression $T(\omega) = j\omega/(j\omega + 10)$ at $\omega = 5, 10, 20, 50, 100$.

Answers: $0.447\angle 63.4°$, $0.707\angle 45°$, $0.894\angle 26.6°$, $0.981\angle 11.3°$, $0.995\angle 5.71°$

APPENDIX D

BUTTERWORTH AND CHEBYCHEV POLES

Transfer functions of low-pass filters have the form

$$T(s) = \frac{K}{q_n(s)} \tag{D–1}$$

where $q_n(s)$ is an nth-order polynomial whose roots define the poles of $T(s)$. The gain response of the filter is found by setting $s = j\omega$ and forming the magnitude $|T(j\omega)|$. The purpose of this appendix is to illustrate how the poles of $T(s)$ are determined for Butterworth and Chebychev gain responses.

Finding the poles of $T(s)$ involves finding the poles of the squared gain response $|T(j\omega)|^2$. The function $T(-j\omega)$ is the conjugate of $T(j\omega)$. The product of any function and its conjugate is the square of its magnitude, in this case $|T(j\omega)|^2$. Since $T(j\omega)$ is found by setting $s = j\omega$, we can write

$$|T(j\omega)|^2 = T(j\omega)T(-j\omega) = T(s)T(-s)|_{s=j\omega} \tag{D–2}$$

Combining Eqs. (D–1) and (D–2), we obtain

$$|T(j\omega)|^2 = \frac{K^2}{q_n(s)q_n(-s)|_{s=j\omega}} \tag{D–3}$$

The denominator of Eq. (D–3) indicates that the poles of $|T(j\omega)|^2$ are either the roots of the polynomial $q_n(s)$ or $q_n(-s)$. The roots of $q_n(s)$ must all lie in the left half plane for $T(s)$ in Eq. (D–1) to be the transfer function of a stable circuit. If $s = \alpha + j\beta$ is any root of $q_n(s)$, then the real part α must be negative. The corresponding root of $q_n(-s)$ is $s = -\alpha - j\beta$, whose real part $-\alpha$ must be positive. In other words, the roots of $q_n(-s)$ all lie in the right half plane. Thus, we can find the poles of $T(s)$ by finding the left half plane poles $|T(j\omega)|^2$.

The square of the normalized ($\omega_C = 1$) Butterworth gain response in Eq. (14–28) is

$$|T(j\omega)|^2 = \frac{K^2}{1 + \omega^{2n}} \tag{D–4}$$

The s-plane poles of $|T(j\omega)^2|$ occur when the denominator of Eq. (D–4) is zero. The gain response involves setting $s = j\omega$. Solving for ω gives $\omega = s/j$, which means the poles of $|T(j\omega)|^2$ occur when $1 + (s/j)^{2n} = 0$. Combining this result with the denominator of Eq. (D–3) gives us

$$q_n(s)q_n(-s) = 1 + (s/j)^{2n} = 0 \tag{D–5}$$

Thus, the poles of a normalized ($\omega_C = 1$) Butterworth low-pass filter are the left half plane roots of Eq. (D–5).

In what follows, we find the left half plane roots of this equation for consecutive values of the integer n. We do not describe the root-finding process itself since the widespread availability of math solvers like Mathcad or MATLAB makes polynomial root finding a routine computation.

When $n = 1$, Eq. (D–5) yields the polynomial $1 + (s/j)^2 = 1 - s^2 = 0$, whose left half plane root is $s = -1$. When $n = 1$, the poles of $T(s)$ are defined by

$$q_1(s) = s + 1$$

For $n = 2$, Eq. (D–5) yields the polynomial $1 + (s/j)^4 = 1 + s^4 = 0$, whose left half plane roots are $s = -0.707 \pm j0.707$. When $n = 2$, the poles of $T(s)$ are defined by

$$\begin{aligned} q_2(s) &= (s + 0.707)^2 + (0.707)^2 \\ &= s^2 + 1.414s + 1 \end{aligned}$$

For $n = 3$ we get $1 + (s/j)^6 = 1 - s^6 = 0$, whose left half plane roots are $s = -1$ and $s = -0.5 \pm j0.866$, and the poles of $T(s)$ are defined by

$$\begin{aligned} q_3(s) &= (s + 1)[(s + 0.5)^2 + (0.866)^2] \\ &= (s + 1)[s^2 + s + 1] \end{aligned}$$

For $n = 4$ we get $1 + (s/j)^8 = 1 + s^8 = 0$, whose left half plane roots are $s = -0.9239 \pm j0.3827$ and $s = -0.3827 \pm j0.9239$. In this case the poles of $T(s)$ are defined by

$$\begin{aligned} q_4(s) &= [(s + 0.9239)^2 + (0.3827)^2][(s + 0.3827)^2 + (0.9239)^2] \\ &= [s^2 + 1.848s + 1][s^2 + 0.7654s + 1] \end{aligned}$$

These calculations illustrate a procedure for finding the poles of a normalized low-pass transfer function with Butterworth gain characteristics for any value of n. In general, when n is odd, there is one real pole and $n - 1$ pairs of complex conjugate poles. When n is even, the poles are all complex conjugates. The polynomials $q_1(s)$, $q_2(s)$, $q_3(s)$, and $q_4(s)$ derived above are the first four entries in Table 14–1. The entries for $n = 5$ and $n = 6$ are derived using the algorithm given above. That is, by finding the left half plane roots of Eq. (D–5) for $n = 5$ and $n = 6$.

The derivation of the poles of a Chebychev low-pass filter follows a process similar to the Butterworth method described above. We begin with the square of the normalized ($\omega_C = 1$) Chebychev gain response in Eq. (14–31).

$$|T(j\omega)|^2 = \frac{K^2}{1 + C_n^2(\omega)} \tag{D–6}$$

where $C_n(\omega) = \cos[n \cos^{-1}(\omega)]$ is an nth-order Chebychev polynomial. When $n = 0$, we have $C_0(\omega) = \cos[0] = 1$, and for $n = 1$ we have $C_1(\omega) = \cos[\cos^{-1}(\omega)] = \omega$. Higher-order polynomials are obtained using the recursion relationship $C_n(\omega) = 2\omega C_{n-1}(\omega) - C_{n-2}(\omega)$. Using these results, the next five Chebychev polynomials are found to be

$$\begin{aligned}
C_2(\omega) &= 2\omega C_1(\omega) - C_0(\omega) = 2\omega^2 - 1 \\
C_3(\omega) &= 2\omega C_2(\omega) - C_1(\omega) = 4\omega^2 - 3\omega \\
C_4(\omega) &= 2\omega C_3(\omega) - C_2(\omega) = 8\omega^4 - 8\omega^2 + 1 \\
C_5(\omega) &= 2\omega C_4(\omega) - C_3(\omega) = 16\omega^5 - 20\omega^3 + 5\omega \\
C_6(\omega) &= 2\omega C_5(\omega) - C_4(\omega) = 32\omega^6 - 48\omega^4 + 18\omega^2 + 1
\end{aligned} \tag{D–7}$$

As in the Butterworth procedure, our goal is to find the poles of $|T(j\omega)|^2$. According to Eq. (D–6), these poles occur when $1 + C_n^2(\omega) = 0$. As previously noted, the gain response involves setting $s = j\omega$. Solving for ω gives $\omega = s/j$, which means these s-plane poles occur when $1 + C_n^2(s/j) = 0$. Combining this result with the denominator of Eq. (D–3) produces

$$q_n(s)q_n(-s) = 1 + [C_n(s/j)]^2 = 0 \tag{D–8}$$

Thus, the poles of a normalized ($\omega_C = 1$) Chebychev low-pass filter are the left half plane roots of Eq. (D–8). We now show how to find these roots for successive values of the integer n.

For $n = 1$ we insert $C_1(\omega) = \omega$ into Eq. (D–8) to get the polynomial

$$1 + [C_1(s/j)]^2 = 1 + [s/j]^2 = 1 - s^2 = 0$$

whose left half plane root is $s = -1$. When $n = 1$, the poles of $T(s)$ are defined by

$$q_1(s) = s + 1$$

For $n = 2$ we insert $C_2(\omega)$ from Eq. (D–7) into Eq. (D–8) to get the polynomial

$$\begin{aligned}
1 + [C_2(s/j)]^2 &= 1 + [2(s/j)^2 - 1]^2 \\
&= 4s^4 + 4s^2 + 2 = 0
\end{aligned}$$

whose left half plane roots are $s = -0.3218 \pm j0.7769$. When $n = 2$, the poles of $T(s)$ are defined by

$$\begin{aligned}
q_2(s) &= (s + 0.3218)^2 + (0.7769)^2 \\
&= s^2 + 0.6436s + (0.8409)^2
\end{aligned}$$

For $n = 3$ we use $C_3(\omega)$ from Eq. (D–7) in Eq. (D–8) to produce the polynomial

$$\begin{aligned}
1 + [C_3(s/j)]^2 &= 1 + [4(s/j)^3 - 3(s/j)]^2 \\
&= 16s^6 - 24s^4 - 9s^2 + 1 = 0
\end{aligned}$$

whose left half plane roots are $s = -0.298$ and $s = -0.149 \pm j0.9037$. When $n = 3$, the poles of $T(s)$ are defined by

$$q_3(s) = (s + 0.298)[(s + 0.149)^2 + (0.9037)^2]$$
$$= (s + 0.298)[s^2 + 0.298s + (0.9159)^2]$$

For $n = 4$ we use $C_4(\omega)$ from Eq. (D–7) in Eq. (D–8) to get the polynomial

$$1 + [C_4(s/j)]^2 = 1 + [8(s/j)^4 - 8(s/j)^2 + 1]^2$$
$$= 64s^8 + 128s^6 + 80s^4 + 16s^2 + 2 = 0$$

whose left half plane roots are $s = -0.2052 \pm j0.392$ and $s = -0.08501 \pm j0.9464$. When $n = 4$, the poles of $T(s)$ are defined by

$$q_4(s) = [(s + 0.2052)^2 + (0.392)^2][(s + 0.08501)^2 + (0.9464)^2]$$
$$= [s^2 + 0.4104s + (0.4425)^2][s^2 + 0.17s + (0.9502)^2]$$

These calculations illustrate a general procedure for finding the poles of a normalized low-pass transfer function with Chebychev gain characteristics for any value of n. The algorithm involves finding the left half plane roots of Eq. (D–8) using the appropriate $C_n(\omega)$ from Eq. (D–7), or the recursion formula. When n is odd, there is one real pole and $n - 1$ complex conjugate poles. When n is even, the poles are all complex conjugates.

The polynomials $q_1(s)$, $q_2(s)$, $q_3(s)$, and $q_4(s)$ derived above produce the same poles as the corresponding entries in Table 14–2 but do not have the property $q_n(0) = 1$. To obtain this property, we factor out the constant term (coefficient of s^0) from each of the first- and second-order factors in $q_n(s)$.

ANSWERS

TO SELECTED PROBLEMS

CHAPTER ONE

1–1 (a) 52 kW; (b) 925 kHz; (c) 20 μs; (d) 2.5 kV

1–3 11.88 MC

1–5 (a) 10^{-6}; (b) 10^{3}; (c) 10^{-3}; (d) 10^{6}

1–7 3 mA

1–9 $i(0) = 0$ A; $i(1) = 6$ A; $i(3) = 18$ A

1–11 $q(10) = 75$ C

1–13 (a) 16.7 A; (b) 6 hr

1–15 $q(t) = 1.5(1 - e^{-2t})$ C for $t > 0$

1–17 at $v = -2$ V, $p = 19.7$ W, absorbing
at $v = 2$ V, $p = -5.22$ W, delivering
at $v = 3$ V, $p = 30.3$ W, absorbing

1–19 $i_{max} = 33.3$ mA

1–21 (a) $p = -72.6$ W, transfer from B to A
(b) $p = 14.4$ mW, transfer from A to B
(c) $p = 1.5$ W, transfer from A to B
(d) $p = 645$ mW, transfer from A to B

CHAPTER TWO

2–1 $v = 60$ V

2–3 $G = 0.25$ mS

2–5 $R_x = 5$ kΩ, $p = 0.5$ W

2–7 $v_{max} = 50$ V

2–11 (a) nodes A, B, C; loops 1, 2;2, 3, 4;1, 3, 4
(b) series 3 & 4; parallel 1 & 2
(c) KCL: node A $-i_1 - i_2 - i_3 = 0$;
node B $i_3 - i_4 = 0$;
node C $i_1 + i_2 + i_4 = 0$
KVL: loop 1, 2 $-v_1 + v_2 = 0$;
loop 2, 3, 4 $-v_2 + v_3 + v_4 = 0$;
loop 1, 3, 4 $-v_1 + v_3 + v_4 = 0$

2–13 (a) nodes A, B, C, D;
loops 1, 3, 2; 2, 4, 5; 3, 6, 4; 1, 5, 6; 2, 3, 6, 5
(b) series none; parallel none
(c) KCL: node A $-i_2 - i_3 - i_4 = 0$;
node B $-i_1 + i_3 - i_6 = 0$;
node C $i_1 + i_2 + i_5 = 0$;
node D $i_4 - i_5 + i_6 = 0$
KVL: loop 1, 3, 2 $-v_1 + v_2 - v_3 = 0$;
loop 2, 4, 5, $-v_2 + v_4 + v_5 = 0$;
loop 3, 6, 4 $v_3 - v_4 + v_6 = 0$

2–16 $v_1 = 20$ V; $v_5 = 0$ V

2–17 $i_1 = 11$ A; $i_5 = -9$ A

2–21 $v_x = 44$ V; $i_x = 4$ A

2–23 $v_x = 500$ V

2–27 $R_{EQ} = 150\ \Omega$

2–29 open, $R_{EQ} = 190\ \Omega$; closed, $R_{EQ} = 150\ \Omega$

2–31 $R_{AB} = 90\ \Omega$; $R_{AC} = 60\ \Omega$; $R_{AD} = 40\ \Omega$; $R_{BC} = 130\ \Omega$; $R_{BD} = 110\ \Omega$; $R_{CD} = 40\ \Omega$

2–33 $R_{EQ} = 15$ kΩ is not possible

2–35 $R_S = 50\ \Omega$, $i_S = 100$ mA

2–41 $v_L = \dfrac{R_L v_S}{R + 2R_L}$

2–43 $i_x = 0.5$ A

2–45 $0 \leq v_O \leq 4$ V

2–47 $R_1 = 40/19\ \Omega$; $R_2 = 10/19\ \Omega$

2–51 $i_x = i_S/4$; $v_x = 3Ri_S/4$

2–53 $v_x = 2.273$ V

2–55 $i_x = 50$ mA

CHAPTER THREE

3–1 (b) $v_x = 20$ V; $i_x = 0.333$ A

3–3 (b) $v_x = -20$ V; $i_x = -6.67$ mA

3–5 (b) $v_x = -4$ V; $i_x = -0.8$ mA

3–7 (b) $v_A = v_B = 4$ V; $v_C = 0$

3–9 (b) $v_x = -3$ V; $i_x = 1$ mA

3–11 (b) $v_x = 2.5$ V; $i_x = 0.25$ A

3–13 (b) $v_x = 4$ V; $i_x = 3.35$ mA;
(c) power delivered by S1 is -40.2 mW

3–15 (b) $v_x = 25$ V; $i_x = -6.25$ mA;
(c) total power delivered is 581 mW

3–21 $K = \dfrac{R_1R_3}{R_1 + R_2 + R_3}$

3–23 $K = \dfrac{R_1R_4 + R_3R_4}{R_1 + R_2 + R_3 + R_4}$

3–25 $i_O = 120$ mA

3–27 $i_O = 44$ mA

3–29 $i_O = 923$ μA

3–31 $v_O = R(i_1 - 2i_2)/4$

3–37 $R_T = 8\ \Omega$; $v_T = 16$ V;
$v_L = 6.15$ V for $R_L = 5\ \Omega$;
$v_L = 8.89$ V for $R_L = 10\ \Omega$;
$v_L = 13.8$ V for $R_L = 50\ \Omega$

3–39 $R_T = 60$ kΩ; $v_T = 4$ V;
$p_L = 66.1$ μW for $R_L = 50$ kΩ;
$p_L = 47.3$ μW for $R_L = 200$ kΩ

3–41 $v_x = 5$ V

3–43 $v = 10$ V

3–45 $R_L \geq 225\ \Omega$

3–47 $R_L = 12.5$ kΩ

3–49 $v = 2.58$ V; $i = 4.84$ mA

3–51 $R_L = 116.6\ \Omega$ or $R_L = 3.431\ \Omega$

3–53 $R_T = 30\ \Omega$; $v_T = 10$ V

3–55 $p_{MAX} = 1.125$ W for $R_L = 5$ kΩ

3–57 impossible since $v_L = 80 > v_{OC} = 75$ V

CHAPTER FOUR

4–1 $K_V = -6.4$; $K_I = -1.6$

4–3 $K_V = 12$; $K_I = 24$

4–5 $K_I = -500$

4–7 $v_O = 5$ V

4–9 $K_V = \dfrac{gR_SR_O + R_O}{gR_SR_O + R_O + R_S}$

4–11 $K_V = \dfrac{r}{R_S + r}$

4–13 $R_{IN} = R(\beta + 1)$

4–15 $i_N = i_S$; $R_N = R_O/(1 + gR_O)$

4–16 for $v_S = 2$ V; $i_C = 2.6$ mA; $v_{CE} = 7.2$ V;
for $v_S = 5$ V; $i_C = 5$ mA; $v_{CE} = 0$ V

4–18 for $v_S = 0.8$ V; $i_C = 0.5$ mA; $v_{CE} = 3.75$ V;
for $v_S = 2$ V; $i_C = 1$ mA; $v_{CE} = 0$ V

4–19 $i_C = 1.41$ mA; $v_{CE} = 0.794$ V

4–21 $v_O = -4.5\, v_S$

4–23 (a) $v_O = 4\, v_S$; (b) $i_O = 0.6$ mA

4–25 pos. no. 1, $K_V = 10$; pos. no. 2, $K_V = 5$; pos. no. 3, $K_V = 1$

4–27 (a) $R_1 = 50\,k\Omega$; $R_2 = 25\,k\Omega$; $R_3 = 10\,k\Omega$;
(b) $-11 \leq v_1 \leq 19$ V

4–29 $v_O = -2v_{S1} + 0v_{S2}$

4–32 $v_O = v_S - Ri_S$

4–35 $v_O = v_{S1} + (1 + R_1/R_2)(v_{S3} - v_{S2})$

4–37 $v_O = 3\, v_1 + 30$

4–39 $v_O = -0.3\, v_S$

4–51 full-scale output $v_O = 7.75$ V; LSB change is 250 mV

4–52 for (0, 1, 0, 1, 0, 1), $v_O = 3.5$ V; LSB change is 167 mV

4–57 for $v_S = 3$ V, $v_O = 0$ V;
for $v_S = -3$ V, $v_O = 0$ V;
for $v_S = 6$ V, $v_O = 15$ V

4–60 for $V_{REF} = 6$ V;
$v_S = 3.5$ V $\rightarrow$ (0, 0, 1, 1, 1);
$v_S = 2.3$ V $\rightarrow$ (0, 0, 0, 1, 1);
$v_S = 5.3$ V $\rightarrow$ (1, 1, 1, 1, 1)

CHAPTER FIVE

5–4 (a) $v_1(t) = 2u(1 - t) - 5u(t - 1) + 5u(t - 2)$;
(b) $v_2(t) = -4r(t) + 6r(t - 2) - 2r(t - 6)$

5–5 (a) $v_1(t) = 5u(t + 1) - 10u(t - 1) + 5u(t - 3)$ V;
(b) $v_2(t) = 5u(t) - 2.5r(t) + 2.5r(t - 2)$ V

5–6 (a) $dv_1/dt = 5\delta(t + 1) - 10\delta(t - 1) + 5\delta(t - 2)$ V/s;
(b) $dv_2/dt = 5\delta(t) - 2.5u(t) + 2.5u(t - 2)$ V/s

5–7 $v(t) = 5r(t) - 5r(t - 2) - 10u(t - 4)$ V

5–9 $-10e^{-20t}u(t)$ V/s; $25(1 - e^{-20t})$ mV-s

5–10 $T_C = 7.21$ ms

5–11 2.45 V

5–14 (a) $T_0 = 4$ ms; $f_0 = 250$ Hz; $V_A = 14.1$ V; $T_S = -0.5$ ms; $\phi = -0.785$ rad;
(b) $T_0 = 1$ ms; $f_0 = 1$ kHz; $V_A = 36.1$ V; $T_S = -0.594$ ms; $\phi = 3.73$ rad

5–16 $v(t) = 20\cos(1000\pi t - 90°)$ V; $\phi = -90°$; $T_S = 0.5$ ms

5–17 (a) $a = -20$ V; $b = 0$; $f = 250$ Hz; $\omega = 500\pi$ rad/s;
(b) $a = 0$; $b = 10$ V; $f = 250$ Hz; $\omega = 500\pi$ rad/s

5–18 $v_1(t) + v_2(t) = -20\cos(500\pi t) + 10\sin(500\pi t)$V

5–22 (a) $v_{max} = 30$ V; $v_{min} = 10$ V;
(b) $v_{max} = 17.9$ V; $v_{min} = -10$ V

5–23 $v_{max} = 1.47$ V at $t = 200$ ms

5–24 $V_A = 12$ V; $V_B = 7$ V; $\alpha = 112\ s^{-1}$

5–25 $v(t) = -1.5 - 3.5\cos(1000\pi t)$ V

5–26 $v_{max} = 16$ V; $v_{min} = -4$ V; $\beta = 1.571$ krad/s

5–28 $v_{max} = 16.05$ V at $t = 20.1$ ms

5–31 (a) $V_p = 14.1$ V; $V_{pp} = 28.2$ V; $V_{avg} = 0$ V; $V_{rms} = 10$ V;
(b) $V_p = 36.1$ V; $V_{pp} = 72.1$ V; $V_{avg} = 0$ V; $V_{rms} = 25.5$ V

5–33 $V_p = 3$ V; $V_{pp} = 3$ V; $V_{avg} = 1.5$ V; $V_{rms} = 1.87$ V

5–35 $V_p = 2$ V; $V_{pp} = 4$ V; $V_{avg} = 0$; $V_{rms} = 1.155$ V

5–37 $V_{avg} = 2V_A/\pi$; $V_{rms} = V_A/\sqrt{2}$

CHAPTER SIX

6–1 $i_C(t) = -24e^{-2000t}u(t)$ mA; $p_C(t) = -72\, e^{-4000t}\, u(t)$ mW

6–3 $v_C(t) = 10 - 20\cos(2000t)$ V

6–5 $C = 1$ nF

6–7 0 V; 5 V; 5 V

6–9 $p_L(t) = 90(t - 2)$ W; $w_L(t) = 45(t - 2)^2$ J; both

6–11 $i_L(t) = 100\sin(1000t) + 13.3\cos(3000t) - 13.3$ mA

6–13 $v_L(t) = (0.2 - 200t)e^{-1000t}$ V; both

6–15 $i_C(t) = -0.8e^{-4000t}$ A; $p_C(t) = 8e^{-4000t} - 16e^{-8000t}$ W; both

6–17 $i_L(t) = 40e^{-1000t} - 20$ mA;
$p_L(t) = -0.32e^{-2000t} + 0.16e^{-1000t}$ mW; both

6–21 $v_O(t) = 10e^{-500t}$ V

6–23 13.5 s

6–25 $v_O(t) = -2.5e^{-50t}$ V

6–27 $V_A \leq 15$ V

6–29 $v_O(t) = 2v_S(t) + RC\dfrac{dv_S(t)}{dt}$

6–36 $C_{EQ} = 3.5\ \mu$F; $L_{EQ} = 1.049$ mH

6–38 $L_{EQ} = 2.4\,\text{mH}$

6–39 $C_{EQ} = 2\,\mu\text{F}$

6–40 $C_{EQ} = 3\,\mu\text{F}$; $L_{EQ} = 2\,\text{H}$; connected in parallel

6–41 (a) 41.7 nF; (b) 75 nF

6–42 four strings of three 20-μF capacitors

CHAPTER SEVEN

7–1 $y(t) = 10e^{-5t}\ t \geq 0$

7–3 C1: $T_C = 0.3$ ms; C2: $T_C = 0.2$ ms

7–5 $v_C(t) = 30[e^{-4000t}]u(t)$ V; $i_O(t) = 2[e^{-4000t}]u(t)$mA

7–7 $v_C(t) = 5[e^{-250t}]u(t)$ V

7–9 $v_O(t) = I_A R\,[1 - e^{-t/RC}]u(t)$

7–11 $v_O(t) = 0.5V_A[e^{-t/2RC}]u(t)$

7–13 $v_C(t) = [6 - 3e^{-200t}]u(t)$ V

7–15 $v(t) = [-0.5e^{-10t} + 0.5\cos(20t) + \sin(20t)]u(t)$ V

7–17 $i_L(t) = [-0.45e^{-15t} + 0.45\cos(5t) + 0.15\sin(5t)]u(t)$ A

7–19 $v_C(t) = [5 + 4e^{-200t}]u(t)$ V

7–23 (a) $C = 100$ nF; $i_C(t) = -e^{-1000t}u(t)$ mA;
(b) $w_C(t) = 91.6$ nJ at $t = 0.2$ ms

7–25 (a) $v_S = 5$ V; $R = 500\ \Omega$; $i_C(t) = 20e^{-2000t}\ u(t)$ mA;
(b) $w_C(t) = 0$ at $t = \ln(2)/2$ ms

7–27 (a) $v_S = 0.5$ V; $R = 100\ \Omega$; $L = 50$ mH;
(b) $w_L(t) = 0$ at $t = \ln(2)/2$ ms

7–31 $v(t) = 10[te^{-5t}]u(t)$ V

7–33 $v(t) = [2 + 3e^{-5t}\cos(10t) + 4e^{-5t}\sin(10t)]u(t)$ V

7–35 $v_C(t) = [20e^{-1000t} - 10e^{-2000t}]u(t)$ V;
$i_L(t) = [10e^{-1000t} - 10e^{-2000t}]u(t)$ mA; overdamped

7–37 $v_C(t) = e^{-1000t}[60\cos(3000t) - 20\sin(3000t)]u(t)$ V;
$i_L(t) = 50e^{-1000t}\sin(3000t)u(t)$ mA; underdamped

7–39 $v_C(t) = [20e^{-500t} - 5e^{-2000t}]u(t)$ V;
$i_L(t) = 12.5[-e^{-500t} + e^{-2000t}]u(t)$ mA; overdamped

7–41 $v_C(t) = 10^5\,te^{-5000t}u(t)$ V;
$i_L(t) = [5 - 5e^{-5000t} - 25000te^{-5000t}]u(t)$ mA; critically damped

7–43 $v_C(t) = -8e^{-100t}\sin(700t)u(t)$ V;
$i_L(t) = [-6 + 7e^{-100t}\cos(700t) + e^{-100t}\sin(700t)]u(t)$ mA;
underdamped

7–45 $v_C(t) = [5 - 5e^{-2000t} - 10^4te^{-2000t}]u(t)$V;
$i_L(t) = [20te^{-2000t}]u(t)$ A; critically damped

7–47 $v_O(t) = 20e^{-400t}\sin(2800t)u(t)$ V

7–49 $\omega_0 = \frac{1}{\sqrt{LC}}$; $\zeta = \frac{R}{4}\sqrt{\frac{C}{L}}$

7–51 (a) $R = 200\ \Omega$; $L = 2\ \text{H}$;
(b) $i_L(t) = 20e^{-50t}\sin(350t)u(t)$ mA

7–53 $R = 1\ \text{k}\Omega$; $L = 2\ \text{H}$; $C = 5\ \mu\text{F}$

7–55 $v_C(t) = -10[e^{-10t}\cos(20t) + 2e^{-10t}\sin(20t)]u(t)$ V

CHAPTER EIGHT

8–1 $\mathbf{V}_1 = 125 + j216$ V; $\mathbf{V}_2 = 100 - j150$ V;
$v_1(t) + v_2(t) = 234.6\cos(\omega t + 16.47°)$ V

8–3 (a) $v_1(t) = 10\cos(10^4 t - 30°)$ V;
(b) $v_2(t) = 60\cos(10^4 t - 220°)$ V;
(c) $i_1(t) = 5\cos(200t + 90°)$ A;
(d) $i_2(t) = 2\cos(200t + 270°)$ A

8–5 $451\cos(25t + 124°)$ V/s

8–7 $v_2(t) = 320\cos(250t + 141.3°)$ V

8–9 $v_1(t) = 0.928\cos(100t - 21.8°)$ V

8–11 $Z = 20 + j0 = 20\angle 0°\ \Omega$

8–13 $Z = 54 - j20 = 57.6\angle -20.3°\ \Omega$

8–15 $Z = 675 - j175\ \Omega$

8–17 $Z = 29.4 - j2.35\ \Omega$

8–19 $R = 833.3\ \Omega$; $C = 1.6$ nF

8–21 $i(t) = 0.354\cos(2000t - 45°)$ A

8–23 $i_R(t) = 212\cos(2500t - 45°)$ mA;
$i_C(t) = 212\cos(2500t + 45°)$ mA

8–25 $i_R(t) = 10\cos(2000t - 90°)$ mA;
$i_C(t) = 10\cos(2000t)$ mA

8–27 $v_x(t) = 126.5\cos(2500t + 18.4°)$ V

8–29 $\mathbf{I}_x = 0.25 - j0.75$ A

8–31 $v_x(t) = 17.9\cos(1000t - 26.6°) + 5.66\cos(2000t - 135°)$ V

8–33 $\mathbf{V}_x = -j19.2$ V

8–35 $\mathbf{V} = 44.7\angle -58.8°$ V; $\mathbf{I} = 0.2\angle -122°$ A

8–41 $v_x(t) = 76.5\cos(4000t - 82.2°)$ V

8–43 $\mathbf{V}_A = 141 - j35.3$ V; $\mathbf{V}_B = -23.5 + j106$ V

8–45 $\mathbf{I}_A = -138 + j140$ mA; $\mathbf{I}_B = 245 - j119$ mA

8–47 $\mathbf{I}_{OUT} = 3.45 + j0.431$ A

8–49 $\mathbf{I}_{IN} = 18.1 + j13.1$ mA; $\mathbf{V}_O = -39.0 + j6.24$ V

8–51 $\mathbf{V}_S = 166 + j138$ V

8–53 $K = 0.847 - j0.0281$; $Z_{IN} = 21.3 - j9.71$ kΩ

8–57 $P_L = 31.3$ mW

8–59 (a) $P_L = 144$ mW; (b) $P_{MAX} = 450$ mW; (c) $Z_L = 1 - j2$ kΩ

CHAPTER NINE

9–1 $F(s) = \dfrac{-A(s + 2\alpha - \gamma)}{(s + \gamma)(s + \alpha)}$

9–3 $F(s) = \dfrac{-A(s + \beta)(s - \beta)}{s(s^2 + \beta^2)}$

9–5 $F(s) = \dfrac{As^2}{(s + \beta)^2 + \beta^2}$

9–7 (a) $F(s) = \dfrac{-5(s - 10)}{(s + 5)(s + 20)}$; (b) $F(s) = \dfrac{12s(s^2 + 15^2)}{(s^2 + 10^2)(s^2 + 20^2)}$

9–9 (a) $F(s) = \dfrac{(s + 25)(s + 75)}{(s + 50)^2}$; (b) $F(s) = \dfrac{1000(s + 10)}{s(s + 20)(s^2 + 10^2)}$

9–17 (a) $f_1(t) = [1 + e^{-10t}]u(t)$;
(b) $f_2(t) = \delta(t) + [5e^{-5t} - 20e^{-10t}]u(t)$

9–19 (a) $f_1(t) = [1 - \cos(\beta t) + \sin(\beta t)]u(t)$;
(b) $f_2(t) = \delta(t) + [\beta\cos(\beta t) - \beta\sin(\beta t)]u(t)$

9–21 (a) $f_1(t) = [2e^{-\alpha t} - 3e^{-2\alpha t} + e^{-4\alpha t}]u(t)$;
(b) $f_2(t) = [2e^{-\alpha t} + 4e^{-4\alpha t}]u(t)$

9–23 (a) $f_1(t) = \left[\dfrac{8}{3} - \dfrac{3}{2}e^{-2t} - \dfrac{1}{6}e^{-6t}\right]u(t)$;
(b) $f_2(t) = [1 + \cos(t) + \cos(2t)]u(t)$

9–25 (a) $f_1(t) = [1 + e^{-4t}\cos(4t) - 3e^{-4t}\sin(4t)]u(t)$;
(b) $f_2(t) = [2 + e^{-40t} - 2e^{-10t}]u(t)$

9–31 (a) $y(t) = [10e^{-5t}]u(t)$;
(b) $y(t) = [2 - 12e^{-20t}]u(t)$

9–33 (a) $2 \times 10^{-3}\dfrac{di_L(t)}{dt} + i_L(t) = 0.05u(t) \quad i_L(0) = 0$;
(b) $i_L(t) = 50[1 - e^{-500t}]u(t)$ mA

9–35 (a) $\dfrac{1}{5000}\dfrac{dv_C(t)}{dt} + v_C(t) = 12e^{-2500t}u(t) \quad v_C(0) = 0$;
(b) $v_O(t) = 15[e^{-2500t} - e^{-5000t}]u(t)$ V

9–37 $y(t) = [20\,e^{-5t} - 10\,e^{-10t}]u(t)$

9–39 (a) $6.25 \times 10^{-6}\dfrac{d^2 i_L(t)}{dt^2} + 5 \times 10^{-3}\dfrac{di_L(t)}{dt} + i_L(t) = 0$

$i_L(0) = 20$ mA; $\dfrac{di_L(0)}{dt} = 0$

(b) $i_L(t) = 20[(1 + 400t)e^{-400t}]u(t)$ mA

9–41 (a) $12.5 \times 10^{-6} \dfrac{d^2 v_C(t)}{dt^2} + 7.5 \times 10^{-3} \dfrac{dv_C(t)}{dt} + v_C(t) = 20u(t)$

$v_C(0) = 0; \dfrac{dv_C(0)}{dt} = 0$

(b) $v_C(t) = [20 - 40e^{-200t} + 20e^{-400t}]u(t)$ V

9–45 (a) $f_1(0) = 0$; $f_1(\infty) = 0$
(b) $f_2(0) = 2$; $F_2(s)$ has j-axis poles

9–47 (a) $f_1(0) = 2$; $f_1(\infty) = 0$;
(b) $f_2(0) = 10$; $F_2(s)$ has right half plane poles

9–49 (a) $f_1(0) = 2$; $f_1(\infty) = 1$;
(b) $f_2(0) = 1$; $f_2(\infty) = 2$

CHAPTER TEN

10–1 $Z_{EQ}(s) = 2R\left[\dfrac{RCs + 1}{3RCs + 1}\right]$;
pole at $s = -1/3RC$;
zero at $s = -1/RC$

10–3 $Z_{EQ}(s) = \dfrac{4000(s + j5000)(s - j5000)}{(s + 5000)^2}$;
double pole at $s = -5$ krad/s;
zeros at $s = \pm j5$ krad/s

10–5 $Z_{EQ}(s) = \dfrac{Ls(2Ls + 3R)}{Ls + R}$;
pole at $s = -R/L$;
zeros at $s = -3R/2L$ and $s = 0$

10–7 $Z_{EQ}(s) = \dfrac{(2RCs + 1)(RCs + 2)}{Cs(2RCs + 3)}$;
poles at $s = -3/2RC$ and $s = 0$;
zeros at $s = -1/2RC$ and $s = -2/RC$

10–11 $I_L(s) = \dfrac{V_A}{R} \dfrac{1}{s + R/L}$; $i_L(t) = \dfrac{V_A}{R}[e^{-Rt/L}]u(t)$

10–13 $I_C(s) = \dfrac{-2V_A}{R} \dfrac{1}{s + 1/RC}$; $i_C(t) = -\dfrac{2V_A}{R}[e^{-t/RC}]u(t)$

10–15 $I_L(s) = \dfrac{V_A}{s(2Ls + R)}$; $i(t) = \dfrac{V_A}{R}(1 - e^{-Rt/2L})u(t)$

10–17 (a) $I_L(s) = \dfrac{-V_A C}{LCs^2 + R_2Cs + 1}$; (b) $i_L(t) = -[60te^{-1000t}]u(t)$ A

10–19 $I_R(s) = \left[\dfrac{LCs^2 + 1}{LCs^2 + RCs + 1}\right] I_1(s)$

10–21 $V_{Czs}(s) = \dfrac{LsV_S(s)}{RLCs^2 + Ls + R}$; $V_{Czi}(s) = \dfrac{RLCsV_0}{RLCs^2 + Ls + R}$

10–23 $v_O(t) = [-10e^{-400t} + 15e^{-100t}]u(t)$ V;
forced pole at $s = -100$ rad/s; natural pole at $s = -400$ rad/s

10–25 $I(s) = \left[\frac{1}{2RCs + 1}\right]I_S(s)$

10–27 $Z_T(s) = \frac{2000(s + 1000)}{3s + 1000}$ Ω; $V_T(s) = \frac{100}{3s + 1000}$ V-s

10–31 (b) $I_2(s) = \frac{R_1CsV_1(s)}{(R_1 + R_2)LCs^2 + (L + R_1R_2C)s + R_1}$;
(c) $i_2(t) = [-50e^{-500t} + 50e^{-1000t/3}]u(t)$ mA

10–33 (b) $V_2(s) = \frac{R_1R_2C_1C_2s^2V_1(s)}{R_1R_2C_1C_2s^2 + (R_1C_1 + R_2C_2 + R_1C_2)s + 1}$;
(c) $v_2(t) = [40e^{-200t} - 10e^{-50t}]u(t)$ V

10–35 (b) $V_2(s) = \left[\frac{R_1R_2C_1C_2s^2 + (R_1C_1 + R_1C_2)s + 1}{R_1R_2C_1C_2s^2 + (R_1C_1 + R_2C_2 + R_1C_2)s + 1}\right]V_1(s)$

10–37 $i_A(t) = 33.3[e^{-500t} + 2e^{-2000t}]u(t)$ mA
$i_B(t) = 66.7[e^{-500t} - e^{-2000t}]u(t)$ mA

10–39 (a) $\Delta(s) = [(RCs)^2 + (3 - \mu)RCs + 1]/R^2$;
(b) $\mu = 1.586$; $RC = 2 \times 10^{-4}$

10–41 $\Delta(s) = \left[s^2 + \left(\frac{1}{RC} + \frac{R}{L}\right)s + \frac{2}{LC}\right]\frac{C}{s}$;
$L = 0.4$ H; $C = 200$ nF; $R = 1000\sqrt{2}$ Ω

10–43 (a) mesh A and B are dependent: $I_B(s) = \frac{\beta}{\beta - 1}I_A(s)$

(b) mesh A: $\left(R_1 + \frac{Ls}{1 - \beta}\right)I_A - R_1I_C = V_S$;

mesh C: $-\left(R_1 + \frac{\beta}{\beta - 1}R_2\right)I_A + \left(R_1 + R_2 + \frac{1}{Cs}\right)I_C = 0$

10–45 $V_O(s) = \frac{4000}{s^2 + 3000s + 8 \times 10^5}$;
$v_O(t) = 1.661(e^{-295.8t} - e^{-2704t})u(t)$ V

CHAPTER ELEVEN

11–1 $Z_{IN}(s) = \frac{(R_1 + R_2)Cs + 1}{Cs}$; $T_V(s) = \frac{R_2Cs + 1}{(R_1 + R_2)Cs + 1}$

11–3 $Z_{IN}(s) = \frac{R_1}{R_1Cs + 1}$; $T_V(s) = -\frac{R_2}{R_1}(R_1Cs + 1)$

11–5 $T_V(s) = \frac{R}{2RLCs^2 + (L + R^2C)s + R}$

11–7 $T_V(s) = \dfrac{RCs + 2}{RCs + 1}$; any combination that produces $RC = 1/500$ s

11–9 $T_V(s) = \dfrac{10^5}{(s + 100)(s + 1000)}$

11–11 $v_2(t) = [500e^{-2000t}]u(t)$ V

11–13 $v_2(t) = 0.5[1 - \sqrt{2}e^{-500t}\cos(500t - 45°)]$ V

11–15 $v_2(t) = \delta(t) + [10e^{-50t}]u(t)$ V

11–17 $g(t) = [-1 + 2e^{-500t}]u(t)$

11–19 $h(t) = [-150e^{-10t} + 450e^{-30t}]u(t)$

11–21 $y(t) = 5[t - 10^{-3}(1 - e^{-1000t})]u(t)$

11–26 $v_{2SS}(t) = 4.77\cos(300t - 17.44°)$ V;
$v_{2SS}(t) = 8.07\cos(600t - 36.19°)$ V

11–28 $i_{2SS}(t) = 1.99\cos(500t + 174°)$ mA;
$i_{2SS}(t) = 1.414\cos(5000t + 135°)$ mA

11–30 $v_{2SS}(t) = 35.4\cos(5000t + 45°)$ V;
$v_{2SS}(t) = 11.2\cos(2500t + 63.4°)$ V

11–31 $y_{SS}(t) = 6.86\cos(200t - 30.96°)$

11–32 $y_{SS}(t) = 9.70\cos(2000t + 14.04°)$

11–34 $y_{SS}(t) = 1.246\cos(400t + 131.6°)$

11–38 $y(t) = \begin{cases} 0 & t \leq 1 \\ t - 1 & 1 < t \leq 2 \\ 1 & 2 < t \end{cases}$

11–40 $y(t) = \begin{cases} 0 & t \leq 1 \\ 0.5(t^2 - 2t + 1) & 1 < t \leq 2 \\ 0.5 & 2 < t \end{cases}$

11–41 $y(t) = [1 - e^{-(t-2)}]u(t - 2)$

11–42 $y(t) = [te^{-t}]u(t)$

11–43 $y(t) = [t - 1 + e^{-t})]u(t)$

11–49 $y(t) = [5t - 1 + e^{-5t})]u(t)$

CHAPTER TWELVE

12–1 $T_V(s) = \dfrac{500}{s + 1000}$
(a) $|T_V(0)| = 0.5$; $|T_V(\infty)| = 0$; $\omega_C = 1$ krad/s; low pass;
(c) $|T_V(0.5\omega_C)| = 0.447$; $|T_V(\omega_C)| = 0.354$; $|T_V(2\omega_C)| = 0.224$

12–3 $T_V(s) = \dfrac{800}{s + 200}$
(a) $|T_V(0)| = 4$; $|T_V(\infty)| = 0$; $\omega_C = 200$ rad/s; low pass;
(c) increase 30-kΩ resistance to 90 kΩ

12–5 $T_V(s) = \frac{2s}{s + 500}$

(a) $|T_V(0)| = 0$; $|T_V(\infty)| = 2$; $\omega_C = 500$ rad/s; high pass;
(b) decrease 100-nF capacitance to 10 nF

12–7 13.9 dB at $\omega = 200$ rad/s; 17.5 dB at $\omega = 400$ rad/s; 19.2 dB at $\omega = 800$ rad/s

12–9 (a) high pass; $\omega_C = 2$ krad/s; $K = 200$;
(b) $T_{SL}(1000) = 100$; $T_{SL}(2000) = T_{SL}(4000) = 200$

12–14 $\omega_{C1} = 1/R_2C_2$; $\omega_{C2} = 1/R_1C_1$

12–15 $\omega_{C1} = 1/R_3C_2$; $\omega_{C2} = 1/R_2C_1$

12–16 $\omega_{C1} = 1/R_2C$; $\omega_{C2} = R_1/L$

12–17 $T(s) = \left(\frac{s}{s + 100}\right)\left(\frac{2000\sqrt{10}}{s + 2000}\right)$

12–19 $T(s) = \left(\frac{500}{s + 500}\right)\left(\frac{10 \times 2000}{s + 2000}\right)$

12–21 $L = 6\ \mu$H; $C = 104$ pF; $\omega_{C1} = 38.05$ Mrad/s; $\omega_{C2} = 42.05$ Mrad/s

12–23 (a) $C = 2\ \mu$F; (b) $Q = 1.25$; $B = 4$ krad/s

12–25 $L = 3.18\ \mu$H; $C = 7.96$ nF

12–27 $L = 1.65$ H; $C = 4.26\ \mu$F

12–28 $B = 1$ Mrad/s; $Q = 1$; $\omega_{C1} = 0.618$ Mrad/s; $\omega_{C2} = 1.618$ Mrad/s

12–31 $T_V(s) = (s + 10)/(s + 110)$; $A_{SL} = 2.73$ V; $A_{ACTUAL} = 2.77$ V

12–32 $T_V(s) = (s + 110)/(s + 10)$; $A_{SL} = 36.7$ V; $A_{ACTUAL} = 36.1$ V

12–33 low pass; $\omega_C = 0.4$ rad/s; passband gain = 0 dB

12–35 high pass; $\omega_C = 100$ rad/s; passband gain = 28 dB

12–37 high pass; $\omega_C = 2.5$ rad/s; passband gain = 34 dB

12–39 $T(s) = 100(s + 200)/(s + 20)$

12–41 $T(s) = 250{,}000s/[(s + 100)(s + 50{,}000)]$

12–46 $T(s) = 300(s + 40)/[(s + 10)(s + 100)]$; low pass; $\omega_C = 10$ rad/s; passband gain = 22 dB

12–47 $T(s) = 3(s + 100)^2/[(s + 20)(s + 500)]$; bandstop; bandpass gain = 10 dB

12–48 $T(s) = 7800s/[(s + 50)(s + 2000)]$; bandpass; passband gain = 12 dB

12–49 $g(t) = 10 - 9e^{-20t} \quad t > 0$

CHAPTER THIRTEEN

13–1 (b) $a_n = 0$, $b_n = -V_A/n\pi$ for all n

13–3 $a_n = 0$, $b_n = 2V_A/n\pi$ for all n

13–5 $v(t) = 12.7\sin(2\pi\,500t) + 4.24\,\sin(2\pi\,1500t) + 2.55\cos(2\pi\,2500t) + 1.82\,\cos(2\pi\,3500t)$ V

13–7 $v(t) = 140 - 93.4\cos(2\pi 100t) - 18.7\cos(2\pi 200t) - 8.0\,\cos(2\pi\,300t) - 4.44\,\cos(2\pi\,400t)$

13–9 $f_0 = 1$ kHz; $g(t) = 10 + 4.05\,\cos(2\pi f_0 t) + 0.45\,\sin(2\pi\,3f_0\,t) + 0.162\,\sin(2\pi\,5f_0\,t)$

13–11 (a) shifted rectangular pulse with amplitude V_A and duration $T = T_0/3$

(b) $a_0 = \dfrac{V_A}{3}$; $a_n = (-1)^n \dfrac{2V_A}{n\pi}\sin\left(\dfrac{n\pi}{3}\right)$; $b_n = 0$ for all n

13–17 $v_O(t) = 14.5\cos(200t - 117°) + 0.999\cos(600t + 34°) + 0.241\cos(1000t - 168°) + 0.091\cos(1400t + 16°)$ V

13–19 $i(t) = 9\,\cos(250t - 225°) + 5.69\,\cos(500t - 243°) + 4.03\,\cos(750t - 252°) + 3.09\,\cos(1000t - 259°)\,\mu$A

13–21 fundamental reduced more than the harmonics by a high-pass circuit with $\omega_C = 100$ rad/s

$n\omega_0$	INPUT AMPLITUDE	OUTPUT AMPLITUDE
40	4.78	1.77
80	2.38	1.49
120	1.59	1.22
160	1.19	1.01

13–25 second harmonic at 400 rad/s dominates

$n\omega_0$	INPUT AMPLITUDE	OUTPUT AMPLITUDE
200	5	0.806
400	2.5	2.5
600	1.67	0.484
800	1.25	0.207

13–27 $P = 8.5$ W; $V_{rms} = 20.62$ V

13–29 $V_{rms} = V_A/\sqrt{6}$; $P = V_A^2/6R$; 37.5% by dc component, 62.5% by ac components

13–31 $Y_{rms} = 15.9$ V

13–33 first three terms give $V_{rms} = 1.165$ V, first six terms give $V_{rms} = 1.193$ V, estimate $V_{rms} = 1.2$ V.

13–35 23.1 mW

CHAPTER FOURTEEN

14–4 select R_1, then calculate $R_2 = 4\zeta^2 R_1$ and $C = 1/(\omega_0\sqrt{R_1R_2})$

14–5 select C_1, then calculate $C_2 = 2\zeta C_1$, $R_1 = 1/(\omega_0 C_1)$ and $R_2 = R_1/(2\zeta)$

14–6 select C_1, then calculate $C_2 = 4\zeta^2 C_1/9$ and $R = 1/(\omega_0\sqrt{C_1C_2})$

14–7 select R_2, then calculate $R_1 = 4\zeta^2 R_2$ and $C = 1/(\omega_0\sqrt{R_1R_2})$

14–11 $T(s) = \dfrac{2.5 \times 10^6}{s^2 + 1000s + 2.5 \times 10^5}$

14–13 $T(s) = \dfrac{1000s}{s^2 + 200s + 10^6}$

14–15 $T(s) = \dfrac{10(s^2 + 10^6)}{s^2 + 200s + 10^6}$

14–26 $T(s) = \dfrac{1}{(s/460 + 1)^4}$

14–27 $T(s) = \dfrac{1}{(s/200 + 1)[(s/200)^2 + s/200 + 1]}$

14–28 $T(s) = \dfrac{0.707}{(s/168.2)^2 + 0.7654(s/168.2) + 1}$

14–41 $T(s) = \dfrac{10s^2}{s^2 + 2828s + 2000^2}$

14–42 $T(s) = \dfrac{s^3}{(s + 2000)(s^2 + 2000s + 2000^2)}$

14–43 $T(s) = \dfrac{0.707s^2}{s^2 + 1820s + 2378^2}$

CHAPTER FIFTEEN

15–1 (a) $v_1(t) = 10 \times 10^{-3}\dfrac{di_1(t)}{dt} + 7 \times 10^{-3}\dfrac{di_2(t)}{dt}$

$v_2(t) = 7 \times 10^{-3}\dfrac{di_1(t)}{dt} + 5 \times 10^{-3}\dfrac{di_2(t)}{dt}$;

(b) $v_2(t) = 140\sin(100t)$ V

15–3 (a) see 15–1(a) above;
(b) $v_S(t) = -5\cos(1000t)$ V

15–5 $v_1(t) = 1.25\cos(1000t)$ V; $i_1(t) = 0.75\sin(1000t)$ A

15–7 $v_x(t) = -100\sin(1000t)$ V

15–9 $k = 0.894$

15–11 (a) $n = 1/5$; (b) $v_1(t) = 120\cos(2\pi 60t)$ V;
$i_1(t) = 6\cos(2\pi 60t)$ mA

15–13 $R_{EQ} = 3\ \Omega$

15–15 $i_S(t) = -8\sin(400t)$ A; $i_L(t) = 4\sin(400t)$ A

15–17 $n = 4.472$

15–19 $R_{IN} = 0.353\ \Omega$

15–21 $\mathbf{V}_2 = 65.3 + j25.2 = 70.0\angle 21.1°$ V; $Z_{IN} = 70.8 + j64.6\ \Omega$

15–23 $Z_{IN} = 27.1 + j20.8\ \Omega$

15–25 $\mathbf{V}_1 = 44.3 - j34.7$ V; $\mathbf{V}_L = 13.2 - j5.57$ V; $Z_{IN} = 40.1 + j131\ \Omega$

15–27 $Z_{IN} = 2.26 + j1.19$ kΩ; $i(t) = 78.4\cos(2500t - 27.8°)$ mA; $P_{IN} = 6.94$ W

15–29 $\mathbf{V}_1 = -29.4 - j9.34$ V; $\mathbf{V}_2 = -147 - j46.7$ V

15–31 $L_1 = 0.398$ H; $L_2 = 3.62$ H; $M = 0.796$ H

CHAPTER SIXTEEN

16–1 (a) $P = -130.7$ W; $Q = -1494$ VAR; delivering;
(b) $P = 82.1$ W; $Q = 465$ VAR; absorbing

16–3 (a) $P = 1.24$ kW; $Q = 4.64$ kVAR; pf $= 0.259$; lagging;
(b) $P = 260$ W; $Q = -1.48$ kVAR; pf $= 0.174$; leading

16–5 (a) $S = 435 - j116$ kVA; (b) $S = 4 - j3$ kVA

16–7 $P = 24$ kW; $Q = 18$ kVA; $I_{rms} = 12.5$ A; $Z_L = 154 + j115\ \Omega$

16–9 pf $= 0.894$; $Z_L = 22.2 + j11.1\ \Omega$

16–11 $S = 436 + j247$ VA

16–13 $R_L = 112\ \Omega$; $L_L = 198$ mH

16–15 $I_L = 6.26 + j1.96$ A; $S_L = 2.58 - j1.29$ kVA

16–17 $S = 325 + j150$ VA; pf $= 0.908$

16–19 (a) $I_A = 20.5 - j6.36$ A; $I_B = -16.4 + j1.82$ A; $I_N = -4.09 + j4.55$ A;
(b) $S_{upper} = 2250 + j700$ VA; $S_{lower} = 1800 + j200$ VA

16–21 $\mathbf{V}_S = 2.57 + j0.148$ kV; $S_S = 25.4 + j25.1$ kVA

16–22 $\mathbf{V}_S = 4.56 + j0.157$ kV; $S_S = 51.7 + j34.6$ kVA; pf $= 0.831$

16–23 $|\mathbf{V}_S| = 1.91$ kV; $|\mathbf{V}_L| = 1.62$ kV

16–24 (a) $\mathbf{V}_S = 4.49 + j0.063$ kV; $\mathbf{I}_L = 5.28 - j3.62$ A;
(b) $S_S = 23.5 + j16.6$ kVA;
(c) $\eta = 98.9\%$

16–25 $|\mathbf{V}_S| = 429$ V; $|\mathbf{V}_L| = 372$ V

16–26 32.9 μF

16–31 (a) $\mathbf{V}_{AN} = 277\angle 0°$ V; $\mathbf{V}_{BN} = 277\angle -120°$ V; $\mathbf{V}_{CN} = 277\angle -240°$ V; $\mathbf{V}_{AB} = 480\angle 30°$ V; $\mathbf{V}_{BC} = 480\angle -90°$ V; $\mathbf{V}_{CA} = 480\angle 150°$ V

16–33 (a) $\mathbf{I}_A = 6.06\angle 0°$ A; $\mathbf{I}_B = 6.06\angle 120°$ A; $\mathbf{I}_C = 6.06\angle -120°$ A; $\mathbf{I}_1 = 3.5\angle -30°$ A; $\mathbf{I}_2 = 3.5\angle 90°$ A; $\mathbf{I}_3 = 3.5\angle -150°$ A

16–35 (a) $\mathbf{I}_A = 42.8 - j6.5$ A; $\mathbf{I}_B = -15.7 + j40.3$ A; $\mathbf{I}_C = -27 - j33.8$ A; (b) $S_L = 28.1 + j22.5$ kVA

16–37 $|\mathbf{I}_L| = 80.25$ A; pf $= 0.938$

16–39 (a) $|\mathbf{I}_L| = 36.1$ A; (b) $Z = 17.3 + j15.2\ \Omega$

16–41 (a) $|\mathbf{I}_L| = 7.22$ A; (b) $Z = 144 + j127\ \Omega$

16–43 $|\mathbf{V}_{L2}| = 2.38$ kV; $S_2 = 15.8 + j7.88$ kVA

16–45 $|\mathbf{I}_L| = 48.1$ A; $|\mathbf{V}_{L1}| = 7.68$ kV; $S_1 = 487 + j416$ kVA

16–47 $|\mathbf{I}_L| = 6.56$ A; $|\mathbf{V}_{L1}| = 460$ V

16–49 $S_L = 876 + j342$ kVA

APPENDIX W1

W1–1 $F(\omega) = \dfrac{A}{j\omega}(1 - e^{-j\omega})$

W1–3 $F(\omega) = A\left[-4\pi\dfrac{\cos(\omega)}{4\omega^2 - \pi^2}\right]$

W1–5 $f(t) = 10\left[\dfrac{\cos(t) - 1}{t}\right]$

W1–7 (a) $f_1(t) = 20[e^{-20t} - e^{-40t}]u(t)$; (b) $f_2(t) = [-e^{-20t} + 2e^{-40t}]u(t)$

W1–13 (a) $F_1(\omega) = \dfrac{2j\omega + 4}{-\omega^2 + 4j\omega + 20}$; (b) $F_2(\omega) = \dfrac{j\omega + 6}{(j\omega + 2)(j\omega + 4)}$

W1–15 (a) $f_1(t) = 2 + 4\cos(2t)$; (b) $f_2(t) = 2 + \text{sgn}(t)$; (c) $f_3(t) = 2u(t)$

W1–17 (a) $f_1(t) = 2 + \text{sgn}(t - 2)$; (b) $f_2(t) = (2e^{-2(t-2)})u(t - 2)$; (c) $f_3(t) = \text{sgn}(t + 2) + \text{sgn}(t - 2)$

W1–19 $g(t) = 0.25(1 - 2e^{-2t} + e^{-4t})u(t)$

W1–27 $v_2(t) = 8\,[e^{5t}u(-t) + e^{-20t}u(t)]$

W1–29 $v_2(t) = 10e^{-100t}u(t)$

W1–31 $y(t) = 0.5\,[u(-t) + e^{-2t}u(t)]$

W1–33 $y(t) = 4e^{-t}u(t) - \text{sgn}(t)$

W1–35 system is an ideal high-pass filter with $\omega_C = \pm\beta$

W1–41 $W_{1\Omega} = A^2/2\alpha$

W1–43 $W_{1\Omega} = A^2/4\alpha$; $W_\alpha = 0.182\,W_{1\Omega}$

APPENDIX W2

W2–1 $z_{11} = 400\ \Omega$; $z_{12} = z_{21} = 100\ \Omega$; $z_{22} = 167\ \Omega$

W2–2 $y_{11} = 2.94$ mS; $y_{12} = y_{21} = -1.77$ mS; $y_{22} = 7.06$ mS

W2–7 $I_1 = 18$ mA; $I_2 = -12$ mA

W2–8 $T_V = -4.17$

W2–9 $V_2 = 9$ V; $I_1 = 30$ mA; $I_2 = -6$ mA

W2–10 $Y_{IN} = 25 + j20$ mS

W2–13 $h_{11} = 0$; $h_{12} = 1$; $h_{21} = -1$; $h_{22} = Y$

W2–15 $h_{11} = R_1 + R_2$; $h_{12} = 0$; $h_{21} = \beta$; $h_{22} = G_3$

W2–18 $A = 1$; $B = R_2$; $C = G_1$; $D = R_2G_1$

W2–19 $V_T = 6$ V; $R_T = 250\ \Omega$

W2–21 (a) $R_{IN} = 800\ \Omega$; (b) $R_{IN} = 400\ \Omega$; (c) $R_{IN} = 600\ \Omega$

INDEX

ac (alternating current) signal, 233
ADC, *see* Analog-to-digital converter
Active circuit, 138
Active device, 138
  OP AMP, 157
  transistor, 152
Active filter, 651
Admittance, 359, 451
  parallel connection, 359, 455
  transfer, 500
Aliasing, 643
Ampere (unit), 5
Amplifier:
  differential, 170
  Inverting, 166
  noninverting, 160
  summing, 168
  voltage follower, 165
Amplitude, 210, 214, 219
Amplitude spectrum, 625, W-1
Analog-to-digital converter, 193
  flash, 195
  full-scale input, 195
  resolution, 195
Angular frequency, 221
Apparent power, 730
Applications:
  ac power system, 757
  analog switch, 17
  autotransformer, 713
  band-limited differentiator, 266
  binary current divider, 45
  cellular osmosis, 437
  decibels, 556
  digital clock detection, 294
  digital clock skew, 542
  digital clock waveform, 236
  ECG waveform, 238
  gain-bandwidth product, 562
  impedance bridge, 357
  loading, 122
  potentiometer, 41
  residential ac power, 742
  resonance, 362
  sample-and-hold circuit, 252
  signal bandwidth, 640
  signal sampling, 641
  step response descriptors, 515
  s-plane geometry, 471
  time-limited integrator, 265
Average power, 386, 728
  of a periodic signal, 638
Average value, 233

Balanced three-phase, 744
  Y-   connection, 750
  Y-Y connection, 747
Bandpass filter, 554
  cascade design, 690–692
  narrow band, 579, 662

Bandpass filter (*cont.*)
  parallel *RLC*, 583
  second order, 660
  series *RLC*, 577
  using first-order circuits, 570
  wide band, 579, 662
Bandstop filter, 554
  parallel design, 692–694
  second order, 662
  series *RLC*, 581
  using first-order circuits, 570
Bandwidth, 578, 640, 662
Bode diagram, 556
  complex poles and zeros, 597–604
  real poles and zeros, 586–597
  using Mathcad, 601
  using MATLAB, 602
Bridge circuits:
  impedance, 357
  Maxwell, 358
Bus, 753
Butterworth:
  high-pass response, 684
  low-pass response, 670
  polynomials, 671

Capacitance, 246
Capacitive reactance, 355
Capacitor, 246
  average power, 386
  dc response, 268
  energy, 248
  impedance, 352, 450
  i–v relationships, 246–247
  parallel connection, 266
  series connection, 267
Cascade connection, 507
Cascade design, 529, 651
Causal waveform, 231
Center frequency, 578, 662
Chain rule, 507
Chapter Objectives, 3
Characteristic equation:
  first order, 279
  second order, 311, 319, 331
Charge, 5
Chebychev:
  high-pass response, 688
  low-pass response, 677
  polynomials, 677
Circuit, 2, 20
Circuit analysis, 2
  computer aided, 51
  phasor domain, 349–352
  resistive, 26–32
  s domain, 446–458
  using Fourier series, 630–635
Circuit design, 2
  ac voltage divider, 356
  amplifier, 161
  bandpass filter, 573
  Bode plot, 592
  complex poles, 473
  high-pass filters, 683–690
  input-output relationship, 264
  low-pass filters, 665–680
  interface circuits, 114–122, 189
  network function, 529–539
  OP AMP circuit, 180–184
  RLC circuits, 330, 333, 580, 583
  second-order active filters, 652–665
  step response, 539
  summer, 169
  transistor inverter, 156
Circuit determinant, 470
Circuit reduction, 45
Circuit theorems:
  maximum power transfer, 112, 387
  proportionality, 91, 365, 458
  superposition, 95, 366, 460
  Thévenin/Norton, 98, 370, 464
Comparator, 192
Complex frequency, 315
Complex numbers, A-12
  Arithmetic operations, A-14
  conjugate, A-13
  exponential form, A-13
  imaginary part, A-12
  real part, A-12
Complex power, 730
  and load impedance, 732
  conservation of, 733
Computer tools, 51
  Electronics Workbench, 178, 327, 541, 575
  Excel, 215, 305, 471, 635
  MATLAB, 86, 382, 414, 516, 602, 691, 693, 739
  Mathcad, 256, 425, 601, 620, 633
  Orcad Capture, 51, 124, 149, 283, 369
Conductance, 15, 359

Connection constraints, 19–26
  phasor domain, 349
  s domain, 446
Convolution, 511, 523, W-22
  graphical interpretation, 526
  integral, 523, W-23
Corner frequency, 559
Coulomb (unit), 5
Coupled inductors, 701–703
  energy, 706
  dot convention, 704
  i–v relationships, 703
Coupling coefficient, 707
Cramer's rule, A-3
Critically damped, 317
Current, 5
  line, 745
  mesh, 81
  short circuit, 100
  phase, 750
Current division, 42, 359, 456
  two path, 43
Current gain, 138
Current source, 18
  dependent, 138
Cutoff frequency, 553, 578

-connection, 744
-Y transformation, 750

DAC, *see* Digital-to-analog converter
Damped ramp, 228
Damped sine, 228
Damping, 316
Damping ratio, 324
dc (direct current) signal, 208
Decade, 556
Decibel, 554
Delay time, 515
Dependent source, 138
Design, *see* Circuit design
Determinant, A-3
Device, 15
Differential amplifier, 170
Differential equation:
  first order, 278
  second order, 309, 319, 324
  solution by Laplace transforms, 426–434
Differentiator, 260
Digital-to-analog converter, 185
  full-scale output, 185
  resolution, 185
Dirichlet conditions, 618, W-2
Dot convention, 703–706
Driving point impedance, 499
Duality, 258, 584

Element, 15
Element constraints, 15–19
  phasor domain, 350
  s domain, 446–449
Energy, 5
  capacitor, 248
  coupled inductors, 706
  inductor, 255
Equivalent circuits, 32
  capacitance, 267
  dc, 268
  inductance, 268
  resistance, 33
  sources, 35
  summary of, 38
  Thévenin/Norton, 98, 370, 464
  transformer, 719
Euler's relationship, A-13
Evaluation, 2
  bandpass circuits, 581
  cascade connections, 509
  high-pass filters, 568
  interface circuits, 119
  low-pass filters, 681–683
  OP AMP circuits, 183, 187, 192
  pressure transducers, 607
  RC circuits, 299
  step response designs, 539
  transducer interface circuits, 192
  transfer function designs, 538
Even symmetry, 628
Exponential waveform, 214
  double, 230
  properties of, 216

Farad (unit), 246
Filter, 553
  active, 651
  bandpass, 554
  bandstop, 554
  high pass, 554
  low pass, 554

Filter (*cont.*)
passive, 651
tuned, 579
Final conditions, 295
Final-value property, 434
First-order circuit, 278
design with, 570–573
differential equation, 278
frequency response, 557–567
RC and *RL* circuits, 278
sinusoidal response, 301–307
step response, 287–293
zero-input response, 279, 280
zero-state response, 293
Flux linkage, 253
Forced pole, 470
Forced response, 287, 482
Fourier series, 617
alternative form, 625
coefficients, 618
in circuit analysis, 630–635
table of, 623
Fourier transforms, W-1
Frequency:
angular, 221
center, 578, 662
complex, 315
corner, 559
critical, 412
cutoff, 553, 578
fundamental, 230, 618
harmonic, 230, 618
natural, 315
notch, 664
radian, 315
resonant, 362
Frequency response, 553
and step response, 604–607
Bode diagrams, 554
first order, 557–567
RLC circuits, 576–586
terminology, 554
Fundamental frequency, 230, 618

Gain function, 553
Ground, 9

Half-wave symmetry, 628
Harmonic frequency, 230, 618
Henry (unit), 253
Hertz (unit), 221
High-pass filter, 554
Butterworth, 684–687
Chebychev, 688–690
first order, 564
second order, 655

Ideal models:
current source, 18
OP AMP, 160
switch, 17
voltage source, 18
Ideal transformer, 708–713
i–v relationships, 710
input resistance, 712
Impedance, 351, 449
input, 499
magnitude scaling, 535
series connection, 355, 455
transfer, 500
Impulse, 211
Impulse response, 510
and convolution, 523, W-22
from network function, 510–512
from step response, 514
Inductance, 253
mutual, 702
self, 702
Inductive reactance, 355
Inductor, 253
average power, 386
dc response, 268
energy, 255
impedance, 352, 450
i–v relationships, 253–254
parallel connection, 268
series connection, 268
Initial conditions, 295
Initial–value property, 434
Input, 90
Input impedance, 499
Instantaneous power, 727
three-phase, 756
Integrator, 259
Interface, 2
Interface circuit, 115, 188
Inverting amplifier, 166

Joule (unit), 5

Kirchhoff's laws, 19
current (KCL), 20
phasor domain, 349
s domain, 446
voltage (KVL), 22

Ladder circuit, 45
Lagging power factor, 731
Laplace transforms, 401
inverse, 403, 416–421
table of pairs, 410
table of properties, 410
uniqueness property, 403
using Mathcad, 425
using MATLAB, 414
Leading power factor, 731
Linear circuit, 2, 90
Linear element, 15
Line current, 747
Line spectrum, 625
Line voltage, 745
Load-flow problem, 733, 737, 753
Loading, 123
s domain, 508
Lookback resistance, 106
Loop, 20
Low-pass filter, 554
Butterworth, 669–674
Chebychev, 675–680
first order, 557
first-order cascade, 666–669
second order, 652

Magnitude scaling, 535
Matrix A-5
adjoint, A-6
algebra A-7
and linear equations A-8
determinant, A-6
inverse A-7
Maximum signal transfer, 111
Mesh current, 81
Mesh-current analysis, 81–90
by inspection, 84
fundamental property, 81
phasor domain, 375
s domain, 475–481
summary of, 90
with current sources, 86
with dependent sources, 146
Mutual inductance, 702

Natural frequency, 315
Natural pole, 470
Natural response, 287, 482
Netlist, 52
Network function, 459, 496
design of, 529–542
driving-point impedance, 499
finding, 500–507
transfer function, 499
Node, 20
Node voltage, 67
Node-voltage analysis, 67–80
by inspection, 70
fundamental property, 67
phasor domain, 375
s domain, 467–475
summary of, 80
with dependent sources, 142
with OP AMPs, 175
with voltage sources, 75
Noninverting amplifier, 160
Nonlinear element, 15, 108
Norton equivalent circuit, 98
phasor domain, 370
s domain, 464
Notch frequency, 664
Nyquist rate, 643

Octave, 556
Odd symmetry, 15, 628
Ohm (unit), 15
Ohm's law, 15
OP AMP, 157
dependent source model, 159
effect of finite gain, 162
ideal model, 160
in the *s* domain, 448
notation, 157
operating modes, 158
OP AMP circuits:
bandpass filter, 661
bandstop filter, 664
differential amplifier, 170
differentiator, 260
high-pass filter, 657
integrator, 259
inverting amplifier, 166
low-pass filter, 653

OP AMP circuits: (*cont.*)
noninverting amplifier, 160
subtractor, 170
summary of, 172, 261
summing amplifier, 168
voltage follower, 165
Open circuit, 16
Open-circuit voltage, 100
Operational amplifier, *see* OP AMP
Output, 90
Overdamped response, 316
Overshoot, 515

Parallel connection, 25
admittances, 359, 455
capacitors, 266
inductors, 268
resistors, 33
Parseval's theorem, W-26
Partial fraction expansion, 416
of improper rational functions, 421
with complex poles, 418
with multiple poles, 423
Passband, 553, 665
Passive sign convention, 9
Passive filter, 651
Peak-to-peak value, 232
Peak value, 232
Period, 219
Periodic waveform, 231
rms value of, 234, 637
Phase angle, 219
Phase current, 750
Phase function, 553
Phase sequence, 745
Phase spectrum, 625, W-1
Phase voltage, 745
Phasor, 344
diagram, 344
domain, 353
properties, 345–346
rotating, 345
Phasor circuit analysis, 349–353
device constraints, 350
connection constraints, 349
impedance concept, 351
Planar circuit, 81, 375
Pole, 412
forced, 470
multiple, 423
natural, 470
number of, 484
simple, 416
stable, 482
Pole-zero diagram, 413
Port, 114, W-37
Potentiometer, 41
Power, 5
apparent, 730
average, 386, 638, 728
complex, 730
maximum, 112, 387
reactive, 728
three phase, 748, 751
Power factor, 730
angle, 730
correction, 737
Power triangle, 731
Primary winding, 708
Proportionality, 91, 365, 458

Quality factor (Q), 579, 662

Ramp, 212
Rational function, 416
Reactance, 355
Reactive power, 728
Reciprocal spreading, 605
Residue, 416
Resistance, 15, 355
lookback, 106
standard values A-1
Resistor, 15
average power, 386
impedance, 351, 450
i–v relationships, 15
power, 16
Resonance, 362
Resonant frequency, 362
Response:
ac steady state, 304
dc steady state, 268, 304, 513
critically damped, 317
forced, 287, 482
frequency, 553
impulse, 510
natural, 287, 482
overdamped, 316
sinusoidal steady state, 304, 518
step, 287, 323, 513

  - underdamped, 316
  - zero input, 279, 293, 315
  - zero state, 293, 460
- Rise time, 515
- *RLC* circuits:
  - frequency response, 576–586
  - design, 330, 333, 580, 583
  - parallel, 318, 583
  - series, 308, 577, 581
  - zero-state response, 312–315
- Root-mean-square (rms) value, 234, 637
- Rotating phasor, 345

- Sampling, 641
  - aliasing, 643
  - Nyquist rate, 643
  - theorem, 642
- Secondary winding, 708
- Second-order circuit, 308
  - differential equation, 309, 319, 324
  - parallel *RLC*, 318, 583
  - series *RLC*, 308, 577
  - step response, 323–331
  - zero-input response, 315
- Self inductance, 702
- Sensor, *see* Transducer
- Series connection, 25
  - capacitors, 267
  - impedances, 355, 455
  - inductors, 268
  - resistors, 33
- Short circuit, 16
- Short-circuit current, 100
- Siemens (unit), 15
- Signal, 2
  - ac, 233
  - dc, 208, 233
  - composite, 226
  - exponential, 214
  - impulse, 211
  - ramp, 212
  - signum, 226
  - sinusoidal, 218
  - step function, 209
- Single phase, 742
- Sinusoidal steady-state response, 304, 518
  - from network functions, 518–522
  - using phasors, 349–352
- Sinusoidal waveform, 218
  - damped, 228
  - properties of, 223
  - phasor representation, 344
- Source:
  - current, 18
  - dependent, 138
  - equivalent, 35
  - practical, 19
  - three phase, 744
  - voltage, 18
- Source transformation, 36
  - phasor domain, 371
  - *s* domain, 465
- Spectrum, 618, W-1
- Stable, 482
  - and pole locations, 482
  - asymptotic, 511
  - marginally, 483
- Standard form, 324, 331
- State variable, 248, 255, 278
- Steady-state response, 304, 497
- Step function, 209
- Step response, 513
  - descriptors, 515
  - first-order circuit, 287
  - from frequency response, 604–607
  - from impulse response, 513
  - from network function, 513–517
  - second-order circuit, 323–331
  - using MATLAB, 516
- Stopband, 553, 666
- Summing amplifier, 168
- Supermesh, 87
- Supernode, 76
- Superposition, 95, 366, 460
- Susceptance, 359
- Switch, 16
  - analog, 17
  - ideal, 17

- Tank circuit, 585
- Thévenin equivalent circuit, 98
  - phasor domain, 370
  - *s* domain, 464
  - with dependent sources, 150
  - with nonlinear loads, 108
- Three-phase power analysis, 747–752
  - load flow, 753
  - single-line diagram, 753
  - Y- connection, 750
  - Y-Y connection, 747

Time constant, 215
  RC circuit, 280
  RL circuit, 281
Time shift, 210, 219
Transconductance, 138
Transducer, 188
  interface circuit, 188
Transfer function, 499
  admittance, 500
  current, 500
  impedance, 500
  voltage, 500
Transformer, 708
  equivalent circuit, 719–723
  sinusoidal steady state, 714–718
  step down, 709
  step up, 709
  turns ratio, 709
Transition band, 666
Transistor, 147, 149, 152
  operating modes, 152
Transresistance, 138
Tuned filter, 579
Turns ratio, 709
Two-port network, 114, W-37
  parameters, W-38

Undamped natural frequency, 324
Underdamped response, 316
Unit output method, 92, 365
Units, table of, 4

VA (unit), 730
VAR (unit), 729
Volt (unit), 5
Voltage, 5
  line, 745
  node, 67
  open circuit, 100
  phase, 745
Voltage division, 39, 355, 455
Voltage follower, 165
Voltage gain, 138
Voltage source, 18
  dependent, 138

Watt (unit), 6
Waveform, 208
  partial descriptors, 231–233
Waveform symmetries, 627–630
Weber (unit), 253

Y-connection, 744

Zero, 412
Zero–input response, 279, 293, 315, 460
Zero–state response, 293, 460

BASIC LAPLACE TRANSFORMATION PROPERTIES

PROPERTIES	TIME DOMAIN	FREQUENCY DOMAIN
Independent Variable	t	s
Signal Representation	$f(t)$	$F(s)$
Uniqueness	$\mathcal{L}^{-1}\{F(s)\}(=)[f(t)]u(t)$	$\mathcal{L}\{f(t)\} = F(s)$
Linearity	$Af_1(t) + Bf_2(t)$	$AF_1(s) + BF_2(s)$
Integration	$\int_0^t f(\tau)d\tau$	$\frac{F(s)}{s}$
Differentiation	$\frac{df(t)}{dt}$	$sF(s) - f(0-)$
	$\frac{d^2 f(t)}{dt^2}$	$s^2 F(s) - sf(0-) - f'(0-)$
	$\frac{d^3 f(t)}{dt^3}$	$s^3 F(s) - s^2 f(0-) - sf'(0-) - f''(0-)$
t-Translation	$[f(t-a)]u(t-a)$	$e^{-as}F(s)$
s-Translation	$e^{-\alpha t}f(t)$	$F(s+\alpha)$
Scaling	$f(at)$	$\frac{1}{a}F\left(\frac{s}{a}\right)$
Final Value	$\lim_{t\to\infty} f(t)$	$\lim_{s\to 0} sF(s)$
Initial Value	$\lim_{t\to 0+} f(t)$	$\lim_{s\to\infty} sF(s)$